American Institute of Chemical Engineers

Separations Division

Presentations at the
2007 AIChE Annual Meeting

November 4-9, 2007
Salt Lake City, Utah, USA

Volume 1 of 2

Printed from e-media with permission by:

Curran Associates, Inc.
57 Morehouse Lane
Red Hook, NY 12571
www.proceedings.com

ISBN: 978-1-60560-000-0

Some format issues inherent in the e-media version may also appear in this print version.

ISBN: 978-1-60560-000-0

American Institute of Chemical Engineers
Proceedings
Three Park Avenue
New York, NY 10016-5991
Phone: 212-591-8100

www.aiche.org

American Institute of Chemical Engineers

Separations Division
2007

TABLE OF CONTENTS

Volume 1

Volume 2

24a Development Of A Universal Solvent Selection Methodology For In-Situ Extraction

C. van den Berg[1], P.J.Th. Bussmann[2], C.P.M. Roelands[1], and L.A.M. Van der Wielen[3]. (1) Separation Technology, Dutch Institute for Applied Science (TNO), Laan van Westenenk 501, Apeldoorn, Netherlands, (2) Food and Biotechnology innovations, Dutch Institute for Applied Science (TNO), Laan van Westenenk 501, Apeldoorn, Netherlands, (3) Department of Biotechnology, Delft University of Technology, Julianalaan 67, 2628 BC Delft, Delft, NL-2628 BC

At TNO a Pseudomonas Putida S12 strain was developed which produces phenol from glucose as a carbon source[1]. However, this strain was severely hindered by product inhibition. To alleviate product inhibition from fermentation broths the product must be removed from the environment of the micro-organism. One way of achieving product removal is in-situ extraction. In this project so-called solvent impregnated resins (SIRs) are developed which are able to selectively remove phenol from fermentation broths. An extractant is impregnated in a carrier polymer which separates the solvent from the fermentation broth. These SIRs are then applied inside the fermentor to remove the product from the fermentation broth into the extractant phase of the particle. This new application of SIRs puts extra constraints on the solvent selection, requiring a novel approach. The selection of an extractant for a certain product should include parameters for fermentation biocompatibility, product separation and extractant regeneration. Toxicity, melting/boiling point, biodegradability, logPo/w, solubility, charge, and hydrogen bond acceptor/donor ability [2] are therefore important factors to take into account for both the product and extractant. We have constructed a database including more than 180 extractants which are a wide range of alcohols, amines, phosphates, aromatics, alkanes, alkenes, thiols, aldehydes and other extractant families. A two-step approach is employed that includes a decision model to obtain a short-list of extractants and an experimental part that consists of measuring and predicting phenol partition coefficients. A quick first selection out of all possible extractants can be made using a 2-step model. For making a final decision between several extractants we use an automated solvent screening system which is able to measure partition coefficients of aromatics over any two-phase system. Using the combination of known physical properties, molecular interaction predictions and measured data an optimal solvent choice can be made. This solvent selection methodology will be developed towards all other uncharged solutes. Results for the model compound phenol will be presented.

1. Wierckx N.J.P., et al., Engineering of solvent-tolerant Pseudomonas Putida S12 for bioproduction of phenol from glucose. Applied and environmental microbiology, 2005. 71(12): p. 8221-8227. 2. Abraham M.H. and Platts J.A., Hydrogen bond structural group constants. Journal of organic chemistry, 2001. 66: p. 3484-3491.

Web Page: www.b-basic.nl

24b Estimation Of Phospholipid Solubility In Conventional Solvents And Supercritical Fluids

Rebecca K. Toghiani, Chemical Engineering, Dave C. Swalm School of Chemical Engineering, Mississippi State University, 330 Swalm, Mississippi State, MS 39762

The extraction of phospholipids from biomass, in particular from waste sewage sludge, has the potential to provide an additional feedstock for biodiesel production. One challenge in developing a competitive, economical recovery process is the identification of a suitable solvent and the optimum temperature and pressure for the recovery operation. Use of predictive computations provides a means to narrow the scope of conditions (temperature and pressure) as well as solvent composition that must be examined experimentally. For most substances, vapor pressure data as well as critical point information are readily available in the literature. However, for many phospholipids, these data are not readily available and would be prohibitively expensive to obtain experimentally due to the cost and limited availability of purified phospholipids. What is readily available for each phospholipid is its chemical structure. The ultimate goal of this effort is to provide a basis for predictive screening of solvent mixtures and identification of operating conditions suitable for economical recovery of mixed phospholipids from a complex matrix.

The prediction of the solubility of select phospholipids in solvents of differing compositions will be examined, including supercritical solvent mixtures. This effort combines the use of chemical structure for physical property estimation with the use of COSMO-Rs for the prediction of pure component vapor pressure of phospholipids and phase equilibria of phospholipids in conventional solvents. Prediction of phospholipid solubility in supercritical solvents is accomplished using the Peng-Robinson equation of state, utilizing estimated vapor pressure estimates for the phospholipids. Validation efforts and predictions for three model phospholipids, distearoyl phosphatidylcholine, dipalmitoyl phosphatidylcholine and dipalmitoyl phosphatidylethanolamine, will be presented.

24c Preliminary Investigation Of Cellular Lipid Extraction Using Electroporation As An Enhancement Technique

Robert G. McComas[1], Mark Bricka[1], and W. Todd French[2]. (1) Dave C. Swalm School of Chemical Engineering, Mississippi State University, PO Box 9595, Mississippi State, MS 39762, (2) Chemical Engineering, Mississippi State University, Box 9595, Mississippi State, MS 39762

Recently Mississippi State University (MSU) has been conduction research to increase the ability of bacteria to accumulate oil. This work was ultimately targeted towards providing an alternative method of producing oil in contrast to that of chemical liquid extraction which is the primary method practiced today. The purpose of this oil is to serve as a feed material for biodiesel (or other oil based fuels).

To ensure that the developing oil accumulation technology is economically completive new methods are required to release the oil from the cell membrane. One technology being considered is Electroporation (EP). EP is a technology that has traditionally been utilized to as a way to introduce substance into cells (such as loading cells with molecular probes). Pores are formed as voltage is passed across the membrane and it exceeds its dielectric strength. In contrast to introducing substances it is hypothesized that these voltage induced pores will provide an outlet for the oil from the cell membrane. This should improve the recovery of the hyper-accumulated lipids by the cell.

This paper will present describe the apparatus being currently being utilized in the laboratory to conduct electroporation investigations as well as the results generated to date. Preliminary results are quite promising indicating that solutions up to 30% oil can be generated.

24d Extraction And Identification Of Julibroside Saponins From The Bark Of *albizia Julibrissin*

Abigail S. Engelberth[1], D. J. Carrier[2], and E.C. Clausen[1]. (1) Chemical Engineering, University of Arkansas, 3202 Bell Engineering Center, Fayetteville, AR 72701, (2) Biological and Agricultural Engineering, University of Arkansas, 3202 Bell Engineering Center, Fayetteville, AR 72701

The co-generation of value added products is vital for the integration of biobased products into the current economy. One such product, saponins, can be found in the bark of Albizia julibrissin (mimosa). Saponins are a natural detergent and are be found in a variety of plants, ranging from the Yucca schildigera plant in Mexico to the Quilaja saponaria plant in Chile. The major types of saponins found in plants have either a steroid nucleus, found in the Yucca plant, or a triterpenoid structure as in the Quilaja. Saponins contain both a water-soluble and a fat-soluble component. Some saponins have been shown to have antiprotozal activity in ruminals and to have been able to reduce blood cholesterol levels in mammals. Triterpenoidal saponins, julibrosides, are known to be present in Albizia julibrissin [1], and have previously been extracted using organic solvents, separated by RP-HPLC and analyzed with mass spectrometry. Unfortunately, the julibrosides are not available commercially as reference compounds. Hence, the first objective of this research project is to extract the saponins from mimosa bark in order to secure reference material. The next objective is to explore the use of centrifugal partition chromatography (CPC) to separate and purify the extracted saponins. The overall goal of this project is to determine the feasibility of extracting saponins from mimosa biomass prior to energy conversion.

[1] Zou, K., Tong, W., Linag, H., Cui, J., Tu, G., Zhao, Y., Zhang, R., Diasteriomeric saponins from Albizia julibrissin. Carbohydrate Research. 2005, 340, 1329-1334

Sub- and Supercritical Fluid Extraction of Bioactive Compounds From *Ganoderma lucidum*

ASKIN, Ruhan[1)], SASAKI, Mitsuru[1)], GOTO, Motonobu[1)*]

[1)] Department of Applied Chemistry and Biochemistry, Kumamoto University

2-39-1 Kurokami, Kumamoto 860-8555, Japan

*Corresponding author, TEL: +81-96-342-3664, FAX: +81-96-342-3679, e-mail: mgoto@kumamoto-u.ac.jp

ABSTRACT

Ganoderma lucidum, a polypore group fungus, provides bioactive compounds in which several triterpenoids and polysaccharides are core structures that claim to posses cancerous and immunomodulatory properties. These substances may be useful as starting materials for the development of chemical therapeutic agents in cancer treatment and for other ailments. Up to now, numerous methods of extraction have been developed to obtain extracts with higher yields and lower costs. Such is the case of extraction with organic solvents, such as methanol, ethanol, and acetone. Recently, supercritical CO_2 ($SCCO_2$) and pressurized hot water or sub-critical water has become of great interest as an alternative solvent for extraction of natural active compounds. The objective of this work has been to establish a method of extraction from grinded form of *Ganoderma lucidum* with water in sub-critical conditions and with $SCCO_2$ at distinct temperatures to obtain extracts rich in water-soluble organic compounds (WSOC), including mostly beta glucans and triterpenoids including ganoderic acids and alcohols respectively. $SCCO_2$ experiments were carried out at pressures of 10, 20 and 30 Mpa and temperatures ranging from 40 to 60 ^{o}C with and without modifier. Ethanol was used as modifier at flow rates of 0.1, 0.2 and 0.4 mL/min. The CO_2 flow rate was maintained at 4 mL/min for 2 hours extraction time. The extracts were analyzed by HPLC. The temperatures for experiments that were carried out by batch-scale sub-critical water extraction kept constant at different time intervals starting from 373 up to 573 K. For semi-continuous scale experiments, samples were extracted once with water at 373, 423, 448 and 473 K, respectively, working at pressure around 10 Mpa to keep the water in the liquid state. The flow rate used was 1 mL/min. The extracts were analysed by TOC for water soluble organic compound recovery. The highest yield obtained at 473 K as 78.1 % and as 57.4 % for batch- and semi continuous scale experiments respectively. It was demonstrated that the modified $SCCO_2$ extraction is suitable for the extraction of Ganoderma lucidum. The advantage of modified supercritical extraction over non-modified supercritical extraction was in the polarity component extraction and enhancement of the fluidity of extracts. It was also seen that molecular weights of extracts became smaller for high temperatures having increasing values for $SCCO_2$ extraction in comparison to hydrothermal treatment.

Keywords: *Ganoderma lucidum*, $SCCO_2$, beta-glucans, triterpenoids, sub-critical water, water soluble organic compounds

1. INTRODUCTION

Extracting flavor essences, fats, oils, antioxidants, polysaccharides, triterpenoids and other food ingredient substances from natural plants constitutes a fundamental chemical engineering processes, featuring time-proven methods [1-3]. Many solvents used in traditional extraction are considered harsh, even hazardous. Being a polypore group fungus, *Ganoderma lucidum* provides bioactive compounds in which several triterpenoids and polysaccharides are core structures that claim to posses cancerous and immunomodulatory properties [4]. Supercritical fluid extraction using carbon dioxide as a solvent has provided an excellent alternative to the traditional solvent extraction of natural products over the last two decades. Supercritical CO_2 with its clean, safe, inexpensive, nonflammable, nontoxic, environment-friendly, nonpolluting solvent properties was found to be selective in the separation of desired compounds without leaving toxic residues in extracts and without the risk of thermal degradation of processed products [5]. Water is an interesting medium and reagent for various conventional chemical reactions, purification and separation processes, because it is non-toxic, non-flammable, cheap and easily available. Similarly, the extraction and isolation of compounds from solid matrices can also be performed efficiently with water at elevated temperatures and pressures. The objective of this work has been to establish a method of extraction from grinded form of *Ganoderma lucidum* with water in sub-critical conditions and with $SCCO_2$ at distinct temperatures to obtain extracts rich in water-soluble organic compounds (WSOC), including mostly β-glucans and triterpenoids including ganoderic acids and alcohols respectively.

2. EXPERIMENTAL

2.1 Material

Grinded form of dried *Ganoderma lucidum* in 3 mm of particle sizeused in this study were supplied by Refarmer Co., Ltd., Japan. *Ganoderma* contains 6.52 % hydrogen, 1.20 % nitrogen, 50.26 % oxygen and 42.02 % carbon (values are given by Kumamoto University Elemental Analysis Center, Kumamoto- Japan). They were stored at room temperature keeping away from humidity until they used and did not undergo any further pretreatment. Polystyrene standard kits were acquired from Shodex Company (Shodex STANDARD S Series- SM 105, Shodex Ltd., Tokyo- Japan). GPC-grade solvent, tetrahydrofuran (THF) was purchased from Wako Chemicals (Japan). CO_2 was acquired from Uchimura Sanso with a purity of 99.9%. Ethanol (EtOH) used as a modifier and rinsing solvent, and as a modifier in

$SCCO_2$ extraction and HPLC-grade solvents, including acetonitrile (CH_3CN) and acetic acid (AcOH) were purchased from Wako Chemicals (Japan).

2.2 Methods

2.2.1 Sub-critical Water Extraction (SWE)

The selectivity of sub-critical water extraction allows for manipulation of the composition of the extracts by changing the operating parameters. Two different extraction mechanisms were used named as batch type and semi-continuous type sub-critical [6] water extraction. These kinds of water treatments were done in order to emphasize the effect of water mobility in the sub-critical extraction process [7, 8].

In this study, our objective was by using hydrothermal treatment to recover water-soluble organic compounds-polysaccharides in order to determine optimum extracting conditions.

So, as for operating conditions, water soluble and insoluble polysaccharide fractions were isolated from *Ganoderma lucidum* by extraction with sub-critical water in batch- and semi-continuous scales.

2.2.1.1 Batch Scale Sub-critical Water Treatment

Experiments were done in reaction chamber (electric furnace) with a single cap stainless steel tube reactor (SUS316 batch-type reactor, 5 ml). 7 experiments were done for each of temperatures of 100 °C, 150 °C, 200 °C, 250 °C and 300 °C. 0.05 g of *Ganoderma lucidum* (1.25 w/v %) was inserted into the reaction tube for decided time durations starting from 5 minutes up to 60 minutes in 10-minute intervals for individual temperature values and argon gas was sprayed as a purging gas. After reaction, the treated sample was taken from the tubular extractor by washing totally 5 ml of pure water, homogenized using ultrasonic cleaner for 15 minutes and centrifuged for 30 minutes. As a following step, filtration (0.45 μm pore size) was done and liquid fractions were separated in order to use for the analyses. After filtration, the water soluble portion was analyzed with Total Organic Carbon Analyzer (TOC) (TOC-5050A, Shimadzu Corporation, Kyoto-Japan) and Gel Permission Liquid Chromatography (GPC), residue from filtration was given to vacuum drying for analytical weight determination.

2.2.1.2 Semi-continuous Scale Sub-critical Water Extraction

A schematic diagram of a laboratory built subcritical water extraction system is presented in Figure 1. The system consists of two HPLC pumps (PU-2080-100 MPa, Jasco Co., Japan), an oven, a stainless steel extraction cell (10 ml Vessel, Thar Designs, Inc., USA), and a collecting flask. Water was de-oxygenated for 30 min using ultrasonic cleaner (Honda, W-211) prior to the extraction. With the HPLC pump, the water was then delivered at a constant flow rate to the extractor and the extraction cell was completely filled with plant material (4.46 g of ground *Ganoderma lucidum*) and mounted vertically in the oven with water flowing from top to bottom. The water was brought to a set temperature by means of the preheating coil inside the oven before entering the extractor. The extraction pressure was controlled by adjusting the back-pressure regulator (AKICO) connected to the outlet coil.

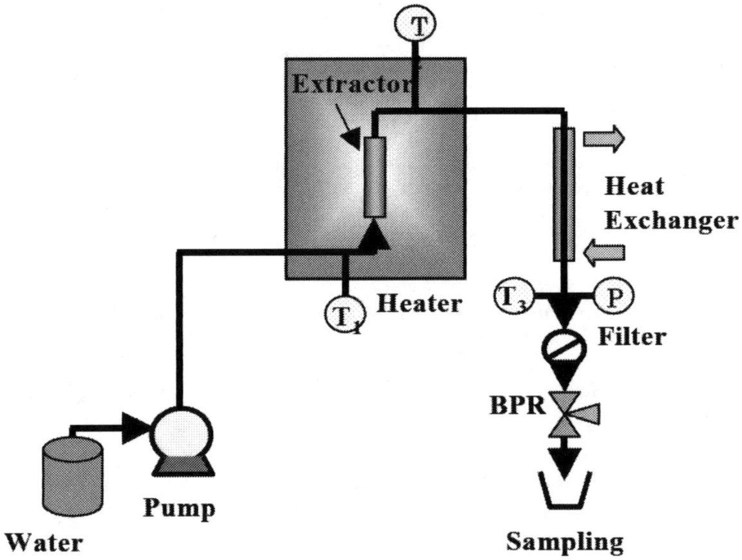

Fig. 1. Subcritical water extraction flow diagram.

A chiller (Sibata Co., Japan) was used to cool the extract from the oven to a constant temperature close to 298 K, thus avoiding losses of the products caused by the hot water. The second pump connected to the outlet coil was used to deliver ethyl alcohol to flush through any organic compounds that were precipitated in the coil as the temperature of the water was cooled. The extract was collected in fractions in collecting flasks.

Two different extraction procedures have been used depending on whether individual extractions (at a chosen temperature). The extraction experiments were carried out mainly to determine the effect of temperature holding the pressure and water flow rate constant. The effect of temperature in the extraction process was studied. For this purpose, three different assays were carried out. Among them, samples were extracted once with water at 100, 150 and 200 °C, respectively, working at pressure around 10 Mpa to keep the water in the liquid state. The flow rate used was 1 ml/min. All runs were performed at least in duplicate. Any doubtful results were checked and the experiments were repeated up to five times.

2.2.2 Supercritical Carbon Dioxide Extraction ($SCCO_2$)

The extraction of ganoderic acids and alcohols from reishi with $SCCO_2$ were carried out using experimental diagram given in Figure 2.

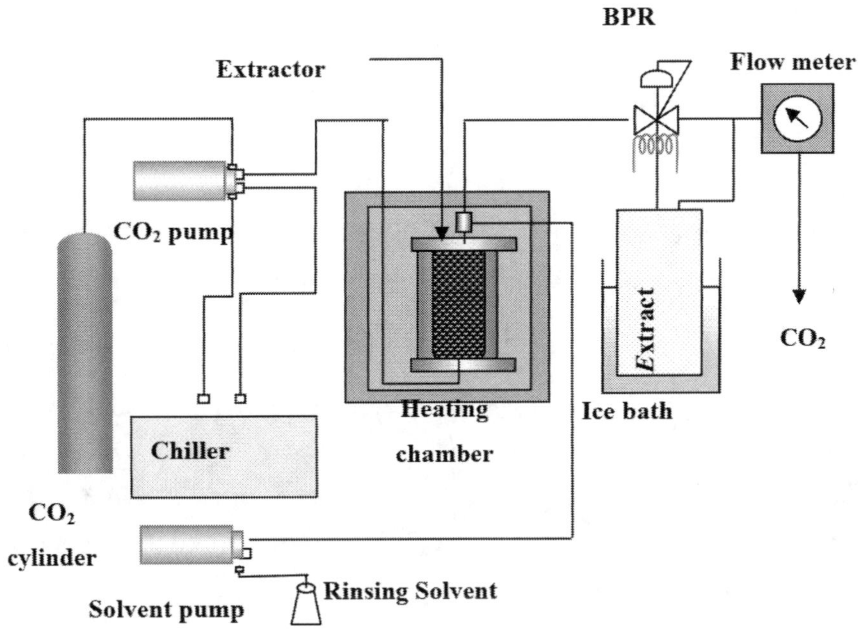

Fig. 2. Diagram of the supercritical fluid extraction apparatus.

Dried and crushed *Ganoderma lucidum* is placed into the extractor column for the extraction run. The total volume of the extraction vessel was well filled with dry ganoderma lucidum ground particles and glass beads as the fixed bed formation. Filters were placed at the top and bottom of the vessel to provide continuous flow of CO_2. EtOH was used as entrainer solvent during extraction and as used for washing the system after extraction at the connection point illustrated in Figure 2 in order to square with the solvent injected into the HPLC column and the mobile phase used in the analysis. The extraction runs were carried out in a pressure range of 10-40 MPa, and the CO_2 flow rate was kept at approximately $4*10^{-6}$ m3/min, a modifier flow rate range of 0.1-0.4 mL/min. To avoid degradation of temperature-sensitive material, the temperatures of 40, 50 and 60 °C were used for extraction. The extracted materials were weighted after the ethanol was evaporated.

2.3 Analysis

The total amount of the total extractable water-soluble organic compounds (WSOC) was determined after extractions in both batch and semi- continuous scale sub-critical water extractions. The recovery efficiency for WSOC was determined by Total Organic Carbon (TOC) analysis. System was calibrated using glucose solutions prepared at various concentrations before each analysis.

The average molecular weights of the polysaccharide compounds extracted from *Ganoderma lucidum* were estimated by Gel Permeation Chromatography (GPC) equipped with UV-visible detector (UV-970, Jasco, Japan) with TSK Gel GMH_{XL}-L column (Tosoh Corporation) (7.8 mm (ID)x30.0 cm (L)) in reference to calibration curve that was formed using polystyrene standards of known molecular weights (Shodex STANDARD S Series- SM 105, Shodex Ltd., Tokyo- Japan). THF was used as the mobile phase with a flow of 1.0 ml/min (10 μl injection volume) at 298 K column temperature in 25 min detection time; the detection was set at 254 nm for both polysaccharide compounds and polystyrene standards where the molecular weights were previously known.

For determination of triterpenoid constituents, including ganoderma alcohols and ganoderma acids in the products of *Ganoderma lucidum*, an analytical system was developed using HPLC with an ODS column. The mobile phase was composed of 1 % $AcOH/H_2O$-CH_3CN and 2 % $AcOH/H_2O$-CH_3CN, and the elution profile was monitored at 243 and 250 nm for ganoderma alcohols and acids respectively.

Particle structures of the powder microcapsules were evaluated by JEOL JSM-5200 model (Tokyo, Japan) scanning electron microscope. Powders were attached to SEM stubs using a 2-sided adhesive tape and left in desiccator. Then they were coated with a fine layer of gold through Sputter Coating Attachment of JEOL JFC- 1100 E (Jeol, Tokyo, Japan), in vacuumed evaporators before examination. For observation, a Scanning Electronic Microscopic JEOL JSM-5310LV (Jeol, Tokyo, Japan) working with a voltage of 15 kV. The microphotographs were carried out with a camera coupled to the microscopic. The samples were systematically observed with 1500 of magnification.

3. RESULTS AND DISCUSSION

3.1 Hydrothermal Treatment

In case of hydrothermal experiments, batch scale hydrothermal treatment of *Ganoderma lucidum* was carried out by contacting the sample material in tubular reactor for a specific contact time.

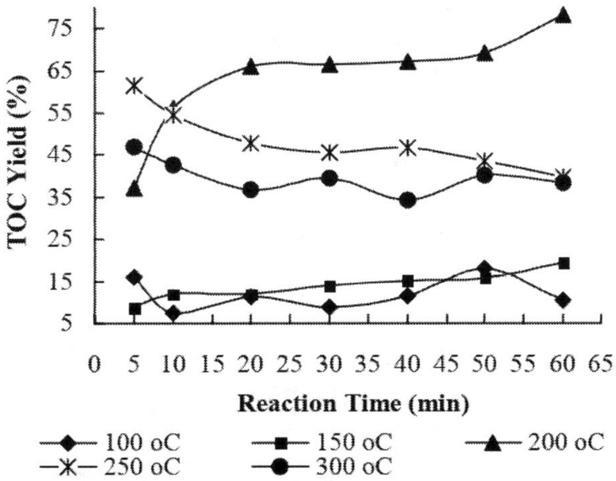

Fig. 3. Percent recovery of WSOC at various temperatures.

After treatment process, it was seen that the yield of recovery for total water-soluble organic carbon (WSOC) was not changed with sample amount, it was greatly effected by temperature. Lowest recovery was recorded as 7.52 % at 373 K. Figure 3 shows temperature effect on WSOC recovery efficiency depending on time. Temperatures up the 200 °C were gradually increased and showed positive effect on the recovery efficiency having highest values of 18.05 %, 19.37 % in individual temperatures for 100 °C and 150 °C respectively.

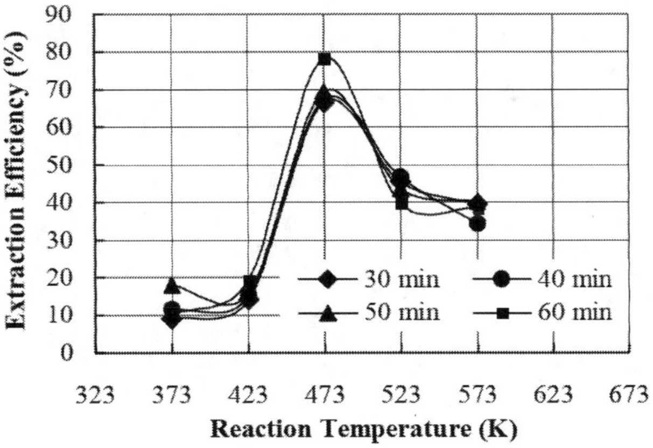

Fig. 4. Relation between extraction efficiency and reaction temperature

After 200 °C, the efficiency became lower again. The highest yield obtained at 200 °C as 78.11 % and decreased to 61.59 % and 46.82 % for 250 °C and 300 °C respectively. The relation between the extraction efficiency and temperature can be neatly observed from Figure 4.

The semi-continuous scale extraction of water-soluble polysaccharides was carried out by contacting the *Ganoderma lucidum* with a continuous flow of pure de-oxygenated water brought to sub-critical state. the extracted amount of total water-soluble organic carbon (WSOC) recovery including polysaccharides increased with an increase in temperature. The highest yield of WSOC extracted with sub critical water at 100, 150, and 200 °C were 30.25, 74.72 and 241.05 mg WSOC g^{-1} of dry sample respectively. This temperature dependence of the yield is due to the increased solubility of water-soluble organic compounds in sub critical water as the water temperature increases. The highest extraction yield (57.37%) was achieved in extracts obtained by individual extractions at 200 °C, which are the sum of the yields obtained in sequential extractions at 200 °C with a total duration of 130 min (Figure 5).

The results suggest that pressurized hot water extraction is greatly affected by extraction temperature in case of WSOC recovery (Figure 6) as 473 K has been selected for most efficient extraction condition for WSOC in both batch and semi-continuous modes.

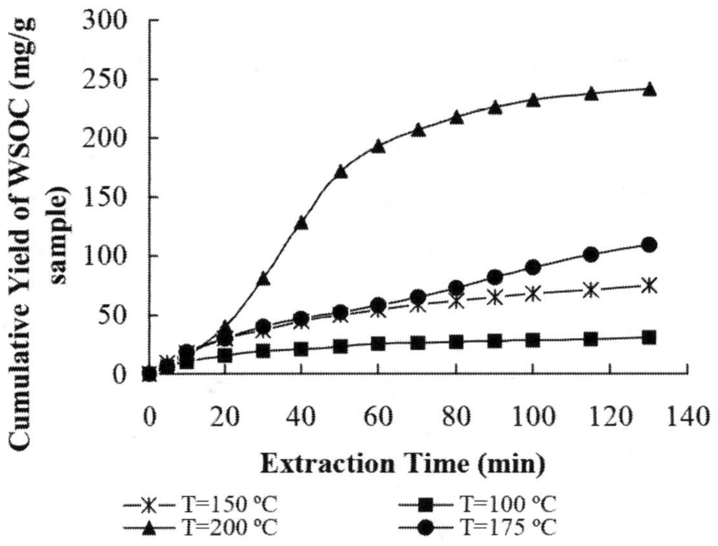

Fig. 5. Effect of temperature on WSOC extraction curve, P=10 MPa, F=1 ml/min

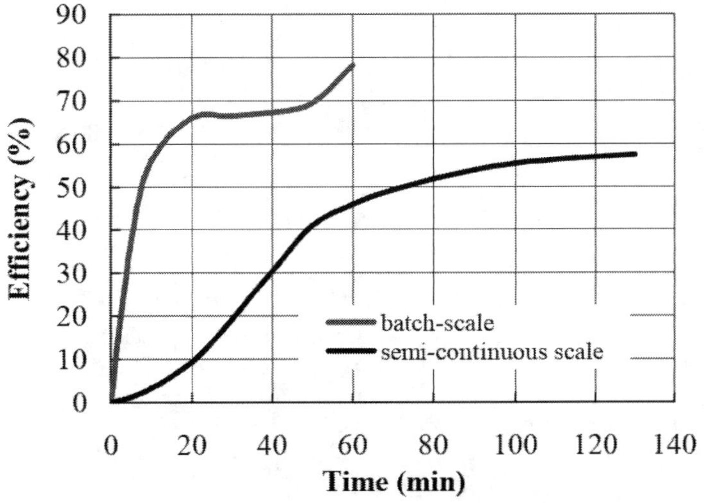

Fig. 6. Comparison of extraction efficiencies for batch- and semi-continuous scale treatments, 200 °C.

Referring to the Gel Permeation Chromatography (GPC) was done after both experiments. For the estimation of average molecular weights of water soluble organic compounds, calibration curve was formed using polystyrene standard kits including 7 portions having the molecular weights from 1.24×10^3 up to 3.44×10^6 taking the concentration as 0.05 % for all injected samples.

All analysis conditions were same for the polystyrene standards and the sample material. By the end of analysis it was seen that the elution volume for the samples were too small comparing the values detected for polystyrene standards, changing around 9-10 ml for all extracted samples in both batch and semi-continuous scale extractions where the elution volume was at least 7.25 ml for the smallest size polystyrene standard. This indicates that the long chained structure of the water soluble polysaccharides divided into very small portions in assistance of the high temperature effect so that the exact numerical detection could not be done referring to reference calibration curve. So, it can be said that the molecular weights for the extracted water-soluble organic compounds, which are mainly composed of polysaccharides including mostly beta glucans, were very small. In addition to the case explained, it was also seen that the amount of extract was little bit increased with high temperature getting smaller in structural composition.

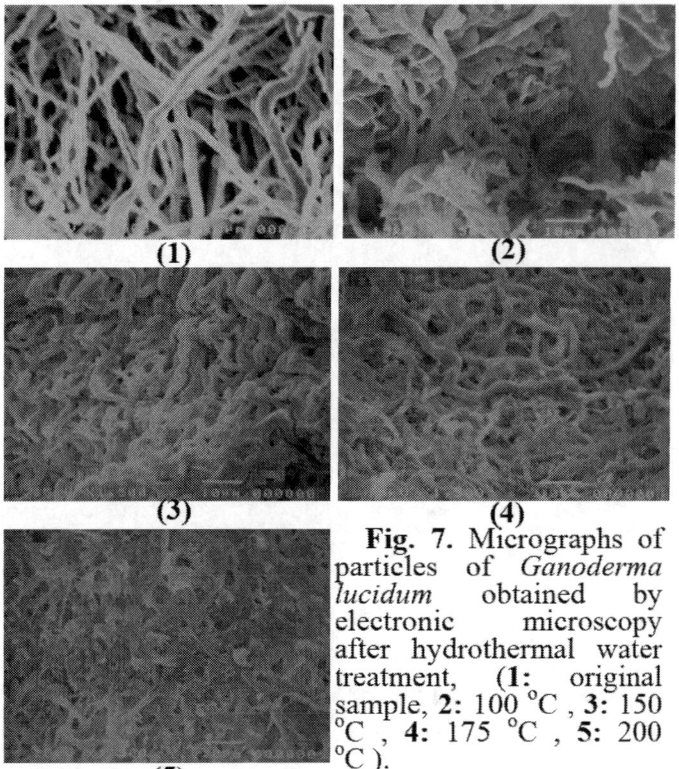

(1) (2) (3) (4) (5)

Fig. 7. Micrographs of particles of *Ganoderma lucidum* obtained by electronic microscopy after hydrothermal water treatment, (**1:** original sample, **2:** 100 °C , **3:** 150 °C , **4:** 175 °C , **5:** 200 °C).

Under all operating conditions where the details were given above, SEM photos of Ganoderma samples were taken in order to investigate the effect of temperature on physical structure of extract powders. Representative SEM micrographs, taken at magnification of 1500 are shown in Figure 7 that there are distinctive differences between the molecular structures being getting smaller as a result of treatments at higher temperatures.

3.2 SCCO$_2$ Extraction

The extractions of Ganoderma lucidum were carried out by contacting the sample with a continuous flow of SCCO$_2$. Extraction efficiency as a function of extraction temperature, pressure at specific modifier flow rates. By the end of extraction experiments, it is obvious to have a look at extraction efficiencies in the most comprehensive means. So, Figure 8 illustrates the extraction efficiencies for overall experimental conditions. It can be seen that triterpenoid components were best extractable at 40 °C having the maximum efficiency of 13, 38 %.

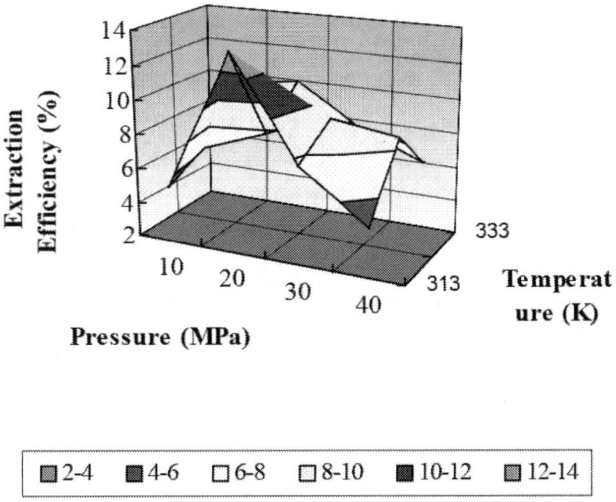

Fig. 8. $SCCO_2$ extraction efficiencies for temperatures 40- 60 °C, pressures 10-40 MPa at EtOH flow rates of 0-0.4 ml/ min

To avoid the degradation of temperature-sensitive materials (like triterpenoid components), the modified supercritical extraction was carried out at lower temperature. A comparison of extraction under 40 and 50 °C is shown in Figure 9, which depicts that not always the higher temperature had the higher extraction yield. This could be explained as the effects of specific temperature values on diffusion and fluidity of extracts from the matrix to the supercritical medium.

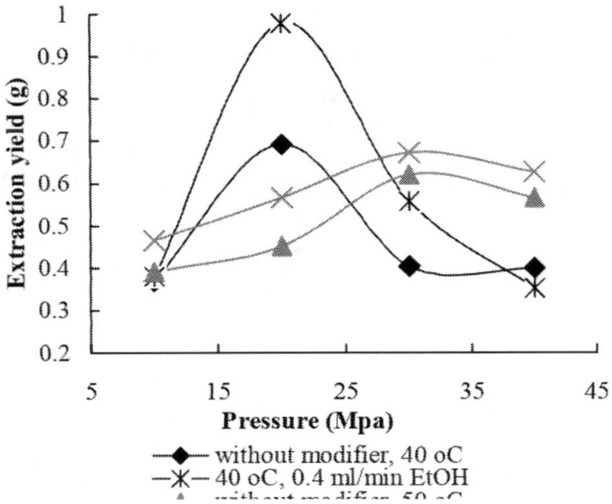

Fig. 9. Temperature and flow rate effects on extraction

4. CONCLUSION

The results of this study indicate that operating conditions (temperature, pressure, co-solvent flow rate, time, etc.) are crucial points in the $SCCO_2$ of triterpenoids from *Ganoderma lucidum*. The optimum extraction condition for triterpenoids was determined to be CO_2 at a density of 0.84 g/ml and 40 °C having 13, 38 % efficiency with a co-solvent of 10 % EtOH with the most influential variable on the extraction being the nature of the co-solvent, percentage of co-solvent followed by CO_2 density, extraction temperature and pressure. For hydrothermal treatment, increase in temperatures showed positive effect on the recovery efficiency for water-soluble organic compounds. The highest yield obtained at 200 °C as 78.1 % and as 57.4 % for batch- and semi continuous scale experiments respectively.

REFERENCES

1. M. S. Shao, "Triterpenoid Natural Products in Fungus Ganoderma lucidum", J. Chin. Chem. Soc., 39 (1992), p. 669.
2. Y. M. Komoda, M. Shimizu, Y. Sonoda, Y. Sato, "Ganoderic Acid and its Derivatives as Cholesterol Synthesis Inhibitors", Chem. Pharm. Bull., 37 (1989), p. 531.
3. T. S. Mizuno, "Bioactive Biomolecules of Mushrooms: Food Function and Medicinal Effect of Mushroom Fungi", Food Rev. Int. 11 (1995), pp. 151-166.
4. T. Sakai and G. Chihara, "Health foods and medicinal usages of mushrooms", Food Reviews International, 11 (1995), pp. 69-81.
5. W. F. Sye, "Improvement Method of Extraction and High-Performance Liquid Chromatographic Separation of Ganoderic Acids from Ganoderma Lucidum", J. Chin. Chem. Soc., 38 (1991), p. 179.
6. King, J. W., Fundamentals and applications of supercritical fluid extraction in chromatographic science. J. Chromatogr. Sci., 27 (1989), pp. 355–364.
7. T. Mizuno, M. Goto, A. Kodama, T. Hirose; Supercritical water oxidation of a model municipal waste, Ind. Eng. Chem. Res., 39 (2000), pp. 2807-2810.
8. CL Phelps, NG Smart, CM Wai. Past, present, and possible future applications of supercritical fluid extraction technology. J. Chemical Education, 73 (1996), pp. 1163-1168.

24f Supercritical Carbon Dioxide Extraction Of Capsaicinoids From Poblano Pepper (Capsicum Annuum L.) Using A Dynamic Method

Octavio Elizalde-Solis and Luis A. Galicia-Luna. Lab. de Termodinamica-ESIQIE, INSTITUTO POLITECNICO NACIONAL-ESIQIE, Lab. Termodinamica, UPALM, Edif. Z, 1er Piso, Sec. 6, Mexico, 07738, Mexico

Supercritical fluid technology is the most recommended method used for obtaining natural extracts, particularly on those where thermolabile compounds are contained. In Mexico a great variety pepper fruits are cultivated. In this study poblano green pepper (Capsicum annuum L.) was used to establish a supercritical extraction method to obtain capsaicinoids (capsaicin, dihydrocapsaicin, and nordihydrocapsaicin); which are responsible of the pungency in peppers.

Supercritical extractions were carried out by using a SFE-500 extraction device (Thar) via a dynamic method with a carbon dioxide as solvent at constant flow. The main components of this equipment are the extractor and separator vessels, each of 500 ml; a coriolis flow meter; and an automated back-pressure regulator. Pepper extracts were collected from the separator vessel and were analyzed by an HPLC method. A DAD detector was used to quantify capsaicinoid compounds.

The influence of temperature and pressure on the extraction yields and capsaicinoid extracts was studied in the ranges of 313 - 333 K and 10 - 30 MPa.

Web Page: www.termo.esiqie.ipn.mx

28a Adsorption Of Light Gases In Large-Pore Isoreticular Metal Organic Frameworks

Youn-Sang Bae[1], Andrew Nelson[2], David Dubbeldam[1], Houston Frost[1], Joseph T. Hupp[2], and Randall Q. Snurr[1]. (1) Department of Chemical and Biological Engineering, Northwestern university, 2145 Sheridan Road, Evanston, IL 60208, (2) Department of Chemistry, Northwestern university, 2145 Sheridan Road, Evanston, IL 60208

Metal-organic frameworks (MOFs) have emerged as promising materials for separations, gas storage, catalysis, and chemical sensing. One class of widely-studied MOFs are the isoreticular metal-organic frameworks (IRMOFs), discovered by Yaghi and co-workers. Although there have been many studies of IRMOFs, most of the studies to date concentrated on the IRMOFs -1, -6, and -8. Therefore, little is known about the separation and storage characteristics of the other IRMOFs. Especially, to our knowledge, there are few experimental studies on IRMOFs -10, -12, -14, and -16, which are the non-interpenetrating counterparts of IRMOFs -9, -11, -13, and -15, respectively. This may be due to the difficulties in synthesizing these materials and the instabilities of such large-pore MOF structures. Moreover, it can be quite difficult to effectively remove all of the solvent molecules (used in the synthesis procedure) from the pores.

In this study, several large-pore IRMOFs were synthesized by the solvothermal method developed by Yaghi et al. Two solvent removal procedures were compared: the original method involving chloroform exchange suggested by Yaghi and a new procedure developed in our labs. We investigated the adsorption properties of light gases such as H2 and CO2 on IRMOFs from both procedures at cryogenic and room temperatures. In addition, grand canonical Monte Carlo (GCMC) simulations of light gases, with and without co-adsorbed solvent molecules, were used to help understand and interpret the experimental results.

28b The Role Of Open Metal Sites In Adsorption Of Polar And Nonpolar Molecules In Cu-Btc

Krista S. Walton and J. Reddy Karra. Department of Chemical Engineering, Kansas State University, 1005 Durland Hall, Manhattan, KS 66506

Metal-organic frameworks (MOFs) have emerged as a new class of adsorbents with the potential for making an important impact in adsorption applications. MOFs are synthesized by a self-assembly process in which organic linkers coordinate with metal centers or clusters. Cu-BTC is a prototypical MOF constructed from copper and benzene tri-carboxylic acid. This material possesses a zeolitic interconnected network of pores in which copper atoms are built into the pore walls in a regular fashion. These metal sites are also coordinatively unsaturated, and thus, available for interaction with guest molecules. Traditional adsorbents such as zeolites and impregnated carbons are well known to possess selective adsorption and catalytic properties that can be attributed to the availability of open metal sites within the pores. It follows then that designing MOFs with open metal sites is a viable strategy for tailoring the adsorptive selectivity of these materials. In this work, we explore the role of open metal sites in the adsorption of carbon dioxide, carbon monoxide, and methane in Cu-BTC. Cu-BTC was synthesized and characterized in our lab. Adsorption equilibrium data were measured at 298 K. Grand canonical Monte Carlo simulations were also performed in conjunction with experiments for examining preferred adsorption sites and binding energies for the three molecules. In this paper, we will present these results and will discuss the role of the open metal sites in the framework compared to surface area/ void volume considerations in affecting the adsorption behavior of pure gases and mixtures.

ANALYSIS OF TRACER ZLC DESORPTION FOR SINGLE-FILE SYSTEMS

Abduljelil Iliyas[1], *Mladen Eić* [1*], *M. Hassan Zahedi-Niaki*[2] *and Sergey Vasenkov*[3]
[1]Dept. of Chemical Engineering, University of New Brunswick, Fredericton, Canada
[2]Department of Chemical Engineering, Laval University, Québec, Canada
[3]Chemical Engineering Dept., University of Florida, Gainesville, USA

*Corresponding author: meic@unb.ca

1. Introduction

The growing interest in the application of one-dimensional (1D) molecular sieves both as solid catalysts and selective adsorbents has stimulated considerable attention in the study of diffusive transport of guest molecules in the 1D pores. It has been established that normal Fickian diffusion is not necessarily applicable to describe the diffusion mechanism of guest-host interaction in 1D microporous solids, particularly when the diameter of the diffusion molecule is greater than channel radius [1]. For these systems, single-file diffusion mechanism has been employed to interpret the data from experimental studies using techniques such as PFG NMR [1] and QENS [2] to reasonable degree of correlation.

Recently, we have compared the tracer exchange curves obtained with 3-D toluene/β-zeolite with those of 1-D toluene/ZSM-12 system under similar conditions [3]. A non-linear behavior was obtained for the long-time asymptote in the 1-D system. This behavior can be explained by the existence of a distribution over the channel lengths in spherical particles of ZSM-12. The objective of this study is to provide evidence of the suitability of tracer ZLC technique in observing molecular transport under conditions of anomalous, single-file diffusion. More specifically, we aim to observe the center-of-gravity diffusion mechanism, which is expected for sufficiently large diffusion times in single-file systems of finite length [4, 5]. In relation to this, the tracer exchange profiles of large probe molecules, i.e., toluene having kinetic diameter of 0.61nm in a host circular channeled SAPO-5 (0.73nm dia.) and elliptical channeled ZSM-12 (0.56 x 0.6 nm dia.) is compared to the corresponding tracer exchange profiles of smaller guest molecules, e.g., water ($d \approx 0.26$nm) and/or acetylene ($d \approx 0.33$nm).

2. Theory

The mean square displacement (MSD) for one-dimensional diffusion can be defined, in accordance with the Einstein equation , as [1]

$$< z^2(t) > = 2tD_o \quad (1)$$

where D_o is the effective self-diffusivity. In contrast to above, MSD of diffusing species in single-file systems of an infinite channel length is given by:

$$< z^2(t) > = 2F\sqrt{t} \quad (2)$$

However, at large times, when the length scale of correlations of molecular movements becomes comparable to the file-length, the laws of normal diffusion:

$$< z^2(t) > = 2D_{sf}t \quad (3)$$

is applicable, but with a greatly reduced diffusivity given by

$$D_{sf} = \frac{d}{L}\left(\frac{1-\theta}{\theta}\right)D_o \quad (4)$$

where *d* denoting size of a molecule and L is the length of a channel. The diffusion in the latter case is often referred to as the center-of-gravity diffusion. Refs. [4] and [5] report theoretical expressions for the diffusion time and the corresponding MSD for the transition from the single-file to center of gravity diffusion for the case when the diffusion process can be described as random walk of solid particles with exclusion interaction in narrow, one-dimensional channels.

3. Experimental

The detailed synthesis procedure for ZSM-12 and SAPO-5 samples used in this study is described elsewhere [6]. Textural and morphological characterizations were performed using BET, SEM and XRD methods. The tracer exchange profiles of acetylene, water and toluene were carried out using TZLC technique [7] and continuously monitored with a quadrupole mass spectrometer (Dycor Dymaxion Quadrupole MS).

4. Results and Discussion.

The SEM, XRD and N_2-adsorption results confirmed that the as-synthesized samples are of very good quality with high crystallinity. Toluene molecules in SAPO-5 and ZSM-12 exhibit a significant deviation in the shape of the measured tracer exchange curves from monoexponential behaviour (Fig.1). The monoexponential behaviour is expected in the limit of large times for normal, i.e. Fickian diffusion in channel systems with one and the same channel length. The results in Fig.1 are in agreement with the expectation of the existence of the distribution over the channel lengths in sorbent particles. For the case of the center-of-gravity diffusion, which we expect for toluene, the influence of the distribution over the channel lengths on the measured curves should be further enhanced in comparison to the normal diffusion. Such an enhancement is expected because the center-of-gravity diffusivity is inversely proportional to the channel length (see Eqn 4). Hence, the distribution over the channel lengths should lead to a larger gap between the rates of tracer exchange in channels with different lengths for the case of the center-of-gravity (single file) diffusion than for the case of normal diffusion.

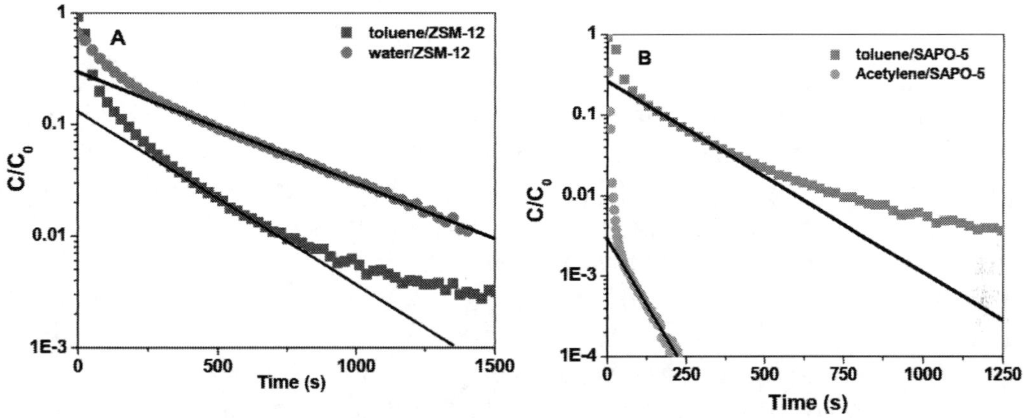

Fig. 1. Experimental tracer ZLC response curves (solid symbols) and theoretical Fickian curves (solid lines) for (A) water/ZSM-12 (dark green), toulene/ZSM-12 (blue), at 0.71 torr, 60°C, 50ml/min; (B) toluene/SAPO-5 (pink) at 0.71 torr, 30°C, 50ml/min, and acetylene/SAPO-5 (orange) at 285 torr, 30°C, 10ml/min.

To verify the existence of such additional distribution over diffusivities leading to an enhancement of deviations in the measured curves form the monoexponential behavior, tracer exchange experiments with water/ZSM-12 and acetylene/SAPO-5 were carried out. The results are also presented in Fig. 1A and B for water/ZSM-12 and acetylene/SAPO-5 systems, respectively in the same range of concentration decrease as toluene for direct comparison. It is evident from both figures that water/ZSM-12 and acetylene/SAPO-5 systems exhibit a long-time behavior close to mono-exponential, as expected for Fickian diffusion mechanism. However, it is somewhat surprising that no significant deviations form the monoexponential behavior due to channel length distribution was observed in this case. Nevertheless, an observation of larger deviations from the monoexponential behavior for toluene/ZSM-12 and toluene/SAPO-5 than for water/ZSM-12 and acetylene/SAPO-5 is in agreement with an assumption of the center-of gravity diffusion in the former two systems.

5. Conclusion

This study provides first-time TZLC evidence of the occurrence of anomalous, single-file diffusion in microporous sorbents.

6. References

[1] Hahn, K., J. Karger, and V. Kukla, Phys. Rev. Lett., , 76 (1996) 2762-2765.
[2] Jobic, H., et al., J. Phys. Chem. B, 101 (1997) 5834-5841.
[3] Iliyas, A. Zahedi-Niaki., M. H.; Eic, M., AIChE Annual Meeting, 2006.
[4] Hahn, K.; Kärger, J. J. Phys. Chem., 102 (1998) 5766-5771.
[5] Nelson, P. H.; Auerbach, S. M., J. Chem. Phys., 110 (1999) 9235-9243.
[6] Iliyas, A. Zahedi-Niaki., M. H.; Eic, M.; Kaliaguine, S. Micro. Meso. Mater., 102 (2007) 171-177.
[7] Brandani, S., Z. Xu, and D. Ruthven. Micro. Mater., 7 (1996) 323-331.

Applicability of water-activated carbon isotherm models to water-carbon nanotube isotherms

Pyoungchung Kim,[1] *Yijing Zheng,*[1] *Sandeep Agnihotri*[1,*]

[1]Environmental Engineering, Department of Civil and Environmental Engineering, University of Tennessee, Knoxville, TN. 37996-2010. *Corresponding author. Email: sagnihot@utk.edu. Phone: (865) 974–7728. Fax: (865) 974–2669

Abstract

The applicability of established water-activated carbon interaction theory to water-carbon nanotube interactions remains to be fully explored. We have conducted sensitive experiments of adsorption kinetics and equilibrium of water on purified and treated single-walled carbon nanotubes (SWNTs) procured from 5 commercial suppliers. Experiments were conducted using a customized set-up that consisting a gas generation system, high sensitivity gravimetric balance (detection limit = 0.1 μg) and a data acquisition system. Adsorption was measured in an open-system in which a carrier gas of a known quantity of vapor continuously flows over the sample while the sample weight is being monitored in real time. Isotherms for each sample (3 – 4 mg) were obtained point-by-point at approximately 20 intermediate concentrations between 2% to 95% relative humidity (RH) by administering adsorption kinetics and then extracting equilibrium capacity at each data point. Triplicate runs were conducted for statistical purposes. The isotherms for most samples were type V in shape. This is similar to water-activated carbon isotherms. This isotherm characteristic arises from water adsorbing onto functional groups and water adsorption by conventional pore filling. We have applied several isotherm models, such as the D.S. model and D.D Do model to isotherms obtained for all SWNT samples. These models account for the semi-chemisorption characteristic of water, and were originally developed for water adsorption in activated carbon. The D.D.Do describes water adsorption by deconvoluting experimental adsorption into separate contributions from adsorption on functional groups and micropore filling. The micropore volume by water adsorption was identical to that by standard N_2 (77 K) adsorption for all samples. It is emphasized that the application of LDF model to kinetics and D.D.Do model to water adsorption in carbon nanotubes is not tractable in the present literature. These are common water-activated carbon models and their applicability to water-nanotube system remains unexplored. Additionally, due to the employment of high-sensitivity analytical techniques this study can provide insightful examination on the realism of the methodology and inter-molecular potentials commonly used in molecular modeling water and nanotubes, which has become the most common tool for studying adsorption of water.

Introduction

Developing a fundamental understanding of water adsorption mechanisms in carbon nanotubes is essential. So far, several theoretical and experimental studies of water adsorption on CNTs have predicted interesting results, such as freezing of water inside nanotubes, reduced hydrogen bonding of water confined in the nanospace of nanotubes, filling and emptying of nanotubes by sequential addition or removal of a single-file chain of water molecules, and gating mechanism of water transport in carbon nanotubes, effects of different microporosities and degree of surface functionalities on water adsorption in single-walled carbon nanotubes (SWNTs). Experimental measurements of water-nanotube interactions have been carried out by nuclear magnetic resonance technique, Raman spectroscopy, Fourier Transform infrared spectroscopy (FT-IR). In contrast, experimental studies reporting water adsorption in carbon nanotubes by direct measurement of adsorbed mass (i.e., gravimetric estimation) are rare. Of most significance, Kaneko and co-workers reported gravimetric adsorption of water vapor in carbon nanohorns. They suggested a possibility of cluster-mediated mechanism for pore filling as opposed to the more conventional layer-by-layer adsorption. Recently, Kanellopoulos and coworkers also used an intelligent gravimetric analyzer: a combination of volumetric and gravimetric techniques, to obtain water vapor adsorption kinetics and equilibrium on SWNTs. Such analytical studies are essential to generate experimental data to complement theoretical results to determine, for example, the role of surface chemistry on adsorption capacity and reliability of kinetics parameters for transport of water through nanopores. In this work, we are reporting the development and application of an experimental setup that allows measurement of water uptake during adsorption-desorption equilibrium and kinetics with a resolution of 0.1 µg for commercially available carbon nanotubes (SWNTs) and activated carbons. It is necessary for an assessment of carbon nanotubes in competition with more versatile and less expensive activated carbons.

Methodology

SWNT samples selected for this study are commercially produced and were purified by the manufacturers. The sample description is as follows: EA95 (95% SWNTs, MER corp.); CS70 and CS80 (70% and 80% SWNTs, Carbon Solutions); AC (Calgon activated carbon) and ACF10 (non-woven activated carbon fiber, American Kynol).

Development and Experimentation in High-sensitivity Microbalance The apparatus used to measure the adsorption isotherms consists of a gravimetric balance (Thermo Cahn, Digital Recording Balance, DRB-200), a gas generation system and a data acquisition system in Figure 1. This equipment has 0.1 µg limit of detection and is capable of measuring weight in an open-system with continuous flow of gas. The system is operated in an environmental chamber with controlled temperature (± 0.5 °C) and relative humidity (± 1%), and vibration reduction designed especially for the balance. The gas generation system consists of ultra high purity N_2 in which water is injected by bubbling the gas through nanopure water, the concentration of which is measured by a relative humidity meter. Prior to each test, 200 sccm of UHP-N_2 flow is established in the balance. The balance is calibrated and zeroed. The adsorbent sample (2-3 mg) is placed on the sample pan and will be heated to over 140 °C to clean its surface off any volatile materials physically adsorbed during storage. The sample is cooled and its dry weight is recorded. The vapor laden gas is continuously passed through the balance such that the sample gains mass upon adsorption. The data acquisition system gathers mass, time and temperature data every 2 seconds during a test. Equilibrium is assumed when no change in mass is detected (mass change in 30 min ≤ 1 μg). The gain in sample mass is normalized with respect to the dry weight and will be reported as sample's adsorption capacity at the particular vapor concentration. The entire adsorption isotherm is thus obtained point-by-point by progressively increasing the concentration of water vapor in the gas stream and by measuring the equilibrium adsorption capacity at each vapor concentration. Desorption isotherm is also obtained in a similar manner, but with decreasing concentration of water vapors.

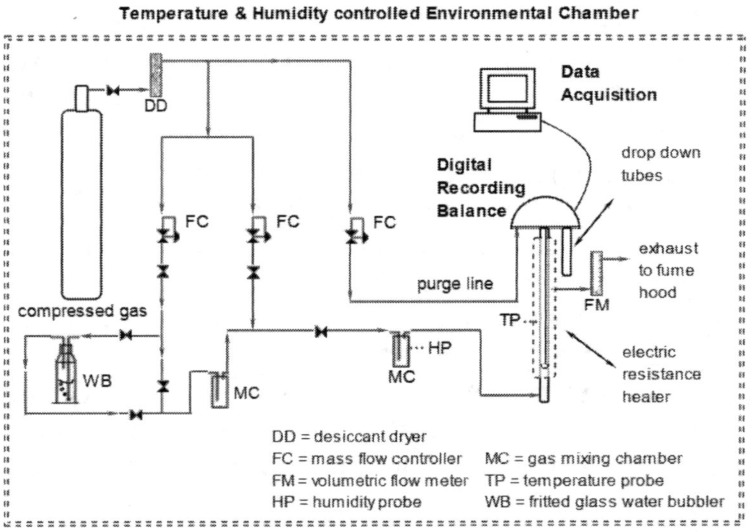

Figure 1. Experimental setup for adsorption of water vapor in carbon nanotubes.

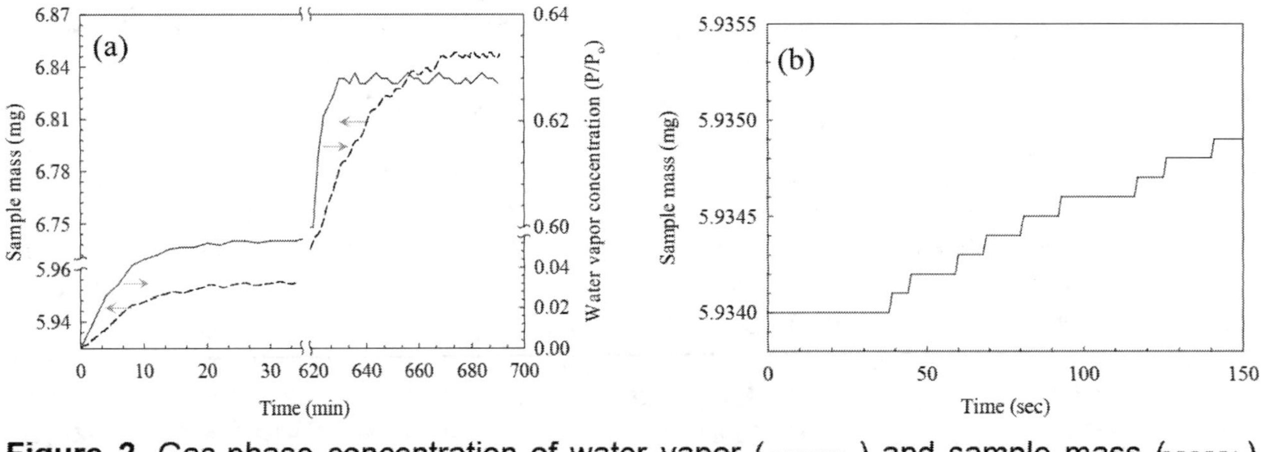

Figure 2. Gas-phase concentration of water vapor (———) and sample mass (- - - - -) measured as a function of time in a typical adsorption experiment.

Figure 2a represents a typical experiment in which both, sample mass as well as water vapor, is monitored as a function of time. It was observed that higher vapor concentrations (P/P_o > 0.15) reach the desired constant value much faster than lower concentrations. Since adsorption requires a constant gas phase concentration, this transient behavior of the water vapor concentration in carrier gas itself is taken into account in the adsorption kinetics analysis presented here. Figure 2b shows that the adsorption kinetics is measured with a 0.1 µg resolution after careful adjustments in the experimental protocol related to humidity in the environmental chamber housing the experimental setup. In the following text, we are reporting application of our experimental setup to determine water adsorption equilibrium and kinetics properties of carbon nanotubes, and comparison of those with the same of activated carbons. Since great amount of diversity exists in both carbon types, we have focused our discussion on identifying similarities and differences in adsorption trends for a better understanding of water-nanotube interactions as well as the practicality of information needed.

Results and Discussion

Kinetics of Water Adsorption. Adsorption kinetics is most useful in designing and assessing adsorbent beds. The LDF model was applied to the kinetics of water uptake at all equilibrium data points presented as following figure for water adsorption isotherms. The LDF model is described as shown in right. Here, M_t is the mass uptake in time t, M_e is the equilibrium uptake for the given pressure increment, and k is the rate constant. As a representation of the data fitting, figure 3 compares the experimentally measured mass versus time data with that calculated by the LDF model for both, adsorption as well as desorption. A near perfect fit to the experimental data is observed in all cases. It should also be noted that the experimental data presented here is complementary to that presented in figure 2 where the relevance of high-sensitivity kinetic measurements with less than 0.5 µg fluctuations is apparent. Adsorption and desorption kinetics in figure 3 shows that the LDF model is applicable to the data with approximately 95% accuracy and prediction of sample mass within ± 0.0035 mg of experimental data. This demonstrates the applicability of the LDF model to carbon nanotubes, and the accuracy of kinetics parameters calculated from the adsorption data obtained from the experimental setup.

$$\frac{M_t}{M_e} = 1 - e^{-kt}$$

The rate constant, k, for adsorption as well as desorption for all test samples was calculated by the LDF model. Representative results for sample EA95 and ACF10 are presented in figure 4. Sample EA95 showed no kinetic rate constant values up to 0.4 P/P_o due to little to no uptake of water, followed by sharply increasing rate constants up to 0.6 P/P_o most likely due to capillary condensation and then decreased up to 0.94 P/P_o due to pore filling. In contrast, the general trends in adsorption rate constants were much smoother in sample ACF10 with clear distinction of two peaks representing adsorption on primary sites and capillary condensation. This indicates that of all samples tested, ACF10 is relatively a more homogenous adsorbent than carbon nanotubes. The more random dependence of k on P/P_o for carbon nanotubes might be attributed to the impurities in nanotube samples. We expect the contributions of impurities to be lumped into the data provided here. Our experimental technique is not capable of differentiating water adsorption on nanotubes from that on impurities in nanotube samples; however, we can predict that adsorptive contributions of impurities might be separated by modeling impurities as non-porous carbon the water adsorption characteristics of which will obviously be much different from those of porous carbons.

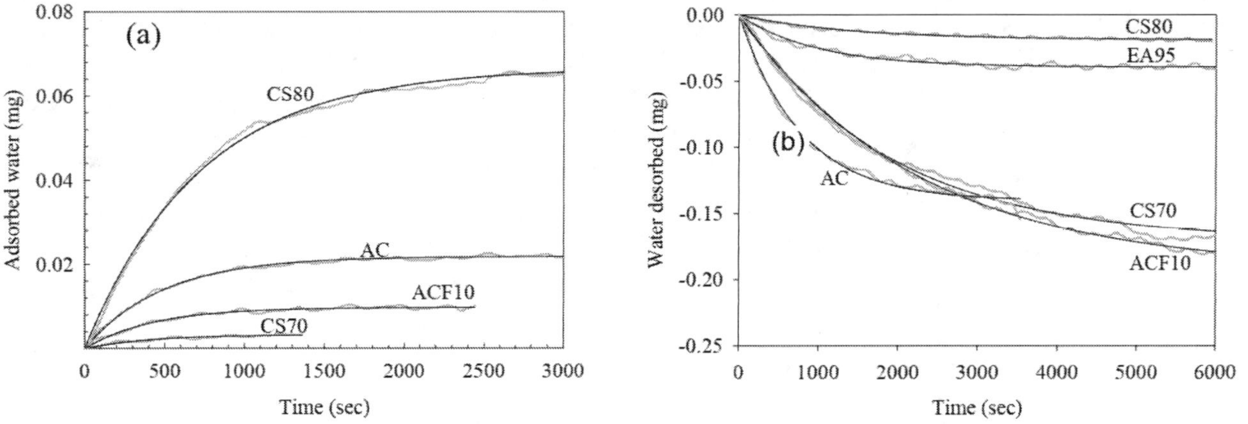

Figure 3. Kinetics of water vapor (a) adsorption at P/P_o = 0.05 and (b) desorption at P/P_o = 0.5 from a samples equilibrated at P/P_o = 0.6. The experimental data is presented as grey lines. Smooth line represents the kinetics modeled by LDF equation.

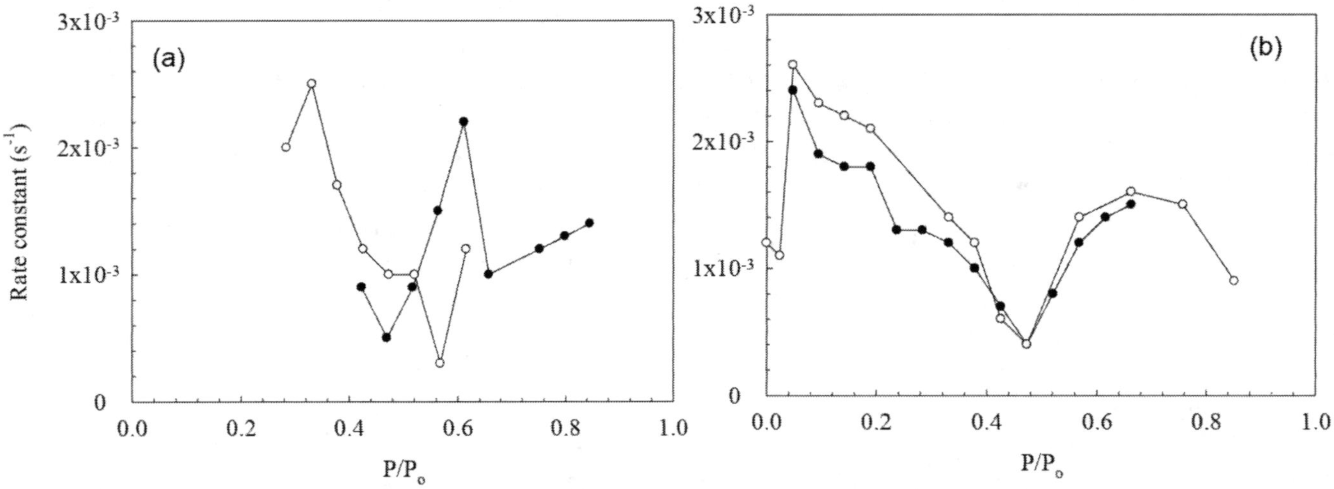

Figure 4. Rate constants, *k*, of water adsorption (•) and desorption (○) calculated from the LDF model fitted to the adsorption kinetics at 20 intermediate points from 0 to 0.94 P/P_o. Sample (a) EA95 (b) ACF10.

Water adsorption equilibrium. Isotherms of water vapor adsorption and desorption on carbon nanotube and activated carbon samples were measured at P/P_o between 0 and 0.94 (Figure 3). Sample EA95 showed a slow and an almost insignificant rise in adsorption capacity until 0.4 P/P_o, a sharp increase between 0.4 and 0.6 P/P_o followed by a slow rise in adsorption capacity from 0.6 to 0.94 P/P_o. Therefore, majority of water vapor uptake by this sample would take place in the 40% to 60% relative humidity range (relative humidity = 100 x P/P_o). The activated carbon sample ACF10 exhibited a similar isotherm characteristic with a slow increase until 0.3 P/P_o, a shape increase between

0.3 and 0.6 P/P_o followed by only a marginal increase in adsorption. This adsorption isotherm is a type V characteristic, which is typical for water adsorption on microporous carbon adsorbents. Samples CS70 and AC also exhibited type V characteristics. However, sample CS80 showed an abrupt increase at the initial low relative pressure ($P/P_o < 0.05$) and a monotonous increase in adsorption capacity up to 0.94 P/P_o with no apparent sharp uptake of water. This is either a type IV or type V characteristics. The morphology of nanotubes in this sample is similar to that of sample CS70. It has been reported that functional groups can affect the starting relative pressure and shape of the isotherms in the carbon adsorbents. Therefore, the ambiguity of adsorption type for this sample is most likely due to a much higher concentration of functional groups that would shift the isotherm rise to a lower P/P_o and change the characteristics to that of an increasingly steeper type IV isotherm.

Adsorption isotherms were fitted to the Dubinin-Serpinski (DS) model that is developed for the interpretation of water uptake data (eq. 2). Here, S_o (mmol/g) is the number of primary adsorption centers, which increases with the degree of oxidation of carbon surface (typical $0.05 < S_o < 5$ mmol/g), c (unitless) is a ratio of the desorption rate to the adsorption rate and k (g/mmol) is a parameter representing the loss of secondary sites upon water adsorption. It affects the maximum amount of water that can adsorb on a sample; it does not, however, affect the initial adsorption behavior. DS equation was selected as a model appropriate starting point for comparison of nanotubes with activated carbons because of lesser number of parameters to be fitted in data modeling, and the ability of the model to *quantify* the surface chemistry relevant to water adsorption.

$$\frac{P}{P_o} = \frac{C_\mu}{c(1 - kC_\mu)(S_o + C_\mu)}$$

The D.D.Do model describes water adsorption isotherm based on the growth of the water cluster at the functional groups and the penetration of water clusters into the micropore in the form of pentamer (5 molecules, approx. 1 nm), which has sufficiently high dispersion energy to remain in the micropore. The D.D.Do model deconvolutes total water adsorption isotherm into adsorption by functional groups (first term on rhs) and that by micropore filling (second term on rhs). In this model, the BET equation describes the adsorption of water by functional groups. S_o is the functional group concentration (mmol/g), $C_{\mu s}$ is the saturation concentration in the micropore (mmol/g), K_f and $K\mu$ are the equilibrium rate constants for, respectively, chemisorption of water on functional groups and water filling of micropore, and n is the average number of molecules in the main cluster (unitless).

$$C_\mu = S_o \frac{K_f \Sigma_{i=1}^{n} i * \left(\frac{P}{p_o}\right)^i}{1 + K_f \Sigma_{i=1}^{n} \left(\frac{P}{p_o}\right)^i} + C_{\mu s} \frac{K_\mu \left(\frac{P}{P_o}\right)^6}{K_\mu \left(\frac{P}{P_o}\right)^6 + \left(\frac{P}{P_o}\right)}$$

All isotherms were fitted to both models. The fitting parameters are presented in Table 2 and the fit to the experimental data is presented as continuous lines in figure 5. It was observed that the D.D.Do model satisfactorily fitted the experimental isotherms for all samples, both nanotubes and activated carbons. This model could also deconvolute total isotherm into adsorption on functional groups and in the pores. It was noted that adsorption on functional groups does not saturate, rather water continues to simultaneously adsorb on functional groups while it is filling the micropores. The D.S. model also fit all sample; however, it appeared less compatible with SWNT sample EA95 and activated carbon sample ACF10. A closer examination of fitting parameters (S_o) showed an extremely hydrophobic surface for these samples, which would explain their incompatibility with the surface-chemistry mediated theory of water adsorption.

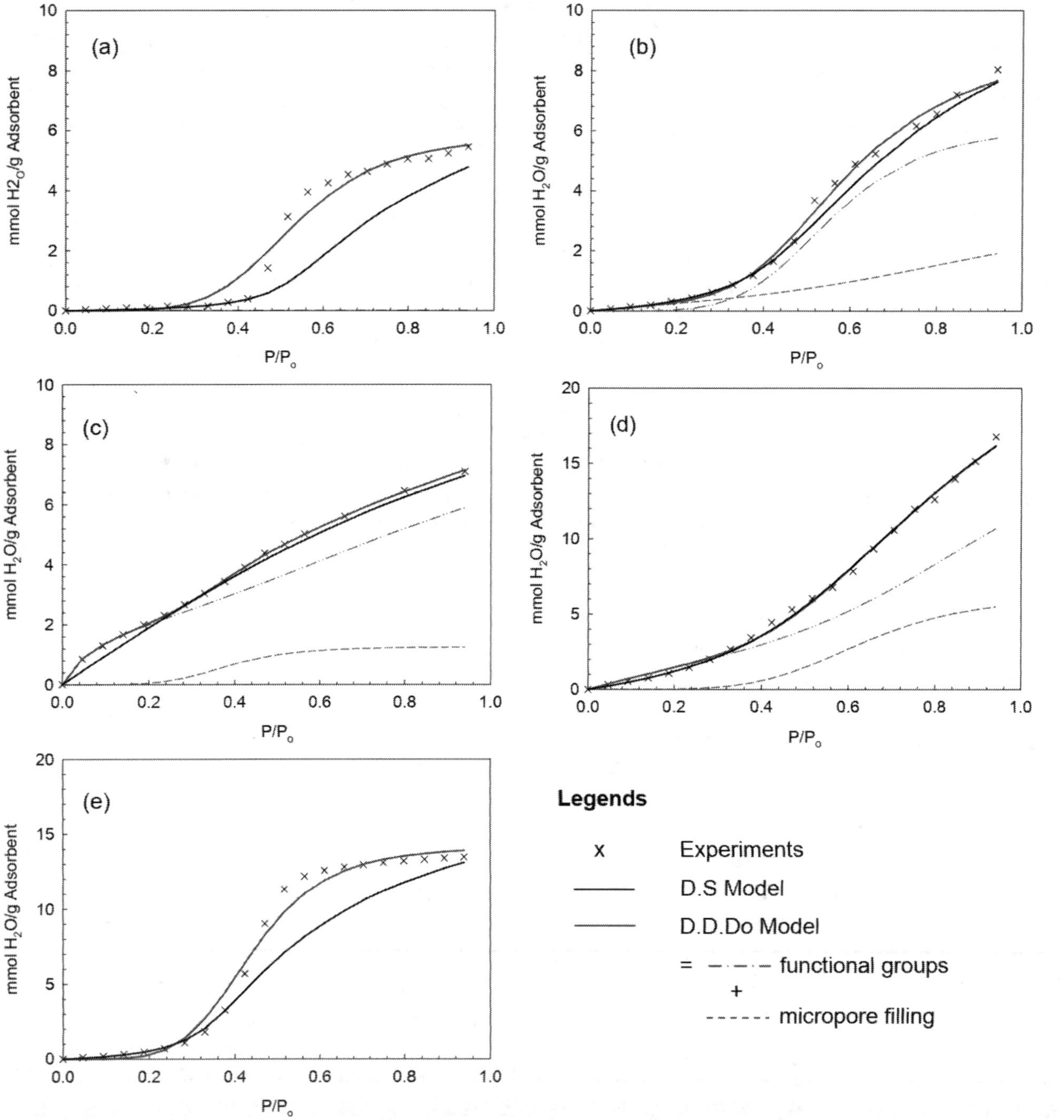

Figure 5. Water vapor adsorption isotherms and fits to common isotherm models for carbon nanotube samples (a) EA95, (b) CS70, (c) CS80, and activated carbon (d) AC and (e) ACF10.

Interestingly, the values of S_o calculated for all samples was nearly the same from both DS and D.D.Do models. This is an important observation because the former has only 3 parameters while the latter has 6 parameters which should affect their values obtained from the nonlinear-regression method used in curve fitting. Thus, it could be said that the D.D.Do model can be used for quantification of sample surface chemistry with credibility. Furthermore, the micropore volume of samples by water adsorption ($C_{\mu s}$) was less than or equivalent to that determined by standard N_2 adsorption at 77K, which implies that in SWNTs water would adsorb only inside the nanotubes and not on the external surface of the bundles. Exception was observed for sample CS80 due mainly to that fact that this sample has a much higher concentration of functional groups (manufacturer provided information) which was also apparent from the shape of its water adsorption isotherm. For the same sample, it was observed that the average size of water molecules cluster (parameter n) was smaller than that for sample CS70. It is emphasized that the cluster size depends on the average pore size and the concentration of hydrophilic functional groups. The diameter of SWNTs in both samples, CS70 and CS80, are exactly the same as they are extracted from the same as-produced materials. Thus, it could be said that increasing the concentration of functional groups would result in formation of smaller average cluster sizes on the primary sites.

Table 2. Fitting parameters for water isotherm models.

Sample	$C_{\mu s}$ (cm^3/g)		S_o (mmol/g)		K_f	K_μ	n	m
	Water	N_2	DS	D.D.Do				
EA95	0.10	0.16	0.07	0.07	0$^\Psi$	22.10	6	5
CS70	0.11	0.15	0.51	0.55	2.10	20.14	8	5
CS80	0.02	0.00	>5	2.10	13.27	114.70	5	5
AC	0.11	0.18	2.75	2.70	2.96	9.73	8	5
ACF10	0.26	0.24	0.44	0.40	0$^\Psi$	60.39	5	5

$^\Psi$ K_f for these samples was almost negligible (10^{-10} to 10^{-15})

The trends in surface chemistry estimated by fitting water adsorption isotherms were also observed directly from surface analysis by Raman spectroscopy (Figure 6). The spectra were obtained by Nicolet 6700 FT-IR-Raman by Thermo Fisher. The excitation λ is in the near infrared region (λ= 946nm). The instrument is fitted with a liquid N_2 cooled high sensitivity Ge detector. It was operated at a power rating of 0.06 W (less than 5% of maximum power) to minimize sample heating which could be one major drawback of a non-dispersive Raman technique. Samples were prepared by grinding approximately 0.05% (w/w) adsorbent in KBr powder. Spectra were collected at 10 different points on the pellet. The spectra were corrected by subtracting blanks. Average of 10 spectra is presented in figure 6. It is known that for carbon samples the peak originating at approximately 1,350 cm^{-1} Raman shift (D peak) is related to the sp^2-sp^3 covalently bonded C and that at 1,580 cm^{-1} Raman shift (G peak) represents the pristine sp^2-sp^2 graphitic structure. The I_D/I_G intensity ratio is commonly used to estimate the degree of functionalization. Sample EA95 exhibited an extremely low I_D/I_G ratio among all samples, similar to the much lower values of S_o estimated from water vapor modeling. The general trend in S_o by water adsorption were observed to be similar to that in the I_D/I_G Raman ratio for both, CNTs as well as activated carbon.

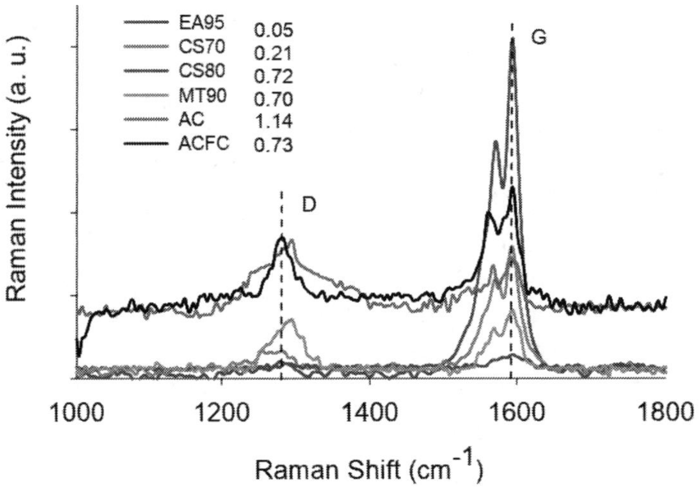

Figure 6. Raman spectra of carbon nanotubes and activated carbon samples. Shoulder peak at 1550 cm^{-1} is due to resonance from metallic nanotubes. It is not included in the I_D/I_G ratio presented in the legend.

Another interesting similarity between the characterization results obtained from the two non-complementary techniques of water adsorption and Raman scattering is noted when comparing the first point of inflexion in isotherms with the I_D/I_G ratio. Each isotherm has two points of inflexion which divides the isotherm into three regions: water uptake on primary adsorption sites, simultaneous water adsorption by capillary condensation and functional groups, and pore saturation. The first point of inflexion occurs when adsorption on hydrophilic functional group is just saturated enough that further increase in water vapor concentration would result in water clusters migrating into the pores i.e., initiation of capillary condensation. It can be noticed that P/P_o at which this phenomenon occurs is inversely related to the I_D/I_G ratio. This is because increasing I_D/I_G value also represents increasing hydrophilicity. This would exhibit significant adsorption at lower vapor concentration; thus, reaching the point of inflexion earlier in the isotherm. This also explains why the adsorption isotherm type appears ambiguous for sample CS80, which is an extremely hydrophilic sample. The I_D/I_G ratio represents total functionalization, i.e., hydrophilic and hydrophobic. Therefore, the similarity of trends in S_o and P/P_o at inflexion does indicate the application of a common water-carbon adsorption model, such as D.D.Do equation, to satisfactorily quantify the nanotubes' surface chemistry of relevance to water adsorption.

Summary and Conclusions

Adsorption equilibrium and kinetics of water vapor were measured gravimetrically on several commercially available carbon nanotubes and activated carbon. An experimental set up is developed that allows real-time measurement of water vapor adsorption kinetics with a resolution of 0.1 μg. Water adsorption in SWNTs followed typical type V adsorption isotherms. The quantification of hydrophilicity of samples was performed by fitting the isotherms to the Dubinin-Serpinski model. The information extracted could be related to sample surface characterization by Raman scattering. The kinetics data was fitted to the linear-driving force model. It is emphasized that application of LDF model and D.D.Do model to water adsorption in carbon nanotubes is not tractable in the present literature. These are common water-activated carbon models and their applicability to water-nanotube system remains unexplored. Here, we found that water adsorption characteristics of nanotubes are similar to that of activated carbons, and the common theoretical models for water-carbon theory can be extended to water-carbon nanotubes. Precise measurement of water vapor adsorption on small sample sizes, as reported here, is attractive especially due to the availability of novel adsorbent materials in limited quantities. Additionally, due to the employment of high-sensitivity analytical techniques this study can provide insightful examination on the realism of the methodology and inter-molecular potentials commonly used in molecular modeling water and nanotubes, which has become the most common tool for studying adsorption of water.

28e Carbon-Silica Composite Adsorbent: Characterization And Adsorption Of Light Gases

T. Grant Glover[1], Kamini I. Dunne[2], Robert J. Davis[2], and M. Douglas LeVan[1]. (1) Department of Chemical Engineering, Vanderbilt University, VU Station B 351604, 2301 Vanderbilt Place, Nashville, TN 37235-1604, (2) Department of Chemical Engineering, University of Virginia, 102 Engineers' Way, P.O. Box 400741, Charlottesville, VA 22904-4741

A novel carbon-silica composite has been synthesized from MCM-41 and polyfurfuryl alcohol and characterized via X-ray diffraction and nitrogen adsorption. The adsorption and X-ray diffraction studies of the parent MCM-41 and the novel material are contrasted with a carbon derived from pure polyfurfuryl alcohol. The novel material has a high surface area, 850 m^2/g, and a narrow pore size distribution centered at approximately 5 angstroms. The novel material is unique when compared to traditional activated carbons. The pores contained in the novel material are assembled using templating techniques and are half the size of pores contained in traditional carbons. Room temperature adsorption isotherms for the novel material and MCM-41 are presented for nitrogen, carbon dioxide, methane, and ethane.

28f Experimental And Theoretical Studies Of Gas Adsorption In $Cu_3(Btc)_2$

Jinchen Liu[1], Karl Johnson[1], Jeffery Culp[2], Sittichai Natesakhawat[3], Bradley Bockrath[2], S. G. Sankar[4], Brian Zande[4], and Giovanni Garberoglio[5]. (1) Department of Chemical Engineering, University of Pittsburgh, 1249 Benedum Hall, Pittsburgh, PA 15261, (2) NETL, 626 Cochrans Mill Road, Pittsburgh, PA 15236, (3) National Energy Technology Laboratory, P. O. Box 10940, Pittsburgh, PA 15236, (4) Advanced Materials Corporation, 850 Poplar Street, Pittsburgh, PA 15220, (5) Dipartimento di Fisica dell', Universita di Trento, Trento, Italy

Metal-organic frameworks (MOFs) are interesting materials for studying gas adsorption because they can be made in high purity, are highly crystalline, and have a very narrow range of pore sizes with virtually all of their pore volume falling within the IUPAC microporous regime. CuBTC [$Cu_3(BTC)_2$, where BTC=1,3,5- benzenetricarboxylate] (also known as HKUST-1) has garnered a good deal of attention since it was first reported by Chui et al.[1] We have improved the activation process for CuBTC by extracting the DMF solvated crystals with methanol; we identify material activated in this way as CuBTC-MeOH. This improvement allowed the activation to be performed at a much lower temperature, thus greatly mitigating the danger of reducing the copper ions. A review of the literature for H_2 adsorption in CuBTC shows that the preparation and activation process has a significant impact on the adsorption capacity, surface area, and pore volume. CuBTC-MeOH exhibits a larger pore volume and H_2 adsorption amount than any previously reported results for CuBTC. We have performed atomically-detailed modeling to complement experimentally measured isotherms. Quantum effects for hydrogen adsorption in CuBTC were found to be important at 77 K. Simulations that include quantum effects are in good agreement with the experimentally measured capacity for H_2 at 77 K and high pressure. We have studied the effects of framework partial charges on H_2 adsorption isotherms. Inclusion of charge-quadrupole interactions improves the agreement between the simulations and experiments at low pressures. We have compared the adsorption isotherms from simulations with experiments for H_2 adsorption at 77, 87, 175, and 298 K, nitrogen adsorption at 253 and 298 K, and argon adsorption at 298 and 356 K. Reasonable agreement was obtained in all cases. We have calculated the self and transport diffusivities of several different gases in CuBTC. The computed diffusivities are similar to those observed for gases in silicalite.

[1] Chui, S.S.Y.; Lo, S.M.F.; Charmant, J.P.H.; Orpen, A.G.; Williams, I.D. *Science* **1999**, *283*, 1148.

28g Length Exclusion in the Adsorption of Chain Molecules on Chabazite Type Zeolites

Inge Daems[1], Ranjeet Singh[2], **Gino V. Baron**[1], and Joeri F.M. Denayer[1]. (1) Dept. Chemical Engineering, Vrije Universiteit Brussel, Pleinlaan 2, Brussel, B-1050, Belgium, (2) Chemical Engineering, Monash University, Clayton, VIC 3800, Australia

In the adsorption of linear C-1-C-8 alkanes, alkenes and alcohols on zeolite chabazite, molecules smaller than 6.7 angstrom are adsorbed in significant amounts, whereas longer chains are almost fully excluded from the micropores.

35a Membranes for Energy Efficiency and Sustainability

Keith Murphy, Prism Membranes, Air Products and Chemicals Inc., 11444 Lackland Road, Saint Louis, MO 63146

Membranes applied to separation and purification have and can contribute significantly to more efficient processes in a wide variety of industries. In addition, significant opportunities and challenges remain, where improved membrane technologies could make further contributions to energy efficiency and sustainability. This tutorial will use a range of examples to demonstrate utility and potential.

35b Electrolytes For Low And Intermediate Temperature Fuel Cells: Their Properties And Interplay With Other Components

Thomas Zawodzinski and Hossein Ghassemi. Chemical Engineering, Case Western Reserve University, A.W.Smith Building, 10900 Euclid Avenue, Cleveland, OH 44106

Polymer electrolytes are the central feature of the eponymous PEM fuel cells. However, a variety of system and other aspects play into what is required of these materials. As systems are developed, specific, application-driven requirements have revealed shortcomings of existing materials. In the face of these shortcomings, at least two possible tacks can be taken: one can develop new and improved materials or one can 'engineer around' the properties of existing materials.

This tutorial will provide an overview of the features of several available electrolyte types together with a critical discussion of their chemistry and physics. These will be put into the context of the requirements of the cell and system to provide additional guidance on electrolyte choices. In particular, we will consider several case studies.

In addition to a critical 'taxonomy' of electrolytes, some topics that will be touched on include an evaluation of the relative importance of electrolyte conductivity to cell performance, limitations due to cross-over of reagents, system start-up, mechanical requirements and material durability in use. Emphasis will be placed on presenting a feeling for the trade-offs inherent in engineering fuel cell systems.

Finally, recently proposed electrolyte types will be critically reviewed in light of all of the above.

35e New Membrane Separations in Biotech Processes

Willem Kools, Millipore Corporation, 900 Middlesex Turnpike, Billerica, MA 01821

Membranes are used throughout the manufacturing process of monoclonal antibodies. This presentation in the membrane tutorial section will highlight some recent developments in which new membranes are explored to improve process economics and address regulatory demands.

First, high performance tangential flow filtration will be discussed as an example of how a standard method of seperation can be enhanced by membrane modification.

Secondly, the use of membranes as chromatography media will be discussed. Rather than relying on size exclusion as is typically done for membrane operations, the convective flow through pores is used to more effectively utilize all the surface area of membranes. Some new devlopments will be highlighted here.

35d Novel Hybrid Distillation-Membrane Analysis Using Thermally Coupled Columns

Mark Peters[1], Shehzaad Kauchali[2], Diane Hildebrandt[3], and David Glasser[3]. (1) Wits University, Private Bag x3, PO Wits, Johannesburg, 2050, Johannesburg, 2050, South Africa, (2) Centre of Materials and Process Synthesis, Wits University, Private Bag x3, PO Wits, Johannesburg, 2050, Johannesburg, 2050, South Africa, (3) Centre of Materials and Process Synthesis,, Wits University, Private Bag x3, PO Wits, Johannesburg, 2050, Johannesburg, 2050, South Africa

A continuous membrane column involves the recycle of some permeated material, and operates in a counter-current manner. The assumption of constant molar overflow is not applicable here. To date, membrane columns have yet to find their place in industrial practice, even for simple configurations. The most probable reason for this is that the design of a membrane column is complex. Any membrane column, no matter how complex its configuration, can be divided into column sections. The Difference Point Equation (DPE) has previously been derived for constant molar overflow counter-current column sections. A more general form of the DPE has been derived to describe to retentate composition down the length of a section. For demonstration purposes, a simple Knudsen membrane is assumed. Using this as the permeation model, in conjunction with the DPE, allows one to graphically plot a column profile for a given initial condition. This allows one to not only configure simple columns, but complex configurations too. As an example, a thermally coupled membrane column will be discussed, using the same membrane throughout. Through simplifying assumptions, the system can be reduced to coupled column sections. Material balance constraints have to be satisfied in order for the configuration to be feasible. It was found that a thermally coupled membrane column is in fact not possible under any operating conditions. Although not a positive one, this is a very useful result, since both time and money can be saved by avoiding any further investigations. Another more useful and practical design is that of a membrane thermally coupled to a distillation column. In much the same way as before, this hybrid process was analyzed. Under certain conditions, there exists a region which makes such a configuration viable. An understanding into the behavior of both the membrane as well as the distillation profiles will allow one to avoid poor initial decisions and designs. Furthermore, this method introduces a novel approach to analyzing hybrid processes.

35c Carbon Nanotube Membranes

Bruce J. Hinds, Chemical and Materials Engg., University of Kentucky, 177 Anderson Hall, Lexington KY, KY 40506-0046

Carbon nanotubes have three key attributes that make them of great interest for novel membrane applications 1) atomically flat graphite surface allows for ideal fluid slip boundary conditions 2) the cutting process to open CNTs inherently places functional chemistry at CNT core entrance and 3) CNT are electrically conductive allowing for electrochemical reactions and application of electric fields gradients at CNT tips. Towards this goal, a composite membrane structure containing vertically aligned carbon nanotubes passing across a polystyrene matrix film have been fabricated. Fabrication steps, material characterization and ionic diffusion transport properties are described. Plasma oxidation during the fabrication process introduces carboxylic acid groups on the CNT tips that are modified using carbodiimide mediated coupling between carboxylic acid on the CNTs and accessible amine groups of the functional molecule. The entrances to CNT's cores were thus functionalized with aliphatic amines of different lengths, charged dye molecule and an aliphatic amine elongated by spacers containing poly-peptides and the simultaneaous permeation of two differently sized but equally charged molecules was studied. The relative selectivity of ruthenium bi-pyridine [Ru-(bipy)3+2] and methyl viologen [MV+2] was seen to vary from 1.9 to 3.6 as a function of tip-functionalization chemistry Anionic charged functional groups are seen to sharply increase flux of cationic permeates. This effect is reduced at higher solution ionic strength consistent with shorter Debye screening length screening attractive charge at the CNT core entrance. Using a hindered diffusion to model observed selectivities was consistent only with a geometry of only CNT tip functionalization, not along the length of CNT core. Bio-chemical gating of CNTs is also seen by tethering desthiobiotin to CNT tips with the reversible binding to streptavidin. The complete ATP cycle (phosphylation/dephosphylation) can be performed on CNT tips with corresponding modulation of flux across CNT membrane. Strong electrostatic effects of binding protein are seen with enhanced cationic flux seen for the relativel open anioic protein binding at CNT tip entrance. The functional density of tethered charge molecules can be substantially increased by the use of electrochemical grafting of diazonium salts. Functionality can be forced to occur at the CNT tip entrances by fast fluid flow of an inert solvent through the core during electrochemical functionalization. The selectivity between Ru(bi-pyridine)32+ and methyl viologen 2+ flux is found to be as high as 23 with -130mV bias applied to the membrane with tethered anionic dye molecule. Changes in the flux and selectivity support a model where charged tethered molecules at the tips are drawn into the CNT core at positive bias hindering/gating flux across the membrane. Application towards controlled transdermal drug delivery are discussed. In general, the transport mechanisms through CNT membrane are a) ionic diffusion is near bulk expectation with no enhancement from CNT b) gas flow is enhanced by ~1-2 order of magnitude due to specular reflection off of flat graphitic surface c) and pressure driven flux of a variety of solvents (H2O, hexane, decane ethanol, methanol) are 4-5 ORDERS OF MAGNITUDE FASTER than conventional Newtonian flow due to atomically flat graphite planes inducing nearly ideal slip conditions. This workshop will focus on the current state of the art in the reported literature on Carbon Nanotube membranes, fundamental transport mechanisms, and chemical modification design strategies for separation applications.

Improving Efficiency and Operation of Heat and Mass Transfer Equipment

Andrei Merenov (asmerenov@dow.com), Dan Tirtowidjojo (DTirtowidjojo@dow.com)

The Dow Chemical Company

Freeport, TX

The integration of heat and mass transfer within one unit operation is an efficient way to design a process. This approach leads to significant capital and energy savings. However, design and operation of such unit operations is not very well understood. In some cases, these unit operations are designed under the assumption of a "limiting factor" when either mass transfer or heat transfer alone is considered. Two cases from industrial experience will be discussed. In both cases neglecting simultaneous heat and mass transfer led to product losses and operating problems.

In the first case, an absorber for a strongly non-ideal liquid was operated without considering mass transfer limitations. The operating efforts were focused solely around removing the significant heat of absorption. A model taking into the account simultaneous heat and mass transfer in the absorber was developed. This approach allowed us to improve operation of the absorber, prevented significant material losses and eliminate potential environmental liability.

In the second case extensive tray damage was observed during operation of a stripping column with a built-in condenser. A dynamic model for the column and the built-in condenser operation was developed. The model identified a problem with the column start up procedure that did not take into account the built-in condenser dynamics. An improved start up procedure considering combined dynamics of the column and condenser was developed which resolved the problem.

These cases demonstrated that quite often the complex nature of the unit operations dealing with simultaneous heat and mass transfer can not be ignored.

Case 1 – Absorption of Strongly Non-Ideal Mixture in a Falling Film Absorber

Background

A vapor – liquid absorption was performed in an industrial-size falling film absorber. The absorbed component was used as a homogeneous catalyst in a downstream process. The falling film absorber flow diagram is presented in Fig.1.1. The vapor stream is introduced at the top of the absorber. Fresh absorbent and recycled absorbent are also introduced at the top of the absorber. The vapor liquid mixture travels downwards co-currently in the tubes of the falling film absorber. The thermodynamic equilibrium in the system is well studied and highly non-ideal. The significant heat of the absorption is removed using a cooling liquid circulating in the falling absorber shell. The non-absorbed vapor is vented to a scrubber (not shown) where the component is recovered by the means of chemical absorption. The rich absorbent accumulates in a storage tank and is recycled to the falling film absorber inlet. According to the downstream process requirements, the concentration of the absorbed component in the absorbent should be in a certain limits. This defines the fresh absorbent/vapor ratio for the process.

A relatively large amount of the component was not absorbed in the unit. This led to the raw material losses, the losses of chemical absorbent in the down stream scrubber and created a bottleneck in the operation of the plant. Several attempts to increase absorber efficiency were made. The most successful process change was an increase in process pressure to its maximum allowable value. Although this improved the process performance the efficiency of the absorption unit did not reach the design value.

Several hypotheses were developed to explain the low performance: high amount of inert components in the vapor feed, liquid maldistribution in the falling film absorber, fouling of the tubing. The analysis of the inlet stream confirmed the presence of less than 4% inert components; thus negating the first hypothesis. In order to verify the second and the third hypothesis the plant shutdown would be required which would be extremely expensive. Before proceeding with the verification of these two hypotheses a decision to investigate alternative hypothesis was made.

Modeling of Falling Film Absorber

An alternative hypothesis is that the poor mass transfer in the absorber is the problem. The absorber was designed using rigorous heat transfer model available in commercial HTRI^{TM} software. The model was used under the assumption that the mass transfer is not a limiting step in the process. In order to test our hypothesis, an ASPEN Plus model was developed using standard blocks available in the software. The flow diagram of the model is shown in Fig.1.2. fresh absorbent, F-ABS, vapor containing component to be absorbed and inert components, VAPOR, and recycling absorbent, RCCM. The mixer block, MIX1, combines the resulting vapor-liquid mixture, FEEDCMM, is going to heater block ABSHM. The heater block is modeled using heat transfer coefficient obtained from HTRI software. The cooled vapor liquid mixture, ABSCOLDM, is sent to the flash block, ABSCM, where vapor and liquid are separated at the adiabatic conditions. The separated liquid, L-S1CM, is sent to separator block EFF. In this block part of the absorbed components separate. This block takes into the account the mass transfer efficiency of the absorber. The calculations used in this block will be discussed below. The components resulting from the separation in the block are sent to mixer MIX2 where they mixed with vapor stream from flash ABSCM, V-S1CM. The rest of the rich absorbent, R-ABS, is sent to splitter RSPLIT where part of the stream is recycled back – stream RCM. Stream RCM is sent to heater block ENVHT which takes into account heat transfer with environment. Finally the recycled absorbent, RCCM, returns to the beginning of the process.

The model includes the effect of heat transfer by using ASPEN Plus HTRX model, ABSHM, and mass transfer efficiency ASPEN Plus model SEP, EFF. The following calculations were implemented in ASPEN Plus calculator block and applied to block EFF:

$$\eta = \frac{Y_{out} - Y_{in}}{Y_{eqv} - Y_{in}}$$

$$Yout = Y_{in} + \eta(Y_{eqv} - Y_{in})$$

$$M'_{HCl} = \frac{Y_{out} \quad M^{*}_{total} - M^{*}_{A}}{1 - Y_{out}}$$

$$Fraction \quad to\ return = \frac{M'_{A}}{M_{A\_L}}$$

Where η is mass transfer stage efficiency, Y_{in} is component A fraction in incoming vapor, Y_{eqv} is equilibrium component A vapor fraction, M^*_{total} is mass of equilibrium vapor flow, M^*_A is mass of component in equilibrium vapor flow, M'_A is mass of A to return to vapor phase, and M_{A_L} is mass of absorbed component A.

Plant Data Analysis

These calculations allowed us to introduce the concept of mass transfer stage efficiency in the ASPEN Plus heat transfer calculations using available software models. Five representative plant data sets were used to validate the model. The results of the validation are shown in Table 1.1. The results of calculations using just ASPEN flash and ASPEN heater block are also presented for comparison. As it can be seen from the data, the new model predicts the amount of the vent flow from the falling film absorber significantly better than the models just based on equilibrium calculations.

This demonstrated that mass transfer plays significant role in the performance of the falling film absorber. This conclusion brought several new opportunities to improve the performance of the absorption process. One of them is to increase the absorption factor by increasing liquid to gas ratio L/G. The amount of fresh absorbent is defined by the specification on the concentration of absorbed component and can not be varied. On the other hand, the increase of the L/G ratio can be achieved by an increase of the absorbent-rich recycle. The historical plant data at high pressure were analyzed and recommended new increased L/G ratio. The fraction of absorbed feed vapor stream versus L/G ratio at the historical and recommended plant operation regimes is shown in Fig 1.3. At the new regime, the vent stream contains primarily inert components and only traces of the component to be absorbed. This simple change in the operating parameter solved the problem.

Conclusion

The consideration of simultaneous heat and mass transfer in the falling film absorber led to identification of the performance problem of the unit operation. It also facilitated the identification of the problem solution. This example illustrated the positive results of such approach to the analysis of existing falling film absorbers. It would be also interesting to consider the expansion of the approach to the design of new units.

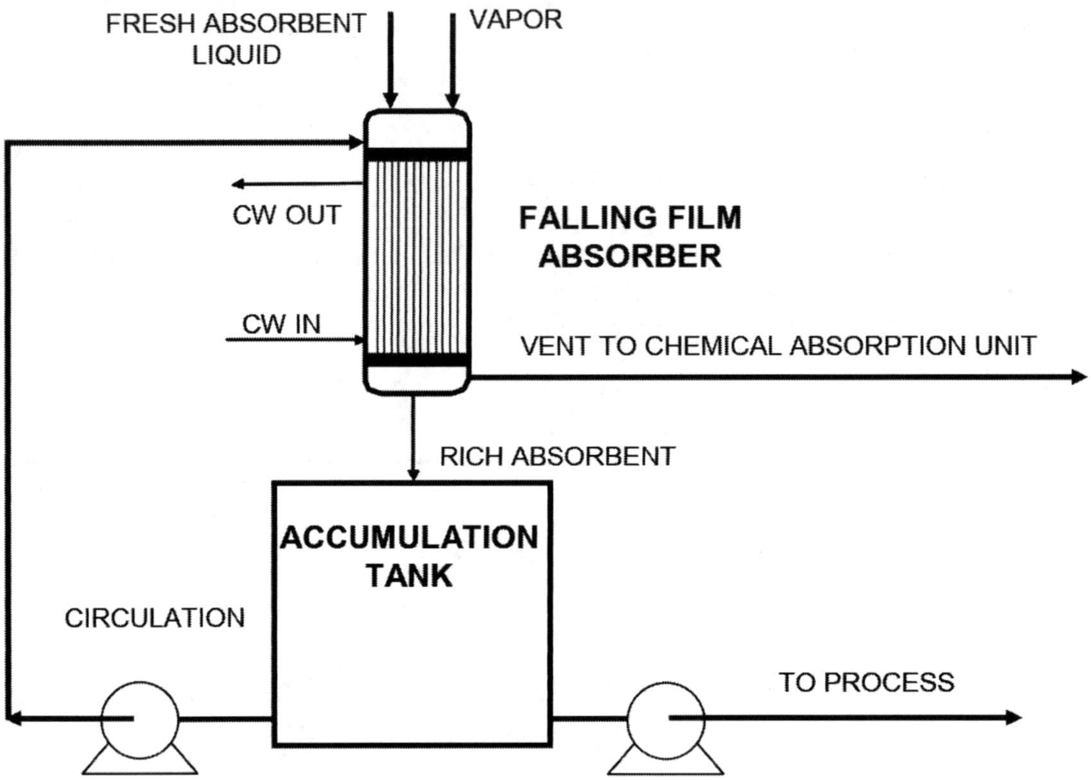

Figure 1.1. Falling film absorber flow diagram.

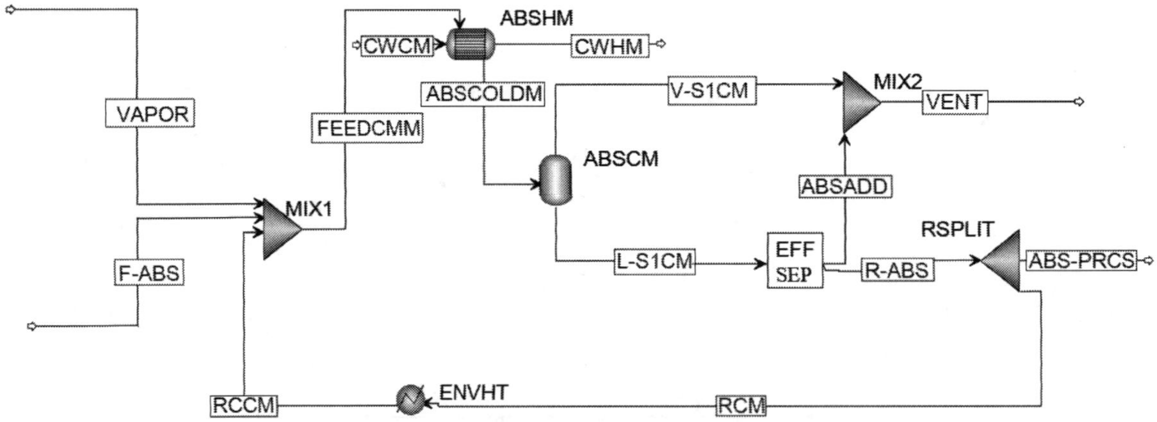

Figure 1.2. Flow diagram ASPEN Plus model of falling film absorber.

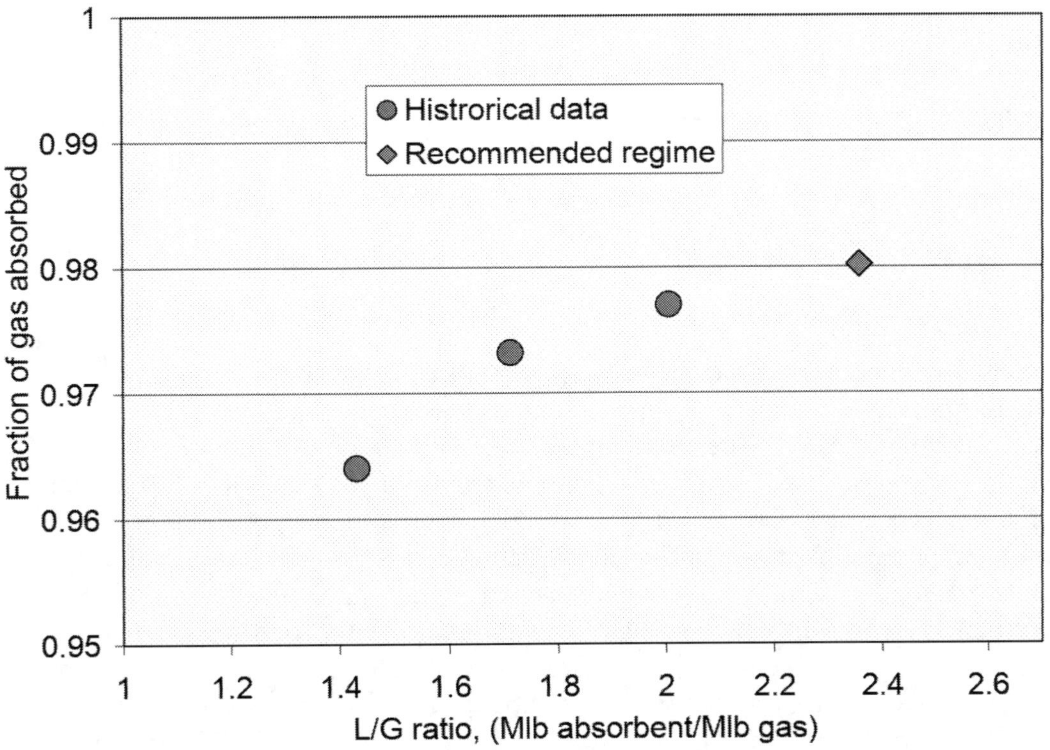

Figure 1.3. Fraction of gas absorbed vs. liquid/gas ration.

Table 1.1. Comparison of model predictions with plant data.

	PLANT	ASPEN FLASH	ASPEN Heater	Model/ 50% Stage efficiency	Pressure
Data set	Normalized Vent	Normalized Vent	Normalized Vent	Normalized Vent	
1	1.00	0.33	0.33	1.07	Low
2	0.66	0.26	0.27	0.79	High
3	1.96	0.52	0.59	1.83	Low
4	0.50	0.20	0.21	0.60	High
5	0.30	0.08	0.09	0.32	High

Case 2 – Tray Damage due to inert-gas absorption

Background

Extensive tray damage was observed during operation of a stripping column with a built in condenser. The trays were mechanically damaged especially in the upper section of the column close to the condenser. The purpose of the column is to separate wide boiling mixture. The design temperature of the bottoms is higher than 180 C. The design temperature in the condenser is less than 50 C. The process pressure is significantly higher than atmospheric. The column is a part of commercial operations and has more than 30 trays.

One of the indications of the tray damage was inability of the column to achieve the desired purity of the light key component. The temperature at the top of the column was almost 30 °C higher than the design temperature. The out of specification quality of the light key was also confirmed by the results of analytical measurements. The higher than design temperature was observed immediately after the initial start up of the column. Inspection of the column internals confirmed the damage of trays. At that time the hypothesis was developed that the tray damage occurred during the installation. The trays were repaired and the column was started again. After the second start up the similar high temperatures at the top of the column were observed. The gamma-ray scan of the column showed similar tray damage in the upper section of the column. The visual inspection of the trays during the plant shut down confirmed the results of the gamma-ray scan. This led us to conclude that start up procedure of the column should be investigated.

Investigation of the Column Start-up Procedure

The simplified schematic of the column is presented in the Fig. 2.1A. The operating features of the column include the pressure control at the top of the column by manipulating the flow rate of the top product – light key component. A relatively small flow of inert component is added to reduce the fluctuations of the pressure due to potential upsets in the built in condenser operations. The major steps of the start up procedure are outlined in Table 2.1.

In order to study the column behavior during the start up procedure, we developed an ASPEN Dynamics model of the column. A steady state ASPEN Plus model was developed and validated against the plant data. Special attention was paid to the modeling of control valves and the pressure drop across them. We also modeled pressure drop across the trays in the column using ASPEN built in correlation. We were able to achieve satisfactory results with "update pressure profile" option in the ASPEN Radfrac block. The steady state model was converted to the ASPEN pressure driven dynamic simulation. The configuration of the partial built in condenser model in the dynamic simulation is shown in Table 2.2. Under this configuration the model calculated the heat duty of the condenser at given temperature of the cooling liquid and the built in condenser geometry.

The start up procedure described in the Table 2.1 was modeled using the ASPEN Dynamics package. The flow diagram of the ASPEN Dynamics model is shown in Fig 2.1B. The model includes ASPEN Radfrac block to model the column, the valves, V1 and V2, on the product lines. In order to model the column start up, the three controllers were included in the model: liquid level controller LC1 to control amount of liquid in the condenser, liquid level controller LC2 to control amount of liquid in column sump and the pressure controller PC1 to control the column pressure. The dashed lines in the diagram indicate the elements of the model manipulated by the controllers. The liquid level controller LC1 manipulates the internal reflux of the column.

The temperature and pressure profiles in the column were monitored and the values of the controllable and manipulated variables in the controller models. The most dramatic results were observed in the behavior of the pressure controller PC1, as shown in Fig. 2.2. During the step 7 of the start up procedure, ~ 12 min almost immediately the pressure at the top of the column dropped by 19 psi. The pressure controller in attempt to compensate for the dramatic pressure reduction totally closed valve V1 and stopped the flow of top product. This caused the pressure at the top of the column collapsed very rapidly practically to the full vacuum.

The behavior of column bottom temperature confirms that of the column top pressure, as shown in Fig.2.3. The collapse of the pressure occurs almost immediately after the bottom temperature recovery.

Analysis of the other variables in the model allowed us to understand the behavior of the column leading to the pressure collapse. As the heat duty of the reboiler is ramping up the vapor from the reboiler reaches the condenser and the vapor from the reboiler condenses in the built in condenser, a certain amount of inert component absorbs in the condensate. This leads to the pressure decrease. The returning relatively cold reflux decreases the bottom temperature and decreases amount of vapor going to the condenser. Consequently, this creates a short unstable equilibrium period when column operates at decreased pressure. As the heat duty of the reboiler is ramping up the temperature of the reboiler increases. This leads to the increase of the amount vapor going to the top of the column and the increase of condensate that acts as inert-gas absorbent. At a certain moment the model predicts that inert gas will be totally absorbed at the top of the column. This moment corresponds to the total pressure collapse. Eventually the condensate is saturated with the absorbed inert component and the inert component is desorbed at the reboiler and builds up in the top of the column. The build up of the inert component corresponds to the pressure recovery at the top of the column. As one of the consequences of the pressure collapse there is a prediction of pressure drop of almost 50 psi between the top and the bottom of the column. The pressure drop of such magnitude is capable to damage column internals.

Conclusions

The modeling work suggested a hypothesis that the major cause of the tray damage in the column is the brief pressure collapse during the start up. A very simple remedy to prevent this is an increase amount of the inert component feeding to the column top above absorption limits. The required increase was estimated using equilibrium flash calculations and verified this amount using dynamic simulations. The suggested changes in start up procedure were implemented during the next plant start up. The column temperature profile after the start up is shown in Fig.2.4. A column temperature profile observed in the plant before is shown for the comparison. The top column temperature decreased almost 30 C and reached the design value. The purity of the top product – light key improved and reached the design specification. The follow up inspection confirmed that the internal of the column are intact and not damaged. This allowed us to make a conclusion that the hypothesis considering pressure collapse was correct.

The column start up was defined by operating features of the column: separation of wide boiling mixture at relatively high pressure, and industrial rates. This created an interesting situation where the dynamic behavior of the built in condenser as a mass transfer device led to the troubled start up. Therefore it is important to consider modeling such dynamic behavior to insure smooth start up.

Lessons Learned

The simultaneous consideration and modeling of mass transfer and heat transfer can help resolve problems with chemical plant equipment.

The heat transfer and mass transfer effects should be considered and modeled in the whole operating range including start ups, normal operation and shut downs.

Results from different commercially available tools can be combined to obtain reliable answers.

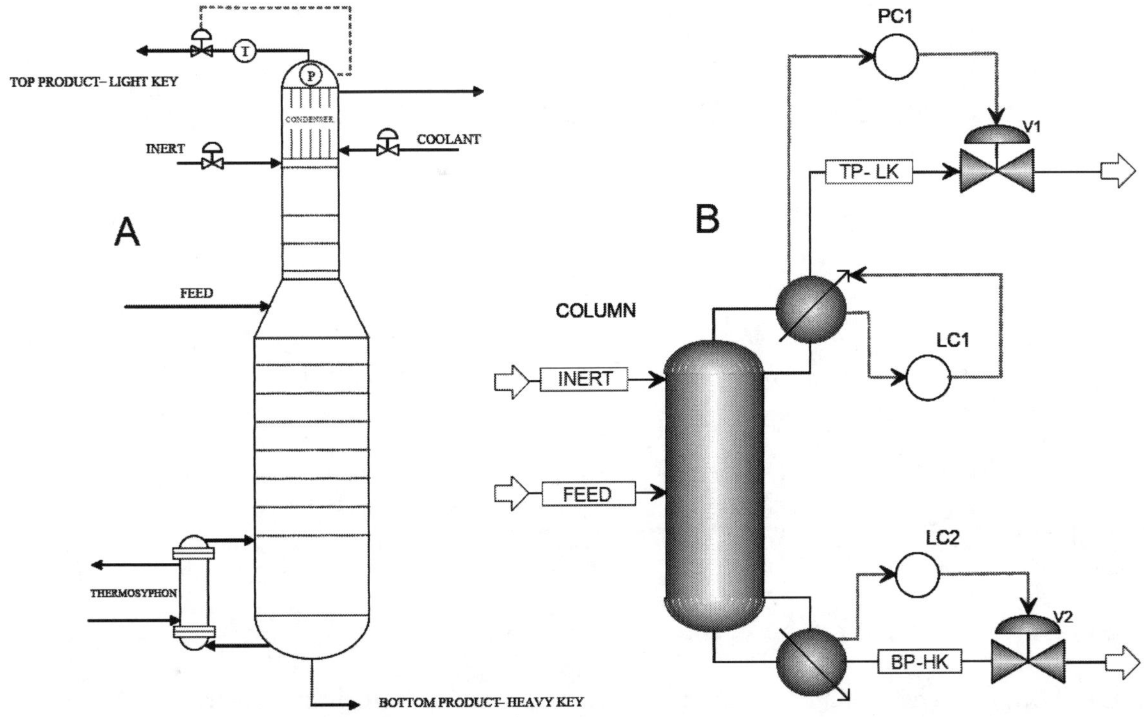

Figure 2.1. Simplified schematic of the column:
A – general column layout.
B - flow diagram of dynamic simulation

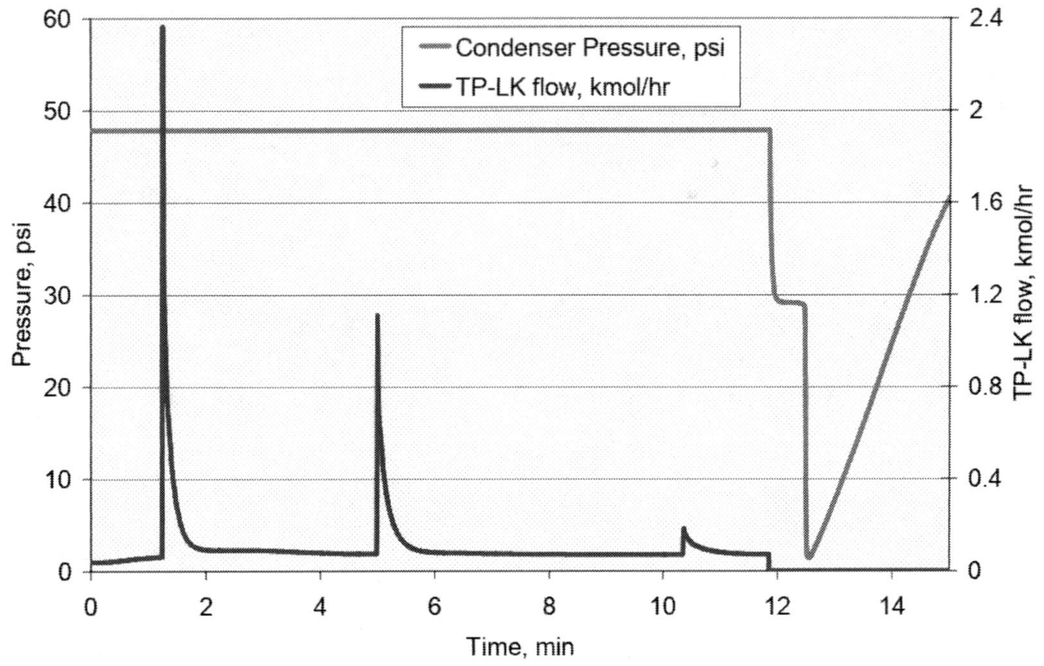

Figure 2.2. The behavior of the pressure controller PC1during the start up. Modeling results.

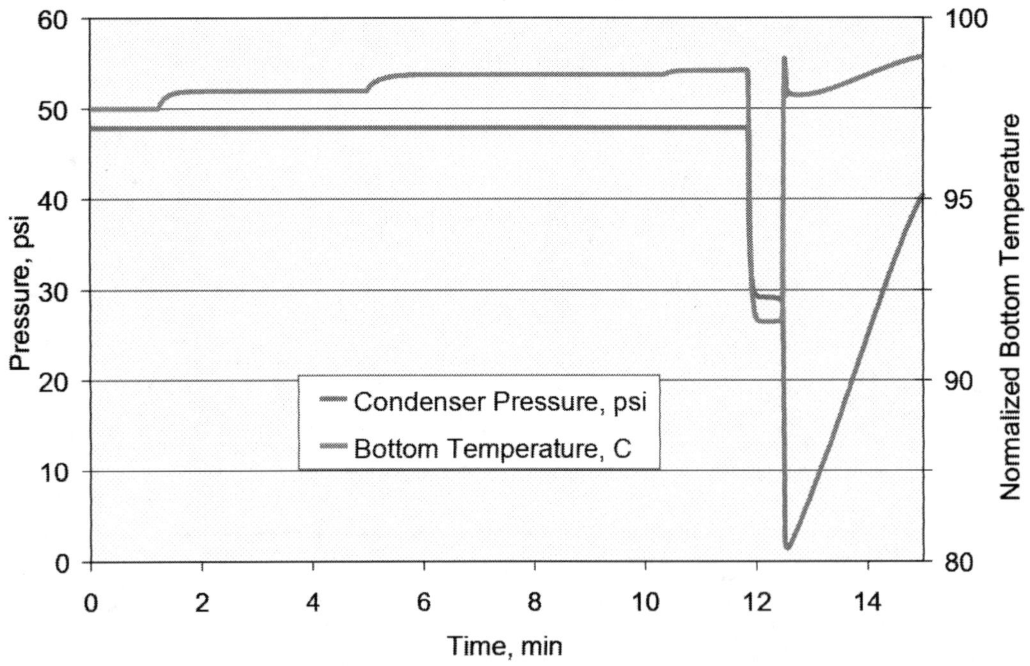

Figure 2.3. The behavior of column bottom temperature in comparison with column top pressure during start up. Modeling results.

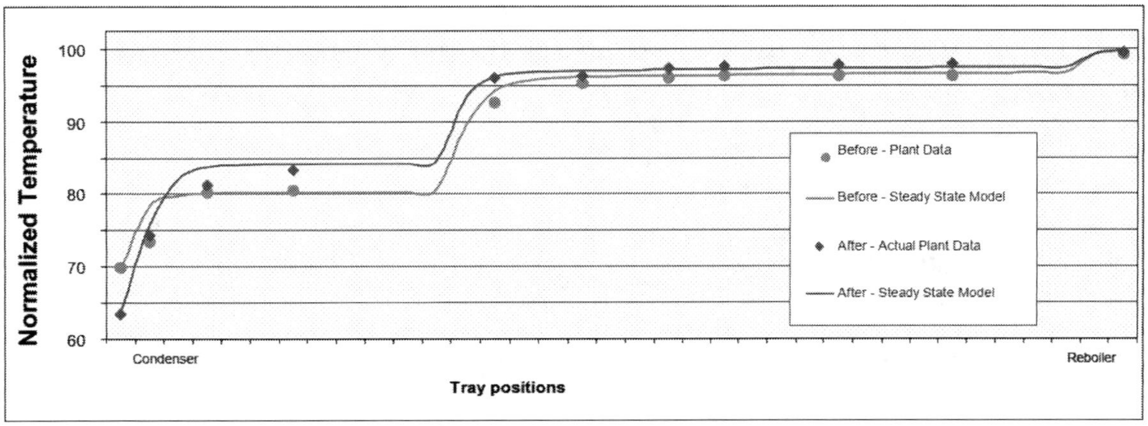

Figure 2.4. The column temperature profile before and after the modified start up.

Table 2.1. The major start up steps.

1	Empty column with 14.7 psia of nitrogen pressure.
2	Bring column pressure from 14.7 psia to operating pressure with inert flow at the top.
3	Control column pressure by the manipulating the flow at the top of the column
4	Fill column sump to 50% level with feed mixture. Continue to level control by manipulating feed flow as the mixture in the sump is heated and vaporized.
5	Start flow of coolant to the built in condenser
6	Start hot oil to the thermosiphon
7	Ramp thermosiphon heat duty to the design value by manipulating the hot oil flow
8	Continue pressure control at the top of the column by manipulating the flow of the top product
9	As the temperature profile builds in the column. Start bottom flow and switch to the normal operating control strategy

Table 2.2. Configuration of partial built in condenser in ASPEN Dynamics model. The names of the variables are given according to the standard convention of ASPEN Dynamics and ASPEN Custom modeler.

Variable	Description	Spec/Value
FeedStages("INERT")	Feed stage numbers	1†
FeedConven("INERT")	Feed conventions	CONDENSER†
condenser(1).QR	Specified duty	Free‡
condenser(1).Tmed_in	Condenser cooling medium inlet temperature	Fixed‡
condenser(1).Fl_med	Medium flow rate	Fixed‡
condenser(1).UA	Overall heat transfer coefficient x area	Fixed‡

† - value
‡ - spec

46b Revamping Crude Units: Pre-Heating Trains and Pumparound Load Adjustnment

Miguel J. Bagajewicz[1], Kitipat Siemanond[2], and Warapon Sripayap[2]. (1) Chemical, Biological and Materials Engineering, University of Oklahoma, 100 E. Boyd St., T-335, Norman, OK 73019, (2) Chulalongkorn University, Chula Soi 12, Phyathai Rd, Bangkok, 10330, Thailand

Crude fractionation units are energy intensive. Therefore any measures taken to heat integrate these units saves operating expenses.

These units make use of so-called pump-around circuits (liquid extracted from the column, cooled down and sent to a few trays above), as means of distributing the reflux of the condenser. These circuits were originally installed aiming at "balancing the traffic of liquid". It was later found, after the energy crisis, that they are also effcient in reducing energyu consumption because they could be effectively used to pre-heat the crude fed to the column.

In a recent string of papers (Bagajewicz et al, IECR) a new methodology to design crude fractionation columns has been presented. The methodology makes use of process simulation to determine fractionation needs in conjunction with heat demand supply diagrams. This allowed the appropriate determination of pump around circuits loads to minimize energy consumption. This is a targeting method that uses the well-know "plus-minus" principle of pinch technology. The technique provided good targets for subsequent heat exchanger network designs.

In a 2006 AICHe conference paper we presented a method where we replaced the demand-supply diagrams by HEN,MILP-based, synthesis methods (Barbaro and Bagajewicz, Computers and Chemical Engineering, 2005), under conditions in which we simulatneously vary the pumparound rates thus capturing the level of pumparound loads that lead to most appropriate balance between capital and energy costs.

In this presentation we will focus on retrofit under similar conditions. Retrofit is based on adding exchangers as well as altering the effective area of existing ones.

Characteristics of Energy-Saving Distillation Through Internal Heat Exchange (Hidic)

Shuzo Ohe, Science University of Tokyo
1-3 Kagurazaka, Shinjuku-ku, Tokyo, Japan 162-8601

INTRODUCTION

Nowadays, the world is faced with significant energy problems. High oil prices are forcing a need for energy reduction. Global warming must be headed off by reducing CO_2. For these and many other reasons, energy saving is a practice that must be pursued constantly by everyone.

In distillation too, energy conservation has long been a key problem. Separating mixtures by distillation requires a great deal of heat energy. In conventional systems, many approaches have been tried, but with limited results. Using recent energy-saving technology, the industry has been able to reduce the amount of heat energy consumed in a large-scale chemical plant, compared to earlier operations. However, the capability of that technology has reached its limits.

In March 2006, the Japanese government concluded the HIDiC project, a four-year project on energy saving in distillation. HIDiC stands for Heat Integrated Distillation Column. This innovative system offers some strong and unusual advantages. As chairman of the project review committee, I would like to provide here a brief summary of the system's principles, as well as some of the project's goals and results.

The HIDiC system employs a breakthrough concept in the distillation process that can deliver a new, higher capability for saving energy. As a result, HIDiC can offer more cost savings than any previous system. (For example, the overall energy savings target for the project was 30%, but the actually demonstrated savings were 60% for the petrochemical plant and 40% for the air separation plant.)

The striking difference between HIDiC and predecessor systems is in the way that HIDiC handles heat. In the conventional distillation system, the waste heat at the top of the enriching section is not recovered. This makes the heat utilization of the conventional system comparatively low. The HIDiC system, in contrast, makes significantly more utilization of waste heat.

HIDiC makes use of the heat from the enriching section to heat the stripping section. This is made possible through the HIDiC system's distinctive longitudinal partitioning of the distillation column into its enriching and stripping sections. In this approach pressure is applied to the vapor, compressing it and causing the temperature to rise, and the resultant heat is thus transferred to the stripping section.

As a result, unlike conventional distillation systems, HIDiC requires no external reflux. That is because, by transferring heat from the enriching section to the stripping section, the HIDiC system can make its own internal reflux.

1. PRINCIPLES

HIDiC adapts a principle which Mah et al. originally presented about thirty years ago as SRV (Secondary Reflux and Vaporization) distillation. (The SRV column is also cited in Perry's Handbook.) HIDiC manifests this concept industrially to achieve heat transfer from the enriching section to the stripping section. It does this by applying SRV and eliminating reflux at the top of the enriching section. The stages of the process are as follows.

1. Vapor from the stripping section does not go directly to the enriching section. Instead, it goes there after compression raises the vapor temperature.

2. Then heat effectively transfers from the vapor in each enriching section tray to the liquid in the matching stripping section tray.

3. The pressure in the enriching section is higher than in the stripping section. After the pressure of the liquid from the enriching section is reduced, the liquid is then fed to the stripping section.

4. Then the vapor rate in the enriching section decreases progressively as it approaches the top. At the same time, the liquid rate in the enriching section increases progressively as it approaches the bottom of the section.

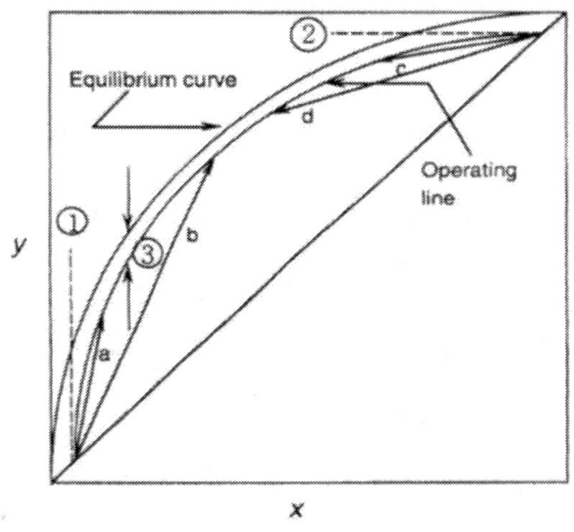

Figure 1. Conditions Required for Internal Heat Exchange[1]

[1]Takamatsu, B., M. Nakaiwa, T. Nakanishi, *Kagaku Kogaku Ronbunshu*, **22**, 985 (1996)

The SRV principle originated by Mah et al. was the springboard for the HIDiC principles developed by Takamatsu, Nakaiwa, and Nakanishi. Their HIDiC concept is illustrated here in an x-y diagram. They asserted that the implementation of a HIDiC system requires all three of the following conditions:

1. The operating line in the *stripping* section is *vertical* at the *bottom* of the column.
2. The operating line in the *enriching* section is *horizontal* at the *top* of the column.
3. The distance between the operating line and the equilibrium curve is an appropriate value throughout the column.

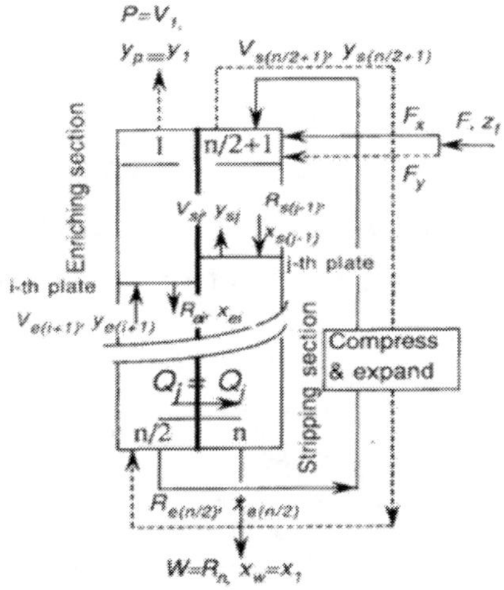

Figure 2. Principles on Which HIDiC is Based[1]

[1]Takamatsu, B., M. Nakaiwa, T. Nakanishi, *Kagaku Kogaku Ronbunshu,* **22**, 985 (1996)

Figure 2 shows the relationships of calculations for flow rate and composition involved in the HIDiC system. The feed, F, enters at the top of the stripping section, as marked by the left-facing arrow, and goes to the stripping section. The vapor, V, from the stripping section is compressed and fed to the enriching section. The top product, P, exits the top of the enriching section.

The mass balances are given by equations (1) through (3). One of them can also define the q-value. Based on the amount of Q transferred from the enriching section, equation (4) can calculate the amount of V in the stripping section.

Mass Balances

$$F = P + W, \quad Fz_f = Py_p + Wx_w \tag{1}$$

$$W / F = q \tag{2}$$

$$z_f = (1-q)y_p + qx_w \tag{3}$$

Action of Internal heat exchange

$$R_{e(n/2)} = V_{s(n/2+1)} - P = V_{s(n/2+1)} - F_y$$
$$= \sum_{i=1}^{n/2} Q_i / \lambda = \sum_{j=n/2+1}^{n} Q_j / \lambda = Q / \lambda \tag{4}$$

The operating line of the enriching section is given by equation (7). In this equation, R_{ei} is smaller, so the slope $R_{ei}/R_{e(i+1)}$ becomes smaller. This means that a condenser is not needed. However, internal reflux is needed. Therefore internal heat exchange is required for reflux of each plate.

i-th plate in enriching section

$$V_{e(i+1)} = R_{ei} + P \qquad (5)$$

$$V_{e(i+1)} y_{e(i+1)} = R_{ei} x_{ei} + P y_p \qquad (6)$$

$$y_{e(i+1)} = \frac{R_{ei}}{V_{e(i+1)}} x_{ei} + \frac{y_p}{V_{e(i+1)}} \qquad (7)$$

The operating line of the stripping section is given by equation (10). In equation (10), $V_{s(j-1)}$ is smaller, so the slope $R_{s(j-1)}/V_{s(j-1)}$ becomes larger. This means that a reboiler is not needed. However, internal vaporization is needed. Therefore internal heat exchange is required for vaporization of each plate.

j-th plate in stripping section

$$R_{s(i-1)} = V_{sj} + W \qquad (8)$$

$$V_{s(i-1)} y_{sj} = R_{s(j-1)} x_{sj} - W x_w \qquad (9)$$

$$y_{sj} = \frac{R_{s(j-1)}}{V_{s(j-1)}} x_{s(j-1)} - \frac{x_w}{V_{s(j-1)}} \qquad (10)$$

The operating line equations (11) and (12) are shown here. The internal reflux ratio and internal reboiling ratio are also given.

Enriching section

$$y_{e(i+1)} = \frac{re_i}{1+re_i} x_{ei} + \frac{y_p}{1+re_i} \quad \text{Internal reflux ratio}: re_i = \frac{R_{ei}}{P} \qquad (11)$$

Stripping section

$$y_{sj} = \frac{1+rs_j}{rs_j} x_{s(j-1)} - \frac{x_w}{rs_j} \quad \text{Internal vaporization ratio}: r_{sj} = \frac{V_{sj}}{W} \qquad (12)$$

The internal reflux ratio is different at each plate. The ratio is expressed as the product of *i* and dr_e, in equation (13) or the product of (n-j+1) and dr_s in equation (14). *The values for* dr_e and dr_s are determined from the amount of internal heat transfer.

Enriching section

$$y_{e(i+1)} = \frac{idr_e}{1+idr_e} x_{ei} + \frac{y_p}{1+idr_e} \quad re_i = i \times dr_e \qquad (13)$$

Stripping section

$$y_{sj} = \frac{1+(n-j+1)dr_s}{(n-j+1)dr_s} x_{s(j-1)} - \frac{x_W}{(n-j+1)dr_s} \quad rs_j = (n-j+1)dr_s \qquad (14)$$

$$dr_e = Q_k / \lambda / P, \qquad dr_s = Q_k / \lambda / W \qquad (15)$$

2. PILOT PLANT PROJECT

HIDiC is evaluated highly as a chemical process in the international effort to combat global warming. In Japanese industry as a whole, 15% of all energy is expended in the chemical industry. In the chemical industry, separation by distillation expends 40% of total energy. The basic research on HIDiC has already yielded promising results. However, significant issues remain before profitability can be assured. Much additional funding is required for its development, but support from private enterprise cannot be expected. Hence, the project must be conducted by NEDO, an agency of the Japanese government.

For the basic study, a project was performed (1993-2000), with encouraging results. The energy saved was calculated as more than 30% over that of the same kind of plant operating with a conventional system. Moreover, the specifications were maintained without reflux – a condition essential for ordinary distillation – for five full hours.

The project focused on the tray, structured packing and plate fin column systems in distillation processes of the petrochemical and air separation industries. The term was four years (from April 2002 to March 2006) and the budget was 1,400,000,000 yen (about US $12,730,000 at that time).

The project was broken down into five portions. For each portion, highly competent leaders were selected from companies with reputations for expertise in research and development in the subject areas.

The budget was distributed according to the objective of each portion. One third of the budget was allocated to the construction and operation of the pilot plant. The distributions for the three types of HIDiC distillation systems (investigating shell and tube, tray, and plate fin) were also sufficient, taking into consideration the cost of making the test equipment.

The four-year schedule was divided into two terms of two years each. The first term was used for basic study and the second term was used for pilot plant construction and testing to evaluate the project. Special care was given to designing a cost-effective and expedient project schedule.

Between the two terms, evaluation of the competing systems was performed by an independent committee comprised of members from industries and academia. This evaluation showed that the principle of competition was promoting efficient use of budget and time.

To optimize the HIDiC design, especially for large-scale distillation columns, the following items will require further R&D.

1. Achievement of both homogeneous distribution of vapor and liquid flow and effective heat transfer through the wall.
2. Development of technology for multi-component systems (ternary and greater).
3. Design of a structure that precludes the complex problems that can arise when the up and down flow rates differ greatly.
4. Development of precise pressure control technology to enable practical application of the azeotropic mixture.

The HIDiC pilot plant (Fig. 3) is located in the factory of Maruzen Petrochemical Company, Limited, near Tokyo, Japan.

Figure 3. HIDiC Pilot Plant

Figure 4 shows the compressor for pressurizing the vapor from the stripping section before it enters the enriching section.

Figure 4. Compressor for the Pilot Plant

Outline of the Project

The basic study, HIDiC 1, extended from 1993 to 2000. Two major results were achieved: energy savings of more than 30%, and maintaining of process specifications for five hours without reflux. The basic project focused on the tray, structured packing, and plate fin column systems in distillation

HIDiC 2, the advanced project, ran from 2002 to 2006. The budget was one trillion, 400 billion yen (about 14 million U.S. dollars). The project was broken down into five portions. One third of the budget was allocated to the construction and operation of the pilot plant.

Project Budget and Time Schedule

Table 1 shows the budget and schedule of the project. The project started in 2002 and concluded at the end of the 2005 Japanese fiscal year i.e., the end of March 2006. The budget is shown for each year, in millions of yen and millions of dollars. The total budget used was $1,360,000, at a rate of 100 yen to the dollar.

Table 1. Project Time Schedule and Budge

Schedule	Fiscal year	2002	2003	2004	2005	
	Petro-Chemical	—	—	—	—	
	Air Separation	—	—	—	—	
	Multi-component	—	—	—	—	
	Report				—	
Budget	Fiscal year	2002	2003	2004	2005	Total
	Million yen	220.0	343.2	572.4	232.2	1367.7
	Million dollar	2.200	3.432	5.724	2.322	13.677

Project Team

Figure 5 is the project's organization chart. The project was organized by a Japanese government entity, NEDO (the New-Energy and Industry Technology Development Organization). The project leader was Dr. Nakaiwa, who belongs to AIST, a Japanese government technology institute. There were two group leaders. One was Mr. Iuchi, director of Maruzen Petrochemical Company. The other was Mr. Kawakami, manager of Taiyo Nissan. Five companies took part in this project.

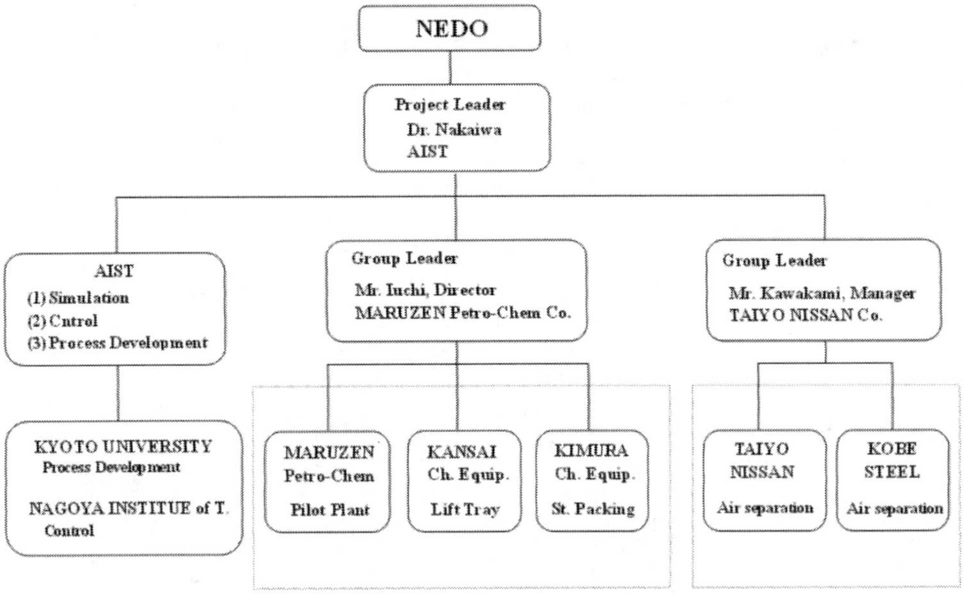

Figure 5. Project Organization Chart

Review Committee

The six-member independent Review Committee listed in Table 2 was composed of representatives from three industrial companies and professors from three universities. At the conclusion of the project, the Committee's results were presented at a meeting of 47 individuals from Japanese government agencies (METI, the Ministry of Economy, Trade and Industry; and NEDO), and the companies that received government funding and took part in the project.

Table 2. Review Committee Members

Position	Name	Title, Affiliation
Chairman	Shuzo Ohe	Professor, Science University of Tokyo
Vice Chairman	Yuji Naka	Professor, Tokyo Institute of Technology
Member	Shigeru Koyama	Professor, Kyushu University
Member	Yoshitaka Saito	Manager, Alcohol Association
Member	Kazuo Matsuda	Manager, Chiyoda Engineering Company
Member	Shizuo Midori	Consultant, Yokaichi Petrochemical Company

3. PILOT PLANT CONSTRUCTION AND PERFORMANCE

HIDiC Pilot Plant Specifications

The structure of the HIDiC is that of a shell and tube heat exchanger. The shell side is used as the enriching section, while the tube side is used as the stripping section. Both sides are filled with structured packing. Dimensions are shown in this table. The column diameter is 1.4 m and the inner tube diameters are 267 mm, 216 mm and 165 mm. The size difference is based mainly on the difference in the vapor rate in the tube as heat is transferred to the shell side.

Table 3. Comparison of the Pilot Plant with a Commercial Plant

		HIDiC	Conventional
1	Column	HIDiC	Conventional
	Height	27 m	31 m
	Diameter	1.4 m	1.0 m
	Internal	Structured Packing 13.5 m H.	Bubble Cap 55 trays
2	Compressor	45 KW	-
3	Reboiler	7.4 m^2	23 m^2
4	Preheater	4.8 m^2	-
5	Condenser	23.5 m^2	87 m^2

HIDiC Structural Details

Figure 6 shows the internal structural details of the HIDiC unit. Each tube is an independent column. The enriching section is in the tube side and the stripping section is in the shell side.

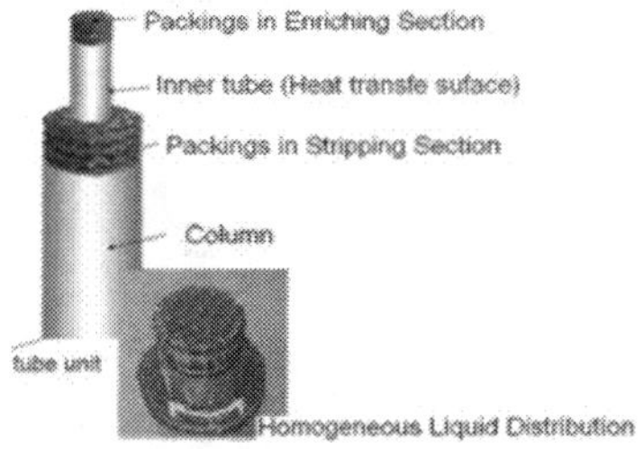

Figure 6. HIDiC Internal Structural Details

The distillation efficiencies were satisfactory for both sides or sections. Figure 7 shows the improvement in efficiency after the project. The NSTM number of theoretical plates per unit height of packing bed, originally ranging from one to two point five, increased to a constant of three.

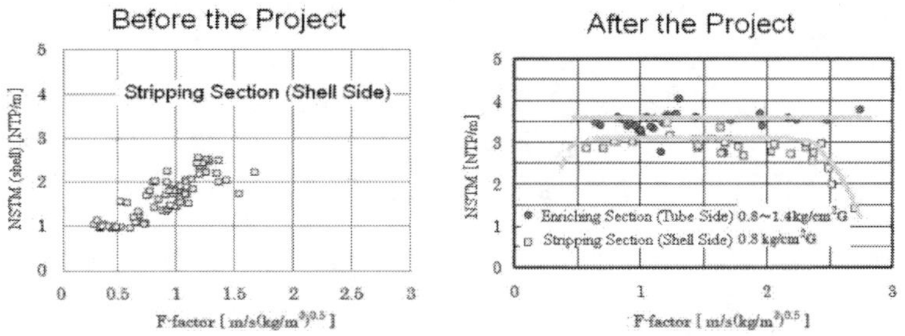

Initial efficiency After modification

Figure 7. Improved Efficiency after Modification

Testing confirmed the capability of continuous operation for 1000 hours. Table 4 shows the results for three tests. While the average showed slightly more than 60% improvement in energy savings, the individual test results were each considerably higher.

Table 4. Energy Savings Results

Energy Saving Result

Date	Feed Rate (kg/hr)	External Reflux Ratio	Energy Saving (%)
Nov. 11, 2005	1900	0	63.9
Feb. 2, 2006	1700	0	67.0
Mar. 10, 2006	2100	0	63.5
Average (year)	1714	1.5	61.2

The chart in Fig. 8 shows the control performance. After the feed rate was changed, steady states were achieved within 3 hours. A steady state was maintained for each flow rate — reflux, bottom, and distillate. This demonstrates HIDiC's exceptional stability in operating performance.

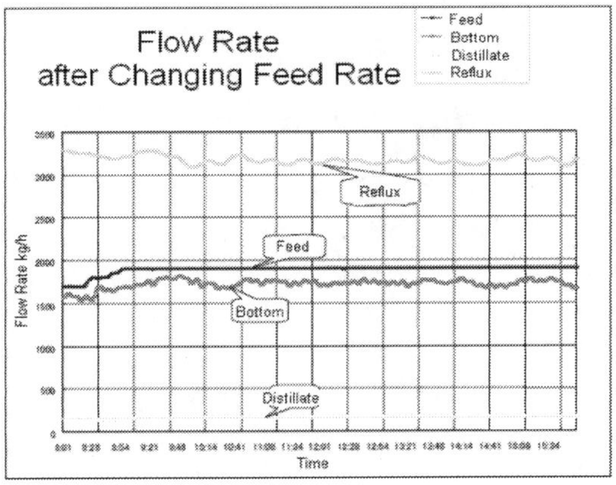

Figure 8. Flow Rate After Hanging Feed Rate

4. DISCUSSION

HIDiC's innovative design principles eliminate the need for external reflux. Here we will discuss the reasons for this.

NTP of Ideal HIDiC

First, let us examine the calculation of the number of theoretical plates (NTP). Figure 9 is reproduced with the permission of the original authors[1]. Note that at the top, the operating line is almost horizontal, whereas at the bottom it is vertical.

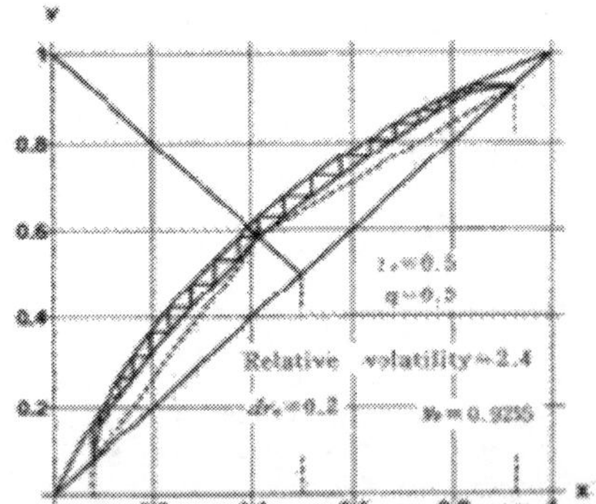

Fig. 6 A numerical example of No. of theoritical plates calculation in the ideal HIDC for benzene-toluene system.

Figure 9. NTP of Ideal HIDiC

[1]Takamatsu, B., M. Nakaiwa, T. Nakanishi, *Kagaku Kogaku Ronbunshu*, **22**, 985 (1996)

Using bubbling point calculations, we calculated the NTP for the differing pressures of the stripping and enriching sections. These results are shown in Fig. 10.

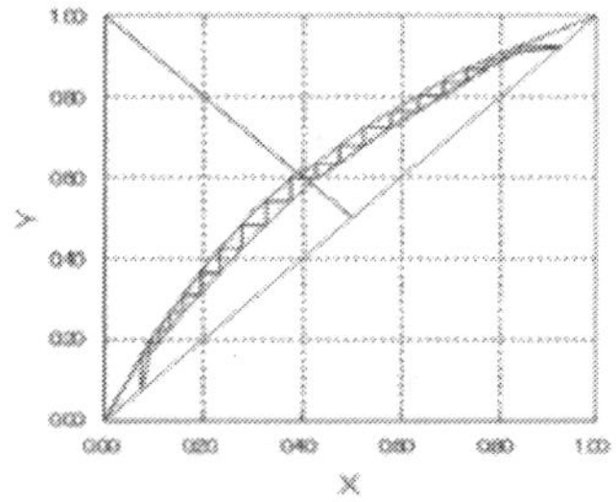

NTP Calculation of Ideal HIDiC (Bubble and Dew point Calculation)

Figure 10. NTP for the Differing Pressures

Figure 11 is the temperature profile for the differing pressures in the stripping section (760 mm Hg) and the enriching section (1520 mm Hg). The difference of temperature between the sections is about 20°C.

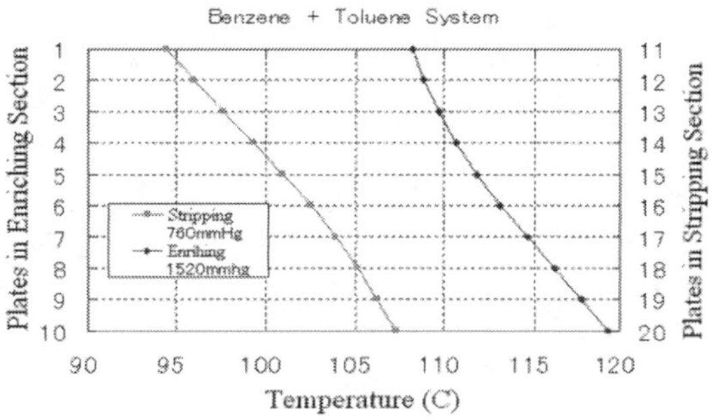

Figure 11. Temperature Profile

Figure 12 shows the vapor and liquid flow rate for each section. In contrast to the conventional single vertical column, the plates of the enriching section and of the stripping section are mounted side by side in parallel. As the vapor rises in both sections, the vapor rate in the enriching section decreases, while that in the stripping section increases.

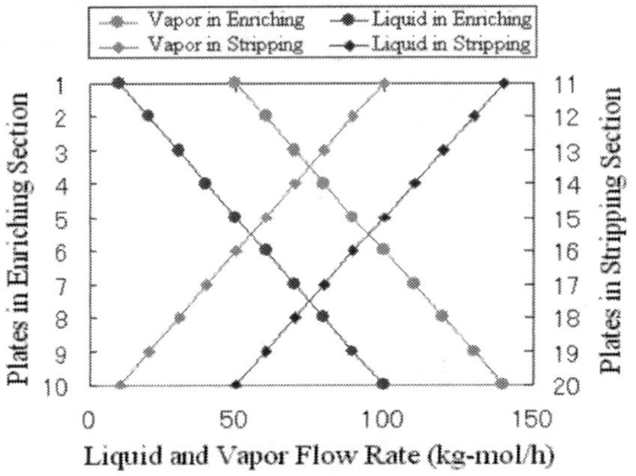

Figure 12. Flow Rates in HIDiC Columns

Figure 13 compares the flow rate of the HIDiC to that of a conventional system. In the conventional system the flow rate is constant, whereas in the HIDiC, the flow rate changes. The figure also shows the HIDiC's "average flow rate." From this flow rate, we calculated the reflux ratio, which was 0.9. This is less than the minimum reflux ratio.

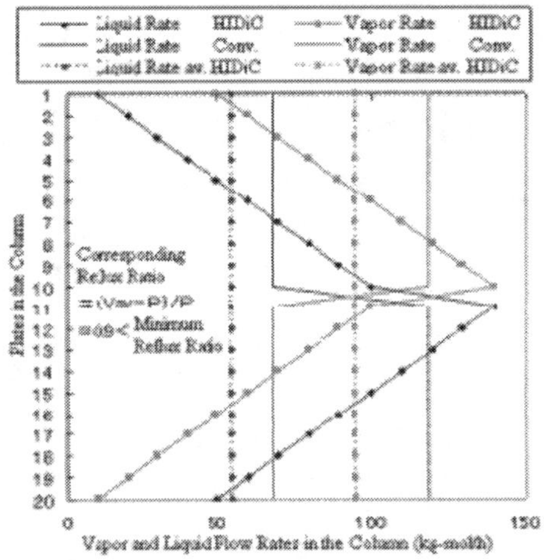

Figure 13. Flow Rates in HIDiC and Conventional Columns

Feasibility of HIDiC for Ethanol + Water System

To examine the feasibility of using HIDiC, we considered an ethanol + water system. Table 5 shows the effect of pressure on the azeotropic composition of an ethanol + water system. Fortunately, we found that the azeotropic composition of ethanol decreases as the pressure increases.

Table 5 Effect of Pressure on Azeotropic Composition of Ethanol + Water System

Ethanol + Water System					
Composition	Pressure	Temp.	Composition	Pressure	Temp.
Mole Fraction	(mmHg)	(℃)	Mole Fraction	(atm)	(℃)
0.9820	100	34.25	0.8940	1.00	78.3
0.9863	100	34.20	0.8887	1.41	87.26
0.9687	130	39.35	0.8874	1.91	95.5
0.9338	200	47.97	0.8820	3.40	112.6
0.9419	250	53.00	0.8740	6.81	136.7
0.9041	500	68.00	0.8620	13.61	164.2
0.8957	740	76.90	0.8520	20.41	182.6
0.8940	760	78.15			
0.8945	760	78.39			

By making use of this characteristic, absolute ethanol may be separated from its mixture with water. Figure 14 shows the feasibility of using the HIDiC process. The first column is used for distillation of a lower composition of ethanol than that of azeotropic composition at reduced pressure. HIDiC is applied to the second column for a higher composition of ethanol. The feed from the top of the first column is pressurized and fed to the enriching section and separates the pure ethanol and azeotropic mixture.

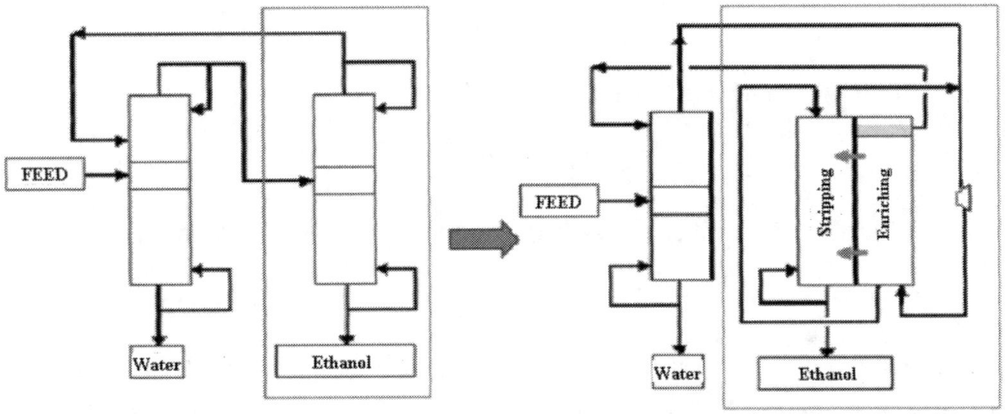

Figure 14. Feasibility of HIDiC for Ethanol + Water System

Figure 15 shows the distillation calculation for the first column. Using this calculation, the mixture is separated into water at the bottom and azeotropic mixture at the top. Figure 16 shows that it is possible to separate pure ethanol at the increased pressure.

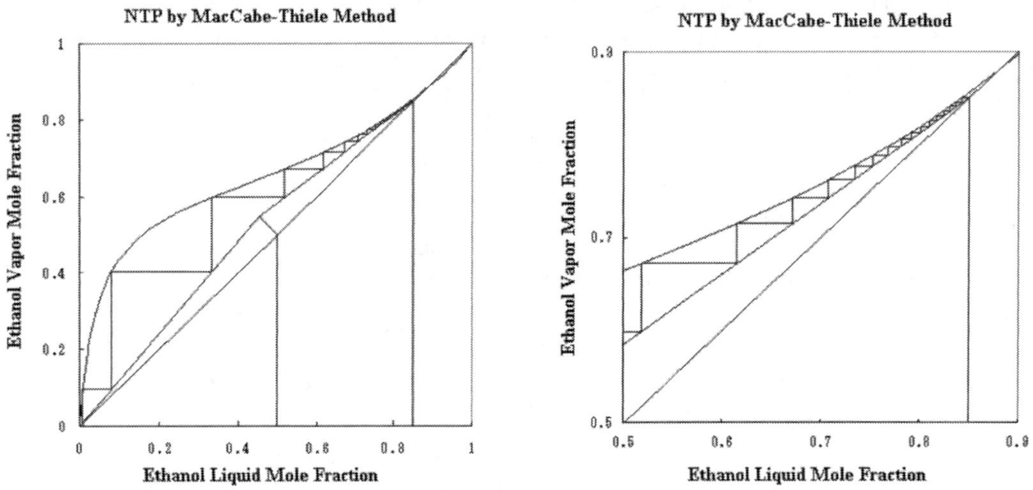

Figure 15. Conventional Distillation for Ethanol + Water System (1)

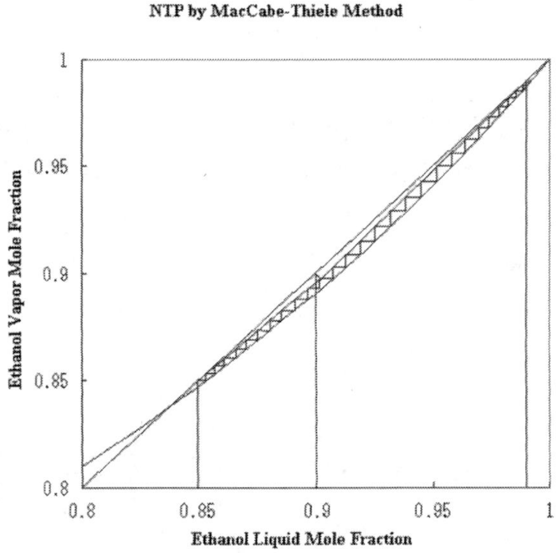

Figure 16. Conventional Distillation for Ethanol + Water System (2)

Figure 17 shows the results of applying HIDiC, as well as the operating parameters. The operating pressure for the stripping section is 101.3 kPa and for the enriching section, 200 kPa. Figure 18 shows the temperature difference between HIDiC's enriching and stripping sections. A sufficient temperature difference is regarded as that of more than five degrees centigrade.

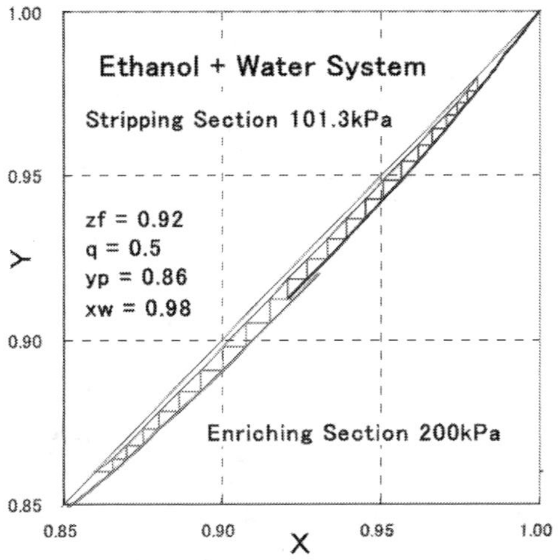

Figure 17. Application of HIDiC for Ethanol + Water System (2)

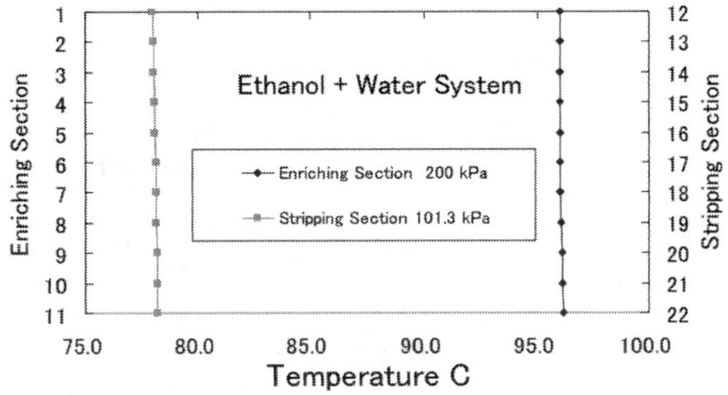

Figure 18. Temperature Difference between HIDiC's Enriching and Stripping Sections

However, in the case of a methanol + water system, the difference is zero at some plates, even if the pressure in the enriching section is increased to as much as three times that of the stripping section (Figure 19). Figure 20 shows the temperature difference between HIDiC's enriching and stripping sections for a methanol + water system when the pressure of the enriching section is increased more than about 3.5 times than that of the stripping section.

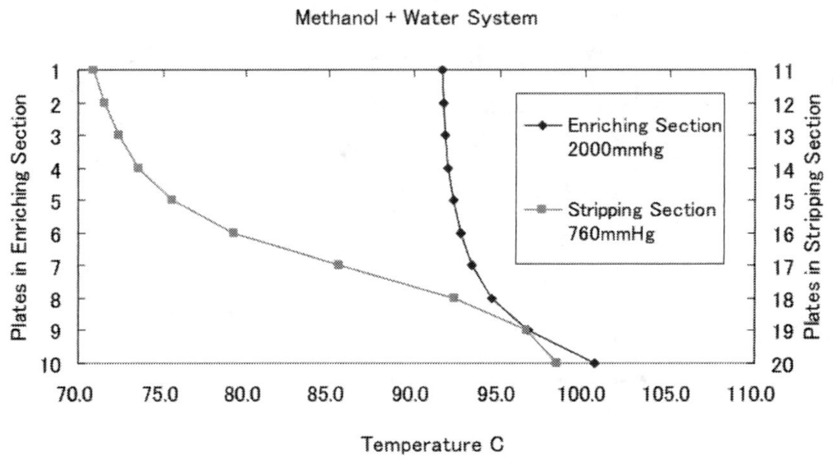

Figure 19. Temperature Difference between HIDiC's Enriching and Stripping Sections (1)

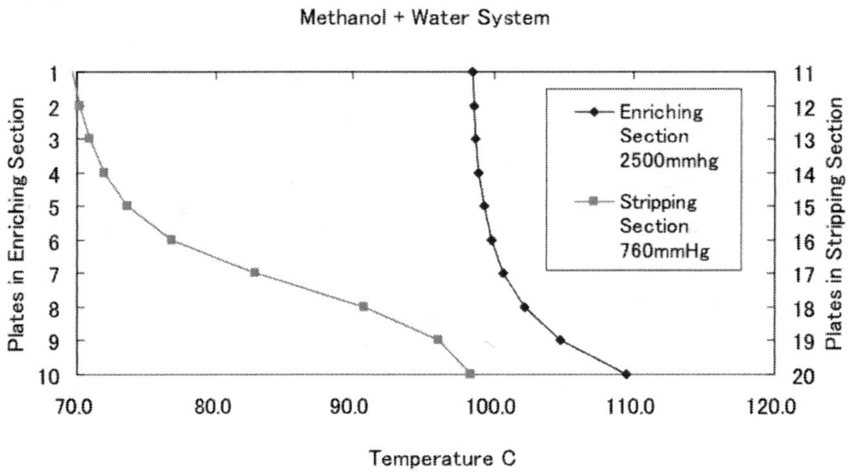

Figure 20. Temperature Difference between HIDiC's Enriching and Stripping Sections (2)

We then compared the temperature profile of the distillation of methanol + water with that of a benzene + toluene system. Figures 21 and 22 show that the change of the methanol + water temperature profile is more complicated than that of the benzene + toluene system.

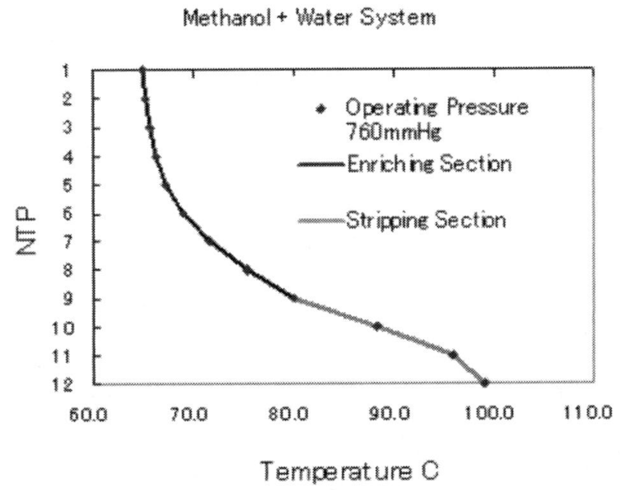

Figure 21. Temperature Profile in Conventional Distillation

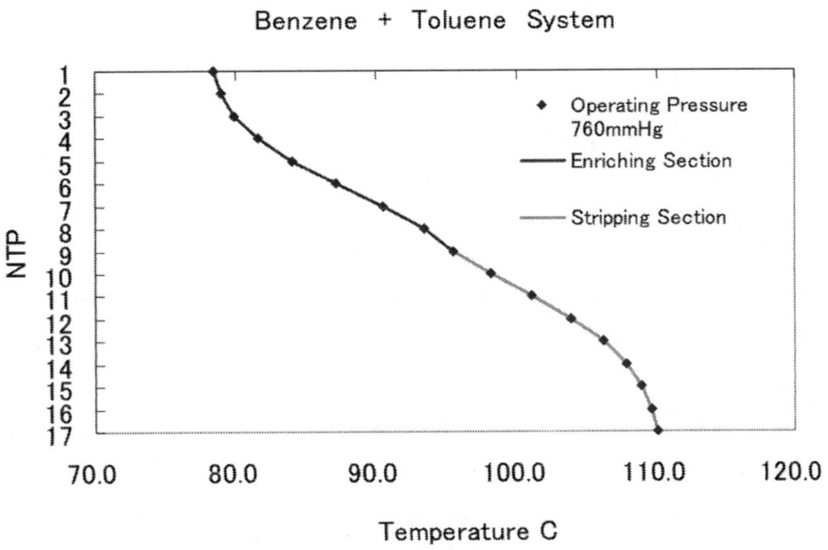

Figure 22.Temperature Profile in Conventional Distillation

The characteristics of the bubble point temperatures in the two mixtures are shown in Figure 23. From these characteristics, we can predict the difference in temperature between HIDiC's stripping and enriching sections. The bubble point change is more complex for the methanol + water system than for the benzene + toluene system.

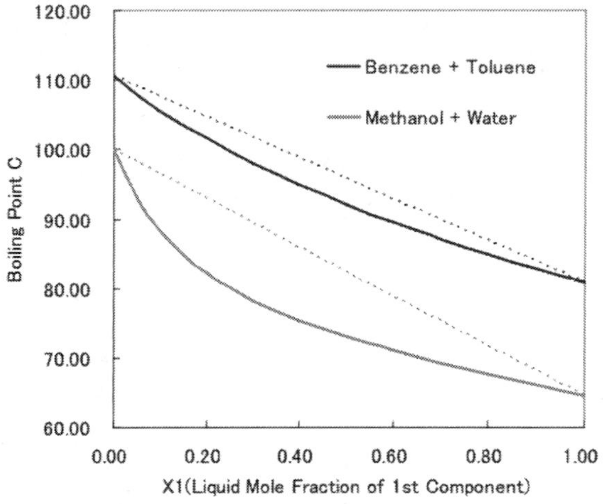

Figure 23. Characteristics of Bubble Point Temperature in Mixture

5. CONCLUSIONS

We have presented the ground-breaking design of HIDiC and efficient operation of the HIDiC pilot plant. We have demonstrated why HIDiC has such a high capability for saving energy — over 60% greater than that of conventional systems. HIDiC also offers some unusual and innovative features. Among them is the fact that HIDiC does not require external reflux, because it can produce internal reflux through internal heat exchange. Moreover, the internal reflux ratio with HIDiC is less than the conventional minimum reflux ratio.

We have considered aspects of the feasibility of using HIDiC for the separation of an azeotropic ethanol + water mixture. Fortunately, the azeotropic composition decreases under pressure. Consequently we can say that the combination of pressurized phase equilibria with HIDiC makes an excellent fit for the separation of an ethanol + water system.

Acknowledgments

We gratefully acknowledge the following persons, agencies and companies.

HIDiC Project: Dr. M. Nakaiwa, Project Leader

HIDiC Review Committee: Prof. Y. Naka, Prof. S. Koyama; Mr. Y. Saito, Mr. K. Matsuda, Mr. S. Midori

Government representatives of the Ministry of Economics, Trade and Industry (METI) and New-Energy Development Organization (NEDO)

Companies: Maruzen Petrochemical Co. (pilot plant site host), Kimura Chemical Equipment Co., Kansai Chemical Equipment Co., Taiyo Nissan Co., Kobe Steel Co

CO_2 CAPTURE PROCESS WITH MEA

Rosa-Hilda Chavez, Instituto Nacional de Investigaciones Nucleares, Estado de México, México.
Javier de J. Guadarrama, Instituto Tecnológico de Toluca, Estado de México, México.
Jaime Klapp, Instituto Nacional de Investigacion es Nucleares, Estado de México, México.

Abstract

Carbon dioxide is considered to be the major source of greenhouse gas es responsible for global warming; man-made CO_2 contributes approximately 63.5% to all greehouse gases. Amine absorption technology, in particular that based on the Monoethanolamine (MEA) process, is considered to be viable for low pressure flue gas CO_2 capturelllll because of the MEA-CO_2 fast reaction rate. MEA absorption processes are associated with high capital and operating cost bec ause a significant amount of energy is required for solvent regeneration and severe operating problems are present such as corrosion and solvent loss and degradation. Comparisons among cases were performed to determine the best operating conditions for CO_2 capture. An analysis of the lean loading and recovery percent were carried out as well as the different absorber and striper combinations by using the chemical processes simulator Aspen Plus™.
Keywords: CO_2 capture; MEA; process modeling.

Introduction

Climate change is one of the greatest and probably most challenging of environmental, social and economical threats facing the world. The issue of climate change is not abou t air quality or smog but about global warming. Human activities have altered the c hemical composition of the atmosphere through the buildup of significant quantities of greenhouse gases, which remain in the atmosphere for long periods of time and intensify the natural greenhouse effect. Increasing concentrations of greenhouse gases are likely to accelerate the climate change rate. Concerns are growing about how increases in atmospheric greenhouse gases caused by human activities are contributing to the natural greenhouse effect and raising the earth average temperature.

Our atmosphere is a complex mixture of gases that trap the sun ́s heat near the earth surface, similar to how the glass of a greenhouse traps the sun warmth. A number of gases are involved in the greenhouse effect, among which the main are: Carbon dioxide (CO_2), Water vapor (H_2O), Methane (CH_4), Nitrous oxide (N_2O), tropospheric Ozone (O_3) and Halocarbons (CFCs, HFCs, etc.). Carbon dioxide is the most important of the greenhouse gases released by human activities. It is the main contributor to climate change because of the q uantities released [1-4].

On a worldwide basis, CO_2 emissions generated from anthropogenic activities are known to be relatively small and in comparison with the gross carbon fluxes from natural systems, they represent only a small fraction (~2%) of total global emissions. However, evidence suggests that they account for most of the observed CO_2 accumulated in the atmosphere [5-8]. On the basis of global emissions information, the primary sources of CO_2 generated from anthropogenic activities are fossil fue l combustion from industry and industrial processes such as cement production, transportation, space heating, electricity generation and cooking, vegetation changes in natural prairie, woo dland and forested ecosystems [9].

A Mini-Plant flue gas CO_2 capture using amine-based CO_2 capture technology requires huge amounts of energy, mostly in the form of heat. The overall objective of this study is to evaluate the feasibility of obtaining the heat required for amine absorption for a particular recovery of carbon dioxide for a given set of equipment specifications and operating conditions in a Mini-Plant, and to develop a model that simulates the removal of CO_2 using Monoethanolamine (MEA) absorption from a Mini-Plant flue gas, and using the chemical processes simulator Aspen Plus™, (simulator of process modeling software).

Process Description

A continuous scrubbing system is used to separate CO_2 from the flue gas stream. As illustrated in Figure 1, the system consists of two main elements: an absorber where CO_2 is removed, and a regeneration column (stripper or desorber), where CO_2 is released (in concentrated form) and the original solvent is recovered. Table 1 shows the description of each parameter used in Figure 1, as well as its temperature and flow values. The pressure remains constant at 1 bar throughout the CO_2 capture and MEA regeneration. In the heat exchanger 1 the stream L1 cools before entering the absorber. The heat exchanger 2 warms the stream L7 before it enters the stripper. The heat exchanger 3 cools L12 leaving the stripper. The mixer combines L6 and L14 in order to consider the MEA losses from evaporation. At the scrubber 1 the temperature of the G1 stream is reduced, whereas scrubber 2 and 3 retain MEA from evaporation.

A key feature of amine systems is the large amount of heat required to regenerate the solvent, this heat is typically drawn from the steam cycle and significantly reduces the net efficiency of the power plant. Substantial electrical energy is also needed to compress the captured CO_2 for pipeline transport to a storage site. The overall energy penalty of this process has a major impact on system performance and cost.

From a multi-pollutant perspective, there are also important interactions between the CO_2 capture system and the emission control of other air pollutants, especially SO_2 and NOx. Acid gases like SO_2 and NO_2 react with MEA to form heat-stable salts that reduce the CO_2 absorption capacity of the solvent. Thus, very low concentrations of these gases (on the order of 10 parts per million) are desirable to avoid excessive loss of (costly) solvent.

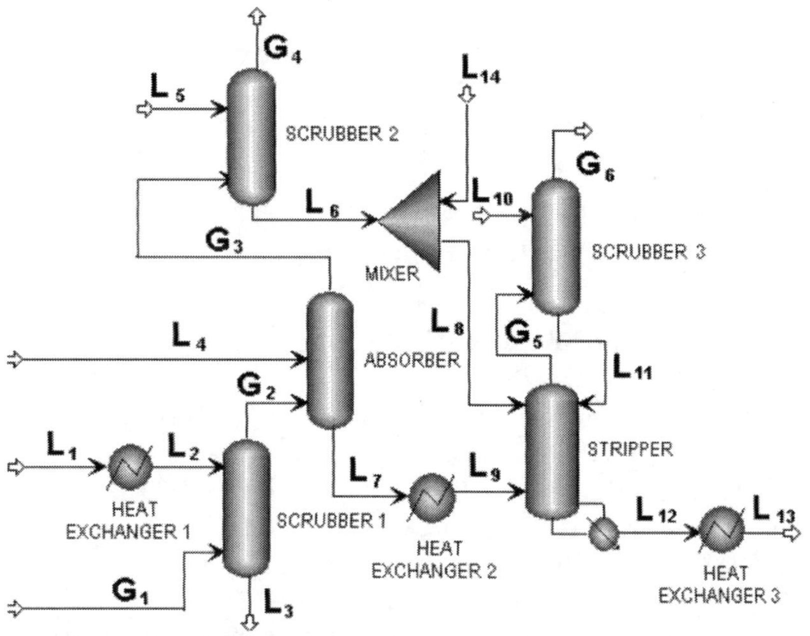

Figure 1. Amine-based system flowsheet for flue gas CO_2 capture.

Table 1. Temperature and Flow Characteristics of the Different Streams.

Stream	Descriptions	T °C	Flow kmol/h	Stream	Descriptions	T °C	Flow kmol/h
L_1	scrubber 1 hot stream	49.25	17.21	L_{11}	stripper inlet stream	92.89	1.11
L_2	scrubber 1 cold stream	35.00	17.21	L_{12}	stripper outlet hot stream	102.32	15.33
L_3	scrubber 1 outlet stream	49.89	17.48	L_{13}	stripper outlet cold stream	35.00	15.33
L_4	absorber inlet stream	35.00	15.33	L_{14}	mixer inlet stream	30.00	0.022
L_5	scrubber 2 inlet stream	48.34	0.83	G_1	CO_2 contaminated hot stream	130.00	3.60
L_6	scrubber 2 outlet stream	48.84	0.82	G_2	absorber inlet cold stream	62.42	3.33
L_7	absorber outlet stream	59.57	15.69	G_3	absorber outlet stream	49.92	2.97
L_8	stripper inlet stream	48.02	0.02	G_4	CO_2 free stream	48.80	2.98
L_9	stripper inlet stream	90.00	15.60	G_5	stripper outlet stream	96.83	1.49
L_{10}	scrubber 3 inlet stream	35.00	1.02	G_6	CO_2 rich stream	92.20	1.40

The elementary steps for the reaction are represented by the following equilibrium reactions:

$$2H_2O \quad H_3O^{+}+OH \tag{1}$$

$$CO_2+2H_2O \quad H_3O^{+}+HCO^{-}_3 \tag{2}$$

$$HCO^{-}_3+H_2O \quad H_3O^{+}+CO^{-2}_3 \tag{3}$$

$$MEAH^{+}+H_2O \quad H_3O^{+}+MEA \tag{4}$$

$$MEACOO^{-}+H_2O \quad HCO_3^{-}+MEA \tag{5}$$

Simulation Results and Discussion

The absorber CO_2 molar gas fraction YCO_2 is presented in Figure 1 with respect to the CO_2 molar liquid fraction XCO_2. As predicted by theory, the absorption operation line is located above the equilibrium one when gas is absorbed by the liquid, and the opposite occurs in the stripper when gas is desorbed by the liquid. The feeding (G1) and leaving (G4) CO_2 absorption column concentrations are 0.09265 and 0.00336, respectively. The CO_2 capture percentage is 96% and the purity percent (G6) in the regeneration column is 98 % with $YCO_2 = 0.9391$.

Figure 2 shows the temperature T, XCO_2 and YCO_2 against the number of mass transfer stages (Nmts, it is the height of packing which gives a change in compositions equivalent to one transfer unit) at the absorber. The temperature increases from 60 to 74°C from stages 1 to 4, then remains approximately constant for stages 4 to 8 and by the end decreases to 50°C. The YCO_2 value decreases from 0.0832 (stage 1) to 0.0111 (stage 5), and then slowly keeps decreases to 0.00081437 (stage 10) with an average value of 0.00598361. The XCO_2 value decreases from 0.047 to 0.026.

Figure 3 describes the YCO_2 behavior at the outlet stream G3 with respect to the number of design stages (Nds). The outlet YCO_2 values are 0.021 (stage 5), 0.016 (stage 10) and 0.0125 (stage 20). From stages 5 to 10, the change is significant, but then until stage 20 the variation is small, and so the inclusion of more stages is not economically justified.

In Figure 4 we present the CO_2 absorption percentage at the inlet and outlet streams as a function of Nds. The absorption percentage has a moderate increase until Nds = 10 for all values of the lean load (it is defined as the CO2/MEA mole ratio), but for greater Nds, the increase slows down for 0.20 and maintains and almost constant value for 0.20. The absorption percentage is lower for 0.25 . This suggests that the absorption column should be operated with low values in order to have high absorption percentages. The YCO_2 value increases moderately until Nds=10 but then there is no further significant increase. Hence, it is not justified an Nds greater than about 10.

Figure 5 presents the absorption percentage with respect to the mass flow stream L4. It is interesting that the absorption percentage is rather insensitive upon Nds. The absorption percentage increases from 60% to roughly 95% as L4 goes from 200 to 360 kg/h. Then it remains approximately constant up to 500 kg/h. The maximum absorption of about 98% occurs at Nds=10 at which L4 is 360 kg/h.

The enriched load in L7, is depicted in Figure 6 as a function of the lean load in L4. In here, the enriched load is also rather insensitive upon Nds. The maximum enriched load in L7 is 0.46 mol CO_2/mol MEA with a lean load of 0.26 mol CO_2/mol MEA in L4.

Figure 7 presents T, XCO_2 and YCO_2 against Nmts at the stripper. The values of XCO_2 and YCO_2 increase with Nmts. At the bottom (stage 1) the CO_2 concentration is small because at this stage CO_2 stripping ends. At the top of the column (stage 7) we have the highest CO_2 gas concentration because this is the point at which the liquid stripping begins. The YCO_2 value is 92% at the dry base in the top the column, while the XCO_2 value at the bottom is 2.5%.

Figure 8 is similar to Figure 7 but for a simulation carried out with 15 Nds. For the first 8 stages, the T, XCO_2 and YCO_2 values remains almost constant and then in the next 7 stages, that is, from stages 8 to 15, the behavior is similar to that showed in Figure 7, which means that only 7 stages are required for the efficient operation of the stripper.

Figure 9 shows the T and YCO_2 values at stripper with respect to Nmts. The temperatures decreases while YCO_2 increases with Nmts. The pressure goes from 1 to 1.5 bars. Greater pressure produces higher CO_2 captures. If the operation pressure increase, less energy is required for CO_2 removal in the stripper. For example, for 50% CO_2 removal, we require 50kW and 100 kW for 1.5 and 1 bars, respectively. It is then not economically justified to operate at 1.5 bars unless it is desired to reduce the heat load provision in the boiler.

We then studied the variation of the temperature with Nds for load boiler values at the stripper from 15 to 50 kW as is shown in Figure 10. The temperature for all cases remains almost constant up to stage 7 and then slowly drops from about 103 C to much lower values. An interesting featur e in the simulation is that there is an inflection point at stage 7. This result is unrealistic and so we can only trust these results up to stage 7.

Figure 11 presents XCO_2 and YCO_2 against Nds for load boiler values at the stripper from 15 to 50 kW. The liquid concentrations have a significant decrease during the first two stages, and then we obtain a plateau where all concentrations remain almost constant until about stage 9. The behavior then changes and the concentrations increase by a significant amount until stage 15. On the other hand, the gas concentration remains constant during the first 8 stages and then increases significantly, following a similar evolution as the liquid phase.

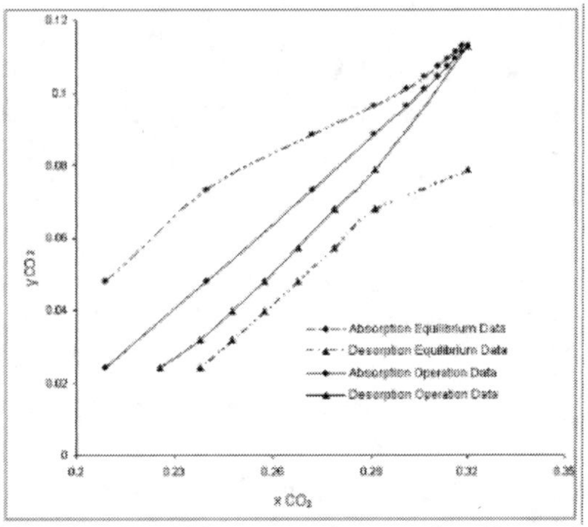

Figure 1. Equilibrium and operation lines of absorption and stripper columns.

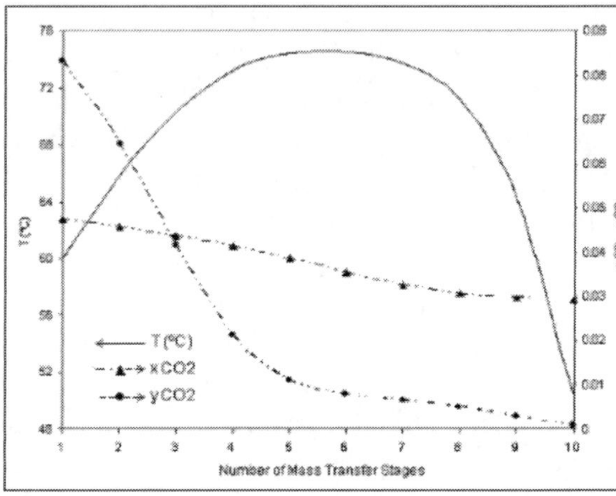

Figure 2. Temperature and CO_2 concentration profile at absorber with respect to the number of mass transfer stages.

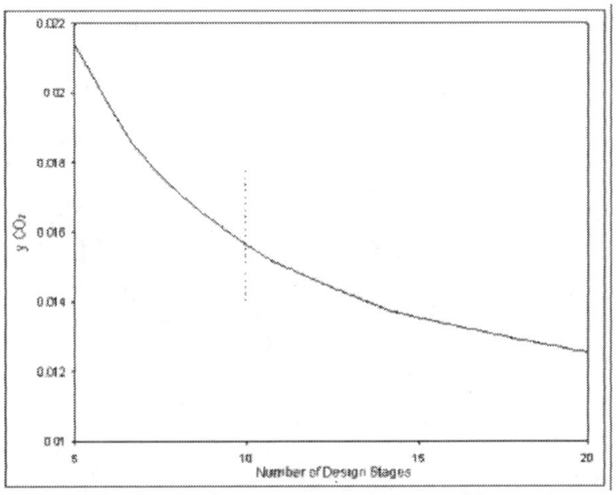

Figure 3. Gas CO_2 concentration profile at absorber with respect to design mass transfer stages.

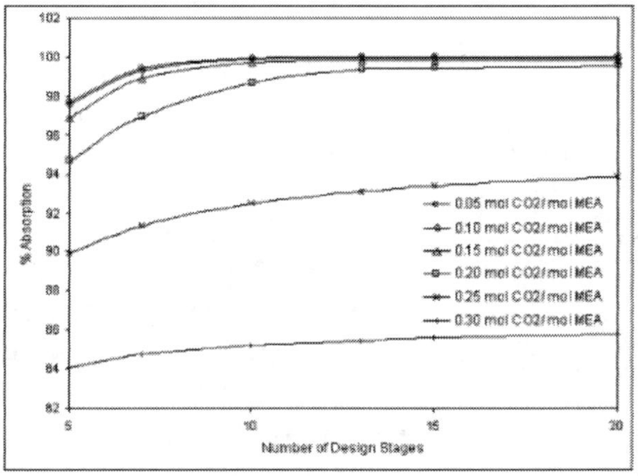

Figure 4. CO_2 absorption percentage with respect to the number of design stages, at different lean load L4.

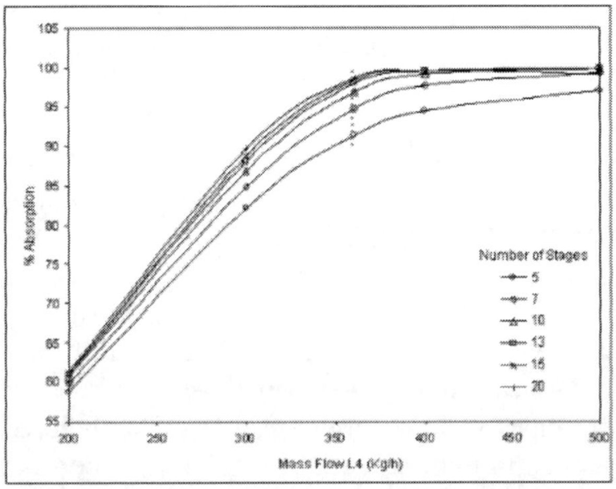

Figure 5. CO_2 absorption percentage profile with respect to lean load L4 at different number of design stages.

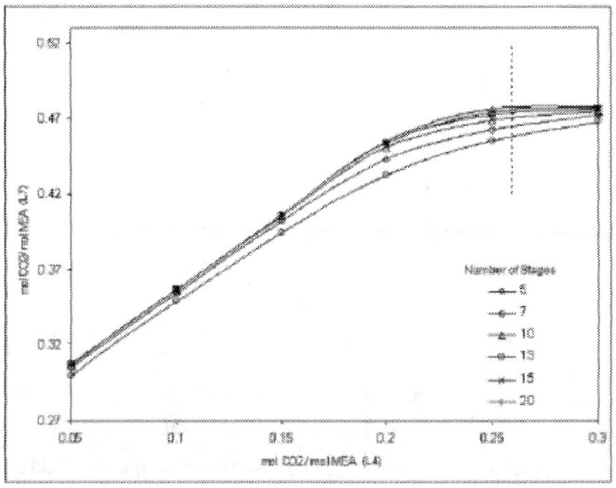

Figure 6. The rich load L7 with respect to lean load L4 at different design stages.

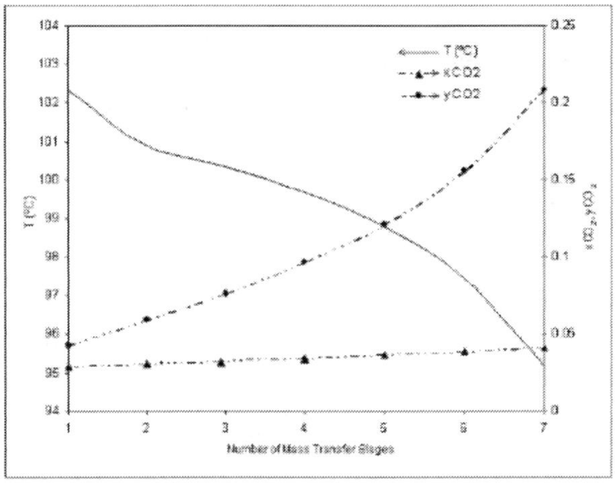

Figure 7. Temperature and concentration profiles at stripper with respect to the number of mass transfer stages.

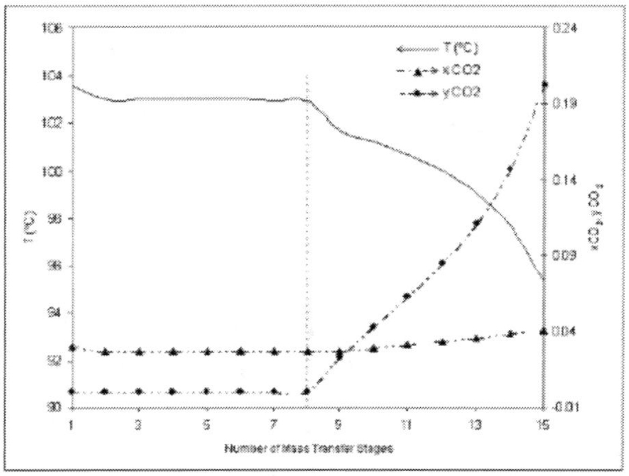

Figure 8. Temperature and CO_2 concentration at the stripper with respect to the number of mass transfer stages.

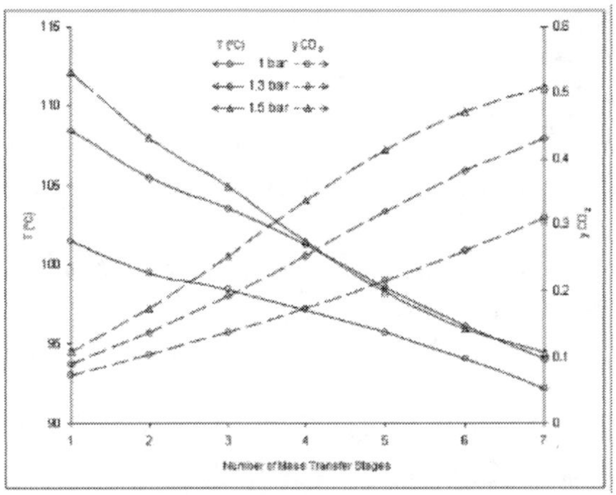

Figura 9. Temperature and CO_2 concentration profiles at the stripper with respect to the number of mass transfer stages from 1 up to 1.5 bar.

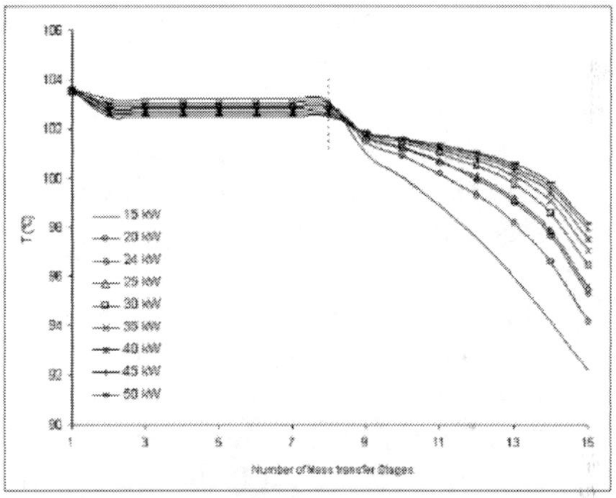

Figure 10. Temperature profiles at stripper with respect to the number of mass transfer stages at different boiler load.

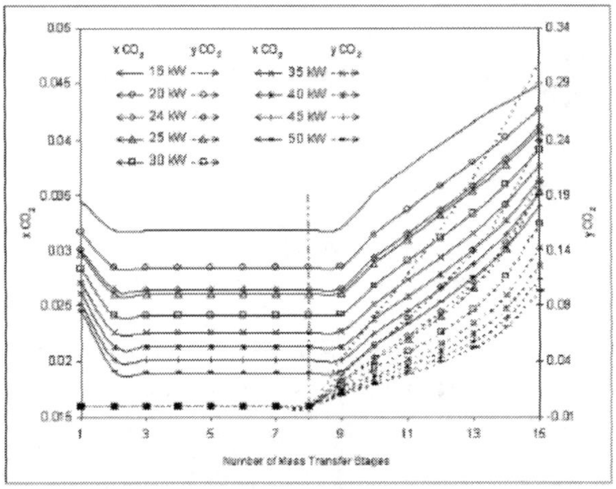

Figure 11. Liquid and gaseous CO_2 concentration profiles at stripper for 15 mass transfer stages at different boiler loads.

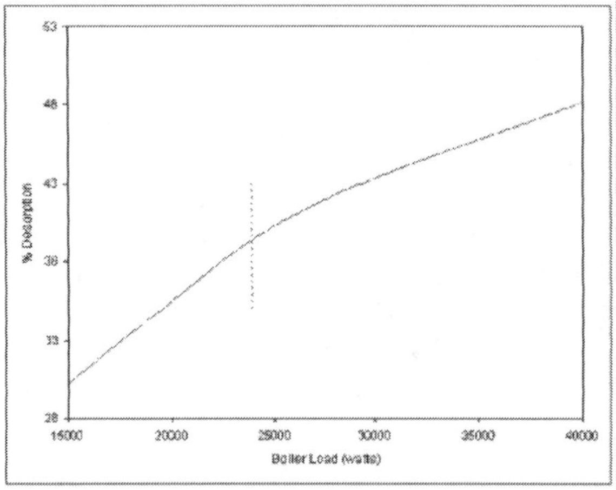

Figure 12. Desorber percentage with respect to boiler load.

Finally, in Figure 12 we present the desorption percentage with respect to the boiler load. We obtain an almost linear increase with boiler load with an inflection point at 24 kW. At this point the percentage of CO_2 removal is 39%.

Conclusions

In this work we have the evaluation of obtaining the heat required for amine absorption for a particular recovery of carbon dioxide for a given set of equipment specifications and operating conditions in a Mini-Plant, and to develop a model that simulates the removal of CO_2 using MEA absorption from flue gas by using the chemical processes simulator Aspen Plus™.

The main conclusiones are:

- The CO_2 capture percentage is 96% and the purity percent in the regeneration column is 98 %. It occurs at the number of design stages of 10 at which absorber intet stream is 360 kg/h.
- From stages 5 to 10 at absorption column, the gas and liquid CO_2 concentration change is significant, but then until stage 20 the variation is small, and so the inclusion of more stages is not economically justified.
- The absorption column should be operated with low values in order to have high absorption percentages. The YCO_2 value increases moderately until the number of design stages of 10 but then there is no further significant increase. Hence, it is not justified an the number of design stages greater than about 10.
- The optimum parameters have been determined for this size of the Plant, which are: 10 stages at absorber and 7 stages at stripper, both operated at 1bar.
- The maximum enriched load in L7 is 0.46 mol CO_2/mol MEA with a lean load of 0.26 mol CO_2/mol MEA in L4.
- In the desorption column, at 50% CO_2 removal, it requires 50kW and 100 kW for 1.5 and 1 bars, respectively. It is then not economically justified to operate at 1.5 bars unless it is desired to reduce the heat load provision in the boiler.
- In the desorption column, it is obtained an almost linear increase with boiler load with an inflection point at 24 kW. At this point the percentage of CO_2 removal is 39%. This process requires 24000kW boiler load in order to regenerate the solvent.
- The developed model that simulated the removal of CO_2 using MEA absorption lets simulate others internals (Mellapak) as the one developed at the ININ, and the use of new solvents with known characteristics.

Acknowledgment

Financial partial support for this work was provided by the Consejo Nacional de Ciencia y Tecnologia (CONACyT), project I36297-U, by the Consejo del Sistema Nacional de Educación Tecnológica (COSNET), project 643.02-PA and the Secretaría de Relaciones Exteriores of México. The authors credit the help of Miss Jazmin Cortez and Mr. Cesar Garcia to edit the Figures.

References

1. Afshin, M et al. (August 2004), Canada's Greenhouse Gas Inventory 1990 -2002, Greenhouse Gas Division, Environment Canada, Ottawa, Ontario, K1A 0H3 .
2. R. Hilda Chávez T., "Study of CO2 Captur e Process by Reactive Absorption", Technical Report 66/2006 at Institute Technische Thermodynamic and Thermische Verfahrenstechnik (ITT) University Stuttgart, Germany ", 30 April 2006 and IT.G.C.AMB/DFR/09/2006 at Instituto Nacional de Investigaciones N ucleares, July 2006.
3. Rosa-Hilda Chavez, Javier de J. Guadarrama and Jaime Klapp, " CO_2 Capture for Atmosphere Pollution Reduction", Towards a Clener Planet, Energy for the Future, Environmental Science and Engineering, Subseries: Environmental Science, Springer, Germany, p99-111, ISSN 1863-5520, 2007.
4. Information on Greenhouse gas sources and sinks, Environment Canada, http://www.ec.gc.ca/pdb/ghg/overview_e.cfm , September 10, 2002.
5. IPCC (2001b), Climate Change 2001: The Scientific Basis, Contributio n of Working Group 1 to the Third Assessment Report of the Intergovernmental Panel on Climate Change, Cambridge University Press, U.K. 944 pp.
6. Pidwirny, M, Fundamentals of Physical Geography, Chapter7: Introduction to the Atmosphere, Department of Geogr aphy, Okanagan University College, B C. http://www.physicalgeography.net/fundamentals/7h.html, August 22, 2004.
7. Sullivan, K. M. (1990), Coal and the greenhouse issue, Paper 90 -141.5, presented at the 83rd Annual Meeting of the Air and Waste Management A ssociation, Pittsburgh, Pennsylvania, June 24–29.
8. Thompson, A. M., K.B. Hogan, and J.S. Hoffman (1992), Methane reductions: Implications for global warming and atmospheric climate change, Atmospheric Environment, 26A (14): 2665–2668.
9. United Nations Environmental Program (UNEP) and UN Framework Convention on Climate Change (2001): Climate Change Information Kit.

EVALUATION OF THE ABSORPTION OF SULFUR DIOXIDE IN WATER USING THREE TYPES OF STRUCTURED PACKING

Rosa-Hilda Chavez, Instituto Nacional de Investigaciones Nucleares, Estado de México, México .

Abstract

In order to determine the height of a separation colu mn, hydrodynamic and mass transfer models are necessary to evaluate the pressure drop and the height of the global mass transfer u nit. Starting from experimental values, such adjusted models have shown a high level of accuracy, confirming the validity and merit of different systems and packings. The present work evaluates SO_2 absorption in a water-based process. The absorption column uses three kinds of mesh corrugated structured packing, one developed in the Mexican National Institute of Nuclear Research (ININ, for its acronym in Spanish) and other similar packings, named Sulzer BX and Mellapak. The optimization was made applying a Two Film model, which is based on the number of global mass transfer units, involving the separation efficiency as a function of the height of a global transfer unit. A comparison of the height of the mass transfer unit on gas film and liquid film was made with respect to the equilibrium line slope, for the three packings considered. The results for the pressure drop, mass transfer coeficients and heigth of the mass transfer unit show a high influence of the estimated value on the separation efficiency from the slope equilibrium line. The deviation of the estimated values from the experimental data is 10%. The Sulzer BX packing showed higher separation efficiency than the ININ and Mellapak packing. These results allow the modification of several parameters of the structured packing, in order to achieve a fitting geometric specification for a very specific s eparation-process application.

Keywords: structured packing; absorption; sulfur dioxide

Nomenclature

Symbol	Process data	Unit
a	surface area per unit packed volume	m^2/m^3
a_g	geometric area of the packing	m^2/m^3
a_{Ph}	interfacial area per unit packed volume	m^2/m^3
B	length of the flow channel	m
C	constant	
d_h	hydraulic diameter	m
D	diffusion coefficient	m^2/s
g	gravitational constant	m/s^2
h_L	liquid holdup	m^3/m^3
HTU	height of a mass transfer unit	m

Symbol	Process data	Unit
L	mass flow of liquid	kg/h
m	slope of the equilibrium line	
n	exponent	
n_t	number of gauze threads per square foot	
NTU	number of transfer units	
S	side length	m
u	superficial load	m/s
$\tilde{u}$	average effective velocity	m/s
V	mass flow of gas	kg/h
Z	column total packed height	m

Symbol	Process data	Unit
a	surface area per unit packed volume	m^2/m^3
a_g	geometric area of the packing	m^2/m^3
a_{Ph}	interfacial area per unit packed volume	m^2/m^3
B	length of the flow channel	m
C	constant	
d_h	hydraulic diameter	m
D	diffusion coefficient	m^2/s
g	gravitational constant	m/s^2
h_L	liquid holdup	m^3/m^3
HTU	height of a mass transfer unit	m

Symbol	Process data	Unit
L	mass flow of liquid	kg/h
m	slope of the equilibrium line	
n	exponent	
n_t	number of gauze threads per square foot	
NTU	number of transfer units	
S	side length	m
u	superficial load	m/s
$\tilde{u}$	average effective velocity	m/s
V	mass flow of gas	kg/h
Z	column total packed height	m
α	corrugated angle with respect to vertical axis of the column	

Greek symbols		
β	mass transfer coefficient	m/s
ε	void fraction	m^3/m^3
θ	corrugated angle	°
μ	dynamic viscosity	$kg/m\,s$
λ	stripping factor	
ξ	resistance coefficient	
ρ	density	kg/m^3

Subscripts		
Fl	flooding point	
L	liquid	
o	overall	
S	loading point	
V	gas	
w	water	

Introduction

Atmospheric pollution in cities with a large population is a significant problem. A percentage of this pollution comes from the burning of heavy oil, a primary source of energy, with high pollution effects due to exhaust products (gases and particles) that contaminate the atmosphere. Even though new laws seek to control atmospheric pollution there are no solutions adapting to each country characteristics. Heavy oil is used in diverse productive activities such as electric power generation, cements, ceramics, glass and bricks for the construction industry , etc. [1].

The main purpose of this paper is to present the experimental evaluation of the absorption of sulphur dioxide in water with high efficiency packings, made in the *Mexican National Institute of Nuclear Research* (ININ, for its acronym in Spanish), using hydrodynamic and mass transfer models. The behaviour of the models were compare using adjusted parameters of ININ structured packing and the reported parameters of other similar structured packings: Sulzer BX and Mellapak.

Materials and Methods

The methodology was divided in two parts: evaluation of adjusted parameters of ININ packi ng, using hydrodynamic and mass transfer models in order to determine liquid hold up, recovery percentage and height of the global mass trasfer, respectively [2, 3] for each packing and the camparison with other similar packings [4].

Hydrodynamic and mass transfer models

In the model the bed flow is upward for the gas and downward for the liquid. It is assumed that under stationary conditions, the inertial forces are neglected, and the shear and gravity forces at the surface of the film are in equilibrium with the shear forces in the gas stream. Operating parameters are the liquid load and the gas velocity, which also affects the liquid hold -up. As expected, the differential equation solution depends upon the flow pattern.

Up to now, the only equations that were known for calculating the mass transfer during a two-phase countercurrent flow in packed columns were those that apply to the range extending up to the loading point [5].

The gas and liquid streams flow separately through the column below but not ab ove this point. Above it, the shear stress in the gas stream supports an increasing quantity of liquid in the column, with the result that the liquid holdup greatly increases. Finally, at the flood ing point, the liquid accumulates to such an extent that co lumn instability occurs. Mass transfer in this upper loading range can be described if these fluid dynamic relationships are taken into consideration.

Hydrodynamic model for hazardous and structured packings

A model that describes the fluid dynamic relat ionships in packed columns with countercurrent flow of the gas and liquid phases was developed in a previous work by Billet and Schultes [8]. This model allows describing the flow conditions up to the flood point. This model assumes that the void fraction in a bed of packing could be represented by a multiplicity of vertical channels through which the liquid flows downwards in the form of a film countercurrent to the ascending gas stream. This model also calculates the mass transfer in the loading range up to the flood point [5].

The liquid hold-up $h_{L,S}$ and the model flow factor ξ_S [6] have been used for the packed columns prediction. The load point is described by the equations:

$$u_{V,S} = \sqrt{\frac{1}{\xi_S}}\left(\varepsilon - h_{L,S}\right)\sqrt{\frac{h_{L,S}}{a_p}A} \qquad (1)$$

$$h_{L,S} = \left(12\,\frac{1}{g}\,\frac{\mu_L}{\rho_L}\,u_{L,S}\,a_p^{\,2}\right)^{1/3} \qquad (2)$$

$$\xi_S = \frac{g}{C_S^2\left[\frac{L}{V}B\left(\frac{\mu_L}{\mu_V}\right)^{0.4}\right]^{2n_S}}, \qquad \text{where: } A = \sqrt{\frac{\rho_L}{\rho_V}}, B = \sqrt{\frac{\rho_V}{\rho_L}} \qquad (3)$$

The liquid holdup at the flooding point $h_{L,Fl}$ must be determined iteratively from Eq. (5) for the mass flow ratio $\frac{L}{V}$ that relates to the specific problem. In this case, the only values of physical significance are those within the range $\frac{\varepsilon}{3} \leq h_{L,Fl} \leq \varepsilon$.

$$u_{V,Fl} = \sqrt{2}\sqrt{\frac{g}{\xi_{Fl}}}\frac{(\varepsilon - h_{L,Fl})^{3/2}}{\varepsilon}\sqrt{\frac{h_{L,Fl}}{a_p}A} \tag{4}$$

$$h^3_{L,Fl}(3\,h_{L,Fl} - \varepsilon) = \frac{6}{g}a_p^{\,2}\;\varepsilon\;\frac{\mu_L}{\rho_L}\;\frac{L}{V}\;B\;u_{V,Fl} \tag{5}$$

$$\xi_{Fl} = \frac{g}{C^2_{Fl}\left[\frac{L}{V}B\left(\frac{\mu_L}{\mu_V}\right)^{0.2}\right]^{2\,n_{Fl}}} \tag{6}$$

The gas stream exerts a strong effect on the liquid holdup above the loading point. The liquid holdup $h_{L,S}$ up to the loading point can be calculated from Eq. (2); and that at the flood point $h_{L,Fl}$ from Eq. (5).

Eqs. (7-11) determine the liquid holdup at the loading point, the loading capacity, liquid holdup at the flooding point, between loading and flooding points. The increase in h_L with the gas load can be expressed by Eq. (11).

$$u_V = 70\%Flooding\;\;u_{V,Fl} \tag{7}$$

$$u_L = \frac{L}{V}\;\frac{\rho_V}{\rho_L}u_V \tag{8}$$

$$h_{L,S} = \left(12\;\frac{1}{g}\;\frac{\mu_L}{\rho_L}\;u_L\;a_p^{\,2}\right)^{1/3} \tag{9}$$

$$h^3_{L,Fl}(3\,h_{L,Fl} - \varepsilon) = \frac{6}{g}a_p^{\,2}\;\varepsilon\;\frac{\mu_L}{\rho_L}\;\frac{L}{V}\;A\;u_V \tag{10}$$

$$h_L = h_{L,S} + (h_{L,Fl} - h_{L,S})\left(\frac{u_V}{u_{V,Fl}}\right)^{13} \tag{11}$$

Mass transfer model for structured packing.

The two-resistance model [6] is used, with the assumption of thermodynamic equilibrium at the phase interface. This is useful for either the rate-based or the equilibrium stage-based computational routines. The basic parameters of the model are the gas (or vapor) and liquid phases mass transfer coefficients β_V and β_L, respectively, and the effective interfacial area a_{Ph}:

$$\frac{a_{Ph,S}}{a} = 1.5\left(a_p d_h\right)^{-0.5} \mathrm{Re}_L^{-0.2} We_L^{0.75} Fr_L^{-0.45}, \text{ where: } \mathrm{Re}_L = \frac{u_L\, d_h}{\mu_L},\ We_L = \frac{u_L^2\, \rho_L\, d_h}{\sigma_L},\ Fr_L = \frac{u_L^2\, \rho_L\, d_h}{\sigma_L} \quad (12)$$

$$\frac{a_{Ph,FI}}{a} = 7\left(\frac{\sigma_L}{\sigma_W}\right)\frac{a_{Ph,S}}{a} \quad (13)$$

$$\frac{a_{Ph}}{a} = \frac{a_{Ph,S}}{a} + \left(\frac{a_{Ph,FI}}{a} - \frac{a_{Ph,S}}{a}\right)\left(\frac{\mathrm{u_V}}{\mathrm{u_{V,FI}}}\right)^{13} \quad (14)$$

$$\tilde{u}_L = \frac{\mathrm{u_L}}{h_L}\left\{1 - \left(\frac{\mathrm{u_V} - \mathrm{u_{V,S}}}{\mathrm{u_{V,FI}} - \mathrm{u_{V,S}}}\right)^2\right\}, \text{ If } \mathrm{u_{V,S}} \le \mathrm{u_V} \le \mathrm{u_{V,FI}} \quad (15)$$

$$\tilde{u}_L = \left(\frac{g\, \rho_V^2\, U_V^2}{12\, \mu_L\, a_p\, \rho_L}\right)^{1/3}\left(\frac{L}{V}\right)^{2/3}\bar{C}, \text{ where } \bar{C} = 1 - \left(\frac{u_V - u_{V,S}}{u_{V,FI} - u_{V,S}}\right)^2 \quad (16)$$

$$\beta_L a_{ph} = C_L\ 12^{1/6}\, \tilde{u}_L^{1/2}\left(\frac{D_L}{d_h}\right)^{1/2} a\, F \quad (17)$$

$$\beta_V a_{ph} = C_V \frac{1}{(\varepsilon - \mathrm{h_L})^{1/2}} \frac{a^{3/2}}{d_h^{1/2}} D_V G^{3/4} E^{1/3} F, \text{ where: } F = \frac{a_{Ph}}{a}, G = \frac{\mathrm{u_V}}{a\, \mu_V}, E = \frac{\mu_V}{D_V} \quad (18)$$

All the constants were determined by the experiments reported in this work, using an absorption column with a diameter of 0.252 m, with stainless steel gauze corrugated sheet packing, named ININ, by means of the air-water and SO_2-water systems.

The two-film model is based on the number of gas and liquid resistance transfer global units, NTU, that are related to the efficiency in terms of the height of a transfer global unit HTU [7].

The column total packed height Z for the gas and liquid are:

$$Z_V = HTU_{OV} * NTU_{OV} \quad (19)$$

$$Z_L = HTU_{OL} * NTU_{OL} \quad (20)$$

The application of the two-film model is frequently used to relate the height of the transfer global unit (HTU_{OV} or HTU_{OL}) with the height of the gas HTU_V and liquid HTU_L transfer units to the absorption:

$$HTU_{OV} = HTU_V + \lambda\, HTU_L \quad (21)$$

$$HTU_{OL} = HTU_L + \frac{1}{\lambda} HTU_V \quad (22)$$

The height of the transfer global units HTU_{OV} is determined through the expression:

$$HTU_{OV} = HTU_V + \lambda NTU_L, \text{ where } HTU_V = \frac{u_V}{\beta_V a_{Ph}} \text{ and } NTU_L = \frac{u_L}{\beta_V a_{Ph}} \quad (23)$$

The quality λ is given by $\lambda = m\frac{V}{L}$ (24)

where m is the ratio of the slope of the equilibrium line to the operation line, and $\frac{L}{V}$ is known as the removed factor. The absorption factor is the inverse of λ.

Use of Structured Packing

Table 1 shows the geometric characteristics of three structured packing.

Table 1. Geometric Characteristic of the Structured Packings [3, 4].

Packing	**ININ**	**Sulzer BX**	**Mellapak**
Material	Stainless Steel	Stainless Steel	Grooved Perforated Plastic sheet
α	35	30	45
n_t	36	60	-
B (m)	0.0165	0.012	0.023
S	0.012	0.009	0.017
θ (°)	45	45	45
ε	0.9633	0.90	0.95
a_g m^2/m^3	418	498	249.18

Results and Discussion

Figures 1 and 2 describe the behavior of the liquid hold up, SO_2 recovery percentage and the HTU_{OV}, respectively, for the SO_2-air/water system at 304.16 K and a pressure of 69.864 kPa. A column of 2.52 m of packed bed height and 0.252m diameter was used. The SO_2 fraction mol of the feeding gas was 0.0012035.

Tables 2 and **3** show adjustment parameters for the ININ packing at loading and flooding points, and the mass transfer behavior between those regimens. The deviation of the estimated values from the experimental data is 10%.

Figure 1 shows the liquid holdup as a funtion of the gas velocity of the three types of structured packing. The liquid holdup has a moderate increase gas velocity from 0.1 to 2 m/s. For greater velocities, the liquid holdup values are rather sensitive. The ININ packing let manage higher gas flow rates than the others two packings. This result is a consequense of the lower pressure drop of ININ packing.

Figure 2 shows SO_2 recovery percentage and the overall height of a mass transfer unit predicted values with respect to packed height. The SO_2 recovery percentage increases as the packed height does. Between 1 to 2 m of packed height, the SO_2 recovery percentage is almost a constant value for ININ 18 and Sulzer BX packing, and between 2 to 3.5 m/s for Mellapak packing, at SO_2 recovery of 66%. At SO_2 recovery of 59%, we obtained 2, 2.4 and 3 m of packed height for Sulzer BX, ININ18 and Mellapak Y250 packing, respectively.

At 70% of the flooding point, the overall height of a mass transfer unit was determined to be 3.33 m. Increasing such percentage up to 90%, decreases the HTU_{OV} to 0.98m, thus achieving lower heights in the stripping of SO_2.

Table 2. Results for the Loading and Flooding Regimen Adjustment Parameters.

$u_V = m/s$	$u_L = m/s$	C_S	n_S	C_{Fl}	n_{Fl}
	0.0234777	20	-0.723	15	-0.708
	0.0221484	18.2	-0.723	13.5	-0.708
	0.0183244	16.8	-0.723	12.5	-0.708
1.014285673	0.0159110	16	-0.723	11.7	-0.708
	0.0136070	14.7	-0.723	10.7	-0.708
	0.0103570	9.4	-0.326	9	-0.194
	0.0073521	8.7	-0.326	8.5	-0.194

Table 3. Results For the Mass Transfer Adjustment Parameters.

$u_V = m/s$	$u_L = m/s$	C_L	C_V
	0.0234777		
	0.0221484		
	0.0183244		
1.014285673	0.0159110	1.963	1.02
	0.0136070		
	0.0103570		
	0.0073521		

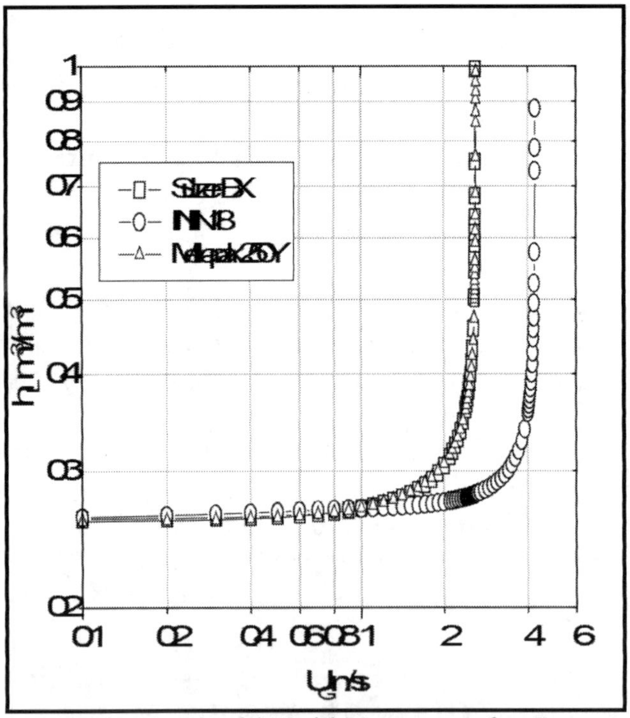

Figure 1. Liquid Hold Up versus the Gas Velocity of Sulzer BX, ININ 18 y Mellapak Y250.

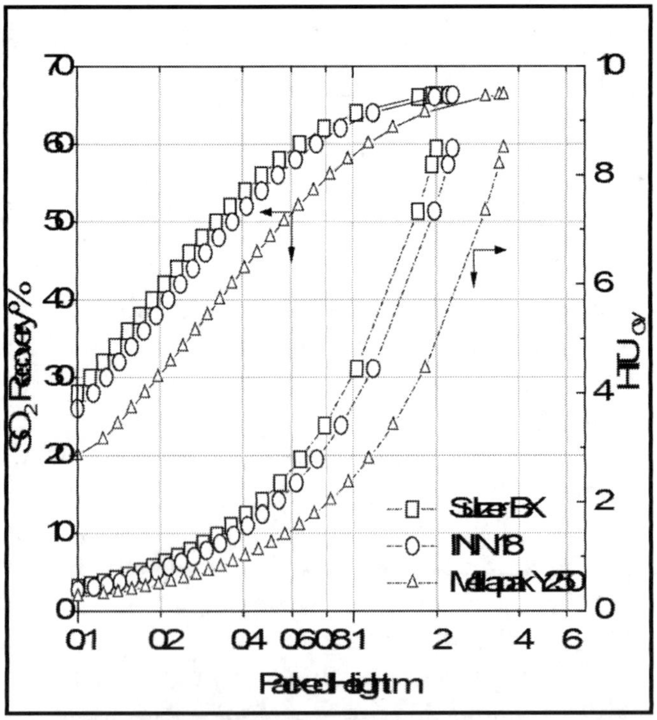

Figure 2. SO_2 recovery percentage and HTU_{OV} versus the column packed height of Sulzer BX, ININ 18 & Mellapak Y250, at liquid and gas velocities of 0.023 and 1.014 m/s, respectively.

Conclusions

In this work the evaluation of obtaining the experimental evaluation of the absorption of sulphur dioxide in water with high efficiency packings by using hydrodynamic and mass transfer models has the main conclusions:

- The fluid dynamics and mass transfer description takes into account the great diversity in the geometry, structure and materials for packings in industrial columns, which normally entails differences in the fluid dynamics under operation conditions and thus in the packing performance.
- The deviation of the estimated values from the experimental data is 10%.
- Based on the present results it seems like the recommended packing for SO_2 recovery is the ININ packing, however more experimental data is necessary.
- At 70% of the flooding point, the overall height of a mass transfer unit was determined to be 3.33 m. Increasing such percentage up to 90%, decreases the HTU_{OV} to 0.98m, thus achieving lower heights in the stripping of SO_2.

Acknowledgment

Partial financial support of this work was provided by Mexican Foreign Relations Office. The author credit the help of Mr. Omar medina and Mr. Cesar García-Atanacio to edit Figures.

References

1. Chávez, T.R.H. y Rivera, D.D., Especificaciones del horno tabiquero de prueba, en el Municipio de San Bartolomé Tlaltelulco, Metepec, Estado de México, IT,CA -DEA-9817, ININ 1998.
2. Chávez R.H. y Guadarrama J.J., "Ajuste de un Modelo Hidrodinámico para una Columna con Empaques Estructurados usando Valores Experimentales", Información Tecnológica - Revista Internacional – La Serena –Chile, Vol. 15, No. 2, 2004, pp. 19-24.
3. Fair, J.R., Seibert, A.F., Behrens, M., Saraber, P.P. and Olujic, Z., "Structured packing performance-Experimental evaluation of two predictive models", Ind. Eng. Chem. Res., Vol. 39, 2000, pp. 1788-1796.
4. Olujic, Z., Seibert, A.F. and Fair, J.R., "Influence of corrugation geometry on the performance of structured packings: an experimental study", Chemical Engineering and Processing, Vol. 39, 2000, pp. 335-342.
5. Billet R. and Schultes, "Fluid Dynamics and Mass Transfer in the Total Capacity Range of Packed Columns up to the Flood Point", Chem. Eng. Technol. Vol. 18, 1995, pp. 371 -379.
6. Billet R. "Contribution to desing and scale-up of packed columns", Fat Sci. Techonol., No. 9, 1990, pp. 361-370.
7. Sherwood, T.K., Pigford, R.L. and Wilke C.R., "Mass transfer", International Student Edition, Japan, 1975.

MULTIPLE STEADY STATES IN CATALYTIC DISTILLATION FOR DEEP HYDRODESULFURIZATION OF DIESEL

A. Rosales[a], *F. D. Vargas-Villamil*[a]
[a]*Programa de Matemáticas Aplicadas y Computación*
Instituto Mexicano del Petróleo, México, D.F.

Abstract

Catalytic distillation (CD) is a technology that has the potential to reduce the capital and operating costs of the oil fraction deep hydrodesulfurization (DHDS) process. However, integrating the reaction and separation in situ within a single unit leads to complex nonlinear interactions. As a result, multiple steady states (MSS) and complex dynamics may arise. Thus, MSS must be taken into consideration during design, operation, and control. This work deals with the steady state multiplicities for DHDS via catalytic distillation of a real light gas oil (LGO) fraction. First, a CD cost optimal design is obtained, then bifurcation plots for both the Dibenzothiophene (DBT) and 4,6-DiMethyl Dibenzothiophene (4,6-DMDBT) conversions are presented against the reboiler duty, bottom flow rate, reflux ratio, and catalyst loading. The classical S-shape is obtained for the reboiler duty, reflux ratio, and catalyst loading for the DBT conversion but not for the 4,6-DMDBT conversion. When the bottom flow rate is the continuation parameter, *isola* branches arise for both the DBT and 4,6-DMDBT conversions. Input-output multiplicities exist for both DBT and 4,6-DMDBT when the bottom flow rate is the continuation parameter. A process with input-output multiplicities may have a zero gain which makes it uncontrollable, while *isolas* may lead to a region were the zero dynamics are unstable. To ease operation and control, the operation point must be in a region characterized by high total sulfur conversion and no multiplicity.

Introduction

The hydrodesulfurization (HDS) of oil fractions is a key process in the refinery industry. In recent decades, it has caught the attention of both academia and industry due to more stringent environmental regulations. Worldwide, current legislations allow 15 to 50 ppm of sulfur in diesel fuel and the tendency is to reduce it to less than 15 ppm to produce ultra low sulfur fuels [1]. These regulations have become a serious challenge since they require the elimination of the refractory sulfur compounds, which may be eliminated at more severe operating conditions. In addition, the production of heavy crude oils, with high sulfur concentration, has been increasing since light oil reservoirs are scarce. To comply with the new laws, refineries will have to make important investments to improve their current HDS technology.

Catalytic distillation (CD) is a viable alternative to the conventional technology because it uses the reactants, the energy, and the services more efficiently. Hidalgo-Vivas and Towler [2] propose the use of CD for the HDS of diesel and present several column configurations. CD has been proposed for light gas oil HDS since it can reduce capital and operating costs when compared to the conventional

HDS [3, 4] (Fig. 1). In industry, CDTech claims to have the technology for gas oil HDS, but this information is proprietary [5].

The *in-situ* integration of the reaction and separation leads to a non linear interaction between the phase equilibriums and chemical kinetics in a countercurrent flow. As a result, multiple steady states (MSS), hysteresis, and a complex dynamic behavior may arise [6]. MSS have a significant impact during the design, operation, and control. Operating problems, as well as lower conversions or product purity, may be a result of the existence of MSS. This work deals with the steady state multiplicities for the DHDS via catalytic distillation of a real light gas oil (LGO) fraction. Bifurcation plots for catalyst mass, reflux ratio, heat duty, and bottom flow are presented.

LGO fraction and kinetics

Oil fractions are complex mixtures and often represented by pseudocompounds. They represent boiling ranges with a cut temperature that can be defined by the true boiling temperature and ASTM-D86 distillation curves. The boiling point range of the fraction used in this work is 180-370 °C [7]. Three pseudocompounds are defined, which represent the light, medium, and heavy gas oil fractions. Sulfur organic compounds are represented by DBT and 4,6-DMDBT. Both compounds are among the most refractory molecules and the most abundant in the gas oil. The reaction kinetics of both the DBT and 4,6-DMDBT have two reaction paths: hydrogenolysis and hydrogenation.

$$DBT + 2H_2 \rightarrow BiPh + H_2S \quad (1)$$
$$DBT + 5H_2 \rightarrow CHB + H2S \quad (2)$$

$$4,6\text{-}DMDBT + 2H_2 \rightarrow 3,\ 3'\text{-}DMBPH + H_2S \quad (3)$$
$$4,6\text{-}DMDBT + 5H_2 \rightarrow 3\text{-}MCHT + H_2S \quad (4)$$

Where BiPh is biphenyl, CHB is cyclohexylbenzene, 3,3'-DMBPH is 3,3'- Dimethyl biphenyl, and 3-MCHT is 3-Methylcyclohexyltoluene. In this work, the reaction rates proposed by Vanrysselberghe and Froment [8] and Vanrysselberghe et al. [9] are used for DBT and 4,6-DMDBT respectively. The reactions are carried out on a Ni-CoMo/Al_2O_3 catalyst.

Catalytic distillation column cost optimal design

The hydrodesulfurization catalytic distillation column described in Fig. 2 is optimized. The aim is to find the optimal value for the reflux ratio (RR), bottom flow rate (B), column diameter (D), condenser temperature (T_1), reboiler duty (Q_r), and hydrogen feed ratio ($F_{H2,1}/F_{H2,2}$) such that the Total Annual Cost (TAC) and the sulfur concentration ($C_{S,Diesel}$) are minimized. A compromise must be obtained between the TAC and sulfur concentration since they have opposite effects. This results in the following multi-objective optimization problem:

$$\min_{u=\{RR,B,D,M_{cat},Q_r,T_1,F_{H2,1}/F_{H2,2}\}} J(u) \quad (5)$$

$$J(u_k) = [w_1 \; w_2]\begin{bmatrix} TAC \\ C_{S,Diesel} \end{bmatrix} \quad (6)$$

s.t.

$$L_{j-1} - L_j + V_{j+1} - V_j + F_j + \sum_{i=1}^{i=n_c} R_{i,j} = 0 \quad (7)$$

$$L_{j-1}x_{i,j-1} + L_j x_{i,j} + V_{j+1}y_{i,j+1} - V_j y_{i,j} + F_j z_{i,j} + R_{i,j} = 0 \quad (8)$$

$$L_{j-1}H^L_{j-1} + V_{j+1}H^v_{j+1} - V_j H^v_j - L_j H^L_j + F_j H^F_j = 0$$

$$R_{i,j} - \sum_{k=1}^{k=n_r} M_{cat,j}\, v_{i,k}\, r_{k,j} = 0 \quad (9)$$

$$y_{i,j} - K_{i,j}x_{i,j} = 0$$

$$\sum_{i=1}^{i=n_c} x_{i,j} - 1 = 0 \quad (10)$$

$$\sum_{i=1}^{i=n_c} y_{i,j} - 1 = 0 \quad (11)$$

$$T_{bed} \le 648\ K \quad (12)$$

$$T_{NT} \le 673\ K \quad (13)$$

where,

$$TAC = Operation\ cost + \frac{Capital\ cost}{Payback}, \quad (14)$$

$$Operation\ cost = J_f + J_w + J_{cat} + J_{H_2}, \quad (15)$$

$$Capital\ cost = J_C + J_t + J_c + J_R, \quad (16)$$

J_f is the annual reboiler fuel cost, J_w the annual condenser cooling water cost, J_{cat} the annual catalyst cost, J_{H2} the annual hydrogen consumption cost, J_C the column cost, J_t the tray cost, J_c the condenser cost, and J_R the reboiler cost. A three year payback and a catalyst life of two years are assumed. The temperature in the reaction section must remain bound to avoid hydrocracking and to protect the catalyst while the bottom temperature must remain bound to avoid thermal cracking in the bottom section (400°C). The phase equilibria was modeled using the Soave-Redlich-Kwon equation of state.

Figure 3 shows the Pareto optimal front and the column optimal design. Reducing the sulfur content requires more severe working conditions and more catalyst, which increases de TAC. Therefore, a compromise must be achieved (Fig. 3a). Fig. 3b shows the optimal design when the sulfur concentration is 15 ppm. The optimization problem was solved using ASPEN Plus.

Multiple steady states

Next, bifurcation diagrams were constructed for the catalyst mass, reflux ratio (RR), heat duty (Q_r) and bottom flow rate as a function of the DBT and 4,6-DMDBT conversions. In these, input, output, and input-output multiplicities appear. Systems with multiplicities may be difficult to operate when the operating point is in the region where the multiplicity is, unless it is a stable point and its attraction region is large enough. So, it is recommended that the operating point be in an interval with only one solution.

Figure 4 shows the bifurcation plots of the mass loading, reboiler duty (Q_r), and reflux ratio (RR) with respect to the DBT conversion (x_{DBT}). Output multiplicities are found having the classical S-shape, which is characterized by having ignition and extinction regions that are stable and an intermediate conversion region that is unstable. These systems are difficult to control since one input results in more than one output. To control these systems more than one control input is required. Figure 5 shows the bifurcation plots of the mass loading, reboiler duty (Q_r), and reflux ratio (RR) as a function of the 4,6-DMDBT conversion ($x_{4,6\text{-}DMDBT}$). Input-output multiplicities having a crisscross shape are found. One input results in more than one output while one output may be produced by more than one input. A process with input-output multiplicities may have a zero gain which makes it uncontrollable. Figures 6 and 7 show the bifurcation plots for the DBT and 4,6-DMDBT as a function of the bottom flow rate. Similarly, input-output multiplicities are found; however, an isolated solution branch appears. In this region, the zero dynamics may be unstable. A classical isola branch appears for the DBT, while for the 4-6-DMDBT an intricate crisscross isola pattern with up to nine solutions arises. To our knowledge, only Chen et al. [10] have reported *isola* branches for reactive distillation.

Conclusions

A catalytic distillation column for the light gas oil hydrodesulfurization is studied. A Pareto optimal front of a multi-objective hydrodesulfurization catalytic distillation column is obtained. Then, bifurcation diagrams are obtained for the main operating variables, i.e., mass loading, RR, B, and Qr. Input multiplicities and input-output multiplicities are found. An *isola* branch, with up to nine solutions, appears for the bottom flow rate and crisscross shapes are found for the 4,6-DMDBT.

Acknowledgment

A. Rosales would like to thank the Instituto Mexicano del Petróleo and CONACyT for their financial support.

References

1. Mendoza-Canales, J., Perez-Castellanos, A., Galina-Hidalgo, S. and Romo-Rico, D. (2006). "Roadmap for Mexico's clean fuels". *Hydrocarbon Processing*, 85(2), pp. 57-62.
2. Hidalgo-Vivas, A. and Towler, G.P.. AIChE Annual Meeting, 1998, Miami, Florida.
3. Vargas-Villamil, F.D., Marroquin, J., de la Paz, C., and Rodriguez, E. (2004). "A catalytic distillation process for light gas oil hydrodesulfurization." *Chemical Engineering and Processing*, 43, pp. 1309-1316.
4. Rosales, A., Vargas-Villamil, F.D., and Arce-Medina, E. (2007). "Sensitivity analysis of light gas oil hydrodesulfurization column via catalytic distillation." *To be published in Catalysis Today*, 2007.
5. http://www.cdtech.com/.
6. Taylor, R. and Krishna, R. (2000). "Modeling of a reactive distillation." *Chemical Engineering Science*, 55, pp. 5183-5229.
7. De La Paz-Zavala, C., Rodríguez, E., and Vargas-Villamil, F. (2004). "An algorithm for pseudocompound delumping and lumping into homologous groups." *Petroleum Science and Technology*, 22(1-2):4560.
8. Vanrysselberghe, V. and Froment, G. (1996). "Hydrodesulfurization of dibenzothiophene on a CoMo/Al2O3 catalyst: Reaction network and kinetics." *Ind. Eng. Chem. Res.*, 35, pp. 3311-3318.
9. Vanrysselberghe, V., Le Gail, R., and Froment, G. (1998). "Hydrodesulfurization of 4-methyldibenzothiophene and 4,6-dimetyldibenzothiophene on a CoMo/Al_2O_3 catalyst: Reaction network and kinetics." *Ind. Eng. Chem. Res.*, 37(4), pp.1235-1242.
10. Chen, F., Huss, R., Doherty, M., and Malone, M. (2002) "Multiple steady states in reactive distillation: kinetic effects." *Computers & Chemical Engineering*, 26(1), 81-93.

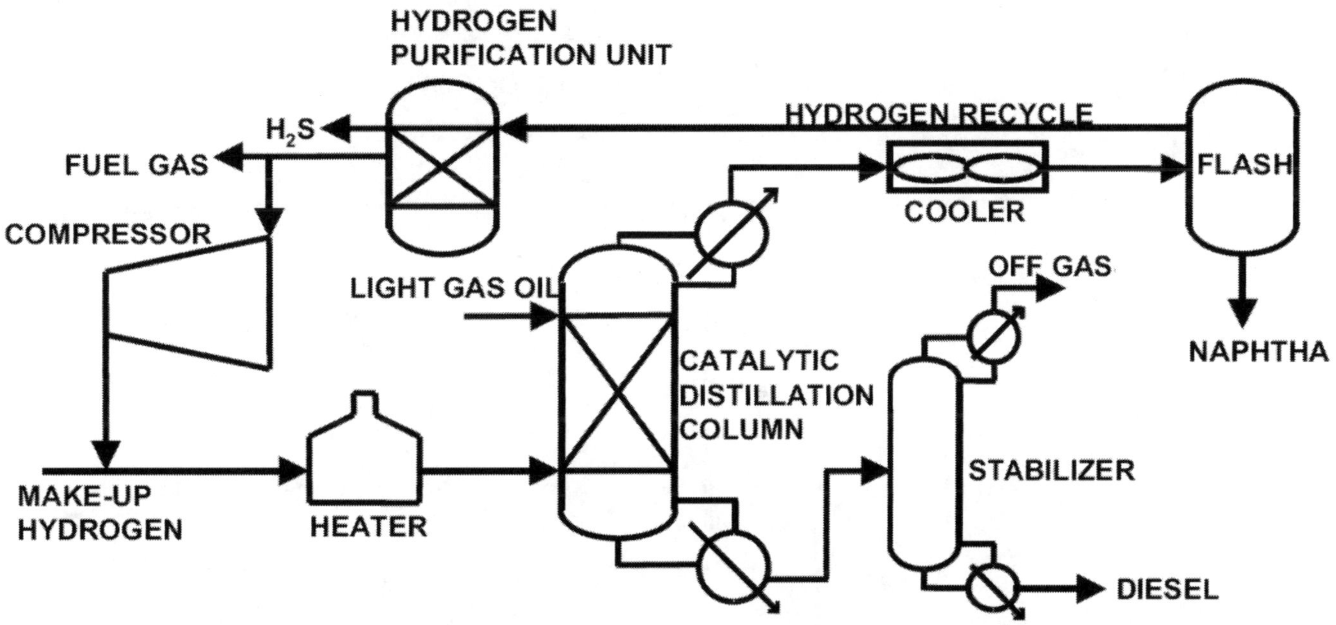

Figure 1. Hydrodesulfurization of light gas oil by catalytic distillation.

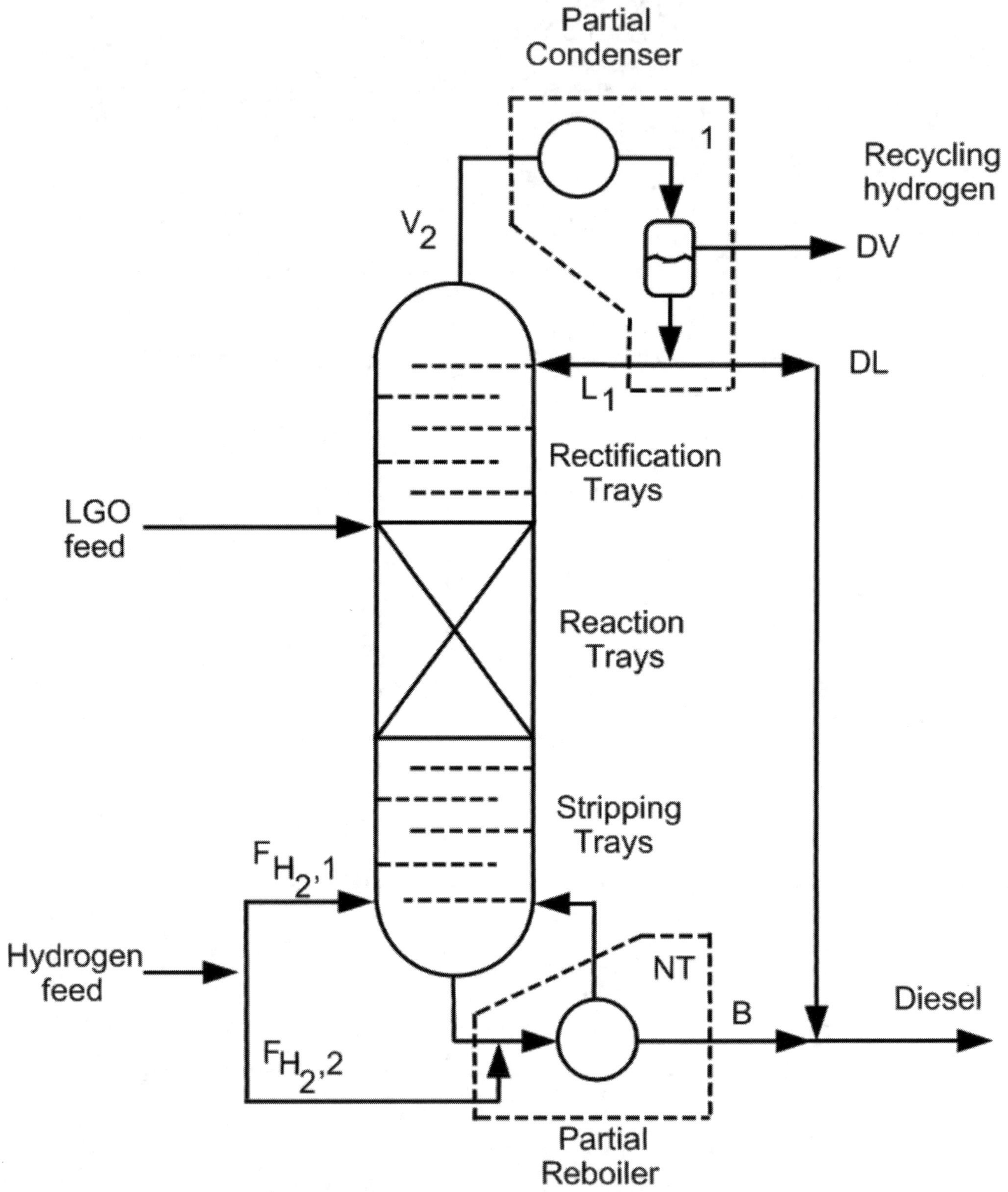

Figure 2. Hydrodesulfurization catalytic distillation column.

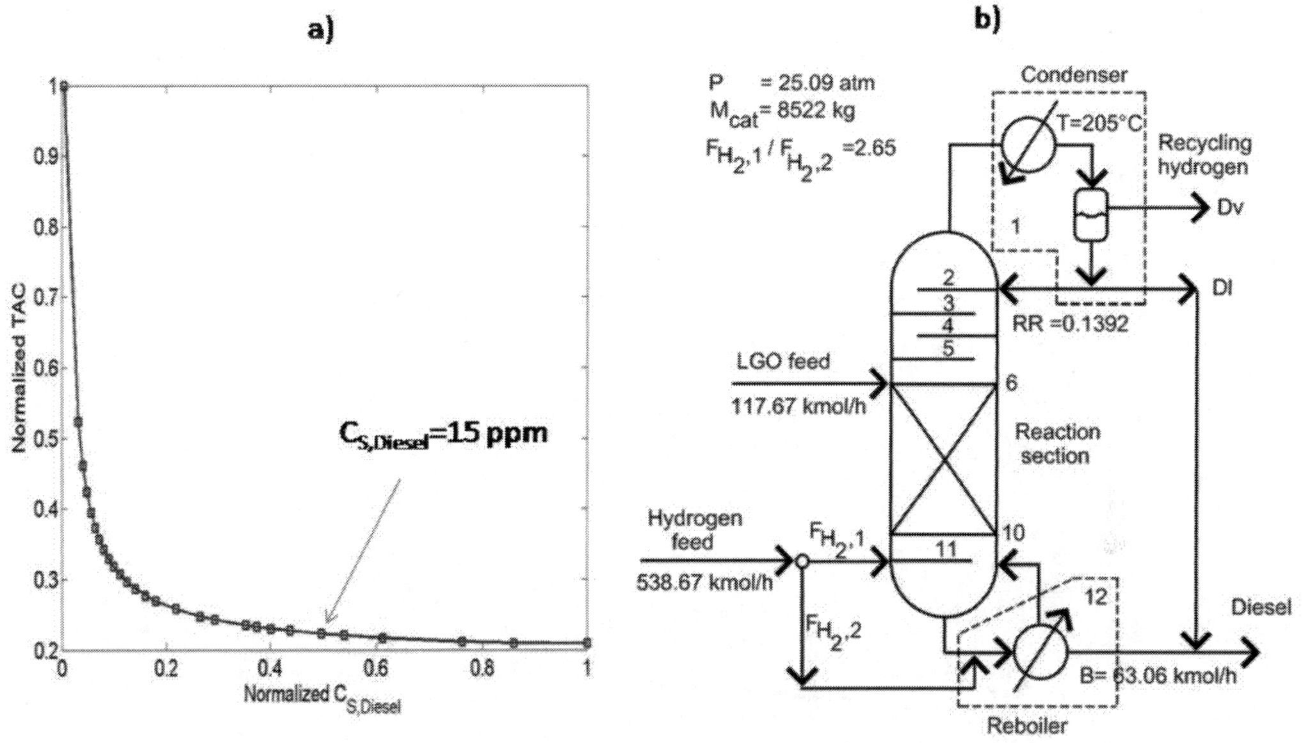

Figure 3. a) Pareto plot. b) Cost optimal design at $C_{S,Diesel}$=15. ppm.

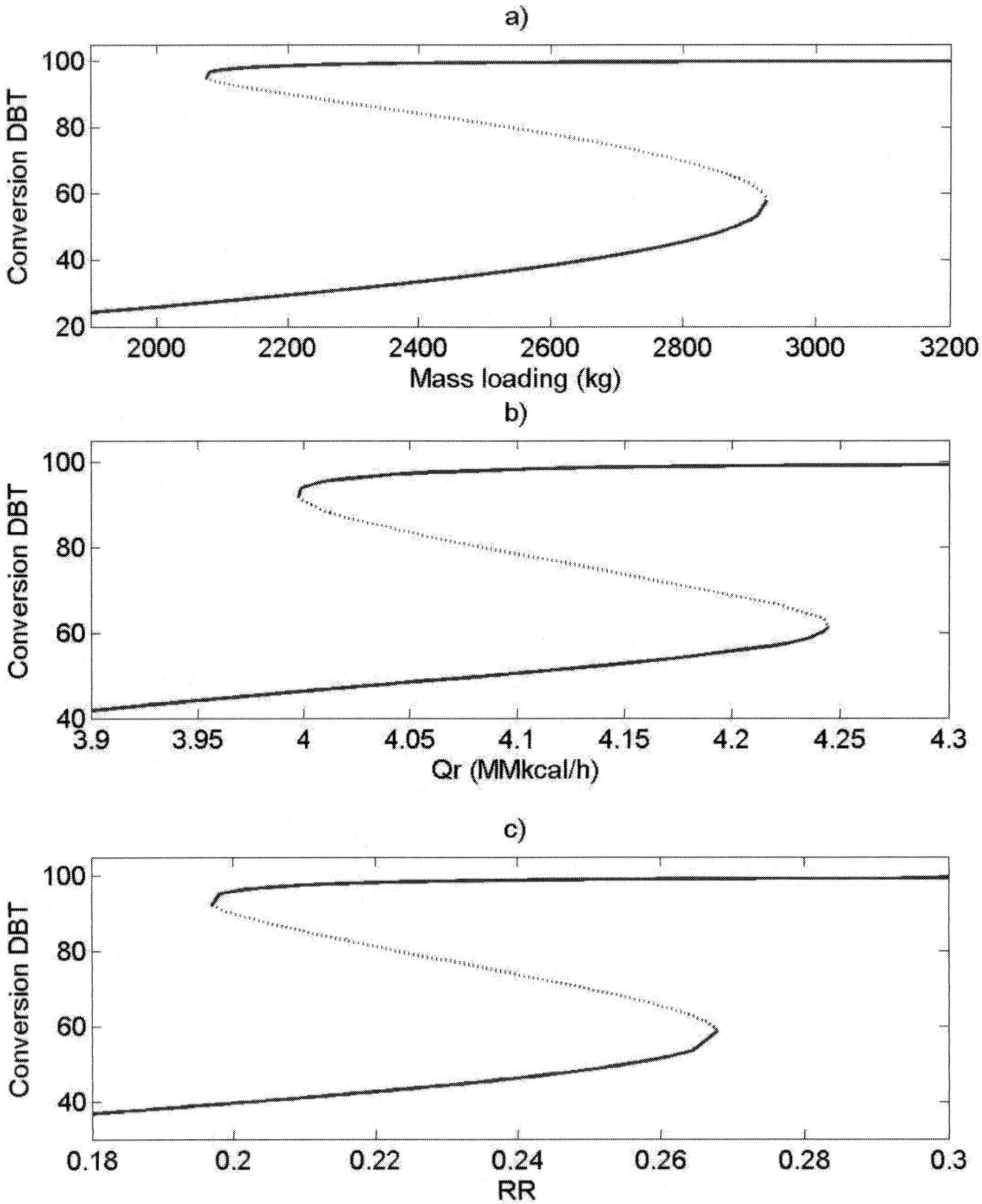

Figure 4. Bifurcation plots for DBT.

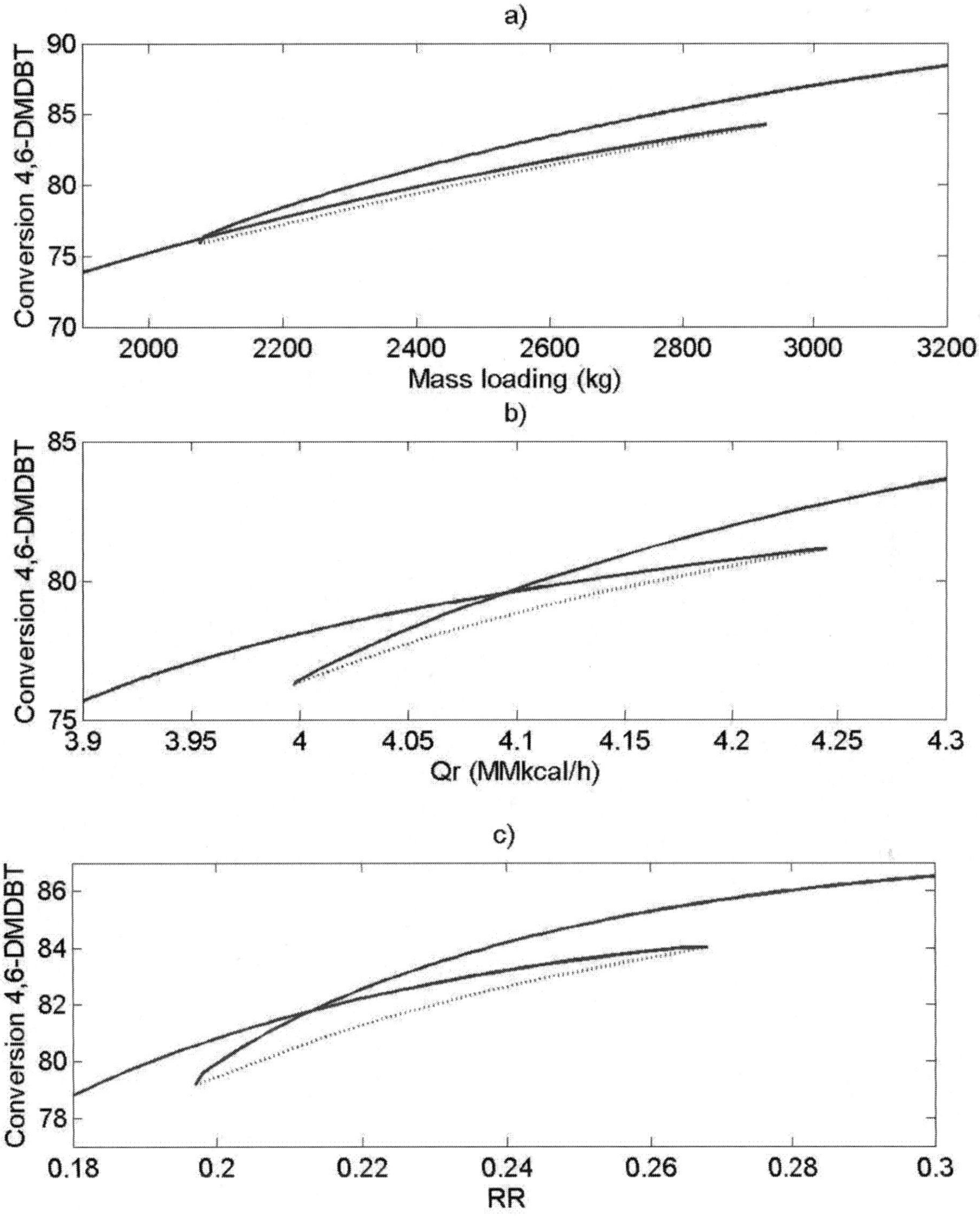

Figure 5. Bifurcation plots for 4,6-DMDBT.

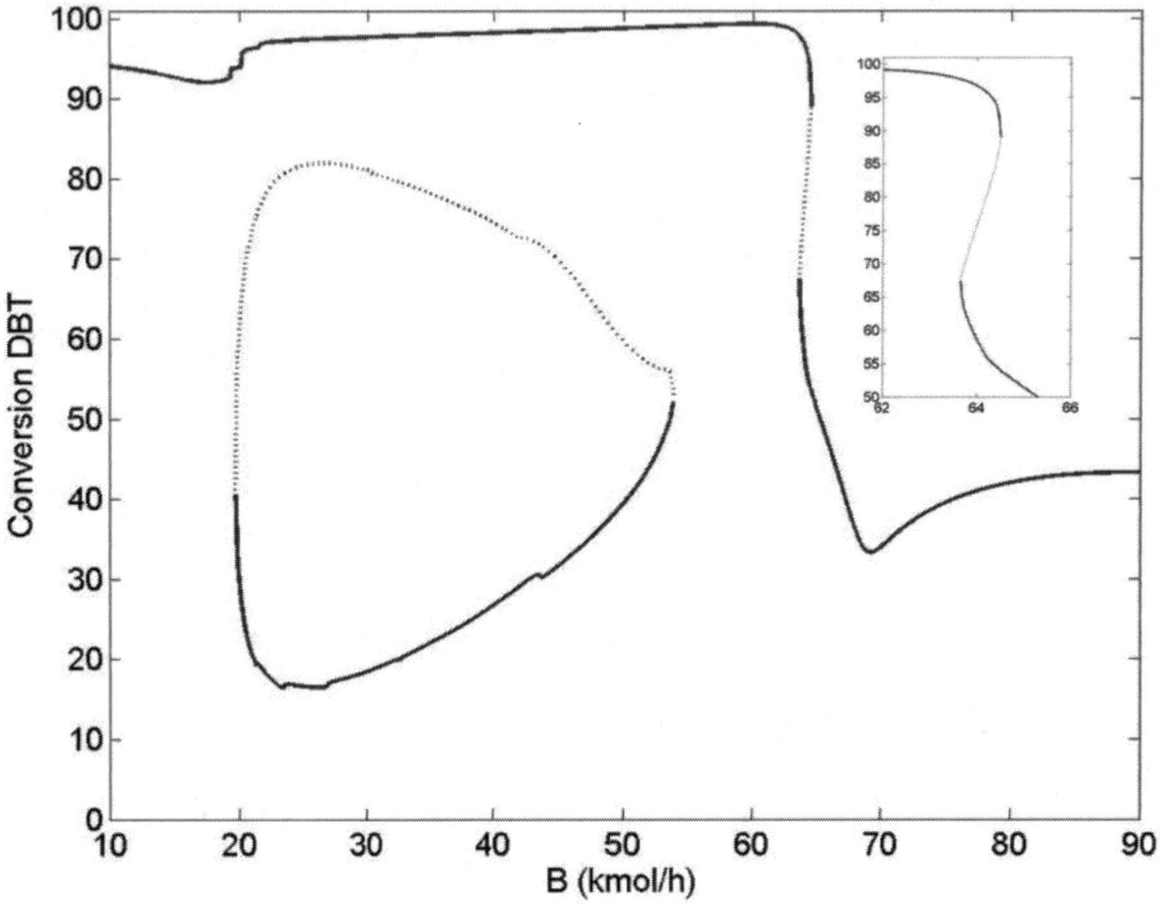

Figure 6. Bifurcation plot for DBT.

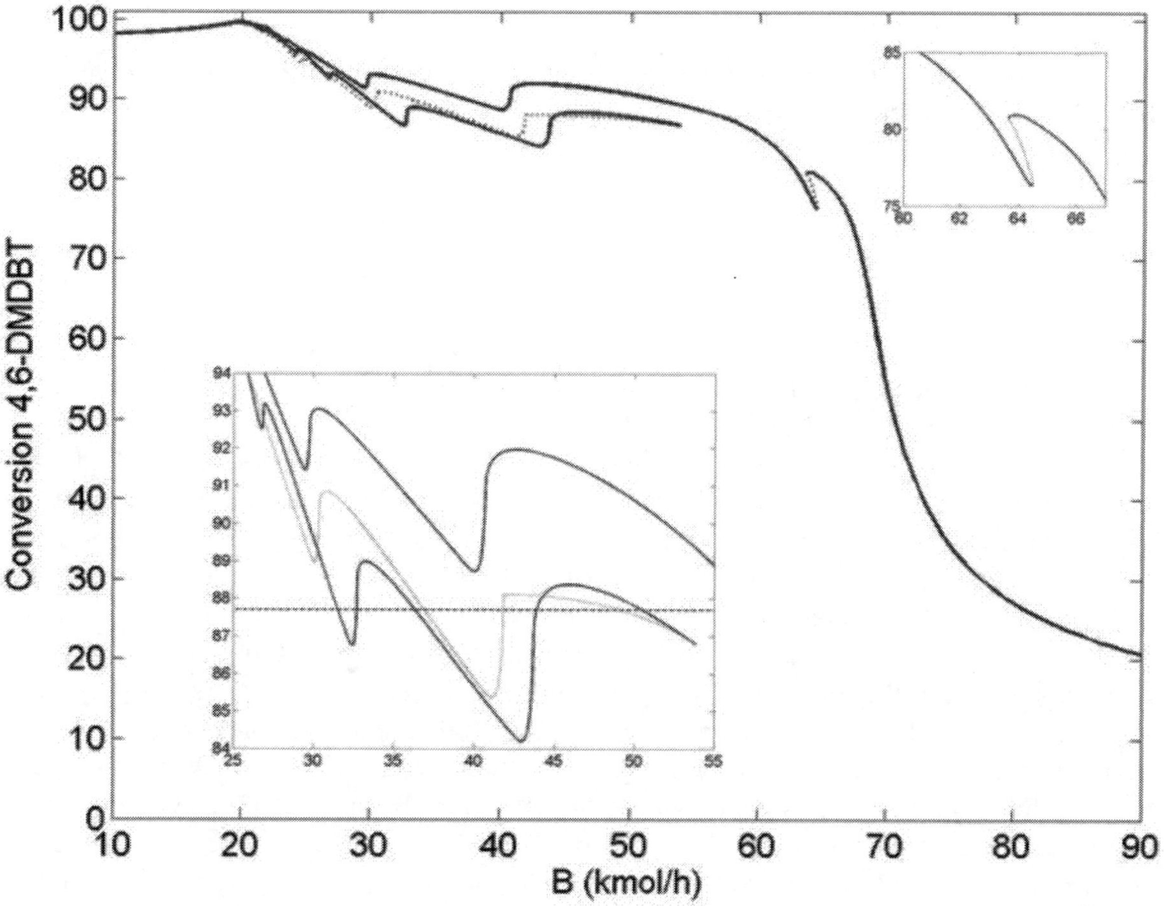

Figure 7. Bifurcation plot for 4,6-DMDBT.

48a Operation of an Industrial Smb Unit for P-Xylene Separation Accounting for Adsorbent Ageing Problems

Pedro Sá Gomes, Mirjana Minceva, and Alirio E. Rodrigues. Laboratory of Separation and Reaction Engineering, University of Porto - Faculty of Engineering, Rua Dr. Roberto Frias s/n, 4200-465 Porto, Porto, Portugal

Among the first and relevant applications of the Simulated Moving Bed (SMB) concept, as patented by Brougthon and Gerhold (1961), was the separation of p-xylene from C8 isomers mixtures at the petrochemical aromatic sites. With Parex® process from UOP (Broughton et al., 1970), and the later introduction of the Eluxyl® process from IFP (Ash et al., 1994), it is possible that around 60 % of the total world p-xylene production is now performed by the application of this kind of adsorptive technology combined with an isomerisation reactor (Minceva and Rodrigues 2007). Stressing out the all importance that p xylene represents nowadays as one of the biggest world commodities, used as a raw material in the manufacture of polyethylene terephthalate (PET) in production of polyester fibbers, molded plastics, films, and blown beverage bottles, it can be perceived that although p-xylene separation has been one of the first is still quite prominent in the SMB separation field. 50 years of experience on SMB application to p-xylene separation, given it the rank of "Old" application, nevertheless, and despite all the work published or patented innovations related with this technology, just a few refer to the adsorbent ageing problems among this type of units (Minceva and Rodrigues, 2002). The information considering this issue remains restricted to the plant operators and technology owners and mainly particle size re-distribution and capacity decline are mentioned as a result of adsorbent ageing leading to process performance decrease. It is from the common wisdom the use of guard beds, as in the gas processing industry, avoiding impurities to enter in the separation process, or more elaborated methods such as online adsorbent removal (Pilliod et al., 2006a,b) and innovative fluid distribution apparatus (Frey et al., 2006), to mitigate fines production. Nevertheless, ageing problems such as "parasite" reactions with the solid or liquid phase, and other attrition effects, leading to fine and channelling continue to appear as time goes by during these units operation. By treating adsorbent ageing as a disturbance it is possible to compensate it by applying SMB online controllers as pointed out by Engell or Morari research groups (Natarajan and Lee, 2000; Abel et al., 2004, or in Engell, 2007). However, the implementation of model predictive control methods should also be based on robust ageing models as compensative measures. Additionally, when the problem is an all complete new unit design the study and analysis of adsorbent ageing factors can have a significant weight on the financial management choices involved on these projects. The possibility of extended life to one of the most predominant economical factors on these units design, the adsorbent cost, can have considerable influence on all unit complete design, if not it the unit viability. Additionally, concepts as sustainable development (Brundtland et al., 1987), the problematic of resources depletion and eco-efficient techniques have to be taken into account in the 21st century innovation path. Extending the sieve life is also a way of driving this kind of separation technology throughout the sustainable lane. A general study on SMB adsorbent ageing has recently been presented by Rodrigues in last AIChE meeting (Sá Gomes et al., 2006), or the more detailed study submitted to publication (Sá Gomes et al., 2007), but still modelling-based research is required to define the ageing problem consequences and indicate strategies that can be used to mitigate capacity loss or other deactivation effects on this particular industrial separation, as involving the use on the so-called non-conventional operating modes as Varicol®, as pointed out in Sá Gomes et al., 2007 for ageing in chiral separations case. Therefore, is objective of this work the presentation of a complete analysis to the adsorbent ageing effect on the performance parameters of an industrial SMB unit for p-xylene separation as well as the development of some strategies to overcome this problem.

References: Broughton, D.B., Gerhold, C.G., 1961. Continuous Sorption Process Employing Fixed Bed of Sorbent and Moving Inlets and Outlets. US Patent no 2 985 589. Broughton, D.B., Neuzil, R.W., Pharis, J.M., Brearley, C.S., 1970. The Parex Process for Recovering ParaXylene. Chem. Eng. Progress 66, 70-75. Ash, G., Barth, K., Hotier, G., Mank, L., Renard, P., 1994. ELUXYL: A New Paraxylene Separation Process Rev. Inst. Fr. Petr. 49(05), 541-549. Minceva, M. and Rodrigues, A. E., 2007. Understanding and Revamping of Industrial Acale SMB units for p-xylene Separation. AIChE J. 53, 138-149. Minceva, M., Rodrigues, A.E., 2002. Modelling and Simulation of a Simulated Moving Bed for the Separation of p-Xylene. Ind. Eng. Chem. Res. 41, 3454-3461. Brundtland, G.H. et al., 1987. Report of the World Commission on Environment and Development – "Our Common Future", General Assembly of United Nations 42th session item 83(e). Frey, S.J., Sechrist, P. A., Kauff, D. A., 2006. Fluid Distribution Apparatus, WO Patent no 055222 A1. Pilliod, D.L., Randall, K.C., Harding, E.M., 2006a. A process for Hydrocarbon Conversion with On-Line solid Particulate Material removal. WO Patent no 118739 A2. Pilliod, D.L., Randall, K.C., Harding, E.M., 2006b. An Adsorption Process with On-Line Adsorbent Removal. WO Patent no 118740 A1. Natarajan, S., Lee, J.H., 2000. Repetitive model predictive control applied to a simulated moving bed chromatography systems, Computers & Chemical Engineering 24, 1127–1133. Abel, S., Erdem, G., Mazzotti, M., Morari, M., Morbidelli, M., 2004. Optimizing control of simulated moving beds—linear isotherm, Journal of Chromatography A 1033, 229–239. Engell, S., 2007. Feedback Control for Optimal Process Operation. Journal of Process Control, article in Press. Sá Gomes, P., Minceva, M., Rodrigues, A.E., 2006. Operation of Simulated Moving Bed in Presence of Adsorbent Ageing, 2006 AIChE Annual Meeting, paper 148c. Sá Gomes, P., Minceva, M., Rodrigues, A.E., 2007. Operation of Simulated Moving Bed in Presence of Adsorbent Ageing. Submitted to publication in Chemical Engineering Science.

48b Separation Of Dilute Binary Gases By Simulated-Moving Bed With Pressure-Swing Assist: The Smb/psa Process

Kyle P. Kostroski and **Phillip C. Wankat**. Chemical Engineering, Purdue University, Forney Hall of Chemical Engineering, 480 Stadium Mall Drive, West Lafayette, IN 47907

Simulated-moving bed (SMB) processes have been studied for both gas and liquid phase separations. However, when both gas and liquid phase SMB processes are feasible for a given separation, liquid operations have been commercialized since they appear to be more economical. For cases when liquid operation is difficult, gas phase SMB processes are of interest. In the current work, we have developed a new hybrid simulated-moving bed process with pressure-swing assist ("SMB/PSA"). As developed, SMB/PSA has two advantages over the standard gas SMB process: i.) regeneration at low pressure and ii.) internal generation of desorbent (carrier gas). SMB/PSA takes advantage of gas expansion at low pressure as well as the desorbent present in the feed stream to decrease overall molar desorbent requirements. However, the addition of a pressure-swing adds complexity to a standard gas SMB process. Both 4-zone and 5-zone clean-in-place SMBs (Abel et al., J. Chromatogr. A. (2004) 1043, 201; Paredes et al., Ind. Eng. Chem. Res. (2004) 43, 6157; Xie et al., Ind. Eng. Chem. Res. (2005) 44, 9904) are easy to retrofit with a pressure-swing if only the pressure of the extract-producing zone is varied, but there are restrictions on operating parameters.

The relatively difficult gas phase separation of R- and S- enflurane enantiomers was selected as a model to determine the feasibility of the SMB/PSA process. Prior work on enantiomeric enflurane separations using a gas chromatographic simulated-moving bed (GC-SMB) is reported in the literature (Juza et al., J. Chromatogr. A. (1998) 813, 333; Biressi et al., Chem. Eng. Sci. (2000) 55, 4537; Biressi et al., J. Chromatogr. A. (2002) 957, 211). The GC-SMB studied previously was configured as a 4-zone SMB with a feed of dilute enflurane in a nitrogen carrier gas. In the current work, simulations were done with Aspen ADSIM: the pressure ratio was PH/PL = 4 bar/0.2 bar = 20, the feed temperature was 25°C, and the other operating conditions were chosen based on the literature values for the 4-zone GC-SMB and adjusted as necessary. Productivity was fixed at the same value for all of the SMB and SMB/PSA processes studied, thereby allowing for a fair comparison based on purity. Biressi et al. (J. Chromatogr. A. (2002) 957, 211) showed that the 4-zone GC-SMB with 2 beds per zone had relatively large molar desorbent requirements. For example, products with an average purity of 0.98 were obtained with a molar desorbent to feed ratio (D/F) of 54. Our simulation results for the 5-zone SMB showed some reduction in molar desorbent requirements since desorbent was added at low pressure. An average purity of 0.92 was attained at a molar desorbent to feed ratio (D/F) of ~ 8 and a volumetric purge to feed ratio (gamma) of 5.0 for the extract-producing zone. To further increase average purity, high pressure desorbent was also added and an average purity of 0.97 was attained at a molar desorbent to feed ratio (D/F) ~ 22.

The practical limitations associated with adding a pressure swing to the 4- and 5-zone SMB processes are eliminated when the SMB/PSA concept is applied to the 2-zone SMB process (Jin and Wankat, Ind.

Eng. Chem. Res. (2005) 44, 1565) since both beds can be easily regenerated at low pressure; thus, the 2-zone SMB/PSA is the focus of the current work. By adding the necessary depressurization and repressurization steps, the 2-zone SMB process is converted to the 2-zone SMB/PSA for gas separations. The four-step cycle: i.) feed, ii.) depressurization/production, iii.) regeneration/production, and iv.) repressurization with carrier gas is followed by a port switch. The bed configuration in the 2-zone SMB-PSA process can consist of a single 2-zone SMB/PSA train with a tank or two 2-zone SMB/PSA interacting trains that are out-of-phase with each other. The latter configuration resembles a typical PSA process and is continuous.

Without adding any external carrier gas (D/F = 0), the decoupled 2-zone SMB/PSA with one train and a tank and one bed per zone (i.e. the tank – (1,1) configuration) achieved an average purity of 0.84 with a recycle ratio (RR: the relative amount of material being circulated between zones I and II) of 0.8, a volumetric purge to feed ratio (gamma: the volume ratio of purge gas to feed gas) of 2.0 per bed, and a bed purge ratio (BPR: the relative amount of total desorbent used to regenerate the raffinate-producing zone) of 0.4. Replacing the tank with a second train (i.e. the decoupled (1,1) – (1,1) configuration) increased the average purity to 0.86 at the same values of recycle rate and bed purge ratio. This indicated that the decoupled (1,1) – (1,1) configuration was more favorable since it eliminated mixing effects caused by the tank. Further manipulation of bed purge ratio at a recycle rate (RR) of 0.8 and a volumetric purge to feed ratio (gamma) of 2.0 per bed with no added desorbent demonstrated that either raffinate or extract products (but not both) could be produced with purity greater than 0.95. However, due to the competing effects of recycle ratio and volumetric purge to feed ratio, average purities greater than 0.86 could not be achieved by the decoupled (1,1) – (1,1) SMB/PSA without adding external desorbent.

In an effort to further increase average purity, the recycle ratio was increased. This allowed more internally generated desorbent to be recycled and less to be used for regeneration, thereby effectively decreasing the volumetric purge to feed ratio. Consequently, external desorbent (pure nitrogen) was added; this allowed the volumetric purge to feed ratio to be manipulated independently of the recycle ratio. At a recycle ratio (RR) of 0.97 and a molar desorbent to feed ratio (D/F) of 5.5, the decoupled (1,1) – (1,1) system produced a product with average purity of 0.92 at the optimal bed purge ratio (BPR) of 0.35 and a volumetric purge to feed ratio (gamma) of 2.0 for each bed. Subsequent increases in the molar desorbent to feed ratio (D/F) and volumetric purge to feed ratio (gamma) at a recycle rate (RR) of 0.97 and concomitant optimization of bed purge ratio (BPR) resulted in continual increases in maximum average purity up to 0.99 at D/F ~ 15.0 for the decoupled (1,1) – (1,1) SMB/PSA.

Increasing the number of beds per zone to two caused some improvement by further lessening molar desorbent requirements. Under the same operating conditions, the (2,2) – (2,2) configuration achieved an average purity of 0.99 at a molar desorbent to feed ratio (D/F) of 12. Other decoupled bed configurations were found to perform between the (1,1) – (1,1) and (2,2) – (2,2) limits, with the (2,1) – (2,1) configuration outperforming the (1,2) – (1,2) configuration. Bed coupling during regeneration was also investigated: at D/F values of 0.0, 5.5, and ~ 15.0, the coupled (1,1) – (1,1) SMB/PSA achieved average purities 0.87, 0.94, and 0.99, respectively. After a comparison with the results for the decoupled (1,1) – (1,1) SMB/PSA process reported previously, it appears that bed coupling is only marginally favorable.

48c Separation Of D-Psicose And D-Fructose From The Binary Mixture Using Simulated Moving Bed Chromatography

Thai Hoang Le[1], Van Duc Long Nguyen[2], Il-Yong Kim[2], Jin-Il Kim[2], Ju-Weon Lee[3], Jeung-Kun Kim[1], and Yoon-Mo Koo[2]. (1) Center for Advanced Bioseparation Technology, Incheon, South Korea, (2) Deparment of Biological Engneering, Inha University, Incheon, South Korea, (3) Deparment of Chemical Engineering, Inha University, Incheon, South Korea

D-psicose monosaccharide, a rare sugar used as a valuable pharmaceutical, has been synthesized by converting from D-fructose in an enzymatic reaction. The D-psicose has been studied to isolate from the post-reaction mixture using Simulated Moving Bed technique. The post-reaction mixture of monosaccharides was treated to remove the contaminants such as buffers, proteins, and organic materials by adsorption and ion exchange resins before it was used as feed solution in SMB process. Four columns packed with Dowex 50WX4-Ca2+ 400 mesh ion exchange resin were used for separation of D-psicose and D-fructose in a design of the four-zone SMB with one column per zone. The single-step frontal analysises for binary mixture were carried out to estimate the isotherm parameters for each D-psicose and D-fructose. The operating conditions of SMB process were determined based on the Triangle theory. As the simulation result of SMB process showed that purity and yield of extract product were 99.315% and 99.99%, respectively and those of raffinate product were 99.352% and 99.963%, respectively.

48d Production Of Pure Enantiomers By Integrating (Continuous) Chromatography And Racemisation Reactions

Malte Kaspereit[1], Javier García Palacios[1], Tania Meixus Fernandez[1], Tuomo Sainio[2], Andreas Seidel-Morgenstern[3], and Achim Kienle[4]. (1) Process Synthesis and Process Dynamics, Max Planck Institute for Dynamics of Complex Technical Systems, Magdeburg, Germany, (2) Laboratory of Industrial Chemistry, Lappeenranta University of Technology, Lappeenranta, Finland, (3) Physical and Chemical Foundations of Process Engineering, Max Planck Institute for Dynamics of Complex Technical Systems, Magdeburg, Germany, (4) Chair for Automation/Modelling, Otto von Guericke University, Magdeburg, Germany

Chemical processes most generally consist of a reaction step and a subsequent separation of the target product from unconverted educts and side products. This holds also for the production of enantiomers (i. e., optical isomers with a special importance for pharmaceuticals, flavours, etc.).

However, since usually only one enantiomer has the intended physiological effect, it is highly desirable to combine enantioseparations with an interconversion (racemisation) of the undesired enantiomer. Ultimately, this would increase the maximum overall yield of the desired species from 50% to 100%.

Racemisation reactions can be triggered, for example, by increased temperature, adjusting pH value, or by enzymes, homogeneous and heterogeneous catalysts, respectively. Furthermore, several chromatographic techniques are available where the classical four-zone SMB process might represent a sub-optimal choice. Interesting alternatives are here SMB schemes with a different number of zones and single column processes like one-column SMB analogs, closed loop and steady state recycling, etc.

Obviously, the manifold options for reaction and separation lead to a multitude of possible process schemes that combine enantioseparation and racemisation of the undesired enantiomer. In this work, different integrated processes will be studied systematically, ranging from fully integrated chromatographic reactors (SMB and single column) to flowsheet-integrated schemes with racemisation in a recycle stream. Investigations are performed both model-based as well as experimentally for a model substance. The presentation will conclude with a comparison of the different processes and the derivation of design guidelines.

48e Optimization Of Steady-State Recycling Chromatography Under Reduced Purity Constraints

Tuomo Sainio, Laboratory of Industrial Chemistry, Lappeenranta University of Technology, Skinnarilankatu 34, Lappeenranta, 53850, Finland and Malte Kaspereit, Max Planck Institute for Dynamics of Complex Technical Systems, Magdeburg, Germany.

Simulated moving-bed chromatography (SMB) has become an industry standard for binary chromatographic separations. As has been demonstrated in multiple difficult separations, the SMB method usually offers lower solvent consumption and often higher productivity than batch elution chromatography. On the other hand, recently interesting alternatives have been proposed based on single-column schemes with recycling (for example, one-column SMB anologs and steady state recycling). Due to their technological simplicity, such processes allow for lower high investment costs and higher versatility than SMB units.

Steady-state recycling chromatography (SSR) is a discontinuous single-column process that combines the aforementioned benefits: simple process configuration and hardware, high specific productivity, and low eluent consumption. The SSR concept has recently been commercialized by Novasep, France, (CycloJetTM) and licensed for example by CYBA Technologies Inc. (SteadyCycleTM) for pharmaceutical applications.

In the SSR process, overloaded injections are applied to the column. Two product fractions are obtained from the leading and tailing parts of the outlet profile, whereas the unresolved part is recycled. Several recycling schemes have been proposed. In closed-loop recycle, the recycled fraction from the column outlet is fed directly into the column inlet, followed or preceded by an injection of the fresh feed. In a mixed recycle mode, the recycled fraction is first mixed with the fresh feed prior to feeding into the column.

While some design procedures for SSR processes have been developed, a systematic methodology for arbitrary purity requirements is missing. In this work, the optimal design of SSR chromatography for the general case is investigated. The limiting case of infinite column efficiency is analyzed in the framework of equilibrium theory. Furthermore, a rapid optimization method, based on integration of the governing equations in the hodograph plane, is presented.

The influence of column efficiency is elucidated by means of numerical simulations. It is shown that the equilibrium theory based optimization can be used as an initial guess for optimization of SSR chromatography with reduced column efficiencies. Calculation results will be supported by experimental investigations of a model system.

Finally, a comparison of the performance of separations by SMB and SSR chromatography will be presented.

48f Two- and Three-Column Compact Smb Processes for Ternary Separation

Rui C. R. Rodrigues[1], João M. M. Araújo[1], and J. P. B. Mota[2]. (1) Requimte/CQFB, Chemistry Department, Faculdade de Ciências e Tecnologia, Universidade Nova de Lisboa, Caparica, 2829-516, Portugal, (2) Departamento de Química, Faculdade de Ciências e Tecnologia - Universidade Nova de Lisboa, Quinta da Torre, 2829-516 Caparica, Portugal

Over the last years there has been intensive research on the simulated moving bed (SMB) process and, as a result, several new cyclic operating schemes have emerged. These include, among others, the cyclic modulation of internal flow rates [1] and asynchronous port switch [2]. These new schemes give rise to highly efficient setups, making it possible, for example, to operate an SMB unit with less than four columns.

The SMB process was originally designed for binary separation, although several alternatives have been suggested to achieve a ternary separation using SMB technology, such as tandem [3] or parallel [4] SMB schemes, and the JO process [5], among others.

In this work, SMB units with only two and three columns are used to achieve a ternary separation. The productivity and solvent requirement for each combination of process are optimized numerically based on a detailed column model. The dynamic optimization problem is formulated using a full-discretization approach [6,7] and solved using an efficient interior-point solver [8]. From the optimization results, the most favorable process for a given separation problem can be selected. Optimized configurations for several schemes are validated experimentally for the ternary separation of nucleosides by reversed phase.

References [1] E. Kloppenburg, E.D. Gilles, A new concept for operating simulated moving-bed processes, Chem. Eng. Tech. 22(1999) 813. [2] O. Ludemann-Hombourger, R.M. Nicoud, M. Bailly, Separation Science and Technology 35(2000) 1829. [3] Y. Xie, S.-Y. Mun, J.-H. Kim, N.-H.L. Wang, Standing wave design and experimental validation of tandem simulated moving bed process for insulin purification, Biotech. Prog. 18 (2002) 1332. [4] P.C. Wankat, Simulated moving bed cascades for ternary separations, Ind. Eng. Chem. Res. 40 (2001) 6185. [5] T. Masuada, T. Sonobe, F. Matsuda, M. Horie, Process for the fractional separation of multicomponent fluid mixture, US Pat. 5 198 120 (1993). [6] J.M.M. Araújo, R.C.R. Rodrigues, J.P.B. Mota, Use of single-column models for efficient computation of the periodic state of a simulated moving bed process, Ind. Eng. Chem. Res. 45 (2006) 5314. [7] J.M.M. Araújo, R.C.R. Rodrigues, J.P.B. Mota, Optimal design and operation of a certain class of asynchronous simulated moving bed processes, J. Chromatogr. A (2006) in press. [8] Y. Kawajiri, L. T. Biegler, Optimization strategies for simulated moving bed and powerfeed processes, AIChE J. 52 (2006) 1343.

48g Standing Wave Design And Simulation Of Simulated Moving Beds Based On Monolithic Columns

Javier Nieves Remacha and **Nien-Hwa Linda Wang**. Chemical Engineering, Purdue University, FRNY, 480 Stadium Mall Drive, West Lafayette, IN 47907-2100

Monolithic silica columns are continuous porous and permeable materials that have a bimodal porous structure. Macropores (0.5-8 micrometers) and high bed porosity (0.6-0.9) allow fast convection, while thin silica skeletons (0.3-5 micrometers) allow rapid diffusion. The high mass transfer efficiency of monolithic columns has been used in analytical batch chromatography to achieve fast separation and high resolution. Only feasibility studies have been carried out for SMB applications using silica monoliths. The objective of this study is to explore the potential of SMB based on monolithic columns.

A new design method for monolithic SMB is developed in this study based on the concept of Standing Concentration Waves (SWD). A general rate model for monoliths is developed for SMB systems (VERSE-MONOLITH). The design based on standing waves is validated with the simulations using VERSE-MONOLITH.

This study shows that the maximum productivity of a monolithic SMB based on the commercial Prep-Chromolith columns is pressure limited. Permeability has to be improved to increase productivity by increasing bed porosity or skeleton diameter. Wave spreading due to mass transfer effects is an order of magnitude lower than in packed bed SMB, leading to much lower desorbent requirement for the same productivity. Axial dispersion is the major mechanism for wave spreading in the monolithic SMB.

48h Analysis Of Separation Of Dilute Binary Gas Mixture By Thermal Swing Adsorption Of Honeycomb Rotor Adsorbent

Tsutomu Hirose, Advanced Adsorption Technology, 19-7, 4 Chome, Miwadai, Higashi-ku, Fukuoka, Japan

Thermal swing adsorption has been developed as honeycomb rotor adsorbers and applied to dehumidification of air, removal of VOC and so on. It was limited to a system of single component to purify the bulk inert gas. In the present paper, we discuss a binary mixture where a more adsorbable component P and a less adsorbable component N diluted in the bulk inert gas are to be separated from each other. A honeycomb adsorbent rotor is divided into four sections as shown in Fig.1 and experiences four successive steps: Adsorption of P-component (PA step), Adsorption of N-component (NA step), Desorption of P-component (PD step) and Desorption of N-component (ND step). The P-rich and N-rich products, respectively, are obtained in the P- and N-stream after the PD- and PA-steps. A preliminary discussion is made on a continuous moving bed for a simple case of instantaneous thermal response and independent linear isotherms. The slope of operating line in PA step has to be greater than the adsorption coefficient of NA isotherm to achieve the preferential adsorption of P-component in the PA step by depressing the N-component adsorption. Similar discussion leads to relationships between the slope of operating lines and adsorption coefficients for successful separation of P and N. This criterion was found to be applied to the cross flow moving bed that is an operation mode of the usual honeycomb rotor adsorbent. Thus, the optimal operating condition was derived for circulating flow rate Q and step time. The same idea will be extended to more complicated cases such as additional withdrawal of inert gas and nonlinear isotherm.

Extraction of LCST-Type Hydrogen-Bonding Organics from Aqueous Solution

Timothy C. Frank
The Dow Chemical Company
Engineering & Process Sciences Laboratory
Midland, Michigan USA

AIChE National Meeting, Salt Lake City, Utah
Extractive Separations Session 02C02, Paper 82a
November 4-9, 2007

Compared to distillation, liquid-liquid extraction often is better suited to recovering hydrogen-bonding organic solutes from dilute aqueous solutions. This paper reviews distillation and extraction options for the class of hydrogen-bonding organics characterized by the presence of a lower critical solution temperature (LCST) in the organic + water phase diagram, including glycol ethers, polyglycols, and certain nonionic surfactants. Compounds of this type often are employed in bioprocessing applications as well as other applications in which organics are mixed with water (For example, see Allen *et al.*, *Biotechnol. Prog.* **2007**, *23*(5), in press). General rules for the temperature-dependence of relative volatility and partition ratios have been developed from representative activity-coefficient data (Frank *et al.*, *Ind. Eng. Chem. Res.* **2007**, *46*(11), 3774-3786). 1) For distillation, relative volatility with respect to water is likely to increase with increasing temperature for stripping volatile LCST-type compounds from dilute aqueous solution, provided the pure-component vapor pressure relative to water also increases or stays about the same; and 2) For extraction of LCST-type compounds from aqueous solution, the partition ratio is likely to increase with increasing temperature provided mutual solubility between phases is low. The application of MOSCED and UNIFAC activity coefficient prediction methods has been shown to be an effective approach to screening useful extraction solvents (Frank *et al.*, *Ind. Eng. Chem. Res.* **2007**, *46*(13), 4621-4625). Example calculations show good correlation between calculated activity coefficients and *K* data.

Introduction

Liquid-liquid extraction is a versatile technology with a wide range of commercial applications. Numerous process schemes, solvent systems, and equipment options have been described in the literature (*1*). Compared to distillation or steam stripping, liquid-liquid extraction often is better suited to recovering hydrogen-bonding organic solutes from dilute aqueous solution because it can allow a significant reduction in energy consumption (*2*). In other cases involving non-volatile or low-volatility solutes, distillation simply is not a feasible option. Although extraction is a fairly mature technology, little information has been published concerning general trends in the temperature dependence of the partition ratio (*K*) for non-reactive extractions. Furthermore, many of the *K* data reported in the literature were generated at or near room temperature with relatively little data reported at other temperatures. This paper briefly describes efforts at The Dow Chemical Company to develop new uses for glycol ethers, and reviews the temperature-sensitivities that were observed during the course of developing distillation and extraction options for recovering glycol ether residues from aqueous process streams (*2*). The results of this work suggest that organics with similar hydrogen-bonding behavior including polyglycols and aliphatic amines are likely to exhibit the same general trends.

Glycol Ethers and Similar LCST-Type Hydrogen-Bonding Organics

Glycol ethers are oxygenated organic solvents typically made from reacting ethylene oxide (EO) or propylene oxide (PO) with an alcohol (Figures 1 and 2). Those glycol ethers derived from EO are called ethylene glycol ethers or E-series glycol ethers, and those derived from PO are propylene glycol ethers or P-series glycol ethers. The resulting compounds have the general structure $R'\text{-}O\text{-}[CH_2\text{-}CH(R)\text{-}O]_n\text{-}H$, where R′ is an alkyl or aryl group, R is H for ethylene glycol ethers or CH_3 for propylene glycol ethers, and n = 1, 2, or 3. The common naming convention refers to monoethylene, diethylene, and triethylene glycol alkyl (or aryl) ethers, or the analogous propylene glycol ethers. The prefix mono-, di-, or tri- reflects the value of n; that is, the number of oxyalkylene repeating units in the molecule. The alkyl (or aryl) name refers to the R′ group arising from the alcohol used in its preparation. The properties of selected glycol ethers are summarized in Figures 3 - 5. Additional information is available from the literature (*2-5*) and from supplier web sites.

The chemical structures of glycol ethers are related to those of polyglycols and nonionic surfactants commonly used in bioprocessing operations. In general, polyglycols may be viewed as higher molecular weight analogs of glycol ethers, normally containing up to about 100 oxyalkylene repeating units. All of these compounds possess both hydrophilic and hydrophobic functionality in the form of hydrophilic hydroxyl and ether groups and hydrophobic alkyl or aryl groups. Furthermore, most of these compounds exhibit a lower critical solution temperature (LCST) in mixtures with water. Below the LCST, the compound is completely miscible with water in all proportions. Above the LCST, the mixture is only partially miscible, and in many cases the solubility of the organic compound in water decreases with increasing temperature (commonly termed inverse solubility). This behavior is due to the ability of the organic compound to hydrogen bond with water at temperatures below the LCST, resulting in an attractive interaction and complete miscibility. At temperatures above the LCST, hydrogen bonding is disrupted and hydrophobic interactions begin to dominate as temperature increases. This inverse solubility behavior corresponds to the lower temperature range of the two-liquid-phase envelope on a temperature-composition diagram. Figure 6 shows such a diagram for ethylene glycol *n*-butyl ether (EB) + water to illustrate these general features. Diagrams for other glycol ethers and for polyglycols are available elsewhere (*5-10*). The LCST values of selected glycol ethers are summarized in Figure 7. Other compounds that typically exhibit LCST behavior in mixtures with water include aliphatic amines and certain water-soluble polymers.

The LCST value is a measure of the hydrophilic/hydrophobic nature of the organic compound in aqueous solution; a higher LCST value indicates a more hydrophilic compound. A cloud-point temperature often is reported for polypropylene glycols, copolymer polyglycols, and nonionic surfactants, instead of a LCST value. Typically, the cloud point is defined as the temperature at which a 1 wt% solution of polyglycol or surfactant dissolved in water becomes cloudy on heating (or clear on cooling). The cloud point will always be somewhat higher than the LCST. Some compounds may be viewed as being completely miscible with water because the LCST is greater than 100 °C and outside the range of normal process temperatures. For example, 1-ethoxy-2-propanol (common name propylene glycol ethyl ether or PE) and most of the polyethylene glycols behave this way. Relatively hydrophobic compounds that exhibit a LCST below 100 °C include 2-butoxyethanol, 1-butoxy-2-propanol (common name propylene glycol n-butyl ether or PnB), polypropylene glycols, and most of the copolymer polyglycols.

Because of their hydrogen-bonding properties, glycol ethers are widely used in aqueous-organic product formulations including coatings and cleaners. These and other applications are listed in Figure 8. Some of the newer applications include their use for selective release of recombinant proteins from within bacterial cells, for lower cost production of therapeutic proteins and industrial enzymes (*11*), and production of chiral glycol ethers for use as intermediates to chiral pharmaceuticals (*12*). These applications are illustrated in Figures 9 – 11. Global production of glycol ethers was over 1 million metric tons in 1999 (*13*). Most glycol ethers are biodegradable and many toxicology studies have been published (*14-18*).

Temperature-Sensitive Activity Coefficients

Tables 1 – 3 summarize relative volatilities ($\alpha^{\infty}_{i,aqueous}$) and infinite-dilution activity coefficients ($\gamma^{\infty}_{i,aqueous}$) for three representative glycol ethers dissolved in water and brine: propylene glycol *n*-butyl ether (PnB), propylene glycol *n*-propyl ether (PnP), and dipropylene glycol *n*-butyl ether (DPnB). The data were generated using the Raleigh distillation apparatus shown in Figure 12, as described in ref. 2. Activity coefficients were determined from measured relative volatilities using the relationship

$$\alpha^{\infty}_{i,water} = \gamma^{\infty}_{i,aqueous} \frac{p_i^{SAT}}{p_{water}^{SAT}} \tag{1}$$

where $p_i^{SAT} / p_{water}^{SAT}$ is the ratio of pure component vapor pressures. In this case, $\gamma^{\infty}_{i,aqueous}$ values are significantly greater than unity, indicating significant hydrophobic interactions. Note that the ratios of pure component vapor pressures are significantly less than unity, so repulsive hydrophobic interactions are responsible for the strippability of these compounds. The results also show a significant increase in the value of $\gamma^{\infty}_{i,aqueous}$ as temperature increases from 50 to 80 °C. This may be explained in terms of the disruption of hydrogen bonding interactions between glycol ether and water in solution due to increasing molecular motion, such that hydrophobic interactions become more and more significant. Thus, for these compounds, relative volatility increases with increasing temperature in part because the aqueous liquid phase becomes a more hostile environment, resulting in an increase in the tendency for the glycol ether to enter the vapor phase (*2*). As a general rule, relative volatility should be greater than α = 30 or so to allow for moderate energy consumption and an economical stripping operation (*2*), so steam stripping these compounds is not a particularly attractive option.

Table 4 summarizes partition ratio data for extraction of PnP from aqueous solution using a variety of hydrophobic organic solvents. The data were generated over a range of temperatures from 40 up to 95 °C and show a significant driving force for extraction. The *K* values, expressed in terms of solute mole fractions in each phase, are directly related to solute activity coefficients:

$$K(\text{mole fraction basis}) = \frac{y_i}{x_i} = \frac{\gamma_{i,aqueous}}{\gamma_{i,organic}} \quad (2)$$

where y_i denotes solute concentration in the extract phase and x_i denotes solute concentration in the aqueous raffinate phase, at equilibrium. Equation 2 allows calculation of $\gamma_{PnP,organic}$ from experimental values for K and $\gamma_{PnP,aqueous}$ (obtained from Table 2) using the expression $\gamma_{PnP,organic} = \gamma_{PnP,aqueous} / K$. These values are listed in Table 4. The $\gamma_{PnP,aqueous}$ data from Table 2 were used without adjustment for PnP concentration because the concentration of PnP in the aqueous phase was always very dilute and similar to that used for the relative volatility experiments (less than 0.003 mole fraction), so extrapolation to a different concentration was not needed. Furthermore, the aqueous phase contained very little dissolved organic solvent from the mutual solubility of extract and raffinate phases (less than 0.0001 mol fraction), so any effect due to the presence of solvent dissolved in the aqueous phase could be neglected. To account for small differences in temperature, the $\gamma^{\infty}_{PnP,aqueous}$ data in Table 2 were slightly adjusted by extrapolation from 50 °C to 40 °C and from 80 °C to 95 °C (*2*). In every case, the values of $\gamma_{PnP,organic}$ obtained in this way decrease with increasing temperature, indicating that PnP/solvent interactions in the organic phase become somewhat more attractive as temperature increases. This trend is at least qualitatively consistent with regular solution behavior, where γ_i decreases with temperature according to $RT\ln\gamma_i$ = constant.

General Rules for LCST-Type Hydrogen-Bonding Organics Mixed with Water

Relative Volatility. The limiting activity coefficients summarized in Tables 1 – 4 allow a glimpse into the molecular interactions occurring in solution for representative LCST-type hydrogen-bonding organic + water mixtures. It appears that the presence of a LCST in the phase diagram is generally indicative of an increase in $\gamma^{\infty}_{i,aqueous}$ with increasing temperature due to disruption of hydrogen-bonding, at least over the range of temperatures within the lower region of the two-liquid-phase envelope (as illustrated in Figure 6). Furthermore, this indicates that infinite-dilution relative volatilities should increase with increasing temperature, provided the pure-component vapor pressure relative to water ($p_i^{SAT} / p_{water}^{SAT}$) increases or stays about the same, as indicated in Equation 1.

Partition Ratios. The trends in the activity coefficient data in Tables 1 – 4 also have implications for the temperature dependence of partition ratios for extraction of LCST-type organics from water. They indicate that K should increase with increasing temperature due to disruption of hydrogen bonding between the organic solute and water in the aqueous phase (resulting in an increase in the activity coefficient for the solute dissolved in the aqueous phase), as well as increasingly attractive interactions between the organic solute and the extraction solvent in the organic phase (and a corresponding reduction in the solute's activity coefficient in the organic phase). Although the relative magnitude of these trends is difficult to predict, the important concept to emphasize here is that both trends are likely to contribute to some degree to an increase in K, and there is no obvious mechanism for a decrease in K with increasing temperature. Table 5 lists K data for a number of LCST-type organics, indicating that this trend appears to be generally observed.

Screening Extraction Solvents Using MOSCED or UNIFAC

The application of MOSCED (*19,20*) and UNIFAC (*21,22*) activity coefficient prediction methods has been shown to be an effective approach to screening useful solvents for extraction of glycol ether residues from water (*23*). Estimates of limiting activity coefficients for PnP dissolved in the organic phase ($\gamma^{\infty}_{PnP,organic}$) obtained using MOSCED and UNIFAC are highly correlated with *K* data for a variety of hydrophobic organic solvents including various alcohols, ketones, ethers, chlorinated hydrocarbons, aromatics, and aliphatic hydrocarbons. The results obtained using MOSCED are shown in Figure 13. Similar results were obtained using UNIFAC.

This approach to ranking candidate solvents assumes that any effect due to differences in the solubility of water in the solvent phase can be neglected, and that the activity coefficient for solute dissolved in the aqueous phase is essentially constant as a function of extraction solvent. In many cases, these assumptions yield valid first-order approximations that allow rapid assessments. The methodology avoids the difficulties and uncertainties involved in accurately predicting the magnitude of *K* or the effect of temperature; instead, the approach focuses on determining the relative order of limiting activity coefficients in the organic phase, which the representative example indicates is an excellent indicator of relative *K* values and is not in itself a strong function of temperature (*23*).

Acknowledgements

Many of my colleagues at The Dow Chemical Company have contributed to the technology described in this review, including Felipe Donate, John (Joe) Anderson, Andrei Merenov, Grant Von Wald, Tom Thyne, Jim Olson, Jeff Allen, Anant Patkar, Sol Resnick, and Barbara Alstad. They are co-authors of a number of the articles and patents cited herein. I also wish to acknowledge Prof. Charles Eckert at Georgia Institute of Technology for valuable discussions.

References

1. Frank, T. C.; Dahuron, L.; Holden, B. S.; Prince, W. D.; Seibert, A. F.; Wilson, L. C. Liquid-Liquid Extraction and Other Liquid-Liquid Operations and Equipment, Section 15 in *Perry's Chemical Engineers' Handbook*, 8th Edition, D. W. Green, Editor-in-Chief; McGraw-Hill: New York, 2007.
2. Frank, T. C., Donate, F. A.; Merenov, A. S.; Von Wald, G. A.; Alstad, B. J.; Green, C. W.; Thyne, T. C. Separation of Glycol Ethers and Similar LCST-Type Hydrogen-Bonding Organics from Aqueous Solution Using Distillation or Liquid-Liquid Extraction. *Ind. Eng. Chem. Res.* **2007**, *46(11)*, 3774-3786.
3. Flick, E. W. Glycol Ethers. Chapter 10 in *Industrial Solvents Handbook*, 5th Ed; Noyes Data Corp.: Westwood, New Jersey, 1998.
4. Cheremisinoff, N. P. *Industrial Solvents Handbook*, 2nd Ed.; Marcel Dekker: NY, 2003.
5. Christensen, S. P.; Donate, F. A.; Frank, T. C.; LaTulip, R. J.; Wilson, L. C. Mutual Solubility and Lower Critical Solution Temperature for Water + Glycol Ether Systems. *J. Chem. Eng. Data* **2005**, *50*, 869-877.
6. Johansson, H.; Karlstroem, G.; Tjerneld, F. Experimental and Theoretical Study of Phase Separation in Aqueous Solutions of Clouding Polymers and Carboxylic Acids. *Macromolecules* **1993**, *26*, 4478-4483.

7. Saraiva, A.; Persson, O.; Fredenslund, A. An Experimental Investigation of Cloud-Point Curves for the Poly(Ethylene Glycol)/Water System at Varying Molecular Weight Distributions. *Fluid Phase Equil.* ***1993***, *91*, 291-311.
8. Saeki, S.; Kuwahara, N.; Nakata, M.; Kaneko, M. Upper and Lower Critical Solution Temperatures in Poly(Ethylene Glycol) Solutions. *Polymer* **1976**, *17*, 685-689.
9. Prange, M. M.; Hooper, H. H.; Prausnitz, J. M. Thermodynamics of Aqueous Systems Containing Hydrophilic Polymers or Gels. *AIChE J.* **1989**, *35*, 803-813.
10. Persson, J.; Nystrom, L.; Ageland, H.; Tjerneld, F. Purification of Recombinant Proteins Using Thermoseparating Aqueous Two-Phase System and Polymer Recycling. *J. Chem. Technol. Biotechnol.* **1999**, *74*, 238-243.
11. Allen, J. R.; Patkar, A. Y.; Frank, T. C.; Donate, F. A.; Chiu, Y. C.; Shields, J. E.; Gustafson, M. E. Use of Glycol Ethers for Selective Release of Periplasmic Proteins from Gram-Negative Bacteria, *Biotechnology Progress* **2007**, *23(5)*, in press.
12. Resnick, S. M.; Donate, F. A.; Frank, T. C.; Thyne, T. C.; Foley, P. Enzymatic Resolution of Propylene Glycol Alkyl (or Aryl) Ethers and Ether Acetates, US 7264960, Sept. 4, 2007, assigned to The Dow Chemical Company.
13. Chinn, H.; Anderson, E.; Yoneyama, M. *Chemical Economics Handbook Report: Glycol Ethers*; SRI International: Menlo Park, CA, 2000.
14. *ECOTEC Technical Report No. 95. The Toxicology of Glycol Ethers and Its Relevance to Man*, 4th Ed., vols. I and II; European Centre for Ecotoxicology and Toxicology of Chemicals: Brussels, Belgium, 2005.
15. Bingham, E.; Cohrssen, B.; Powell, C. H. *Patty's Industrial Hygiene and Toxicology*; Wiley: New York, 5th Ed., pp. 73-395, 2001.
16. Cowan, R.; Kwon. Aerobic Biodegradation of Ethylene Glycol Ethers. *J. Hazardous Ind. Wastes* **1999**, *31*, 273-282.
17. Staples, C. A.; Bostman, R. J.; Cano, M. L. Ethylene Glycol Ethers: An Environmental Risk Assessment. *Chemosphere* **1998**, *36*, 1585-1613.
18. Gonsior, S. J.; West, R. J.; Biodegradation of Glycol Ethers in Soil. *Environ. Toxicol. Chem.* **1995**, *14*, 1273-1279.
19. Thomas, E. R.; Eckert, C. A. Prediction of Limiting Activity Coefficients by a Modified Separation of Cohesive Energy Density Model and UNIFAC. *Ind. Eng. Chem. Proc. Des. Dev.* **1984**, 23, 194-209.
20. Lazzaroni, M. J.; Bush, David; Eckert, C. A.; Frank, T. C.; Gupta, S. K.; Olson, J. D. Revision of MOSCED Parameters and Extension to Solid Solubility Calculations, *Ind. Eng. Chem. Res.* **2005**, *44(11)*, 4075-4083.
21. Fredenslund, A.; Gmehling, J; Michelsen, M. L.; Rasmussen, P; Prausnitz, J. M. Computerized Design of Multicomponent Distillation Columns Using the UNIFAC Group Contribution Method for Calculation of Activity Coefficients. *Ind. Eng. Chem. Proc. Des. Dev.* **1977**, *16*, 450-462.
22. Wittig, R.; Lohmann, J.; Gmehling, J. Vapor-Liquid Equilibria by UNIFAC Group Contribution. 6. Revision and Extension. *Ind. Eng. Chem. Res.* **2003**, *42*, 183-188.
23. Frank, T. C.; Anderson, J. J.; Olson, J. D.; Eckert, C. A. Application of MOSCED and UNIFAC to Screen Hydrophobic Organic Solvents for Extraction of Hydrogen-Bonding Organics from Aqueous Solution, *Ind. Eng. Chem. Res.* **2007**, *46(13)*, 4621-4625.

Tables

Table 1. Rayleigh Distillation Results – Stripping PnB from Water and Brine (*2*)

Solutes		PnB	PnB	PnB + 3% NaCl	PnB + 3% NaCl
Pressure	mm Hg	95	370	95	370
Temperature	°C	50	80	50	80
Average PnB concentration in the pot	mole fraction	0.0035	0.0028	0.0029	0.0028
PnB material balance accountability	%	99.98	95.1	91.4	94.0
Relative volatility, $\alpha^{\infty}_{PnB,aqueous}$		5.3 ± 0.6	10.6 ± 1.2	7.1 ± 1.2	14.0 ± 1.5
Vapor press. ratio, $p^{SAT}_{PnB} / p^{SAT}_{water}$		0.065	0.082	0.065	0.082
Activity coefficient, $\gamma^{\infty}_{PnB,aqueous}$		81 ± 9	130 ± 16	109 ± 20	170 ± 19

Table 2. Rayleigh Distillation Results – Stripping PnP from Water and Brine (*2*)

Solutes		PnP	PnP	PnP + 3% NaCl	PnP + 3% NaCl
Pressure	mm Hg	95	365	95	365
Temperature	°C	50	80	50	80
Average PnP concentration in the pot	mole fraction	0.0016	0.0020	0.0016	0.0020
PnP material balance accountability	%	97.4	92.6	100.9	98.1
Relative volatility, $\alpha^{\infty}_{PnP,aqueous}$		5.5 ± 0.7	9.7 ± 1.5	7.3 ± 0.7	12.3 ± 1.3
Vapor press. ratio, $p^{SAT}_{PnP} / p^{SAT}_{water}$		0.120	0.153	0.120	0.153
Activity coefficient, $\gamma^{\infty}_{PnP,aqueous}$		46 ± 5	64 ± 10	61 ± 6	81 ± 7

Table 3. Rayleigh Distillation Results – Stripping DPnB from Water and Brine (*2*)

Solutes		DPnB	DPnB	DPnB + 3% NaCl	DPnB + 3% NaCl
Pressure	mm Hg	93	356	93	356
Temperature	°C	50	80	50	80
Average DPnB concentration in the pot	mole fraction	0.0010	0.0008	0.0010	0.0010
DPnB material balance accountability	%	96.4	100.8	97.6	99.8
Relative volatility, $\alpha^{\infty}_{DPnB,aqueous}$		1.1 ± 0.1	3.0 ± 0.3	1.5 ± 0.2	3.6 ± 0.3
Vapor pressure ratio, $p^{SAT}_{DPnB} / p^{SAT}_{water}$		0.0049	0.0083	0.0049	0.0083
Activity coefficient, $\gamma^{\infty}_{DPnB,aqueous}$		224 ± 25	360 ± 30	306 ± 30	434 ± 30

Table 4. Values of $\gamma_{PnP,organic}$ Obtained by Analysis of Stripping and Extraction Data (*2*).

Organic Solvent	Temp.	*K* (exp'l)	x_{PnP} (exp'l)		y_{PnP} (exp'l)		$\gamma_{PnP,aq}$	$\gamma_{PnP,org}$
	°C	mol/mol	wt%	mol%	wt%	mol%		
2-ethylhexanol	40	66	0.75	0.115	8.48	7.63	41	0.62
	95	210	0.23	0.035	8.63	7.30	74	0.35
1-octanol	50	62	0.73	0.112	7.95	6.95	46	0.74
	95	106	0.40	0.061	7.92	6.50	74	0.70
methyl isobutyl ketone (MIBK)	40	33	1.23	0.191	8.08	6.23	41	1.24
	95	74	0.56	0.085	8.57	6.36	74	1.00
methyl heptyl ketone (MHK)	40	28	1.89	0.293	7.54	8.21	41	1.46
	95	80	0.69	0.106	8.06	8.45	74	0.93
toluene	50	23	1.78	0.275	7.97	6.24	46	2.00
	95	45	0.97	0.149	8.65	6.75	74	1.64

Table 5. Selected Partition Ratio Data as a Function of Temperature (*2*)

Organic Solute	Extraction Solvent	LCST or UCST for organic solute + water binary	Partition Ratio (wt% solute in organic phase / wt% solute in aqueous phase)
Glycol Ethers			
propylene glycol methyl ether	dichloromethane	LCST > 100 °C	0.37 at 22 °C 0.82 at 50 °C
dipropylene glycol methyl ether	dichloromethane	LCST ~100 °C	1.9 at 22 °C 5.5 at 50 °C
propylene glycol n-propyl ether	dichloromethane	LCST = 32 °C	7.4 at 25 °C 10.2 at 35 °C
propylene glycol methyl ether	2-ethylhexanol	LCST > 100 °C	0.42 at 22 °C 0.90 at 50 °C
Aliphatic Amines			
trimethylamine	toluene	LCST > 160 °C	0.16 at 0 °C 0.38 at 20 °C 0.78 at 40 °C
propylamine	toluene	LCST = 55 °C	0.23 at 18 °C 0.38 at 32 °C
diethylamine	toluene	LCST = 48 °C	0.7 at 20 °C 2.8 at 58 °C
Other Compounds			
tetrahydrofuran	2-methylfuran	LCST = 72 °C	2 at 20 °C 5 at 55 °C
2-butanol	2,4,4-trimethylpentene	LCST ~ 5 °C UCST = 115 °C	3 ± 1 at 20 °C 3 ± 1 at 60 ° C
methyl ethyl ketone	hexane	LCST = -20 °C	1.5 at 25 °C 2.0 at 38 °C
quinine	ethyl ether	hydrogen-bonding solid (for comparison)	18 at 0 °C 45 at 20 °C

Figures

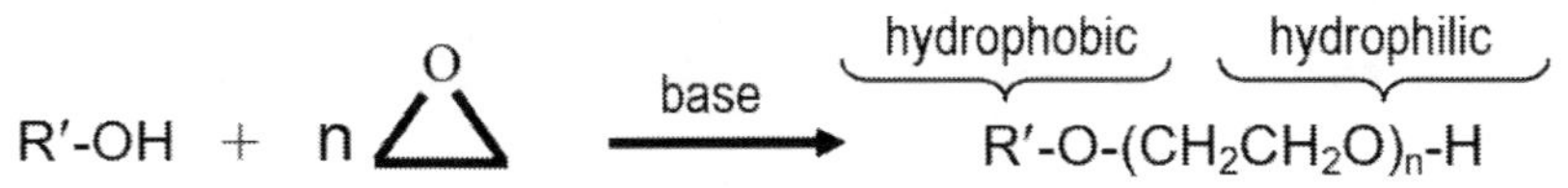

R′ = methyl, ethyl, n-propyl, n-butyl, n-hexyl, phenyl

n = 1	Ethylene glycol alkyl (or aryl) ether (2-alkoxyethanol)	$R'\text{-O-}CH_2CH_2\text{-OH}$
n = 2	Diethylene glycol alkyl ether (2-(2-alkoxyethoxy)ethanol)	$R'\text{-O-}(CH_2CH_2\text{-O})_2\ H$
n = 3	Triethylene glycol alkyl ether (2-(2-(2-alkoxyethoxy)ethoxy)ethanol)	$R'\text{-O-}(CH_2CH_2\text{-O})_3\ H$

Compounds with n > 3 normally are called polyethylene glycols

Figure 1. Ethylene glycol ether structures and nomenclature (E-Series).

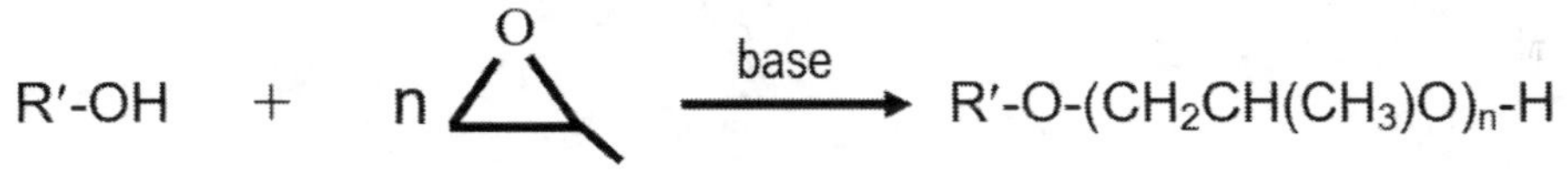

n = 1	Propylene glycol alkyl (or aryl) ether (1-alkoxy-2-propanol)	$R'\text{-O-}CH_2CH(CH_3)\text{-OH}$
n = 2	Dipropylene glycol alkyl ether (1-(2-alkoxy-2-methylethoxy)-2-propanol)	$R'\text{-O-}(CH_2CH(CH_3)\text{-O})_2\ H$
n = 3	Tripropylene glycol alkyl ether (1-(2-(2-alkoxy-1-methylethoxy)-1-methylethoxy)-2-propanol)	$R'\text{-O-}(CH_2CH(CH_3)\text{-O})_3\ H$

Compounds with n > 3 normally are called polypropylene glycols

Figure 2. Propylene glycol ether structures and nomenclature (P-Series).

	E-Series		
Alkyl Group	1 EO	2 EO	3 EO
methyl	----	194	249
ethyl	----	202	256
n-butyl	171	230	280
n-hexyl	208	259	----
phenyl	244	----	----
	P-Series		
Alkyl Group	1 PO	2 PO	3 PO
methyl	120	190	243
n-propyl	149	213	256
n-butyl	171	230	274
phenyl	243	----	----

Figure 3. Normal boiling points (°C) of typical glycol ethers. Colors indicate isomers.

	E-Series		
Alkyl Group	1 EO	2 EO	3 EO
methyl	----	3.5	6.2
ethyl	----	3.6	6.8
n-butyl	2.9	4.9	7.9
n-hexyl	4.5	6.2	----
phenyl	22	----	----
	P-Series		
Alkyl Group	1 PO	2 PO	3 PO
methyl	1.7	3.7	5.5
n-propyl	2.4	3.9	6.3
n-butyl	2.8	4.9	7
phenyl	25	----	----

Figure 4. Viscosity (cP at 25 °C) of typical glycol ethers. Colors indicate isomers.

	E-Series		
Alkyl Group	1 EO	2 EO	3 EO
methyl	----	1.02	1.05
ethyl	----	0.99	1.02
n-butyl	0.90	0.95	0.99
n-hexyl	0.87	0.93	----
phenyl	1.11	----	----
	P-Series		
Alkyl Group	1 PO	2 PO	3 PO
methyl	0.92	0.95	0.97
n-propyl	0.88	0.92	0.94
n-butyl	0.88	0.91	0.93
phenyl	1.06	----	----

Figure 5. Specific gravity of typical glycol ethers at 25 °C. Colors indicate isomers.

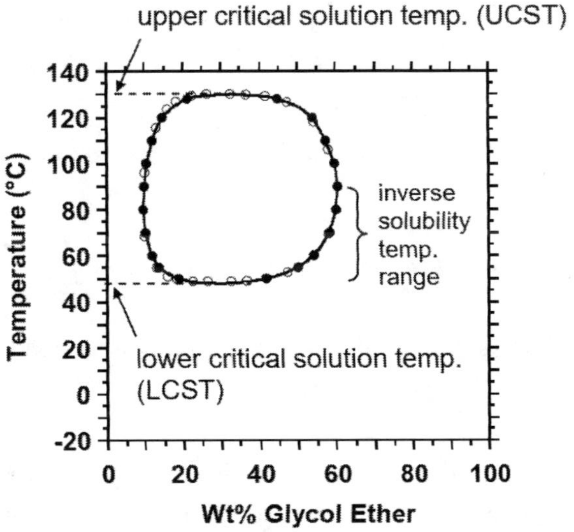

Figure 6. Temperature-composition diagram for ethylene glycol *n*-butyl ether + water showing LCST behavior (*5*).

E-Series			
Alkyl Group	1 EO	2 EO	3 EO
n-butyl	48	> 100	> 100
n-pentyl	< 20	40	----
2-Me-1-butyl	< 20	30	----
n-hexyl	~ -10	12	45
n-heptyl	----	~ -10	29
P-Series			
Alkyl Group	1 PO	2 PO	3 PO
n-propyl	32	10	2
n-butyl	~ -10	-10	----
P-Series Mixture (Equimolar)			
A	Propylene glycol n-propyl ether		
B	Dipropylene glycol n-propyl ether		
A+B	LCST = 18 °C		

Figure 7. Summary of LCST Values for Glycol Ethers (5).

- **Solvents and Co-solvents in Product Formulations**
  - coatings, inks, adhesives, latex formulations, cosmetics, hard-surface cleaners, agricultural chemical formulations, …
  - water-borne coating formulations
    - use less organic solvent
    - lower VOC emissions
- **Industrial Fluids**
  - electronics manufacturing, textile processing
  - automotive brake fluids, equipment cleaners, …
- **Reaction Media**
  - solvents for free radical acrylic monomer polymerization
  - reaction intermediates
- **Bioprocessing**
  - Primary release of periplasmic proteins from bacterial cultures
  - Extraction of hydrophilic compounds from aqueous solutions
- **Chiral Glycol Ethers (P-Series) as Intermediates to Chiral Pharmaceuticals**

Figure 8. Uses for glycol ethers.

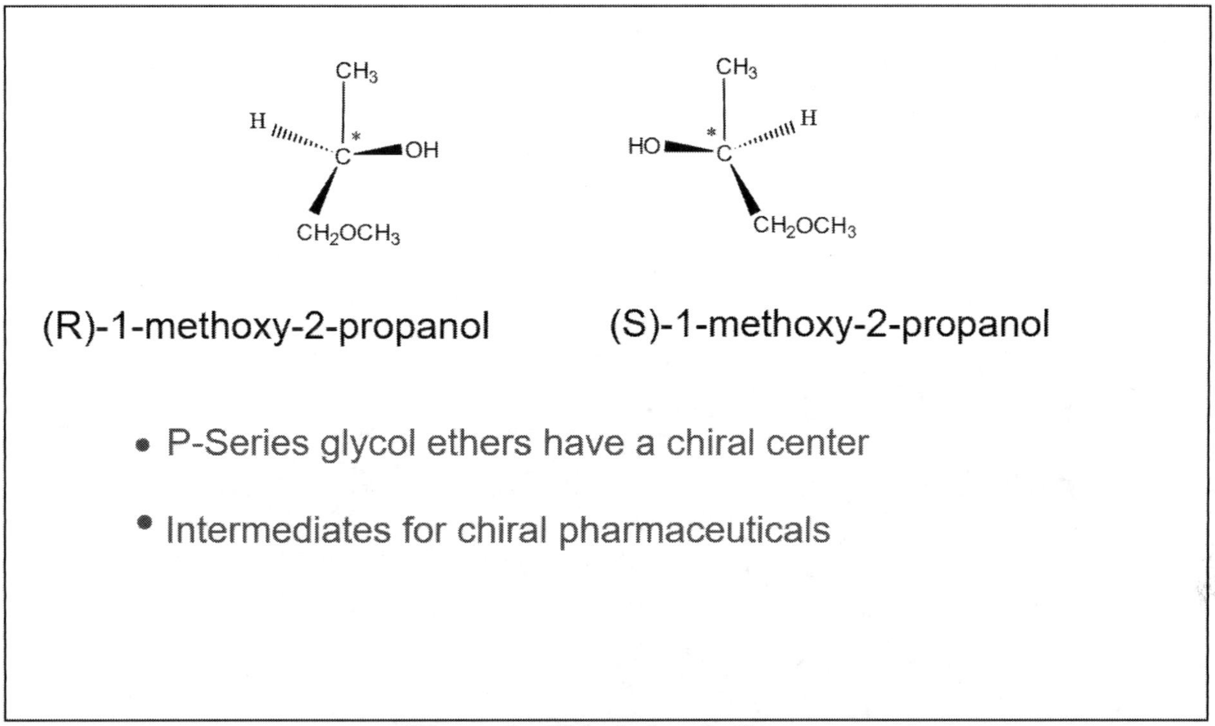

Figure 9. Chiral glycol ethers.

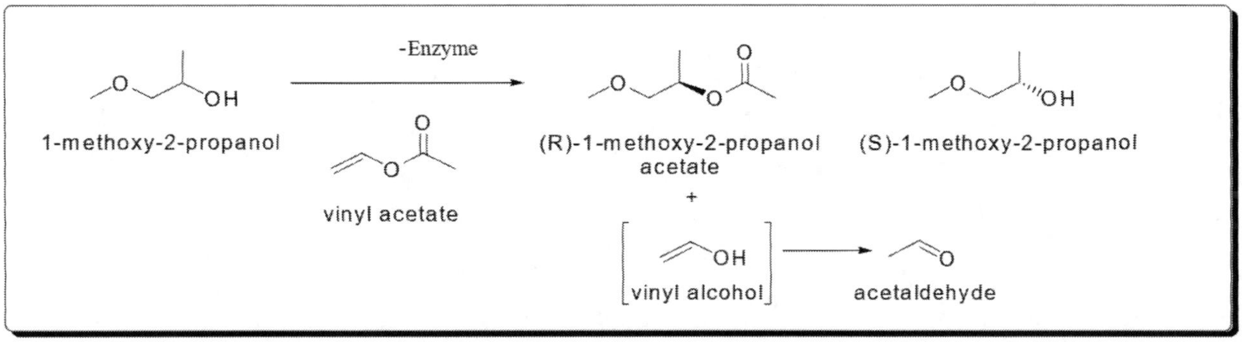

- Chemistry proceeds at high substrate loading
- Various P-series glycol ethers resolved in >98% ee
- Enzyme recycle demonstrated (immobilized CAL-B Chirazyme L-2)
- Experimental quantities of various enantiomers are available
- Felipe Donate (fadonate@dow.com)

Figure 10. Enantioselective enzymatic transesterification route to chiral glycol ethers. See US patent 7264960 (*12*).

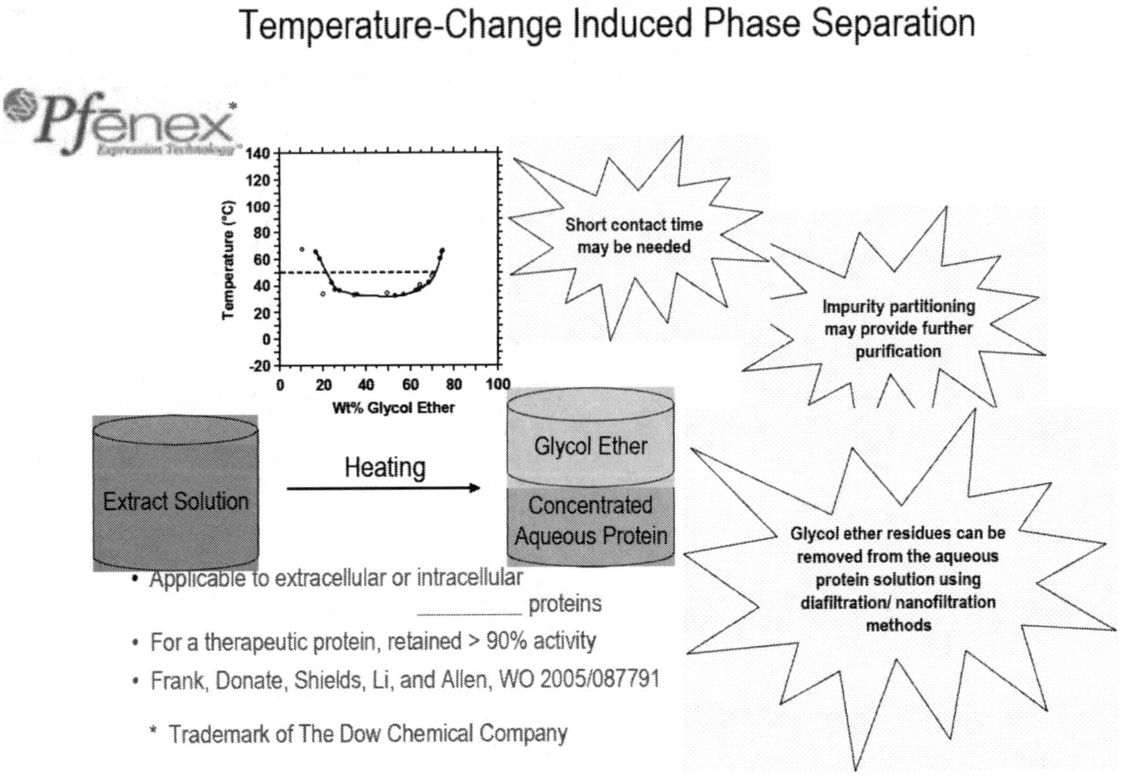

Figure 11. Release and concentration of intracellular proteins using glycol ethers (*11*).

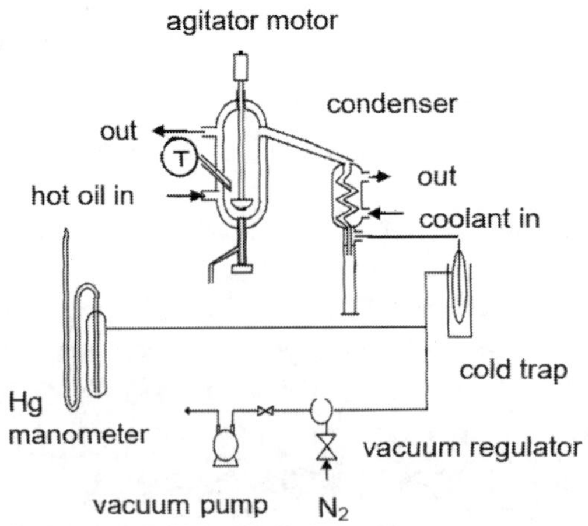

Figure 12. Rayleigh distillation apparatus (*2*).

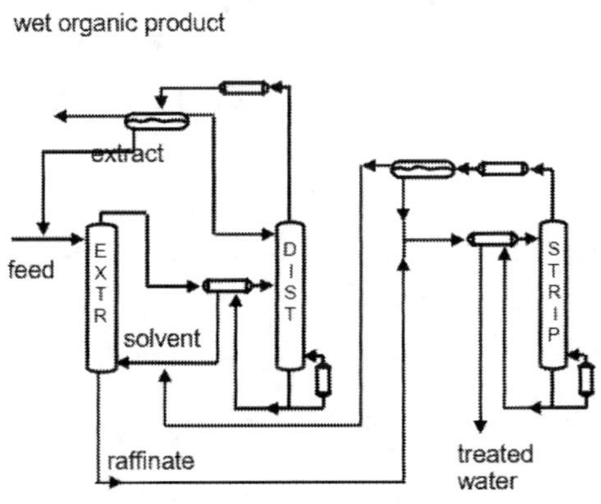

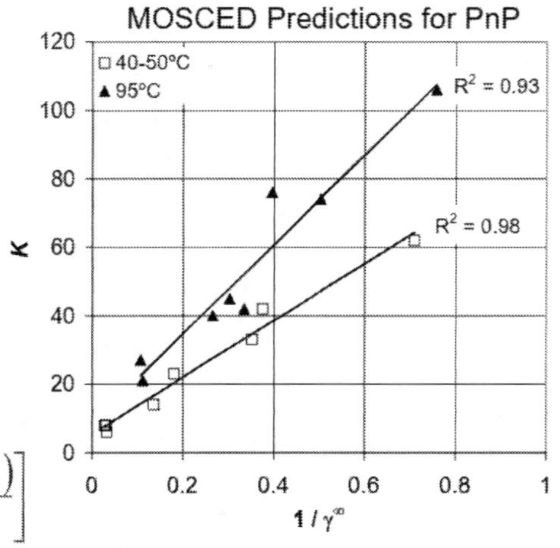

$$K(\text{mol fraction basis}) = \frac{y_i}{x_i} = \frac{\gamma_{i,aqueous}}{\gamma_{i,organic}}$$

$$\ln\gamma_{2,1}^{\infty} = \frac{v_2}{RT}\left[(\lambda_1-\lambda_2)^2 + \frac{q_1^2 q_2^2(\tau_1-\tau_2)^2}{\psi_1} + \frac{(\alpha_1-\alpha_2)(\beta_1-\beta_2)}{\xi_1}\right]$$

- Extraction of glycol ether residues from aqueous effluent streams
- Using a steam-strippable hydrophobic organic solvent

Figure 13. Removal of glycol ether residues from aqueous solution by extraction into a more hydrophobic organic solvent (*23*).

82b Extraction-Distillation or Extractive Distillation?

Jose Bravo and Don Foulk. Shell Global Solutions (US), PO Box 4327, Houston, TX 77210

The Extraction-Distillation (Stripping) combination is a common one in industry. With the avaialbility of powerful analysis tools and simulation programs, as well as understading of some fundamentals, engineers can analyze conventional Extraction flowsheets to determine if Extractive Distillation is an economic alternative. This paper will explore some of this fundamental concepts.

EXTRACTION OF METHANOL AND GLYCERIN FROM CRUDE BIODIESEL

Authors:

Donald J. Glatz
Lori Mason

Koch Modular Process Systems, LLC.

Prepared for Presentation at

AIChE National Meeting

Salt Lake City, Utah
November 5-9, 2007

INTRODUCTION

The biodiesel industry is a rapidly expanding industry and optimized unit operations are required in order to insure economical processes. One of the process steps involves the removal of methanol and glycerin from crude biodiesel feed, immediately after transesterification and phase separation. This step can be accomplished by liquid-liquid extraction (LLE) using water as the solvent to extract methanol and glycerin from the crude biodiesel feed. Many of the grass-roots biodiesel plants have used an inefficient mix/decant batch process for this unit operation. A counter-current, multi-stage LLE column should offer a process improvement resulting in higher efficiency and lower operating cost. This paper covers the evaluation of three (3) different types of extraction columns; the KARR® Reciprocating Plate Column, the SCHEIBEL® Column, and a static column using structured packing.

The feedstock studied was an unrefined, biodiesel produced from refined, bleached soybean oil. The test procedure used for this evaluation is outlined as follows:

- Generate liquid-liquid equilibrium data for the biodiesel feed and water. This establishes the equilibrium curve for this application.
- Test the biodiesel washing process in three (3) pilot scale columns (KARR®, SCHEIBEL®, and packed) at different operating conditions. The inlet and outlet data will establish the operating line for each run.
- Select the optimal results for each column and determine the efficiency (stages per meter) via graphical solution.
- Plot the optimal data for each column on a Stichlmair type plot[1] and use this as the basis of comparison.

The biodiesel feed contained 2.55 wt. % methanol and 6000 ppm glycerin. The target raffinate concentrations were less than 0.1% methanol and less than 200 ppm glycerin.

LIQUID-LIQUID EQUILIBRIUM DATA

Liquid-liquid equilibrium data was generated using a standard "Shake Test" procedure[2]. A 2-liter, round bottom reactor-type flask, with half moon agitator (Figure 1) was used for mixing and phase separation. A series of five (5) mix/decant runs were performed using a solvent to feed (S/F) ratio of 0.05, "as is" weight basis.

The results from the shake tests revealed that the glycerin level was reduced to less than 200 ppm after only 2 mix/decant runs. A total of 4 mix/decant runs were required to achieve less than 0.1% methanol. Based upon this finding, it was decided that methanol would be the key for evaluating the various column options. Therefore, glycerin data has not been presented in this paper.

The liquid-liquid equilibrium curve generated for the methanol extraction is shown on Figure 2.

PILOT PLANT EQUIPMENT SET-UP AND PROCEDURE

The equipment set-up for the pilot plant test is shown on the attached Figure 3. Essentially, this set-up consists of the following:

- A glass shell extraction column for visual observation of the countercurrent phase behavior in the column. The actual column configurations for the KARR®, SCHEIBEL®, and packed columns are shown in Figure 4. A manifold system for the water inlet location was used for the agitated columns (shown in Figure 4) to allow testing at several agitated heights.

- The organic feed and water solvent were received in 55-gallon drums. These were charged to the pilot extraction column directly from the drums using a dip tubes and variable speed metering pumps (FMI pumps). Flow rates were monitored by mass flow meters and the pumps were adjusted as required to maintain the desired flowrates.

- The bottoms flow rate (aqueous extract phase) and interface were controlled via an FMI pump with manual adjustment. Overhead (organic raffinate phase) was allowed to flow by gravity to receiver.

Test Procedure

The following procedure was used during the test program:

- Initially, the crude biodiesel feed was spiked to target concentrations of 2.9% methanol and 0.62% glycerin.

- The organic, biodiesel feed and water were charged directly from 55-gallon drums into the extraction column. The operation was conducted at room temperature.

- For startup each day, the column was simultaneously filled with both the biodiesel feed and water and the agitator set at a slow, initial speed. During this time, the flowrates were adjusted and set at the desired values.

- Once the column was full, the bottoms pump was turned on and adjusted to establish an interface in either the bottom or top disengaging chamber of the column. The bottom take-off rate was continuously monitored and adjusted to maintain a constant interface level.

- The agitator speed was then adjusted to the desired set point.

- The test supervisor specified conditions for each run. Feed and solvent rates were confirmed every 30 minutes throughout each run. At the completion of a run, raffinate and extract flow rates were measured and samples of each phase were taken for analysis.

- During startup each day, a total of five (5) column turnovers were performed before the initial samples were taken. A column turnover is defined as the total column volume divided by the combined feed and solvent rates. After the first run, and following adjustment of the variables, a total of three (3) turnovers were performed before additional sampling.

TEST RESULTS AND DISCUSSION

Operating conditions for all pilot plant runs are shown in Table 1. For each run, a plot was prepared using the equilibrium curve and the operating line based upon analytical results for the samples. The efficiency was determined by stepping off the number of theoretical stages using the McCabe-Thiele method. A description of the results achieved by each column follows:

KARR® Column

All runs in the KARR® Column were made with the aqueous phase dispersed (interface at the bottom of the column). Distinct dispersed phase droplets were produced and a clear, sharp interface was maintained in the lower disengaging chamber. No coalescence of the dispersed phase droplets was observed on the PTFE plates and shaft spacers.

KARR® and SCHEIBEL® are registered trademarks of Koch-Glitsch, LP.

Initially, the KARR® Column was operated with constant 2" plate spacing per the standard design. However, it became obvious after the first few runs that this spacing arrangement was not optimal. By visual observation, most of the column operated with very low dispersed phase hold-up and the column flooded near the bottom when attempts were made to increase the agitation speed above 90 SPM.

The plate stack was modified to allow for larger plate spacing at the bottom of the column with progressively smaller plate spacing moving up the plate stack. As can be seen by the data in Table 1, this allowed for higher agitation speeds and from a visual observation more disperse phase hold-up in the column.

The column efficiency was optimized for three (3) capacities, generating the following results:

Capacity (M^3/M^2 hr)	Maximum Efficiency Theo. Stages per Meter
20.3	1.0
24.4	1.1
28.5	1.2

SCHEIBEL® Column

All runs in the SCHEIBEL® Column were made with the water phase dispersed. Distinct dispersed phase droplets were produced and a clear, sharp interface was maintained in the lower disengaging chamber. A minimal amount of dispersed phase coalescence occurred on the metal baffles; however this was easily overcome by the agitation to produce the discrete aqueous droplets.

Based upon the operational results from the runs in the KARR® Column, the SCHEIBEL® Column was set up with variable impellor sizes, i.e. smaller impellors at the bottom of the column with progressively larger impellors moving up the column.

Visually, the SCHEIBEL® Column operated with smaller dispersed phase particle size and more hold-up than the KARR® Column. The only capacity tested during this test program was a specific throughput of 16.3 M^3/M^2 hr (scale-up basis).

The optimal column efficiency was 5.1 theoretical stages per meter.

Packed Column

The packed column was initially tested with the aqueous phase continuous (interface at the top) at a capacity of 20.4 M^3/M^2 hr. During the testing it was difficult to see and maintain the interface in the top chamber. There was no clear interface, rather a large emulsion band. Attempts to run at higher capacity were unsuccessful due to flooding.

The packed column was also tested with the organic phase continuous (interface at the bottom) at capacities of 20.4 and 28.5 M^3/M^2 hr. The dispersed phase droplets were large, typically ≥ ¼" diameter. Again, there was no clear interface, rather a large emulsion band in the lower disengaging chamber.

The optimal results achieve in the packed column were:

Capacity (M^3/M^2 hr)	Maximum Efficiency Theo. Stages per Meter
24.4	0.4
28.5	0.3

Stichlmair Type Plot

Figure 5 shows efficiency data presented by Stichlmair[1] (shaded regions) and Cusack, Holmes, and Karr[3] (curves). This data was all generated for the toluene-acetone-water system, which is the standard test system as suggested by Stichlmair.

The original Stichlmair curves for the SCHEIBEL® Column, KARR® Column and packed column are shown on Figure 6 along with data points generated during this test program with Biodiesel. It is obvious from Figure 6 that the water-methanol/glycerin-biodiesel system is a more difficult system because the efficiencies for each column are significantly lower for the Biodiesel system than for the standard test system.

CONCLUSION

The results from this test program demonstrate that agitated columns offer significant operational and efficiency improvements as compared to a column packed with structured packing for the Biodiesel feed tested (unrefined, biodiesel produced from refined, bleached soybean oil). Both the KARR® and SCHEIBEL® Columns were operationally sound, producing discrete, dispersed phase droplets and a clean, sharp interface. The SCHEIBEL® Column produced the highest efficiency.

RECOMMENDATION

The results from this study were based upon one feedstock only. Biodiesel can be produced from a broad range of feeds, both plant oils and animal fats. In order to select the best column type and provide an optimized design, we recommend testing each individual feed in pilot column(s) before finalizing a production design.

LITERATURE CITED

1. Stichlmair, J. "Performance and Cost Comparison of Various Equipment Designs for Liquid-Liquid Extraction", Chemie-Ingenieur-Technik, 52(3) 1980.

2. Glatz, D., Parker, W., "Extraction Enhancement, One Step at a Time", Chemical Engineering, November 2004

3. Holmes, T., Karr, A., Cusack, R., "Performance Characteristics of Packed and Agitated Extraction Columns", AIChE Summer National Meeting, August 17, 1987.

Table 1 - Pilot Plant Test Data

KARR Column Runs

Test #	Plate Stack Ht. (feet)	Capacity (GPH/sq.ft)	S/F "as is"	Agitation (SPM)	Flow Rate (g/min) Feed	Water	Raffinate	Extract	Note
K1	10	700	0.1	90	197	10	196	14	1
K2	10	700	0.05	80	196	10	202	10	1
K3	8	700	0.05	90	196	10	198	15	1
K4	6	700	0.05	90	196	10	196	12	1
K5	6	700	0.1	90	187	20	188	24	1
K6	10	700	0.1	120	187	20	185	22	2
K7	10	700	0.1	140	186	20	194	11	2
K8	10	700	0.05	80	186	10	185	16	2
K9	10	700	0.05	100	186	10	Flooded		2
K10	8	600	0.06	100	167	10	170	8	2
K11	6	700	0.05	80	196	10	202	12	2
K12	8	700	0.05	80	196	10	196	13	2
K13	10	700	0.11	140	185	21	181	24	2
K14	10	600	0.11	140	161	18	158	19	2
K15	10	600	0.11	160	160	18	Flooded		2
K16	10	500	0.11	140	133	15	132	14	2
K17	10	700	0.11	140	186	21	Flooded		2
K18	8	700	0.11	100	186	20	186	24	2

SCHEIBEL Column Runs

Test #	Agitated Stages	Capacity (GPH/sq.ft)	S/F "as is"	Agitation RPM	Flow Rate (g/min) Feed	Water	Raffinate	Extract	Note
S1	40	400	0.11	300	200	21	199	22	3
S2	40	400	0.1	375	201	21	199	22	3
S3	40	400	0.05	375	201	11	195	15	3
S4	30	400	0.05	375	201	11	196	15	3

Packed Column Runs

Test #	Packing Ht. (ft)	Capacity (GPH/sq.ft)	S/F "as is"		Flow Rate (g/min) Feed	Water	Raffinate	Extract	Note
P1	10	500	0.1		1240	124	1223	143	4
P2	10	500	0.05		1241	62	1239	74	4
P3	10	700	0.1		1760	176	1778	177	5
P4	10	500	0.1		1241	124	1236	127	5

Notes:
1. First 5 KARR Column runs used the standard, constant plate spacing for plate stack
2. Remainder of KARR Column runs used variable plate spacing for plate stack
3. Capacity specified for SCHEIBEL Column is based upon scale up rate
4. First 2 runs in packed column were with organic phase dispersed (interface on the top)
5. Last 2 runs in packed column were with aqueous phase dispersed (interface on bottom)

Figure 1

Equipment Set-up for Generating Liquid-Liquid Equilibrium Data

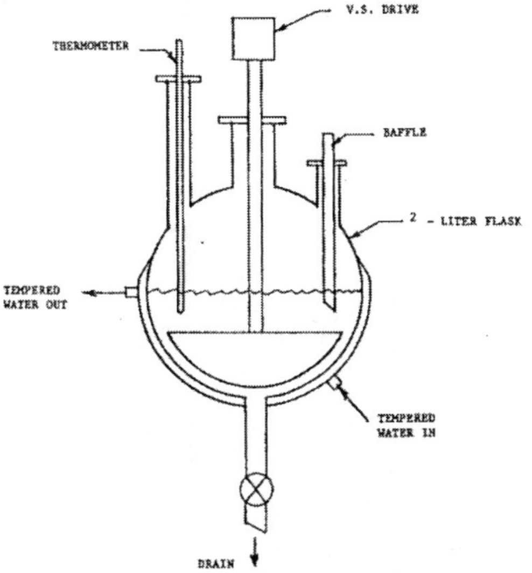

KOCH MODULAR PROCESS SYSTEMS, LLC

EXTRACTION TECHNOLOGY GROUP

Figure 2 - Biodiesel Equilibrium Curve

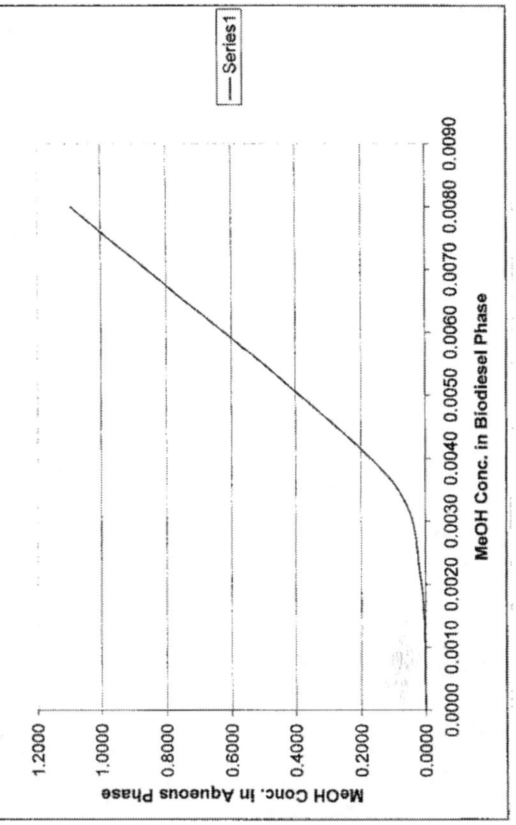

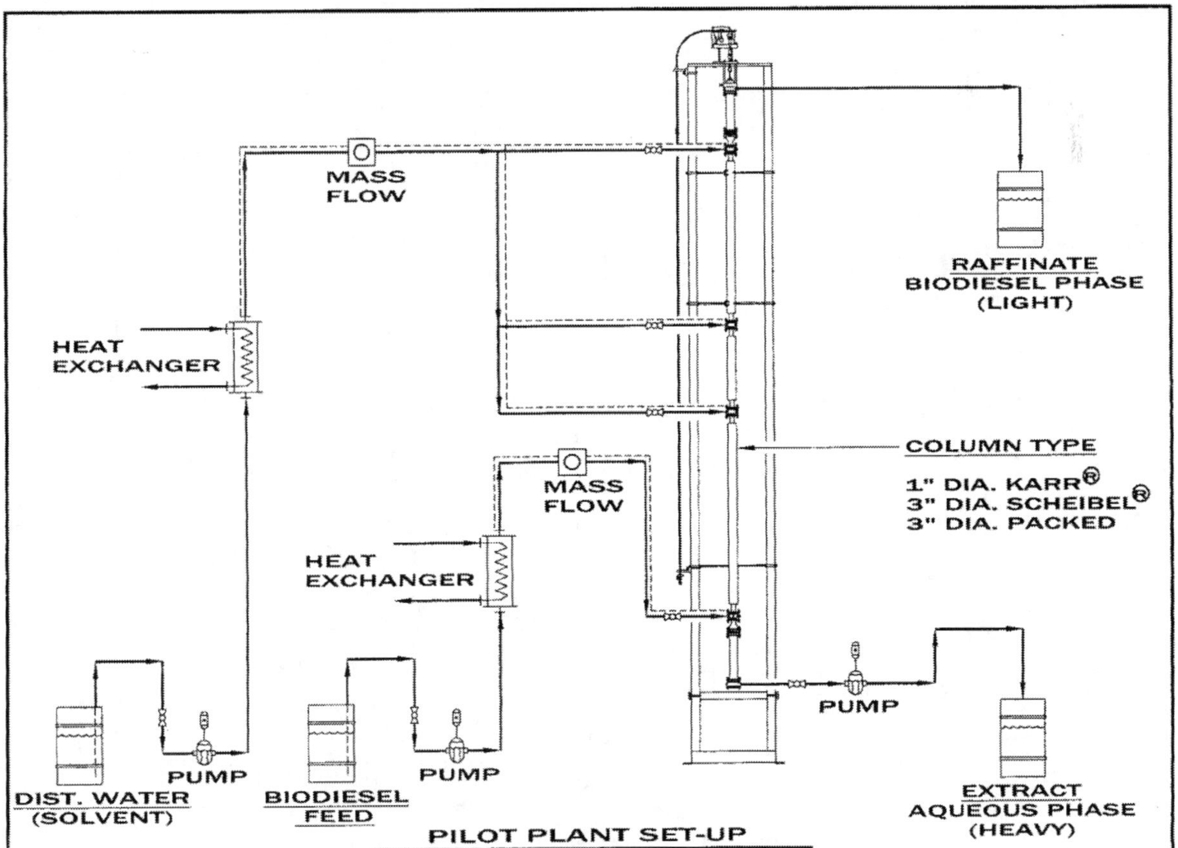

PILOT PLANT SET-UP

FIGURE - 3

KOCH MODULAR PROCESS

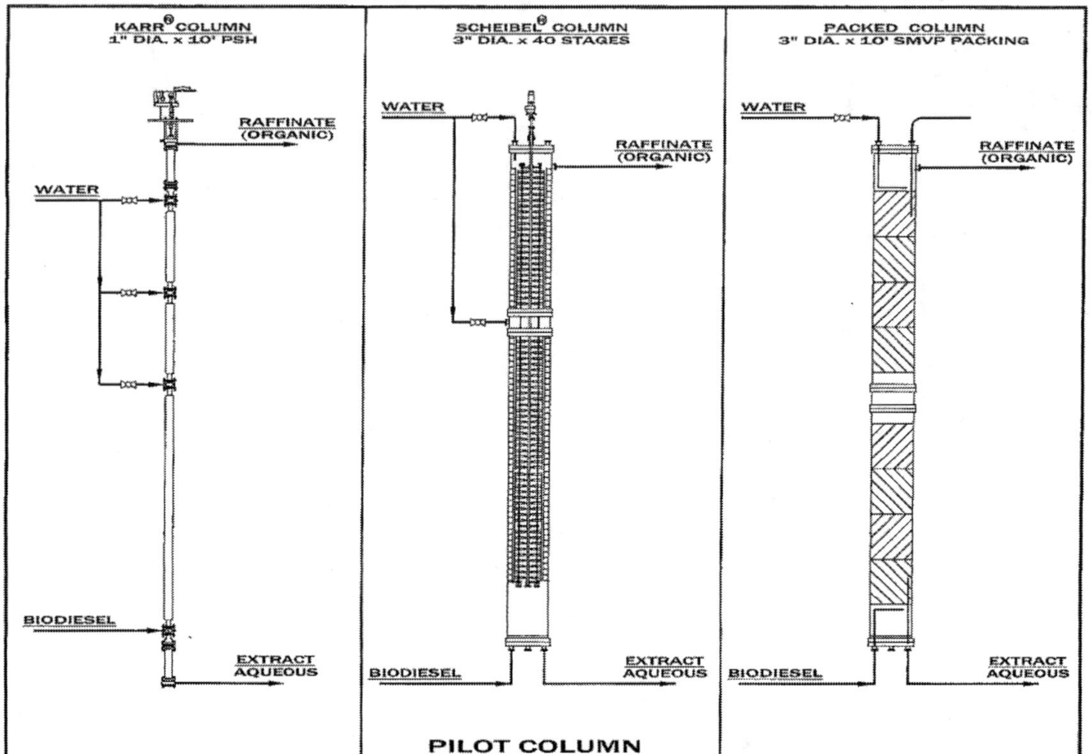

PILOT COLUMN CONFIGURATION

FIGURE - 4

KOCH MODULAR PROCESS

Figure 5 – Stichlmair Type Plot

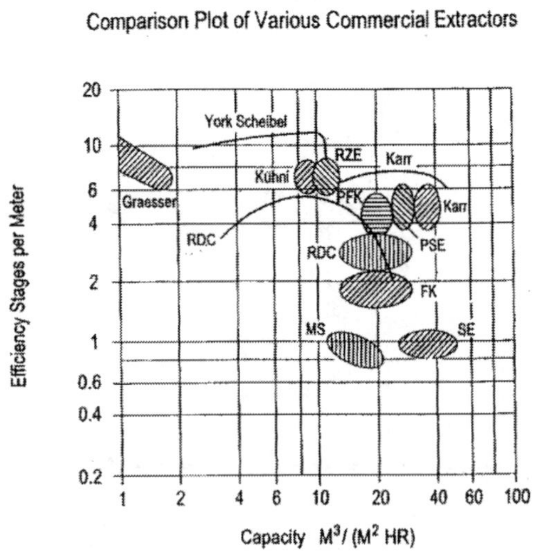

Key:

Graesser	= Raining Bucket Extractor	RDC	= Rotating Disc Contactor
MS	= Mixer Settler	RZE	= Agitated Cell Extractor
SE	= Sieve Plate Extractor	Karr	= Karr Reciprocating Plate Column
FK	= Random Packed Column	Kuhni	= Kuhni Extractor
PFK	= Pulsed Packed Column	York-Scheibel	= Scheibel Column
PSE	= Pulsed Sieve Plate Column		

Source: Shaded Area =
Curves =

(1) J. Stichlmair, University of Essen, Wiesbaden, Germany, "Performance and Cost Comparison of Various Equipment Designs for Liquid-Liquid Extraction", Chemie-Ingenieur-Technik 52(3),1980

(2) T. Holmes, R. Cusack, A. Karr, The Otto H. York Company, AIChE Summer National Meeting, August 17, 1987

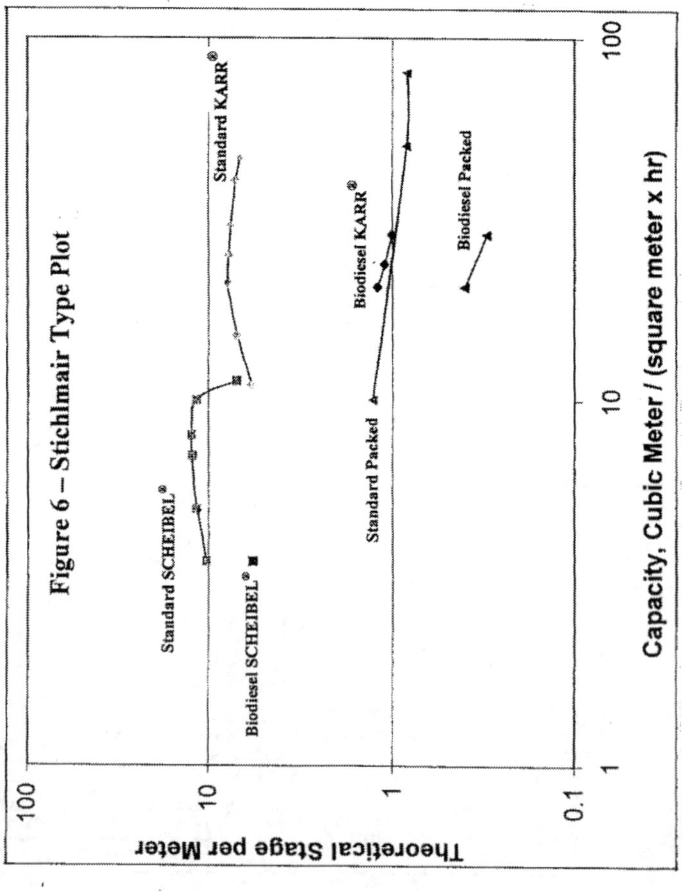

Figure 6 – Stichlmair Type Plot

82d New Approach for Modelling Mechanically Agitated Extraction Columns

Frank Seibert, J. Christopher Lewis, and Timothy Rochelle. Separations Research Program, University of Texas at Austin, 10100 Burnet Road, Bldg 133/CEER, Austin, TX 78758

A new modeling approach for the Karr reciprocating plate extractor is being evaluated which applies fundamental static models with an added correction factor to account for the mechanical energy input. Correction factors for drop diameter, slip velocity, film mass transfer coefficient and axial mixing are used to determine the flooding velocity and apparent mass transfer efficiency. The initial results appear very promising based on data obtained using the toluene/acetone/water system (a high interfacial tension system). Future work is planned to check the proposed modeling approach with the butanol/succinic acid/water system (a low interfacial tension system).

82e Membrane Composition and Osmosis in Liquid Membrane Permeation

Marlene Fritz, Matthäus Siebenhofer, and Rolf Marr. Department of Chemical Engineering and Environmental Technology, Graz University of Technology, Inffeldgasse 25/C/II, Graz, A-8010, Austria

The system Zn2+/DEHPA is a standard test system in reactive extraction [1].From application in emulsion liquid membrane permeation significant water transport from the aqueous feed phase across the membrane due to osmosis has been reported. Present research activities focus on the investigation of supported liquid membranes with the test system Zn2+/DEHPA. For modelling purposes water flux due to osmosis has been subject of investigation. The experiments were carried out in a u-tube membrane cell. DEHPA concentration in the membrane phase was varied within 0-30 [wt%]. Water flux is expected, because osmotic pressure between phase I (stripping phase, cH2SO4 = 1 [mol/L]) and phase III (aqueous feed phase, cZn = 400 [mg/L]) is not in balance. The difference in osmotic pressure between phase I and phase III is deltap = 48 [bar]. Unexpectedly osmosis driven water flux drops with increasing DEHPA concentration. Beyond DEHPA concentration of 5 [w%] no water flux is observed. Transmembrane water flux is blocked by the ion exchanger DEHPA which forms an osmotic sink.

[1] H.-J. Bart, M.J. Slater, Standard Test system for reactive extraction - Zink/D2EHPA, European Federation of chemical engineering working party on distillation, absorption and extraction, 2001

82f Prediction of the Partitioning Coefficients with Cosmo-Rs

Michael Diedenhofen, COSMOlogic GmbH & Co. KG, Burscheider Str. 515, Leverkusen, Germany and Andreas Klamt, COSMOlogic GmbH&CoKG, Burscheider Str., Leverkusen, 51381, Germany.

Because of the increasing number of solvents that can be suitable for a certain extraction problem, theoretical methods that are able to give a fast prediction of the solubility of a substance in a solvent, are valuable tools for the preselection of a set of interesting candidates. Especially ionic liquids, which have attracted enormous attention during the last decade, represent a broad an fast growing field of new solvents. The COSMO-RS model, which is a quantum chemical based statistical thermodynamic procedure, has been used successfully for the prediction of solubilities and partitioning coefficients.1 Although the method was originally designed and parameterized for neutral compounds, it has been shown, that it is applicable to ionic liquid solutions as well. In this talk we will give an overview of the basic COSMO-RS theory and the applications on the field of extraction with "classical" solvents and ionic liquids.

1) F. Eckert, A. Klamt, AICHE Journal, Volume 48, Issue 2, 369-385.

85a Fundamental Study of the Adsorption of Chiral Solutes on Polysaccharide Based Chiral Stationary Phases for Chiral Separation Applications

Rahul B. Kasat, N. H. Linda Wang, and **Elias I. Franses**. School of Chemical Engineering, Purdue University, 480 Stadium Mall Drive, West Lafayette, IN 47907-2100

Even though polysaccharide based chiral stationary phases are used widely in pharmaceutical industry for chiral drug separations, the chiral recognition mechanisms of these phases are not yet elucidated. Recently, we have studied these biopolymers in dry state, and also upon interactions with solvents and simple solutes in detail [1-3]. In this study, the adsorption of a chiral solute on the polysaccharide based chiral stationary phases are studied in detail using HPLC, attenuated total reflection infrared spectroscopy (ATR-IR), X-ray diffraction (XRD), and molecular modeling. The capacity factors and selectivities of enantiomers of norephedrine vary significantly with different adsorbents and solvents. The changes observed in the wavenumbers and the intensities of amide I and amide II bands of these polymers in the ATR-IR spectra upon adsorption of each of the enantiomer are stereo-specific and correlate strongly with the selectivities in HPLC. The IR wavenumbers, and the H-bonding interaction energies of the polymer side chains with each enantiomer in different configurations predicted using the DFT/B3LYP/6-311+g(d,p) level of theory are in agreement with the capacity factors observed in HPLC. The changes in the polymer crystallinity upon absorption of each enantiomer are characterized with XRD. We conclude that the H-bonding states of the polymer change significantly upon adsorption of each enantiomer and usually result in an increase in the polymer crystallinity. These results have implications in understanding the fundamental molecular recognition mechanisms of these polymers. The results should generate general guidelines for selecting solvents, and sorbents, to optimize retention times, selectivities, productivity, and solvent consumption in chiral chromatography for both analytical separation applications and large-scale production of single enantiomers.

1. R. B. Kasat, Y. Zvinevich, H. W. Hillhouse, K. T. Thomson, N.-H. Linda Wang, and E. I. Franses, Journal of Physical Chemistry B, 110, 2006, 14114-14122.

2. R. B. Kasat, C. Y. Chin, K. T. Thomson, E. I. Franses, and N.-H. Linda Wang, Adsorption, 12, 2006, 405-416.

3. R. B. Kasat, N.-H. Linda Wang, and E. I. Franses, Biomacromolecules, 8, 2007, 1676-1685.

85b Separation of Reboxetine Enantiomers Using Compact Simulated Moving Bed Schemes

João M. M. Araújo, Rui C. R. Rodrigues, and José P. B. Mota. Requimte/CQFB, Chemistry Department, Faculdade de Ciências e Tecnologia, Universidade Nova de Lisboa, Caparica, 2829-516, Portugal

Reboxetine, (RS)-2-[(RS)-α-(2-ethoxyphenoxy)benzyl]-morpholine is a antidepressant drug which belongs to the so-called norepinephrine reuptake inhibitors (NRI drugs), because its selectivity inhibits the uptake of this important neurotransmitter from the presynaptic neurones. Reboxetine has two asymmetrical carbons, thus it can exist as two enantiomeric pairs; however due to the stereoselectivity of the synthesis only the (R,R)-, (S,S)-pair is present as a racemic mixture in the active principle and the commercial formulations. Recent studies support the hypothesis that the (S,S)-enantiomer is a more potent norepinephrine reuptake inhibitor than the (R,R)- and that its is responsible for the vasomotor and cardiac side effects of reboxetine, which features the necessity to have reliable methods for the separation and analysis of reboxetine enantiomers.

Simulated moving bed (SMB) chromatography due to its many advantages with respect to discontinuous batch chromatography, such as higher product purity, less solvent consumption, and higher productivity per unit stationary phase; is an upright tool for the separation of reboxetine enantiomers. Thanks to recent developments in cyclic operation policies, a number of possibilities for improving SMB performance have emerged through variation of parameters during the switching interval. These include, among others, the nonsynchronous shift of the inlet and outlet ports within a global switching period, and cyclic modulation of feed flow and feed concentration. These advances are pushing the trend toward the use of units with smaller number of columns, because less stationary phase is used and the setup is more economical.

Two distinct compact SMB schemes with two and three columns were developed for the binary separation of reboxetine enantiomers on Chiralpak AD using as solvent a mixture of Hexane-Ethanol-DEA, for both linear and nonlinear isotherms. The developed compact schemes include a two-zone SMB with feed flow modulation, and a three-column four-zone SMB with asynchronous port switching and feed flow modulation, that were optimized using a full-discretization method implemented on AMPL and solved using a efficient interior-point solver.

Product quality control is mandatory for chiral chromatographic separations, and requires robust on-line monitoring of absolute enantiomer concentrations. In this work, our recent developed on-line monitoring system consisting of a high-speed HPLC setup with two UV detectors was implemented to monitor the concentration profile of each reboxetine enantiomer in the SMB unit. This monitoring scheme does not require the use of a polarimeter.

Keywords: enantiomers; reboxetine; chromatography; simulated moving bed; flow modulation; asynchronous port switching.

85c Theory Of Electrodeionisation. A Link Between Theories Of Ion Exchange Columns And Electrodialysis

Georges Grévillot, François Lapicque, and Laurence Muhr. Chemical Engineering Science Laboratory-CNRS, 1, rue Grandville, Nancy, France

Electrodeionisation is the combination of electrodialysis and ion exchange processes where ion exchangers are disposed between the membranes. The main application actually is water deionisation. We develop a theory for electrodeionisation based on the well-known equilibrium theory of ion exchange columns. In the differential mass balance equation for ion exchange columns, a term is added accounting for mass transport through the membrane. The second equation is the ion exchange isotherm. The system is solved by the method of characteristics. Several cases are considered depending of the membrane flux law. Transient and steady states are studied. During transient state an ion exchange front develops and propagates through the bed with a downstream plateau of non-constant concentration and an upstream concentration varying exponentially. An interesting result is shown: at steady state, there is no influence of the ion exchange isotherm. The electrical properties of the ion exchanger only influence the process. Several aspects of the model are discussed: non-linear isotherm, transversal concentration gradient, axial dependant flux.

85d An Extended Theory of Adsorption in An Ion Exchange Resin

Ann Marie Hardin and C. F. Ivory. Chemical Engineering, Washington State University, PO Box 642710, Dana 114, Spokane St, Pullman, WA 99164-2710

Ion exchange chromatography is commonly used in the purification of proteins and other ionic species. Understanding the relationship between the intraparticle conditions and those of the surrounding buffer is crucial for predicting the optimum operating conditions for ion exchange. It is additionally necessary to consider interactions between adsorbed and mobile species. We present a new theoretical model of ion exchange adsorption that incorporates consideration of these interactions as well as the intraparticle conditions and their effect on the diffusing species. This model considers three dimensions of interaction. The ability of the protein to enter to pore is first considered and is affected by the relationship between the molecule hydrodynamic radius and the open pore space. The foot print, or shadow, of a molecule is also a function of the hydrodynamic radius and affects the amount of a particular mobile species that may be adsorbed onto the internal surfaces of the resin. The final area of consideration is the distance of closest approach between the adsorbing species and the charged sites on the resin surface. This parameter is again a function of the molecule's radius as well as the molecular charge and the surrounding conditions. The distance of closest approach indicates the strength of the interaction between the mobile species and resin surface. Our model will incorporate all three of these parameters to predict the adsorption of a protein onto an ion exchange resin. The model is developed from first principles and validated using direct indication of the resin internal pH and adsorption experiments.

Web Page: che.wsu.edu/~ahardin/Research.htm

85e Adsorption Of Ethoxylated Nonionic Surfactants On Various Interfaces

Abhay Ladhe, Dept. of Chemical and Materials Engineering, University of Kentucky, Lexington, KY 40506 and Dibakar Bhattacharyya, Chemical and Materials Engineering, University of Kentucky, Lexington, KY 40506.

Ethoxylated nonionic surfactants are used in many industrial applications, which involve their adsorption on various interfaces. Fundamental understanding and quantification of the adsorption behavior is therefore important. In the present study, adsorption behavior of pure ethoxylated nonionic surfactants on cotton and polyester surface in hydrophobic environment was quantified. The role of ethylene oxide group variation in surfactant-polymer interaction was established using pure surfactant with same alkyl chain length but varying ethoxylate chain lengths. It was observed that surfactant with more ethylene oxide groups per molecule, being more hydrophilic, interacts favorably with cotton than polyester in the hydrophobic environment. The adsorption of the pure surfactants on model gold surface was also established using Quartz Crystal Microbalance with Dissipation monitoring (QCM-D) system. Effect of ethylene oxide chain length and surfactant concentration on the extent of adsorption was precisely quantified and adsorption isotherms were constructed. Adsorption was studied from both aqueous solutions and solutions in hydrophobic solvent of the surfactants. Information about thickness and structure of adsorbed layer was calculated from the QCM-D frequency-dissipation data.

MASS TRANSPORT MODEL FOR THE REMOVAL OF PARA-CHLOROPHENOL BY COLUMN ADSORPTION USING ACTIVATED CARBON

Michael F. F. Sze, The Hong Kong University of Science and Technology, Hong Kong
Vinci K. C. Lee, The Hong Kong University of Science and Technology, Hong Kong
Gordon McKay, The Hong Kong University of Science and Technology, Hong Kong

Abstract

This project aims at purifying para-chlorophenol contaminated industrial effluents by column adsorption using Chemviron Filtrasorb 400 (F400) activated carbon. Two column adsorbers, namely, a constant diameter column (0° tapered) and a tapered column (5° tapered), have been operated under same conditions in the adsorption study. These two columns are operated with the same volumetric flowrate and activated carbon loading, but different values of initial concentration of para-chlorophenol. The para-chlorophenol concentration time curve (breakthrough curve) from each column run has been analyzed, and hence the characteristic of each column adsorber with different configuration has been evaluated.

To predict the breakthrough curves for the para-chlorophenol/F400 carbon system, the Homogeneous Surface Diffusion Model (HSDM), which includes the effect of the external mass transfer and the effect of the intraparticle mass transfer by the homogenous surface diffusion, is used. The SIPS equilibrium isotherm model has been incorporated with this model to describe the adsorption equilibrium at the interface of adsorbent solid and solution. This HSDM applies the Crank-Nicholson implicit finite difference method which was proposed by Mathews and Weber [1] for solving diffusion equations. In literature, it has hitherto been applied to fixed-bed adsorption systems on activated carbon [2-5].

The HSDM is modified to describe the dynamic adsorption process in tapered fixed bed column adsorbers. Due to the unique configuration of the tapered columns, the interstitial velocity of adsorbate flow is changing along the column, and hence gives a variable external mass transfer coefficient (k_f), which has been used to represent the laminar boundary layer surrounding the F400 carbon particle at different bed heights. The surface diffusion coefficient (D_s) is assumed to be independent of the surface coverage of the F400 carbon solids in this project. A constant surface diffusion coefficient is used to represent the migration of para-chlorophenol molecules on the surface or pore walls of the F400 activated carbons.

From the equilibrium time test, the para-chlorophenol/F400 carbon system has been shown to attain equilibrium after 20 days under constant agitation. The adsorption pilot plant column study, however, is a dynamic process, in which the F400 activated carbon did not possess a full adsorbing capacity in the duration of the column run. As a result, the bed utilization rate from the concentration time curve (breakthrough curve) is calculated to reflect the effective bed adsorption capacity towards para-chlorophenol in this dynamic process. With the introduction of this factor, the HSDM could provide a good fit to the experimental data. This new modified HSDM could be applied successfully to evaluate the dynamic adsorption behavior of para-chlorophenol/F400 activated carbons in two different kinds of columns – constant diameter columns and tapered columns with different inclinations.

1. Introduction

The adsorbent particle is considered as a homogeneous solid sphere in which adsorbate is transported by surface diffusion. Adsorbates diffuse through an external film to the outer surface of a particle. Once reaching the particle surface, adsorption takes place instantaneously and equilibrium is established between the adsorbate fluid and the particle surface.

The developed mathematical model started with the overall mass balance around the fixed bed (neglecting the axial dispersion), which is given as

$$v\left[\frac{\partial C}{\partial Z}\right]_t + \left[\frac{\partial C}{\partial t}\right]_Z + \rho_p\left(\frac{1-\varepsilon}{\varepsilon}\right)\left[\frac{\partial q}{\partial t}\right]_Z = 0 \qquad (1)$$

Corresponding initial and boundary conditions

For t=0 and z=0, $C=C_o$

For t=0 and z>0, C=0 and q=0

The rate of accumulation of solute in the solid phase, and hence could be described by the rate of solute transfer across the stagnant liquid film.

$$\rho_p\left[\frac{\partial q}{\partial t}\right] = \frac{3k_f}{R_p}(C - C_s) \qquad (2)$$

The adsorption rate of solute by spherical particle is governed by the surface diffusion and hence represents as

$$\frac{\partial q}{\partial t} = \frac{1}{r^2}\frac{\partial}{\partial r}[D_s r^2 \frac{\partial q}{\partial r}] \qquad (3)$$

Boundary Condition 1 at the center of the particle: $\left.\frac{\partial q}{\partial r}\right|_{r=0} = 0$

Boundary Condition 2 at the surface of the particle: $\frac{k_f}{\rho_p}(C - C_s) = D_s \left.\frac{\partial q}{\partial r}\right|_{r=R}$

Initial conditions: 1) q = 0 for all r at t=0 and 2) $q = q_s(t)$ for all R_p at t=t

In order to predict the breakthrough curve, the numerical solutions for the fluid and solid phase PDEs are coupled and solved by using a finite difference method. By implementing the Crank-Nicolson method to the Eq. (2), the solutions to the solid phase PDE could provide an estimate of solid phase concentration for the fluid phase PDE solution. With the respective boundary conditions associated with the solid phase PDEs, the solid phase PDE could be transformed into a matrix and solved at each time step. With the SIPS isotherm equation and solid phase concentration, the fluid phase concentration can be solved at the current time level and the whole breakthrough curve could be obtained by increasing the time step.

The velocity term in the new derived equation is no longer constant along the tapered column, in fact, it varies with bed height. By simple geometry (shown in Figure 1), the linear velocity of liquid (v) in different bed heights (Z) could be re-defined as a function of bed height. Linear velocity of solute (v) in tapered column at bed height Z is calculated as $v = \frac{Q}{A} = \frac{Q}{\pi r_{col}^2}$, where $r_{col} = r_{col,1} + Z\tan\theta$.

The velocity term is incorporated in the k_f evaluation by the Olive and Lacoste (1979) external mass transfer correlation, which means k_f values would also change with bed height. The Olive and Lacoste expression comprises dimensionless numbers, in which the Sherwood number (Sh) is a function of the Reynolds number (Re) and the Schmidt number (Sc), and is given as $Sh = 4.3Re^{0.35}Sc^{1/4}$, with the validity of $0.1<Re<3$. In order to develop an applicable HSDM for tapered systems, two terms (velocity and the external mass transfer coefficient) have to be expressed as a function of bed height and incorporated into the external film balance calculation in the simulation program.

2. Results and Discussion

The adsorption column experiment is dynamic process, the fixed bed of activated carbon could not be fully saturated. Incorporating the equilibrium adsorption capacity to the HSDM did not give a good fit of the experiment observation. Figures 2 and 3 show the original HSDM results for removing 300mg dm^{-3} para-chlorophenol in 0° degree and 5° degree tapered columns, both operating at 200mL/min.

The effective bed adsorption capacity is introduced to the modified HSDM and hence provides good fit to the experiment results. Actually, this factor represents the extent of saturation in adsorption columns. It could be determined by analyzing the breakthrough curve of each system, which is the ratio of bed saturation in the duration of column operation to that at equilibrium. Table 1 summarized the operating conditions, effective bed adsorption capacity, optimized surface diffusivity (D_s) and the average external mass transfer coefficient (k_f) values of each column system. From the simulation program, the average external mass transfer coefficients for 0° degree and 5° degree tapered columns are 8.53E-03cm/s ± 4.26E-04cm/sand 1.44E-02cm/s ± 6.99E-03cm/s respectively.

The discrepancy of the simulated HSDM and experiment results are greatly reduced after the introduction of the effective bed adsorption capacity. Figures 4 and 5 show the improvement of introducing the effective bed adsorption capacity to the HSDM, which could evaluate the dynamic adsorption behavior of para-chlorophenol/F400 activated carbons successfully in two different kinds of columns – constant diameter columns and tapered columns.

3. Conclusion

A Homogeneous Surface Diffusion Model (HSDM) using the Crank-Nicholson implicit finite difference method is used to solve the diffusion equations in the para-chlorophenol/F400 carbon systems. This model is modified by using the Olive and Lacoste correlation to evaluate the variable external mass transfer coefficient (k_f) values in tapered columns and introducing the effective bed adsorption capacity to each column run. After these modifications, the breakthrough curves of the para-chlorophenol/F400 carbon systems operated in constant diameter columns and tapered columns could be successfully predicted.

Nomenclature

A	=	Cross-sectional area of column, m^2
C	=	Adsorbate concentration in liquid phase, mg dm^{-3}
C_s	=	Liquid phase adsorbate concentration at adsorbent external surface, mg dm^{-3}
D_s	=	Surface diffusivity, m^2 s^{-1}
k_f	=	External film mass transfer coefficients, m s^{-1}
q	=	Solid phase adsorbate concentration at time *t*, mg g^{-1}
q_s	=	Solid phase adsorbate concentration at the surface of adsorbent particle, mg g^{-1}
Q	=	Pollutant volumetric flowrate, dm^3 hr^{-1}
r	=	Radial position within adsorbent particle, m
r_{col}	=	Radius of adsorption column, m
$r_{col,1}$	=	Radius of adsorption column at bed height of zero meter, m
R_p	=	Radius of adsorbent particle, m
Re	=	Reynolds number, dimensionless
Sc	=	Schmidt number, dimensionless
Sh	=	Sherwood number, dimensionless
SSE	=	Sum of the squares of the error, dimensionless
t	=	Service time/operating time of the adsorption column, hr
v	=	Linear velocity of fluid, m hr^{-1}
Z	=	Bed height, m
θ	=	Tapered angle, degree
ρ_p	=	Density of adsorbent particle, kg m^{-3}
ε	=	Bed voidage, dimensionless

Acknowledgements

The authors would like to thank the Research Grants Council (RGC) of Hong Kong SAR for the financial support for this research project.

References

1. Mathews, A. P. and Weber, W. J. J. (1976), "*Effects of external mass transfer and intraparticle diffusion on adsorption rates in slurry reactors*", American Institute of Chemical Engineers, Symposium Series 73(166): p.91-107.
2. Chang, S., Waite, T. D. and Fane, A. G. (2005), "*A simplified model for trace organics removal by continuous flow PAC adsorption/submerged membrane processes*", Journal of Membrane Science, 253(1-2): p.81-87.
3. Guo, W. S., Shim, W. G., Vigneswaran, S. and Ngo, H. H. (2005), "*Effect of operating parameters in a submerged membrane adsorption hybrid system: Experiments and mathematical modeling*", Journal of Membrane Science, 247(1-2): p.65-74.

4. Thacker, W. E., Crittenden, J. C. and Snoeyink, V. L. (1984), "*Modeling of adsorber performance: Variable influent concentration and comparison of adsorbents*". Journal of the Water Pollution Control Federation, 56(3 I): p.243-250.
5. Wilmanski, K. and Van Breemen, A. N. (1990), "*Competitive adsorption of trichloroethylene and humic substances from groundwater on activated carbon*". Water Research, 24(6): p.773-779.

Table Caption

Table 1 Operating conditions, effective bed adsorption capacity, optimized surface diffusivity (D_s) and average external mass transfer coefficient (k_f) of para-chlorophenol/F400 column systems

Figure Caption

Figure 1 Transverse section of a tapered column

Figure 2 Breakthrough curve of adsorption of para-chlorophenol onto F400 carbon in 0° degree tapered column (300mg dm^{-3} + 200mL/min)

Figure 3 Breakthrough curve of adsorption of para-chlorophenol onto F400 carbon in 5° degree tapered column (300mg dm^{-3} + 200mL/min)

Figure 4 Simulated results by HSDM (dash line) and modified HSDM (solid line) for the adsorption of para-chlorophenol onto F400 carbon in 0° degree tapered column (300mg dm^{-3} + 200mL/min)

Figure 5 Simulated results by HSDM (dash line) and modified HSDM (solid line) for the adsorption of para-chlorophenol onto F400 carbon in 5° degree tapered column (300mg dm^{-3} + 200mL/min)

Table 1

Tapered degree	Initial concentration (mg dm^{-3})	Flowrate (mL/min)	Effective bed adsorption capacity (%)	D_s (cm^2/s)	k_f (cm/s)	SSE
0	150	200	66.4	1.14E-08	8.53E-03	0.248
0	200	200	72.9	1.49E-08	8.53E-03	0.255
0	300	200	75.0	1.73E-08	8.53E-03	0.125
5	150	200	76.5	1.01E-08	1.44E-02	0.171
5	200	200	80.3	1.22E-08	1.44E-02	0.247
5	300	200	81.7	1.89E-08	1.44E-02	0.143

Figure 1

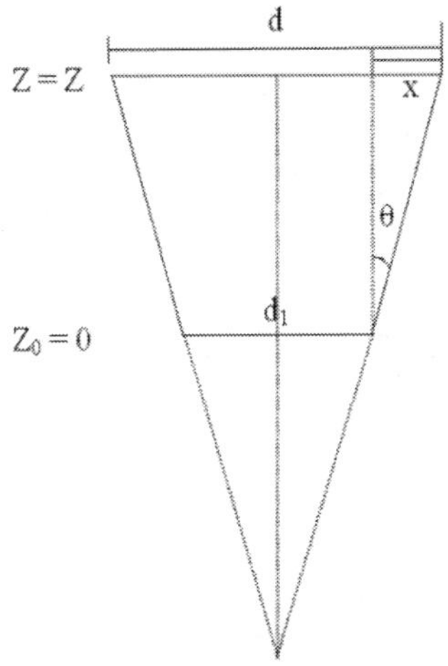

Figure 2

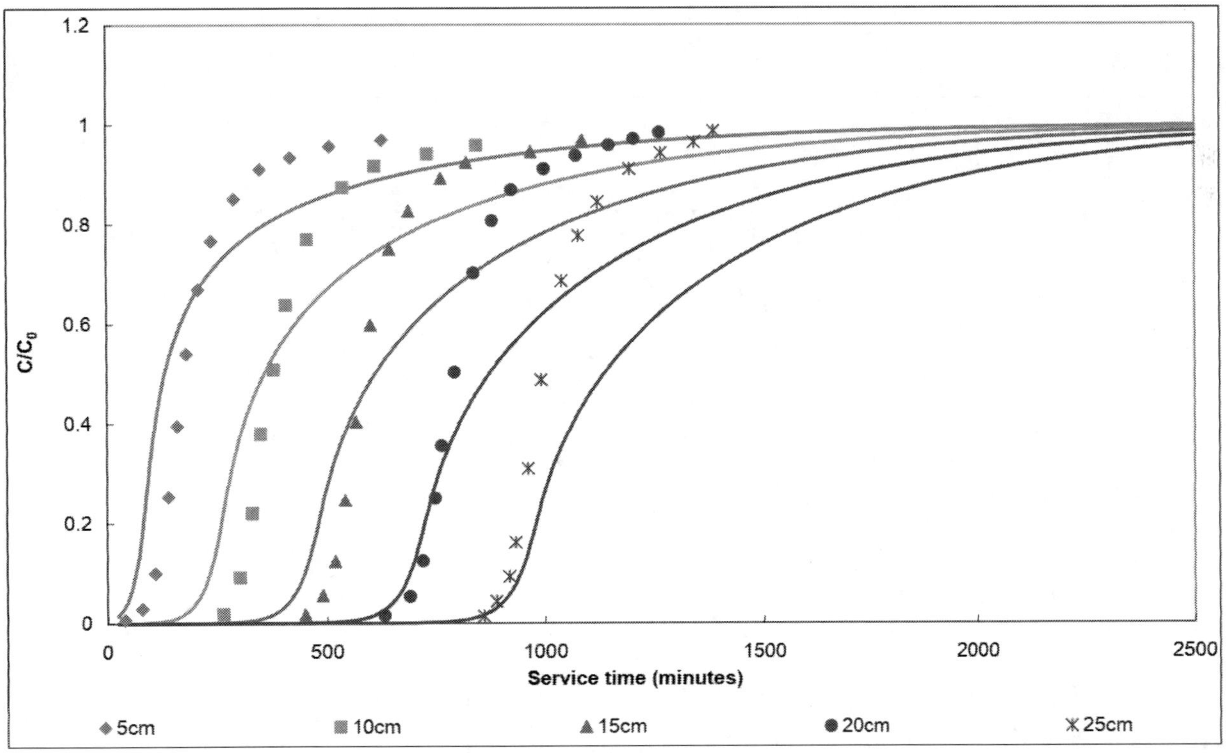

Figure 3

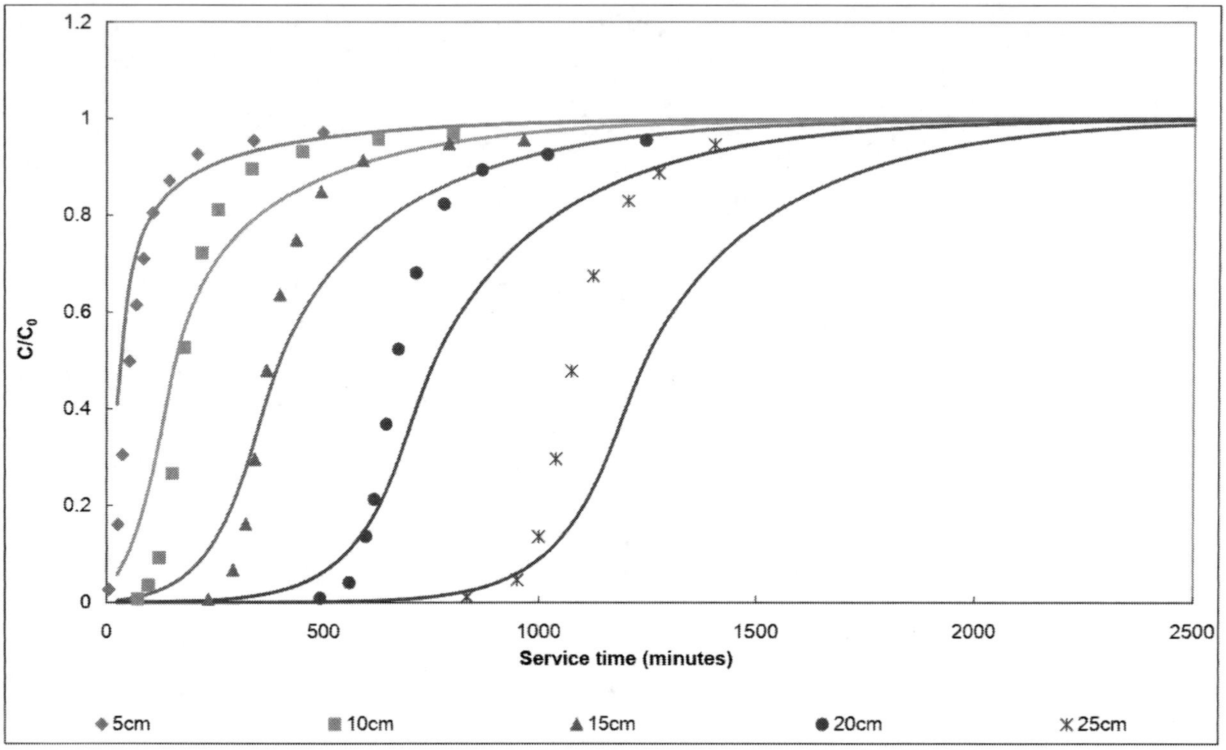

Figure 4

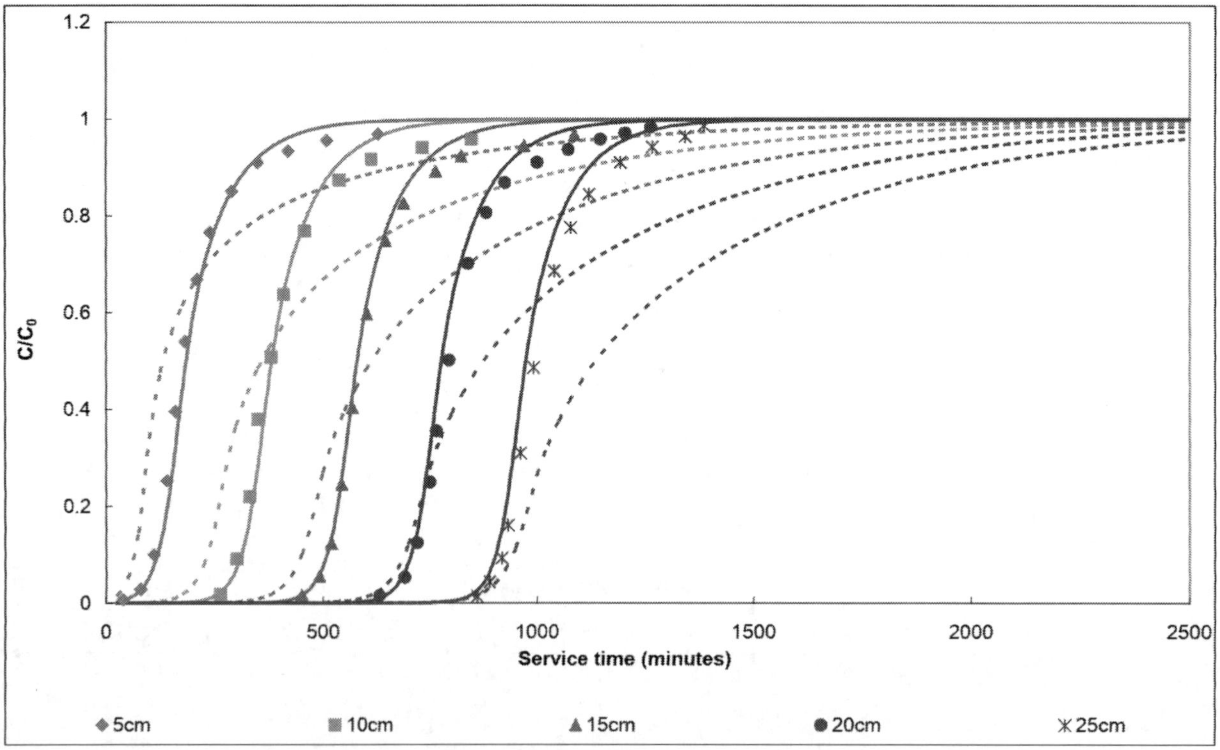

Figure 5

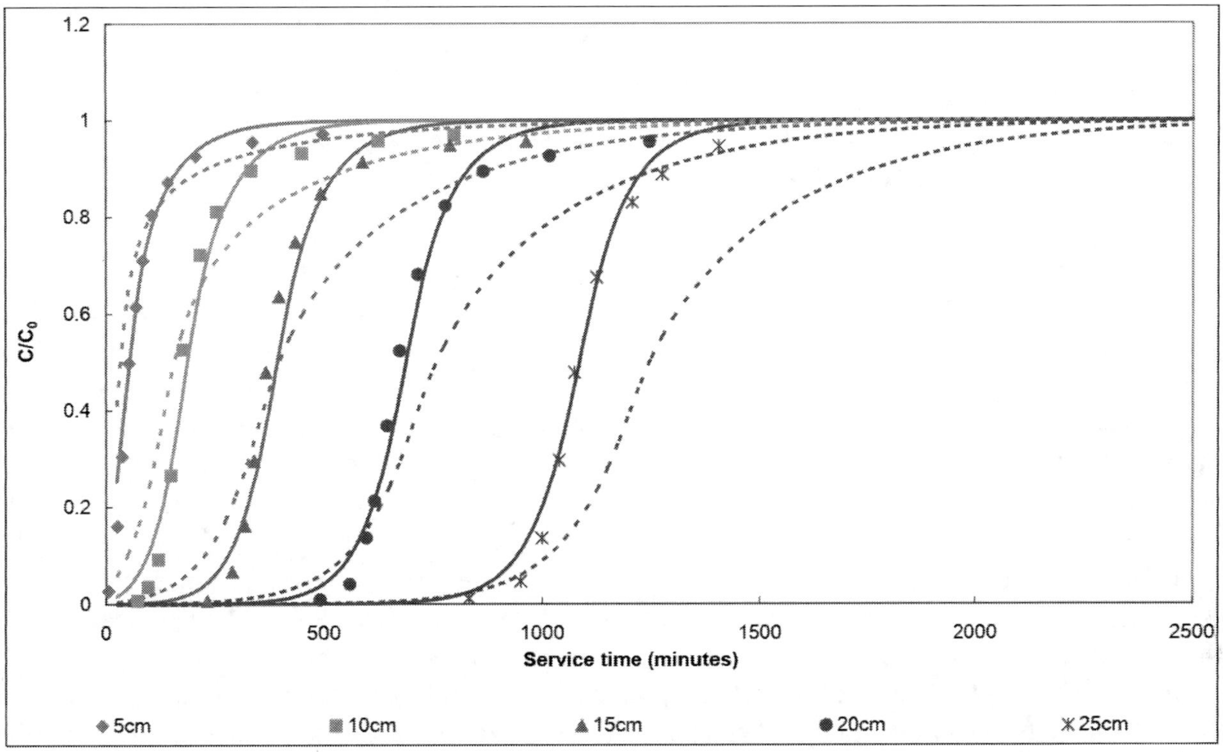

1.2
1
0.8
0.6
0.4
0.2
0
C/C₀
0
500
1000
1500
2000
2500
Service time (minutes)
5cm
10cm
15cm
20cm
25cm

85g Adsorption/desorption Studies Of CO2, Isobutane And 2,2,4-Trimethylpentane In Beta-Zeolite And Usy-Zeolite Using A Tapered Element Oscillating Microbalance (Teom)

Kening Gong[1], Yao A. Houndonougbo[2], Subramanya Nayak[3], Brian B. Laird[2], Bala Subramaniam[4], M. P. Dudukovic[3], and P. A. Ramachandran[5]. (1) Chemical Engineering, The University of Kansas, 1530 W.15th Street, Rm 4132, Learned Hall, Lawrence, KS 66045, (2) Chemistry, University of Kansas, 1251 Wescoe Hall Drive, Lawrence, KS 66045, (3) Chemical Engineering, Washington University in St Louis, St Louis, MO 63130, (4) Center for Environmentally Beneficial Catalysis/Chemical Engineering, University of Kansas, 1501 Wakarusa Dr., Suite A-110, Lawrence, KS 66047, (5) Chemical Engineering, Washington University in St. Louis, Box 1198, One Brookings Drive, St. Louis, MO 63130

In developing stable and durable solid acid catalysts for 1-butene+isobutane alkylation, reliable understanding of the adsorption and desorption rates of reactants, products and solvents in catalysts is essential to screen potential solid acid catalysts. Toward this end, we are currently investigating the adsorption/desorption phenomena of model compounds (olefins and paraffins) and potential solvents (i. e. CO_2 media have been demonstrated to enhance pore-accessibility and catalyst stability) on various solid acid catalysts and supports. Complementary theoretical models are implemented to reliably interpret the experimental data and obtain fundamental parameters. Equilibrium adsorption isotherms and adsorption/desorption rates of CO_2, isobutane and 2,2,4-trimethylpentane in β-zeolite and USY-zeolite have been experimentally investigated by using a tapered element oscillating microbalance (TEOM) which provides excellent sensitivity, allowing mass changes as little as 10 μg to be detected. It was found that CO_2 and isobutane adsorption and desorption rates on β-zeolite and USY-zeolite are fast (at 303-373 K, under CO_2 or isobutane partial pressure 0-1.4 bar). At all temperatures investigated, saturation adsorption is not achieved even up to 1.4 bar for CO_2. The adsorption uptakes of 224-TMP in β-zeolite or USY-zeolite change dramatically (the time to approach equilibria can differ by 2-3 orders of magnitude) depending on the operating conditions. The adsorption/desorption dynamics of 224-TMP in both zeolites are much slower at ambient temperatures preferred for alkylation. The significant buildup of 224-TMP can impose significant transport limitations inside the β-zeolite and USY-zeolite pores. The accumulation of 224-TMP and other heavy compounds inside the pores will result in catalyst deactivation. A GCMC (Grand Canonical Monte Carlo) simulation program has been employed to understand the relationship between molecular-level structure and observable macroscopic properties in this adsorption system. A meso-scale model has been developed to estimate the effective diffusivity and adsorption/desorption parameters.

113a Dynamic Mean Field Theory Of Adsorption And Desorption In Porous Materials

P. A. Monson, Univ. of Massachusetts, Dept. of Chem. Eng., Amherst, MA 01003

Density functional theories have been an important tool in understanding the molecular level behavior of fluids confined in porous materials[1, 2]. These theories allow us to calculate the free energy and density distribution for fluids inside porous materials for equilibrium states of the system. They also allow us to study metastable states associated with hysteresis loops in adsorption/desorption isotherms. It is also of interest to study the dynamics in the approach to equilibrium and metastable equilibrium for fluids in pores. This has recently been done using molecular dynamics[3, 4] and dynamic Monte Carlo simulations[5]. In this paper we present a theoretical approach to studying the dynamics that makes use of a lattice gas treatment of the system. The theory yields the time evolution of the density distribution in the system. The equilibrium or metastable equilibrium states from mean field density functional theory emerge as the long time limit in the dynamics. A similar approach has been used to study phase separation in metal alloys[6] and diffusion in porous materials[7].

To illustrate the utility of this approach we present applications to two systems. A slit pore in contact with bulk fluid and an inkbottle pore geometry in contact with bulk fluid. We study the dynamics of the adsorption and desorption in the systems and compare the system behavior during reversible and irreversible adsorption and desorption processes.

1. Evans, R., Fluids Adsorbed in Narrow Pores - Phase-Equilibria and Structure. Journal of Physics-Condensed Matter, 1990. 2(46): p. 8989-9007. 2. Gelb, L.D., K.E. Gubbins, R. Radhakrishnan, and M. Sliwinska-Bartkowiak, Phase separation in confined systems. Reports on Progress in Physics, 1999. 62 (12): p. 1573-1659. 3. Sarkisov, L. and P.A. Monson, Hysteresis in Monte Carlo and molecular dynamics simulations of adsorption in porous materials. Langmuir, 2000. 16(25): p. 9857-9860. 4. Sarkisov, L. and P.A. Monson, Modeling of adsorption and desorption in pores of simple geometry using molecular dynamics. Langmuir, 2001. 17(24): p. 7600-7604. 5. Woo, H.J. and P.A. Monson, Phase behavior and dynamics of fluids in mesoporous glasses. Physical Review E, 2003. 67(4): p. 041207. 6. Gouyet, J.F., M. Plapp, W. Dieterich, and P. Maass, Description of far-from-equilibrium processes by mean-field lattice gas models. Advances in Physics, 2003. 52(6): p. 523-638. 7. Aranovich, G.L. and M.D. Donohue, Diffusion equation for interacting particles. Journal of Physical Chemistry B, 2005. 109(33): p. 16062-16069.

THE EFFECT OF PARTICLE SIZE OF CHITOSAN ON ADSORPTION OF ARSENATE AND CHARACTERIZATION

Katrina C. M. Kwok, The Hong Kong University of Science and Technology, Hong Kong
Vinci K. C. Lee, The Hong Kong University of Science and Technology, Hong Kong
Gordon McKay, The Hong Kong University of Science and Technology, Hong Kong

Abstract

The degree of deacetylation of chitosan with different particle sizes was determined from the ratio of nitrogen-to-carbon with the elemental analyzer (EA) at around 82 % indicating the total amount of deacetylated groups are similar. The swelling capacity of chitosan was investigated as a function of contact time and their particle sizes. It is found that the polymer of the chitosan particles expanded to certain extents in contact with aqueous medium which leads to the opening of new sorption sites in the polysaccharide structure. The compositions and the surface properties of the chitosan with varying particle sizes were characterized with the FT-IR analysis and the BET analysis. The sorption properties of chitosan are subject to both its chemical and physical properties. The characterization of chitosan is important in the investigation of the adsorption mechanism and the adsorption capacity of As(V) ions and chitosan. The adsorption of As(V) ions and chitosan with varying particle sizes in batch kinetic systems were studied and explained on the basis of the properties of chitosan.

Arsenic contamination of groundwater and surface water is a world-wide problem which is caused by the dissolution of naturally present arsenic-containing minerals and a variety of human activities posing severe heath threats during contact and utilisation [1].

Chitosan, (poly-β(1-4)-2-amino-2-deoxy-D-glucose), a natural, non-toxic, biodegradable polysaccharide copolymer of 2-glucosamine and N-acetyl-2-glucosamine, is derived by deacetylation of chitin, a major component of crustacean shells of prawn, crab, shrimp or lobster, received increasing attention as one of the renewable materials for extensive applications in the pharmaceutical and biomedical industries for enzyme immobilization and purification and in chemical plants for the treatment of metal contaminated water and wastewater. Its main attributes correspond to its polycationic nature and the protonation of amine functional groups in aqueous acid environments enabling ion exchange with metal ions and basic dyestuffs forming salts with the negatively charged acid dyes [2, 3] and adsorption of arsenic [4, 5] or similar anions like perrhenate ion [6] .

The batch kinetic studies were conducted with three typical particle sizes of chitosan ranging from 355 – 500 μm, 500 – 710 μm and 710 – 1000 at the initial concentration of As(V) solution of 3000 μg l^{-1} and initial pH of 3.50. It was found that the rate of adsorption of As(V) ions on chitosan increased with decreasing the particle size of the adsorbent, indicating that the particle size influenced the adsorption kinetics of As(V) ions. Faster adsorption uptake kinetics can be achieved with smaller particle size (355 – 500 μm) in contact time of 6 hours. The experimental data have been modelled using a new kinetic model, the pseudo-first order reversible model.

1. Introduction

One of the reasons for manipulating particle size is to facilitate the adsorption by controlling the surface-to-volume ratio of the particles. Larger surface-to-volume ratios could be obtained with smaller particle size, which in turn could increase the adsorption capacity of the adsorbent and/ or increase the rate of adsorption. The adsorption capacities of As(V) ions and chitosan is greatly dependent on the amount of free amine group ($-NH_2$) of chitosan, the extent of protonation of the amine group in acidic aqueous medium to its protonated form ($-NH_3^+$) and the physical properties of chitosan, namely, its surface area, pore size distribution, pore volume, crystallinity and the elemental composition which influence the accessible number of amine group for the sorption of metal ions.

2. Results and Discussion

2.1 Physical properties of chitosan

2.1.1 Surface area of chitosan

The variations of surface area (m^2 g^{-1}) of the chitosan flake with four discrete particle size ranges were determined by the BET surface area analyzer and shown in Table 1. From the results, the surface areas of chitosan increased with decreasing particle sizes, implying higher adsorption capacity of adsorbate on chitosan with smaller particle size (250 – 355 μm). The surface areas of chitosan decreased from 1.119 m^2 g^{-1} to 5.859x10^{-2} m^2 g^{-1} with increasing particle size from 250 – 355 μm to 500 – 710 μm.

2.1.2 Pore volume of chitosan

The variations of total pore volume of the adsorbent, chitosan with four discrete particle size ranges were determined with the BET analyzer and shown in Table 1. The total pore volume of the chitosan also decreased from 2.688 x10^{-2} cc g^{-1} to 9.138 x10^{-3} cc g^{-1} with increasing particle size. It is believed that the chitosan with smaller particle size has higher surface area and higher total pore volume. Thus, faster uptake kinetics of arsenate on chitosan with smaller particle size was achieved as shown in Figure 1.

2.1.3 Effect of swelling property of chitosan

The extent of the swelling property of chitosan was studied by contacting 0.250g of the chitosan with 50 ml of deionized water adjusted pH_i = 3.50 ± 0.05 for the time intervals of 6, 24 and 48 hours. The particle size of the swollen chitosan, assumed as spherical particle was determined using a particle size analyzer. The sorption capacity of chitosan can be directly influenced by its crystallinity [7] which can be affected by the particle size and the extent of swelling of chitosan in water, which is correlated with the kinetics of diffusion and sorption of adsorbate. These parameters influence both the sorption capacity of chitosan at equilibrium and the uptake kinetics. The sorption capacity is mainly controlled by the amount of accessible sorption sites while the uptake kinetics is governed by the adsorbate diffusion and the swelling kinetics of the particles.

It is noticed that the crystalline region of the chitosan is not accessible to both water molecules and metal ions. Hence, the decrease in crystallinity of the chitosan will lead to increase in accessible sites. In the study of the swelling in water, it is found that the chitosan particle with the smallest particle size range of 250 – 355 μm has the highest ratio of surface area and volume of the chitosan particles of 2.56×10^{-2} μm^{-1}. The variation of the surface area-to-volume ratio was plotted against the particle size ranges of chitosan upon the changing contact times of 6, 24 and 48 hours as shown in Table 2. It is found that the differences in the surface area-to-volume ratio with the four discrete particle size ranges are larger in a shorter contact time of 6 hours and the differences in the ratio decrease when the contact time of chitosan in water increases to 48 hours. This phenomenon might lead to a faster sorption kinetic rate of As(V) and chitosan with smaller particle size in the initial period of batch kinetic studies and a slower desorption rate of As(V) ionsi in the later.

2.1.4 Elemental composition of chitosan

The elemental compositions of the chitosans with different particle size ranges in term of weight percentage (%) of nitrogen (N), carbon (C), sulphur (S) and hydrogen (H) were analyzed in Table 3. The degree of deacetylation of the chitosans was determined with the carbon-to-nitrogen ratio with using Eq. (1) and shown in Table 3. It can be seen that both parameters, the weight percentage of nitrogen (N%) and the degree of deacetylation (DDA%), of the chitosan increases slightly with increasing particle size range from 250 – 355 μm to 710 – 1000 μm, indicating the total number of amine group in the chitosan with large particle size range is greater. However, the amine groups ($-NH_2$) and the acetyl groups ($-COCH_3$) in chitosan are randomly distributed in the particles. The sorption capacity and the uptake kinetics can be influenced by the surface area of the chitosan particles as well as the number of accessible amine groups on the particle surface.

$$\frac{N(\%)}{C(\%)} = \frac{14}{96 - 24DDA(\%)} \quad (1)$$

2.1.5 FT-IR analysis of chitosan

The chemical structures of the chitosan were examined with FT-IR analyzer. Table 4 shows the FT-IR spectra of chitosan with four discrete particle size ranges and the adsorbed chitosan with absorbance against the wavenumber (cm^{-1}) at 1655 cm^{-1} (N-H stretching) and 3450 cm^{-1} (O-H stretching). The degree of deacetylation of the chitosan with different particle size ranges are calculated using Eq. (2) [8]:

$$DDA = [1 - (A_{1655} / A_{3450})/1.33] \times 100\% \quad (2)$$

From the result, it is found that the degree of deacetylation of chitosan decreases with increasing particle size, indicating that the total amount of accessible amine functional groups on the particle surface decreases. The decrease in the accessible amine groups in the larger particles of chitosan lead to a decrease in the adsorption rate in the batch kinetic studies.

2.2 Effect of particle size on batch kinetic studies

The effect of particle size of chitosan with the initial As(V) concentration of 3000μg l^{-1} at initial pH_i = 3.50 in batch kinetic study was investigated as shown in Figure 1. Three particle size ranges of 355 – 500 μm, 500 – 710 μm and 710 – 1000 μm were used in this study. It is found that the batch kinetic systems of arsenate ions and chitosan are mainly composed of two sections, a rapid initial sorption within the first 30 minutes and a slow steady desorption after 30 minutes contact time.

3. Kinetic modelling development

Conventional kinetic models, such as the pseudo-first order and the pseudo-second order models [9, 10] have been used to simulate the sorption of arsenate ions and chitosan However, these two kinetic models cannot correlate well in the slow desorption section of the batch kinetic results in this study. A novel kinetic model, namely, the pseudo-first order reversible model has been developed to describe the sorption and desorption of the batch kinetic systems of As(V) and chitosan with different particle size ranges simultaneously. The new kinetic model was developed based on the introduction of the reversible reaction (desorption) between the arsenate ions and chitosan into the pseudo-first order model.

The rate law for the sorption section can be described by Eq. (3).

$$\frac{dq_t}{dt} = k_{+1}(q_{\max} - q_t) \quad (3)$$

The rate law for the desorption section can be represented by Eq. (4).

$$\frac{dq_t}{dt} = -k_{-1}q_t \quad (4)$$

By combining Eq. (3) and Eq. (4), it becomes:

$$\frac{dq_t}{dt} = k_{+1}(q_{\max} - q_t) - k_{-1}q_t \quad (5)$$

With the initial condition q_t = 0 with t = 0, $t \rightarrow \infty$, $q_t \rightarrow q_e$, the solution of Eq. (5) becomes:

$$q_t = \frac{k_{+1}q_{\max}}{k_{-1} - k_{+1}}\left(e^{-k_{+1}t} - 1\right) - \left(q_e + \frac{k_{+1}q_{\max}}{k_{-1} - k_{+1}}\right)\left(e^{-k_{-1}t} - 1\right) \quad (6)$$

In this model, the values of the equilibrium solid phase concentration, q_e, were determined in the previous sorption studies. The values of the model parameters in different particle size ranges are shown in Table 5. It can be seen that the rate constants of the sorption step, k_{+1}, decrease with increasing particle size of the chitosan from 355 – 500 µm to 710 – 1000 µm. The rate constants of the desorption step, k_{-1} are fairly constant for the desorption of As(V) from the adsorbed chitosan.

4. Conclusion

The surface properties and the chemical compositions of the chitosan with varying particle sizes were determined by the BET analysis, elemental analysis (EA) and FT-IR analysis. It is demonstrated the variations of particle size can significantly influence the number of accessible amine groups in the chitosan, which governs its sorption capacity and uptake kinetics. The characteristic swelling property of chitosan can affect its surface area-to-volume ratio of chitosan with varying particle sizes which can be used to explain the phenomenon of rapid initial sorption and slow steady desorption of arsenate ions and chitosan in batch kinetic studies. A novel pseudo-first order reversible model has been developed to describe the sorption and the desorption of the batch kinetic systems of As(V) and chitosan simultaneously.

Nomenclature

C_0	= Initial liquid-phase concentration, $\mu g\ l^{-1}$
C_e	= Equilibrium liquid-phase concentration, $\mu g\ l^{-1}$
$DDA\%$	= Degree of deacetylation of chitosan, %
k_{+1}	= Rate constant of sorption in Eq. (3)
k_{-1}	= Rate constant of desorption in Eq (4)
pH_i	= Initial pH
q_e	= Equilibrium solid-phase concentration, $\mu g\ g^{-1}$
q_{max}	= Maximum sorption capacity, $\mu g\ g^{-1}$
q_t	= Solid-phase concentration at time t, $\mu g\ g^{-1}$
SSE	= Sum of the squares of the error, dimensionless
t	= Contact time of solid phase and aqueous phase, min

Acknowledgements

The authors (K.C.M., Kwok and V.K.C., Lee) would like to thank Hong Kong RGC for the provision of support throughout this research study.

References

1. Berg, M., Tran, H.C., Nguyen, T.C., Pham, H.V., Schertenleib, R., and Giger, W., (2001), "Arsenic contamination of groundwater and drinking waiter in Vietnam: A human health threat," *Environmental Science & Technology*, 35(13) pp. 2621-2626.
2. McKay, G., Blair, H.S., Gardner, J.R., and McConvey, I.F., (1985), "Two resistance mass transfer model for the adsorption of various dyestuffs onto chitin," *Journal of Applied Polymer Science.*, 30 pp. 4325-4335.
3. Ng, J.C.Y., Cheung, W.H., and McKay, G., (2002), "Equilibrium studies of the sorption of Cu(II) ions onto chitosan," *Journal of Colloid and Interface Science*, 255(1) pp. 64-74.
4. Dambies, L., Vincent, T., and Guibal, E., (2002), "Treatment of arsenic-containing solutions using chitosan derivatives: uptake mechanism and sorption performances," *Water Research*, 36(15) pp. 3699-3710.
5. Kartal, S.N. and Imamura, Y., (2005), "Removal of copper, chromium, and arsenic from CCA-treated wood onto chitin and chitosan," *Bioresource Technology*, 96(3) pp. 389-392.

6. Kim, J.C., Lee, H.Y., Kim, M.H., Lee, H.J., Kang, H.Y., and Kim, S.M., (2006), "Preparation and characterization of chitosan/gelatin microcapsules containing triclosan," *Colloids and Surfaces B: Biointerfaces*, 52(1) pp. 52-56.
7. Guibal, E., (2004), "Interactions of metal ions with chitosan-based sorbents: A review," *Separation and Purification Technology*, 38(1) pp. 43-74.
8. Khan, T.A., Peh, K.K., and Ch'ng, H.S., (2002), "Reporting degree of deacetylation values of chitosan: the influence of analytical methods," *Journal of Pharmacy and Pharmaceutical Sciences*, 5(3) pp. 205-212.
9. Lagergren, S., (1898), "About the theory of so-called adsorption of soluble substances," *Kungliga Svenska Vetenskapsakademiens. Handlingar*, Band 24(No. 4) pp. 1-39.
10. Ho, Y.S. and McKay, G., (1999), "Pseudo-second order model for sorption processes," *Process Biochemistry*, 34(5) pp. 451-465.

Table Caption

Table 1 Surface area and total pore volume of chitosan

Table 2 Swelling properties of chitosan

Table 3 Elemental composition and degree of deacetylation of chitosan

Table 4 FT-IR analysis and degree of deacetylation of chitosan

Table 5 Summary of pseudo-first order reversible model parameters

Figure Caption

Figure 1 Simulation of the pseudo-first order reversible models for the batch kinetic studies of As(V) ions and chitosan with varying particle size of adsorbents

Table 1. Surface area and total pore volume of chitosan

Particle size (μm)	250 – 355	355 - 500	500 - 710	710 - 1000
Surface area (m^2/g)	1.12E+00	7.26E-01	5.86E-02	2.98E-01
Total pore volume (cc/g)	2.69E-02	1.93E-02	9.14E-03	1.94E-02

Table 2. Swelling properties of chitosan

Particle size (μm)	Surface area-to-volume ratio (μm^{-1}) Contact time		
	6-hours	24-hours	48-hours
250 - 355	2.56E-02	1.77E-02	1.77E-02
355 - 500	1.46E-02	1.01E-02	1.11E-02
500 - 710	1.01E-02	6.33E-03	6.33E-03
710 - 1000	5.76E-03	5.25E-03	4.78E-03

Table 3. Elemental composition and degree of deacetylation of chitosan

Particle size (μm)	N%	C%	S%	H%	DDA%
250 - 355	7.49	40.93	0.00	9.68	81.26
355 - 500	7.47	40.82	0.00	9.70	81.15
500 - 710	7.73	42.15	0.00	10.12	81.92
710 - 1000	7.78	42.32	0.12	9.84	82.76

Table 4. FT-IR analysis and degree of deacetylation of chitosan

Particle size (μm)	Absorbance A_{1655}	A_{3450}	DDA%
250 - 355	0.89	1.52	55.83
355 - 500	1.57	1.95	39.51
500 -710	2.17	2.22	26.64
710 - 1000	1.47	1.47	24.76

Table 5. Summary of pseudo-first order reversible model parameters

Particle size (μm)	355 - 500	500 - 710	710 - 1000
k_{+1}	2.05E-01	1.51E-01	1.06E-01
k_{-1}	3.42E-04	2.57E-04	2.52E-04
q_{max}	5.13E+03	4.90E+03	4.03E+03
q_e	9.12E+02	5.23E+02	2.66E+02
Total SSE	7.71E+05	1.72E+06	7.10E+05

Figure 1

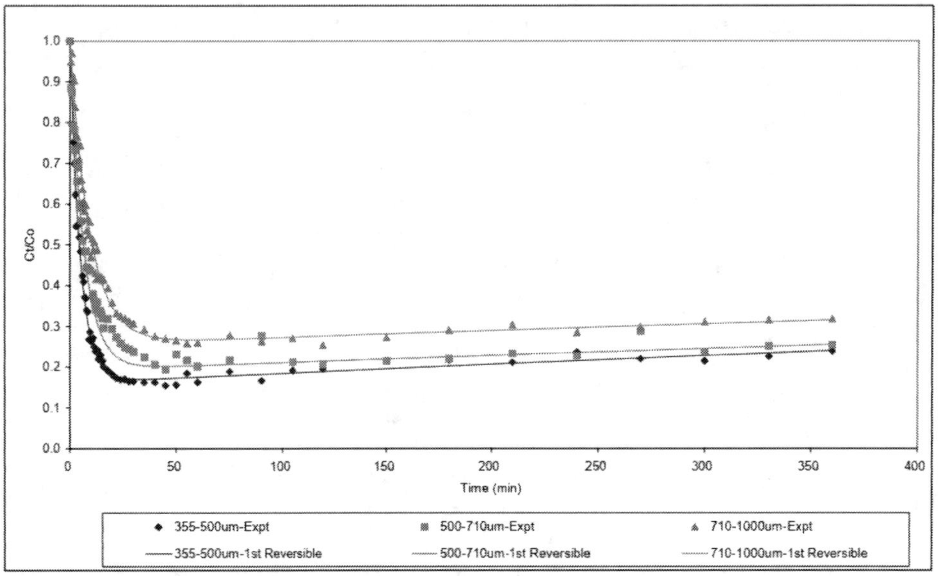

Ct/Co
1.0
0.9
0.8
0.7
0.6
0.5
0.4
0.3
0.2
0.1
0.0
0
50
100
150
200
250
300
350
400
Time (min)
355-500um-Expt
500-710um-Expt
710-1000um-Expt
355-500um-1st Reversible
500-710um-1st Reversible
710-1000um-1st Reversible

113f Molecular Modeling Of Carbon Replica Obtained From Silica Template Using Grand Canonical Monte Carlo Simulations

Surendra K. Jain, Department of Chemical Engineering, North Carolina State University, North Carolina State University, Raleigh, NC 27695, Keith E. Gubbins, Department of Chemical and Biomolecular Engineering, North Carolina State University, Raleigh, NC 27606, and Roland J.-M. Pellenq, CNRS, Campus de Luminy, Centre de Recherche en Matière Condensée et Nanosciences, Marseilles, 13288, France.

We present a Grand canonical monte carlo study to obtain carbon replica from a silica template. Many experimental studies have been reported on obtaining carbon replica using different templating materials like SBA-15, MCM-41, Zeolites etc. The formation of these templated carbon is a two fold process. A carbon rich precursor in vapor or liquid phase is first deposited inside the porosity of the matrix template. The adsorbed precursor is then carbonized by heating at a high temperature in an oxygen free environment to produce a carbon-matrix composite. Finally, the matrix is removed by treating with an acid and the carbon replica is obtained. In this work, we obtain Carbon rods and carbon pipes by doing carbon adsorption in a model MCM-41 pore using grand canonical monte carlo simulations. The carbon-carbon interaction is modeled using the bond order potential developed by Brenner [1]. The matrix-carbon interaction is modeled using the PN-Traz potential [2]. These carbon rods and carbon pipes are then used to obtain simple models for CMK-1, CMK-3 and CMK-5. We further study the adsorption of argon in these CMK models at 77 K using grand canonical monte carlo simulations.

[1] D. W. Brenner, Phys. Rev. B 1990; 42(15):9458. [2] R. J.-M. Pellenq and D. Nicholson, J. Phys. Chem. 1994; 98:13339.

113e The Effect Of Local Defects On Water Adsorption In Silicalite-1 Zeolite

Alain H. Fuchs[1], Fabien Caillez[2], Anne Boutin[3], and Isabelle Demachy[3]. (1) ENSCP, Paris Graduate School of Chemical Engineering, Paris, France, (2) University of Paris-Sud, batiment 349, Orsay, 91405, France, (3) Chemistry, University of Paris-Sud, Batiment 349, Orsay, France

We report a joint experimental and molecular simulation study of water condensation in silicalite-1 zeolite. A sample was synthesized using the fluoride route, and was found to contain essentially no defects. A second sample synthesized using the hydroxide route was found to contain a small amount of silanol groups. The thermodynamics of water condensation was studied in these two samples, as well as in a commercial sample, in order to understand the effect of local defects on water adsorption. The molecular simulation study enabled to qualitatively reproduce the experimentally observed condensation thermodynamics features. A shift and a rounding of the condensation transition was observed with an increasing hydrophilicity of the local defect, but the condensation transition was still observed above the water saturation vapor pressure. Both experiments and simulations agree on the fact that a small water uptake can be observed at very low pressure, but that the bulk fluid does not develop in the gas phase. The picture that emerges from the observed water condensation mechanism is the existence of a heterogeneous internal surface that is overall hydrophobic, despite the existence of hydrophilic "patches". This heterogeneous surface configuration is thermodynamically stable in a wide range of reduced pressures (from P/P0 = 0.2 to a few thousands), until the condensation transition takes place.

113d Extension of the Quenched Solid Density Functional Theory to Adsorption on Carbons

Peter I. Ravikovitch, Present address: ExxonMobil Research and Engineering Co, Annandale, NJ 08801 and **Alexander V. Neimark**, Department of Chemical and Biochemical Engineering, Rutgers University, 98 Brett Road, Piscataway, NJ 08854-8058.

Recently, we have suggested a model of adsorption in pores with amorphous and microporous solids walls, named the quenched solid density functional theory (QSDFT) that had been successfully applied to modeling adsorption in pores of siliceous materials [1]. QSDFT represents a multicomponent non-local density functional theory (NLDFT), in which the solid is treated as a quenched component with a fixed spatially distributed density. In this paper, QSDFT approach, which takes into account roughness and corrugation of pore walls, is extended to carbon adsorbents.

[1] Ravikovitch P.I. and Neimark A.V., Langmuir 22, (2006) 11171

113g Determination Of Pore Accessibility In Disordered Nanoporous Carbons

Thanh X. Nguyen and **Suresh K. Bhatia**. Division of Chemical Engineering, The University of Queensland, Brisbane, Australia

The understanding of pore accessibility of gases in disordered microporous materials is a long standing problem, having significance for accurate prediction of adsorption equilibrium and dynamical properties as well as explanation of hysteresis phenomena in such materials. In recent work, we have proposed a novel algorithm, which enables one to determine pore accessibility in an atomistic structural model of the disordered nanoporous material. This algorithm has been successfully utilized to determine pore accessibility of nitrogen and argon in the atomistic structural model of the sacharose char obtained using the hybrid reverse Monte Carlo technique (HRMC) and further validated using transition state theory (TST).

In this work, we investigate pore accessibility of carbon dioxide and methane in the sacharose char over a wide temperature range. The results calculated using TST show that there is no pore accessibility problem for carbon dioxide at 273 K with its very short crossing time (~10-5 second) while methane indicates slow diffusion even at high temperature (above 300 K) with its crossing time within practical time scale (seconds). Such significantly slower diffusion of methane at high temperature practically leads to underprediction of carbon dioxide adsorption in the same carbon, when using the pore size distribution extracted from supercritical methane adsorption, due to the latter's lower accessibility. From our recent work, the crossing time of methane at 300 K and argon at 87 K are rather similar and within practical time scale (several seconds). However, it can be inferred that the crossing time of methane above 300 K significantly exceeds that of argon at 87 K if the pore mouth becomes slightly smaller due to the former's larger molecular size. Consequently, it can be seen that accessibility of methane, argon, and carbon dioxide increases in that order.

Finally, The findings above are validated by prediction of high pressure methane and carbon dioxide adsorption measurements in several carbons over a wide range of temperatures using structural characteristics of these carbons, obtained from interpretation of experimental measurements of argon adsorption at 87 K by our finite wall thickness (FWT) and infinite wall thickness (IWT) DFT-based models. Lowest pore accessibility of methane in the PCB carbon is clearly evident with significant overestimation of experimental methane adsorption data but correct prediction of carbon dioxide.

Hybrid Ion Exchange/ Nanofiltration (HIX-NF) Process for Energy Efficient Desalination of Brackish and Seawater

Sudipta Sarkar and Arup K. SenGupta
Fritz Engineering Laboratory, Department of Civil and Environmental Engineering, Lehigh University, Bethlehem, PA 18015. Ph: 610-758-3534. Email:arup.sengupta@lehigh.edu

Introduction

Desalination of sea and brackish water is one of the ways of meeting increased water demand around the globe in near future. However, till today production of potable water from saline water sources is not popular due to it's significantly higher cost compared to the conventional water treatment processes from surface or non-saline ground water sources. According to the Desalination and Water purification Roadmap (DWPTR) released by the Department of Interior of the US, the present state-of-the-art desalination technologies are not energy efficient to gain wide-spread uses [**1**]. Thus there is a need for the development of next-generation cost-effective desalination processes for sea and brackish water. Reverse osmosis, the most widely used process of desalination suffers from the three major shortcomings, namely, high energy consumption, durability of reverse osmosis membrane and susceptibility of membrane to fouling [**2-4**]. Forward Osmosis (FO) and membrane distillation processes, which are being pursued to be candidate technologies for the future, like reverse osmosis, are also membrane driven processes [**5, 6**]. The present research endeavors in the scientific world is mostly concerned about development of advanced quality membranes with low fouling and higher flux. In this communication we present a different approach towards achieving an energy-efficient process of desalination. The new hybrid ion exchange assisted nanofiltration (HIX-NF) process of desalination involves partial alteration of feed water chemistry through a reversible ion exchange process prior to feeding to a membrane. Such an alteration of feed water chemistry allows for reduction in osmotic pressure of the saline water as well as makes desalination possible by the use of nanofiltration (NF) membrane in place of energy intensive reverse osmosis (RO) membrane. As a result, desalination at much lower energy consumption is possible using HIX-NF process.

Hybrid Ion Exchange-Nanofiltration (HIX-NF): Underlying Concept

Three major underlying scientific tenets of the proposed HIX-NF process are:

- Ion exchange processes work on exchange of equivalents of ions while the osmotic pressure of an aqueous solution is governed by molar concentrations of the ions or solutes.
- At identical feed electrolyte concentrations with divalent ions, NF membranes offer significantly greater permeate water flux for the same solute rejection compared to RO membranes, all other conditions remaining identical.
- Ion exchange selectivity between divalent sulfate (SO_4^{2-}) and monovalent chloride (Cl^-) ions is tunable i.e., it can be manipulated by tailoring the amine functional groups of anion-exchange resins and accompanying selectivity reversal effect at high salt concentrations.

The process protocol of the hybrid process can be illustrated as follows:

Considering sodium chloride (i.e., Na^+ and Cl^-) to be the primary constituent of brackish or seawater, the osmotic pressure of the brackish/sea water can be greatly reduced by using ion exchanger in pre-saturated forms of choice, namely, sulfate and/or magnesium, as illustrated in **Figure 1** (situations **A** and **B**). Also, the ion exchange process is reversible; therefore, sodium chloride can be reproduced from Na_2SO_4 or $MgSO_4$ by reversing the flow direction as shown by the dashed lines in **Figure 1**. For relative ease of implementation of the process as a

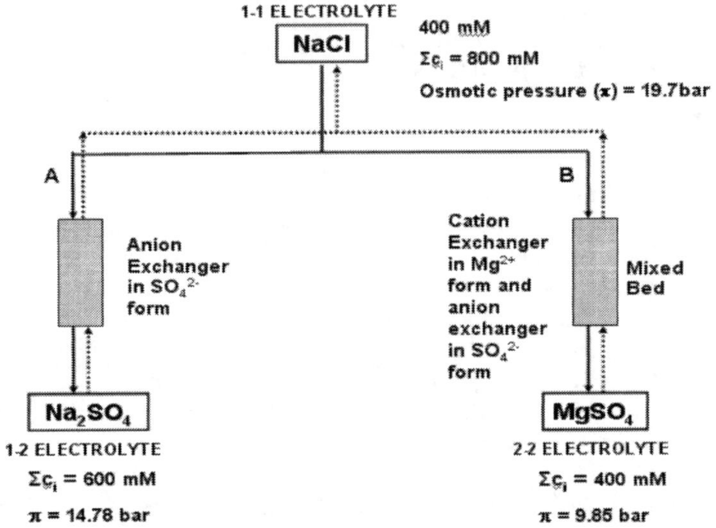

Figure 1. Reduction in osmotic pressure of sodium chloride following passage through ion exchangers pre-saturated with divalent ions.

viable engineered process, we consider the situation B where NaCl (1-1 electrolyte) is converted to Na_2SO_4 (1-2 electrolyte) with an anion exchanger in sulfate form. Two trains of anion exchange columns and one NF membrane are needed and the process schematic is illustrated in **Figure 2**. Individual process steps and underlying scientific rationale are presented hereunder.

Step 1. Incoming brackish/sea water containing NaCl is passed through an anion exchanger in sulfate form.

$$\overline{(R_4N^+)_2SO_4^{2-}} + 2NaCl \Leftrightarrow \overline{2(R_4N^+)Cl^-} + SO_4^{2-} + 2Na^+ \qquad (1)$$

The resulting Na_2SO_4 (1-2 electrolyte) has 30% lower osmotic pressure but the equivalent electrolyte concentration remains the same. Overbar denotes the solid exchanger phase while R_4N^+ represents its quaternary ammonium functional group.

Step 2. Now the sodium sulfate solution is fed under pressure through a sulfate or divalent ion-rejecting nanofiltration (NF) membrane (DOW SR-90 or equivalent). Since the sulfate passage through the NF membrane is extremely low and the osmotic pressure is reduced due to the conversion of 1-1 NaCl to 1-2 Na_2SO_4, very high water flux is obtained at low transmembrane pressure. In this regard, earlier studies have shown that electrolyte passage through RO and NF membranes during pressure-driven processes is essentially a coupled transport process [**7, 8**]. Since sodium ions are coupled with sulfate in the feed, sodium passage through the SR-90 membrane would also be minimal and controlled by sulfate.

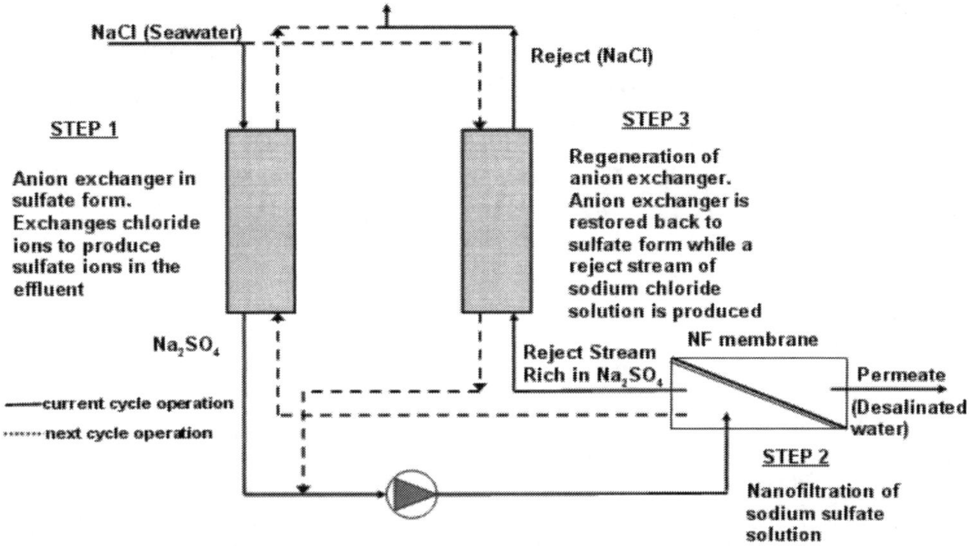

Figure 2. Schematic of the HIX-NF desalination process delineating three major operational steps.

Step 3. The reject stream from the NF membrane, extremely rich in sulfate content, now passes through the previously exhausted anion-exchange column (now in chloride form) from Step 1. The column is transformed back into sulfate form and the resulting effluent contains NaCl (1-1) electrolyte.

$$2\overline{(R_4N^+)Cl^-} + Na_2SO_4 \Leftrightarrow \overline{(R_4N^+)_2SO_4^{2-}} + 2Cl^- + 2Na^+ \qquad (2)$$

No external regenerant is required and the column is now ready for operation in Step 1. From a theoretical perspective, the operation is self-sustaining, i.e., anion exchangers switch back and forth between sulfate and chloride forms while nanofiltration offers high water flux at low transmembrane pressure.

Results

Runs With NF and RO

Figure 3A shows a plot of permeate flux vs. transmembrane pressure for both NF and RO membranes for a feed solution of 280 mM sodium sulfate (Na_2SO_4). Experiments were carried out using GE Osmonics falt-leaf cell apparatus. Permeate flux for NF membrane (DOW SR 90) was nearly two to four times greater than the RO membrane (DOW SWHR RO) at identical transmembrane pressure. It is to be recognized that 280 mM Na_2SO_4 is obtained following the passage of 560 mM NaCl (≈30,000 mg/L TDS) through an anion exchanger in sulfate form. For the same set of membrane runs, **Figure 3B** shows the percentage salt rejection versus transmembrane pressure. Note that well over 98% salt rejections were achieved for both NF and RO membranes.

In a similar vein, **Figures 3C and 3D** show plots of water fluxes and % salt rejection for RO and NF membranes respectively at different transmembrane pressures for 140 mM feed solutions. Again, while salt rejections are comparable and consistently greater than 98% for both NF and RO membranes, permeate or water fluxes for NF membrane are over three times greater than that for RO.

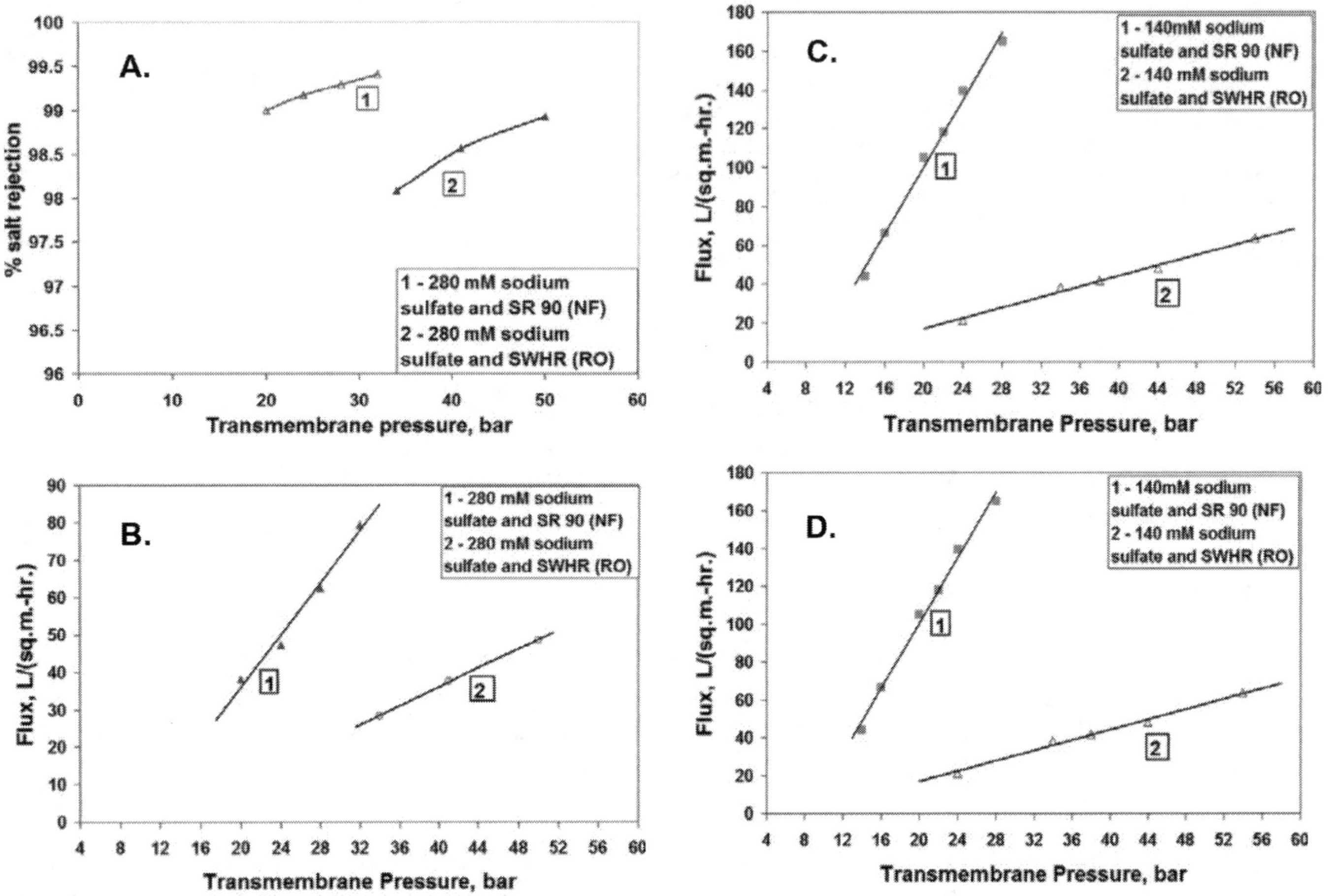

Figure 3. Comparison of salt rejections (**A** and **C**) and permeate fluxes (**B** and **D**) for RO and NF membranes with two different feed sulfate concentrations (280 mM and 140 mM).

Ion Exchange Column Runs

A fixed bed anion exchange column in sulfate form (Purolite A850 strong base anion exchange resin) was fed with 560 meq/L NaCl solution; **Figure 4** shows the effluent history. Note that sulfate concentration at the exit of the column was exactly at 560 meq/L and it remained consistently at that concentration until chloride broke through sharply after three bed volumes. Very sharp sulfate and chloride profiles also suggest that at such high concentrations, resistances from the liquid phase film- and intra-particle diffusion are insignificant. For all practical or design purposes, the equilibrium model holds barring minor non-ideality resulting from axial dispersion. Osmo-molaltites of the solutions were determined using a Vapro 5520 Osmometer (Wescor Inc.). Note that osmotic pressure of the solution dropped from 24 bar in the feed (NaCl) to 15 bar in the effluent (equivalent Na_2SO_4) following anion exchange.

In order to confirm the reversibility of the anion exchange step in the HIX-NF process, we considered a brackish water of 2500 mg/L NaCl (≈ 40 meq/L) with 60% product recovery or yield. The reject stream shall contain approximately 100 meq/L Na_2SO_4 (**Figure 5A**). We fed a strong base anion exchanger (Purolite A-850 originally in sulfate form) with 10 bed volumes of 40 meq/L NaCl and subsequently regenerated counter-currently with four bed volumes of 100 meq/L Na_2SO_4. **Figure 5B** shows y_S and y_{Cl} (equivalent fractions of sulfate and chloride in the exchanger phase) after each complete cycle following countercurrent regeneration. Note that the bed is sufficiently sulfate loaded (y_S>0.95) after four cycles confirming insignificant loss of sulfate from the system. Also, **Figure 5C** shows that sulfate predominates over chloride (i.e., $x_S >> x_{Cl}$) in the water being fed to the NF membrane after anion exchange. Thus, the process is sustainable at the feed concentration of about 2350 mg/l NaCl.

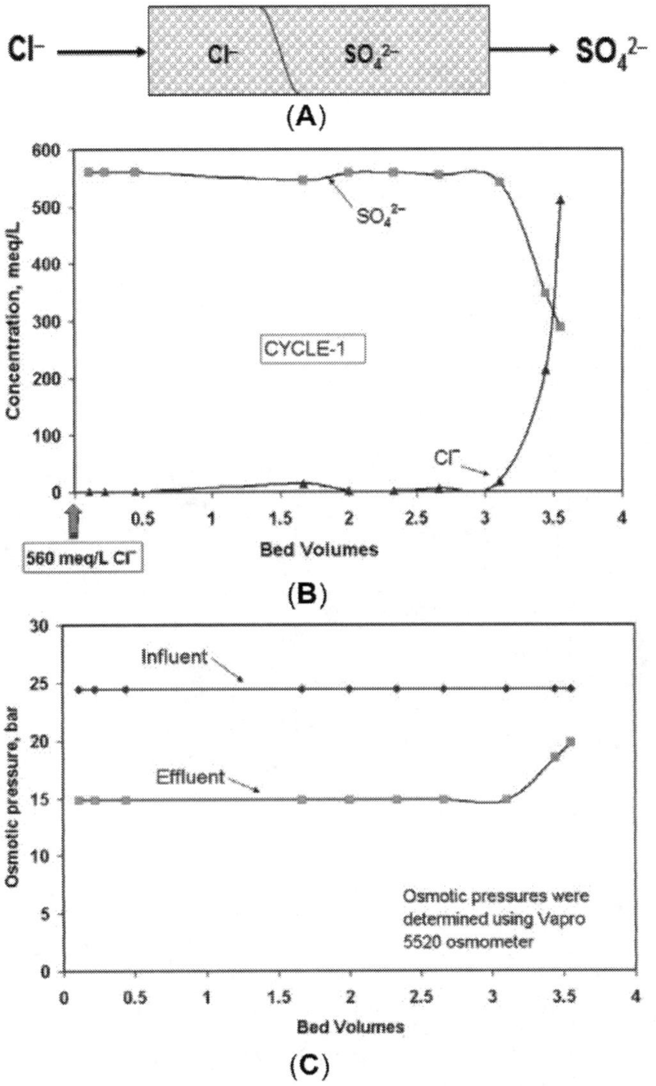

Figure 4. **A.** A schematic showing exchange of chloride with sulfate in an anion exchanger. **B.** Sulfate elution profile with a feed chloride solution of 560 meq/L. **C.** Evidence of osmotic pressure reduction through chloride-sulfate exchange.

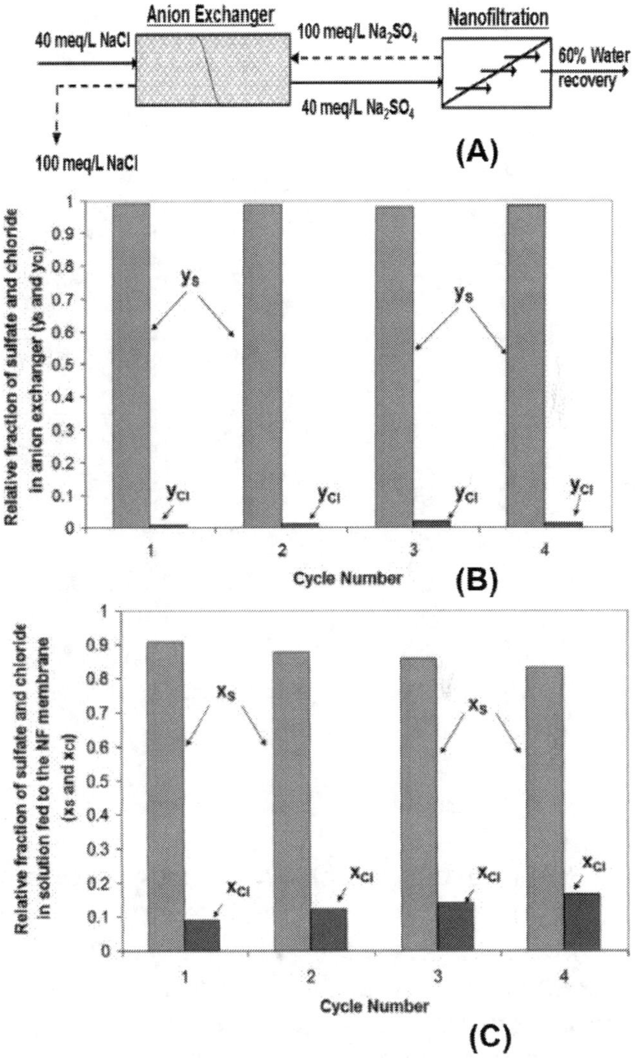

Figure 5. A schematic of HIX-NF for a brackish water with 2350 mg/L TDS (**A**). Relative fractions of sulfate and chloride remaining in the anion exchanger (**B**) and in the solution fed to the NF membrane at each cycle (**C**)

Flux and Energy Comparison

Nanofiltration membranes offer far better flux compared to RO and FO membranes. The reversible anion exchange step in the HIX-NF process helps to attain better flux for desalination by making it possible to use the NF membranes for rejecting only divalent anions. **Table-1** shows a comparison of permeate fluxes reported when saline water containing NaCl were subjected to different desalination processes. It may be noted that for HIX-NF process the flux value is way greater than any other processes.

Table-1. Comparison of permeate fluxes among different membrane processes*

Flux (L/sqm.hr)	Pressure (bar)	Solution strength	Membrane type	Reference
30.96	**–**	**0.05M**	CTA (Forward Osmosis)	9
33.83	**–**	**0.2 M**	CTA (Forward Osmosis)	9
23.04	**–**	**0.5M**	CTA (Forward Osmosis)	9
14.4	**15**	**0.13M**	RO TFC Polyamide, Koch	10
13.52	**55**	**0.55 M**	RO TFC Polyamide, Filmtech	11
38	**–**	**0.01M**	PTFE, Direct Contact Membrane distillation	12
32	**40**	**0.35M**	RO, Hydraunautics, 25M100AI	13
22	**40**	**0.51M**	RO, Hydraunautics, 25M100AI	13
62	**16**	**0.28 M**	HIX-NF	Unpublished Lehigh University work
80	**32**	**0.56M**	HIX-NF	

* For salt rejection values greater than 98%.

We have run different concentrations of saline water in our laboratory in both RO (Dow SWHR membrane) and HIX-NF (Dow SR-90 NF membrane) processes using the flat-leaf membrane test set-up and calculated the energy consumption for the membrane process from the data on flux and transmembrane pressure. **Table-2** provides a comparison of energy requirement of the two processes. It may be noted that desalination using HIX-NF process consumed significantly much less energy compared to the RO process.

Table 2. Comparative consumption of energy between RO and HIX-NF processes

Concentration of Saline water (as mg/L NaCl)	Energy required to produce one cum of desalinated water (kWh)	
	RO	**HIXNF**
30,000	1.68	0.62
15,000	1.16	0.48

Conclusions

Attributes

The key attributes of the HIX-NF process are: (**1**) RO membranes which are more expensive, more susceptible to fouling and less energy efficient can be replaced altogether with NF membranes; (**2**) Sulfate-rejecting NF membranes with high water flux at low transmembrane pressures greatly reduce the energy requirement, and (**3**) Reduced osmotic pressure of the feed and enhanced permeate flux with NF will yield significantly higher product water recovery. Thus, the disposable reject brine will attain greater dissolved solids concentration and consequently, the reject volume will diminish.

Note that additional capital costs of ion exchange units for the HIX-NF process will be greatly offset by the reduced costs of NF membranes and decreased replacement needs. Also, the sizes of the anion exchange columns can be small corresponding to less than five minutes of detention or empty bed contact time.

Challenges

It is already demonstrated that HIX-NF process is sustainable for solutions with NaCl concentrations up to 2350 mg/L using a strong base anion exchange resin (A-850). However, all ion exchange processes experience a phenomenon known as selectivity reversal which occurs at high ionic strength of a solution. According to **Figure 6,** selectivity reversal for A-850 takes place around a concentration of 100 meq/L (i.e. near the diagonal) where separation factor or $\alpha_{S/Cl}$ value is unity and it decreases further at higher concentration (e.g. 400 meq/L). Thus, for reject concentrations beyond 100 meq/L, sustainability of HIX-NF process is affected due to loss of significant amount of sulfate after each cycle. The challenge ahead is to make HIX-NF process sustainable for a wide range of salinity, up to the level of sea water. Resins for which selectivity reversals take place at a much higher concentration would be an ideal in this situation. Several research investigations on monovalent-divalent anion selectivity have clearly demonstrated that the relative degree of substitution of hydrogen in amine functionality of the anion exchange resin by an alkyl group is strongly correlated to divalent-monovalent

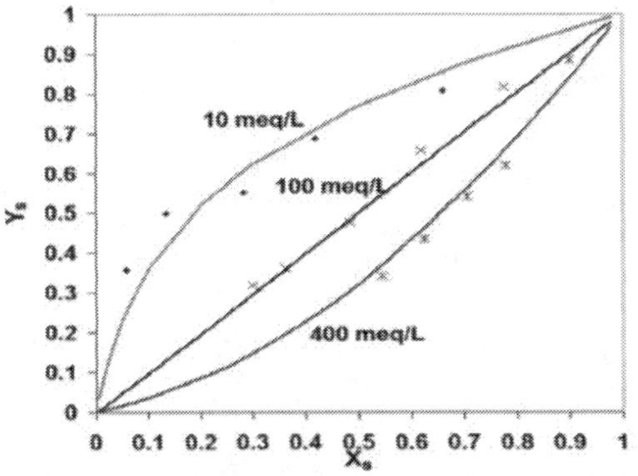

Figure 6. Sulfate-chloride isotherms for anion exchange resin A-850 at three different total electrolyte concentrations showing selectivity reversal above 100 meq/L.

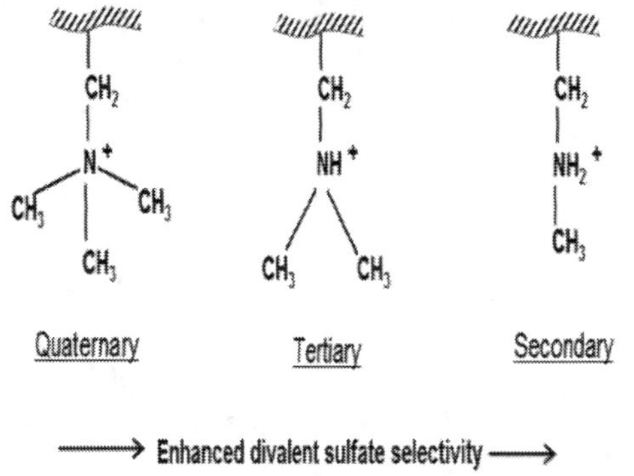

Figure 7. An illustration depicting an enhancement of sulfate selectivity from quaternary amine to tertiary and secondary amine functional groups.

selectivity [**14-18**]. **Figure 7** illustrates how sulfate-chloride selectivity is enhanced from quaternary amine to secondary amine functional groups. Our calculations on experimentally available data [**17**] show that resins with polyamine functionality (Duolite A-7, Rohm and Haas Co.) will exhibit $\alpha_{S/Cl}$ value greater than unity at concentrations as high as 1000 meq/L or 60,000 mg/L TDS. We are presently working on development of a mixed bed process for different anion exchange resins that will sustain the reversibility of sulfate-chloride exchange without significant loss of sulfate from the system.

References

1. USBR and SNL (United States Bureau of Reclamation and Sandia National Laboratories). (2003) "Desalination and Water Purification Technology Roadmap: A Report of the Executive Committee. Desalination & Water Purification Research & Development Report #95", Denver, CO: U.S. Department of the Interior, Bureau of Reclamation, Water Treatment and Engineering Group.
2. Semiat, R. (2000) "Desalination: Present and Future", *Water Int.*, 25(1), 54-65.
3. Ettouney, H.M., El-Dessouky, H.T., Faibish, R.S., et. al. (2002) "Evaluating the Economics of Desalination", *Chemical Engineering Progress*, 32 – 39.
4. Service, R.F. (2006) "Desalination Freshens Up," *Science*, Vol. 313, August 25th.
5. Cath, T.Y., Childress, A.E., Elimelech, M., "Forward Osmosis: Principles, Applications, and Recent Developments", *J. Membr. Sci.*, 281 (1-2), (2006), 70-87.
6. Cath, T.Y. and Childress, A.E. (2005) "Membrane Contactor Processes for Wastewater Reclamation in Space. II. Combined Direct Osmosis, Osmotic Distillation, and Membrane Distillation for Treatment of Metabolic Wastewater", *J. Membr. Sci.*, 257, 111–119.
7. Mukherjee, P. and SenGupta A.K. (2006) "Some Observations about electrolyte permeation mechanism through reverse osmosis and nanofiltration membranes", *J. Membr. Sci.*, 278, 301-307.
8. Mukherjee, P. and SenGupta A.K. (2003) "Ion exchange selectivity as a surrogate indicator of relative permeability of ions in reverse osmosis processes", *Env.l Sci.Technol.,* 37, 1432 - 1440.
9. McCutcheon, J.R., McGinnis, R.L., and Elimelech, M. (2006) "Desalination by a Novel Ammonia-carbon dioxide Forward Osmosis Process : influence of draw and feed solution concentrations on process performances ," *J. Membr. Sci.*, 278, 114-123.
10. Tay, K.G., Song, L., Ong, S.L., and Ng, W. J. (2005), "Nonlinear behavior of permeate flux in full scale Reverse Osmosis Process," *J. Env. Engineering*, 131 (11), 1481-1487.
11. Dow Chemicals, Catalogue of Filmtec seawater RO elements for membrane systems, Catalogue no. 609-00377-703.
12. Cath, T.Y., Adams, V.D., and Childress, A.E. (2004) "Experimental Study of Desalination using Direct Contact Membrane Distillation: a new Approach to Flux Enhancement," *J. Membr. Sci.*, 228.
13. Goosen, F.A.M., Sablani, S.S., Al-Maskari, S.S., Al-Belushi, R.H. and Wilf, M.(2004), "Effect of feed temperature and flow rates on permeate flux in spiral wound reverse osmosis systems" at Latin American and Caribbean Conference for Engineering and Technology (LACCEI), 2-4 June, Miami, Florida.
14. Clifford, Dennis, Weber, Jr., Walter J. (1983). "The determinants of divalent/ monovalent selectivity in anion exchangers," *Reactive Polymers, Ion Exchangers, Sorbents* 1(2): 77-89.
15. SenGupta, A.K., Roy, T. and Jessen, D., "Modified anion-exchange resins for improved chromate selectivity and increased efficiency of regeneration," *Reactive Polymers*, (1988), 9, 293-299.
16. Guter, G. (1995), "Nitrate removal from contaminated groundwater by anion exchange, "in *Ion Exchange Technology: Advances in Pollution Control*, SenGupta, A.K. (Ed.), Technomic Publishing Co., Inc., Lancaster, PA.
17. Liberti, L., Petruzzelli, D., Helfferich, F.G. and Passino, R. (1987) "Chloride/Sulfate ion exchange kinetics at high solution concentration," *Reactive Polymers*, 5, 37-47.
18. Pohl, C.A., Stillian, J.R., Jackson, P.E. (1997) "Factors controlling ion-exchange selectivity in suppressed ion chromatography", *J. Chromatography A*, 789, 29-41.

148b Synthesis And Characterization Of Polycaprolactone Nanoporous Membranes For Liquid-Phase Controlled Release

Chi Yen, Hongyan He, L. James Lee, and W.S. Winston Ho. Chemical and Biomolecular Engineering, The Ohio State University, 140 WEST 19TH AVE, Columbus, OH 43210

In recent years, implantable drug delivery devices have been promising therapeutic systems for the treatment of chronic diseases. Chronic patients suffer from health problems as a result of permanent illness and/or organ failure. Therefore, medical methodology which can be effective for a long time has attracted a lot of attention. Implantable drug delivery devices are able to function in the human body for a long period. However, an improper amount of drug released from the device would cause uncomfortable side effect. To optimize the pharmacotherapy, drug release has to be well controlled in accordance with the remedy during a treatment course. An ideal drug delivery device is capable of releasing the desirable amount of drug to the targeted site for a required period of time. In order to achieve this goal, biodegradable polycaprolactone nanoporous membranes may be expected to control the drug release rate via tuning the porous structure.

In this study, biodegradable and biocompatible polycaprolactone (PCL) nanoporous membranes have been successfully prepared via the combination of thermally and nonsolvent induced phase separation techniques. The polymer solution with a specific composition was cast on a suitable plate. Then, the cast film of the solution was immersed into a lower temperature coagulation bath, i.e., both thermally and nonsolvent induced phase separations took place simultaneously. In the membrane preparation, the thermally induced phase separation could enhance porosity or avoid the formation of a dense top-layer. On the other hand, the nonsolvent induced phase separation could play a role on the formation and the dimension/size of nanoscale pores. Our results showed that the pore size reduced and the porosity increased as the coagulation bath temperature decreased. Also, the release rate of the model drug compound, lysozyme, into an aqueous solution was higher as a lower coagulation bath temperature was used. As a result, the drug release rate can be well controlled via adjusting the parameters for the membrane preparation, including coagulation bath temperature and casting solution composition. In addition, the nanoporous membrane provides a barrier for preventing the passage of cytotoxic cells, macrophages, antibodies and other immune substances.

148c Evaluation Of Solvent Resistant Nanofiltration As An Alternative To Distillation In The Isolation Of An Active Pharmaceutical Intermediate

Jennifer M. Pollard, Kent E. Goklen, and David J. Roush. BioPurification Development, Merck & Company, Inc., 126 E. Lincoln Ave, P.O. Box 2000, Rahway, NJ 07065

The isolation of fermentation derived secondary metabolites is often required for the synthesis of semi-synthetic anti-infective drugs. Typically, secondary metabolites have a molecular weight range of 300-2000 kD and potentially can be thermally labile. Single stage distillation is widely used during multi-step organic solvent-based processes in order to concentrate product streams, switch solvents for the next processing step or recover solvent for subsequent reuse. However, this form of distillation is not very efficient when processing with high boiling solvents such as isobutyl alcohol or solvent switching to a lower boiling solvent. It also does not provide any purification as typical fermentation impurities are nonvolatile.

Solvent resistant nanofiltration (SRNF) provides a novel alternative to distillation for the concentration and solvent switch of an active pharmaceutical intermediate. SRNF provides the potential advantages of increased efficiency, purification and selective solvent removal. This paper will present a comparison of a distillation based process versus a nanofiltration process, in terms of yield, purity, and productivity (kg/hr). The effect of molecular weight cut-off (MWCO), pressure, membrane loading, and diavolumes on the yield and purity of the product for two steps will be presented. SRNF was used to replace distillation for the initial concentration of an isobutyl alcohol whole broth extract. A 12 wt% increase in purity was observed vs. distillation based process - a purity increase which was carried through the subsequent isolation. A second distillation step of a mixed solvent composition (78% IPAc, 10% IBA, 7% MeOH & 5% water) was also replaced with SRNF. The concentration and diafiltration of the isopropyl acetate extract provided a clearance of IBA to the target specification of < 1% which increased the purity of the downstream precipitation by 10%. Overall both NF steps increased the productivity by 10 fold.

148d A Countercurrent Cascade Of Crossflow Hollow Fiber Membrane Modules For Direct Contact Membrane Distillation: Experimental Studies And Modeling

Hanyong Lee[1], Liming Song[2], Jack Gilron[1], and Kamalesh K. Sirkar[1]. (1) Chemical Engineering, New Jersey Institute of Technology, Warren Street, Newark, NJ 07102-1982, (2) Chemical Engineering Department, New Jersey Institute of Technology, Membrane Separations, 161 Warren St, Newark, NJ 07102

Direct contact membrane distillation (DCMD) is a thermally-driven process for desalination of highly saline water using low-grade energy without the corrosion problem. We have recently developed a novel crossflow module of highly porous hydrophobic polypropylene hollow fibers having a porous coating of silicone-fluoropolymer to mitigate temperature polarization and pore wetting in DCMD. Crossflow, large fiber wall thickness, large fiber bore and the porous coating on these hollow fiber membranes have yielded a very high water vapor flux, high thermal efficiency, low temperature polarization and no distillate contamination [1, 2]. The competitiveness of such a promising technology relies, in part, on the maximization of energy efficiency. Recovering energy from the hot distillate product to heat up the cooled brine feed to be recycled using high packing density polymeric hollow fiber heat exchangers can minimize the net thermal energy input required for distillate production. Integration with crossflow DCMD modules in a countercurrent cascade may allow the DCMD process an opportunity to maximize simultaneously both water production and heat recovery [3]. A model for a cascade of DCMD modules is essential to search for optimal number of modules in a cascade and operating conditions. This study focuses on the development and experimental verification of a model for such a process. A mathematical simulation program for our crossflow DCMD modules in a countercurrent cascade mode was developed by integrating coupled heat- and mass-transfer models of membrane modules. The developed simulation program allows one to predict temperatures of effluents, water production rate, thermal efficiency and gained output ratio (GOR) for a given number of modules (stages) in the cascade, inlet temperatures, and flow rates. The simulation results were verified with the experimental results from cascades consisting of 2 to 8 stages. The membrane mass transfer coefficient (km) is not known. A series of experiments were also done to determine km. Results presented will illustrate the development of the simulation program, experimental verification of simulation results and the application of the simulation for heat recovery as well as water production. How the GOR varies with membrane area and other quantities is of great interest.

[1] B. Li and K. K. Sirkar, I & EC Res., 43 (17), 5300-5309 (2004). [2] L. Song, B. Li, K.K. Sirkar and J. Gilron, I & EC Res., 46, 2307-2323 (2007). [3] J.Gilron, L. Song and K.K. Sirkar, I & EC Res., 46, 2324-2334 (2007).

148e Dehydration of Hydriodic Acid and Sulfuric Acid by Pervaporation

Frederick F. Stewart and Christopher J. Orme. Chemical Sciences, Idaho National Laboratory, P.O. Box 1625, Idaho Falls, ID 83415-2208

The ability to remove water from acidic solutions has numerous potential industrial applications. A particular application of interest to INL has been the development of acid concentration processes to enable the Sulfur-Iodine (S-I) thermochemical cycle for hydrogen production. A thermochemical cycle is simply a chemical process for splitting water in which only oxygen and hydrogen are the only effluents. The S-I cycle is a variant based on concurrent iodine and sulfur oxidation and reduction cycles. Pervaporation is a membrane separation technique that uses solubility selectivity to remove one component and reject another. In pervaporation, a feed is circulated at low pressure across the upstream side of the membrane, while a vacuum is applied downstream. Selected permeants sorb into the membrane, transport through it, and are vaporized from the backside. Thus, a concentration gradient is established, which provides the driving force for transport. Common applications for pervaporation are in the removal of water from organic streams, such as ethanol. For the application of acid concentration, other factors must be considered; first and foremost is the chemical stability of the membrane. Thus, durable materials must be selected. Recent efforts at the INL have focused on applying pervaporation of water through Nafion-117 and Nafion-112 membranes for the removal of water from HI/water and HI/Iodine/water, sulfuric acid/water feedstreams. In this work, separations have been performed at temperatures as high as 134 degrees Celsius. Transmembrane fluxes of water are commercially competitive (~5000 g/m2h) and separation factors have been measured as high as 8000, depending on the membrane and the water content. In these experiments, the common trade off in membrane performance is observed in that as flux is increased, separation factor decreases. Nafion-112, a thinner membrane, exhibited much higher fluxes than the Nafion-117 with no significant loss in separation factor indicating that the permeability of iodine and HI through Nafion materials can be low depending on experimental conditions. Data for the sulfuric acid concentration suggests performance similar to the HI experiments. All membranes studied for the HI, HI/iodine and sulfuric acid feeds exhibited no degradation in membrane performance during use. Particular attention in the paper will be paid to the performance dependence on the pressure gradient across the membrane.

155a The Development Of A Macro-Scale Process To Release Periplasmic Proteins Using Osmotic Shock

Anant Patkar, **Subrata Sen**, Mike Chappell, and Yuk Chiu. Bioprocess R&D, The Dow Chemical Company, 5501 Oberlin Dr., San Diego, CA 92121

The conventional laboratory approach of small scale batch osmotic shock to release proteins from microbial cells consists of mixing and equilibration of cell paste with high osmolarity buffer, followed by dilution in a low molarity buffer. This method is time-consuming, inefficient, difficult to control and scale up. This presentation will describe development of a novel approach which contacts in a continuous flow, high osmolarity fermentation broth with an appropriate low osmolarity shock buffer through a static mixer coupled to a continuous centrifuge for simultaneous solid-liquid separation. This method has a high throughput, the mixing time is easier to control, and scale-up is straightforward and quantitative. We successfully demonstrated the applicability of this operation using human growth hormone expressed in P. fluroescens as a model system. Under best conditions, high step yields (~90% and above) and purity comparable to the batch protocol were achieved. This development is a successful example of an integrated approach to biological product recovery and solid-liquid separation.

155b Optimizing Filtration Of Phytochlorin Compounds

Phillip Blake, Nicholas Kamerath, and Keith Roper. Chemical Engineering, University of Utah, Salt Lake City, UT 84102

Phytochlorin and its derivatives are used in photodynamic therapy to create singlet oxygen species that lyse cancer cells. Preparation of phytochlorin from plant extracts uses dead-end filtration after salt and solvent precipitation to recover viscous phytochlorin solids from mother liquor. We quantitate effects of particle size, pH, mixing, filter media, and batch-to-batch consistency on filtration rate and membrane capacity. Data show a characteristic particle settling time aids in filtering the sodium salt form of the phytochlorin compounds. Use of high energy mixing to prepare the phytochlorin precipitation increases filtration rate. Improvements to phytochlorin processing increased production rates of a photosensitive drug intermediate.

Flocculation of Yeast Suspension by a Cationic Polymer: Characterization of Flocculent-Cell Interaction

Subrata Mondal[1], Ranil Wickramasinghe[1], Jong Leng Liow[2], and Y. K. Leong[3]. (1) Department of Chemical and Biological Engineering, Colorado State University, Fort Collins, CO 80523-1370, (2) Department of Mechanical Engineering, James Cook University, Townsville, Australia, (3) Chemical Engineering, James Cook University, Townsville, Australia

Abstract: Addition of a cationic polymer can lead to the flocculation of yeast cells thus forming macroscopic flocs [1]. Flocculation occurs by two main mechanisms (a) formation of macromolecular bridges between the particles, and (b) surface potential and charge reduction due to the adsorption of highly charged polyelectrolytes on oppositely charged particles. The efficiency of flocculation is determined by the structure of polymer layers formed on the surface, i.e., the spacial distribution of adsorbed segments near the interface, the adsorption layer thickness, number and length of loops and tails protruding from the surface into the solution per unit area [2]. In studying the flocculation behavior of microbial cells, many interaction forces need to be considered, such as, gravitational force, van der Waals attractive force, electrostatic repulsive force and specific short-range force [3]. In this study, we have investigated the flocculation of yeast cell suspensions by a cationic polymeric flocculent. The flocculation behavior of yeast cell suspensions at different times during the flocculation process is explained in terms of floc size, physicochemical surface characteristics of yeast and cell surface electric potentials (zeta potentials), dynamic mobility and Kappa value measurements. The flocculent particle size was visualized by a video camera, SILICON VIDEO® 2112 CCD. The initial rate of flocculation and growth of floc particles were dependant on the flocculent dosage. Particle size increased with increasing flocculent dosage up to certain point and then decreased. The magnitude of negative value of the zeta potential decreased and finally reached a positive plateau with increasing flocculent dose. However, during the flocculation process zeta potential was negative and magnitude decreased with the increasing flocculent dose. Our results indicate that homoflocculation occurs at the beginning of flocculation process. However, aggregation occurs at higher flocculent doses as the flocculation proceeds.

Materials and Methods

Flocculation experiments were conducted by using baker's yeast cells. DWT 767-PH, a cationic polyelectrolyte with molecular weight of 15-20 $\times 10^6$ and charge density (percentage of monomer units that are charged) of 40 % and pH 4-6 was obtained from Cytec Industries (Stamford, CT, USA). The baker's yeast cells were washed and suspended in distilled water containing glucose. The zeta potential of yeast flocculations was measured by using a colloidal dynamics ZetaProbe analyzer, USA. The calculation of the zeta potential was assisted with software wizards. The particle size during the flocculation process was observed under a microscope with the given magnification from the mm scale on the ruler. The video camera was a SILICON VIDEO® 2112 CCD with the pictures frame grabbed into the computer. The frame rate given in the document file was about 8 per minute.

Brief Results and Discussion

The influence of time on floc size of the initial flocculation stages is shown in Figure 1. The flocculation rate was increases from 0.19 ppm to 3.98 ppm flocculant dosage and then decreases. This phenomenon could be explained in this way that the flocculation takes place due to the charge neutralization as well as hydrogen bonding. Mainly, the hydrogen bonding takes prominent place as the cataionic polymers were not fully ionized at lower pH (Figure 1). When the concentration exceeds a certain value, the hydrogen bonding is no longer favorable due to marked repulsion and these results in the stabilization of the suspension [4].

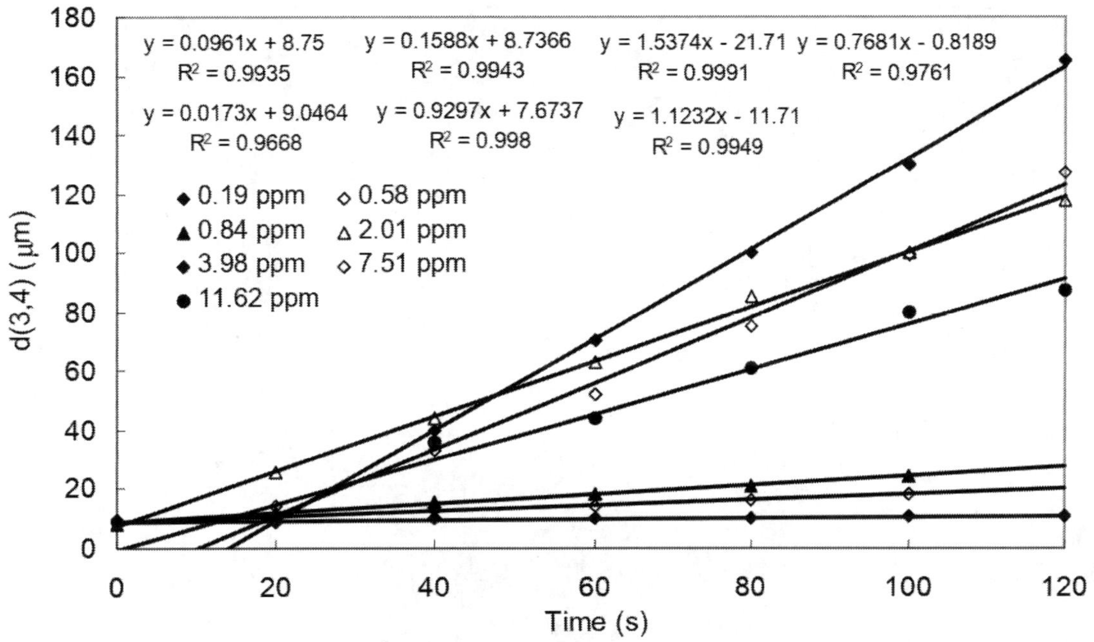

Figure 1: Calculation of the initial flocculation rate from the initial flocculation stages. Measurements are conducted out every 20 s. Experimental conditions: Flocculant DWT767-PH, dosage 7.51 ppm, added after 40 s; yeast concentration 0.117 g/L, stirring speed 300 rpm.

To visualize the interactions between polymer and yeast cell and to verify results obtained by flocculation process, images were taken with the microscope. Images were taken at different time interval during flocculation process. The micrographs at different time interval are shown in Figure 2. As seen in Figure 2, the particle size was increases with increasing the flocculation time, hence supported our previous data during the flocculation process. We observed the homoflocculation initially between yeast and cationic polymers. As the flocculation proceeds, the increasing in flocculant ionization results in the instability of the individual particles, causing them to aggregate. This suggests that at the beginning of flocculation, the flocculation behavior is derived from a more random, interactions of the particles.

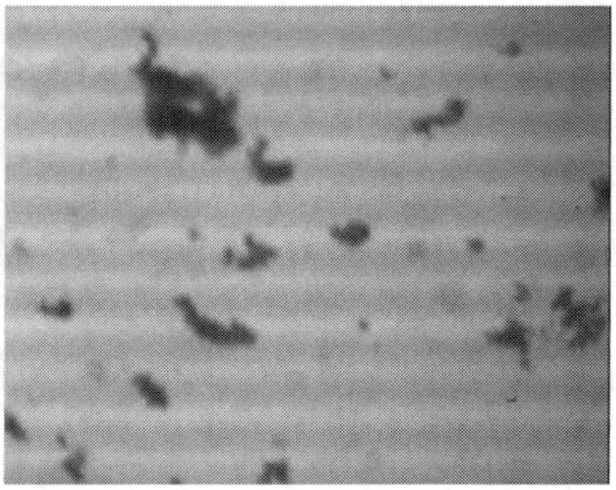

80 sec 1456 sec

3024 sec

Figure 2: Micrographs of yeast flocculation at different time intervals

References

1. M. Ginovart, D. López, A. Giró and M. Silbert, Flocculation in brewing yeasts: A computer simulation study, Biosystems, 83 (1), 2006, 51-55.
2. S. Barany and A. Szepesszentgyörgyi, Flocculation of cellular suspensions by polyelectrolytes, Advances in Colloid and Interface Science, 111 (1-2), 2004, 117-129.
3. R. C. W. Berkeley, J. M. Lynch, J. Melling, P. R. Rutter, and B. Vicent, Microbial Adhesion to Surfaces, Ellis Horwood, Chichester, 1980.
4. Gregory, J. (1973) Rates of flocculation of latex particles by cationic polymers. J. Colloid Interf. Sci. 42, 448-456

164a Methodology For Screening Adsorbents For CO2 Capture

Jun Zhang and **Paul A. Webley**. Chemical Engineering, Monash University, Wellington Road, Clayton, VIC 3800, Australia

As is increasingly recognised world wide, global warming is mainly caused by emission of greenhouse gases (mostly CO_2) from the burning of fossil fuels. Correspondingly, adsorbents for CO_2 capture have attracted significant research efforts recently. To understand the impact of improvement in adsorbent characteristics on the CO_2 capture process and economics, an assessment tool is needed. Various such tools (experimental and theoretical) have been used previously. Theoretical tools include use of isotherms (isothermal working capacity, adiabatic working capacity and selectivity) and experimental tools include thermal gravimetric analysis (TGA). However, the actual operation of a pressure swing adsorption process is not readily captured reliably with these methods. In this study, different adsorbent screening approaches were applied to CO_2 capture, using hypothetical adsorbents, NaX, NaY and Activated Carbon, and the results from different methods were analysed and compared. It was found that isothermal approaches are not suitable for the screening of adsorbents for CO_2 capture, because of the great amount of heat of adsorption for CO_2. A new index, "Capture Figure of Merit" (CFM) is a better assessment tool for developing adsorbents and it can also be readily converted to CO_2 capture costs. Robust numerical simulation and experimental process data have confirmed the validity of CFM. In addition, it is shown that heat of adsorption plays a very important role in cyclic adsorption process and such thermal effect should be given sufficient consideration in adsorbent design and assessment.

164f Design of a Hyperbranched Pei/ Mesoporous Silica Hybrid High-Capacity Sorbent for CO2 Capture from Flue Gas

Jason C. Hicks[1], Daniel J. Fauth[2], McMahan L. Gray[2], and **Christopher W. Jones**[1]. (1) Chemical & Biomolecular Engineering, Georgia Institute of Technology, 311 Ferst Dr., Atlanta, GA 30332, (2) U.S. Department of Energy, National Energy Technology Laboratory, 626 Cochrans Mill Road, P.O. Box 10940, Pittsburgh, PA 15236-0940

Carbon dioxide capture and sequestration will likely become an important technology in the next decades, as concern about climate change drives governments to consider regulating green-house gas emissions. Undoubtedly, one of the largest sources of human-generated CO2 is from coal-fired power plants. Thus, there has been a recent emphasis on assessing current CO2 separation technologies for application to dilute CO2 streams with large water contents, as well as developing new technologies for this separation. One of the most promising is adsorptive separation using amine-modified organic or inorganic polymer sorbents. Unfortunately, a significant drawback with sorbents of this type has been low CO2 sorbent capacities, requiring very large masses of (usually expensive) adsorbent.

In this talk, we will describe a new, rationally designed hybrid sorbent comprised of PEI covalently tethered to silica. This material has the highest adsorption capacity reported in the open literature under realistic flue gas conditions. The material is low cost, owing to the simple method of synthesizing the adsorbent. In addition, the adsorbent can be used with effectively no change in capacity over multiple cycles (10 or more times), demonstrating the potential to have the durability needed for practical application.

164c Materials For Reversible High Temperature Chemisorption Of CO2

Michael G. Beaver, Ki Bong Lee, Hugo S. Caram, and Shivaji Sircar. Chemical Engineering, Lehigh University, 111 Research Drive, Mountaintop Campus, Bethlehem, PA 18015

Results of experimental tests into the equilibrium and column dynamic data for the chemisorption of CO_2 on two materials has identified the materials as potential candidates for the capture of CO_2. The first of the materials is a K_2CO_3 - promoted hydrotalcite that displays good sorption capacity in 400-500 °C range. The second is a Na_2O -promoted alumina that has shown good sorption capacity in 250-400 ° C temperature range. The two materials both exhibited Langmuirian behavior in the low pressure region, but deviated substantially in higher pressure regions. A new analytical model that simultaneously accounts for Langmuirian chemisorption and an additional surface complexation reaction between gaseous and sorbed CO_2 has been proposed to describe the measured equilibrium data for both materials. Experimental breakthrough tests showed fast kinetics and narrow mass transfer zones for CO_2 adsorption. The isosteric heats of chemisorption and heats of additional complexation reaction on both materials were estimated to be low, indicating that desorption of CO_2 from both materials could be achieved with relative ease. Tests have confirmed that both materials show stable sorption capacity after several sorption-desorption cycles. These characteristics make them attractive candidates for use in cyclic processes for the capture of CO_2. The Na_2O promoted alumina shows promise as a candidate for capture of CO_2 from flue gas of a coal fired power plant, while the K_2CO_3 promoted hydrotalcite will be a better candidate for the Sorption Enhanced Reaction process to simultaneously produce fuel cell grade H_2 and high purity CO_2 at feed gas pressure.

164d Molecular Modeling Of The Adsorption Of CO2 In Metal-Organic Frameworks

A. Ozgur Yazaydin[1], Randy Snurr[1], and John J. Low[2]. (1) Chemical and Biological Engineering, Northwestern University, 2145 Sheridan Road E136, Evanston, IL 60208, (2) Advanced Characterization, UOP LLC, 50 East Algonquin Road, Des Plaines, IL 60017

Metal-organic frameworks (MOFs) are nanoporous materials with potential applications in adsorption separations, catalysis, and gas storage. They are synthesized in a "building-block" approach by self-assembly of metal or metal-oxide vertices interconnected by rigid organic linker molecules. Several MOFs have been shown to exhibit interesting CO2 adsorption behavior and have high storage capacity. On the other hand, the CO2 adsorption performance of these materials above room temperatures and at low pressures, which are typical for flue gases, has not been studied well.

We have used atomistic grand canonical Monte Carlo simulations to investigate the selective adsorption of CO2 over other gases such as N2, CH4 and H2 in several classes of MOFs to explore the possibilities for the removal of CO2 from flue gas. Comparisons of predicted single-component isotherms and heats of adsorption will be made with experimental data from the literature. Predictions for mixtures will be used to make suggestions for possible separations applications.

164e Atomistic Simulations Of CO_2 And N_2 Adsorption And Diffusion In Zeolites

Daniela Kohen, Jameson Dahlin, David Selassie, and Disan Davis. Chemistry, Carleton College, 1 North College St, Northfield, MN 55057

To efficiently separate CO_2 from multi-species gas streams, sorbents must have high CO_2 selectivity, and CO_2 mobility within the materials must not be a limiting factor. Zeolites are one attractive option as sorbents due to their variety of structures, availability, and thermal and chemical stability. To investigate the influence of pore geometry on adsorption and selectivity, we have used Grand Canonical Monte Carlo to study the adsorption of CO_2, N_2, and CO_2/N_2 mixtures onto three zeolites with identical chemical composition (SiO_2,) but different structures; silicalite, ITQ-3, and ITQ-7. These zeolites preferentially adsorb CO_2 over N_2 in both single component and mixture adsorption. The CO_2/N_2 selectivities observed in the three siliceous zeolites studied vary strongly as the adsorbent's crystal structure changes, with the selectivity in ITQ-3 being the largest. We have thus concluded that pore geometry strongly affects adsorption in the different zeolites. We present here the complementary studies of non-isotropic CO_2 diffusion within these materials. Molecular Dynamics simulations have been used to compute CO_2, N_2 and CO_2/N_2 mixtures self-diffusion, corrected diffusion, and transport diffusion within ITQ-3, ITQ-7, and silicalite. Significant diffusion occurs, and carbon dioxide molecules explore large regions in all three zeolites. Rates of diffusion vary between zeolites, decreasing with increased pore loading in ITQ-7 and silicalite but showing a non-monotonic dependence in ITQ-3. In a complementary study we have also identified the preferred CO_2 and N_2 sites within silicalite, ITQ-3, and ITQ-7.

164b Adsorption Of Carbon Dioxide And Nitrogen On Carbonaceous Sorbents

Maciej Radosz, Xudong Hu, and Youqing Shen. Chemical and Petroleum Engineering, University of Wyoming, 1000 E. University Ave, Laramie, WY 82071

Experimental sorption of CO2 and N2 up to 10 bar and up to 130°C on carbon-rich (carbonaceous) sorbents greatly depends on the properties of the sorbent material. In general, the CO2/N2 sorption selectivity is not very sensitive to temperature, but it decreases with increasing pressure. This suggests that an isobaric thermal recovery is a more efficient approach than an isothermal pressure-swing recovery. The CO2 sorption and desorption rates are found to be high. The time needed for the carbonaceous sorbents to become saturated with CO2 varies from about 15 to 45 minutes, and it is not very sensitive to temperature and pressure. The temperature cycles do not affect the CO2 sorption capacity, its rate, and the CO2/N2 sorption selectivity. These sorption characterization data are used to design a low-pressure carbon filter for removing CO2 from flue gas and other gas mixtures.

156a CO2 Separation from Fuel Gas by Surfactant-Assisted Gas-Hydrate Process

Guochang Zhang[1], **Rudy E. Rogers**[1], Ni Liu[2], Shangmin Xiong[1], and William T. French[1]. (1) Swalm School of Chemical Engineering, Mississippi State University, P.O.Box 9595, Mississippi State University, MS 39762, (2) Institute of Refrigeration Technology, College of Power Engineering,, University of Shanghai for Science and Technology, 516 Jungong Road, Shanghai, 200093, China

Separation of methane and CO2 by means of gas hydrates promoted by surfactants is reported. In a surfactant-water-CO2-hydrate system, CO2 enriched the solid hydrate phase during hydrate formation, where thermodynamic system equilibrium ultimately determined CO2 separation extent. Synthetic surfactants and biosurfactants significantly promote gas hydrate formation from nonpolar hydrocarbon gases, but for polar gases, such as CO2, no surfactant to promote CO2 hydrates has previously been reported. In this study, CO2-philic surfactants were developed to selectively promote CO2 hydrate formation and assist separation of a CO2-fuel gas. In the presence of surfactants developed in the study, CO2 was further enriched in the solid hydrate phase compared to hydrate formation from distilled water without surfactant. In the experiments, commercially-available surfactants and laboratory-modified surfactants were screened and tested to selectively promote CO2 hydrates in both quiescent and stirred systems. Separation experiments were conducted on mixtures of 50% CO2 and 50% methane initial gas content. The effects of surfactant concentration, solution pH, pressure, temperature, volume ratio of water to gas, as well as THF additive on the separation are reported. While using laboratory-modified surfactants, gas hydrate formation rates sharply increased when pH values rose to 4.0. The CO2 concentration in the gas phase after hydrate formation with surfactants decreased to as low as 24.53%. From visual observations, surfactants not only increased the formation rate of CO2 hydrate, but also helped accumulate CO2 hydrate on special surfaces and influenced the morphology of the solid hydrates.

156b Carbon Dioxide Capture Via Oxy-Fuel Combustion And Integrated Pollutant Removal

Stephen J. Gerdemann, Thomas Ochs, Danylo Oryshchyn, and Cathy Summers. National Energy Technology Laboratory, U.S. Department of Energy, 1450 Queen Avenue, SW, Albany, OR 973212152

Oxy-fuel combustion uses off-the-shelf technology and denitrified air for combustion in a coal-fired power plant to produce a relatively pure CO2 flue gas. Denitrification of the combustion air reduces the flue-gas volume by as much as 80%, resulting in a post combustion gas containing mostly CO2 and water. This flue gas can be cooled, condensing water, and compressed to produce a CO2 stream suitable for sequestration. Heat can be recovered from the cooling step and integrated into the thermal cycle of the power plant to minimize the energy cost of CO2 capture. This paper describes a model of the material and energy flows through a CO2-capturing oxy-fired power plant incorporating NETL's integrated pollutant-removal (IPR) process.

156c Evaluation Of Ionic Liquid Based Green Separation Agents For Carbon Combustion Flue Gas Stream Purification

Samuel A. Morton III and Debra Shlossman. Chemical Engineering, Lafayette College, 266 Acopian Engineering Center, Easton, PA 18042

If carbon sequestration approaches that employ biological methods are to be made viable for the capture of carbon from fossil fuel combustion waste gas streams, a separation technique that is both economically and environmentally green is needed. One concern is the potential negative consequences of certain gases, such as sulfur dioxide, on aqueous-based sequestration approaches. This study experimentally determined the ability of room-temperature ionic liquids and ionic liquid-based mixtures to absorb combustion-related acid gases, carbon dioxide and sulfur dioxide, from gas streams. Ionic liquids are novel organic salts that are liquids at or near room temperature. These solvents have a negligible vapor pressure, and have been called green "designer" solvents due to the vast number of anion-cation combinations that are possible. This work has directly measured the uptake of target gases by the ionic liquid utilizing a quartz-crystal microbalance (QCM), a technique that has demonstrated sensitivity on the nanogram scale. The acid gases studied in this experiment were tested on a selection of ionic liquids of the alkyl-imidazolium family as well as mixtures containing ionic liquids. 1-hexyl-3-methylimidazolium bis(trifluoromethylsulfonyl)imide, [hmim][Tf2N], was used as a benchmark due to the presence of accepted literature data for certain of the above acid gases. Additionally, special attention will be paid to the variation of results with respect to ionic liquids structural and functional selection. New developments will also be discussed.

156d Modeling Carbon Dioxide Solubility In Poly(Ionic Liquid)S With Cosmotherm

Karl Johnson, University of Pittsburgh, 1249 Benedum Hall, Department of Chemical Engineering, Pittsburgh, PA 15261, **Yang Wang**, Chemical Engineering, University of Pittsburgh, 1249 Benedum Hall, Department of Chemical Engineering, Pittsburgh, PA 15261, and Steve R. Lustig, Central Research & Development, The DuPont Company, Experimental Station, Route 141, Wilmington, DE 19880-0356.

Poly(ionic liquid)s (PILs) are solid polymers synthesized by cross-linking ionic liquids. Recent experimental work has shown that PILs have the very high affinity for CO_2, much higher than the corresponding ionic liquid monomers. We use COSMOtherm to model the solubility of CO_2 in various PILs and their corresponding monomers. We compare different methods of generating the sigma profiles required for the COSMOtherm calculations and compare our results to experiments.

156e CO2 Sequestration Via Brine/bauxite Residue Mixture

Yee Soong[1], R. M. Dilmore[1], S. W. Hedges[1], C. Griffith[1], D. E. Allen[2], and C. Zhu[3]. (1) National Energy Technology Laboratory, U.S. Department of Energy, Pittsburgh, PA 15236-0940, (2) Salem State College, Salem, MA 01970, (3) Indiana University, Bloomington, IN 47405

In recent years, a great deal of concern has been expressed with regard to global climate change and its link to growing atmospheric concentrations of carbon dioxide. In order to decrease the impact of anthropogenic CO2 on global climate, several strategies are under development that will potentially remove CO2 from the atmosphere or decrease CO2 emission. One such strategy involves the capture of CO2 from large point sources (such as fossil fuel-fired power plants) and the long-term storage of CO2 underground. Carbon dioxide may be sequestered in various geological formations, including depleted oil and gas reservoirs, unmineable coal seams, basalt formations, and deep saline aquifers. However, there are additional options for CO2 sequestration that should be considered including the use of industrial waste products. One of the proposed strategies under investigation by the National Energy Technology Laboratory (NETL) is sequestration of CO2 from the atmosphere through promotion of solubility and mineral trapping in Ca, Mg, and Fe-bearing caustic waste streams.

Currently 20-30 billion barrels of saline wastewater are produced annually with the production of oil and gas in the USA. About 35% of this acidic wastewater (pH 3 to 5) is treated and discharged as surface water. Many of the gas/oil-field brines have high concentrations of dissolved Ca, Mg, and Fe ions. Carbonate minerals could be formed in the presence of CO2 under certain conditions. This suggests the possibility of using Ca, Mg, and Fe ions from saline wastewater for CO2 sequestration via the formation of carbonate minerals.

Over 70 million tons of bauxite residues are generated annually when aluminum is extracted from the principal ore, bauxite. The residue is mostly comprised of iron and titanium oxides, silica, calcium carbonate, and unrecoverable alumina and soda. The pH of the liquid is as high as 13.5, and the solids and solid surfaces also contain high alkalinity. The caustic nature of the residue causes concerns associated with long-term environmental liability. Worldwide, there are about 200 million tons of bauxite residues that are mostly stored in tailings ponds. For decades, the aluminum industry has been investigating options for treating, disposing, and using bauxite residue. However, environmentally and economically sound methods and processes are still elusive. Addition of CO2 to this caustic material will not only result in carbonate mineral formation but will also serve to decrease the pH of the residue, making disposal more environmentally sound. Moreover, bauxite residue (pH 13.5) can serve as an effective source of caustic material to treat oil and gas field wastewater brines (pH 3 to 5) for mineral carbonation. The use of bauxite residue/brine to capture and store CO2 will serve to not only help mitigate the impact of anthropogenic CO2 on global warming but will also help neutralize caustic industrial waste for safe storage and reuse.

NETL have explored a novel concept that achieves full neutralization of bauxite residue through reaction with acidic oil and gas field wastewater brines with subsequent carbonation. Experiments are conducted to determine the ability of bauxite residue brine reactant mixtures to absorb and sequester CO_2 both from a concentrated stream and as a component of a mixed flue gas. The use of bauxite residue/brine mixtures to capture and store CO_2 can serve to not only help mitigate the impact of anthropogenic CO_2 on global warming but also serve to achieve complete neutralization of the caustic industrial waste for safe storage and potential reuse.

156f Membrane And Solid-Sorbent Alternatives To Amine Absorption For Separating Carbon Dioxide From Flue Gas

Maciej Radosz, Xudong Hu, and Youqing Shen. Chemical and Petroleum Engineering, University of Wyoming, 1000 E. University Ave, Laramie, WY 82071

Novel membrane and sorbent materials are synthesized, characterized and evaluated for separating carbon dioxide from power-plant flue gas relative to an amine-absorption benchmark process. The cost of recovered CO2 using the amine process is about $40-50/ton. Examples of alternative separation materials are ionic liquids, ionic polymers, PEG-grafted ionic polymers, brominated polyphenylene oxide (BPPO), silica-impregnated BPPO nanocomposites, zeolites, and carbon-rich solid sorbents. These materials are characterized with respect to their capacity and selectivity relative to nitrogen. These characterization data are used to estimate the cost of recovered CO2 for a variety of separation approaches. By and large, the approaches that involve compression, high sorption heat, and expensive materials lead to high CO2 recovery costs.

Synthesis of Multilayer Layered Double Hydroxide (LDH)/Bentonite Membranes for CO_2 Separation

Liangxiong Li [a] *, *Ning Liu*[a], *Brian McPherson*[a,b], *Robert Lee*[a]

[a] Petroleum Recovery Research Center, New Mexico Institute of Mining and Technology, Socorro, NM 87801, US

[b] Department of Civil and Environmental Engineering, University of Utah, Salt Lake City, US

Abstract

Layered double hydroxides (LDHs), with a typical formula of $Mg_6Al_2(OH)_{16}CO_3{\cdot}4H_2O$, have a reversible Mg-(OH)-Mg and Mg-(OH)-Al layered structure with interlayer spacing occupied by carbonate ions (CO_3^{2-}) and hydrated water. The octahedral crystal structure evolves to amorphous mixed solid oxide upon heat treatment at high temperatures. Specifically, the interlayer CO_3^{2-} will decompose and release CO_2 by heating to above 400°C. At presence of CO_2 and water vapor, the microporous structure of layered double hydroxides will be recovered, assigning the LDHs unique properties for membrane fabrication, particularly for application in high temperature CO_2 separation.

---

* To whom correspondences should be addressed. Phone: (505) 835-6721; Fax: (505) 835-6031; Email: Li@prrc.nmt.edu.

In this work, supported layered double hydroxide (LDH) membranes were synthesized by depositing oppositely charged LDH nanoparticles and bentonite colloids onto porous α-alumina support. The monodispersed layered double hydroxide nanoparticles were synthesized by direct hydrothermal coprecipitation. Uniform thin film (~1.0 μm) of layered double hydroxide on porous α-alumina support was further formed by dip the porous substrate into a homogeneous suspension containing spherical LDH nanoparticles. To minimize the interparticle pores, a second coating with bentonite colloids was performed. The synthesized LDH/bentonite membranes have a thickness of ~2μm. Different techniques including X-ray diffraction (XRD), scanning electron microscope (SEM), high-resolution transmission electron microscope (HRTEM), and nitrogen adsorption isotherm (BET) have been applied to study the membrane properties such as crystallinity, membrane morphology, and micropore size distribution. Binary gas permeation of equimolar N_2 and CO_2 were performed at elevated temperatures ranging from 50°C to 400°C. The membranes showed CO_2 permselective with a separation factor of 1.39 and CO_2 permeance of 7.68×10^{-8} $mol.m^{-2}.s^{-1}.Pa^{-1}$. The permeation behavior was explained by preferable chemisorption of CO_2 on layered double hydroxide and micro-, mesoporous structure of the membranes.

Key words: Layered double hydroxide, LDH nanoparticle, CO_2 separation, membrane, thin film.

1. Introduction

Layered double hydroxides (LDHs), also known as hydrotalcite, are microporous lamellar compounds consisting of positively-charged brucite-type layers and interlayer hydrated anions for charge compensation, such as NO_3^-, Cl^- or CO_3^{2-}. The positive charge of brucite-like layers arises from the substitution of part of its divalent ions (i.e. Mg^{2+}) with trivalent ones such as Al^{3+}. The interlayer spacing ranges from 0.3 nm to 3.0 nm when different anions are intercalated into the interlayer galleries.[1]

A typical LDH with a formula of $Mg_6Al_2(OH)_{16}CO_3 \cdot 4H_2O$ has a reversible Mg-(OH)-Mg and Mg-(OH)-Al layered structure with interlayer spacing occupied by carbonate ions (CO_3^{2-}) and hydrated water. The octahedral crystal structure will evolve to amorphous mixed solid oxide upon heating to high temperatures. Specifically, the interlayer CO_3^{2-} will decompose and release CO_2 upon heating to above 400 °C.[2] Such an unique property makes LDHs superior for micro- or mesoporous membrane fabrication and high temperature CO_2 separation. By fabricating micro- or mesoporous LDH membranes, CO_2 separation at high temperatures could be achieved at a mechanism of surface adsorption and diffusion because of the strong bonding of CO_2 with material surface and high stability of LDHs at high temperature and water vapor.

Recently, a lot of efforts have focused on how to prepare LDH thin films for application in membrane separation and sensor development. Zn-Al layered double hydroxide films with high transparency were synthesized on glass substrate by a sol-gel process followed with hot water treatment for crystal conversion.[3] Okamoto[4] reported the synthesis of highly-oriented LDH films on glass surface through a process of decarbonation, anion-exchange, delamination, and casting. Other techniques used for LDH film preparation include surface reaction[5], precipitations[6], and self-assembling.[7] The LDH films reported so far are either formed on dense supports such as glass and aluminum metal surfaces[4,6] or by self-supporting.[8] To apply LDH films for practically gas separation, supported thin LDH films on porous supports must be synthesized. However, very few publications have reported the synthesis of supported LDH membranes because the LDS suspensions generally have a two-dimensional crystal structure and the membranes can not be fabricated by slip casting. The 2-D crystal structure of LDH platelets have a large aspect ratios ranging from 10 to 1000, resulting in the formation of house-of-cards structure and large interparticle pores.[9] Recently, direct precipitation and two-step crystallization have been deployed to synthesize LDH nanoparticles with small aspect ratios (<5) which shows promising application in membrane fabrication.[10]

In this work, supported LDH membranes on porous alumina supports were synthesized by a two-step coating in which positively-charged nanocrystalline LDH suspension and negatively-charged bentonite colloids are coated layer by layer. Furthermore, gas permeations of N_2/CO_2 were tested and the permeation behavior was explained by sorption-diffusion model and microstructure of the LDH membranes synthesized therein.

2. Experimental

2.1 Materials and apparatus

Thin film of LDHs was formed on surface of disc-shaped α-alumina substrates. The porous supports were made in the laboratory from commercial α-alumina powder (dia.=0.44 μm, Alcoa) with a diameter of 28mm and thickness of 3.0mm. Chemicals used in this study include aluminum nitrate ($Al(NO_3)_3 \cdot 9H_2O$) (Reagent Grade, Fisher Scientific), Magnesium nitrate ($Mg(NO_3)_2 \cdot 6H_2O$) (Reagent Grade, Fisher Scientific), urea (A.C.S., Fisher Scientific), sodium hydroxide (>99.9%, Sigma-Aldrich), hydroxypropyl cellulose (HPC, M_w=100,000, Sigma-aldrich), and deionized (DI) water. Gases used in the gas permeation test include N_2 (Trigas, >99.9%), and CO_2 (Trigas, >99.9%).

Nitrogen adsorption isotherms at 77K were determined using ASAP 2020 MicroPore System (Micromeritics, USA). The pore size distribution was obtained by applying BJH and HK models to the adsorption isotherms. Particle size and membrane morphology were studied by scanning electron microscopy (SEM, JEOL 5800LV) and transmission electron microscopy (JEOL 2010, HRTEM). The crystal structure of the porous materials was confirmed by X-ray diffraction pattern (XRD, Rigaku® Geigerflex diffractometer using Cu Kα radiation). All the samples for the X-ray test were dried at 120°C for 24 hr.

2.2 Synthesis of bentonite clay colloid and LDH nanoparticles

Bentonite colloid Na-bentonite colloids were separated from commercial Wyoming bentonite (Wyo-ben hydrogel) by a standard sedimentation and separation process.[11] The bentonite colloid was further transferred into a membrane tubing (MWCO: 12-14,000, Spectrum/Por®) for ion leaching in large amount of DI water. Figure 1 shows the XRD pattern of purified bentonite clay colloid. The inset shows the TEM images of bentonite clay colloid purified therein. The benonite clay suspension has an average particle size in the range of 50 to 100 nm.

Mg-Al-LDH nanoparticle Well-dispersed Mg-Al-LDH nanoparticles were prepared with a two-step crystallization process reported in the literatures.[9,10] Typically, 3.85g $Mg(NO_3)_2 \cdot 6H_2O$ and 1.86g $Al(NO_3)_3 \cdot 9H_2O$ were dissolved into 50mL DI water. The mixture was poured into a 200ml 0.15M NaOH solution under vigorous stirring in a Teflon beaker. After stirring at 45°C for 30 min, the solution was then separated by centrifuge. The precipitate was redispersed into 200 ml DI water by vigorous shaking. The suspension prepared therefore was placed in a preheated oven for aging at 100°C for 10 hr. The resultant homogenous suspension has a light blue color and was confirmed to be LDH nanoparticles by XRD and TEM, as shown in Figure 2. The broadening XRD diffraction peak was applied to the Scherrer equation, obtaining the particle thickness down the c-axis of ~7.5 nm. The lateral size in the basal plane of LDH nanoparticle is in the range of 20 to 30 nm, giving an aspect ratio of ~4. The suspension is very stable and no aggregation was observed after sitting at 5 °C for three months.

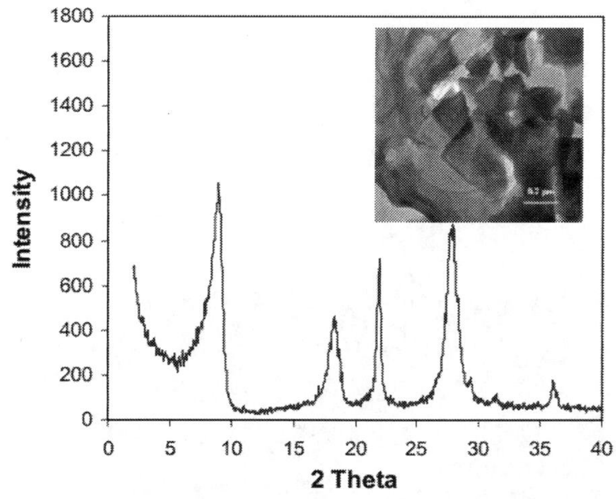

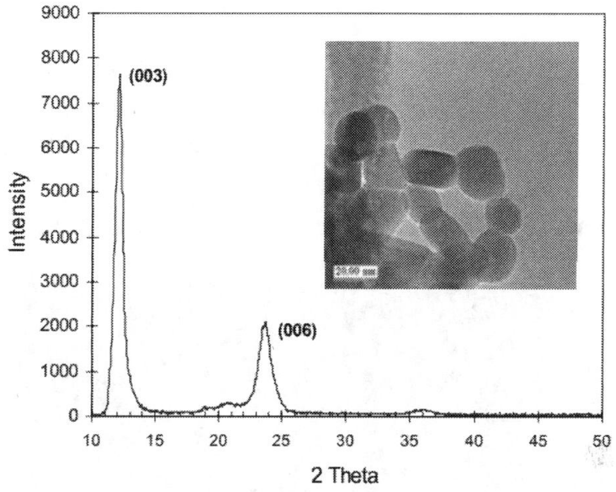

FIG. 1. XRD pattern of purified bentonite clay colloid. Inset is the TEM image.

FIG. 2. XRD pattern of the Mg-Al-LDH nanoparticles. Inset is the TEM image.

2.3 Membrane fabrication and characterization

Thin LDH membranes were prepared by coating multilayer LDH and bentonite thin layers with opposite charge. First, a LDH colloidal suspension was prepared by mixing LDH nanoparticles, HPC polymer binder and DI water in a Teflon beaker. After ultrasonic dispersion for 5 min, a polished substrate was dipped into the suspension for 2 to 3 sec. Under capillary pressure, liquid will be sucked into the porous substrate and the suspended nanoparticles will accumulate on the surface, forming a very thin layer of LDH. Since the synthesized LDH suspension has a positive zeta potential while the alumina substrates are negatively charged in aqueous solution, no pH adjustment was needed. The coated LDH membrane was then dried at 45 °C at controlled humidity for 48 hr. After drying, the membranes were calcinated in air at 550 °C for 8 hr with a heating rate of 0.5 °C/min and cooling rate of 1.0 °C/min. Following calcinations of LDH thin film, negatively-charged bentonite layer was coated to cover the interparticle pores. The bentonite colloid suspension was prepared by dispersing 0.5wt% purified bentonite suspension and 0.04% HPC into DI water. The bentonite coating process was the same as that of the LDH coating. Figure 3 shows the SEM images of cross-section on the membranes coated with LDH layer and LDH/Bentonite multilayer. The thin LDH layer formed by single coating has a thickness of ~1μm. The second coating of clay colloid with opposite charge increases the membrane thickness to ~2μm as shown in Figure 3(b).

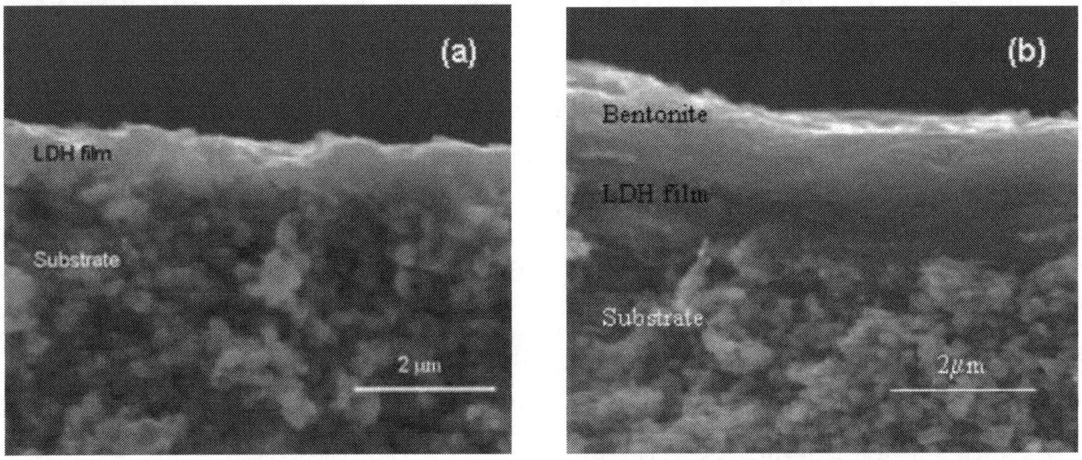

FIG. 3. Cross-section of LDH/Bentonite membranes synthesized by layer-by-layer deposition.

It was observed that the solid concentration, binder concentration and coating time all play important roles in the formation of continuous LDH membranes. LDH films with good surface coverage were formed from a suspension containing solid contents ranging from 0.3wt% to 1.2wt%. Membrane preparation from concentrations out of that range results either particle penetration into the substrate or loose coating film due to excessive gelation on surface. The optimized composition for LDH film formation is: (0.4–0.8 wt%) LDH + 0.05wt%HPC + DI water. The alternate clay layer was formed from a suspension containing 0.4 wt% bentonite and 0.04wt% HPC.

The remaining coating suspension was dried and calcinated at the same condition as the membrane preparation to simulate the film formation process. The resultant powder sample was analyzed for N_2 adsorption isotherm at 77k. Further, the Horvath-Kawazoe (HK) method, and the Barret, Joyner and

Halenda (BJH) model were applied on the isotherm plot for pore size distribution analysis in the range of micropores and mesopores respectively.

2.4 Measurement of CO_2/N_2 permeation

Gas separation performances of the membranes were investigated using equimolar dual gas permeation (N_2/CO_2) at varying temperatures. The permeation system was schematically shown in Figure 4. A gas mixture with equimolar N_2 and CO_2 was flowing across the membrane surface while the membrane downstream was swept by helium at a flow rate of 15.0mL/min. The permeate composition was monitored by an on-line GC analysis. The composition of the feed stream was recalibrated every 24 hours. The stabilized molar percentage (x_i) of N_2 and CO_2 at each temperature was measured and the gas permeance (P_i) and separation factor (S_i) were calculated from the following equation,

$$S_i = \frac{[x_i/(1-x_i)]_{permeate}}{[x_i/(1-x_i)]_{feed}} \qquad (1)$$

$$P_i = \frac{Q_i}{A \cdot \Delta t \cdot P^0} \qquad (2)$$

Where $Q_i / \Delta t$ is molar flow rate, mol/s; A is membrane surface area, m^2; and P^0 is partial pressure of component i in feed stream.

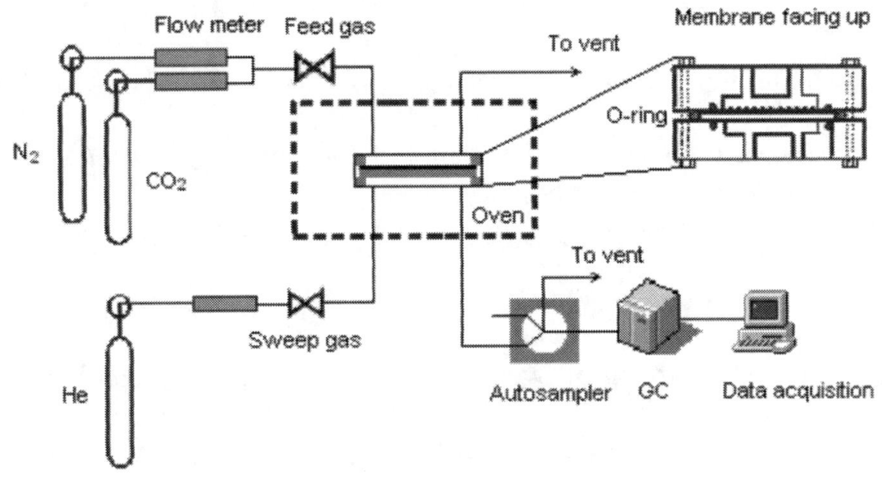

FIG. 4. Schematic diagram of the N_2/CO_2 separation system.

3. Results and discussion

3.1 Multilayer coating for membrane fabrication

Figure 5 shows the SEM images of LDH thin film prepared by coating LDH suspension with particle sizes of ~20nm and ~60nm, respectively. After calcinations at 550°C for 8 hr, high density of cracks were observed on the membranes fabricated from suspensions containing ~20nm nanoparticles. The membranes synthesized from a suspension of ~60nm LDH particles show good surface coverage but large interparticle pores as shown in Figure 5 (b).

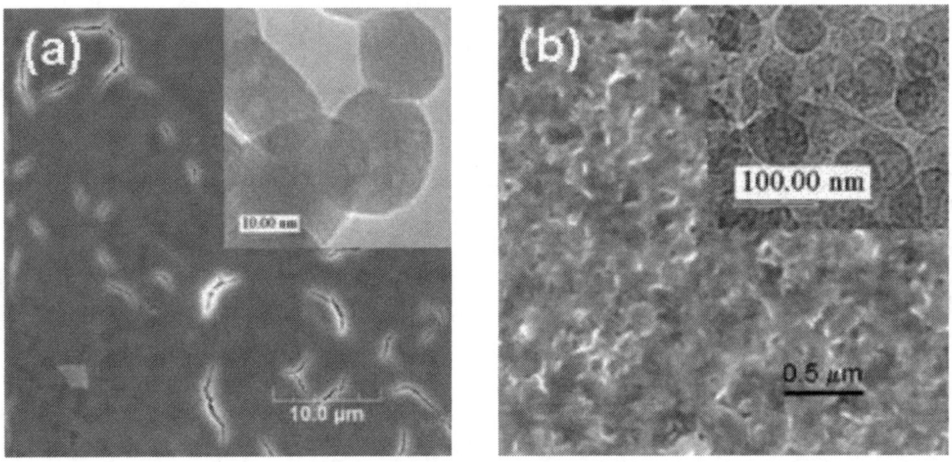

FIG. 5. Surface SEM image of LDH thin film synthesized from suspension with different particle sizes, (a) ~20nm, (b) ~60nm. The insets are TEM images of the nanoparticles.

Water and carbonate ions usually present in the interlayer spacing of LDH nanoparticles. Deep dehydration and decarbonization during calcinations process cause significant shrink of thin LDH layer.[12] The incompatibility of thermal expansion between the thin film and substrate gives the explanation of the crack formation during membrane calcination. For the membranes synthesized from LDH particles with larger size (i.e., ~60nm), no defects were observed by random surface searching of SEM images. Unfortunately, the loosely packing of two dimensional LDH crystals and house-of-cards structure makes the membrane less selective. Multiple coating will be necessary to minimize the interparticle microdefects. Repeated coating of LDH thin layers was first tested and the second layer was peeled off easily upon calcination. Insufficient capillary pressure and slowing drying during second coating may result the loosely packing and instability of material structure. Thus, a second coating of bentonite colloid was deployed to form a layer-by-layer structure.

In the second coating, a LDH membrane was dipped into a bentonite suspension containing negatively-charged bentonite nanosheets. Under electric charge and capillary pressure, a thin bentonite layer forms on the surface of LDH film. Figure 6 gives the surface SEM image of the LDH/bentonite membrane and TEM images of bentonite nanosheets used for the membrane fabrication. The bentonite nanosheets have a particle size of ~100 nm. The alternate thin layers of LDH and bentonite with opposite charges form a layer-by-layer structure that is very stable at elevated temperatures (i.e., 550°C).

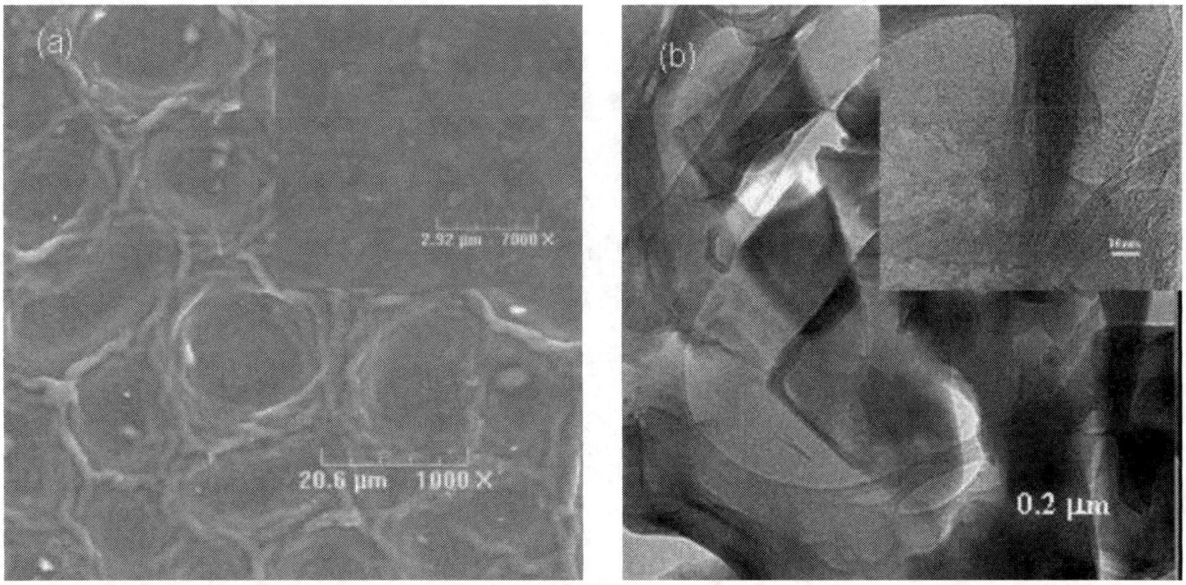

FIG. 6. SEM images of LDH membranes formed by layer-by-layer coating, (a) surface SEM image, (b) TEM image of bentonite nanosheet.

3.2 Packing style, texture and their influence on membrane pore structure

Layered double hydroxide nanocrystallines have a two dimensional structure. The LDH nanoparticles deployed for membrane fabrication in this work have a lateral size 20 to 60 nm and aspect ratio of 4~10.

In aqueous solution, these nanosheets are generally organized by face-to-face, edge-to-edge, and edge-to-face aggregation. Some idealized packing styles during a membrane fabrication process are shown in Figure 7. The hydrated water and carbonate ions are localized in the interlamellar space between two successive sheets.

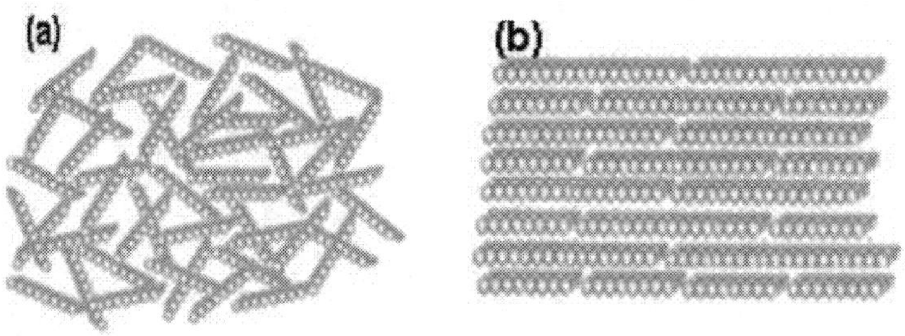

FIG. 7. Packing style of LDH membranes (a) house-of-cards, (b) (001) plane parallel to the surface, (c) (001) plane perpendicular to the surface.

A fast coating and drying process forms a highly porous structure due to the formation of house-of-cards structure by the primary edge-to-face packing of 2-D LDH nanosheets.[5] Reducing the particle size is generally effective for minimize the interparticle pore size. Even with the random packing in the house-of-cards structure, LDH nanoparticles tend to oriented with (001) plane parallel to the surface as confirmed by the XRD pattern

FIG. 8. Edge-to-face structure of LDH film with (001) plane perpendicular to the surface.

shown in Figure 2. For a comparison, LDH film formed by a direct precipitation is shown in Figure 8, which indicates a combination of edge-to-edge and edge-to-face aggregation. In both situations, large interparticle micropores will form and lower the membrane separation efficiency.

Figure 9 gives the N_2 adsorption isotherm and pore size distribution results. The LDH membranes obtained by coating LDH nano- sheets have a large surface area of 331.5 m^2/g and micropore volume of 0.287 cm^3/g. The inset of Figure 9(b) is the BJH pore size distribution derived from adsorption branch of the isotherm, revealing a bimodal pore structure of the membrane with pore size distribution in the range of ~0.7nm and ~4.0nm respectively which are corresponding to the interlayer micropores and interparticle nanopores. The interlayer micropore size is determined by the counter ion species while the interparticle nanopores are governed by the particle size and packing style.[9,13] For platelet materials, natural slow precipitation result in oriented stacking and minimized interparticle spacing while freeze-drying and fast deposition result the house-of-cards structure with large volume of mesopores.[14,15]

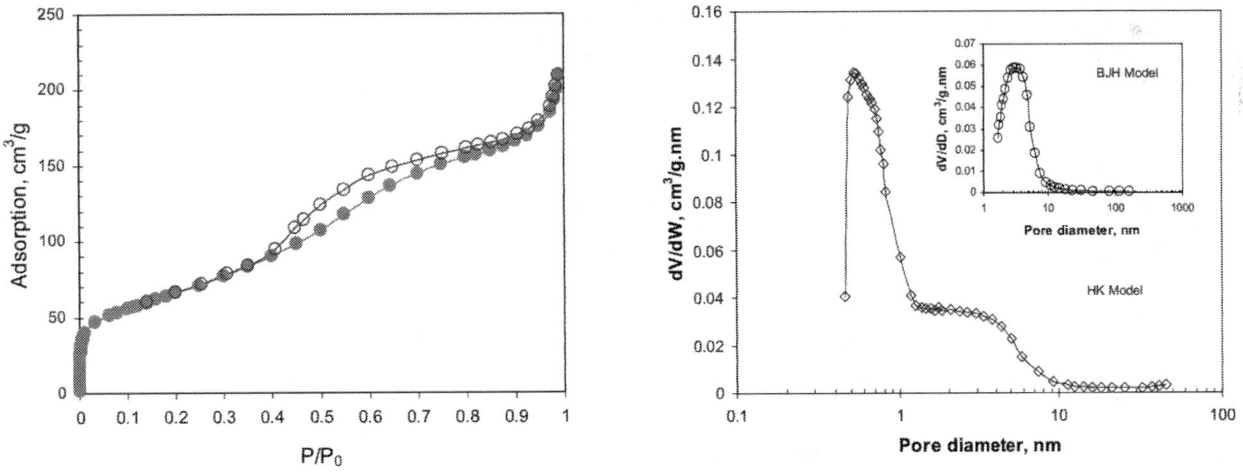

(a) N_2 adsorption isotherm at 77K (b) pore size distribution of LDH film

FIG. 9. Microporous structure of LDH film synthesized by dip-coating of suspended LDH nanoparticles.

3.3 Gas permeation

Figure 10 shows the permeance of N_2 and CO_2 and the separation factor as a function of operation temperature for a LDH/bentonite membrane. The permeance of CO_2 decline slightly when temperature was increased from 50 to 100 °C. Above that temperature, both permeances of N_2 and CO_2 increase with increasing temperature. Molecular sieving is not expected to affect the separation of this mixture because the two components have similar molecular sizes, both of which are much smaller than the pore size of the membrane. The dynamic sizes of N_2 and CO_2 are 0.33 nm and 0.37 nm, respectively. The membrane has pore sizes of 0.7 nm and 4.0 nm. The CO_2-N_2 separation factor was greater than 1.0 in the entire tested temperature range. This is opposite to the behavior of Knudsen diffusion, which is expected from an essentially mesoporous membrane and gives a CO_2/N_2 separation factor of 0.80 (<1).

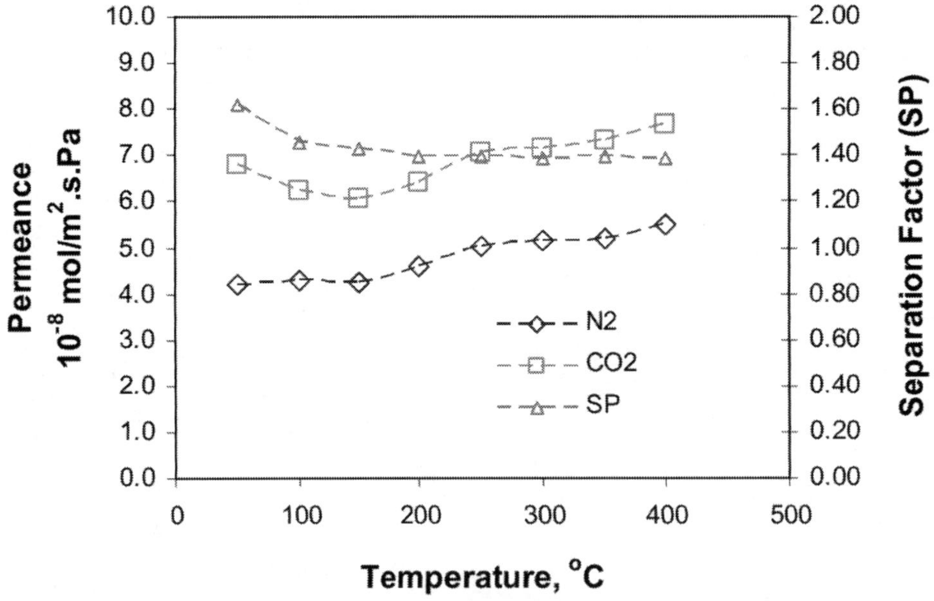

FIG. 10. Gas permeation on a supported LDH membrane.

4. Conclusions

Supported LDH membranes were synthesized on porous α-alumina support by coating LDH nanocrystalline and alternate bentonite colloid. The coating of nanocrystalline LDH and bentonite with opposite charges creates a layer-by-layer structure that is very stable at elevated temperatures. Particle size of LDH was found to be critical for the formation of continuous thin films. Reducing the particle size and controlling the orientation were expected to be effective for reducing the interparticle pores. Unfortunately, large quantities of micro-cracks were found on the LDH membranes synthesized from nanosheets with particle size at ~20nm. Decarbonization and dehydration upon heating caused incompatible thermal expansion and shrinkage. As a result, high density of micro-cracks formed on membrane surfaces during calcination. Membranes fabricated from LDH suspension containing ~60 nm LDH nanosheets show continuous surface coverage and high gas permeance due to the house-of-cards structure and loosely packed LDH nanoparticles.

The supported LDH membranes have a mesoporous structure with bimodal pore sizes in the range of ~0.7nm and ~4.0nm. The pore size distribution is affected by stacking style of LDH nanosheets and crystal orientation. LDH films formed by direct precipitation from reaction solution have an edge-to-face and edge-to-edge structure and large interparticle pore size. A second coating of bentonite colloid is effective to reduce the interparticle pore size and enhance the permselective of LDH membranes. LDH membranes synthesized by the two-step coating show CO_2 permselective at elevated temperatures up to 400°C.

Acknowledgement:

This research was sponsored by the U.S. Department of Energy through the National Energy Technology Laboratory (grant # DE-FC26-04NT15548) and Southwest Carbon Sequestration Partnership (grant # DE-FC26-05NT42591).

Reference:

(1) Goswamee, R.L., P. Sengupta, K.G. Bhattacharyya, and D.K. Dutta, *Appl. Clay Sci.*, 13 (1998), 21-34.

(2) Kim, Y., W. Yang, P.K.T. Liu, M. Shimi, and T.T. Tsotsis, *Ind. Eng. Chem. Res.* 43 (2004), 4559-4570.

(3) Yamaguchi, N., T. Nakamura, K. Tadanaga, A. Matsuda, T. Minami, and M. Tatsumisago, *Crystal Growth & Design,* 6 (2006), 1726-1729.

(4) Okamoto, K., T. Sasaki, T. Fujita, and N. Iyi, *J. Mater. Chem.*, 16 (2006), 1608-1616.

(5) Gao, Y.F., M. Nagai, Y. Masuda, F. Sato, W.S. Seo, and K. Koumoto, *Langmuir*, 22 (2006), 3521-3527.

(6) Liu, J., X. Huang, Y. Li, K.M. Sulieman, X. He, and F. Sun, *J. Phys. Chem. B*, 110 (2006), 21865-21872.

(7) Geraud, E., V. Prevot, J. Ghanbaja, and F. Leroux, *Chem. Mater.* 18 (2006), 238-240.

(8) Wang, L, C. Li, M. Liu, D.G. Evans, and X. Duan, *Chem. Commun.*, 2007, 123-125.

(9) Gursky, J.A., S.D. Blough, C. Luna, C. Gemez, A.N. Luevano, and E.A. Gardner, *J. Am. Chem. Soc.* 128 (2006), 8376-8377.

(10) Xu, ZP., G.S. Stevenson, C. Lu, G.Q. Lu, P.F. Bartlett, and P.P. Gray, *J. Am. Chem. Soc.* 128 (2006), 36-37.

(11) Li, L.X., T.M. Whitworth, and R. Lee, *J. Membr. Sci.* 217 (2003), 215-225.

(12) Climent, M.J., A. Corma, S. Iborra, K. Epping, and A. Velty, *J. Catal.* 225 (2004), 316-326.

(13) Adachi-Pagano, M., C. Forano, J.P. Besse, *J. Mater. Chem.*, 13 (2003), 1988-1993.

(14) Molinard, A., E.F. Vansant, *Adsorpt.-J. Int. Adsorpt. Soc.*, 1 (1995), 49-59.

(15) Pacula, A., E. Bielanska, A. Gawel, K. Bahranowski, and E.M. Serwicka, *Appl. Clay Sci.*, 32 (2006), 64-72.

DEVELOPMENT OF PRACTICAL SUPPORTED IONIC LIQUID MEMBRANES: A SYSTEMATIC APPROACH

David R. Luebke[1], Jeffery B. Ilconich[2], Christina R. Myers[1], and Henry W. Pennline[1]
(1)United States Department of Energy, National Energy Technology Laboratory,
(2) Parsons/RDS, Pittsburgh, PA[1,2]

Introduction

Current technology has taken several approaches to address greenhouse gas emissions, but a critical component of any strategy remains the capture and sequestration of carbon dioxide (CO_2). There are several ways to reduce CO_2 emissions which include: increasing process efficiency; the use of lower-carbon content fuels such as natural gas and biofuels including ethanol from corn and cellulose; and decreasing fuel consumption. These methods are essential, but not sufficient, to meet the goal of ending the rise in atmospheric CO_2 concentration. It is also necessary to address the capture and sequestration of CO_2. The reason is that fossil fuels, while an important energy source, are also the principal green house gas source. A significant portion of our future energy production will rely on the ability to produce zero emission power from fossil fuels, where the majority of expense in overall emissions reduction will be associated with CO_2. The optimum place to start is the capture of CO_2 from point sources such as power plants.

Investigative studies and research have closely examined and are beginning to address the problems associated with CO_2 capture for various types of plants: i.e. a gasifier cannot be retrofit to any existing system. Holding the most promise for the future is the Integrated Gasification Combined Cycle (IGCC), a clean coal technology and a significant component of future power generation which allows efficiency enhancement, offers the potential to benefit the environment and could provide low carbon fuel by producing hydrogen, electricity, and liquid hydrocarbons. The possibility for hydrogen (H_2) and/or CO_2 separation prior to combustion in the IGCC makes it even more promising. The greatest CO_2 capture efficiency in IGCC systems is achieved by separating concurrently with the equilibrium limited water gas shift reaction. By removing CO_2, a product, the reaction is forced to completion allowing a smaller reactor and the use of less steam.

Several technologies have been investigated for co-production of high purity hydrogen with recovery of CO_2 ready for sequestration. Focus has turned towards membrane technology for its simplicity, modular design, and ease of scaling. Membranes have already demonstrated the capability to separate H_2 and CO_2 in various applications and types of systems but must overcome elevated temperatures and pressures as well as degradation by contaminant gases and particulates [1].

Supported liquid membranes (SLMs) are a class of materials that allow researchers to utilize the wealth of knowledge available on liquid properties to optimize membrane performance. These membranes also have the advantage of liquid phase diffusivities, which are higher than those observed in polymers and grant proportionally greater permeabilities. The primary shortcoming of the supported liquid membranes demonstrated in past research has been the lack of stability caused by volatilization of the transport liquid. Ionic liquids, which may possess high CO_2 solubility relative to light gases such as H_2, are excellent candidates for this type of membrane. They are stable at elevated temperatures and have negligible vapor pressure.

A study has been conducted evaluating the use of a variety of ionic liquids in supported ionic liquid membranes for the capture of CO_2 from streams containing H_2. In a joint project, researchers at the University of Notre Dame synthesized and characterized ionic liquids, while the researchers at the National Energy Technology Laboratory incorporated candidate ionic liquids into supports then evaluated the membrane performance for the resulting materials.

Experimental

The ionic liquids [hmim][Tf_2N], [NH_2C_3mim]Tf_2N], and [NH_2C_3mpy][Tf_2N] were synthesized at the University of Notre Dame using standard procedures [2,3]. Polyethersulfone, HT Tuffryn® (polysulfone) and Biodyne® A (crosslinkable nylon) polymeric porous substrates from Pall Corporation were used as supports for ionic liquid membranes. The support discs were 25 mm in diameter with an active membrane area of 2.2 cm^2.

The SILMs were made by placing the polymeric support in a container and slowly depositing the ionic liquid drop wise on top of the membrane with a pipette. Enough ionic liquid was added to completely cover the surface of the substrate and the membrane was allowed to absorb the ionic liquid for at least eight hours. The SILMs were then removed from the container and the excess ionic liquid was removed by blotting the SILM.

The membrane was placed on the permeate side of a Millipore® stainless steel filter holder and an unmodified substrate identical to the one used to prepare the membrane was then placed against the membrane on the feed side to reduce the mechanical stress. The feed gas flowed at approximately 30 ml/min and consisted of ~20% carbon dioxide, ~20% hydrogen and the balance argon with one mixture containing ~1% hydrogen sulfide. A sweep gas, argon, was used on the permeate side with a flow between 1.0 to 2.0 ml/min. The flows and pressures were controlled by a Cole Parmer® flowmeter and Honeywell® pressure transducers and digital readout, respectively. The pressure was approximately 108 kPa for the feed and the permeate pressure was less then 102 kPa. Between experiments, the system was purged overnight with either UHP helium or UHP argon and then allowed to stabilize with the desired gas before testing began.

Testing was performed in a flow system in which the permeate and retentate gas compositions were determined using either a HP 5890 or a Perkin Elmer Clarus 500 gas chromatograph with twin TCD detectors and Alltech Hayesep D 100/120 packed columns. The gas analysis was collected and stored using Total Chrom Navigator® software. Flows were measured with the OPTIFLOW 520 Digital Flowmeter, Humonics®. Temperature was measured by a Type K thermocouple in contact with the surface of the testing cell. The temperature was controlled by a Thermo Electron Corporation furnace controller. Thermal control was carried out by ramping to the desired temperature at 1°C/minute (small temperature gaps i.e. 15°C) or 2°C/minute (larger temperature gaps i.e. >20°C) and holding for several hours or longer during the testing.

Results and Discussion

Several steps have been taken in the development of practical ionic liquid membranes. Proof-of-concept was established by showing that ionic liquids could be used as the transport media in SLMs. A base ionic liquid [HMIM][Tf2N] was chosen to be placed on polysulfone (PSF) and polyethersulfone (PES) supports which were then tested over the temperature range 37-135°C. Both sets of membranes showed similar performance at room temperature, but increasing the temperature to 50°C caused the PES supported membranes to fail. The PSF had a lower glass temperature when not in contact with the ionic liquid; however with the

IL its glass transition temperature was only slightly depressed as opposed to that of PES which was greatly affected. The results for PSF are shown in Figure 1.

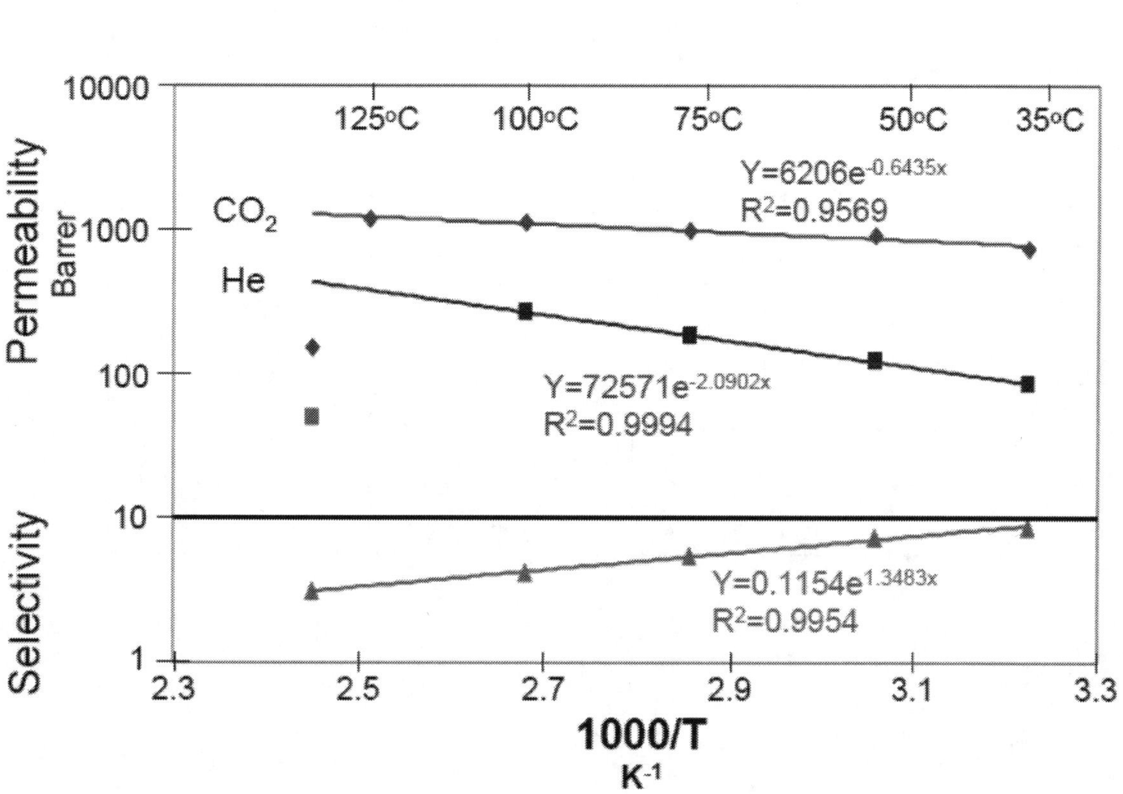

Figure 1. CO_2 (♦) and helium (▪) permeability and the selectivity (▲) as a function of temperature.

CO_2 permeability is higher than that of helium, used as a surrogate for H_2 in the initial study. Selectivity and permeability for both He and CO_2 show the Arrhenius dependence expected for solution diffusion up to 125°C. Above 125°C, the Arrhenius dependence of the permeabilities fail, but that of selectivity remains intact. This behavior is an indication that pore collapse occurs within the support restricting effective membrane area without changing the transport mechanism [4].

The results of the proof of concept study made it clear that support improvements would be necessary in order to operate at the temperatures of interest for CO_2/H_2 separations. Several supports were examined using the base ionic liquid [HMIM][Tf2N] over the temperature range 37-300°C. The supports that showed the most promise and stability were Biodyne® A (crosslinkable nylon) polymeric porous substrate supports from Pall Corporation. The difference between the two supports, Biodyne®1 and Biodyne®2, was pore size; Biodyne®1 had larger pores. Table 1 and Figure 2 show how the different pore sizes affected the permeability of CO_2 and H_2 and the selectivity over the temperature range 37-300°C but more importantly that both are stable to 300°C.

Table 1. CO_2 and H_2 permeability and selectivity from 37-300°C. **[HMIM][Tf₂N]-** Biodyne®

	Temp. °C						
	37	50	100	150	200	250	300
CO_2 (Barrer)	*average*						
BIODYNE®1	417	446	508	606	699	767	825
BIODYNE®2	502	571	772	861	1009	942	1165
H_2 (Barrer)	*average*						
BIODYNE®1	43	50	72	122	199	323	567
BIODYNE®2	54	67	123	205	367	535	918
CO_2/H_2 Selectivity							
BIODYNE®1	9.72	9.03	6.84	4.72	3.13	2.13	1.44
BIODYNE®2	9.30	8.47	6.24	4.18	2.74	1.76	1.27

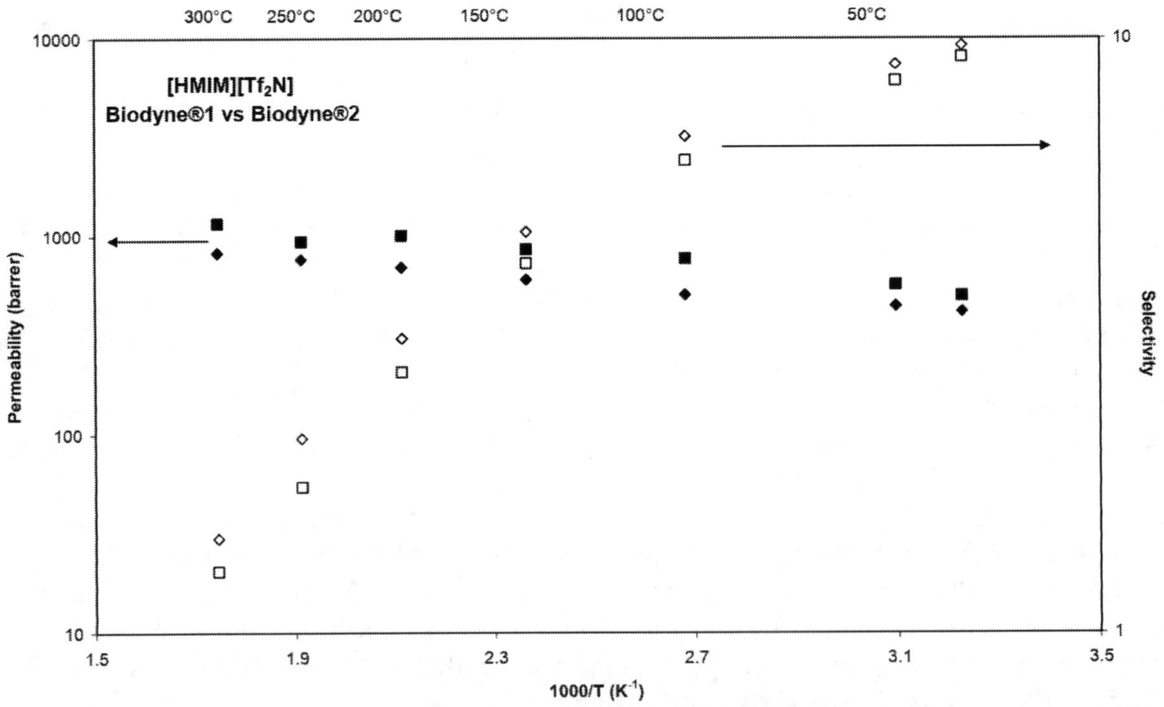

Figure 2. CO_2 permeability (solid figures) and selectivity (open figures) of BIODYNE®1(♦) and BIODYNE®2 (▪) as a function of temperature.

Preliminary testing showed that PALL supports had stability over the greater temperature range, but selectivity was still low at elevated temperature. Now that the supports could operate at high temperatures while maintaining their properties, it was necessary to investigate other ionic liquids in order to increase selectivity. It was hypothesized that the use of other ionic liquids including those that form chemical complexes with CO_2 based on amine interactions could improve performance at elevated temperatures by allowing a facilitated transport mechanism to become dominant.

Two complexing ionic liquids were placed on the Biodyne®2 support, which was chosen since its overall performance was higher and tested in the temperature range 37-300°C. Test results for the complexing ionic liquid membranes were compared to the base ionic liquid [HMIM][Tf2N] on the BIODYNE®2 support. The first amine-based IL tested on the BIODYNE®2 support was [HN$_2$PMIM][Tf$_2$N] as shown in Figure 3.

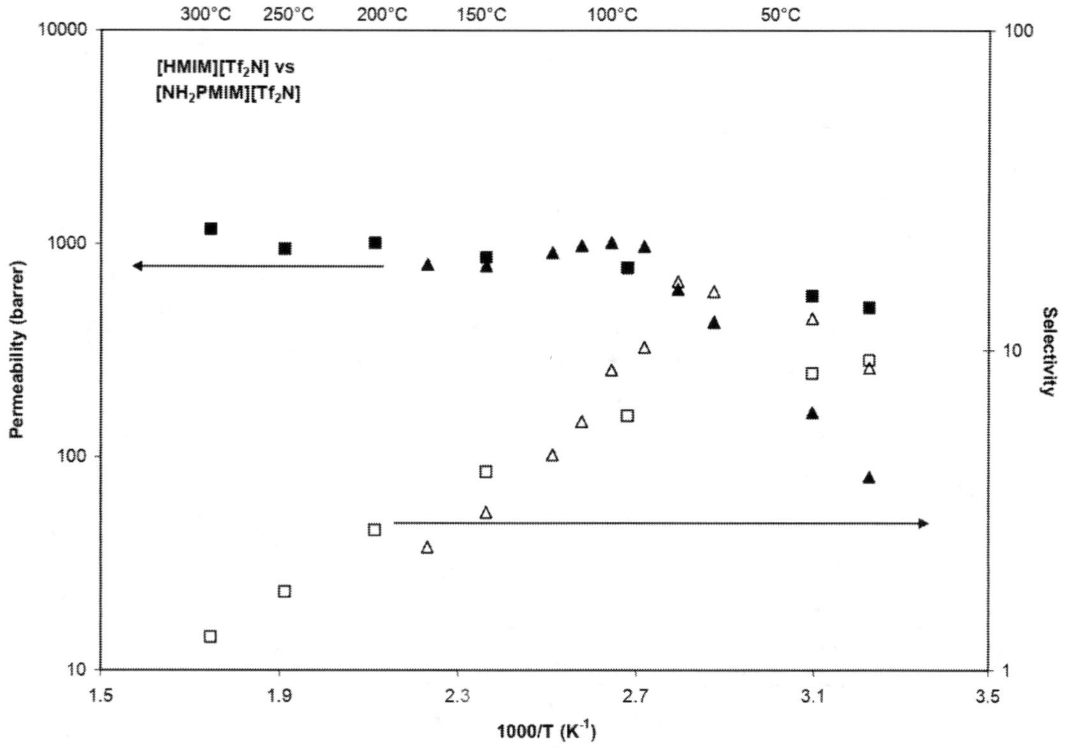

Figure 3. CO_2 permeability (solid figures) and selectivity (open figures) of [HMIM][Tf2N] (▪) and [NH2PMIM][Tf2N] (▲) as a function of temperature.

At the lower temperature, the CO_2 permeability of the complexing ionic liquid membrane increases rapidly compared to the base ionic liquid, but as the temperature increases, the complexing IL peaks then begins to taper off similarly to the base ionic liquid. The peak observed in selectivity is not observed in solution diffusion, but rather is the effect of facilitated transport, in which chemically complexed CO_2 results in stronger interactions at higher temperatures and a change in rate limiting step from decomplexing of CO_2 to diffusion. The next complexing ionic liquid, [NH2PMPY][Tf2N], showed similar behavior but with slightly diminished performance as shown in Figure 4.

The CO_2 permeability and selectivity of [NH2PMPY][Tf2N] is slightly lower than [NH2PMIM][Tf2N], and [NH2PMIM][Tf2N] reached its maximum selectivity at 75°C while the peak for [NH2PMPY] [Tf2N] appeared at 50°C. Since it is probable that stronger complexes lead to peak selectivity at a higher temperature and increased selectivity, it is possible to suggest that [NH2PMIM][Tf2N] forms a more stable complex with CO_2. Table 2 below shows a summary of the base ionic liquid [HMIM][Tf2N] and the two complexing ionic liquids.

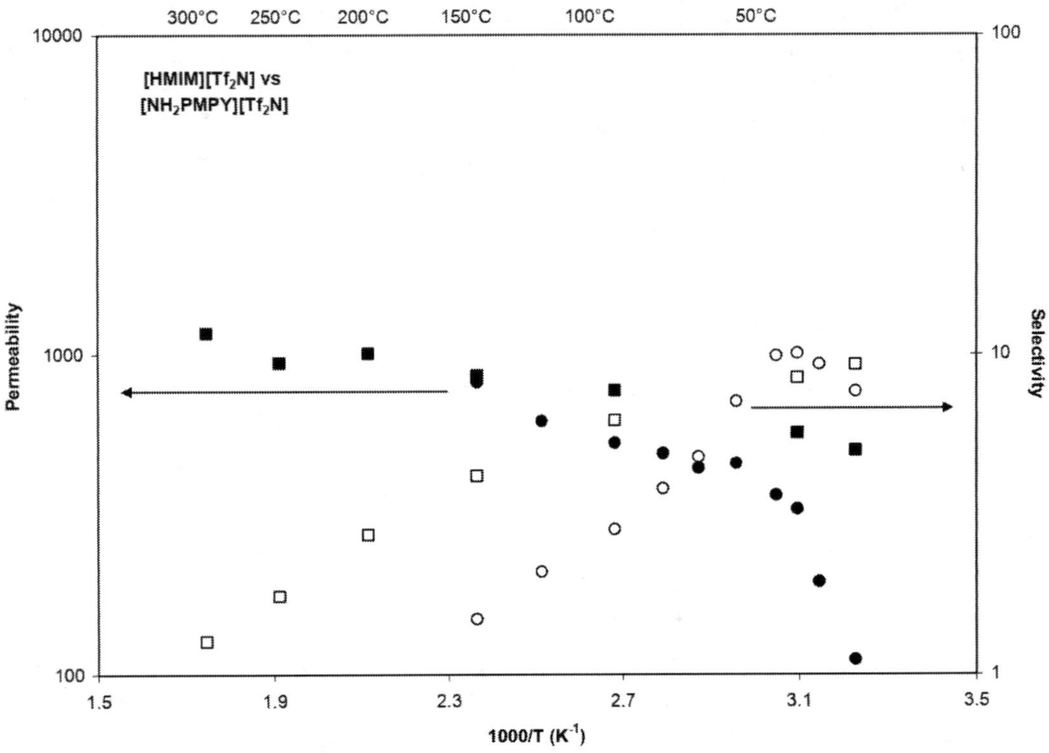

Figure 4. Results of CO_2 permeability (solid figures) and selectivity of [HMIM][Tf2N] (•) and [NH2PMPY][Tf2N] (°) as a function of temperature.

Table 2. Summary of base and complexing ionic liquids with Biodyne®2 support.

[HMIM][Tf$_2$N]	37°C	300°C
CO_2 (barrer)	502	1165
H_2 (barrer)	54.0	918
Selectivity CO_2/H_2	9.30	1.27
[NH$_2$PMIM][Tf$_2$N]	37°C	75°C
CO_2 (barrer)	80.5	429
H_2 (barrer)	10.8	28
Selectivity CO_2/H_2	9.58	15.3
[NH$_2$PMPY][Tf$_2$N]	37°C	50°C
CO_2 (barrer)	102	329
H_2 (barrer)	14.4	32.5
Selectivity CO_2/H_2	7.69	10.11

Table 3 shows the base ionic liquid on Biodyne®2 in a commercial application of CO_2 selective membranes, natural gas sweetening. Interesting results for CO_2/H_2 separation led us to question the usefulness of these membranes in other applications.

Table 3. Natural Gas Sweetening

[HMIM][Tf_2N]-	37°C	60°C
CO_2 (barrer)	378	416
CH_4 (barrer)	41.0	59.0
α (CO_2/CH_4)	9.30	7.09

The data in Table 3 are comparable to the commercial gas sweetening membranes, but the results were not as high as expected. It is believed that methane solubility, presumably much greater than H_2 solubility, leads to unexpectedly low selectivity.

Conclusions

We have demonstrated that CO_2 selective membranes can be fabricated from ionic liquids. These membranes are capable of operating at high temperatures. Further development will include ionic liquid improvements to increase complex strength and performance, identification of the effect of contaminants, work to increase transmembrane pressure tolerance, and scale up to modules.

Acknowledgements

The authors gratefully recognize the Joan Brennecke and Ed Maginn research groups at the University of Notre Dame. Their efforts in the synthesis and characterization of the ionic liquids along with their invaluable expertise have been instrumental in the development of this exciting new technology. We look forward to continued rewarding collaboration.

Disclaimer

Reference in this report to any specific commercial product, process, or service is to facilitate understanding and does not necessarily imply its endorsement or favoring by the United States Department of Energy.

References

1. Ritter, James A., Ebner, Armin D., (2007), "State-of-the-Art Adsorption and Membrane Separation Processes for Hydrogen Production in the Chemical and Petrochemical Industries," *Separation Science and Technology,* 42, pp. 1123-1193.
2. J.M. Crosthwaite, S.N.V.K. Aki, E.J. Maginn, J.F. Brennecke, (2004) "Liquid phase behavior of imidazolium-based ionic liquids with alcohol," *Journal of Physical Chemistry B,* 108, p. 5113.
3. J.M. Crosthwaite, S.N.V.K. Aki, E.J. Maginn, J.F. Brennecke, (2005) "Liquid phase behavior of imidazolium-based ionic liquids with alcohols: effect of hydrogen bonding and non-polar interactions," *Fluid Phase Equilib.*, 228, p 303.

4. Ilconich, Jeffery, Myers, Christina, Pennline, Henry, Luebke, David (2007), "Experimental investigation of supported ionic liquid membranes for CO_2/He separation at temperatures up to 125°C," *Journal of Membrane Science,* 298, pp. 41-47.

156i CO_2 and N_2 Diffusion in Zsm-5 Zeolites with Varying Si/ Al Ratios

David A. Newsome, Delft University of Technology, DelftChemTech, Julianalaan 136, 2628 BL, Delft, Netherlands, Anton J. Dammers, Delft University of Technology, DelftChemTech, Julianalaan 136, Delft, 2628 BL, Netherlands, and Marc-Olivier Coppens, The Isermann Department of Chemical and Biological Engineering, Rensselaer Polytechnic Institute, 110 8th Street, Troy, NY 12189.

Zeolites are crystalline inorganic microporous materials, with pores that are of a size comparable to that of many small molecules. Therefore, they can selectively discriminate between molecules with relatively small differences in size and shape. This feature makes zeolites ideal candidates for membrane separations, catalysis, and sequestration. One important environmental application for zeolites is the separation of CO_2 from gaseous mixtures containing N_2 or CH_4.[1-3] Experimental studies have been performed to analyze the adsorption and diffusion of these mixtures in zeolite membranes, a well-established example being ZSM-5.[4, 5] This zeolite contains a two-dimensional network of intersecting channels running in the *x* and *y*-directions, and having a size of ~5.5 Å. ZSM-5 preferentially adsorbs CO_2 over N_2 due to the strong quadrupole moment of CO_2.[1] The CO_2-favored selectivity of ZSM-5 depends significantly on the Si/Al ratio that defines the zeolite composition. The presence of Al substitutions into the framework requires the presence of charge-balancing cations, such as Na^+ and Rb^+. The cation and CO_2 strongly interact, which causes increased adsorption and decreased diffusion.[6] We have performed molecular simulations of CO_2 and N_2 in ZSM-5 to study the impact caused by the presence of Na^+ on diffusion and adsorption. The roles of temperature and Si/Al ratio are considered.

1 – Katoh, M. et. al. *J. Colloid Int. Sci.* **2000**, *226*, 145-150.

2 – Harlick, P.; Tezel, F. *Micro. Meso. Mat.* **2004**, *76*, 71-79.

3 – Siriwardane, R. et. al. *Energy and Fuels* **2001**, *15*, 279-284.

4 – Shin, D. et. al. *Micro. Meso. Mat.* **2005,** *85*, 313-323.

5 – Zhu, W. et al. *Ind. Eng. Chem. Res.* **2006**, *45*, 767-776.

6 – Wirawan, S. and Creaser, D. *Micro. Meso. Mat.* **2006**, *91*, 196-205.

168a Balancing Distillation Column Diameter with Vapor Feeds

Phillip C. Wankat, Purdue University, Forney Hall of Chemical Engineering, 480 Stadium Mall Drive, West Lafayette, IN 47907-2100

Methods for balancing distillation vapor loads to achieve smaller, more uniform calculated column diameters are developed for columns with vapor feeds and a large calculated column diameter in the enriching section. The column diameter for distillation is usually designed to be a set percentage of flooding, typically in the range of 60 to 80 %. The diameter calculated at different locations usually varies. Typically, the column is designed at the location with the largest column diameter and built with a constant diameter unless the calculated diameters at different locations differ by a large amount. When this design procedure is followed, the percentage of flooding will be less, often significantly less, than the nominal design percent of flooding everywhere in the column except for the location with the maximum calculated diameter. In cases with extreme variations in calculated diameter, the column is built in two sections with different diameters. In this research we consider a large difference in the calculated diameters as potential to either reduce the column volume or as potential to process more material in a retrofitted column. If the calculated diameter is high in the enriching section, we reduce the vapor flow rate in this section. To do this we reduce the reflux ratio from the condenser and do additional condensing either with an intermediate condenser or with a condenser on all or part of the feed.

The calculated diameter can be reduced by: 1. cooling the entire feed to produce a two phase feed or a liquid, 2. condensing vapor inside the column with an intermediate condenser or 3. cooling or condensing a part of the feed while leaving the other part vapor with a two-enthalpy feed system. The first two methods are well-known in the literature while the third is not. To retain the same product purities it may be necessary to increase the energy loads to the column or increase the number of stages.

First, applications are illustrated for the main distillation column in an ethanol biorefinery that receives the vapor from the beer still as feed. In this case the use of the two-enthalpy feed system has an advantage since a considerable amount of vapor can be condensed without changing the minimum boilup rate. For example, for distillation of a saturated vapor feed of 10 mole % ethanol and 90 mole % water the use of two-enthalpy feed resulted in a 53 % reduction in column volume with the same purity and no increase in energy use if 11 % more stages were added. Reductions in volume as large as 76 % are predicted but with a significant increase in reboiler duty. This method can also be used for debottlenecking existing columns that were constructed with a large diameter.

Additional calculations are illustrated for the separation of vapor feeds of methanol and water and for vapor feeds of propane and n-butane. Since the diameter balancing method is a processing technique for distillation, it is generally applicable to any type of equipment used for distillation. Thus, the methods developed in this paper can be applied to different types of trays and different random and structured packings.

168b The Long and the Short of Energy Consumption in Distillation

Angelo Lucia, University of Rhode Island, Dept of Chem. Engineering, Kingston, RI 02881

We all feel it every time we go to the gas pump, or pay an electric bill, or a home heating bill. Energy costs are continuously rising – and there seems to be no relief in sight. The global demand for energy has and will continue to grow at an unprecedented rate as the US becomes more addicted to oil and gas and as China and India become industrialized nations. We are all also aware of the impact that burning fossil fuels has on the environment – putting large amounts of carbon dioxide into the atmosphere and potentially contributing to global warming and climate change. This rather bleak picture has prompted renewed interest in energy conservation throughout the private and industrial sectors in the US since using less fuel has the two-fold advantage of reducing consumption and carbon dioxide emissions. One place where significant energy can be saved is in the chemical process industries where reaction and separation abound to produce all of the modern conveniences we now take for granted (e.g., the synthetic fibers, plastics, and polymers). Distillation is a widely used and versatile separation process that is used throughout the petroleum, chemical, pharmaceutical, and other industries to separate products, by-products, solvents, and raw materials. However, it is also a very large energy consumer with an estimated 40,000 distillation columns in the US that use approximately 24 % of the energy consumed by the manufacturing sector.

In this talk, I will give an overview of the recent unifying approach to energy consumption and energy efficiency proposed by Lucia and co-workers as it relates to distillation and, more generally, to multi-unit chemical processes that involve distillation. This novel methodology is based on differential geometry and the view that longest and shortest metrics (i.e., distances, surface areas, and volumes) associated with distillation paths give insight into the energy required for a given set of separation tasks. I will outline what can be demonstrated through example as well as what can actually be proved rigorously. I will show how this new approach encompasses all existing methods for determining minimum energy requirements for distillation – such as the McCabe-Thiele method, Underwood's method, boundary value methods, and so on. I will also show that it provides a rigorous and implementable way of finding energy efficient solutions that other methods cannot find – such as non-pinched minimum energy solutions. Throughout this talk, I will emphasize the facts that this new methodology can be applied to mixtures with any number of components, that it is applicable to multi-unit processes, and that it has a rigorous theoretical foundation rooted in differential geometry. Finally I will use many geometric illustrations to elucidate key features of this new approach to energy consumption in the synthesis and design of chemical processes.

168c Process Analysis Of Fully Thermally Coupled Distillation Columns Based On Experimental Pilot Plant Data And Dynamic Simulation

Gerit Niggemann and Georg Fieg. Process and Plant Engineering, Hamburg University of Technology, Schwarzenbergstr. 95c, Hamburg, Germany

Distillation remains the most commonly used separation process in the chemical industry. The separation of a homogenous multicomponent mixture into more than two products is usually realized in series of distillation columns, with the amount of energy used in especially these column arrangements being considerable. In that respect, the application of fully thermally coupled distillation columns, such as dividing-wall distillation columns, is a promising alternative to conventional distillation towers. Especially the aspect of getting high value-added products in merely one distillation column with simultaneous savings in both, energy and capital costs, attracts the investors of distillation plants. Furthermore, a growing demand for high-purity products in fine and commodity chemistry within the last years increased the interest of operating dividing-wall columns. Although several theoretical studies have been executed on dividing-wall columns, to the best of our knowledge, an extensive set of experimental data is still missing. In this contribution we present real pilot plant data of a dividing-wall distillation column in combination with simulation results obtained from a mathematical process model. For that purpose a lab scale dividing-wall distillation column has been recently setup in our laboratory. The separation of an industrially relevant multicomponent mixture in products of high purities is examined. Extensive and high quality measurement devices allow for a detailed analysis of the steady state and the unsteady state behaviour. In earlier studies a rigorous process model has been developed, which allows for a detailed analysis of the column startup, whereas startup implies the transition from ambient conditions to the steady state operting point. The model considers several features, which are characteristic for a dividing-wall distillation column, such as the self adjusting vapour split and the heat transfer across the dividing wall. The presentation focuses on the setup of the pilot plant and the experimental data, whereas the profiles of characteristic process variables are analysed and discussed. After introducing the rigorous process model, the simulation results are compared with the pilot plant data for validation purposes. The extensive validation considers the steady state, the dynamic and the startup characteristics of the dividing-wall column. This comprehensive validation is indispensable, as a model-based design of startup strategies demands for a reliable process model. Finally, some results of practically feasible startup strategies will be presented, which enable large savings in startup time. The extensively validated process model is a promising basis for developing time-optimal startup strategies and process control concepts in the near future. Successfully designed control approaches, which are tested by simulation, will be implemented and applied at our pilot plant. These examinations will be the subject of further conference contributions.

168d A Rate-Based Model Approach To Internally Heat Integrated Distillation Columns (Hidic)

Koichi Iwakabe[1], Keigo Matsuda[2], Masaru Nakaiwa[3], Kinpei Horiuchi[4], Toshinari Nakanishi[5], Akio Horiguchi[6], Kazuya Kubo[6], Takao Ohmori[1], Takuji Yamamoto[1], and Sho Kataoka[1]. (1) Energy-Efficient Chemical Systems Group, Research Institute for Innovation in Sustainable Chemistry, National Institute of Advanced Industrial Science and Technology (AIST), Japan, AIST Tsukuba Central 5, 1-1-1, Higashi, Tsukuba-shi, Ibaraki, 305-8565, Japan, (2) Department of Chemistry and Chemical Engineering, Graduate School of Science and Engineering, Yamagata University, 4-3-16, Jonan, Yonezawa, 992-8510, Japan, (3) Research Institute for Innovation in Sustainable Chemistry, National Institute of Advanced Industrial Science and Technology (AIST), Japan, AIST Tsukuba Central 5, 1-1-1, Higashi, Tsukuba-shi, Ibaraki, 305-8565, Japan, (4) Project Development Department, Maruzen Petrochemical Co., Ltd., 3, Goi, Minami-kaigan, Ichihara-shi, Chiba, 290-8503, Japan, (5) Kimura Chemical Plants Co. Ltd., 2-1-2, Terajima Kuise, Amagasaki-shi,, Hyogo, 660-8567, Japan, (6) R&D Division, Mitsubishi Chemical Group Science and Technology Research Center, Inc., 1, Tohocho, Yokkaichi-shi, Mie, 510-8530, Japan

Distillation is the most widely-used but the most energy-consuming separation process in chemical industries. The internally Heat-Integrated Distillation Columns (HIDiC) are one of the most promising technologies for the energy saving of conventional distillation processes. HIDiCs have similar structure to a heat exchanger. The rectifying section of HIDiCs is contacted with the stripping section through walls for heat exchange between these two sections. If temperatures of the rectifying section are higher than those of the stripping section, the residual heats in the rectifying section can be utilized in the stripping section. If the coupling of these two sections is appropriate, it will work like well-controlled side heaters/coolers. The HIDiC technology attracts a lot of industrial and academic interests since the energy savings performance of the HIDiC has been proved by the pilot plant in Japan. Although a lot of operation data of the HIDiC pilot plant have been already obtained, the mass and heat transfer phenomena in the HIDiC are still unclear. In order to analyze the phenomena, a simulator with rate-based model has been developed for HIDiCs. Simulation results showed that the separation performance of the HIDiC is affected by the internal heat exchange. The results also suggested that the internal heat exchange might suppress the mass transfer. Here we report significant outcomes from the study.

168e Synthesis Of Optimal Distillation Networks

Arun Giridhar, Chemical Engineering, Purdue University, 480 Stadium Mall Drive, West Lafayette, IN 47907 and **Rakesh Agrawal**, School of Chemical Engineering and Energy Center at Discovery Park, Purdue University, 480 Stadium Mall Drive, West Lafayette, IN 47907.

The synthesis of optimal distillation networks for a given application is a challenging problem in process design. One difficulty is that the space of all possible configurations for a given multicomponent feed is combinatorially large, with a large fraction of these configurations using more columns than necessary, without compensating with better energy consumption. Our prior work in this area classifies and reduces the search space to a set of basic configurations, all of which use (n-1) columns for a given n-component feed. We have shown that most of these basic configurations are more efficient than the sharp-split configurations commonly used in industry, when considered on the basis of first law heat duty alone. A source of design complexity still remains unsolved in that while the overall heat duty can be reduced for a distillation network, more heating or cooling might be needed at temperatures farther removed from ambient temperatures than in the original network, and therefore be more expensive on a second law basis. In this work, we present a framework to deal with the effect of utility temperatures and heat duties on the optimal distillation configuration. We present case studies where a large saving in overall heat duty was achieved, but the new design required significantly more cooling at the lowest temperature, affecting the cost. We show how this can be taken into account both by minimizing a cost function that takes the nature of the utility into account, as well as by calculating the thermodynamic exergy loss. We present ways to maintain the saving in heat duty while reducing the exergy loss.

168f The Attainable Region For Binary Distillation Processes

Diane Hildebrandt[1], David Glasser[1], and Germaine Ntunka[2]. (1) Centre of Materials and Process Synthesis,, Wits University, Private Bag x3, PO Wits, Johannesburg, 2050, Johannesburg, 2050, South Africa, (2) COMPS, Wits University, Private Bag x3, PO Wits, Johannesburg, 2050, Johannesburg, 2050, South Africa

The Attainable Region (AR) was defined by Horn for reactor systems. The definition can however be broadened to include distillation processes. In this context we define the Attainable Region for a distillation column as the set of all possible outputs (and corresponding inputs) for a given feed, distillate composition and VLE thermodynamic model. The fundamental processes considered are VLE based separation and mixing.

We consider the variables of the AR to be composition and a variable that we call "kappa" that arises naturally from the system equations and is analogous to residence time in reacting systems. The kappa variable includes energy and number of stages in a single variable.

We will develop the theory and demonstrate it by considering various examples. We will show that we can find the AR for binary disitllation systems and most importantly that both the optimal structure and operating conditions come out from the solution of the AR. This allows us to determine targets for performance to which we can compare conventional columns to see what room for improvement there is in the performance of these. Furthermore, the solution also gives us insight into the way in which the operation of these conventional systems should be changed in order to get closer to the target performance.

188a Use of Free-Solution Conjugate Electrophoresis (Fsce) as a Highly Sensitive Method to Differentiate Genetically Engineered Protein Polymers with Slight Differences in Length or Charge Distribution

Jennifer S. Lin[1], Jennifer A. Coyne[1], Xiaoxiao Wang[2], and Annelise E. Barron[1]. (1) Chemical and Biological Engineering, Northwestern University, 2145 Sheridan Road, Evanston, IL 60208, (2) Department of Biochemistry, Molecular Biology & Cell Biology, Northwestern University, 2205 Tech Drive, Hogan 2-100, Evanston, IL 60208

We have used genetic engineering to create a series of non-natural, repetitive polypeptides ("protein polymers") that are water-soluble and adopt random-coil conformations in solution. These protein polymers were produced by expression in E. coli bacteria, with molecular weights ranging from 9 kDa to 36 kDa and with designed sequences consisting of Gly, Ala, Thr, Ser, and positively charged Arg amino acids.

Here we demonstrate that Free-Solution Conjugate Electrophoresis (FSCE) in fused silica capillaries is a highly sensitive technique that can discriminate between protein polymers that have only slight differences in length and charge. Commonly used methods to assess protein purity after expression, such as RP-HPLC, MALDI-TOF, and SDS-PAGE have not shown sufficient resolution to reliably separate these protein polymers. In FSCE, a protein polymer is conjugated to a DNA oligonucleotide and analyzed by free-solution electrophoresis. The hydrodynamic drag of the protein polymer causes a mobility shift in the migration velocity of the conjugated DNA compared to free DNA. Polydisperse samples are easily profiled, with high resolution, using this method.

The same proteins were expressed in a vector containing an N-terminal decahistidine tag for affinity purification and a separate expression vector encoding a C-terminal octahistidine tag. Although all proteins appeared pure using the standard techniques mentioned above, we show that using FSCE we can identify multiple truncated protein products that resulted from incomplete expression of the full-length protein when expressed with only the N-terminal affinity tag. We also demonstrate that FSCE is able to distinguish between protein polymers with identical overall charges but with the charges in different positions along the chain, due to "end-effects" lending more weight to charges at the ends of the protein.

188b Modeling A Preparative-Scale Dynamic Field Gradient Focusing Instrument

Noah I. Tracy and Cornelius F. Ivory. Chemical Engineering, Washington State University, PO Box 642710, Pullman, WA 99164

Separating biologically active polymers often proves difficult due to the similarity between polymers with different biological activities. The small differences in charge and or size between polymers can give them different electrophoretic mobilities. This difference in mobility serves as a means for separation using dynamic field gradient focusing (DFGF). Separation and concentration by DFGF occurs in free-solution within a vertical, annular separation chamber filled with a low-conductivity buffer. Unlike isoelectric focusing, no pH gradient or carrier ampholytes are used in the separation.

Dynamic field gradient focusing uses two opposing forces to focus a molecule. One force, due to an electric field, is proportional to the molecule's electrophoretic mobility and decreases in intensity as the molecule moves down the chamber. Pumping buffer into the bottom of the separation chamber pushes the molecule up the chamber in opposition to the force of the electric field. The molecule will then focus at the height in the chamber where the opposing forces are equivalent. Thus, DFGF separates molecules by electrophoretic mobility and concentrates them in one step.

We use COMSOL Multiphysics, a finite element differential equation solving package, to numerically solve equations for mass and current conservation. A 2D axially symmetric mathematical model of the separation process in the preparative DFGF instrument confirms prior experimental results using Hb and BSA. The model results indicate that the preparative DFGF instrument functions as designed. The model also serves as a guide for optimizing the instrument.

188c Identification Of Conformation Differences Between Protein Isoforms For The Development Of Resolute Separation Methods

Madhukar Patel[1], Joshua Waitzman[1], Matthew B. Kerby[1], Anuj Chauhan[2], and Anubhav Tripathi[1]. (1) Engineering, Brown University, Providence, RI 02912, (2) University of Florida, 419, ChE Bldg., Gainesville, FL 32611

Development of approaches to separate protein isoforms is propelled by the central role that these macromolecules have been found to have in the pathological states of disease and mutations. Protein isoforms are protein molecules that vary from each other due to differences in a small number of amino acid residues. Since isoforms are alike in their make-up, they often have similar physical characteristics that make them very difficult to identify and separate. The proposed approach aims to perform separation of protein isoforms under conditions where separation of close mobility molecules can be made extremely resolute. Such conditions will provide an environment in which differences in electrophoretic mobility and/or adsorption kinetics become more discernable and exploitable.

Surface Characterization of Blended Chitosan- Poly(Vinyl Alcohol) Membranes Using Atomic Force Microscopy

MN Hyder, RYM Huang and P. Chen
Department of Chemical Engineering, University of Waterloo, Waterloo, ON, Canada, N2L 3G1

Abstract

In Pervaporation, high permeability and selectivity are always desirable for successful separation. Polymers have a variety of intrinsic chemical, physical, mechanical and morphological properties. By blending of two different polymers, the advantages of each polymer can be combined for newer and improved materials for separation [1]. Now a day's researchers have focused on blended membrane of different polymers for their improved film forming ability, high mechanical strength and chemical resistance and excellent separation performance [2]. In this study, blended membrane of poly(vinyl alcohol) and Chitosan were prepared in different composition and characterized using Atomic Force Microscopy (AFM) to study the physiochemical structure of the membranes and their effect on pervaporation. Tapping mode AFM surface scans showed that the membrane surfaces were rough on a nanometer scale. Water contact angle measurements were also performed to determine the surface hydrophilicity of the membranes. Compared to non-crosslinked membranes, the crosslinked membranes were shown to change in chemical structure by FTIR and become less hydrophilic by water contact angle measurement. Hydrophilicity of the blended membranes did not change significantly compared to pure poly (vinyl alcohol) membrane or pure Chitosan membrane. Dehydration of isopropanol-water mixtures were conducted over a range of isopropanol concentrations (40-95 wt%) in the feed solution. The present study shows that compared to the total flux, the selectivity is significantly affected by the blending of the two polymers. The physiochemical results were then quantitavely correlated to the separation performance.

The blended membranes were crosslinked with glutaraldehyde and characterized by FTIR and DSC. Results from the FTIR shows the reduction in –OH groups and appearance of –COC groups after crosslinking. Results of the thermal scans from DSC experiment shows the melting point of the blended membrane is in between the melting points of poly(vinyl alcohol) and chitosan.

Scanning electron micrographs of the blended membrane shows a thick and dense structure without any pores. Atomic Force Microscope (AFM) was used to determine the topography of the composite membrane surfaces [3]. After image acquisition, roughness of the membrane surface was determined by a program in the AFM image processing toolbox [3-4].

Figure 1 shows the 3D and top scan AFM images of the surface of a blended membrane over a scan area of $6 \times 6\ \mu m^2$. The membrane surface is rough on a nanometer scale. The roughness slightly increases as the amount of chitosan increases in the blend. PVA itself has a surface roughness of 8 nm and chitosan to be 30 nm. In the blended membrane, the highest surface roughness of 80 nm was found for CS-PVA (75:25) composition. These rough surfaces will be in contact with the feed solution during the pervaporation process. The hydrophobicity of the rough PVA surface was quantified using contact angle measurement.

To evaluate the hydrophobicity of the membrane surface, sessile drops of water were used to measure the static contact angle (θ)[3]. The fresh (not crosslinked) CS-PVA composite membrane shows a static contact angle of 22.7^0. Crosslinking of PVA increases the contact angle making them less hydrophilic. Crosslinked blended membranes with higher amount of Chitosan content show higher contact angles and therefore less hydrophobic, e.g., 61^0 on the CS-PVA(75:25), 47^0 on the CS-PVA(50:50) and 43^0 for CS-PVA(25:75).

The total flux and selectivity were increased after the blending the polymers. Among the blended membranes, CS-PVA(75:25) shows the highest flux and selectivity of 0.86 kg/m^2/hr and 412 respectively. The surface roughness and hydrophobicity were also found to be highest for this particular membrane.

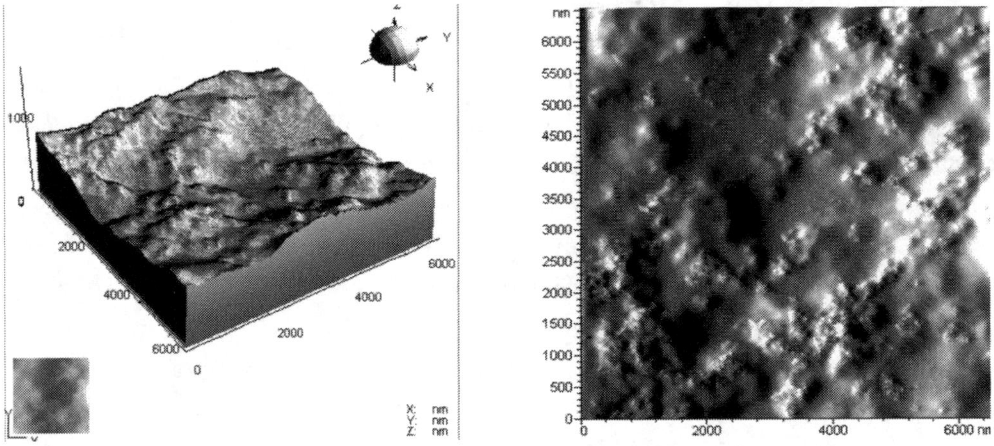

Figure 1: 3D and topographic surface scan of CS-PVA blended membrane.

References

[1] Huang, R. Y. M., *Pervaporation Membrane Separation Processes*, Elsevier, Amsterdam, (1991)

[2] Smitha B, Suhanya D, Sridhar S and Ramakrishna M, *Separation of organic-organic mixtures by pervaporation-a review*, J Membrane Sci, **241**, 1 (2004).

[3] Hyder M.N., Huang R.Y.M. and Chen P., *Correlation of physicochemical characteristics with pervaporation performance of poly(vinyl alcohol) membranes,* J Membrane Sci., **283,** 281 (2006).

[4] Zeng, Z., Xiao X., Gui Z. and Li, L., *AFM study on surface morphology of* $Al2O_3$-SiO_2-TiO_2 *composite ceramic membranes*, J. Membrane Sci., **136,** 156 (1997).

189b Synthesis & Characterization Of Thin Film Metal Oxide Membranes

Kyle P. Kostroski[1], **Makonnen M. Payne**[2], Benjamin A. McCool[3], David C. Calabro[3], Harry W. Deckman[3], and Andrew E. Riley[3]. (1) Chemical Engineering, Purdue University, Forney Hall of Chemical Engineering, 480 Stadium Mall Drive, West Lafayette, IN 47907, (2) Chemical Engineering, The City College of New York and The Graduate School and University Center of the City University of New York, Steinman Hall (T-305), 140th Street and Convent Avenue, New York, NY 10031, (3) ExxonMobil Research and Engineering, 1545 Route 22 East, Annandale, NJ 08801

Membrane-based separations are becoming increasingly more important in a variety of applications such as those in the biotechnology and energy production areas. As a result, the demand for high quality, defect-free membrane materials has grown, fostering a synergistic relationship between membrane synthesis and characterization. Within this relationship, membrane characterization has become a "feedback control," thereby providing vital information about membrane quality that can be incorporated back into membrane synthesis. In this work, a series of metal oxide membranes were synthesized and characterized in an effort to optimize the synthesis process and gauge the performance of these membranes for potential ultrafiltration applications.

The metal oxides were synthesized using sol-gel chemistry and the selective metal oxide layers were deposited on commercially available symmetric and asymmetric gamma-alumina supports via spin-coating. During the first-pass synthesis process, gas permeation testing results showed both Knudsen and Poiseuille flow contributions to overall membrane permeance. This indicated the presence of defects within the metal oxide active layer. After several successive depositions of additional metal oxide layers on the gamma-alumina support, the Poiseuille flow contribution to total membrane permeance decreased. In general, the membrane effective pore size seemed to approach its nominal pore size as the number of metal oxide layer depositions increased. These results indicate defect healing with successive layer addition. Transient porosimetry analysis appeared to agree with these findings, indicating that the average pore size decreased with the addition of successive metal oxide layers. The porosimetry results also showed some narrowing of the pore size distribution with increasing number of metal oxide layers.

Overall, the deposition of additional active layers has proven to be effective for healing defects and reducing defect flow contributions to total membrane permeance. Further, membrane characterization proved its worth as a feedback control: the information garnered from gas permeation testing and transient porosimetry analysis proved to be instrumental for the future synthesis of quality membrane materials.

189c Multi-Stage Reactive Separations Using Three-Dimensional Porous Catalytic Polymer Materials

Jung Hun Song, Department of Chemical Engineering, Graduate Center and City College of City University of New York, 140th St. at Convent Ave., New York, NY 10031 and Ilona Kretzschmar, Department of Chemical Engineering, The City College of City University of New York, 140th St. at Convent Ave., New York, NY 10031.

Three-dimensional porous catalytic polymer materials (PCPMs) have enormous potential for the field of reactive separations. Controlled pore sizes, porosity, and selective catalyst material deposition in PPMs have shown to be essential characteristics in the field.[1] Our work is concerned with the fabrication of three-dimensionally periodic PPMs that have well-defined sections with different pore sizes and different catalyst materials using colloidal templating.

Catalyst coated colloids (CCCs) of different sizes are convectively assembled inside capillaries. A liquid polymer is used to fill the interstitial spaces formed by the colloidal assembly. Subsequently, the polymer is cured to form a solid CCC-polymer matrix. An inverse opal structure forms upon dissolution of the colloids from the matrix, leaving behind a catalyst embedded multi-sectioned PCPM. We have assembled 0.5, 2.4, and 9.6 μm sized sulfate-polystyrene (PS) colloids in 50 μm inner-diameter polymethylmethacrylate (PMMA) capillaries with various sequences. The hexagonally close-packed crystals formed have been infiltrated by a UV curable prepolymer, which has been cured subsequently under a long-wave UV light (365 nm). Treatment of the cured CCC-polymer matrix with organic solvents leads to porous catalytic polymeric materials with a controlled catalytic material and pore size distribution.

We have investigated the structure of the colloidal assemblies and the resulting PCPMs using scanning electron microscopy and EDAX elemental analysis. Further, knowledge of properties and characteristics of these PPMs are critical in practical applications, therefore, a preliminary tensile strength analysis will be also presented.

(1) Ulbricht, M.; Advanced functional polymer membranes. Polymer 2006, 47, (7), 2217-2262.

189d Membrane Fouling, Surface Characterization, And Transport Models In Water Treatment Applications

Shih-Chieh Tu[1], Varadarajan Ravindran[1], Wonho Song[2], and **Massoud Pirbazari**[1]. (1) University of Southern California, Department of Civil and Environmental Engineering, Los Angeles, CA 90089-2531, (2) Civil and Environmental Engineering, University of Southern California, Los Angeles, CA 90089-2531

Membrane processes in water treatment often experience permeate flux decline due to membrane fouling attributed to natural organic matter (NOM). Membrane transport models were developed for prediction and simulation of permeate flux, and two different models were compared for ultrafiltration and nanofiltration membranes. The first model essentially considers surface fouling due to deposition of organic foulants. The second model involves both surface sorption and pore diffusion to address organic fouling due to surface deposition and internal pore blocking. In these studies, tannic acid and dextran are used as model compounds to simulate organic fouling. Additionally, a real source water is employed for model calibration and verification.

The membrane transport models considered resistances due to membrane film, concentration polarization, gel formation, and internal pore fouling. Membrane filtration tests were conducted to evaluate permeate flux decline patterns due to organic fouling, as well as solute concentration profiles in the permeate and concentrate streams. The pore diffusion model exhibited good predictive capability for permeate fluxes and solute rejections for both nanofiltration and ultrafiltration membranes. The pore diffusion model (second model) provided better predictions was superior to the surface fouling model (first model) in predicting the permeate fluxes for the ultrafiltration membranes. However, no significant differences in permeate flux predictions between the two models were observed for nanofiltration membranes.

Model simulation and sensitivity studies were conducted to evaluate the effects of various input parameters pertaining to operating conditions and fluid-dynamic regimes. These parameters include the mass transfer coefficient, solution dynamic viscosity, membrane resistance, and resistance per unit gel layer thickness. These studies provide a well-founded theory based on the dependence of transport resistance components and membrane performance on input parameters. The studies illustrate the boundary layer effect on gel layer formation and the overall mass transfer resistance.

Furthermore, the study used several surface characterization techniques on virgin, fouled, and cleaned membranes under different conditions, to evaluate membrane fouling potentials and to elucidate membrane-solution and membrane-foulant interactions. This analysis involved membrane surface charge characteristics, functional group properties, hydrophobicity, foulant affinity, micro-morphology, micro-topography, and foulant compositions. Routine evaluations such as dynamic adsorption tests, zeta-potential measurements, and scanning electron microscopy were well supported by atomic force microscopy (tapping mode), X-ray photoelectron spectroscopy, and Fourier transform infrared (FTIR) spectroscopy. Membrane filtration tests were conducted by varying operating parameters such as solute concentrations, trans-membrane pressures, and reject flow rates. These tests showed that the nanofiltration membrane composed of polypiperazine amide and the ultrafiltration membrane composed of polyether sulfone were more susceptible to organic fouling by tannic acid than the nanofiltration membrane made of cross-linked aromatic polyamide. These observations were well supported by surface potential measurements that demonstrated higher negative surface charges and greater hydrophilic property for the aromatic polyamide membrane.

Key Words: Membrane transport model, fouling potential, membrane surface characterization techniques, pore diffusion, internal pore fouling; surface fouling, concentration polarization, gel formation, ultrafiltration; nanofiltration

189e The Impact Of Surface Topology On Scaling Propensity Of Polyamide Membranes

Hsiao-yu Nancy Lin[1], Gregory T. Lewis[1], Wen-Yi Shih[2], and Yoram Cohen[1]. (1) Chemical and Biomolecular Engineering, University of California, Los Angeles, 5531 Boelter Hall, Los Angeles, CA 90095, (2) University of California at Los Angeles, Dept. of Chemical Engineering, Los Angeles, CA 90095-1590

Membrane mineral salt scaling is one of the major factors that limit recovery in reverse osmosis desalination of brackish groundwater. Mineral salt scaling leads to flux decline and shortening of membrane useful life. The primary mineral salt scalants that are typically of concern in inland water desalination are calcium carbonate, calcium sulfate and barium sulfate, typically in the calcite, gypsum and barite forms, respectively. In order to develop predictive models of mineral scale development on RO membranes, there is a need to first assess the importance of membrane topology and surface chemistry on mineral salt scaling. Such information could also be beneficial to developing optimal criteria for designing RO/NF membranes of low scaling propensity. Accordingly, a systematic surface crystallization study was carried out, with gypsum as the model scalant, to evaluate the kinetics of surface crystallization on polyamide membranes of different surface topologies. The study was carried out using a quartz crystal microbalance (QCM) with polyamide membranes that were interfacially polymerized onto the surface of quartz crystals. With the developed methodology one can create differing surface topologies for the same polyamide membrane surface chemistry, thereby enabling isolation of the impact of membrane topology on surface crystallization. To examine the impact the surface topology, membrane surface roughness was varied through various pH adjustments of the underlying polyelectrolyte adhesion layer. The kinetics of surface crystallization was then studied in a QCM flow cell in which the accumulation of surface mineral mass was continuously monitored. The study was carried out with supersaturated feed solutions (with respect to gypsum), prepared by mixing the ionic ingredients just prior to delivery to the QCM flow cell in order to avoid bulk crystallization in a standing feed solution. The feed solution (in all experiments) was maintained at a supersautarion index of ~1.5 for which the bulk crystallization induction time was of the order of 9 hr. Therefore, bulk crystallization was avoided in this system given the much shorter convective residence time of the solution in the system (< 1 min), relative to the bulk crystallization induction time. As a consequence, crystallization was restricted to surface crystallization in the QCM flow cell. Imaging of the QCM surfaces by SEM and AFM provided detailed information on the surface crystal number density, crystal size and morphology. This information was used to examine the correlation between surface topology and gypsum surface crystallization kinetics. The kinetics of surface crystallization (i.e., time evolution of gypsum surface scale) was also analyzed using a single crystal growth model along with a population growth model. This information provided, for the first time, a direct measure of surface crystallization kinetics on RO membranes. The results of the study demonstrated that surface topology can have a significant impact on surface crystallization suggesting that there is merit in developing approaches of optimization membrane surface topology so as to reduce scaling propensity.

Effect of Gas Temperature on Integrity Testing: A Comprehensive Study of Gas Temperature inside Filter Housing during Pressurization and Depressurization

Jibin Hao, Bioprocess R&D, Millipore Corporation, Bedford, MA

Abstract

An automatic integrity tester is routinely used in pharmaceutical production to test the integrity of sterilizing filters. Among other factors, test accuracy is affected by gas temperature changes on the upstream side of the filter. During the integrity test, the upstream gas temperature changes considerably when pressurizing or depressurizing the gas in the filter housing, which is part of the integrity test protocol.

Thermocouples were placed inside a filter housing to monitor the gas temperature changes during pressurization and depressurization. It was found that the air temperature inside the filter housing increased from 22°C to over 70°C when pressurized from 0 to 60 psig. The air temperature inside the housing dropped from 22°C to -45°C when depressurized from 56 to 0 psig. The theory of gas temperature changing with pressure was studied and the experimental values of gas temperature changes were well within the limits projected from adiabatic models. Understanding the correlation between pressure and instantaneous temperature changes is critical for the design of automated test equipment.

Introduction

In sterile filtration, the regulatory agencies require the integrity of membrane filters to be tested before and after their use. The test procedure typically includes three steps: loading filter(s) into the filter housing; wetting the filter(s) with manufacturer's suggested liquid; and pressurizing the upstream side of the filter(s) with the test gas and measuring the integrity properties of the filter(s), such as the diffusion flow rate and/or bubble point. Ideally the gas diffusion flow rate and bubble point should be measured at the downstream side of the filter(s).Currently all the commercial automatic integrity testers measure the gas flow rate or/and bubble point at the upstream side of the filter(s). This practice eliminates the potential contamination to the filtration products. But it may lead to inaccuracy of the measurement, that is, the discrepancy between upstream and downstream measurements.

To accurately measure the gas flow rate at the upstream side of the filter(s) is a challenging task. This is because the upstream gas temperature is not steady and thus leads to fluctuation in the upstream flow rate. For example, integrity testing of a 10", 0.1μm Millipore filter (Catalogue Number: CVVL71TP3) is typically conducted at 56 psig with 1.8 liter upstream volume. According to ideal gas law, a change of upstream air temperature at 0.1°C/min will result in a 2.7 ml/min upstream gas flow rate fluctuation. This flow rate fluctuation is large in comparison with 10 ml/min, which is the typical air diffusion flow rate of the filter. Upstream gas temperature has a considerable impact on the accuracy of commercial integrity testers.

The key issue to integrity testing is to keep the upstream gas temperature constant. This is not easy because there are three temperatures interacting with one another during integrity testing, that is, upstream gas temperature, wetting liquid temperature, and environment temperature where the filter housing is located. If these three temperatures are not the same, upstream gas temperature will change with time. This temperature change has almost no impact on the gas flow rate at the downstream but affects the upstream gas flow rate significantly. Only when these three temperatures are the same, is the gas flow rate at upstream equal to that at downstream. The environmental temperature and the wetting liquid temperature can be addressed in the integrity test procedure but the gas temperature in the filter housing can not be controlled. This is because the upstream gas temperature will increase whenever there is an upstream pressure increment, which is a part of the integrity test protocol. The objective of this study is to understand, experimentally and theoretically, the impact of pressure change on the upstream gas temperature in integrity testing.

Theory

Pressurization

Figure 1 is a schematic diagram for pressurizing a filter housing. The tank volume of the gas source is much larger than the volume of the filter housing and thus the pressure and temperature in the gas source can be treated as constant. Theoretically, the pressurization process can be assumed as adiabatic, but not isentropic (reversible adiabatic). As shown in the Appendix, the gas temperature in the filter housing for adiabatic process can be expressed as follows [1-4].

$$T = \frac{r}{\left(1-\frac{p_0}{p}\left(1-r\frac{T_s}{T_0}\right)\right)}T_s \qquad (1)$$

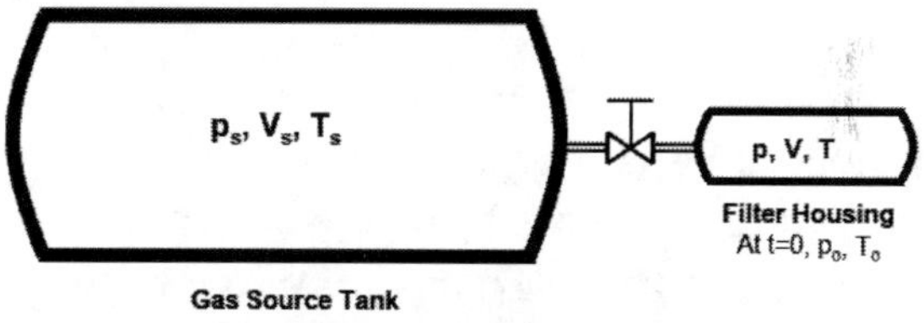

Figure 1. Schematic diagram of pressurizing a filter housing from a high pressure source tank
Assuming $V_s >> V_h$ and constant p_s and T_s

Equation 1 indicates that gas temperature in the filter housing is dependent on the temperature of the source gas but not its pressure. It also shows that the gas temperature in the filter housing is dependent on the ratio of the initial and final gas pressure in the housing (p_0/p).

Depressurization

Figure 2 is a pressurized housing with a tube connected to the environment. If there is no heat exchange between the housing and environment and the tube is frictionless, the depressurization can be treated as isentropic (reversible adiabatic) process. Thus we have [1]:

$$\frac{p}{\rho^r} = \frac{p_0}{\rho_0^r} \qquad (2)$$

By combining above equation with ideal gas law, the gas temperature in the housing can be written as follows.

$$T = \left(\frac{p}{p_0}\right)^{\frac{r-1}{r}} T_0 \qquad (3)$$

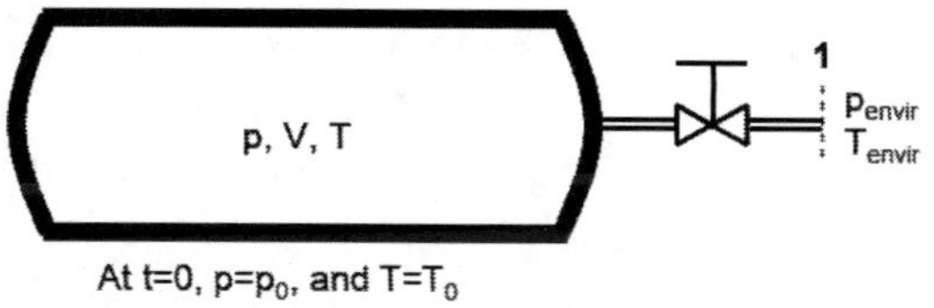

Figure 2. Schematic diagram of depressurizing a high pressure housing to environment

If $\frac{p_{envir}}{p} < \left(\frac{r+1}{2}\right)^{r/(1-r)}$, the gas flow inside the tube is choked flow [1-4] and its velocity reaches the speed of sound , which is $\sqrt{\frac{rRT_1}{M}}$. At this condition, the gas pressure at surface 1 is different from environment pressure and is expressed as $p_1 = \left(\frac{r+1}{2}\right)^{r/(1-r)} p$ ($p_1 \approx 0.53p$ for air).

Experimental

In this study, the gas temperatures in the filter housing are measured constantly while the filter housing is pressurized and depressurized. The pressure and temperature data are recorded with the aid of a data acquisition system. As shown in Figure 3, the equipment used in the test consisted of a 10" Millipore filter housing (thermocouples inside), a pressure transducer Setra-370 (Setra Systems, Boxborough, MA), DaqBook/260 installed with DBK 19 card (IOtech, Cleveland, OH), and Laptop computer loaded with DASYLAB software (DASYTEC USA, Amherst, NH). The pressure transducer and DaqBook/260 (converting thermocouple's voltage signal to temperature) were calibrated prior to the experiment.

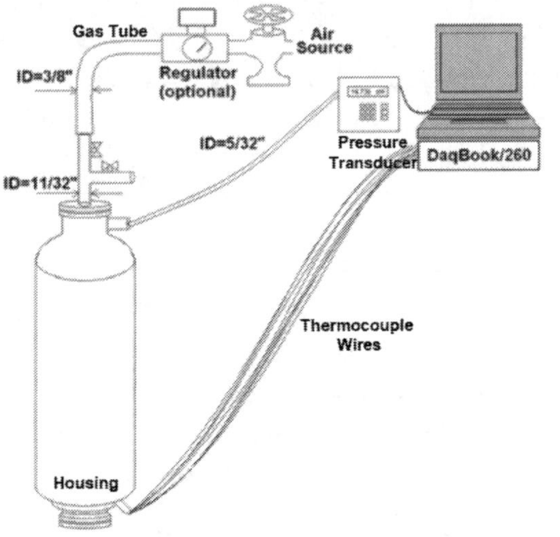

Figure 3. Experiment set-up for pressurization and depressurization

As shown in Figure 4, ten J-type thermocouples (Omega Engineering Inc, Stamford, CT), five 30 gauge (0.25 mm in diameter) and five 40 gauge (0.076 mm in diameter), were used to measure the gas temperatures inside filter housing. The temperature response time of 40 gauge thermocouple is much faster than 30 gauge because of its smaller diameter. Prior to the experiment, the accuracy of these thermocouples was checked against two NIST-traceable glass thermometers (one ranging between -38 and 2 ^{0}C and the other between -1 and 101 ^{0}C) from -30 ^{0}C to 80 ^{0}C.

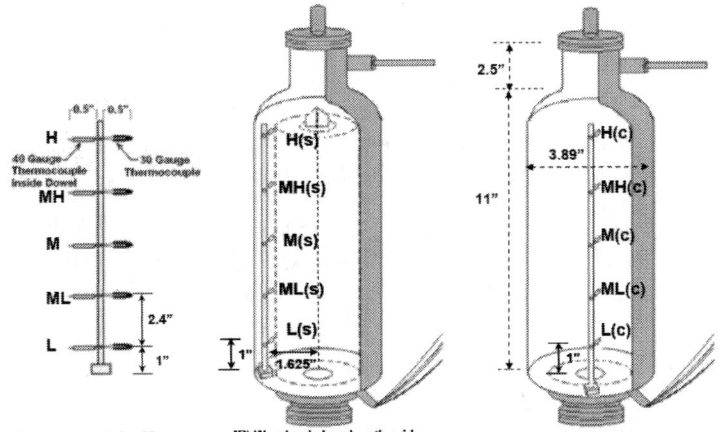

4A) Wood rod with five 30-gauge and five 40-gauge thermocouples

4B) Wood rod placed on the side of the base. The housing is either empty or loaded with 10" or 5" filter

4C) Wood rod placed on the center of the base of the empty housing

Figure 4. Schematic diagram of thermocouple locations with a 10" Millipore housing
Symbols in Figure: L=low; M=Middle; H=high; (s)=side; (c)=center

The filter housing used in this study was a 10" Millipore housing (Catalogue number: H4117USAC), which consisted of a dome and a base. As shown in Figure 4, the thermocouples were evenly positioned on a wood rod of 3/16" diameter. The wood rod was placed either on the side of the filter housing (Figure 4B) or on the center of the base (Figure 4C). In the former case, the rod was at the center between the outer surface of the filter and inner surface of the dome, that is, 1.625" from the center of the base. The membrane filter shown in the Figure 4B was optional. The thermocouple wires were introduced to the DaqBook/260 through the hole located on base of the filter housing.

Results and Discussion

Gas Temperature in Housing during Pressurization

When pressurizing a filter housing, the gas temperature inside the housing increases, then decreases, and finally stabilizes at room temperature. Gas temperature inside the housing is not uniform and there is a temperature distribution across the filter housing, vertically and horizontally. These are discussed below in detail.

A. Gas Temperature Profile inside Housing during Pressurization

Figure 5 shows air temperature inside a 10" empty housing when the housing is pressurized from 0 psig to 60 psig. The gas temperatures, which were measured by the 30-guage and 40-gauge thermocouples at ML(c) position (see Figure 4C), are shown on the plots. As shown in Figure 5, air temperature in the housing increases sharply from room temperature (~22°C) to over 70°C when the compressed air is introduced to the housing. Then air temperature drops with time and will reach the room temperature eventually. How fast the temperature drops depends on the configuration of the test system, which will be discussed below in section C.

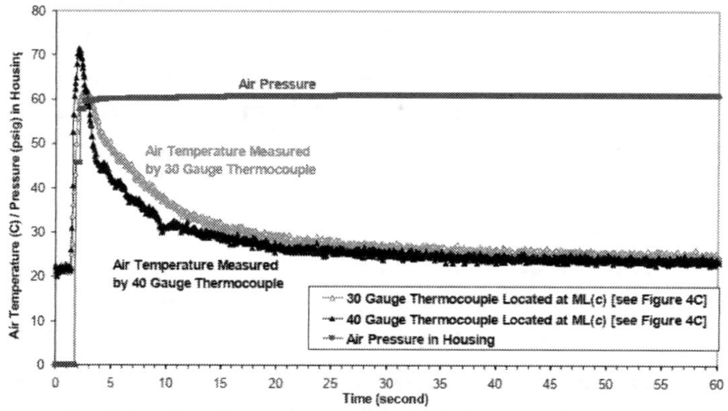

Figure 5. Comparison of air temperature profiles measured by 30 and 40 gauges thermocouples located at ML(c) [see Figure 4C] in 10" empty housing when pressurized from 0 to 60 psig

Figure 5 indicates that 40 gauge thermometer responds to temperature changes much faster than 30 gauge thermometer. For 40 gauge thermocouple, it reaches the peak temperature at 0.6 second. It takes 30-gauge thermocouple 1.3 second to reach its peak. Also the peak temperature measured by 40 gauge thermocouple is 71.4°C, which is much higher than 62.2°C measured by 30 gauge thermocouple. This is due to the fast temperature response of 40 gauge thermocouple. Selecting the right thermocouples is the key in measuring temperature changes during pressurization. It should be noted that temperatures measured by 40 gauge thermocouple are not necessarily the true temperature inside filter housing, but it is much closer than 30 gauge thermocouple.

B. Temperature Profile inside Housing

The temperature inside the filter housing is not uniform because heat transfers between the gas inside filter housing and environment. To better understand the temperature profiles inside the housing during pressurization, the wood rod mounted with thermocouples was first placed at the center of the empty housing (see Figure 4C) for the pressurization experiment. Then the wood rod was moved to the side (see Figure 4B) of the filter housing and the same experiment was repeated. Figure 6 shows the temperature profile measured by 40 gauge thermocouples across the empty housing when the filter housing is pressurized from 0 psig to 20 psig. All temperatures shown on Figure 6 are the peak temperatures recorded by the 40 gauge thermocouples during the pressurization.

As shown in Figure 6, air temperatures are lower at the bottom and top, and higher at middle. This temperature profile is anticipated because at the top and bottom heat transfers not only to the side but also the top or bottom of the filter housing. At the middle, heat mainly transfers to the side of the filter housing and thus the highest temperature is observed at somewhere between the top and bottom. Horizontally, the plots

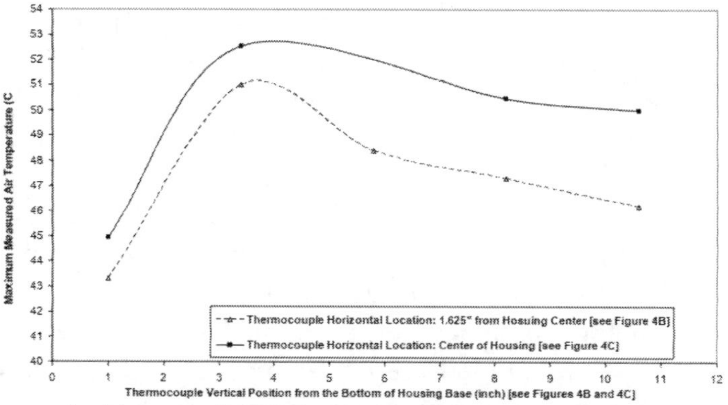

Figure 6. Gas temperature distribution in 10" empty housing when pressurized from atmosphere to 20 psig
Air temperatures were measured by 40 gauge thermocouples; initial air temperature in housing was ~22 °C

show that the gas temperature at the center of the housing is much higher than that at the side. This is expected because the direction of heat transfer is from the gas inside housing to the environment and it only transfers from high temperature to low temperature.

C. Effect of Wetted Filter on Temperature Profile

Figure 7 compares the gas temperatures inside the filter housing with three different configurations: filter housing loaded with 5", 10" or no filter. The filters (5" or 10") in the experiment were completely wetted with room temperature water and the temperatures in Figure 7 were measured with 40 gauge thermocouple located at H(s) position in the housing (see Figure 4B). As shown in Figure 7, the upstream gas temperature inside the empty housing reaches as high as 61°C. After 60 seconds, the gas temperature inside the housing is 24.5°C, about 2.5°C higher than the room temperature. For the filter housing loaded with 10" wetted filter, the observed peak temperature is 41.1°C, lowest among the three settings. This configuration takes the least amount of time, about 50 seconds, for the upstream gas to stabilize at room temperature. The temperature of 10" filter housing loaded with 5" filter, as expected, is in-between. The peak temperature during pressurization is 54.9°C. The gas temperature inside the housing is about 23.6°C after 60 seconds, which is still 1.6°C above room temperature.

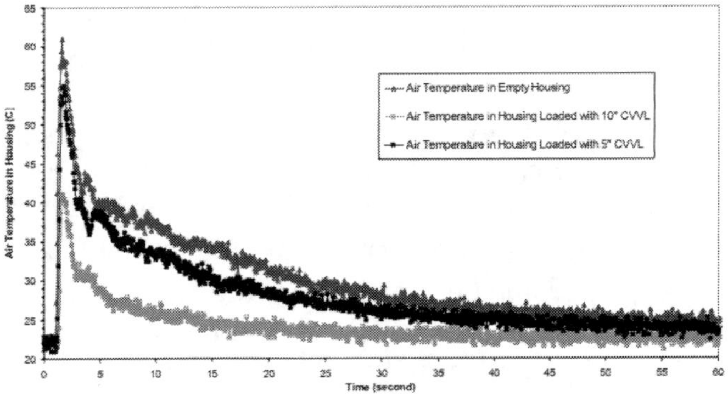

Figure 7. Effect of wetted filter on air temperature in 10" housing when pressurized from 0 psig to 56 psig
Air temperature was measured by 40 gauge thermocouple located at H(s) position [see Figure 4B]

Figure 8 shows the gas temperature profile inside the 10" filter housing loaded with 5", 10", or no filter. The temperatures in plots are the peak temperatures recorded by the 40-gauge thermocouples located at the side (see Figure 4B) of the housing. As shown in Figure 8, the gas temperature is highest for the empty housing. For the filter housing loaded with 5" or 10" filter, the temperatures are almost the same at the lower part of the filter housing. However, at the upper part of the filter housing the gas temperature for the filter housing loaded with 5" filter is much higher than that for the housing loaded with 10" filter. This is because 5" filter is only about the half the length of the 10" filter and thus the upper half of the filter housing is empty for this configuration.

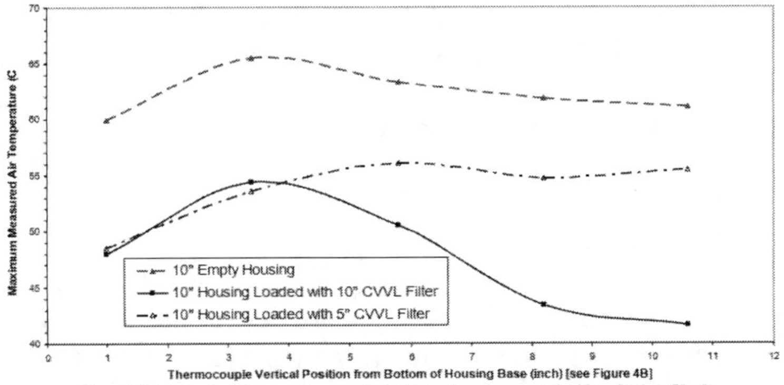

Figure 8. Effect of wetted filter on air temperature in 10" housing when pressurized from 0 psig to 56 psig
Air temperatures were measured by 40 gauge thermocouples located at the side the housing base [see Figure 4B]
Initial air temperature in the housing was ~22 °C

The wetted filter functions as a heat sink and absorbs the heat generated during the pressurization. Consequently gas temperatures inside the housing during the pressurization are largely dependent on the ratio of the wetted membrane area to upstream volume. Higher ratio of membrane area to upstream volume leads to faster heat transfer to the wetted membrane filter and thus takes less time for upstream gas temperature to reach stable (equilibrium) temperature. The membrane filtration systems at pharmaceutical production vary considerably and the ratio of the membrane area to upstream volume differs considerably from test system to test system. In testing the integrity of the membrane filters in these systems, the key is to adjust the stabilization time so that the upstream gas temperature is stable before measuring the upstream gas flow rate. If measuring the gas flow rate before the upstream temperature is stabilized, the measured upstream gas flow rate will be much different from the real flow rate at the downstream, regardless of what type of commercial integrity testers used in the test.

D. Comparison of Experimental Data and Theoretical Predication

Figure 9 shows the peak temperature recorded by 40 gauge thermocouple (located at ML(s) and ML(c), see Figure 4) in the empty housing and the projected temperature from the adiabatic model. As expected, the thermocouple temperature recorded at center is a little higher than that at the side.

The adiabatic model shown in Figure 9 is the ideal situation and assumes that there is no heat transfer between the gas and the environment through the empty filter housing. In reality, heat does transfer to the environment during the pressurization. Also the temperature response time of the thermocouples is not zero and thus the temperature measured by thermocouple is lower than the true temperature. Consequently the gas temperature recorded by the thermocouple inside the filter housing is lower than the temperature projected from the adiabatic model. The magnitude of the temperature difference between the adiabatic theory and experiment is largely dependent on the time to pressurize the housing to the desired pressure, which is controlled by the diameter of gas inlet tube. If the time to pressurize the housing is short, the gas temperature inside the housing is close to that projected from the adiabatic model. If the pressurization time is very long, the gas temperature inside the housing will be close to the environment temperature because of constant heat transfer.

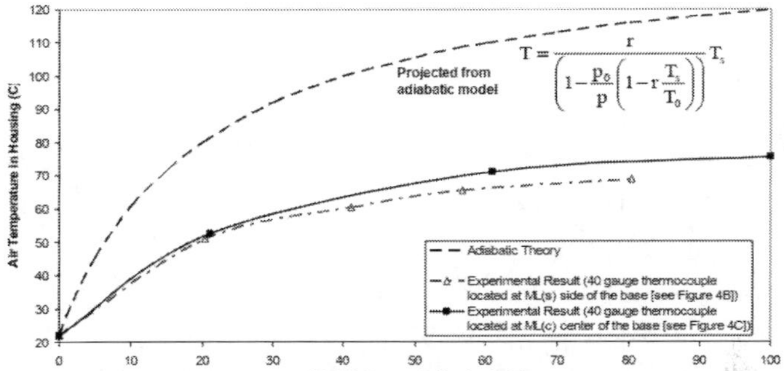

Figure 9. Air temperature in a 10" empty housing when pressurized from atmosphere
Initial air pressure (p_0) and temperature (T_0) in housing: 0 psig and ~22 °C;
Temperature of air source (T_s): ~22 °C r=Cp/Cv (r=1.4 for air)

Gas Temperature in Housing during Depressurization

The gas temperature in the housing drops very significantly during depressurization. Figure 10 shows the gas temperature profiles of 30- and 40-gauge thermocouples at H(c) position (see Figure 4C) when the gas pressure in the housing is depressurized from 56 psig to 0 psig. The initial gas temperature in the housing is at room temperature (~22°C). During the depressurization, the dip temperature observed from 40 gauge thermocouple in the housing is -45.9°C, which is significantly lower than -25.4°C observed by 30 gauge thermocouple. In addition, the dip temperature recorded by 40-gauge thermocouple is about 1 second earlier than 30-gauge thermocouple, 0.7 second for 40-gauge thermocouple vs. 1.7 seconds for 30-gauge thermocouple. As discussed before, this is due to the response time of 40-gauge thermocouple being much faster than that of 30-gauge thermocouple.

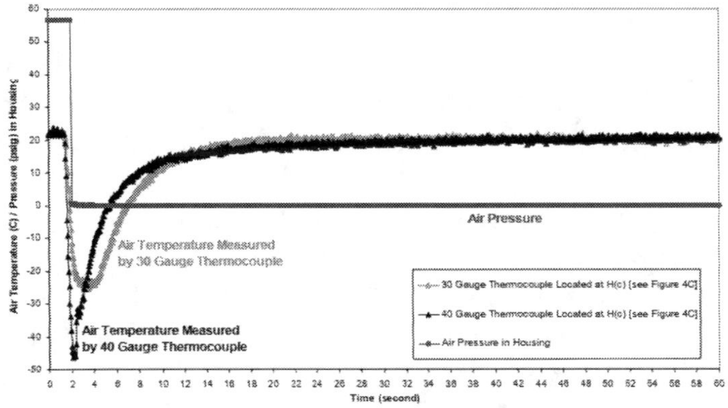

Figure 10. Comparison of air temperature profiles measured by 30 and 40 gauges themocouples located at H(c) [see Figure 4C] in 10" empty housing during depressurization

Figure 11 compares the observed dip temperatures by 40-gauge thermocouple in the empty housing with the projected temperature from isentropic (reversible adiabatic) model. In the experiment, it is found that the lowest temperature is observed at the top H(c) position (see Figure 4C) when the wood rod is at the center of the base plate. If the wood rod is placed at the side of the base plate, the lowest temperature is recorded at MH(s) (see Figure 5B).

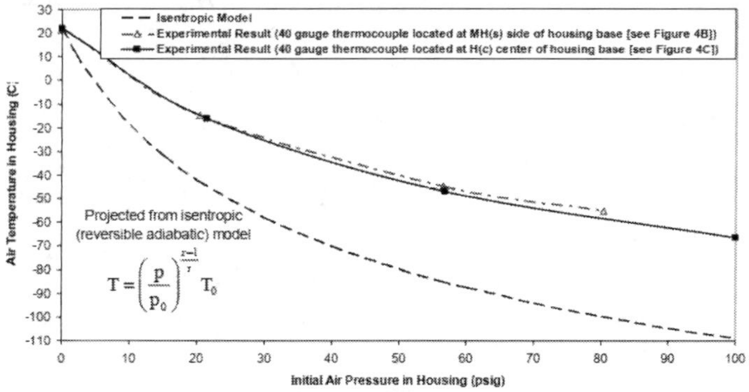

Figure 11. Air temperature in a 10" empty housing during depressurization to atmosphere
Final air pressure (p_f) and initial air temperature (T_0) in housing: 0 psig and ~22 ^{0}C
r=Cp/Cv (r=1.4 for air)

The plots in Figure 11 show that gas temperature observed by 40-gauge thermocouple in the 10" empty housing is much higher than that projected from the isentropic (reversible adiabatic) model. As we discussed above, this is because the housing is not adiabatic and there is heat transfer between the environment (through filter housing) and the gas in the housing. How close to the adiabatic model is mainly determined by how fast the gas is depressurized. The shorter it takes to depressurize, the closer it approaches to the adiabatic model. The plots of Figure 11 indicate that the gas temperature at the center is lower than that at the side. This is expected because of the distance difference to the outer surface of the housing.

Conclusions

In the integrity testing, the filter housing is gas pressurized as part of test protocols. The upstream gas temperature increases significantly during pressurization and then cools off with time. The temperature change has a great impact on the accuracy of upstream flow rate measurement, which is the typical method used in all commercial integrity testers. To get accurate upstream flow rate, it needs to wait until the upstream gas temperature is stable. The waiting (stabilization) time is largely dependent on the ratio of membrane area inside the housing to the upstream volume. At higher ratio of membrane area to the upstream volume, the time for upstream gas temperature to reach stable is shorter. At lower ratio of membrane area to the upstream volume, the opposite is true. The wetted membrane inside filter housing functions as a heat sink during the gas pressurization and it is the key to determine the stabilization time. Understanding how temperature affects the accuracy of integrity testing is fundamental in designing and operating the automatic integrity testers for pharmaceutical industry.

The gas temperature inside filter housing can be modeled as adiabatic during pressurization and isentropic (reversible adiabatic) during depressurization. In reality, the filter housing is not adiabatic and thus there is gap between theoretical projection and experimental data. This gap can be minimized if the pressurization or depressurization process is instant, that is, the time for pressurization and depressurization is approaching zero.

Acknowledgements

Jacob Andrews, who performed most of the tests in this study
Michael Joens, who helped to set up the Data Acquisition Systems

Symbols

p: gas pressure
ρ: gas density
T: gas temperature
r=Cp/Cv (Cp: heat capacity at constant pressure; Cv: heat capacity at constant volume)
t: time
M: molecular weight
ω: mass flow rate
Û: internal energy per unit of mass
Ĥ: enthalpy per unit of mass

References

1. R. Byron Bird, Warren E. Stewart, and Edwin N. Lightfoot (2002), "Transport Phenomena", second edition, New York, NY, John Wiley & Sons, Inc.
2. Patrick H. Oosthuizen and William E. Carscallen (1997), "Compressible Fluid Flow", New York, NY, McGraw-Hill Companies, Inc.
3. Michel A. Saad (1993), "Compressible Fluid Flow", second edition, Englewood Cliffs, New Jersey, Presentice Hall.
4. J. Craig Dutton and Robert E. Coverdill (1997), "Experimements to Study the Gaseous Discharge and Filling of Vessels", Int. J. Engng Ed. Vol. 13, No. 2, pp. 123-134.

Appendix

As shown in Figure 1, the gas temperature and pressure at the source (T_s and p_s) are assumed to be constant. The mass flow rate from gas source tank to filter housing can be written as follows.

$$\omega = \frac{d}{dt}(\rho V) \qquad \text{(A.1)}$$

If the pressurization is adiabatic and the kinetic energy in gas resource and filter housing is neglected, the energy equation can be written as

$$\frac{d}{dt}\left(\rho V \hat{U}\right) = \omega \hat{H}_s \qquad \text{(A.2)}$$

$$\hat{C}v\, T V \frac{d\rho}{dt} + \rho V \hat{C}v \frac{dT}{dt} = V \frac{d\rho}{dt} \hat{C}p\, T_s \qquad \text{(A.3)}$$

$$\frac{dT}{T - rT_s} = -\frac{d\rho}{\rho} \qquad \text{(A.4)}$$

$$\frac{T - rT_s}{T_0 - rT_s} = \frac{\rho_0}{\rho} \qquad \text{(A.5)}$$

Using ideal gas law to replace the density term in equation A.5, one has

$$T = \frac{r}{\left(1 - \frac{p_0}{p}\left(1 - r\frac{T_s}{T_0}\right)\right)} T_s \qquad \text{(A.6)}$$

197a A High Temperature Adsorption Process For Air Separation

Arvind Rajendran, School of Chemical and Biomedical Engineering, Nanyang Technological University, Nanyang Drive 62, Singapore, Singapore and **Shamsuzzaman Farooq**, National University of Singapore, 4 Engineering Drive 4, Singapore, 117576, Singapore.

With the current oil price at a record high, the importance of conserving non-renewable energy sources and harnessing energy from renewable alternatives cannot be over-emphasized. Most industrial heating processes generate required energy by combusting hydrocarbon fuel, such as oil or natural gas, with air as the oxidant. In many cases, there is room to enhance the combustion process by using either a pure or an enriched oxygen stream. High purity oxygen is also an essential requirement in many of the emerging technologies such as solid oxide fuel cells (SOFCs) and methane to syngas production. Hence, economics of oxygen generation technologies is intimately related to the optimum utilization of fossil fuel in an environmentally friendly manner. Perovskites, also known as ABO3 type mixed metal oxides (where A and B are exchangeable metal ions) are known to produce high degree of oxygen deficiency in their structure at high temperature. We have completed a preliminary investigation where it has been shown that the aforementioned perovskite property can be effectively utilized to develop a high temperature adsorbent for air separation with practically infinite oxygen selectivity. A couple of promising compositions have been identified, and equilibrium and kinetics of oxygen adsorption in these perovskite samples have characterized. Potential of these samples for air separation by pressure swing adsorption is currently being explored and the findings will be presented in the proposed presentation.

197b Novel Thermal Swing Sorption Enhanced Reaction Process For Simultaneous Production Of Fuel-Cell Grade H2 And Compressed CO2 From Synthesis Gas

Ki Bong Lee, Michael G. Beaver, Hugo S. Caram, and Shivaji Sircar. Department of Chemical Engineering, Lehigh University, Iacocca Hall, 111 Research Drive, Bethlehem, PA 18015

Hydrogen is a potential clean energy and hydrogen production technologies are important research topics that can lead into a new energy era. Coal is a cheap and abundant energy resource compared to expensive and limited petroleum. Synthesis gas produced from coal gasification can be used to produce hydrogen through water gas shift (WGS) reaction (CO + steam ↔ CO2 + H2) and purification procedures. The conventional route includes pressure swing adsorption (PSA) processes to produce a high purity hydrogen product gas. PSA processes can also be designed to produce a CO2 by-product at ambient pressure. This study discusses the "Thermal Swing Sorption Enhanced Reaction (TSSER)" which simultaneously carries out the WGS reaction and the separation of CO2 from the reaction zone by using a novel CO2 chemisorbent in a single unit operation. The new process produces a fuel-cell grade hydrogen (< 10 ppm CO) and compressed CO2 as a by-product gas. The simultaneous removal of CO2 during WGS reaction circumvents the thermodynamic limitation and shifts equilibrium to the right resulting in the enhancement of the extent of the forward reaction. As a result, a pure hydrogen stream can be directly produced, and the process is simpler and potentially more economical than the conventional WGS. Sorption enhanced WGS reaction was tested experimentally and confirmed to produce high purity hydrogen. The thermal swing concept was used for column regeneration and production of compressed CO2. A five-step continuous process was tested by simulation. High net conversion of CO to hydrogen product, high CO2 recovery, and efficient column regeneration was achieved.

197c Extreme Configurations For Heavy Reflux Psa Cycles

James Ritter, Amal Mehrotra, Steven P. Reynolds, and Armin D. Ebner. Department of Chemical Engineering, University of South Carolina, Columbia, SC 29208

Using an adsorption process simulator, Reynolds et al. [1-4] have been studying twelve different stripping pressure swing adsorption (PSA) cycles over a wide range of process conditions for concentrating CO2 from a stack gas effluent at high temperature using K-promoted HTlc. These twelve stripping PSA cycles have ranged from light reflux (LR) only cycles like the Skarstrom cycle that utilizes a low pressure purge step [1,2], heavy reflux (HR) only cycles that utilize a high pressure rinse step in lieu of a LR step [2-4], and LR and HR, i.e., dual reflux (DR), cycles that utilize both the low pressure purge and high pressure rinse steps [2-4]. In addition, a co-current depressurization step has been studied with the LR only cycles [2], and recovery (REC) and feed plus recycle (F+R) steps have been studied with the HR only and DR cycles [4].

In general, they found that the addition of the LR step resulted in the best process performance, the addition of the REC step made the process performance slightly worse and required an additional column, and the addition of the F+R step made the process performance slightly better and did not require an additional column. The best cycle based on overall performance was a 5-bed 5-step stripping PSA cycle with LR, HR from countercurrent depressurization (CnD), and F+R (CO2 purity of 98.8%, CO2 recovery of 100.0% and feed throughput of 5.8 L STP/hr/kg) [4]. The next best cycle was a 5-bed 5-step stripping PSA cycle with LR, HR from LR purge, and F+R (CO2 purity of 96.6%, CO2 recovery of 71.3% and feed throughput of 57.6 L STP/hr/kg) [4]. These improved performances were caused mainly by the counterintuitive use of a very small purge to feed ratio (0.02) for the former cycle and a larger one (0.50) for the latter cycle. The former cycle also was good at producing CO2 at high purities and recoveries but at lower feed throughputs, and the latter cycle was good at producing CO2 at high purities and feed throughputs but at lower CO2 recoveries. The best performance of a 4-bed 4-step stripping PSA cycle with HR from CnD (with and without REC or F+R steps) was disappointing because of low CO2 recoveries (CO2 purity of 99.2%, CO2 recovery of 15.2% and feed throughput of 72.0 L STP/hr/kg) [4]. This last result was also counterintuitive and revealed that the CO2 recoveries of this cycle would always be much lower than the corresponding cycles with a LR step, no matter the process conditions, and that the LR step was very important to the performance of these HR cycles for this particular application.

In general, they also found that the HR step is as interesting as it is flexible when it comes to cycle configuration. In a HR only cycle, the HR must come from the CnD effluent, with some fraction taken as reflux and the remaining fraction taken as heavy product. This constitutes one possible cycle configuration for a HR only cycle. In contrast, for a DR cycle, there are seven possible cycle configurations, because all the HR can come from the CnD, LR or both steps. This is why these DR cycles are so interesting. The seven possible cycles are comprised of the following configurations.

1) All the CnD effluent can be used as HR, with the heavy product obtained from all the LR effluent.

2) All the LR effluent can be used as HR, with the heavy product obtained from all the CnD effluent.

3) All the CnD effluent can be used as HR, with some fraction of the LR effluent also used as HR, with the remaining fraction taken as heavy product.

4) All the LR effluent can be used as HR, with some fraction of the CnD effluent also used as HR, with the remaining fraction taken as heavy product.

5) All the CnD effluent can be taken as heavy product, with some fraction of the LR effluent also taken as heavy product, with the remaining fraction taken as HR.

6) All the LR effluent can be taken as heavy product, with some fraction of the CnD effluent also taken as heavy product, with the remaining fraction taken as HR.

7) Some fraction of both the CnD and LR effluents can be used as HR, with the remaining fraction of both effluents taken as heavy product, with the two fractions not necessarily being equal, especially since the flow rates of the CnD and LR effluent streams tend to be vastly different.

Moreover, in every one of these HR cycle configurations, an option exists for where to put the light gas effluent produced during the heavy reflux step. It can be taken as light product or it can be recycled back into the process. There are two ways to recycle the light gas effluent back into the process. The first way is simply to blend this light gas effluent with the feed during the feed step. This method does not add another step to the cycle and is referred to as the F+R step. The second way is to feed this light gas effluent to the heavy end of a bed just after the feed step. This method does add another step to the cycle and hence bed, and is referred to as the REC step.

Only a fraction of these cycle configurations has been studied so far [2-4]. The objective this presentation is to introduce these relatively new ways to operate the HR step and to provide an overview of the results obtained from them. In all cases, the system of interest is the concentration of CO2 from a stack gas effluent at high temperature using K-promoted HTlc. The goal is to determine if these different configurations can improve the process performance even beyond that already achieved.

1. S. P. Reynolds, A. D. Ebner and J. A. Ritter, New Pressure Swing Adsorption Cycles for Carbon Dioxide Sequestration," Adsorption, 11, 531-536 (2005).

2. S. P. Reynolds, A. D. Ebner, and J. A. Ritter, "Stripping PSA Cycles for CO2 Recovery from Flue Gas at High Temperature Using a Hydrotalcite-Like Adsorbent," Ind. Eng. Chem. Res., 45, 4278-4294 (2006).

3. S. P. Reynolds, A. D. Ebner and J. A. Ritter, "Carbon Dioxide Capture from Flue Gas by PSA at High

Temperature using a K-Promoted HTlc: Effects of Mass Transfer on the Process Performance," Environmental Progress, 25, 334-342 (2006).

4. S. P. Reynolds, A. Mehrotra, A. D. Ebner and J. A. Ritter, "Heavy Reflux PSA Cycles for CO2 Recovery from Flue Gas. Part I. Performance Evaluation," Adsorption, submitted (2007).

197d Development Of Computer Aided Design Tools For Psa Processes

Giovanna Fiandaca[1], Eric Fraga[1], and Stefano Brandani[2]. (1) Department of Chemical Engineering, University College London, Torrington Place, London, WC1E 7JE, United Kingdom, (2) School of Engineering and Electronics, University of Edinburgh, Sanderson Building, The King's Buildings, EH9 3JL, Edinburgh, United Kingdom

In order to provide the reduction of greenhouse gas emissions needed to attempt to stabilise atmospheric CO_2 concentrations, significant changes in the way energy is produced will have to be implemented and many different options will have to be pursued. Since fossil fuel will remain the main energy source for a number of decades, development of CO_2 capture and storage (CCS) becomes essential. According to the estimate of the Intergovernmental Panel on Climate Change (IPCC), in most CCS systems the cost of capture (including compression) is the largest cost component. As a result, efforts are needed to provide more efficient and cost effective separation options. Recently, Pressure Swing Adsorption (PSA) has attracted attention as a valuable technology for CO_2 capture since it is a more economic and effective alternative to other separation options such as cryogenic, membrane and absorption processes. The aim of our project is the development of computer aided design tools for PSA processes. Our focus is on fast cycle operations which are needed to reduce capital and operating costs. However, both modelling and optimisation of such operations are challenging. The simplest model for a PSA process is based on the assumption of a linear driving force (LDF) to mass transfer between the fluid and the solid adsorbent. It simplifies the model by eliminating the need to describe concentration profiles in the solid. Unfortunately, the LDF model fails to capture the dynamics of fast cycle operations. As a result, we use more detailed diffusion models. These involve coupled nonlinear partial differential and algebraic equations in time and one or two space dimensions. The detailed diffusion models imply more difficulties for optimisers, especially gradient based methods. We are currently investigating the efficiency of direct search methods for the optimization of a PSA process described by the detailed model. The eventual goal is the development of multi-criteria optimization methods for this problem, in order to achieve high efficiency both in terms of costs of the operation and purity/recovery of the products.

197e Simulation Of Unique Pressure Changing Steps And Situations In Psa Processes

Armin D. Ebner[1], Amal Mehrotra[1], James C. Knox[2], M. Douglas LeVan[3], and James A. Ritter[1]. (1) Department of Chemical Engineering, University of South Carolina, Columbia, SC 29208, (2) Marshall Space Flight Center, Huntsville, AL 35812, (3) Department of Chemical Engineering, Vanderbilt University, VU Station B 351604, 2301 Vanderbilt Place, Nashville, TN 37235-1604

A more rigorous cyclic adsorption process simulator is being developed for use in the development and understanding of new and existing PSA processes. Unique features of this new version of the simulator that Ritter and co-workers have been developing for the past decade or so include: multiple absorbent layers in each bed, pressure drop in the column, valves for entering and exiting flows and predicting real-time pressurization and depressurization rates, ability to account for choked flow conditions, ability to pressurize and depressurize simultaneously from both ends of the columns, ability to equalize between multiple pairs of columns, ability to equalize simultaneously from both ends of pairs of columns, and ability to handle very large pressure ratios and hence velocities associated with deep vacuum systems. These changes to the simulator now provide for unique opportunities to study the effects of novel pressure changing steps and extreme process conditions on the performance of virtually any commercial or developmental PSA process.

This presentation will provide an overview of the cyclic adsorption process simulator equations and algorithms used in the new adaptation. It will focus primarily on the novel pressure changing steps and their effects on the performance of a PSA system that epitomizes the extremes of PSA process design and operation. This PSA process is a sorbent-based atmosphere revitalization (SBAR) system that NASA is developing for the new Crew Exploration Vehicle (CEV).

This SBAR system consists of a 2-bed 3-step 3-layer system that operates between atmospheric pressure and the vacuum of space, evacuates from both ends of the column simultaneously, experiences choked flow conditions during pressure changing steps, and experiences a continuously changing feed composition, as it removes metabolic CO2 and H2O from a closed and fixed volume, i.e., the CEV cabin. Important process performance indicators of this SBAR system are size, and the corresponding CO2 and H2O removal efficiencies, and N2 and O2 loss rates. Results of the fundamental behavior of this PSA process during extreme operating conditions will be presented and discussed.

197f Axial Dispersion And The Danckwerts Boundary Conditions

Stefano Brandani, School of Engineering and Electronics, University of Edinburgh, Sanderson Building, The King's Buildings, EH9 3JL, Edinburgh, United Kingdom

In many papers reporting models of adsorption processes, for historical reasons, the axial dispersion term is often either written in an incorrect form or the boundary conditions applied are not consistent. We review the correct formulation of these terms and show why the Danckwerts boundary inlet condition should be used in preference to the assigned inlet concentration. We also present a new formal proof of the exit boundary condition for open-closed systems and extend the results to open-open systems, which are practically relevant when solving adsorption models using finite element numerical schemes.

Investigation of protein imprinting using miniemulsion polymerization through the epitope approach

Tan Chau Jin and Tong Yen Wah

Department of Chemical and Biomolecular Engineering
4 Engineering Drive 4
National University of Singapore
SINGAPORE 117576

Keywords: molecular imprinting, Ribonuclease A, Cytochrome C, nanoparticles, protein, miniemulsion polymerization, protein, epitope

Prepared for Presentation at the 2007 Annual Meeting AIChE

November 2007

Unpublished

Abstract

Molecular imprinting is a state-of-the-art technique that can be applied for the preparation of polymeric materials with antibody-like affinity for pre-determined target molecules, thus offering an array of possible applications in the field of analytical chemistry and separation. Nevertheless, imprinting of protein macromolecules has only been met with limited success due to its incompatibility with the conventional molecular imprinting system. In this study, it was aimed to achieve successful protein imprinting through the epitope approach in a miniemulsion polymerization system. The peptide epitopes employed were obtained through direct chemical proteolytic digestion (for RNAse A) or synthesized using the solid-phase peptide synthesis (for cytochrome C). Epitope-imprinted nanoparticles (about 50 nm) were prepared. The nanoparticles were morphologically characterized through direct microscopy and nitrogen sorption method. In order to elucidate their recognition property, they were subjected to batch adsorption runs in single-protein and competitive environments. Preliminary results showed that the RNase A epitope-imprinted nanoparticles failed to exhibit the desired affinity towards the protein. The remaining characterization studies are still in progress.

1.0 Introduction

Molecular imprinting has been widely recognized as a promising technique for imparting pre-determined molecular recognition property onto synthetic materials such as polymers. For the past few decades, the interest and attention shown towards this field has been increasing at an amazing pace. This is mainly attributed to the potential wide applications of molecularly imprinted polymers (MIPs) in the fields of separation[1,2], catalysis[3,4], analytical chemistry[5,6] and biosensing[7,8]. Compared to its biological counterparts like enzymes and antibodies, MIPs can not only display comparable molecular selectivity, they are also more robust, reusable and most of all, easy and cheap to prepare. Therefore, MIPs represent a new class of materials that could mimic and possibly replace their biological equivalents. However, to date, with the conventional approach, most of the successes in molecular imprinting have been based on small target ligands whereas the imprinting of macromolecules like proteins has been limited. One of the major difficulties faced by these large molecules for the imprinting application lies in the diffusion limitations which restricts the ease of template removal as well as uptake. In response to such limitation for protein imprinting, the epitope approach has been proposed as a viable strategy. This approach was first proposed and demonstrated by Rachkov et al[9]. In this pioneering effort, it was shown that by imprinting a short peptide, the imprinted polymer displayed preferential uptake of the associated bigger peptide, oxytocin. This work first illustrated the applicability of the epitope approach. In subsequent years, there were few other investigative attempts via the epitope approach for peptide[10,11] and protein[12] imprinting and the target molecules under study include dengue virus NS1 protein, cytochrome C (Cyt C), alcohol dehydrogenase (ADH) and bovine serum albumin (BSA). These attempts primarily focused on the preparation of epitope-imprinted film. However, more illustrations of the epitope approach for protein imprinting have been largely absent.

We had previously prepared protein surface-imprinted nanoparticles through miniemulsion polymerization system[13]. However, in our following study on the system mechanism, it was found that there are some requirements on the inherent properties of the template proteins for successful imprinting. As such, in this work, we aimed to investigate the applicability of the epitope approach for protein imprinting using miniemulsion polymerization in the hope of making the system amenable to all different proteins. Methyl methacrylate (MMA) and ethylene glycol dimethacrylate (EGDMA) had been used as the functional and cross-linking monomer respectively. The peptide epitopes used in investigation were obtained either from chemical proteolytic digestion or through solid-phase peptide synthesis. The proteins under study were ribonuclease A (RNase A) and Cyt C. Based on our understanding of the miniemulsion polymerization system for molecular imprinting, we had also made slight modifications like including a short sequence of hydrophobic amino acid resides in the peptide epitope in an attempt to improve the imprinting efficiency. The preliminary results obtained for the RNase A epitope-imprinted nanoparticles were rather discouraging due to the lack of observed affinity towards the protein. The remaining investigative work is currently being carried out in the laboratory.

2.0 Experimental

The peptide epitope used for RNase A was the N-terminal 13-mer peptide (K-E-T-A-A-A-K-F-E-R-Q-H-M). The epitope was obtained through direct proteolytic digestion of RNase A using cyanogen bromide (CNBr). To carry out the chemical digestion, 48.5 mg of RNase A and 0.6 g (400 molar excess) of CNBr were weighed and added into 15 mL of 70% formic acid solution. The solution was magnetically stirred and the reaction was allowed to proceed in dark for 24 hours. Upon completion, the solvent was removed through rotary evaporation and yellowish gel-like product was obtained. It should be noted that cyanides in the solvent must be oxidized to harmless waste before disposal. Commercial bleach was used in excess to decompose cyanide. The product gel was then dissolved in DI water for subsequent preparative HPLC purification. Matrix-assisted laser desorption/ionization (MALDI) time-of-flight (TOF) mass spectrometry (MS) was employed to identify the correct fraction (the peptide epitope) to be collected.

On the other hand, the peptide epitope used for Cyt C was the C-terminal peptide (A-Y-L-K-K-A-T-N-E-G). The epitope peptide was synthesized through solid-phase peptide synthesis. In addition, a modified peptide epitope with 5 terminal hydrophobic amino acid residues (alanine) was also synthesized as the template (A-Y-L-K-K-A-T-N-E-A-A-A-A-A-G). Similarly, the synthesized epitope peptides were purified using HPLC and analyzed with MALDI-TOF-MS.

The epitope-imprinted nanoparticles were prepared using the miniemulsion polymerization modified from Miller's work[14]. The first aqueous phase was prepared by mixing 57.7 mg sodium dodecyl sulfate (SDS), 46.9 mg sodium bicarbonate and 194 mg cetyl alcohol (CA) in 20 mL of DI water. To prepare the second aqueous phase, 0.2 g of SDS were dissolved in 400 ml of DI water. Following that, 0.8 mL of MMA and 4.2 mL of EGDMA were mixed, forming the oil phase. The monomer mixture oil phase was then slowly added into the first aqueous phase and followed by homogenization at 24,000 rpm with a homogenizer for 40 seconds. Subsequently, the stipulated amount (as shown in Table 1) of the epitope peptide was added into the miniemulsion and mixed magnetically for 30 minutes to allow effective monomer-template interaction. The miniemulsion was then added into the second aqueous phase and transferred into a 1 L 3-neck round-bottomed flask. The flask was mechanically stirred at 300 rpm, and slowly heated to 40 $^{\circ}$C. When the pre-polymerization miniemulsion reached 40 $^{\circ}$C, the reaction vessel was pre-purged with nitrogen gas for 15 minutes to displace oxygen and finally, sodium bisulfite (0.230 g), followed by ammonium persulfate (0.252 g), were added into the mixture for polymerization up to 24 hours.

Table 1. *The amount of template epitope peptide added for the imprinting polymerization.*

Epitope peptide	Amount added (mg)
K-E-T-A-A-A-K-F-E-R-Q-H-M	2.80
A-Y-L-K-K-A-T-N-E-G	2.71
A-Y-L-K-K-A-T-N-E-A-A-A-A-A-G	2.04

Upon completion, the polymeric nanoparticles were washed five times each with DI water and a solution of SDS: acetic acid (10 w/V%: 10 V/V%) to remove the template peptide epitopes, four times with excess ethanol to remove the surfactant or any unreacted monomer and initiator, and six times with DI water. The washed imprinted nanoparticles were suspended in water under room temperature. Non-imprinted nanoparticles were prepared in a similar manner as above, except without the addition of the template epitope.

In the morphological characterization of the MIP nanoparticles, field-emission scanning electron microscopy (FESEM) and nitrogen sorption method (in progress) were used. The swelling ratios of the nanoparticles were also determined (in progress). In examining the selectivity of the nanoparticles, batch and competitive rebinding tests were carried out with lysozyme (Lys) as the competitive assay.

3.0 Results and Discussion

As seen in the microscopic images of Figure 1, RNase A epitope-imprinted nanoparticles that sized around 50 nm were prepared. The nanoparticles were regularly shaped, monodispersed and under direct observation, there were no significant morphological differences between the imprinted and the non-imprinted nanoparticles.

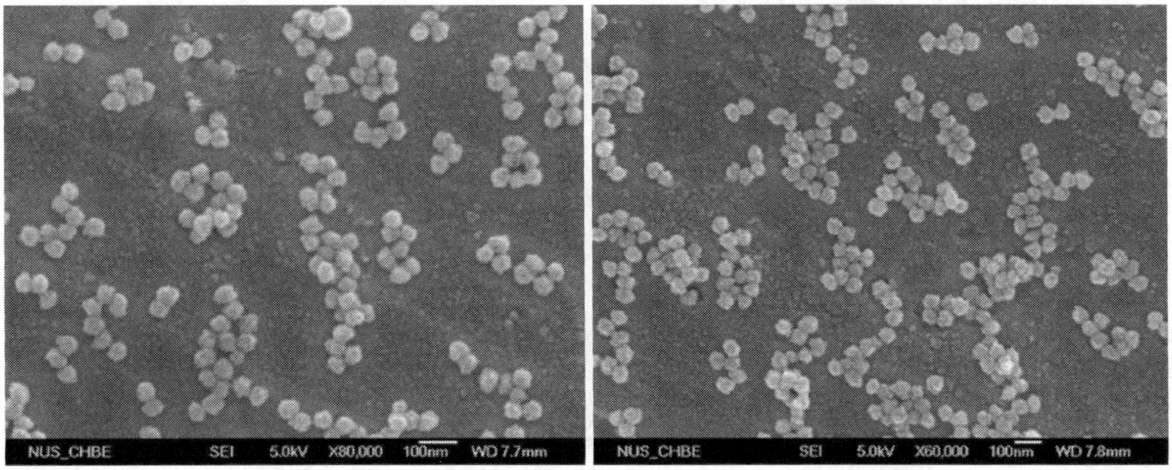

(a) **(b)**

Figure 1. *FESEM images of (a) non-imprinted (eNIP) and (b) RNase A epitope-imprinted (eMIP) nanoparticles.*

When the nanoparticles were subjected to single-protein batch adsorption runs, it was surprising to us that the eMIP nanoparticles did not display any preferential uptake of RNase A molecules. The results are shown in Figure 2. While the remaining investigative work is still in progress, at this point of time, the exact reason for the observation remains unknown to us. Nevertheless, there are 3 possibilities. First of all, it could be that the N-terminal peptide sequence chosen for RNase A may not be significant enough to function as an epitope. Secondly, miniemulsion polymerization may not be amenable to the epitope approach of protein imprinting; and thirdly, the imprinting system was not well-optimized for achieving good imprinting efficiency. More results and discussions will be presented on the conference.

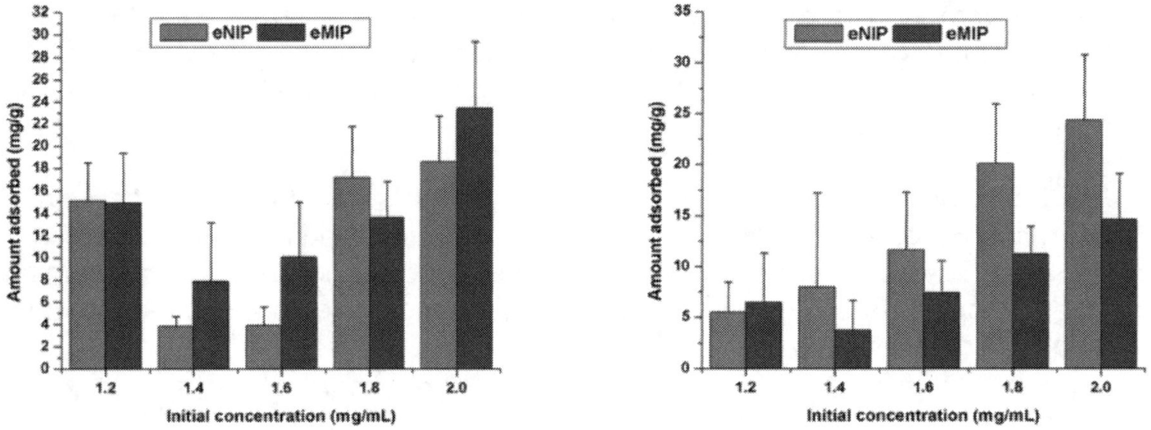

(a) (b)

Figure 2. *Results of (a) Lys and (b) RNase A batch-rebinding tests.*

4.0 References

(1) Ulbricht, M. *J. Chromatogr. B* **2004**, *804*, 113-125.
(2) Da Costa Silva, R. G.; Augusto, F. *J. Chromatogr. A* **2006**, *1114*, 216-223.
(3) Cheng, Z.; Li, Y. *J. Mol. Catal. A-Chem.* **2006**, *256*, 9-15.
(4) Pasetto, P.; Maddock, S. C.; Resmini, M. *Anal. Chim. Acta* **2005**, *542*, 66-75.
(5) Hawkins, D. M.; Trache, A.; Ellis, E. A.; Stevenson, D.; Holzenburg, A.; Meininger, G. A.; Reddy, S. M. *Biomacromolecules* **2006**, *7*, 2560-2564.
(6) Bossi, A.; Piletsky, S. A.; Piletska, E. V.; Righetti, P. G.; Turner, A. P. F. *Anal. Chem.* **2001**, *73*, 5281-5286.
(7) Li, X.; Husson, S. M. *Langmuir* **2006**, *22*, 9658-9663.
(8) Liu, K.; Wei, W. Z.; Zeng, J. X.; Liu, X. Y.; Gao, Y. P. *Anal. Bioanal. Chem.* **2006**, *385*, 724-729.
(9) Rachkov, A.; Minoura, A. *B. B. A.* **2001**, *1544*, 255-266.
(10) Titirici, M. M.; Sellergren, B. *Anal. Bioanal. Chem.* **2004**, *378*, 1913-1921.
(11) Tai, D. F.; Lin, C. Y.; Wu, T. Z. ; Chen, L. K. *Anal. Chem.* **2005**, *77*, 5140-5143.
(12) Nishino, H.; Huang. C. S.; Shea, K. J. Angew. *Chem. Int. Ed.* **2006**, *45*, 2392-2396.
(13) Tan, C. J.; Tong, Y. W. Langmuir **2007**, *23*, 2722-2730.
(14) Miller, C. M.; Sudol, E. D.; Silebi, C. A.; El-Aasser, M. S. *Macromolecules* **1995**, *28*, 2754-2764.

209b Investigation Of Chemical Selective Displacers Using Robotic High Throughput Screening, Spr, NMR And Md Simulations

Steven Cramer, Chris Morrison, Scott McCallum, Rahul Godawat, James Moore, and Shekhar Garde. Dept. of Chemical Engineering, RPI, 110 8th Street, Troy, NY 12180

High throughput screening was employed in concert with several analytical techniques to identify and evaluate the behavior of chemically selective displacers for protein purification in ion exchange systems. A robotic liquid handling system was adapted to efficiently carry out this parallel batch screen of selective displacers on multiple protein pairs. The results identified potential selective displacers and important functional group chemistries and also indicated that this selectivity was due primarily to the selective binding between the displacer and targeted proteins. Nuclear Magnetic Resonance was then conducted on several protein/displacer mixtures verifying the binding of the selective displacers to targeted proteins and the location of the binding event. Surface plasmon resonance experiments and molecular dynamic simulations were also carried out to corroborate the NMR results. This proof of concept study shows that more specific selectivities may also be possible by utilizing affinity based selective displacers for explicit protein systems.

209c Microscopic Analysis Of Adsorption Of E. Coli And Staphylococcus Aureus On Single-Wall Carbon Nanotube Aggregates

Shuguang Deng[1], Venkata Upadhyayula[1], Goef Smith[2], and Martha Mitchell[1]. (1) Chemical Engineering Department, New Mexico State University, P.O. Box 30001, MSC 3805, Las Cruces, NM 88003, (2) Biology Department, New Mexico State University, P.O. Box 30001, MSC 3805, Las Cruces, NM 88003

Adsorption of a Gram-positive bacterium, Staphylococcus aureus (S. aureus) and Gram-negative bacterium, Escherichia coli (E. coli) on carbon nanotube aggregates was investigated in shaker experiments and microscopic analyses for biosensor applications. Carbon nanotube-based biosensors have the potential of rapid concentration and detection of pathogens in drinking water systems. The signal strength of the biosensors is directly proportional to the immobilized analyte concentration on the carbon nanotube probes and is highly dependent on how well the microorganisms are associated with the nanotubes. Both adsorption kinetics and equilibrium of the bacterial species on carbon nanotubes were determined in the shaker experiments. Microscopic analyses including SEM, TEM with and without ultramicrotomy, and confocal microscopy were performed to visualize the association of the bacteria with carbon nanotubes. It was found from the adsorption measurements that S. aureus have 12-100 times stronger and closer association with the single-wall nanotube aggregates than the E. coli. The small size S. aureus diffuse 25 times faster than E. coli in the carbon nanotube aggregates. Microscopic analyses have qualitatively confirmed the adsorption results. The preliminary results from this work suggest that single-wall carbon nanotubes are good candidate for biosensors.

209d Generation and Use of Single Component Internal pH Gradients for Separation of Protein Isoforms

Natarajan Ramasubramanyan[1], Alan Hunter[1], Timothy M. Pabst[2], and Giorgio Carta[3]. (1) Global Biologics, PGRD, Pfizer Inc, 700 Chesterfield Pkwy West, Chesterfield, MO 63005, (2) University of Virginia, Charlottesville, VA 22904, (3) Chemical Engineering, University of Virginia, 102 Engineers' Way, Charlottesville, VA 22904-4741

Proteins biologics often contain several isoforms or variants, arising due to the differences in expression, post-translational modifications, protease action or chemical modification. These isoforms are best understood by evaluating their biological activity using bioassays that measure their biochemical or biological function. Availability of preparative amounts of these variants is essential for these assays. Many of the analytical techniques that are capable of separating these isoforms either do not provide material that can used readily or are not amenable to fractionation or scale-up.

In this study we report successful separation and/or enrichment of protein isoforms that arise due to differences in aggregation and sialation using preparative, high resolution pH based separation methods. Shallow gradients of over 20 column volumes over one pH unit provide a high resolution power, even under relatively high loads (2-5 mg/ml resin). Representative data from a couple of representative protein molecules that have been successfully separated using these techniques is presented.

209e Protein Refolding By Means Of Gradient Chromatography

Esteban Jaramillo Freydell[1], Ottens Marcel[2], Wielen Luuk[2], and Michel Eppink[3]. (1) Department of Biotechnology, Delft University of Technology, Julianalaan 67, Delft, 2628 BC, Netherlands, (2) Biotechnology, Delft University of Technology, Julianalaan 67, Delft, 2628 BC, Netherlands, (3) DSP, Diosynth Biotechnology Europe, Oss, Netherlands

PROTEIN REFOLDING BY MEANS OF GRADIENT CHROMATOGRAPHY

The over expression of recombinant proteins in Escherichia coli leads in most cases to their accumulation in the form of insoluble aggregates referred to as inclusion bodies. The inclusion bodies are composed of partially folded intermediates, which in most cases have a native-like secondary structure. In order to release the product from them, the inclusion bodies must be solubilised in a buffer composition that most of the times include aggregation suppressors and redox agents. To obtain the native and bioactive product the concentration of the chemicals used for the solubilization have to be changed preferably in a controllable manner, minimizing the product lost through the competitive reactions, namely misfolding and aggregation. The industrial process generally used is dilution, which is an economical process but limited by the formation of native product and the equipment size required to cope with the product demand. Recent developments have indicated that application of chromatography for refolding may be successful. In this work we will experimentally and theoretically show column refolding in batch as well in continuous mode, making use of an SMB system, including urea and redox gradients in order to promote refolding of a fusion protein trapped in inclusion bodies.

Web Page: www.bt.tudelft.nl

209f Size Exclusion Chromatography Applied To Recovery, Purification And Characterization Of Microbial Hyaluronic Acid

Diana C. S. Azevedo, Anayla S. Sousa, Artemízia P. Guimarães, Ivanildo J. Silva Jr., and Célio L. Cavalcante Jr. Chemical Engineering, Universidade Federal do Ceará, Campus do Pici, bl. 709, Fortaleza, 60.455-900, Brazil

This work reports the use of size exclusion chromatography (SEC) applied to recovery, purification and characterization of hyaluronic acid (HA) obtained from Streptococcus zooepidemicus by submerged and solid state fermentation. The presence of proteic contaminants was identified by electrophoresis in polyacrylamide gel under denaturant conditions (SDS-PAGE) and quantified by the Bradford method. The recovery and pre-purification of HA from the proteic contaminants were performed by subsequent operations of precipitation with ethanol and dissolution with saline solution (performed up to four times). After four purification steps, for both submerged and solid state fermentation, the total protein concentration was reduced about 90 % and the HA concentration decreased about 16%. The characterization of molar mass distribution of the HA samples was performed by size exclusion chromatography (SEC) in analytical scale using Shodex OHPak SB806M HQ column, refractive index detector, 0.1 M sodium nitrate solution as mobile phase, flow rate of 0.8 mL.min-1. Experiments were also performed in semi-preparative scale for protein separation and molar mass fractionation using Superose 6 column, refractive index and UV/Vis (in wavelength of 280 nm) detector, 0.1 M sodium nitrate solution as mobile phase and flow rate of 0.8 mL.min-1. The injection volumes were 20 μL in analytical scale and 400 μL in semi-preparative scale. All runs were performed at room temperature for both scales. Molar mass (MM) distribution of HA samples was assessed by SEC using pullulan markers (polymer of similar hydrodynamic volume as compared to HA) as molar mass standards over the range of 5.8 x 103 Da to 8.53 x 105 Da. The MM of samples of HA obtained from submerged fermentation were found to be in the range of 103 to 107 Da, after being submitted to four pre-purification steps. The purification steps not only decreased proteic contamination but also increased the average molar mass of the samples. For solid state fermentation, the MM distribution was found to be in the range of 103 to 105 Da. In the runs with the semi-preparative Superose 6 column, it was possible to separate HA fractions, free of proteic contaminants, with molar mass above 105 Da.

211a New And Novel Batch Distillation Processes

Rakesh Agrawal, Purdue University, 480 Stadium Mall Drive, West Lafayette, IN 47907 and **Girish S. Joglekar**, Batch Process Technologies, Inc., 1291 Cumberland Avenue, West Lafayette, IN 47906.

A number of new and novel batch distillation column solutions are developed to distill multicomponent mixtures in multiple pure product streams. An important feature of these new and novel solutions is that they retain the flexibility of the original conventional batch distillation column and yet provide large reduction in energy consumption. These configurations also lead to large reduction in batch cycle time. Depending on the mixture to be distilled, the time reduction can be as large as 50%. This paper will introduce all the new features and the operating modes.

The new batch distillation processes are expected to find application in a typical pharmaceutical, specialty chemical and agricultural industrial facility that utilizes a wide range of solvents. Another important feature of the new processes is that they can potentially eliminate slop cuts (i.e. waste solvent streams) and thereby remove the need to send waste solvents to environmentally hazardous incinerators.

211b Non-Pinched Minimum Energy Solutions in Distillation

Amit Amale[1], Ross Taylor[2], and Angelo Lucia[1]. (1) Chemical Engineering, University of Rhode Island, Dept of Chem. Engineering, Kingston, RI 02881, (2) Clarkson University, Department of Chemical Engineering, Potsdam, NY 13699

Distillation is still an important unit operation for the separation of chemicals, and due to its versatility, it will continue to be used in the forseeable future. Very good techniques for finding minimum energy requirements for distillation, like the McCabe-Thiele and Underwood methods, have existed for quite some time. More recently, methods like the boundary value of Doherty and co-workers, the Vmin method of Halvorsen and Skogestad, and the rectifying body method of Urdaneta et al. have been proposed to handle multicomponent mixtures that exhibit azeotropes and multiple column configurations. For the most part, these methods exclusively find minimum energy requirements for distillations with pinched column profiles. There are no methods that can find non-pinched minimum energy designs for distillations in a rigorous and systematic way – either for single distillation columns or for multi-unit processes in which distillation is a part. In this presentation, non-pinched solutions of single distillation columns and multi-unit chemical processes are studied. It is shown that the recent approach proposed by Lucia and co-workers based on the shortest stripping line distance can be used to find non-pinched solutions. This new methodology consists of a Mixed Integer NonLinear Programming (MINLP) strategy in which pinched solutions are determined by 1) First solving a NonLinear Program (NLP) for the boil-up ratio that gives the feasible column with the shortest stripping line distance from the desired bottoms composition to the stripping pinch point curve. 2) Subsequently solving an integer program (IP) for a finite number of stages to further reduce the stripping line distance under the conditions that the column is feasible and the boil-up ratio is fixed at the solution provided by the NLP. A number of interesting example problems are presented that show that the shortest stripping line methodology is capable of finding minimum energy requirements with non-pinched distillations. Many geometric illustrations for three and four component mixtures are used to highlight key attributes of the shortest stripping line approach.

211c Process Intensification For The Distillation With Heat-Integration

Keigo Matsuda[1], Koichi Iwakabe[2], Takeshi Higuchi[1], Masayuki Toda[1], and Masaru Nakaiwa[3]. (1) Department of Chemistry and Chemical Engineering, Graduate School of Science and Engineering, Yamagata University, 4-3-16, Jonan, Yonezawa, 992-8510, Japan, (2) Project Development Department, Maruzen Petrochemical Co., Ltd., Energy-Efficient Chemical Systems Group, Research Institute for Innovation in Sustainable Chemistry, AIST Tsukuba Central 5, 1-1-1, Higashi, Tsukuba-shi, Ibaraki, 305-8565, Japan, (3) National Institute of Advanced Industrial Science and Technology, Central 5, 1-1-1, Higashi, Tsukuba, 305-8565, Japan

The proposed of the present work is to discuss a method for "Process Intensification" of the distillation process of separation. The model of an internally heat-integrated distillation column (HIDiC) with concentric type and compact heat exchanger type were developed by a rate-based model and the separation of multicomponent mixture was simulated. This has been implemented in the simulator which is capable of predicting dynamic conditions. The developed model has been used as concentric and compact heat exchanger types for HIDiC, and their separation and operation performance was discussed. In the results, The HIDiC compact heat exchanger process can be also obtained for zero reflux and/or no reboil operation. This process improved the function and performance of conventional distillation separation process and could make it a miniature. In addition, the HIDiC compact heat exchanger showed highly energy saving performance compared with the conventional distillation processes.

211d Opportunities For Reduction Of Energy Consumption In Batch Distillation Processes

Nana Yaw Asiedu, Diane Hildebrandt, David Glasser, and Brendon Hausberger. SCHOOL OF CHEMICAL AND METALLURGICAL ENGINEERING, UNIVERSITY OF THE WITWATERSRAND,, JOHANNESBURG, South Africa

Batch distillation is one of the most important unit operations for small- scale production of speciality and fine chemicals.It is also used in pharmaceutical and perfume industries and as well as for laboratory studies. Distillation of any form (Batch,continuous,complex batch,etc)is an energy intensive process and this has a significant effect on the overall cost of production.This paper investigates ways of reducing the energy consumption of a simple conventional binary batch distillation process by finding the Attainable Region method for a binary batch boiling process. The fundamental processes considered are boiling and mixing.It is shown that there is a lower limit to the amount of energy required to produce for a given composition of material from a specified initial material. This limit can be achieve by use of a specified mixing policy where this policy is obtained from the Attainable Region approach. It is shown that using this approach, the binary batch distillation requires significant less energy compared to the conventional batch processes to achieve a particular distillation product.

211e Dynamic Modelling And Simulation Of Distillation Columns To Overcome The Initialization Issue

Ivan Dones and Heinz A. Preisig. Chemical Engineering, NTNU, S. Saelandsveg 6, Trondheim, 7491, Norway

Modelling of chemical processes is today the foundation to almost any operation including design, control, planning, retrofitting. As varied as the use of models is its structure: models are built for a particular purpose, to meet the requirement of the application. Though there is obviously a need for dynamic models, at least in chemical engineering the most common models are steady-state models mainly used in designing and retrofitting, but also in some control and planning applications. With distillation being the most common separation unit, its models have been subject of studies for decades. The computation of steady-state distillation models is a notorious problem as it consists of a large number of equations which proliferates mainly with the number of components and number of trays. Solving such a model implies to solve this large-scale algebraic problem for a single set of roots representing the equilibrium state of the distillation. The numerical methods being used for this purpose have usually a very small convergence radius, which gave rise to a lot of research (Komatsu and Holland 1977, Holland 1981, Taylor et al. 1996, Rabeau et al. 1997, Grossmann et al. 2005). Thus whilst computing has improved tremendously, little progress has been made in solving the initialization problem. The authors' idea for approaching the initialisation problem is to use a dynamic description (Rademaker et al. 1975, Skogestad 1992) that mimics the behaviour of a distillation column close enough so as to simulate a dynamic path starting with a very simple generic initial state and integrating up to the desired operating condition. Thereafter one may switch to a steady state description of the plant and continue the computations, if so required. Earlier studies have shown that bringing two isolated phases into contact will converge monotonically to the steady-state under the condition that the heat transfer is significantly faster than the diffusional mass transfer through the phase boundary (Asbjørnsen 1973, Preisig 2004). This is taken as a starting point to suggest a sequence of physical operations which will guide the dynamic model to converge to the desired steady-state condition. Thus by switching from a steady-state simulation to a dynamic simulation one substitutes the complex initialisation problem by a simple one plus an integration over a feasible physical path. Mathematically this implies that one solves a set of Differential Algebraic Equations (DAE) instead of a large set of algebraic equations. With DAE solvers having advanced tremendously over the past decades, using a dynamic model may even be competitive to using a steady state model. The model is constructed from simple components: each tray, boiler and condenser are considered as a dynamic flash whereby each flash is contained in a fixed volume, tray volume, boiler volume and condenser volume respectively. Staking up such two-phase containments, the column becomes a tower of dynamic flash "rums". The transfer rates for mass and heat were chosen empirically so as to satisfy above-mentioned conditions. The internal streams are driven by the heat source and sink in the boiler and the condenser, respectively. The procedure representing a feasible physical path leading to the desired steady state starts with a set of isolated containments forming pairs of gas and liquid both in isolation. Those are first brought in contact with each other, pair by pair thereby fixing the respective joint volume to represent the volume of the boiler, each stage and condenser, respectively. The individual phases are initialised within the domain of operation. Since a priori the two phases are not in equilibrium, they can exchange energy (in the form of heat and volumetric work) and mass both being diffusional processes through the common phase boundary. The conductive heat stream, driven by the different temperatures of liquid and gas phase, is significantly faster than the mass diffusion, which makes the temperatures in the two phases to converge quickly. The mass exchange driven by the chemical potentials, accordingly to fundamental transfer laws, is slower. The mass stream induces also convective energy transfer as mass carries internal energy. The heat transfer in the liquid phase is assumed to be very fast (event dynamics). This makes the temperature on the boundary identical to the liquid bulk temperature. The energy associated with the phase change is thus solely associated with the liquid phase. This simplifies also the computations as otherwise one has to compute the conditions on each phase boundary, which involves solving for roots of a set of algebraic equations (dimension: number of components + 1 for each boundary). The first step is completed when the two phases in each part of the column have reached an equilibrium. Then it is time for the second step: adjusting the liquid levels. If the liquid level is too little, meaning the fluid does not reach up to the weir height on each stage, liquid is added on the top with the overflows modelled with the weir equation. In case there is too much fluid, one may have to drain some out of the boiler so as to adjust the overall hold-up. With the next step energy is drawn in form of heat from the liquid phase of the condenser driven by the difference between the temperature of the liquid phase and the temperature of a cold reservoir. This procedure excludes anything like drop or film formation. Up to this point the gas phases are not connected. Only at this point this communication of mass driven by the pressure difference is enabled. The transfer model consists of two parts, namely a resistance in the construction, the dry tray resistance, and the hydraulic pressure drop on the tray. Finally heat is supplied to the liquid phase of the boiler driven by the difference in temperature between a warm reservoir and the liquid phase excluding any bubble formation. The complete model consists for each stage of number of components mass balances for the gas phase and the liquid phase, and an energy balance for each liquid and gas phase. The approach was found to be extremely robust. Different initial conditions were tested. Some conditions lead initially to an inverted pressure profile, that is, the pressure on the top was higher than the pressure in the boiler. Adjusting the liquid levels on the trays did not have a large effect on the pressure distribution. The pressure though changed drastically as cooling and heating was enabled. In all cases, the column converted to the correct pressure profile thereby going through a complicated inversion process for the pressure as well as the temperatures converging monotonically towards the final profile during the last phase.

Synthesis of Complex Distillative Separation Sequence

Libin Zhang and Andreas Linninger

Extended Abstract for a presentation at the AIChE 2007 Annual Meeting
in Salt Palace Convention Center, Salt Lake City, Utah, November 4 -9, 2007

Laboratory for Product and Process Design
Bioengineering, University of Illinois at Chicago.
e-mail: {lzhang10, linninge}@uic.edu;

Summary:

Separation processes make up 40%-70% of capital and operating costs of chemical manufacturing. Distillation accounts for more than 60% of the total process energy for the manufacture of commodity chemicals. Complex column configurations are estimated to harness energy savings up to 70%. Therefore, distillation synthesis with complex is a meaningful target for energy improvements on an industry-wide scale. However, design of complex separation system requires (i) structural decisions such as sequencing of distillation columns for solvent recovery, and (ii) determination of associated operating conditions (e.g. distillate/.bottoms compositions, reflux, ratio etc). For computer tools this process synthesis constitutes a very formidable challenge. Technically MINLP incorporating binary as well as continuous variables can address synthesis problem. However, infeasible operation following from ineffective structural decision may jeopardize the robustness and convergence of MINLP. In this presentation, we demonstrate a hybrid algorithm combining stochastic with gradient-based search techniques. A evolutionary algorithm will construct automatically structurally different complex separation network. Feasibility of the design will be delegated to an advanced feasibility tests based on temperature collocation of finite elements (Zhang and Linninger, 2004, 2006). Our search algorithm is capable of solving complex distillative synthesis problem robustly and reliably without user interactions. Application will demonstrate the algorithms performance in complex column sequencing problems for solvent recovery.

Scope:

We propose a novel bi-level hybrid algorithm for the synthesis of complex column sequences. The master problem will make structural decisions using evolutionary algorithms, while the subproblem will rigorously assess the feasibility of each of these structures using our novel MInimum Bubble Point DIstance (MIDI) Algorithm. The minimum distance algorithm computes the rectifying and stripping temperature profiles by uses orthogonal finite-element collocation technique.

Each specification e.g. flowsheet structure, reflux ratio, distillate and bottoms composition, can be represented by a novel chromosome. At every generation, the chromosomes are cultivated by crossover and nonuniform dynamic mutation. In order ensure feasibility of the chromosomes, our MIDI algorithm computes the minimum bubble point distance for each specification, i.e. a chromosome. This minimum bubble point distance is obtained by solving a gradient based NLP optimization problem. A minimum distance of zero ensures feasibility, whereas positive values indicate infeasible specifications. Penalizing the genetic algorithm's objective for infeasible specifications ensures that the chromosomes are feasible after a few generations.

Significance:
The major innovation of the proposed approach is an evolutionary program combined with a novel temperature collocation algorithm to systematically build and optimize complex column configurations. Massive problem size reductions due to temperature collocation ensure the realistic composition profiles of each column in the network without sacrificing the computational and thermodynamic rigor. The evolutionary algorithm constructs automatically different separation sequences; feasibility of the synthesized sequences is ascertained by means of the MIDI algorithm as a subroutine to the master GA. In contract to existing approaches, this method synthesizes several clusters of designs solutions, each one corresponding to regions of local optimality.

Zhang, L. and Linninger, A. A., "Towards Computer-aided Separation Synthesis", AIChE J.,52, 1392-1409, 2006.

Zhang, L. and Linninger, A. A., "Temperature Collocation Algorithm for Fast and Robust Distillation Design", Industrial and Engineering Chemistry Research, 2004, 43, 3163-3182.

222a CO2 Capture By Adsorption Processes – Vacuum Swing Adsorption

Paul A. Webley[1], Jun Zhang[2], Penny Xiao[2], and Gang Li[2]. (1) Monash University, Wellington Road, Clayton, VIC 3800, Australia, (2) Chemical Engineering, Monash University, Wellington Road, Clayton, VIC 3800, Australia

Adsorption processes for CO2 capture from a variety of process streams have received increasing attention as the concern over climate change becomes more urgent. Work started around the world in earnest in the 1990's to understand the capabilities and obstacles to successful operation of adsorption processes for CO2 capture. Most of these efforts examined adsorbent development with much less attention devoted to process development. Over the last three years we have studied the CO2VSA (vacuum swing adsorption) process intensely using both our pilot scale VSA and our simulator. Through cycle design and economic evaluation, we have determined the operating region and performance window of the VSA process for 13X zeolite. Overall, a purity in excess of 95% is achievable with recoveries ranging from 70-80% depending on process parameters. In this study we report on how to exploit the adsorbent – process synergy to produce a viable VSA design. Initial economic evaluation indicates promise for the technology although several obstacles still remain such as the presence of water and impurities in the stream and the large scale of operation. However, the great advances in materials development coupled to careful adaptation of process conditions means that there will be a role for adsorption processes in CO2 capture in the future.

222b Comparison Of Vacuum Swing Adsorption And Electrothermal Swing Adsorption For Carbon Dioxide Recovery From Fossil Fuel Power Plant Flue Gas

Nabil Ettlili, Cécile Vallières, and **Georges Grévillot**. Chemical Engineering Science Laboratory-CNRS, 1, rue Grandville, Nancy, France

The capture and the recovery of CO2 from fossil fuel power plant flue gas are a very important technical challenge. Adsorption processes can be applied to recover CO2 from such dilute and voluminous flue gas (more than 106 m3 h-1 for 750 MWe). In Pressure Swing Adsorption (PSA) the gas mixture flows through a packed bed of adsorbent at elevated pressure and the bed is regenerated by reducing the pressure. In Temperature Swing Adsorption (TSA), the bed is regenerated by raising its temperature. TSA is mainly used in VOC capture from air. The desorption is performed even by flowing hot air, in case of VOC desorption prior incineration or by flowing steam, in case of VOC recovery and recycling. These two processes do not seem adapted for CO2 recovery. The first one needs the compression of a very large volume of gas. TSA based on air heating will produce diluted CO2.

In this work, we study ESA (Electrothermal Swing Adsorption),VSA (Vacuum Swing Adsorption) and their combination.

Three adsorption processes were examined experimentally and theoretically: Electrothermal Swing Adsorption (ESA), Vaccum Swing Adsorption (VSA) and coupled ESA-VSA. Different adsorbents were tested (activated carbon, carbon molecular sieves, polymer adsorbents, zeolites...) and zeolite 5A was chosen due to its high adsorption capacity and selectivity. The adsorption cell (column: diameter = 35 mm, height = 100 mm) contained a thread in contact with the adsorbent. Carbon dioxide and nitrogen concentrations were recorded using an IR spectrometer and a gas chromatograph. Three solutions were tested for desorption:

1- ESA: at the end of the adsorption step, the entrance of the column is closed and an electrical current is applied to the thread. Carbon dioxide desorbs as the temperature increases and flow out the column. This recovered CO2 is very pure (its concentration reaches 99.9 %). More than 70 % of the total adsorbed CO2 is recovered at 200 °C.

2- Coupled ESA-VSA: after ESA, the pressure is reduced. With the two steps together, more than 98 % of the total adsorbed CO2 is removed from the column. Its purity is higher than 99 %Vol.

3- VSA: regeneration with vacuum only was also studied. The duration of the regeneration step is reduced. Not only the regeneration is performed without heating the bed but also the temperature of the adsorbent decreased of more than 10 °C. The most part of the adsorbed CO2 is recovered with a high purity.

A theoretical investigation was done by numerical solving of equations using COMSOL Multiphysics software.

222c Parametric Analysis of CO2 Capture by Temperature Swing Adsorption Using 5A Zeolite

Marc Clausse, Jérôme Mérel, and Francis Meunier. Cnam - IFFI (EA 21), case 331, 292 rue Saint-Martin, 75141 Cedex 03, Paris, France

To deal with the Greenhouse Gases (GHG) emission increase, CO2 capture and storage is one of the possible way to lower emissions due to fossil fuel use for electricity production. Different technologies for CO2 capture are currently under development: membrane separation, physical/chemical absorption, adsorption. If the separation is easier in the pre-combustion case as the CO2 molar fraction is high (typically about 30%), post combustion capture is the main way to lower the emission of existing power plants and future power plants where CO2 is produced during the combustion. The major difficulty of post combustion capture is to produce a highly concentrated CO2 stream matching the purity requirement for transportation while CO2 is diluted in the flue gas. Its concentration ranges between 4% for Natural Gas Combined Cycle and 14% for pulverised coal for example. Of course, the energy consumption has to be low while keeping high CO2 recovery. Chemical absorption is today reference process with an energy cost ranging from 1.2 to 4 MJ/kgCO2 depending on the amine type and process configuration. Adsorption processes were first not really considered as competitive for CO2 capture when compared to absorption. Nevertheless, recent work has proven the interest of PSA/VSA processes for this application showing low energy consumptions. However the consumed energy is of mechanical work type whereas heat is directly used in chemical absorption. For these reasons, we have chosen to evaluate the potential of post combustion capture by Temperature Swing Adsorption (TSA). Contrarily to PSA/VSA, TSA can be directly heat driven. However, the main drawbacks of TSA are its low productivity, which results in large adsorbent, amount and adsorbate dilution during desorption because of the regeneration by hot gas. To avoid these drawbacks, an indirect TSA process developed in our laboratory is used. The originality of this process comes from the indirect heating during the regeneration step of the process using an internal heat exchanger. Thanks to this heat exchanger, the adsorption step is cooled as well.

In this paper, a numerical parametric study of our adsorption process is presented. The model describes mass, energy and momentum balance coupled with equations for equilibrium, equation of state and thermodynamic or transport properties. The following input parameters were used : feed gas CO2 composition, temperature and flowrate, desorption temperature, and purge flowrate. Furthermore, a pre-cooling step influence was addressed as well. To qualify the performances, CO2 capture rate, purity, productivity and specific energy consumption were used. If very high CO2 purities can be achieved (>95%) a trade-off has to be found with productivity. Concerning energy consumption, the net amount is considered but the temperature level of the requested heat source is discussed as this aspect will be of major importance for process integration.

222d Temperature Dependent Non-Equilibrium Kinetic Model That Describes The Reversible Adsorption And Desorption Behavior Of CO2 In A K-Promoted Htlc

Armin D. Ebner, Hai Du, and James A. Ritter. Department of Chemical Engineering, University of South Carolina, Columbia, SC 29208

Ritter and co-workers [1] recently showed that an initially fast adsorption or desorption phenomenon was followed by a state of extremely slow uptake or release of CO2 in K-promoted HTlc that requires hours perhaps days to reach equilibrium. The CO2 loadings at such an equilibrium state were also considerably different than those attained during the initial stages of adsorption or desorption. Then, they developed a model to describe this behavior in terms of a Langmuir-Hinshelwood type of kinetic mechanism [2].

That model was a based on a non-equilibrium kinetic approach that describes the reversible adsorption and desorption behavior of CO2 on a K-promoted HTlc. It combined adsorption, diffusion and reaction together with a Langmuir-Hinshelwood mechanism to describe the uptake and release processes of CO2 on a K-promoted HTlc. The results from that work convincingly revealed that the adsorption and desorption behavior of CO2 on a K-promoted HTlc is associated with complex, highly coupled, completely reversible adsorption, diffusion and reaction phenomena. It also revealed that a true equilibrium state might never be achieved in this material or that it might take an inordinate amount of time to achieve, with the claimed equilibrium states reported in the literature perhaps only being associated with the initial fast stages of adsorption or desorption observed by Ritter and co-workers [1,2].

The objective here is to report on the development of a temperature dependent form of the non-equilibrium kinetic model that describes the reversible adsorption and desorption behavior of CO2 on a K-promoted HTlc also as a function of temperature. This model combines adsorption, diffusion and reaction together with a Langmuir-Hinshelwood approach to describe the uptake and release processes of CO2 on a K-promoted HTlc but now over a broad range of temperatures. Experimental results and corresponding model predictions will be presented in terms of the CO2 loadings in a K-promoted HTlc during repeated CO2 adsorption and desorption cycles with a half cycle time of 60 min and for temperatures ranging between 300 and 440 oC.

1. A. D. Ebner, S. P. Reynolds and J. A. Ritter, "Understanding the Adsorption and Desorption Behavior of CO2 on a K-Promoted HTlc through Non-Equilibrium Dynamic Isotherms," Ind. Eng. Chem. Res., 45, 6387-6392 (2006).

2. A. D. Ebner, S. P. Reynolds and J. A. Ritter, "Non-Equilibrium Kinetic Model that Describes the Reversible Adsorption and Desorption Behavior of CO2 in a K-Promoted HTlc," Ind. Eng. Chem. Res., 46, 1737-1744 (2007).

222e Multifunctional Carbon Filter Process For CO2, Nox, Sox, And Mercury Flue-Gas Clean-Up

Maciej Radosz, Xudong Hu, and Youqing Shen. Chemical and Petroleum Engineering, University of Wyoming, 1000 E. University Ave, Laramie, WY 82071

A Multifunctional Carbon Filter Process (MCFP) is developed and evaluated for separating not only carbon dioxide but also other pollutants of power-plant flue gas, such as NOx, SOx and mercury. Relative to an amine-absorption benchmark, the cost of the recovered CO2 using a carbon-rich sorbent at ambient pressure can be reduced by a factor of 2 or more. Such a sorbent is selective at near ambient temperatures, easy to keep at constant sorption temperature, and easy to recover by direct steam heat because its heat of sorption is very low, which will be illustrated with laboratory data. Furthermore, MCFP is relatively insensitive to flue-gas moisture, which poses a serious problem for more hydrophilic sorbents such as zeolites. The MCFP technology can be integrated with a coal power plant toward producing an Enhanced Oil Recovery (EOR) grade CO2, which will be illustrated with laboratory and economic data for a Wyoming power plant.

POST-COMBUSTION AND PRE-COMBUSTION CO_2 CAPTURE SOLID SORBENTS

Ranjani V. Siriwardane, U.S. Dept. of Energy, National Energy Technology Laboratory, Morgantown, WV
Robert W. Stevens, Jr., Parsons, National Energy Technology Laboratory, Morgantown, WV
Clark Robinson, Parsons, National Energy Technology Laboratory, Morgantown, WV

Introduction

Fossil fuels supply more than 98 percent of the world's energy needs. However, the combustion of fossil fuels is one of the major sources of the greenhouse gas, carbon dioxide. It is necessary to develop technologies that will allow us to utilize the fossil fuels while reducing the emissions of greenhouse gases. Commercial CO_2 capture technology that exists today is expensive and energy intensive. Improved technologies for CO_2 capture are necessary to achieve low energy penalties. Pressure swing adsorption/sorption (PSA/PSS) and temperature swing adsorption/sorption (TSA/TSS) are some of the potential techniques that could be applicable for removal of CO_2 from both high- and low-pressure gas streams.

NETL research group has developed a novel solid sorbent utilizing a liquid-impregnation technique. The sorbent is capable of capturing CO_2 in the presence of steam at ambient temperatures and can be regenerated at temperatures below 80 °C. Recent systems analysis conducted at NETL indicate that this process is economically favorable. The liquid-impregnated solid sorbent showed higher capacities during both low and high pressure tests.

According to a 1999 IEA report, the PSA and TSA processes may be more suitable for CO_2 capture from gasification processes. The system analysis conducted in the Netherlands shows that the PSA and TSA systems would be even more energy efficient for integrated gasification combined cycle (IGCC) systems if the sorbents were operational at warm gas temperatures (250–350 °C).[1] To address this problem, researchers at NETL developed a novel magnesium-based sorbent that can capture CO_2 at 200–315 °C and are regenerable at 375 °C. The capture process with this novel warm gas CO_2 removal sorbents involves a chemical reaction. The sorbent has a high CO_2 sorption capacity at 200-400 °C, which is considerably higher than that of the Selexol process.

NETL researchers have developed a novel sodium-based sorbent at that can capture CO_2 at 600 °C. The sorbent is regenerable at 900 °C and 1 atm. This sorbent may be suitable for more severe capture conditions as well as sorbent-enhanced water gas shift reaction.

The objective of this work is to develop regenerable sorbents that have high selectivity, efficient regenerability, and high adsorption capacity for CO_2 over wide temperature ranges. These properties are critical for the success of the PSA/PSS and TSA/TSS processes.

Experimental

The liquid-impregnated solid sorbent pellets suitable for fixed bed applications were prepared at Süd-Chemie utilizing the procedure provided by NETL.[2] Competitive gas adsorption studies were conducted in a bench-scale fixed-bed reactor (2 in. diameter with 6 in. bed height) at 1 atm (~1.01 x 10^5 Pa) and at 30 °C using a gas mixture with a composition that represented flue gas (15 percent CO_2, 82 percent nitrogen, and 3 percent oxygen saturated with water vapor) . The

---

[1] I. Smith, "CO_2 Reduction Prospects for Coal," IEA research report, ISBN 92-9029-336-5, 1999.
[2] R.V. Siriwardane, "Solid Sorbents for Removal of Carbon Dioxide from Gas Streams at Low Temperatures," U.S. Patent 6,908,497 B1, June 21, 2005.

regeneration was conducted at 80 °C with steam and N_2 mixture. A liquid impregnated solid sorbent (<700 microns) suitable for other reactor bed applications such as fluidized reactor bed was also prepared. These sorbents were also tested in a lab scale reactor with simulated flue gas at 30-60°C and were regenerated at 80 °C.

Testing was conducted with the magnesium-based solid sorbent at 200–250 °C in flow reactors at 10-30 atm and was regenerated at 375 °C at both 1 atm and 10.2 atm with steam and N_2. The space velocity used for the test was 250-1000 h^{-1}.

Testing was conducted with the sodium-based solid sorbent at 600°C in a flow reactor at 1 atm. Regeneration was performed at 885°C at 1 atm with steam and N_2. The gas hourly space velocity for both capture and regeneration was 500 h^{-1}. The sorbent was also employed during water gas shift reaction testing to evaluate its ability to enhance the reaction. Sorbent-enhanced water gas shift testing was performed at 300-600°C at 1-20 atm with a reactor feed consisting of 15 percent CO, 18 percent H_2O, and 67 percent He.

Results and Discussion

Atmospheric Pressure Tests with the Liquid-Impregnated Solid Sorbent

When the sorbent was tested at an atmospheric pressure of 30 °C, it was able to remove CO_2 from 15–17 percent to parts per million (ppm) levels from a gas mixture containing 15–17 percent CO_2, 3 percent O_2, and 82 percent N_2 saturated with water. The removal efficiency was 99 percent at 1,000 hr-1 space velocity. Multi-cycle tests conducted with the fixed bed sorbent showed that the sorbent reactivity was stable indicating that the regeneration of the sorbent at 80 °C with steam did not affect the sorbent's performance. The sorbent CO_2 sorption capacity was 1.5–2.0 moles per kg. Preliminary systems analysis conducted at NETL indicated that the CO_2 removal utilizing the solid sorbent is more economical than the process with liquid amines since the regeneration of the solid sorbent requires less energy.[3] Since steam does not affect the sorbent's performance, regeneration in the presence of steam would be a reasonable choice. The ability to regenerate the sorbent at a lower temperature and the resistance to steam are advantages of this novel, ambient-temperature CO_2 removal sorbent. The sorbent (< 600 microns) prepared for other reactor bed showed 99% removal efficiency and CO_2 capacity of about 1.5 moles per kg at 40 °C. This is very close to the acceptable CO_2 capture capacities identified by the reactor design studies.

Atmospheric Pressure Tests with the Liquid-Impregnated Solid Sorbent Suitable for Fluidized-Bed and Isothermal-Bed Reactors

The sorbent was prepared with a particle size less than 600 μm as required for fluidized-bed and isothermal-bed reactors. The reactor design studies conducted at NETL indicated that heat of regeneration of the sorbent is strongly depended on the specific heat capacity of the sorbent, the CO_2 capture delta capacity (difference between CO_2 sorption capacity at the absorption temperature and the CO_2 sorption capacity at regeneration temperature) and the heat of sorption of CO_2. For the liquid impregnated clay sorbent, the required minimum CO_2 delta capacity is estimated to be about 1.6-1.9 mol/kg for isothermal-bed reactors. However, this requirement may be lower for the fluidized-bed reactors. The sorbent (<600 μm) was tested with simulated flue gas; results of the testing yielded a 99 percent removal efficiency of CO_2 and a capture capacity of approximately 2.1 mol CO_2/kg of sorbent at 40 °C as shown in Figure 1. Since the particle density of the sorbent is ~2 g/cm3, the corresponding CO_2 capture capacity (volume basis) is approximately 4.6 mol/liter of sorbent. The sorbent can be regenerated at 80-100 °C; since steam will likely be utilized during regeneration, the

[3] T. Tarka, "System Analysis of NETL Solid Sorbent," internal report, U.S. Department of Energy, NETL, 2005.

majority of the regenerations were conducted at 100 °C. The CO_2 capture delta capacity for sorption at 40 °C and regeneration at 100 °C was determined to be 1.6 mol/kg (3.2 mol/liter) and 1.3 mol/kg (2.6 mol/liter) for sorption at 60 °C. The value obtained at 40 °C is very close to the acceptable CO_2 capture delta capacities identified by the reactor design studies. A 10-cycle test performed with capture at 60 °C and regeneration at 100 °C showed stable reactivity. The sorbent preparations were also made in collaboration with Süd Chemie, Inc. to develop materials with more attrition resistance.

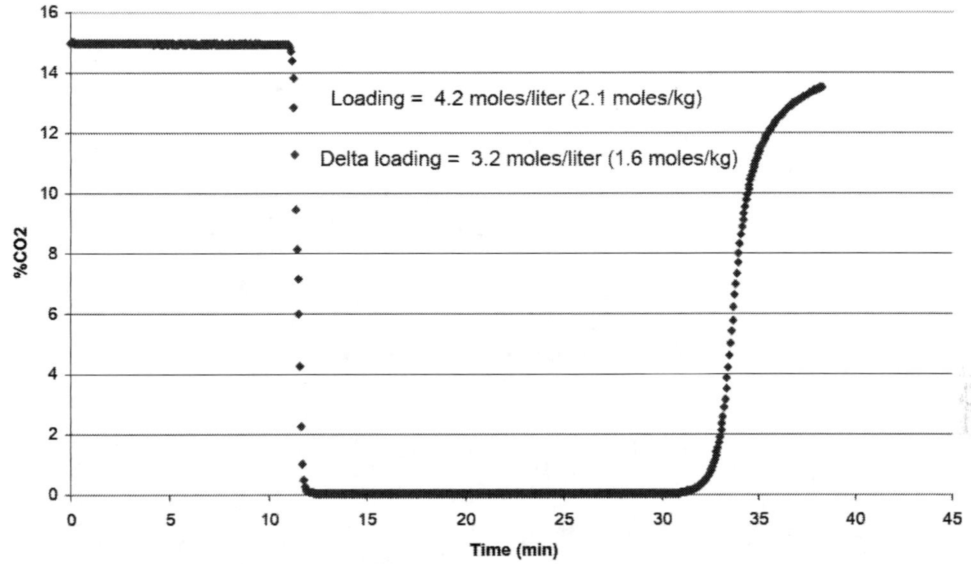

FIGURE 1. Lab-scale flow reactor test data with sorbent (particle diameter < 600 μm) suitable for fluidized-bed/isothermal-bed reactors at 1 atm; 40 °C; GHSV ~ 500 h^{-1}; Feed composition = 16% CO_2, 3% O_2, 82% N_2, saturated with water vapor.

Warm Temperature Sorbent

The CO_2 capture sorbents that operate at 200–315 °C would be very useful for IGCC applications. The water gas shift reactor converts the gas to CO_2 and H_2 at 200–350 °C. A pure hot H_2 stream can be obtained if CO_2 is removed at 200–315 °C. Magnesium-based sorbents prepared at NETL were tested in a lab-scale reactor at 200–250 °C at 10.2 atm with 28 percent CO_2 in He saturated with steam. Regeneration was conducted at 375 °C and 10.2 atm. A 10-cycle flow reactor test conducted at 250 h^{-1} space velocity with about 3 grams of magnesium-based sorbent indicated that the sorbent has a very high CO_2 capture capacity (approx. 4 moles per kg) as shown in Table 1.

Further testing was conducted at a bench scale to evaluate process variables (i.e., temperature, pressure, H_2O feed content, etc.). Testing was conducted at 200-250 °C, 10-30 atm, with H_2O content between 10-30 percent and with 10 percent CO_2. Testing revealed that an increase in pressure and/or H_2O concentration resulted in improved CO_2 capture capacity. The test data at 10.2 atm and 15 percent H_2O is shown in Figure 2.

Regenerable sorbents at 200–315 °C with high CO_2 capture capacities are not reported in the literature, and these novel sorbents offer great promise for IGCC applications. The high capacities will contribute to low regeneration cost and small vessel size. The regeneration temperature of the sorbent is 375 °C, and the temperature swing from absorption to regeneration is very low. Preliminary system analysis conducted at NETL showed promising data with restricted use of steam during regeneration.

Table 1. Test conditions and capture capacities during the 10-cycle test of the magnesium-based warm gas temperature CO_2 removal sorbent.

	CO_2 Sorption			Regeneration	
Cycle	T (°C)	P (atm)	Capacity (mol/kg)	T (°C)	P (atm)
1	200	6.8	1.47	375	1.0
2	250	6.8	2.11	375	1.0
3	250	10.2	2.13	375	1.0
4	200	10.2	2.84	375	1.0
5	200	6.8	2.40	375	1.0
6	200	10.2	3.46	375	10.2
7	200	10.2	4.09	375	10.2
8	200	10.2	4.14	375	10.2
9	200	10.2	4.27	375	10.2
10	200	10.2	3.22	-------	-------

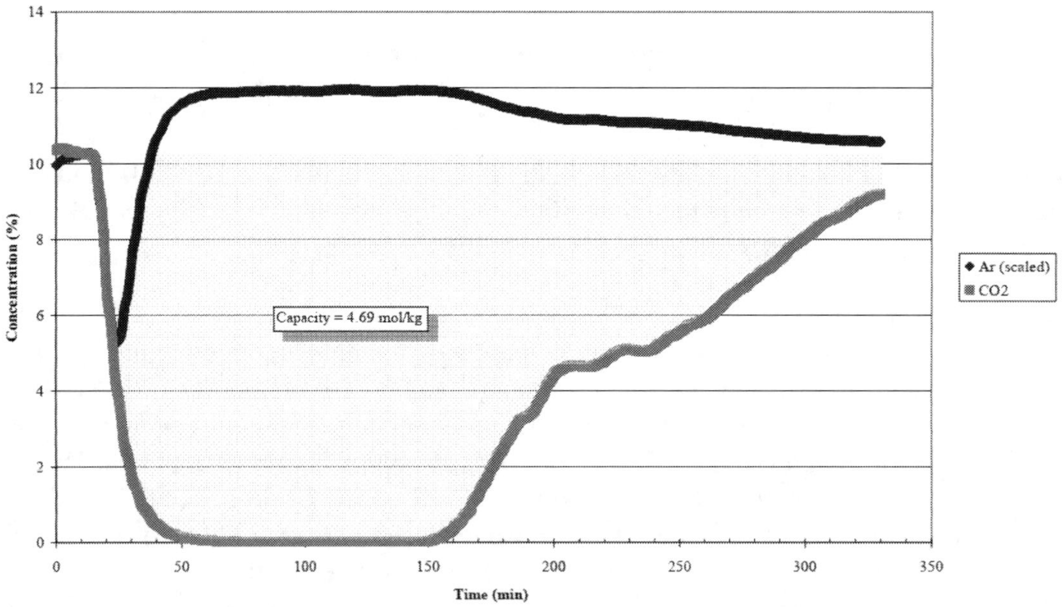

Figure 2. Bench-scale testing with magnesium-based sorbent. Test performed at 200°C, 10.2 atm, 250 h^{-1}. Feed composition = 10% CO_2, 15% H_2O, 10% Ar, bal. N_2.

High Temperature Sorbent

Sorbents that operate at temperatures higher than 300 °C would be useful for more severe capture requirements. Sodium-based sorbents prepared at NETL were tested in a bench-scale reactor at 600 °C and 1 atm with a feed composition of 10 percent CO_2, 30 percent H_2, 15 percent H_2O, and 45 percent He at a space velocity of 500 h^{-1}. Regeneration was performed at 885 °C and 1 atm in flowing steam and He. CO_2 capture capacity of this sorbent was determined to be very high (5-7 moles per kg). A multi-cycle test was conducted; capture capacity of the final capture cycle was greater than the first capture cycle. This sorbent also enhanced the water gas shift reaction and converted CO/H_2O to H_2 without an additional catalyst as shown in Figure 3.

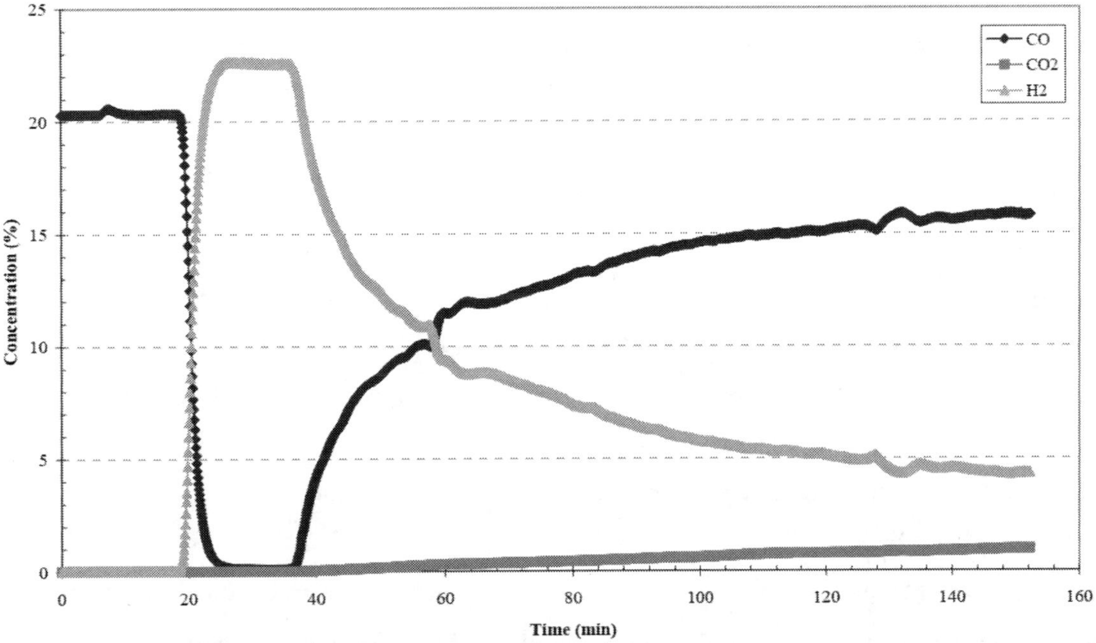

Figure 3. Bench-scale sorbent-enhanced water gas shift reaction testing with sodium-based sorbent. Test performed at 600°C, 1 atm, 500 h^{-1}. Feed composition = 15% CO, 18% H_2O, 67% He.

Summary

The novel liquid-impregnated solid sorbent showed stable CO_2 sorption performance at ambient temperature and atmospheric pressure during a multi-cycle test and regenerability below 80 °C. The sorbent's performance was not affected due to the presence of steam. The capture capacity of the liquid-impregnated solid sorbent is significantly higher at 20 atm. Acceptable sorbent capacity and delta loading was achieved for 40 °C sorption and 100 °C regeneration.

The novel magnesium-based sorbent was able to capture CO_2 (4 moles per kg) at 200–250 °C and 10.2 atm, and it was possible to regenerate the sorbent at 375 °C and 10.2 atm. The performance was stable during a 10-cycle test.

The novel sodium-based sorbent was able to capture CO_2 (5-7 moles per kg) at 600 °C and 1 atm; it was possible to regenerate the sorbent at 885°C and 1 atm. The performance was stable over a multi-cycle test. This sorbent was also able to enhance the WGS reaction to produce H_2 without a catalyst.

Acknowledgements

This technical effort was performed in support of the National Energy Technology Laboratory's on-going research in carbon capture under the RDS contract DE-AC26-04NT4181. Assistance provided by Frank Thomas, Michael Thomas, and Al Batton during testing is greatly appreciated.

SEGREGATION OF SOLIDS IN A BATCH FLUIDIZED BED

Vaka Murali Mohan1[a], Nageswara Rao V2[a], Rajendra Prasad, P3[a*], Sujatha, V4[a*]*
and
Ramappa Siddi 5[b]

a Department of Chemical Engineering, Andhra University, Visakhapatnam – 530 003 , A.P, India; e-mail: prprasad_au@yahoo.co.in
b Department of Biotechnology, ANITS, Visakhapatnam, A P, India

ABSTRACT

Fluidization has many industrial applications. One of the applications of the fluidization is segregation of particles. The segregation of the particles occurred in a gaseous fluidized bed when the particle difference in their size and density. This paper reports the influence of size difference, particle density, gas flow rate and weight proportions of particles on their segregation and correlates the effect of various parameters on the separation.

Keywords : Segregation, fluidization, modeling, size difference, particle density.

INTRODUCTION

Fluidization is the technique by which solids are transformed in to fluid like state through contact with the fluid, either gas or liquid. The technique has been tried since 1910 and a commercial Winkler gasified came into existence in 1921. Since then fluidization is being in use in several physical operations like heat and mass transfer, drying, coating and for chemical operation like combustion, cracking, and chemical synthesis. Several industrial gases like so_2, co_2 etc is being obtained by fluidized bed roasting of ores. The importance of this technique in industrial applications has led many investigators and institutions to work on it and several books have appeared like [2-4,6,7,9,10,16]. Enhanced research interest on fluidization has led to organize many symposia and conferences at different places like those of M.I.T. in 1948, in London in 1951 [5] and at kharagpur in 1964. The meeting in 1955 at Brooklyn polytechnic institute received process data. in 1956 French association for the study of fluidization conducted a meeting at which many papers on fundamental and process were presented. In December 1957, a special symposium on fluidization was conducted in Chicago by AICHE. Despite its serious drawbacks the compelling advantages of overall economy of fluidized contacting have been responsible for its successful use in industrial operations. This is possible due to the efforts of research made to overcome the above mentioned draw backs by making modifications in the bed. One such modification was by sectioning the bed using horizontal or vertical baffles, known as multistage fluidized bed [8, 12-15]. Multistage fluidized beds with horizontal baffles involves [12-14] careful down comer design while those with vertical baffles [1,15] involve careful design of gas distributor. Other investigations shows that the efficiency of fluidized bed can be improved by use of packing inside the bed use of fines which act as lubricants in a bed of coarse particles and use of vertical internal like a spiral internal [11]in a single state fluidized bed.

All the above investigations have been made aiming at reduced gross irregularities and instability of the bed induced by gas flow so that the advantageous characteristics prevail and can be made use of it to a maximum possible extent. But all these modifications increase complexity of the bed rendering control and operation more difficult.

Hence the best way can be the use of a single stage fluidized bed with minimum modification in the geometry and reducing the spread in residence times in a different way. The present work has been carried out in this line as mentioned above the fluidized bed with uniform solids has the same usual drawbacks. But these can be reduced by using a material of which the major portion wills another material. This reduces the spreads of solids residence time compared to the bed of uniform particle size.

EXPERIMENTAL SET UP

A schematic of the experimental setup used in the present study is shown in Fig. 1. The experimental setup consists of a fluidized bed of 69 mm internal diameter and a height of 60 cm made of acrylic column. A perforated distributor plate with a free area of 13% (2 mm perforation diameter with 5 mm triangular pitch) is used for uniform distribution of gas. A fine mesh is fixed to the perforated plate to prevent the particle flow through the distributor plate. A bottom discharge pipe with a provision to vary different diameters is used. This pipe is directly connected to the bottom product collection tank. An over flow pipe of 25 mm diameter is provided at a height of 23.5 cm from the distributor plate to withdraw the top product. Air as a fluidizing medium is introduced at the bottom of the column from the compressor, pressure regulator and calibrated rotameters. For uniform distribution of air, glass beads of 2-3 mm diameter are packed between the distributor plate and inlet of the air. The binary heterogeneous solid mixture to be separated is fed continuously into the bed at the top of the column under gravity flow from a solid feed tank (hopper). Desired flow rates of solid can be obtained by moving a scale (No. 9 in Fig. 1) horizontally with different sizes of apertures. The steadiness of solid flow rate was achieved by maintaining sufficient height of solids in the hopper. To measure the pressure drop across the bed, two pressure taps are provided, one below the distributor plate and the other at the top of the column. Pressure drop is continuously recorded by a differen- data-acquisition unit consists of a PC equipped with 12-bit A/D data acquisition board at a rate of ten samples per second. A digital camera is placed near the fluidized bed to record the growth of jetsam layer as a function of time at a rate of one frame per second, which is connected to the computer through pressure transducer connected to data-acquisition unit.

PROCEDURE

The separation of solids may be carried for the solids having different sizes, densities or shapes. Either incipient fluidization technique or particulate fluidization technique was used for separation of solids in a fluidized bed. Depending on the process industries any one of the above method was used. Presently it was difficult to separate the solids by incipient fluidization method, because after separation it needs drying of solids and also it was difficult to separate the solids which were coming along the fluid. So the other method, particulate fluidization was preferred for segregating the solids of different sizes. It was therefore preferérably suggested to particulate fluidization, because the difference between the density of fluid and material was more comparable to that of incipient fluidization. Initially various bed heights Ho=4, 8, 12 cm of binary mixtures were fed to the column and known amount of fluid was passed through the distributor zone. Before the fluid enters into the fluidization zone, it was conformed that the distribution in the column was completed. Based on the difference in size the flotsam particles moves at the top and the jetsam particles will settle at bottom. Every batch of solids was fluidized for 3-4 min and the samples were collected through the sampling pores situated at various heights H=15, 30, 45 cm respectively. The collected samples were analyzed by means of sieve analysis and the fraction of fines was estimated. During the process the pressure drop along the column was noted down. The fine fractions were calculated for various heights. The same procedure was repeated for the various feed compositions (70:30, 50:50, 30:70) of coarse and fine for various bed heights.

The present study confined to the study on the effect of feed composition and the effect of bed heights on fine separation of solids. As many as thirty experiments considering various alternatives have been carried.

FITTING OF THE CORRELATION

From the obtained results of these experiments a correlation was developed with the help of a numerical method called least square analysis by using least square analysis. Various criteria might be used to estimate the coefficients in a model from experimental data. In present section it was assumed that $y = u(x)$ and that N discrete values of y are known along with a corresponding number of x. these may be designed as $y_1, y_2, \ldots.. y_N$ and $x_1, x_2,$.

. . . x_N. No particular spacing on the x values was assumed, and some may even repeated. These values may result from measuring the response of a system or process to an input variation, or they may be specific values from transcendental function.

The suitable approximation was here in taken as a linear combination of (m+1)

Linearly independent functions $u_1(x), u_2(x), \ldots u_m(x)$. Thus

$$\Phi_m(x) = a_0u_0(x) + a_1u_1(x) + \ldots + a_mu_m(x) \qquad [1]$$

Where the $a_0, a_1 \ldots a_m$ are constants to be determined from the fit of [1] to the specified values $y_1, y_2, \ldots y_N$.

There are a number of possible linearly independent analytical functions which can be in [1]. If, however, N>(m+1), an obvious conclusion is that the approximation cannot, in general, coincide with all the data. In this case the means of determine the coefficients in[1] such that the resulting expression is the 'best' fit to the data. When the values $y_1, y_2, \ldots y_N$ are exact or are subject to random errors, it can shown that the least-square error criterion yields the best estimation for $a_0, a_1 \ldots a_m$. This criterion states merely that if the approximation error at each point is given by

$$R_i = |\, y_i - \Phi_m(x_i)\, | \qquad i = 1, 2, 3, \ldots N \qquad [2]$$

Then one desires that

$$R_1^2 + R_2^2 + \ldots + R_N^2 = \text{minimum} \qquad [3]$$

The sum of the squares of the residuals is given by

$$S_m^2 = \sum R_i^2 >= 0 \qquad [4]$$

That is necessary that

$$\partial S_m^2 / \partial x_i = 0 \qquad i = 1, 2, 3, \ldots N \qquad [5]$$

Equation [5] leads to the minimum form

$$\partial S_m^2 / \partial x_i = 0 = 2\sum (y_i - \sum a_k x_i^k)(x_i^j) \qquad k = 0, 1 \ldots N \qquad [6]$$

By simplifying the nomenclature and Expanding this equation results in

$$P_j = \sum a_k n_{jk} \qquad j = 0, 1 \ldots m \qquad [7]$$

$$P_0 = a_0n_{00} + a_1n_{01} + \ldots + a_mn_{0m}$$

$$P_1 = a_0n_{10} + a_1n_{11} + \ldots + a_mn_{1m}$$

$$\ldots\ldots\ldots\ldots\ldots\ldots \qquad [8]$$

$$P_m = a_0n_{m0} + a_1n_{m1} + \ldots + a_mn_{mm}$$

These equations were called the normal equations for the least square formulation. The unknown can be determined, at least theoretically, by any of the matrix methods.

The derivation of [8] can be obtained in an almost trivial way by using matrix notation as follows

$Y = [\, y_1, y_2 \,..\, y_N]^T$

$a = [\, a_0, a_1 \,....\, a_m]^T$

$$X = \left\{ \begin{matrix} x_1^0\, x_1^1 \,..\, x_1^m \\ x_2^0\, x_2^1 \,....\, x_2^m \\ \cdots\cdots\cdots \\ x_N^0\, x_N^1 \,...\, x_N^m \end{matrix} \right\} \quad [9]$$

where X is a matrix of order of N times (m+1). Premultiplying X_a and y by X^T and setting the results equal to each other yields

$$X^T X_a = X^T y \qquad [10]$$

$$\text{Then } a = (X^T X)^{-1}(X^T y) \qquad [11]$$

This was done with the help of a code developed in C language. Comparison of Experimental U_{mf} with Theoretical U_{mf} was shown in Table 1.

Theoretical U_{mf} was calculated using Chetester et al (1984) equation,

$$U_{mf} = (\mu / d_p\rho_g)\, [\, (28.7)^2 + 0.0494\, (d_p^3 \rho_g (\rho_s - \rho_g) g) / \mu^2 \,]^{0.5} - (\mu / d_p\rho_g)\, (28.7) \qquad [12]$$

Where U_{mf} = minimum fluidization velocity, μ = Viscosity of air at room temperature d_p = average diameter of the sand particle, ρ_g = density of air, ρ_s = density of sand

RESULTS AND DISCUSSION

In particulate fluidization some of the following parameters may effect the separation of fines from coarse particles. Various feed compositions of solids, Various bed heights, Density difference in feed material, Fluid flow rate. The calculation of empty tower pressure drop Vs superficial velocity of air was shown in fig 3 and reveals as the velocity increases the pressure developed in the empty column. The calculation of Minimum fluidization Velocity for various size sand particles for L/D = 1 was shown in fig 4 of various size solid particles and reported the velocity increased up to a maximum and then remains constant. The effect of diameter of particles on terminal velocity for various densities was shown in fig 5 and reports as the density of the particle increased, the effect of the diameter of the particle increased with terminal velocity.

The effect of Bed height on fine segregation, for various feed composition of solids fed to the column were shown in fig 6. For a particular Ho, the separation of fines increased with the increase of fine composition in the feed mixture, this may be due to the increase of fines which fluidizes more at a fixed velocity compare to that of coarse particles. The individual plots for fixed feed compositions shows that the fines concentration increased with the decrease of bed heights this may be the reason due to decrease of coarse particles in the solid mixture fed to the column, i.e. having a change of fluidizing only fine particles more compare to coarse particles.

The effects of feed composition on the fine separation of solids were shown in fig 9. For a particular bed height and for a particular collecting sample pore height the separation of fines increased with the increase in the ratio of coarse particles to the fine particles, because fine particles fluidizes more than the coarse particles. The individual plots shows that the collection of fine particles increased with the increase in of height from the distributor plate, because the jetsam particles always tends to settle down at the bottom and the floatsam particles always tends to move at the top. The effect of Umf for various size and various samples was shown in fig 12 and reveals as Umf increased with the diameter of the particles, as the density of the particles increased.

CONCLUSIONS

Experimental investigation of solids separation aspects using a binary solid mixture using different bed heights and different feed compositions were done. Fitting the results with a suitable model and estimation of the parameters. Measurement of pressure drop due to solids. velocity increases the pressure developed in the empty column. The effect of the diameter of the particle increased with terminal velocity. the fines concentration increased with the decrease of bed heights this may be the reason due to decrease of coarse particles in the solid mixture fed to the column

REFERENCE

1. Beran, Z and Lutchaa, J., `Optimising Particle Residence Time in a Fluidized bed Drier`, The Chem. Engg., 678 (1975).
2. Botterill, J.S.M., `Fluid Bed Heat Transfer`, Academic Press London (1975).
3. Cheremisinoff, Nicholas P., `Hydrodynamics of Gas Solid Fluidization`, Gulf Publishing Company, Huston (1984).
4. Davidson, J.F. and Harrison, D., `Fluidization`, Academic Press, London (1971).
5. `Fluidization Symposium`., London, J. Appl.Chem. (London), 2, Suppl.Issue 1(1952).
6. Grace, J.R and Matsen, J.M., `Fluidization`, Plenum Press, London(1980).
7. Keairns, D.L., `Fluidization Technology`, Hemisphere Pub. Crio., Washington (1975).
8. Krishnaiah. K and Varma. Y.B.G., `Pressure Drop, Solids concentration and mean holding time in multistage fluidization,' can.j.chem.engg.,60,346[1982].
9. kunil.D and Levenspiel.O., 'Fluidization Engg.'., John Wiley & sons, New York (1969)
10. Leva.M, 'Fluidization' Mc Graw Hill Book Company, New York (1959)
11. Pydi Setty, Y., 'Axial Mixing of Solids in Spiral Fluidiuzed Beds'., Ph.D Thesis, IIT Madras, 1983.
12. Raghuraman, J and Varma, Y. B. G., 'Residence Time Distribution of Solids in Multistage Fluidized beds', Chem.Engg.Sci., 28, 305 (1973a).

13 Raghuramanrma, y. b g., a model for residence time distribution, in multistage system with cross-flow betweenactive and dead regions, chem..engg.sci.,1973.
14 Raghuraman,j and verma,Y.B.G., An experimental investigation of the residence time distribution of solids in multistage fluidition, chem.engg.sci.,30,145,[1975].
15.Reay,d., particle residence time distribution in plug flow fluid bed dryers, proc. first int. symp.drying, montreal, Canada 1978.
16.Reboux P the phenomena of fluidization, association francaise de fluidization,paris[1954].
17.Shirai, T., review of fluidization literature Tokyo iit, tokyo,1954.
18.Vanecek,V., Markvart, M., Drbohlav, R., fluidized bed drying, leonard hill, london[1966].
19. Zenz,f.a and Othmer, fluidization and fluid particle systems,pub.corp., new york[1960]
20.Rowe, P.N., Nienow, A.W.,&Agbim,A.J.Traqns.Instn. Chem.Engrs, 1972.
21 Othmer.D.F., 'Fliuidization', Reinhold publishing Corporation, New York (1956).
22. Goldstein. A.A, N.Levine,and J.B.Hereshoff: J.Assoc. Comput. Mach., 1957,4,43.

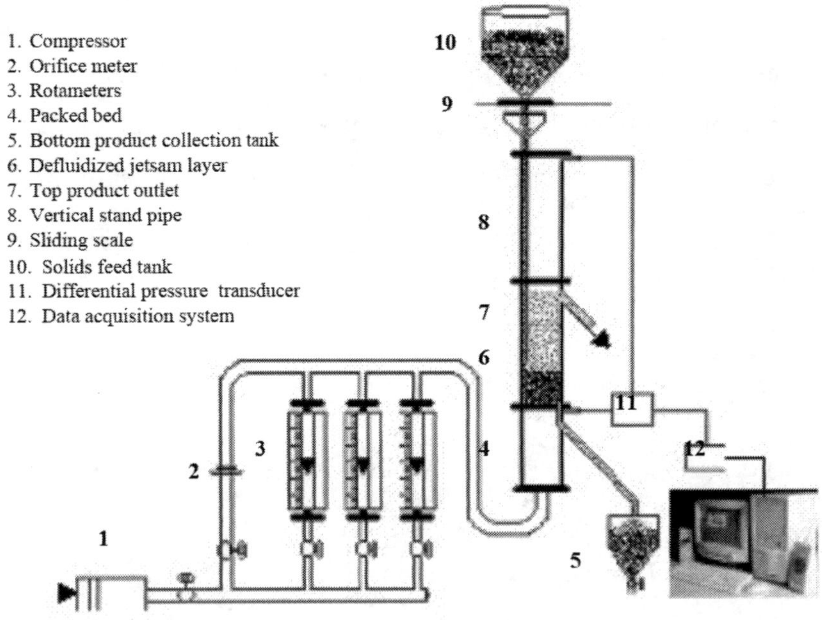

Fig. 1. Schematic of experimental setup.

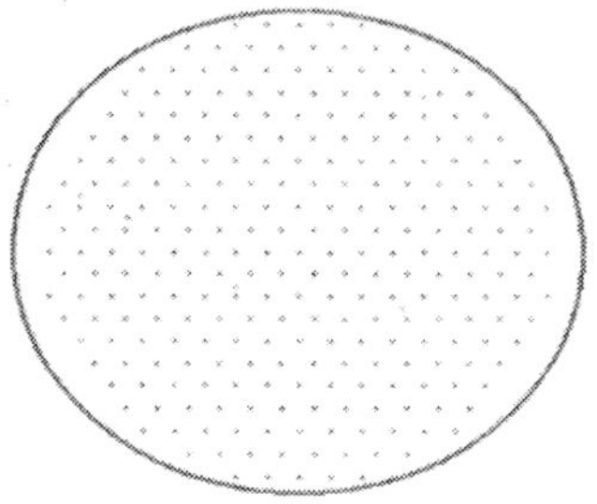

fig 2. DISTRIBUTOR PLATE DESIGN for triangular pitch of 5mm and dorr of 2mm

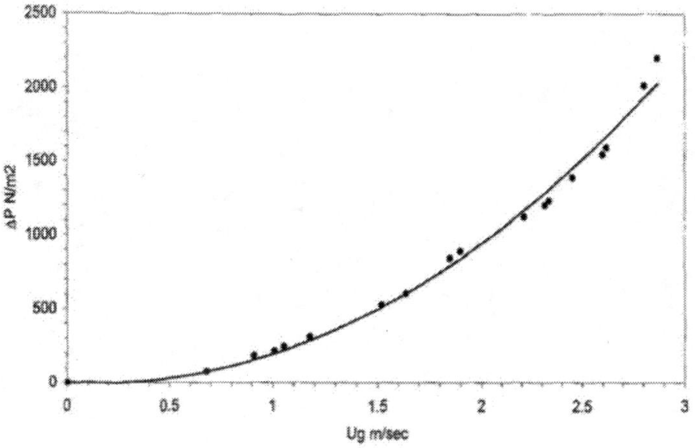

fig 3. **Calculation of empty column pressure drop at various gas velocities.**

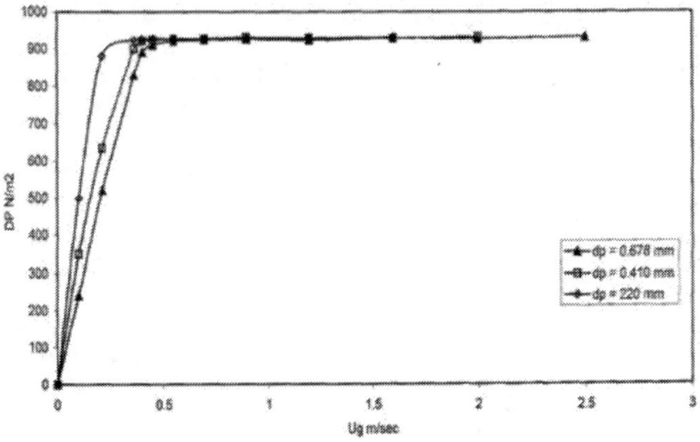

fig 4. Calculation of minimum fluidization velocity for various size sand particles for L/D = 1(nearly 350 grams of sample taken)

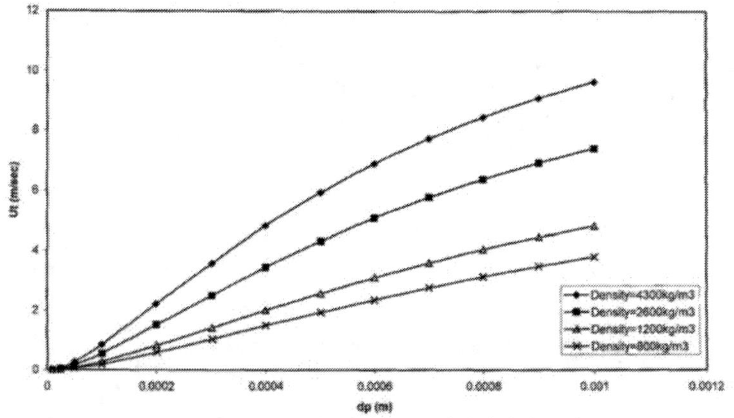

fig 5. **Terminal Velocity vs Dia. of the particle (Ut vs dp)**

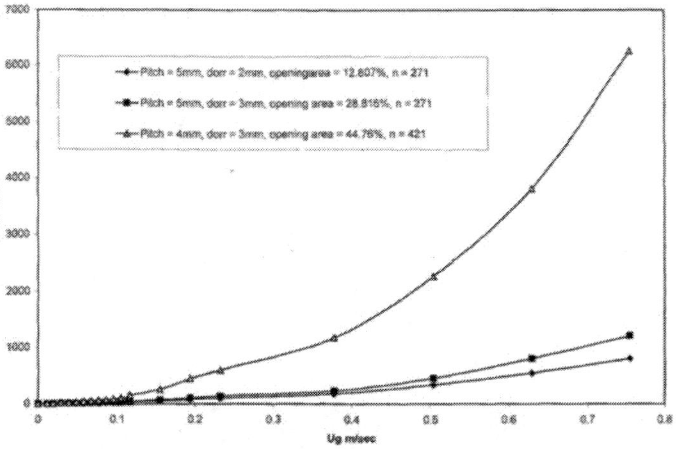

fig 12. **Calculation of Pressure drop vs empty tower velocity for various distributor plates**

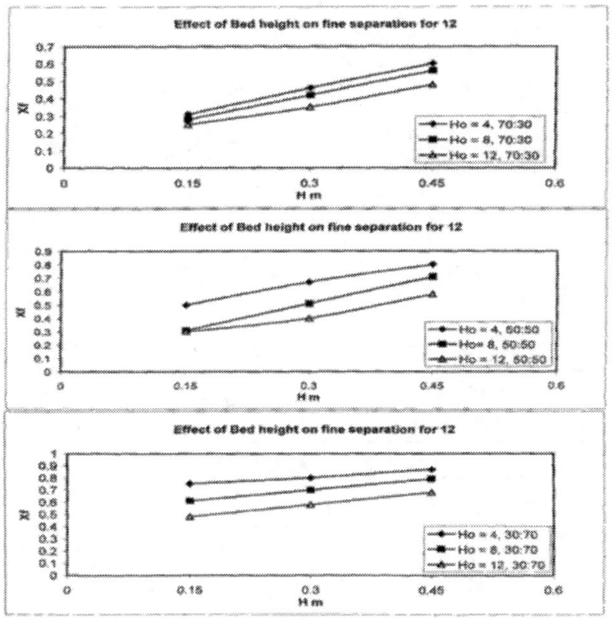

fig 6. Effect of Bed height on fine separation for 687 & 410 microns

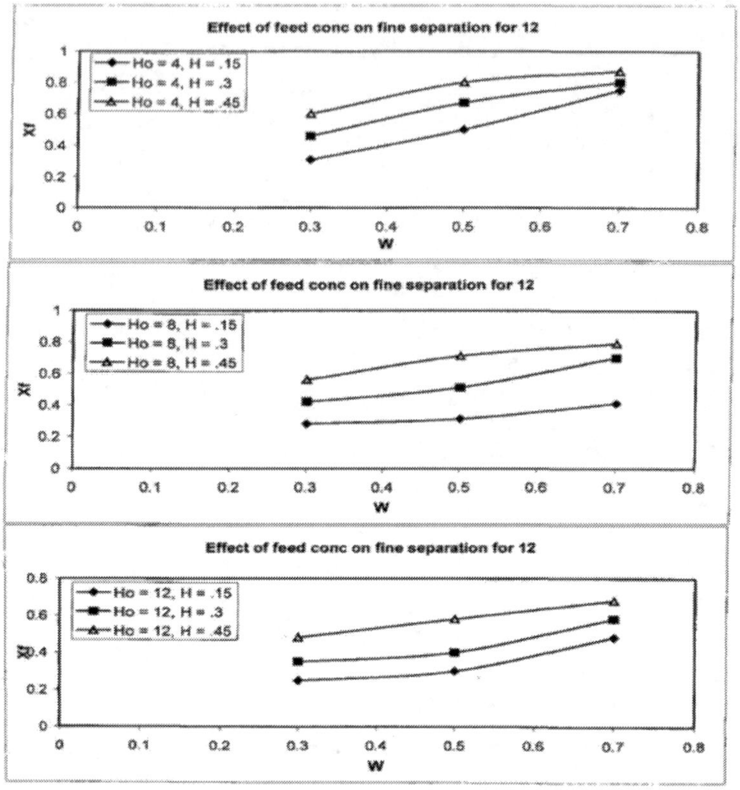

fig 9. Effect of feed conc on fine separation for
687 & 410 microns

Table 1 Comparison of Experimental U_{mf} with Theoretical U_{mf}

S.No	Size of the sample, μ	Experimental U_{mf}, m/s	Theoretical U_{mf}, m/s
1	687	0.454	0.458
2	410	0.38	0.392
3	223	0.25	0.261

CERAMIC NANOFIBER FILTER MEDIA

S. J. Park, B. Raghavan, S. Bhargava, and G.G. Chase
The University of Akron
Akron, OH 44325

Abstract

Ceramic nanofibers are synthesized by electrospinning polymer – metal organic mixtures. These fibers are combined with microfibers to form ceramic composite filters. Such filters can be used to efficiently filter submicron sized particles from high temperature gases. The fiber surfaces can be augmented with catalysts to make the filters chemically react with gases and particulates. Because the ceramics are inert at high temperatures the filters can be regenerated by heating to remove combustible particles. This paper will discuss recent developments to make ceramic nanofiber filter media.

1. Introduction

Ceramic nanofibers have a remarkable range of applications as support materials, high temperature filtration, composite reinforcements and biomedical membranes. The advantage of nanofibers is the large surface area to volume ratio. Electrospinning is a simple technique to produce submicron diameter ceramic nanofibers from polymer and ceramic precursor mixtures.[1] Calcination of polymer from the electrospun nanofibers results in ceramic nanofibers. Thus we can obtain chemically inert fibers for high temperature filtration.

The nanofibers can be mixed with microfibers to form ceramic composite filters. The combination of nano and microfiber forms filter media that can capture a wide range of particle sizes with an acceptable pressure drop. The nanofibers enhance the filter media by efficient capture of sub-micron particles gases. The ceramic nanofibers can also be augmented with catalyst particles for gas phase catalytic conversion. These catalytic filters combine filtration with catalysis.[2] They are capable of simultaneously removing sub-micron sized particulates from combustion gases and reducing chemical pollutants by catalytic reaction. Example combustion gases include those from waste incinerators, pressurized fluidized-bed coal combustors, diesel engines, boilers, and biomass gasifiers.

In this work we have synthesized alumina ceramic nanofibers augmented with Palladium catalyst. These nanofibers were mixing with commercially available alumina microfibers to from the filter media. Nanofibers are expected to improve performance of the catalyst particles because of high surface-area per unit area[3,4]. Properties of the filter media with nanofibers were measured using several techniques to characterize porosity, permeability, and the presence of catalyst. Further catalytic properties of these ceramic nanofibers will be studied. Also a theoretical modeling of ceramic fibrous filter medium was model to estimating the pressure drop and quality factor with respect to rise in temperature with some common assumptions.

2. Experimental

2.1. Preparation of electrospun fibers and characterization

Polyvinylpyrrolidone (PVP, Aldrich, Mw: 1,300,000) of 7wt % was prepared by dissolving in ethanol (AAPER alcohol). PVP solution was mixed with aluminum acetate (Aldrich), water and ethanol. Palladium chloride (Sigma Aldrich, 60 % Pd) was dissolved in the above solution in the ratio of 5 wt% $PdCl_2$ with respect to aluminum acetate. The solution is mixed overnight at 40 °C.

Nanofiber mats of palladium doped alumina nanofibers were prepared by applying voltage at 20 kV (Gamma High Voltage Research Inc.) and with 3.0 µl/min flow rate using the syringe pump (World Precision Instruments SP1011). The needle was positioned about 20 cm above the grounded collector consisting of aluminum foil collector. The nanofibers mat was calcined at 600 °C for 2hrs. These samples were characterized using Transmission Electron Microscope (TEM) and X-ray Photoelectron Spectroscopy (XPS).

2.2. Construction of filter media

The slurry for the filter media was made of alumina microfiber, binder and ceramic nanofibers. Fibers and binder were mixed in water and the slurry was vacuum molded under a vacuum of less than 5 psi to form filter disks of 2.5 cm diameter. Porosity, permeability and surface morphology of filter media with and without palladium oxide particle doped nanofibers were measured using a pycnometer, Frazier air permeability test, transmission electron microscopy (TEM) and scanning electron microscopy (SEM), and by X-ray Photoelectron Spectroscopy (XPS).

2.3 Model and assumption of fibrous filter media

The Objective of this modeling to find out the how the ceramic microfiber filter behaves with respect to rise in temperature. A hot gas (air) at 20 psi is passed through a ceramic filter at specific conditions with some common assumptions listed below.

The porosity of the filters prepared in our laboratory have porosities in the range of 0.95 ~0.98. Thus, we considered porosity to be 0.95 and filtration of spherical particles of 150 nm is considered for our modeling study. Operating conditions which are assumed for these modeling are the mass flow rate of air through the filter to be 7.25 g/s. Capture efficiency of 99% is considered for this study.

Some of the common assumptions which are taken into account for this modeling are:

1. Steady state
2. Isothermal process
3. One dimensional flow, only in Z – direction, at the volume averaged scale
4. Flow is evenly distributed among the fibers
5. The permeability of the media are uniform and isotropic
6. Incompressible filter media
7. Effects of binder materials are neglected
8. The model only considers clean filter media. Performance does not consider particle loading of the filter media
9. Captured particles do not affect pressure drop, flow rate and capture efficiency
10. All fibers are identical having uniform and identical diameters

Single fiber efficiency theory was applied to find out how the Pressure drop and Quality factor of ceramic filter changes with respect to fiber diameter for varying fiber diameters. In this model, we considered direct interception and brownian diffusion acting together for capture of sub-micron particles. Considering all above mentioned mechanisms, quality factor and pressure drop across the filter is given by Eqs. (1) and (2).[5]

$$\Delta P = \frac{16\mu L V_Z^G (1 + 1.996Kn)\varepsilon_i}{d_f^2\left[(-0.5\ln(\varepsilon_i) - 0.75) + \varepsilon_i - \frac{\varepsilon_i^2}{4} + 1.996Kn\left(-0.5\ln\varepsilon_i - 0.25 + \frac{\varepsilon_i^2}{4}\right)\right]} \quad (1)$$

$$QF = (1-\varepsilon)E_{RD}\left(\frac{4\pi V_Z{}^G \mu(1+1.996Kn)\varepsilon_i}{d_f\left((-0.5\ln(\varepsilon_i)-0.75)+\varepsilon_i-\frac{\varepsilon_i^2}{4}+1.996Kn\left(-0.5\ln(\varepsilon_i)-0.25+\frac{\varepsilon_i^2}{4}\right)\right)}\right)^{-1} \quad (2)$$

Where, μ – Viscosity of the fluid, $V_Z{}^G$ – Gas phase velocity, L - Length of the filter media, Kn – Knudsen number, d_f – Diameter of fiber, E_{RD} – Overall single fiber efficiency ε_i – Porosity, ΔP – Pressure drop across the filter, QF – Quality factor for fibrous filter

3. Results and discussion

Scanning Electron Microscope (SEM) of alumina nanofibers after heating 600 °C showed the diameter of around 100 nm. Transmission Electron Microscope (TEM) images of the fiber and palladium oxide particles on the surface of alumina nanofibers heated at 600 °C are shown in Figure 1 (b). The palladium oxide doped electrospun alumina nanofibers have diameters of around 110 nm. Particles of palladium oxide of around 20 nm size were well distributed on the surface of the supported alumina nanofibers.

(a) (b)

Figure 1. SEM image of alumina nanofibers (a) and Transmission Electron Microscope (TEM) images of palladium oxide doped alumina nanofibers (b) after heated 600 °C for 2hrs.

X-ray Photoelectron Spectroscopy (XPS) of electrospun alumina nanofiber and palladium oxide doped alumina nanofibers after heating at 600 °C for 2hrs are shown in Figure 2. Figure 2 (b) shows the chemical state of palladium from the chemical shift of the Pd ($3d_{5/2}$) peak: (1) before hydrogenation, and (2) after hydrogenation. It was located at 336.8 eV before hydrogenation and shifted 1.1 eV toward lower binding energies after hydrogenation[6]. This suggests that the palladium is in its oxide form before hydrogenation and in a metallic form after hydrogenation.

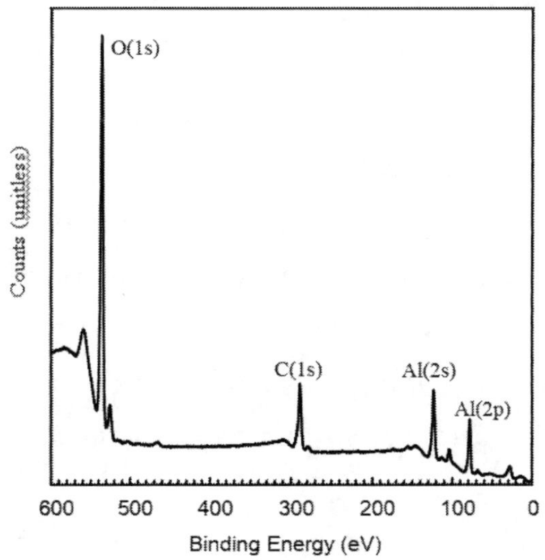

(a)

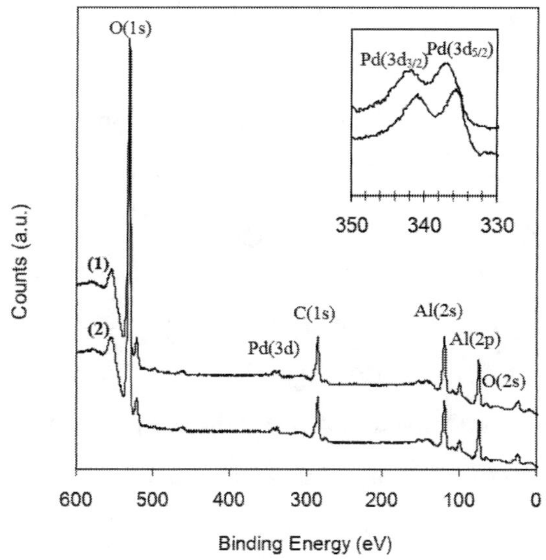

(b)

Figure 2. X-ray Photoelectron Spectroscopy (XPS) of (a) electrospun alumina nanofiber (b) alumina nanofibers doped with palladium calcined at 600 °C.

Filter media were formed from the slurry of alumina microfibers using vacuum molding apparatus. The top view of an alumina microfiber and nanofiber filter disk after heating at 600 °C for 2hrs is shown in Figure 3 (a) and image of alumina micro and nanofibers shown in (b). Ceramic nanofibers such as alumina and alumina supported palladium oxide catalysts have been successively prepared by electrospinning techniques. These nanofibers have diameters of about 110 nm. Particles of palladium oxide of around 20 nm size were well distributed on the surface of the supported alumina nanofibers.

(a) (b)

Figure 3. Filter media (a) top view of alumina microfibers filter media after heating at 600 °C for 2hrs, (b) scanning electron microscopy image of the alumina micro and nanofibers.

Table 3.1 lists the charateristics of the filters for several commercially available microfibers. The fibers in the filter are held together using commercially available binder materials. The binders typically use starch as part of the binding process; the amounts of starch used to from the media are indicated. Shot are small spherical particles that are present in the micro fibers supplied to us from manufacture.

Table 3.1. Characterization of filter media.

		Porosity				**Permeability x 10^{-11} (m^2)**			
Microfibers type		ASBF1	ALBF		SAFFILHA	ASBF1	ALBF		SAFFILHA
Binder type		Megasol	Megasol	Alumina	Alumina	Megasol	Megasol	Alumina	Alumina
Potato starch	0.01g			0.97				3.46	
	0.02g		0.9				2.98		
	0.03g	0.99		0.94				1.50	
	0.06g	0.85							
Corn Starch	0.03g				0.99				1.77
					0.98				1.07
Binding condition		good	good	good	good	good	good	good	good
Shot		No	Little	Little	No	No	Little	Little	No

- Microfibers: ASBF-1(Zircar Ceramics, 50% alumina, 50% silica), ALBF(Zircar, 97% alumina, 3% silica), SAFFIL HA (Thermal ceramics, 97% alumina, 3% silica)
- Binders: Megasol (silica binder), Alumina(Zircar Ceramics, alumina rigidizer)

Ceramic filter media prepared as above can be used for hot gas filtration. For this particular application. The model described previously was used to estimate the pressure drop and quality factor for varying fiber diameter at same mass flow rate, capture efficiency, particle size, diameter of filter for normal temperature, 21°C and at higher temperature of 500°C.

Figure 4 shows the length of the media and pressure drop of fibrous filters that are 99% efficient in particle capture for two different temperatures. The plot shows that the length of the fibrous filter is the least for fiber diameters less than 100 µm for both temperatures. For filters of diameters greater than 100 µm the filter lengths are not practical in order to attain 99% capture efficiency. Also it is apparent that length of filter media, pressure drop across media increases with the rise in temperature, while keeping all other conditions similar.

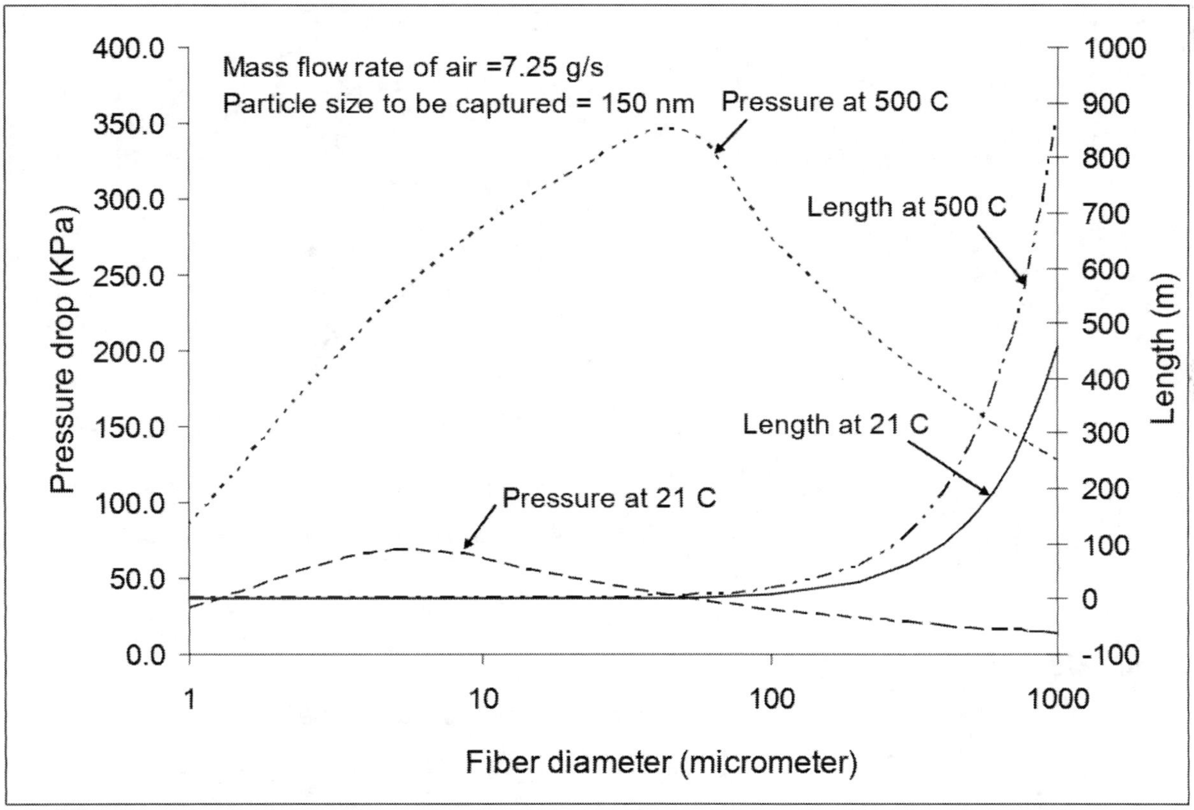

Figure 4. Pressure drop and length of the media with varying the fiber size at two different temperatures. The plots compare filters having the same 99% capture efficiency.

From Figure 5, it is apparent that Quality factor for fibrous filter at 21 °C is higher than the 500 °C, because of lower pressure drop in fibrous filter. Quality factor of the media at a temperature decreases initially and then increases back. This is because of the fact that Brownian diffusion dominants the capture mechanism at lower temperature and direct interception is dominant mechanism at higher temperatures.

Quality factor of fibrous filters decrease and then increases as fiber diameter is increased. Quality factor at 1 μm is almost similar to Quality factor at 100 μm. Thus, the choice is between finer fiber diameters with thicker fiber diameter. Thicker fiber diameter costs us in thickness of filter media. Fabrication of filter media with finer fiber diameter looks promising.

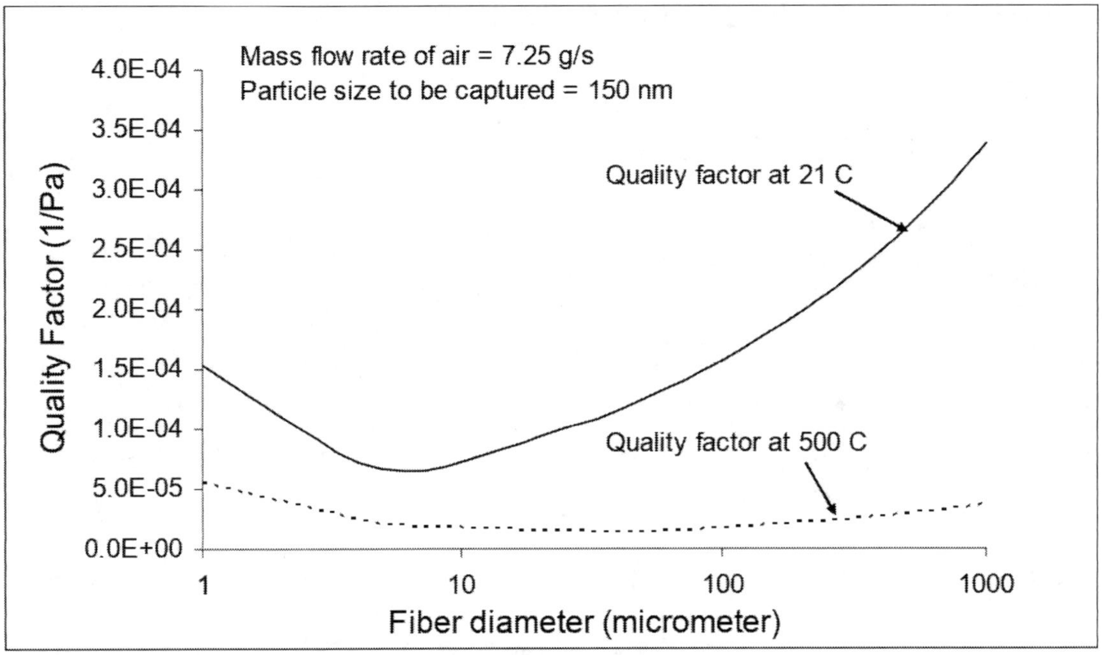

Figure 5. Quality factor of fibrous filters with varying the fiber size at two different temperatures.

4. Conclusion

Filter media of alumina micro and nanofibers combined were prepared and characterized. Filter performance of pressure drop, capture efficiency and quality factor has been modeled for ambient and elevated temperature conditions. The model results show these filters may be useful for hot gas filtration.

This work is still in progress. Future work will include testing of filters at elevated temperatures and testing of catalyst performance.

Acknowledgments

This work was supported by the Coalescence Filtration Nanomaterials Consortium: Ahlstrom Paper Group, Donaldson Company, Cummins Filtration, Hollingsworth and Vose, and Parker Hannifin. This work is also supported by the National Science Foundation grant STTR IIP 0637539.

REFERENCE

1. W. Kataphinan, R. Teye-Mensah, E.A. Evans, R.D. Ramsier, D.H. Reneker, D.J. Smith, *J. Vac. Sci. Technol. A* **21**, 1574-1578 (2003).
2. A. Cybulski, J.A. Moulijn, *Structured catalysts and reactors,* 417-424 (1998).
3. E.T. Bender, P. Katta, A. Lotus, S.J. Park, G.G. Chase, R.D. Ramsier, *Chemical Physics Letters* **423,** 302–305 (2006).
4. R. Teye-Mensah, V. Tomer, W. Kataphinan, J.C. Tokash, N. Stojilovic, G.G. Chase, E.A. Evans, R.D. Ramsier, D.J. Smith, D.H. Reneker, *J. Phys. Condens. Matter* **16,** 7557 (2004).
5. R.C. Brown, Pergamon Press, Oxford (1993).
6. M.C. Militello, S.J. Simko, *Surface Science Spectra* **3**, 395-401 (1997).

243a Predicting Morphology For Membranes Formed Via Liquid – Liquid Tips

Patrick L. Hanks and Douglas R. Lloyd. Chemical Engineering, The University of Texas at Austin, 1 University Station C0400, Austin, TX 78712

A deterministic approach was taken to predict membrane morphologies resulting from the matrix solidification step that occurs in the formation of microporous membranes via liquid – liquid thermally induced phase separation (L–L TIPS). Many studies have examined the growth rate of droplets in the coarsening stage of membrane formation, but few have attempted to extend this information into the subsequent processing steps of matrix solidification, diluent extraction/exchange, and drying. In the research presented here, the approach taken to model matrix solidification utilized Monte-Carlo routines to provide quantitative information on cell size and cell size distribution for a representative polymer – diluent system. The model accounted for shrinkage due to crystallization of the polymer-rich phase, diluent expulsion from the polymer-rich phase, and continued droplet growth during the solidification process, but did not seek to include blooming effects in which diluent is expelled to the surfaces of the membrane. The predicted structures were compared to experimentally formed membranes. The effect of extraction on membrane morphology was shown experimentally to be negligible and therefore was not included in the modeling. The drying stage was simulated using finite element analysis to investigate the effects of capillary pressures in deforming the final membrane structure.

243b Investigation of Corrugation Phenomenon in the Inner Contour of Hollow Fibers during the Non-Solvent Induced Phase-Separation Process

Sina Bonyadi, Chemical & Biomolecular Engineering, National University of Singapore, 10 Kent Ridge Crescent, Singapore, Singapore, Tai Shung Chung, Department of Chemical and Biomolecular Engineering, National University of Singapore, 10 Kent Ridge Crescent,, Singapore, 119260, Singapore, and William B. Krantz, Biomolecular and Chemical Engineering, National University of Singapore, 10 Kent Ridge Crescent, Singapore, 117602, Singapore.

In this paper by proposing a novel mechanism, we reveal one of the most controversial issues in the hollow fiber fabrication process regarding the instability leading to the deformed crosssection of fibers fabricated through nonsolvent induced phase separation. We have analyzed possible instability mechanisms based on our experimental observations and then postulated that the principal instability occurs in the external coagulation bath where the pressure induced in the nascent fiber outer layer as a result of diffusion/convection, precipitation, densification and shrinkage buckles the rigid precipitated polymer shell in the dope and bore fluid interface. In addition, the effect of some spinning conditions such as air-gap distance, bore fluid composition, take-up speed, external coagulant and dope concentration on the final shape of the fiber cross-section have been investigated. The proposed mechanism is in good qualitative agreement with all our observations.

243c Silicalite-1/amorphous Silica Mixed-Matrix Membranes By Layer-By-Layer Deposition

Joshua A. Sheffel, Department of Chemical Engineering and Materials Science, Univeristy of Minnesota, 421 Washington Avenue SE, Minneapolis, MN 55455-0132, Jungkyu Choi, Chemical Engineering and Materials Science Dept., University of Minnesota, Department of Chemical Engineering and Materials Science, 421 Washington Avenue SE, Minneapolis, MN 55455-0132, and Michael Tsapatsis, Department of Chemical Engineering and Materials Science, University of Minnesota, 421 Washington Avenue SE, Minneapolis, MN 55455-0132.

Recent work in the area of zeolite membrane fabrication has demonstrated the feasibility of thin, continuous, oriented zeolite films for gas and vapor separations. However, the development of simple fabrication techniques that offer facile scale-up for industrial use is a work in progress. Mixed-matrix composite membranes represent a compromise between separation performance and ease of membrane formation whereby some of the performance characteristics of the zeolite are forfeited in order to gain the processability associated with a chosen matrix material. In negotiating this compromise, balancing the matrix adsorption and diffusion characteristics with those of the zeolite is crucial. We will discuss a model that addresses this material design problem and apply it to the case of a silicalite-1 composite for organic vapor separations. Guided by insights into the optimal matrix material for this system, we will present results on the fabrication of silicalite-1/amorphous silica composites by a layer-by-layer deposition method. The films consist of alternating layers of amorphous silica and b-oriented, plate-like silicalite-1 particles with near monolayer coverage. Characterization by XRD, SEM, and vapor permeation will be presented.

243d Fundamental Understanding of the Morphology and Macrovoid Evolution of the Single-Layer P84 and Dual-Layer Ultem/P84

Natalia Widjojo, Tai-Shung Chung, and William B. Krantz. Biomolecular and Chemical Engineering, National University of Singapore, 10 Kent Ridge Crescent, Singapore, 117602, Singapore

Single-layer P84 and dual-layer Ultem/P84 hollow fiber membranes with different morphology have been fabricated simultaneously via a dry-jet wet-spinning process. A preliminary study of the effects of spinneret dimension and its coupling effect with air- gap distance on macrovoid evolution in the single-layer P84 hollow fibers show that the number of macrovoids considerably increases with an increment in spinneret annulus gap thickness. In addition, the number of inward-pointed macrovoids increases while the number of outward-pointed macrovoids decreases with an increase in air-gap distance. Based on the study of single-layer P84 hollow fiber membranes, the desired dual-layer Ultem/P84 hollow fiber membrane morphology with a sponge-like outer layer and free-macrovoid inner layer structures for pressure-driven process applications can be tailored by determining a proper combination of air gap distance, first external coagulant, and bore-fluid chemistry. It is found that dual-layer hollow fiber membranes spun with a longer air-gap and a first coagulant bath using higher alcohol series, i.e. ethanol or 2-propanol are more prone to a closed-cell structure formation as compared to those spun with a lower air-gap and a first coagulant bath using water or methanol. Furthermore, the number of macrovoids in the inner layer increases with an increase in air-gap distance which is similar to the phenomenon found in the single-layer hollow fiber membranes. In addition to the air-gap length, the type of the first coagulation bath and bore-fluid also has a significant impact to the formation of macrovoids in the inner layer.

243f Evolution of Macrovoids and Critical Factors to Form Macrovoid-Free Hollow Fiber Membranes

Na Peng, Tai-Shung Chung, and Kaiyu Wang. Chemical and Biomolecular Engineering, National university of Singapore, 10 Kent Ridge Crescent, Singapore 119260, Singapore, Singapore

The origins of macrovoid and the ways to eliminate it have received great attention and heavy debates during the last five decades, but no convincing and agreeable comprehension has been achieved. This work will systematically investigate the key factors to form macrovoid-free hollow fiber membranes. We have discovered there should be critical values of polymer concentration, air gap distance and take-up speed, only above all of which the macrovoid-free hollow fibers can be successfully produced. This observation was confirmed for hollow fibers spun from different materials such as polysulfone, P84 and cellulose acetate, and may be even universally applicable for other polymers. The major mechanisms why these critical parameters can effectively suppress macrovoids have been elaborated. The concept of acceleration of stretch was also proposed. A quantitative relationship was observed to relate it with the number of macrovoids at the critical velocity and critical air gap distance: $n=C1*a^{1/2} + C2$(where C1 and C2 are materials and processing related constants which possibly are functions of dope formulation, rheology, precipitation and spinning conditions). Even though this mathematical description is still open and requires further investigation due to the complexity of spinning process, this model does provide some fundamental understanding of material related and process related tendency of macrovoid formation.

243e Growth of ETS-10 Films on Gallium Arsenide for Photocatalytic Applications

Mariam N. Ismail, Juliusz Warzywoda, and Al Sacco Jr. Department of Chemical Engineering, Center for Advanced Microgravity Materials Processing (CAMMP), Northeastern University, 360 Huntington Ave, 147 Snell Engineering Center, Boston, MA 02115

Engelhard titanosilicate ETS-10 has attracted much interest for photocatalytic applications due to its monatomic semiconductor …-Ti-O-Ti-O-Ti-… chains and its band gap energy of 4.3 eV. Preparation of ETS-10 films on nonporous GaAs substrates was investigated using gel-forming synthesis mixtures with molar composition 5.2Na2O : 0.5K2O : 1.0TiO2 : 5.5SiO2 : 113H2O, and initial pH≈10.5. Hydrothermal syntheses of ETS-10 in the presence of the GaAs substrates were carried out at 503 K for 2-6 days. After 2-day synthesis, very few ETS-10 crystals appeared to be electrostatically attached onto the GaAs substrates, with the bulk product being fully crystalline. After 6-day synthesis, the same amount of crystals appeared to be present on the substrate as after 2-day synthesis. This indicated that direct hydrothermal synthesis was not the route to consider in ETS-10 film growth on GaAs. Use of covalent linkers in the supported zeolite and zeotype film growth has been investigated by several groups. Yoon et al. [1] proposed that binding of zeolite A crystals onto XP-coated (where P is propylsilyl, and X is: C for chloro, B for bromo, or I for iodo) glass plates proceeded via direct nucelophilic substitution of the terminal halide in the glass-bound XP groups by the hydroxyl groups located on the zeolite A external crystal surface. In order for covalent linkage to occur, the CP and the hydroxyl groups on the glass plate and zeolite, respectively, must be present, as shown by the following schematic 1, where G is the glass substrate, and Z is zeolite A: Utilizing this methodology of covalent linkers a polysiloxane photoconductive polymer (e.g., hexamethyldisiloxane) was used to coat GaAs substrates, and then ETS-10 crystals were attached via refluxing under argon, or using a dip coating technique. In order to utilize a similar method to that of Yoon et al. [1], modification of ETS-10 is necessary in order to increase the number of terminal hydroxyl groups. Several groups investigated post treatment using acids to modify the ETS-10 crystal structure, which enhances the photocatalytic activity of ETS-10 due to the increase in the amount of TiOH sites created during the acid treatment [2]. Lv et al. [3] found that acid treatment changes the octahedral titanium to penta- and/or tetra-coordinated titanium, and generates terminal TiOH groups. This enhances the photocatalytic activity of ETS-10. Therefore, it can be hypothesized that acid treatment may be beneficial not only for photocatalysis, but also for formation of intact ETS-10 films. Preliminary results on glass substrates will be presented. [1] Kwang Ha, Yun-Jo Lee, Han Ju Lee, Kyung Byung Yoon, Adv. Mater. 2000, 12, No. 15, 1114. [2] Llabrexsi Xamena, F. X., Calza, P., Lamberti, C., Prestipino, C., Damin, A., Bordiga, S., Pelizzetti, E., Zecchina, A. J. Am. Chem. Soc. 2003,125, 2264. [3] L. Lv, J. K. Zhou, F. Su, X. S. Zhao, J. Phys. Chem. C 2007, 111, 773-778.

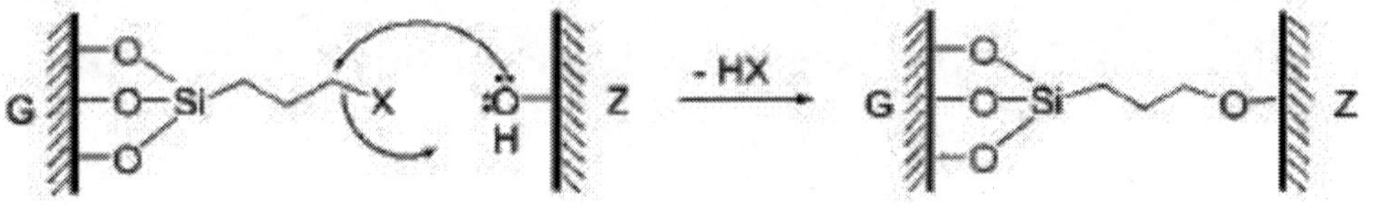

(1)

Porous PVDF Membranes Containing Amphiphilic Comblike Polysiloxane Formed Via Phase Inversion Process

Baoku Zhu*, Yanling Qian, Jianhua Wang, Youyi Xu

Institute of Polymer Science, Zhejiang University, Hangzhou, 310027, China

(zhubk@zju.edu.cn)

Abstract: Amphiphilic comb-like polysiloxane containing polyether side chains (ACPS) is presented for modification of PVDF membrane prepared via phase inversion process. The stability of ACPS in phase inversion process, the effects of ACPS on morphology, chemistry structure, crystallinity, mechanical property, hydrophilicity and filtration performance were discussed to extend. It was found that not only the ACPS molecules stay stable in membrane formation process and their side PEG chain enrich on membrane surface, but also these stability and enrichment can be reserved in membrane application process. Via supposing the schematic model, the mechanisms relating to structure and membrane performance are interpreted. From the observed results, it is concluded the ACPS should contribute to the potential candidate materials for high performance filtration membranes.

Keywords: Amphiphilic polysiloxane, comb-like, poly (vinylidene fluoride) (PVDF), porous membrane, hydrophilication, phase inversion

Introduction

As one of the most popular materials for filtration membranes, poly (vinylidene fluoride) (PVDF) has earned lots of attention due to its comparative mechanical properties, thermal and chemical stability, radiation resistance etc. Whereas, PVDF membranes are inclined to be fouled by proteins, oil, sewage, and may bring high driving force for water filtration for the intrinsic hydrophobic property of PVDF [1-2]. Usually, blending with hydrophilic material as the third component to the casting solution [3-5] is an effective methods to improve the hydrophilicity of membranes.

Previous researches revealed that the linear hydrophilic polymers, such as poly (vinyl pyrrolidone) (PVP) and poly (ethylene glycol) (PEG) etc have acted as pore inducing components instead of hydrophilication component in membranes [6-9] due to the fact that the linear polymer incline to be diffuse out either in phase inversion process or in real application. In comparison, introduction of come-like amphiphilic copolymer could endow the more stable improved hydrophilicity for membrane [10-13]. Principally, the strong attraction between hydrophobic segments of the amphiphilic copolymer and hydrophobic matrix (PVDF), as well as the tangling of comblike molecule can increase stability of the copolymer, while hydrophilic side chains endow the blend membrane with super hydrophilicity.

To explore the efficient additives for PVDF filtration membrane, an easily obtained amphiphilic comb-like polymer, polysiloxane containing polyether side chains (ACPS, Fig 1) were proposed as hydrophication reagent for PVDF membrane in this work. Via a phase inversion process, ACPS was introduced into PVDF membrane matrix. Basing the structure and property investigation, it was expected that stable and applicable hydrophic PVDF/ACPS membranes should be realized.

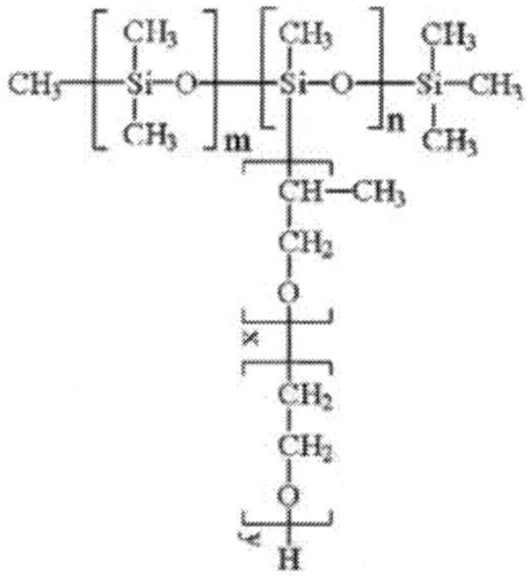

Figure 1. Schematic structure of ACPS

Experimental

Materials: PVDF (Mw=1,000,000), purchased from Shanghai 3F new materials Co., LTD (China). N, N-dimethyl formamide (DMF), Bovine serum albumin (BSA) (Mw=67,000) and Poly (ethylene glycol) (PEG, $M_n = 2\times10^4$) was purchased from Sinophorm Co. ACPS was synthesized by a hydrosilylation coupling reaction with unsaturated poly (ethylene oxide)-block-poly (propylene oxide) utilizing a platinum catalyst in toluene under nitrogen at 85 °C. The molecular weight and component of ACPS was, M_n=2.5×10^4, M_w=3.6×10^4, n : (n +m) =0.4 (corresponding to Figure 1). The PPO and PEO composition was 38.7wt% and 44.0wt% respectively.

Preparation of Membranes: PVDF/ACPS blend membranes were prepared by typical phase separation method. In preparation, PVDF and ACPS were dissolved in DMF at 80° to yield a homogeneous casting solution. After degassing under vacuum at 60°C, the solution was cast into film on glass plate and immersed into 60°C coagulation bath of pure water to form the solid blend PVDF/ACPS membrane. By leaching the membrane thoroughly and drying in vacuum at 60°C, the membrane was used for further characterization. As a reference sample, the porous PVDF membrane using PEG20000 as the additive (the weight ratio of PEG/PVDF was 20/80) was also prepared following the similar processes.

Characterization of blend membranes: The gellation kinetics during membrane formation was monitored by light transmission instrument. The morphology of the membranes was observed using the scanning electronic microscope (SLRI10N SEM, FEI co., Holland). The porosity was measured by mercury porosimeter (Autopore IV Demo 9500, micromeritics). Water contact angle of membranes was measured by a contact angle goniometer (Dataphysics, OCA20, Germany) to testify its surface hydrophilicity. The surface and bulk chemical composition of blend membranes were measured by X-ray photoelectron spectroscopy (XPS, VG Multi lab 2000) and element analysis instrument (vario EL, Elementar Co., Germany) respectively. The water permeability of membranes was measured with cross-membrane pressure in 0.1MPa at 25°C. By being shake the membranes immersed in 60°C water, and characterizing the water contact angle and flux corresponding membrane, the stability of ACPS polymers in membrane was investigated.

Results and discussion

Forming kinetics and morphology of membrane

For casting solution with ACPS/PVDF mass ratio in 0/100, 5/95, 10/90, 20/80 and 30/70, the corresponding changes of light transmission, as well as the inferred precipitation time (t, characterized as the time from immersing liquid membrane into the coagulation bath to the sharp decrease of light transmission), during the membrane formation were illustrated in Figure 2. When the ACPS content increased, the delayed time increased. The observation suggested that the phase separation of casting solution shifted from an instantaneous demixing to a delayed demixing. From the membrane structure in Figure 3, it could be found that the instantaneous demixing process resulted in fingerlike voids in the sublayer of PVDF/ACPS membrane, whereas the delayed demixing yielded spongelike pore voids with smaller size.

For casting solution containing PEG20000, instead of ACPS, the phase inversion behavior with delayed precipitation similar to PVDF/ACPS was found in membrane formation process. However, there was difference in membrane structure between two systems. While PEG20000 was used, the formed membrane has higher porosity, larger finger-like pores. Moreover, many sphere crystals were formed on the membrane surface when only PEG20000 was used as additive. In principle, larger void might yield the decrease in mechanical strength, and crystallization would lead lower permeability of flux of the corresponding membrane.

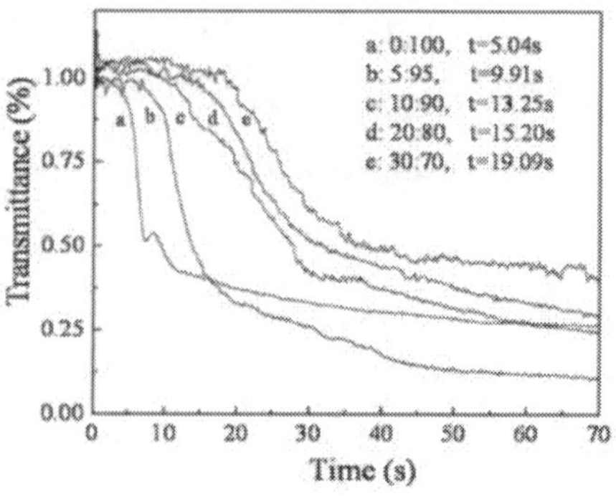

Figure.2 Changes of light transmission during formation of membrane using the castiong solution with ACPS/ PVDF mass ratio in (a) 0/100; (b) 5/95; (c) 10/90; (d) 20/80; (e) 30/70

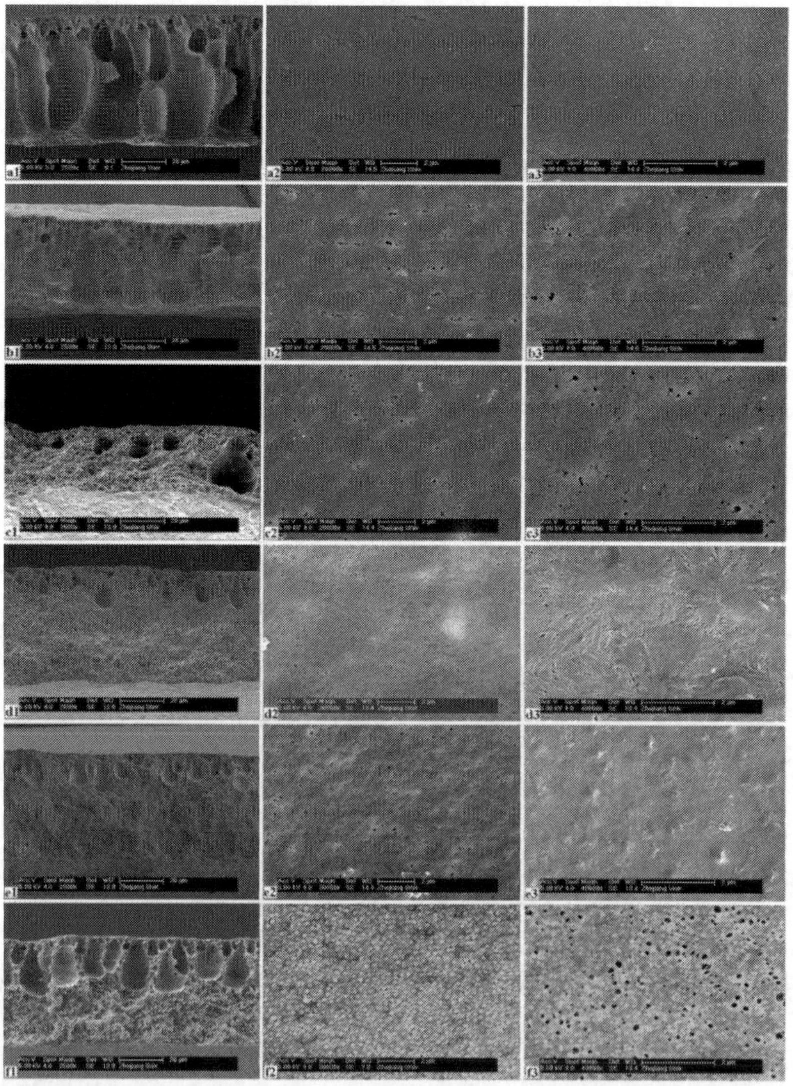

Figure 2. SEM images of membranes prepared from the casting solution with ACPS/PVDF mass ratio of (a) 0/100, (b) 5/95, (c) 10/90, (d) 20/80 and (e) 30/70, and (f) PEG/PVDF in 20/80.

(1) cross section (2) top surface (3) bottom surface

Chemical composition and hydrophilicity of membrane

Concerning ACPS is water-soluble polymer, it is reasonable that ACPS molecules would dissolve into water bath in membrane formation process. And thus, the mass ratio of ACPS/PVDF might be much smaller than that in casting solution. To reveal the reservation of ACPS, the chemical compositions in whole membrane matrix and in surface layer were measured by elemental analysis and XPS respectively. For the membrane prepared from casting solution with mass ratio of ACPS/PVDF in 30/70, the molar atom ratio of O/F ratio in membrane bulk is observed as 0.25. This value is perfectly close to 0.27, which was calculated basing the composition in casting solution. This result indicated that ACPS could be reserved to a very high degree. In other words, the ACPS molecules were not inclined be dissolved out off the membrane in precipitation process. In mechanism, the reservation should come from not only the intense hydrophilic interaction between the backbone of ACPS with PVDF, but also the obvious tangling effect of comblike structure of ACPS, which prevented the ACPS diffusing out of the solidified membrane. This case is quite different to that using linear water soluble polymer (e.g. PEG or PVP) as additives. The higher reservation degree of hydrophilic additive is very beneficial to stability of chemistry structure and hydrophicity of membrane.

In XPS spectrum, the emission peaks at 284.7eV for C (1s), 685.3eV for F (1s), 534.2eV for O (1s), 101.3eV for Si (2p), respectively. By calculation from this spectrum, the O/F ratio in surface layer was found as 0.34, obviously larger than that in whole bulk membrane (0.25) or the value in casting solution (0.27). Such observation clearly inferred the enrichment of ACPS in membrane surface layer. This enrichment could be reasonable explained as followed. In phase inversion process, the hydrophilic poly(ethylene oxide) side chains in ACPS tend moving to the interface between membrane and water bath. At the same time, the both the strong hydrophobic interaction between ACPS backbone and PVDF, and the entangling effect of comblike structure of ACPS prevent the whole ACPS diffusing out off the membrane. As a result, a structure in that the PEO side chain extended extending toward to coagulation bath (also stretching out off membrane surface) was formed and fixed. According to the relationship between chemical structure on surface and hydrophilicity of membrane, theoretically, this enrichment of ACP would endow the surface hydrophilicity of prepared membrane more efficiently. In fact, such enrichment not only occurred in membrane surface layer, but also on wall of voids or tunnels inside the membrane matrix. In brief, the dispersion structure of ACPS in blend membrane could be proposed as illustrated in Fig 4(a).

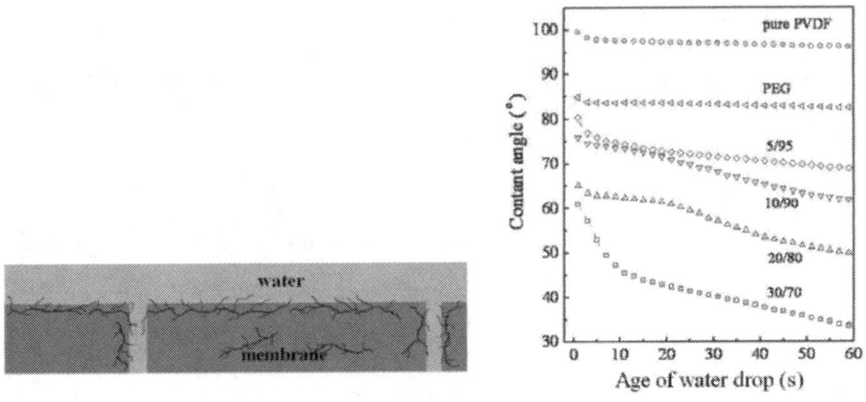

(a) (b)

Figure 4. (a) Supposed illustration schematic of ACPS in the membranes and (b) water contact angles on the blend membrane prepared from different ACPS/PVDF ratio

As shown in Figure 4(b), the hydrophilicity of PVDF/ACPS membrane was performed. With the increase of ACPS content, the hydrophilicity increased gradually. In comparison, the membrane using PEG20000 as additive performed the typical hydrophobicity due to the higher loss degree of PEG in membrane formation. Obviously, the amphiphilic comblike ACPS is more efficient in improving the hydrophilicity of blend PVDF membrane.

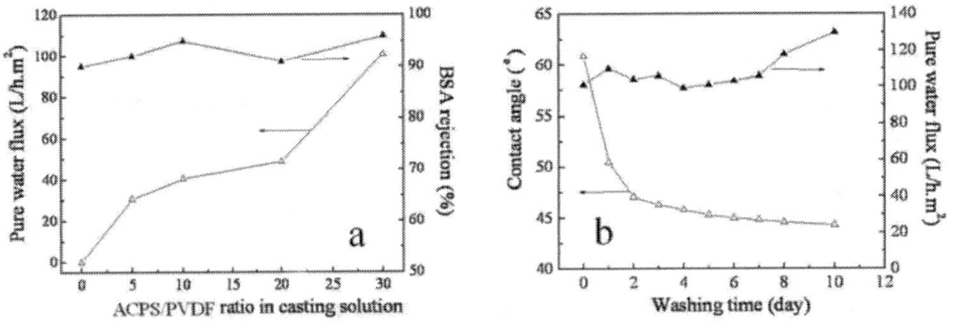

Figure 5. (a) Water flux and BSA rejection of membranes prepared from casting solution with different ACPS/PVDF ratio. (b) Changes of water flux and initial contact angle with leaching time for the blend membrane prepared from the casting solution with ACPS/PVDF ratio of 30:70

Membrane filtration property and stability

Pure water flux and BSA rejection were tested to study the filtration and rejection properties of the prepared PVDF/ACPS membranes. The pure PVDF membrane and PEG/PVDF membrane didn't allow water to pass through under a pressure of 0.1MPa without additional treatment, indicating total hydrophobicity of these two membranes. After being wetted by ethanol, their water flux increased to 12.5 $L/h.m^2$ and 31.4 $L/h.m^2$, while the BSA rejection of both membranes was 90% and 81.7% respectively. For the PVDF/ACPS membrane without additional pre-wetting, the water flux increase to 100 $L/h.m^2$ for the membrane with ACPS/PVDF ratio in 30/70 (Fig 5a). The BSA rejection of all PVDF/ACPS membranes was higher than 90%, indicating ultrufiltration characteristics of these blend membranes.

The stability of structure and performance is very important in practical application of membrane. To simulate the practical filtration application in this work, the PVDF/ACPS membrane (ACPS/PVDF mass ratio: 30/70) was immersed in 60°C water and continuously shaken for various time. By monitoring the changes in water contact angle and water flux, the stability of ACPS in membrane and filtration property could be inferred. With prolonging shaking time, as shown in Fig 5b, the static water contact angle decreased gradually to a constant value, while the water flux could increase about 20%. The observations that the hydrophilicity didn't become larger meant the relative stability of ACPS in membrane. The decrease in contact angle water with shaking time might be caused by the further enrichment of ACPS still in shaking process. Considering the pore structure of blend membrane was not affected by shaking process, the increasing flux should be the result of increasing hydrophilicity. This performance approved the reliability of PVDF/ACPS membrane in application process. With further work on controlling the pore structure, ACPS could used as practical hydrophilication reagent for PVDF ulfiltration membrane.

In conclusion, the amphiphilic comb-like polysiloxane containing hydrophilic PEO side (ACPS) chains was successfully blended into PVDF ultrafiltration membranes via phase inversion route. The ACPS could not only reserve stably in membrane, but also enrich in membrane surface in membrane formation process. The enrichment of ACPS in membrane surface layer endowed the blend membrane with efficient hydrophilicty, which could improve the antifouling and permeation property of membrane. Moreover, the stable reservation of ACPS in filtration process made ACPS a potential additive for high performance PVDF ultrafiltration membrane.

Acknowledgement Financial support form National "863" Program (Granted No. 2006AA03Z233) of China is gratefully acknowledged.

References

1. Yamato N, Kimura K, Miyoshi T, Watanabe Y. Difference in membrane fouling in membrane bioreactors (MBRs) caused by membrane polymer materials. J. Membr. Sci. 2006; 280: 911-919.

2. Fang HHP, Shi XL. Pore fouling of microfiltration membranes by activated sludge. J. Membr. Sci. 2005; 264: 161-166.
3. Fontananova E, Jansena JC, Cristianoa A, Curcio E, Drioli E. Effect of additives in the casting solution on the formation of PVDF membranes. Desalination, 2005; 192: 190-197
4. Wang YQ, Wang T, Su YL, Peng FB, Wu H, Jiang ZY. Remarkable Reduction of Irreversible Fouling and Improvement of the Permeation Properties of Poly(ether sulfone) Ultrafiltration Membranes by Blending with Pluronic F127. Langmuir, 2005; 21:11856-11862
5. Yeow ML, Liu YT, Li K. Preparation of porous PVDF hollow fibre membrane via a phase inversion method using lithium perchlorate (LiClO4) as an additive. J. Membr. Sci. 2005, 258:16-22
6. Kim JH, Lee KH. Effect of PEG additive on membrane formation by phase inversion. J. Appl. Polym. Sci. 1998; 138: 153-163
7. Kim IC, Lee KH. Effect of poly(ethylene glycol) 200 on the formation of a polyetherimide asymmetric membrane and its performance in aqueous solvent mixture permeation. J. Membr. Sci. 2004; 230:183-188
8. Chen NP, Hong L. Surface phase morphology and composition of the casting films of PVDF–PVP blend. Polymer, 2002; 43:1429-1436
9. Qin JJ, Wong FS, Li Y, Liu YT. A high flux ultrafiltration membrane spun from PSU/PVP (K90)/DMF/1, 2-propanediol. J. Membr. Sci. 2003; 211: 139-147
10. Park JY, Acar MH, Akthakul A, Kuhlman W, Mayes AM. Polysulfone-graft-poly (ethylene glycol) graft copolymers for surface modification of polysulfone membranes. Biomaterials, 2006; 27:856-865
11. Ma XL, Su YL, Sun Q, Wang Y Q, Jiang Z Y. Preparation of protein-adsorption-resistant polyethersulfone ultrafiltration membranes through surface segregation of amphiphilic comb copolymer. J. Membr. Sci. 2007; 292: 116-124
12. Hester JF, Banerjee p, Won YY, Akthakul A, Acar MH, Mayes AM. ATRP of Amphiphilic Graft Copolymers Based on PVDF and Their Use as Membrane Additives. Macromolecules. 2002; 35: 7652-7661.
13. Hester JF, Banerjee P, A.M. Mayes. Preparation of Protein-Resistant Surfaces on Poly(vinylidene fluoride) Membranes via Surface Segregation. Macromolecules. 1999; 32: 1643-1650.

244a High Performance Protein Separation by Ion Exchange Membrane Partitioned Free Flow Isoelectric Focusing System

Jiuhua Cheng[1], Yi Li[2], Tai Shung Chung[2], and Shing-Bor Chen[2]. (1) Department of Chemical & Biomolecular Engineering, National University of Singapore, 10 Kent Ridge Crescent,, Singapore, 119260, Singapore, (2) Department of Chemical and Biomolecular Engineering, National University of Singapore, 10 Kent Ridge Crescent,, Singapore, 119260, Singapore

The objective of this study is to investigate a combination of Ion Exchange Membranes and Free-Flow Isoelectric Focusing (FFIEF) technology for high-selectivity and high-flux protein separation, in which ion exchange membranes are used as the separation media. A FFIEF device has been successfully designed and extensive experiments have been conducted to prove its effectiveness in enhancing the protein separation performance. Three types of membranes were employed in this work and they were commercial microfiltration (MF) ion exchange membranes, commercial neutral UF cellulose membranes, and home-made ultrafiltration sulfonated polysulfone (UF SPSf) ion exchange membranes. The protein separation results show that home-made UF SPSf membranes have the superior selectivity and flux to other membranes. This is due to the fact that the pH gradient between two sides of membranes as well as the interaction between the protein molecules and membrane surface plays an important role in the high-performance protein separation. This phenomenon is very similar to the native mass transfer system in human body. By applying a semi-batch separation process and optimizing various experimental conditions, a high-purity (> 90%) and concentrated target protein is obtained at the permeation side of home-made UF SPSf membranes with a high flux. More over, it is possible to separate multi-components mixture into high purity fractions continuously through one multi-chambers IEM-FFIEF process.

244b The Contribution of Protein-Solvent Interaction in Protein Ultrafiltrations: The Prediction of Limiting Flux through the Free-Solvent Model for Osmotic Pressure

Yiheng Wang and Victor G. J. Rodgers. Department of Bioengineering, The University of California, Riverside, A220 Bourns Hall, Riverside, CA 92521

During protein ultrafiltration (UF), high protein concentration results at the surface of the membrane. The concentrated protein contributes largely to the osmotic pressure of the system, which has direct effects to the ultrafiltration performances such as permeate flux. This osmotic pressure contribution might also be the direct cause of the limiting flux phenomenon, which is an important phenomenon that imposes the performance limitation of ultrafiltration.

The free-solvent model for osmotic pressure recognizes that the protein-solvent interactions, protein hydration and protein-ion binding, to be specific, can have a tremendous impact on the osmotic pressure in concentrated protein solutions. This model well predicts the osmotic pressure of protein with high concentration (on the order of 100g/L), so it is an ideal model for protein ultrafiltration process.

Ultrafiltration parameters, such as permeate flux and protein permeate concentration, are predicted using free-solvent model for both single protein and binary protein ultrafiltrations. The results show reasonably good agreements to the experimental data. More importantly, the limiting flux phenomenon is predicted and characterized using the free-solvent model. It is well demonstrated that the limiting flux phenomenon is the natural consequence of the osmotic pressure contribution from concentrated protein at the membrane surface.

244c Steric and Electrostatic Interactions Govern Nanofiltration of Amino Acids

Shankar Chellam, Department of Civil and Environmental Engineering, Department of Chemical Engineering, University of Houston, 4800 Calhoun Rd., Houston, TX 77204-4003 and Yongki Shim, Civil and Environmental Engineering, University of Houston, 4800 Calhoun Rd., Houston, TX 77204-4003.

Crossflow nanofiltration experiments were performed to investigate the factors influencing the removal of amino acids by a polymeric thin-film composite membrane.

The first part of the presentation will focus on the removals of five monoprotic (Ala, Val, Leu, Gly, and Thr), one diprotic (Asp), and one dibasic (Arg) amino acids in a range of permeate fluxes, feed pH values, and ionic strengths. Transport of these hydrophilic amino acids was analyzed using the phenomenological model of Kedem and Katchalsky. At any given pH and ionic strength, reflection coefficients (rejection at asymptotically infinite flux) of monoprotic amino acids increased with molar radius demonstrating the role of steric interactions on their removal. Additionally, consistent with Donnan exclusion, higher reflection coefficients were obtained when the membrane and the amino acids both carried the same nature of charge (positive or negative). In other words, both co-ion repulsion and molecular size determined amino acids removal. Importantly, the removal of effectively neutral amino acids were significantly high than neutral sugars and alcohols of similar size demonstrating that even near their isoelectric point, zwitterionic characteristics preclude them from being considered as strictly neutral.

The second part of the presentation will demonstrate the importance of sorption in determining the dynamic nanofiltration of two hydrophobic amino acids, viz. phenylalanine and tryptophan. Batch tests showed that these amino acids having hydrophobic side chains were substantially absorbed on the polymeric membrane, which was quantitatively modeled using the Freundlich isotherm. Crossflow nanofiltration revealed that convective flow makes adsorption sites more accessible to amino acids. Hence, the permeate concentration changed more rapidly at the initial stages of filtration compared with batch tests.

In summary, co-ion repulsion and steric interactions govern hydrophilic amino acids removal by "tight" nanofiltration membranes whereas sorption plays an important role during the initial filtration stages for hydrophobic amino acids.

244d Membrane Chromatography: Surface-Initiated Atom Transfer Radical Polymerization of 2-Dimethylaminoethyl Methacrylate

Bharatkumar V. Bhut and Scott Husson. Chemical and Biomolecular Engineering, Clemson University, 112 Earle Hall, Dept. of Chemical and Biomolecular Engineering, Clemson University, Clemson, SC 29634-0909

Protein separation by ion-exchange chromatography remains a focus for many researchers working in the field of bioseparations. When used as chromatography media in filtration module, synthetic microporous or macroporous membranes offer advantages over resin-based media, such as low pressure drop, high production rate, and facile scale up and set up. The objective of this research was to surface modify commercially available regenerated cellulose membrane by atom transfer radical polymerization to produce high-capacity ion-exchange membranes for protein chromatography. The monomer 2-dimethylaminoethyl methacrylate was polymerized from cellulose membranes to convert them into ion-exchange membranes. Physicochemical properties and performance properties were studied as a function of polymerization time with various analytical measurement techniques; these include average membrane pore size, pore-size distributions, membrane surface chemistry, and ion-exchange capacities. Protein adsorption experiments were done with an Atka purifier to evaluate dynamic capacities of our surface modified membranes.

244e Ultrafiltration of Parvovirus

Ranil Wickramasinghe and David Grzenia. Colorado State University, Department of Chemical Engineering, Fort Collins, CO 80523-1370

Ultrafiltration is frequently used in the biotechnology industry for protein purification. The main applications of ultrafiltration are for protein concentration and buffer exchange. This paper focuses on ultrafiltration for purification of parvovirus. The feasibility of using ultrafiltration to remove contaminating host cell proteins from the virus particles has been investigated. Purification of virus particles and virus vectors for clinical applications of gene therapy and in the manufacture of viral vaccines is a major large scale separations problem. Today, parvoviruses, such as adeno associated virus, are being extensively investigated as gene therapy vectors. Consequently, development of robust purification operations will be essential.

Tangential flow ultrafiltration and high performance tangential flow filtration of Aedes Aegypti densonucleosis virus has been investigated using flat sheet membranes with a nominal molecular weight cut off of 30, 50, 100 and 300 kD. Virus particles were detected in the permeate of the 300 kD membrane for both modes of operation. In tangential flow filtration no virus particles were detected in the permeate from the 100 kD membrane. However, during high performance tangential flow filtration significant passage of virus particles through the membrane was observed. The results obtained here are in general agreement with results obtained in previous studies of high performance tangential flow filtration for protein purification. Optimization of the operating conditions of high performance tangential flow filtration may result in a highly selective unit operation for purification of virus particles and virus vectors.

244f Buffer Effects on Viral Filtration of Mab Solutions with Hollow Fiber Membranes

Bruno F. Marques, Sianny Christanti, C.S. Brian To, Michael E. Laska, and David J. Roush. BioPurification Development, Merck & Co., PO Box 2000, Rahway, NJ 07065

Current large-scale monoclonal antibody (MAb) purification processes typically include a viral filtration (VF) step after primary recovery and chromatography, but prior to final formulation. The ability to perform VF after final formulation could have significant economic (i.e. less membrane area due to lower filtrate volume), as well as regulatory (i.e. reduced risk of adventitious viral contamination) advantages. This scenario would require filtration of highly-concentrated feed streams (20-50 g/L) in a variety of buffer matrices. In previous work, we have demonstrated efficient filtration of MAb solutions up to 20 g/L, with hollow fiber membranes, in a system with two primary buffering species. Here, we discuss experimentation with different buffer conditions (i.e. conductivity, pH, multiple components) to allow for filtration of MAb solutions containing product at > 20 g/L. Results from dynamic light scattering (DLS) measurements will be employed to evaluate the impact of MAb-specific self-association behavior that significantly affects VF performance. Along with DLS data, we will use an experimental method outlined by Syedain *et al* (*Biotechnol. Prog.* **2006,** *22,* 1163) to study the localized concentration profile of several MAbs on the surface of hollow fiber membranes, as well as possible MAb-membrane interactions. The data from these different studies could eventually be used to create a theoretical/predictive model for VF operation.

247a 3D Microdevices for Particle Separation

Christina Randall[1], Timothy Leong[2], and **David H. Gracias**[2]. (1) Department of Biomedical Engineering, Johns Hopkins University, Baltimore, MD 21218, (2) Department of Chemical and Biomolecular Engineering, Johns Hopkins University, 125 Maryland Hall, 3400 N Charles Street, Baltimore, MD 21218

We describe a new class of three dimensional microdevices for separation of particles from fluids based on size and chemical affinity.

Web Page: www.jhu.edu/chbe/gracias/

247b Preparation of Uniform Microcapsules with Silicone Oil as Continuous Phase in a Microchannel Device

Xingchu Gong, **Yangcheng Lu**, and Guangsheng Luo. Department of Chemical Engineering, The State Key Laboratory of Chemical Engineering, Tsinghua University, Beijing, 100084, China

Microcapsules have very interesting characteristics and the technology of microcapsulation has been used in various fields such as medicine, agriculture, foods and the chemical industry. Solvent evaporation method is widely used in preparation of microcapsules. This technique neither requires special equipments nor phase separation-inducing agents. Controlled particle sizes in the nanometer to micrometer range can be achieved. But careful selection of encapsulation conditions and materials is needed to yield high encapsulation efficiencies. So far, most researchers use dichloromethane or chloroform as the evaporated solvent. As a result, only polymers which are soluble in the chlorohydrocarbons mentioned above can be used as a shell material. Because species of shell material greatly influence the mass transfer performance of microcapsules, it will be very helpful to produce perfect microcapsules if more shell materials can be used. Poly(dimethylsilicane) (PDMS), also named silicone oil when it is in liquid state. Silicone oil has very good chemical stability, thermal stability, and physiological safety. The solubility of polar compounds or macromolecular compounds in silicone oil is often very small. This paper presents an improved solvent evaporation method with silicone oil as the continuous phase to make more polymer solvents available. A T-shaped microchannel device was used to produce emulsions instead of mechanical stirring to prepare the mono-dispersed microcapsules. Lower evaporation temperature and shorter evaporation time were required in the new process. Uniform polyacrylonitrile (PAN) microcapsules containing Aliquat 336 (ALQ) have been successfully prepared. N,N-dimethylformamide (DMF), a widely used solvent with very good dissolution ability but lower evaporation ability, was applied as the polymer solvent. The prepared microcapsules have rough surfaces and homogeneously internal structures. Microcapsules' mean diameters and loading ratio were influenced by the composition of polymer solution and evaporation temperature. By changing the two-phase flow rate, the mean size of microcapsules can be easily controlled. When more ALQ was added in polymer solution, the loading ratio of microcapsules increased. The mass transfer performance and stability were determined by extraction of Cr2O72- ions from its aqueous solution. The mass transfer rate are enough fast. After three times of repeated extraction and stripping, the microcapsules kept almost the same extraction ability, which indicated that the microcapsules have very good stability.

247c Functionalized Carbon Coated Cobalt As A Novel Magnetic Nano-Devices In Separation Technology

Robert N. Grass, Evagelos K Athanassiou, Fabian M. Koehler, and Wendelin J. Stark. Chemistry and Applied Biosciences, Swiss Federal Institute of Technology (ETH Zurich), Wolfgang-Pauli-Str. 10, ETH Hönggerberg, Zurich, 8093, Switzerland

The miniaturization of sampling volumes in analytical chemistry and medical diagnostics requires fast and reliable separation methods. Based on the energy demand and difficulties in miniaturization, standard unit operations are often not applicable and novel separation methods have to be introduced. Magnetic separation using magnetic beads has often been proposed for a wide range of such applications. However, due to the large cost of the magnetic materials, large particle sizes (>> 1 μm), relatively poor magnetic properties, and very limited binding capacities, the wide spread use of this technology has been limited to high-value biotechnological applications.

In this contribution we present an improved magnetic nanobead material – carbon coated cobalt – which can be manufactured at large scale and has highly beneficial properties for applications in miniaturized separation technology. The material can be described as nanoparticles with a size ranging from 20 – 40 nm and consisting of a metallic cobalt core which is coated with some layers of graphene carbon. The use of metallic cobalt as core material offers optimal magnetic properties and the carbon layer protects the metallic core from oxidation in a wide range of solvents, pH ranges and temperatures of up to 200°C.

The carbon layer can be covalently functionalized with amino- or chloro- groups, allowing a modification of the surface chemistry and the binding and de-binding of specific ligands. We further demonstrate the application of the material in separation technology with biotechnological relevant ligands.

The Figure shows a photograph of a suspension of carbon coated cobalt nanoparticles in water which are removed from the liquid by the use of a commercial neodymium magnet (photograph taken ~5 seconds after placing the magnet)

References:

[1] R. N. Grass, E. K. Athanassiou, W. J. Stark, *Angew. Chem. Int. Ed.* **2007**, in print.

[2] R. N. Grass, W. J. Stark, *J. Mater. Chem.* **2006**, 16, 1825.

[3] R. N. Grass, M. Dietiker, R. Spolenak, W. J. Stark, *Nanotechnology* **2007**, 18, 035703.

247d Protein Separation By Isoelectric Focusing Coupled With Isotachophoresis On A Microfluidic Device

Huanchun Cui[1], Talukder Zaki Jubery[2], Prashanta Dutta[2], and Cornelius F. Ivory[1]. (1) Chemical Engineering, Washington State University, Pullman, WA 99164, (2) Mechanical Engineering, Washington State University, Pullman, WA 99164

This paper reports the experimental application of isoelectric focusing coupled with isotachophoresis on a networked microfluidic chip, which has an ITP channel and an IEF channel. Those two channels share a channel segment, where the transition from IEF to ITP takes place. A mixture of three fluorescent proteins was first concentrated and segregated into adjacent ITP zones. After the ITP zones migrated into the channel segment, current was switched from ITP channel to IEF channel. A pH gradient was established by ampholytes pre-filled in the IEF channel and the proteins in the segment were separated by IEF. More IEF channels will be added along the ITP channel to separate protein samples in the ITP train in future. As the future work is completed, protein will be first segregated by ITP according to electrophoretic mobility and then separated by IEF based on pI.

247e Microfluidic Laminar-Flow Extraction Of Functional Proteins From Cell Lysate

Robert J. Meagher, Yooli K. Light, and Anup K. Singh. Sandia National Laboratories, P.O. Box 969, MS 9292, Livermore, CA 94551

Advances in small-scale cell culture have made it possible to perform highly parallel experiments in which hundreds of proteins or protein variants can be expressed simultaneously. These techniques can be useful for fields as diverse as drug discovery, fundamental proteomics, or engineering of enzymes with enhanced or novel activities. Traditional bench-scale methods such as chromatography for purifying proteins are laborious and do not scale well, and would be impractical for purifying small amounts of protein from many cultures for high-throughput applications.

To this end, we have developed a microfluidic technique for isolating specific proteins from cell lysate. The process is based on continuous-flow aqueous two-phase extraction using a PEG-potassium phosphate system. The low interfacial tension of the two-phase system enables stable side-by-side laminar flow of the two phases over a wide range of flow conditions. The aqueous two-phase system is mild, preserving protein structure, allowing us to recover intact, functional proteins and enzymes from the process. Aqueous two-phase systems are highly tunable, allowing specific recovery of nearly any protein of interest while rejecting undesired proteins and cell debris. We have also demonstrated simple engineering of proteins to modify their partitioning behavior and enhance recovery. The objective of the separation is typically to remove unwanted or interfering proteins or activities; the extraction can also serve as an initial, crude purification prior to a second orthogonal purification technique. The fluid mechanics and mass transport of the two-phase flow are interesting and varied, and we present simple modeling and experiments to predict the performance of the process.

The microfluidic extraction approach has several advantages. The process operates in a continuous-flow, rather than batch mode, enabling automation. No packed particles, chromatography supports, or other heterogeneous materials are required, which greatly simplifies chip setup and operation. The small scale of the microfluidic device allows us to processes microliter and sub-microliter samples in minutes, eliminating wasteful, oversized bench scale processing. Parallel processing can be enabled by simply adding additional microchannels. Finally, the process can be integrated into a multi-step lab-on-a-chip platform as an intermediate purification step for true high-throughput analysis.

247f Characterizing The Long-Term Performance And Stability Of A Micro Gas Chromatograph

Mathew Luebbers[1], Adarsh D. Radadia[1], Byunghoon Bae[2], Junghoon Yeom[2], Jea-Hyeong Han[3], Ilwhan Oh[4], Mark A. Shannon[5], and Richard I. Masel[6]. (1) Chemical and Biomolecular Engineering, University of Illinois, 600 S. Mathews St., Urbana, IL 61801, (2) Department of Mechanical and Industrial Engineering, University of Illinois at Urbana-Champaign, 1206 W. Green St., Urbana, IL 61801, (3) Mechanical and Industrial Engineering, University of Illinois at Urbana-Champaign, 1206 W. Green St., Urbana, IL 61801, (4) Department of Chemical and Biomolecular Engineering, University of Illinois at Urbana-Champaign, 294 RAL, MC-712, 600 S. Mathews Ave., Urbana, IL 61801, (5) Mechanical Science and Engineering, University of Illinois at Urbana-Champaign, 1206 W. Green Street, Urbana, IL 61801, (6) Chemical Engineering, University of Illinois at Urbana-Champaign, 600 S Mathews, Urbana, IL 61801

For reasons of homeland security as well as a wide variety of other applications, the development of a small, accurate, and portable micro gas chromatograph has become a topic of great interest. Such a device could be used to monitor air samples and separate individual components which allows for several possible applications such as monitoring for a hazardous chemical species. However, to make implementing the device practical, one of the most important demands is that the device components must all behave properly for long periods of time and work extremely consistently over millions of cycles. This is extremely demanding the case of this device because the entire chromatogram detected is expected to take around 4 seconds with peak widths on the order of tens of milliseconds. Any malfunction or degradation in the performance of a component could lead to peak shifting or peak dispersion which if severe enough would make the device unusable.

In this project, the micro gas chromatograph is form by integrating an array of detectors with a microfabricated pump, a microfabricated column, and a microfabricated preconcentrator assembly with microfabricated valves. For the purposes of characterizing the device performance, a set of tests must be run to determine a false-positive rate. The amount of jitter, or peak shifting, seen during the repetition of thousands of testing cycles will be indicative of the expected rate of false-positive responses.

This paper will report on the repetition jittering testing of a number of testing setups. Initially we will test with a commercially available sampling valve and a standard fused silica column. A sample containing a trace vapor will be injected over the course of many thousands of cycles and the peak position will be tracked over all trials. The microfabricated valves and column will then be integrated into this test setup and the same experiment will be repeated. In addition, tests will be performed in which multiple components are separated and tracked over repeated cycles, thus giving a solid feasibility test for the long term operation of the device.

247g Recent Advances In On-Chip High Speed Gas Chromatograph With Carbon Nanotube Sensors

Adarsh D. Radadia[1], Mark A. Shannon[2], Richard I. Masel[3], Michael S. Strano[4], and Keith Cadwallader[4]. (1) Chemical and Biomolecular Engineering, University of Illinois, 600 S. Mathews St., Urbana, IL 61801, (2) Mechanical Engineering, University of Illinois, 600 S. Mathews St., Urbana, IL 61801, (3) Chemical Engineering, University of Illinois, 600 S. Mathews St., Urbana, IL 61801, (4) University of Illinois, 600 S. Mathews St., Urbana, IL 61801

A micro-GC allows for small, handheld, low power equipments with on-chip devices for environmental sampling. In order to use a micro-GC as a fast chemical warfare agent scanner in highly industrious places like airports, there is a definite need for reduction in analysis times from the current state of the art. Current technology cannot scan (separate and detect) air samples in relatively short times (~ 4 seconds). High speed analysis GC requires 1) a continuous sampling part , 2) a preconcentration part, 3) micro-columns with high separation efficiency and resolution ability and 4) a detector with fast analysis time.

We use standard microfabrication techniques to create each of the components and test individual performance. Integration of the fabricated parts will not be demonstrated. A MEMS pump with a fluctuating polyimide membrane is fabricated to form the continuous sampling component of the GC. Preconcentration principle is demonstrated in a microfluidic processor with valves using metal-organic framework molecules as the adsorbing species. Micro-columns are fabricated in silicon using DRIE technique. Micro-columns are coated with modified polydimethylsiloxane films to perform the gas phase separation. High speed separation on chromatography has been targeted to build a highly sensitive and specific sensor. High resolution columns are fabricated after studying effects of 1) various micro-channel bend geometries and 2) micro-channel configurations on chip. The detector side effort is focused on building a carbon nanotube based chemi-sensors. Carbon nanotubes are highly promising materials for sensing chemical species like thionyl chloride and dimethylmethylphosphonate, a simulant for chemical warfare agent sarin. Results of individual operational components are evaluated using an ordinary modified GC.

250a Modeling Adsorption And Desorption In Ordered Mesoporous Materials

Bradd Libby and **P. A. Monson**. Univ. of Massachusetts, Dept. of Chem. Eng., Amherst, MA 01003

The field of porous materials research has been revolutionized by the synthesis of more ordered mesoporous silica materials via templated self-assembly processes[1]. Materials such as MCM-41, MCM-48, SBA-16 etc. exhibit a relatively high degree of geometrical order on the mesopore length scale and are thus excellent test beds for theories of adsorption, particularly those based on molecular concepts. Many studies of these systems have been made using density functional theory [2] and Monte Carlo simulations [3] for molecular models with geometries representative of the real materials. Such studies reveal important insights into the molecular level behavior and provide a basis for new methods for isotherm analysis and characterization. However, it could be argued that none of the calculations made thus far provides truly quantitative agreement with experiment over the full range of the adsorption and desorption isotherms - including any hysteresis present. In this work we explore some of the issues that underlie this and describe how much improved agreement with experiment can be obtained from Monte Carlo simulations of molecular models. We focus on the cases of MCM-41, based on the cylindrical pore geometry, and MCM-48 and SBA-16, based on minimal surface pore geometries.

While these materials display a substantial degree of geometric order, we have found that the inclusion of some disorder into the models is necessary to reproduce the behavior seen in experiments. In our work we do this by including: i) variations in pore size across the sample through averaging over a pore size distribution; ii) surface roughness in the pore walls; iii) constrictions within the pores. We describe an accurate method for interpolating adsorption and desorption isotherms, which makes it possible to efficiently average Monte Carlo simulation results over pore sizes sampled from a distribution.

1. Beck, J.S., et al., J. Am. Chem. Soc., 1992, 114, 10834; Zhao, D.Y., et al., J. Am. Chem. Soc., 1998. 120, 6024. 2. Ravikovitch, P.I., A. Vishnyakov, and A.V. Neimark, Phys. Rev. E, 2001, 64, 011602. 3. Schumacher, C., et al., J. Phys. Chem. B, 2006, 110, 319; Coasne, B., et al., Langmuir, 2006. 22, 194.

250b A Monte Carlo Study of Gas Adsorption on Atomistic Models of Templated Mesoporous Silicas

Francisco R. Hung[1], Supriyo Bhattacharya[2], Benoit Coasne[3], Matthias Thommes[4], and Keith E. Gubbins[2]. (1) Department of Chemical Engineering, Louisiana State University, Baton Rouge, LA 70803, (2) Center for High Performance Simulation and Department of Chemical and Biomolecular Engineering, North Carolina State University, Raleigh, NC 27606, (3) Institut Charles Gerhardt Montpellier, CNRS (UMR 56253), University Montpellier 2, and ENSCM, Montpellier cedex 5, France, (4) Quantachrome Instruments, 1900 Corporate Drive, Boynton Beach, FL 33426

Templated mesoporous silica materials MCM-41 and SBA-15 consists of hexagonal arrays of cylindrical pores with diameters between 1.5 and 20 nm, narrow pore size distributions and negligible pore networking. These and other properties make these materials suitable for a number of applications in catalysis, adsorption, optics and electronics [1]. Detailed knowledge of the pore size and pore volume is crucial in order to optimize these applications. Argon adsorption is a key standard measurement for characterization of porous materials [2]. Due to its low vapor pressure, krypton adsorption at 77 K is often used to measure very low surface areas [2,3], and is currently also being explored for the pore size analysis of thin silica films [4].

In this work we report molecular simulations of argon and krypton adsorption on atomistic models of templated mesoporous silicas. These models [5] add atomistic levels of detail to mesoscale representations [6] of these porous materials; the latter were originally generated from lattice Monte Carlo simulations mimicking the synthesis process of templated mesoporous silicas. Three atomistic porous samples were considered: (A) a model of SBA-15 including both mesopores and micropores with mean diameters of 5.4 nm and 1.1 nm, respectively; (B) a model of MCM-41, which is similar to (A) but without microporosity; and (C) a regular cylindrical pore model. Our results [7] for Ar and Kr adsorption indicate that, among the three pore models considered, model A provides the best agreement with experimental data of gas adsorption on SBA-15 samples. Our calculations also suggest that fine-tuning the microporosity and the surface chemistry of model A can lead to better agreement with experiments. Marked differences were observed in the simulated adsorption isotherms, isosteric heat curves, pore filling mechanisms, and structure of the confined phases inside the three pore models. These results suggest that pore surface roughness, microporosity and other morphological features such as constrictions, play an important role in the mechanism of gas adsorption and filling of the mesopores.

[1] G. J. de A. A. Soler-Illia, C. Sanchez, B. Lebeau and J. Patarin, Chem. Rev. 102, 4093 (2002); F. Schüth and W. Schmidt, Adv. Mater. 14, 629 (2002). [2] F. Rouquerol, J. Rouquerol and K. Sing, Adsorption by powders and porous solids, Academic Press (1999); S. Lowell, J. E. Shields, M. A. Thomas, and M. Thommes, Characterization of porous solids and powders: surface area, pore size and density, Springer (2004). [3] C. Y. Chiu, A. S. T. Chiang and K. J. Chao, Microp. Mesop. Mater. 91, 244 (2006); A. Zukal, Microp. Mesop. Mater. 92, 220 (2006). [4] M. Thommes, N. Nishiyama and S. Tanaka, Stud. Surf. Sci. Catal., in press (2007). [5] B. Coasne, F. R. Hung, R. J.-M. Pellenq, F. R.

Siperstein and K. E. Gubbins, Langmuir 22, 194 (2006); S. Bhattacharya, B. Coasne, F. R. Hung and K. E. Gubbins, Langmuir, to be submitted (2007). [6] F. R. Siperstein and K. E. Gubbins, Mol. Sim. 27, 339 (2001); Langmuir 19, 2049 (2003); S. Bhattacharya and K. E. Gubbins, J. Chem. Phys. 123, 134907 (2005). [7] F. R. Hung, S. Bhattacharya, B. Coasne, M. Thommes and K. E. Gubbins, Adsorption, submitted (2007); F. R. Hung, S. Bhattacharya, B. Coasne, M. Thommes and K. E. Gubbins, Langmuir, to be submitted (2007).

250c Computational Investigation Of The Influence Of Surface Disorder On Wetting Behavior

Eric M. Grzelak[1], Vincent K. Shen[2], and Jeffrey R. Errington[1]. (1) Dept. of Chemical and Biological Engineering, University at Buffalo, 303 Furnas Hall, Buffalo, NY 14260, (2) NIST, Physical and Chemical Properties Division, 100 Bureau Dr., MS 8380, Gaithersburg, MD 20899-8380

The wetting properties of a fluid in contact with an attractive substrate depend significantly on the properties of that substrate, including its roughness. This relationship has helped scientists create novel materials with applications ranging from superhydrophillic surfaces used in film spreading to biomimetic self-cleaning materials. In this presentation we describe our recent computational efforts aimed at understanding the link between wetting properties and the molecular-level roughness of a surface. Grand canonical transition matrix Monte Carlo simulations are used to obtain the surface free energy of a fluid in contact with an atomistically detailed surface. This information is subsequently used to calculate spreading coefficients at conditions below the wetting temperature and prewetting saturation points in the partially wetting regime. Extrapolations of these properties allow us to pinpoint the temperature of the wetting transition. To study wetting on rough surfaces, we use a Lennard-Jones fluid interacting with a substrate comprised of static particles. A collection of substrates with varying degrees of molecular disorder are considered, including various orientations of the perfect face-centered cubic lattice, thermally disordered crystalline configurations, and amorphous structures. A Voronoi tessellation is used to characterize the surface topology of a given structure, which provides a means to quantify the molecular roughness of the substrate.

250d Packing Induced Selectivity During Adsorption Of Long Chain N-Alkane Mixtures In Zeolites Under Liquid Phase Conditions

Sudeep Punnathanam and Randall Snurr. Chemical and Biological Engineering, Northwestern University, 2145 Sheridan Road, Room E136, Evanston, IL 60208

Liquid phase adsorption differs from gas phase adsorption due to the high density of the adsorbates inside the pores. This results in the intermolecular forces among adsorbate molecules playing a significant role in the adsorption mechanism. Although in principle grand canonical Monte Carlo (GCMC) simulation of an adsorbed phase in contact with a liquid or gas phase is similar, there are several challenges in practice. Because of the high density in the pores under liquid-phase conditions, insertions and deletions of molecules become inefficient in convention GCMC simulations, resulting in poor sampling averages. Previously [1-3] we have shown that through the use of advanced techniques such as configuration biasing and identity swaps, one can perform GCMC simulations for adsorption of n-alkanes with chain lengths up to 10 C atoms in zeolites under liquid phase conditions. However, simulations of adsorption of alkanes with chain lengths exceeding 10 C atoms, especially of mixtures, are still inefficient.

In this study we make use of the parallel tempering technique to improve the efficiency of simulations for long-chain alkane mixtures in zeolites. These simulations helps to us to understand the molecular origins of phenomena such as azeotrope formation and selectivity inversion observed during experiments of adsorption of alkane mixtures.

1. K. M. A. DeMeyer, S. Chempath, J. F. M. Denayer, J. A. Martens, R. Q. Snurr and G. V. Baron, "Packing effects in the liquid-phase adsorption of C5-C22 n-alkanes on ZSM-5.", J. Phys. Chem. B, 107, 10760-10766, 2003

2. S. Chempath, J. F. M. Denayer, K. M. A. DeMeyer, G. V. Baron and R. Q. Snurr, "Adsorption of liquid-phase alkane mixtures in silicalte: simulations and experiment", Langmuir, 20, 150-156, 2004

3. I. Daems, G. V. Baron, S. Punnathanam, R. Q. Snurr and J. F. M. Denayer, "Molecular cage nestling in the liquid-phase adsorption of n-alkanes in 5A zeolite", J. Phys. Chem. C, 111, 2191-2197, 2007

250e Study On Hexane Adsorption In Zeolite Itq-29 By Molecular Simulation

Miguel A. Granato[1], Thijs J. H. Vlugt[2], and Alírio E. Rodrigues[1]. (1) Department of Chemical Engineering, LSRE - Laboratory of Separation and Reaction Engineering - Faculty of Engineering - University of Porto, Rua Dr. Roberto Frias, Porto, 4200-465, Portugal, (2) Condensed Matter and Interfaces, Utrecht University, P. O. Box 80.000, Utrecht, 3508 TA

Adsorption isotherms and isosteric heat of adsorption of hexane in Zeolite ITQ-29 were simulated using the Configurational Bias Monte Carlo technique in the grand-canonical (μ VT) ensemble and compared with results published by Gribov, Sastre and Corma (*J. Phys. Chem. B.*, **109**, 23794-23803. **2005**) which reports studies on hexane adsorption in zeolite ITQ-29, by IR spectroscopy. The fractional loadings of hexane in ITQ-29 are presented in units from integral intensities of the absorption bands [u.a.]. In this work we present the simulation loadings of hexane in ITQ-29 converted to fractional coverages and compared to the experimental results. The simulations were performed using a united atom force field that explicitly distinguishes silicon and aluminum atoms through the different types of oxygen framework atoms. Our simulation results are in good agreement with the experiments, although an overestimation of the adsorption capacities has been found in the lowest pressure region.

250f On the Crossing of Low Temperature Adsorption Isotherms of Argon on Graphite Surface

David Nicholson and **Duong Do**. Chemical Engineering, University of Queensland, Department of Chemical Engineering, Brisbane, 4072, Australia

Low temperature adsorption isotherms of argon on graphite are studied using Grand Canonical Monte Carlo simulation. The goal of this investigation is to see whether the phenomenon of crossing of isotherms (when plotted against the reduced pressure) is reproduced by the simulation, and to gain insight into the microscopic reasons for the crossing. Simulations were carried out with adsorption of argon on a structureless surface at 71.11, 73.52, 74.96 and 78.47 K. The results were plotted against the pressure scaled against the GEMC simulated vapour pressure. If surface mediation is not accounted for, the simulated isotherms do not cross; however they do when surface mediation is taken into consideration. This is another support for the possible existence of the surface mediation, a phenomenon that has been shown to be necessary for the correct description of adsorption isotherms of noble gases and other gases on graphitized thermal carbon black.

250g The Binding Of Flue Gas Pollutants On Pd, Au, Ag, Cu And Their Alloys

Erdem Sasmaz and Jennifer Wilcox. Chemical Engineering, Worcester Polytechnic Institute, 100 Institute Road. Goddard Hall, Department of Chemical Engineering, Worcester, MA 01609

With the reduction of natural gas and petroleum sources and subsequent increases in their costs, energy from coal is becoming increasingly popular. However, coal is not a clean technology, so with this demand comes a subsequent demand to make this energy source more environmentally-friendly. Trace elements, such as mercury, arsenic, and selenium, are highly volatile and are known to escape into the atmosphere from coal combustion flue gas.

Many sorbent materials such as activated carbon, metal oxides, metal sulfides and pure metals can remove mercury under different conditions. Among these sorbent materials activated carbon is one of the most-studied sorbents. In spite of the high mercury removal capacity of activated carbon, it has been reported that activated carbon prevents concrete from meeting the freeze-thaw requirements in the recycling step involving the addition of particulate matter for concrete manufacturing. It would be more efficient to find a concrete-friendly sorbent material while simultaneously maintaining a high mercury removal capacity.

Density Functional Theory calculations have been performed to predict the binding mechanism of flue gas pollutants such as SO2, Hg0, HgCl2, HgO and SeO2 on pure Pd (001), Au (001), Ag(001), Cu(001) and their binary alloys. The vast majority of the mercury released from coal combustion is elemental (Hg0) and its binding on PdAu (111), PdAg (111) and PdCu (111) has been investigated. It has been found that the Pd surface is more reactive than the Au, Ag and Cu surfaces to the pollutants of coal combustion flue gas. Calculations for Hg0 binding on binary alloys have shown that small additions of Au, Ag and Cu to Pd increase the binding energy by improving the surface reactivity of the alloy.

255b Adsorption Equilibrium And Kinetics Of Hydrogen In Metal Organic Framework (Mof)-5

Dipendu Saha, Zhiguan Yang, and Shuguang Deng. Chemical Engineering Department, New Mexico State University, P.O. Box 30001, MSC 3805, Las Cruces, NM 88003

The short life of fossil fuel along with severe pollution to environments puts the traditional automobiles in the backburner as the reliable means for future-day transport systems. Fuel cell, driven by hydrogen gas, has become a promising member as the source of alternative energy, but the lack of economic, efficient, endurable and safe on-board hydrogen storage system provides a mammoth task for present day technologist to put fuel-cell driven cars into the competitive position with the traditional ones. A worldwide effort to bring up effective means for storing hydrogen gas has resulted in several methodologies and materials, but none of them could suffice to the actual need. Until recently, a new kind of porous crystalline material, known as Metal Organic Framework (MOF) has gained huge potential as a hydrogen storing material. The flexible and case-specific tailoring of basic molecular architecture and functional groups, large pore size, very high specific surface area and selective uptake of small gas molecules have provided MOF a huge benchmark as promising hydrogen storage material. The so called reticular synthesis (logical construction of networks from molecular building blocks) has resulted in six 'catenated' MOFs (isoreticular MOF, IRMOF) reported in literature, amongst which MOF-5 (or IRMOF-1) has become most interesting member of the family.

The present study focuses the characteristics and hydrogen adsorption criteria of MOF-5 (IRMOF-1). With little modification based on the published synthesis procedures, the newly synthesized MOF-5 exhibits Langmuir surface area above 4000 m2/gm with dihydrogen uptake capacity around 1.45 wt% up to one atmosphere pressure and 77K. Characterized by type-1 adsoprtion, sharp increase in the isotherm curve up to the atmospheric pressure indicates that the material is hardly saturated and is expected to have more uptakes at the elevated level of pressure. It is the highest storage capacity ever reported by a crystalline microporous material at low pressure level.

255c Removal Of Mercury From Flue Gases Using Ionic Liquid Coated Chelating Adsorbents

Lei Ji[1], Stephen W. Thiel[2], and Neville G. Pinto[2]. (1) Chemical & Materials Engineering, University of Cincinnati, P.O. Box 210012, Cincinnati, OH 45221-0012, (2) Chemical &Materials Engineering, University of Cincinnati, P.O. Box 210012, Cincinnati, OH 45221-0012

Thermally robust chelating adsorbents for the direct capture of vapor phase mercury have been developed. In these adsorbents, chelating ligands are bonded to a mesoporous substrate and coated with an ionizing surface nanolayer. The effectiveness of these adsorbents for capture of oxidized mercury at elevated temperature (160°C) has been demonstrated in fixed-bed experiments; oxidized mercury capacities as high as 58 mg/g adsorbent have been observed. The capacity of the fully functionalized adsorbent is believed to be the sum of the capacities of the ionic coating, uncovered ligands, and covered ligands. Pore size does not have a significant effect on mercury adsorption. Mercury diffusion through the ionic coating layer is believed to be the rate limiting step for mercury capture. The chelating adsorbent technology can also be used to capture elemental mercury at elevated temperature by proper selection of ionic liquid coating; elemental mercury capacities as high as 15.6 mg/g adsorbent were observed for one unoptimized prototype adsorbent, which is quite high compared to the 2-3 mg/g adsorbent obtained using commercially available activated carbon. These results suggest that the simultaneous removal of elemental and oxidized mercury at elevated temperatures is feasible.

255d Rapid Pressure Swing Adsorption System (Rpsa) For Oxygen Enrichment Of Air : A Comprehensive Model & Analysis

Pallavi Chitta and Bruce J. Tatarchuk. Center for Microfibrous Materials Manufacturing ,Chemical Engineering, Auburn University, 208 Ross Hall, Auburn, AL 36830

Oxygen-enriched air helps increase reaction conversion as well as efficiency in any combustion related process. The primary objective of the research is to design and develop an oxygen enrichment system. The paper presents a comprehensive mathematical model and a system level analysis of a pressure swing (PS), temperature swing (TS) and temperature-aided pressure swing (PTS) adsorption process. The parameters required for the model are generated empirically. The sorption experiments are carried out in a fixed bed reactor (2cm dia).The sorbents consist of molecular sieves and other cation exchanged zeolites primarily including Li-LSX, Li-X. The results of sorbent screening as well as process screening studies are presented in detail. All the experimental results are presented in terms of nitrogen and oxygen breakthrough curves. The experimental feasibility study has demonstrated that the Li-LSX zeolite combined with a short cycle time Rapid PSA shows good potential for the production of oxygen enriched air of low-medium purity. A parametric analysis is presented for particular case of Oxygen enrichment of air by Rapid PSA.

255e Hybrid Adsorption-Distillation System For Olefin-Paraffin Separations

Ramiel Nassara and **F. Handan Tezel**. Department of Chemical Engineering, University of Ottawa, 161 Louis Pasteur, Ottawa, ON K1N 6N5, Canada

The objective of this study has been to look at the possibility of combining adsorption and distillation processes to reduce the energy demand for the separation of olefins and paraffins.

Two adsorbents are studied in this work for ethylene/ethane separations. The first, AgNO3/SiO2, has been considered in previous work to be a very promising adsorbent for olefin/paraffin separations in industry. The second, CuCl/SiO2, has been created based on the high affinity of Cu+ ions for olefins among the π-complexation metal group. Previous adsorbents had been used with Cu+ ions impregnated within zeolites or activated carbons, but there was nothing prevalent on the impregnation of silica gel with copper ions. This is mostly due to the difficulty of finding a method that will successfully implement the impregnation. Wet impregnation was used for the preparation of AgNO3/SiO2, while wet impregnation and thermal monolayer dispersion were used for the CuCl/SiO2 adsorbent preparation to see which case was more effective in impregnating the silica gel.

Volumetric experiments were conducted with the two adsorbents for ethylene and ethane at temperatures of 34°C and -73°C. Langmuir, Toth, and TD-Toth isotherms were used to model the data and binary adsorption behaviour predictions were made accordingly using the Extended-Langmuir model. An economic analysis was conducted for the addition of a Pressure Swing Adsorption system to a cryogenic distillation column.

255f Gas Separation Cycles Using Multiple Adsorption Units

Alberto I. LaCava, Computer Information Systems, Felician College, 262 S Main Street, Lodi, NJ 07644

Adsorption technology has produced very useful separation units that are widely used in industrial practice. Some of the better known separations include Pressure Swing Adsorption (PSA) and Temperature Swing Adsorption (TSA). In the quest to expanding the usefulness of adsorption separations, cycles have been produced that look at the adsorption process in a different way. These include the "inverted PSA" process and several processes that tend to improve the separation of the most adsorbable component at higher levels of purity. As the need of separate components with higher levels of purity increases, the option of using multiple adsorption units becomes an important option. These options are illustrated for cases such as hydrogen recycle in Solid Oxide Fuel Cell cycles and the concentration of oxygen from very low levels to high purities for special applications.

255g Enhancement Of Bio-Ethanol Production By Adsorption

Rudy Jones, **F. Handan Tezel**, and Jules Thibault. Department of Chemical Engineering, University of Ottawa, 161 Louis Pasteur, Ottawa, ON K1N 6N5, Canada

The transportation sector is a major contributor to green house gas (GHG) emissions. It is widely accepted that immediate action is required before irreversible damage is caused to our environment. There is a need to progressively move to renewable energy sources. Biomass, which typically uses approximately the same amount of CO2 during growth as the CO2 generated during its consumption, can advantageously serve as a renewable source of energy. Bio-ethanol produced from cellulosic biomass can mitigate the net CO2 released for the transportation sector. Biomass can potentially be produced from energy crops and other agricultural materials.

During the fermentation process, it is not feasible to go above 5-6 % (w/w) ethanol when fermenting xylose to ethanol with the currently available technology. In the research described herein, it is proposed to continually remove the ethanol from the fermentor by adsorption, to shift the fermentation towards the more rapid production of ethanol to reduce the load of the distillation column downstream in the bio-ethanol production.

Adsorption of ethanol from aqueous solution by activated carbon and ZSM-5 adsorbents was performed in a column-recirculation system to determine the effective adsorption isotherm. The measured capacity in these experiments was lower than experiments performed in previous well mixed batch systems over the same concentration range. The effect of adsorbent attrition is thought to play a role in the capacity discrepancy as particle breakdown was noted in the well mixed batch systems. The effect of the particle size is also discussed.

A model based on finite difference method is presented for the purpose of mass transfer parameter estimation in batch systems and in breakthrough column studies. The applicability of breakthrough simulations based solely on batch kinetic experiments and column-recirculation isotherms is discussed.

A rough economic analysis of the process based on the preliminary data is also presented.

260a Direct Nanoflux in Ptmsp and Composites Thereof

Jason P. Killgore and Rene M. Overney. Chemical Engineering, University of Washington, Box 351750, Seattle, WA 98195

Poly(1-trimethylsilyl-1-propyne) (PTMSP) has garnered significant research interest as the most permeable poymer yet discovered. This extremely high permeability is most apparent in the transport of organic vapors, where sorption of the gas in the polymer plays a dominant role. Morphologically, PTMSP exhibits very high free volume and therefor exhibits poor size selectivity with regards to diffusion dominated permanent gases. This combination of high sorption selectivity and low diffusive selectivity leads to reverse selective separations and has prompted consideration for a number of industrial processes where it is desirable to remove a larger species from a mixed gas while maintaining pressurization of the smaller species.

To date, all investigations of PTMSP have relied on bulk measurements of permeability, while generally ignoring local heterogeneities. Here, we invoke Flux Lateral Force Microscopy (F-LFM), to explore local gas flux in PTMSP. By rastering a scanning probe microscope (SPM) and observing the lateral deflection of the cantilever, it is possible to determine local friction at the polymer surface. Much like an air hockey table, when the substrate is exposed to a gas stream, a corresponding decrease in friction is observed. In addition to providing local flux information, the technique is unique in that it provides access to both diffusive and rheological behaviors, allowing study of their coupled influence on polymer transport.

While suitable for the study of bulk homogoneous polymers, the technique is most valuable in the study of heterogenous or anisotropic membranes whose permeability varies spatially on micron and nanometer length scales. To this end, we have studied the presence of silica interfaces in PTMSP, which have been shown to significantly improve permeability and selectivity of the membrane, particularly when the interfaces originate from high surface area to volume nanoparticles. This enhancement is unique given that traditional Maxwell type models suggest that permeability should decrease volumetrically with the addition of an impermeable phase. Here, flux in the vicinity (~100nm) of the nanoparticles has been explored and compared to the bulk to determine the extent of the interfacial modification.

260b Natural Gas Purification Using High Performance Crosslinked Hollow Fiber Membranes: Effects of High Pressure CO2 and Toluene Feed

Imona C. Omole[1], Stephen J. Miller[2], and William Koros[1]. (1) Chemical and Biomolecular Engineering, Georgia Institute of Technology, 311 Ferst Drive, N.W., Atlanta, GA 30332-0100, (2) Chevron Energy Technology Company, 100 Chevron Way, Richmnond, CA 94802

Natural gas is one of the fastest growing primary energy sources in the world today. The increasing world demand for energy requires increased production of high quality natural gas. For the natural gas to be fed into the mainline gas transportation system, it must meet the pipe-line quality standards. Natural gas produced at the wellhead is usually "sub-quality" and contains various impurities such as CO2, H2S, and higher hydrocarbons, which must be removed to meet specifications.

In natural gas purification, membranes are attractive alternatives to the amine absorption processes, due to their lower cost, lower energy requirement, reduced environmental impact, and reduced size. In this work, it is desired to remove CO2 from natural gas using highly permselective hollow fiber membranes. Hollow fiber modules offer high surface area/volume for increased productivity compared to spiral wound and plate-in-frame modules. Carbon dioxide is usually the largest impurity in natural gas feeds and high CO2 partial pressures in the feed can lead to plasticization and loss of some methane product. Moreover, the presence of highly sorbing higher hydrocarbons in the feed can further reduce membrane performance. For polymer membranes to be used in these applications, they must be robust, resistant to plasticization, easily formed into asymmetric membrane forms and also exhibit adequate separation properties.

Covalent crosslinking has been shown to increase plasticization resistance in dense films by suppressing the degree of swelling and segmental chain mobility in the polymer, thereby preserving the selectivity of the membrane. This research focuses on extending the dense film success to asymmetric hollow fibers. The crosslinked hollow fibers reported here show the high selectivity and fluxes required for industrial application and are tested with model natural gas feeds at high pressures. The performance of the crosslinked hollow fiber membrane in the presence of high CO2 partial pressures and toluene (a highly sorbing contaminant) is investigated.

260c Sulfonated Polyethersulfone-Based Membranes with Different Ion Forms for High-Performance Gas Separation

Yi Li and Tai-Shung Chung. Department of Chemical and Biomolecular Engineering, National University of Singapore, 10 Kent Ridge Crescent,, Singapore, 119260, Singapore

Amorphous polymeric materials, which are cost-effective with sufficient selectivity and good processability, are one type of dominating material in the membrane separation technology. It offers the greatest promise for both industrial application and fruitful academic research. However, an issue, namely "upper bound trade-off curve" raised by Robeson may constrain the further application of polymeric materials in the membrane separation technology. To further expand membranes for industrial applications, one must find ways to improve the gas permeation flux (productivity) and permselectivity by synthesizing and developing high-performance membrane materials.

The sulfonation of polymeric membrane materials is found to be an effective method for advancing membrane-based gas separation performance. Two types of sulfonated polyethersulfone (SPES) polymers with different degrees of sulfonation (DS) were synthesized by adjusting the used amount of chlorosulfonic acid and confirmed via elemental analysis and X-ray photoelectron spectroscopy (XPS) in this work. The SPES-H membranes exhibited a decrease in the permeability and an increase in the selectivity with increasing the degree of sulfonation because of the effect of electrostatic crosslinking. Novel transition metal counterions were applied to conduct the ion exchange treatment of SPES-H polymer samples to change the affinity properties of some specific penetrants in the membrane. The SPES-Zn membrane exhibited the best gas separation performance in this study due to a combination of the strong electrostatic crosslinking and the affinity between Zn^{2+} and some specific gas molecules. Compared with flat dense PES membranes, the SPES-Zn membrane displayed impressive permselectivity increments of approximately 140%, 86%, 18% and 57% for He/N_2, H_2/N_2, O_2/N_2 and CO_2/CH_4 gas pairs, respectively. These newly developed SPES membranes with different ion forms showed high resistance to the CO_2-induced plasticization and a comparably high CO_2/CH_4 selectivity in both pure gas and mixed gas measurements, which demonstrate that these SPES membranes are a type of potential and excellent membrane material for gas separation in industrial applications.

260d Supercritical Carbon Dioxide ($scCO_2$) Separations Using Torlon® Based Asymmetric Hollow Fiber Membranes

Madhava R. Kosuri and William J. Koros. School of Chemical and Biomolecular engineering, Georgia Institute of Technology, 778 Atlantic Dr. NW, Atlanta, GA 30332

Pushing the limits of polymeric membrane technology to supercritical carbon dioxide ($scCO_2$) separations is of particular interest for fluoropolymer synthesis. Developing a robust membrane that can withstand the plasticization effect of such high pressure CO_2 is a key challenge. Plasticization results in increased segmental mobility of the polymeric chains and thus reduces its separation capacity.

We have chosen Torlon®, a polyimide-amide polymer, as the polymeric material of choice for this separation. It is believed that the polyimide-amide forms intra/inter chain hydrogen bonding which can provide resistance to plasticization. We further aim to decrease the plasticization by incorporating molecular sieves in the polymer matrix. In the case of enhanced polymer zeolite interactions, it is believed that the chains near the sieve interface become rigidified and thereby provide resistance to plasticization.

A defect free asymmetric hollow fiber membrane has been successfully formed from Torlon® using a dry-wet spinning process. This defect free membrane provides stable selective separations up to 1100psi of CO_2. Details of the separation performance will be disclosed. Preliminary studies in making a composite membrane with a good polymer zeolite interaction will also be presented.

260e Enhanced Gas Separation Performance of 6fda-Oda-Nda/psf Dual-Layer Hollow Fiber Membranes Via Substructure Resistance Reduction Using the Mixed Matrix Approach

Natalia Widjojo, Shaohua David Zhang, and Tai-Shung Chung. Biomolecular and Chemical Engineering, National University of Singapore, 10 Kent Ridge Crescent, Singapore, 117602, Singapore

In the past decades, the dual-layer hollow fiber membrane technology has become an emerging research area for gas separation due to its high applicability and superior saving on expensively functional materials compared to those single-layer ones. By one step co-extrusion, it may fabricate a composite hollow fiber membrane consisting of an ultra-thin dense selective skin as the function layer and a porous substrate as the supporting layer. As a result, the dual-layer hollow fiber configuration may potentially possess the most practical way that maximizes the function of a highly expensive material or brittle material for gas separation.

In this work, copoly (4,4'-diphenyleneoxide/1,5-naphthalene-2,2'-bis(3,4-dicarboxyl phenyl) hexa fluoro propane diimide) (6FDA-ODA-NDA) / polysulfone (PSF) dual-layer hollow fiber membranes with enhanced gas separation performance with minimum substructure resistance have been fabricated. Prior investigation in the interface morphology of the as-spun dual-layer fibers and its relation with gas separation performance indicate that the inner layer provides a substantial resistance to the permeating gases at various spinning conditions, thus causing the gas separation performance to be predominantly determined by the substructure layer rather than the dense outer skin layer.

To overcome the substructure problem, three novel strategies have been proposed in this work: (1) the addition of Al_2O_3 nanoparticles in the inner layer dope solution; (2) the introduction of early convective premixing with the aid of an indented dual-layer spinneret; and (3) the combination of methods (1) and (2).

The advantages, disadvantage and limitations of each approach have been determined, and will be discussed and revealed at the conference. In summary, dual-layer hollow fibers spun with separation performance above 90% of the intrinsic selectivity value of 6FDA-ODA-NDA dense film have been achieved when combining the first and second approaches.

260f The Effects of Spinneret Dimension and Hollow Fiber Dimension on Gas Separation Performance of Ultra-Thin Defect-Free Torlon® Hollow Fiber Membranes

Na Peng and Tai-Shung Chung. Chemical and Biomolecular Engineering, National university of Singapore, 10 Kent Ridge Crescent, Singapore 119260, Singapore, Singapore

Defect-free hollow fiber membrane with an ultra-thin dense layer is the most desirable apparatus in gas mixture separation due to its high separation performance and good resistance to plasticization effect or heavy hydrocarbon contamination. Its formation is also the most challenging task because of the trade-off between the formation of ultra-thin selective layer and the generation of the defects. However, we have for the first time successfully produced defect-free Torlon® hollow fiber membranes with an ultra-thin dense layer of around 540 • from only a one polymer/one solvent binary system. The O2/N2 permselectivity is much higher than the intrinsic value of Torlon® dense film. The formation mechanism has been studied and will be elaborated in the conference. The key strategy to enhance the permselectivity herein is to achieve finer monodisperse interstitial chain space by inducing an optimum elongation stretch. The effects of spinneret dimension and hollow fiber dimension on gas separation have also been investigated and reported for the first time. It is found that spinnerets with different annular gaps will result in different membrane morphology and separation performance. Torlon® poly (amide-imide) is employed as the membrane material because of its good potential for further treatments to stand against plasticization in nature gas separation.

270a Complex Hydride Hydrogen Storage Vessel Design

Yongfeng Wang, Armin D. Ebner, and **James A. Ritter**. Department of Chemical Engineering, University of South Carolina, Columbia, SC 29208

Hydrogen is being considered very seriously as an alternative, renewable energy carrier for use in internal combustion engines or fuel cells [Ritter et al, 2003]. However, despite some recent advancement, hydrogen storage is still a major concern. In addition to cost, safety and achieving sufficient gravimetric and volumetric densities, a hydrogen storage system, for both stationary and mobile applications, requires suitable thermal integration to facilitate optimal charging and discharging.

Among the commercially viable solid matrix methods of storing hydrogen, the complex hydride, NaAlH4, when doped with Ti, is one of the few, perhaps even the best, option for both stationary and mobile applications. The excitement about this system has been due to its relatively high gravimetric density and its reversibility at relatively low temperatures. However, a paucity of information is available on the design and development of hydrogen storage vessels containing complex hydrides.

The primary objective of this work is to further develop an existing 2-D mathematical model [Gadre et al, 2005; Wang et al, 2007a,b] to describe the dynamic behavior of a complex hydride hydrogen storage vessel during uncontrolled charge operations. This objective will be facilitated by drawing on the previous work by the authors on the modeling of more conventional metal hydride hydrogen storage vessels [Gadre et al., 2003; Jiang et al, 2005; Gadre et al, 2005; Wang et al, 2007a,b]. A secondary objective of this work is to study different ways of thermally integrating the hydrogen storage vessel, ranging from internal water or air heat exchangers, to external water or air heat exchangers. Again, this objective will be facilitated by drawing on the previous work by the authors on internal water heat exchanger vessel designs for metal hydride hydrogen storage vessels [Gadre et al., 2003; Jiang et al, 2005; Gadre et al, 2005; Wang et al, 2007a,b].

It is the contention here that the models developed by Ritter and co-workers for metal hydrides can be easily modified to explore the use of Ti-doped NaAlH4 as the reversible hydrogen storage media. Physical, thermodynamic and kinetic properties of Ti-doped NaAlH4, typically as a function of temperature and pressure, are needed. Some of this information is available in the literature, such as conductivity and heat capacities [Lesch et al., 2004; Wang et al., 2004], hydrogen storage capacities [Gross et al., 2002], charge and discharge kinetics [Bellosta von Colbe et al., 2005]. Some of it might have to be estimated. It is also the contention here that these models can be easily modified to explore different heat transfer modes. Different geometries for internal and external fluid heat exchangers will be readily programmed into the Comsol Physics software package, and different heat transfer coefficient correlations will be examined for the internal and external air heat exchanger configurations, such as finned geometries. This presentation will provide an up to date account of this model development effort being carried out for the design of a complex hydride hydrogen storage vessel.

Bellosta von Colbe, J. M., M. Felderhoff, B. Bogdanovic, F. Schueth, C. Weidenthaler, "One-Step Direct Synthesis of a Ti-Doped Sodium Alanate Hydrogen Storage Material, Chemical Communications, 37, 4732-4734 (2005).

Gadre, S. A., A. D. Ebner, S. A. Al-Muhtaseb and J. A. Ritter, "Practical Modeling of Metal Hydride Hydrogen Storage Systems," Ind. Eng. Chem. Res., 42, 1713-1722 (2003).

Gadre S.A., A.D. Ebner and J.A. Ritter, "Two Dimensional Model for the Design of Metal Hydride Hydrogen Storage Systems," Adsorption, 11, 871-876 (2005).

Gross, K.J., G.J. Thomas and C.M. Jensen, "Catalyzed Alanates for Hydrogen Storage", J. Alloys Compd. 683-690 (2002)

Lesch, D A., Lewis G.J., Low, J.J., Sachtler, J.W.A., "Discovery of Novel Complex Metal Hydrides for Hydrogen Storage through Molecular Modeling and Combinatorial Methods" DOE FY 2004 progress report III.C.4

Jiang, Z., R.A. Dougal, S. Liu, S.A. Gadre, A.D. Ebner and J.A. Ritter, "Simulation of a Thermally-Coupled Metal Hydride Hydrogen Storage and Fuel Cell System," J. Power Sources, 142, 92-102 (2005).

Ritter, J. A., A. D. Ebner, J. Wang and R. Zidan, R., "Implementing a Hydrogen Economy," Materials Today, 9, 18-23 (2003).

Wang, J. et al., "Hydride Development for Hydrogen Storage", DOE FY 2004 progress report III.C.2

Wang Y., A.D. Ebner and J.A. Ritter, "Thermal Characterization of a Metal Hydride Hydrogen Storage System with a Two Dimensional Model during Controlled Discharge," Int. J. Hydro. Energy, to be submitted (2007a)

Wang Y., A.D. Ebner and J.A. Ritter, "Thermal Characterization of a Metal Hydride Hydrogen Storage System with a Two Dimensional Model during Rapid Charging," in preparation (2007b).

270b A Thermodynamic Model For Hydrate Equilibrium In Sediments

Patricia Taboada-Serrano and Costas Tsouris. Oak Ridge National Laboratory, P.O. Box 2008, Oak Ridge, TN 37831-6036

Clathrate hydrates are solid crystalline structures in which water hydrogen-bonded cages enclose gas molecules in a unit volume equivalent to one hundred and sixty unit volumes of gas at standard conditions. The conditions for the formation and stability of gas hydrates involve high pressures and low temperatures, a few degrees above the freezing point of water. Natural gas hydrates can be found in coastal sediments or the permafrost, and their exploitation constitutes an attractive future source of energy. In the case of marine sediments, natural gas hydrate can be present as disseminated particles, hydrate lenses, or massive structures. A thermodynamic model developed to describe the equilibrium of gas hydrates in sediments is discussed in this work. The model is analogous to the analysis performed to predict frost damage in soils. Effects of the porous medium on hydrate equilibrium are included in the model in terms of surface work and grain stress. The objective is to better describe the hydrate stability zone within marine sediments (i.e., depths at which natural gas hydrate is stable), and to predict the characteristics of natural-gas hydrate deposits. Accurate prediction of natural gas hydrate equilibrium in sediments is important for the design of future gas-hydrate exploitation strategies.

270c Adsorption Heat Storage For Solar Energy Applications

Dan Dicaire and **F. Handan Tezel**. Department of Chemical Engineering, University of Ottawa, 161 Louis Pasteur, Ottawa, ON K1N 6N5, Canada

Solar radiation is an abundant renewable source of energy. It has been used in the past for different applications as an energy source. The problem with it has been the loss of the excess energy that can be harvested, but can not be utilized, due to lack of proper energy storage system. If a storage system can be designed so that the excess energy can be stored and utilized at a later time, the efficiency, and therefore the economic benefit of the solar energy system will increase drastically. This project will look at a possible scenario to solve this problem using reversible adsorption. Combining this technology with solar collectors, energy stored during the day can be used for heating during the night and the heat stored from the summer can be used for heating during winter months.

Adsorption is exothermic which is complemented by desorption which is endothermic. This cycle has the potential to be used for physical thermal energy storage in which the heat can be stored for an indefinite period of time without deterioration. The ultimate application for this technology is to store solar radiation harvested from solar concentrators and use the charged adsorbent beds as a clean renewable source of heating for commercial and residential use. An adsorbent storage system has been designed, built and tested at the University of Ottawa. The system has proven that thermal energy can be stored and released from the adsorbent beds and that saturated beds can be regenerated and used again. The study has focused on the screening of adsorbents (such as Zeolite 13X and Silica gel) to determine the optimal working pair for thermal energy storage, the conditions for an optimal regeneration cycle, as well as the optimal flow rate, humidity and regeneration parameters to extract the maximum amount of heat from the adsorbent beds.

270d Adsorption Equilibrium And Kinetics Of Fluoride On Alumina-Based Adsorbents

Arely Torres, Shuguang Deng, Daniel Candelaria, Laura Nunez, and Ryan McCool. Chemical Engineering Department, New Mexico State University, P.O. Box 30001, MSC 3805, Las Cruces, NM 88003

One of the most serious problems along the U.S.A.-Mexican border is the lack of access to adequate quality drinking water. The groundwater from wells in Columbus, NM and Palomas, Mexico border region contains significantly high levels of fluoride and arsenic that could potentially cause many serious diseases for the residents, especially young children in this border region. There is an urgent need for purifying the drinking water supplied by the municipals to protect the health of the residents in the border region. MnO2 and CaO were used to modify the sol-gel derived alumina to enhance its adsorption for fluoride. Both adsorption equilibrium and kinetics on pure alumina and MnO2 and CaO doped alumina adsorbents were investigated in batch adsorption experiments. The MnO2-AA adsorbent showed significant improvement on fluoride adsorption capacity when the fluoride concentration in water is less than 4 ppm, while the CaO-AA adsorbent exhibits higher adsorption capacity at fluoride concentration above 4 ppm. The adsorption of fluoride on the sol-gel derived adsorbent can be described by a Freundlich type of adsorption isotherm equation. The fluoride adsorption kinetics is slightly faster on MnO2-AA than CaO-AA adsorbent. The estimated effective pore diffusivities of fluoride in the MnO2-AA and fluoride in the CaO-AA are 4.5x10-7 cm2/s and 4.3x10-7 cm2/s, respectively. These values are about 10 times higher than the effective pore diffusivities of fluoride in the pure sol-gel derived alumina.

Mesoporous Carbon Impregnated with Iron Oxide Nanoparticles for Arsenic Removal from Aqueous Solution

Xinhua Li, Liangxiong Li, and Robert Lee*

Petroleum Recovery Research Center, New Mexico Institute of Mining and Technology, Socorro, NM 87801, USA

Abstract

Heavy metals (i.e. As^{3+}) in wastewater or groundwater streams have caused worldwide environmental concerns because of the health risk posed by their presence. Nanostructured iron oxides show excellent arsenic removal efficiency due to the strong interaction between nano-sized rust particles and arsenic species. Unfortunately, the subsequent removal of rust nanoparticles offers new challenges for water purification.

In this work, mesoporous carbon impregnated with iron oxide nanoparticles has been synthesized by colloidal imprinting and iron nitrate decomposition. First, silica-sucrose composite material was synthesized by dispersing colloid silica into a sucrose solution, forming a composite structure of well-dispersed silica nanoparticles in a sucrose matrix. Secondly, the sucrose was carbonized at 550°C in N_2 atmosphere. Finally, the dispersed silica nanoparticles were removed by chemical dissolution to form a mesoporous structure. The silica/sucrose ratio was varied to form a stable, 3-dimentional, and ordered

* Author to whom correspondences should be addressed. Phone: (505) 835-6507; Fax: (505) 835-6031 Email: Li@prrc.nmt.edu.

mesoporous structure. Scanning electron microscopy (SEM), micropore surface analysis (BET), and electron microscopy were used for material characterization. The mesoporous carbon synthesized at a sucrose/silica ratio of 0.8 has a bimodal pore structure with large pores at 2μm and uniform mesopores at ~19.9 nm. The surface area of the mesoporous carbon varies from 672 to 1276.2 m^2/g as the sucrose/silica ratio declines from 2.4 to 0.8. The mesoporous carbon was further treated by pore filling and decomposition of iron nitrate to form a mesoporous carbon with encapsulated iron oxide nanoparticles.

The removal of As(V) from an aqueous solution was undertaken through bulk adsorption experiments. Encapsulation of iron oxide in the mesoporous carbon is essential to enhance As(V) adsorption capacity. The high adsorption capacity suggests that the use of mesoporous carbon with encapsulated iron oxides is promising for heavy metal removal from water.

Key words: Mesoporous, carbon, Colloidal-imprinting, Iron oxide, As(V) removal, water treatment.

1. Introduction

Micro- or mesoporous carbons have great potential for application in catalytic and environmental cleaning since they usually have large surface areas and custom-designed pore geometry for removing large-molecule contaminants. One efficient method for porous carbon synthesis is a process calling hard-templating, in which ordered mesoporous silicas (i.e., SBA-15 or MCM-48) are employed as hard templates.[1] The general synthesis route includes pore filling of a carbon precursor, carbonization, and dissolution of the hard template, resulting in an inverse replica of the mesoporous silica.[2,3] The pore size and geometry of the synthesized carbon is a reverse replica of the mesoporous silicas. For example, the mesoporous carbon CMK-3 has a hexagonally ordered mesopore with pore size of 8.4 nm, similar to the pore wall structure of the hard template of SBA-15.[4] Recently, mesoporous carbon with larger pore sizes ranging from 10 nm to 100 nm has been synthesized using colloid silicas as templates.[2] The mesoporous carbon with larger pore sizes is an ideal host material for postsynthetic doping or grafting because the large surface areas and pore sizes are accessible to dopants and large organic molecules.

In this work, a nanostructured mesoporous carbon was synthesized by using colloid silica as a template. Nanoparticles of iron oxide were further impregnated into the spherical pores of the mesoporous carbon by pore filling and decomposition. Environmental application of the nanostructured mesoporous carbon was investigated for As(V) removal from an aqueous solution.

2. Experimental

2.1 Chemicals and instrumental

The chemicals used in this experiment include sucrose (ACS, Fisher), colloidal silica (40 wt%, Aldrich), H_2SO_4 (93%), NaOH (ACS, Aldrich), alcohol blend (100%, Stephens), and iron nitrate nonatrydrate, ($Fe(NO_3)_3 \cdot 9H_2O$, p.a.,Acros), sodium arsenate dibasic heptahydrate ($Na_2HAsO_4 \cdot 7H_2O$, Sigma-Aldrich), and distilled water (DI).

A gravity convection oven (Model DVS 400) and a Ney Vulcan 3-550 were used to dry and pyrolysize the carbon samples. A stirrer/ hot plate (Cole- Parmer Instrument Company, Model 4658) was used to prepare the synthesis solution. The surface area and pore structure were measured by a surface area and porosity analyzer (Model ASAP

2020). The surface morphology and pore structure were studied by field emission scanning electron microscopy (SEM, HITACHI). The As(V) concentration was analyzed with atomic absorption spectrometers (Zeeman AA, Varian).

2.2 Synthesis of mesoporous carbon impregnated with iron oxide

The synthesis of colloid imprinting mesoporous carbon includes three steps: (1) preparation of well-dispersed colloid silica into a sucrose solution, (2) carbonization of sucrose to form a composite material with silica nanoparticles embedded into a sucrose matrix, and (3) dissolution of silica nanoparticles ot form a porous structure with uniform spherical pores in the range of 10 to 30 nm.

The synthesis of mesoporous carbon was performed using sucrose as the carbon source and colloid silica as the hard template. First, 25 wt% sucrose solution was prepared by adding 4 g sucrose into 16 g DI water. A small amount of H_2SO_4 (0.2 ~ 0.3g) was added to catalyze the carbonization reaction of sucrose. The mixture was stirred at room temperature for 5 min. Different amounts of colloid silica solutions were then added to obtain sucrose/silica ratios of 2.4, 1.9, 1.2, 0.8, and 0.4. After stirring at room temperature for 30 min, the temperature was increased to 60°C for water evaporation. Stirring was continued at 60°C for about 12 hr; about 50% water was removed by evaporation, forming a very viscous, light yellow slurry. The viscous slurry was then transferred to a gravity conventional oven (Model DVS 400) for drying at 80°C for 5 hr. The oven temperature was then increased to 100°C for 3 hr and 160°C for another 3 hr. The precarbonized sucrose with embedded silica nanoparticles was further heated at 500°C for 6 hr under N_2 atmosphere. The carbon-silica mixture obtained therein has a structure with well-dispersed silica nanoparticles embedded in the carbon matrix. Silica dissolution by NaOH solution was performed with a procedure reported in the literature.[5] The silica–carbon composite was treated with 1.0 M NaOH solution (50% alcohol–50% H_2O) at 120°C for 10 hr. After washing with ethanol, the dissolution was repeated until no trace silica could be detected by EDS. The final mesoporous carbon was washed with a large quantity of DI water (1000 mL/g carbon) and dried at 120°C for 8 hr.

The impregnation of iron oxide into the mesoporous carbon was prepared by a process of infiltration of iron nitrate solution and decomposition. In a typical synthesis of iron oxide-impregnated, mesoporous carbon, 0.3 g carbon was dispersed into 10 ml iron nitrate solution and stirred for 6 hr to impregnate the mesopores with iron nitrate solution. The iron nitrate solutions had concentrations of 0.10 M, 0.5 M, and 1.0 M. After filtration and drying at 60°C for 12 hr, the temperature was increased to 90°C and the solution was held for 3 hr under vacuum. The encapsulated iron nitrate was further decomposed to iron oxide by treating at 350°C for 5 hr under N_2 atmosphere.

2.3 Material characterization

Surface area and pore size distribution of the mesoporous carbon with and without iron oxide infiltration were investigated by N_2 adsorption and desorption at 77 K (BET, ASAP 2020). The pore size distribution was derived from a BJH model on the adsorption branch of the N_2 adsorption-desorption isotherm. The surface morphology and pore geometry of mesoporous carbon was studied by scanning electron microscopy.

2.4 Adsorption experiments

The As(V) removal experiments were carried out in a batch reactor. The As(V) stock solution with concentration of 1000 mg/L was prepared by dissolving $Na_2HAsO_4.7H_2O$

into DI water. The test solution containing 1.0 mg/L As(V) was prepared by diluting the stock solution with DI water. The batch experiments were carried out in a plastic bottle containing 200 ml solution with initial As(V) concentration of 1.0 mg/L. Then 0.1 g mesoporous carbon was added under continuous stirring and the samples were collected at definite intervals through a syringe filter (0.22μm, Fisher). The concentration of As(V) was analyzed immediately after sample collection. The amount of As(V) adsorbed was calculated from the difference between the initial and final solutions.

3. Results and Discussion

Figure 1 gives the surface area of mesoporous carbons synthesized from solutions containing different sucrose/silica ratios. During the sucrose-silica composite synthesis, the sucrose is the continuous phase and the silica nanoparticles are embedded in the sucrose matrix. Increasing the sucrose/silica ratio from 0.4 to 0.8 results in an increase in surface area from 1226.8 to 1276 m^2/g. A low concentration of sucrose (i.e., sucrose/silica=0.4) is not sufficient to form a continuous carbon phase. As a result, the synthesized mesoporous carbon has the morphology of fine powder and a relatively small surface area. An appropriate sucrose/silica ratio is critical to form a continuous matrix of carbon with large surface area. As the sucrose/silica was increased from 0.8 to 2.4, the surface area of the synthesized mesoporous carbon declined gradually from 1276 m^2/g to 612 m^2/g. The chemical composition of the synthesis solution also shows some influence on the pore size distribution of the resulting mesoporous carbon. The BJH pore size shifted from 24 nm to 18.5 nm as the sucrose/silica ratio declined from 2.4 to 0.4, as shown in Fig. 2. The pore shrinkage of mesoporous carbon is caused by high stress occurring during the sucrose carbonization process. The mesoporous carbon synthesized with a high sucrose content have a stronger carbon framework and show less contraction during the silica dissolution in a strong NaOH solution. Mesoporous carbon synthesized from the composite with a sucrose/silica ratio of 0.8 shows a large surface area and thus is used for the posttreatment for As(V) removal.

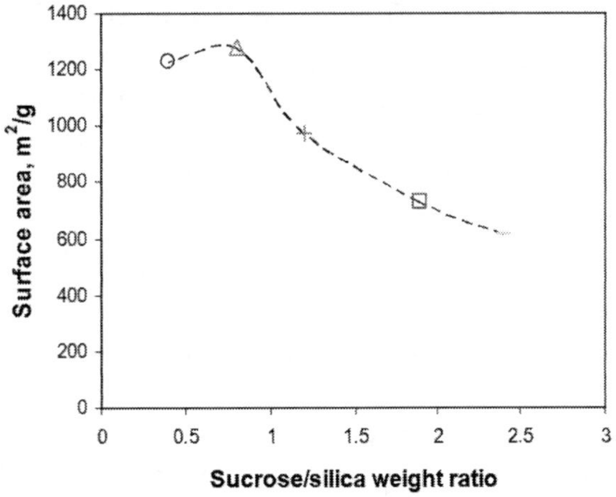

FIG.1. Surface area as a function of sucrose/silica ratio.

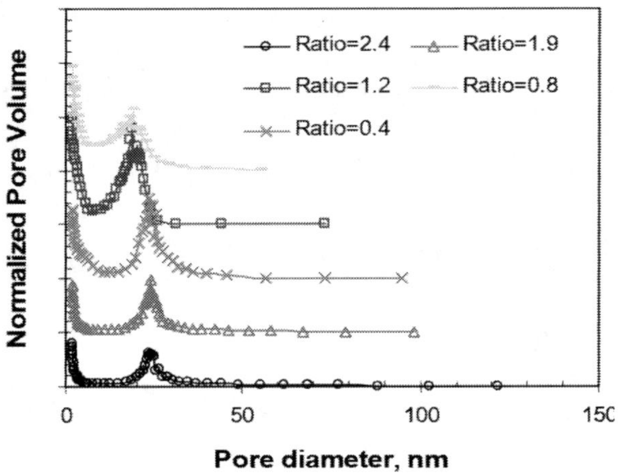

FIG. 2. Pore size distribution of porous carbon synthesized from solution containing different sucrose/silica ratios.

Figure 3 gives SEM images of the mesoporous carbon, showing a bimodal pore structure. The large pores with pore size at 0.1–2.0 μm is formed when the sucrose solution dries while the uniform nanopores are derived from the dissolution of silica nanoparticles. Such a bimodal pore structure is superior for application in water treatments because the large pores provide an interconnected mass transport channel while the nanopores enhance the adsorption capacity.

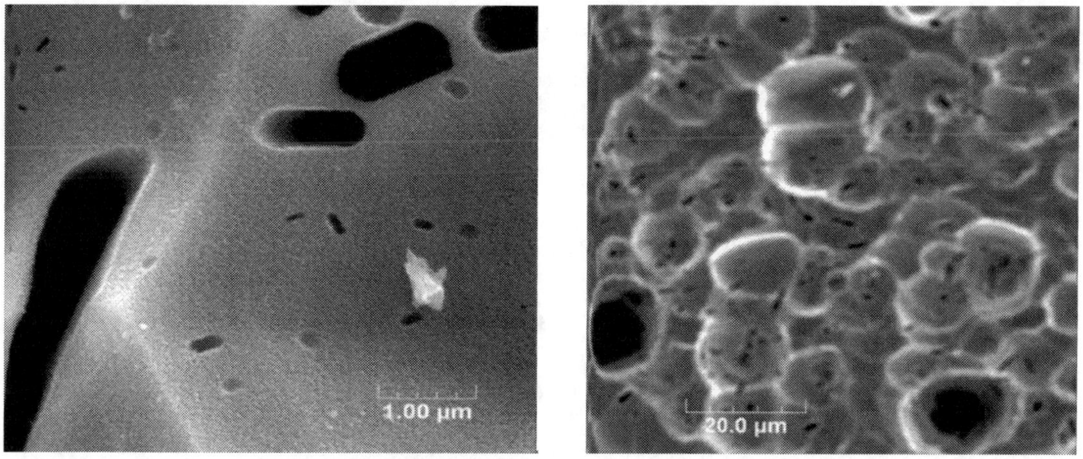

FIG. 3. SEM images of mesoporous carbon at different magnifications.

The synthesized mesoporous carbon was further treated by iron oxide filling for arsenic removal. Figure 4 gives the N_2 adsorption isotherm and pore size distribution derived from the BJH model on the adsorption branch.

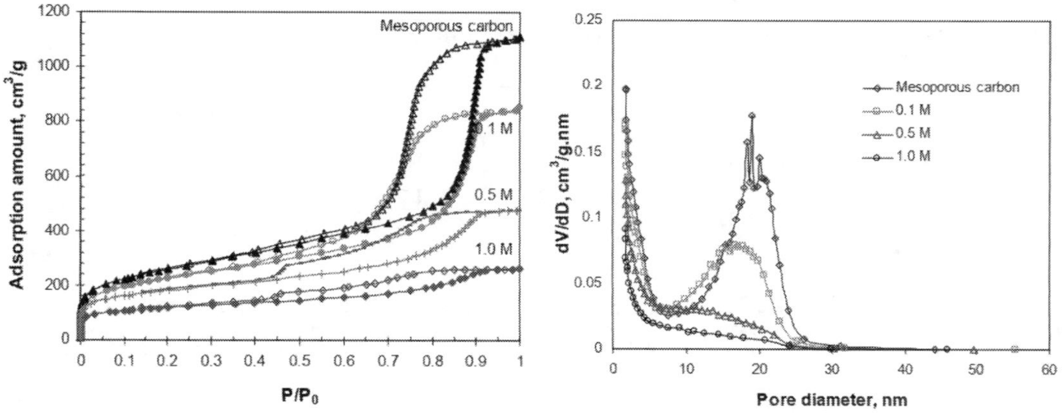

(a) N_2 adsorption isotherm (b) BJH pore size distribution

FIG. 4. N_2 adsorption isotherm and pore size distribution of mesoporous carbon with different amounts of pore filling by iron oxide.

The N_2 adsorption isotherm shows typical mesopore characteristics, as illustrated in Figure 4(a). Increasing iron nitrate concentration reduces pore volume. It is interesting to observe that the hysteresis loop occurs at nearly the same relative pressure (P/P_0=0.45), suggesting the same curvature of the pore radius even though the pores were partially filled with iron oxide. A possible explanation is that the iron oxides exist as nanoparticles instead of coating the internal surface of mesoporous carbon, but more evidence (i.e., transmission electron microscopy imaging) is needed. The same conclusion was derived from the pore size distribution analysis as shown in Figure 4(b). Broadening of pore size distribution is also observed that is attributed to the bimodal characteristics of synthesized carbon. A substantial quantity of iron nitrate solution entrained in the large pore cavities will be concentrated during the drying process. The concentrated solution will fill adjacent mesopores upon further drying and decomposition, resulting in variations in quantities of iron oxide filling pores. Table 1 summarizes characteristics of mesoporous carbon impregnated with iron oxide synthesized by infiltration and decomposition.

Table 1. Langmuir Surface Area and Pore Size Distribution

No.	Pore filling	$S_{Langmuir}$, m^2/g	Pore size, nm	Pore volume, cm^3/g
1	Mesoporous carbon (sucrose/silica=0.8)	1276.2	19.9	1.70
2	0.10 M $Fe(NO_3)_3$	1101.8	17.2	1.30
3	0.46 M $Fe(NO_3)_3$	866.5	14.4	0.73
5	0.86 M $Fe(NO_3)_3$	558.6	13.1	0.41

The iron oxide-impregnated mesoporous carbon synthesized from 0.10M $Fe(NO_3)_3$ solution was tested for As(V) adsorption by bulk experiments. The time-dependence of As(V) adsorptions for the mesoporous carbon with and without impregnating iron oxide is shown in Figure 5. The As(V) adsorption increases rapidly at an initial two hours and approaches equilibrium with elapsed time. The impregnation of iron oxide plays a critical role in As(V) adsorption capacity increasing that of mesoporous carbon from 0.23 mg/g to 1.25 mg/L.

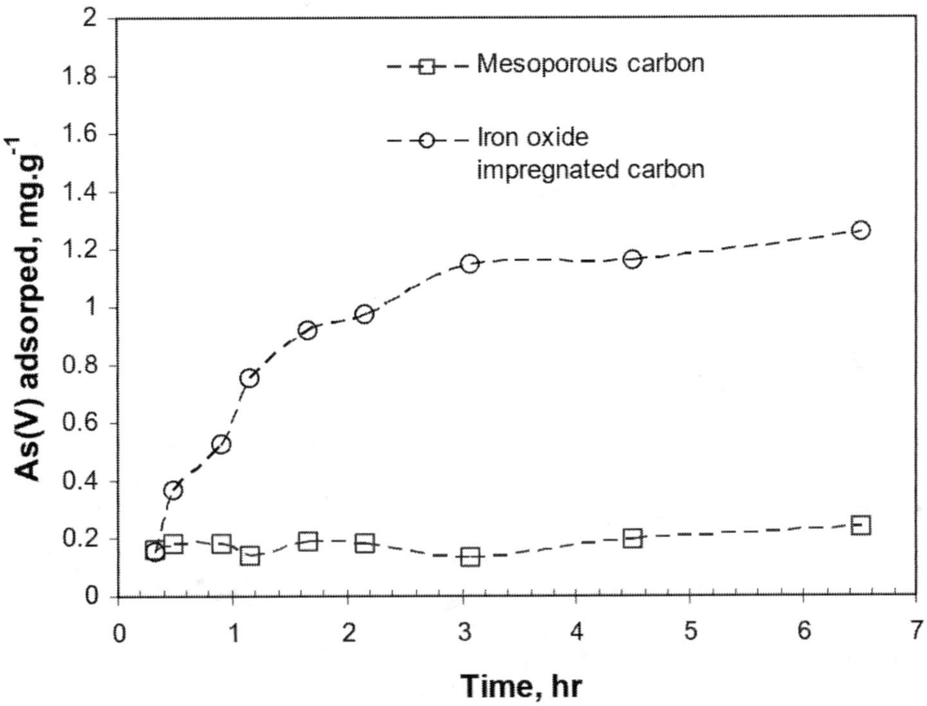

FIG. 5. The effect of contact time on the uptake of As(V).

Table 2 compares the arsenate removal efficiency of synthesized mesoporous carbon and other reported materials. The iron-impregnated mesoporous carbon shows a significantly higher adsorption capacity.

Table 2. Comparison of Mesoporous Carbon with Other Reported Iron-Oxide Coated Materials

Adsorbent	Dosing, g/L	Initial As(V) concentration	Removal (%)	Adsorption capacity, mg/g	Reference
Iron oxide-coated sand	10	1.0	80	0.065	Joshi et al.[6]
Iron-impregnated sand	10	1.0	98.5	0.098	Vaishya et al.[7]
Iron oxide-coated cement	3	2.0	95.5	0.64	Kundu et al.[8]
Mesoporous carbon	0.5	1.0	10.7	0.23	This work
Iron oxide-impregnated carbon	0.5	1.0	57.1	1.25	This work

4. Conclusion

Mesoporous carbon impregnated with iron oxide was synthesized by filling mesoporous carbon with iron nitrate solution, followed with drying and decomposition. The mesoporous carbon with a large surface area and uniform spherical pore structure was synthesized through the colloid imprinting process in which colloid silica nanoparticles embedded in a sucrose matrix were selectively removed by carbonization and dissolution. A typical mesoporous carbon was synthesized from a solution containing 4 g sucrose + 10 mL colloid silica + 16 g DI water. The mesoporous carbon has a large surface area of 1276 m^2/g with pore size of 19.9 nm.

Postsynthesis treatment by pore filling with iron oxide formed a composite structure with iron oxide nanoparticles encapsulated in the mesopores of the carbon host. The concentration of the filling solution plays a critical role ing the pore structure and pore volume of the resulting material. The iron-impregnated mesoporous carbon shows a high As(V) adsorption capacity and can be potentially used as a novel adsorbent for water treatment.

Acknowledgement:

This research was sponsored by the U.S. Department of Energy through the National Energy Technology Laboratory (grant # DE-FC26-04NT15548) and Southwest Carbon Sequestration Partnership (grant # DE-FC26-05NT42591). The authors would like to thank Ms. Liz Bustamante for editing the manuscript.

References

[1] Li, Z., and M. Jaroniec, *J. Am. Chem. Soc.*, 123 (2001), 9208-9209.

[2] Roggenbuck, J., and M. Tiemann, *J. Am. Chem. Soc.,* 127 (2005), 1096-1097.

[3] Ryoo, R., S.H. Joo, and S. Jun, *J. Phys. Chem. B.*, 103 (1999), 7743-7746.

[4] Jun, S., et al., *J. Am. Chem. Soc.*, 122(2000), 10712-10713.

[5] Vinu, A., M. Miyahara, V. Sivamurugan, T. Mori, and K. Ariga, *J. Mater. Chem.*, 15 (2005), 5122-5127.

[6] Joshi, A., M. Chaudhuri, *J. Environ. Eng.-ASCE.*122 (1996), 769-776.

[7] Vaishya, R.C., S.K. Gupta, *J. Environ. Eng.-ASCE.* 129 (2003), 89-93.

[8] Kundu, S., and A.K. Gupta, *Colloids and Surface A: Physicochem. Eng. Aspects* 273 (2006), 121-128.

270f Adsorption of Frying Oil Polar Components by Magnesium Silicate and Activated Carbon

Ozgul Ozcan, Chemical Engineering Dept., Istanbul Technical University, Maslak, Istanbul, Turkey and **Sevil Ozgul-Yucel**, Bioengineering Dept., Y ıld ız technical University, Davutpa°a kampüsü, Istanbul, Turkey.

ABSTRACT

This study covers the laboratory studies on the adsorption of polar compounds of frying oil by synthetic magnesium silicate and activated carbon produced from the same origin rice hull. The polar components adsorption capacity of the produced magnesium silicate was 644 mg polar comp./g adsorbent. This value was higher compared to the capacities of the industrial magnesium silicate Magnesol XL, an industrial activated carbon and laboratory produced activated carbon which have the values of 616, 552 and 368 mg polar compound/g adsorbent, respectively. Magnesium silicate was produced with a very simple reaction between magnesium chloride and the sodium silicate solution which was obtained before by the alkaline treatment of the industrial rice hull ash of Thrace region in Turkey. The product magnesium silicate had a surface area of 680 m2/g and a particle size around 40 ì. Laboratory producd activated carbon however had a 43 m2/g surface area and 150 ì particle size.

272a "Performance of Liquids Recovery Plant Demethanizer"

Glenn Shiveler, Chemical-Gas Group, Sulzer Chemtech USA, 4019 South Jackson, Tulsa, OK 74145 and Derrick Oneal, Thomas Russell Co., 7050 South Yale, Suite 210, Tulsa, OK 74136.

Abstract: The performance of a gas plant with a cryogenic Demethanizer equipped with several sections of modern random packing and tower internals is examined to show the mass transfer separation performance, ethane recovery and energy usage under various modes of operation. This study compares the performance of the plant in GSP operation with full refrigeration versus partial refrigeration during plant start-up. Comparison is made between plant data and process simulation. The packing hydraulics is studied and compared to vendor hydraulic models. The packing mass transfer efficiency is compared to hydraulic load parameters, separation factor and the stripping factor.

272b Mass Transfer Characteristics Of Packed Beds Containing Catalytic Structured Packing: Model Vs. Experiment

Zarko Olujic[1], Marcel Behrens[2], and Peter Jansens[1]. (1) Separation Technology, Delft University of Technology, Leeghwaterstraat 44, Delft, 2628 CA, Netherlands, (2) ET GEO-Europe, Air Products and Chemicals, Hersham Place, Molesey Road, Hersham, KT12-4RZ, United Kingdom

Combining heterogeneously catalysed chemical reaction with thermal separation in one unit appears to be an effective way to intensify industrial processes, where applicable. Further significant improvements in this direction are expected from a new generation of structured packing based modular contactors which allow certain flexibility with respect to the variation in the reaction to separation requirements in a single unit. However, little is known about hydraulic and mass transfer performance characteristics of such complex contactor-separator configurations. Total reflux experiments have been performed at atmospheric pressure using pilot plant facility available at TU Delft with Sulzer KatapakSP 12 packing, in conjunction with two test systems, an aqueous system and an organic mixture. The tests were performed first with a bed comprising the structured packing MellapakPlus 752.Y, then with a bed comprising only KatapakSP 12, and finally with a bed simulating actual case with three elements of KatapakSP 12 placed in between short beds of MellapakPlus 752.Y on upper and lower side. As expected the pressure drop and mass transfer performance of hybrid bed were in between those of two homogeneous beds. Taking a predictive model developed originally for common corrugated sheet structured packings, which was later extended to so-called high capacity packings, as a building block, a new, parallel channel model was developed that makes a distinction between catalyst filled pockets in the reaction section and the open and/or closed channels in separation section. With this simple approach it appeared possible to account properly for specific geometric features of modular catalytic structured packing, without using any empirical, packing type specific constant. The new model proved to be capable to predict the mass transfer efficiency around the loading (design) point.

272c Dry Tray Pressure Drop Of Sieve Trays

Daniel R. Summers, Sulzer, 4019 South Jackson Avenue, Tulsa, OK 74107

The sieve tray has been in the market place as a distillation device for many decades. It has been used extensively on distillation tray equipment worldwide as a highly efficient vapor/liquid-contacting device. Many people have examined dry tray pressure drop data for sieve tray decks and there have been several attempts to correlate this data. This paper will be a supplement to the author's AIChE March 2001 paper on the dry tray pressure drop of movable and fixed valves. A correlation for sieve tray pressure drop will be presented that incorporates tray thickness, hole diameter and open area. This correlation will be compared to data. It is felt that a definitive dry tray pressure drop correlation would be of great interest to those working in Column Design.

272d High Capacity Co-Current Flow Distillation Internals

Zhanping Xu, Paul Sechrist, Greg Wisniewski, and Gavin P. Towler. UOP, 175 E Park Dr, Tonawanda, NY 14150

The SimulFlow™ device is a next generation mass transfer device capable of achieving a significant capacity increase while maintaining or improving product purities.

Testing in the UOP 8-foot diameter air-water column has indicated that the SimulFlow device hydraulic capacity is ~235% (constant liquid/vapor ratio) that of a conventional valve tray, at a pressure drop which is significantly lower. Further testing in an F.R.I. 4-foot diameter distillation column with iC4/ nC4 at 100 psia (6.9 bar) demonstrated a potential capacity which exceeds the calculated F.R.I. system limit at a low, 16-inch (406-mm) stage spacing. The operating flexibility of the device was established at F.R.I. by consistently achieving a stage efficiency of ~70% across an operating range of more than 2 to 1.

At these test conditions, the SimulFlow device at a 16-inch (406-mm) stage spacing has the same height equivalent to a theoretical plate (HETP) as a tray at a typical 24-inch (640-mm) tray spacing with an efficiency of 105%. Therefore a column containing SimulFlow devices can generate the same number or more theoretical plates in a given height with up to 235% of the flow capacity.

This capacity increase is achieved by co-current liquid and vapor flow against gravity within a contacting channel, followed by efficient vapor/liquid phase separators. Each stage is comprised of individual linear modules of varying horizontal length to accommodate a circular column area. Successive stages are rotated by 90 degrees in order to ensure uniform fluid distribution throughout the column. These features form a unit cell structure, allowing predictable results at various column diameters with high confidence.

CFD STUDY FOR CHIMNEY TRAY DESIGN IN CRUDE FRACTIONATORS

Karolline Ropelato[(1)], *Ricardo Rodrigues*[(1)], *Andre C. Bellote*[(2)], *Gunter Schnaibel*[(2)], *Glaucia A. S. Torres*[(3)], *Maria E. P. L. Marsiglia*[(3)], *Carlos N. Fonseca*[(4)], *Antonio V. S. Castro*[(4)]

(1) ESSS – Engineering Simulation and Scientific Software, Florianopolis/SC, Brazil.
(2) REPLAN – Paulinia Refinery, São Paulo/SP, Brazil.
(3) PETROBRAS Headquarters, downtown, Rio de Janeiro/RJ, Brazil.
(4) CENPES – PETROBRAS Research and Development Center, Rio de Janeiro/RJ, Brazil.

ABSTRACT

Liquid distribution makes the difference between good and bad equipment performance in petroleum refining, so its understanding is a major issue with regards to technology applied to distillation columns. The present work considers a Computational Fluid Dynamics (CFD) study applied to a real chimney tray. This study represents the base case to analyze the liquid maldistribution phenomenon and proposes enhancements on the design concept, since the liquid maldistribution affects directly the operational performance. CFD studies gave support in the comparison of a real chimney tray behavior with different arrangements at three levels of mass flow rate. A three-dimensional (3D) transient, non homogeneous and free surface model (Eulerian-Eulerian approach) has been designed for the flow and geometry improvements of chimney trays in a fractionation tower. Results show the importance of employing the Eulerian-Eulerian model to predict the flow patterns of liquid and gas phases.
KEYWORDS: Chimney tray, distillation column, multiphase flow, CFD.

INTRODUCTION

Computational fluid dynamics (CFD) is a strong tool applied by process engineers in design, investigation and development of deep industrial knowledge basis. It is especially useful for complex geometries and under conditions which human access is not allowed with measuring equipment. CFD technology allows its users to investigate and precisely evaluate flow variables throughout numerical optimization before building expensive experimental prototypes. This results in better-designed technology that can be put into operation quicker and more cost effective. As a powerful tool, the CFD technique was chosen to understand the collectors pan multiphase phenomena.

The formation of inoperative or stagnation zones must be avoided in the columns, because the mass-transfer processes take place in the effectively wetted regions so it is fundamental that the distribution of the liquid in contact with the gas be homogeneous. The uniformity of fluid flow characterizes the performance of liquid distributors Heggemann *et al.* (2007). Collector pans establish the condition of liquid feed to distributors as they are in the previous section of the tower. In petroleum refining its understanding is a major issue with regards to technology applied to distillation columns. One of recent works in this approach relies on the work of Pizzo *et al.* (1998), who analyzed liquid distribution in a packed column, at pilot scale, checking efficiency values and showing that when there is a tiny distance between the distributor and the packing surface, the initial distribution has a better flow profile. Olsson (1999) analyzed liquid distribution cases to evaluate details about design, manufacture, testing, installation, and inspection of distributors. Spiegel (2006) proposed a method to assess liquid distribution quality. Heggemann *et al.* (2007) validated CFD results with fluid flow experimental

data on liquid distributors determining local flow rates through orifices including the analysis of discharge coefficients of extensive devices in a feasible expenditure of time. A study comparing different distributor designs has been performed by Lehner *et al.* (2006).
According to Kister *et al.* (2001) less attention has been paid on chimney trays, which are considered the second most problematic distillation internal, just after packed-tower distributor type.
The collector pan study has been concerned on how the geometrical arrangement influences the liquid distribution at downboxes. The purpose of this analysis is to study the liquid maldistribution that affects directly the operational condition, aiming to propose changes to the conceptual design.

Model set-up

Figure 1 represents the base case or original conception of collector pan with near 7 meters of diameter. Blue color indicates the downboxes, red color the chimney arrangement and green color the sump. The symmetrical conditions of the collector pan allowed a reliable simplification of the mathematical model.

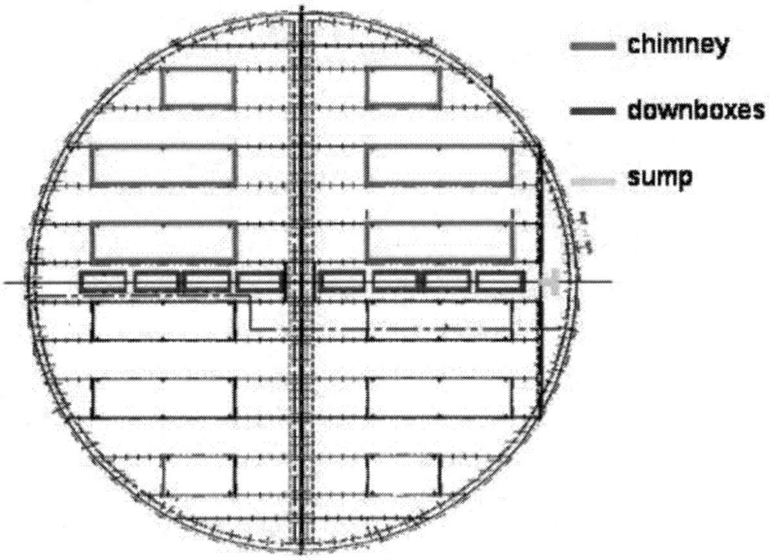

Figure 1: Base condition of collector pan.

The study has considered three different geometries based on the chimney arrangement: original, shifted (the chimneys are shifted 200 mm away from the downboxes) and 90º rotated. Figure 2 presents the geometrical conditions, blue color represent the water inlet, red color the downboxes and green color the sump. Downboxes and sump represent the liquid outlets. The liquid (water) feeds the system and flows around chimneys to the downboxes and the sump. The mass flow rate at sump withdrawal presented on Table 1 guarantees the liquid draw. The downboxes mass flow is the response that will be affected by chimney arrangement or withdrawal condition.

CFD analysis aids in comparing the real chimney performance with different arrangements at three levels of mass flow rates at sump withdrawal. Liquid distribution depends on applicable principles which are dependent on criteria such as amount of liquid to be distributed and its operating range, as shown on Table 1, the internal reflux represents the flow through downboxes.

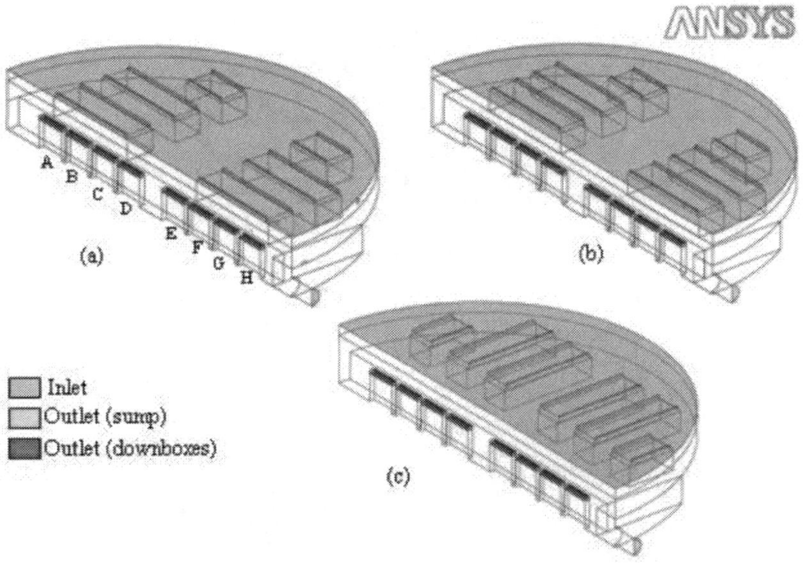

Figure 2: Boundary conditions: (a) original, (b) shifted and (c) 90º rotated.

Table 1: Evaluated conditions.

	Inlet (kg s^{-1})	Outlet (kg s^{-1})	
		Internal reflux	Withdrawal (sump)
Low	95	79.16	15.84
Medium	95	47.5	47.5
High	95	15.84	79.16

Numerical details

ANSYS-CFX discretizes the governing equations using an element-based finite volume method. This numerical method has the advantage of combining geometric flexibility of the finite volume method with strict properties conservation of the finite volume method. A polyhedral control volume is constructed around each mesh vertex. The collector pan geometry consists of a tetrahedral mesh of prismatic elements with the concept of extruded elements.

Different geometries have been studied with roughly the same number of nodes ($\cong 3.10^5$) to be consistent when comparing results under the same conditions. Some important details had been considered to simulate a multiphase flow like extruding the mesh in the region close to the continuous phase (air). This is important to represent the flow interface with enough refinement close to equipment wall. The inlet condition is characterized by droplets and this region was numerically discretized by a coarse mesh as the numerical results in this region are not important for this study.

The model has been generated as three-dimensional (3D), transient and the continuous phase (air) has been modeled using the k-ε turbulence model, which is the most prominent turbulence model implemented to the present condition. It has been proved that this model is stable and numerically robust providing a well established regime of predictive capability with accuracy (Wilcox, 2000). A high resolution scheme was used to model the advection terms of the momentum equation. A second order advection scheme based on flux limiters was applied in the momentum equation for the volume fraction equations.

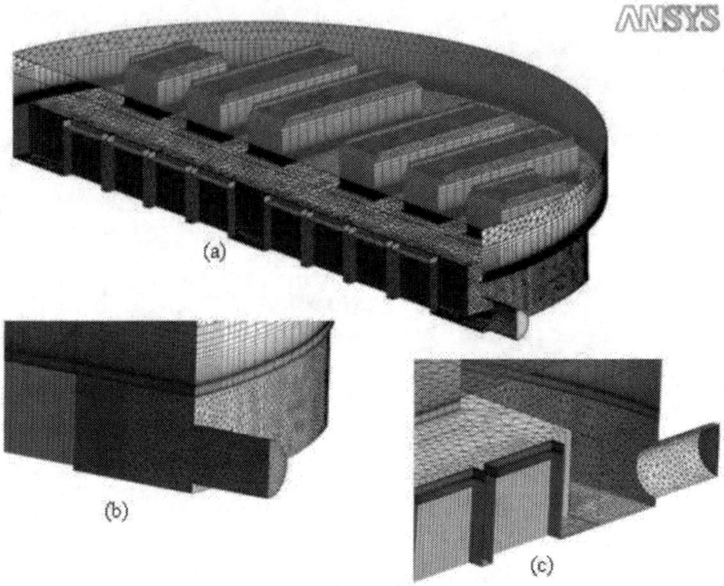

Figure 3: Numerical mesh, (a) Rotated mesh, (b) Symmetry and interface refinement in detail, (c) Mesh details inside.

GOVERNING EQUATIONS

An Eulerian-Eulerian non-homogeneous model has been applied to analyze the mass and momentum behavior. Free surface flow is a special class of multiphase flow in which it is often appropriate to assume a single velocity field at each point in the space (Zwart, 2007). Although a non-homogeneous model has to be applied in this case, the free surface flow allows the two phases to separate. This condition is required because the dispersed phase (water) is entrained into continuous phase (air) and after that, the two phases will separate again. This condition does not fit in the homogeneous model approach.

In the non-homogeneous model each fluid has its own flow field and the fluids interact via interphase transfer terms and there is only one solution field for each separate phase. Transported quantities also interact via interphase transfer terms, e.g. two phases may have separate velocity and temperature fields, but there will be a tendency for them to come to equilibrium through interphase drag.

A transient, multiphase method has been modeled to represent the flow on the collector pans. Considering the α phase, the continuity and momentum equations are given by Equations (1) and (2) respectively:

$$\frac{\partial(r_\alpha \rho_\alpha)}{\partial t} + \nabla \bullet (r_\alpha \rho_\alpha u_\alpha) = 0 \tag{1}$$

$$\frac{\partial}{\partial t}(r_\alpha \rho_\alpha u_\alpha) + \nabla \bullet (r_\alpha \rho_\alpha u_\alpha u_\alpha) - \nabla \bullet (r_\alpha \mu_\alpha (\mathbf{T}_\alpha + \mathbf{T}_\alpha^T)) = -r_\alpha \nabla p + r_\alpha \rho_\alpha g + \mathbf{M}_\alpha \tag{2}$$

The total sum of volume fraction is the unity:

$$\sum_{\alpha} r_{\alpha} = 1 \qquad (3)$$

ANSYS-CFX applies the volume fraction conservation equation as in the Equation (3) to generate a discretized pressure equation and solve pressure and velocity fields coupled.

The form presented in Equation (4) is applied to interphase drag force model acting on α phase due to β phase:

$$\mathbf{M}_{\alpha} = \frac{C_D}{8} A_{\alpha\beta} \rho_{\alpha} \left| \mathbf{u}_{\beta} - \mathbf{u}_{\alpha} \right| \qquad (4)$$

Interfacial transfer of momentum however is directly dependent on the contact surface area between the two phases (air and water) and a free surface model has to be considered to analyze collector pans fluid dynamics. The free surface model attempts to solve the interface between the fluids. Considering the two phases, with air and water, the following equation is used for interfacial area density.

$$A_{\alpha\beta} = \left| \nabla r_{\alpha} \right| \qquad (5)$$

The interfacial area density, shown by Equation (5), represents the gradient module of volume fraction. As schematically presented on Figure 4, it represents the fluid flow characteristics of the dispersed phase (water in blue color) and the continuous phase (air in yellow color). A condition of high values of interfacial area density in a volume means that the interphase definition is well done. At initial condition the water layer was defined as the same height of downboxes wall and it has been observed to decrease with simulation time. In practical terms it can be assumed that a more uniform downbox liquid entrance presents better feeding uniformity to the following section.

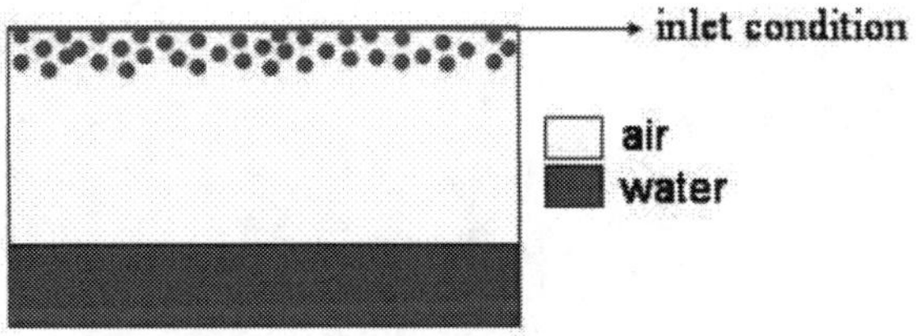

Figure 4: Schematic representation of fluid flow characteristics.

***Liquid distribution* characterization**

The coefficient of variation (C_V) is the most important variable to analyze the standard deviation of velocity, is a measure of dispersion of a probability distribution, represent in this case the quality of gas/liquid distribution in a cross-section area (Billingham, J. F. *et al.* (1996) and Mohamed *et al.*, 2001).

When considering a multiphase study, the C_V applied to a continuous flow profile can be described by Equation (6).

$$C_V = \left[\frac{1}{a_t} \sum_{i=1}^{N} a_i \left(\frac{v_i - \bar{v}}{\bar{v}} \right)^2 \right]^{0.5} \quad (6)$$

And the average gas velocity ($\bar{v}$) can be defined by Equation (7).

$$\bar{v} = \frac{1}{a_t} \sum_{i-1}^{N} a_i v_i \quad (7)$$

Where N is the total number of cells; a_t is the total column cross sectional area; a_i is the cross sectional area of a cell; v_i is the local gas velocity.

RESULTS AND DISCUSSION

The flow field that could be represented by stremlines The condition of medium liquid withdrawal was chosen to characterize the streamlines as shown on Figure 5. Under others conditions such as low and high liquid withdrawals, the general fluid dynamic is the same with local differences that can be noticed at each downbox mass flow rate. The original chimney arrangement had a severe design problem: chimneys excessively close to downboxes causing a significant difference on their feed, as can be seen comparing the original and shifted cases, Figures 5(a) and 5(b) respectively. This kind of arrangement makes the feed to downboxes “F” and “G” non-effective and lead downbox “H” to receive more liquid. This can be explained basically by the fact that the chimneys position partially blocks liquid flow.

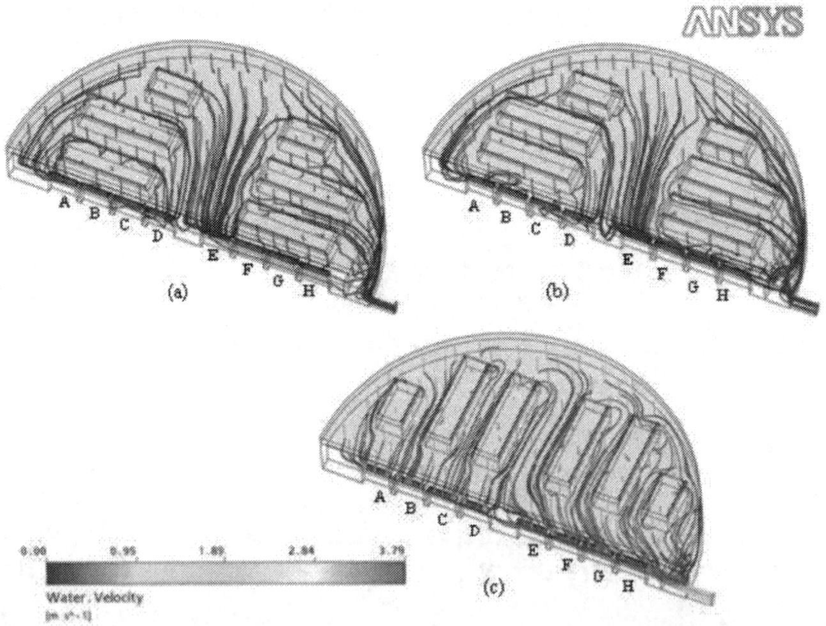

Figure 5: Water streamline, (a) original, (b) shifted and (c) 90º rotated.

The condition of 90º rotated chimneys shows significant differences in the flow field, mainly this chimney arrangement does not block up downboxes.

A consistent analysis of collector pans must consider different conditions of withdrawal and chimney arrangement. The individual downbox mass flow has been chosen as the second variable of this study. As there were concerns about transient conditions, the mass flow non-variance on time was considered as permanent condition.

CFD results indicate that the collector pan is not considerably affected under low withdrawal condition for the evaluated geometries as shown on Figure 6. Nevertheless the 90º rotated chimneys layout presents the best result under this withdrawal rate.

On Figure 7 can be observed that the increase of mass flow withdrawal directly affects liquid distribution profile and provides difficulties to select the best chimney arrangement.

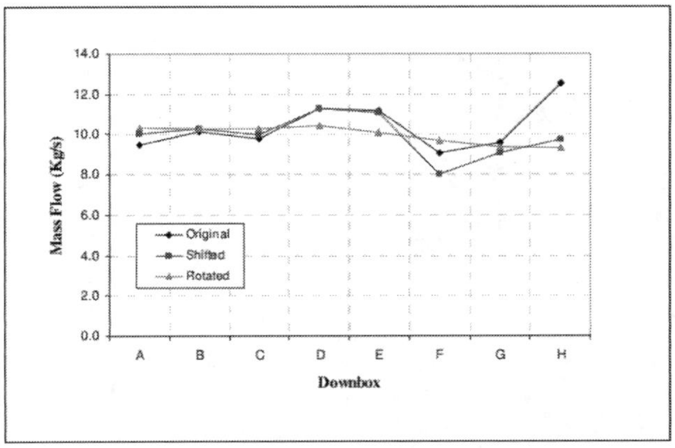

Figure 6: Low mass flow withdrawal.

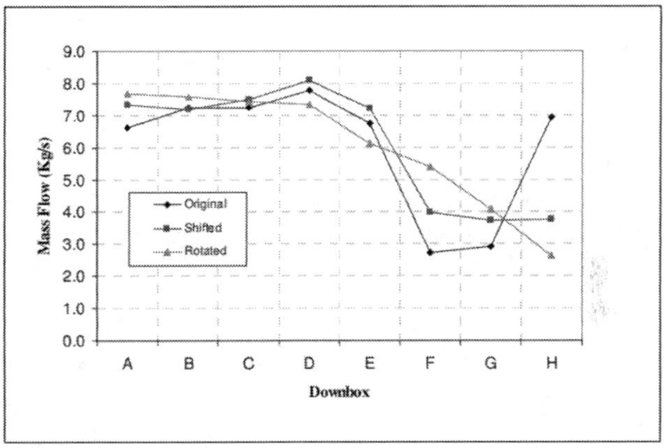

Figure 7: Medium mass flow withdrawal.

An interesting phenomenon is observed under high mass flow withdrawal. As shown on Figures 8 and 9, the increase on withdrawal rate completely disturbs downboxes feed. These conditions suggest that a second draw off nozzle, in the opposite side of the column, would be required to minimize disturbance.

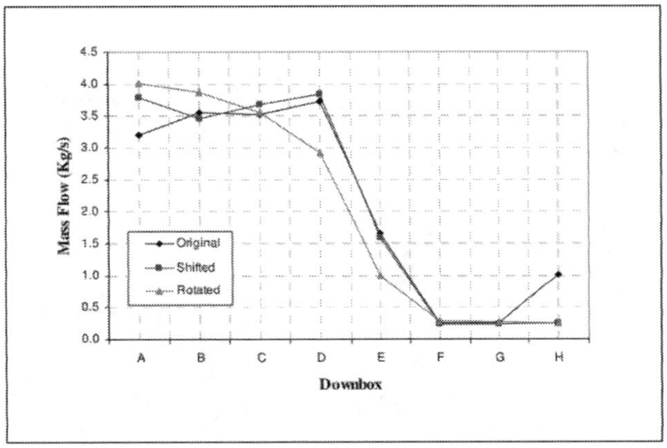

Figure 8: High mass flow withdrawal.

The application of Equations (6) and (7) to the liquid distribution gives the value of C_V. However, C_V by itself is not sufficient to define the best distribution and this variable should be analyzed together with stream lines to provide fluid dynamics characteristics and mass flow fields on each downbox.

Low values of C_V can be noticed on the 90º rotated geometry, which represent the best condition under low withdrawals as pointed by the stream lines of Figures 5c and 6. The medium condition of withdrawal has presented non-significant C_V difference among the geometric layouts studied, even computing the relation of C_V analysis, even so the 90º rotated geometry shows best results. This can be observed at Figure 9. Values of C_V at high withdrawal condition indicate better results for the original arrangement, however for high withdrawal the C_V values are far from the ideality of fluid dynamic behavior so for this condition a geometrical symmetric condition must be considered, as for example, including second sump.

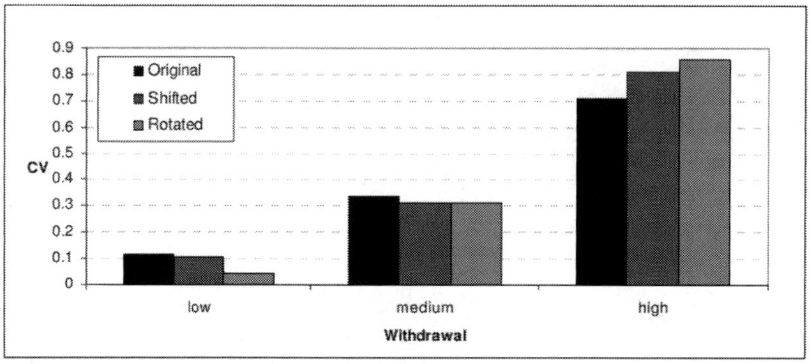

Figure 9: C_V analysis for different withdrawal conditions.

When considering the high values of withdrawal the importance of symmetrical condition can be verified on Figure 10. At this condition, the C_V comparing one and two sumps shows that the use of two sumps decreases the C_V, for the same geometry. It is important to be noticed for this condition on each sump, the relation of each withdrawal to internal reflux is different when comparing the 1 sump condition, because for 2 sumps the mass flow condition at the withdrawals is shared by two. This explains why we observe for 1 sump better results for original and shifted geometries and for 2 sumps the best result is obtained for 90º rotated geometry.

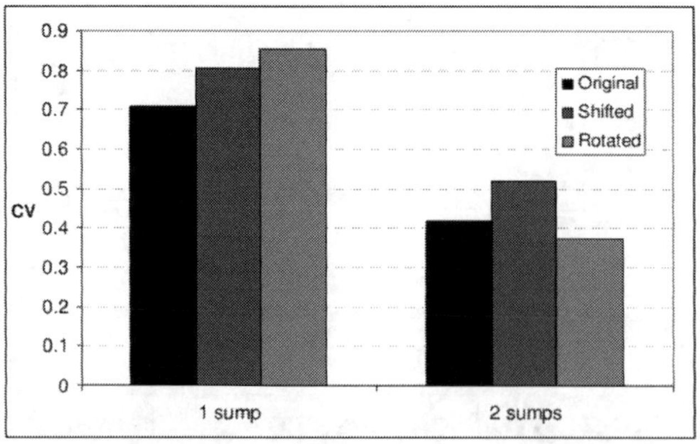

Figure 10: C_V analysis for high withdrawal comparing one and two sumps.

Table 2 presents the results overview that proposes a rule of thumb of how should be chosen the geometry of the chimney arrangement and the number of withdrawals (sumps). The hachure squares in the table represents the region of better results, and the cross is the best result among the results of one condition.

Table 2: Results overview.

	1 Sump			2 Sumps		
	Original	Shifted	Rotated	Original	Shifted	Rotated
Low			X			
Medium		X	X			
High						X

CONCLUSIONS

Despite the simplicity of the model presented in this study it has been proved to be useful in determining the collector pan behavior. According to presented results, CFD methodology provides the characterization of liquid distribution and the influence of the chimney arrangement and sump, draw off rates (withdrawal conditions) in association with streamlines, mass flow and C_V analysis.

The results of this study pointed the need of an engineering analysis for column internals selection, taking into account general operation condition, especially for packed columns. The optimum performance and desired operating ranges can only be achieved when tower packing/internals are precisely designed and carefully selected for individual cases.

Other studies are being carried out analyzing the effect of uneven flow rates from downboxes as the inlet condition to distributors, modeling the mass and momentum behavior.

NOMENCLATURE

a_i ≡	Cross section area of a cell [m^2]
a_t ≡	Total collector pans cross section area [m^2]
A ≡	Interface area density [m^2]
C_D ≡	Drag force [-]
C_V ≡	Coefficient of variation [-]
M ≡	Interfacial forces [ms^{-2}]
N	Total number of cells [-]
p ≡	Pressure [Pa]
r ≡	Volume fraction [-]
t ≡	Time [s]
u ≡	Velocity Vector [$m\ s^{-1}$]
v_i	Local gas velocity [$m\ s^{-1}$]
$\bar{v}$	Average gas velocity [$m\ s^{-1}$]
T and T^T ≡	Viscous and Turbulent stress tensors [$kg\ m^{-1}\ s^{-2}$]
α and β ≡	Phases [-]
ρ ≡	Density [$kg\ m^{-3}$]
μ ≡	Dynamic viscosity [$kg\ m^{-1}\ s^{-1}$]
$\nabla \bullet \mathbf{u}$ ≡	Divergence of u $\left[\nabla \bullet u = \frac{\partial u_x}{\partial x} + \frac{\partial u_y}{\partial y} + \frac{\partial u_z}{\partial z}\right]$

$\nabla r \equiv$ Gradient of r $\left[\nabla r = \frac{\partial r_i}{\partial x} + \frac{\partial r_j}{\partial y} + \frac{\partial r_w}{\partial z}\right]$

REFERENCES

1) Billingham, J. F., Bonaquist, D. P., Lockett, M. J. (1996) *Characterization of the Performance of Packed Distillation Column Liquid Distributors*. Praxair Technology Inc. p 841-851.
2) Olsson, F. R. (1999) *Detect Distributor Defects Before they Cripple Columns.* Chem. Eng. Progress. Pp. 57-61.
3) Heggemann, M., Hirshberg, S., Spiegel, L., Bauchmann, C. (2007) *CFD Simulation and Experimental Validation of Fluid Flow in Liquid Distributors.* IChemE, 85, pp. 59-64.
4) Kister, H. Z., Rosenzwelg, T. (2001) *Troubleshoot Chimney Trays effectively*. Hydrocarbon Processing, pp. 101-107.
5) Lehner, M., Verfahrenstechnik, R. (2006) *Distributors Designs Liquid distributor design criteria.* Hydrocarbon Engineering, 11, pp. 55-58.
6) Mohamed Ali, A., Jansens, P., Olujic, Z. (2001) *The Use of Computational Fluid Dynamics to Model Gas Flow Distribution in Packed Columns.* 6th World Congress of Chemical Engineering Melbourn, Australia 23-27 September.
7) Pizzo, S. M., Jr. Moraes, D., Fernandes, F. A. N., Kobayasi, M. S. Pazini. R. J. (1998) *Analysis of Liquid Distribution in a Packed Column on a Pilot Scale*. Ind. Eng. Chem. 37, pp. 2844-2849.
8) Wilcox, D. C. (2000) *Turbulence Modelling for CFD.* DCW Industries, pp. 314.
9) Zwart, P. J., Burns, A. D., Galpin, P. F. (2007) *Coupled Algebraic Multigrid for Free Surface Flow Simulations*. Proceedings of OMAE, 1-10.

A case study of packed column liquid maldistribution and its impact on separation efficiency

Pankaj S. Gautam, Wesam El Ramahi, Andy Grahn, Don Paciorkowski, Mohan Khadilkar
SABIC Innovative Plastics, Mt. Vernon, USA

1. Abstract

This paper presents a case study of a distillation column used to purify a process stream consisting of 2 components (a major component and a trace component '*B*'). The separation efficiency of the distillation column declined after startup coming out of shutdown such that the desired separation was no longer achievable under design conditions. Use of gamma scans indicated liquid maldistribution in the bottom packed section of the column. The gamma scans also indicated the malfunctioning of the nuclear level transmitter in the column receiver, which was recalibrated. The lower column efficiency was partially offset by increasing reflux rate that allowed column to function for a few additional months till the scheduled shutdown. The visual observations made after opening up the column indicated severe damage to the bed limiter on top of the bottom packed section as the root cause of column malfunction.

2. Introduction

Packed tower liquid maldistribution was ranked sixth on the list of top ten malfunctions in the last 50 years[1]. Most of these problems have been a result of liquid redistributor malfunction – plugging, overflow, internal damage, poor irrigation quality, and fabrication/installation defects etcetera[2]. The case below demonstrates how the liquid maldistribution in a packed section of the distillation column results in lower separation efficiency. In the study below, we have provided a brief description of the distillation tower schematics followed by the problem description. This is followed by a description of the root cause investigation process.

3. Case description

3.1 Background

The distillation column in our case is used to separate product from trace levels of a compound that shall henceforth be referred to as '*B*', by taking advantage of a large difference in the boiling points of the two components. Fig. 1 shows a schematic of the distillation column. The feed to the distillation tower consists of more than 95% product along with a trace component *B* and other inerts. The column overheads are condensed using cooling water in a condenser fitted with a boot. The liquids are withdrawn from the condenser boot and the reflux is pumped back to the column while the distillate liquid is sent over to the storage tank. The uncondensed vapor stream is further cooled in a series of heat exchangers to recover product, which is sent to the product storage tanks. The column is equipped with a thermosiphon reboiler using steam as the heating medium. The column level at the bottom of the tower is controlled by a level controller that manipulates steam to the reboiler to achieve the desired set point. The column bottoms go to the receiver that uses a combination of electric power and steam to further flash off the residual product. The column bottoms are drawn off through a control valve that operates in a batch mode. The *B*-rich liquid from the receiver is discharged after going through a series of processing steps.

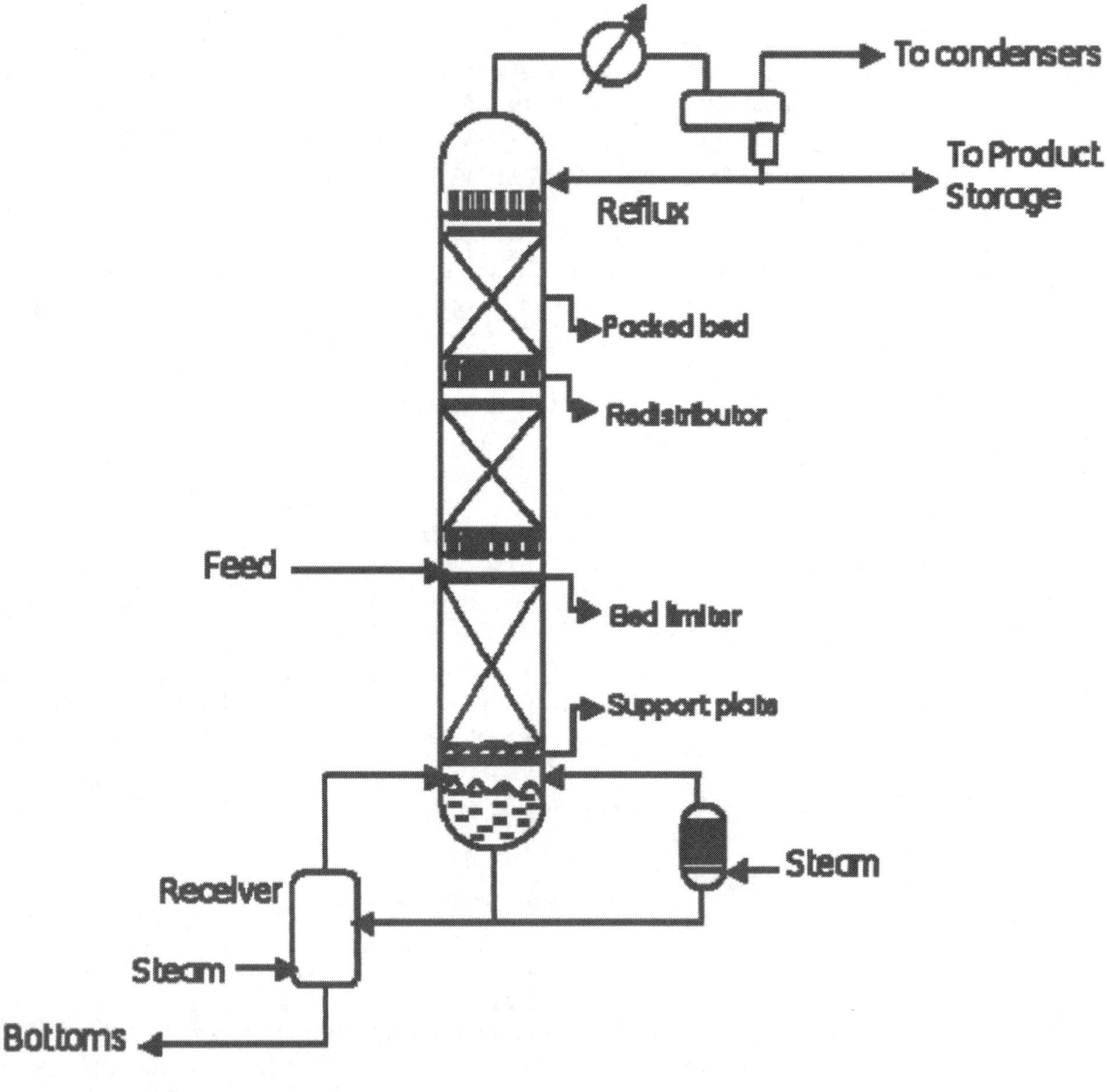

Fig. 1: A schematic of product distillation tower

The column is a 3'-0'' ID tower designed by Koch-Glitsch®. Fig. 1 shows some of the relevant details of the column design. It consists of three packed beds- two packed beds in the rectifying section and one packed bed in the stripping section. The stripping section is packed with #25 IMTP packings while the rectifying section is packed with #40

IMTP packings. There is a gas injection support plate at the bottom of each packed section and a bed limiter plate on top. There are a total of three redistributors- one liquid redistributor at the top for the reflux and two redistributors in the middle separating the packed sections.

3.2 Problem description

The distillation column could not achieve the desired separation after startup following Turnaround (TA). A plot of *B* vs. feedrate before and after shutdown revealed a few interesting features as shown in Fig. 2:

i. The concentration of *B* in the overhead product vapor from the column was higher at a given feedrate post-TA indicating a drop in separation efficiency of the column and,

ii. High concentration of *B* in the overhead product vapors was observed at low feedrates, which is counterintuitive.

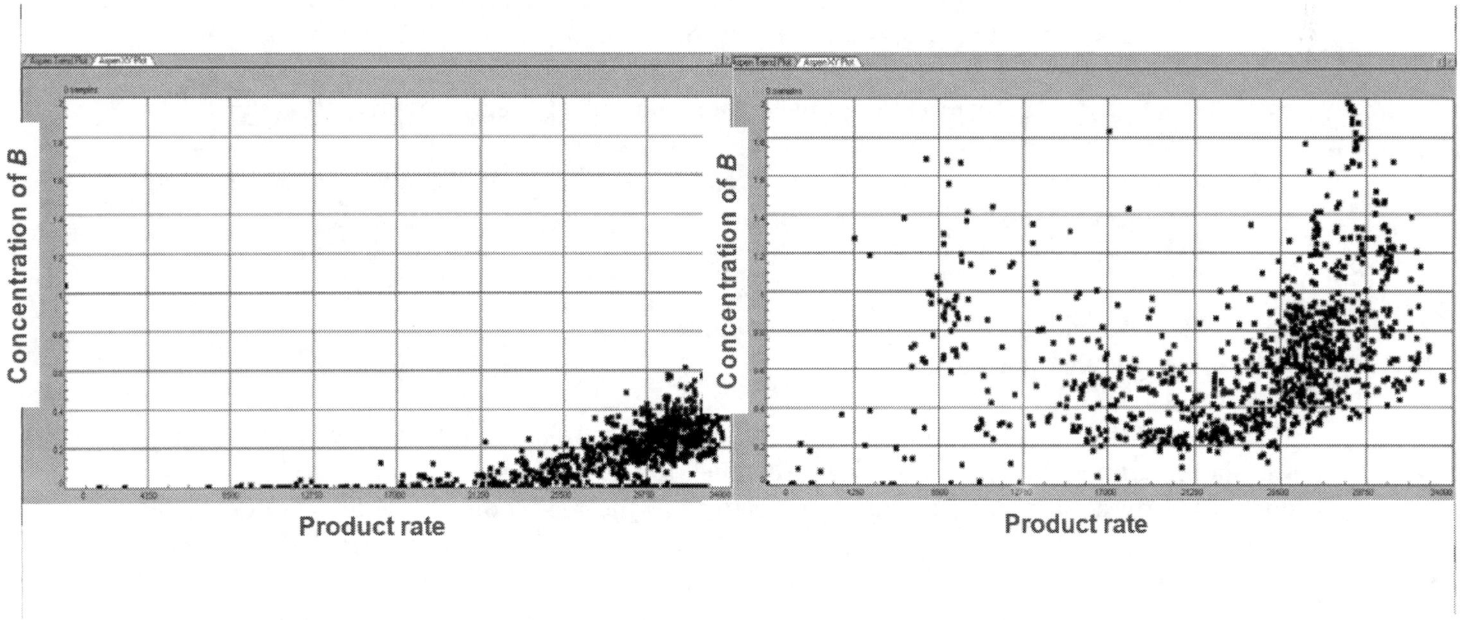

Fig. 2: A comparison of column performance before and after TA

It should also be pointed out that the second observation above may not be a true steady state response as the feedrate to the column was often brought down in response to a high concentration of *B* in the overhead vapors. However, it is certain that separation efficiency of the column went down post TA, especially at high feedrates. A gamma scan of the column was carried out at different reflux and feedrates. The reflux rate on the column was increased from 36 gpm to 39 gpm to offset some of the loss in separation efficiency. The column performance improved and a reasonable degree of separation could be obtained for 3-4 months before the column performance deteriorated sharply. A gamma scan was again performed on the column and the reflux rate was raised once again from 39 gpm to 43 gpm following gamma scan to improve separation efficiency. The separation efficiency improved but stayed below pre-TA level. The column operation was continued in this fashion till next scheduled TA when the column was eventually opened up and examined to determine the root cause.

4. Root Cause Investigation

4.1 FMEA and ASPEN simulations

A failure modes and effects analysis was used as a troubleshooting tool to generate hypotheses. Based on FMEA, a number of testable hypotheses were generated and a series of experiments were designed to test them. Of these, some of the important hypotheses that were either ruled out or deemed less plausible by a combination of experimentation, field walk-downs, and simulation are as follows:

i. The feed to the distillation column contains higher amounts of *B* due to malfunctioning of one of the reactors upstream.
ii. Solid catalyst particles from the reactor got entrained with the feed stream into the distillation tower and plugged up the packings resulting in channeling inside the column.
iii. The feed composition contains higher amount of inerts lowering the separation efficiency.
iv. The measurement system for concentration of *B* is malfunctioning and reading incorrect values.
v. Miscalibrated instrumentation on the tower such as level transmitters, pressure gauges, and flow meters etcetera.
vi. Condenser and/or reboiler malfunctioning.
vii. Misvalving during TA.

A number of other hypotheses were also generated that could not be tested. ASPEN simulation model was setup to study the impact of parameters such as higher concentration of *B* in feed, higher inert loading in the feed, condenser duty, and reboiler duty etcetera on the separation efficiency of the column. The simulations also indicated the aforementioned factors cannot satisfactorily explain the column behavior if the separation efficiency of the column is kept constant. In other words, the simulations demonstrated a decrease in separation efficiency as the most plausible explanation for the column malfunction. The decrease in separation efficiency could be caused due to a variety of factors, prominent ones being partial flooding, liquid/vapor maldistribution inside the tower, and a change in the packing wettability due to contaminated feed stream. The loss of separation efficiency could not be verified by temperature or pressure difference between the top and the bottom of the column. This is because a change in concentration of the component *B* in the overheads or the bottoms in the range of interest does not result in a measurable difference in the temperature of the overheads or bottoms, *B* being a trace component

4.2 Gamma Scans

Gamma scan technology is a well-established tool for troubleshooting column problems; especially, problems involving flooding, liquid/vapor maldistribution and damage to the column internals[3,4,5]. The results from one of the column scans are shown in Fig. 3. It consists of three equal chord scan lines of the column located symmetrically alongside the column circumference. SE denotes southeast direction, NE denotes northeast direction and W stands for the west direction. Two gamma scans of the column were carried out at different times after the upset. Both are qualitatively similar although the liquid maldistribution seems to be more pronounced in the one shown in Fig. 3. The authors consulted with Henry Kister regarding interpretation of gamma scans and some of the inferences are summarized below:

i. Liquid holdup in the SE side of the column in the packed section below the feed is consistently lower than the NE side.
ii. Liquid holdup in the SE side of the column in the packed section at the top of the column below the redistributor appears to be somewhat lower compared to NE and W sides.
iii. The NE chord in the region just below the feed shows the packing height of the bottom section extending about 1' in the vertical direction above what is indicated by SE, W chords. It should be noted that there might be some interference with the manway here causing scanline change. Besides, the NE scanline never touches the clear vapor bar. One interpretation could be the packings may have blown towards the NE side through the bed limiter in some plant upset.
iv. Within the typical range of operating parameters, no evidence of flooding was observed in any section of the column. This observation is based on gamma scans of the column at various operating conditions similar to Fig. 3 but not shown here.

The gamma scans of the receiver were also carried out to test if the nuclear level transmitter was functioning correctly. The gamma scans showed the receiver level was considerably lower than the value measured by nuclear level transmitters. This raised the possibility of the steam coils in the receiver being exposed due to a low liquid level, thus resulting in *B* flashing off back into the column causing poor separation. The nuclear level transmitter on the receiver was recalibrated following this discovery. This factor may also have contributed to the improved performance of the column in addition to the reflux rate increase from 39 gpm to 43 gpm.

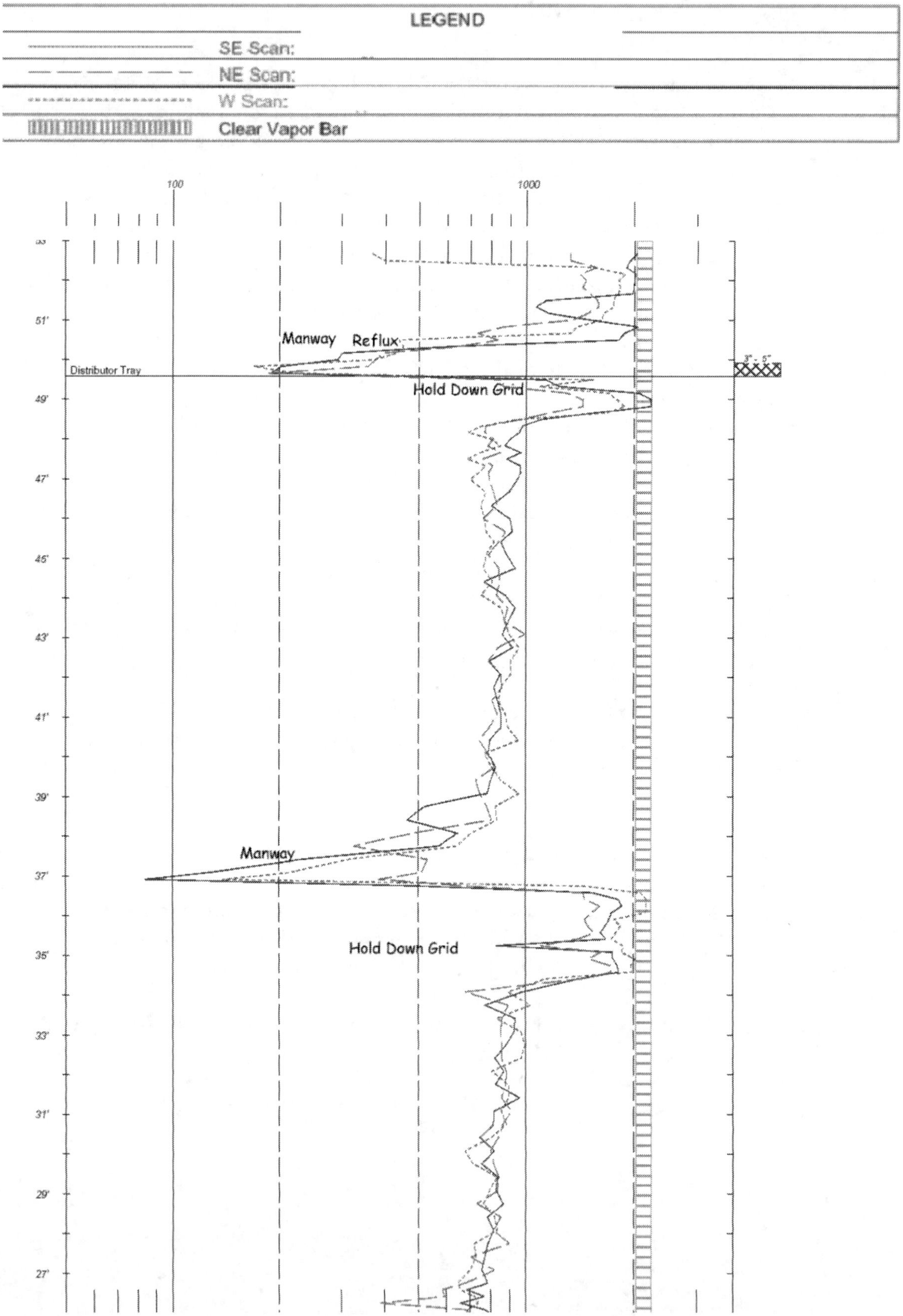
LEGEND
SE Scan:
NE Scan:
W Scan:
Clear Vapor Bar
100
1000
51'
Manway
Reflux
Distributor Tray
Hold Down Grid
49'
47'
45'
43'
41'
39'
Manway
37'
Hold Down Grid
35'
33'
31'
29'
27'

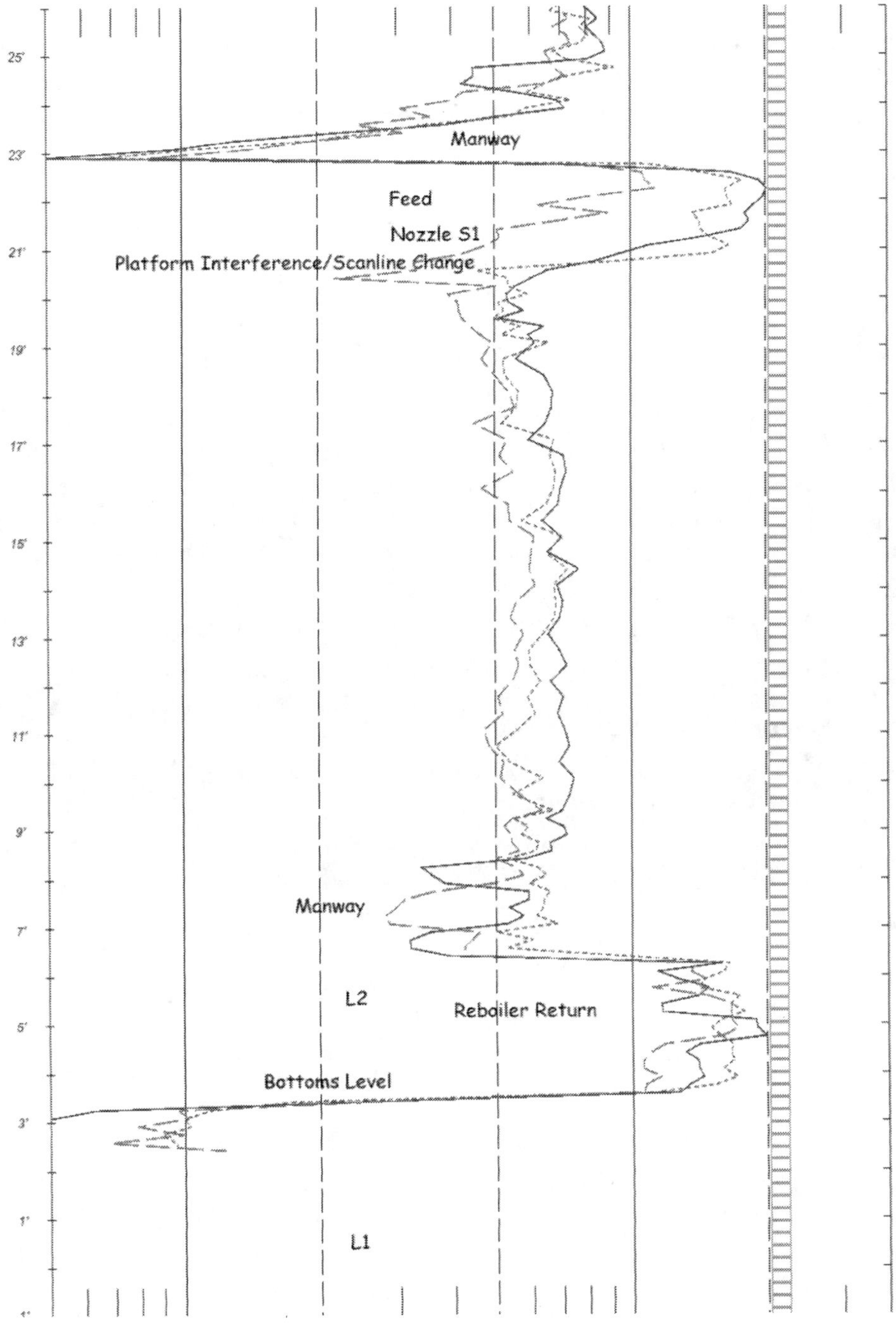

Fig. 3: Gamma scans of the column operating at a reflux rate of 39 gpm

4.3 Inverted U response

Fig. 4 shows a 'inverted U' shaped relationship between the concentration of *B* in overheads vs. reflux ratio that was observed after the reflux rate was increased following gamma scan. The separation efficiency was high both at low and high reflux ratios but low at the intermediate reflux ratios. It should be noted that the different reflux ratios for a single reflux rate essentially reflect the different feedrates. We do not have a satisfactory explanation for this kind of column behavior although it might have to do with the damaged bed limiter as described in the section below.

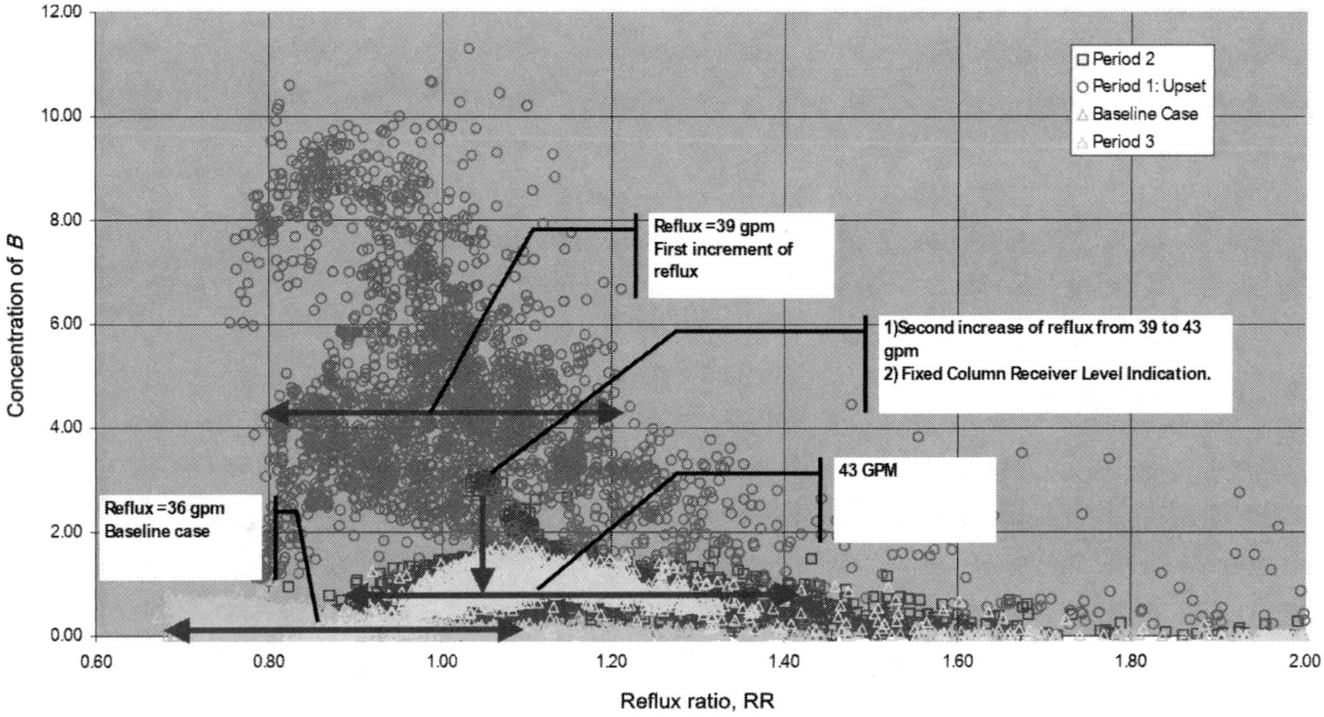

Fig. 4: Overhead concentration of *B* vs. reflux ratio (RR)

4.4 Shutdown inspection

The distillation column was opened up during the shutdown after the upset to find evidence of damaged internals. It was found that the bed limiter plate of the bottom packed section below the feed had lifted off and was sitting loose at a slanted angle. The packings in the bottom section had come out and spilled over into the space between the redistributor and bed limiter. Fig. 5 shows a photograph of the torn bed limiter plate sitting loose on top of the bottom bed. A photograph of the redistributor (Fig. 6) just above the feed inlet nozzle suggested some of the packings might have plugged distributor holes further worsening the problem of liquid maldistribution in the lower packed section of the column. It appears that the support plate above the redistributor is also misaligned. These findings are in agreement with the observations from the gamma scans that showed high density in the NE chord below the distributor on top of the bottom packed section. In-depth analysis of the root cause revealed that it was a result of high base level damage to the packings and bed limiter while trying to startup the column after the shutdown. Apparently, during column startup before the upset, the base level transmitter failed allowing the liquid in the bottom of the column to rise into the bottom packed section. Vapor returning from the reboiler slugged through this liquid level, possibly lifting off the bed limiter on top of the bottom packed section and spilling the packings into the space between the redistributor and the bed limiter. It is sobering to note that tower base level and reboiler return are number 2 on top 10 malfunctions in the distillation towers in the last 50 years[1,6].

Fig. 5: The bed limiter plate on top of the bottom packed bed just below the feed inlet

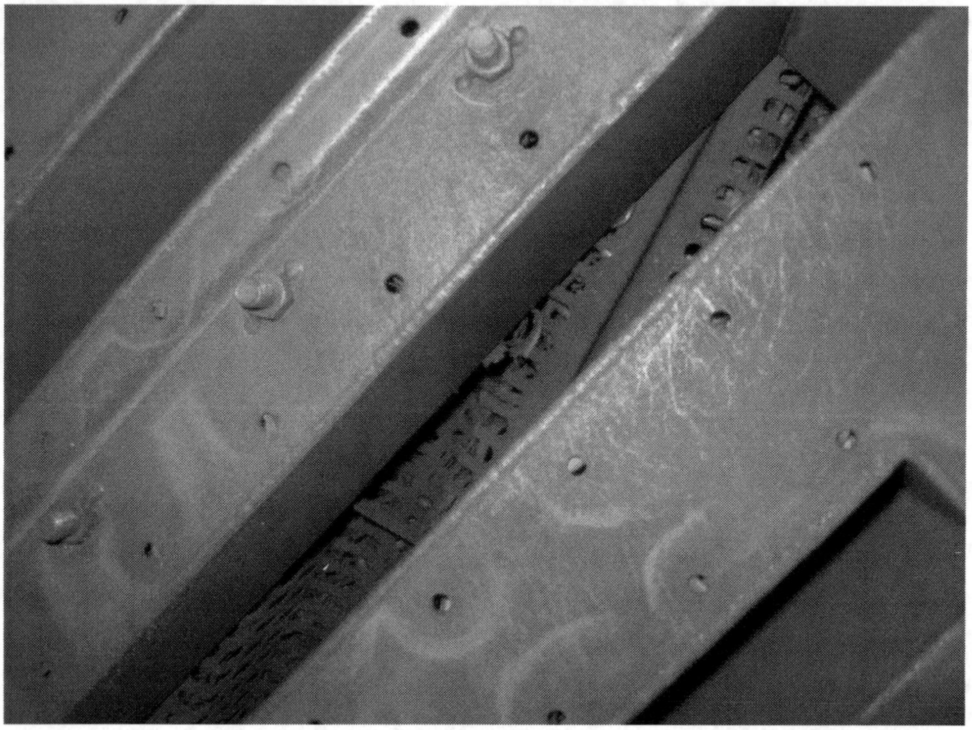

Fig. 6: The redistributor below the second packed bed right above the feed inlet

5. Summary

Gamma scans are a very valuable tool in the analysis of the column malfunctioning. In the case discussed here, gamma scans of the column helped in the identification of the problem with the liquid level transmitter in the receiver. The gamma scans correctly diagnosed the problem of liquid maldistribution in the bottom section of the packed column. The gamma scans also hinted at the possibility of the bed limiter lifting off though it was confirmed only after the column was opened up. One of the key learning from this case study is to install redundant base liquid level measurements at the bottom of the column and to establish administrative controls during column startup to eliminate the high-base-level damage incidents.

Acknowledgement

The authors wish to thank Henry Z. Kister for offering his advice and suggestions on interpreting gamma scans and column performance data and educating us on the column malfunctions in general. We also want to acknowledge the contributions of Dave Albright, especially for his suggestions on the reflux rate and overall project leadership.

References

1. Kister, Henry Z., "What caused tower malfunctions in the last 50 years?" Trans. IChemE, 81, Part A, Jan 2003, p. 5.
2. Kister, Henry Z, "Distribution Troubleshooting," John Wiley & Sons, Inc., 2006, ISBN 0-471-46744-8.
3. Xu, S.X., Mixon, W., "Diagnosing maldistribution in towers," Chemical Engineering Progress, May 2007, p. 28-35.
4. Vasquez, Pablo A.S., Costa, F.E., Rela, P.R., Calvo, W.A.P., LeRoux, G.A.C. and Hamada, M.M., "Gamma scan evaluation for random packed column," 2005 IEEE Nuclear Science Symposium Conference Record, 2005, p. 519-523.
5. Marchot, P. et al., "Investigation of liquid maldistribution in packed columns by X-ray tomography," Transactions of the Institute of Chemical Engineers, v77 (6), 1999, p. 511-518.
6. Kister, Henry Z., "Distillation troubleshooting," John Wiley & Sons, ISBN 0-471-46744-8, 2006.

290a Characterization Of Oxygen Scavenging Membranes Based On Polybutadiene

Hua Li[1], Keith Ashcraft[1], Scott Matteucci[1], Kevin Tung[1], Benny Freeman[1], Mark Stewart[2], and Jason Jenkins[2]. (1) The University of Texas at Austin, 10100 Burnet Road, Building 133, Center for Energy and Environmental Resources, Austin, TX 78758, (2) Global PET Technology, Eastman Chemical Company, P.O. Box 2002, Kingsport, TN 37662

There is considerable interest in the use of reactive membranes in oxygen scavenging systems. One reason for this interest is that conventional packaging materials, such as PET, do not provide sufficient barrier resistance to oxygen for certain food and electronics packaging applications.

Polybutadiene (PB) is an important synthetic rubber that readily oxidizes upon exposure to even low levels of atmospheric oxygen. The addition of a transition metal salt catalyst to PB markedly increases its ability to oxidize. This study presents results related to oxygen scavenging materials prepared from commercial polybutadiene doped with low levels of cobalt neodecanoate. The resulting membranes react with oxygen at room temperature, and they sorb up to 20 weight percent oxygen as a result.

This presentation will discuss the influence of preparation and reaction conditions on oxidation of 1,4-polybutadiene. Oxygen uptake is measured using an analytical balance and a non-invasive oxygen sensor based system, and the results from these two measurements techniques are compared. The effect of cobalt concentration, film thickness and oxygen partial pressure on oxygen uptake rate and amount were determined. Crosslinking was observed during the oxidation process, so the effect of crosslinking on oxygen uptake was also studied. Oxidation proceeds faster at higher oxygen partial pressure. Permeation tests using uncatalyzed and catalyzed 1,4-polybutadiene membranes show that O_2 and CO_2 permeability through such membranes decreases by more than two orders of magnitude as a result of oxidation.

Commercial 1,2 and 1,4-polybutadiene contain UV stabilizer packages that can be removed prior to preparing oxygen scavenging membranes. To remove the inhibitors, PB is first dissolved in toluene and then precipitated in methanol. Oxygen sorption experiments also demonstrate that oxygen uptake and oxidation rate depend on the number of purification steps (i.e., the extent of inhibitor removal). Oxygen uptake and oxidation rate are stable after a sufficient number of purification steps, suggesting that inhibitors can be removed to levels that do not influence oxygen scavenging properties.

Measurement of diffusion and permeation coefficients of oxygen in polymers based on luminescence quenching

Sanchari Chowdhury [(1)], Justin Dodson [(1)], Venkat R. Bhethanabotla [, (1)] and Rajan Sen [(2)]*
Department of Chemical Engineering [(1)], Department of Civil Engineering [(2)]
University of South Florida, Tampa, Fl 33620-5350

INTRODUCTION

Over the past decade, polymer supported luminescence oxygen sensors are increasingly being used for different applications ranging from the measurement of pressure distribution of air on wing of an aircraft using pressure sensitive paint [1] to the study of oxygen diffusion properties in polymers [2-7]. These technologies are developed based on the concept that the luminescence intensity of some luminophores quenches in the presence of oxygen. In the present work main objective is to measure diffusion properties of oxygen in polymers. A number of cases are reported in the literature where luminescence quenching by oxygen is used for measuring diffusion coefficients of the oxygen in polymers. In most of the cases, the luminophore was directly dispersed in the polymer and average oxygen concentration change in polymer was monitored by studying the average intensity change or life time change of luminophore using a spectrofluorometer [2-4, 6-7]. Our research work focused on the measurement of permeability and diffusivity of oxygen in polymers using luminescence sensor with fluorescence microscope. Our technique is an improvement over past reported methods and successfully applied for different polymers.

Yekta *et al.* [3]were first to develop an appropriate model combining Fick's law of diffusion and the SV equation to extract diffusion coefficients from experimental data for both "diffusion in" and "diffusion out" experiments. This model was based on the concept that the intensity change of the luminophore corresponds to the average oxygen concentration within the polymer. They also assumed that the change in luminescence intensity with oxygen concentration follows linear SV equation. Their model was extensively used later to find diffusion coefficients for different luminophores containing polymer films which follow linear SV equation [2-7]. However, most times it is difficult to get oxygen sensors of linear response. Heterogeneity of luminescence sensors occurs due to incompatibility of the polymer and luminophore. Influence from unquenchable emission from aggregates of luminophore and also sometimes the unquenchable background emission may contribute to the nonlinear behavior. Influence of dual or continuous gas sorption in the polymer may also results in nonlinearity [8-13].

Several models were developed for describing the nonlinear response of the sensors. One multi-site quenching model was developed considering the sensor to consist of a number of independently emitting species with different single-exponential lifetimes and relative contributions[11-13].

Researchers proposed different models depending on several quenching mechanisms: surface quenching, a two-site model, and a combination of surface quenching and a two-site model [12]. Based on Froster-type energy transfer Draxler *et al.* [8, 9] proposed a modified model to describe non-linear SV plots of ruthenium dyes in polymer matrixes. Another model was

[*] Author for correspondence at venkat@eng.usf.edu

developed considering the Langmuir adsorption with Henry's law of solubility following nonlinear solubility model [10].

However, it is highly complicated to derive analytical models combining different nonlinear SV models with the Fick's law subjected to different set of boundary conditions. Recently researchers reported microscopy study of heterogeneity in luminescence sensor [14-16]. Fluorescence microscopy technique allowed them to study the SV response of the luminescence sensors with microscopic spatial resolution. They reported homogeneous regions of sensors show better response to oxygen concentration than regions where the dye is aggregated.

Diffusion measurement methods where the luminophore is dispersed in the polymer were successfully used for several polymers, but this technique is only applicable for transparent polymers. To overcome this limitation Rharbi *et al.* [5] reported a different method for measuring the oxygen diffusion and permeation coefficients in the polymer. They measured the oxygen permeation and diffusion coefficients of free standing polymer using the accumulation in volume technique and the diffusion coefficients of viscous liquid polymer by the film on sensor method.

In the present work, we report the use of fluorescence microscopy to investigate spatial distribution of SV response of the sensor at different oxygen concentrations at the microscopic level. Fluorescence microscopy allowed us to detect relatively homogeneous regions. The responses from these regions were analyzed to calculate the oxygen diffusion coefficients. This method avoids the complexity of including nonlinear SV equations in the analytical model. Additionally, in the present study we used the film on sensor method and the accumulation in volume technique to investigate oxygen diffusion behavior in polymers.

Oxygen diffusion and permeation coefficients in Teflon and PDMS were measured to validate our new technique. Then the technique was used to measure the diffusion coefficient of a HP polymer (black silicone elastomer) and PDMS containing different weight percentage of zeolite. We combined the SV equation with Fick's law of diffusion to extract the diffusion coefficient from experimental data.

EXPERIMENTAL SECTION:

Sensor films and polymer:

The oxygen sensing material was prepared by dispersing the 1.3283e-004 mol luminophore platinum tetrakis (pentafluorophenyl) porphyrin (PtTFPP) (Frontier Scientific, Inc., Logan UT) in 1 liter solution of polystyrene (Sigma-Aldrich; Milwaukee, WI,USA, Avg. M_w 280,000 by GPC)/Toluene (0.24 $g \cdot l^{-1}$). This solution was spin-coated on cleaned 19 mm diameter glass cut from 1 mm thick microscope slide. Then these were cured at room temperature for 1 hour and at 120 °C for 5 hours.

The permeation and diffusion coefficient of DuPont's Teflon FEP film were measured using accumulation in volume technique. The thickness of Teflon film was 25 µm. The diffusion coefficients of Sylgard 184 a common poly(dimethyl-siloxane) (PDMS) and 3-6265 HP polymer (silicone elastomer) (Dow Corning) were measured by film on sensor technique. In this case, the polymer was directly cast on a 19 mm luminescent sensor. The PDMS film was cured at

120 °C for 1 hour. The 3-6265HP polymer film was cured at 100 °C for 35 minutes. To disperse zeolite ($Ca_{/n}Na_{12-2n}[(AlO_2)_{12}(SiO_2)_{12}] \cdot xH_2O$, Molecular sieves, 5 Å, beads, 4-8 mesh, Sigma Aldrich, Milwaukee, WI) into PDMS the zeolite was first ball-milled in SPEX 8000 Mixer/Mill with 8 mm stainless steel ball for 2 hours. Different weight percentages of zeolite were dispersed in PDMS solution. These solutions were cast on the sensors, and cured for 1 hour at 120 °C.

The Leica DMI 4000b inverted research fluorescence microscope equipped with Leica DFC340 FX CCD Camera was used in this study. The specimen was only exposed to illumination while taking images and the lamp intensity was maintained at 10 % of its total intensity. The shutter was open while acquiring an image. Thus, photobleaching of the fluorescent sensors was minimized. For observing the SV behavior of sensor, Images of the sensor were acquired at different concentrations of oxygen and analyzed for intensity using Image Pro-plus.

Stainless steel cells were constructed for diffusion measurements of the polymers. For the accumulation in volume experiment there were two chambers in the diffusion cells (figure 1). The polymer films were placed between the chambers using viton gaskets to prevent leakage. Upstream section of polymer film was continuously flashed with nitrogen. In this experimental configuration the concentration of oxygen inside the volume of the downstream chamber changes, with commensurate change in the luminescence intensity of the sensor.

For the film on sensor experiment the polymer film attached to a sensor was placed in a stainless steel chamber (figure 2). The surface of this film was continuously purged by nitrogen. In all cases, the polymer film was first equilibrated with air; then exposed to a lower concentration of oxygen. Luminescence intensity changes measured the changes in oxygen concentration at the sensor/polymer film boundary in the film-on-sensor technique. The diffusion cell was mounted on the stage of the microscope in such a way that the sensor faced the illumination source and detector. The image of the sensor film was captured through the 1 mm thick glass on which the sensor film was coated.

ANALYTICAL MODELS:

In this work, diffusion of gas through the polymer material is described by Fick's law. The solution of this equation depends on the boundary conditions at the edges of the film. We combined the linear SV equation with the diffusion model to extract the diffusion and permeation coefficients from the experimental data.

Combining the diffusion model[17,18] with the SV equation, the luminescence intensity is related to the partial pressure of oxygen:

$$\left(\frac{I_{air}}{I}-1\right)=\left(\frac{I_{air}}{I_0}-1\right)\left(1-\sum_k \frac{2\left(\beta_k^{\,2}+\eta^2\right)\sin\beta_k \frac{x}{l}}{\beta_k\left[\beta_k^2+\eta^2+\eta\right]} e^{-\frac{D\beta_k^2 t}{l^2}}\right) \qquad (1)$$

From this model both the diffusion and permeation coefficients of oxygen in a polymer can be calculated.

Film on sensor model:

For the film on sensor experiment where the upstream of the film was maintained with pure nitrogen and initially the film was conditioned with air, the solution for Fick's second law is given by Crank [18]. Combining that equation with the SV final equation can be written as [5]:

$$\frac{\left(\frac{I_{air}}{I}-1\right)}{\left(\frac{I_{air}}{I_0}-1\right)}=\left(1-\frac{4}{\pi}\sum_{n=0}^{\infty}\frac{(-1)^n}{2n+1}\exp\left(\frac{-D(2n+1)^2\pi^2 t}{4l^2}\right)\right) \quad (2)$$

RESULT AND DISCUSSION:

Characterization of sensor:

Spatial distribution of the SV response of fluorophore PtTFPP in luminescence sensor was demonstrated by using a fluorescence microscope. A representative image is shown in figure 1. From this image it may be seen that there are some bright fluorescent spots in a nearly homogeneous background region of low intensity. This shows the dye was not homogeneously dispersed in the sensor. Bright spots (blue and green regions in sensor) are due to micro-crystal formation of the luminophore due to its incompatibility with the polystyrene matrix. Four different regions of different intensity in the sensor were investigated for their SV response at nine different oxygen pressures.

It may be seen from the results (Table 1) that SV constants are high for nearly homogeneous low-intensity regions (red and orange)) (k_{sv}=0.9 psi $^{-1}R^2 = 0.998$); whereas the high-intensity microcrystals region shows poor SV response (K_{sv} = 0.33 psi^{-1}, R^2 = 0.986). Past investigations on heterogeneity of different sensors also have shown that microcrystalline area show greater intensity but less quenching by oxygen [14]. In the present work average intensity over whole image also showed very good SV response. This can be attributed to the fact that the microcrystalline area occupying a small percentage of the entire sensor. In comparison, in the homogeneous region the SV response from the background dominates the response from heterogeneous region.

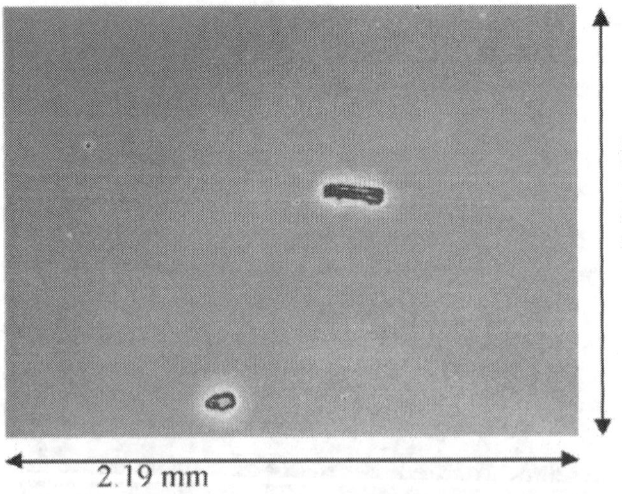

Figure 1: Pseudo-colored image of luminescence sensor (PtTFPP/PS) (intensity increases from red to blue)

Table 1. SV constants of different microscopic regions of luminescence sensors

Region	Fresh Sensor		After 30 minutes photobleaching	
	K_{SV} (psi $^{-1}$)	R^2	K_{SV} (psi $^{-1}$)	R^2
1 (Red)	0.85	0.999	0.77	0.998
2 (Orange)	0.90	0.998	0.79	0.998
3 (Green)	0.61	0.999	0.54	0.996
4 (Blue)	0.33	0.986	0.25	0.958
Average	0.86	0.999	0.77	0.998

Based on the SV analysis of different regions of the sensor film, the intensity change of regions which follow the linearity of SV equation and have high SV constants were examined for evaluation of oxygen diffusion parameters.

Measurement of diffusion and permeation coefficients:

Diffusion coefficient of oxygen in Teflon was measured using accumulation in volume technique. Teflon film was tightly placed in diffusion cell. Figure 2 shows intensity ratio vs. time data for Teflon. The diffusion model (Equation 1) was fit to experimental data to extract both permeation coefficient and diffusion coefficient. For 25 μm Teflon, from experimental data diffusion coefficient and permeation coefficient calculated were 1.12×10^{-11} m^2/sec and 1.55×10^{-10} mol·m^2/m^3sec·atm respectively. These data compare with reported value of diffusion coefficient 1.84×10^{-11} m^2/sec and permeation coefficient 1.62×10^{-10} mol·m^2/m^3sec·atm. [19]. The reason for little variation between calculated data and reported data may be the diffusion and permeation coefficients in literature were measured at higher temperatures. We extrapolated these data at room temperature (25 °C) to compare with our experimental findings.

Film on sensor method was first used to measure diffusion coefficient of PDMS for validating the technique. Experimental data fitted with diffusion model (equation 2) are given in figure 3. Our data for PDMS fall in the range of data reported in literature [20].

We measured the oxygen diffusion coefficient in 3-6265 HP polymer. The color of this polymer is black. Experimental data fit well with diffusion model (figure 4). New data for this polymer are given in Table 2 which is the same order of magnitude for silicon elastomers.

This technique is proven to be suitable for the polymer for which it is difficult to prepare uniform free standing thin films. Another advantage of this technique is experimental time is comparatively less.

The diffusion coefficients of oxygen in the polymer containing zeolites were measured using film on sensor technique. It is shown in figure 5 that experimental data fit well to the Fickian diffusion model as described by equation 2. Diffusion coefficients reduce as the weight percentage of zeolite increases in PDMS. This trend also agrees with literature [21]. So it can be predicted that presence of zeolite in PDMS introduces more free volume as well as more pores which enhance oxygen diffusion.

Table 2: Oxygen diffusion coefficients for various polymers

	Zeolite weight percentage	Diffusion coefficient (10^{-9} m^2/sec)
3-6265 HP polymer (silicone elastomer).	0.0	4.52
PDMS	0.0	0.82
PDMS + zeolite	2.5	1.02
PDMS + zeolite	10.0	1.17
PDMS + zeolite	20.0	1.22
PDMS + zeolite	30.0	1.26

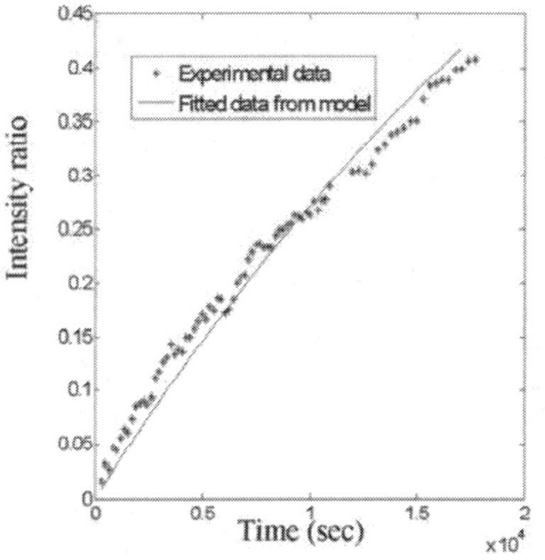

Figure 2: Experimental data and fitted data for the 0.025 mm thick Teflon film.

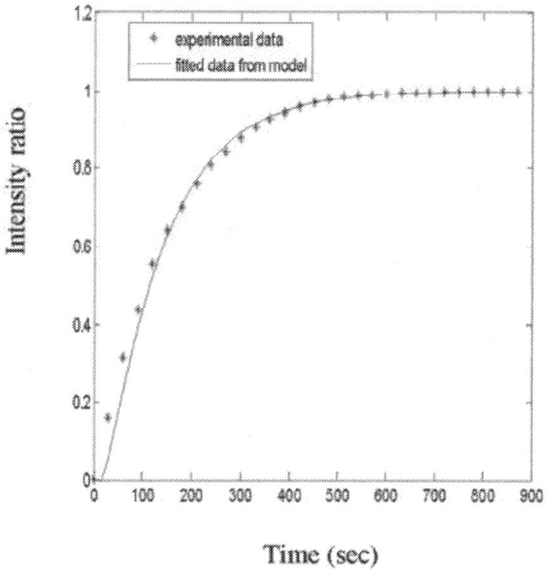

Figure 3: Experimental data and fitted data for the 0.8 mm thick PDMS film.

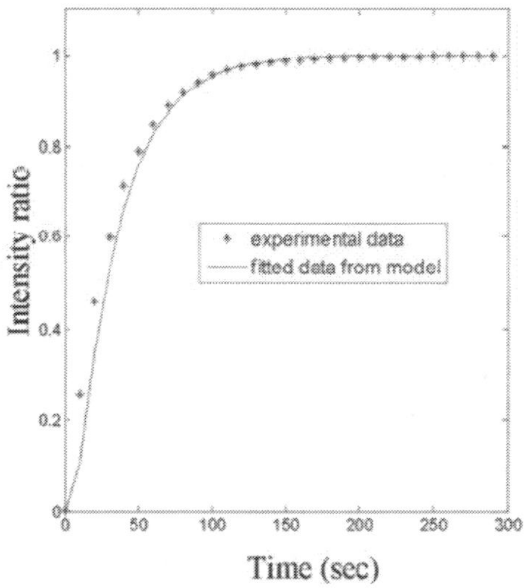

Figure 4: Experimental and fitted data for 0.55 mm 3-6265 HP polymer film (silicone elastomer)

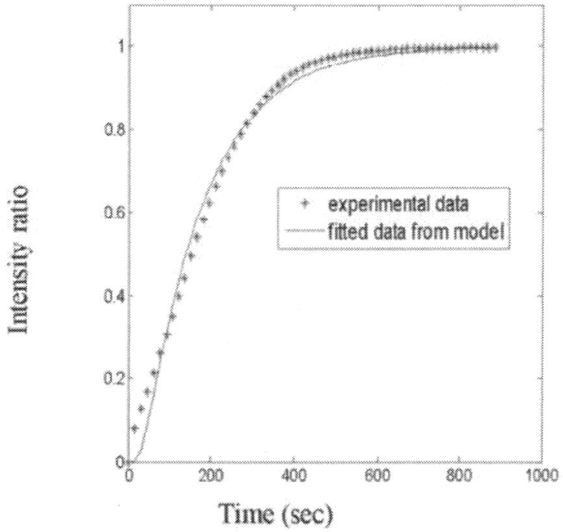

Figure 5: Data for the 0.65 mm thick PDMS film containing 10% zeolite.

CONCLUSIONS:

A fluorescence microscopy technique was developed to measure diffusion and permeation coefficients of oxygen in polymers. Using this technique the microscopic level SV responses of heterogeneous sensor are measured. This technique allows for the distinction of the responses of background region (nearly homogeneous regions) from the regions of aggregated luminophores. As the nearly homogeneous regions show better response to oxygen concentration and follow the linearity of the SV equation, studying the response of these, one can eliminate the complexity of combining the nonlinear SV equation with a diffusion model. With this method data for Teflon and PDMS were obtained and compared well with literature values. New data for 3-6265 HP polymer (silicone elastomer) and PDMS containing zeolite are on an expected order of magnitude with comparable materials.

We developed a new, simple quasi-steady model for describing diffusion phenomena in the case of accumulation in volume technique. Photostable luminophore is essential for these kinds of experiments. Minimizing photobleaching of the luminophore is a challenge for this method that was overcome by shuttering techniques. The method developed here can be applied for measuring oxygen diffusion properties in different polymers ranging from transparent to opaque subjected to the condition that no component is present in the polymer interferes with the response of sensor to oxygen concentration. This method is also suitable for polymer composites.

REFERENCES

1. Hubner, J. P.; Carroll, B. F.; Schanze, K. S.; Ji, H. F. *Experiments in Fluids* **2000**, *28*, 21.
2. Bowyer, W. J.; Xu, W.; Demas, J. N. *Anal. Chem.* **2004**, *76*, 4374.
3. Yekta, A.; Masoumi, Z.; Winnik, M. A. *Can. J. Chem.* **1995**, *73*, 2021.

4. Schappacher, G.; Hartmann, P. *Anal. Chem.* **2003** *75*, 4319.
5. Rharbi, Y.; Yekta, A.; Winnik, M. A. *Anal. Chem.* **1999**, *71*, 5045.
6. Mills, A.; Chang, Q. *Analyst* **1992**, *117*, 1461.
7. Kneas, K. A.; Demas, J. N.; Nguyen, B.; Lockhart, A.; Xu, W.; DeGraff, B. A. *Anal. Chem.* **2002**, *74*, 1111.
8. Draxler, S.; Lippitsch, M. E.; Klimant, I.; Kraus, M.; Wolfbeis, O. S. *J. Phys. Chem.* **1995**, *99*, 3162.
9. Draxler, S.; Lippitsch, M. E. *Anal. Chem.* **1996**, *68*, 753.
10. Demas, J. N.; Deora, B. A.; Xut, W. *Anal. Chem.* **1995**, *67*, 1377.
11. Carraway, E. R.; Demas, J. N.; DeGraff, B. A.; Bacon, J. R. *Anal. Chem.* **1991**, *63*, 337.
12. Carraway, E. R.; Demas, J. N.; DeGraff, B. A. *Langmuir* **1991**, *7*, 2991.
13. Carraway, E. R.; Demas, J. N.; DeGraff, B. A. *Anal. Chem.* **1991**, *63*, 332.
14. Kneas, K. A.; Demas, J. N.; DeGraff, B. A.; Periasamy, A. *Microscopy and Microanalysis* **2000**, *6*, 551.
15. Bowman, R. D.; Kneas, K. A.; Demas, J. N.; Periasamy, A. *Journal of Microscopy* **2003**, *211*, 112.
16. Bedlek-Anslow, J. M.; Hubner, J. P.; Carroll, B. F.; Schanze, K. S. *Langmuir* **2000**, *16*, 9137.
17. Paul, D. R.; Dibendetto, A. T. *J. Polym. Sci.C* **1965**, *10*, 17.
18. Crank, J. *The Mathematics of Diffusion* second ed.; Oxford University Press, 1975.
19. Koros, W. J.; Wang, J.; Felder, R. M. *J. Appl. Polym. Sci.* **1981**, *26*, 2805.
20. Shiku, H.; Saito, T.; Wu, C.; Yasukawa, T.; Yokoo, M.; Abe, H.; Matsue, T.; Yamada, H. *Chemistry Letters* **2006**, *35*, 234.
21. Jia, M.; Peinemann, K.-V.; Behling, R.-D. *J. Membr. Sci.* **1991**, *57*, 289.

290c Direct Measurement Of The Chemical Potential Driving Force Behind The Evaporation-Induced Diffusion Of Water Through Hydrogels

Tobias D. Wheeler and Abraham D. Stroock. Chemical and Biomolecular Engineering, Cornell University, 120 Olin Hall, Ithaca, NY 14853

Hydrogel materials are used in many technologies, including biomaterials for contact lenses and wound dressings, and membranes for pervaporation processes and fuel cells. The evaporation-induced diffusion of water through the hydrogel material is a significant factor in determining the performance of hydrogel materials in these applications. For this reason, it is important to characterize the mass transfer properties of hydrogels with respect to the transport of water. We report a simple, accurate experimental technique that makes this characterization possible. The technique measures the chemical potential drop across a hydrogel membrane with one side exposed to a vapor phase and the other to a liquid water reservoir or to a second vapor phase with higher water vapor content than the first. The chemical potential drop is determined by using a heat and mass balance of the multi-phase system and a psychrometric calibration of mass transport in the vapor phase above the membrane. We determine the response of the hydrogel membrane to changing chemical potential drops by measuring the steady-state flux of water vapor through the membrane. This response is summarized by global mass transfer functions. We also calculate the diffusivity of water within the membrane using the measured steady-state flux and chemical potential drop together with Flory-Rehner polymer-solvent mixture thermodynamics and various transport models. Experiments are performed with poly(hydroxyethyl methcacrylate) membranes. We test membranes of varying thickness contain varying amounts of cross-linking agent to determine the effects of these attributes on transport properties. Experimental results reveal that the model-dependent diffusivity of water in the material decreases as the driving force for diffusion is increased. The decreasing diffusivity is explained by a water desaturation-induced glass transition of the polymer network constituting the solid fraction of the hydrogel, the occurrence of which is confirmed using Differential Scanning Calorimetry at controlled relative humidity.

290h Molecular Dynamics Simulation Of Diffusion And Sorption Of Water In Conducting Polyaniline

Muhammad Sahimi, Mayur Ostwal, and Theodore Tsotsis. Mork Family Department of Chemical Engineering and Materials Science, University of Southern California, 925 W Bloom Walk, HED 216, Los Angeles, CA 90089

Energy minimization and molecular dynamics simulation are used to develop atomistic models of HCl- and HBr- doped conducting polyaniline, in order to study diffusion and adsorption of water vapor in these polymers. Various morphological properties of the polymers are computed, including their pair correlation functions that are found to be in good agreement with the experimental data, and their accessible free volumes. The sorption isotherms and self diffusivity of water vapor in these polymers are also computed at various temperatures. The computed sorption isotherms are in good quantitative agreement with the experimental data, while the diffusivities are within an order of magnitude of the data. The computed activation energy compares well with that published in literature for water sorption in doped polyaniline.

290e Water And Ethanol Counter-Diffusion In Ion Exchange Membranes

John Pellegrino[1], Peter McCormick[1], and Giulio Sarti[2]. (1) CEAE Department, University of Colorado, ECOT411, Boulder, CO 80309-0428, (2) Chemical Engineering, Mining and Environmental Technology, University of Bologna, viale Risorgimento 2, Bologna, Italy

Measurements of counterdiffusion of water, ethanol, and NaCl through several commercial ion exchange membranes have been conducted in the context of evaluating the suitability of using ethanol as the "draw" agent for forward-osmosis-based water recovery from highly "impaired" water sources. The water versus ethanol selectivity (ratio of permeances) was determined for a range initial water salinities (all greater than seawater) and ethanol mole fractions. Differential swelling at the two membrane interfaces provides for a maximum in the observed selectivity with respect to the relative driving forces. The transport is modeled with a classical solution-diffusion model with consideration of variations in the molar densities at the two interfaces.

Diffusion of cyclohexane in styrene block copolymers, relationship between mass transport and viscoelastic properties

F. *Bacchelli* [(1)], *S. Coppola* [(1)], *M. Giacinti Baschetti* [(2)], *M. G. De Angelis* [(2)], *F. Doghieri* [(2)]

[(1)] *Centro Ricerche Elastomeri, Polimeri Europa S.p.A., via Baiona, 107 – 48100 Ravenna - Italy*
[(2)] *Dip. Ingegneria Chimica, Mineraria e delle Tecnologie Ambientali - Università di Bologna - Viale Risorgimento, 2 - 40136 Bologna - Italy*

Abstract

The mass transport properties of cyclohexane in different phase separated styrene block-copolymer, namely polystyrene-block-polybutadiene-block-polystyrene [SBS], polystyrene-block-polyisoprene-block-polystyrene [SIS] and polystyrene-block-poly(ethylene butylene)-block-polystyrene [SEBS] were investigated. The solubility and diffusivity in films samples have been measured, as function of concentration, at 60°C by means of sequences of differential vapor sorption experiments run in a pressure decay apparatus. The effects of both molecular architecture (linear and star-branched for SBS) and microstructure of the elastomer mid-segment (polybutadiene vs poly(ethylene butylene) and polyisoprene) were considered. The sorption results were interpreted and modeled in terms of contributions from rubbery and glassy phases. Comparison between the different data in the binary system as function of solute concentration revealed that friction factor for solute molecules in SEBS and SIS is significantly higher than in SBS.

Such results were also confirmed by linear viscoelastic characterization of the different polymers that has been carried out performing small angle oscillatory shear experiments in a wide range of temperatures between the soft block Tg and room temperature. The latter experimental data were analyzed by Time-Temperature Superposition. The relaxation spectrum has been determined and the monomeric friction factor has been extracted. This value has been correlated with results from diffusion experiments leading to a substantial agreement.

Introduction

Styrene-diene block copolymers are widely used in many industrial applications due to their processability and to their mechanical properties; these material indeed are known to produce phase segregated structures in which glassy styrenic and rubbery dienic domains can be found [1-3]. As a consequence these thermoelastic polymers have mechanical properties similar to those of crosslinked polymers when used at low temperatures, but high flexibility and good processability when brought at high temperatures where also polystyrene domains behave like a rubber .

In these years many studies have dealt with characterization of such materials; the mechanical and physical properties, the rheology as well as the structure of these materials have been widely investigated by several authors [1-7], and many data can be found in the open literature. Nonetheless such copolymers are considered as complex materials whose behavior is still far from being completely understood.

In this concern the present works focused the attention on the mass transport properties of three styrene based block copolymers, with the idea that such studies could help in completing the characterization and help the understanding of such materials behavior. Indeed to date only a few studies have considered such topics and a limited amount of data have been presented concerning the solubility ad diffusivity of gases and vapors in styrene-diene block-copolymers [2,8-10].

In particular the attention will be focused on the sorption and diffusion process of cyclohexane in SBS, SIS and SEBS block copolymer with a styrene mass fraction of about 30%. Several differential sorption run, made with a pressure decay apparatus, were collected which allowed the calculation of both solubility and diffusivity values at 60°C and at activity values ranging from 0 to 0.8.

Moreover, a linear viscoelastic characterization was carried out performing small angle oscillatory shear experiments in a wide range of temperatures between the soft block Tg and room temperature. The latter experimental data were then processed and analyzed by Time-Temperature Superposition [11] in order to obtain the master curves and the shift factors. The relaxation spectrum was then determined with the Rheoplus® software (Anton Paar-Physica), and in particular the monomeric friction factor which can be compared with similar quantities obtained from the vapor sorption experiments.

Experimental

Mass transport experiments have been carried out using the pressure decay apparatus showed in Fig.1; it is a simple single cell device in which the data are obtained monitoring the pressure change in the sample compartment, where the specimens is placed, once it has been filled with the desired amount of vapor penetrant; the latter is obtained directly from vaporization of the liquid solvent contained in the penetrant reservoir. The apparatus is connected to a vacuum pump and a nitrogen trap for system evacuation and solvent recovery after each experiment. During measurements, the temperature of the sample cell was kept at 60°C with a water-jacket, connected to a thermostatic bath through a recirculating pump.

Several differential sorption tests were performed for the different materials, the sample used were rectangular slabs with dimensions of about 50x10x3 mm. The data were processed through an algorithm which allows the solution of the second Fick's law with the appropriate boundary condition, in particular time dependent boundary conditions were considered to account for pressure variations in the sample cell during the experiments. The following problem was considered and solved:

$$\frac{\partial \omega_{pen}}{\partial t} = D\frac{\partial^2 \omega_{pen}}{\partial x^2}$$

$$\omega_{pen} = \omega_{pen}^{I} \quad \forall\, x \in I \ \ \forall\, t > 0$$

$$\omega_{pen} = \omega_{pen}^{0} \quad \forall\, x \qquad t = 0$$

where D is the diffusion coefficient and ω the penetrant concentration in the polymer. The superscripts 0 and I refers to the initial time and to the interface position, respectively, and as said before ω_{pen}^{I} is itself a function of time due to the pressure change during sorption.

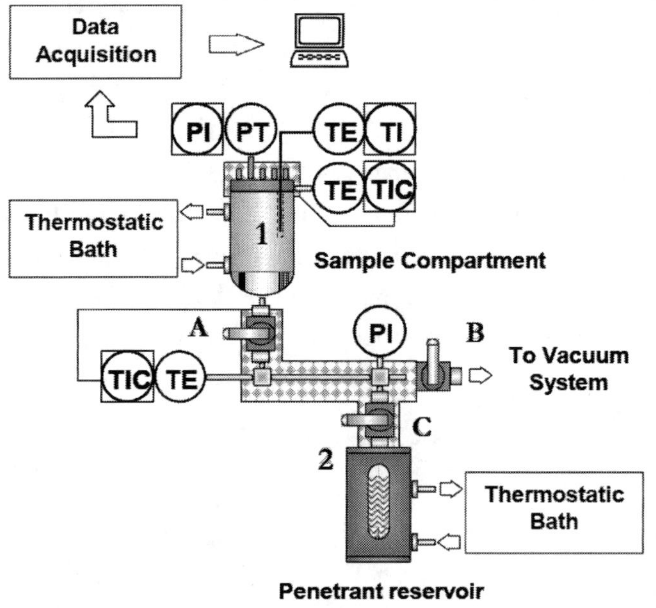

Fig.1 Schematic of the pressure decay apparatus used for sorption experiments.

The linear viscoelastic properties were tested through a Dynamic Mechanical Analysis (DMA) performed on the different materials using a Strain Controlled Rheometer (ARES – Rheometrics). With such rheometer torsional rectangular and parallel plates geometries were used and dynamic frequency sweep tests were performed in the range 100 to 0.01 rad/s. The different experiments were carried out at a number of temperatures around the soft block Tg (≈-55°C for SEBS and SIS , ≈-90°C for SBS) which were reached using liquid nitrogen as cooling agent.

Table 1 Charasteristic of the block-copolymers considered in the present work

	Styrene content (wt%)	$\mathbf{M_w}$	**Polydispersity**	**Diene(1,2)**
SBS (radial)	34	234000	1.2	9%
SIS (linear)	26	131000	1.14	3%
SEBS (linear)	34	54000	1.02	25%

The materials tested were provided by Polimeri Europa S.p.A. and were three different monodisperse styrene based block-copolymers namely SIS, SEBS and SBS with similar amount of glassy polymer. Their different properties are reported in Table 1.

The penetrant was reagent grade cyclohexane purchased from Sigma-Aldrich and used without any further purification.

Results and Discussion

The diffusivity data for the three polymers considered are presented in Fig. 2, where also data for linear polybutadiene (PB) are shown for comparison. As it can be seen, the block copolymer diffusion coefficients lie below the homopolymer data and are slightly increasing with penetrant content. Among the different copolymers, SBS has the highest diffusivity, followed by SEBS and SIS respectively. In particular, those two materials show similar diffusion coefficients at low penetrant content, while SEBS diffusivity increases more markedly with cyclohexane concentration and its value, at the maximum concentration inspected, results to be about 30% higher than what obtained for SIS in the same conditions.

Interestingly, the copolymer behavior leads to the conclusion that the branched SBS system, with double bonds in the backbone chain, has higher diffusion coefficient with respect to linear block-copolymers with higher degree of hydrogenation. This result suggests that the higher chain flexibility, that should be characteristic of linear SEBS and SIS, is not the key factor in determining the diffusion coefficient and other contributions, as for example steric hindrance or different phase distribution, can play a major role.

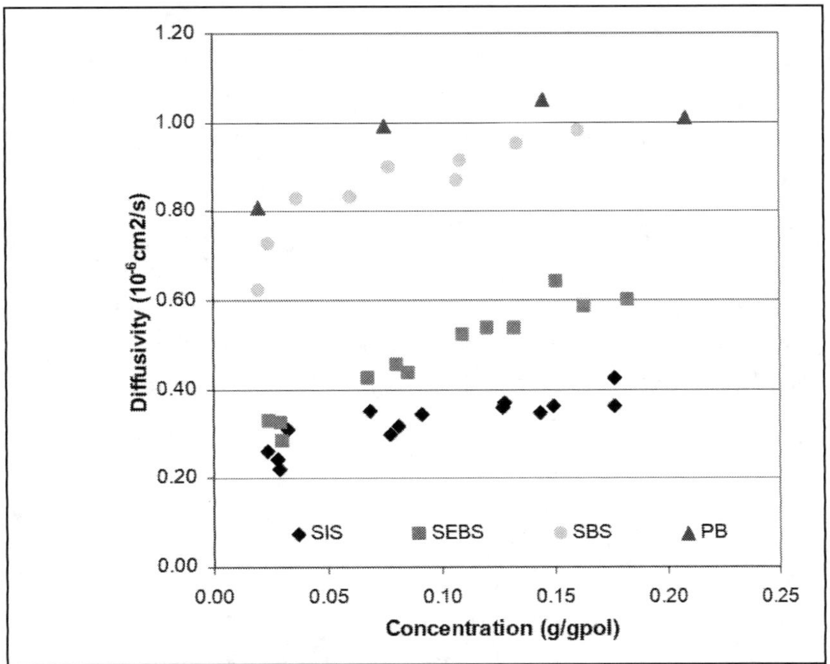

Fig.1 Cyclohexane diffusion coefficient calculated as at 60°C for the different polymers inspected.

Due to the fact that the diffusion coefficient of the styrenic glassy phase, which is about two orders of magnitude lower that that obtained for the present materials, the styrenic domains can be considered as impermeable to the penetrant in the duration of the experiment. As a consequence, they may be considered as "obstacles" for the diffusing molecules and their effect on diffusion can be accounted for through a tortuosity factor that can be roughly related to their mass fraction in the different copolymers. On the other hand, the three copolymers have similar styrene content and seems to have also similar structure, with sharp interfaces between rubbery and styrenic domains so that the differences observed in the diffusivity are probably not related to presence of the glassy phase or to its distribution inside the materials.

It must also be noticed that the diffusion coefficients obtained from sorption experiments are directly related to polymer short range mobility, which, in a free volume approach to diffusion, is directly related to the probability of the penetrant molecule to find an accessible hole in its neighborhood suitable for jump.

Similar conclusions are obtained by dynamic mechanical analysis: such method operates at temperatures close to the T_g of soft segments, thus allowing, to focus the attention on the viscoelastic properties of the dienic phase only, while the styrenic domains behave like elastic solids.

In Fig. 3 the results of the rheological characterization of the different materials are reported. The data represent the master curve of PB and of the different block copolymers at -55°C, in particular the storage and loss modulus of PB and SBS were obtained at -90°C and then extrapolated to –55°C using WLF relationship, that is:

$$\log a_T = c_1 \frac{(T - T_g)}{c_2 + (T - T_g)}.$$

where a_T is the ratio between the value of the inspected property at temperature T and the corresponding value at T_g, while c_1 and c_2 are empirical constants that can be obtained by fitting the experimental data; in the present case, the fitted values are -11.75 and 30.76°C respectively, for both PB and SBS.

From Fig.3 the different behavior of these copolymers with respect to PB can be noticed, in particular in all the copolymers inspected one can notice a shift in the storage modulus plateau towards higher values, that is associated to the presence of rigid styrenic domains.

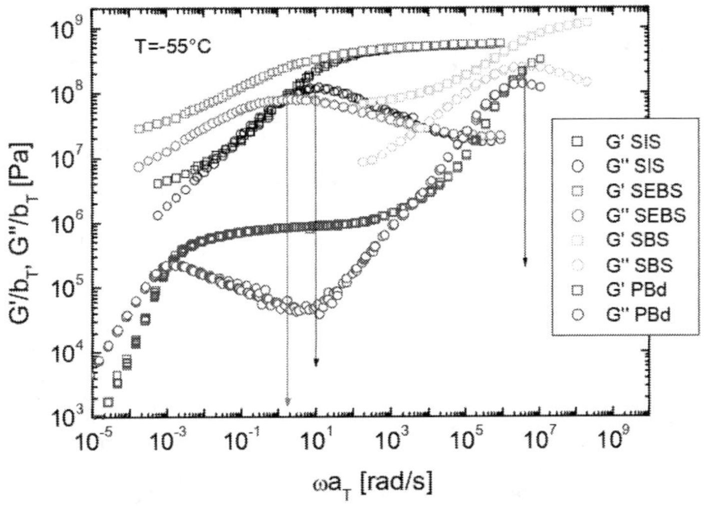

Fig.3 Master curve of storage G' and loss (G") modulus for the different polymer inspected at -55°C. The arrows indicates the frequency corresponding to a the maximum in the loss modulus.

In DMA analysis, short range mobility can be associated, at least qualitatively, to the maximum of loss modulus which is observed at high frequency, in particular, the higher the frequency, the lower is the relaxation time associated and the higher is the chain mobility.

After these considerations, therefore, one can conclude that SBS has an higher intrinsic mobility of chain segments with respect to SIS and SEBS, that show very similar behavior. The fact that SBS and PB have similar mobility confirms the assumption initially made, that in the present analysis styrene domains do not affect the dynamic mechanical behavior, therefore the DMA data refer only to the dienic phase.

When considering SIS and SEBS DMA data, their behavior differs from what obtained from sorption data. Indeed, SIS shows lower diffusivity but higher mobility with respect to SEBS. However, it should be noticed that the rheological characterization has been carried out on the penetrant-free copolymer, and should therefore be compared with diffusivity data in the limit of vanishing penetrant content. In this range indeed SIS and SEBS show very similar values of the diffusion coefficient and an opposite trend with respect to those followed at high concentration is not to be excluded.

In conclusion, it can be said that sorption experiments and rheological characterization showed to be in good agreement and both indicated that linear SIS and SEBS block-copolymers are endowed with a lower soft segment chain mobility with respect to SBS. In addition, the experimental results suggest that mass transport properties obtained by sorption experiments can be, with much care, used to estimate viscoelastic properties at temperatures hardly accessible by means of rheology and viceversa. The present results suggest that the different behaviors observed, both in the case of diffusion and of mechanic behavior, are not related to the structure of the copolymers or to the presence or distribution of the styrenic domains, but concern only the structure and properties of soft segments.

Conclusions

In the present study the mass transport properties of different styrene-diene block-copolymers, namely SIS, SBS and SEBS, have been tested using cyclohexane as a probe. The three polymer had similar styrene content, about 30% and were linear (SIS and SEBS) or star-branched (SBS). The experiments were performed with a pressure decay technique at a temperature of 60°C and up to an activity of 0.8, and allowed the calculation of the cyclohexane diffusion coefficient for the three copolymers considered. The diffusivity showed an increasing trend with penetrant concentration for all the polymers considered and was higher for SBS with respect to SIS and SEBS, which are linear and have a higher degree of hydrogenation. Such results, although counterintuitive, are confirmed by rheological characterization performed through DMA analysis, which shows that SBS has a short range chain mobility substantially higher than that of the other two copolymers considered.

References

1. Aggarwal, S. L.; Livigni, R. A. Polymer Preprints, Division of Polymer Chemistry, American Chemical Society, v 18, n 1, Mar, 1977, p 212-220
2. Chiang, Kwan T.; Sefton, Michael V. Journal of Polymer Science, Polymer Physics Edition, v 15, n 11, Nov, 1977, p 1927-1935
3. LaMonte A.J.; Graessley W.W.; Register, R.A. *Macromolecules*, 27, 1994, p 6026-6032
4. Carreras E.S., Piau J.-M.,* El Kissi N., Pignon F., Panine P., *Journal of Rheology*, 50 (2006), pp. 803–829,
5. Rivera-Gastelum M.J.; Robles-Vazquez, O.; Puig, J.E.; Garcia-Garduno, M.J.; Castano, V.M; Monroy, V.M.: *Materials Letters*, 17, 1993, p 84-90
6. N.R. Legge, G. Holden, H.E. Schroeder, *Thermoplastic elastomer – A comprehensive review*, Hanser publisher, Munich - New York, 1987
7. G.V. Vinogradov , V.E. Dreval, A.Ya. Malkin, G. Yanovsky, V.V. Barancheeva, E.K. Borisenkova, M.P. Zabugina, E.P. Plotnikova, O. Yu. Sabsai, *Rheol. Acta*, 17 (1978) 258-268
8. W.R. Lau, C.J. Glover e J.C.J. Holste, *J.Appl.Polym.Sci.*, 27, 1982, 3067-3077
9. Wang, J.; DeRocher, J. P.; Wu, L.; Bates, F.S.; Cussler, E.L. *Journal of Membrane Science*, 270, 2006, p 13-21
10. Giotto, M.; Lin, G.; Canfield, A.; Jones, A.A. *Macromolecules*, 38, 2005, 9904-9905
11. G. Strobl, *The Physics of Polymers*, Springer-Verlag, Berlin-Heidelberg-NewYork (1997)

Spinodal Decomposition in Ternary Systems with Significantly Different Component Diffusivities

Abdulrahman A. Alfarraj and E. Bruce Nauman

The Isermann Department of Chemical and Biological Engineering

Rensselaer Polytechnic Institute, Troy, New York 12180

Extended Abstract

Most simulations of diffusion in ternary systems have been limited to systems where the component diffusivities are equal or else where there is one dominant component. The case of unequal diffusivities has proved more difficult because the standard approach to multi-component diffusion can give non-physical results that violate a material balance constraint. Presented here is a solution technique that is similar computationally to the standard method but ensures that the material balance is satisfied despite dramatically different diffusion coefficients and near-zero concentrations of some components.

The results outlined in this paper will appear in detailed form together with appropriate literature citations in two separate papers, one in *Macromolecular Theory & Simulations* and one in *Polymer.* The composition and gradient dependent chemical potentials are given by are given by:

$$\nabla \mu_i = (1-\phi_i)\nabla\left(\frac{\delta G}{\delta \phi_i}\right) - \sum_{\substack{j=1 \\ j\neq i}}^{N} \phi_j \nabla\left(\frac{\delta G}{\delta \phi_j}\right) \qquad i,j = 1,\ldots,N \tag{1}$$

It is the chemical potential gradients that drive diffusion. We assume

$$-j_i = \phi_i D_i \nabla \mu_i \qquad i = 1,\ldots.,N \tag{2}$$

Note that any cross diffusion terms have been ignored. Also, the inclusion of ϕ_i as a pre-multiplier allows the system to reduce to Fickian diffusion when the mixture is ideal. The traditional approach at this point is to derive the component continuity equations for N - 1 of the components, giving modified Cahn-Hilliard equations. Instead, we apply Equation 2 to each of the components and satisfy a material balance constraint (such as constant density) by forcing the sum of fluxes into a point to be zero:

$$\sum_{i=1}^{3} j_i = 0 \qquad (3)$$

This is done by summing the incoming and outgoing fluxes for the various components and then scaling the larger in magnitude of the incoming and outgoing so that Equation 3 is satisfied, the scaling being done proportionately for the individual species.

Simulation Results

Here we give only a few representative results. Details are provided in the forthcoming publications. One simulation was for a ternary system with a dual discrete particle distribution where the minor components have distinctly different diffusivities. The new result predicts the bimodal particle size distribution, which was not possible previously.

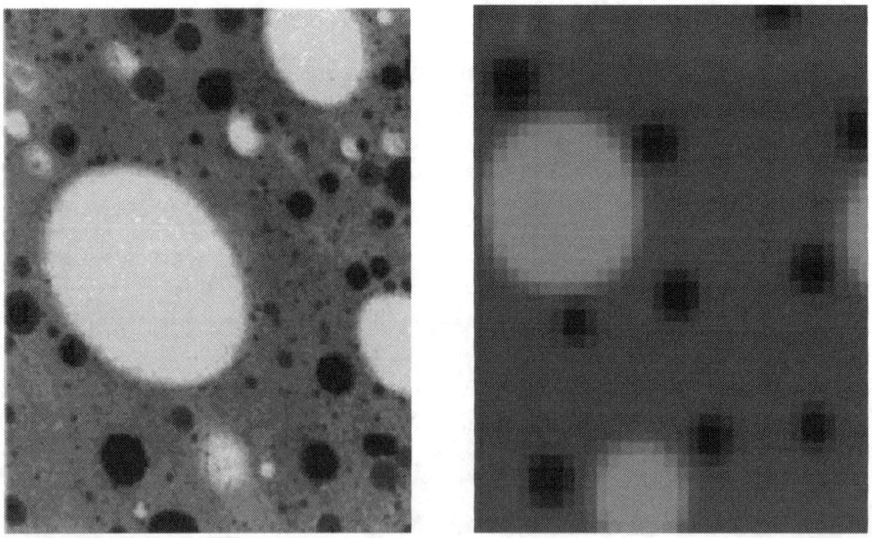

Comparison of (a) Experimental and (b) Simulated PS/PMMA/PB (77/15/8)

An interesting and unexplained result is that the scaling exponents for the two discrete phases are quite different, 0.32 for the larger particles and 0.12 for the smaller particles.

Another simulation treated reaction induced phase separation. The simulated system was styrene / polystyrene formed by gradual polymerization / and polybutadiene. The simulation results for an agitated batch system predict the occlusion morphology observed in impact-modified polystyrene.

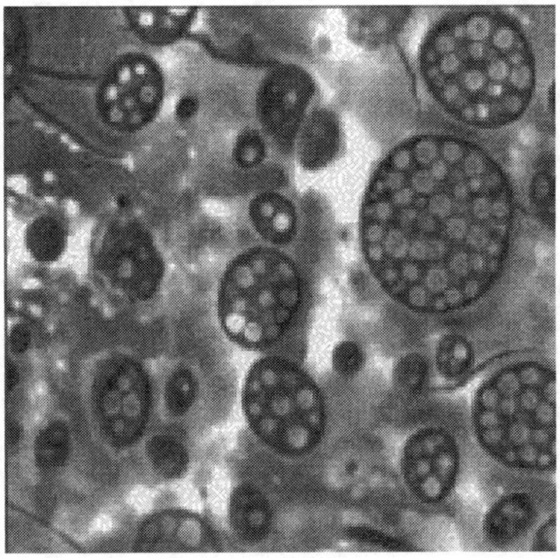

(a)

Experiment

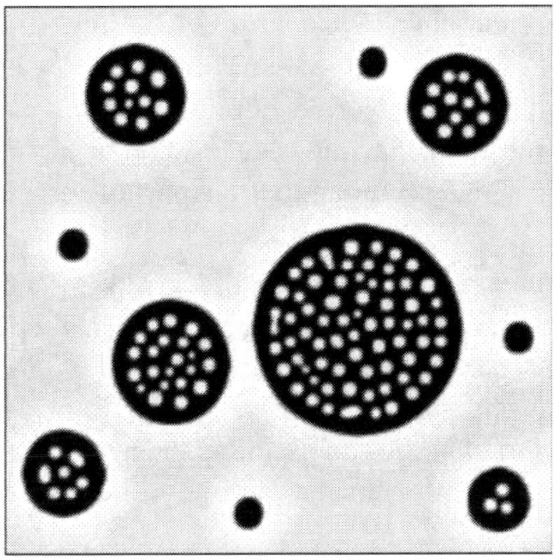

(b)

Simulation

Simulation of a quiescent batch polymerization exhibited phase inversion contrary to the early patent literature that claimed agitation was necessary to produce a particulate rubber phase. The discrepancy is attributed to crosslinking of the rubber that occurs during the polymerization of styrene that stabilizes the rubber phase and maintains cocontinuity.

290d Water Transport In Nafion® Using Time-Resolved Ftir-Atr Spectroscopy

Daniel T. Hallinan Jr. and Yossef A. Elabd. Chemical and Biological Engineering, Drexel University, 3141 Chestnut Street, Philadelphia, PA 19104

The hydrogen polymer electrolyte membrane (PEM) fuel cell is an innovative, environmentally-benign alternative power source with potential for large-market applications, such as transportation. Its success depends on increasing its efficiency, which in large part is limited by properties of the PEM. Current PEMs, such as Nafion®, require high water contents to maintain conductivities above 10 mS/cm. However, other factors necessitate hydrogen fuel cell operation at temperatures above 120°C and humidities below 20% relative humidity (RH). A PEM with high conductivity at these hot, dry external conditions will require either decoupling proton conductivity from membrane water content or enhancing the ability of the membrane to retain water. To achieve either of these goals, a fundamental understanding of multicomponent transport of molecules and ions in PEMs is necessary.

In this study, the transport of water in Nafion® was investigated using time-resolved Fourier-transform infrared, attenuated total reflectance (FTIR-ATR) spectroscopy. This technique not only provides molecular-level contrast between diffusants and polymers in real time, but can also determine chemical interactions via shifts in the infrared spectra. In other words, the diffusion of molecularly distinct species in the polymer membrane can be independently and simultaneously measured in the presence of a concentration gradient, while at the same time monitoring changes in intermolecular interactions.

In these experiments, water diffusion in Nafion® was examined as a function of external relative humidity using time-resolved FTIR-ATR spectroscopy. Regions of the infrared spectrum associated with the stretching and bending vibrations of the hydroxyl bond (O-H) were deconvoluted. Infrared bands associated with different states of water (hydronium ions and several states of bulk water) and anhydrous sulfonic acid in Nafion® were identified and their rates quantified in real time. The ability to monitor separate molecular diffusing species reveals new underlying transport mechanisms. Specifically, the dynamic and equilibrium state of each species and its relationship to water sorption, water diffusivity, and proton conductivity will be discussed. These results not only provide new insights into understanding water management in current PEMs, but also could have a significant impact on the development of new PEMs for high temperature, low humidity fuel cell operation.

292a Protein Adsorption Kinetics in Chromatographic Particles Visualized by Optical Microscopy

Melani C. Stone, David M. Schlientz, and **Giorgio Carta**. Chemical Engineering, University of Virginia, 102 Engineers' Way, Charlottesville, VA 22904-4741

We have recently developed a new method to image transient patterns of protein adsorption in individual spherical chromatographic particles under strong binding conditions. The method takes advantage of the difference in refractive index between the protein-free and protein-saturated adsorbent matrix. When the particles are viewed with an ordinary microscope using white light illumination, the adsorption front appears as a bright ring that moves in time from the surface of the particle to its center. Experimental data were previously obtained for a limited number of proteins with the cation exchanger SP-Sepharose-FF. Sharp rings were observed confirming that protein mass transfer within the particles occurs via a shell-progressive diffusion mechanism. Quantitative analysis based on the shrinking core model provided an accurate and precise way of determining the intraparticle diffusivity for individual particles. In this work we extend these measurements to investigate (a) the effect of ionic strength on protein adsorption in IEX resins, (b) the adsorption of different proteins in agarose-based HIC resins, and (c) the adsorption of IgG in Protein A adsorbents.

292e Effect Of Detector Response On Dynamic Column Breakthrough Experiments

Arvind Rajendran, School of Chemical and Biomedical Engineering, Nanyang Technological University, Nanyang Drive 62, Singapore, Singapore, Shamsuzzaman Farooq, National University of Singapore, 4 Engineering Drive 4, Singapore, 117576, Singapore, and Vinay Kariwala, Division of Chemical and Biomolecular Engineering, Nanyang Technological University, Singapore, Singapore.

Dynamic Column Breakthrough (DCB) experiments, routinely used to complement adsorption and diffusion studies at the particle scale, constitute an important step in the development and verification of a dynamic process model. A typical experimental set-up consists of: a mass flow controller to control the feed flow rate; pressure measuring and regulating devices; a mass flow meter to monitor the exit flow; valves to appropriately direct the flow; and (preferably) an on-line detector to analyze the exit flow composition. The tubing, fittings and accessories from the exit of the mass flow controller to the inlet of the adsorber and from the adsorber exit to the detector (including the detector volume) constitute the dead volumes. Our experience with laboratory-scale breakthrough experiments suggests that the tubing, fittings and accessories primarily contribute a delay to the response of the adsorber. However, the detector, in addition to a delay from its internal volume, may also have some degree of spread depending on the operating volumetric flow rate. A common practice is to measure a blank response under the same flow, pressure and temperature conditions as the actual experiment by simply bypassing the adsorption column with a tube (or a connector) of negligible volume and this blank response is then subtracted point by point from the combined response (i.e., including the adsorption column). The underlying assumption here is that blank and column responses are linearly additive, both in terms of mean residence time and spread.

In the proposed presentation, it will be shown that the delay and spread contributed by the dead volumes of a laboratory-scale breakthrough rig may be effectively represented as a combination of a plug flow reactor followed by several CSTRs acting in series. This model for representing the dead volumes will then be used to evaluate the conditions under which the linear additivity assumption behind the aforementioned method of point by point correction is valid. It will be shown that at other conditions the mixing effect in the detector, in fact, sharpens the response received from the adsorption column and hence the point by point correction leads to over-compensation. A rigorous procedure to calculate the breakthrough response at the column outlet from the combined breakthrough and blank responses will be presented with detailed illustration.

292c Is Differential Volumetric Technique More Accurate

Dr. Orhan Talu, Chemical and Biomedical Engineering, Cleveland State University, 1960 East 24th street SH 455, Cleveland, OH 44115 and Asli Ertan, Chemical and Biomedical Engineering Department, Cleveland State University, 2121 Euclid Avenue, Stilwell Hall 455, Cleveland, OH 44115.

There have been several reports indicating higher accuracy of the differential volumetric technique for isotherm measurements. The basic premise is the fact that the differential pressure transducers are inherently more accurate at high pressure than absolute pressure transducers resulting in ten-fold increase in accuracy. This approach is especially promoted for hydrogen storage applications. In this project, we constructed a differential volumetric apparatus with great care to increase its accuracy. The apparatus also provide data for standard “absolute” volumetric measurements. The PT diagram of palladium hydride is measured in this system. (PT diagram of metal hydrides also characterize gas uptake by solids just like isotherms). By comparing the data calculated both ways, differential and absolute volumetric approaches, the accuracy issues are addressed directly.

292d Diffusion Of Solvent Vapors In Nanoporous Adsorbents: Single Particle Studies Using Concentration Swing Frequency Response

M. Douglas LeVan, Yu Wang, and T. Grant Glover. Department of Chemical Engineering, Vanderbilt University, VU Station B 351604, 2301 Vanderbilt Place, Nashville, TN 37235-1604

A new frequency response method using concentration variation has been extended to investigate diffusion of solvent vapors in nanoporous adsorbents. The method is based on periodically time-varying species feed concentrations with a constant total molar inlet flow rate. A single particle with a mass of 1 to 4 milligrams is used to study the vapor diffusion rates in the adsorbents. The method requires only a small amount of adsorbent, which is advantageous for new materials development. For analysis, a mathematical model is formulated considering nanopore diffusion, surface barrier resistance, external film resistance, and axial dispersion.

The system was used to examine the diffusion of water in silica gel and hexane in activated carbon. The water vapor diffusion mechanism is described by the linear driving force model, but the hexane vapor is controlled by nanopore diffusion. The diffusion rates for different concentrations have been measured to study the concentration dependence. The examples considered show that the novel concentration-swing frequency response apparatus is capable of accurately measuring mass transfer rates for adsorption of vapors.

292b Determination of the Competitive Isotherms of Enantiomers by a Hybrid Inverse Method Using Overload Band Profiles and the Periodic State of the Smb Process

João M. M. Araújo, Rui C. R. Rodrigues, and José P. B. Mota. Requimte/CQFB, Chemistry Department, Faculdade de Ciências e Tecnologia, Universidade Nova de Lisboa, Caparica, 2829-516, Portugal

In the present work a numerical method was developed and tested to determine adsorption isotherms in simulated moving bed (SMB) chromatography. The numerical parameters of prescribed adsorption isotherm model are derived using a hybrid inverse method, that incorporates overload band profiles of the racemic mixture, which includes breakthrough data from a single frontal analysis to overcame the shortcoming arising from the underestimation of the saturation capacity, once the concentration range sampled by the bands is diluted with regards to concentration injected; and the cyclic steady state (CSS) concentration profile of an SMB experiment. Each competitive adsorption isotherm model is coupled with an axially dispersed flow model with finite mass transfer rate to describe the experimental band profiles. The numerical constants of the isotherms models are tuned so that the calculated and measured band profiles match as much as possible.

This method was applied to determine the binary adsorption isotherms of the two enantiomers of Trögers base on a system made of pure methanol as the mobile phase and of Chiralpak AD. We show that this numerical inverse method can be applied even without the knowledge of the individual band profile of the pure solutes. The results indicate that the hybrid inverse method offers a reliable and quick approach to determine the competitive adsorption isotherms for a specific SMB separation.

Keywords: competitive isotherms; chiral separations; inverse method; band profiles; cyclic steady state; simulated moving bed chromatography.

MATHEMATICAL MODELING AND VISUALIZATION OF GAS TRANSPORT IN A ZEOLITE BED USING A SLICE SELECTION PROCEDURE

Michel Petryk
Modélisation du Transfert de Masse en Milieu Hétérogène et Nanoporeux, Université Technique d'Etat Ivan Pul'uy, Ternopil, Ukraine
Sebastien Leclerc, Daniel Canet
Méthodologie RMN, Nancy University, Vandoeuvre-les-Nancy Cedex, France
Jacques Fraissard
Physique Quantique, ESPCI, Paris Cedex 05, France

Abstract

We present the analytical solution of the equations of gas diffusion in a heterogeneous zeolite bed. The problem is handled by assuming that the bed consists of a large number of very thin layers of solid, perpendicular to the direction of propagation of the gas. Mass transfer by diffusion in such a material is determined by a system of differential equations with boundary and interface conditions. The results allow the theoretical determination of the time dependence of the concentration profiles and the inter- and intra-crystallite diffusion coefficients of a gas in each layer of the bed. A numerical application concerns the diffusion of benzene in a cylindrical bed of ZSM5 displaced vertically and rapidly, step by step, inside the NMR probe. Thus we can obtain the time dependence of the concentration of gas absorbed at the level of each slice. These coupled investigations give a better understanding of the diffusion process in this multilayer material.

Introduction

Mathematical modeling of mass transfer by diffusion in zeolite beds or other solids and the methods for finding the analytical solutions have been extensively studied by several authors [1-19]. Molecular transport in a bed of zeolite involves two processes: diffusion in the macropores formed by the space between the crystallites (intercrystallite diffusion) and diffusion in the micropores within the crystallites (intracrystallite diffusion). To determine the contribution of each of these processes to the overall diffusion process, one has to know the values of certain parameters, such as the pressure in the gas phase, the dimensions of the crystallite bed, the adsorption isotherms, etc. In the present work we assume that the bed consists of a large number of very thin layers of solid perpendicular to the direction of propagation of the gas.

As an application of the theoretical results we present the results of a study dealing with the penetration of a gas (benzene) in a zeolite catalyst (ZSM5) bed using the technique described in [20]. The proposed experimental method relies on the vertical displacement, step by step, of the bed during the adsorption of the gas, the NMR detection using a very thin coil. In this way, the region effectively probed is limited to a very thin slice. Thus we can obtain the variation as a function of time of the concentration of gas absorbed at the level of each slice. This dual approach enables us for the first time to determine the diffusion coefficient profiles at every position in the bed. But we have to say even now that the main objective of this study concern the co-diffusion of several gases in a complex microporous solid.

Experiments

NMR apparatus

Our aim is to determine the evolution of the NMR spectrum along the length of a heterogeneous sample in the z direction of the magnetic field B_0. This original method of imaging has been presented in ref. 20.The idea is to select successively several sections at different positions of the sample along the z axis (Fig. 1). In this way one obtains for each position a spectrum corresponding to the composition of the sample at this point. In order to perform these experiments a device was designed to displace the sample step by step in the magnet.

Fig. 1: Principle of experiment

Blue: sample; Red: zone of excitation

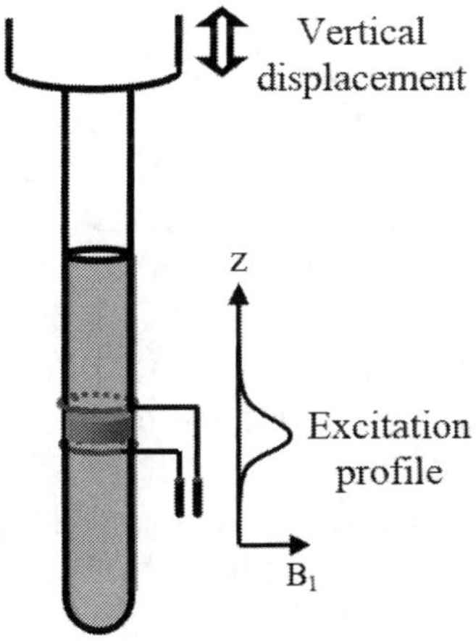

Adsorption measurement

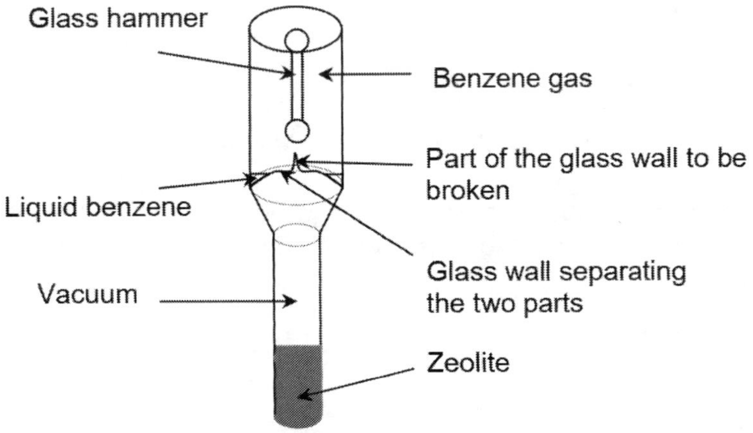

Fig. 2: Experimental set-up

The gas to be adsorbed is in equilibrium with the liquid, in the upper part of the ampule (Fig. 2). It begins to diffuse into the zeolite bed when the wall separating the two parts is broken by means of the glass hammer. In accordance with the principle mentioned above, the zeolite sample is displaced vertically in the magnet during the adsorption, a few microns at a time. The signal of the gas is detected for each position, rapidly so as to obtain the best possible precision. When the sample reaches the uppermost or lowest position it starts again in the opposite direction, and continues in this way until adsorption equilibrium is reached. By this means one obtains a curve of gas adsorption as a function of time for each level in the adsorbent.

In the present case we have studied the adsorption of benzene at 25°C in a 15 mm-deep bed of ZSM5 zeolite. In the upper part of the ampule the vapor pressure of benzene corresponds to the saturation vapor pressure in equilibrium with the liquid at 25°C.

Mathematical modeling

The model developed here is analogous to the bipore model developed in references [2-12, 15-19, 21,22] . A cylindrical bed of microporous zeolite crystallites, assumed to be spherical (radius R), is exposed to a constant concentration of adsorbate in the gas phase (Fig. 2 and 3). One face of this bed is permeable to the gas. In this case one can consider that gas diffusion is linear in the macropores (z direction along the height, l, of the bed) and radial in the micropores. We assume that the zeolite bed consists of a large number, n, of very thin layers of solid, of thickness $\Delta l_k = l_{k+1} - l_k$, perpendicular to the propagation of the gas in the *z* direction. We have made the following assumptions: *i*) during the evolution of the system towards equilibrium there has to be a concentration gradient in the macropores and/or in the micropores; *ii*) the effect of heat is negligible; *iii*) diffusion occurs in the Henry's law region of the adsorption isotherm.

The mathematical model of gas diffusion kinetics in the zeolite bed (which we consider to be a heterogeneous and multilayer porous medium) is defined by the solutions of the system of differential equations:

$$\varepsilon_{inter,k}\frac{\partial c_k}{\partial t} = D_{inter,k}\cdot\varepsilon_{inter,k}\frac{\partial^2 c_k}{\partial z^2} - \frac{3\left(1-\varepsilon_{inter,k}\right)}{R}\cdot\left(D_{intra,k}\frac{\partial q_k}{\partial r}\right)\Big|_{r=R}; \quad (1)$$

$$\frac{\partial q_k}{\partial t} = D_{intra,k}\left(\frac{\partial^2 q_k}{\partial r^2} + \frac{2}{r}\frac{\partial q_k}{\partial r}\right) \quad (2)$$

in the domains $I_n = \left\{ t > 0, r \in (0,R), z \in \bigcup_{k=1}^{n+1}(l_{k-1}, l_k); l_0 = 0; l_{n+1} \equiv l < \infty \right\}$

corresponding to the mass balance in the macropores (c_k) and in the micropores (q_k) for each *k* layer. The meaning of all notations is given in the Nomenclature Section.
With the initial conditions:

$$c_k(t=0,z)=0, \quad q_k(t=0,r,z)=0; \quad (3)$$

the boundary conditions:

$$c_{n+1}(t,z=l) = c_{\infty\, n+1}, \quad q_k(t,r=R,z) = K_k \cdot c_k(z,t); \quad (4)$$

$$\frac{\partial c_1}{\partial z}(t,z=0) = 0, \quad \frac{\partial q_k(t,r=0,z)}{\partial r} = 0 \quad (5)$$

and the interface conditions between adjacent layers along the z axis:

$$\left[c_k(t,z)-c_{k+1}(t,z)\right]\Big|_{z=l_k}=0;\ \left[\frac{\partial}{\partial z}c_k(t,z)-\frac{\partial}{\partial z}c_{k+1}(t,z)\right]\Big|_{z=l_k}=0,\ k=\overline{1,n} \qquad (6)$$

where $K_k=\frac{q_{\infty_k}}{c_{\infty_k}};k=\overline{1,n+1}$.

Using dimensionless quantities (except time),

$$C_k=\frac{c_k}{c_{\infty k}};\qquad Q_k=\frac{q_k}{q_{\infty k}};\qquad Z=\frac{z}{l};\qquad X=\frac{r}{R},\qquad L_k=\frac{l_k}{l}$$

in Eqs. (1) and (2) the problem becomes: to find the solution of the partial differential equations:

$$\tau_{M_k}\cdot\frac{\partial C_k}{\partial t}=\frac{\partial^2 C_k}{\partial Z^2}-\Gamma_k\frac{\partial Q_k}{\partial X}\Big|_{X=1}; \qquad (7)$$

$$\tau_{m_k}\cdot\frac{\partial Q_k}{\partial t}=\frac{\partial^2 Q_k}{\partial X^2}+\frac{2}{X}\frac{\partial Q_k}{\partial X}; \qquad (8)$$

in the domain: $D_n=\left\{t>0, X\in(0,1), Z\in\bigcup_{k=1}^{n+1}(L_{k-1},L_k); L_0=0; L_{n+1}\equiv 1\right\}$

with the initial conditions:

$$C_k(t=0,Z)=0,\qquad Q_k(t=0,X,Z)=0; \qquad (9)$$

the boundary conditions:

$$C_{n+1}(t,Z=1)=1;\qquad Q_k(t,X=1,Z)=C_k(t,Z); \qquad (10)$$

$$\frac{\partial C_1}{\partial Z}(t,Z=0)=0;\quad \frac{\partial Q_k}{\partial X}(t,X=0,Z)=0 \qquad (11)$$

and interface conditions:

$$\left[C_k(t,Z)-C_{k+1}(t,Z)\right]\Big|_{Z=L_k}=0;\ \left[\frac{\partial}{\partial Z}C_k(t,Z)-\frac{\partial}{\partial Z}C_{k+1}(t,Z)\right]\Big|_{Z=L_k}=0;k=\overline{1,n}; \qquad (12)$$

here: $\Gamma_k=\frac{3}{P_{inter_k}}\cdot\frac{\tau_{M_k}}{\tau_{m_k}};\qquad \tau_{M_k}=\frac{l^2}{D_{inter_k}};\qquad \tau_{m_k}=\frac{R^2}{D_{intra_k}};$

$$P_{inter_k}=\frac{\varepsilon_{inter_k}\cdot C_k}{\varepsilon_{inter_k}\cdot C_k+\left(1-\varepsilon_{inter_k}\cdot C_k\right)\cdot Q_k}\approx\frac{\varepsilon_{inter_k}}{\varepsilon_{inter_k}+\left(1-\varepsilon_{inter_k}\right)\cdot\frac{Q_k}{C_k}},k=\overline{1,n+1}.$$

The adsorption constant K_k of molecules (for example, hydrocarbons) in zeolites is usually very high. Therefore ε_{inter} is much smaller than $(1-\varepsilon_{inter})$ K and can be neglected in the expression of $P_{inter,,k}$ giving the relative amount of gas in the intercrystallite space ΔL_k.

$$P_{inter_k}=\frac{\varepsilon_{inter_k}}{\left(1-\varepsilon_{inter_k}\right)\cdot K_k},k=\overline{1,n+1}.$$

If $\tau_{m,k} >> \tau_{M,k}$, $\Gamma_k << 1$, the process in the layer ΔL_k is controlled by intracrystallite diffusion. It is the opposite if $\tau_{M,k} >> \tau_{m,k}$ ($\Gamma_k >> 1$). If $\Gamma_k \approx 1$, inter- and intracrystallite diffusions should control the overall diffusion in such a layer [9].

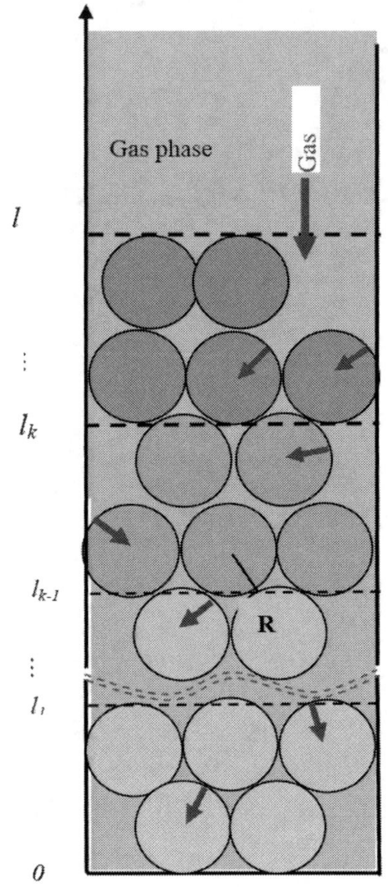

Diffusion in macropores

Length of the bed: l

Characteristic position of the layer: l_k, , $k=\overline{1,n+1}$

Thickness of layer k: $\Delta l_k = l_{k+1} - l_k$, $k=\overline{1,n+1}$

Intercrystallite diffusion coefficient in k layer: $D_{inter,k}$;

Corresponding characteristic time: $\tau_{inter,k}$

In the theoretical part, l is the top of the bed and 0 is the bottom.

Diffusion in micropores

Crystallite diameter: R

Intracrystallite diffusion coefficient in the k layer: $D_{intra,k}$

Corresponding characteristic time: $\tau_{intra,\,k}$

In the theoretical part, r = 0 corresponds to the center of the sphere.

$l >> R$; $D_{inter,k} >> D_{intra,k}$; $k=\overline{1,n+1}$

Fig. 3. The process of diffusion in the zeolite bed

Putting $Q_k = \dfrac{N_k}{X}$, $k=\overline{1,n+1}$ into Eqs. (7)-(12), the system becomes:

$$\tau_{M,K}\frac{\partial C_K}{\partial t}=\frac{\partial^2 C_K}{\partial Z^2}-\Gamma_k\left(\frac{1}{X}\frac{\partial N_k}{\partial X}-\frac{1}{X^2}N_k\right)_{X=1}\equiv\frac{\partial^2 C_k}{\partial Z^2}-\left[\frac{\partial N_k}{\partial X}-N\right]_{X=1}, \tag{13}$$

$$\tau_{m,K}\frac{\partial N_K}{\partial t}=\frac{\partial^2 N_K}{\partial X^2}, \tag{14}$$

in the domain: $D_n=\left\{t>0, X\in(0,1), Z\in\bigcup_{k=1}^{n+1}(L_{k-1},L_k); L_0=0; L_{n+1}\equiv 1\right\}$

with the initial conditions:

$$C_k(t=0,Z)=0; \quad N_k(t=0,X,Z)=0; \tag{15}$$

the boundary conditions:

$$C_{n+1}(t,Z=1)=1; \qquad N_k(t,X=1,Z)=C_k(t,Z); \tag{16}$$

$$\frac{\partial C_1}{\partial Z}(t,Z=0)=0, \qquad N_k(t,X=0,Z)=0 \tag{17}$$

and the interface conditions:

$$\left[C_k(t,Z)-C_{k+1}(t,Z)\right]_{Z=L_k}=0;\left[\frac{\partial}{\partial Z}C_k(t,Z)-\frac{\partial}{\partial Z}C_{k+1}(t,Z)\right]_{Z=L_k}=0; k=\overline{1,n} \tag{18}$$

Analytical solution: concentration profiles

The analytical solutions of the problem described by Eqs. (13)-(18) can be obtained using the Laplace integral transformation and the Cauchy function methods. The calculations are too complex and long to be reported here. They are developed in [23].

Results

Benzene adsorption curves

Fig. 4 represents the variation with time of the benzene concentration for each level of the zeolite bed. It must be noted that the z coordinate is determined relative to the bottom of the tube. For example, z = 15 mm (dimensionless Z = 1) corresponds to the surface of the bed directly in contact with the gas phase.

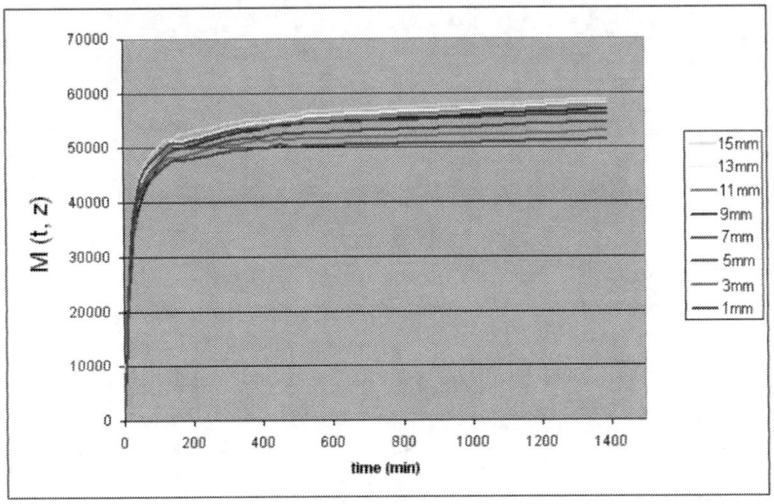

Fig. 4. Amount (arbitrary units) of adsorbed benzene against time in each layer *k* of the bed, from z = 15 mm (layer in contact with the gas phase) to z = 1 mm (layer near the bottom of the bed).

The amount of adsorbed benzene at the equilibrium of adsorption is 1.25 mmol.g^{-1}, in agreement with the literature.

Numerical simulation and analysis

A program in the Microsoft Visual C++ environment was especially developed to provide the solutions to the previous equations:

- Using the experimental curves of Fig. 4, the intercrystallite D_{inter} (macropores) and the intracrystallite D_{intra} (micropores of the crystallites) diffusion coefficients;
- Using these diffusion coefficients, the concentration distribution in the intercrystallite space $C(t,Z)$, in the intracrystallite space $Q(t,X,Z)$ and the integral mass $M(t,Z)$.

Fig. 5 shows the computed variations of D_{inter} and D_{intra} versus time at four different distances z (in mm) from the bottom of the bed: z = 14; z = 10; z = 7; z = 3. D_{inter} and D_{intra} decrease approximately exponentially with time increase, which is normal since the crowding increases in parallel with the concentration of adsorbed molecules. D_{inter} decreases from 1.4 m^2/s at the beginning of adsorption to 0.001 m^2/s near the equilibrium state. In the same way, D_{intra} decreases from 10^{-12} to $5{\cdot}10^{-14}$ m^2/s. These results are in good agreement with the literature data [24, 25].

It may appear surprising that the beginning of the curve corresponding to z = 14 mm presents a lower diffusion coefficient than the others for the same time. This is due to the fact that there is not a perfect concordance between the moment the solid comes into contact with the benzene and the acquisition of the first NMR signals. The first signals correspond then to the upper layers which have already adsorbed a significant amount of gas, making thus diffusion more difficult.

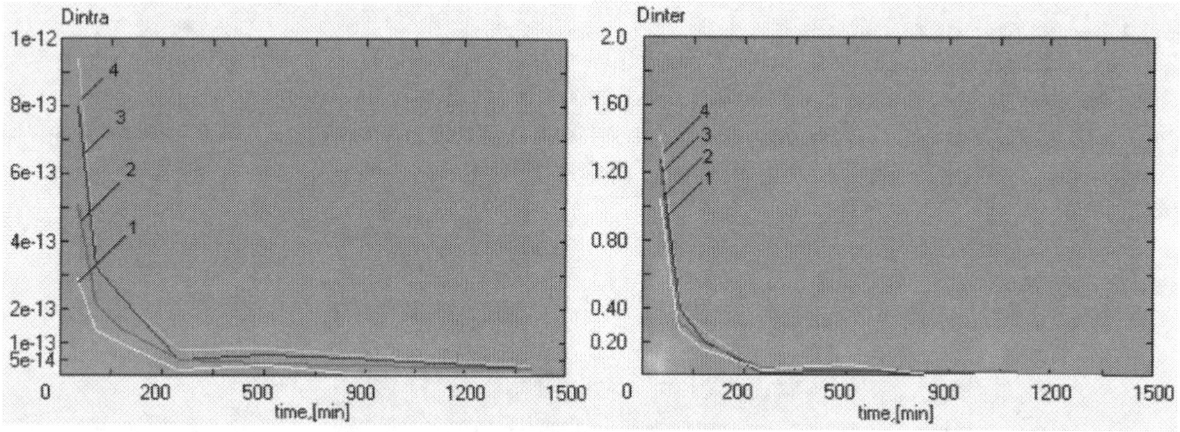

Fig. 5. Benzene diffusion coefficient profiles D_{intra} and D_{inter} (m^2/s) against time t (min)
Curves: 1) z = 14 mm; 2) z = 10 mm; 3) z = 7 mm; 4) z = 3 mm.

Using the values of D_{inter} and D_{intra} we can determine the concentration distributions in the intercrystallite space $C(t,z)$, in the intracrystallite space $Q(t,X,z)$ and the total adsorbed mass $M(t,z)$, for each time and position in the bed (z) and in the crystallites (z,r). For example, Fig.6 shows the calculated and experimental kinetic curves of dimensionless total adsorbed mass $M(t,z)$ versus time *t* for four positions z (mm) from the bottom of the bed: z = 15 mm; z = 11 mm; z = 7 mm; z = 1 mm. The calculated variations agree satisfactorily with the experimental ones.

Fig.7 shows the concentration profiles of C_6H_6 in intercrystalline space C(t,z) for four positions z. Fig. 8 presents the calculated profiles of C_6H_6 concentrations in crystallites, $Q(t,X,z)$, as a function of dimensionless distance from the crystallite center X for different times *t* (min) and four distances from the bottom of the crystallite bed z (mm).

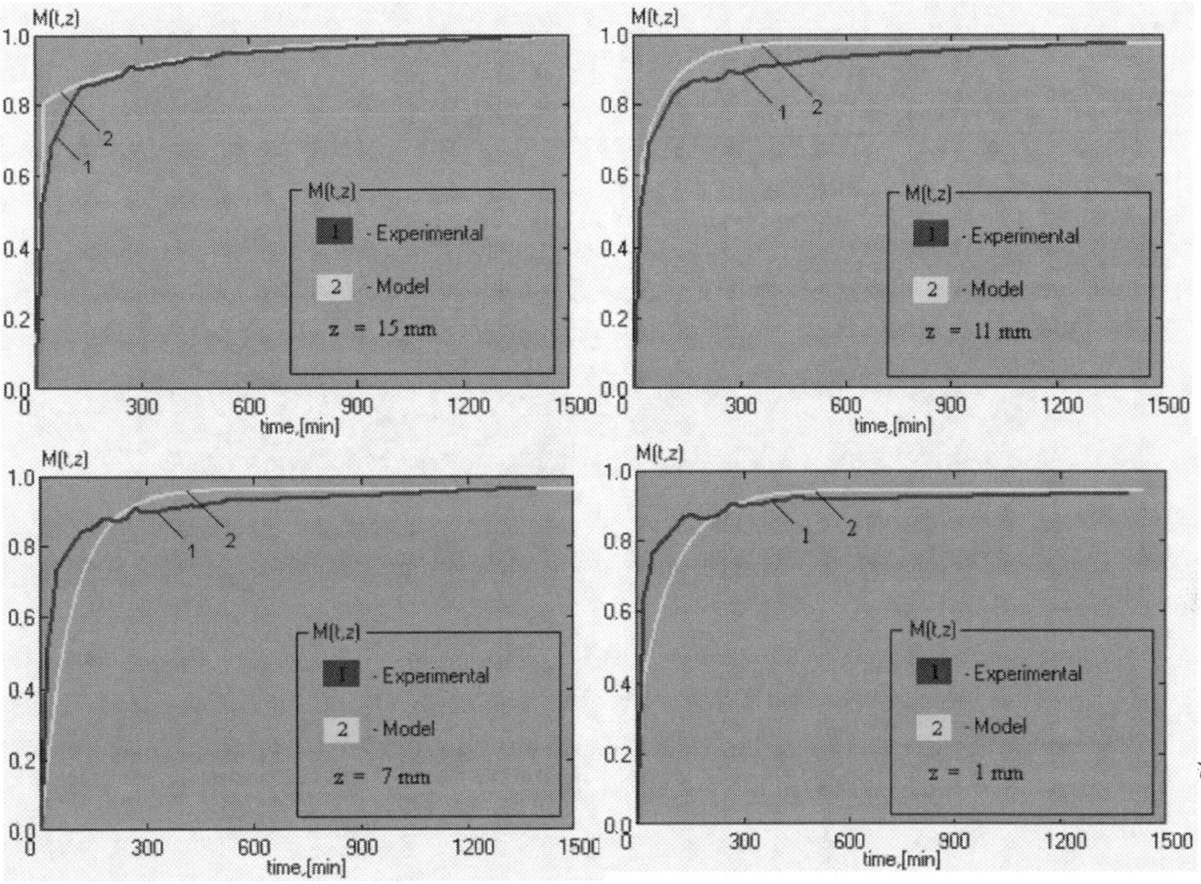

Fig.6. Model and experimental kinetic curves of dimensionless total adsorbed mass versus time *t* (min) for different values of distance (mm) from the bottom of the zeolite bed *z*.

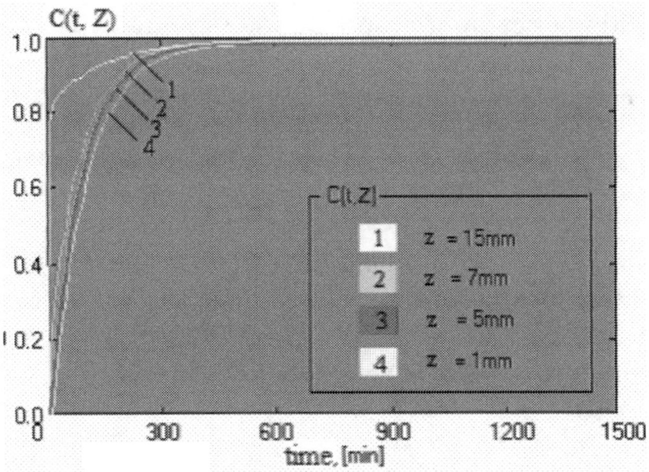

Fig. 7. Intercrystallite concentration profiles versus time t (min) for different values of the distance z (mm) from the bottom of the bed.

a), b)

c),d)

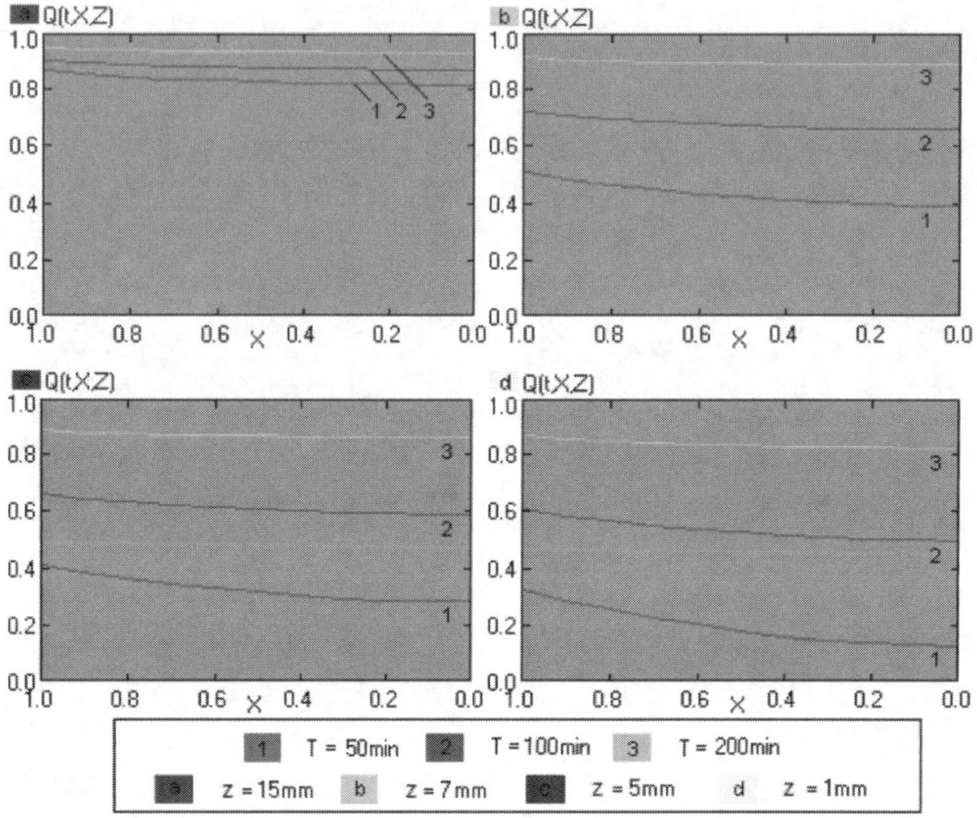

Fig. 8. Concentration profiles, *Q(t,X,z)*, in zeolite crystallites located at different positions *z* (*mm*) in the bed, and for different times t (min.): a) z = 15 mm; b) z = 7 mm; c) z = 5 mm; d) z = 1 mm.

Conclusion

We have shown the interest of the analytical solution of the equations of gas diffusion in a zeolite bed, assuming it to consist of a large number of very thin layers of solid, perpendicular to the direction of propagation of the gas. The solution is based on the integral Laplace transformation and Cauchy methods. We have applied the analytical solution to the study of benzene diffusion in the macropores and micropores of a ZSM5 zeolite bed.

We have been able to calculate the inter- and intracrystallite diffusion coefficients, D_{inter} and D_{intra} of benzene, respectively, as a function of time and the crystallite position in the bed. For each position inside the micro- and macropores of a ZSM5 bed we have determined the variation of the concentration of this gas against time.

This technique should prove to be very interesting for studying the co-diffusion of several gases in a catalytic bed, since it should make it possible to detect the characteristic spectra of each gas at every moment and position in the bed. We have presented here the first step of a general study.

Nomenclature

$k=\overline{1,n+1}$: layer number. Subscript *k* will be added to all the following symbols to specify that they are characteristic of the *k* layer.

c : adsorbate concentration in macropores.

c_∞ : adsorbate equilibrium concentration in macropores.

$C = c/c_\infty$: dimensionless adsorbate concentration in macropores.

D_{inter} : macropore diffusion coefficient, m^2/s
D_{intra} : micropore diffusion coefficient, m^2/s
K : adsorption equilibrium constant
l : bed length, mm.
$\Delta l = l_k - l_{k-1}; k = \overline{1, n+1}$: layer thickness (we assume that all layers have the same thickness)
L: dimensionless bed length
M_m : micropore uptake.
M_M : macropore uptake.
M_t : total uptake at time t.
M_∞ : total uptake at equilibrium.
p_{inter} : relative adsorbed gas in macropores in equilibrium condition.
p_{intra} : relative adsorbed gas in micropores in equilibrium condition.
q : adsorbate concentration in micro- and nanopores.
q_∞ : equilibrium adsorbate concentration in micropores.
$Q = q/q_\infty$: dimensionless adsorbate concentration in nanopores.
r : distance from crystallite center, mm.
R : mean crystallite radius, mm (we assume that the crystallites are spherical).
$X = r/R$: dimensionless distance from crystallite center.
z : distance from bottom of crystallite bed, mm.
$Z = z/l$: dimensionless distance from bottom of the crystallite bed.

Greek letters
ε_{inter} : bed porosity.
τ_M : diffusion time required for a molecule to penetrate the crystallite bed from top to bottom, s.
τ_m : diffusion time required for a molecule to penetrate a crystallite from the external surface to the center, s.
τ_{inter} : time constant of diffusion in intercystallite space, s .
τ_{intra} : time constant of diffusion in intracrystallite space, s .
Γ : dynamic parameter.

References

1- Lykov A.V. and Mykhaylov Y.A., *The Theory of Mass Transport,* State Energy Publishing, Moscow (1963).
2- Ruckenstein E., Vaidyanathan A.S. and Youngquist G.R., *Sorption by Solids with Bidisperse Pore Structures,* Chem. Eng. Sci., **26**, 147-152 (1971).
3- Barrer R.M., *Diffusion and Flow in Porous Zeolite, Carbon or Ceramic Media, Characterization of Porous Solids*, Society of Chemical Industry, London (1979).
4- Kärger J. and Ruthven D., *Diffusion in Zeolites and Other Microporous Solids*, John Wiley & Sons, New York (1992) and references therein.
5- Chen N.Y., Degnan T.F. and Smith M.C., *Molecular Transport and Reaction in Zeolites: Design and Application of Shape Selective Catalysis*, VCH, Weinheim and New York (1994).
6- Lee L.K., *The kinetics of sorption in a biporous adsorbent particle*, AIChE J., **24**, 531-533 (1978).
7- Magalhães F.D., Laurence R.L., Conner, W.C., Springuel-Huet M.A., Nosov A. and Fraissard J., *Study of molecular transport in beds of zeolite crystallites: semi-quantitative modeling of ^{129}Xe NMR experiments,* J. Phys. Chem. B, **101**, 2277-2284 (1997).

8- Springuel-Huet M.A., Nosov A., Kärger J. and Fraissard J., *^{129}Xe NMR study of bed resistance to molecular transport in assemblages of zeolite crystallites*, J. Phys. Chem., **100**, 7200-7203 (1996).
9- N'Gokoli-Kekele P., Springuel-Huet M.A. and Fraissard J., *An Analytical Study of Molecular Transport in a Zeolite Crystallite Bed.* Adsorption, **8**, 35-44 (2002)
10- Petryk M., Shabliy O., Leniyk M. and Vasylyuk P., *Mathematical Modeling and Research for Diffusion Process in Multilayer and Nanoporous Media.* Fluid Transport in Nanoporous Materials. (W.C. Conner and J. Fraissard, eds.), NATO Science Series, Series II: Mathematics, Physics and Chemistry, Vol. 219. Springer (Amsterdam) 685-655 (2006).
11 - Kärger, J., *A study of fast tracer desorption in molecular sieve crystals,* AIChE J., **28**, 417-423 (1982).
12- Kärger, J. and Heink W., *The propagator representation of molecular transport in microporous crystallites*, J. Magn. Reson., **51**, 1 (1983).
13- Kärger J. and Pfeifer H., *NMR self-diffusion studies in zeolite science and technology*, Zeolites, **7**, 90-107 (1987).
14- Kärger J., Pfeifer H. and Heink W., *Advances in Magnetic Resonance.* (J.S. Waugh, ed.), Vol. **12**, p. 1, Academic Press, San Diego, 1988.
15- Koresh, J. and Soffer, A.J., *Molecular Sieve Carbons, Part 3 – Adsorption kinetics according to a surface barrier model*, Chem. Soc. Faraday Trans., **77**, 3005-3018 (1981).
16- Ma, Y.H. and Ho S.Y., *Diffusion in synthetic fine powder and pellets,* AIChE J., **20**, 279-283 (1974).
17- Ma, Y.H. and Lee T.Y., *Transient diffusion in solids with a bipore distribution*, AIChE J., **22**, 147-152 (1976).
18- Ruthven, D.M., *Principles of Adsorption and Adsorption Processes*, Wiley, New York, 1984.
19- Younquist, G.R., *Diffusion and flow of gases in porous solids*, Ind. Eng. Chem., **62**, 52-63 (1970).
20- Leclerc S., Trausch G., Cordier B., Grandclaude D., Retournard A., Fraissard J. and Canet D., *Chemical Shift Imaging (CSI) by precise Object Displacement,* Magn. Reson. Chem., **44,** 311-317 (2006).
21- Petryk M. and Vorobiev E., *Mass Transfer from Liquid containing Spherical Particles During The Pressing of Biological Porous Material*, Proceedings of the 2nd W. European Conference on Filtration and Separation, Compiègne (France), Vol. 2, 763-766 (2000).
22- Petryk M., Petryk N and Matiega V., *The Intensification of Process of Vibrofiltration in non Regular Cylindrical Media.* Emerging Solid/Liquid Separation Technologies ICEST-2002, Compiègne (France), **1,** 143-149 (2002).
23- Petryk M., Leclerc S., Canet D. and Fraissard J., "Diffusion-Fundamental.org" on line, **4,**11.01-11.23 (2007).
24- Foger K. , Sanders J.V. and Seddon D., *Channel arrangements and activity of some ZSM zeolites.* Zeolite, **4,** 337-345 (1984).
25- Springuel-Huet M.-A., *^{129}Xe NMR for diffusion of hydrocarbons in zeolites and ^{1}H NMR imaging for competitive diffusion of binary mixtures of hydrocarbons in zeolites*
Fluid Transport in Nanoporous Materials (2006)
NATO Science series II ¨Mathematics, Physics and Chemistry, Vol.219, pp 315- 332.
Wm. Curtis Conner and Jacques Fraissard, Edit., Springer , Dordrecht, Publisher.

Computer Program for Filter Media Design Optimization

S. Dharmanolla and G. G.Chase

Microscale Physiochemical Engineering Center
The University of Akron, Akron, OH, 44325-3906
330-972-7943 gchase@uakron.edu

This paper presents a computational approach for designing, mixed fiber filter media for depth filtration. The goal is to find an optimum solution, given the range, for a set of design parameters: thickness of the media, diameter of microfiber, and diameter of nanofiber, surface area ratio of nanofiber to microfiber and mass of microfiber. The objective of this work is to develop a software program that can be used as an aid to accelerate filter design and reduce the number of laboratory experiments.

This program applies a Genetic Algorithm to search for an optimum filter media design based on Quality Factor which quantifies the filter performance. A user friendly computer program is developed that provides inputs, outputs and controls to design a filter media. The program provides a starting point for constructing filter media for testing and design for particular applications.

THEORY AND BACKGROUND

In filtration, efficiency of the media can be quantified by defining certain parameters, namely; Filter Coefficient, Single Fiber Efficiency and Quality Factor (R.C.Brown, 1993). In FiL2RO we use the Quality Factor to characterize the filter performance. To determine the Quality Factor we must solve the species and momentum balances to determine the concentration and pressure profiles. The model is based upon the following assumptions:

- Media fibers are rigid and stationary.
- The media is incompressible.
- Isothermal process.
- Unidirectional flow in z direction.
- The particles are captured by the fibers.
- The permeability of media is uniform and isotropic
- The flow rate and porosity of media are constant.
- Binder materials to hold the fibers together are neglected.
- Clean filter (not loaded with particles).

Species Balance

The number of particles captured (*N*) by a medium layer of thickness (Δz) are proportional to the number incident (*N*) to the thickness (Δz) and to a 'constant', known as the filter coefficient, α

$$\Delta N = -\alpha N \Delta z \tag{1}$$

It is formulated as:

$$\alpha = E_{RD}\, \alpha_{100\,\%} \tag{2}$$

where E_{RD} is the overall single fiber efficiency taking into account mechanisms of direct interception, and Brownian diffusion. The single fiber efficiency is the ratio of the number of particles actually removed relative the number that would be removed if particle capture was proportional to the projected area of the fibers. The projected area is used to define the 100% filter coefficient as

$$\alpha_{100\%} = \frac{4(1-\varepsilon)}{\pi d_f}$$ where ε_i is the porosity of the medium. (3)

If the filter medium is made of two fiber sizes then α is calculated by

$$\alpha = \sum E_i \alpha_i \tag{4}$$

where E_i is the single fiber efficiency for the fiber of size d_i and α_i is calculated from

$$\alpha_i = \frac{4\varepsilon_i}{\pi d_i} \tag{5}$$

where ε_i is the volume fraction of the medium occupied by the d_i size fiber.

The filter coefficient is used in the species balance to determine the particle concentration in the gas stream. The one dimensional species balance at steady state has the form

$$\frac{dC}{dz} = -\alpha C \tag{6}$$

where C is the local particle concentration. This integrates to obtain the concentration profile as

$$C(z) = C_{in} \exp(-\alpha z) \tag{7}$$

Momentum Balance

The steady state one-dimensional momentum balance reduces to the form

$$\varepsilon \frac{dP}{dz} = F_z \tag{8}$$

where F_z is the drag force due to fluid flow past the fibers. For a two fiber filter medium the drag force is related to the fiber size, volume fraction, and force per unit length of fiber by the expression

$$F_z = \frac{4\varepsilon}{\pi} \sum \left(\frac{\varepsilon_i f_i}{d_i^2} \right) \tag{9}$$

The f_i values are determined from correlations relating pressure drop to flow rate for flows through fibrous filters. The pressure drop is obtained by integrating the momentum balance to obtain

$$\Delta P = \frac{4L}{\pi} \sum \left(\frac{\varepsilon_i f_i}{d_i^2} \right) \tag{10}$$

Quality Factor (QF):

The Quality Factor (QF) is defined as the ratio of negative logarithm of penetration to pressure drop. The penetration is defined as the ratio of the outlet concentration to the inlet concentration.

$$QF = \frac{-\ln\left(\frac{C_{out}}{C_{in}}\right)}{\Delta p} \tag{11}$$

In filtration, pressure drop is related to energy expenditure hence QF is a measure of separation performance achieved against energy expended.

Sample model calculations are compared in Figure 1. The quality factor and parameter values are normalized by dividing by a base case condition. The effects of changing parameters on the quality factor are plotted. The results show the nanofiber diameter, face velocity, and particle diameter have the biggest effects on performance, while holding other quantities constant. The base case is taken at temperature of 210C, 20 Psi pressure, face velocity of 2.1 m/sec, particle diameter of 150nm, 2 micron microfiber diameter, 200nm nanofiber diameter, 1 cm thickness of media, 3gms of microfiber and area ratio of nanofiber to microfiber as 0.1. The quality factor calculated for the base case is 0.135 (1/Kpa). The variations in quality factor due to fiber amounts, fiber sizes and other design parameters suggest that a computer program may be written to determine optimum design values where the filter has maximum performance.

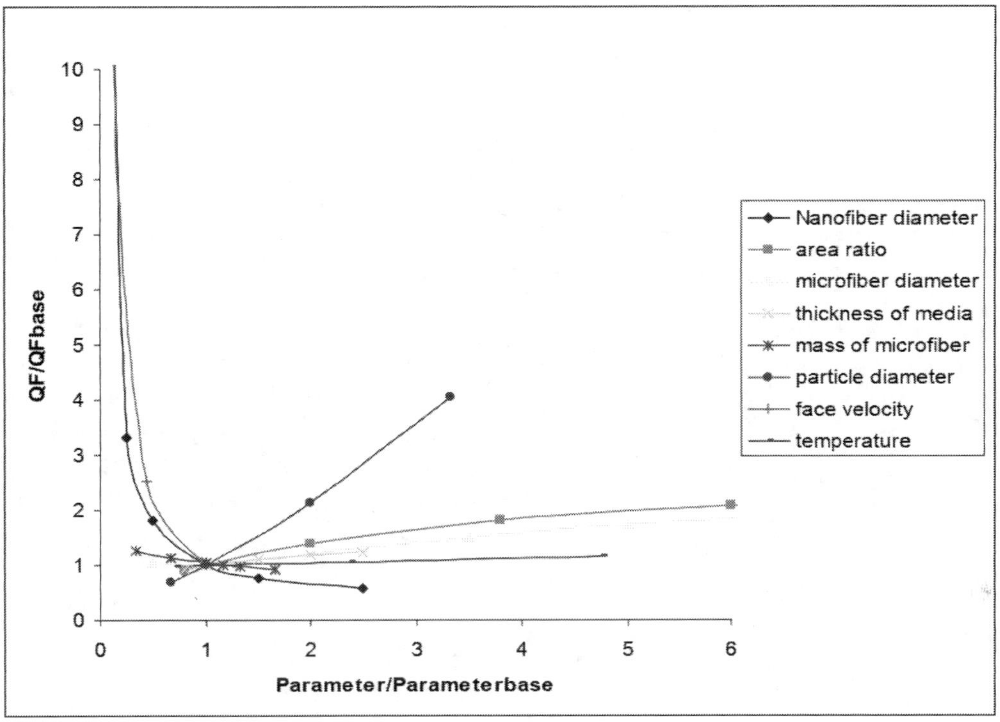

Figure 1. Effect of parameters on quality factor

GENETIC ALGORITHM

Genetic Algorithms (GA) are robust search routines that enable use to search complex solution space to find optimum solutions. The GA is used here to search out the optimum size and amounts of nanofiber and microfiber and other design parameters for composite filter media.

Figure 2 shows the flow chart for the program calculations. The GA starts with a population of chromosomes. Each chromosome represents one possible solution. The chromosomes are stored in binary form in the population so that the GA can manipulate the individual bits. The fitness of each chromosome is determined by calculating the Quality Factor for that particular solution. The solutions having higher Quality Factors are “more fit” than those with lower Quality Factors. The GA has several operations that must occur.

- Reproduction: Reproduction of a population is based on survival of the fittest. The better solutions are reproduced more in the next population than the less optimal solutions.
- Crossover: This operation is performed by exchanging parts between two selected individual parent chromosomes to form new chromosome solutions in the new population analogous to the exchange of chromosomes in offspring in a two-parent biological sense.
- Mutation: This operation is used to prevent premature convergence to local optima by randomly sampling new points in the search space which is carried out by flipping bits at random. Mutation rate should be very low.
- Stopping Criteria: The stopping criteria are used to stop calculating new generations. In our program the calculations stop when the after a sufficient number of new generations the optimum answer has converged to a steady value.

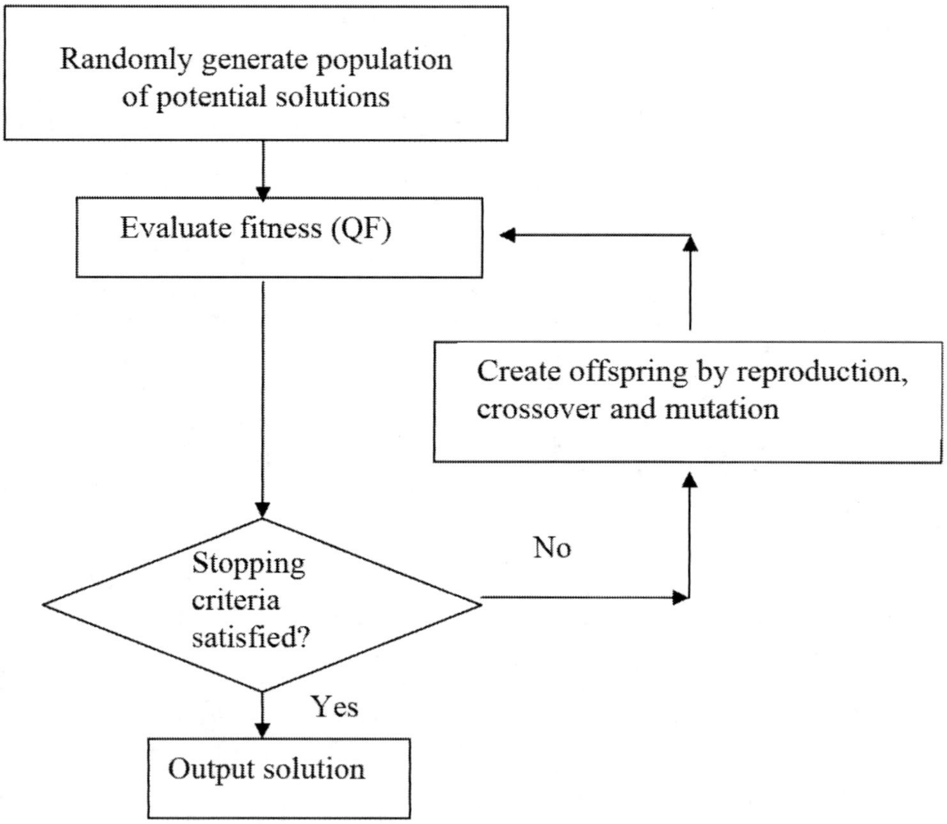

Figure 2. Flow chart for Genetic Algorithm calculations.

COMPUTER PROGRAM

The computer program, called FiL2RO, has different modules or sections shown in the window. These sections are explained here.

- *Process parameters*

User specify operating conditions such as pressure, temperature , gas face velocity and provide information for diameter of filter media , particle size and density of microfiber with units.

- *Optimized parameters*

In this module the user can select any of these five parameters (thickness of media, diameter of microfiber, diameter of nanofiber, area ratio of nanofiber to microfiber and mass of microfiber) to optimize by checking the appropriate box. The user must give a range for all these variables using a hyphen "-" between values.

- *Genetic Algorithm Parameters*

In this module the user gives the number of iterations (generations) in which the program runs and searches for the solution in the given range. The population is set to 1000. Mutation probability range is from 0 to 1 with a default value of 0.05. Depending on the convergence tolerance, the program stops and shows the result. If convergence tolerance is not achieved, the number of generations can be increased and the program can be re-run. A convergence tolerance of less than 10^{-8} is recommended to get better solutions.

- *Result*

This window shows the optimum design parameters for the filter media with iterations (generation number), convergence and quality factor. It also calculates the porosity of media, pressure drop and permeability of media with units.

- *Plot*

Generation vs. convergence plot and Quality factor vs. Design parameters can be plotted after the program has converged.

- *Control buttons*

After the user has entered the data, clicking on start button runs the program. After the program is executed we can see the results by clicking on view results. Using save results data can be saved into an external text file by specifying your own path. View plot will show the generation vs. convergence plot. The default value button will reset all the parameters to the default values. Using load file we can see the saved data with results by selecting a previously saved file. About and Help button give details about the program. Exit closes the program.

Other features of the FiL2RO software include:

- Interactive presentation of data to enable user to input interpretations of the measured data.
- This program can be used as a starting point for experimental preparation of filter media.
- It can reduce an effort and time for laboratory testing and development of media.
- Warns the user if data is incorrect or data is missing.
- Conversion of units according to your choice.
- Checking of input data to protect against possible incorrect data entry.
- Output data in graphical and tabular format.
- The user selects the range for optimum parameters and the program looks for solution in that range and shows the output.
- If the convergence is not achieved it warns the user to increase the iterations or increase the convergence.
- User can view input parameters and graphical interpretation of final result.
- Results can be saved into a file to be viewed later.
- All data required for filter design are shown.

CONCLUSIONS

The FiL2RO program calculates optimum design parameters of mixed fiber filter media. The program uses a genetic algorithm to search for optimal solutions within the user specified parameter ranges.

ACKNOWLEDGEMENT

This work was supported by the Coalescence Filtration Nanomaterials Consortium: Ahlstrom Paper Group, Donaldson Company, Cummins Filtration, Hollingsworth and Vose, and Parker Hannifin. This work is also supported by the National Science Foundation grant numbers CTS-0310429.

REFERENCES:

1. R.C. Brown, *Air filtration*, An integrated approach to the Theory and Applications of Fibrous Filters, Pergamon, New York, 1993
2. Bird, R.B., Steward W.E., Lightfoot E.N., *Transport Phenomena*, (1960), Wiley, New York
3. Goldberg David E., Genetic *algorithms in search, optimization and Machine learning*, 1989, Addison-Wesley, New York.

293b 2-D Simulation Of Continuous Separation Of Magnetic Particles In Single-Wire And Multi-Wire Arrays

Fei Chen, Kenneth A. Smith, and T. Alan Hatton. Department of Chemical Engineering, Massachusetts Institute of Technology, 77 Massachusetts Avenue, Cambridge, MA 02139

High gradient magnetic separation (HGMS) has long been used as an efficient solid/liquid separation technique in the mining and metallurgical industries. Its potential applications in food, biological and pharmaceutical areas have recently attracted increasing attention. The design HGMS process suitable for emerging applications in these areas has become an important topic around the world. In this work, the trajectories of magnetic particles are simulated in separation column with combined magnetic and flow fields. The fields and forces are calculated with finite element method (FEM). Based on the particle trajectories, the capture performance of single round wire, single non-round wire and multi-wire array are evaluated. In order for our model to be scalable, all governing equations are non-dimensionalized and three dimensionless numbers (Reynolds number, dimensionless characteristic magnetic force and magnetic force based Grashof number) are used for the analysis of particle trajectories. The influences of three dimensionless numbers and other parameters on the capture efficiency are investigated. The dynamic process of particle buildup growth and its effect on flow field and magnetic field are also addressed in this work. With clear physical and mathematical descriptions, a continuous separation concept of simultaneously collecting and removing magnetic particles is proposed and evaluated. All methods and conclusions presented here can be extended to novel magnetic separator design and help us understand the entire separation process.

293e Computer Aided Analysis in Mechanical Operations – Filtration

Asha Immanuel Raju Chaduvula[1], Ravi Prasad Andra[1], Jagannadha Rao Patruni[2], Srinivasa Babji Josyula[2], Kiran Kumar Nalla[3], Murali Yugandhar Nikku[4], and Israel Yedluri[5]. (1) Chemical Engineering, Andhra University, Andhra University, Visakhapatnam - 530 003, AP, INDIA, Visakhapatnam, 530003, India, (2) Chemical Engg, Andhra University, Andhra University, Visakhapatnam - 530 003, AP, INDIA, Visakhapatnam, 530003, India, (3) Chemical Engg, Center for Energy Engg, Room No: 109, Andhra University, Visakhapatnam, 530003, India, (4) Chemical Engineering Department, Andhra University, Andhra University, Visakhapatnam - 530 003, AP, INDIA, Visakhapatnam, 530003, India, (5) Department of Geophysics, Andhra University, Visakahpatnam, 530003, India

ABSTRACT

Today's Era is commonly known as computer Era. The computations involving time of washing, optimum cycle time, total area of filtrate surface, total cycle time, the cake thickness, minimum number of frames, filtration constants, dimensions of filter press and amount of wash water required in filtration are estimated at a more speed with the help of the computer aided analysis in which a computer flow diagram will be prepared. A computer flow diagram is simply nothing but the flow sheet illustrating all the steps of the computation process and the detailed procedure followed in each step. In the present paper numerous problems are taken, coded in 'C' language, tested with the manual and computer calculations and the results are compared and discussed provided with the application of numerical methods like Interpolation, extrapolation, Simpson's, Euler methods etc. The application of numerical methods is also incorporated for example for calculating the area under curve. The main reason for the computer aided analysis in mechanical operations (Filtration) is the repetition of the calculations.

293d A Predictive Pressure Drop Model For A Multi-Filter Array Filtration System

Ryan A. Sothen, Chemical Engineering, Auburn University, 0322A Haley Center, Auburn, AL 36849 and Bruce Tatarchuk, Center for Microfibrous Materials Manufacturing, 230 Ross Hall, Dept. of Ch. E., Auburn, AL 36849-5127.

Adsorbent or catalyst entrapped media, such as microfibrous materials engineered at Auburn University, provides a novel method to efficiently entrap particles as small as ten microns within a sinter-locked matrix of fibers. The small adsorbent diameters greatly increase the contacting efficiency of the media; however, the particles also increase the pressure drop across the media due to additional drag. The increase in pressure drop is a limiting factor of the media's capacities. If a thick media is designed to increase adsorbent loading, the resulting rise in pressure resistance will drastically increase energy consumption and could potentially overload the air handler.

Pleating the media is a simple approach to increase the capacity and reduce the pressure drop of the filter. A second technique to greatly increase the capacity and further reduce the pressure drop is to employ a filtration array composed of multiple pleated filters. To fully utilize the potential of the filter array, a comprehensive model of pressure drop factors is constructed. By employing various fundamental fluid dynamics equations and empirical data, the research establishes a predictive model capable of estimating pressure resistance for flow through a multi-filter array. The contributions of filter count, pleat count, filter depth, and media type on total pressure drop across the system is assessed.

Using this model, the capacity and contacting performance of a multi-filter array is compared to a packed bed at a fixed pressure drop and volumetric flows. Through the use of novel material and packaging techniques, the amount of adsorbent present in a filtration unit can dramatically increase while maintaining a lower pressure drop than commercially available adsorbent units.

293c Design Characteristics Of Pleated Filters And Their Corresponding Effect On Pressure Resistance

Ryan A. Sothen, Chemical Engineering, Auburn University, 0322A Haley Center, Auburn, AL 36849 and Bruce Tatarchuk, Center for Microfibrous Materials Manufacturing, 230 Ross Hall, Dept. of Ch. E., Auburn, AL 36849-5127.

Pressure resistance plays a fundamental role in determining air handler specifications and is a key component in determining the energy needed to flow air through the system. Pressure drop within a HVAC system is created by a combination of resistances such as flow in the duct, around bends, across baffles, and through filtering elements. Numerous filtration systems are commercial available; however, pleated filters are one of the more common styles. To increase the available filtration area, a pleated filter uses a highly folded media. The additional area reduces the air velocity through the media, thus reducing the pressure drop across the filter. The extra media also extends the filter's useful life. If the pressure drop across a pleat filter can accurately be predicted, then the system's energy consumption and removal efficiency can be optimized.

For a pleated filter, the pressure resistance curve possesses a “U” shaped curve when plotted against the pleat count. At low pleat counts, the primary contributor to total pressure drop is the resistance in the media. As the pleat count increases, the resistance due to the media decreases; however, the viscous resistances begin to rise. Thus, each pleated filter passes through an optimal pleat count corresponding to the minimal pressure drop.

This study's objectives are to identify the design parameters and assess their consequential pressure drops. The research proposes a model capable of making accurate predictions of the pressure drop across a pleated filter based solely on physical properties of the media. The design parameters are determined as follows: filter dimensions (height, width, and depth), media type (thickness and permeability), filter housing, and pleat count. An empirical model is created by systematically altering the filter's parameters and observing the resulting pressure changes. Experimental data is collected from twenty filters with various media types, filter depth, and pleat counts. The data is utilized in conjunction with Darcy's law, Bernoulli's equation, and the equation of continuity to model the filter as a series of individual resistances.

293f Cfd Modeling Of A Candle Filter Vessel: Ash Bridging Effects

Ali R. Mazaheri[1], **Isaac K. Gamwo**[2], and Goodarz Ahmadi[1]. (1) Department of Mechanical and Aeronautical Engineering, Clarkson University, Potsdam, NY 13699, (2) National Energy Technology Laboratory, U.S. Department of Energy, Pittsburgh, PA 15236-0940

Pilot power plants such as Pressurized fluidized-bed combustion (PFBC) and integrated coal gasification combined cycles use hot gases from coal to drive a gas turbine. Candle filters are used to remove hot-gas particulates that erode and corrode gas turbine blades. The Reynolds stress turbulence model of FLUENT was used to model the gas flow behavior inside the filter vessel. Particle equations of motions are employed, transport and deposition of the micron-size aerosols are studied. Computational results predict that ash bridging leads to a non-uniform temperature distribution along the ceramic candle filters in the bridging region. The analyses of ash bridging deposition on the internal surfaces of the filter show that the absence of ash bridging tends to promote the deposition of particle of 10 ìm on the surfaces.

293g Effect of Colloid-Colloid Interactions on Deposition in Porous Media

Huilian Ma and William P. Johnson. Geology and Geophysics, University of Utah, Salt Lake City, UT 84112

Colloid-colloid interactions have been largely neglected in the past studies of particle deposition in porous media. Except for the complexity involved in accounting for such interactions, this neglect may largely result from limitations of classic filtration theories, which predict no attachment under unfavorable surface interactions (i.e., in the presence of repulsive energy barriers to deposition due to like-charged surfaces). Recent advances in understanding the mechanisms of colloidal deposition in the presence of energy barriers (e.g., wedging at grain-grain contacts and retention in secondary energy minima) provide a basis for inclusion of colloid-colloid interactions in predictions of particle deposition. Here, effects of colloid-colloid interactions on particle retention in porous media are investigated through a Lagrangian particle trajectory model that has been parallelized to allow particle-particle interaction. Results show that despite their mutual electrostatic repulsion, colloid-colloid interactions can mitigate or enhance retention in porous media depending on the fluid velocity and chemical conditions. In addition, parameters characterizing repulsive colloidal interactions, pore geometry are explored in developing a correlation equation for predicting particle deposition in porous media in the presence of energy barriers.

305a Fouling Development on Reverse Osmosis Membranes

Thuy Tran[1], Brian Bolto[1], Stephen Gray[2], and Eddy Ostarcevic[3]. (1) Manufacturing & Materials Technology, CSIRO, Gate 5 Normanby Road, Clayton, Victoria, 3168, Australia, (2) Institute of Sustainability and Innovation, Victoria University, Hoppers Lane, Werribee, Melbourne, Victoria, 8001, Australia, (3) GWMWater, 24 Darlot Street, Horsham, Victoria, 3402, Australia

Like other membrane filtration processes, fouling is a major obstacle that prevents efficient operation of reverse osmosis (RO) systems, causing deterioration of both the quantity and quality of treated water, and consequently resulting in higher treatment cost. Numerous autopsy studies have been carried out to better understand the physico-chemical processes governing the fouling [see, for example, 1-5]. Methods commonly used in many of these studies such as ICP-MS, GC-MS, FTIR and XRD provide only information on the average composition of the surface deposits. Studies that employ direct observation methods including optical and electron microscopy often focus on the top surface deposits, but not on the underlying deposit layers. These deficiencies lead to an incomplete understanding of the deposition kinetics of various foulants, and therefore of the fouling mechanisms, particularly where thicker deposits have been developed.

In addition, many mechanistic studies on RO membrane fouling often focus on one foulant type for the purpose of simplicity. Whilst this approach is useful, it is very important to understand the effects of interactions between various foulant types on the fouling mechanisms. For instance, it has recently been reported that the enhanced concentration polarization of salt ions within the colloidal cake layer may result in an increase in osmotic pressure and rapid flux decline during cake layer development [6,7]. As well, the interactions between colloidal and organic foulants has been found to give rise to considerable synergistic effects, as manifested by a significantly higher flux decline compared to the additive effects of colloidal fouling and organic fouling alone [8].

This study reports the autopsy results of a spiral wound RO membrane after nearly one year of service in a brackish water treatment plant using optical and electron microscopic methods, FTIR and ICP-AES. Both the top surface and the cross-section of the fouled membrane were analysed to provide further insights into the development of the fouling layer.

It has been found that the extent of fouling was uneven across the membrane surface with regions underneath or in the vicinity of the strands of the feed spacer being more severely affected. The fouling appeared to have developed through different stages. In particular, it consisted of an initial thin fouling layer of an amorphous matrix with embedded particulate matter. The amorphous matrix comprised organic-Al-P complexes and the particulate matter was mostly aluminium silicates. The use of aluminium sulphate as coagulant and phosphonate-based antiscalant is suggested to contribute to the high levels of Al and P in the fouling layer. As the fouling layer reached a thickness of about 5 to 7 μm, further amorphous material, which is suggested to include extracellular polymeric substances such as polysaccharides, started to deposit on top of the existing fouling layer. This secondary amorphous material did not seem to contain any particulate matter nor any inorganic elements within it, but acted as a substrate upon which aluminium silicate crystals grew exclusively in the absence of other foulants, including natural organic matter (NOM).

[1] Butt, F. H., Rahman, F. and Baduruthamal, U. (1997) Characterization of foulants by autopsy of RO desalination membranes. Desalination 114, 51-64.

[2] Speth, T. F., Summers, R. S. and Gusses, A. M. (1998) Nanofiltration foulants from a treated surface water. Environ. Sci. Technol. 32, 3612-3617.

[3] Sahachaiyunta, P., Koo, T. and Sheikholeslami, R. (2002) Effect of several inorganic species on silica fouling in RO membranes. Desalination 144, 373-378.

[4] Vrouwenvelder, J. S. and van der Kooij, D. (2002) Diagnosis of fouling problems of NF and RO membrane installations by a quick scan. Desalination 153, 121-124.

[5] Gwona, E. M., Yu, M. J., Oh, H. K. and Ylee, Y. H. (2003) Fouling characteristics of NF and RO operated for removal of dissolved matter from groundwater. Water Research 37, 2989-2997.

[6] Hoek, E. M. V. and Elimelech, M. (2003) Cake-enhanced concentration polarization: a new fouling mechanism for salt-rejecting membranes. Environ. Sci. Technol. 37, 5581-5588.

[7] Lee, S., Cho, J. and Elimelech, M. (2005) Combined influence of natural organic matter (NOM) and colloidal particles on nanofiltration membrane fouling. J. Membr. Sci. 262, 27-41.

[8] Li, Q. and Elimelech, M. (2006) Synergistic effects in combined fouling of a loose nanofiltration membrane by colloidal materials and natural organic matter. J. Membr. Sci. 278, 72-82.

305b Studies On Scaling Of Membranes In Desalination By Dcmd

Fei He[1], Jack Gilron[2], Hanyong Lee[1], Liming Song[1], and Kamalesh K. Sirkar[1]. (1) Chemical Engineering, New Jersey Institute of Technology, Warren Street, Newark, NJ 07102-1982, (2) Zuckerberg Institute for Water Research, Ben Gurion University, POB 653, Beer Sheva, 84105, Israel

Membrane distillation (MD) whether of the DCMD (direct contact membrane distillation) or VMD (vacuum membrane distillation) variety can have a role to play in desalting highly saline waters that have considerable osmotic pressures where reverse osmosis (RO) operation becomes more expensive and problematic. Using MD in this way would allow increased recovery and help reduce the problem of concentrate disposal vexing inland desalination. To realize this promise, MD must show itself to be more resistant to scaling than RO and thus not limited by it in the way that RO is. An analysis of the scaling potential in hollow fiber membrane-based crossflow DCMD is presented in terms of the saturation index profiles throughout the hollow fiber membrane module as a function of the location in the module for the sparingly soluble salt, CaSO4 and CaCO3, individually or mixed together. Modeling shows that the highest scaling potential is to be found at the high temperature end of the module both due to the high brine temperature and concentration polarization associated with high local fluxes. Concentration effects are far more important than temperature, although concentration polarization estimated in crossflow hollow fiber DCMD units is lower than that in spiral wound modules in RO for similar flux values. Experimental results at high saturation indices show that even when the precipitation rate is fast in the mixed CaSO4-CaCO3 system at elevated temperatures or high concentrations, no significant loss in water vapor permeation is observed. Possible explanations for this resistance to scaling are presented and discussed in terms of crossflow and the nature of the coating on the surface.

305c Rapid Characterization of Membrane Fouling by Proteins in Constant Flux Microfiltration and Development of Mathematical Model for Fouling Prediction

Xinghua Sun and Raja Ghosh. Chemical Engineering, McMaster University, 1280, Main Street West, Hamilton, ON L8S 4L7, Canada

Fouling of microfiltration membranes by proteins is a major challenge in bioprocessing. It is affected by a diverse range of factors: operating condition such as permeate flux and system hydrodynamics, protein solution properties such as pH, ionic strength and protein concentration, and membrane properties such as hydrophilicity, roughness, pore size and distribution. Knowledge of protein fouling mechanisms is essential for prediction, control and prevention of fouling. We discuss a technique for rapid assessment and characterization of fouling during constant flux microfiltration of protein solutions. This technique is based on the pulsed sample injection technique for fouling assessment, earlier developed in our group for studying fouling of ultrafiltration membranes. Different stages of fouling are identified form the transmembrane pressure-time (TMP-time) profiles in microfiltration tests. The initial fouling rate is found to be slow from which there is transition to a rapid fouling stage followed by a slow fouling stage. A multi-mechanism mathematical model which takes into account pore blockage, cake formation and pore constriction are discussed. The model predictions are compared with experimental results.

305d Kinetics Of Ro Membrane Scaling In The Presence Of Antiscalants

James Au[1], Myung-man Kim[1], Anditya Rahardianto[1], Eric Lyster[2], and Yoram Cohen[1]. (1) Chemical and Biomolecular Engineering, University of California, Los Angeles, 5531 Boelter Hall, Los Angeles, CA 90095, (2) Chem Eng, UCLA, 5531 Boelter Hall, Los Angeles, CA 90095

Most inland brackish water sources contain precursor ions of sparingly soluble mineral salts (e.g. calcium, sulfate, bicarbonate, etc.) that can lead to membrane scaling in RO desalting operations and thus impair plant productivity (i.e. permeate flux), product water quality, and membrane operational life time. Currently, membrane scaling is typically mitigated using commercial polymeric formulations, commonly known as antiscalants, to retard the onset and growth of mineral salt crystals in the bulk and onto the membrane surface [1,2]. The use of antiscalants can add up to 15% of the total water production cost and thus optimization of antiscalant use can be beneficial to lowering the cost of RO desalting. Current understanding of antiscalants action is primarily inferred based on permeate flux decline studies and studies of mineral salts precipitation in bulk solutions [3]. Direct measurements of the kinetics of mineral salt scaling on RO membranes are needed in order to provide data needed for the development of predictive scale formation models, optimize RO operating conditions and antiscalant choice and dosage. While a number of studies have been recently published on direct observation of scale formation, there is lack of much needed information on the direct influence of antiscalants on mineral scale crystallization on membrane surfaces.

In the present study, the influence of antiscalants on retarding the onset and growth of mineral salt crystals on RO membranes was quantified and elucidated using a novel approach for direct (during high pressure RO process) optical imaging of surface mineral scale growth [4]. RO membrane scaling tests were conducted, in a novel transparent RO cell, using model solutions representative of field samples of brackish water from typical locations in California. The study focused on gypsum scale which is the most problematic scalant in inland water desalination in most areas in the U.S. A number of commercial antiscalants were first evaluated with respect to their effectiveness in suppressing gypsum bulk crystallization. The antiscalants were then tested in membrane RO experiments, at varying doses, to directly measure their impact on the time-evolution of crystal site density and the growth rate of individual surface crystals. An approach to quantify antiscalant effectiveness was employed based on determination of kinetic growth parameters using a simple surface scale formation model.

The results of the study have shown, for the first time, that different commercial membranes differ in their response to antiscalant scale suppression treatment,. Specifically, it was found that surface nucleation was affected by both the specific antiscalant and membrane used. In general, the ranking of antiscalant effectiveness based on bulk crystallization induction time correlated with membrane surface scale coverage and flux decline measurements in membrane RO studies. Membrane surface roughness and hydrophobicity (characterized by contact angle measurements) were shown correlate to some degree with scaling propensity of the membranes and with antiscalant effectiveness in suppressing scale. The present approach to evaluating antiscalant effectiveness provides a rationale approach to antiscalant and membrane selection for brackish water desalting where propensity for mineral scaling can be an impediment for successful process implementation.

[1] Rahardianto, A., W. Shih, R. Lee, Y. Cohen. Diagnostic characterization of gypsum scale formation and control in RO membrane desalination of brackish water. J. Mem. Sci. 279 (2006) 655.

[2] Wen-Yi S., A. Rahardianto, R. Lee, Y. Cohen. Morphometric characterization of calcium sulfate dihydrate (gypsum) scale on reverse osmosis membranes. J. Mem. Sci. 252 (2005) 253.

[3] Amjad, Z. Kinetics of crystal growth of calcium sulfate dihydrate. The influence of polymer composition, molecular weight, and solution pH. Can. J. of Chem. 66 (1988) 1529.

[4] Uchymiak, M., A. Rahardianto, E. Lyster, J. Glater, Y. Cohen. A novel RO ex situ scale observation detector (EXSOD) for mineral scale characterization and early detection. J. Mem. Sci. 291 (2007) 86.

305e Potential Fouling Resistant Coating Materials Synthesized By UV Induced Polymerization

Yuan-Hsuan Wu and Benny Freeman. Chemical Engineering, University of Texas at Austin, 10100 Burnet Rd. Building 133, Austin, TX 78758

The lack of sufficient amounts of pure water is one of the world's most serious concerns. Consequently, water reuse and management is becoming more and more important. Produced water is the single largest waste stream in oil and gas production, often containing salts, heavy metals, emulsified oil and other organics. If the organic content and salinity of produced water could be reduced to acceptable limits, produced water could represent a potential new water source with a wide variety of uses. Membranes may be an effective tool for treating water produced during oil and gas production. However, membrane fouling is a serious problem and limits the efficiency of water purification in this application.

The initial objective of this research was to examine potential fouling-reducing coating materials and determined fundamental structure/property relationships governing water transport, rejection of model foulants such as emulsified oil droplets or protein aggregates, and fouling characteristics. A systematic series of hydrogel-based polymers was prepared, and their permeation and fouling properties were studied and correlated with their structure. In a future study, composite membranes prepared via coating of commercial UF membranes will be performed.

In this presentation, the effects of UV intensity, water content in the prepolymer solution, and crosslinker content on film morphology, water transport and rejection properties are presented. Crosslinked free-standing films were prepared by UV-induced free radical polymerization at different UV intensities, using N,N'-methylene bisacrylamide (MBAA) as the crosslinker and N-vinyl-2-pyrrolidone (NVP) as the monomer. The water content in the prepolymer solution was varied from 30 wt % to 70 wt%, and the MBAA content of total solids was varied from 15 wt% to 30 wt%. The crosslinked NVP films show polymerization induced phase separation (PIPS). These PIPS films contain pores that scatter visible light and render them opaque. The films exhibited a porous morphology in CryoSEM studies. Liquid water uptake increased from 100 to 350 g water/100 g dry polymer as water content in the prepolymer solution increased. Dead end filtration of deionized water, oil/water solutions (1500 ppm), and bovine serum albumin (BSA) solutions (1g/l), was performed to explore the fundamental transport and fouling properties of these materials. Different internal structures of crosslinked NVP films at the same composition were made by polymerization at different UV intensities and characterized by molecular weight cutoff (MWCO) measurements. Polymerization under high intensity UV light (~3000 μW/cm2) leads to high water permeability, up to ~3000 L μm/(hr m2 bar) in film containing 70 wt% water and 30 wt% solids (NVP/MBAA: 85/15) and high MWCO, ~10,000 kDa. At the same composition with polymerization at low intensity (~1100 μW/cm2), the water permeability was ~1300 L μm/(hr m2 bar) and the MWCO was also lower, ~650 kDa. The total permeate flux of pure water was almost equal to that of the BSA containing solution and oil/water emulsion containing solution, indicating that crosslinked NVP films resist fouling by oil droplets and BSA under the conditions studied. The oil rejection was ~97% in all samples, while the protein rejection depended on the prepolymer water content, crosslinker concentration, and UV intensity.

DEVELOPMENT OF NOVEL HYDROPHILIC BACTERICIDAL MEMBRANES

*Ramamoorthy Malaisamy[1], David Berry[2], Diane Holder[2], Lutgarde Raskin[2], and Kimberly L. Jones[1]**
Howard University, Washington, DC, USA.
[1]Corresponding and Presenting (Email: malaisamy@gmail.com)
*[1]*Corresponding (Email: kljones@howard.edu)*
[2]University of Michigan, Ann Arbor, MI, USA

Introduction

Microfiltration membranes have been used in water and wastewater treatment for integration in membrane bioreactors, to remove particulates, and as pretreatment for reverse osmosis or nanofiltration membranes. Although membranes can produce very high quality effluent, fouling limits more widespread usage and significantly increases the cost of a membrane based water treatment system. Inorganic, organic and colloidal fouling can be significantly reduced by removing the foulant particles or molecules; however biofouling is much more difficult to control. Microorganisms grow as long as there is a food source available, so even trace levels of the microorganisms would result in the eventual growth and formation of a biofilm. Consequently biofouling will follow[1].

Research has been conducted during the last decade on improving the bactericidal qualities of membranes in order to reduce biofouling[2-5]. Grafting of quaternary ammonium polymers onto membrane surfaces[6,7] has yielded promising results in resolving this issue. In addition to increasing bactericidal properties, modifying membrane surfaces by grafting quaternary ammonium polymers may also reduce organic fouling. This study will address additional issues related to this type of membrane modification such as the exact mechanisms for long term biofouling control and the effect of the modifications on other performance variables such as organic fouling, rejection and flux.

Goal, Objectives and Tasks

Our goal is to create membrane surfaces that will be resistant to bacterial growth and biofouling. Our approach is to surface modify commercially available microfiltration membranes by UV assisted graft-co-polymerization of an antimicrobial monomer. The modified antimicrobial membranes should not be significantly less permeable or more prone to organic fouling than the virgin membranes. We framed the following objectives and tasks to meet this goal.

Objective 1: Identify optimal conditions and compositions of monomer ([2-(Acryloyloxy)ethyl] trimethyl ammonium chloride - AETMA and Acrylic Acid - AA) for the surface grafting on polyethersulfone membranes, by UV-assisted graft-co-polymerization.

Objective 2: Evaluate the anti-biofouling properties of the modified membranes.

Objective 3: Compare the overall performance properties of modified and unmodified membranes.

Methodology

We have chemically modified commercially available microfiltration polyethersulfone polymer membranes using UV-initiated graft-co-polymerization of a reactive quaternary ammonium derived acrylic acid monomer, [2-(Acryloyloxy)ethyl] trimethyl ammonium chloride and acrylic acid.

Membrane surface modification by UV-initiated grafting

In order to carry out the reaction, commercial PES membranes of 0.1μM pore diameter were cut into coupons of 44.5 mm diameter and soaked in pure water overnight to remove wetting agents applied by the manufacturers. The membranes were dried in air oven at 37± 3°C for several hours and weighed. Monomer type and irradiation times were varied on an attempt to optimize the modification conditions for a high degree of grafting to reduce bacterial growth without negatively impacting flux.

- Wavelength of light : 300nm
- Concentration: AETMA: 1M; AA:0.3M
- Irradiation time: AETMA - 5, 10 and 15 minutes; AA – 8 minutes

Degree of grafting

After the polymerization reaction was complete, the membranes were thoroughly rinsed in water and dried at a temperature controlled air oven for 4-6 hours at 37± 3°C. The initial weight before the modification and the weight after the modification were measured to determine the degree of grafting. A duplicate of all the experiments were done and the values averaged.

FTIR Spectroscopy

The modified dried membranes were scanned on an ATR crystal in an ATR-FTIR spectrophotometer in order to identify the specific functional groups of the monomers and to confirm the graft modification.

Permeate flux measurements

The modified and unmodified membrane coupons were loaded into a stirred cell and hydrolytically compacted at 40 psi for 1 h. The applied pressure was then reduced to the operating pressure, 10 psi. Permeability was measured at each pressure value.

E. coli suspension, filtration and assays

EGFP tagged *Escherichia coli* K12 was grown overnight in Luria Bertani (LB) broth and was re-suspended in PBS buffer and diluted to a concentration of approximately 10^6 cells/mL and filtered through the membranes. After a prescribed concentration of *E.coli* was filtered, the water flux through the membrane before and after *E. coli* filtration was measured in order to determine the flux recovery. Following the filtration experiments, fluorescence microscopy was used to determine the presence of GFP labeled *E. coli* on the membranes. Pure water flux was measured after rinsing (with pure water) and chemical cleaning (chlorine solution of 10 ppm free chlorine) following the *E.coli* filtration [8].

Results and Discussion

Membrane modifications

Figure 1 shows the degree of grafting of AETMA monomer on the PES membrane support with different irradiation times and monomer concentrations. When the time of irradiation was at 5 min, the degree of modification was 44 μg/cm^2. Upon increasing the irradiation time to 7, 10 and 15 minutes, the degree of grafting increased almost linearly to 133, 257 and 415 μg/cm^2, respectively.

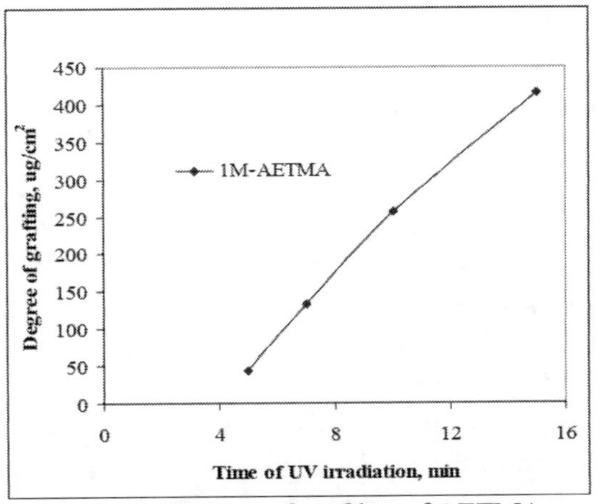

Figure 1. Degree of grafting of AETMA on PES membranes. Irradiation was at 300 nm.

We observed a similar linear trend in acrylic acid system, with an increase in the monomer concentration. Although the reaction appears to be linear with respect to concentration and time, this trend may be only in the lower initial monomer concentration range. We will confirm this behavior in a wider initial concentration range.

FTIR Spectroscopy

FTIR spectroscopy further confirmed the membrane modification with the monomer. Figure 3 is a spectra showing the peaks corresponding to the functional group (carbonyl, =C=O) of the AETMA monomer at 1740 cm^{-1}.

Permeate flux

The legends on the graph indicate the different stages of the filtration series. They are: (1) pure water flux of a pre-compacted membrane. (2) *E. coli* suspension flux. (3) pure water flux of the *E. coli* fouled membrane after rinsing in water. (4) pure water flux of *E. coli* fouled membrane after cleaning in 10ppm chlorine solution for 30 minutes.

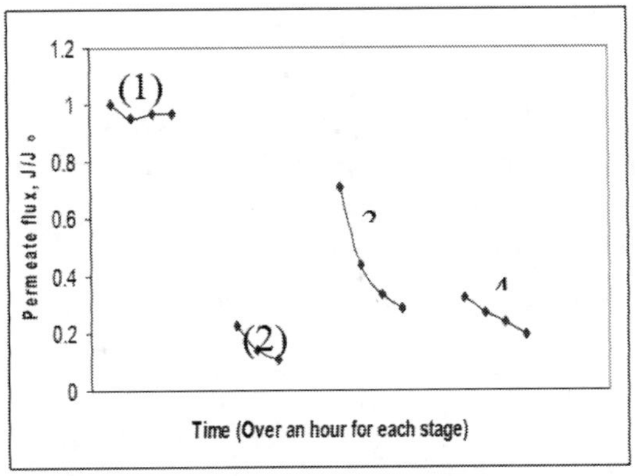

Fig. 2. Various stages of filtration for unmodified PES at 10 psi.

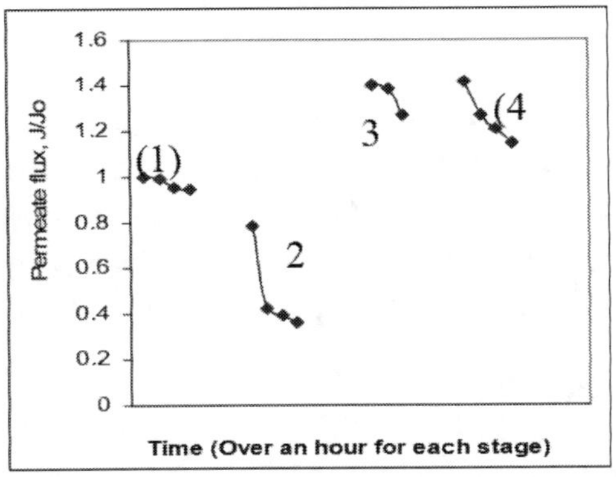

Fig. 3 Various stages of filtration for 1M AETMA grafted PES (10 min) at 10 psi

The *E.coli* suspension flux was only 10 % of the steady state pure water flux for unmodified membrane (Fig.2). Upon rinsing with water we were able to recover the flux by removing the surface-adsorbed cells to some extent, however the chlorine cleaning was not useful. Thus, the irreversible fouling is relatively higher for the unmodified membrane.

For modified membranes with relatively shorter irradiation time (5 min), the *E.coli* flux is considerably lower than the pure water flux and there is only a minimal flux recovery upon water rinsing and chemical cleaning, which is sensible because the amount of modified polymer (AETMA) is insufficient to act on the bacterial cells. However, when the irradiation time was increased to 10 minutes (Fig.3), the degree of grafting increased to about 250 $\mu g/cm^2$ and the *E.coli* suspension flux was higher than for the 5 min irradiation time. In addition, the flux recovery was remarkable and surpassed the original pure water flux. Though there is no considerable effect of chemical cleaning on the flux, it is exciting to note that the pure water flux of the membrane improved.

At 15min irradiation time, the *E.Coli* suspension flux increased over 3 fold from that of the original pure water flux. The other interesting finding with this modification is that the water rinsing step and chlorine cleaning step have almost no effect, which means the fouling may be negligible. In the following months, we will repeat these experiments and determine which surface properties of the modified membranes affect flux, rejection and fouling.

We also performed studies on the modification of PES membranes with acrylic acid monomer with various concentrations for comparison to the AETMA results. The results show that the *E.coli* suspension filtration is only about 20% of the pure water flux. However subsequent water rinsing and chemical cleaning have remarkably recovered the flux to 50%. When compared, the membranes modified with AETMA are better at maintaining flux following modification.

Anti-BioFouling Analysis

The live/dead staining showed a low fraction of dead cells for unmodified membranes (which was expected) and for 0.3M AA modified membranes. There was an increased proportion of dead cells in all modified membranes. Cell death percent maximum of about 75% was achieved.

Conclusion summary

We have successfully identified the optimal conditions for graft modification of PES membranes under UV light and modified the membranes with both AETMA and AA monomers on PES membranes and confirmed the modification. Preliminary results for the antimicrobial properties of the modified membranes are promising, however we need to reproduce the results and determine why flux recovery varies with degree of grafting and type of monomer.

Acknowledgement

This material is based upon work partially supported by the Center of Advanced Materials for the Purification of Water with Systems, a National Science Foundation Science and Technology Center, under Award No. CTS-0120978

References

1. Flemming, H-C. Schaule, G. Griebe, T. Schmidt, J. and Tamachkiarowa, A. (1997), *Desalination*, 113, pp. 215-225.
2. Pieracci, J. Crivello, J.V. and Belfort, G. (1999), *J.Membr.Sci.*, 156, pp. 223-240
3. Hyun, J. Jang, H. Kim, K. Na, K. and Tak, T. *J.Membr.Sci* (2006), 282, pp. 52-59
4. Xu, W. and Chellam, S. *Env.Sci Technol* (2005), 39, pp. 6470-6476.
5. Kim, S.H. Kwak, S.-Y. Sohn, B.-H. and Park, T.H. *J.Membr.Sci* (2003), 211 pp. 157-165
6. Hilal, N. Kochkodan, V. Al-Khatib, L. and Levadna, T.I. *Desalination* (2004), 167, pp. 293-300.
7. Kochkodan, V.M. Hilal, N. Goncharuk, V.V. Al-Khatib, L. and Levadna, T.I. *Colloid J.*, (2006), 68, pp. 267-273.
8. Herzberg, M. and Elimelech, M. *J.Membr.Sci* (2007), 295, pp. 11-20

Novel Yeast & Oil Drop Microfiltration Equipment

Michael T. Stillwell, Wiwit Sumritwatchasai, Richard G. Holdich, Serguei R. Kosvintsev

Introduction

The conventional microfiltration of yeast and oil is problematic due to irreversible fouling on the membrane surface and within the internal pore structure. These problems result in the need for high shear at the membrane surface, which entails:

- Higher operating costs
- Periodic replacement of deteriorated filters
- Cell or drop damage, affecting product quality and filtration performance

A new type of filter is commercially available, which is a true surface filter with no internal structure, where each pore forms a direct channel of uniform size from one side of the membrane to the other. These membranes can be used in an oscillating filtration system, which provides a high peak shear directly at the membrane surface. The benefits of this system are:

- High permeate fluxes
- Low operating pressures
- A gentle process not involving excessive shear forces
- Greater membrane longevity
- Lower operating costs

The oscillating microfiltration system has recently been tested in a consortium project involving three large oil companies. The aim was to filter Produced Water from oil rigs and reduce the oil concentration in the water from over 1000 ppm to 30 ppm in order to comply with the new environmental pollution limits. The system was capable of meeting the 30 ppm legal requirement and provided all of the above advantages including being a compact lightweight unit that operated at low noise levels.

In this work, filtration tests have been carried out to validate the system at different shear rates by varying the frequency and amplitude of oscillation for calcium carbonate, yeast and oil challenge suspensions. The membrane rejection was determined at different shear rates for each challenge suspension. It was found that high shear rates provide higher critical fluxes because the membrane surface is kept cleaner. However, the rejection decreases at higher shear rates because more of the smaller particles can pass through into the permeate. Experimental data relating the permeate pressure to the filtration time were compared to a predictive pore blocking model for constant rate filtration. The model was found to fit the experimental data well at higher shear rates but was poor at the low shear rates. It is suspected that cake filtration occurs at the lower shear rates, as there is not enough shear stress to keep the membrane surface clean. Simultaneous development work has been performed to develop a composite membrane coating, which is designed to improve the membrane rejection for the smaller particle sizes, whilst retaining the advantages of the system stated previously.

The Equipment

The oscillating system was provided by Micropore Technologies Ltd., Leicestershire, UK. A tubular metal membrane with active area length 65 mm and diameter 14 mm was used with a slotted pore geometry of 400 microns length and 6 microns width (Figure 1).

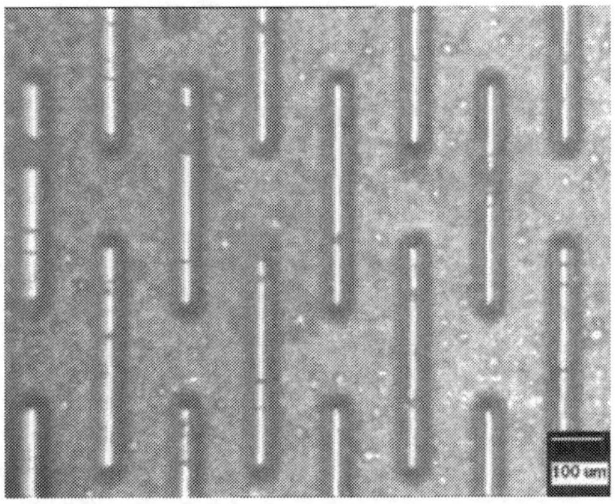

Figure 1 Microscope image of the tubular membrane used in the oscillating filtration system with pore dimensions of 400 microns length and 6 microns width.

The membrane had a durable low surface free energy coating in order to minimise fouling during filtration. The membrane was attached to an electrically driven oscillating unit, which could be controlled for frequency and amplitude (Figure 2).

The membrane was inserted into a beaker containing the challenge suspension, which was agitated using a magnetic stirrer. The permeate was drawn through the membrane pores using a peristaltic pump. Flux measurements were made using a measuring cylinder and stopwatch, and pressure measurements were taken from the gauge located upstream of the pump. For each challenge suspension, the shear rate was increased from 249 s^{-1} to 7761 s^{-1} by altering the frequency and amplitude.

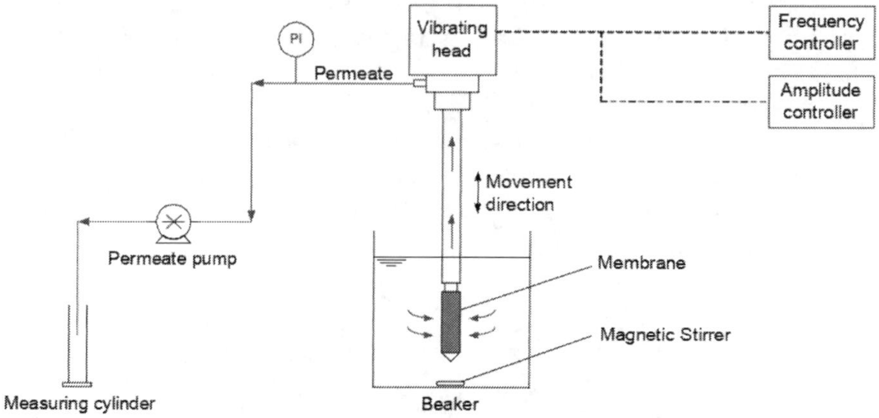

Figure 2 The experimental equipment to measure the filtration performance of the oscillating membrane.

The feed and permeate size distributions were measured using a Coulter Multisizer. This device allows the mass of particles in each size grade to be calculated and thus the rejection can be calculated according to the following equation:

$$rejection = \left(1 - \frac{mass\ of\ particles\ in\ permeate\ grade}{mass\ of\ particles\ in\ feed\ grade}\right) \times 100$$

Grade Efficiency Results

Calcium carbonate suspensions, yeast suspensions and dispersions of crude oil in water were prepared using reverse osmosis water, the details of which are contained in Table 1.

Table 1 Data for each challenge suspension.

Suspension	**Mean Size, microns**	**Concentration, g/litre**
Calcium Carbonate	2.1	1.0
Crude Oil	3.7	1.0
Yeast	3.9	0.1

The feed and permeate from each experiment were analysed using a Coulter Multisizer to determine the rejection efficiency as a function of particle size.

For calcium carbonate and yeast particles larger than 6 microns, 100% rejection was obtained for all shear rates, which is expected for a 6 micron pore width. However, the affects of oil droplet deformation at the membrane surface were observed as the rejection was only 97% for droplets 6 microns in diameter at the maximum and minimum shear rates. For an additional ten fold increase in the permeate pump setting, the rejection of oil droplets at 6 microns decreased to 90% at the maximum and minimum shear rates, which provides further evidence of droplet deformation. A comparison of other rejection values for each suspension at a particle size of 2.5 microns is shown in Table 2 for the maximum and minimum shear rates.

Table 2 Summary of the rejection data for each suspension at a particle size of 2.5 microns for the maximum and minimum shear rates.

Suspension	Rejection, %	Shear rate, s-1
Calcium Carbonate	45	7761
	85	249
Crude Oil	77	7761
	88	249
Yeast	0	7761
	90	249

Table 2 shows that membrane rejection for the smaller particles is dependent upon the shear rate. For calcium carbonate particles sized 2.5 microns, only a 45% rejection was obtained at the highest shear rate of 7761 s^{-1}, whereas a rejection of 85% was obtained at the lowest shear rate of 249 s^{-1}. This trend occurs also for the crude oil and yeast suspensions. This demonstrates that the membrane surface is kept relatively cleaner at the highest shear rate, which prevents a cake layer from forming and allows more of the smaller material into the permeate. This is demonstrated quite clearly by the 0% rejection of yeast particles at the highest shear rate of 7761 s^{-1}, which had a low feed concentration of 0.1 g/litre and contained only 18% of the total particles in the calcium carbonate suspension.

Pore Blocking Model

A pore blocking model has been proposed for constant pressure filtration (Filippov et al. 1994), which has been modified to represent a constant rate process. The model is based on a sieve mechanism that takes into account the membrane pore size and particle size probability distribution functions. By assuming that the permeability at time infinity, k_∞ = 0, the model can be written as:

$$J(t) = \frac{k_o}{\frac{1}{\Delta P} + k_o \cdot y \cdot c \cdot t}$$

Where k_o is the initial permeability (m^3/s.kPa), "y" is the blocking area (empirical constant), c is the concentration of the suspension (kg/m^3) and "t" is the time (s).

The model predictions were compared to experimental permeate pressure measurements for a challenge suspension containing 0.2 g/litre calcium carbonate and 0.457 g/litre yeast. The model predictions were obtained by adjusting the empirical value "y" for each shear rate until the best fit was obtained. An example at the highest shear rate of 7761 s^{-1} is shown in Figure 3.

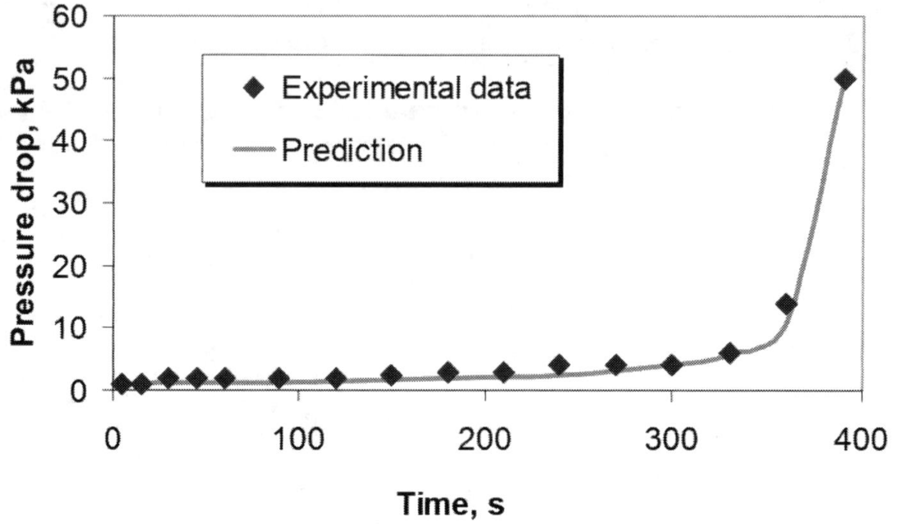

Figure 3 Pore blocking model predictions compared to experimental results indicating a good agreement at high shear rate ($7761 s^{-1}$).

Figure 3 shows that there is a good agreement between the model and experimental data, which is also obtained for the next two highest shear rates of 6468 s^{-1} and 5174 s^{-1}. At a shear rate of 4225 s^{-1}, the fit begins to deteriorate and the agreement becomes poor. This indicates that pore blocking occurred at the highest shear rate, but at low shear rates the fouling mechanism probably occurred via a cake filtration mechanism.

Membrane Coating Development

The minimum slotted pore width is currently 4-5 microns due to manufacturing limitations, which means that it is not possible to achieve a 100% particle rejection below 5 microns. The data shown previously also suggests that as the shear rate at the membrane surface increases, the rejection decreases because a cleaner membrane surface allows more of the smaller particles through into the permeate. In order to address these problems, a composite membrane has been developed which involves attaching a thin layer of particles onto the surface of the filter to create a permanent secondary membrane. Filtration is therefore performed entirely by the secondary membrane and the metal filter acts purely as a support to the coating. To demonstrate this concept, a metal filter with relatively large slot dimensions of 35 microns width and 800 microns length was used in filtration studies.

Commercially available glass micro-spheres shown in Figure 4a were deposited onto 42 mm diameter flat disc membranes and sintered under pressure to create a thin and robust porous coating (Figure 4b).

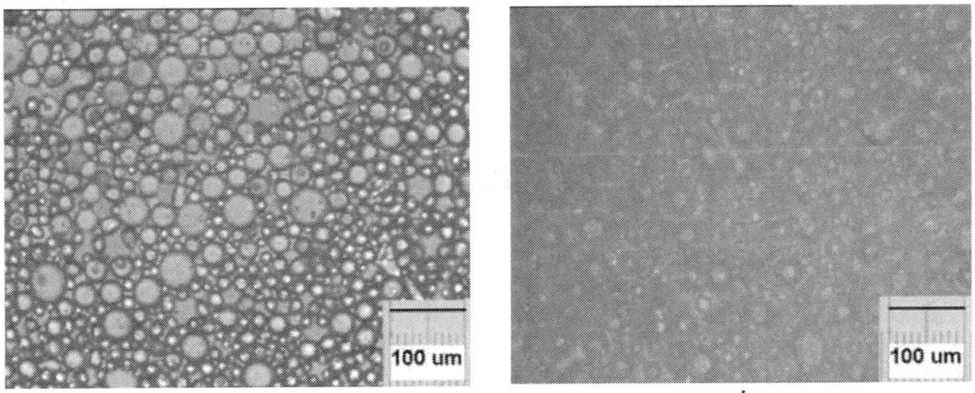

a b

Figure 4 (a) the commercially available glass microspheres used to form the top coating and (b) an example of a sintered coating ready for filtration tests.

The filtration performance of the sintered discs were tested using a dead-end filtration device known as a stirred cell. Two types of calcium carbonate challenge suspension and a yeast challenge suspension were prepared using reverse osmosis water, the details of which are contained in Table 3.

Table 3 Details of the challenge suspensions used with the composite membranes.

Suspension	Mean Size, microns	Concentration, g/litre
Calcium Carbonate (Large)	7.5	0.5
Calcium Carbonate (small)	3.0	0.5
Yeast	4.3	0.1

Low concentrations were used to minimise fouling in order to determine the approximate pore size of the membrane sample. A high shear rate of 5500 s^{-1} was used in each experiment in order to minimise fouling. The permeate pressure and flux trends are shown in Figures 6a and 6b.

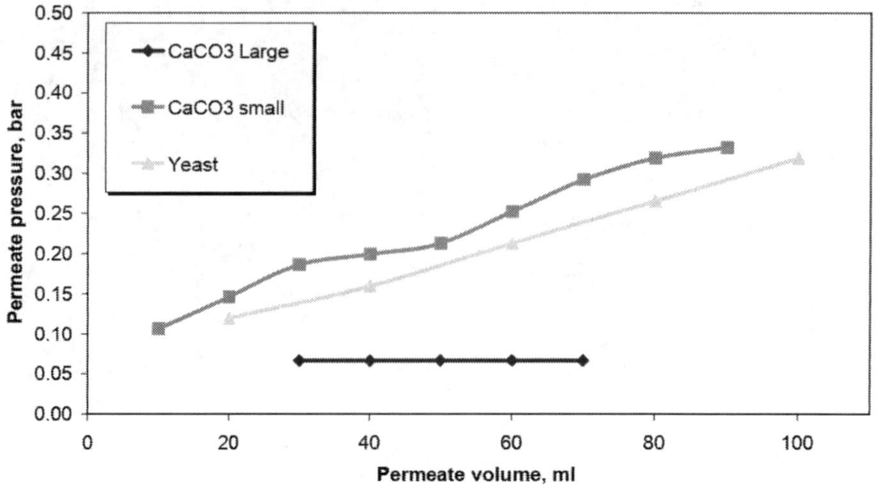

a

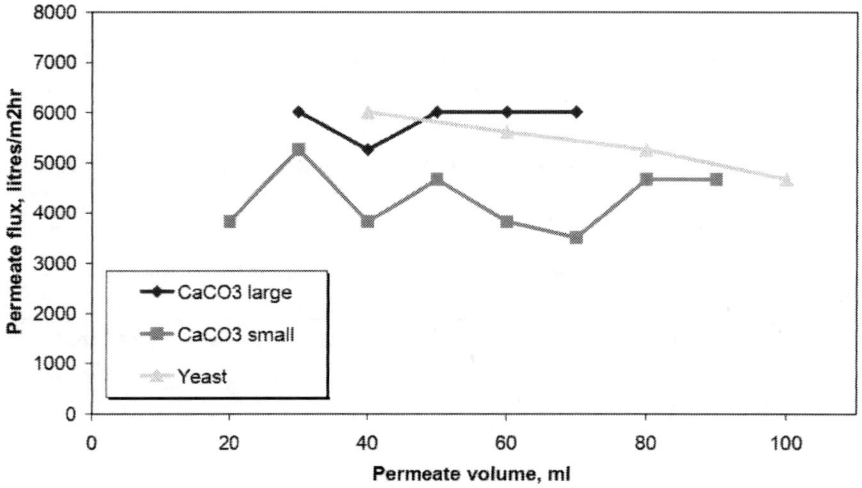

b

Figure 5 (a) Permeate pressure and (b) permeate flux trends as a function of permeate volume for a composite sintered membrane using different challenge suspensions.

Figures 6a and 6b show that the large calcium carbonate suspension had a constant permeate pressure of less than 0.07 bar and a constant permeate flux of 6000 litres/m^2.hr. This indicates that a cake layer was not formed on the membrane surface during the filtration. The permeate was clear and a microscope analysis could not detect any particles indicating a 100% rejection of suspended material.

Under the same conditions of pump speed setting and shear rate, the permeate pressure for the small calcium carbonate suspension gradually increased to 0.3 bar. The permeate flux was lower than for the large calcium carbonate and showed a slight decline over time with a value of approximately 4500 litres/m^2.hr. This indicates that particles in the feed became trapped inside the internal pore structure of the sintered membrane coating during the filtration. The measured rejection was approximately 90% for particle sizes down to 2.7 microns, but the feed particles deposited inside the sintered pore structure would have contributed to this rejection value. This result still provides useful information about the internal pore size of the sintered coating prepared under the particular set of bake conditions including temperature, time and pressure. Composite coatings that were not sintered to the same extent as the one reported here could easily pass all of the small calcium carbonate particles in the feed and provided no resistance to their flow.

The permeate pressure and flux trends for yeast are similar to those of the small calcium carbonate, but they also indicate that less internal blocking of the sintered coating occurred compared to the small calcium carbonate. This could be attributed to the deformable nature of the yeast under the high flux conditions, which pulled more yeast particles into the permeate. The measured rejection was approximately 85% for particle sizes down to 2.7 microns.

These results suggest that the concept behind the composite coating is working, because the internal pore structure of this particular membrane sample could trap particles down to 2.7 microns using a 35 micron slotted pore substrate as a support. Work is currently underway to reduce the sintered pore size further in order to carry out more extensive tests on yeast filtration, which is known to be a problematic challenge suspension even for certain types of surface filter (Chandler, Zydney 2006).

Conclusions

This investigation has shown that high shear rates are successful at minimising surface fouling, but this also leads to a decrease in the membrane rejection. It is also possible for deformable material larger than the slotted pore width to enter the permeate. At high shear rates, a pore blocking model for constant rate filtration can accurately represent the membrane fouling. At low shear rates, membrane fouling is attributed to a cake filtration mechanism. Initial development work to overcome the problems of membrane rejection has focused on the membrane coating. A composite coating that involves forming a permanent secondary membrane over the top of the metal surface filter shows good signs for future development work.

References

Chandler, M. & Zydney, A. 2006, "Effects of membrane pore geometry on fouling behavior during yeast cell microfiltration", *Journal of Membrane Science,* vol. 285, no. 1-2, pp. 334-342.

Filippov, A., Starov, V.M., Lloyd, D.R., Chakravarti, S. & Glaser, S. 1994, "Sieve Mechanism of Microfiltration", *Journal of Membrane Science,* vol. 89, no. 3, pp. 199-213.

306a Exploration of Dual-Layer Sulfonated Polyethersulfone/ polyethersulfone Hollow Fiber Membranes in Protein Separation

Yi Li and Tai-Shung Chung. Department of Chemical and Biomolecular Engineering, National University of Singapore, 10 Kent Ridge Crescent,, Singapore, 119260, Singapore

Proteomics is the large-scale study of proteins - the ultimate products of genes. By understanding proteins, researchers hope to understand the biochemical mechanisms of disease, leading to the development of drugs with an improved safety profile and a greater probability of clinical trial success. To gain a comprehensive understanding of protein function and regulation, researchers must separate proteins, then identify and characterize the expressed proteins. Unfortunately, the protein separation is a difficult process due to the complexity of proteins selves and their biological environments. The proteins are composed of numerous amino acids and the sequence of amino acids in each protein is unique to that protein; therefore, each protein has its own particular 3-D shape or native conformation. An amino acid is electrically neutral at its isoelectric point (pI) and has a positive or negative charge at lower or higher pH, respectively. To obtain a completely purified protein, many separation processes have been developed based on different protein properties, and they include chromatography, electrodialysis, electrophoresis, membrane chromatography and ultrafiltration (UF). Among them, the membrane-based protein separation processes may be potentially attractive ones.

The sulfonated polyethersulfone (SPES) material with a negative charge was applied to separate a bovine serum albumin (BSA) and hemoglobin (Hb) mixture through a electrostatic interaction mechanism in this work because these two proteins possess the very similar molecular sizes (66kDa vs. 65kDa) but different pI values (4.8 vs. 6.8). The dual-layer hollow fiber configuration was utilized in this work because SPES materials with a high degree of sulfonation (DS) (i.e. a high charge density) become highly hydrophilic and are normally difficult to precipitate and solidify in water and other common non-solvents, such as methanol, ethanol and isopropanol. Through continuous endeavor in developing the co-extrusion technology, the delamination-free dual-layer hollow fiber membranes consisting of a selective SPES outer layer and a porous supporting PES inner layer were successfully fabricated. These newly developed dual-layer SPES/PES hollow fiber membranes exhibit the best and most durable protein separation performance among all spun fibers in this work. They have a BSA/Hb separation factor of above 10 and last for almost 1 day, which is very attractive for practical applications of membrane technology in protein separation because the membrane separation performance normally decreases rapidly with a function of time induced by some reasons such as fouling and saturation of binding sites.

306b Purification Of Equine Immunoglobulin G (Igg) From Horse Serum Using An Enhanced Membrane Based Hybrid Bioseparation Technique

Lu Wang and Raja Ghosh. Chemical Engineering department, McMaster University, 1280 Main Street West, Hamilton, ON L8S 4L7, Canada

Equine immunoglobulin G (IgG) is widely used as antivenom. It is also used anti-rabies and anti tetanus agents. Purification of equine IgG is challenging due to the complexity of serum which contains large amounts of albumin. We discuss an enhanced hybrid bioseparation technique which is based on a standard hybrid bioseparation technique developed earlier in our group for the purification of human IgG. The technique gives high resolution and high throughput by combining ammonium sulphate induced protein precipitation, microfiltration and hydrophobic interaction based membrane adsorption. The enhanced technique overcomes some of the limitations of the standard hybrid bioseparation technique, i.e. antibody dilution and membrane fouling. The equine IgG purity and recovery were both significantly higher than those obtained with currently used commercial techniques

DYNAMIC CHARACTERIZATION OF AFFINITY MEMBRANES FOR MONOCLONAL ANTIBODY PURIFICATION

Simone Dimartino, Cristiana Boi and Giulio C. Sarti
Department of Chemical Engineering, Mining and Environmental Technology, University of Bologna, viale Risorgimento 2, Bologna, Italy,

Introduction

Membrane affinity chromatography is a novel and attractive protein purification technique that is receiving increasing attention for the downstream processes. It has been developed as an alternative to conventional chromatography in order to overcome several limitations due to packed bed columns, such as long process time, mass transport controlled by diffusion and high pressure drops [1-3].

In this work, the adsorption of human IgG onto SartoAIMs Protein A membranes has been studied. Complete chromatography cycles, including adsorption, washing and elution, have been experimentally analyzed both in batch and dynamic experiments, and the results are discussed.

Experimental

Materials and method

SartoAIMs Protein A membranes (Sartorius Stedim Biotech, Göettingen, Germany) are stabilized reinforced cellulose membranes coupled with Protein A; these membranes have a nominal pore size of 0.45 µm, an average thickness of 183 µm and a void fraction of 0.78.

Phosphate Buffered Saline, PBS pH 7.4, was used as equilibration and washing buffer; to obtain the IgG samples, PBS was also used to dilute a concentrated human polyclonal antibody solution purchased as Gammanorm from Octapharma (Stockholm, Sweden). A solution of 0.1 M glycine pH 3.5 was used as elution buffer.

Batch experiments have been performed by soaking the membranes in a IgG solution kept in gentle agitation using an orbital shaker. The protein concentration at the beginning and at the end of each experiment was measured by UV absorption at 280 nm by using a UV-Visible Spectrophotometer Shimadzu UV-1601 (Shimadzu Italia, Milan, Italy). The dynamic experiments have been performed using an FPLC AKTA Purifier 100 controlled by Unicorn software (GE Healthcare, Milan, Italy). In both experiments the membranes were cut into circles of 2.5 cm in diameter, for the dynamic runs a layered stack of membranes was housed in a flow cell.

Batch experiments

In order to check the adsorption behavior of SartoAIMs Protein A membranes, several batch adsorption experiments have been performed prior to the dynamic runs. The static binding capacity, SBC, was evaluated from a mass balance by calculating the difference between the initial value of protein concentration in solution and the value at equilibrium. The experimental data have been fitted with the Langmuir adsorption model [4], obtaining the values of maximum binding capacity, q_{max} = 5.46 mg/ml, and the Langmuir dissociation constant, K_d = 0.13 mg/ml. The experimental results together with the Langmuir adsorption isotherm are shown in Figure 1. These new SartoAIMs affinity membranes are characterized with a maximum binding capacity more than double with respect to previous Protein A based affinity membranes [5].

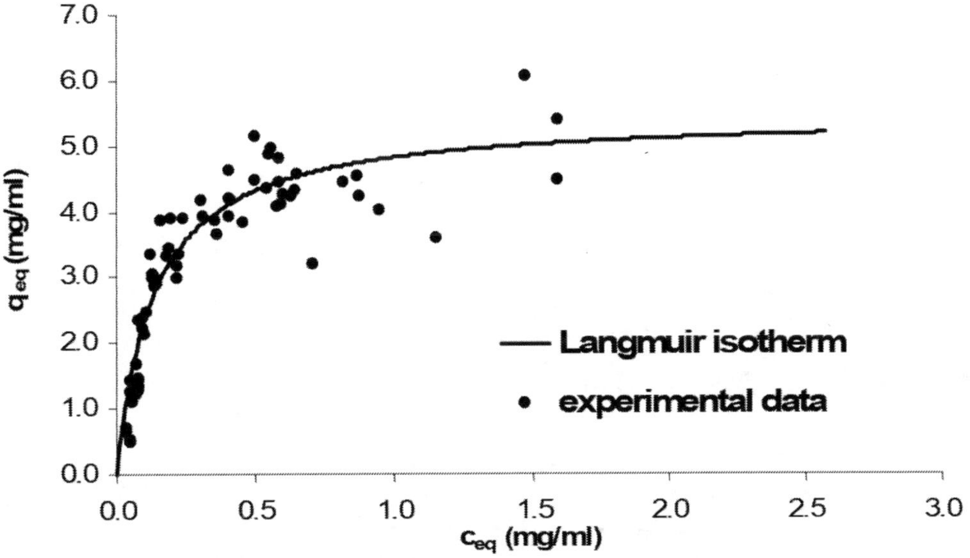

Figure 1. Equilibrium adsorption isotherm of human IgG on SartoAIMs Protein A membranes.

Dynamic behavior

The dynamic runs were performed by using pure Gammanorm solutions and consist of complete adsorption, washing and elution cycles. The membranes were regenerated after every fifth cycle, the relevant operating conditions are summarized in table 1. All cycles have been performed using a layered stack of ten membranes.

Table 1. Experimental conditions of dynamic experiments.

	Volume (ml)	Solution
Equilibration	10	PBS pH 7.4
Adsorption	Depending on the feed IgG concentration	human IgG solution diluted in pBS
Washing	20	PBS pH 7.4
Elution	10	Glycine 0.1 M pH 2.8
Re-Equilibration	10	PBS pH 7.4
Regeneration	30 (every 5th cycle)	1 M NaCl, 50 mM NaOH

A complete dynamic cycle at a flow rate of 2 ml/min and feed concentration of 0.3 mg/ml is shown in Figure 2. The breakthrough curve obtained shows a sudden increase in concentration for a short period, followed by a slower increase as the membranes approach saturation. Similar behaviors, with broadened breakthrough curves have been frequently reported in the literature [1-3].

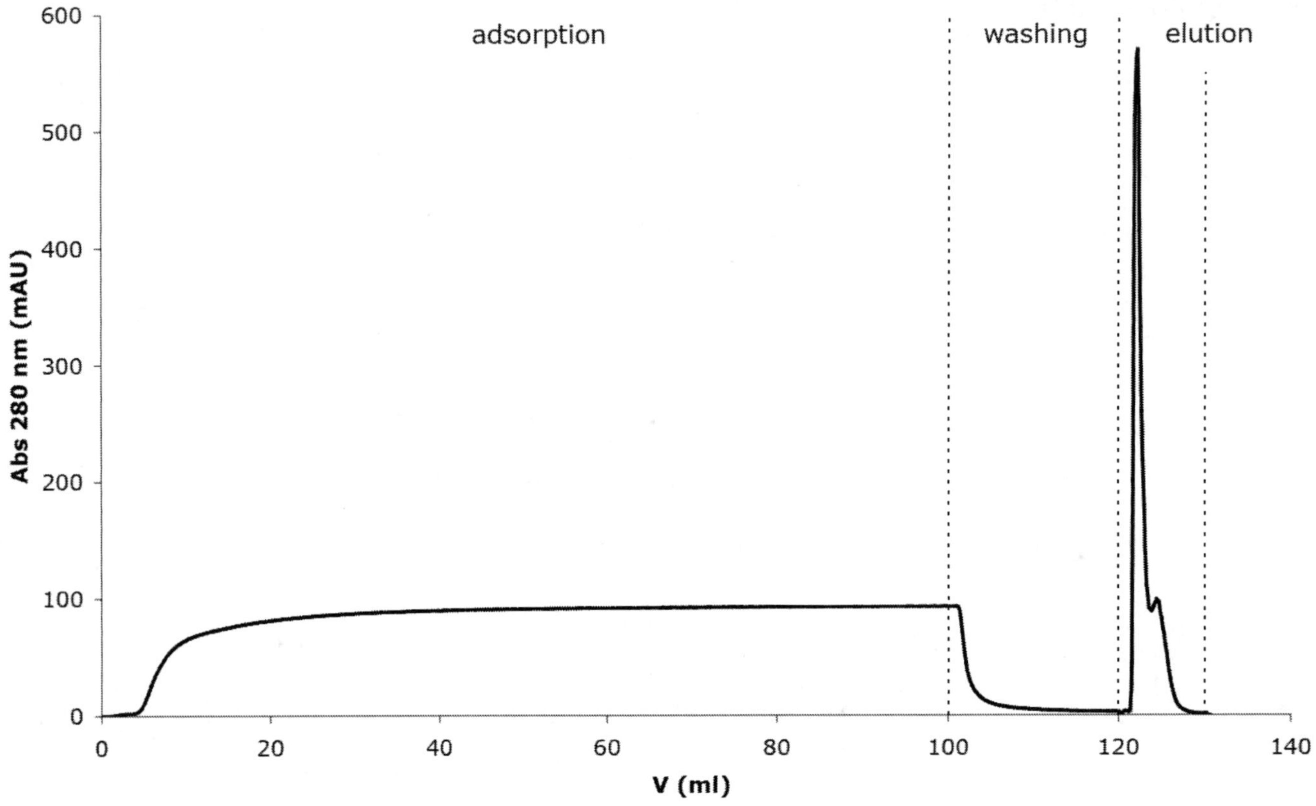

Figure 2. Complete adsorption cycle of polyclonal human IgG onto SartoAIMs Protein A membranes at 2 ml/min and 0.3 mg/ml.

Figures 3 – 5 show detailed examples of the different chromatographic steps of the cycles performed at two different operating conditions, 2 ml/min with $c_{0,IgG}$ = 0.3 mg/ml and 1 ml/min with $c_{0,IgG}$ = 0.5 mg/ml.

At the onset of breakthrough, a small increase in the absorbance signal was recorded up to the first 4 ml of feed volume. That is likely due to the presence of IgG_3 in the feed solution, an IgG subclass that is weakly retained by Protein A, and therefore it exits immediately from the membrane module while the other subclasses are still adsorbed.

The washing step has a fundamental role in this kind of separation processes, since during this stage the membranes are cleaned from non specifically adsorbed proteins. However, in order to economize buffer consumption, the washing step was ended when the absorbance signal reduced to 3% of the maximum value recorded in the adsorption step. The experimental results show that, for the experimental system under investigation, a washing volume of 20 ml is more than enough. That washing volume is definitely much smaller than the feed volume required for membrane saturation.

The elution step shows a first well defined peak, followed by a second broad peak. The reasons for this behavior are not simple and are currently under investigation.

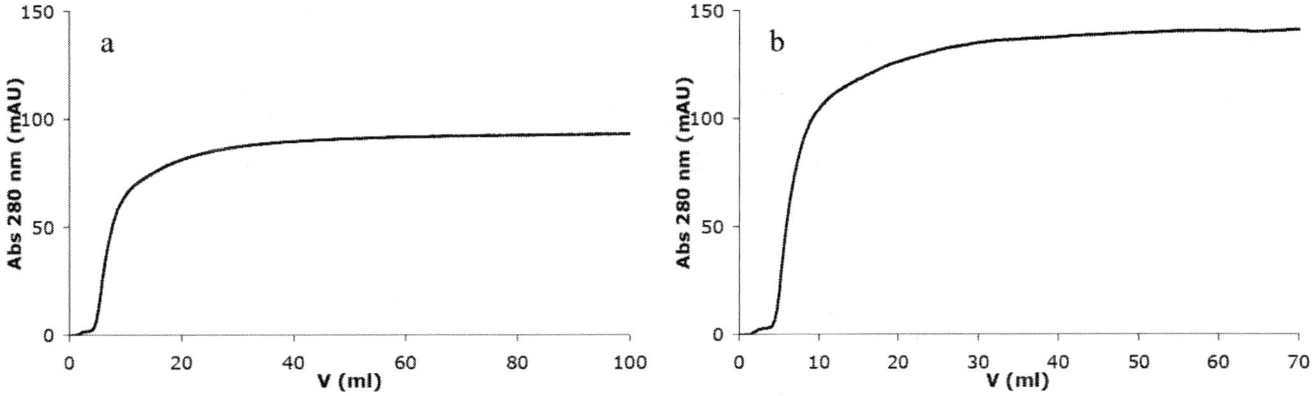

Figure 3. Loading step for the system human IgG - SartoAIMs Protein A membranes: (a) 2 ml/min and 0.3 mg/ml; (b) 1 ml/min and 0.5 mg/ml.

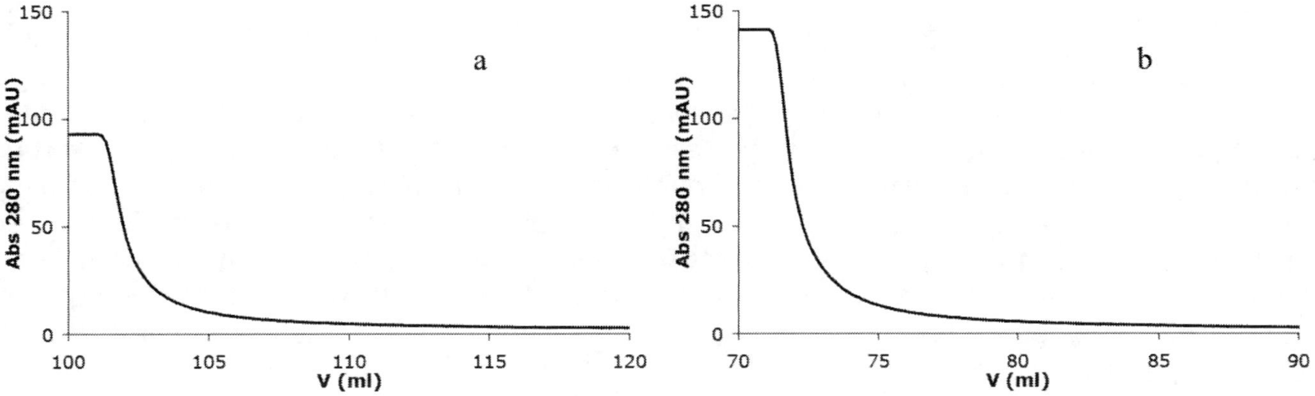

Figure 4. Washing step for the system human IgG - SartoAIMs Protein A membranes: (a) 2 ml/min and 0.3 mg/ml; (b) 1 ml/min and 0.5 mg/ml.

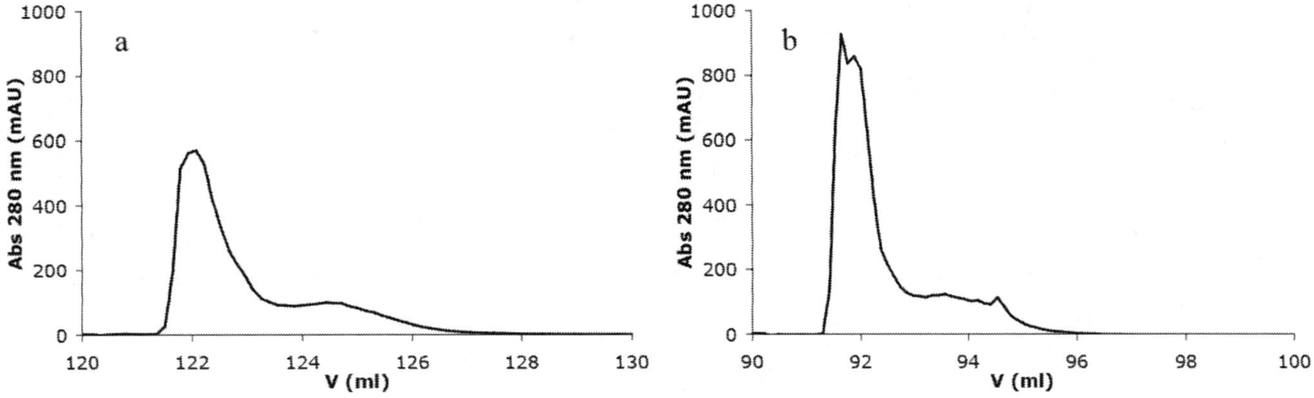

Figure 5. Elution step for the system human IgG - SartoAIMs Protein A membranes: (a) 2 ml/min and 0.3 mg/ml; (b) 1 ml/min and 0.5 mg/ml.

A typical mass balance for two different experimental runs is summarized in Table 2. The error between the amount of antibody adsorbed and eluted is below 6%, i.e. within experimental error. One can thus affirm that SartoAIMs Protein A membranes allow a recovery of nearly 100% under mild elution conditions, and no IgG is retained in the membranes after elution.

Table 2. Mass balance on the adsorption cycles.

Flow rate (ml/min)	$c_{0,IgG}$ (mg/ml)	Amount adsorbed (mg)	Amount eluted (mg)	Error %
2	0.3	2.76	2.87	-2.73
1	0.5	3.57	3.31	5.36

The dynamic binding capacities, DBC, and the relevant process yields are reported in Table 3. Chromatographic separations are generally operated up to the breakthrough point, defined as the point in which the outlet protein concentration equals 10% of the inlet concentration. The bed utilization is defined as the ratio between the amount of IgG adsorbed at the breakthrough point and the maximum amount adsorbed at the same operating conditions, namely $DBC_{10\%}$ / $DBC_{100\%}$. For the two cases investigated in this study, the bed utilization is around 44%, indicating that the membrane stack is far from being completely exploited at the breakthrough point. Nevertheless, the process yield at 10% breakthrough, i.e. the ratio between the protein adsorbed at 10% breakthrough and the protein fed at 10% breakthrough, results to be around 98%.

The static binding capacities that correspond to equilibrium concentrations of 0.3 and 0.5 mg/ml are 2.53 and 2.88 mg/ml, respectively. At the same concentration values the dynamic binding capacity is more than 1.5 times larger than the static binding capacity, indicating an increased accessibility of the Protein A active sites in dynamic conditions.

Table 3. Dynamic binding capacities and process yield of the adsorption cycles.

Flow rate (ml/min)	$c_{0,IgG}$ (mg/ml)	$DBC_{10\%}$ (mg/ml)	Yield @10% BT %	$DBC_{100\%}$ (mg/ml)
2	0.3	1.74	98	4.19
1	0.5	2.54	98	5.55

Conclusions

New affinity membranes with immobilized Protein A have been tested for adsorption of human IgG in batch and dynamic experiments. These SartoAIMs Protein A membranes are endowed with a maximum static binding capacity at least double than that of commercially available affinity membranes for human IgG purification.

An elution buffer at pH 3.5 allows complete recovery of the IgG immobilized on the membranes.

The dynamic binding capacity at 100% breakthrough is always higher than the static binding capacity measured at the same values of equilibrium concentration. In flow conditions a higher proportion of the membrane internal area is accessible to the IgG solution and ligand-protein interactions are more likely to occur.

Due to breakthrough broadening, the bed utilization at 10% breakthrough results around 44%; future work will be devoted to understand the reasons for such behavior and to obtain sharper breakthrough curves. However, it is worth noticing that at the breakthrough point the process yield is nearly 98%.

Acknowledgements

The "Advanced Interactive material by Design" (AIMs) project for funding under the Sixth Research Framework Programme of the European Union, and in particular Sartorius AG as project partner.

MIUR, Italian Ministry of Education, University and Research and the University of Bologna for partial financial support.

References

[1] C. Boi, S. Dimartino, G. C. Sarti, J. Chromatogr. A 1162 (2007) 24.
[2] H. Yang, M. R. Etzel, Ind. Eng. Chem. Res. 42 (2003) 890.
[3] O. P. Dancette, J. L. Taboureau, E. Tournier, C. Charcosset, P. Blond, J. Chromatogr. B 723 (1999) 61.
[4] I. Langmuir, J. Am. Chem. Soc. 40 (1918) 1361.
[5] C. Boi, G. C. Sarti, Development and Characterization of Affinity Membranes for Immunoglobulin Purification, in press Separation Science and Technology 2007.

306d Affinity Based Separation Of Proteins Using Functionalized Microfiltration Membranes

Saurav Datta[1], D. Bhattacharyya[1], and A. Nath[2]. (1) University of Kentucky, 177 Anderson Hall, Lexington, Kentucky 40506, (2) Johns Hopkins University, Department of Neurology and Neuroscience, Baltimore, MD 21287

Functionalized microfiltration (MF) membranes are becoming increasingly popular in biotechnology and pharmaceutical industries. Affinity based separation using functionalized MF membranes are emerging as a potential alternative to conventional bead-based column chromatography due to their hydrodynamic advantages. Functionalization of MF membranes offer choices of attaching the active groups or moieties on different types of materials (hydrophilic, hydrophobic), environments (external surface, internal pore surface) and dimensions (pore size varying from 0.05 to 10 *f*Ým).This study deals with functionalization of microfiltration (MF) membranes within the pore structures and their applications in the field of bio-separations. Wide pore structure of MF membranes allows easy access of the active groups/moieties to the target molecules, under convective flow condition. Therefore, the mass transfer resistance, and hence the processing time per unit volume of feed is less. This study describes functionalization of MF membranes with active groups. Once the pore surface of the membrane is activated, further attachments of groups, moieties or molecules (ligands) are carried out. In this research work, avidin is attached as a ligand. Then, biotin-tagged protein is separated from mixture, purified and isolated from the matrix. The quality of the isolated pure protein is analyzed and compared with that obtained by conventional bead-based column chromatographic separation. Characterizations of the functionalized membranes are also presented in this study. Characteristic fouling of the membrane matrix associated with bio-separations is also analyzed critically. Some, simplified mathematical model is presented to describe the transport of bio-molecules within the membranes.

306e Transport Of Proteins In A Nanopore

Leili Javidpour[1], M. Reza Rahimi Tabar[1], and **Muhammad Sahimi**[2]. (1) Department of Physics, Sharif University of Technology, Tehran 11365, Tehran, Iran, (2) Mork Family Department of Chemical Engineering & Materials Science, University of Southern California, Los Angeles, CA 90089

Transport of protein through nanoporous membranes is important to several problem, ranging from their purification to many biomedical applications. Molecular simulation of this phenomenon can shed light on the underlying physics of the problem and, hence, provide us with insight for practical applications. We have carried out such simulations in a slit pore, using the discontinuous molecular dynamics (DMD) technique, in order to study the stability and folding of the proteins in the pore. Using a novel combination of the DMD and the Langevin equation, we also study transport of proteins in nanopores. A wide range of temperature was considered, the effect of the pore's size on the folding temperature was studied, and both attractive and repulsive interaction potentials between the proteins and the pore's walls was considered. We have utilized an intermediate-resolution protein model, the so-called PRIME model.

306f High Sensitivity Binary Gas Integrity Test For Porous Membranes

Sal Giglia and Mani Krishnan. Millipore Corporation, 80 Ashby Road, Bedford, MA 01730

A novel non-destructive integrity test for microporous and ultraporous membranes has been developed that is far more sensitive to detecting defects than a conventional gas-liquid diffusion test. In a gas-liquid diffusion test, the measured diffusive flow rate through a wetted membrane is compared to a theoretical or empirically established flow rate for an integral membrane. A measured flow rate value that exceeds the specification for an integral filter signals the presence of one or more defects. However, the sensitivity of the diffusion test is compromised by the fact that the diffusive flow rate for an integral membrane spans a range that can, sometimes, be large when compared to the flow contributed by a defect, resulting in a relatively poor signal-to-noise ratio.

The test developed here uses a binary gas mixture and is based on the principle of differing gas permeabilities through the liquid layer that results in a concentration enhancement of the faster permeating gas. In an integral membrane, the permeate composition can be predicted based on the transport properties of the gases permeating through the liquid layer and the known operating conditions. A deviation from the expected concentration is an indication of the presence of a defect or open pores. Unlike the gas-liquid diffusion test, the binary gas test has low sensitivity to membrane porosity, liquid layer thickness, and membrane area. Consequently, integral devices will exhibit a relatively a narrow range of integral values resulting in a superior defect signal-to-noise ratio. In this study, a binary gas integrity test method was developed and applied to a newly developed virus clearance filter. The binary gas test was demonstrated to provide a significantly higher level of virus retention assurance compared to the air-water diffusion test, and can be implemented by the filter vendor as an additional quality assurance test prior to shipping the product to end users

AUTHOR INDEX

AUTHOR INDEX

AUTHOR INDEX

AUTHOR INDEX

AUTHOR INDEX

AUTHOR INDEX

AUTHOR INDEX

AUTHOR INDEX

AUTHOR INDEX

AUTHOR INDEX

AUTHOR INDEX

AUTHOR INDEX

AUTHOR INDEX

American Institute of Chemical Engineers
3 Park Avenue
New York, New York 10016-5991

ISBN: 978-1-60560-000-0

American Institute of Chemical Engineers

Separations Division

Presentations at the
2007 AIChE Annual Meeting

November 4-9, 2007
Salt Lake City, Utah, USA

Volume 2 of 2

American Institute of Chemical Engineers

Separations Division

Presentations at the
2007 AIChE Annual Meeting

November 4-9, 2007
Salt Lake City, Utah, USA

Volume 2 of 2

Printed from e-media with permission by:

Curran Associates, Inc.
57 Morehouse Lane
Red Hook, NY 12571
www.proceedings.com

ISBN: 978-1-60560-000-0

Some format issues inherent in the e-media version may also appear in this print version.

ISBN: 978-1-60560-000-0

American Institute of Chemical Engineers
Proceedings
Three Park Avenue
New York, NY 10016-5991
Phone: 212-591-8100

www.aiche.org

American Institute of Chemical Engineers

Separations Division
2007

TABLE OF CONTENTS

Volume 1

Volume 2

Author Index

Removal of Nanoparticles from Copper Chemical Mechanical Planarization Wastes via Novel Separation Methods

Eduardo V. Carrera, Kimberly Ogden* and Moses N. Bogere
Department of Chemical Engineering, University of Puerto Rico at Mayaguez, Mayaguez, Puerto Rico, 00681; *Department of Chemical and Environmental Engineering, University of Arizona, Tucson, AZ 85721

Abstract

Chemical mechanical planarization (CMP) wastewater emanating from semiconductor processing contains many compounds including copper (II) ions and alumina and/or silica particles abrasives with an average size of 100 nm. Recycle of CMP wastes requires treatment to remove copper, while disposal of treated wastewater requires that all potential pollutants including copper be removed. Coagulation and filtration were studied in order to identify the most viable and optimal way to employ coagulants using common methodologies for wastewater treatment to primarily remove nanoparticles.

It is difficult to separate nanoparticles from CMP wastewater through settling and conventional filtration. Chemical aggregation of nanoparticles was found to enhance the settling and filtration characteristics. The efficiency of four coagulant reagents was evaluated, and reagents were based on aluminum and iron salts, natural coagulant and polyelectrolytes with different electrical charge. Optimal operating conditions such as coagulant dosage, pH, type of coagulant, and sedimentation time were established. The parametric studies demonstrated the viability of coagulants to reduce the amount of particles in wastes. Over 95% turbidity reduction was achieved and less than 5 NTU (the recommended value for good quality water) was realized. Filtration characteristics of aggregated particles were also established.

1. Introduction

Chemical mechanical planarization (CMP) is a critical step in integrated circuit device manufacturing and involves the process of planing and smoothing the wafer surface by chemical and mechanical forces (Stanley & Ogden, 2003). CMP slurries are needed in microelectronics industry because wafer polishing depends on them to clean its surface and remove imperfections. Due to rapid growth, this industry consumes large volumes of ultra pure water in CMP slurries and post-polishing cleaning (Golden *et al.*, 2000). CMP slurries once used in the polishing processes pick up inorganic ions such as copper which is toxic to flora and fauna in aquatic systems.

Environmentally benign processing demands that used water and slurry such as CMP wastes be treated and recycled if possible. CMP wastewater contains nanoparticle abrasives such as alumino-silicates, copper and many other compounds (Golden *et al.*, 2000). There is a concerted effort to treat CMP wastes via biosorption (Stanley, 2003) and other novel filtration/biotreatment schemes (Padilla et al., 2006). Stanley established that efficiency of biotreatment/biosorption is significantly reduced due to fouling of biosorption bed by the nanoparticles. It has also been established (Padilla et al., 2006) that under certain conditions copper adheres to the nanoparticles also. A detailed discussion of other methods including electrocoagulation, crossflow filtration and ion exchange are provided elsewhere (Carrera et al., 2007).

This work focuses on removal of nanoparticles from cu-CMP wastewater by employing methods based on novel separation techniques such as coagulation and filtration. The main objective was to study the efficiency of five chemical coagulants and establish the operating parameters for efficient removal of nanoparticles before the biotreatment step. In this way, it is important to establish a strategy to reduce water consumption, explore the possibility of recycling water in non-process applications and comply with environmental regulations (U.S EPA) in order to completely degrade pollutants and lead to environmentally benign compounds as end products. The results of the present study may lead to a rapid, efficient, and economical method of treatment of Cu-CMP wastewater.

2. Experimental procedure

CMP wastewater samples were prepared from commercially procured iCue® 5001 slurry obtained from Cabot Microelectronics (Wisconsin, USA). Some samples were obtained from University of Arizona tool room. The raw slurry and different dilutions made of this slurry were characterized according to turbidity, total solids and pH using the standard EPA methods. Fresh CMP slurries in dilution did not contain copper II ions, so a standard cupric nitrate hydrate aqueous solution (Cu^{2+}) was added to these diluted samples in order to formulate different concentrations of this ion in the Cu-CMP waste samples. The wastewater turbidity (in terms of NTU) was measured using a LaMotte turbidimeter (model 2020 e/i). Turbidity was determined before and after chemical coagulation and coagulation/filtration experiments. An Orion 410A pH-meter was used to determine and adjust pH of the wastewater to be analyzed. A Phipps & Bird six-paddle stirrer (model 7790-400) was used to conduct jar test experiments and determine the optimal operating conditions for each coagulant.

The CMP wastewater samples were prepared by diluting raw commercial samples from using deionized water. A waste sample of 500 mL was obtained, yielding suspension turbidities of approximately 200 and 900 NTU. Three equal samples of these solutions were analyzed to determine total solids (TS) with gravimetric method EPA 160.3 and to measure turbidity (in NTU).

Five chemical coagulants were used to remove nanoparticles and copper by means of chemical aggregation (coagulation) in order to enhance the settling and filtration characteristics. Aluminum sulfate, ferric sulfate, chitosan, polyacrylamide and an anionic polymer (all supplied by Sigma-Aldrich) were employed in the coagulation tests. Chemical coagulation was conducted using a jar test apparatus. The apparatus consisted of a paddle stirrer with six 1-L glass beakers. 500 mL of CMP waste sample was put in a jar testing apparatus (containers/beakers) and was mixed thoroughly until complete homogeneity. Initial turbidity was measured and pH was fixed for the six samples. Different amounts of coagulant were added to each container and were rapidly mixed at 100 rpm for 1 min. Then the stirring speed was reduced to 25 rpm and continued mixing for 30 min. The pH was adjusted whenever necessary (only for inorganic coagulants which were pH dependent). The mixer was turned off and the samples were allowed to settle down for 30 min. Final turbidity was measured for each sample. This procedure established the optimal dosage of coagulant at a fixed pH and initial turbidity. But by changing these parameters, another optimal dosage was established. The same procedure was then followed for the other coagulants.

After the mixer (in the jar test) was turned off, the samples were filtered in a vacuum Buchner funnel using different types (pore size) of filter papers. Final turbidity was measured for each sample. Whatman # 2 qualitative (8 μm) and # 50 hardened (2.7 μm) filter papers were used for filtration tests as well as a Corning 430049 bottle top sterile filter (0.2 μm, nylon).

3. Results and discussion

3.1 Characterization of CMP wastes

Table 1 lists the main characteristics of some CMP slurries and copper CMP wastewater obtained from the University of Arizona tool room.

Table 1. Characteristics of CMP wastes

	Commercial CMP slurry	Silica CMP slurry	Surrogate CMP slurry	Cu-CMP waste (UA tool room)
Oxide particles	Al_2O_3	SiO_2	Al_2O_3-SiO_2	SiO_2
pH	8.32	8.66	6.88	7.78
TS (mg/L)	54480	380130	67130	5050
Turbidity (NTU)	>13500	1200	70	240
Copper (ppm)	-	-	20	185-190
Density (g/mL)	1.02	1.20	0.99	1.00
Color	Milky	Milky	Colorless	Faint blue

Slurry chemistries vary from process to process. However, for the studies undertaken, commercial CMP slurry (an alumina-based suspension; Ihnfeldt & Talbot, 2006), silica CMP slurry, surrogate CMP waste and Cu-CMP wastewater (University of Arizona tool room) were characterized Padilla (Padilla et al., 2006). Commercial CMP slurry contained fine oxide particles mainly aluminum oxide (alumina, Al_2O_3) and had a milky color. Silica CMP slurry was composed of fine suspended oxide particles mainly silicon dioxide (silica, SiO_2), and had a milky appearance. Surrogate CMP waste prepared from procured laboratory chemicals (Siddiqui et al., 2005) had a clearer appearance and consequently a low turbidity and contained alumino-silicate nanoparticles. Cu-CMP of University of Arizona tool room was different from the other samples, and had a faint blue and milky color which indicated the presence of copper in solution and fine suspended oxide particles.

Commercial and silica CMP slurries had high total solids (TS) contents of 54480 and 380130 mg/L, respectively, being much higher than 5050 mg/L of Cu-CMP waste (UofA tool room). This significant difference was due to the dilution of the CMP slurries with ultra purified water (UPW) and streams that contained organic compounds during wafer washing and cleaning of CMP process. The lower turbidity of Cu-CMP waste (240 NTU) as compared to that of commercial CMP slurry (>13500 NTU) and silica CMP slurry (1200 NTU) was also due to the fact that slurries were diluted during the polishing procedure.

After wafer cleaning and polishing to reduce wafer topological imperfections, CMP slurries dragged copper ions and some organic compounds. For this reason, there was no copper content in original slurries and thus, were different from typical Cu-CMP waste (Fig. 1). Also this practice slightly reduced pH of the suspensions.

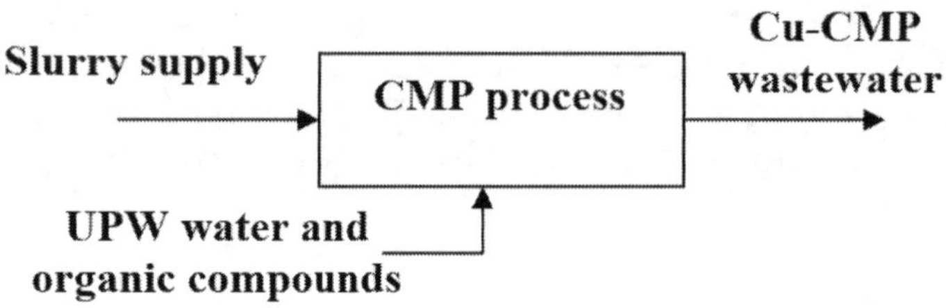

Fig. 1. Schematic diagram of CMP process

Previous experiments (Ihnfeldt & Talbot, 2006) demonstrated that between pH values 6.5 and 9.5 the alumina is negatively charged owing to the measured isoelectric point (IEP) of the iCue® 5001 Cabot alumina-based slurry being pH ~6.5 whereas the IEP of α-alumina has been published as 9.2 in an aqueous dispersion. It means that at low pH values (less than 6.5), alumina particles are positively charged in the slurry.

3.2. Chemical coagulation

In the chemical coagulation process, the nano-sized suspended particles were first aggregated by the addition of chemical coagulants (coagulation step) and then, the aggregated particles formed large, dense and rapid-settling flocs during flocculation step.

It proved difficult to reproduce the initial turbidity for CMP waste samples, therefore, instead of turbidity the results were represented in terms of percentage residual turbidity which was used to represent the amount of nanoparticles present in suspension with respect to the initial amount in the original sample before coagulation.

In order to determine the lowest coagulant dosage at the minimum value of residual turbidity, data was obtained (in graphical form; not shown here but can be found in Carrera et al, 2007) of percentage of residual turbidity versus coagulant dosage alum, ferric sulfate, chitosan, polyacrylamide and anionic polymer. Several replicates of each run were made and the average values reported. The optimum operating conditions established for each coagulant are summarized in Table 2.

Table 2. Optimum operating conditions for coagulation process

Coagulant	Coagulant dosage (mg/L)	pH	Initial suspension turbidity (NTU)	Settling time (min)
Alum	100	5.0	900	18
Ferric sulfate	50	5.0	600	18
Chitosan	1.6	7.0	200	5
Polyacrylamide	315	7.0	190	25
Anionic polymer	75	7.0	190	2

Alum, an inorganic coagulant and widely used in water treatment, effectively removed nano-sized particles at pH of 5.0, 100 mg/L dosage and initial suspension turbidity of approximately 900 NTU with percentage residual turbidity (% R.T.) of 0.79 %. The lowest residual turbidity was established at 100 mg/L dosage at pH=6.5 and ~ 900 NTU as shown in Fig. 2.

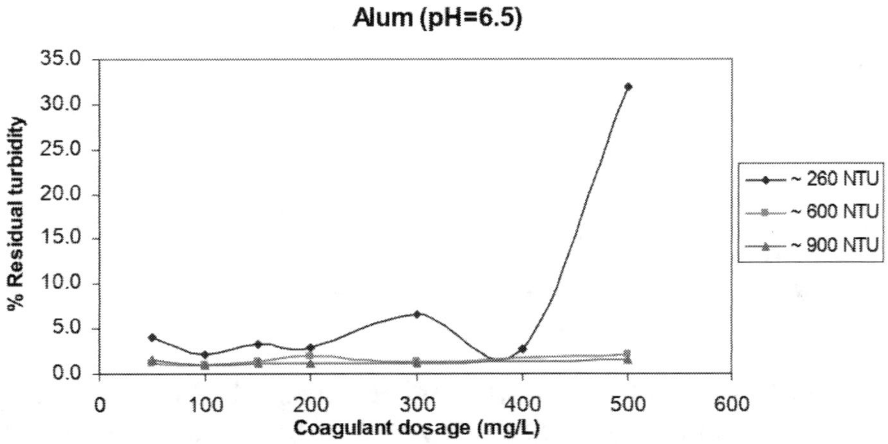

Fig. 2. Effect of aluminum sulfate dosage on the removal of turbidity at pH=6.5 and at different initial suspension turbidities

Since, 100 mg/L dosage and ~900 NTU were the optimal results for alum at pH=6.5 (with %R.T.= 0.88), these conditions were repeated at pH = 5.0 and pH=8.0 yielding a percentage residual turbidity of 0.88 and 47.10 respectively. This explains why a pH of 5 is the optimum one for alum. There were some clear advantages of using alum: ease of handling, not being hygroscopic and not changing color of suspension. On the other hand, alum produced large volumes of sludge since about 100 mg/L was needed.

Ferric sulfate is an inorganic coagulant, which was not as effective as alum. The results established that the lowest residual turbidity was 50 mg/L dosage for pH=6.5 and initial suspension turbidity of ~600 NTU. Ferric sulfate had a good removal of solids with a lower coagulant dosage as compared to alum. On the other hand, it was difficult to handle and humidified rapidly (hygroscopic), and this coagulant changed the color of suspension to yellowish which indicated a poor quality of the treated wastewater.

Chitosan is a natural cationic coagulant. The established optimal conditions were at pH=7.0, 1.6 mg/L dosage and initial suspension turbidity of ~200 NTU with % R.T. = 0.30 %. Experimental results at different initial suspension turbidities are given in Fig. 3. It is clearly observed from the figures that chitosan yields the lowest residual turbidity for the highest colloid concentration but with the highest chitosan dosage. That was due to the fact that the greater the presence of solids in suspension the more chemical coagulant was needed to destabilize the system and then particles were well removed due to the small distance between particles so that they could be enmeshed and swept out.

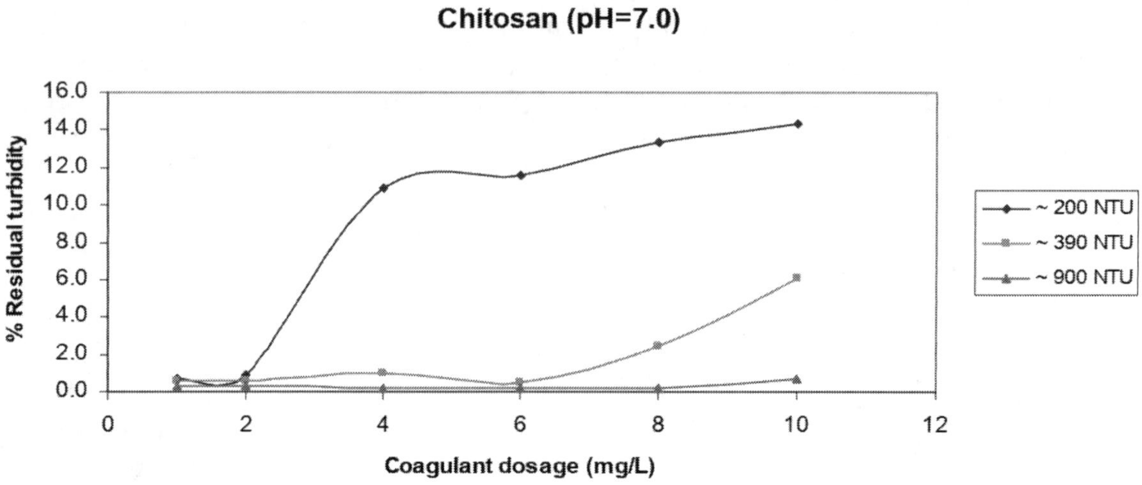

Fig. 3. Effect of chitosan dosage on the removal of turbidity at pH=7.0 and at different initial suspension turbidities

The optimum pH of coagulation for chitosan was found to be 7.0 and % R.T. = 0.3% while at pH 6 and 8 the % R.T. were 6.1 and 0.98 respectively. Chitosan had many advantages such as excellent removal of solids with small coagulant dosages within an acceptable time of sedimentation and it did not change the color of suspension and sludge. At the same time the drawbacks were its virtual insolubility in water if pH is above 6.5 so it had to be dissolved in 1% acetic acid solution, and the fact that it is a costly coagulant.

Polyacrylamide is a nonionic synthetic polymer hence, it did not acquire ionic charge. The optimal operating conditions were not as much effective compared to chitosan and were established to be pH=7.0, 315 mg/L dosage and initial suspension turbidity of ~ 190 NTU with % R.T. = 4.7%. At pH 6 and 8, % R.T. was 5.3 and 17.3 respectively. One advantage of polyacrylamide was that it required a small dosage for coagulation. On the other hand, polyacrylamide had a high viscosity thus, it was difficult to handle and it is also an expensive coagulant. There was neither valuable removal of solids nor satisfactory time of sedimentation. It was thought that since polyacrylamide has a high molecular weight and no ionizable groups, it was not able to effectively destabilize charged particles in CMP slurry. Coagulation would be enhanced, using polyacrylamide as a flocculant aid in conjunction with alum or ferric sulfate.

The anionic polymer, on the other hand, produced very good results. The optimal conditions found were: pH=7.0, 75 mg/L dosage and initial suspension turbidity of ~190 NTU with %R.T. of 0.20 % (Fig. 4.). At pH of 6 and 8, the percentage residual turbidity was 0.24 and 0.89 respectively.

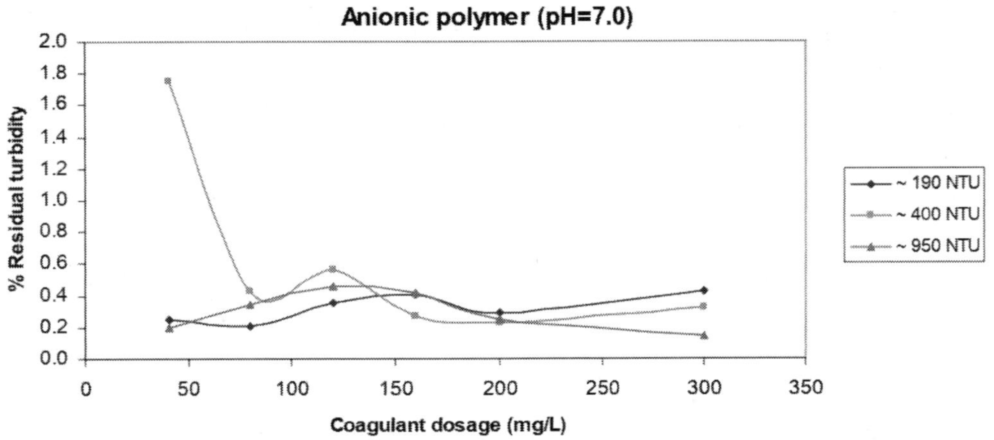

Fig. 4. Effect of anionic polymer dosage on the removal of turbidity at pH=7.0 and at different initial suspension turbidities

Anionic polymer tended to swell and in this way particles were effectively removed and subsequently settled down in an acceptable time frame. Acceptable removal of turbidity and suspended solids was achieved and a clear supernatant solution was obtained. This coagulant is also expensive (even though widely it is used in removing oils from wastewater). The polymer was hygroscopic and difficult to handle. Anionic polymer was also known to double as a flocculant in skimmers for sludge separation from treated water (Melendez, 2002).

3.3. Chemical coagulation and filtration

During coagulation/flocculation process, nanoparticles were stuck together and formed flocs of great size and volume, and subsequently were separated by means of filtration. In this procedure, agglomerates present in the CMP waste after coagulation were removed from the liquid by means of a porous medium which retained the particulate matter but allowed the liquid to pass through. It was reported (Stanley, 2003) that the nanoparticles in CMP slurries were about 200-nm size.

These experiments were conducted at the optimal operating conditions obtained and tabulated in Table 2. Coagulation followed by filtration had as purpose to improve the removal of large size flocs. For filters with pore sizes of 8 μm and 2.7 μm its removal was not as effective as compared with coagulation followed by sedimentation. Nevertheless, when Corning filter (0.2 μm) was utilized, it indeed removed 100% of the flocs and a solution with no turbidity was obtained. In this manner, the agglomerated particles (flocs) that all coagulants formed had particle sizes greater than 0.2 μm. In the case of anionic polymer, their flocs were even greater than 2.7 μm. For the rest of the coagulants, their flocs had varying particle sizes but bigger than 0.2 μm as mentioned.

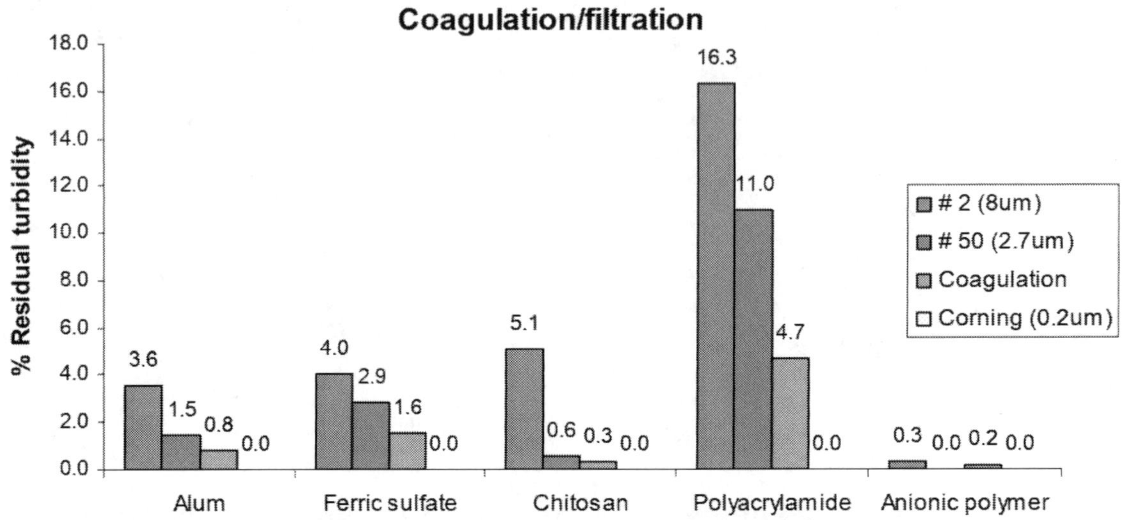

Fig. 5. Effect of coagulants on the removal of turbidity using filters with different pore sizes and a comparison with coagulation

Therefore, suitable results were obtained by means of coagulation in conjunction with filtration using Corning filter. Final turbidity of all the filtrated samples was 0.0 NTU, thus complied with regulations regarding treated wastewater. The motivation for these studies was to demonstrate that high-efficiency depth (bag-type) filters (such as those sold by Ronningen-Petter) can effectively be utilized to remove flocs in high capacity processing of CMP wastes.

3.4. Comparison of results

In order to determine the most effective coagulant to separate nano-sized alumina particles in CMP wastes, a comparison was made on basis of the percentage of residual turbidity and solids removal (as shown in Fig. 6).

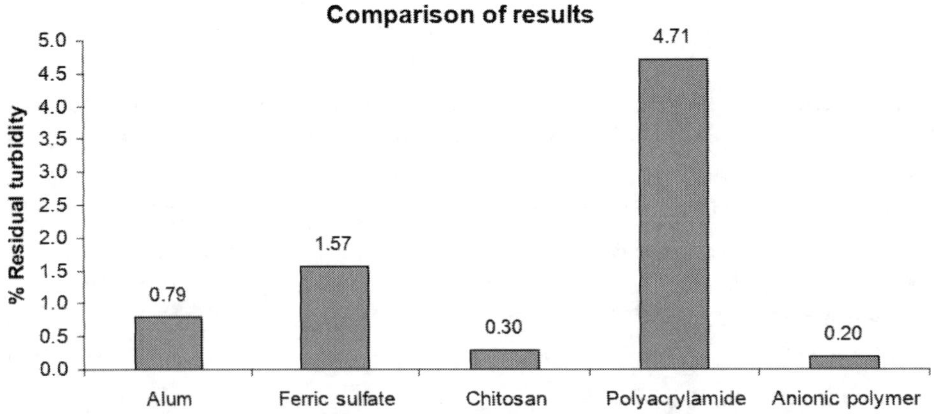

Fig. 6. Comparison of coagulants on the removal of turbidity by means of coagulation (commercial CMP slurry from Cabot Microelectronics iCue® 5001)

As it is shown in Fig. 5 (chemical coagulation followed by filtration) and Fig. 6 (chemical coagulation) either anionic polymer, chitosan and alum, in this order, demonstrated a low residual turbidity and a consequential reduction of nanoparticles in suspension. The anionic polymer produced good results with copper-based CMP samples (results are provided in Carrera et al., 2007) which confirmed its flexibility for distinct operating conditions, versatility and effectiveness to reduce the concentration of nanoparticles in CMP wastes. Therefore, the anionic polymer was the best coagulant to remove copper and nanoparticles from CMP wastes. Further studies are needed to confirm this as CMP slurry chemistry varies.

Flocs formed during coagulation should settle rapidly and be resistant to destruction by shearing forces. Unfortunately, alum produced a light-weight, fragile, slow-settling floc (Fig. 7) consisting primarily of $Al(OH)_3$. Hence, an additional experiment was carried out to improve floc properties and enhance coagulation using alum as primary coagulant (100 mg/L dosage, pH=5.0, initial turbidity ~900 NTU) and anionic polymer as coagulant aid (75 mg/L dosage and added during slow mixing step). That resulted in a residual turbidity of 0.17%, even better than coagulation with alum or anionic polymer alone, and a rapid settling (2 minutes).

Figure 8 shows a great difference between separation by gravity settling compared to coagulation and coagulation/filtration. In this way, the efficiency was improved and this novel separation method was justified.

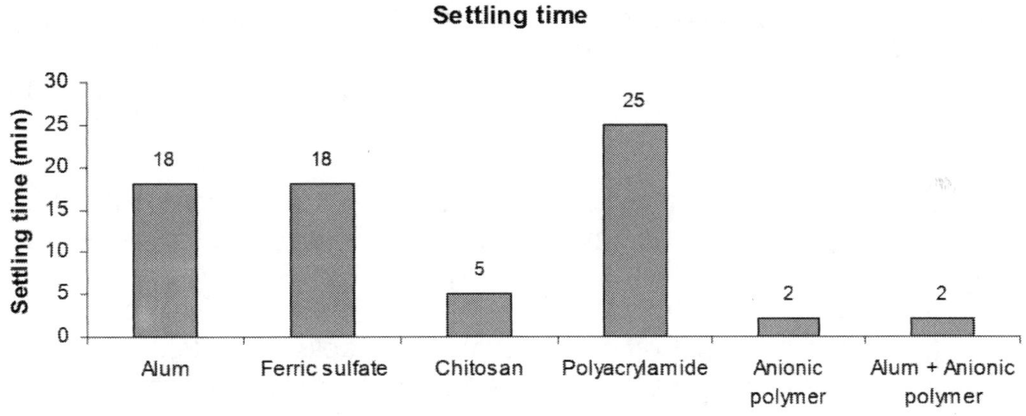

Fig. 7. Settling time for different coagulants at the optimal operating conditions using commercial CMP slurry

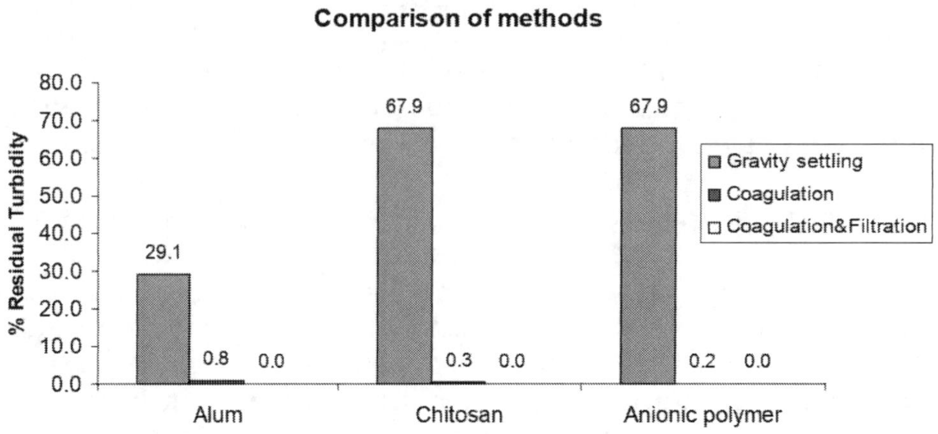

Fig. 8. Comparison of methods on the removal of turbidity from commercial CMP slurry with alum, chitosan and anionic polymer (initial suspension turbidity ~900 NTU, ~200 NTU and ~190 NTU respectively)

4. Conclusions

Removal of nanoparticles from copper chemical mechanical planarization wastewater was studied. The novel treatment method proposed was optimized to remove nanoparticles but provided the benefit to remove copper adsorbed onto the particles (about forty-five percent of the copper ions adhere to the particles). The coagulants provided predictable results for alumina based CMP waste samples. Further work is needed to address the silica nanoparticles.

Based on the experimental results, the following conclusions can be drawn:

1) A commercially available alumina-based CMP slurry was characterized and evaluated in order to establish the optimal operating conditions for removing particulates by means of chemical coagulation followed by either sedimentation and/or filtration. It was found that the original raw slurry had a high total solids content (54480 mg/L) and high turbidity (greater than 13500 NTU) and when it was diluted these values were reduced to within a range of 600-2400 mg/L and 100-1000 NTU, respectively.
2) Comparisons to determine the most effective coagulant were made on basis of the residual turbidity, expressed as percentage of residual turbidity, and settling time. Residual turbidity was used to indicate the relative amount of particles remaining in treated CMP waste. Based on the results it was found that of the five coagulants studied the anionic polymer was the most efficient at removing alumina particles and was also versatile to treat other CMP samples. Its mechanism to destabilize the colloidal system was suspected to be depletion flocculation (Jenkins & Snowden, 1996).
3) Over 95% turbidity reduction was achieved and less than 5 NTU (the recommended value for good quality water) was realized.
4) Removal of copper was evaluated qualitatively and by means of turbidity reduction. Since there was removal of turbidity, insoluble copper-based compounds and adsorbed copper were removed from CMP waste. Copper was suitably removed by the anionic polymer.

5) Chemical coagulation was found to be very effective and viable to reduce the amount of nanoparticles and copper in CMP wastes. In this way, the efficiency of the subsequent biotreatment can be improved. It is the hope that combining coagulation and biosorption will completely solve the problem of cu-CMP waste treatment.

Acknowledgment

Part of this work was funded through an NSF grant. Eduardo acknowledges additional financial support from the department of chemical engineering at UPRM.

References

Aksu S. and Doyle F., The role of glycine in the Chemical Mechanical Planarization of copper. *Journal of the Electrochemical Society*, 149 (6): G352-G361, 2002.

American Quality Water Works Association. Water quality and treatment – *A handbook of community water supplies*. Fifth edition, Chapter 6 by Letterman R., Amirtharajah A., O'Melia C., Mc Graw Hill, 1999.

American Water Works Association and American Society of Civil Engineers. *Water Treatment Plant Design*. Third edition, Mc Graw Hill Professional, 1997.

Asakura S. and Osawa F., On interaction between two bodies immersed in a solution of macromolecules. *Journal of Chemical Physics*, volume 22 (7): 1255-1256, 1954.

Belongia B., Haworth P., Baygents J. and Raghavan S., Treatment of alumina and silica chemical mechanical polishing waste by electrodecantation and electrocoagulation. *Journal of the Electrochemical Society*, 146 (11) 4124-4130, 1999.

Carrera, E. V., Ogden, K. and Bogere M. N., Effective coagulants for primary treatment of chemical mechanical planarization wastewater, *Trans. Semiconductor Manufacturing*, 2007 (to be submitted).

Corlett G., CMP water reduction and waste treatment overview. *Semiconductor Online*, March 17, 1999.

Golden J., Small R., Pagan L., Shang C. and Ragavan S. Evaluating and treating CMP wastewater. *Semiconductor Int.*, http://www.e-insite.net/semiconductor, 2000.

Grassi M., Shi B. and Allen H., Partition of copper between dissolved and particulate phases using aluminum oxide as an aquatic model phase: effects of pH, solids and organic matter. *Journal Brazilian Chemical Society*, Vol. 11 (5): 516-524, 2000.

Ihnfeldt R. and Talbot J., The effects of copper CMP slurry chemistry on the colloidal behavior of alumina abrasives. *Journal of the Electrochemical Society*, 153 (11): G948-G955, 2006.

Jenkins P. and Snowden M., Depletion flocculation in colloidal dispersions. *Advances in colloid and interface science*, 68: 57-96, 1996.

Judd S. and Hillis P., Optimization of combined coagulation and microfiltration for water treatment. *Water Research* 35(12), pp.2895-2904, 2001.

Kawamura S. Effectiveness of natural polyelectrolytes in water treatment. *Journal of the American Water Works Association*, 83 (10): 88-91, October 1991.

Kruilik G., Kramasz K., Golden J., Small R., Shang C. and Ragan L., Copper CMP wastewater chemistry and treatment. *Ultrapure Water* 18 (10) 29-35, 2001.

Meléndez-Colón D., Development of novel dual-function media for treatment and separation of oils & grease from tuna wastewater. M.S. Thesis, University of Puerto Rico, Mayagüez PR, 2002.

Padilla Y., Ogden, K. and Bogere M. N., Development of filtration/biotreatment scheme to recycle CMP wastewater. Paper presented at the AIChE National Meeting, San Francisco Ca., November 2006.

Reker M., *et al.*, Treatment and water recycling of copper CMP slurry waste streams to achieve environmental compliance for copper and suspended solids. *Semiconductor Fabtech* 8th ed.: 141-150.

Rossini M., Garcia Garrido J. and Galluzzo M., Optimization of the coagulation-flocculation treatment: influence of rapid mix parameters. *Water Research* 33 (8), 1817-1826, 1998.

Ruiz A. and Ogden, K., Biotreatment of organic compounds and copper in waste streams from semiconductor manufacturing. NSF/SRC Engineering Research Center for Environmentally Benign Semiconductor Manufacturing, 2004.

Siddiqui J., Castillo D., Kapoor R., Keefover T. and Richards R., Low defectivity product slurry for CMP and associated production method. US Patent 6979252, 2005.

Singh R., Hagerman S., Wargo C. and Roberts W., Effective dispersion of CMP slurry abrasive particles. Paper presented at VMIC Conference, California, 2006.

Stanley L. and Ogden K., Biosorption of copper (II) from chemical mechanical planarization wastewaters. *Journal of Environmental Management*, 69-3: 289-297, 2003.

US EPA., Assessment of high-rate sedimentation processes: microcarrier weighted coagulation jar tests. *Urban Watershed Management Research*. US Environmental Protection Agency, August 1999.

320b Chemical Decontamination of Spills in Water Streams

A. Sadana, University of Mississippi, Chemical Engineering Department, University, MS 38677-1848 and Tingting Zhu, National Center for Computational Hydroscience and Engineering, University of Mississippi, 120 Carrier Hall, University, MS 38655-1848.

A review is presented of the information that is required to help clean up accidental and deliberate chemical spills in water streams. The review includes (a) current EPA, SBIR 9R-NC-07-1055 (closing date May 23, 2007)information that deals with decontamination, drinking water and wastewater security, and emergency response, (b) types of chemical spills--explosives, pathogens/biological agents/proteins, etc., mercury, PCBs, organic chemicals, hazardous chemicals, and metals,(c) methods for treatment of different chemical spills, and benefits of phytoremediation, (d) contaminants suitable for phytoremediation, (e) biological pathogens, (f) devices used for chemical contamination, (f) removal of EDCs (endocrine disrupting chemicals), (g) some clean up-methods (documents available online), (h) some examples of relevant government documents, (i) community's right to know, RCRA, Clean Water Act, etc,(j) relevant databases such as CAMEO, EFDB (Environmnetal Fate Data Base) (k) SQRTs (Screening Quick Reference Tables), (l) other useful databases (m) setting up a data base for a model compound for use by emergency responders and (n) diffusion-limited kinetics and involvement of free-radical mechanisms. The intent of the presentation is to provide a starting point to put together information in one place, so that one can build up a data base for one's personal use or need for chemical decontamination in water streams.

Investigation of the Effectiveness of Various Treatment Methods to Remove Endocrine Disrupting Compounds from Domestic Wastewater

Nitesh Poladia[1], **Robert W. Peters[1]**, and Robert A. Angus[2]
[1]Civil, Construction, and Environmental Engineering, University of Alabama at Birmingham, 1075 13th Street South, Birmingham, AL 35294-4440
[2]Department of Biology, University of Alabama at Birmingham, Birmingham, AL 35294

Abstract

Recently, it has been determined that various synthetic and natural compounds can mimic or interfere with the action of natural hormones and disrupt the endocrine systems of humans and wildlife. These substances, collectively referred to as endocrine-disrupting compounds (EDCs) have been linked to a variety of adverse effects in both humans and wildlife. Numerous EDCs, most of which act as estrogens, have been detected in various surface and groundwaters.

A major source of estrogens in rivers appears to be treated wastewater effluent. Investigations worldwide have detected bioactive estrogens in waters receiving treated wastewater. Municipal wastewater is a complex mixture of natural and synthetic organic chemicals. The most powerful EDCs that are common in treated wastewater include natural hormones 17Beta-estradiol (E2), its breakdown product, estrone (E1), and 17alpha-ethinylestradiol (EE2), the synthetic estrogen used in birth control pills. Other non-steroidal organic chemicals have been shown to possess estrogenic activity, but are much weaker than the steroid hormones. These include the degradation of nonionic surfactants, such as alkylphenolethoxylates and plasticizers. Current technologies for treating municipal wastewaters are only partially successful at removing EDCs. Steroid hormones are especially difficult to remove completely. They contain "steroid rings" that are resistant to degradation by the microorganisms used in wastewater treatment plants. Not only are steroids not efficiently removed from the waste, but those not removed are actually activated by the treatment. As a result, both natural and synthetic estrogens and their degradation products tend to pass through wastewater treatment systems in bioactive forms and can reach concentrations in receiving streams sufficient to produce deleterious biological effects on organisms living in the waters.

This project investigates the effectiveness of various advanced oxidation process (AOP) techniques in degrading the two most potent steroids (E2 and EE2) in aqueous solution, and analyzes water and sediments in rivers in the Birmingham, Alabama area receiving treated wastewater to determine whether the estrogens are present in concentrations sufficient to disrupt endocrine systems of aquatic organisms in the receiving waters. Advanced oxidation processes studied include: ozonation, hydrogen peroxide, sonication, UV light, and combinations thereof. Effectiveness of the treatments are assessed and reported using quantitative chemical analysis using HPLC (indicating the extent to which the amount of the steroid in the spiked sample has been reduced by a given treatment, and also shows the accumulation of breakdown products). A cell-based estrogen screen assay is used to determine whether the treated sample retains biological (estrogenic) activity.

Objectives

The research project has two major objectives: (1) to investigate the effectiveness of various advanced oxidation process (AOP) techniques in degrading the two most potent steroids (E2 and EE2) in aqueous solution, and (2) analyzing water and sediments in rivers in the Birmingham, Alabama area receiving treated wastewater to determine whether the estrogens are present in concentrations sufficient to disrupt endocrine systems of aquatic organisms in the receiving waters.

Introduction and Background

Initial studies have focused on E2 (natural estrogen) and EE2 (a synthetic steroid that is the active ingredient of birth control pills). These steroids have been determined to be the dominant and most potent estrogens in wastewater [Snyder *et al.*, 2001]. They are excreted by women (E2 by all sexually mature pre-menopausal women, and EE2 by women taking birth control pills) and have frequently been detected in treated wastewater [Lee and Peart, 1998; Huang and Sedlak, 2001; Körner *et al.*, 2005; Snyder *et al.*, 2001; Cargouet *et al.*, 2004; Huggett *et al.*, 2003; Kolodziej *et al.*, 2003; Gibson *et al.*, 2005; Tilton *et al.*, 2002; Sarmah *et al.*, 2006; Nakada *et al.*, 2004]. As a result, it is critical that effective and cost efficient methods be developed to remove these endocrine disrupting compounds from domestic wastewater. In our study, we are examining a variety of advanced oxidation processes (AOPs), tested singly or in combination, to assess the extent to which they can degrade E2 and EE2 in water. The AOP treatments being tested include the following:

1. Sonication (employing various acoustic frequencies);
2. Ozone (O_3) [initial concentrations: 10, 100, and 1,000 ppm];
3. Hydrogen peroxide (H_2O_2) [initial concentrations: 10, 100, and 1,000 ppm];
4. Fenton's reagent (iron salts, initial concentrations: 10, 100, and 1,000 ppm); and
5. Ultraviolet light [0, 2, 4, 8, 12, and 16 lamps; 5 W each].

After the individual AOP treatments have been investigated, various combinations will be tested, including:

1. Sonication + H_2O_2:
2. Sonication + O_3;
3. Sonication + UV;
4. UV + O_3;
5. UV + H_2O_2;
6. UV + O_3 + H_2O_2;
7. O_3 + H_2O_2;
8. Sonication + UV + O_3 + H_2O_2.

One of these techniques (sonication) has been shown to be effective at degrading organic compounds [Inazu *et al.*, 1993; Hua and Hoffmann, 1999; Peters *et al.*, 2006]. Effectiveness of the various treatment techniques is evaluated in two ways: quantitative chemical analysis using high performance liquid chromatography (HPLC) to indicate the extent to which the amount of steroid in the spiked sample has been reduced by a given treatment (and may also show the accumulation of breakdown products) and for estrogenic activity using a cell-based estrogenic screen assay to determine whether the treated sample retains biological (estrogenic) activity. For the HPLC analysis, the water samples are prepared using a modified protocol of Jenkins *et al.* [2003]. For each water sample, duplicate replicates of 100-mL are filtered, first through a glass filter with 1.6 μm pore size, followed by a cellulose membrane filter with 0.45 μm pore size. The filtered samples are passed through methanol-washed solid phase extraction (SPE) cartridges which retain the organic compounds in the sample. The SPE cartridges are stored in a freezer and eluted at a later time. Organics are eluted from the SPE cartridges with 5.0-mL HPLC grade methanol, dried under nitrogen, and reconstituted in 50-μL HPLC grade methanol for HPLC analysis. This process concentrates the organics in the original sample by a factor of 2,000. HPLC analysis is performed using a Perkin Elmer HPLC system with water (Solvent A) and 100% methanol (Solvent B) on a Varian, 4.5 cm x 4.0 mm, Microsorb-MV, C-18 column and with a Perkin Elmer UV detector (235 nm). The solvent gradient begins with 40% methanol for 5 minutes; linearly increased to 100% methanol over 20 minutes, and held constant at 100% methanol for another 10 minutes. Quantification of the steroids in water samples is performed by comparing the area under the peak to that of a known standard. Figure 1 shows a typical HPLC chromatogram for an injected sample containing 10.0 nmoles each of E2 and progesterone. Tentative breakdown products of AOP-treated steroids or on compounds found in fractionated river water or sediment samples is performed by HPLC by comparison to retention times of common animal and plant steroids and by relative absorbance of UV light at 210, 240, and 280 nm. Definitive identification of the compounds is conducted using gas chromatograph and mass spectroscopy (GC/MS).

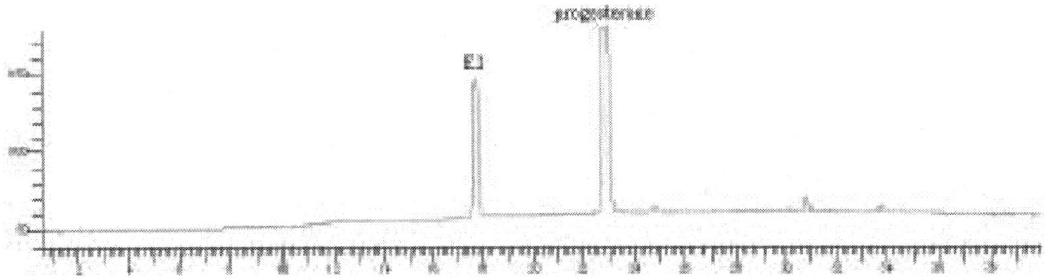

Figure 1. HPLC chromatogram for an injected sample containing 10 nmoles each of E2 and progesterone.

In the second phase of the project, water samples are analyzed using a yeast estrogen screen assay to determine the biological (estrogenic) activity following treatment. The yeast estrogen screen utilizes genetically engineered baker's yeast cells (*Saccharomyces cerevisiae*) containing a constitutively expressed gene for the human α estrogen receptor and an estrogen response element coupled to the E. coli β-galactosidase reporter gene [Arnold *et al.*, 1996]. The assay has two parts: First, cells are incubated overnight in a 96-well cell culture plate in the presence of the compound to be tested. Different wells contain different concentrations (or dilutions) of the text compound, triplicate replicates are done of each. If the compound binds to the estrogen receptor, the complex will in turn bind to the estrogen response element and stimulate expression of the β-galactosidase gene. After the incubation period, the cells are placed in a buffer which lyses the cells and which contains the chromogenic β-galactosidase substrate O-Nitrophenyl β-D-galactopyanoside (ONPG). If the cells have produced β-galactosidase, the enzyme will cleave the OPNG and produce a yellow product. The amount of product is quantified by reading the absorbance at 405 nm. The absorbance is proportional to the estrogenic activity of the compound with which the cells are incubated, and with appropriate controls, can be stated as E2 equivalents. This is the concentration of E2 in water than can have endocrine disrupting effects (<5 ng/L in whole life exposure studies) [Fenske *et al.*, 2005]. Water samples are screened for estrogenic activity with the yeast estrogen screen assay, and will be prepared as for HPLC analyses, except that after drying under nitrogen, the samples are reconstituted in dimethylsulfoxide (DMSO) (since methanol is toxic to living cells). The extraction process concentrates the organics in the water sample by a factor of 2,000. Arnold *et al.* [1996] were able to detect E2 in the yeast estrogen assay at concentrations as low as 0.01 nM. This corresponds to an ability to detect E2 in the original water sample at a concentration of 0.15 ng/L.

Each treatment is being tested for various treatment times (1, 5, 10, 15, and 20 minutes). The waters being treated are spiked with either E2 or EE2; the estrogen concentrations in the water samples are 1 and 10 μg/L. These concentration levels are about 1 to 2 orders of magnitude higher than concentrations typically found in wastewater streams, but they are more amenable to quantification by HLPC methods and permit us to determine treatments that breakdown the steroids most effectively. For example, Clara *et al.* [2005] measured E2 at concentrations from 35 to 125 ng/L and EE2 at concentrations of 3 to 70 ng/L in the influents to various Austrian wastewater treatment plants.

Additionally, river water and sediments downstream of local wastewater treatment plants in the Birmingham, Alabama are being analyzed. Although estrogens have been detected in treated wastewaters in studies worldwide, we are unaware of any published reports investigating the occurrence of endocrine disrupting chemicals in treated wastewater in the State of Alabama. River water and sediments in the Birmingham, Alabama area are being collected and analyzed for steroids using HPLC and for estrogenic activity using the yeast estrogen screen assay.

Preliminary Results:

HPLC analyses are currently being conducted to determine the efficiency of the various AOP treatments as a function of treatment time for destroying the endocrine disrupting compounds. As an example, use of hydrogen peroxide and ozone employing a 20-minute treatment time, removed more than 80% of the E2 initially present in the sample. Results from other treatment conditions will be presented in the oral presentation at the conference.

References Cited:

Arnold, S.F., M.K. Robinson, A.C. Notides, L.J. Guillette, Jr., and J.A. McLachlan, 1996. "A Yeast Estrogen Screen for Examining the Relative Exposure of Cells to Natural and Xenoestrogens", ***Environ. Health Perspect.***, ***104***: 544–548.

Cargouet, M., D. Perdiz, A. Mouatassim-Souali, S. Tamisier-Karolak, and Y. Levi, 2004. "Assessment of River Contamination by Estrogenic Compounds in Paris Area (France)", ***Sci. Total Environ.***, ***324***: 55–66.

Clara, M., N. Kreuzinger, B. Strenn, O. Gans, and H. Kroiss, [2005]. "The Solids Retention Time – A Suitable Design Parameter to Evaluate the Capacity of Wastewater Treatment Plants to Remove Micropollutants", ***Water Res.***, ***39***: 97–106.

Fenske, M., G. Maack, C. Schafers, and H. Segner, 2005. "An Environmentally Relevant Concentration of Estrogen Induces Arrest of Male Gonad Development in Zebrafish, *Danio rerio*", ***Environ. Toxicol. Chem.***, ***24***: 1088–1098.

Gibson, R., M.D. Smith, C.J. Spary, C.R. Tyler, and E.M. Hill, 2005. "Mixtures of Estrogenic Contaminants in Bile of Fish Exposed to Wastewater Treatment Works Effluents", ***Environ. Sci. Technol.***, ***39***: 2461–2471.

Hua, I., and M.R. Hoffmann, 1999. "Kinetics and Mechanism of the Sonolytic Degradation of CCl4 Intermediates and Byproducts", ***Environ. Sci. Technol.***, ***39***(3): 864–871.

Huang, C.H., and D.L. Sedlak, 2001. "Analysis of Estrogenic Hormones in Municipal Wastewater Effluent and Surface Water Using Enzyme-Linked Immunosorbent Assay and Gas Chromatography/Tandem Mass Spectroscopy", ***Environ. Toxicol. Chem.***, ***20***: 133–139.

Huggett, D.B., C.M. Foran, B.W. Brooks, J. Weston, B. Peterson, K.E. Marsh, T.W. La Point, and D. Schlenk, 2003. "Comparison of in vitro and in vivo Bioassays for Estrogenicity in Effluent from North American Municipal Wastewater Facilities", ***Toxicol. Sci.***, ***72***: 77–83.

Inazu, K., Y. Nagata, and Y. Maeda, 1993. "Decomposition of Chlorinated Hydrocarbons in Aqueous Solution by Ultrasonic Irradiation", ***Chem. Lett.***, ***1***: 57–60.

Jenkins, R.L., E.M. Wilson, R.A. Angus, W.M. Howell, and M. Kirk, 2003. "Adrostenedione and Progesterone in the Sediment of a River Receiving Paper Mill Effluent", ***Toxicol. Sci.***, ***73***: 53–59.

Kolodziej, E.P., J.L Gray, and D. Sedlak, 2003. "Quantification of Steroid Hormones with Pheromonal Properties in Municipal Wastewater Effluent", ***Environ. Toxicol. Chem.***, ***22***: 2622–2629.

Körner, W., U. Bolz, W. Sussmuth, G. Hiller, W. Schuller, V. Hanf, and H. Hagenmaier, 2005. "Input/Output Balance of Estrogenic Active Compounds in a Major Municipal Sewage Plant in Germany", ***Chemosphere***, ***40***: 1131–1142.

Lee, H.B., and T.E. Peart, 1998. “Determination of 17β-Estradiol and Its Metabolites in Sewage Effluent by Solid-Phase Extraction and Gas Chromatography/Mass Spectroscopy”, ***J. AOAC Internat.***, ***81***: 1209–1216.

Nakada, N., H. Nyunoya, M. Nakamura, A. Hara, T. Iguchi, and H. Takada, 2004. “Identification of Estrogenic Compounds in Wastewater Effluent”, ***Environ. Toxicol. Chem.***, ***23***: 2807–2815.

Peters, R.W., J.J. Gauthier, and J. Mohammad, 2006. “Final Report: ***Use of Sonication/ Acoustic Cavitation with Advanced Oxidants to Treat Petroleum Hydrocarbons-Contaminated Surface Waters and Groundwaters***”, Final report submitted to the Alabama Water Resources and Research Institute, Auburn, Alabama, [62 pages], (December 20). Report can be accessed at the following web address: http://awrri.auburn.edu/grants/2002_Peters_Final_Report.pdf.

Sarmah, A.K., G.L. Northcott, F.D.L. Leusch, and L. Tremblay, 2006. “A Survey of Endocrine Disrupting Chemicals (EDCs) in Municipal Sewage and Animal Waste Effluents in the Waikato Region of New Zealand”, ***Sci. Total Environ.***, ***355***: 135–144.

Snyder, S.A., D.L. Villeneuve, E.M. Snyder, and J.P. Giesy, 2001. “Identification and Quantification of Estrogen Receptor Agonists in Wastewater Effluents”, ***Environ. Sci. Technol.***, ***35***: 3620–3625

Tilton, F., W.H. Benson, and D. Schlenk, 2002. “Evaluation of Estrogenic Activity from a Municipal Wastewater Treatment Plant with Predominantly Domestic Input”, ***Aquatic. Toxicol.***, ***61***: 211 – 224.

OIL/WATER SEPARATION: EXAMINATION OF VARIOUS TREATMENT TECHNIQUES

TRACY WILLIAMS
Environmental Engineer
Anniston Army Depot
Anniston, AL

ROBERT W. PETERS, Ph.D., P. E.
Professor of Environmental Engineering
Department of Civil, Construction, and Environmental Engineering
University of Alabama at Birmingham
1075 13th Street South
Birmingham, Alabama 35294-4440

M.P. SHARMA, Ph.D.
Professor of Chemical Engineering
Department of Chemical and Petroleum Engineering
University of Wyoming
Laramie, Wyoming 82071

Abstract

Current environmental regulations require that ships can not discharge bilge water containing >15 parts per million (ppm) oil. Defense missions include operations that generate significant amounts of bilge water; a consistently performing treatment system is paramount to sustain these missions and operations. Bilge water is a highly complex wastewater containing various organic and inorganic contaminants in which the oil may exist in several phases, each behaving differently in conventional oil water separation treatment. Due to the nature of this oily phase, a single treatment method can not efficiently treat bilge water to meet these regulatory limits. A series of process steps should be employed to treat each of the oil phases in order to achieve <15 ppm in the final effluent in accordance with international maritime, federal, and defense discharge requirements. A literature review of current technologies and systems revealed a number of potential candidate methodologies. A qualitative assessment of performance criteria, advantages, and disadvantages of these various technologies and systems was conducted to determine theoretical process trains for simulated bilge water types. The qualitative assessment approach was useful as a preliminary screening tool to condense the compilation of technologies to a manageable number for further study, analysis, evaluation, testing, and demonstration.

Introduction

Current environmental regulations require that ships can not discharge bilge water containing >15 parts per million (ppm) oil. The International Maritime Organization (IMO) originally adopted in 1973, and later modified in 1978, provides standards governing the discharge of oily mixtures from seagoing vessels, referred to as MARPOL 73/78. Under these standards, IMO prohibits oily discharges from such vessels unless such discharges are below 15 ppm; discharges are prohibited altogether when vessels are within 12 nautical miles of land [1]. The U.S Code, Title 33, Sections 1901-1911 implements MARPOL 73/78 under the Clean Water Act. The CWA, amended in 1996, authorized the U.S. Environmental Protection Agency and the Department of Defense to jointly develop Uniform National Discharge Standards to control certain discharges incidental to normal operations of Armed Forces vessels [2].

The objective of this research is to develop a theoretical process or processes for the effective onboard treatment of bilge water for the defense community to comply with current regulatory discharge limits of 15 ppm oil. Such theoretical processes should be considered for future evaluation, testing, and demonstration. Research tasks involved: (1) conducting literature review to characterize the nature, concentration, and generation rates of bilge water from defense operations; (2) conducting literature review to identify the various treatment technologies for onboard applications of bilge water treatment; (3) synthesizing the information compiled through the literature review to evaluate the various treatment technologies identified, (4) determining the most promising and eliminating the least suitable technologies, and; (5) developing preliminary recommendations of theoretical treatment options for onboard bilge water applications. The results presented in this paper are based on a qualitative assessment of the operating parameters, performance criteria, and limitations presented in the literature reviewed for this research.

An extensive literature review was conducted, examining scholarly journals, industry and technical journals, textbooks, government reports (U.S government and foreign), patent reports, research and demonstration proposals and reports, and government, vendors, university and other websites to identify the composition of bilge water and the types of treatment technologies and systems currently available for treating bilge water. Approximately 50-75 resources were reviewed, spanning from 1973 to 2006.

A number of operations onboard Armed Forces vessels generate wastewaters containing oily wastes which must be treated prior to disposal, the most significant being bilge water, ballast water, and tank cleanings. Bilge water results from oils, fuels, and other petroleum products leaked, spilled, or otherwise released from ship machinery where it pools in the bottom of the vessel. Ballast water is used for stabilizing ships when fuel levels are low by filling fuel tanks or oil cargo spaces with seawater. Fuel tanks are periodically cleaned, requiring the disposal of the rinsate [3]. While ballast water and tank cleanings pose problems for ships and vessels, bilge water presents complex operational and environmental issues and is the focus of this manuscript.

While housekeeping and maintenance practices can be changed to reduce the frequency of discharges, these steps alone could not completely eliminate the need for discharging bilge water. This is of particular concern to the defense community as its operations generate significant volumes of bilge water and a consistently performing onboard treatment system is needed to ensure minimal impacts to its missions. Such systems are often operated in confined, congested portions of the vessel and can not always be attended to by ship personnel. Residual processing of wastewater and associated wastestreams must be a minimum since onboard space availability is a premium. Finally, shrinking budgets constrain resources to operate and maintain these systems. Consequently, high-performance, high-yield reliable treatment is needed to ensure environmental compliance and mission sustainment.

BACKGROUND

Bilge water is a highly complex wastewater containing various organic and inorganic contaminants in which the oil may exist in several phases, each behaving differently in conventional oil water separation treatment. This 'oily water cocktail' exists in varying composition and concentrations of a wide variety of contaminants including, but not limited to: fuels, oils, detergents, greases, solvents, corrosion inhibitors, surfactants, rusts, paints, insulation material, and other materials and substances which accumulate in the bottom of a ship or vessel. The presence of water and cleaning fluids is the result of cleaning the engine room, and oil products result from leaks from engine room or breaks in engines or oil lines. If there is a crack in the outer hull, seawater or freshwater will enter the bilge. Rust and other deposits are created by the corrosive seawater. Mixing by pumps, and the addition of surfactants, corrosion inhibitors, detergents, and other surfactants result in mechanical and chemical emulsions.

The USEPA and the Navy are jointly working to characterize each of the vessel discharge subject to the UNDS to develop performance standards for devices to control these discharges. Although composition of bilge water varies considerably among vessels, preliminary results presented in a draft EPA report are based grouping vessels fleet-wide according to similarities in engineering and discharge characteristics. For example, **Tables 1 and 2** below summarizes bilge water from surface vessel oil-water separators and submarines, respectively [4, 5].

Table 1. Summary of Composition of Surface Vessel Bilge Water/OWS Discharges

Constituent	Concentration	Annual Mass Loading, (lbs)
Classicals	**(mg/L)**	
Ammonia as nitrogen	BDL – 0.17	67
Nitrate/Nitrite	0.2 – 0.17	202
Total Nitrogen[A]	1.77*	1,304
Total Phosphorus	1.2 – 2.7	1,353
TPH	6 – 16	7,208
Mercury	**(ng/L)**	
Mercury	32.05 – 79.8	0.04
Metals	**(μg/L)**	
Copper		
Dissolved	116 – 201	122
Total	277.5 – 426	255
Iron		
Total	432 – 531	353
Nickel		
Dissolved	109 – 247	132
Total	97.75 – 245	126
Zinc		
Dissolved	511 – 1,260	640
Total	514 – 1,330	657

* – Log normal mean concentration.
[A] – Total nitrogen is the sum of nitrate/nitrite and total Kjeldahl nitrogen.
BDL – below detection limits.

Table 2. Summary of Composition of Submarine Bilge Water Discharges

Constituent	Concentration, (mg/L)	Annual Mass Loading, (lbs)
Cadmium	<0.005 – 0.2	0.75
Chlorine	0.0 – 1.6	7.9
Copper	0.065 – 15	53.4
Cyanide	<0.01 – 0.03	0.15
Heptachlor	0.000005	0.00019
Heptachlor epoxide	0.000003	0.00014
Mercury	<0.0002 – 0.0007	0.0026
Nickel	<0.04 – 11	36.9
Oil	<5 – 820,000	1130
Phenol	<0.01 – 5.4	7.15
Silver	<0.01 – 0.035	0.23
Zinc	<0.02 – 11	51.2

The oil content in bilge water is generally estimated to be 20-25%, although lower concentrations (several parts to million to 0.1%) are not uncommon [6]. The new generation of ships with dryer bilges [generating <1,000 gallons per day (gpd) of bilge water] will produce bilge water with higher concentrations of oil and detergents [7]. Oftentimes, oil in bilge water is not 'pure'; it can exist in several phases, each behaving somewhat differently in an oil/water separator. Also, the bilge water is often contaminated by other materials (e.g., detergents, cleaning solvents, particulate matter, etc.) which are not removed by physical separation and/or can impede gravity settling. The oil can exist as free, dispersed, solubilized, dissolved or chemically emulsified [6]. In an oil-water mixture, the oil particle size may vary considerably, depending on the phase:

- Free - >150 μm
- Dispersed – 20-50 μm
- Emulsions - <20 μm [8]

The volumes of bilge water generated are controlled to a certain extent by ensuring hull tightness and minimizing the amount of water used when cleaning engine rooms. Estimates of bilge water generated range from 0.5-50 m^3/day, and bilge water accounts for about 20% of several tons oily wastewater discharged annually from vessels [9].

Results and Discussion

Treating bilge water by conventional gravity settling requires a long residence time, and consequently large treatment tanks or vessels. Designing an OWS to promote particle size growth (i.e., coalescence) reduces the residence time and vessel size. This improves capture of the oily phases dominated by physical separation (i.e., increasing density differences would promote oil particle rise to the water surface), primarily the free oil, dispersed oil, and some of the emulsified oil. Other treatment methods are required to more effectively treat and remove the solubilized and more stable emulsified phases [10]. Because the oil in bilge water can exist in different states, the water can not be effectively treated by a single treatment method, but rather by using a combination of staged treatment methods. Preferably, the free oil phase is removed first, followed by one or more stages to remove the dispersed and chemically interacting phases (solubilized, dissolved and emulsified phases) [6]. Consequently, a series of process steps should be employed to treat each of the oil phases in order to achieve <15 ppm in the final effluent in accordance with international maritime, federal, and defense discharge requirements.

A literature review of current technologies and systems revealed a number of potential candidate methodologies. This inventory was divided into seven broad categories of technologies and systems:

a. Gravitational methods;
b. Centrifugal technology;
c. Flotation;
d. Membrane methods;
e. Electrochemical advanced oxidation processes;
f. Biological treatment, and;
g. Miscellaneous treatment methods [to include combination/hybrid methods and those methods not meeting the other six categories].

The miscellaneous treatment methods reviewed included: thermal treatment (incineration); filtration utilizing organoclay media; air-sparged hydrocyclone; and sonication; and combination/hybrid systems such as electroflocculation/ ultraviolet treatment; ultrafilitration/reverse osmosis (UF/RO), ultrafiltration/nanofiltration composite ceramic membranes; and microfiltration/ultrafiltration systems.

In order to develop a treatment process train, the nature of the wastestream being treated (physical states, volumes, contaminants, concentrations, etc.) and the conditions under which that wastestream is produced, must be determined. In the absence of analytical data, a waste characterization can be developed based on a process owner's knowledge of the operations producing the waste stream. Using the literature values and data presented herein, two bilge water types were hypothesized: (1) a low-volume, highly concentrated bilge ('dry' bilge water) and (2) a high-volume bilge water containing low concentrations of oil ('wet' bilge water). Both cases can be reasonably encountered inasmuch as older vessels tend to generate large volumes of low-concentration bilge water, while newer vessels are being designed and operated to minimize water generation, thus producing a more concentrated wastestream. For the purposes of this research, the presumptive characteristics of these hypothesized "dry" and "wet" bilge waters are as follows and used as the benchmarks for comparing the various treatment technologies and methods:

Hypothesized 'Dry' Bilge Water

- <1000 gallons of bilge water generated per day
- High oil content - approximately 15%
- Chemically produced emulsions (as a result of cleaning with surfactants)

Hypothesized 'Wet' Bilge Water

- Approximately 100,000 gallons of bilge water generated per day
- Minimal oil content - 0.1% (approximately 1,000 ppm) in primarily dispersed form
- Some settable solids present

Each treatment method or technology was evaluated based on its capability of separately treating each of these components of the respective bilge water type.

In addition to wastestream considerations, the conditions (physical, environmental, and operational) under which these wastewaters are generated must be considered as well. A qualitative assessment of performance criteria, advantages, and disadvantages of the major categories of technologies and systems presented in this paper (gravity/coalescing separators; centrifuges/hydrocyclones; flotation; membrane technology; and electrochemical advanced oxidation processes) was conducted to evaluate for their treatment suitability for each of the proposed hypothesized bilge water types and determine theoretical process trains for these hypothesized bilge water types. The physical, environmental and operational conditions considered in this quality assessment were:

1. Final effluent meets dischargeable limits (<15 ppm);
2. Ambient water temperature (20°C or approximately 68°F) [11];
3. Low salinity environment (3.1.-3.8%) [12];
4. Capable of being installed in a small area;
5. Simple to operate;
6. Minimal maintenance requirements;
7. Minimize residuals (sludge, etc.) produced; and
8. Capable of recovering oil of such a quality that can be recycled or reuse as processed or with minimal reprocessing.

The first seven factors above, as well as the individual components of the hypothesized bilge water types, were used as the basis for the primary selection criteria for the theoretical process train for each wastewater. First, the bilge water must be treated to an acceptable level to discharge – in the case of bilge water, that level (<15 ppm) is defined by regulation. The salinity content and water temperature represent the mean conditions found at sea. Space is a premium in the bilge area of a ship or vessel, so the selected process train must be capable of being installed in a confined space. Furthermore, the system should not be complicated to operate or maintain since operators may not always be able to attend to the treatment system and vessels and ships may be away from port for extended periods of time. The limited onboard space and operator attention also dictate that minimal waste is generated by the process. The last factor was used as a balancing criterion to make the final processing determination in the event more than one potential process train of equivalent performance is identified. A treatment system that can produce a reusable product (in this case, oil) can be very beneficial to DoD ship operations. Effective and efficient recovery of oil from bilge water could produce cost savings and operational efficiency by reducing the amount of virgin oil procured and stored onboard.

Because the oil in bilge water can exist in one or more phases, the sequencing of the process steps is critical to optimal treatment efficiency. As such, it is practical to sequence the processes so that the free oil is removed first, then provide secondary processing to treat the dispersed and chemically interacting phases (solubilized, dissolved, and emulsified phases) [6]. This sequence represents the level of increasing difficulty in treating each phase.

A qualitative assessment of the performance criteria, advantages and disadvantages of the major treatment technologies, and methods reviewed was used to determine the suitability of each technology or method for treating each of the hypothesized bilge water type. The results of this assessment are summarized in **Tables 3 and 4** for the hypothesized 'dry' and 'wet' bilge water types, respectively. This determination was based solely on the literature values and data obtained in this research. Independent verification of the literature values and data presented herein was not performed as this was outside the scope of this research.

Treatment suitability was determined based on the capacity of each treatment method or technology to treat the various bilge water components, achieve the discharge limit, and operate under the prescribed physical, environmental and operational conditions. Based on these factors, a total of eleven criteria were developed to evaluate each treatment technology or method. Each technology or method was assigned a 'pass/fail' rating [equally weighted (an assumed value of unity)] for each of the eleven criteria. The preliminary score was based on the ratings of the primary criteria noted in **Tables 3 and 4**. Those treatment technologies and methods which demonstrated the capability of producing a reusable or recyclable oil were rated slightly higher by weighting that factor slighter higher than the other ten criteria. The final rating in each table reflects a presumed 'degree' of suitability of that process for treating the hypothesized bilge water in question; a value less than 100% demonstrates that an individual process in each column would not completely treat the hypothesized bilge water under all of the physical, environmental, and operational conditions presented. That is, the treatment method or technology would not treat one or more of the bilge water components and/or satisfy one or more of the criteria listed, and as such, additional treatment would be warranted. Hence, the three processes with the highest total 'scores' demonstrated the highest degree of suitability, and consequently the most promising technologies, to develop a theoretical treatment process train for treating the respective bilge water type.

Summary and Conclusions

For the hypothesized 'dry' bilge water (concentrated and chemically-emulsified oil under low-flow regimes), the theoretical process train involved:

- Gravity separation in a coalescing unit, followed by an electrochemical advanced oxidation process (AOP), and membrane technology for final polishing of the effluent. This process train could be further improved by:
  - Air-sparged hydrocyclone (ASH) technology as primary treatment in lieu of gravity separation, and;
  - UF/RO membrane technology contingent upon effluents from upstream processes achieving an oil concentration <400 ppm.

Although centrifuges and hydrocyclones shows a tendency to remove the free oil phase more effectively than gravity/coalescing separators, the latter was selected given hydrocyclones' susceptibility to seawater.

For the hypothesized 'wet' bilge water (low oil concentrations mainly in dispersed phase under high flow regimes); the theoretical process train involved:

- Centrifuges or hydrocyclones for primary treatment, followed by biological treatment, and membrane technology for final polishing of the effluent. Membrane technology offers the added benefit of recovering useable product. This process train could further improved or modified by:
  - Gravity separation in a coalescing unit with subsequent filtration using organoclay filter media as primary treatment in lieu of a centrifuge or hydrocyclone;
  - Incorporation of air-sparged hydrocyclone technology as a more efficient hydrocyclone;
  - Hybrid membrane systems utilizing (1) ultrafiltration/nanofiltration composite membranes, (2) microfilration/ultrafiltration membranes, or ultrafiltration/reverse

osmosis membranes (particularly if upstream effluent concentrations are <400 ppm) for final polishing of the effluent.

- Incineration for complete treatment as this eliminates all the wastewater and generates little residual waste.

Table 3. Qualitative Analysis of Treatment Processes for Hypothesized 'Dry' Bilge Water

Treatment Method/Technology		Gravity/ coalescing OWS	Centrifuge/ hydrocyclone	Flotation	Membrane	Biological	Electrochemical AOP
PRIMARY CRITERIA							
Type of Bilge Water	**Bilge Water Characteristics**						
Dry	<1,000 gpd	P	P	F	P	P	P
	High oil content (~15%)	P	P	P	F	P	P
	Oil primarily in free phase (>150 µm droplet size)	P	F	P	F	P	F
	Chemical emulsions present (<20 µm)	F	P	P	P	F	P
Operational and Other Factors							
Final effluent meets dischargeable limits (<15 ppm)		F	F	F	P	F	F
Ambient temperature (20°C or ~68°F)		P	P	P	P	P	P
Low salinity environment		P	F	P	P	P	F
Capable of being installed in a small area		P	P	P	P	P	P
Simple to operate		P	P	P	P	P	F
Minimal maintenance requirements		P	P	F	P	F	P
Minimize residuals (sludge, etc.) produced		F	P	F	F	F	F
PRELIMINARY SCORE		**72.7%**	**72.7%**	**63.6%**	**72.7%**	**63.6%**	**54.5%**
BALANCING CRTERION							
Capable of recovering oil of such a quality that can be recycled or reused as processed or with minimal reprocessing		F	P	F	P	F	F
TOTAL SCORE		**72.7%**	**81.8%**	**63.6%**	**72.7%**	**63.6%**	**54.5%**

Table 4. Qualitative Analysis of Treatment Processes for Hypothesized 'Dry' Bilge Water

Treatment Method/Technology		Gravity/ coalescing OWS	Centrifuge/ hydrocyclone	Flotation	Membrane	Biological	Electrochemical AOP
PRIMARY CRITERIA							
Type of Bilge Water	Bilge Water Characteristics						
Dry	<1,000 gpd	P	P	F	P	P	P
	High oil content (~15%)	P	P	P	F	P	P
	Oil primarily in free phase (>150 μm droplet size)	P	F	P	F	P	F
	Chemical emulsions present (<20 μm)	F	P	P	P	F	P
Operational and Other Factors							
Final effluent meets dischargeable limits (<15 ppm)		F	F	F	P	F	F
Ambient temperature (20°C or ~68°F)		P	P	P	P	P	P
Low salinity environment		P	F	P	P	P	F
Capable of being installed in a small area		P	P	P	P	P	P
Simple to operate		P	P	P	P	P	F
Minimal maintenance requirements		P	P	F	P	F	P
Minimize residuals (sludge, etc.) produced		F	P	F	F	F	F
PRELIMINARY SCORE		**72.7%**	**72.7%**	**63.6%**	**72.7%**	**63.6%**	**54.5%**
BALANCING CRTERION							
Capable of recovering oil of such a quality that can be recycled or reused as processed or with minimal reprocessing		F	P	F	P	F	F
TOTAL SCORE		**72.7%**	**81.8%**	**63.6%**	**72.7%**	**63.6%**	**54.5%**

Capital, and operational and maintenance (O&M) costs were not included as selection criteria due to the lack of and/or incomplete data for identified in the literature review. It should be noted that cost data is essential in evaluating treatment technologies, and as such any assessment of treatment options would be incomplete without such data. Consequently, the preliminary results presented herein are subject to change, given how cost data will be used in any subsequent assessment and evaluation of the treatment trains proposed in this paper.

The qualitative assessment approach was useful as a preliminary screening tool to condense the compilation of technologies to a manageable number for further study, analysis, evaluation, testing, and demonstration.

References Cited

1. U. S. Navy Military Sealift Command, (1995). "Oil Pollution Prevention: Oily Water Discharges Corrections", *Environmental Issues Newsletter (No.16, 14 April)*, Washington Navy Yard, DC.
2. U. S. Environmental Protection Agency (USEPA), (1999). *Uniform National Discharge Standards (UNDS) for Armed Forces Vessels Factsheet/EPA-821-F-99-009, April*, U.S. EPA, Washington, DC.
3. Harvey, A.; Guzdar, A., *et al.*, (1973). *Centrifugal Coalescer for Separating Oil from Water Discharged from Ships, U. S. Coast Guard Office of Research and Development Project No.430*, Waltham, MS: Foster-Miller Associates, Inc.
4. U. S. Environmental Protection Agency (USEPA), (2003). *DRAFT Characterization Analysis Report Surface Vessel Bilgewater/Oil Water Separator/Section 1.0 – Introduction and Table of Contents/August 2003*, Washington, DC.
5. U. S. Environmental Protection Agency (USEPA), (2003). *DRAFT Discharge Assessment Report: Surface Vessel Bilgewater/Oil Water Separator/June 2003*, Washington, DC.
6. Finger, S. M., and Tabakin, R.B., (1973). *Development of Shipboard Oil/Water Separation Systems". Proceedings for the SAE-ASME-AIAA-ASMA-AIChE Intersociety Conference on Environmental Systems (July 16-19)*, London, England, American Society of Mechanical Engineers.
7. Tompkins, K. and Murphy, L., (2002). *Ultrafiltration Membrane Polishing Systems for Shipboard Treatment of Oily Wastewater*, Raleigh, NC, North Carolina Department of Environment and Natural Resources Division of Pollution Prevention and Environmental Assistance.
8. Cheryan, M., and Rajagopalan, N., (1998). "Membrane Processing of Oily Streams: Wastewater Treatment and Waste Reduction", *Journal of Membrane Science, 15*(1): 13-28.
9. Tomaszewska, M.; Orecki, A.; and Karakulski, K., (2005). "Treatment of Bilge Water Using a Combination of Ultrafiltration and Reverse Osmosis", *Desalination*, *185*(1-2): 203-212.
10. Plat, R., (1994). *Gravitational and Centrifugal Oil-Water Separators with Plate Pack Internals*, The Netherlands, Delft University Press.
11. Wikimedia Foundation, Inc., (2007a). *Sea Surface Temperature*, St. Petersburg, FL.
12. Wikimedia Foundation, Inc., (2007b). *Seawater*, St. Petersburg, FL.

6) Fourier's law was developed from empirical observations at steady state and when used in transient applications is a extrapolation and not adequately confirmed by careful experimentation
7) Over prediction of theory to experiment was found in important industrial systems such as fluidized bed heat transfer to immersed surfaces[13], CPU overheating[14], gel electrophoresis[15], restriction mapping[16], adsorption[17], nuclear fuel rod[18], drug delivery systems[19], when parabolic Fourier model is used indicating another mechanism that has not been considered well.
8) The Casimir limit[20] or during transfer of heat in nanoscale regions the Fourier law is replaced with
Equation of Phonon Radiative Transport. In this limit the flux is described by an expression similar to the one used in radiation heat transfer (Swartz and Pohl[21], Mazumdar[22]. The heat transport, for example, in dielectric crystalline materials is believed to be primarily by atomic or crystal vibrations. These vibrations travel as waves and the energy of the waves quantitated is the phonon, (Kittel[23].)

Boley[24] found that the addition of the second derivative in time of temperature to the governing equation is the only way to remove the singularities found in the solution to parabolic heat conduction equations. The generalized Fourier's law of heat conduction can be written as;

$$q = -k\,\partial T/\partial x - \tau_r\,\partial q/\partial t \qquad (1)$$

This is the damped wave conduction and relaxation equation. When the relaxation time, τ_r is zero Eq. (1) will revert to the Fourier's law of heat conduction. Reference to the use of Eq. (1) can be traced back to Maxwell[25] and Morse and Feshbach[26]. Cattaneo[27,28] and Vernotte[29] postulated this equation independently. Eq.(1) can be used to represent the finite speed of heat and remove the infinite speed implication in the Fourier's law of heat conduction. Reviews of the use of Eq. (1) have been provided by Joseph and Preziosi[30, 31] and Ozisik and Tzou[32]. Tzou[33] has discussed the micro to macro scale behavior. Sharma[8-12] discussed the manifestation of the damped wave transport and relaxation equation in industrial applications and provided bounded solutions well within the constraints of second law of thermodynamics. Some investigators[34] have used the Boltzmann[35] transport equation and derived both the Fourier's law of heat conduction and the damped wave conduction and relaxation equation given by Eq. (1) as special cases. They derive a set of equations for length scales comparable to the mean free path of the molecule.

Tzou[36] attempted to provide the physical significance of the relaxation time in the wave theory of heat conduction. The relaxation time results from the phase-lag between the heat flux vector and the temperature gradient in a high-rate response. Brown and Churchil[37] measured the second sound. Peshkov[38] measured a thermal wave speed of 19 m/s at 1.4 K in Helium. Zehnder and Rosakis[39] measured the temperature distribution at the vicinity of the dynamically propagating cracks in 4340 steel. The relaxation mechanism is fundamental to thermal resonance that cannot be depicted by Fourier's law of heat conduction[40]. For the thermal wave speed around 900 m/s in 4340 steel at 480 ^{0}C the value of relaxation time was found to be of the order of 10^{-11} seconds. A table for relaxation time for materials is not available in the literature. Relaxation times for materials with a non-homogeneous inner structure were presented by Kaminski[41]. For Sodium bicarbonate they report a relaxation time of 29 seconds and 20 seconds for sand and 54 seconds for ion exchange materials. Mitura et. al.[42] claim that the for the falling drying rate period the average relaxation time is of the order of several thousand seconds. For homogeneous substances the relaxation time values range from 10^{-8} to 10^{-10} seconds for gases, 10^{-10} to 10^{-12} seconds for liquids and dielectric solids as concluded by Sieniutycz[43]. Mitra, Kumar et. al.[44] presented experimental evidence of the wave nature of heat propagation in processed meat and demonstrated that the hyperbolic heat conduction model is an accurate representation on a macroscopic level of the heat conduction process in such biological material. They report a relaxation time of the order of 16 seconds.

Some investigators have raised some concerns about the generalized Fourier's law of heat conduction violating the second law of thermodynamics (Bai and Lavine[45] Taitel[46], Barletta and Zanchini[47], Zanchini[48]. Taitel[46] attempted to obtain analytical solution to the governing equations and found that the solution temperature for some values went above the boundary temperature indicating a possible violation of Clausius inequality. Transient instability, including the intrinsic transition from the desirable stability, neutral stability to the ultimate unstable response was investigated by Tzou[49] for a wide spectrum of heating rates. Tzou confirmed that the relaxation time results from the rate equation within the mainframe of the second law in the nonequilibrium, irreversible thermodynamics. Antaki[50] examined the dual phase lag equation that was introduced by Tzou and provided an analytical solution for the case of a semi-infinite medium subject to constant wall flux boundary condition. Sharma[51] derived the damped wave conduction and relaxation equation from free electron theory and Stokes Einstein formulation[52] and by analogy with mass diffusion. The relaxation time was found to be a third of the collision time of the electron and the obstacle. The velocity of heat was found to identical with the velocity of mass derived from kinetic representation of pressure or the Maxwell representation of the speed of molecules. A analytical solution for the case of finite slab subject to constant wall temperature was obtained. The final condition in time as the fourth condition for the second order hyperbolic PDE governing equation for the wave temperature was shown to result in well bounded solutions. This clearly means that care must go into the choice of the conditions used in the boundary of space and initial and final time domains. These have to be physically realistic such as at steady state equilibrium temperature is attained. Only for large relaxation times oscillations were found in the solution for temperature. These oscillations were subcritical and damped. The time conditions used by Taitel are unrealistic from the physical realities of heat transfer and that is why their solution exhibited a overshoot. Thus the equations do not violate the laws of thermodynamics as much as the choice of the space and time conditions as necessary constraints. Baumeister and Hamill[53] presented a analytical solution for the hyperbolic heat equation in a semi-infinite medium subject to a constant wall temperature boundary condition by the method of Laplace transforms. They found interface discontinuity in their solution.

2.0 Separation of Whey Proteins by Radial Chromatography

The sequential separation of whey proteins is often effected by radial chromatography [Mozaffar, Salah, Saxena and Miranda, 2000]. In 1994 the USA exports of dairy products amounted to $717 million. A part of the success is due to the improved cattle breeding and milk processing capabilities. Cheese is made from the milk of various mammals. The basic components of milk contain fat, protein, lactose, ash, enzymes and vitamins. Two major categories of proteins are cassein and whey proteins. Cassein is a colloid and whey is soluble. In the cheese industry, two types of precipitation techniques are most commonly used to separate the total milk proteins separated into caseins and whey proteins, i.e, rennet precipitation and acid precipitation. In rennet precipitation, rennin is added to warm milk at 30-35 ^{0}C, precipitating the caseins and leaving the whey proteins in solution. Whey produced by this method is referred to as "sweet whey." Acid precipitation is carried out at the isoelectric point of milk (i.e., 4.7) through the use of acid to precipitate out casein and leave the whey in solution. Whey produced by this method is referred to as "acid whey." The choice of the method used to make the curd depends on the desired cheese product. Although the whey is discarded during the cheesemaking process, it has significant value as a source of nutrition. Unless whey is used to produce other products (including feed), it is a liability for cheese factories. The high biological oxygen demand (BOD) of whey presents disposal problems. Thus, large quantities of whey are used in candies and special cheese products. In addition, large quantities of dry and concentrated forms of whey are also used in dry and concentrated forms that are mixed with food and feeds. Whey may also be condensed and spray dried by the roller process for other uses. Various whey drinks have been developed, including wines, carbonated beverages, buttermilk substitutes, whey-fruit flavored drinks, and whey-tomato drinks. Whey is also used to produce whey butter, soups, protein hydrolysates, cheeses, processed cheese foods and spreads, bakery products, infant foods, geriatric foods, hydrolyzed lactose syrup, pills, riboflavin concentrates, alcohol (e.g., butyl alcohol), methane, acetone, spirit vinegar, food acidulant, resins, coatings, tanning, acrylic plastics, and biomass. In their undenatured, soluble form, whey proteins are also useful as binders in extruded vegetable or animal protein foods. Whey contains various proteins (e.g., .beta.-lactoglobulin and .alpha.-lactalbumin), lactose, soluble minerals, water-soluble vitamins, and enzymes.

The cheese industry produces large amounts of whey, much of which is used to make whey protein concentrate Commercial-scale fractionation of different whey proteins has been hampered by the lack of an economical fractionation technology. It would be desirable, therefore, to provide a method for the continuous and sequential separation of various proteins from whey in a simple one or two step separation process. None of the existing methods provide a product which can be easily, economically, and efficiently utilized as a supplement for infant formulas, fat substitutes or other commercially important products. A need remains in the industry for economical methods to process whey products, including processes and methods that are designed so as to permit reutilization of buffers and constituents utilized during the purification of whey compounds. In addition, methods are needed to increase the efficiency of whey processing procedures, including methods designed to reduce the time necessary to obtain a final product. Sepragen Corp. San Jose, CA, USA, provides a process for the sequential separation of at least five different proteins from whey and incorporating these separated whey proteins into pharmaceutical and food formulations. The process of the invention is directed to the continuous, sequential separation of whey proteins by chromatography, comprising adsorbing the proteins in liquid whey on a suitable separation medium packed in a chromatographic column and sequentially eluting immunoglobulin (e.g, IgG), .beta.-lactoglobulin (.beta.-Lg), .alpha.-lactalbumin (.alpha.-La), bovine serum albumin (BSA), and lactoferrin (L-Fe) fractions with buffers at suitable pH and ionic strength. Even though both axial and *radial flow chromatography* may be utilized, a horizontal flow column is particularly suitable. The liquid whey is selected from the group consisting of pasteurized sweet whey, pasteurized acid whey, non-pasteurized acid whey, and whey protein concentrate. The spatio-temporal concentration in the radial flow chromatographic column is discussed below.

3.0 Finite Speed Diffusion Effects

The convection effects due to radial flow also have to be taken into account compared with the analysis done in earlier section. The mass balance on a thin cylindrical slice including the radial convective effects can be written as;

$$-1/r\partial(rJ_r)/\partial r - V_r\partial(rC)/\partial r - k'''C = \partial C/\partial t \qquad (2)$$

where k''' is the first order rate constant for irreversible adsorption. V_r is the radial velocity. The damped wave diffusion and relaxation equation can be written is;

$$J_r = -D_{AB}\partial C/\partial r - \tau_{mr}\,\partial J_r/\partial t \qquad (3)$$

In order to eliminate J_r between Eqs. [2-3], Eq. [2] is differentiated wrt to t and Eq. [3] wrt r after multiplying by r

$$1/r\partial^2(rJ_r)/\partial r\partial t - V_r/r\partial^2(rC)/\partial r\partial t - k'''\partial C/\partial t = \partial^2 C/\partial t^2 \quad (4)$$

The damped wave diffusion and relaxation equation can be written is;

$$1/r\partial(rJ_r)/\partial r = -D_{AB}/r\,\partial/\partial r(r\,\partial C/\partial r) - \tau_{mr}/r\partial^2(rJr)/\partial t\partial r \quad (5)$$

Eliminating $1/r\partial^2(rJ_r)/\partial r\partial t$, between Eqs.[4-5],

$$D_{AB}/r\,\partial/\partial r(r\,\partial C/\partial r) - V_r/r\partial(rC)/\partial r - k'''C = (1 + k'''\tau_{mr})\partial C/\partial t + \tau_{mr}(V_r/r\partial^2(rC)/\partial r\partial t + \partial^2 C/\partial t^2) \quad (6)$$

Obtaining the dimensionless form by the following substitutions;

$$X = r/\mathrm{sqrt}(D\tau_r);\quad \tau = t/\tau_r;\quad u = (C/C_s);\quad k^* = k'''\tau_{mr};\quad Pe_m = V_r/\mathrm{sqrt}(D/\tau_{mr}) \quad (7)$$

$$1/X\,\partial/\partial X(X\,\partial u/\partial X) - Pe_m/X\partial(Xu)/\partial X - k^* u = (1 + k^*)\partial u/\partial\tau + Pe_m/X\partial^2(uX)/\partial X\partial\tau + \partial^2 u/\partial\tau^2 \quad (8)$$

The damping term can be removed using a $u = W\exp(-n\tau)$ substitution;

$$\partial u/\partial\tau = -nW\exp(-n\tau) + \exp(-n\tau)\partial W/\partial\tau \quad (9)$$

$$\partial^2 u/\partial\tau\partial X = -n\partial W/\partial X\exp(-n\tau) + \exp(-n\tau)\partial^2 W/\partial\tau\partial X \quad (10)$$

$$\partial(uX)/\partial\tau = -n(WX)\exp(-n\tau) + \exp(-n\tau)X\partial W/\partial\tau + \exp(-n\tau)W$$

$$\partial^2(uX)/\partial X\partial\tau = -n(W)\exp(-n\tau) - nX\partial W/\partial X\exp(-n\tau) + \exp(-n\tau)\partial W/\partial\tau + \exp(-n\tau)X\partial^2 W/\partial X\partial\tau + \exp(-n\tau)\partial W/\partial X$$

$$\partial^2 u/\partial\tau^2 = +n^2 W\exp(-n\tau) + -2n\exp(-n\tau)\partial W/\partial\tau + \exp(-n\tau)\,\partial^2 W/\partial\tau^2$$

Substituting the above, Eq. [3.404] becomes;

$$[1/X + nPe_m - Pe_m/X]\ \partial W/\partial X + \partial^2 W/\partial X^2 - Pe_m/X\partial(XW)/\partial X$$
$$= (1 + k^* - 2n + Pe_m/X)\partial W/\partial\tau + Pe_m/X\ X\partial^2 W/\partial X\partial\tau + W(-n(1 + k^*) + n^2 + k^* + nPe_m/X) + \partial^2 W/\partial\tau^2 \quad (14)$$

For small Pe_m/X Eq. [14] can be written as;

$$[1/X + nPe_m]\,\partial W/\partial X + \partial^2 W/\partial X^2 - Pe_m/X\,\partial(XW)/\partial X$$
$$= (1 + k^* - 2n)\,\partial W/\partial\tau + Pe_m/X\; X\partial^2 W/\partial X\partial\tau + W(-n(1 + k^*) + n^2 + k^*) + \partial^2 W/\partial\tau^2 \tag{15}$$

By choosing $n = (1 + k^*)/2$, Eq. [15] becomes;

$$\partial^2 W/\partial X^2 + 1/X\partial W/\partial X + (1 + k^*)/2Pe_m\partial W/\partial X - Pe_m\partial W/\partial X - Pe_m W/X$$
$$= Pe_m/X\partial^2 W/\partial\tau\partial X - W/4\,|1 - k^*|^2 + \partial^2 W/\partial\tau^2 \tag{16}$$

Eq. [16] can be solved by the method of separation of variables;

Let $W = \phi(X)V(\tau)$,

$$\phi''/\phi + \phi'/\phi\,(1/X) + (1 + k^*)Pe_m/2(\phi'/\phi) - Pe_m\phi'/\phi - Pe_m/X = Pe_m/XV'\phi'/V\phi - 1/4|1 - k^*|^2 + V''/V \tag{17}$$

$$\phi''/\phi + \phi'/\phi\,[(1/X)\{1 - Pe_m - Pe_m V'/V\} + Pe_m(k^* + 1)/2\}] - Pe_m/X$$
$$= V''/V = 1/4|1 - k^*|^2 - \lambda_n^2 \tag{18}$$

The time domain solution can seen to be;

$$V = c_1\exp(+\tau\mathrm{sqrt}(1/4|1 - k^*|^2 - \lambda_n^2) + c_2\exp(-\tau\mathrm{sqrt}(1/4|1 - k^*|^2 - \lambda_n^2)) \tag{19}$$

At infinite time the concentration cannot be infinity and hence c_1 has to be set to zero.

$$V = c_2\exp(-\tau\mathrm{sqrt}(1/4|1 - k^*|^2 - \lambda_n^2)) \tag{20}$$

Then,

$$V'/V = -\mathrm{sqrt}(1/4(|1 - k^*|^2 - \lambda_n^2)) \tag{21}$$

The space domain solution can be obtained

$$X^2\phi'' + \phi'\,[X[1 - Pe_m + Pe_m\,\mathrm{sqrt}1/4(|1 - k^*|^2 - \lambda_n^2] + XPe_m/2\,(1 + k^*)] - \phi\,[X^2(1/4|1 - k^*|^2 - \lambda_n^2) - Pe_m X] = 0 \tag{22}$$

Comparing Eq. [22] with the generalized Bessel equation given in Eq. [A.30],

$$a = 1 - Pe_m + Pe_m\,\mathrm{sqrt}1/4(|1 - k^*|^2 - \lambda_n^2]$$

$$r = 1;\quad b = +Pe_m/4\;(1 + k^*)$$

$$1 - a - r = Pe_m - Pe_m \, sqrt1/4(|1 - k^*|^2 - \lambda_n^2] - 1 = 0$$

$$s = 1/2 \quad ; \; d = Pe_m \; ; \; c = 0$$

Thus λ_n can be chosen from the relation given in Eq. [22]

$$1/4(|1 - k^*|^2 - \lambda_n^2 = (Pe_m - 1)^2 / Pe_m^2 \tag{23}$$

$$[1 + 1/(Pe_m(k^*-1)/2 - 1)]^2 = 1/4(|1 - k^*|^2 - \lambda_n^2)) \tag{24}$$

$$\text{or } \lambda_n = sqrt[\tfrac{1}{4}|1 - k^*|^2 - (Pe_m - 1)^2/Pe_m^2] \tag{25}$$

$$a = 1 - Pe_m + Pe_m (Pe_m - 1)/Pe_m = 0 \tag{26}$$

The order of the Bessel equation is thus; $p = 2/2 = 1$. The solution is given by;

$$\phi = \exp(-X(1 + k^*)Pe_m/4)[c_1 J_1 (2X|Pe_m|^{1/2} + c_2 Y_1 (2X|Pe_m|^{1/2}] \tag{27}$$

The value of b differs between the two coefficients of the Bessel differential equation. The solution is approximately obtained. c_2 can be seen to be zero from the condition that W is finite at X = 0. Thus,

$$\phi = \exp(-X(1 + k^*)Pe_m/4)[c_1 J_1 (2X|Pe_m|^{1/2} \tag{28}$$

$$u = c \exp(-(1 + k^*)\tau/2) \exp(-\tau sqrt(1/4|1 - k^*|^2 - \lambda_n^2)) \exp(-X(1 + k^*)Pe_m/4)[c_1 J_1 (2X|Pe_m|^{1/2} \tag{29}$$

There exists a non-homogeneity at the boundary condition at X = 0. This can be solved by the method of superposition and by expressing the solution as a sum of a transient part and a steady state part. At steady state, Eq. [3.488] becomes;

$$1/X \, \partial/\partial X(X \, \partial u/\partial X) - Pe_m/X\partial(Xu)/\partial X - k^* u = 0 \tag{30}$$

$$\partial^2 u/X^2 + 1/ X\partial u/\partial X - Pe_m u/X - Pe_m \partial u/\partial X - k^* u = 0 \tag{31}$$

For small Pe_m/X

$$X^2 \partial^2 u/X^2 + X(1 - XPe_m)\partial u/\partial X - u[X^2 k^* + X Pe_m] = 0 \tag{32}$$

Comparing Eq. [32] with the generalized Bessel equation
Approximately,

$$a = 1; \quad r = 1; \quad 2b = -Pe_m$$

$$c = 0; \quad s = 1; \quad 1 - a - r = -1; \quad -b(1-a-r) = Pe_m/2 \; ; \quad d = -k^{*1/2}$$

The order of the Bessel solution can be calculated as p = 0. Thus approximately,

$$u = \exp(-Pe_m X/2) \, [c_1 \, I_0 \, (Xk^{*1/2}) + c_2 \, K_0(Xk^{*1/2}) \qquad (33)$$

The value of b differs in the two coefficients of the Bessel differential equation. As a first pass, analysis the variation is neglected. From the condition that the concentration is finite at zero X c_2 can be seen to be zero. For X = 0,

$$1 = c_1$$

Thus at steady state,

$$u = \exp(-Pe_m X/2) \, I_0 \, (Xk^{*1/2}) \qquad (34)$$

Now the transient portion can have the boundary condition at X = 0, u = 0. Eq. [3.425] can be seen to obey the boundary condition at X = 0 for the transient portion of the solution. As the $J_1(x)$ goes to zero at x = 3.8317, beyond a certain distance the concentration will be zero. This critical distance is when,

$$R > 1.916\mathrm{sqrt}(D\tau_r/Pe_m) \qquad (35)$$

The design of the radial flow chromatography column greater than the R calculated by Eq. [35] will not result in any further separation. This is the critical length of diffusion.

References

1 Onsager, L, "Reciprocal Relations in Irreversible Processes", Phys. Review, **37**, 1931, 405-426.

2 Nernst, W., *Die Theoretician Grundalgen des n Warmestatzes*, Kanppe Hall, Frankfurt, 1917.

3 Qiu, T. Q., Tien, C. L., "Heat Transfer Mechanisms during Short Pulse Laser Heating of Metals", *ASME Transactions, Heat Transfer Division*, **Vol. 196**, 1992, 41-49.

4 Xie, Y. S., Yuan, Y. X., Zhang, X. B., "Hyperbolic Heat Conduction Equation and Analytic Solution for Instantaneous Ignition of Solid Propellant", *Bimggmg Xuebao/Acta Aramamentarii,* **Vol. 27**, Issue Supplement, March., 2006, 24-28.

5 Landau, L., and Liftshitz, E. M., *Fluid Mechanics*, Pergamon, UK, 1987.

6 Renganathan, K., "Correlation of Heat Transfer with Pressure Fluctuations in Gas-Solid Fluidized Beds", Ph.D. Dissertation,

West Virginia University, Morgantown, WV, 1990.

7 Byrd, R. B., Stewart, W. E., Lightfoot, E. N., *Transport Phenomena*, Wiley, New York, 2002.

8 Sharma, K. R., *Damped Wave Transport and Relaxation*, Elsevier, Amsterdam, 2005.

9 Sharma, K. R., "Temperature Solution in Semi-Infinite Medium under CWT using Cattaneo and Vernotte for non-Fourier Heat Conduction", *225th ACS National Meeting*, New Orleans, LA, March 23rd - March 28th, 2003.

10 Sharma, K. R., "Finite Speed Heat Conduction - III Relativistic Transformation from Cylindrical Coordinates in Semi-Infinite Medium", *Chemistry Preprint Archive*, Volume 2003, Issue 2, February 2003, 360-377.

11 Sharma, K. R., " Finite Speed Heat Conduction - IV Relativistic Transformation from Spherical Coordinates", *Chemistry Preprint Archive*, Volume 2003, Issue 2, February 2003, Pages 378-395

12 Sharma, K. R., "A Fourth Mode of Heat Transfer called Damped Wave Conduction", *42nd Annual Convention of Chemists, Meeting,* Santiniketan, India, February , 2006.

13 Sharma, K. R., Turton, R., "Mesoscopic Approach to Correlate Surface Heat Transfer Coefficients with Pressure Fluctuations in Gas-Solid Fluidized Beds", *Powder Technology*, **Vol. 99**, (2), 1998, 109-118.

14 Sharma, K. R., "Storage Coefficient of Substrate in a 2 GHz Microprocessor", *225th ACS National Meeting*, New Orleans, LA, March, 2003.

15 Sharma, K. R., " Third Order PDE in Restriction Fragment Length Gel Electrophoresis", *228th ACS National Meeting,* Philadelphia, PA, August 2004.

16 Sharma, K. R., "On the Use of Cutrices in Restriction Mapping of Three Enzymes", *231st ACS National Meeting,* Atlanta, GA, March, 2006.

17 Sharma, K. R., "Adsorber Design for Removal of Acrylonitrile from Water", *AIChE Spring National Meeting,* New Orleans, LA, March /April, 2003.

18 Sharma, K. R. "Critical Radii Neither Greater than the Shape Limit Nor Less than Cycling Limit", *AIChE Spring National Meeting,* New Orleans, LA, March/April 2003.

19 Sharma, K. R., "Bessel Composite Function of the Third Order and First Kind: Solution to the Dissolving Pill Problem", *230th ACS National Meeting,* Washington, DC, August/September, 2005.

20 Casimir, H. B. G., "Note on the Conduction of Heat in Crystals", *Physica,* **5**, 1938, 495 -500.

21 Swartz, E. T., and Pohl R. O., "Thermal Boundary Resistance" *Reviews of Modern Physics*, **61**, 1989, 605 - 668.

22 Majumdar, A., "Microscale Heat Conduction in Dielectric Thin Films", *Journal of Heat Transfer*, **Vol. 115**, 1993, 7-16.

23 Kittel, C., *Introduction to Solid State Physics*, John Wiley, New York , 1986.

24 Boley, A, K. R., *Heat Transfer Structures and Materials*, Pergamon, New York, 1964.

25 Maxwell, J. C., "On the Dynamical Theory of Gases", *Phil. Trans. Roy. Soc.*, **157**, 1867, 49.

26 Morse, P. M., and Feshbach., H., *Methods of Theoretical Physics*, McGraw Hill, New York, 1953.

27 Cattaneo, C., " A Form of Heat Conduction which Eliminates the Paradox of Instantaneous Propagation", *Comptes Rendus*, **247**, 1958, 431- 433.

28 Cattaneo, C., *Sulla Conduzione del Calore*, Societta Tipografica Modenere, 1948.

29 Vernotte, P., "Les Paradoxes de la Theorie Continue de l'equation de la Chaleur", *C. R.. Hebd. Seanc. Acad. Sci. Paris*, 246, **22**, 1958, 3154 - 3155.

30 Joseph, D. D., and Preziosi, L., " Heat Waves", *Reviews of Modern Physics*, **61**, 1989, 41-73.

31 Joseph, D. D., and Preziosi, L., "Addendum to Heat Waves" *Reviews of Modern Physics*, **62**, 1990, 375-391.

32 Ozisik, M. N., and Tzou , D. Y., "On the Wave Theory of Heat Conduction", *ASME J of Heat Transfer*, **116**, 1994, 526- **535.**

33 Tzou, D. Y., *Macro to Microscale Heat Transfer: The Lagging Behavior*, CRC Press, New York, 1996.

34 Chen, G., "Ballistic-Diffusive Heat-Conduction Equations", *Phys. Review Letters*, **Vol. 86,** (11), 2001, 2297 – 2300.

35 Boltzmann, L., *Vorlsungen Uber Kinetische Gastheorie*, J. A. Berth, Leipzing, 1896.

36 Tzou, D. Y., " An Engineering Assessment to the Relaxation Time in Thermal Wave Propagation", *Int. J. Heat Mass Transfer*, **36**, 7, 1993, 1845-1850.

37 Brown, M. A. and Churchill S. W., " Finite-difference computation of the wave motion generated in a gas by a rapid increase in the bounding temperature", *Computers and Chemical Engineering*, **23**, 3, 1999, 357-376.

38 Peshkov, V., "Second Sound in Helium II", *Journal of Physics USSR*, VIII, 1944, 381-386.

39 Zehnder A. T. and Roaskis, A. J., "On the Temperature Distribution at the Vicinity of Dynamically Propagating Cracks in 4340 Steel", *Journal Mech. Phys. Solids*, **Vol. 39**, 1991, 384-415.

40 Tzou, D. Y., "Damping and Resonance Characteristics of Thermal Waves", *ASME Journal Appl. Mech.*, **Vol. 59**, 1992, 862-867.

41 Kaminski, W., "Hyperbolic Heat Conduction Equation for Materials with a Nonhomogeneous Inner Structure" *Journal of Heat Transfer*, **Vol. 112**, 1990, 555 – 560.

42 Mitura, E., Michalowski, S., Kamisnki, W., "A Mathematical Model of Convection Drying in the Falling Drying Rate Period", *Drying Technology*, **Vol. 6**, (10), 1988, 113-137.

43 Sieniutycz, S., "The Variational Principle of Classical Type for Noncoupled Non-Stationary Irreversible Transport Processes ith Convective Motion and Relaxation", *International Journal of Heat and Mass Transfer*, Vol. 20, 2977, 1221-1231.

44 Mitra, K., Kumar, S., Vedavarz, A., Moallemi, M. K., "Experimental Evidence of Hyperbolic Heat Conduction in Processed Meat", *Journal of Heat Transfer*, **Vol. 117**, 1995, 568-573.

45 Bai, C., Lavine, A. S., "On Hyperbolic Heat Conduction and Second Law of Thermodynamics", *Journal of Heat Transfer*, **Vol. 117**, (2), 1995, 256-263.

46 Taitel, Y., "On the Parabolic, Hyperbolic and Discrete Formulation of Heat Conduction Equation", *International Journal of Heat and Mass Transfer*, **Vol. 15**, (2), 1972, 369-371.

47 Zanchini, E., "Hyperbolic Heat Conduction Theories and non-Decreasing Entropy", *Phy. Review B-Condensed Matter and Material Physics,* **Vol. 60**, (2), 1999, 991-997.

48 Barletta, A. and Zanchini, E., "Thermal-Wave Heat Conduction in a Solid Cylinder which undergoes a change of Boundary Temperature", *Heat and Mass Transfer/ Warema-und Stoffuebertragung*, **32**, 4, 1997, 285-291.

49 Tzou, D. Y., "Thermal Control in Solids with Rapid Relaxation", *Journal of Dynamic Systems, Measurement and Control, Trans. of ASME*, **Vol. 125**, (4), 2003, 563 – 568.

50 Antaki, P. J., "Solution for non-Fourier Dual Phase Lag Heat Conduction in a Semi-infinite Slab with Surface Heat Flux", *International Journal of Heat and Mass Transfer*, **Vol. 41**, (14), 1998, 2253-2258.

51 Sharma, K. R., "Manifestation of Acceleration during Transient Heat Conduction", *Journal of Thermophysics and Heat Transfer,* **26**, 4, 799-808.

52 Sharma, K. R., *On the Derivation of an Expression for Relaxation Time from Stokes-Einstein Relation*, 233rd ACS National Meeting, Chicago, IL, March 2007, American Chemical Society, Washington, DC.

53 Baumeister, K. J., Hamill, D., "Hyperbolic Heat Conduction Equation – A Solution to the Semi-infinite Body Problem", *ASME Journal of Heat Transfer,* **Vol. 93**, 1, 1971, 126-128.

326b Recovery Of High Purity Organic Solvents From Waste Photoresist Strippers

Han Sang Oh[1], Dae-Jin Kim[1], Myung-Jun Park[2], **Moon-Kyu Yoon**[3], Moonyong Lee[3], and Kee-Kahb Koo[1]. (1) Department of Chemical and Biomolecular Engineering and Interdisciplinary Program of Integrated Biotechnology, Sogang University, Seoul, 121-742, Korea, (2) Korex Co., Anyang, 431-804, South Korea, (3) School of Chemical Engineering and Technology, Yeungnam University, Kyongsan, 712-749, South Korea

Photoresist strippers used to remove the residue of photoresist in the TFT-LCD or semiconductor process are organic solvent mixtures with some additives. In the present experiments, main organic solvents including NMP and BDG were reclaimed from waste photoresist stripper by using a spinning band distillation column. Purity of final products were obtained to be 99.5%, water content less than 0.05% and most metal contents less than 10ppb, respectively. To extend those experiments further, a computer simulation for a continuous distillation process was made using NRTL and Peng-Robinson equations.

Structural Design and Separation Characteristic of Divided Wall Column by the Shortcut Method

Seunghyun Lee, Mohammad Shamsuzzoha, Moonyong Lee[*]

School of Chem. Eng. & Tech. Yeungnam University, Dae-dong, Kyungsan, Kyungpook, Korea

Abstract

It is suggested an efficient structure design method with interrelation of optimization by the internal separation characteristic for the initial structure design stage of divided wall distillation column. In the structural design of divided wall column (DWC), the main design factors affecting the overall performance are the position of feed tray, divided wall section, and side-stream tray as well as the total number of trays in main and prefractionator sections. In the proposed method, the main factors of structural design in DWC can be easily obtained from a shortcut method by using the Fenske-Underwood equation via employing three conventional simple column's configuration while the existing structural design method requires a lot of efforts for the design of DWC. One of the optimal design factors, the internal-flow in DWC is related to the reflux ratio of the 1st column, and it is also related to each composition of components separated at the top and bottom of the 1st column. This relation clearly has a benefit as the tray position is very simple and efficient by applying on the DWC structure design. Further, the optimal design condition and energy consumption is compared by applying the proposed method on divided wall column and by using the optimal internal-flow via. a case study in simulation. The energy consumption is also studied by changing the composition of intermediate component in feed mixture. The simulation result shows that the energy consumption of DWC using the shortcut method shows an acceptable difference from the global optimal design case. The DWC designed by the proposed method saves energy approximately from 29% to 34% more over the conventional sequential two column system.

Introduction

In the chemical industry, the conventional sequential two column system is the most commonly used to separate the ternary mixture. The control strategy of the product composition for the conventional process is simple although there is a remixing in the 1st column's bottom. The conventional distillation column is energy inefficient and this can be improve by using a thermally integrated column.[1-2] The typical thermally integrated column is a Petlyuk distillation column shown in Figure 1.[3-7] The Petlyuk distillation column contains two columns where one is the prefractionator and the other is the main distillation column. In the Petlyuk distillation column, the low and high boiling components are initially separated at the top and bottom of the prefractionator, respectively. The low and middle boiling components are then fed into the main column through the top section while the high and middle boiling components are fed into the main column through the 1st column's bottom. In the main column, the middle boiling component is withdrawn from the tray of the middle section where the concentration of the middle component is maximum. Although, the Petlyuk distillation column structure is thermally integrated and has a significant advantage in energy saving but also has some difficulties in design and operation. A further step in development of the Petlyuk distillation column is realized by integration of the separation task within one column shell, referred to the DWC shown in Figure 2.[8] Its improved operability makes the DWC commercially attractive over the Petlyuk distillation column. The DWC consists of one column in which the prefractionator is separated from the main column by a divided wall. This structure offers a considerably lower investment as well as operating cost. Furthermore, the balance of pressure and operating problems are naturally solved.

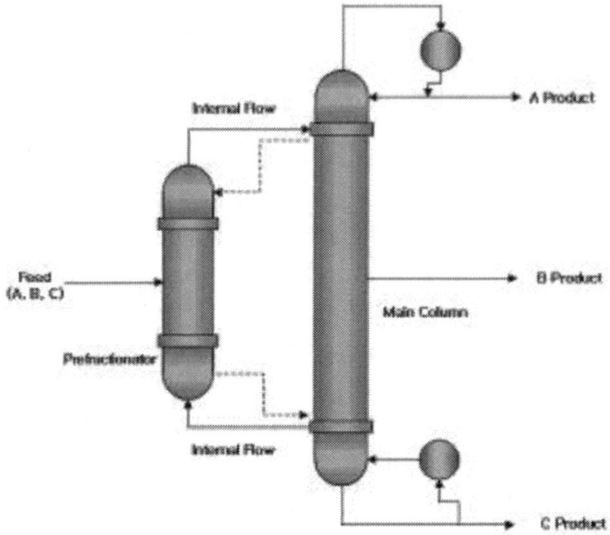

Figure 1. A schematic diagram of the Petlyuk column.

Although the DWC has many advantages, its application in chemical process industries has been still limited. This is due to the lack of proper design and control strategy. Other main difficulty is that the DWC is practically impossible to change the internal flow, which is most important operating variable to determine overall performance, once its structure is fixed because the prefractionator is internally integrated in the main column. Due to this limited flexibility, it is highly required the proper structure design with a rigorous simulation study. Even though the recent literature survey reveals that many studies for design and control of the DWC are available, a simple and efficient method to get the structure in the initial design stage is still rare. In addition, it is important to study the degree of energy saving for the DWC and the relation of main design factor with the pattern of internal flow distribution for optimal design.

In the proposed study, a simple shortcut method is developed to decide the structure of the DWC such as the feed and side-stream locations as well as the divided wall section. The method applies the well-known Fenske-Underwood equation on the equivalent three sequential conventional column configuration. Furthermore, the design characteristic is analyzed using the pattern of the rectifying and stripping flows of the 1st column.

Structure Design

In the DWC, a division wall is installed in the shell and separates it into the two sections, as shown in Figure 2. The remixing effect of the middle boiling component can be minimized with the divided wall.

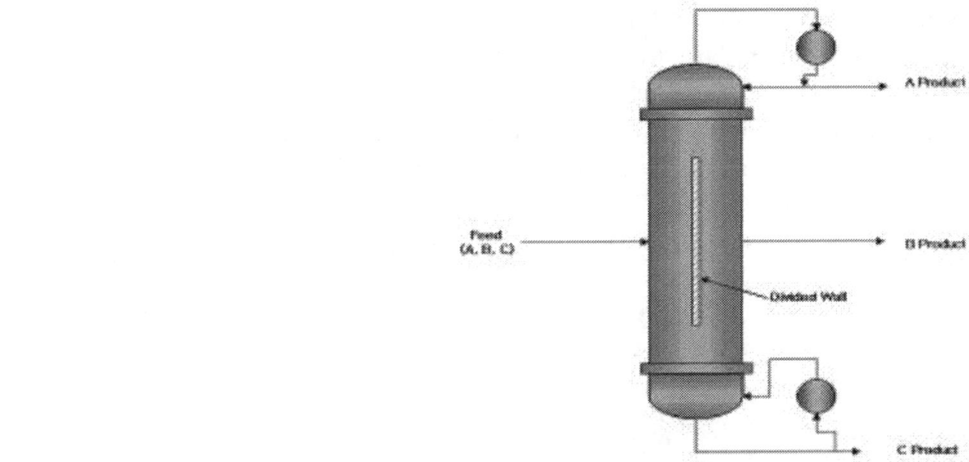

Figure 2. A schematic diagram of the dividing wall column.

The proposed a structure design method which determines the optimal dividing section using the equilibrium distillation curve for the liquid composition profile of the prefractionator and main section and thus finds the theoretical tray and reflux. In this study, however, we propose a new approach to decide the theoretical tray and reflux ratio using the Fenske-Underwood equation.

As shown in Figure 2, the structure of DWC can be divided into four section: the prefractionator section for the feed mixture, the top and bottom sections above and below the divided wall, and the divided wall section. The ternary mixture is feed to the prefractionator and is separated to the low and high boiling component. The middle boiling component is distributed to both the top and bottom of divided wall. The mixture of low and middle boiling components is separated in the upper section of the main column, and the mixture of high and middle components is separated in the lower section of the main column. The DWC structure can be equivalently presented using the conventional column configuration as shown in Figure 3.

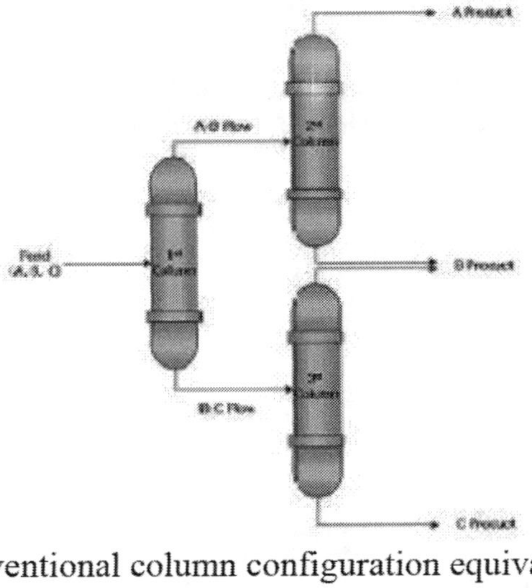

Figure 3. A conventional column configuration equivalent to the DWC.

In this conventional configuration, the 1st column can be matched to the prefractionator in the DWC. The rectifying section of the 2nd column and the stripping section of 3rd column are analogous to the top and bottom section in the DWC. Both the stripping section of the 2nd column and the rectifying section of 3rd column is equivalent to the divided wall section in the DWC. Furthermore, both the bottom stream from the 2nd column and the top stream from the 3rd column is analogous to the side stream of the DWC.

As shown in Figure 2, there are a split of internal liquid flow and that of internal vapor flow at the divided wall position. Therefore, the feed flow should be three in the 1st column because the liquid and vapor internal flow are fed to the prefractionator. In the structural design by the shortcut method, the reflux flow of the 1st column can be considered as the internal liquid flow fed to the prefractionator. Similarly, the boil-up flow of the 1st column is analogous to the splitted vapor internal flow into the prefractionator.

In comparison with the DWC in Figure 2, since the simple column configuration in Figure 3 is not thermally coupled, total energy consumption is likely to be different from each other. The number of trays for each main section in the resulting DWC designed by the proposed method might be more conservative to obtain the specification required. However, it is noted that the relative location of each major section is not much affected by this structural difference between the DWC and the equivalent conventional column configuration.

In this study, HYSYS^{TM} simulation program is used with the Fenske-Underwood equation for the design of column structure. The feed condition of ternary mixture is shown in Table 1.

Table 1. Feed condition of ternary mixture.

		F1	F2	F3
Feed Mixture	A: n-Pentane	A : 0.40	A : 0.33	A : 0.15
	B: n-Hexane	B : 0.20	B : 0.33	B : 0.70
	C: n-Heptane	C : 0.40	C : 0.33	C : 0.15

*1atm, 45kmol/h.

It is assumed the operating pressure is 1atm and the top, side and bottom product purity are required as 98 mol%, 98.5 mol% and 98 mol%, respectively. For the Fenske-Underwood shortcut method, the composition of light key component (n-Pentane) at the bottom and heavy key component (n-Heptane) at the top in the 1st column is set to less than 0.1 mol%. This is based on the fact that the separation efficiency is highest when the low (A) and high (B) boiling component are completely separated in the prefractionator section.

Table 2. Main structure of the equivalent conventional column configuration. (F1 Mixture)

	1st column	2nd column	3rd column
Ext. Reflux Ratio	0.898	1.608	6.86
Total Stage No.	18	14	14
Feed Stage No.	9	5	10
Rectify Vap/Liq(kmol/h)	41.87/19.81	47.75/29.44	36.49/31.85
Stripping Vap/Liq(kmol/h)	41.87/64.81	25.69/29.44	36.49/54.79

*The actual reflux ratio is set as double as the minimum reflux ratio.

The result of HYSYS™ simulation for the F1 mixture is given in Table 2. From the result in Table 2, the number of tray for the prefractionator column is set as that of the 1st column (18 stages) and the number tray for the main column is set as the sum of the 2nd and 3rd column trays (28 stages). As shown in the table, the feed tray of the DWC is the 9th tray and the divided wall is located from the 5th tray (i.e., feed tray of the 2nd column) to the 24th tray (i.e., sum of tray number of the 2nd column and the feed tray of the 3rd column) The side stream is drawn from the 14th tray.

To implement the DWC on the HYSYS™ environment, the Petlyuk column configuration is used because it is thermodynamically equivalent to the DWC. The resulting process flow diagram of the DWC for the HYSYS™ simulation is shown in Figure 4.

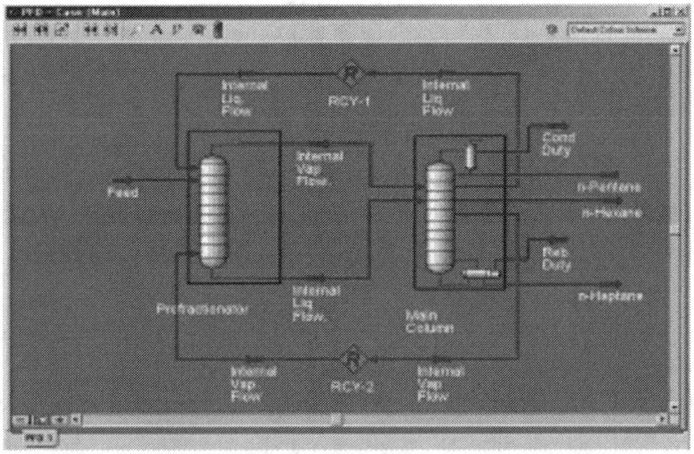

Figure 4. Process flow diagram of the DWC in HYSYS™.

In the simulation study, it is assumed that no heat transfer occurs cross the divided wall section.

Optimal Design of Internal Flow Distribution

Once the column structure is fixed, the distribution of the internal liquid and vapor flows to the prefractionator and main section is the most significant factor to affect the energy consumption and separation efficiency among all other design variables. Furthermore, the internal flow distribution is also the main factor to determine the easiness for realization of the DWC structure. In this study, the internal flow distribution is considered as an optimizing variable. Figure 6 shows the effect of the internal flow on the energy consumption. The figure clearly illustrate that there exists the optimal flow distribution to give the lowest energy consumption and the energy consumption is quite sensitive to the internal flow distribution. In this particular example, the lowest energy consumption is 1.760×10^6kJ/h when the internal liquid and vapor flows are 38kmol/h and 16kmol/h, respectively while the highest energy consumption for a whole searching region is 2.950×10^7kJ/h which is around 16 times larger than the lowest one.

The existence of the optimal internal flow distribution can be explained by the two competing effects of the internal flow as follows: Increase in the internal flow in the prefractionator improves the separation efficiency of the prefractionator and thus leads to the reduced reflux flow of the main column for a given product purity. At the same time, the increase of the internal flow in the prefractionator directly results in the increase of the internal flow in the main column to meet the internal mass balance, which requires more energy in the condenser and the reboiler. These two effects occurs competitively.

The positive effect by separation efficiency improvement dominates up to some internal flow and after then the negative effect by internal traffic increase becomes to be dominant. For this reason, the optimal internal flow distribution essentially appears.

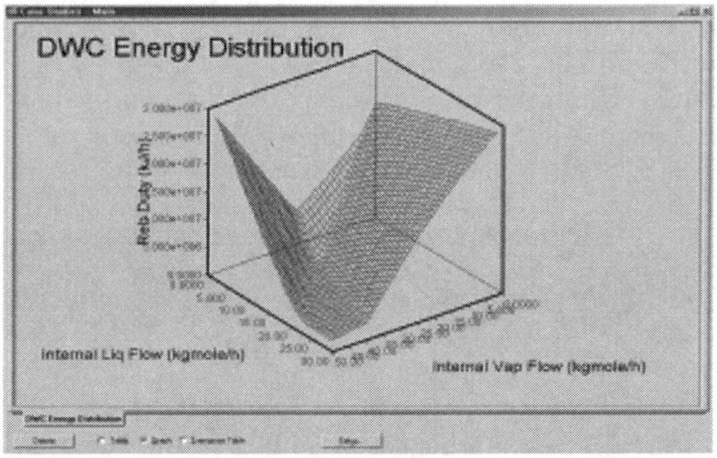

Figure 5. The energy consumption by internal flow distribution

The reflux and boil-up flow of the 1st column in the equivalent conventional column configuration proposed are analogous to the internal liquid and vapor flow in the prefractionator, respectively. In the proposed shortcut method, the reflux and boil-up flow of the 1st column are determined by the product purity of the 1st column which also affects the resulting DWC structure. Depending on the product purity of the 1st column, the internal flows of the main column as well as its structure is accordingly determined. Therefore, in the proposed method, how to select the product purity of the 1st column is very important not only to the optimal structure but also to the optimal internal flow distribution. The optimal internal flow of the DWC by optimization is compared with that by the shortcut method and the result is given in Table 3. As seen from the table, the energy consumption of DWC at the global optimum condition is 1.761×10^6kJ/h while the energy consumption of DWC by the proposed shortcut method is 1.790×10^6kJ/h. Although there is some difference in resulting internal flows, the difference in energy consumption between the two cases is insignificant just as 1.7%.

Table 3. Internal flow and energy consumption of the DWC based on the shortcut method and the optimization (F1 mixture)

	DWC (1st Column)	**DWC (case study)**	**Two sequential column**
Rectify Liq Flow/ Internal Liq Flow	19.81kmol/h	16.00kmol/h	
Stripping Vap Flow/ Internal Vap Flow	41.87kmol/h	38.00kmol/h	
Total Reb Duty (kJ/h)	1.790×10^6	1.761×10^6	2.630×10^6

Figure 6 shows the energy consumption vs. the feed and side-stream tray location in the DWC simulation study. In this simulation study the structure of shortcut column is same as the DWC's separation concept except the difference in the proposal of initial design and the structure of heat integration.

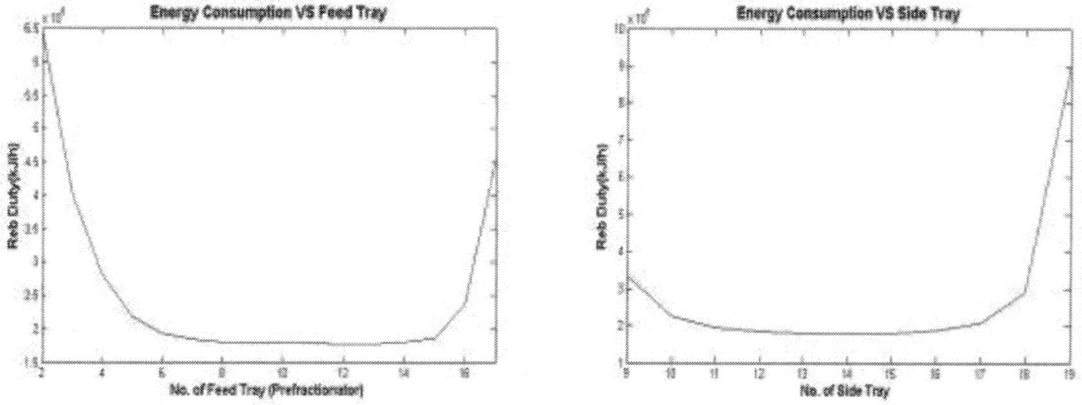

(a) (b)

Figure 6. Energy consumption vs. (a) the feed tray location and (b) the side stream tray location.

The minimum energy consumption is 1.780×10^6kJ/h when the feed tray is located at the 12th tray while that by the shortcut method is the 9th tray with the energy consumption of 1.790×10^6kJ/h. Note that the energy consumption is insensitive to the feed tray location for a wide range. In the case of side stream tray, the minimum energy consumption occurs at the 14th tray both from the shortcut method and the optimization. The above observation demonstrates that the proposed shortcut method is suitable for the optimal DWC structure design.

The separation characteristic of the DWC is closely related with the feed composition. To illustrate the effect of feed composition, we use three feed mixture cases where the composition of the middle boiling component are low, moderate and high. The internal flows with energy consumption of the DWC by the shortcut method and by the optimization are compared. The operating condition and the required product purity are same as in Table 1.

Table 4. Internal flow and energy consumption of the DWC based on the shortcut method and the optimization for different mixtures.

	Shortcut Column		**DWC**		**2 Column**
	Rectify Liq Flow (kmol/h)	Stripping Vap Flow (kmol/h)	Liq Internal flow (kmol/h)	Vap Internal flow (kmol/h)	
F1	19.81	41.87	16	38	
F2	23.18	45.64	15	36	
F3	15.48	30.92	14	35	
Reboiler Duty (10^6kJ/h)	F1 : 1.79 F2 : 2.02 F3 : 2.19		F1 : 1.76 F2 : 1.92 F3 : 2.05		F1 : 2.63 F2 : 2.85 F3 : 3.33

From Table 4, it is apparent that the DWC improves energy efficiency in any case. But the degree of energy saving is changed from 29% to 34%. Improvement of energy efficiency is most significant when the composition of middle boiling component is high. This result seems to be authentic because the energy efficiency by the DWC comes from the prevention of remixing. The difference of energy consumption between the DWC based on the proposed shortcut method and that by the optimization method is only 1.7~6.3%. It indicates the proposed method can be properly used to design the DWC.

Conclusions

This study emphasizes a shortcut method using the Fenske-Underwood equation for the design of DWC structure. The method utilizes the three conventional column configuration equivalent to the DWC to find the proper DWC structure in a simple manner. The DWC designed by the proposed shortcut method shows little difference from that through optimization study in terms of energy consumption. Extensive simulation study illustrates the proposed method is suitable to DWC structure design.

Acknowledgement
This research was supported by the 2006 Energy Resource and Technology Development Program of Korea.

References

1. R. Agrawal (2000), " A Method to Draw Fully Thermally Coupled Distillation Column Configurations for Multicomponent Distillation", *Trans IChemE*, April, pp 454-464.
2. J.G. Segovia- Hernández, V. Rico- Ramírez, A. Jiménez and S. Hernández (2004), "A Comparison of the Feedback Control Behavior Between Thermally Coupled and Conventional Distillation Schemes", *Computers and Chem. Eng*, pp 811-819.
3. A. Jiménez, N. Ramírez, A. Castro and S. Hernández (2003), "Design and Energy Performance of Alternative Schemes to the Petlyuk Distillation System", *Trans IChemE*, May, pp 518-524.
4. I. J. Halvorsen and S. Skogestad (2003), "Minimum Energy Consumption in Multicomponent Distillation. 2. Three-Product Petlyuk Arrangements", *Ind. Eng. Chem. Res*, pp 605-615.
5. E. A. Wolff and S. Skogested (1995), "Operation of Integrated Three-Product (Petlyuk) Distillation Columns", *Ind. Eng. Chem. Res.* pp 2094-2103.
6. I. J. Halvorsen and S. Skogestad (1999), "Optimal Operation of Petlyuk Distillation: Steady-State Behavior", *Journal of Process Control*, pp 407-424.
7. I. J. Halvorsen and S. Skogestad (2004), "Shortcut Analysis of Optimal Operation of Petlyuk Distillation", *Ind. Eng. Chem. Res*, pp 3994-3999.
8. M. I. Abdul Mutalib, A. O. Zeglam and R. Smith (1998), "Operation and Control of Dividing Wall Distillation Columns, Part1: Degrees of Freedom and Dynamic Simulation", *Trans IChemE,* March, pp 308-318.

326d Influence Of Mass Transfer On Conceptual Design Of Distillation Processes

Ross Taylor, Chemical Engineering, Clarkson University, Potsdam, NY 13699 and Angelo Lucia, University of Rhode Island, Dept of Chem. Engineering, Kingston, RI 02881.

In this paper we take a look at the influence of mass transfer on distillation column design. The column design equations incorporating mass transfer effects are similar to the conventional design equations and are as easy to solve numerically.

Lucia and Taylor (2006) have shown that the boundaries in ternary residue curve maps can be defined as the locally longest residue curves in a given region of the composition triangle. Residue curves are equivalent to distillation column profiles under the assumption of equilibrium at every point in a packed column that is operated at total reflux. It is well known, however, that real distillation columns do not operate at equilibrium and the composition profiles are influenced by the rates of interphase mass transfer.

Here we provide evidence that the boundaries in nonequilibrium ternary distillation processes are the longest composition trajectories in a given region of composition space.

We further show that the minimum energy design for a specified separation can be defined by the shortest feasible stripping trajectory (when measured from the bottoms composition to the pinch point). These properties of the distillation trajectories corresponding to boundaries and to minimum boilup/ reflux are the basis for optimization based algorithms that can be used for the determination of energy efficient distillation processes when mass transfer effects are important.

All this means that mass transfer effects can be considered at the conceptual design stage, instead of in later phases of process design. This can be significant in some cases because it is possible, in principle at least, for mass transfer effects to make possible designs that would not normally be considered if a purely equilibrium based model was used as the basis. More importantly, mass transfer effects can render ineffective designs based solely on equilibrium considerations.

326e Energy Efficient Reactive Distillation

Ross Taylor, Chemical Engineering, Clarkson University, Potsdam, NY 13699 and Angelo Lucia, University of Rhode Island, Dept of Chem. Engineering, Kingston, RI 02881.

Residue curves and residue curve maps have been extensively studied for over 100 years. Much of the literature in this field deals with systems that do not react. Chemical reactions can influence residue curve maps in some important ways. For example, it is known that reactions can lead to both the appearance and disappearance of stationary points (azeotropes), and that reactive azeotropes can exist even in systems that otherwise would be considered thermodynamically ideal. It follows that chemical reactions can influence the very existence of separation boundaries and, therefore, the design and synthesis of reactive separation processes.

Lucia & coworkers have shown that distillation boundaries in non-reacting ternary liquid mixtures can be defined as local maxima in the line integral from any unstable node to all reachable stable nodes. In addtion, they have shown that the minimum energy design for a specified separation can be defined by the shortest feasible stripping trajectory (when measured from the bottoms composition to the pinch point). Among other things this approach:

Easily finds minimum energy solutions that do not correspond to separation pinch points.

Is unaffected by the number of components.

Provides knowledge of other solutions that have near minimum energy consumption.

Solves problems other synthesis methodologies cannot.

This paper shows how the geometric approach of Lucia and coworkers can be adapted for the rigorous determination of minimum energy reactive distillation systems. Examples include 4 component systems with a single equilibrium reaction and ternary systems with a kinetically controlled reaction.

326f Problems Arising Out Of Impurities In Steam Turbines

Ehsan Bakhshi, R&D Center of National Petrochemical Company, Tehran, Iran and Fatemeh Abniki, Process Eng.Dept. of National Petrochemical Company,Ghadir Group, Tehran, Iran.

Different impurities are transported into the steam turbine from the Preboiler cycle in the feedwater and the boiler as a total (mechanical plus vaporous) carryover . Their major sources include condenser leak , make up water , and improperly operated condense polishers . The turbine environment is controlled through a control of impurity ingress and various feed water , boiler water , and steam chemistry limits . The corrosiveness of the steam turbine environment is caused by one or more of following : 1) Concentration of impurities from low part per billion levels in steam to percent levels on surfaces and the formation of concentrated aqueous solutions (concentration by deposition or evaporation of moisture) 2) Insufficient pH control (in both acid and alkaline regions) 3) High-velocity , high-turbulence , and low-pH moisture The steam impurities that are of most concern include chlorides , sulfates , fluorides , carbonates , hydroxides , organic and inorganic acids , oxygen , and CO_2 .Their behavior in turbine steam and deposits is well documented . There are strong synergistic effects and interactions with metal oxides . In addition to corrosion during operation , turbines can corrode during : 1-Manufacture (machining fluids and lubricants) , 2- Storage (airborne impurities and preservatives) , 3- Erection (airborne impurities , preservatives , and cleaning fluids) ,4-Nondestructive testing (cleaning and testing fluids) 5- Lay up (deposits plus wet air) Many of the above substances may contain high concentrations of sulfur and chlorine , which could form acids upon decomposition . Decomposition of typical organic for example , carbon tetra-chloride(CCl_4) , occurs at about 150 C . there fore , the composition of all of the above substances should be controlled and most of them should be removed before operation .

Web Page: www.nipc.com

326g Bioseparations Using Magnetic Fluids

Saurabh Tejwani, **Fei Chen**, Kenneth A. Smith, and T. Alan Hatton. Department of Chemical Engineering, Massachusetts Institute of Technology, 77 Massachusetts Avenue, Cambridge, MA 02139

We are developing computational and simulation tools to provide an in-depth understanding of the thermodynamic and dynamic properties of suspensions of nonmagnetic particles immersed in magnetic fluids. A fundamental understanding of these properties is important in the development of magnetic separation systems. The trajectories of magnetic particles in a modified rhombic multi-wire array are simulated in a combined magnetic and flow field. The magnetic field in the array is calculated by taking superposition of single wires and the flow field is obtained by interpolation of a numerical solution of the Navier-Stokes equation. We have explored the effect of the magnetic field on the nonmagnetic particles. The methods and conclusions presented here can be extended to novel magnetic separator design and help us understand the entire separation process.

326h Characteristics Of CO2 Absorption Into Aqueous Ammonia

Kwang Bok Yi[1], You Jeong Kim[2], Chang Hyun Ko[1], Jong-Ho Park[1], and Jong-Nam Kim[1]. (1) Chemical Process Research Center, Korea Institute of Energy Research, 71-2 Jang-dong Yuseong-ku, Daejon, South Korea, (2) Korea Advanced Institute of Science Technology, 374-1, Guseong-dong, Youseng-gu, Daejon, South Korea

Many researches have been performed for CO_2 capture from various industrial gas streams in order to reduce the contribution of CO_2 to global warming. Although being the most effective for CO_2 capture, the chemical absorption process based on amines such as monoethanolamine (MEA) is still too expensive to apply for large CO_2 sources like power plants. In addition, it has been well known that the amine process has inherent problems such as absorbent degradation by acidic compounds and oxygen in gas stream, high energy consumption, and equipment corrosion. Recently, aqueous ammonia has been suggested as a new absorbent which can avoid the above problems. The aqueous ammonia has larger theoretical capacity of CO2 compared to amine solution, low regeneration energy required, low material cost, and potential ability to capture acidic gases in flue gas as well. However, there are only few research reports available in the use of aqueous ammonia for CO_2 capture. Thus, the characteristics of CO_2 absorption into ammonia have not been fully investigated and it requires further in-depth study. In this work, the effects of temperature and concentration of aqueous ammonia on CO_2 absorption capacity were investigated using a semi-batch reactor, and the concentration changes of carbon containing species (ammonium carbonate, ammonium bicarbonate, and ammonium carbamate) during the absorption were also studied using Raman spectroscopy. The bubbling enhanced feature of the reactor reduced the gas-liquid mass transfer resistance remarkably. Hence, the CO2 gas could be absorbed more rapidly into aqueous ammonia. The extremely sharp breakthrough curves were obtained by the reactor and the breakthrough point at given condition was used for interpreting characteristics of CO_2 absorption into aqueous ammonia.

Effect of polymer concentration on surface characteristics and performance of cellulose acetate membranes fabricated for on-chip applications

Jermey N.A. Matthews[a], Hisham Mohamed[b], James N. Turner[b, c], Kimberly L. Jones[a*]

[a]Department of Civil Engineering, Howard University, Washington DC 20059
[b]Wadsworth Center, Albany, NY 12201-0509
[c]Department of Biomedical Sciences, The University at Albany, Albany, NY 12201-0509
*CORRESPONDING AUTHOR EMAIL: kljones@howard.edu; FAX: 202-806-5271

1. Introduction

Membrane filtration processes continue to rank high as efficient and cost-effective means for industrial applications involving treatment and purification of liquids and gases. Recent literature has attempted, with moderate success, to deduce and predict the relationship between membrane property and performance. Experiments and models alike have qualitatively shown a relationship between membrane properties such as pore size, charge density, surface roughness, chemical structure, and hydrophilicity, as it relates to the key performance outputs of flux/permeability, rejection, and fouling. Quantitative data that correlates well with theory is still, however, needed.

Research in our laboratories recently developed a novel method of spin casting (normally used in microelectronics) followed by phase inversion/immersion precipitation for the casting of cellulose acetate polymeric membranes directly unto silicon wafers[1]. Cellulose acetate is a biopolymer with good chemical and mechanical stability that has been used in a vast array of industries and applications[2]. A recent review by de Jong et al.[3] declared the great potential of the casting/phase inversion process for novel microfluidic applications since there is a vast knowledge base for the immersion precipitation method - an established process involving the diffusion of non-solvent molecules through the polymer/non-solvent interface where the dense polymer barrier layer is formed. The solvent exchange leads to subsequent formation of macrovoids below the barrier layer. The porosity has been shown to decrease and skin thickness increase when the initial polymer concentration increases, inhibiting the inward diffusion of the non-solvent[4]. Changing the initial concentration of the polymer is therefore a useful way of controlling porosity, membrane thickness, and surface morphology.

The focus of this work was to investigate the effect of the initial polymer concentration on the surface properties and performance of cellulose acetate membranes fabricated by the SCIP method for on-chip applications. Specifically, the surface properties measured were zeta potential, roughness, hydrophilicity, and effective pore size and the performance parameters were permeability (flux) and salt rejection.

2. Experimental Methodology

Membrane Fabrication

Cellulose acetate solutions were prepared by dissolving cellulose acetate (39.8 wt% acetyl content, Mn=30,000), in N,N–dimethylacetamide, yielding 10, 25, 17.5, and 40% (w/v) solutions. Casting was accomplished using an AC201 Photo Resist Spinner (Headway Research, Inc., Garland, TX) to spin a small volume of the polymer solution for 20 seconds at 2000 rpm across a Celgard™ 2400 microporous membrane. After spinning, the coated Celgard was immersed in a water bath at room temperature for 10 minutes, removed and allowed to dry in(a) air (b) for 3 hours at 50 0C in a vacuum oven. All steps are performed in a clean room.

For the TFC polyamide membrane, the active layer is formed via interfacial polymerization of the amine monomer, piperazine and trimesoyl chloride (TMC) as a cross-linker on a support layer of a hydrophilic poly (ether sulfone) (PES) membrane obtained from GE Osmonics. In the modification steps, the amine monomer is co-polymerized with 2-aminobenzoic acid, which adds more negatively charged functional groups (carboxyl) to the surface.

The dry support PES membrane will be allowed to soak in water for a period of 24 hours then rinsed to remove any preservatives. The PES membrane will then be taped to a glass plate and dipped vertically in the aqueous monomer solution for a period of 5 minutes. The plate is then removed and a rubber roller is used to remove the excess reactant. The glass plate is then dipped into the organic solution and the reaction is allowed to react at the interface for 10 seconds. It is then removed, allowed to drain, and then dried in a vacuum oven for at least 12 hours. The concentration of the cross-linker, TMC, will be altered and the pore size will be calculated from pressure-driven rejection of uncharged solutes and/or AFM.

Flux and Rejection Experiments

The pure water permeability and uncharged solute rejection experiments were performed in a HP 4750 Stirred Cell (Sterlitech, Inc.) which had a capacity of 250 mL and an effective membrane area of 14.6 cm^2. For rejection experiments, the uncharged molecule, Vitamin B_{12} (Sigma-Aldrich) dissolved in ultrapure water was used. Vitamin B_{12} has a molecular weight of 1350 Da and a hydrodynamic radius of 0.85 nm. The feed concentration of vitamin B_{12} was 100 mg/L. Subsequent to pre-compaction with water, the rejection experiments follow the same protocol of the water flux experiments. The permeate concentration was analyzed and measured by a TOC (total organic carbon) Analyzer (Tekmar-Dohrmann) or UV-Vis (Shimadzu Instruments).

Characterization

The membranes were characterized by scanning electron microscopy using a Leo 1550VP FESEM (Leo Electron Microscopy Inc., New York). Fracture surfaces were obtained by freezing the sample in liquid nitrogen prior to breaking and were sputter coated with gold for 1 minute prior to observation.

The zeta potential was measured by streaming potential with the EKA Analyzer (Brookhaven Labs) over a pH range of 3-9 using 1 mM KCl electrolyte.

Contact angle was measured on a goniometer fitted with a light microscope and a scale. The membrane was placed unto the open sample stage, taped down, and water droplets were placed on the membrane. Measurements were done in a clean room and the results of three trials were averaged.

Dry membrane samples were analyzed by ATR-IR (Brucker Vertex 70 FT-IR with a Pike Miracle Diamond ATR accessory) using a variable angle grazer. Results were normalized to untreated Celgard membrane and reported as absorbances.

Samples were imaged by AFM in a PicoPlus fluid cell (Molecular Imaging, Inc.) under Magnetic AC (MAC) mode in 1 mM KCl electrolyte solution. The cantilever was the MAClever Type IV (Molecular Imaging, Inc.) made of silicon nitride with a resonance frequency of 15 kHz and a force constant of 0.02 N/m.

3. Results and Discussion

Cellulose acetate composite membranes with controlled properties were successfully fabricated by a novel technique, SCIP, which integrates spin casting with the immersion precipitation method. The results are summarized in Table 1. The effective pore radius and hydrophilicity decreased with increasing initial polymer concentration, while the effective surface charge density increased. There was no significant change in root-mean-square surface roughness. The pure water permeability of the membranes decreased with increasing initial polymer concentration (Figure 1), whereas the salt rejection increased. Rejection was nearly twice higher for divalent SO_4^{2-} as compared to monovalent Cl^- for all concentration levels (see Figure 2). The feasibility for on-chip applications has already been demonstrated for effective separation the heme molecule (614 Da) from ssDNA[5]. Potential applications also exist for removal of small molecules and charged ions in water treatment.

Acknowledgements

The authors thank Dr. Christopher (Kit) Umbach and Magnus Bergkvist at Cornell University for help with AFM and ATR-IR, respectively, Dr. Hisham Mohamed and Dr. Don Szarowski of the Wadsworth Center (Albany, NY) for fabrication and SEM support, Future Digital Scientific Corporation (Long Island, NY) for help contact angle measurements and analysis, and Celgard, Inc. for sample Celgard membranes. The authors are also grateful to grant support from the Nanobiotechnology Center (NBTC) (Contract ECS-9876771) at Cornell University, Dr. Malaisamy Ramamoorthy for helpful conversations, and the following undergraduate students for experimental support – Zenobia Lewis, Alaina Tarver, and Michael Chestnut.

Table 1. Surface properties of the fabricated cellulose acetate membranes

Polymer Concentration (% CA)	**10 %CA**	**17.5 %CA**	**25 %CA**	**40 %CA**
Surface Roughness, RMS, [nm]	100.0	93.5	102.2	104.1[a]
Normalized Absorbance at 1750 cm^{-1} (C=O bond)	0.94	1.05	1.18	1.19
Normalized Absorbance at 3500 cm^{-1} (O-H bond)	0.11	0.13	0.15	0.16
Zeta Potential at pH 8	-17	-19	-21	-30
Iso-electric Point, IEP	3.5	3.5	3.6	3.7
Contact Angle, θ	34.9 ± 0.6	32.5 ± 1.4	37.0 ± 0.8	48.3 ± 0.8

[a] Performed on a JEOL AFM

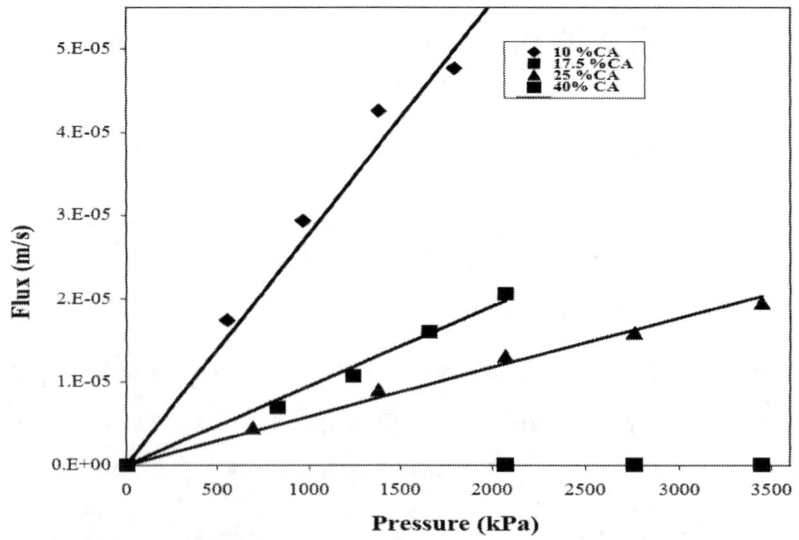

Figure 1. Pure water flux as a function of pressure for CA membranes with different initial polymer concentrations.

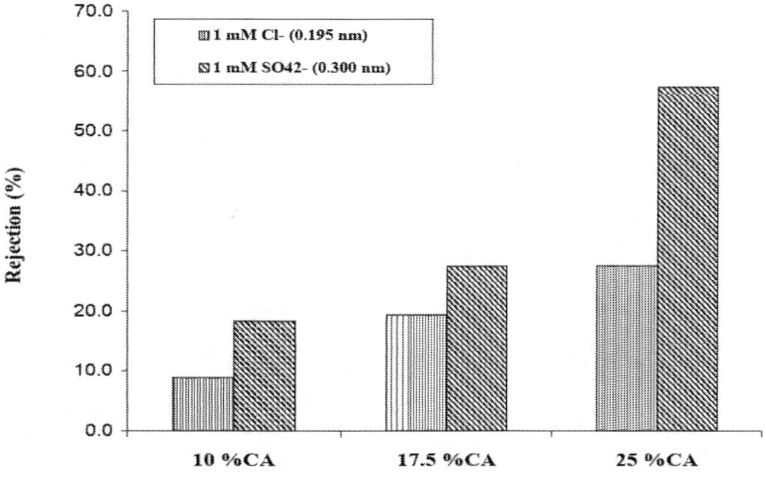

Figure 2. Anion rejection by CA membrane of $CaCl_2$ and $CaSO_4$ solutions.

References

[1] A.P. Russo, S.T. Retterer, A.J. Spence, M.S. Isaacson, L.A. Lepak, M.G. Spencer, D.L. Martin, R. MacColl, J.N. Turner, Separation Science and Technology 39 (2004) 2515.

[2] K.J. Edgar, C.M. Buchanan, J.S. Debenham, P.A. Rundquist, B.D. Seiler, M.C. Shelton, D. Tindall, Progress in Polymer Science 26 (2001) 1605.

[3] J. de Jong, R.G.H. Lammertink, M. Wessling, Lab Chip 6 (2006) 1125.

[4] C.S. Tsay, A.J. McHugh, Journal of Membrane Science 64 (1991) 81.

[5] H. Mohamed, A.P. Russo, D.H. Szarowski, E. McDonnell, L.A. Lepak, M.G. Spencer, D.L. Martin, M. Caggana, J.N. Turner, Journal of Chromatography A 1111 (2006) 214.

Separation of Ink from Recycled Paper Assisted by Electric Field

Akpojotor Shemi, Jeff Hsieh

The behavior of flexographic inks, hydrophilic and submicron particles, in cellulose fiber slurry during flotation de-inking is model. The literature includes model work on the collection of fine ink particles on bubbles in an electrochemical cell. The theory is based on the bubble collection calculated from the particle concentration distribution. The flow around the bubble was described as creeping flow, while the distribution of particles around the pulp was calculated from diffusion equation. But the system studied was homogeneous (i.e. pure ink dispersion). Some experimental work shows that when fiber is added into the system (i.e. heterogeneous system of fiber and inks) the removal of inks by flotation is markedly reduced. The reduced flotation efficiency is due to the deposition of ink on the fiber surface. The ink also get deposited in fiber lumen as well as the interstitial parts of the fiber network during transport. The hydrophilic ink interact well with the hydrophilic fiber.

The fiber network hinders the agglomeration of the inks and reduces the number of collision between the inks and bubbles. Floatability of hydrophobic ink on the other hand is not affected by the presence of fiber in the suspension. Because hydrophobic ink are less likely to deposit on the fiber and repel water, they collects on the rising bubbles. For transport of the hydrophilic ink through the fiber, the size of the pore of the individual fiber and the fiber network is also important. The degree of swelling of the fiber is a function of fiber consistency (i.e. water volume), fiber type and pH of the medium.

The efficiency of the surfactant chemicals is compromise because there is competitive sorption of the chemicals by both the fiber and ink. The flotation efficiency is also reduced in the presence of fiber because of bubble coalesce. For submicron particles, their collection efficiency on bubble decreases with bubble size

The electrochemical cell has been used to float micron size particles. The electrochemical cell is based on the release of ions from the electrodes and the sorption of ions on the particles resulting in the coagulation and agglomeration of the ink on the fiber. The agglomerated inks are then floated by gas bubbles. The gas bubbles are generated by the electrolysis of water which results in the evolution of hydrogen and oxygen bubbles from the electrodes. Modeling the entire system and the intricate chemical interactions is difficult. The model only consider the some important mechanism

The steps in the involves in the model include

1) The transport of ink through the porous fiber. Modeled as diffusion of a single particle. A material balance exist around the fiber.

2) The agglomeration of the ink by the action of metal ions on the inks

Modeled as the two interacting particles. A material balance accounts for coagulated particle and un-coagulated particles

3) The collection of the ink on the fiber. Model as the collection of particle on the surface.

Once all the steps are accounted, the collection efficiency of the single particle is used to determine the collection efficiency of the entire ink concentration.

326k Design Of Reactive Distillation Systems Using A Process Systems Engineering Approach

Alejandro O. Domancich, Chemical Engineering Department, Planta Piloto de Ingenieria Quimica - Universidad Nacional del Sur, Camino La Carrindanga km 7, Bahia Blanca, 8000, Argentina, Nelida B. Brignole, Computing Science Department, Planta Piloto de Ingenieria Quimica - Universidad Nacional del Sur, Camino La Carrindanga km 7, Bahía BLanca, 8000, Argentina, and **Patricia M. Hoch**, Plapiqui - UNS - Conicet, CC 717, Bahía Blanca, Buenos Aires 8000, Argentina.

During the last years, process intensification received a great deal of attention because of the potential cost savings. By process intensification it is meant the significant improvement achieved by a process, at a lower total cost. For example, the entire plant design is upgraded by reducing the number of units involved in a conventional process to obtain the same products. Reactive distillation columns constitute a clear example, because it is possible to use a single piece of equipment in order to perform two tasks: reaction and separation of products, otherwise performed separately.

However, reactive distillation is not suitable for every process where reaction and separation steps occur. Operating conditions, such as pressure and temperature of the reactive and separation processes and perhaps other requirements, must overlap in order to assure the feasibility of the combined process. This limitation can be overcome by fixing adequate operating conditions in the cases where this is possible.

Reactive distillation has been presented as a multifunctional reactor, where the reactive and separation tasks are combined into a single unit, thus reducing investment costs. The advantages of using this configuration have already been reported (Stankiewicz, 2003). A thorough review on the design methods for reactive distillation can be found in Almeida-Rivera et al. (2004). However, a systematic analysis of the performance of reactive distillation columns changing structural variables has not been presented previously. The simplest case exhibits reactive and non-reactive sections with no interrelationship between them, except at the equipment level. Even when the interrelations between sections remain unchanged compared with traditional technology; significant savings (2.5 times as big as the equipment expenditure) can be obtained because of the smaller and cheaper plant needed to perform the same tasks. A clear example is the Urea2000plus™ technology, by Stamicarbon B.V., where a pool reactor is used to combine the urea reactor, the carbamate condenser and the inerts scrubber. However, further improvements can be obtained by really integrating the tasks, on the following items:

• On the reaction: because there is an equilibrium displacement, since the products are being withdrawn

• On the separation: because of the changes in the driving force for mass transfer due to the reaction.

This case is formally presented as a reactive separation, also called a separative reaction. Both tasks occur at the same level. Several disadvantages can be overcome by using these combined schemes, because the selectivity and the reaction yield increase. This makes it possible to avoid thermodynamic constraints, such as azeotropes, thus allowing a considerable reduction not only in the investment costs but also in the operating costs (energy, water and solvents, otherwise required as entrainers). On the other hand, the mathematical model for this piece of equipment is more complex, and the requirements for simulating the combined process are higher. Non-linear coupling of reaction, phase equilibria, and mass and energy transport can give rise to many undesired effects, for example the appearance of reactive azeotropes (called arheotropes) and the multiplicity of steady states.

The performance of reaction with separation in one piece of equipment offers distinct advantages over the conventional, sequential approach (Podrebarac et al, 1998). As advantages of this integration, chemical equilibrium limitations can be overcome, higher selectivities can be achieved, the heat of reaction can be used for distillation, auxiliary solvents can be avoided, and azeotropic mixtures can be more easily separated than in conventionally distillation. This may lead to an enormous reduction of capital and investment costs and may be important for sustainable development due to a lower consumption of resources (Al-Arfaj and Luyben, 2000). Some industrial processes where reactive distillation is used are the sterification processes, such as the synthesis of methyl acetate (Agreda et al, 1990); and the preparation of ethers, like MTBE (Jacobs and Krishna, 1993), TAME and ETBE, used as fuel additives. An explanation for the occurrence of steady-state multiplicity was provided by Hauan et al (1995). A thorough review on the modeling aspects of reactive distillation can be found in Taylor and Krishna (2000), where special emphasis is put on the differences between equilibrium and non-equilibrium based models with their advantages and drawbacks. Doherty and Malone (2001) gave valuable commentaries on future trends and challenges, and a thorough review on the design methods for reactive distillation can be found in Almeida-Rivera et al. (2004).

We present the optimal design of a reactive distillation column, based on a model presented by Almeida-Rivera (2005) and further modified by Domancich et al (2006). The model comprises a combination of pure separation stages and a reactive distillation sector, where both reaction and separation take place at the same time.

For this study, we use a rigorous modeling of the column, with an equilibrium model for the reactive sector. The model of the column includes integer variables for defining the total number of stages, the number of reactive stages and feed locations. The economic optimization of the reactive distillation process is performed using a MINLP algorithm implemented in GAMS (Brooke et al, 2004) and the cost of the resulting scheme is compared to the cost of the conventional process considering that reaction and separation take place in different equipment, showing significant savings.

References

Agreda V.H., L.R. Partin and W.H. Heise (1990). Chemical Engineering Progress 40, 40-46.

Al-Arfaj M. and W.L. Luyben (2000). Industrial and Engineering Chemistry Research 39, 3298-3307.

Almeida-Rivera C.P., (2005) Designing reactive distillation processes with improved efficiency, PhD

Thesis, Technical University of Delft.

Almeida-Rivera C.P., Swinkels P.L.J and J. Grievink, (2004), Designing reactive distillation processes: Present and future, Computers and chemical engineering, 28, (10), pp. 1997-2020.

Brooke A., D. Kendrick, A. Meeraus, R. Raman 2004, GAMS: A Userxs guide.

Doherty M. and M.F. Malone (2001). Conceptual design of distillation systems. McGraw-Hill, USA.

Domancich A.O., A.C. Olivera, N.B. Brignole, P.M. Hoch. 2006, "Performance analysis of reactive distillation columns: adjustment of reactive and non-reactive stages", Proceedings of the XXII Interamerican Congress of Chemical Engineering. ISSN 1850 3535.

Jacobs R. and R. Krishna (1993). Multiple solutions in reactive distillation for methyl tert-butyl-ether synthesis. Industrial and Engineering Chemistry Research 32 (8), 1706-1709.

Stankiewickz, A., (2003), Reactive separations for process intensification: An Industrial Perspective, Chemical Engineering and Processing, 42, pp. 137-144.

Taylor R. and Krishna R, (2000), Modelling reactive distillation, Chemical Engineering Science 55, pp. 5183, 5529.

326l Separation Of Organic Acids From Fermenation Broths Using Electrodeionization: Fundamental Determination Of Wafer Thickness On Transport Properties

Heath Himstedt and Jamie A. Hestekin. Ralph E. Martin Department of Chemical Engineering, University of Arkansas, 3202 Bell Engineering Center, Fayetteville, AR 72701

Bioproccessing offers advantages over traditional processing including fewer by-products and milder operating conditions. Of particular interest is biocatalyst-driven separation of organic acids for applications in industry and medicine. Currently these are quite expensive to produce due to the difficulty of separation from dilute bioprocessing solutions. The proposed technology we will study is wafer enhanced electrodeionization (WE-EDI), which uses ion exchange wafers and an electric field to selectively remove ions from uncharged species even at very low concentrations. WE-EDI is highly pursuable and desirable because of the large binding surface area, concurrent product removal and production, high efficiency and ease of operations compared to loose resins. Current modeling efforts for EDI do not fully take into account water splitting, ion selectivity, and flow rates. We will present some new modeling that will give a better understanding of this technique and suggest applications for future work in this area.

326n Calculation Of Distillation Energy For Life Cycle Inventories

Evan, M. Griffing and Michael R. Overcash. North Carolina State University, Raleigh, NC 27695

Life cycle inventories (LCI) of products and processes can be efficiently created using data from open literature combined with chemical engineering process heuristics. This endeavor requires a simple and reasonably accurate heuristic for each unit operation. We present a distillation model to determine reflux ratios and energy requirements, and compare the results to reflux ratios of real columns in the literature. Distillation is an energy intensive processing step in the chemical industry. It is the most common separation technique for miscible liquids and typically accounts for over half of the process heat requirements of a chemical manufacturing plant (Kunesh et al. 1995). Due to the large energies consumed, an accurate distillation heuristic is important for meaningful inventory calculations. Furthermore, determining the total energy requirement of a distillation column is more challenging than determining the energy of other important unit operations, such as reactors. The basic steps involved in modeling, designing, and optimizing distillation columns are well described in the literature. An economic trade-off of operating and capital costs leads to an economic optimum reflux ratio (Ropt). This optimum reflux is often reported to be 1.2 to 1.3 times the minimum reflux (Rm). Some situations, such as high energy costs, refrigeration, or high material costs cause deviations from this rule. Using the assumption that the actual reflux ratio (R) is a multiple of Rm, we use the following heuristic to determine the energy requirements of a distillation column: (1) Specify feed composition and condition as well as separation efficiency (2) Calculate Rm (3) Determine R from Rm and calculate energy required to condense the flow in the top of the column (4) Assume that energy in the reboiler replaces energy removed in the condenser and supplies specific heat requirements. The level of detail and accuracy in Rm calculations can vary greatly. The distillation model and the vapor liquid equilibrium data both add to the range of modeling choices. A very simple method is to correlate Rm with the boiling point difference of key components. Various short-cut methods and process simulators add complexity, but provide additional accuracy. When a life-cycle inventory is presented, many distillation columns are included. Therefore, transparency requires the model and calculations to be simple and easily verified and understood. Although many texts give the optimum reflux as a multiple of Rm, there are very few real operating columns that are well specified in the literature. The reflux ratios of all distillation columns found in the literature have been tabulated and compared to the values of Rm calculated with a short-cut (Underwood) method. A linear fit produced the ratio R = 1.5* Rm. We have incorporated this method with a simple vapor pressure model into our computational tool for determining LCI data.

326o Extraction Dynamics of Ethyl Ester of Docosahexaenoic Acid with Silver Ion with Slug Flow Prepared by Microreactor

Eiji Kamio[1], Yu Seike[1], Hidekazu Yoshizawa[1], and Tsutomu Ono[2]. (1) Department of Environmental Chemistry and Materials, Okayama University, 3-1-1, Tsushima-naka, Okayama, Japan, (2) Department of Material & Energy Science, Okayama University, 3-1-1, Tsushima-naka, Okayama, Japan

Over the last several decades, N-3 polyunsaturated fatty acids (PUFA) such as docosahexaenoic acid (DHA) and eicosapentaenoic acid (EPA) have been widely noticed for human health. They are contained in fish oils with many other fatty acids. Therefore, purified and concentrated N-3 PUFA which is separated from the other PUFA and saturated fatty acids has to be used for the medical supplies and food additives. For separation and purification of PUFA, liquid-liquid extraction of PUFA ethyl ester (PUFA-Et) with silver ion has attracted attention and has been expanded into more effective systems utilizing supported liquid membrane, emulsion liquid membrane and slug flow prepared by microchip. Among these techniques, we have directed our attention to utilizing slug flow prepared by microchip. Recently, we investigated the extraction of DHA-Et and EPA-Et with the slug flow prepared by T-shape microchip and achieved ultra fast and effective extraction of them. In this study, we carried out a kinetic investigation of DHA-Et extraction with silver ion as the extractant. We used a stirred transfer cell to investigate the extraction rate of DHA-Et. The effect of stirring speed, DHA-Et concentration, silver ion concentration and temperature on initial extraction rate was measured. From the knowledge, we determined the kinetic parameters for complex-formation reaction between DHA-Et and silver ion. In addition, the overall rate equation containing the diffusion resistances of DHA-Et across the liquid laminar film near the interface between aqueous and organic phases was proposed in order to predict the uptake curves of DHA-Et extraction. The calculated uptake curves showed good correlation to the experimental data for the extraction of DHA-Et with the stirred transfer cell and for the ultra fast extraction of it with slug flow prepared by T-shape microchip. By comparing the calculated and experimental uptake curves resulted for the slug flow system, we discuss the characteristic of slug flow and estimate the influential parameter in the ultra fast extraction. The theoretical investigation suggested that the ultra fast extraction realized for the slug flow system is due to the large specific interfacial area of the slug caused by the presence of wall film and the thin liquid laminar film adjacent to the interface caused by the internal circulation.

326p Design of Distillation Process for Separation of Hfp/tfe

Inho Song[1], Hyoseong Ahn[1], Hyunsoo Jeon[1], Houn-Kyu Joung[1], In-Joon Park[2], Soo-Bok Lee[2], and **Yongtaek Lee**[1]. (1) Chemical Engineering, Chungnam National University, 220 Gung-dong, Yuseong-gu, Daejeon, 305-764, South Korea, (2) Interfacial Material Division, Korea Research Institute of Chemical Technology, Jang-dong, Yuseong-gu, Daejeon, 305-764, South Korea

Hexafluoropropylene (C3F6, HFP) is produced 5 to 20 wt% as a byproduct when poly (tetrafluoroethylene) is thermally decomposed to get tetrafluoroethylene (C2F4, TFE) which may be used as a feed gas for the synthesis of another fluoro-compund. Up to now, HFP has been supplied as an inert material together with TFE, resulting in the unstable reaction condition as well as the incomplete separation of the final products. If HFP can be separated from the mixed gases prior to their usage as a feed gas, not only it might help to keep the stable reaction condition but also the separated HFP may be utilized as a raw material to the synthesis of a novel interfacial monomer which could be used for the electronic and the chemical industries. In this study, the Aspen Plus 2004 software is used for the design of distillation process for separation of TFE from the HFP/TFE mixture gas.

Both the optimal theoretical number of plates and the optimal reflux ratio were looked for as the distillation tower was designed using the Aspen Plus 2004. There are several input specifications to be taken into account: the components of the feed gas and their concentrations, the concentration of TFE in the top and bottom products, the type of the condenser, and the thermodynamic equations to describe the states of the mixture. The two steps of computer simulation were carried out: the rigorous simulation by RadFrac following the shortcut calculation by DSTWU. The Peng-Robinson equation of state was used to describe the state of the C2 and C3 gas mixture. Both the complete condenser and the partial condenser were selected by considering the process following the distillation in which HFP is used as an input material. For both types of condenser, the optimal design of the distillation tower was performed in terms of the theoretical number of plates and the reflux ratio at the given operating temperature and the pressure.

For a typical feed gas composed of 6.9 mol% HFP and 93.1 mol% TFE, the input gas is fed to the 10th plate among 22 plates equipped with the partial condenser, which is operated at -10°C and 13.09 atm. With the reflux ratio of 0.6448, the TFE concentration in the top product is obtained to be 99.8 mol% while the HFP concentration in the bottom product is 99.8 mol%.

326q Separation of 1-Butene/n-Butane Using Mesoporous Mcm-41 Containing Cuprous/ferrous Ions Pair

Sung Soo Kim[1], Min Kang[2], Hyung Ik Lee[1], Chang Hyun Ko[3], Jong-Nam Kim[3], and Ji Man Kim[1]. (1) Chemistry, Sungkyunkwan University, Suwon, South Korea, (2) Environment and Development Solutions, Suwon, 442-749, South Korea, (3) Separation Process Research Center, Korea Institute of Energy Research, 71-2 Jang-dong Yusong-ku, Daejon, South Korea

1-butene, which is considerably important monomer in industry, has been usually produced by highly energy consuming separation based on distillation of C4 fractions in petrochemical plant. In this regards, there have been a lot of attempts to develop efficient separation methods of olefin and paraffin. In the present work, the adsorption characteristics of mesoporous MCM-41 containing cuprous, ferrous ions utilizing the g-complexation between the ions and 1-butene for selective adsorption of 1-butene in the mixture of the 1-butene and n-butane were studied. MCM-41 with cuprous/ferrous ions pair exhibit much higher adsorption amount for 1-butene than n-butane, compared to those of MCM-41 with only cuprous ion and microporous 13X zeolite. Especially, the MCM-41 with equal molar ratio of cuprous and ferrous exhibited the highest 1-butene/n-butane adsorption ratio at low pressure.

326r Dynamic Model of a Sfe Plant for the Extraction of Squalene from Olive Oil Residues

João B. Fernandes, Pedro C. Simões, and J. P. B. Mota. Departamento de Química, Faculdade de Ciências e Tecnologia - Universidade Nova de Lisboa, Quinta da Torre, 2829-516 Caparica, Portugal

A dynamic model of a supercritical fluid extraction (SFE) process plant is shown and applied to the fractionation of a binary mixture model of Methyl Oleate and Squalene. The supercritical solvent used in this separation is carbon dioxide. This model combines a mathematic model for each one of the process main pieces of equipment namely: the countercurrent packed extraction column, the separation column for oil recovery and solvent regeneration, the heat exchanger used to pre-heat the solvent stream, the pump used to compress the gas and the solvent make-up. To access the effect of disturbances on main process variables, perturbations were applied on pre-specified operating parameters as: feed composition, solvent mass flow rate, extraction and separation temperatures and pressures. The best set of process parameters was also determined with the help of the Dynamic Optimization Module of gPROMS simulator. A good agreement was obtained between experimental and predicted results. The dynamic model was found to accurately predict the outlet streams compositions and temperatures along the process of all the case studies.

329a Kinetics of Dopant Diffusion into Films for Reducing Radiation-Induced Conductivity in Polymeric Dielectrics

Robert J. Klein[1], John L. Schroeder[1], Shannon M. Lacy[1], Michael E. Belcher[1], Phillip J. Cole[2], and Joseph L. Lenhart[1]. (1) Organic Materials, Sandia National Laboratories, P.O. Box 5800, MS-1411, Albuquerque, NM 87122, (2) NNSA Satellite Programs, Sandia National Laboratories, P.O. Box 5800, MS-1411, Albuquerque, NM 87122

Radiation-induced conductivity in polymeric dielectrics can be significantly reduced by the incorporation of small-molecule traps (Kurtz and Arnold, JAP 57 (1985) 2532). In the appropriate concentration range, small molecules consisting of aromatic rings and strongly electron-withdrawing groups can reduce radiation-induced conductivity by more than 95 percent. Here, we investigate the doping of poly(ethylene terephthalate) films with several fluorenone dopants from solution. The dopant concentrations in the films, measured as a function of time and temperature, were fit by a combined adsorption-diffusion model that was solved analytically. Agreement between experimental data and the model is excellent, and physically meaningful parameters were obtained for all dopants under consideration, including the number of adsorption sites on the film, the equilibrium rate of adsorption, and the rate of diffusion. These parameters are also understood in the context of the chemical makeup of the dopants and the polymer.

Adsorption of Mercury (Hg II) on Coarse Loamy Mixed Hyperthermic Fluventic Hypluostept Soil of Punjab, North-west India

H. S. Hundal, Kuldip Singh, and Dhanwinder Singh.
Department of Soils, Punjab Agricultural University, Ludhiana, India

Introduction

Adsorption is an important process that affects the mobility and fate of contaminants in soil. An understanding of this process and its mechanism is crucial to the assessment of potential groundwater contamination and to the development of cost-effective soil remediation strategies. Generally, soil properties, environmental conditions and the amount of contaminants in soils, affect adsorption (Gold berg and Glaubig 1988; Sposito 1989; Zachara et al. 1989). Remediation processes, usually, also follow the same principle and the most common techniques are based on adsorption and precipitation phenomena. Studies on animals have reported effects such as alterations in testicular tissue, increased resorption rates, and abnormalities in development. The major effect from chronic exposure to inorganic mercury is kidney damage. Therefore, the present investigation was carried out to understand whether an alkaline soil could be used as effective adsorbent material for removal of mercury from contaminant water.

Material and Methods

A Coarse Loamy Mixed Hyperthermic Fluventic Hypluostept Soil of Punjab, North-west India was collected. The soil has pH (H_2O) 8.3 and EC 8.7 dSm^{-1} for a 1:2 extract. It has cation exchange capacity 6.8 cmol (+) kg^{-1}, organic C 2.7 g kg^{-1} and $CaCO_3$ 3.8 %. The soil contained 54 % sand, 28 % silt and 18 % clay. For mercury adsorption, 0.5 gm soil in quadruplicate was equilibrated with 100 ml of 0.01 M $NaNO_3$ solution containing a range of Hg concentration (up to 45µg Hg ml^{-1}) for 20 hours in 200 ml plastic bottles. Different mercury concentration solutions were prepared from $HgNO_3$ of analytical reagent salt and their pH was adjusted to 7.0 by addition of 0.01 M HCl or NaOH.

Adsorption isotherms were determined by incubating soil solution suspension at two different temperature of 278 and 318 ^{0}K. After equilibration period, soil solution suspensions were separated through filtration from Whatman no 42 filter paper. Mercury in the filtrates solution was determined on GBC make Atomic absorption spectrometer (Model Avanta PM) using cold vapours technique on hydride generation assembly. The mercury adsorption was computed from the difference between initial and final concentrations. The Langmuir equation ($C/x = 1/kb + C/b$) was used to interpret the equilibrium adsorption data. Where C is the equilibrium Hg concentration (µg Hg cm^{-3}), x is the adsorption maxima (µg Hg g^{-1} soil) and k is a constant related to the energy of adsorption (cm^3 $µg^{-1}$ Hg). The differential isosteric hest of adsorption, Δ H was obtained by collecting mercury adsorption data for the same soil at 278 and 318^0 K and applying the Clausiues Clapeyron equation to the system. For a given amount of mercury adsorbed, Ø,

$\text{Log } [C_2/C_1] = -\Delta H / 2.303 R [1/T_1 - 1/T_2]$, Where C_1 and C_2 are the equilibrium Hg concentration (µg Hg/cm^3) at temperature T_1 (280^0 K) and T_2 (305^0 K) respectively and R is the molar constant (1.985 E^{-3} Kcal/mol.)

Results and Discussion

Adsorption of mercury on a course loamy mixed hyperthermic Fluventic Hypluostept soil determined at 278 and 318 ^{0}K showed single linear Langmuir plots (Fig.1).

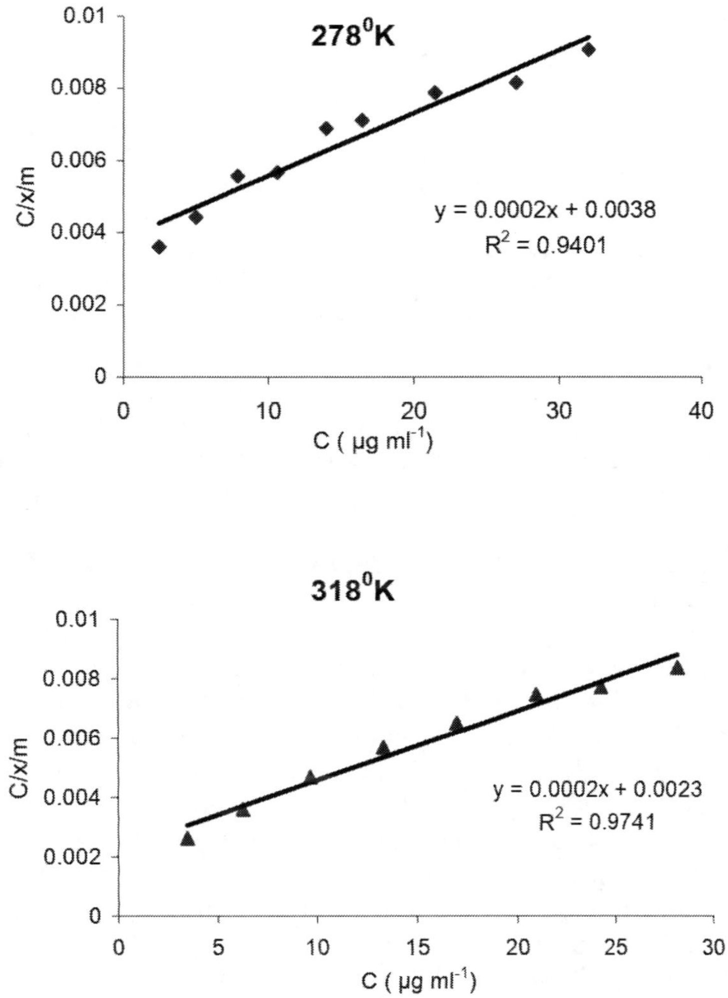

Figure 1 Langmuir plot for mercury adsorptions on coarse loamy mixed hyperthermic fluventic hyplustept soil after equilibration for 20 hours at 278^{0}K and 318^{0}C

The adsorption maxima for Hg adsorption by soils remained constant (5000 µg g^{-1} soil) at both temperature of equilibration of soil: solution suspension at 278 ^{0}K and 318 ^{0}K. The bonding energy of Hg adsorption by soils suspension was higher (0.0526 cm^3 μg^{-1} Hg) at 278 ^{0}K than (0.08695 cm^3 μg^{-1} Hg) at 318 ^{0}K.

The heat released or absorbed during adsorption of mercury is the differential molar heat of adsorption, Δ H. The value of Δ H between 278 and 318 ^{0}K were computed using equation (2) and are plotted as a function of mercury adsorbed (Fig. 2).

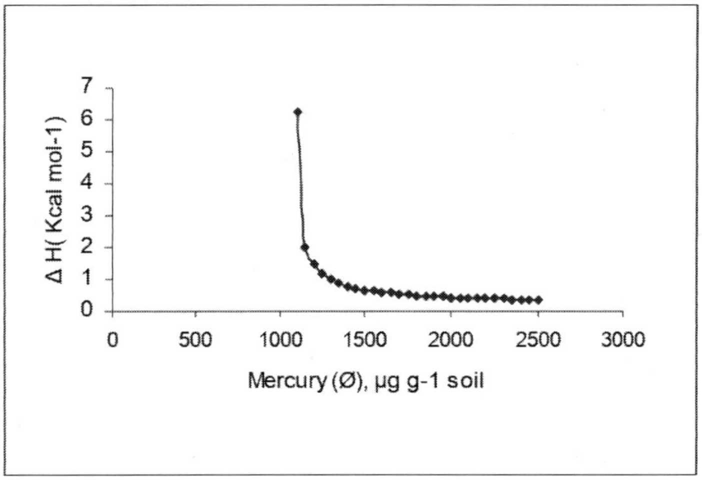

Figure 2. Differential isosteric heat of adsorptions for mercury on coarse loamy mixed hyperthermic fluventic hyplustept soil after equilibration for 20 hours at 5^{0}C and 45^{0}C

The Fig. 2 show that the adsorption process is energy consuming (endothermic), within Δ H varying between 6.26 and 0.36 kcal/mole for a constant surface, Ø, of 1100 to 2500 µg of Hg gm^{-1} soil. This shows that at adsorption process was predominant due to chemisorptions or precipitation of mercury minerals on the surface of soil matrix. The differential isosteric heat of adsorption for Hg adsorption markedly decline from 6.26 to 1.44 $kcalmole^{-1}$ as the amount of Hg adsorbed decreased from 1100 to1200 µg Hg g^{-1} soil. As the amount of Hg adsorbed by soil enhanced from 1200 to 2500 µg Hg g^{-1} soil, the differential isosteric heat of adsorbed change slightly from 1.44 to 0.36 $kcalmole^{-1}$. The higher magnitude of Δ H below 1100 µg Hg g^{-1} soil might be due to the interaction of Hg^{2+} cations with the hydrous oxide site of iron and manganese minerals or clay particles in the soil. When the amount of Hg^{2+} cations adsorbed enhanced from 1100 to 3000 µg g^{-1} soil, there was very little decline of Δ H (1.44-0.58 $kcalmole^{-1}$). This might be due to interaction of adsorbate to adsorbate i.e. between the mercury atoms rather than with the surface sites of soils with mercury atoms. Thus, the surface chemistry of mercury adsorption on the surfaces of soils and soil minerals is different from other heavy metal cations. In the present study, pair interaction between mercury atoms are most likely to dominate on the surface of the adsorbate. Hg_2 is a weekly bound molecules, which require lesser binding energy and eventually very little amount of Δ H is involved during mercury adsorption due to pairing process.

To understand the aqueous species and solid phase of Hg as a function of pe-pH, the Pourbaix diagrams (Fig.3a and b) were drawn for soil- solution containing equilibrated 0.14mM Hg, 30 uM Cl^{-1}, 0.15mM CO_3^{2-} and 0.33 mM SO_4^{2-}. The mercury in soil solution predominantly exits as Hg, $Hg(OH)_2$ and $HgS(SH)^{-1}$ in aqueous forms (Fig. 3a).

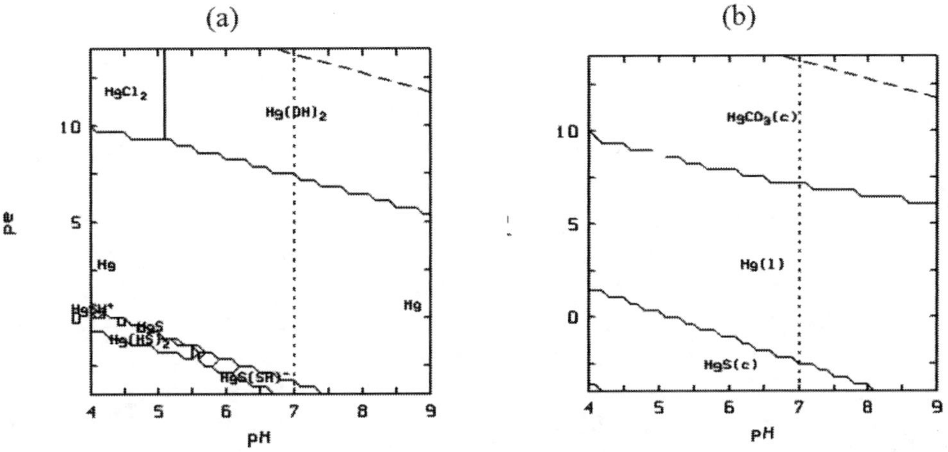

Figure 3. Pourbaix diagram showing the (a) solid phases (b) aqueous species of mercury (II) in soil solution suspension of 0.01 M $NaNO_3$ after equilibration for 20 hours containing 0.14 mM total Hg (II), 30 mM Cl^-, 0.15mM CO_3^{2-} and 0.33 mM

In a pH range of 5 to 9 and at highly oxidized condition, the $Hg(OH)_2$ is predominant aqueous species. When the pH moves below 5 and at highly oxidized condition (pe ≥10), $HgCl_2$ becomes predominant aqueous species. At elevated oxidized condition between pH 4 to 9, the $HgCO_3$(c) was identified as dominant solid phase (Fig. 3b). In moderately reduced to oxidize condition adsorption of mercury is mainly due to pairing of mercury atoms. In reduced environments, removal of mercury from water is mainly due to precipitation of HgS (c).

It is conclusively evident from this investigation that the surface properties of soils available at the sites having elevated mercury ground water can be utilized for remediation work.

References

Sposito, G. (1994). *Chemical Equilibria and Kinetics in Soils.* Oxford University Press,New York.

Zachara, J.M., Fredrickson, J.K., Smith, S.C., and Gassman, P.L. (2001) Solubility of Fe(III) oxide-bound trace metals by a dissimilatory Fe(III) reducing bacterium. *Geochemica et Cosmochimica Acta* 65:75-93

Goldberg S and Glaubig R A. (1988) Anion Sorption on a Calcareous, Montmorillonitic Soil—Arsenic. *Soil sci. Soc. Am. J* 52 :1297-1300

Adsorption of Nickel on Coarse Loamy Mixed Hyperthermic Fluventic Haplustept Soil of Punjab, North-West India.

Kuldip-Singh, Dhanwinder-Singh and H.S.Hundal, Department of Soils, Punjab Agricultural University, Ludhiana, India

Introduction

Nickel (Ni) release into the environment has been increasing continuously as a result of man's industrial activities and posses a significant threat to the environment and public health because of its toxicity, bioaccumulation in the food chain and persistence in nature (Adriano, 1986; Alloway, 1995). Nickel input in the environment is mainly derived from phosphate fertilizers, pesticides, bio-solid sources, nickel electroplating, producing & using nickel catalysts, fabricating parts and structures by welding, flame spraying, cutting, grinding, and polishing of nickel containing alloys, manufacturing nickel cadmium batteries, constructing nickel molds in glass bottle factories and spraying nickel containing paints (Mellis et.al.,2004).). Nickel causes gastrointestinal irritation and lung cancer. Adsorption reactions occur at the solid/liquid interface and are the most important mechanisms for controlling the activity of metal ions in soil solution. In a complex system with amphoteric behavior, the comprehension of the mobility, availability and fate of pollutants in the soil system is crucial for the prediction of the environmental consequences and for development of prevention/remediation strategies. The most common remediation techniques are based on adsorption and precipitation phenomena. The present investigation was undertaken to employ alkaline soil of north-west India as a cheapest adsorbent to remove nickel from the water bodies having elevated nickel content.

Material and Methods

The soil collected from the region for Ni adsorption experiment was coarse loamy mixed hyperthermic Fluventic Hyplustept. The soil has pH (H_2O) 8.3 and EC 8.7 dSm^{-1} for a 1:2 extract. It has cation exchange capacity 6.8 cmol (+) kg^{-1}, organic C 2.7 gkg^{-1} and $CaCO_3$ 3.8 %. The soil contains 54 % sand, 28 % silt and 18 % clay. For Ni adsorption, 0.5 gm soil in quadruplicate was equilibrated for 24 hours in 200 ml plastic bottles containing 100 ml of 0.01 M $NaNO_3$ solution with varying Ni concentration up to 50 µg Ni ml^{-1}. Different Nickel concentration solutions were prepared from $Ni(NO_3)_2$ $6H_2O$ analytical reagent salts and their pH was adjusted to 7.0 by addition of 0.01 M HCl or NaOH. Adsorption isotherms were determined by incubating soil solution suspension at two different temperature of 280 and 318 ^{0}K. After equilibration period, soil solution suspensions were separated through filtration. Nickel in the filtrates solution was determined on atomic adsorption spectrometer. Nickel (II) adsorption was computed from the difference between initial and final concentrations. The Langmuir equation ($C/x = 1/kb + C/b$) was used to interpret the equilibrium adsorption data. Where C is the equilibrium Ni concentration (µg Ni cm^{-3}), b is the adsorption maxima (µg Ni g^{-1} soil) and k is a constant related to the energy of adsorption (cm^3 $µg^{-1}$ Ni). The differential isosteric hest of adsorption, Δ H was obtained by collecting nickel adsorption data for the same soil at 280 and 318^0 K and applying the Clausiues Clapeyron equation to the system. For a given amount of Nickel adsorbed, Ø, $\text{Log } [C_2/C_1] = -\Delta H/ 2.303 R [1/T_1-1/T_2]$, Where C_1 and C_2 are the equilibrium Ni concentration (µg Ni cm^{-3}) at temperature T_1 (280^0 K) and T_2 (318^0 K) respectively and R is the molar constant (1.985 Kcal mol^{-1}).

Result and Discussion

Adsorption of Ni on a coarse loamy mixed hyperthermic soil determined at 280 and 318 ^{0}K showed one break in the slope of Langmuir plots (Fig.1).

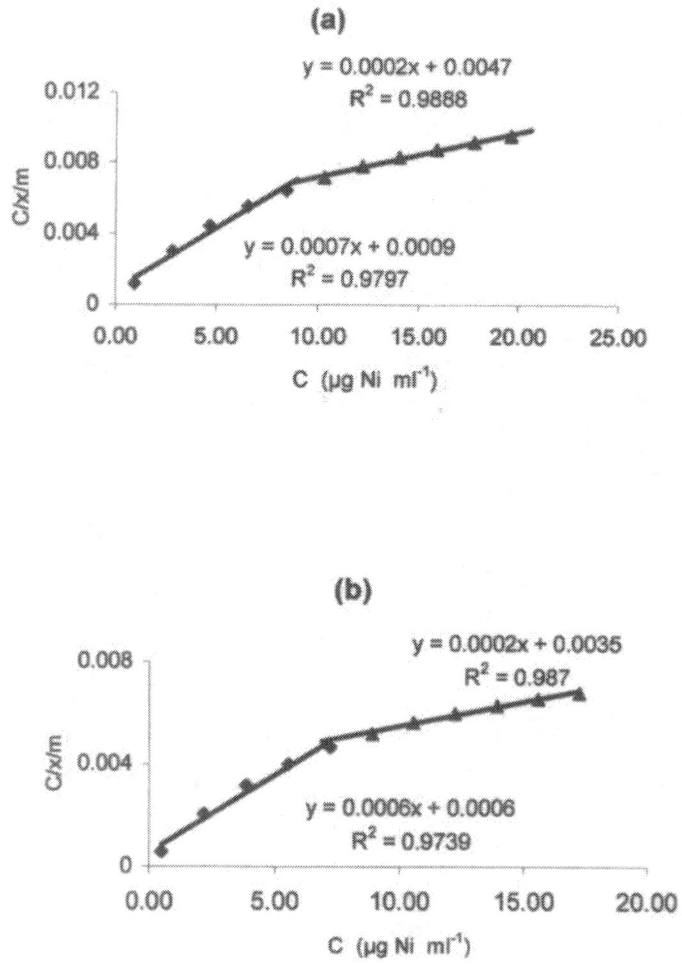

Figure 1 Langmuir plot for Nickel adsorptions on course loamy mixed hyperthermic Fluventic hypluostept soil after equilibration for 24 hours at (a)280^{0}K and (b)318^{0}K

Thus Nickel adsorption on this soil is described by two regions Langmuir isotherm equation, i.e. the plots showed two distinct linear portions. In the Langmuir plot for the region I and II, the bonding energy of nickel adsorption on soil increased with temperature from 280 to 318 ^{0}K (Table1).

Table 1. Langmuir parameters for Nickel adsorption on coarse loamy used mixed hyperthermic Fluventic Hyplustept soil

Adsorption	Region I		Region II	
	k	b	k	b
Soil at 280 ^{0}K	0.78	1429	0.04	5000
Soil at 305 ^{0}K	1.00	1667	0.06	5000

k = bonding energy, $cm^3 \mu g^{-1}$; b = adsorption maxima, $\mu g\ g^{-1}$ soil

The bonding energy for Ni adsorption for region I enhanced from 0.78 to 1.00 $cm^3 \mu g^{-1}$ Ni and for region II enhanced from 0.04 to 0.06 $cm^3 \mu g^{-1}$Ni with increase in temperature of the equilibrated soil solution suspension from 280 to 318 ^{0}K. However, the bonding energy for Ni adsorption onto soil was higher for region I than region II. In region I of Langmuir plot the adsorption maxima for Ni adsorption enhanced from 1429 to 1667 $\mu g\ g^{-1}$ soil with increased in temperature from 280 and 318 ^{0}K. However, the adsorption maxima of 5000 μg Ni g^{-1} of soil in the region II remained unaffected by temperature. The heat released or absorbed during adsorption of Nickel is the differential molar heat of adsorption, Δ H. The value of Δ H between 280 and 318 ^{0}K were computed using Clausiues Clapeyron equation and are plotted as a function of Nickel adsorption (Fig. 2).

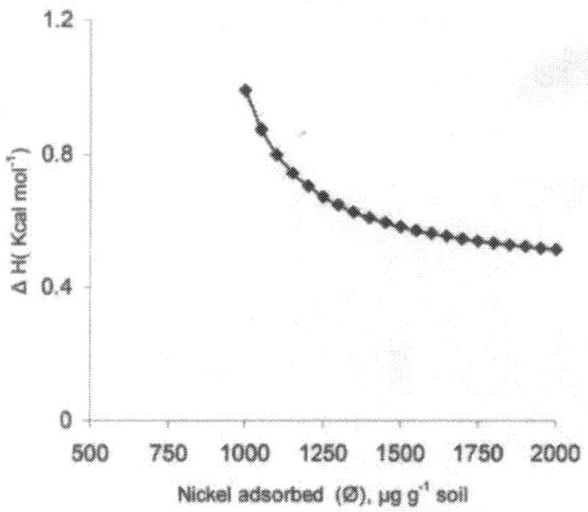

Figure 2. Differential isosteric heat of adsorptions for mercury on course loamy mixed hyperthermic Fluventic hypluostept soil after equilibration for 24 hours at280^{0}C and 318^{0}C.

The Fig. 2 shows that the adsorption process is energy consuming (endothermic), within Δ H varying between 0.52 and 0.99 $kcalmole^{-1}$ for a constant surface, Ø, of 1000 to 2000 μg of Ni gm^{-1} soil. Thus it evident that adsorption process was predominant due to chemisorptions or precipitation of nickel compounds on the surface of soil matrix. It is evident from this investigation that the surface properties of soils available near the polluted water bodies can be utilized for remediation work of the nickel pollution. Mole fraction diagram (Fig.3) illustrate the aqueous species of Nickel (II) in soil solution suspension of

0.01 M $NaNO_3$ after equilibration for 24 hours containing 0.69 mM total Ni (II), 30 mM Cl-, 0.15mM CO_3^{2-} and 0.33 mM SO_4^{2-} . Ni^{2+} aqueous species is predominantly found in equilibrated solution below pH 7.0 When the pH is increased above 7.0, Ni^{2+} species is precipitated in the form of $Ni(OH)_2$ (c) compound and predominantly occurred above pH 7.50.

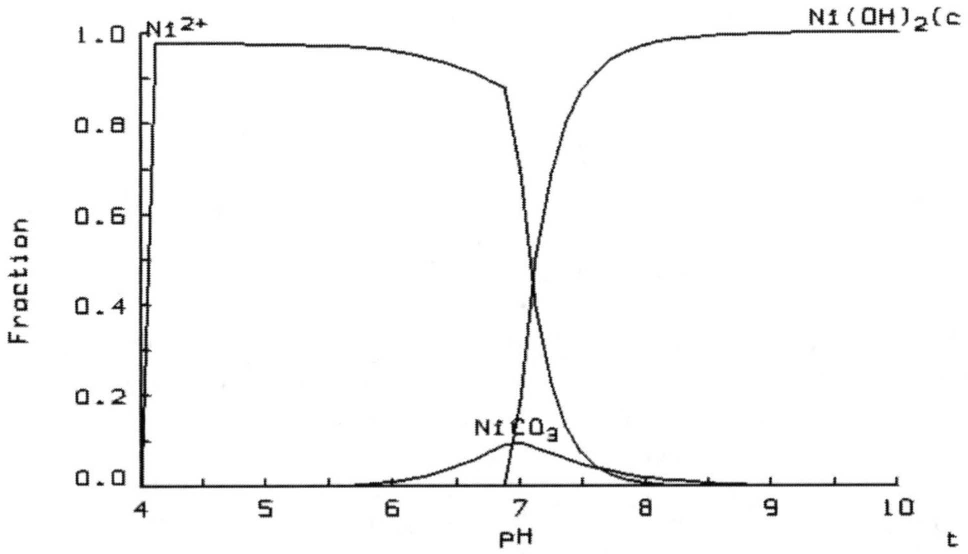

Figure 3. Mole Fraction diagram showing the solid and aqueous species of Nickel (II) in soil solution suspension of 0.01 M $NaNO_3$ after equilibration for 24 hours containing 0.69 mM total Ni (II), 30 mM Cl^-, 0.15mM CO_3^{2-} and 0.33 mM SO_4^{2-}.

Nickel carbonate ($NiCO_3$) in aqueous form was present between pH 6 to 8 and its maximum amount about 10 percent found at pH 7.0, however, its mole fraction decline with increase or decrease in pH from 7.00. The results of the present investigation suggest that local soil could be employed for removal of Ni from contaminated water bodies.

References

Adriano, D.C. (1986) Trace elements in the terrestrial environment. Springer-Verlag, New York.

Alloway, B.J. (1995) The origins of heavy metals in soils. p. 38–57. *In* B.J. Alloway (ed.) Heavy metals in soils. Blackie Academic & Professional, London

Mellis E.V.,. Cruz M.C.P. and Casagrande, J.C. (2004). Nickel adsorption by soils in relation to pH,organic matter, and iron oxides. *Sci Agric* **61**:190-195.

Surface-modified alumina as a high capacity material of $^{99}Mo/^{99m}Tc$ generator column

Jun Sig Lee1[1]*, Jong-Sup Lee[2], Ul-Jae Park, Kwang-Jae Son[1],
Hyon Soo Han[1], and Seung Kon Ryu[2]

(1) Korea Atomic Energy Research Institute
Daedeuk-daero 1045, Yuseong-gu, Daejeon, S. Korea, 305-353
Jlee15@kaeri.re.kr
(2) Dept. of Chem. Eng., Chungnam National University, 220 Gung-dong,
Yuseong-gu, Daejeon, 305-764, South Korea

Introduction

Different types of generators have been developed to extract ^{99m}Tc conveniently [1]. Currently, more than 80 % of the total demand for medical isotopes is on the ^{99m}Tc in the world. This radionuclide is extensively used for the diagnosis of major organs such as the brain, lung, heart, kidney, etc. with appropriate labeling compounds.

^{99m}Tc is produced from the decay of ^{99}Mo as shown below. The radioactivity of ^{99}Mo and ^{99m}Tc in a radioactive molybdenum solution reaches equilibrium approximately 23 hours after a disturbance because of the difference of the decay times. The chemical forms of these nuclides in a solution are generally $^{99}MoO_4^{2-}$ and $^{99m}TcO_4^-$. Based on the chemical property, these nuclides can be separated from one another. Mostly $^{99m}TcO_4^-$ is separated from $^{99}MoO_4^{2-}$ by a chromatographic method, which is called 'generator' technology. From a chromatographic column on which ^{99}Mo is preloaded, ^{99m}Tc is periodically eluted once the radio-activities of the isotopes reach a equilibrium.

$$^{99}Mo \underset{67\ hr}{\overset{\beta^-,\gamma}{\longrightarrow}} {}^{99m}Tc \underset{6\ hr}{\overset{\gamma}{\longrightarrow}} {}^{99}Tc$$

A ^{99m}Tc generator, in general, is composed of a column packed with a ceramic adsorbent, tubing, eluant reservoir or vials, collection vials, and a shielding. The generator technology for the ^{99}Mo/^{99m}Tc is well developed, and the products are available commercially in the case of the fission ^{99}Mo/^{99m}Tc generators because the chemical quantity of ^{99}Mo for a few curries of ^{99}Mo is not a significant amount to load onto a small column. Hence, acidic alumina is normally employed for (n,f) ^{99}Mo/^{99m}Tc generators. However, the current infrastructure of the supply of (n,f) ^{99}Mo is fragile as the research reactors are becoming aged. In this regard, alternative research activities to develop generators with (n,γ)^{99}Mo have been performed by several research groups. To develop a commercially viable (n,γ)^{99}Mo/^{99m}Tc generator, the generator column should have at least a several times higher adsorption capacity for molybdenum than that of generally available alumina. Gel generator [2, 3], PZC [4, 5], and titania columns [6] are good examples to achieve such a high adsorption capacity. Gel generator is normally produced by the irradiation of a mixed oxide of molybdenum and zirconium. Even though a high radioactivity can be achieved by this method, handling radioactive powders without a contamination is a very difficult task. PZC, a zirconium compound, which was developed by a Japanese group has enough capacity for $(n,\gamma)^{99}Mo/^{99m}Tc$ generators. However, this material needs a batch reaction with a ^{99}Mo solution by a boiling for several hours to prepare ^{99}Mo loaded materials. Hence, this method cannot produce the columns easily as it requires a remote handling of a solid-liquid mixture. Hence, these technologies still have restrictions to be applied for the commercial production of $(n,\gamma)^{99}Mo/^{99m}Tc$ generators.

Recently our research group has developed a new adsorbent as a generator column material. This material has more than a 200 mg/g adsorption capacity for molybdenum. Also the elution of ^{99m}Tc from the loaded column is reasonable (~85%). Column preparation is also as easy as that for $(n,f)^{99}Mo/^{99m}Tc$ generators. In this method, a non-radioactive and dried adsorbent is pre-packed into a glass column and then the column is loaded by passing a molybdenum solution through the column just as the process for $(n,f)^{99}Mo/^{99m}Tc$ generators. Hence, this adsorbent can be used for both (n,γ) and (n,f) $^{99}Mo/^{99m}Tc$ generators.

In this paper, the preparation of the adsorbent and adsorption characteristics in a batch and on a column will be described.

Experiments

Preparation of Adsorbent

Detailed synthesis scheme for adsorbents can be found elsewhere [7]. The adsorbent used in this study is synthesized by a sol-gel processing. Aluminum tri-sec-butoxide (97%) and fuming sulfuric acid are purchased from Aldrich chemical Co. and used without further purification. Other necessary chemicals and solvent in the experiments are ACS grades. To functionalize alumina, certain moles of fuming sulfuric acid are added to the aluminum tri-sec-butoxide dissolved in isopropyl alcohol while stirring. After a complete mixing of the solution, certain moles of water are added drop-wise to induce the gelation. The gel is aged at room temperature for 24 hours. Aged gel is dried at 80 °C for 24 hours and then at 130 °C for 8 hours. The gel is crushed and sieved for the particle size of 75 – 150 μm. By this procedure, the adsorbent has 3.4mmol/g of sulfate functional groups.

Adsorption Characterization

Adsorption experiments are performed by using simulated molybdenum solutions in which ^{99}Mo is spiked as a tracer. The molybdenum solution is prepared by dissolving a desired amount of molybdenum oxide (Aldrich Chemical Co.) in a 0.4 N sodium hydroxide (NaOH) solution. The solution pH is adjusted to 10.5 or other specific values. To this solution, a trace amount of ^{99}Mo is added to obtain a 0.3 μCi/ml of the specific activity just before the contact experiments.

Batch loading experiments are performed by shaking 0.5gram of the synthesized adsorbent in 25ml of the solution for 2 hours at room temperature. The radio-activities of the solutions before and after the contacts are measured by using a HPGe γ –ray detector (EG & G Ortec, Model # GEM 10175). The activity difference of the solutions before and after a contact experiment is considered as the fraction of the loaded amount on the adsorbent.

Column experiments are performed by using a column with 1.0cm ID in which 2 grams of adsorbent is pre-packed. Sea sand (~0.5g) is loaded on the top of the adsorbent layer. To this column, 50 ml of a molybdenum solution (10,000mg/L) is introduced at a flow rate of 1.0ml/min and then washed twice with 25 ml of saline solution. The radio-activity difference between the feed and effluent solutions is considered as the fraction of the amount on the column. Elution of ^{99m}Tc is performed by feeding desired amounts of the saline solution (0.9% NaCl) for 23hours after the column is prepared. The activity of the eluted ^{99m}Tc is measured at 140.5KeV by the HPGe γ –ray detector.

Results and Discussion

Synthesis of Adsorbent

The sol-gel processing employed in this study has two steps. The first step is replacing the functional groups of aluminum trialkoxide with sulfate. In the second step, the modified monomers are hydrolyzed and condensed to produce a gel material by the addition of water as shown below:

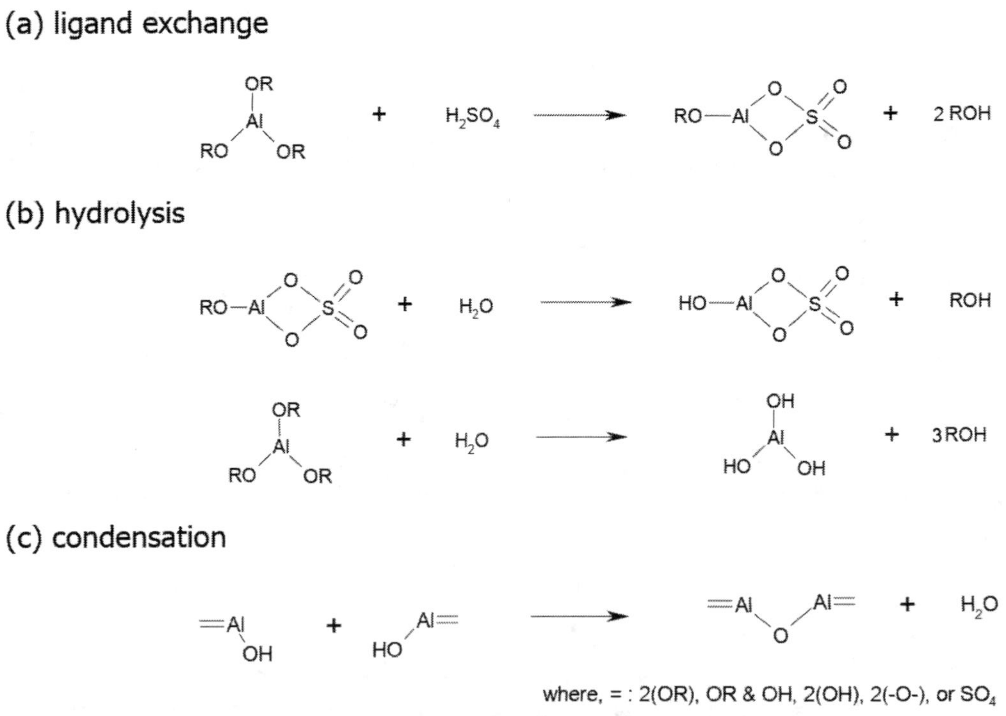

The adsorbent synthesized by this process has 3.4 mmol/g of a sulfate content, which is determined by an elemental analysis. The maximum molybdenum uptake capacity is approximately 217 mg/g and 195 mg/g by a batch and a column experiments, respectively.

Adsorption Isotherm in Batch

The adsorption equilibrium between the adsorbent and molybdenum solution are studied. The aqueous speciation of molybdate is also studied to determine the appropriate pH of the loading solutions as shown in Figure 1. When the solution pH is higher than 5, MoO_4^{2-} is the predominant species, which is a preferable species to produce bidentate molybdenum on the adsorbent. Hence, the solution pH is fixed at 10.5 in this study to produce strongly bounded molybdenum on the adsorbent.

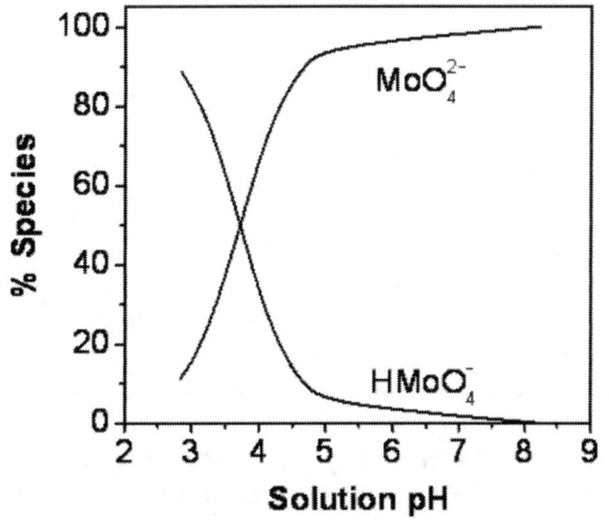

Figure 1. Speciation of Molybdate in an Aqueous Solution

The chemical reaction between molybdate and the sulfate-functionalized surface sites can be described as an ion exchange reaction as shown below:

Chemical Reaction:

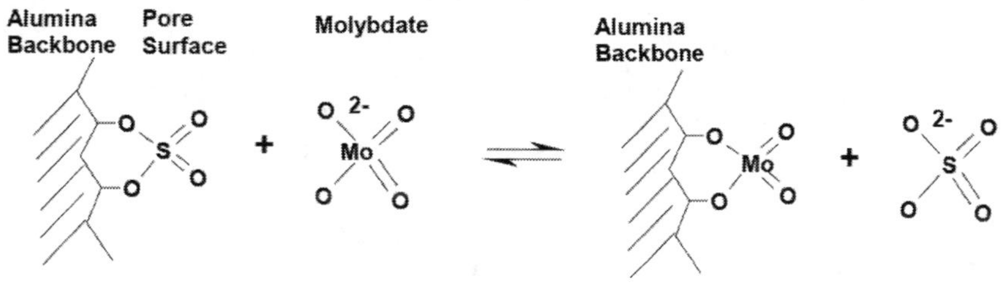

This ion exchange reaction can be described as a Langmuir isotherm presented below:

$$q = \frac{q_{max} \cdot K \cdot C}{1 + K \cdot C} \tag{1}$$

where, q is the equilibrium capacity of the adsorbent (mg/g), q_{max} is the maximum uptake capacity, K is the equilibrium constant, and C is the equilibrium concentration in the solution (mg/L).

As shown in Figure 2, the experimental data can be represented well with the parameters of q_{max} (226 $\pm$ 7 mg/g) and K (0.0054 $\pm$ 0.0007 L/mg).

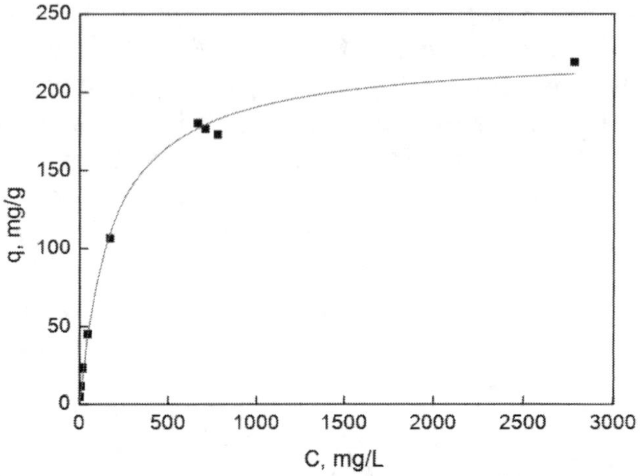

Figure 2. Adsorption Isotherm of Molybdenum by a Batch Experiment (solid line: Langmuir isotherm represented by Equation (1))

Adsorption in Column

Breakthrough experiments are performed to measure the adsorption capacity and the distribution of molybdenum along the column. To determine the breakthrough, the adsorbent is packed into a column without any other backup materials such as acidic alumina. The breakthrough occurs when 25ml of the feed solution is introduced to the column as shown in Figure 3. The total loading capacity on this column is 195 mg/g, which is only 10% less than that from the batch experiment. Considering that the column is not completely saturated, this difference is negligible.

A leaching of molybdenum from the beginning is observed. This leaching is probably due to the existence of fine particles in the column. The sulfated alumina is not washed before packing it into the column to preserve its functionality. Hence, the fine powder may be washed out during the loading process. This leaching of molybdenum is not a problem in the production of a generator when the ratio of $^{99}Mo/^{99m}Tc$ is smaller than the medical requirement for the ^{99m}Tc solution. It is observed from our other experiments that the ratio in the ^{99m}Tc solution is less than the requirement.

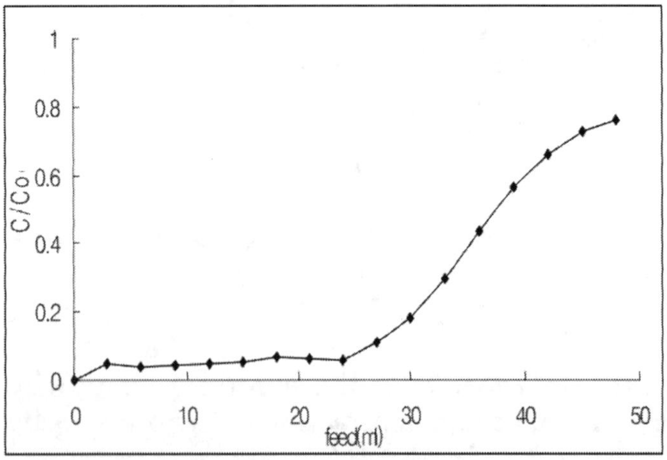

Figure 3. Breakthrough Curve of Molybdenum (feed concentration: 10,000mg/L, feed pH = 10.5, flow rate = 1.0ml/min)

To study the distribution of adsorbed molybdenum along the column, a series of columns are loaded with different quantities of molybdenum. The distribution profile of molybdenum through the column is constructed by discretely measuring the radiation emitted from the adsorbed ^{99}Mo as shown in Figure 4. For this experiment, a TLC scanner (Bioscan, mini-scan B-MS-1000) equipped with a PM tube and a lead collimator with a 1.0 mm slit is used. The measured data are splined for a better representation. From this experiment, it is estimated that only molybdenum is adsorbed at the front of the column when the molar quantity of molybdenum, which can saturate 15% of the functional sites, is introduced. In the case of a near saturation (100%), the molybdenum is distributed throughout the column. This means the reaction between the molybdate and the reactive sites of the adsorbent is not affected by a material transfer such as a diffusion.

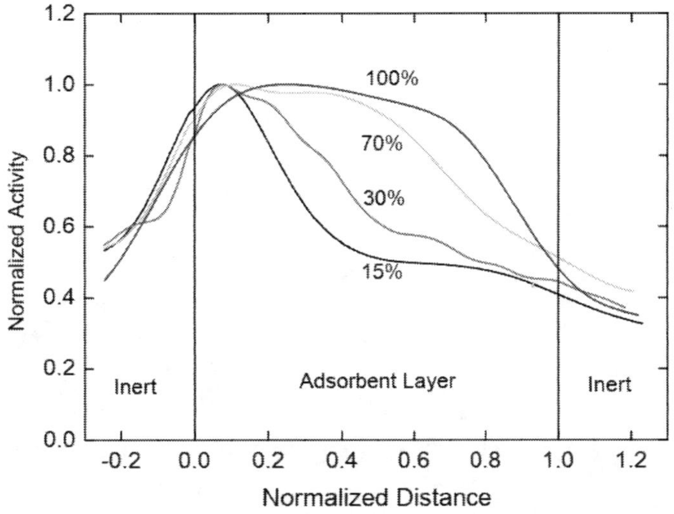

Figure 4. Radiographic Distribution of Adsorbed Molybdenum on the Column

^{99m}Tc Elution

^{99m}Tc is eluted from the loaded column by using a saline solution (0.9% NaCl solution, medical grade). ^{99m}Tc produced by the decay of ^{99}Mo stays on the column as a weakly adsorbed form. Hence, ^{99m}Tc is readily eluted by the saline solution. From the column, ^{99m}Tc is eluted with an efficiency of 67% at the first day, and then approximately 84% for the following 4 days as shown in Figure 5. The elution efficiency with respect to the eluant volume is also measured at each 10 ml fraction as shown in Figure 6. In this experiment, it is observed that the elution efficiency of ^{99m}Tc from the (n,γ) $^{99m}Tc/^{99}Mo$ generator is approximately 10 ~ 15 % less than the efficiency generally reported for (n,f) $^{99m}Tc/^{99}Mo$ generators. Also, the specific activity per volume of the eluant is less than the that from the (n,f) generator, from which more than 95% of ^{99m}Tc is eluted with approximately 10 ml of the saline solution. This diluted elution is due to the longer adsorption layer in the column than that in the (n,f) generator because the adsorbed molybdenum is mostly non-radioactive on the (n, γ)^{99}Mo column. Hence, the solution volume to achieve the same elution efficiency is much larger than that for the (n,f) ^{99}Mo column.

Even though the elution efficiency for the column in this study is lower than that of the (n,f)^{99}Mo generator, it is still competitive when compared to other (n,γ) $^{99}Mo/^{99m}Tc$

generators reported in the literature (i.e. % Elution of gel generator ~ 75% [3]). Also, it is observed from our on-going research that this problem may be solved by altering the pore properties of this column material. The results from this on-going research work will be published in the near future.

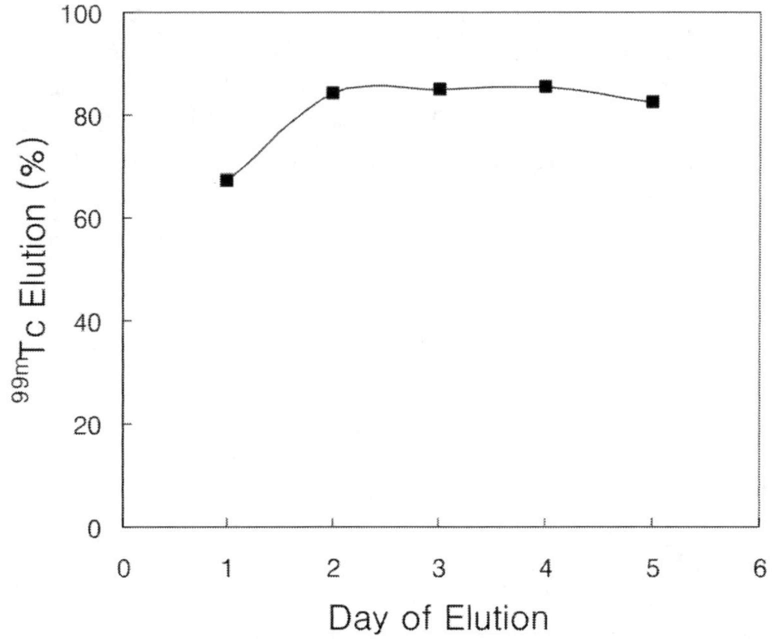

Figure 5. Elution Efficiency of ^{99m}Tc by using the Saline Solution (0.9% NaCl) Each Day

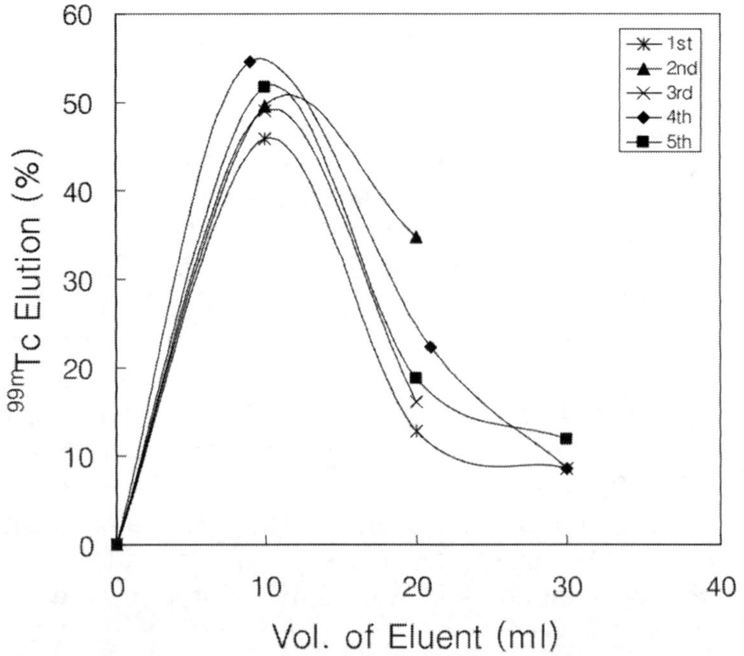

Figure 6. Elution Efficiency of ^{99m}Tc by using the Saline Solution (0.9% NaCl)

Conclusions

In this study, a sulfated alumina material is synthesized to use as a column material for (n,γ) $^{99}Mo/^{99m}Tc$ generators. The material has been tested for its adsorption properties by batch and column experiments. The adsorption capacity for molybdenum in the batch is approximately 217 mg/g. Also, the capacity determined by a column operation is approximately 195mg/g. This material can be packed in a column as a dried non-radioactive powder. Hence, the preparation of a ^{99}Mo loaded column is much easier than any other (n, γ)^{99}Mo columns. The elution efficiency of ^{99m}Tc is approximately 10 ~ 15 % less than that generally obtained from (n, f)^{99}Mo columns. This elution efficiency is still competitive when compared to other reported (n, γ)$^{99}Mo/^{99m}Tc$ generators. Much of work currently being undertaken shows an improvement of the elution efficiency by altering the pore characteristics of the column material in our laboratory.

Acknowledgements

The financial support of Ministry of Science and Technology (MOST) in the Republic of Korea through Mid and Long Term Nuclear R&D Program is gratefully acknowledged.

REFERENCES

1. A. Mushtaq, Journal of Radioanalytical and Nuclear Chemistry, 262, 797 – 810 (2004).
2. O.B. Gurko, E.I. Ravkova, and I.A. Savushkin, Atomic Energy, 90, 390-395 (2001).
3. P. Saraswathy, S. K. Sarkar , G. Arjun, S. K. Nandy, and N. Ramamoorthy, Radiochimica Acta, 92, 259-264 (2004)
4. S. L. Van, Proceedings of the 1997 Workshop on the Utilization of Research Reactor Bandung, Nov. 1997. pp. 6-13
5. Y. Hasegawa, M. Nishino, K. Kurosawa, M. Tanase, and H. Yamabayashi, Proceedings of the 1997, Workshop on the Utilization of Research Reactors, Japan Atomic Energy Research Inst., Tokyo, 1998, p. 3.
6. A. Dash, S. Gandhi, R. Ram, and M.R.A. Pillai, Procedings of BRNS-DAE National Symposium on Nuclear and Radiochemistry, Mumbai, India, 2001, p. 420
7. J.S. Lee, H.S. Han, U.J. Park, K.J. Son, H.Y. Shin, S.B. Hong, K.D. Jang, and J.S. Lee, Absorbents for radioactive element and preparation method thereof, Korean Patent Application Pending, #2006-109983 (Nov. 8, 2006)

Adsorption of Lead on Fine Loamy mixed Typic Dystrudepts Soil, of Himachal Pradesh, North-west India

Dhanwinder Singh, Kuldip Singh and H.S. Hundal, Department of Soils, Punjab Agricultural University, Ludhiana 141004, INDIA.

Introduction

The interaction of contaminants, such as lead (Pb), with minerals and other aquifer materials controls the movement of these contaminants in groundwater. Adsorption/ precipitation are assumed to control the solubility of inorganic compounds. The stronger these interactions, the lesser will be the downwards movement of contaminant. Adsorption/ precipitation are often controlled by the nature of the cations, the nature of the adsorbent, and the specific geochemical conditions (Mckenzie, 1980; Sepos, 2005). In soil, Pb is thought to be mostly bound to the solid phase and, therefore, insoluble. Similar to other heavy metals, Pb reaching the soil may interact with various soil constituents, and eventually affect its solubility and mobility. The present study deals with the adsorption of Pb^{2+}, in an acidic soil, representing Typic Dystrudepts, of Himachal Pradesh, North-west India.

Material and methods

A fine loamy mixed dystrudepts soil of Palampur, Himachal Pradesh, north west India was collected to study the Pb adsorption by it. The soil has pH 6.1 and EC 0.08 dS m^{-1} (1:2 Soil:Water), cation exchange capacity 23.3 cmol (+) Kg^{-1} and organic carbon 9.1 g Kg^{-1}. The soil contained 65% sand, 13% silt and 22% clay. To study Pb adsorption, 2g of soil (in duplicate) was shaken with 20 ml of 0.01 M $Ca(NO_3)_2$ solution containing seven different lead concentrations (40 µg ml^{-1}, 80 µg ml^{-1}, 100 µg ml^{-1}, 120 µg ml^{-1}, 140 µg ml^{-1}, 160 µg ml^{-1} and 200 µg ml^{-1}) for two hours followed by equilibration for 24 hours at 298 ^{0}K. The samples were then centrifuged; the supernatant was decanted and analyzed for Pb on GBC flame atomic absorption spectrophotometer (AAS). Lead adsorption was computed from the difference in initial and final concentrations. Lead adsorption was also studied at 318 ^{0}K. Langmuir adsorption isotherms were derived from Pb adsorption data and adsorption parameters adsorption maxima 'b' and bonding energy coefficient 'k' were calculated. The value of thermodynamic equilibrium constant, K^o for adsorption reaction can be defined as K^o = $a_s/a_e = \gamma_s C_s/ \gamma_e C_e$

Where a_s denotes activity of adsorbed metal, a_e activity of metal in equilibrium solution, C_s milligrams of metal adsorbed per litre of solution in contact with the adsorbent surfaces, C_e milligrams of solute per litre of solution in equilibrium solution, γ_s is the activity coefficient of the sorbed metals and γ_e represents the activity coefficient of metal in equilibrium solution. Since at lower concentration the activity coefficient approaches unity, the equation gets reduced to $K^o = C_s/ C_e$. The values of K^o were obtained by plotting $\ln(C_s/ C_e)$ vs. C_s and extrapolating to zero C_s . The standard enthalpy (ΔH^0) was obtained from integrated form of Vant Hoff equation: $\ln K_2^o/K_1^o = \Delta H^o / R .[1/T_1 -1/T_2]$.

Results and Discussion

The Pb sorption data at 298 ^{0}K and 318 ^{0}K were well described by Langmuir adsorption isotherm (Figure 1). The conformity of the lead adsorption

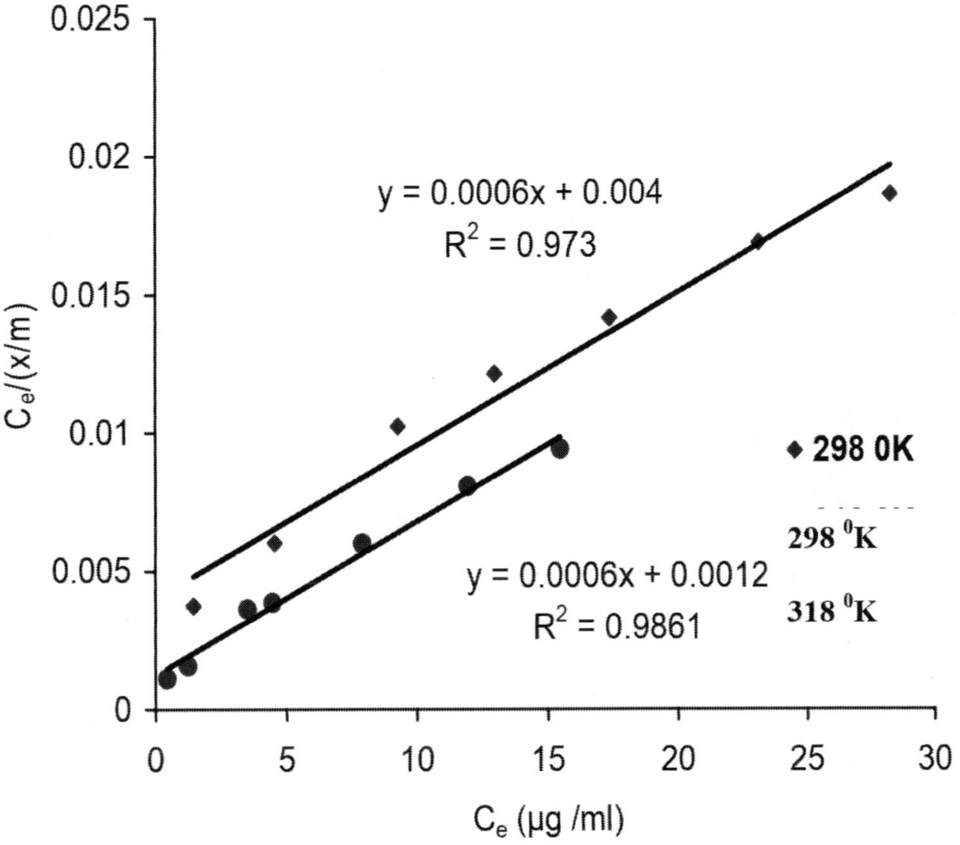

Figure 1. Langmuir plot for Pb adsorption on fine loamy mixed dystrudepts soil after equilibration for 24 hours

data to the linear form of the Langmuir equation was indicated by the highly significant correlation coefficients (r^2 = 0.97 to 0.99) derived from data pairs of $C_e/(x/m)$ and C_e. With the increase in temperature from 298 ^{0}K to 318 ^{0}K, the Langmuir adsorption maxima (b) remained unchanged, whereas, the bonding energy value 'k' increased from 0.15 to 0.50 cm^3 μg^{-1}. Hundal et al. (2003) also observed that bonding energy constant derived from cadmium adsorption isotherms for alkaline soils of Punjab increased with a similar increase in reaction temperature. The heat released or absorbed during adsorption of Pb is differential molar heat of adsorption or Enthalpy of adsorption, ΔH_{ads}. Heat of adsorption provides stability of adsorbed species, strength of adsorbate-substrate bonds and thermodynamic driving forces for surface reactions. The values of ΔH (between 298 and 318 ^{0}K) computed using Vant Hoff equation varied from 5.09 to 1.09 kcal mole^{-1} for a constant surface, Ø, of 650 to 1800 μg Pb g^{-1} soil. Figure 2 shows a plot of ΔH values as a function of of Pb adsorption. The positive values of differential isosteric heat of adsorption (ΔH) on soil between 298 and 318 ^{0}K elucidate that adsorption of Pb is an endothermic process which leads to chemisorption or precipitation of Pb compounds on the surface of soil matrix. It is conclusively evident from this investigation that the surface properties of soils help in retarding the downwards movement of Pb in soil profile and thus preventing the contamination of groundwater. Visual minteq programme was used to compute aqueous species of Pb in soil solution suspension of 0.01 M $Ca(NO_3)_2$ after equilibration for 24 hours containing 0.114 mM total Pb (II), 0.15 mM CO_3^{2-} and 0.50 mM SO_4^{2-}.

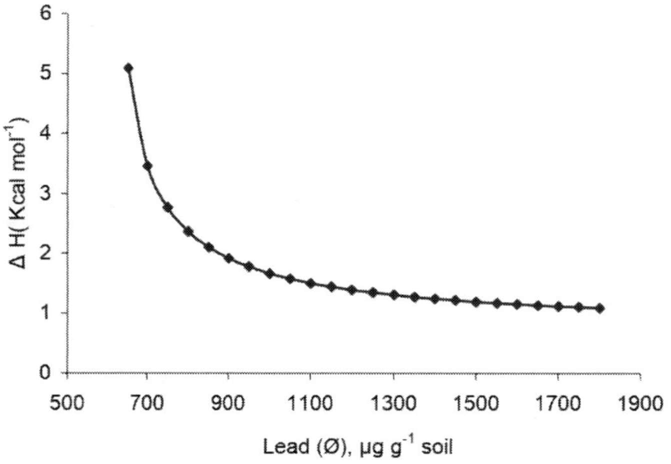

Figure 2. Differential isosteric heat of adsorption for Pb on fine loamy mixed Typic Dystrudepts soil, after equilibration for 24 hours at 298 ^{0}K and 318 ^{0}K.

Mole fraction diagram for the dominant aqueous species of Pb was drawn (Figure 3). Figure shows that, in the pH range of 4-7, the dominant aqueous Pb species include Pb^{2+}(aq), $PbNO_3^+$ and $PbSO_4^0$(aq). Where as in neutral to alkaline pH range, the dominant aqueous species of Pb were $PbOH^+$, $PbCO_3^0$, $Pb_3OH_4^{+2}$ and $Pb(OH)_2$. At its original pH 6.1, the per cent distribution of different lead species in fine Loamy mixed typic dystrudepts soil solution suspension was Pb^{2+} 80.3%, $PbNO_3^+$ 12.2%, $PbSO_4$ (aq) 3.3%, $PbOH^+$ 1.6%, $Pb(NO_3)_2^0$ (aq) 0.29%, $PbCO_3^0$ 0.40% and $PbHCO_3^+$ 1.9%. It can be concluded that depending upon the soil properties, the charged species of Pb could interact with the surface constituents while the ion pairs having zero charge could contaminate the groundwater because of their free downward infiltration.

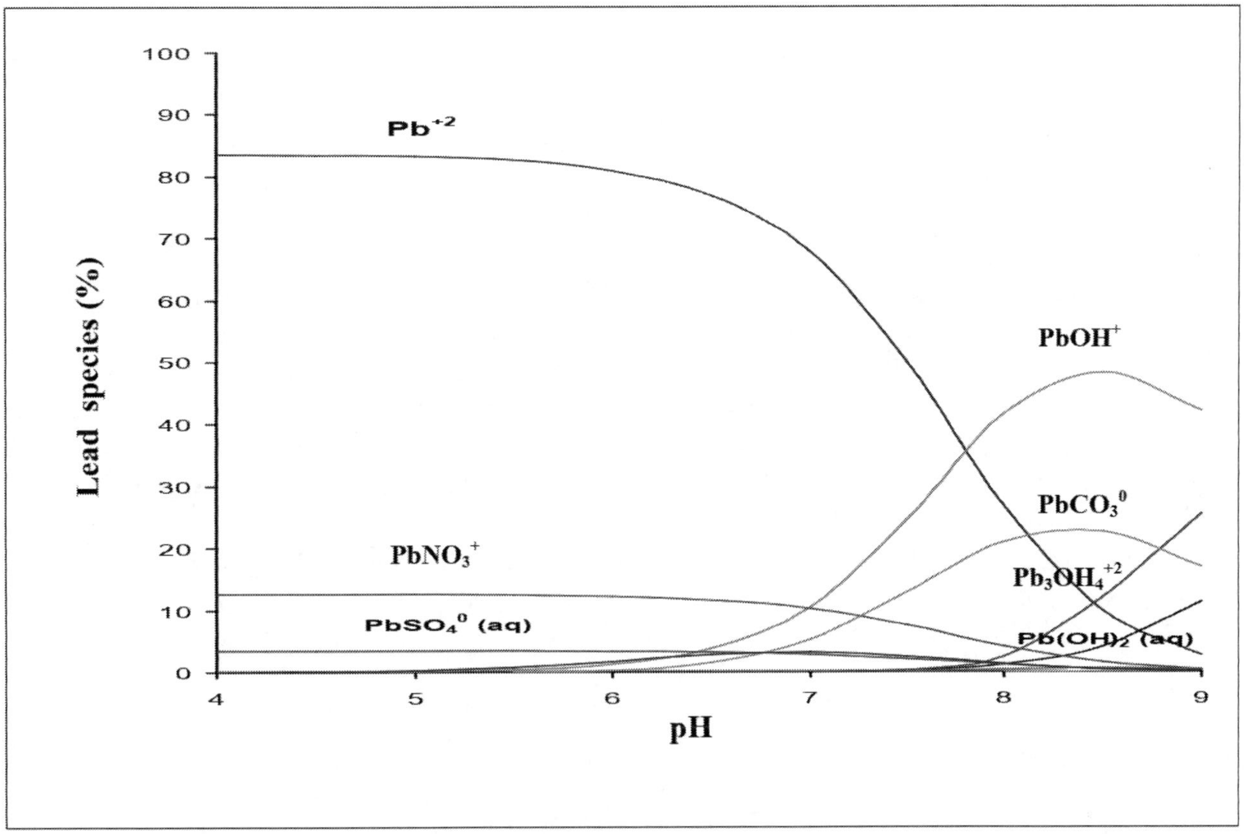

Figure 3. Lead speciation as a function of pH in soil 0.01 M $Ca(NO_3)_2$ system containing equilibrated 0.11mM Pb, 0.15 mM CO_3^{2-} and 0.50 mM SO_4^{2-} concentrations.

References

Hundal, H.S, Raj Kumar and Arora C.L (2003) Cadmium adsorption by some alkaline soils of Punjab state. Agropedology 13: 17-24.

Mckenzie R.M. (1980) The adsorption of lead and other heavy metals on oxides of manganese and iron. Australian Journal of Soil reseach 18: 61-73

Péter Sepos, Tibor Németh, Ilona Mohali and István Dódony (2005) Effect of Soil composition on adsorption of lead as reflected by a study on natural forest soil profile. Geoderma 124: 363-374.

329f Adsorptive Separation of Chloronitrobenzene

Akekasit Leardsakulthong[1], Pramoch Rangsunvigit[1], and Santi Kulprathipanja[2]. (1) The Petroleum and Petrochemical College,Chulalongkorn University, Chula 12,Phyathai road,10330, bangkok, Thailand, (2) UOP,A Honeywell Company, 25 East Algonquin Road,Des Plaines, IL 60017-5016, USA, Des Plaines, IL 60017-5016

Chloronitrobenzenes (CNBs) are used in numerous industries including pesticides, fungicides, pharmaceuticals, preservatives, photochemicals and rubber. There are 2 methods in preparing the CNBs. One is by nitration of chlorobenzene(CB) in the presence of mixed acid, which gives a mixture of about one-third o-CNBs and two-thirds p-CNBs and a small amount of m-CNB. The other method is by chlorination of nitrobenzene. The later method will produce an isomeric mixture containing the meta isomer as the major chloronitrobenzene constituent. To separate p-, o- and m-CNB isomers from each other, a suitable separation process is needed to be developed. This work, an adsorptive process is investigated as a potential CBN isomers separation. Preliminary adsorption results using X and Y zeolites with various cations exchanged are promising. Detail information will be presented and discussed.

Dyes Adsorption and Photo-Degradation Mechanisms on Anatase

Anca Duta and Luminita Andronic.
Transilvania University of Brasov, Centre Product Design for Sustainable Development
Eroilor 29, BRASOV, Romania

Abstract:
TiO_2 - anatase is a wide bad gap semiconductor with photo-catalytic properties, used in organic dyes photo-degradation. Commercial TiO_2 powder (Degussa P25) was used in experiments as dispersions in methyl-orange solutions and as thin films deposited by doctor blade technique. Tests were done in dark (when only adsorption occurs) and under UV irradiation; the efficiency was calculated for different dye concentrations and various ratio solution volume : anatase mass. The adsorption mechanisms are discussed for adsorption and for photo-degradation considering the influence of the catalysts state: dispersed or thin film. The competitive processes adsorption/photo-catalysis are evaluated based on the adsorption parameters. The addition of H_2O_2 in the photo-degradation systems increases the efficiencies when using thin films, at values comparable to those obtained in dispersed systems.
Keywords: photo-degradation, TiO_2, methyl-orange, adsorption

1. Introduction

Organic dyes, particularly azo-dyes, are not likely to be bio-degradable thus the effluents resulted in the textile finishing industry raise real environment problems. Now-a-days, the commonly used technologies imply dyes' coagulation followed by sedimentation; this produces sludge (requesting supplementary treatment, before disposal) while the process is not fully efficient at very low dyestuff concentrations, [1]. Adsorption on activated carbon can be an alternative for advanced removal of organic toxic waste but this partially solves the problem because the substrate re-generation brings supplementary operations in the technology.

The photo-catalytic degradation, on wide band gap semiconductors, offers advantages that justified the large amount of studies published in the past years: complete mineralization of the pollutants forming harmless inorganic compounds, no use of supplementary chemicals and accessible operational parameters (the use of solar or near-UV radiation, room temperature). Recent studies, [2, 3] proved that photo-catalysis is more efficient comparing with adsorption on activated carbon, for diluted solutions of organic dyes and herbicides.

Methylorange, is a high-toxicity product (carcinogenetic), with environmental and health risk, and its large-scale use makes it a priority in wastewater treatment studies, [4].

Photo-degradation of methyl-orange is widely investigated; a recent study, [5], reports on the process efficiency when using different semiconductors (TiO_2, SnO_2, ZnS, CdS) and recommends (as the majority of the studies) the use of TiO_2.

Titanium dioxide is a wide band gap semiconductor, practically insoluble in water and with high chemical stability; these characteristics explains the broad application spectrum in opto-electronic devices such as photo-electrochemical cells and solid state solar cells.

The optoelectronic properties of TiO_2 depend on the crystalline structure. The main two polymorphs have the band gaps of 3.24eV (anatase) and 3.05eV (rutile); due to its band gap, allowing the energy alignment with a broader number of partners in photo-chemical reactions, anatase is reported with a better efficiency in photo-catalytic process, [6].

An amphoteric behavior characterizes anatase in aqueous media and, as result of a variable surface charge, the photo-degradation efficiency is strongly influenced by pH. The point of zero charge depends on temperature, [7] and surface heterogeneity, [8], explaining the different values mentioned in literature; the value corresponding to the standard temperature is $pH_{pzc} = 6.17$, [7]. In acidic media (below the pzc) the surface is positively charged while at pH above pH_{pzc} the surface is negatively charged. Most of the azo-dyes are also ampholites and the photo-degradation efficiency depends on the dye's ionic form and structure.

The photo-degradation process has as first step the dye's adsorption onto the surface, followed by the photo-chemical reaction. Consequently, the pH becomes an important factor, controlling the type of bond between the substrate and the adsorbed specie and, finally, the overall process efficiency.

As an interface process, photo-catalysis is enhanced on substrates with high specific surface. This becomes very important in highly diluted solutions of dyes, when the reactants' diffusion toward the substrate represents a limiting step of the adsorption mechanism. Therefore, the use of nano- or mezzo-structured TiO_2 powders is mainly reported in the studies. Technologically, the use of fine particle dispersion is less advantageous comparing with the immobilized photo-catalyst, with predictable consequences in efficiency. For increasing the photo-degradation efficiency in systems using TiO_2 thin films, advanced oxidation processes (AOP) involving O_2 and H_2O_2 can be an alternative. Studies on the mechanisms, [4], show that the radiation is not directly used for breaking chemical bonds in the pollutant molecule but the catalyst enhanced the $HO^{\cdot}$ formation which reacts and decomposes the adsorbed dye.

This paper presents the results obtained in methyl-orange photo-degradation on TiO_2 powders and thin films. The adsorption mechanisms (in dark) and the overall process (under irradiation) are comparatively discussed in systems with and without H_2O_2.

2. Exerimental

2.1. Materials

Methyl-orange (MO, Merck) solutions were prepared with bi-distilled water, with concentrations up to 0.05mmol/L.

Podwer of TiO_2 (Degussa P25, Germany) was used. The product is a mixture of 75% anatase and 25% rutile, and the morphology involves pores of aprox. 7nm size and a surface area of 50 m^2/g.

A paste of TiO_2 powder, acetylacetone (2,4 pentadione, 99+%, AcAc, Aldrich) in water and nonyl-phenol was smeared on FTO (Libbey Owens) to form a thin film. The film was annealed at 500°C, for 6 hours.

Characterisation of powders and the thin films was done by XRD (Advanced Brucker D8) and AFM (NT-MDT, Ntegra Spectra). The bandgap and the thin film thickness were evaluated using UV-VIS spectroscopy (Perkin Elmer Lambda 25 UV/VIS).

2.2 Adsorption and photo-degradation experiments

Batch experiments were carried out in systems obtained by dispersing amounts of TiO_2 in methyl-orange solutions; two mixing ratios were tested: 0.01g /L and 0.05 g/L. The reactor is a cylindrical flask, of 500mL, open to air.

Based on the thin films dimensions (average thickness 5μm), the volume of the methyl-orange contact solution was estimated, to fit the 0.01 g/L mixing ratio, and the experiments used a photo-reactor of 50mL.

Adsorption test were done in dark, while photo-degradation studies used three F18W/T8 blacklight tubes (Philips), placed annular to the photo-reactor. Each tube emits a broad range of UVA (typically 340-400nm) with λ_{max} = 365 nm. The mean value of the radiation flux, reaching the middle of the reactive system is 3Lx (Luxmeter Mavolux 5032 C/B USM).

Previous studies on MO photo-degradation using anatase (powders and thin films) showed that the efficiency increases with pH, and values above 95% can be expected only at pH>5, [9]. Another group of studies, [10] optimized the photo-catalytic duration at 360 min.

The experiments presented in this study are done at pH = 5.65 (the natural pH of the MO solutions), under stirring, for 360 min, at room temperature (298 K).

3. Results and discussions

3.1. Photo-catalyst characterization

The XRD pattern for the thin films, Fig. 1, shows that the films contain both polymorphs and anatase (A) exceeds rutile (R) in composition. This is not surprising, considering that the transition temperature anatase-rutile is above the annealing temperature. The band gap, calculated based on the optical absorption spectra, is 3.65eV, higher than any of the titania phases, and this can be the result of the oxygen vacancies passivation (during annealing) with the formation of a small amount of amorphous phase.

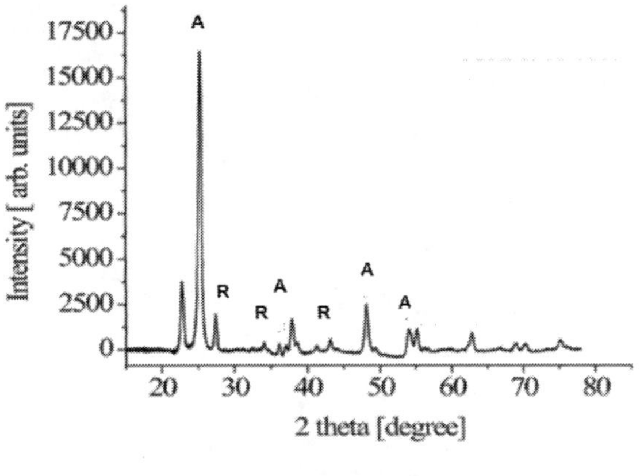

Fig. 1. XRD of TiO_2 thin film (peaks assigned according to JCPDS database)

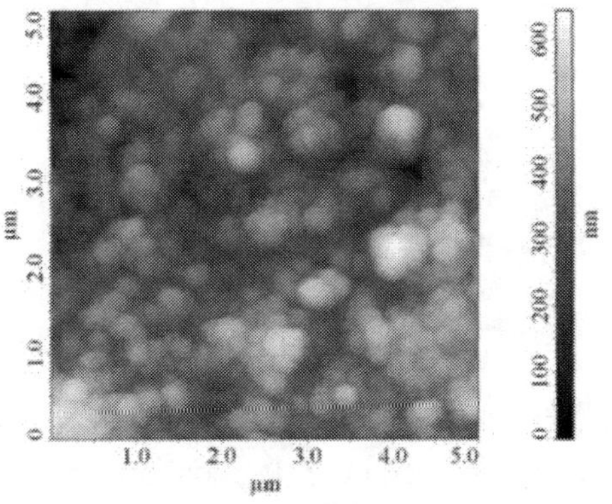

Fig. 2. AFM picture of the TiO2 thin film

The thin film morphology, Figure 2, indicates the formation of aggregated particles, homogenously covering the FTO substrate and resulting in open large pores. This type of morphology insures a large specific area of the film.

3.2. Adsorption and photo-degradation efficiency

The photo-degradation of methyl-orange is started only if the dye is adsorbed on the catalyst surface. At the experimental pH, MO has an anionic structure ($pH_{pzc,\,MO} = 3.4$) and the catalyst is positively charged ($pH_{exp} < pH_{pzc,\,TiO2}$). This favors adsorption based on electrostatic forces, if the preferred orientation of MO towards the substrate is reached. The next step is the photochemical reaction of the active radicals (especially $HO^{\cdot}$ resulted due to the oxygen and/or H_2O_2 reaction with the photo-catalyst) with the anionic dye. Consequently, the photo-degradation efficiency under irradiation can be increased by optimizing the efficiencies of both steps.

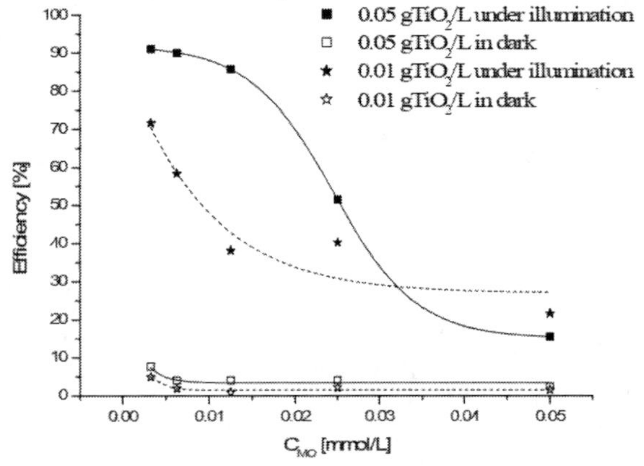

Fig. 3. Adsorption (in dark) and photo-degradation (under iradiation) efficiencies on podwer TiO_2

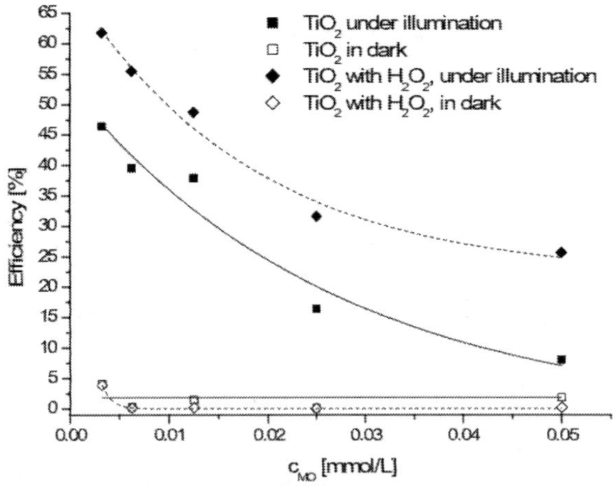

Fig. 4. Adsorption (in dark) and photo-degradation (under iradiation) efficiencies on TiO_2 thin film

As presented in Figure 3 and Figure 4, the photo-degradation overall efficiency is higher in systems containing powder TiO_2. The diluted dispersions (with the TiO_2 concentration of 0.01 g/L) register lower efficiencies, and the slower diffusion of the dye to the substrate can be the explanation. The photo-degradation on thin films has lower efficiencies but the addition of H_2O_2 (as an activator for $HO^{\cdot}$ formation) strongly enhances the efficiency, bringing it at similar values as for the systems with the same concentration of TiO_2 powder.

The adsorption in dark has low efficiencies, below 5.5%. This is a positive aspect, considering the need to form a monolayer, subject of further photo-degradation, and not a stack of multiple layers that can clog the photo-catalyst surface, blocking the radiation access.

3.3. Adsorption and photo-degradation mechanisms

The amount of dye that reaches the mass unit of the photo-catalyst was investigated as function of the pollutant concentration. In dark the result is the well known adsorption isotherm. Under irradiation, the adsorbed dye is photo-decomposed and the by-products are released in the environment, leaving room for the next dye molecules that are adsorbed on the "clean" substrate, etc., so that the results are corresponding to a more complex process that was considered as a "pseudo-adsorption" and was interpreted based on an "adsorption pseudo-isotherm", Figure 5 and Figure 6.

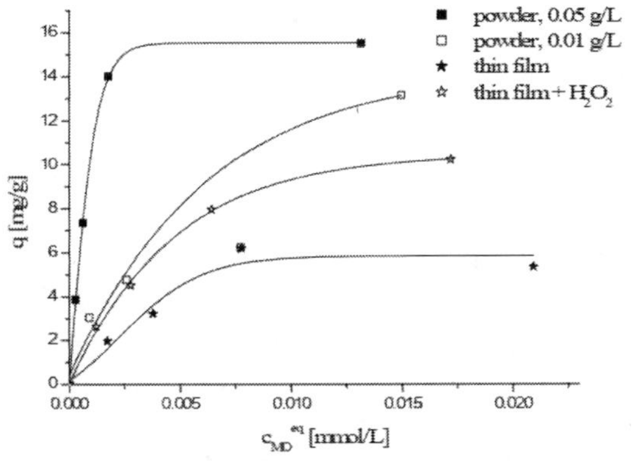

Fig. 5. Pseudo-Adsorption isotherm of MO on TiO_2 powder and film

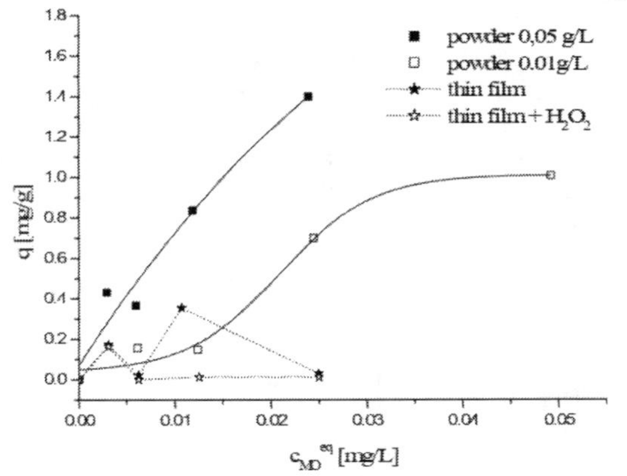

Fig. 6. Adsorption isotherm (in dark) of MO on TiO_2 powder and film

A Langmuir shape characterizes the process under irradiation, for all the investigated systems. The real adsorption, in dark, shows a less clear shape in the powder systems and a random variation on thin films. In accord with the low efficiencies, the shape of the isotherms corresponds to a weak binding of the dye on the substrate. The data obtained for thin films can be explained by the fact that the average dye concentration in the solution is not the same with the dye concentration near the substrate; the system is more sensitive to random variations in concentration (even at long contact times) and batch mixing must be reconsidered. Comparing the results obtained in dark and under irradiation, one can conclude that irradiated TiO_2 exhibits a higher affinity for the dye, active on a space region that allows photo-degradation also on thin films.

The linear form of the Langmuir isotherm was used to calculate the "pseudo-adsorption" parameters:

$$c^{eq} / q = 1/(q_{max} K) + c^{eq} / q_{max} \quad (1)$$

Where: c^{eq} represents the equilibrium concentration of MO, q the adsorption mass coefficient with the maximum value q_{max}, and K is the adsorption equilibrium constant. The linear fit is presented in Figure 7 and the adsorption parameters are presented in Table 1.

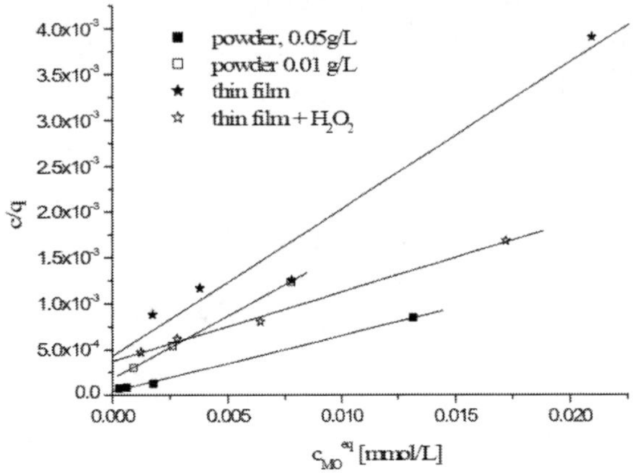

Fig. 7. Linear form of the Langmuir pseudo-isotherm of MO adsorption on TiO_2 powder and film (under irradiation)

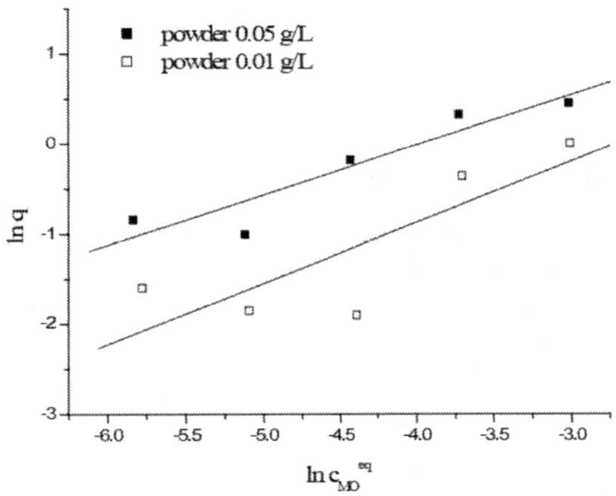

Fig. 8. Linear form of the Freundlich isotherm of MO on TiO_2 powder (in dark)

The absorption in dark, on powder TiO_2 fitted well in the Freundlich isotherm:

$$\ln q = \ln k_F + 1/n \ln c^{eq} \quad (2)$$

Where k and n are the Freundlich parameters, that give information on the relative adsorption capacity (k_F) and on the adsorption intensity (n).
The linearization results are presented in Figure 8 and Table 1:

Table 1 Adsorption Parameters of MO on TiO_2

TiO_2 system	Under irradiation (Langmuir parameters)			In dark (Freundlich parameters)		
	R	K	q_{max} [mg/g]	R	k_F	n
Powder, 0.05g/L	0.99902	1.46E+03	16.37465	0.93412	9.169658	1.797139
Powder, 0.01g/L	0.99992	7.59E+02	7.29501	0.92689	6.328669	1.472971
Thin film	0.97967	3.73E+02	6.228201	-	-	-
Thin Film with H_2O_2	0.9978	2.01E+02	13.28727	-	-	-

The results show that, despite the fact that the same powder is used and the maximum adsorption capacity is actually the same (depending on the specific surface and number of active sites), the dispersed system registers large differences when lowering the powder concentration. This confirms the need to consider all the process steps and parameters, keeping in mind that we are working with low very dye concentrations. This fact is also demonstrated by the results obtained in dark, where the Freundlich equation (assigned usually to heterogeneous substrates) shows good adsorption intensity (n>1) still, no Langmuir fit. It is important to notice the contribution of the H_2O_2 in fastening the photo-degradation process, reflected as an increased overall maximum capacity of the thin film.

For the experimental results obtained under irradiation, thermodynamic calculations were developed to estimate the standard adsorption Gibbs Energy, ΔG, of the photo-degradation process:

$$\Delta G = -RT \ln K_o \qquad (3)$$

Where: R is the gas contestant (8.314 $JK^{-1}mol^{-1}$), T the absolute temperature (K) and K_o the equilibrium adsorption constant.

The equilibrium constant K_o was estimated based on the extrapolation of the ln (c^{ads} / c^{eq}) function vs. c^{ads} to the value $c^{ads} = 0$, [11], Fig. 9 and the results are presented in Table 2.

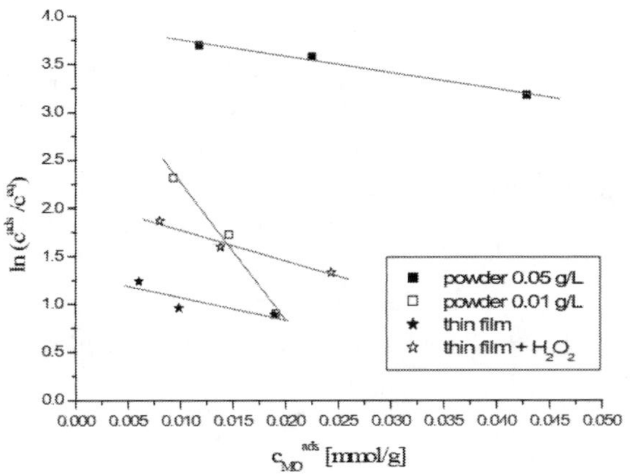

Fig. 9. Approximation of the equilibrium constant for MO photo-degradation on TiO_2

Table 2 Equilibrium constant and ΔG for the MO photodegradation on TiO_2

Powder 0.05 g/L		Powder 0.01 g/L		Thin film		Thin film + H_2O_2	
K_o	ΔG [kJ/mol]	K_o	ΔG [kJ/mol]	K_o	ΔG [kJ/mol]	K_o	ΔG [kJ/mol]
3.92726	-3.3904	3.70916	-3.24879	1.30589	-0.66123	2.09306	-1.83

The ΔG values show that the spontaneous photo-degradation processes could be enhanced by temperature.

4. Conclusions

The MO photo-degradation from highly diluted solutions can be done using TiO_2 powders, with a majoritar anatase content. The dye adsorption process occurs without clogging the substrate and the photo-degradation reaction is responsible for the process efficiency. By lowering the photo-catalyst concentration in the dispersed system, the efficiency decreases and good results (efficiency up to 95%) were registered for systems with TiO_2 concentration of 0.05 g/L.

The use of immobilized photo-catalyst in thin films leads to good photo-degradation efficiencies, especially when H_2O_2 is added to the system, comparable with the efficiency in dispersions with the same amount of catalyst.

The overall photo-degradation process can be described by the Langmuir equation, although it implies a more complex path (adsorption and photo-reaction) while the adsorption is better described by the Freundlich equation.

References

1. Wang, W-Y, Ku, Y., Effect of solution pH on the adsorption and photocatalytic reaction behaviors of dyes using TiO2 and Nafion-coated TiO2, Colloids and Surfaces A, 302 (2007) 261-268
2. Kim, S-H, a.o. Adsorption and photocatalysis kinetics of herbicide onto titanium oxide and powdered activated carbon, Separation and Purification Technology (2007) in press
3. Sharkawy E.A., Soliman, A.Y., Almer, M, Comparative study for the removal of methylene blue via adsorption and photocatalytic degradation, Journal of Colloid and Interface Science 310, (2007), 498-508
4. Tsoukleris, D.S., a.o., Photocatalytic properties of screen-printed titania, Catalysis Today, 124 (2007), 110-117
5. Kansal, S.K., Singh, M., Sud, D, Studies on photodegradation of two commercial dyes in aqueous phase using different photocatalysts, Journal of Hazardous Materials, 141 (2007), 581-590
6. Tayede, R., a.o., Photocatalytic degradation of dyes and organic contaminants in water using nanocrystalline anatase and rutile TiO_2 , Science and Technology of Advanced Materials, (2007), in press
7. Chou, J-C., Liao, L.P., Study on pH at the point of zero charge of TiO_2 pH ion-sensitive field effect transistor made by the sputtering method, Thin Solid Films, 476, (2005), 157-161
8. van Riemsdijk, W.H., a.o., Electrolyte adsorption on heterogeneous surfaces: adsorption models, Journal of Colloid and Interface Science , 109, (1986), 219-228
9. Andronic, L., Duta A., TiO2 thin films prepared by spray pyrolysis deposition and their photocatalytic activity, JOAM, 9 (2007), 1403-1406
10. Andronic, L., Duta, A, TiO2 thin films for dyes photo-degradation, Thin Films, 515 (2007), 6294 – 6297
11. Hasnat M.A., a.o., Adsorption and photocatalytic decolorization of a synthetic dye erythrosine on anatase and ZnO surfaces, Journal of Hazardous Materials, 147, (2007) 471-477

ADSORPTION KINETICS OF CHLORINATED HYDROCARBONS INTO HIGH SILICA ZEOLITE

Kazuyuki Chihara, Shinji Kondo, Takashi Matsumoto,*

Corresponding author: Kazuyuki Chihara, Dept. of Applied Chemistry, Meiji University Higashimita, Tama-ku, Kawasaki 214-8571, Japan,

Introduction

In order to predict the structure of a substance and the physical properties, using a molecular simulation has prospered in recent years. The Grand Canonical Monte Carlo (GCMC) method and Molecular Dynamics (MD) method are simulation method for solving a phenomenon from a microscopic level. It is becoming possible to interpret the adsorption characteristic in a molecule level rationally, and to predict the macroscopic characteristic such as adsorption isotherms [1]. In this study, equilibria and isosteric heat of adsorption for the system of chlorinated hydrocarbons and High-Silica zeolite(USY6.18) were obtained with gravimetric method and chromatographic method. By comparing an experiment result with a molecular simulation result, the validity of forcefield parameters and zeolite model was examined.

Experimental

Gravimetric Method

Gravimetric analysis was used to obtain the adsorption curves plotted for the ratio of volumes of solution (q_t/q_∞) against time (t). Micropore diffusivity was obtained by analyzed adsorption curve. The amount adsorbed was measured corresponding to the pressure of the vapor in the tube. The pressure was measured by pressure sensor at higher pressure range (> 0.013 atm) and baratron at lower pressure range (<0.013 atm). A Time with increment value of adsorption was measured by chart recorder. In this way, the adsorption curves were obtained. In order to compute a diffusion coefficient from an experimental result, it is necessary to create the theoretical curve which performs creation of uptake curve obtained in the weight variation-per-hour measurement experiment, and curve fitting.

Simulation

Dynamics Simulation Module in Cerius2 (Accelrys Inc.) was used throughout MD simulations. Non-bonded interactions that consist of van der Waals and electrostatic terms, were obtained by Mellot et al [2]. As the parameter of the other valence interactions, the parameters of UNIVERSAL1.02 in the database of Cerius2 were used. USY-6.18 has the silica-alumina ratio of SiO_2/Al_2O_3=6.18, so remaining aluminum were considered in zeolite model. The Einstein formulas were used for calculation of a diffusion coefficient.

Result and Discussion

Although chloroform after a simulation existed easily, it moved to the fine pores of adsorbent as calculation progressed. Moreover, in the interface model, the process which the chloroform which exists in the outside of adsorbent moves on the surface of adsorbent by 10ps, and moves to the inside of adsorbent at 50ps was able to be seen. This move process was assumed to be diffusion with a boundary film.
As an example of a result, the result of USY6.18 (FAU type zeolite) and chloroform is shown (Fig.1). The diffusion coefficient and mass transfer coefficient which were obtained by the experiment and the simulation are shown in Table 1 and 2.

Table1 Micropore diffusivity of chloroform into USY6.18.

method	Gravimetric	MD sim.
Temperature [K]	303.15	303.15
D [cm^2/sec]	4.66E-10	2.10E-06

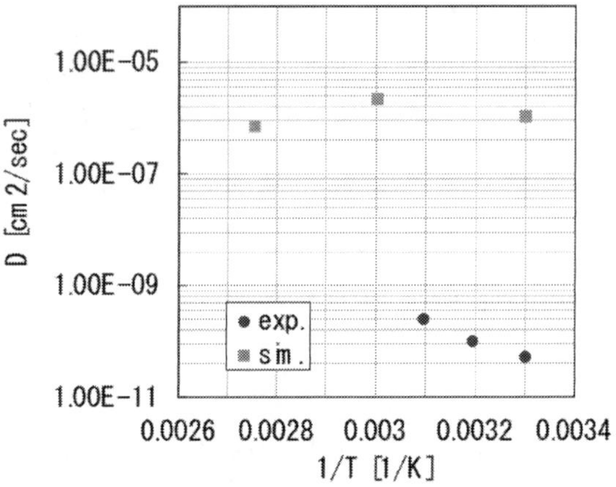

Fig.1Arrhenius plot

Table2 mass transfer coefficient

	exp.303K	exp.313K	exp.323K	sim.303K
$a^2/15D$ [sec]	4.0246	1.6295	1.0048	0.0109
$K*/k_f a_p$ [sec]	4.1984	4.0738	5.0238	0.1359
$1/K_s a_p$ [sec]	8.3571	5.7033	6.0285	0.1468

Conclusions

In ZSM-5 Type, it is necessary to improve an electric charge and some models. The behavior of chloroform, moving from cage to cage, which means diffusion in Y-type zeolite, could be confirmed by MD calculation. As a result of comparing the diffusion coefficients by the MD method with those by chromatograph method, it was suggested that behavior of adsorbates in micropore of zeolite might be estimated by the MD calculation.

References

1. C. F. Mellot, A. M. Davidson, J. Eckert, Adsorption of Chloroform in NaY Zeolite: A Computational and Vibrational Spectroscopy Study, J.Phys. Chem. B, 102, 2530 (1998b)

329j Olefin Selectivity Enhancement Of Silver-Nitrate Impregnated Aluminosilica By Sodium Hydroxide Treatment On Substrate

Chang Hyun Ko, Jince Sebastian, Jong-Ho Park, Kwang Bok Yi, Sang-Sup Han, Soon-Haeng Cho, and Jong-Nam Kim. Chemical Process Research Center, Korea Institute of Energy Research, 71-2 Jang-dong Yuseong-ku, Daejon, South Korea

Olefin/paraffin separation by distillation is an energy-intensive process because the boiling points of olefins are so close to those of their corresponding paraffins. Recently, olefin/paraffin separation using adsorption based on π-complexation has attracted much attention due to its high selectivity and reversible capacity. Silver nitrate ($AgNO_3$) is expected to be one of the most effective materials for olefin/paraffin separation because it showed strong affinity toward double bonds compared to saturated single bonds. To increase the number of effective adsorption sites for olefin in adsorbent, silver nitrate was distributed on high surface area substrates. One of the most effective adsorbent for the separation of olefin from paraffin was silver nitrate supported on aluminosilica. However, aluminosilica has various types of pores such as micropores (< 2nm), mesopores (2 -10 nm), and macropores (> 10 nm). Silver nitrate located inside micropores in aluminosilica is not expected to be active adsorption site for olefin because micropores might be barriers for olefin diffusion toward siliver nitrate. High surface area substrates with minimized portion of micropores or without micropores might be an ideal substrate for olefin separation adsorbents. To achieve this goal, this report investigated the effect of basic solution treatment on the selective removal of micropores in aluminosilica. Although total surface area of aluminosilica decreased after basic solution treatment, 1-butene selectivity to n-butane increased at optimized condition because micropores, which is the barrier for olefin adsorption, was removed. In addition to selectivity, desorption speed was also enhanced for the same reason.

Adsorption Mechanisms Of Heavy Metals On Modified Fly Ash

Maria Visa*, Anca Duta**,
*Chemistry, College For Natural Sciences, Armoniei 13, Brasov, Romania And
**Chemistry, Transilvania University Of Brasov, Eroilor 29, Brasov, Romania.

Abstract
The use of fly ash in wastewater treatment, for heavy metals (Cd, Cu and Ni,) immobilization, represents a topic much addressed to in the past years solving both, the problems of advanced treatment and the intelligent use of fly ash. The study presents the results obtained in Cd^{2+}, Cu^{2+} and Ni^{2+} removal, using fly ash from CHP Brasov (Romania) with modified surface. Studies have compared the adsorption mechanisms of heavy metals (cadmium, copper and nickel) on fly ash with modified surface charge. The fly ash morphology is characterized using AFM while XRD provides composition and crystallinity data. Metal ions solutions ($c = 10^{-4} - 10^{-2}$m) are mixed in a reaction flask with fly ash (powder); samples of fly ash washed in ultra pure water or with modified surfaces after long contact with alkaline (NaOH 1, 2 and 4N) solutions were used in the experiments. Using the optimized adsorption parameters (contact time, wastewater volume: adsorbent mass ratio), the adsorption isotherms are obtained and the adsorption mechanisms are discussed. The experimental data indicate that adsorption of cadmium, nickel and copper ions can be significantly improved by using the fly ash modified with NaOH 2N, when highly efficient chemical adsorption, is registered even at very low heavy metals concentrations.
Keywords: Fly ash, modified fly ash, wastewater treatment, heavy metal removal

1. Introduction

Ash results mainly during burning of pulverized coal in the heat and power stations furnaces. About 85% of the ash is carried out of the furnace into the electrostatic filters (the fly ash, FA), the rest being collected as bottom furnace ash.

The FA has as major components oxides of silicon, aluminum, calcium and iron but lower constituents are also included (magnesia, titania, sodium and potassium oxides) along with trace components (heavy metal oxides, metal sulfates and chlorides, etc.). The dimensions of FA particles range from a few to hundreds of microns; the glassy spherical particles have a rather large specific surface, especially those of small dimensions, [1]. The exact properties of fly ash depend on a complex of factors: composition and dimensions of the coal particles, burning temperature, burning duration, furnace geometry.

The available coal reserves in the world are estimated at more than 400 years' supply but the use of coal in heat and power plants starts to be limited because of the side effects induced by the burning by-products: CO_2, NO_x, SO_x. To fulfill the environment standards, modifications in the burning process are done, by decreasing the burning process temperature (for lower NO_x values) with consequences on the ash composition. An increase in the carbon content (expressed as LOI, loss on ignition) is registered without any supplementary side effects so-far reported, [2].

Fly ash is broadly used in concrete production (mainly in manufacturing aerated concrete blocks) but, large quantities of FA (over 40%) are disposed and stored as stockpiles or in drained lagoons, [3]. They raise supplementary environmental problems, due to the leaches causing underground water pollution. Therefore, identifying novel applications of FA, in large scale processes, represents a topic very much addressed to in the past years.

The FA composition and specific surface recommends it as a possible adsorbent in wastewater treatment. Literature mentions the use of coal fly-ash for heavy metal immobilization and studies were carried out for nickel [4], copper [5-7], and cadmium [5, 7], lead, zinc, etc.

One of the problems that hinders a large scale application of the results, is the properties variation (and adsorption behavior) of the FA from different sources. Therefore, recent studies are targeting the development of FA with controlled surface composition, obtained in efficient and feasible processes. A two steps mechanism can be identified, involving the alkaline dissolution of easily soluble surface components (based mainly on calcium, magnesium, sodium, potassium) followed by re-precipitation of insoluble compounds, with controlled composition. Following these mechanisms, charge controlled surfaces, [8] or zeolite-type surfaces could be obtained, [9, 10] with good adsorption efficiency of copper, cadmium and lead from polluted wastewater. All the results are obtained using highly concentrated alkaline solution for modifying the FA surface, but this lowers the sustainability and increases the costs of the overall process.

This paper presents the results obtained in cadmium, copper and nickel removal from wastewater, using FA modified with alkaline solutions of average concentrations.

2. Materials and Experiments

Heavy metal solutions are prepared using ultrapure water and: $CdCl_2 * 2.5\ H_2O$, (Chemopar); $NiCl_2 * 6H_2O$ (Scharlau); $CuCl_2 * 2H_2O$ (Merck)
The cations concentration was evaluated by AAS(Perkin Elmer, Aanalyst 200), at the relevant wavelengths: $\lambda_{Cd} = 228.8$nm; $\lambda_{Ni} = 341.5$ nm; $\lambda_{Cu} = 324.8$ nm;
Fly ash: Fired coal FA was collected from the Brasov CHP plant, directly from the filters. The ash was washed in ultra pure water for 48 hours, until constant pH of the washing solutions;
FA surface was modified by batch mixing, 48 hours, a dispersion of FA: alkaline solution (1:10):
Alkaline Solutions of NaOH (Merck): c = 1N, 2N and 4N
The FA crystalline structure was analyzed by XRD, using Bruker D8 Advance Diffractometer. The topography of the FA surface is studied using Atomic Force Microscopy (AFM, NT-MDT model BL222RNTE). The images were taken in contact mode with Si-tips (CSG10, force constant 0.15 N/m, tip radius10 nm).
Batch adsorption experiments were conducted, at room temperature, to evaluate the removal efficiency of cadmium, nickel and copper ions from the aqueous solutions. The effect of contact time, ratio FA mass/solution volume, initial metal concentration and FA surface were studied.
As reference, adsorption test, in the same experimental conditions, were done on wet activated carbon powder (Merck).

3. Results and discussions

The Wastewater Discharge Regulations (TM 5-814-8 - developed in US and accepted, almost at the same values, in many countries) set the following values, obtained based on the best available technology in electroplating industry: $BAT_{Cd} = 0.69$ mg/L; $BAT_{Cu} = 3.38$ mg/L; $BAT_{Ni} = 3.98$ mg/L. Obviously cadmium represents the highest risk and imposes the restrictions in a treatment process of wastewater also containing copper and nickel. Therefore, the optimization tests were developed for cadmium removal, using FA with modified surface. The optimized adsorption process was then applied for solutions containing copper or nickel.

Previous studies for cadmium removal, [11], using FA with modified surface based on complexing agents (black eriochrome T, pyrocatechol violet), showed a limited adsorption efficiency of this type of substrate, up to 16%. This proves that the superficial adsorption of a compound (even with high metal affinity) is not enough, and chemical substrate modifications are required. Early tests, [12], involving surface modification using HCl, resulted in very low cadmium adsorption efficiencies, (less than 5%); this result was assigned to the oxides dissolution, not followed by re-precipitation.

Samples of FA were modified by direct contact of the NaOH solutions with the concentrations 1N, 2N and 4N. The XRD patterns of the modified FA are similar, Figure 1 and confirm the existence of carbon (graphite) along with silica and calcium oxide. The analysis confirms the dissolution – precipitation process of silica, solving the tetragonal and forming the orthorhombic polymorph. The magnesium oxide peak is stronger after the treatment, indicating that the process favours the precipitation of Mg^{2+} resulted from the dissolution of soluble compounds; this is not a general behaviour because calcium oxide content and crystallinity is not to affected in the experimental conditions.

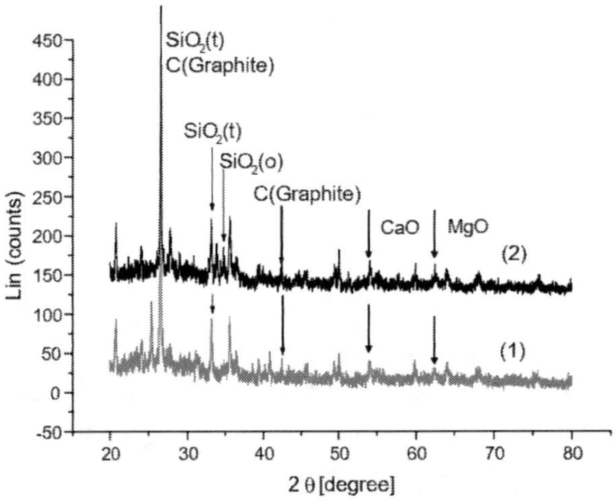

Fig. 1. XRD of raw FA (1) and FA modified with NaOH 2N (2)

The re-precipitation of compounds is confirmed by AFM, showing the crystal growth on the initial spherical morphology, Figure 2.

Fig. 2. AFM picture of the FA modified with NaOH 2N

Literature mentions the affinity of carbon (as activated carbon or resulted after pyrolysis of different organic wastes) in the adsorption process of cadmium from aqueous solutions, [13]. The cadmium adsorption efficiencies on raw, modified fly ash and activated carbon (AC) were preliminary investigated using metal cation solutions (0.01N), at a ratio substrate mass: solution volume of 1:100 g/mL. For a contact time of 10 minutes, the results are presented in Table 1.

Table 1 Cadmium Adsorption Efficiency [%] on different substrates

Time [min]	FA / NaOH 1N	FA/NaOH 2N	FA / NaOH 4N	AC
10	58.17	31.47	99.96	11.62

The results obtained using the FA modified with NaOH 4N are fulfilling one of the main objectives: to identify an efficient substrate for cadmium removal, based on modified FA. But, considering the need to identify an environmental friendly set of process parameters, further experiments were developed, testing the possibility of lowering the concentration of the alkaline solution used for FA modification. The optimization tests were focusing on the contact time and on the modified FA mass needed to adsorb cadmium from 100 mL solution, Table 2:

Table 2 Cadmium adsorption efficiency and optimized parameters on modified FA[*)]

	FA/NaOH 1N	FA/NaOH 2N	FA/NaOH/4N
Contact time, [min]	60	30	10
Adsorbent mass/100 mL solution, [g]	2	4	1
Adsorption efficiency [%]	99.97	99.45	99.96

*) C_{Cd} = 0.01N

Technologically speaking, 10 minutes contact time cannot represent a feasible solution but a contact time of 30 minutes is possible and future test will be developed for studies on immobilized FA, as thin films.

The next experiments were developed using modified FA with NaOH solution, at the optimized parameters, testing mono-ionic solutions of cadmium, nickel and copper, respectively. The adsorption efficiencies, Figure 3, show that the substrate is optimized conditions lead to god removal conditions, in the low concentration range (below 100 ppm) for all the investigated cations. If pre-treatment is done, the residual heavy metal concentrations satisfy the BAT values.

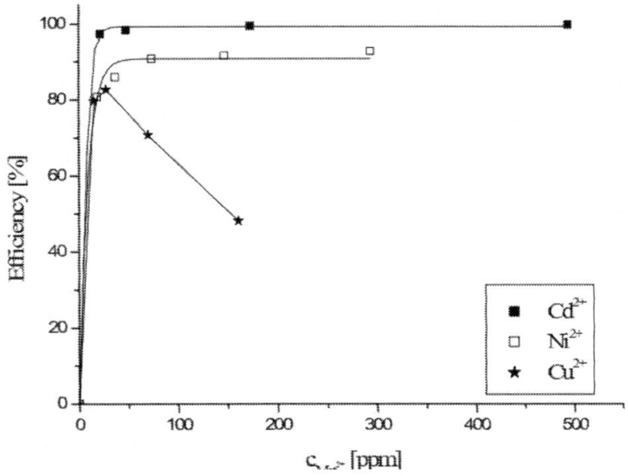

Fig. 3 Adsorption efficiency of cadmium, copper and nickel

The adsorption isotherms in Figure 4 indicate rather weak interactions at low concentrations and fit to the Freundlich linearization fits well to all the adsorption mechanisms, Table 3, confirming the adsorption on a highly heterogeneous substrate. Although the copper adsorption at higher concentration has a lower efficiency (50%) the value of the "n" parameter proves a stronger interaction with the substrate. The contradiction is only apparent, considering that, in aqueous media all the cations are hydrated, with large volumes that have consequences on the diffusion rate towards the substrate. Supplementary, previous tests have shown that adsorption involves the ion without de-hydration but with shape modification and, because of this, more available adsorption sites are blocked by large volume cations leading to a decrease in efficiency even when good affinity with the substrate exists. The reverse situation is registered for nickel and cadmium, with high adsorption efficiencies and low affinity for the substrate.

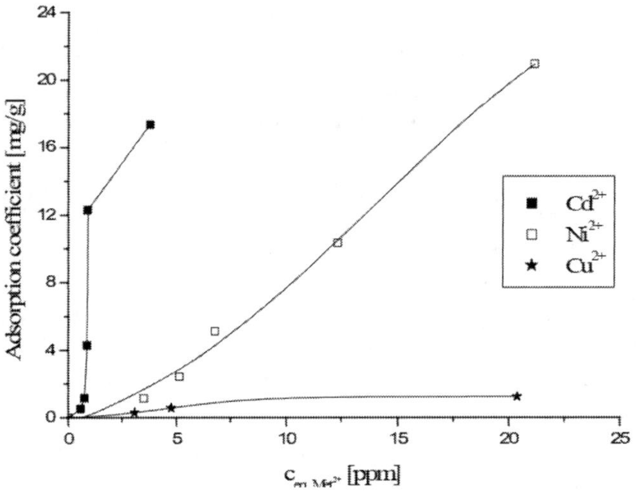

Fig. 4 Adsorption isotherms from mono-ionic solutions of cadmium, copper and nickel

Table 3 *Freundlich adsorption parameters of cadmium, nickel and copper on FA/NaOH 2N from mono-ionic solutions*

Cd^{2+}			Ni^{2+}			Cu^{2+}		
R	k	n	R	k	n	R	k	n
0.9096	15.02928	0.160256	0.9875	$2.4\ 10^6$	0.628931	0.9680	0.209297	1.886792

Further experiments were done on solutions containing equi-molar mixtures of the cadmium, copper and nickel. The adsorption efficiencies, plotted in Figure 5, for cadmium and copper show that in the concentration range below 100 ppm the substrate presents allows advanced removal.

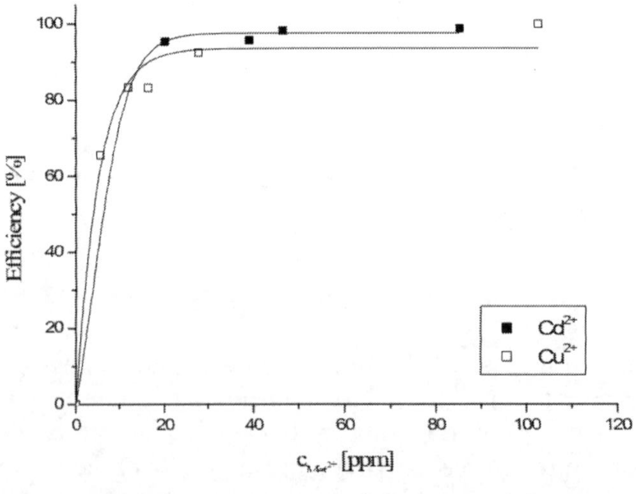

Fig.5 Adsorption efficiency of Cu and Cd from a mixed solution with Ni on FA/NaOH 2N

Similar data were registered for nickel. Due to the different behaviour registered in the mono-ionic solutions, the comparative analysis of copper and cadmium (in the mixed solutions) was considered significant. The final concentrations satisfy the BAT standard for Cu and are very close to the required value for Cd.

The analysis of the adsorption coefficient vs. equilibrium concentration shows a competitive adsorption with very close adsorption values in the experimental conditions (values ranging up to 3mg/g).

4. Conclusions

The surface of fly ash can be modified with average concentration alkali solutions (NaOH 2N) when dissolution and re-precipitation of the superficial compounds leads to an increased specific surface and crystal growth. The modified surface exhibit good adsorption efficiency for copper, nickel and cadmium; this result was registered both in mono-ionic or in mixed solutions. The surface has still a high degree of heterogeneity and the Freundlich isotherm describes well the process.

References

1. Demanet, C.M., Atomic force microscopy determination of the topography of fly-ash particles, Applied Surface Science, Vol. 89, Issue 1, 1995, p. 97-101
2. Hurt, R., Suuberg E., Gao, Y., Unburned carbon in ash: formation, properties, and behaviour, EPRI report, TR-109340, 1997
3. Sear, L. K., Properties and Use of Coal Fly Ash, Thomas Telford Ltd., 2001
4. Serpil, C., Pehlivan E., The use of lfy ash as low cost, environmentally friendly alternative to activated carbon for the removal of heavy metals from aqueous solutions, Colloids and Surfaces A, Vol. 298, Issues 1-2, 2007, p. 83-87
5. Papandreou, C.J., Stournaras, C.J., Panias, D., Copper and Cadmium adsorption on pellets made from fired coal, Journal of Hazardous Materials, (article in psress)
6. Heechan, C., Dalyoung O., Kwanho, K, A study on removal characteristics of heavy metals from aqueous solutions by fly ash, Journal of Hazardous Materials, Vol. 127, Issue 1-3, 2005, p. 187-195
7. Ricou, P., Lecuyer I., Le Cloirec, P., Removal of Cu2+, Zn2+, Pb2+ by adsorption onto fly as hand fly ash/lime mixing, Water Science and Technology, Vol. 39, Issue 10-11, 1999, p. 239 – 247.
8. Hsu, T-C., Yu, C.-C., Yeh C-M., Adsorption of Cu2+ from water using raw and modified coal fly ashes, Fuel, article in press.
9. Shawabeck, R., Al-Harahsheh, a.o., Conversion of oil shale into zeolite for cadmium and lead removal from wastewater, Fuel, Vol. 83, Issue 7-8, 2004, p. 981-985
10. Lee, W.K.W., Deventer, J.S.J., Structural reorganization of class F fly ash in alkaline silicate solutions, Colloids and Surfaces A., Vol. 211, Issue 1, 2002, p. 49-66
11. Visa, M., Duta, A., Advanced Wastewater Treatment for Heavy Metals Immobilisation Using Modified Fly Ash, Materials Science and Engineering, B (in press)
12. Visa, M., Duta, A., Tailoring Fly Ash As Adsorbent For Cadmium, Removal Form Industrial Waste Water, Bulletin of The Transilvania University of Brasov, Vol. 48, p. 142
13. Mohan' D., Pittman, C., a.o. Sorption of arsenic, cadmium, and lead by chars produced from fast pyrolysis of wood and bark during bio-oil production, Journal of Colloids and Interface Science, Vol. 310, Issue 1, 2007, p. 57-73
14. A.W. Marczewski, A practical guide to isotherms of adsorption on heterogenous surfaces, 2002, http://adsorption.org/awm/ads/Ads.htm

DEVELOPMENT OF HIGH TEMPRTATURE DESORPTION SYSTEM IN THE ADSORPTIVE CONCENTRATOR FOR HEAVY COMPONENTS OF VOC (VOLATILE ORGANIC COMPOUNDS)

Ken-ichiro Yamada, Keimei Furuki, Yuuji Fujioka, Hiroshi Okano and Tsutomu Hirose
Seibu Giken Co., Ltd., Fukuoka, Japan

Introduction

Volatile organic compounds (VOC) emitted in a large volume from chemical plants, semiconductor industries, printing and coating processes and so on has been recognized as one of precursors for suspended particle matters and photochemical oxidants. The governmental regulations for the VOC emission are getting more and more severe in many countries in a worldwide movement to the environmental protection. For example in Japan, the air pollution control law revised in 2004 have been enforced in April 2006 and the aim is to reduce the annual emission in 2010 is aimed to be reduced by 30% relative to the total emission of 1.5million tons in 2000 by a combination of the governmental regulations and voluntary efforts of emission sites.

High efficiency units for VOC abatement are required to meet the above target of the air pollution control. VOC of low concentration are destructed finally by direct or catalytic combustion, corona destruction, or ozone decomposition unless the recovery by cooling or compression is economical in the case of high concentration (>1%). VOC-laden air is strongly recommended to pretreated to concentrate VOC and reduce the load of the final destruction units. Preconcentrators are usually based on adsorption on various porous solids, although absorption and membrane processes are not impossible. Remarkable progress in adsorptive VOC treatments in the recent one or two decades may be the introduction of thermal swing honeycomb rotor adsorbers into the market. This type of VOC concentrators were characterized by monolith or honeycomb structure of adsorbent instead of the conventional particles or pellets and by a rapid response to temperature swing.

Seibu Giken Co., Ltd, Fukuoka, Japan, put the VOC concentrators into the market in 1988 as an extension of their corrugation technology developed in honeycomb rotor dehumidifiers and regenerative heat exchangers. The present authors have discussed the efficiency of the thermal swing VOC concentrator in a conference paper by Kuma et al.[1)] and in journals by Mitsuma et al.[2)] and Yamauchi et al.[3)] They described the details of honeycomb rotor adsorbers developed by Seibu Giken Co. Ltd and derived some general principles for design and operation of honeycomb rotor VOC concentrators. The most recent development is discussed in another paper[4)] of this AIChE Annual Meeting.

One of current issues is the deterioration of efficiency due to a long term running. Exhaust gas involves sometimes VOC of high boiling point or VOC easy to polymerize. When these VOCs are processed in the concentrator, they are difficult to be desorbed at the standard temperature and accumulated in the rotor to cause the deterioration of the removal efficiency during a long term operation for years. We developed recently the reactivation technology for the deteriorated adsorbent rotor by a high temperature desorption together with heat proof sealing. The purpose of the present paper is to discuss a new reactivation system to extend the life time of the honeycomb rotor adsorbent.

Honeycomb Rotor Adsorbent and VOC Destruction System

Figure 1 shows the total VOC destruction system combined with the adsorptive concentrator and the thermal oxidizer (combustion) and Fig. 2 shows a 3-D diagram of the concentrator. VOC-laden air is concentrated by a factor of about 10 in a honeycomb rotor concentrator according to the flow sheet given in arrowed lines before it is combusted directly in an oxidizer to destruct it finally. The degree of enrichment is usually sufficient high for combustion of VOC. The direct combustion sometimes is replaced by catalytic combustion or others when VOC concentration in the feed air is too low to concentrate it up to the sufficient concentration. Combustion heat released is utilized through heat exchange to regenerate the honeycomb adsorbent by desorbing the adsorbed VOC at an elevated temperature. Thus a self-sufficient system can be constructed without any external heat source.

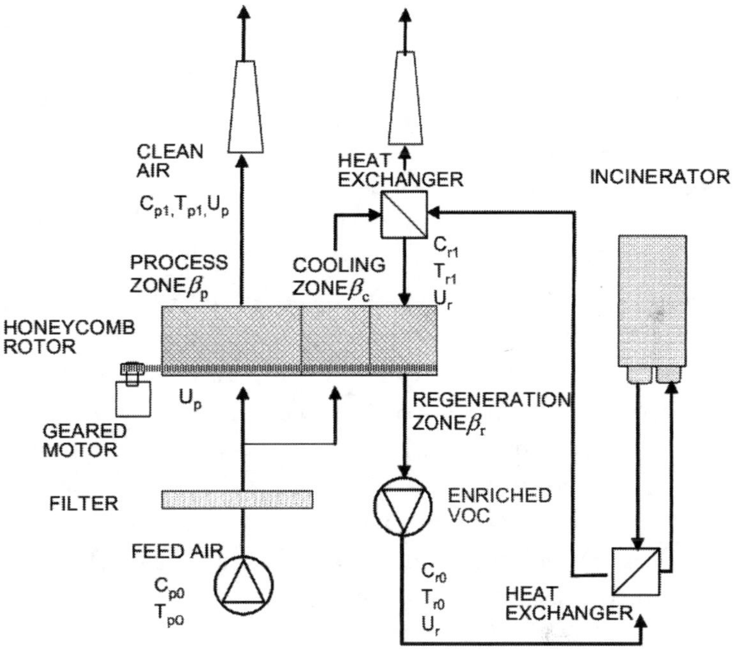

Fig. 1 Integrated system of the VOC destruction

A housing holding a slowly rotating honeycomb adsorbent rotor is subdivided usually into three sectors as shown in Fig. 2. The first is a process zone where VOC vapor was removed from the feed air by adsorption onto the adsorbent surface during passing through narrow channels of the honeycomb structure as shown in Fig. 3. The second is a regeneration zone where the adsorbent rotor is regenerated by hot air streamed countercurrently to the process zone. The last is a cooling zone, which is located between the process and regeneration zones and serves to purge the crude product from the bulk of fine product and to cool the rotor to keep a high adsorption capacity in the subsequent process zone. The rotor was kept airtight to avoid air leakage across each zone by a set of radial and circumferential seals, of which the details are discussed by Mitsuma et al.[5] In typical commercial applications, the ratio of cross-sectional area of process: regeneration: cooling zone is set to 10:1:1 and the regeneration temperature is 180℃.

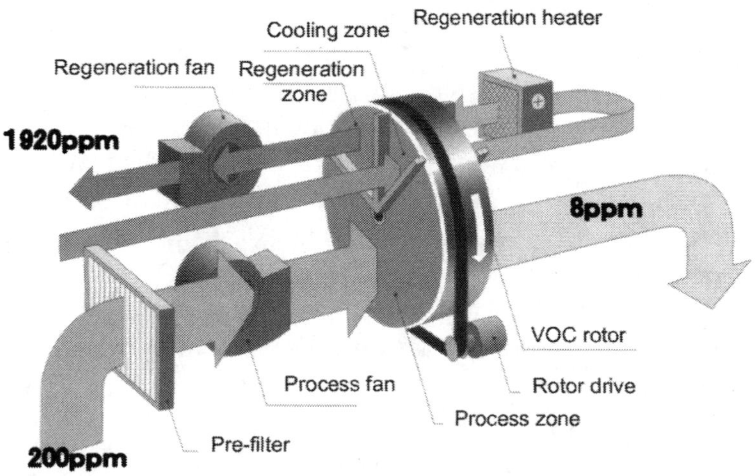

Fig. 2 A 3-D diagram of the VOC concentrator

The key technology of the honeycomb rotor adsorber is the preparation of a high-efficiency adsorbent in a matrix of corrugated sheet; the detailed procedure can be found in patents by Kuma et al.[6-8] A plane strip and a corrugated strip of 0.2mm thick ceramic fiber paper with high voidage were first stacked with heat-resistant adhesive to manufacture a single layer of honeycomb with a typical channel pitch of 3.2mm width x 1.7mm height and the wall thickness of 0.2mm. Then the single-layer honeycomb was rolled up with the same adhesive into the shape of a honeycomb rotor as shown in pictures of Fig. 2 up to a desired diameter while the tension applied to the layer was controlled carefully. The largest commercial size is about 4m in diameter, and a large-diameter unit was manufactured by fixing several precut honeycomb segments in a frame instead of the above single stage rolling-up. The rotor was once calcined at about 500℃ to manufacture a heat-resistant ceramic rotor by removing unnecessary organic matters involved as a binder in the ceramic paper. The ceramic rotor was then impregnated with dilute silica sol in which fine powder of high silica zeolite was suspended. The composite adsorbent of zeolite was solidified in the void of ceramic fiber sheet of the rotor matrix after drying. The adsorbent honeycomb rotor was completed by heating it, cutting it in 450mm width, and finishing the end surface flush. High silica zeolite, i.e. zeolite with high Si/Al ratio, as used here bears a hydrophobic surface to accelerate the adsorbability for organic compounds.

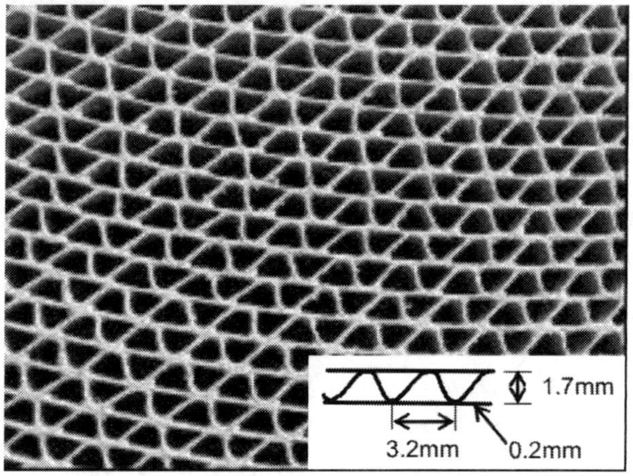

Fig. 3 Cross-sectional view of the honeycomb structure

Many commercial units are operating at the present with the removal efficiency higher than 95% and the concentration ratio of 5-20 under the standard specifications of the rotor diameter of 0.3-3m, the rotor width of 0.4-0.45m, air velocity of 2-4m/s, temperature of regeneration air of about 180 C and co-exiting humidity less than 80%. Scientific aspects of the effect of the operating conditions such as the rotation speed, air velocity etc. on the removal efficiency were discussed by Yamauchi et al.[3].

Deterioration and Reactivation of the Honeycomb Rotor Adsorbent

Exhaust gas from factories related to semiconductors and liquid crystals involves sometimes VOC of low saturated vapor pressure such as DMSO (Dimethyl sulfoxide), MEA (Methyl ethyl amine) etc. or VOC easy to polymerize. When these VOCs are processed in the concentrator incorporated with a hydrophobic zeolite rotor, they are difficult to be desorbed at the standard temperature of 180-200 C and accumulated in the rotor to cause the deterioration of the removal efficiency. In the worst case of miss operation, the accumulated VOC as fuel may burn the rotor down. We arrange the following two measures for this issue; 1) Reactivation by spray washing with water and 2) High temperature reactivation..

Fig. 4 Water works in the water spray washing

Reactivation by Spray Washing with Water.

In the case of water soluble VOC such as DMSO and MEA, the accumulated VOC can be washed out successfully by spraying water on the honeycomb wall as shown in Fig.4 and the performance has been supported by many experiences in the semiconductor factory.

Indications of Adsorption or Adhesion of High Boiling Point Matters

A test piece for the visual observation was sliced from a honeycomb rotor of a VOC concentrator which had operated in a semiconductor factory for 3.5 years. The color of the surface turned to yellow-brown in the used piece from near white in a virgin sample as compared in Fig. 5. It cannot be decolorized by washing with water, implying the adhesion of some substance which is water-insoluble and hardly volatile at the standard regeneration temperature of 180 C. A microscopic observation shows that voids between zeolite crystals were clogged with something and the contour of particles is blurred as shown in Fig. 6. The differential thermal analysis was carried out for the used sample. An endothermic peak and a corresponding weight loss appeared around 300 C due to the desorption or evaporation of probable high boiling point matters, as shown in Fig.7. These preliminary observation imply that the deteriorated adsorbent rotor may be reactivated by exposure of it in a high temperature atmosphere above 300 C.

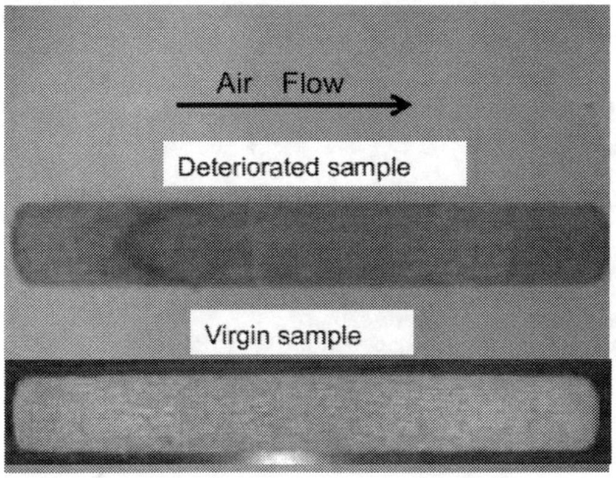

Fig. 5 Color change on the surface

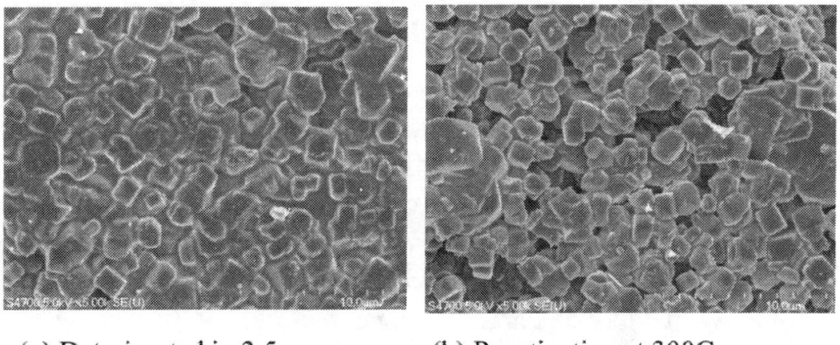

(a) Deteriorated in 3.5 years (b) Reactivation at 300C

Fig. 6 Microscopic observation of the surface of the zeolite honeycomb rotor

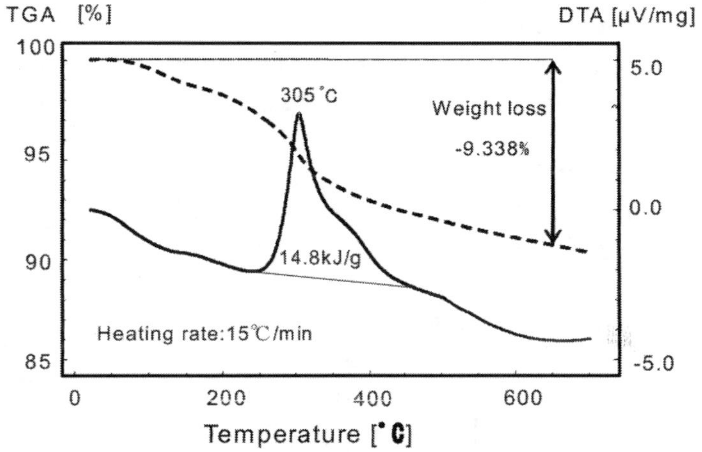

Fig. 7 Thermal analysis of a deteriorated sample of the honeycomb rotor

High Temperature Reactivation System

Static Adsorption Test

The amount adsorbed was compared between a virgin rotor, a deteriorated rotor and a reactivated rotor to confirm the advantage of a high temperature reactivation. The sample was sliced from each rotor in a 10 mm thick disk at a position of 30 mm depth from the inlet end of the feed. The reactivated rotor is the one which was pre treated at 300 C for 1 hour. Saturated vapor of isopropyl alcohol was adsorbed at 25 C in a desiccator. As shown in Fig. 8, the amount adsorbed of the deteriorated rotor had decreased to 61% of the virgin rotor, but it was recovered to 89% of the virgin by the reactivation at 300 C. With the high temperature reactivation at 300 C, the clogging substance disappeared and the particle contour became clear as shown in Fig. 6(b). In the thermal analysis of the reactivated sample, the endothermic peak in Fig. 7 disappeared and the weight loss decreased to 2.8% from 9.3% for the deteriorated one.

Onsite reactivation system

In the case of water insoluble VOC accumulated, spray washing with water is not effective but the reactivation by hot air stream is desirable. Off-site operation, of course, is possible to reactivate by high temperature heating in a separate factory as done with active carbon concentrators. However, it is not a practical process in view of the dismantling expenditure and intermission of operation. The present reactivation at high temperature enabled to reactivate the rotor on site of the VOC concentrator in relatively short time.

A flow sheet of the onsite reactivation system is shown in Fig. 9. A blue line refers to the usual operation (On-line) as a VOC concentrator while a red line refers to the case of Off-line for the high temperature reactivation. In the actual operation of reactivation, the feed air is stopped and the outdoor air is heated to 300 C by a burner after heat exchange to reactivate the deteriorated rotor which rotates very slowly at a few round per hour.

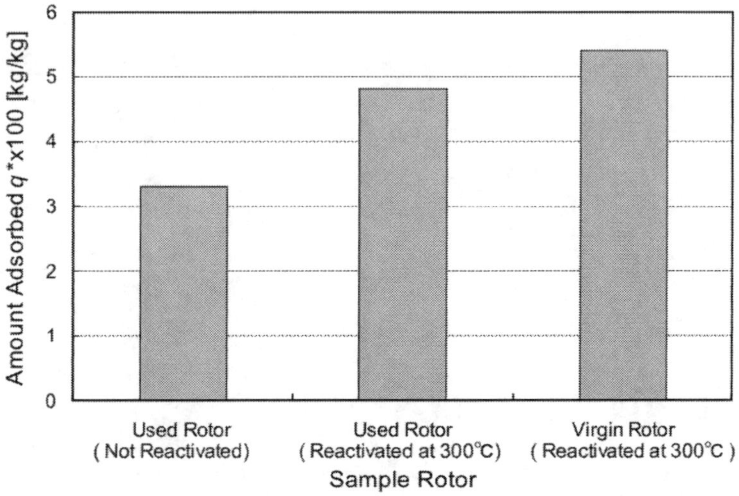

Fig. 8 Static adsorption test to confirm the effect of high temperature reactivation

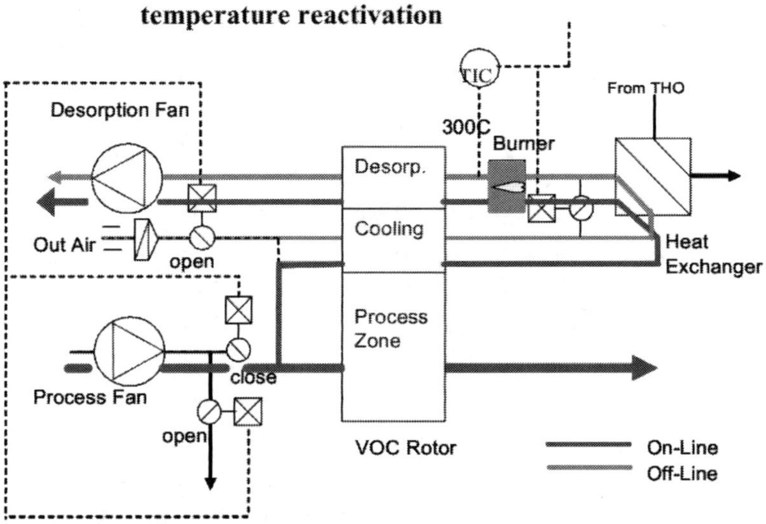

Fig. 9 Flow sheet of the onsite reactivation system by high temperature air stream

Heat Proof Seal

The heat-resistant seal was essential in developing the reactivation system to stand high temperature reactivation. The rotor surface is rubbed usually with flexible seal to prevent the concentrated air from leakage to/from other zones but the heat resistant rubber seal used in the standard specification could not stand a high temperature of 300 C necessary for reactivation. On the other hand, inorganic sealing material is highly heat resistant but has some troubles with easy wear of the rotor surface and less tightness of sealing. We developed in 2005 a hybrid seal mechanism by combining metal seal, Teflon-like seal and heat proof rubber with high seal performance to provide a heat resistant, wear proof, and tight proof seal successfully. Figure 10 shows the seal structure at a boundary between the regeneration zone and the cooling zone. A pair of heat protector plate of Teflon was installed at the tip of a clearance adjuster so that the clearance between the plate tip and the rotor end is kept at a minimum value. Layers of air between the protector plates each other and between the plate and a rubber seal work as thermal insulator to prevent the rubber seal from thermal damage.

Temperature was measured at various points on the rotor surface, i.e. points ①-⑤ in Fig.10, to examine the seal performance. Temperature increases with increasing rotation speed as shown in Fig. 11(a). A rotation speed at 2 rph is slow enough to keep low temperature since the rubber seal is exposed below 200 C as shown in Fig. 11(b).

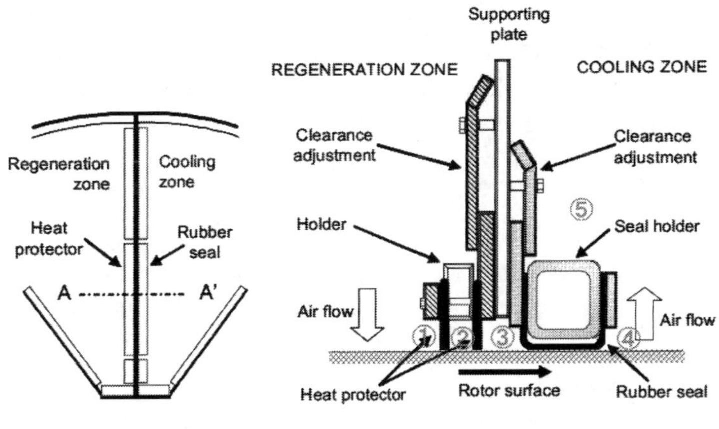

(a) Top view of the rotor (b) A-A' cross-section

Fig. 10 A new mechanism of heat proof seal between the regeneration and cooling zones

Two Staged Reactivation System.

The single operation of either water spray washing or high temperature reactivation cannot be expected to be sufficient for reactivation when both water soluble and water insoluble VOC are accumulated in the honeycomb rotor. A two-staged reactivation system was developed for such a case by carrying out first the water spray washing followed by the high temperature reactivation. This system enabled to treat the exhaust gas having been impossible to be treated so far due to various kinds of hardly desorbed VOCs involved.

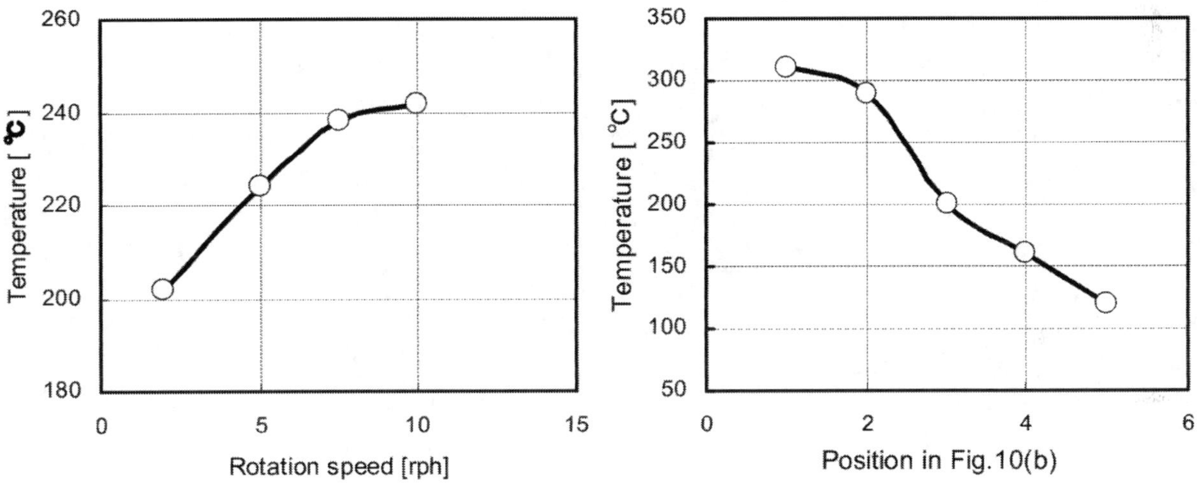

(a) Effect of the rotation speed (b) Temperature profile

Fig. 11 Performance of the new heat proof seal

Concluding Remarks

The performance deterioration due to hardly desorbed compounds and the reactivation by high temperature air stream were discussed and some new findings were obtained.

1) An differential thermal analysis showed that the hardly desorbed compounds can be desorbed at about 300 C. The amount adsorbed of was recovered to about 90% of the virgin rotor when it was treated at 300 C.

2) The onsite reactivation system was developed by passing the hot air of 300 C through the regeneration zone while the process air was stopped.

3) Heat proof seal was developed by the installation of heat protector plates in the regeneration zone to enable the high temperature reactivation.

We are forced to destruct efficiently a large amount of dilute VOC although a powerful technology has been unavailable and thus much dilute VOC was discharged into the atmosphere so far. The VOC concentrator discussed here is a useful tool to solve the problem and a combination with the thermal oxidizer can lead to the energy saving VOC destruction. Much progress is expected in the environmental measures with the VOC concentrator in developing countries as well as developed ones,

References

(1) Kuma, T., Y. Mitsuma, Y. Ota, T. Hirose, "Removal Efficiency of Volatile Organic Compounds, VOCs, by Ceramic Honeycomb Rotor Adsorbents," *Fundamentals of Adsorption (Procedings of the 5th International Conference on Fundamentals of Adsorption)*, **5**, 481(1996).

(2) Mitsuma, Y., Y. Ota, T. Hirose, "Performance of Thermal Swing Honeycomb VOC Concentrators," *J. Chem. Eng. Japan*, **31**, 482(1998)

(3) Yamauchi, H., A.Kodama, T.Hirose, H.Okano, K.Yamada, "Performance of VOC Abatement by Thermal Swing Honeycomb Rotor Adsorbers," *Ind. Eng. Chem. Res.*, **46**, 4316(2007)

(4) Hirose, T., H. Okano, K. Yamada, K. Furuki, Y. Fujioka, "Continuous VOC (Volatile Organic Compounds) Concentrators by Thermal Swing Honeycomb Rotor Adsorbent," *Extended Abstracts of 2007 AIChE Annual Meeting,* #470a (Environmental Applications of Adsorption (09013))

(5) Mitsuma, Y., T. Kuma, H. Yamauchi, T. Hirose, "On Improvement and Scaling-up of Thermal Swing Honeycomb Adsorbent for VOC Concentrators", *Kagakukogaku Rombunshu,* **24**, 248(1998)

(6) Kuma, T., H. Okano, "Active Gas Absorbing Element and Method of Manufacturing", *United States Patent* 4886769, 1989.

(7) Kuma, T., "Method of Manufacturing A Gas Absorbing Element or Catalyst Carrier Having A Honeycomb Structure" , *United States Patent* 5194414, 1993.

(8) Kuma, T., "Gas Adsorbing Element and Method for Forming Same", *United States Patent* 5348922, 1994

329m Adsorption of Water Vapor Onto Microporous Carbons Prepared by Various Oxidation Methods

Takahiro Mikami, Kazunori Kawata, and Yohinobu Otake. Applied Chemistry, meiji university, 1-1-1, Higashi-Mita, Tama-ku, Kawasaki, Japan

Activated carbon has been used as adsorbent for removing pollutants such as volatile organic compounds (VOC). However, adsorption of VOC on activated carbon is hampered by moisture in air, since oxygen functional groups act as adsorption primary sites for water molecules though hydrogen bond. Therefore, in this study, equilibrium isotherms of water vapor on microporous carbons containing oxygen functional groups have been investigated in order to see correlation between amounts of water adsorbed and site concentrations of surface oxygen functional groups. Different types of activated carbons, one activated carbon fiber and two granular activated carbons were oxidized by nitric acid, hydrogen peroxide and air at different conditions. Surface chemistry and pore structure of those carbons were characterized respectively by temperature programmed desorption (TPD) and nitrogen adsorption at 77K. Adsorption of water vapor on the oxidized carbon samples was measured with volumetrically apparatus at 298K. The data acquisition was focused solely on the effect of surface oxygen functional groups, since adsorption of water vapor was affected by both of carbon pore characteristics and surface nature. It was found that amounts of water adsorbed depend on the amount of CO2 producing groups at low relative pressure, regardless of pore structure of the microporous carbons. The isotherms were also analyzed by Dubinin-Serpinski (DS) equation. The use of this equation showed that primary adsorption center a0 attributes to the oxygen functional groups, especially CO2 producing groups.

329n Multi-Scale Simulation Starting at the Molecular Level for N-Hexane Chromatographic Process

Hae-jeong Son[1], Young-il Lim[1], Kyung-soo Nam[2], and Kyung-Seun Yoo[2]. (1) Chemical engineering, Hankyoung national university, 67 soksong-dong, Kyonggi-do, Anseong, South Korea, (2) Environmental engineering, Kwangwoon university, Kwangwoon Road, Nowon-Gu, Seoul, South Korea

Chromatographic adsorption has been used as one of industrial operations for high purity separation. Traditionally, the adsorption process development has been based on heuristic knowledge and expertise, which is time-consuming and costly. Model-based simulation and optimization may help to accelerate the adsorption process development and effectively search optimum operation conditions. This study presents a multi-scale simulation approach starting at the molecular level for the n-hexane chromatographic adsorption process. With the explosive growth of computational power in recent year, molecular simulation has flourished in order to predict the molecular structure of a material and its physical properties. The molecular simulation is performed in Materials Studio v.4.0 (Accerlys Software Inc., USA) with the Forcite and the Sorption modules. The molecular structure of the activated carbon is constructed with periodic boundaries to fit physical properties measured (porosity, specific surface area and density) and is optimized, using the COMPASS (Condensed-phase Optimized Molecular Potentials for Atomistic Simulation Studies) force field including bonding and non-bonding interaction potentials of atoms. A Grand Canonical Monte Carlo (GCMC) method is used for the prediction of adsorption isotherms of n-hexane on the activated carbon structure constructed by utilizing COMPASS. In GCMC simulation, the temperature, volume and chemical potential, which can be directly related to pressure and temperature using the bulk gas equation of state, are specified, while the number of particles and associated configurational energy are allowed to fluctuate. The adsorption isotherms obtained from the molecular simulation are exploited as a model parameter needed for the adsorption process simulation. The process simulation is performed in FAST-Chrom/SMB (Fast and Accurate Simulation Tools for Chromatography and SMB), which developed by our own laboratory. The simulation results are in a good agreement with the experimental results of n-hexane on the activated carbon.

Keywords: Multi-scale simulation, Molecular simulation, Process simulation, Chromatographic adsorption, Activated carbon, COMPASS, Grand canonical Monte Carlo (GCMC) method, Adsorption isotherms.

329o Chemisorption of Hydrogen on Oxidized Activated Carbons after Their Heat Treatment

Atsushi Gabe, Takuma Hirahara, Takanori Baba, and Yoshinobu Otake. Applied Chemistry, Meiji University, 1-1-1,Higashi-Mita,Tama-ku,, kawasaki, 2148571, Japan

Activated carbon was oxidized with HNO3 at 340 K for 3 hours. The samples thus oxidized were then subsequently heat treated(HT) at various maximum temperatures between 400 K and 1273 K for 8 hours under H2 atmosphere in order to prepare H2 -treated samples. This heat treatment simultaneously allowed carbon-oxygen complexes to decompose progressively and hydrogen to be chemisorbed on the oxidized activated carbons. Temperature programmed desorpion(TPD) in a flowing Ar was then performed from 300K to 1273K on the H2-treated samples to estimate the amount of hydrogen chemisorbed on the carbon surfaces. The amount of hydrogen chemisorbed increased with increasing HT temperature up to about 900 K, above which temperature it decreased. This result indicates that hydrogen chemisorption on the carbon surface strongly depends on the HT temperature and the deactivation of sites responsible for hydrogen chemisorption occurs even at the temperature of 900 K.

329p Modeling The Electrokinetic Decontamination Of Concrete By Petrov-Galerkin Finite Element Method

Patrick A. Oglesby, School of Chemical Engineering, Purdue University, Forney Hall, Rm 3007, 480 Stadium Mall Drive, West Lafayette, IN 47907 and Michael T. Harris, Chemical Engineering, Purdue University, Forney Hall of Chemical Engineering, 104, 480 Stadium Mall Drive, West Lafayette, IN 47907.

A one-dimensional, time dependent electrokinetic model has been developed to simulate the transport and sorption of cesium and strontium ions through a concrete disk The sensitivity of the ion concentration in the anolyte and catholyte to the overall mass transfer coefficient and ion diffusion coefficient is investigated. The model predicts the variation in the voltage drop across the disk and the concentrations of all involved species throughout the disk and in the anolyte and catholyte. The equations are solved by using the Galerkin/Finite element method.

329q Comparison Of Two Adsorbents For The Separation Of A Sugar Mixture Based On Smb Process

Se-Hee Jo and Sungyong Mun. Chemical Engineering, Hanyang University, Seoul, 133-791, South Korea

In this study, Dowex and MFG resins are used to separate a sugar mixture using SMB process and the results are compared with each other. The intrinsic parameters in each system with a different resin are obtained by a series of single column experiments. Based on the resultant intrinsic parameters, the numerical simulations are conducted for the four-zone SMB processes based on the two resins. The separation performances of the SMB process with the two resins are then compared in terms of productivity, purity, and recovery.

329r Adsorption Of Acetic Acid Vapour By Protective Packaging Materials Used For Preventive Conservation Of Cultural Heritage Objects

Cécile Vallières, **Georges Grévillot**, and Stéphanie Pacault. Chemical Engineering Science Laboratory-CNRS, 1, rue Grandville, Nancy, France

The harmful effect of pollution on works of art is well known. One solution to protect artefacts is to use mall-size enclosures. This is even more necessary if the object itself off-gasses volatile compounds produced by its natural degradation. Acetic acid is thus produced by decomposition of cellulose of furniture wood or book paper itself. The sorption capacities of several conservation papers towards acetic acid vapour were studied. Adsorbents incorporated in the paper were: zeolite, activated carbon grains and fibres, calcium carbonate, magnesium oxide and mixtures of it. The adsorption capacity was measured by the method of breakthrough curve in a column filled by a large number of small pieces of paper and flowed by a vapour at constant pressure in nitrogen. Isotherms were obtained at ambient temperature and with/without humidity.

329s Simulated Moving Bed Chromatography For Gaseous Applications

Isabel A. A. C. Esteves, Chemistry Department, FCT/UNL, Quinta da Torre, Caparica, 2829-516, Portugal, **J. P. B. Mota**, Departamento de Química, Faculdade de Ciências e Tecnologia - Universidade Nova de Lisboa, Quinta da Torre, 2829-516 Caparica, Portugal, and Mário F. J. Eusébio, Requimte/ CQFB, Chemistry Department, Faculdade de Ciências e Tecnologia, Universidade Nova de Lisboa, Caparica, 2829-516, Portugal.

Keywords. Simulated moving-bed (SMB), gas chromatography, gas separation.

SMB chromatography has been increasingly applied for separations of pure substances in the pharmaceutical, fine chemistry and biotechnology industries. It has many advantages with respect to discontinuous batch chromatography: higher product purity, less solvent consumption, and higher productivity per unit stationary-phase.

The SMB is a practical way of implementing a counter-current chromatographic process. N identical chromatographic columns are connected in series to build a closed loop. By moving the input/ withdrawal ports one bed ahead, in the direction of fluid flow, at fixed intervals, the counter-current contact adsorbent/fluid is simulated. The SMB has been mainly employed for liquids, but the same principle can be applied to separate a gas mixture of two competing adsorbates using a weakly adsorbed gas as eluent. However, SMB gaseous applications have been limited.

Recently, we have shown that the cyclic steady-state of continuous multicolumn gas chromatography, operating under isothermal conditions, can be reproduced experimentally using a single-column setup. The experimental procedure is based on a mimetic single-column chromatographic model in which the part of the outlet stream that is not recovered as product is recycled back to the column with a lag of $(N-1)\tau$ time units, where N is the number of columns of the multicolumn unit and τ is the switching interval. The setup was employed to demonstrate the process feasibility, explore the effect of its major operating parameters and illustrate the performance enhancements that are obtained by asynchronous and time-variable flowrate schemes for gas-phase SMB.

By correctly selecting the step within the cycle for process start-up, the steady periodic state can be achieved in a minimum number of cycles. Hence, the adsorbates and eluent quantities required to experimentally reproduce the periodic state of the SMB process can be greatly reduced. This may be an economic, optimal manner of experimentally testing a set of operating conditions to achieve a given separation performance for new continuous chromatographic applications.

Our implementation was successfully applied to separate CO_2/CH_4 gas mixtures on activated carbon in the Henry's law region using N_2 as purge gas. The work is currently being extended to the nonlinear region of the adsorption equilibrium, where the process operates under non-isothermal conditions with non-negligible changes in fluid velocity across the mass transfer zone. The use of pressure swing to induce both desorption and flow of gas through the columns is also being evaluated.

Acknowledgements. FCT/MCES (Portugal) for Post-doctoral grant SFRH/BPD/14910/2004.

References. 1) Jupke, A., Epping, A., Schimdt-Traub, H. Optimal design of batch and simulated moving bed chromatographic separation processes, Journal of Chromatography A, 944, 93, 2002. 2) Juza, M., Mazzotti, M., Morbidelli, M. Simulated moving bed chromatography and its application to chirotechnology, Trends in Biotechnology, 18, 108, 2000. 3) Mota, J. P. B., J.M.M. Araújo, J. M. M. Single-column simulated-moving-bed with recycle lag, AIChE J, 51, 1641, 2005. 4) Nicoud, R. M. The simulated moving bed: a powerful chromatographic process. LC-GC, 5, 43, 1992. 5) Strube, J., Haumreisser, S., Schmidt-Traub, H., Schulte, M., Ditz, R. Comparison of batch elution and continuous simulated moving bed chromatography, Organic Process Research and Development, 2, 305, 1998. 6) Wankat, P. C., Large Scale Adsorption and Chromatography, CRC Press, Boca Raton, 1986. 7) Mota, J. P. B., Esteves, I. A. A. C., Eusébio, M. F. J. Synchronous and Asynchronous SMB Processes for Gas Separation, AIChE J, 53, 1192, 2007. 8) Mota, J. P. B., Esteves, I. A. A. C., Eusébio, M. F. J. Optimal Design and Experimental Assessment of Time-Variable Simulated Moving Bed for Gas Separation, submitted to Ind. Eng. Chem. Res., 2007.

329t Development of Calcium Oxide Sorbents for Adsorption of Carbon Dioxide

Hong Lu and Panagiotis (Peter) Smirniotis. Chemical & Materials Engineering Department, University of Cincinnati, 2600 Clifton Avenue, Cincinnati, OH 45221

The reversible carbonation and calcination of calcium oxide with CO2 provide potential for capture of CO2 from high temperature gas stream. An important issue of calcium oxide based sorbents is how fast and to what extent carbonation and calcination take place and how they progress in multi cycles. We show here the mechanism determining their behaviors. Structure promoted sorbents were also investigated to try to overcome performance degradation. The CO2 uptake at various temperatures was determined by thermogravimetric analysis. To study the reversibility of the sorbents, multiple cyclic operations were performed at realistic conditions.

329u Alkali Metal Removal From Coal Extracts Using Ion Exchange

Kwang B. Yi, Ji-Won Ha, Chang Hyun Ko, Jong-Ho Park, Si Hyun Lee, and Jong-Nam Kim. Chemical Process Research Center, Korea Institute of Energy Research, 71-2 Jang-Dong, Yuseong-gu, Daejeon, 305-343, South Korea

Restrictions for pollutant emissions are getting more stringent these days due to rising environmental concerns over the world. Immense utilization of fossil fuel over past hundred years has been blamed as a major source of pollutant. However, fossil fuels are still considered as a major energy source. Moreover, coal consumption in the future is expected to increase due to rapid industrialization of developing countries. In order to migrate smoothly to new energy source with long-time visions, development environmentally sound process to utilize fossil fuel is essential. As such efforts, removal of mineral matters and modification of low rank coal recently have been attempted and reported. Ultra Clean Coal project in Australia and Hyper Coal (HPC) project in Japan recently have been demonstrated as efficient demineralization process of coal by alkali solution treatment and organic solvent extraction respectively. Successful development of these processes would enable the direct combustion of cleaned coal powder in gas turbine of power plants which will result in both reduction of CO2 by 20% and increase of efficiency up to 48%. However, both the UCC and HPC also have obstacles for practical operation. The treated coal in the UCC process still contains 0.1 to 0.5% of ash and 60ppm of alkali metals. The ash and alkali metals in the treated coal by thermal extraction (at 360°C and 1MPa) with organic solvent is less than 200ppm. The remained alkali metals in treated coal would cause corrosion on internal parts of the turbine. Therefore, a subsequent process is still needed to reduce alkali metal contents to less than 0.5ppm, which is the currently acceptable level for introduction to gas turbine. It seems that ion exchange method would fit the most for removing trace of alkali metals and the successful development of a feasible ion exchange process would expedite realization of the coal modification process. In this study, removal of alkali metals using ion exchangers was attempted. In order to maximize overall efficiency of the coal modification process, ion exchange must occur in the same condition of thermal extraction(at 360°C and 1MPa). In addition, organic solvents, such as NMP (N-methyl-2-pyrrolidone), are supposed to be in exchange medium. Therefore, inorganic ion exchangers, which can endure such rough condition, were prepared and tested.

329v Sorption Properties Of N2 And H2 On Multi-Walled Carbon Nanotube-Graphite Mixture Modified By Alkaline Hydroxides

Fatma O. Erdogan and **Turkan Kopac**. Department of Chemistry, Zonguldak Karaelmas University, 67100 Zonguldak, Turkey

Carbon has been known as one of the best adsorbents for gases. Since the discovery of carbon nanotubes numerous studies have been conducted on their synthesis, treatment and sorption properties [1]. The physical and chemical activation are well known methods to obtain microporous carbon with high surface area from different precursors [2,3]. Among all the chemical activation agents, alkaline hydroxides as KOH or NaOH are reported to be highly interesting from the performance point of view, allowing activated carbons to be prepared from many kinds of carbonaceous precursors [4]. Hydrogen sorption studies in carbon materials have been of great concern in recent years as hydrogen is regarded as an ideal energy source causing no pollutants or greenhouse gases during combution unlike the fossil fuels [5,6]. In this study, the multiwalled carbon nanotubes-graphite mixture produced as a cathode deposit during the Kratschmer-Huffman arc process modified by alkaline hydroxides (KOH, NaOH) at different temperatures was used to enhance the surface area and the capability of hydrogen sorption. The surface area of the MWCNT-graphite mixture was increased by physical activation in the temperature range of 300-800°C and the highest value of the surface area was obtained at 800°C. KOH and NaOH activations were applied for the sample mixture in a temperature range of 500-800°C. The surface area and the micropore volume of MWCNT-graphite sample mixture was increased by alkaline hydroxides activation. The samples produced by physical and the alkaline hydroxides activations were compared according to their hydrogen sorption characteristics.

1. Yin Y.F, Mays T, McEnaney B, Langmuir, 1999,15, 8714-8718. 2. Lee S.M, Lee S.C, Jung J.H, Kim H.J, Chemical Physics Letters, 2005, 416, 251-5 3. Zhao N, He C, Jiang Z, Li J, Li Y, Materials Letters, 2007, 61,681–5 4. Raymundo-Pi•ero E, Azaïs P, Cacciaguerra T, Cazorla-Amorós D, Linares-Solano A, Beguin F, Carbon, 2005, 786-795. 5. Erdogan O.F, Kopac T, International Journal of Hydrogen Energy, 2007 (in press). 6. Kopac T, Karaaslan, T, International Journal of Hydrogen Energy, 2007(in press).

329w Modeling of Psa Process in Fuel Ethanol Production: Effect of Ethanol Co-Adsorption on the Performance

Marian Simo[1], Daniel M. Leo[2], Mohammadali Safavieh[1], Christopher J. Brown[2], and Vladimir Hlavacek[1]. (1) Chemical and Biological Engineering, University at Buffalo, 303 Furnas Hall, Buffalo, NY 14260, (2) Thermal Kinetics Systems, LLC, 667 Tifft Street, Buffalo, NY 14220

In the production of bio-ethanol, Pressure Swing Adsorption (PSA) process is used to remove water and obtain >99.5 wt% ethanol. The 2-bed 4-step PSA cycle consists of following steps: high pressure feed, counter-current blowdown, vacuum purge and counter-current pressurization with product. Our previously developed PSA simulator for this process assumed that only water is being adsorbed, while ethanol was considered as an inert. However, our recent gravimetric measurements have revealed that ethanol is co-adsorbed as well. Equilibrium data for water-ethanol system were measured and fitted to find reasonable equilibrium model. Adsorption kinetics was approximated by a linear driving force model for both water and ethanol. The effect of ethanol co-adsorption on the separation performance was studied using an updated PSA simulator. Several scenarios were calculated using the binary model and compared to our previous results.

329x Charge and Discharge Cycles in a Storage Vessel for Adsorbed Natural Gas

Marcelo R. Amora[1], A. Eurico B. Torres[2], Diana C. S. Azevedo[3], and **Célio L. Cavalcante Jr.**[1]. (1) Department of Chemical Engineering, Universidade Federal do Ceara, Campus do Pici, Bl. 709, Fortaleza, CE, 60455-760, Brazil, (2) Universiade Federal do Ceara, Campus do Pici, bl. 709, Fortaleza, 60.455-760, Brazil, (3) Chemical Engineering, Universidade Federal do Ceará, Campus do Pici, bl. 709, Fortaleza, 60.455-900, Brazil

Due to the natural gas expanded usage, it is necessary to develop alternative technologies for its transportation in economically viable and safe ways to distant areas from the delivery point in gas pipelines using vessels filled with adsorbent at moderate pressures (about 3.5 to 4.0 MPa) and ambient temperatures. This study intends to accomplish experimental loading and delivery tests of Natural Gas using a prototype storage vessel filled with microporous activated carbon, to investigate the long term behavior of the adsorption capacity with charge and discharge cycles. A larger adsorption capacity was observed for the heavier components, in relation to methane, which decreases the adsorbent efficiency in a long-term usage due to large adsorbed amounts even in low pressures. The relative selectivities of the several constituents of NG in relation to methane were also measured and their variation with pressure was calculated.

329y Modification Methods to Reduce or Eliminate Non-Zeolite Pores for Hydrogen Separations

Moises Carreon, **Dean E. Camper**, Shiguang Li, and Richard Noble. University of Colorado, UCB 424 Department of Chemical and Biological Engineering, Boulder, CO 80309

Various modification methods including swelling have been looked at to reduce or eliminate non-zeolite pores for hydrogen separations.

330a Effect of Concentration Polarization on Flux Decline for Ultrafiltration in Tubular Membranes

Ho-Ming Yeh[1], S. Y. Lin[2], and C. H. Lee[2]. (1) Tamkang University, Department of Chemical and Materials Engineering, Tamsui, 251, Taiwan, (2) Chemical and Materials Engineering, Tamkang University, Department of Chemical and Materials Engineering, Tamsui, 251, Taiwan

The correlation equations for predicting local and average values of the permeate flux in tubular-membrane ultrafilters was derived from mass and momentum balances by the modified resistance-in-series model with the considerations of the increment of concentration polarization and the declines of transmembrane pressure and flow rate, along the membrane tube. Ultrafiltration of dextran T500 aqueous solution in a tubular microporous ceramic module has been carried out under various feed concentrations, transmembrane pressures and feed flow rates, and many experimental data of ten-point local permeate fluxes along the tube were obtained to confirm the correlation predictions. The increment of concentration polarization as well as the decline of permeate flux along the tube was also discussed.

Separation of propane from nitrogen using poly(dimethyl siloxane) thin film composite membranes[1]

Jian-Hua Cao, Membrane Technology and Engineering Research Center, Tsinghua University, Beijing China
Yang-Zheng Lin, Membrane Technology and Engineering Research Center, Tsinghua University, Beijing China
Ji-Ding Li[2], Membrane Technology and Engineering Research Center, Tsinghua University, Beijing China
Cui-Xian Chen, Membrane Technology and Engineering Research Center, Tsinghua University, Beijing China
Shen Lu, Membrane Technology and Engineering Research Center, Tsinghua University, Beijing China

Abstract

Polyvinylidene difluoride (PVDF) membranes coated with poly (dimethyl siloxane) (PDMS) thin film composite membranes were investigated for the separation of propane from nitrogen. An experimental study was performed to evaluate the effects of various operating conditions (i.e., temperature, pressure, feed composition) on the membrane permselectivity with propane/nitrogen mixtures. It was shown that the membrane could be adapted to propane/nitrogen separation, and the membrane selectivity increased with a decrease in temperature and an increase in the feed pressure. The permeability of nitrogen was independent of feed pressure and increased with the temperature increasing. However, the permeability of propane increased with the pressure increasing and the temperature decreasing. The relationship between the permeability of propane and nitrogen and feed temperature can be represented in an Arrhenius equation. At 18℃ and 500 kPa, the membrane exhibited a propane/nitrogen permeability ratio of as high as 10 for feed mixtures containing 6–30 mol% propane.

Instruction

Volatile organic compound (VOC) emissions are one of the major sources of air pollution. As far as the environment pollution problem and significant economic loss are concerned, it is inevitable to recovery VOC. There are a number of separation processes to address such a problem: carbon adsorption, condensation, absorption, membrane fractionation. Of course, membrane vapor separation appears to be attractive. Membranes are required that are much more permeable to organic vapors than to air. Polymer membranes useful for the separation of organic vapors from air should be highly permeable to organic

1 Supported by China Postdoctoral Science Foundation (20060400436), the Major State Basic Research Program of China (No. 2003CB615701), National Natural Science Foundation of China (No. 20576059, No.20676067), CNPC Innovation Foundation (No. 05051143).
2 Corresponding author. E-mail: lijiding@mail.tsinghua.edu.cn (J D Li)
Tel: +86-10-62782432. Fax:+86-10-62770304

solvents, but relatively impermeable to air. PDMS as a candidate of rubbery polymer has been used extensively in the pervaporation [1-3] and gas separation process [4-6].

There is a great amount of data in the literature concerning permeation of gases through polymer membranes, including several recent reviews [7,8]. There are much fewer data on the permeation of organic vapors [9,10], and the factors that affect vapor permeation are less well understood. In this work, the PDMS/PVDF composite membrane prepared and applied in the separation of propane and nitrogen. The effects of operating temperature and pressure on the membrane performance were investigated. Moreover, the permeation of propane/nitrogen gas mixtures through the PDMS/PVDF composite membrane was studied.

Experimental

Materials

Polyvinylidene fluoride (PVDF, Solef®1010, average Mw211 000) was supplied from Solvay. Triethyl phosphate (TEP) and *n*-hexane were of analytic grade and obtained from the Beijing chemical reagent company. PDMS (107, viscosity 10000 Pa s), tetraethylorthosilicate(TEOS), dibutyltin dilaurate (DBTL)were purchased from Beijing Chemical company. Deionized water was used as the non-solvent for the PVDF. Nitrogen and propane (purity, 99.99%) were come from Beijing Huayuan Gas Chemical industry Co., Ltd. All chemicals were used as received.

Preparation of PVDF microporous membrane

The PVDF porous membranes were prepared by means of the phase inversion process. Specific amount of PVDF was dissolved in TEP by mechanical stirrer the viscous solution at 80 ℃ for at least 24h to guarantee complete dissolution of the polymer. The solution was filtrated and degassed under vaccum condition. The transparent and viscous solution was obtained. The solution was cast on a polyester non-woven fiber using a casting knife set a thickness of 180μm. After 15s, the polyester non-woven fiber was immersed in deionized water (21℃). After complete coagulation, the membranes were transferred into a pure water bath, which was refreshed frequently, for at least 24h to remove the trances of solvent. The membrane was dried in oven at 50℃. The white membrane was obtained and the thickness was about 60 μm.

Preparation of PVDF/PDMS composite membrane

PDMS, cross-linking agent TEOS and catalyst DBTL were dissolved in *n*-heptane by magnetic stirrer at ambient temperature. After degassed under vacuum, the solution was cast onto the PVDF membrane. The membrane was first vulcanized under room temperature to evaporate the solvent, and then introduced into a vacuum oven to complete cross-linking. The thickness of the top skin layer could be determined by means of SEM photographs. All experiments in this study were performed with the same membrane.

Characterization of composite membrane

Scanning electron microscopy was used to study the surface and cross sectional morphology of the composite PDMS membrane and to measure the skin layer thicknesses. The composite PDMS membrane was freeze fractured in liquid nitrogen. The sample was coated with a conductive layer of sputtered gold. The cross-section of the composite membrane was investigated using a FLOTU JSM-7401F SEM at 1KV.

Gas permeation test

The experimental setup is schematically shown in Figure 1. The feed gas at a given composition was obtained by two mass flow controllers and the gas composition was determined using a gas chromatograph (GC-14C, Shimadzu) equipped with a TCD. The effective membrane area for permeation was 22.55cm^2. The membrane was cut into circular coupons and fitted in the permeation cell. The permeation cell consisted of two detachable stainless steel parts. A sintered porous stainless steel plate was imbedded in one part of the cell to support the membrane. Double O-rings were used to provide a seal between the membrane and the cell. The residue flow was controlled such that the stage cut was always below 0.05. The pressure of the feed side was controlled by means of a needle valve, while the pressure on the permeate side was kept atmospheric. The permeate flow rate was measured by a bubble flow meter and the permeate composition was analyzed using the gas chromatograph. The permeance of a gas component *i* across the membrane, J_i, was obtained using the following equation:

$$J_i = \frac{Qy_i}{A\Delta P_i} \qquad (1)$$

where J_i is the permeance,which can be expressed customarily in the gas permeation unit GPU(1 GPU=10^{-6} cm^3 (STP)/(cm^2 s cmHg)), Q the overall gas permeation rate, A the effective area of the membrane (cm^2), y_i the mole fraction of component *i* in the permeate, and ΔP_i is the partial pressure difference of component *i* across the membrane. $\Delta P_i = P_h x_i - P_l y_i$, where P_h and P_l are the gas pressures in the feed and permeate, respectively, and x_i is the mole fraction of component *i* on the feed side. The permeability of membrane is equal to the permeance multiplied by the membrane thickness *L* and usually expressed in unit Barrer (1 Barrer = 10^{-10} cm^3 (STP) cm cm^{-2} s^{-1} cmHg^{-1}).

The perselectivity is given by the permeability ratio of the two components

$$\beta = \frac{J_A}{J_B} \qquad (2)$$

where subscripts A and B represent propane and nitrogen, respectively. The separation factor of the gas mixtures was calculated from

$$\alpha = \frac{y_A/y_B}{x_A/x_B} \qquad (3)$$

Results and discussion

The Morphology of PVDF and PDMS/PVDF composite membrane

The surface morphology of the PVDF support layer and the PDMS/PVDF composite membrane was examined using SEM, as shown in Fig. 2 (a and b). It is apparent that there is a significant change in the membrane surface morphology after the PDMS layer was coated onto the PVDF support layer. The composite membrane has a uniform and smooth surface. The asymmetric structure of the membrane was observed by examining the cross-sections of the PVDF support layer and the composite membrane under SEM, as shown in Fig. 2(c and d), respectively. An asymmetric structure was found in the PVDF substrate, which was composed of a porous layer with sponge pores and a denser skin layer (see Fig. 2(c)). In the composite membranes, the top dense layer of PDMS is clearly distinguished from the skin layer of the

PVDF substrate. The adhesion of the top PDMS layer to the PVDF substrate is excellent. The thickness of the PDMS layer can be estimated from the SEM image to be about 15μm.

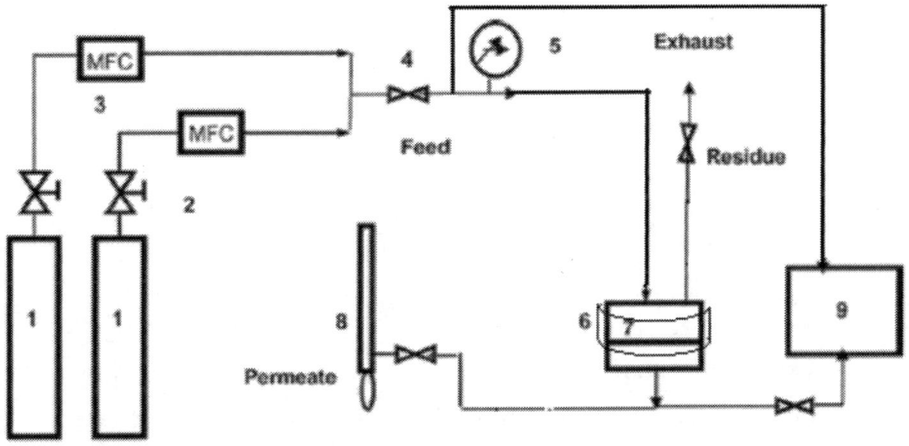

Figure 1 Schematic diagram of experimental setup for vapor permeation: (1) gas cylinder, (2) pressure regulator,(3) mass flow controller, (4) vavle, (5) pressure gauge, (6) heating tape (7) membrane cell, (8) bubble flow meter, (9) gas chromatograph.

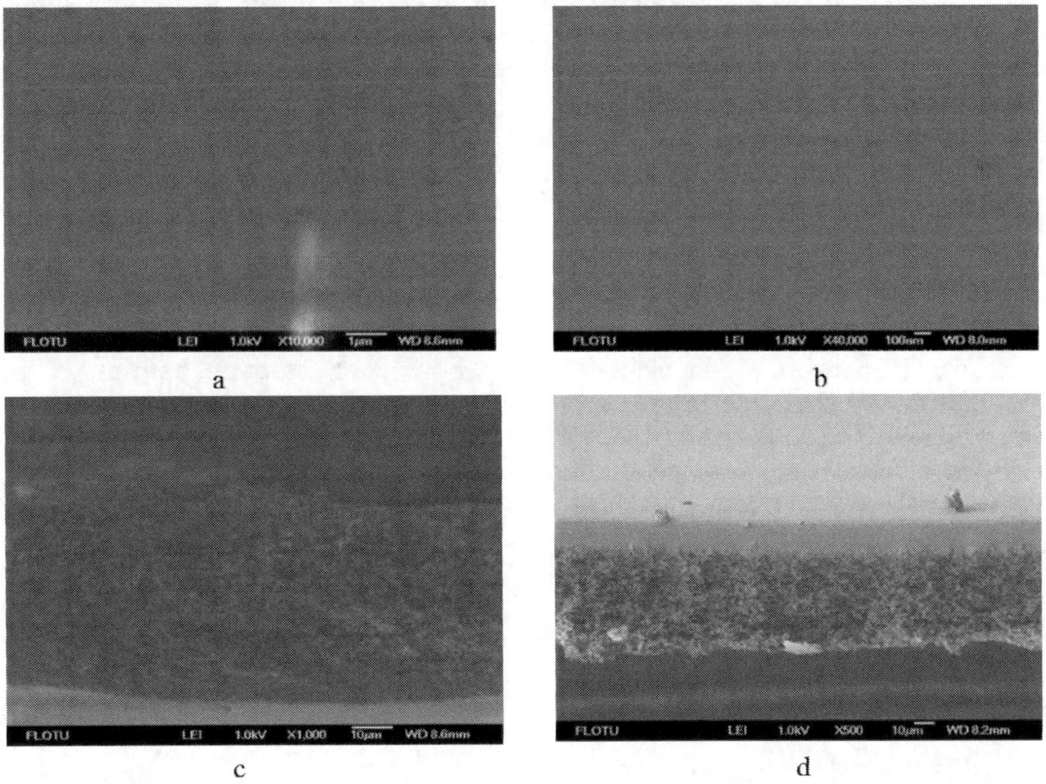

Figure 2 SEM images of membrane morphology.(a, surface of PVDF substrate; b, surface of PDMS/PVDF composite membrane; c cross-section of PVDF membrane; d, cross-section of PDMS/PVDF composite membrane)

Pure gas permeation

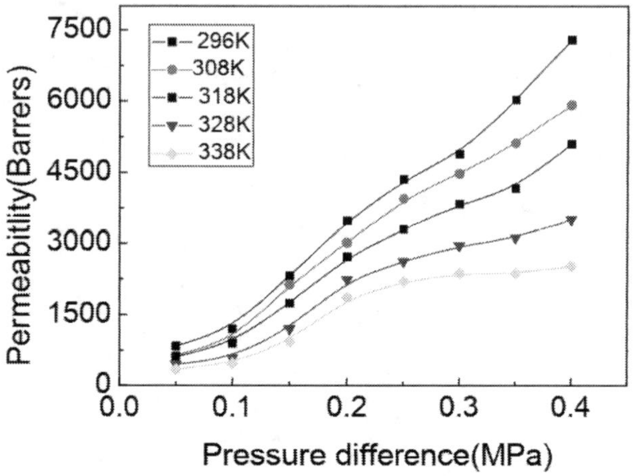

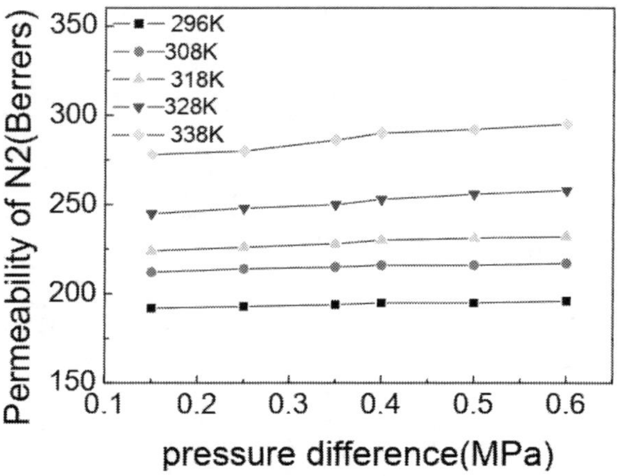

Figure 3. Permeability coefficient of propane at different feed pressures and temperatures.

Figure 4. Permeability coefficient of nitrogen at different feed pressures and temperatures.

To evaluate the permeability and the ideal selectivity of the PDMS, the permeability of pure propane and nitrogen through PDMS/PVDF composite membrane was determined. The investigated pressure was 0.1-0.6MPa and the temperature was 296K-338K. Figure 3 and Figure 4 show the permeability coefficients of pure propane and nitrogen through the membrane at different temperatures and pressures. It can be seen that the permeability of propane increases significantly with an increase in pressure. The permeability of propane decreases with the temperature increasing, especially at high pressure. From Figure4, the permeability of nitrogen is essentially independent of pressure at relatively low temperatures, and it begins to increase slightly with pressure at higher temperatures. With temperature increasing, the permeability of nitrogen increases slightly. For the separation of propane from the nitrogen, important factors like their molecules diameters and critical data must be considered. In general, molecules become more condensable with increasing diameter, hence propane is obviously easily condensable compared to the nitrogen. The solubility of condensable propane is much higher than nitrogen, and propane permeates through the membrane preferentially due to the high solubility-selectivity, in spite of the larger molecular size of propane (and thus relatively lower diffusivity) than nitrogen. An increase in solubility with an increase in pressure and a decrease in solubility with increasing temperature is normally expected. It appears that the solubility plays a major role in the permeation of propane. This results are agreement with the references[11].

To elucidate the effect of temperature on the permeability of propane and nitrogen, the permeability was plotted as a function of reciprocal temperature (Figure 5 and 6). It is clearly that the temperature dependence of permeability of nitrogen and propane followed an Arrhenius-type relation of the form:

$$J = J_0 \exp\left(-\frac{E_p}{RT}\right) \tag{4}$$

where E_p is the activation energy of permeation, J_0 the pre-exponential factor, and R is the gas constant. As shown in Figure 7, the apparent activation energy for nitrogen permeation is positive and unaffected by pressure. The activation energy for propane permeation is negative and dependent upon pressure. It is obvious that the effects of temperature on the permeabilities of nitrogen and propane are opposite.

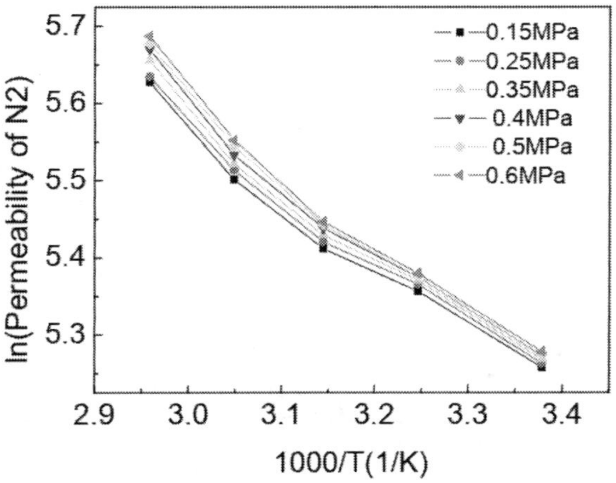

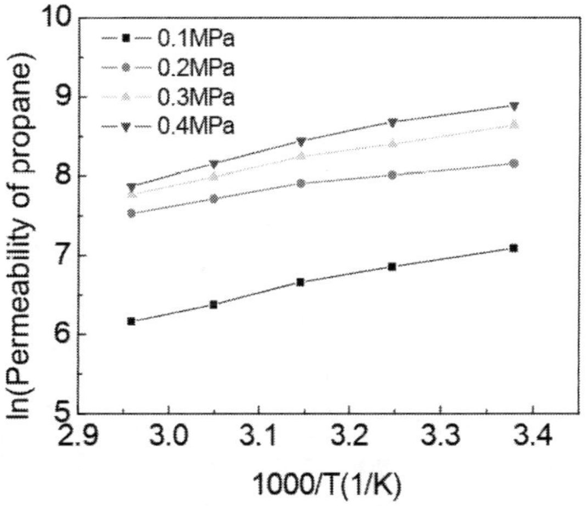

Figure 5 Temperature dependence of nitrogen permeability at different feed pressures.

Figure 6 Temperature dependence of propane permeability at different feed pressures.

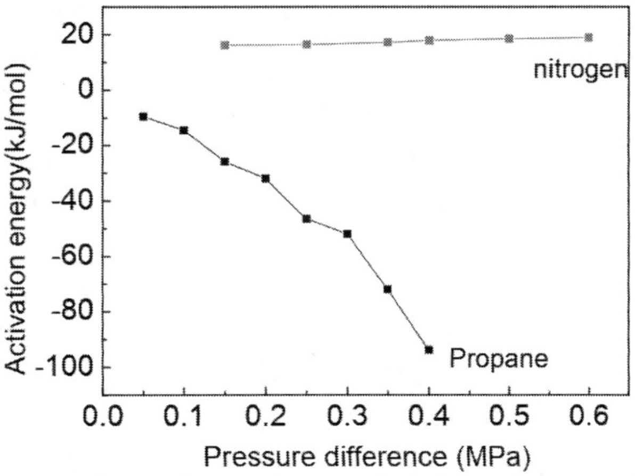

Figure 7 Activation energy of permeation for nitrogen and propane at different feed pressures.

The ideal selectivity of the membrane to propane/nitrogen separation is shown in Figure 8. The selectivity of propane and nitrogen in PDMS will be strongly temperature dependent-decreasing with increasing temperature due to the non-ideal behavior of propane, and this will then influence at which temperature the process should be run. Therefore, a low temperature and a high feed pressure are favorable to the permeation of propane. At 296K and 0.4 MPa, the ideal selectivity of propane/nitrogen is 38.

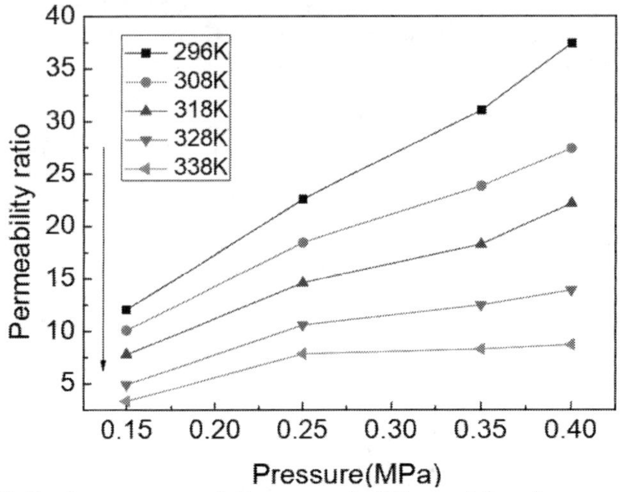

Figure 8 Ideal selectivity for propane/nitrogen at different feed pressures and temperatures based on pure gas permeability ratio.

Propane/nitrogen gas mixture permeation

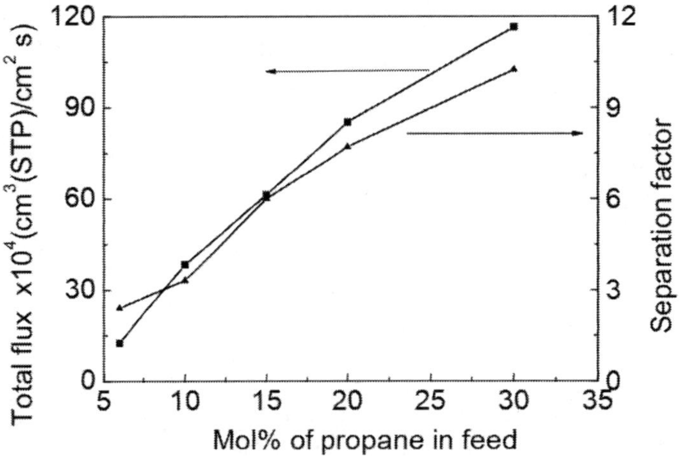

Figure 9 Total permeation flux and separation factor as a function of Mol% of propane in the feed. Operating temperature 18℃, feed pressure 0.5MPa.

In generally, the actual perselectivity of a membrane for gas mixture permeation may be difference from the ideal permselectivity based on pure gas permeation. The separation of propane/nitrogen gas mixtures by the PDMS/PVDF composite membrane was also measured. At a feed pressure of 0.5MPa and 18℃, the total flux and separation factor as a function of the feed propane concentration is shown in Figure 9. Both the total flux and separation factor increase with an increase of the propane concentration in the feed gas mixture when the feed concentration is 6-30% propane. However, the separation factor of mixture gas permeation was lower than the perselectivity of pure gas based on the ratio of permeability of pure propane and nitrogen at the same transmembrane driving force for permeation (i.e. the partial pressure difference. With the feed propane concentration increase, the propane flux increase due to the amount of propane dissolved in the membrane increasing. This resulted in the membrane more swollen by the penetrant and nitrogen to diffuse through the membrane more easily. Consequently, the total flux and the separation factor increased.

Conclusions

The PDMS composite was prepared and can be adapted to propane separation from nitrogen. The propane/nitrogen selectivity can be enhanced by lowering the temperature and increasing the feed gas pressure. The separation factor in the gas mixtures is much lower than the permeability ration of the pure gases at the same operating conditions. At 18 ℃ and 0.5MPa, the separation factor as high as 10 was obtained for feed mixtures containing 6-30mol% propane.

References

1. Garcia Villaluenga J.P. and A. Tabe-Mohammadi, A review on the separation of benzene/cyclohexane mixtures by pervaporation processes, J Memebr Sci, 2000, 169: 159-174.
2. Toraj Mohammadi, Abdolreza Aroujalian and Ali Bakhshia, Pervaporation of dilute alcoholic mixtures using PDMS membrane, Chemical Engineering Science, 2005, 60: 1875 -1880.
3. Lei Li, Zeyi Xiao, Shujuan Tan, Liang Pu and Zhibing Zhang, Composite PDMS membrane with high flux for the separation of organics from water by pervaporation, J Memebr Sci, 2004, 243:177-187.
4. Baker, B. D., J.G. Wijmans, in: D.R. Paul, Y.P. Yampol'skii (Eds.), Polymer gas Separation Membranes, CRC Press, Boca Raton, FL, 1994.
5. Stern, S.A., Shah, V.M., Hardy, B.J., structure-permeability relationships in silicone polymers, J. Polym. Sci. B: Polym. Phys.,1987, 25: 1263.
6. Merkel, T.C., Bondar, V.I., Nagai, K., Freeman, B.D., Pinnau, I., Gas sorption, diffusion, and permeation in poly(dimethylsiloxane), J Polym. Sci. B: Polym. Phys. 2000, 38: 415.
7. Koros W J, Fleming G K, Membrane-based gas separation, J Membr Sci, 1993, 83:1.
8. Maier, G., gas separation with polymer membranes, Angew. Chem. Int. Ed., 1998, 37: 2960.
9. Baker, R.W., Yoshioka, N., Mohr, J.M. and Khan, A.J., Separation of organic vapors from air,J. Membr. Sci. 1987, 31: 259.
10. Jordan, S.M., and Koros, W.J., Permeability of pure and mixed gases in silicone rubber at elevated pressures, J. Polym. Sci. B: Polym. Phys., 1990, 28: 795.
11. Liu L, Chakma A and Feng X S, propylene separation from nitrogen by poly(ether block amide) composite membranes, J Membr, Sci, 2006, 279:645-654.

Synthesis, characterization of polyurethaneimide and pervaporation performances to benzene/cyclohexane

Hong Ye, Jiding Li *, Jian Chen, Cuixian Chen

Department of Chemical Engineering, Tsinghua University, Beijing, 100084, China

ABSTRACT:

Polyurethane-imide (PUI) were successfully synthesized from the soft segment of poly(ethylene adipate)diol (PEA) and hard segments via a two-step method. The hard segment of PUI was from 4, 4'-methylenebisphenylisocyanate (MDI) and pyromelltic acid dianhydride (PMDA). The structures and properties of PUI thus obtained were characterized by means of FT-IR, and XRD. The results showed that the polymers had expected structures with urethane and imide groups,. Furthermore, the membranes of PUI were prepared by solution casting, and the pervaporation (PV) performances were investigated for separating benzene/cyclohexane mixture. And PUI membranes had a specific permeation flux of 15.64 $kg \cdot \mu m \cdot m^{-2} \cdot h^{-1}$ and separation factor of 7.75 in the benzene/cyclohexane mixture at 40°C. With the feed temperatures increasing, the flux of the PUI membranes was enhanced from 15.64$kg \cdot \mu m \cdot m^{-2} \cdot h^{-1}$ to 76.84 $kg \cdot \mu m \cdot m^{-2} \cdot h^{-1}$, but the separation factor was reduced from 7.75 to 4.96.

KEY WORDS Pervaporation Membrane; benzene; Polyurethane; Polyurethaneimide

---

* Corresponding to: Jiding Li, E-mail: lijiding@mail.tsinghua.edu.cn

1. INTRODUCTION

Separation of aromatic/non-aromatic hydrocarbon mixtures is among the most important and difficult processes in petrochemical industry [1, 2]. For example, in the production process of cyclohexane, the unreacted benzene has to be removed for pure cyclohexane recovery. Both benzene and cyclohexane have similar molecular size, polarity and volatility with only 0.6°C difference. Furthermore, they form close boiling or azeotropic mixtures at the entire range of their compositions, which makes this separation very difficult.

Traditionally, extractive distillation and azeotropic distillation are used for this separation. However, these conventional technologies often suffer from complicated processes and high operation costs.

Pervaporation (PV) is a promising alternative to the separation of aromatic/non-aromatic hydrocarbon mixtures. It offers many advantages, such as high separation efficiency, low energy consumption and simple operation [3-6].

Polyurethaneimide (PUI) are attractive polymer materials for their good mechanical property and chemical resistance. The PUI membranes also could be used in pervaporation separation of aromatic from naphtha as reported by the company EXXON [7].

In this paper, novel poly(ethylene adipate) (PEA)-based polyurethaneimide (PUI) were synthesized. Their chemical structures and properties were characterized by FT-IR and XRD. The Pervaporation performances of the PUI membranes were investigated for benzene/cyclohexane mixture.

2. EXPERIMENTAL

2.1 Materials

4,4'-methylene-bis(phenylisocyanate) (MDI, 99.9%) was obtained from Yantai Wanhua Polyurethane Co.Ltd. and used as received. Poly(ethylene adipate) diol (PEA, Mn=2000) was obtained from Yantai Huada chemicals industry Co.Ltd and dried under vacuum at 110°C before used. Pyromellitic dianhydride (PMDA) were purchased from Beijing Chemical Reagents Company and purified by sublimation. N,

N-dimethyl formamide (DMF) was purified by distillation under reduced pressure over calcium hydride and stored over molecular sieves (4Å).

2.2 Synthesis of PUI

The olyurethaneimide (PUI) were synthesized by a two-step method including prepolymerization and chain extension reaction as shown in Fig.1. Initially, the solution of PEA in DMF was placed into a three-neck flask equipped with a reflux condenser and N_2 inlet and then added dropwise with MDI at 80°C. The reaction mixture was stirred intensively for 1.5 h for the synthesis of NCO-terminated PUI prepolymer. Then the solution of PMDA as chain extender was added to perpolymer of PUI. Finally, the reaction mixtures were heated to 85°C and maintained for another 2 h. The molar ratios of PEA, MDI and PMDA were 1:2:1. Thus the solution of PUI was obtained.

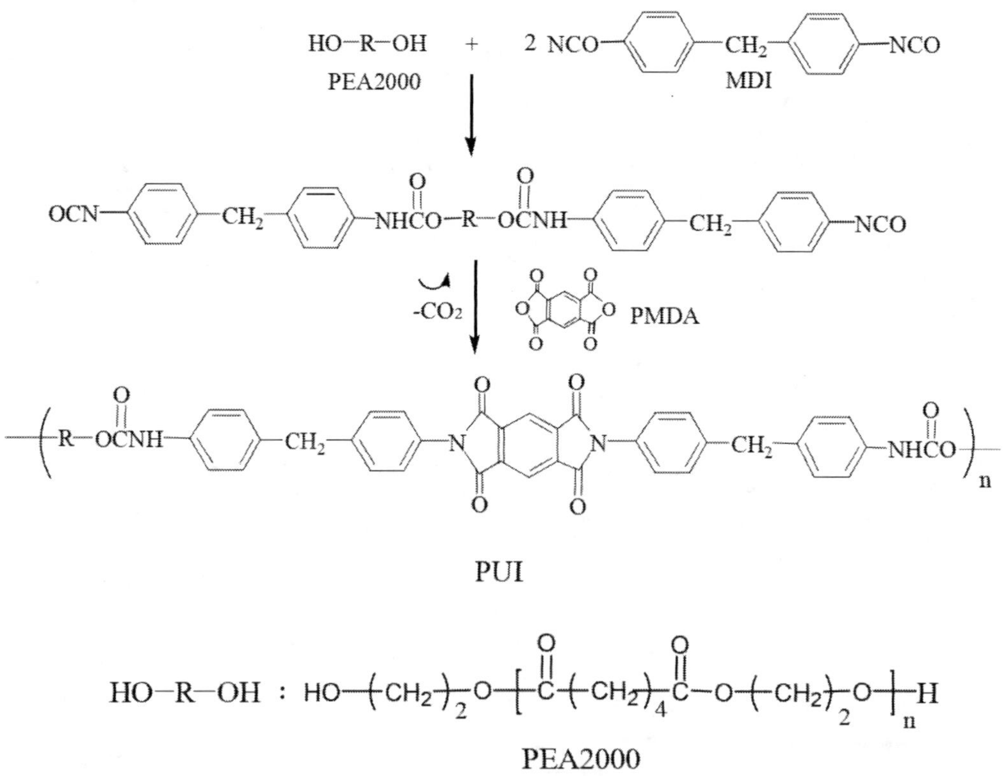

Fig1. Synthesis scheme of PEA-based PUI

2.3 Membrane preparation

The solutions of PUI was diluted to 15wt% in DMF and filtered over a 25 μm metal filter to remove impurities. Then the membranes of PUI were prepared by casting their solutions in a Teflon mould. After the solvent evaporation (60°C, 1.5h), the PUI membranes were further cured at 150°C for the complete formation of imide group. Finally the resulting membranes were stripped from the Teflon mould. The thickness of PUI membranes was 80-110 μm.

2.4 Characterization of PEA-based and PUI membranes

The structure of PUI was characterized by FT-IR spectra (Nicolet IR560) with polymer film. The morphologies of the polyimides were characterized by a Germany Bruker D8 Advance X-ray diffractormeter (XRD) using CuKα.

2.5 Pervaporation experiments

The pervaporation performances of and PUI membranes were tested by using an apparatus developed in our laboratory as shown in Figure 2.

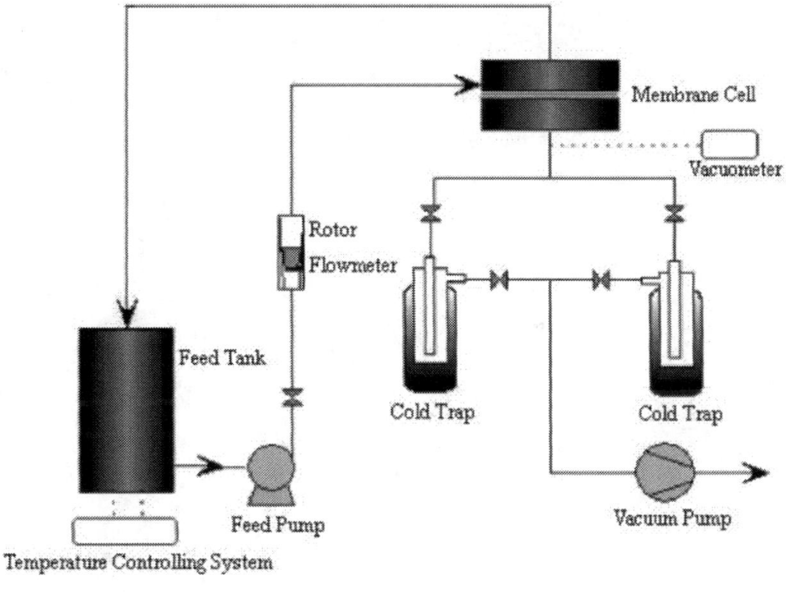

Figure 2 Pervaporation performances test apparatus

3. RESULTS AND DISCUSSION

3.1 Structure characterization of PUI membranes

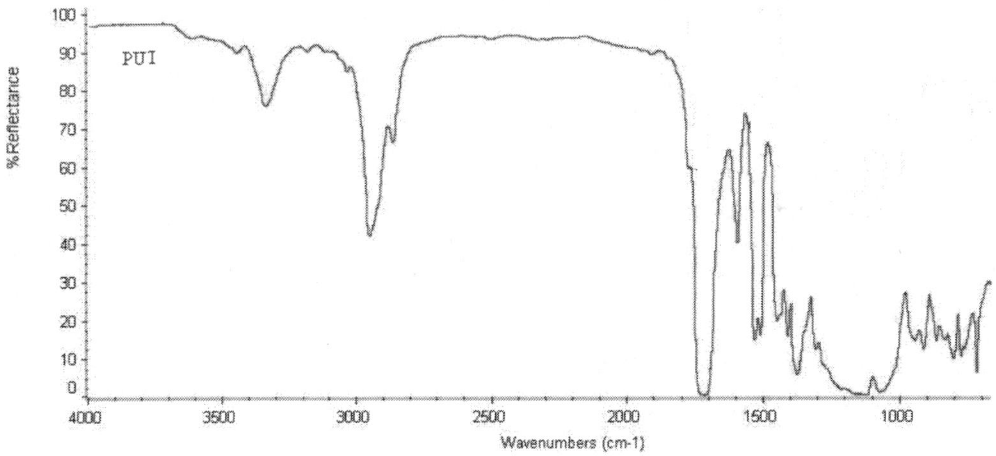

Figure 3 FI-IR spectra of PEA-based PUI

The structure of PEA-based PUI synthesized in this study was shown in Figure 1. The characteristic peaks of PUI were tested by FT-IR and shown in Figure 3. In this figure, FT-IR spectras of PUI demonstrated the characteristic peaks of PEA, urethane group and imide group:

PEA: 2940 cm^{-1} (ν C-H stretch vibration), 1720 cm^{-1} (ν CO), 1220 cm^{-1} (ν COC); urethane: 3300 cm^{-1} (N-H stretch vibration), 1720 cm^{-1} (amide I), 1535 cm^{-1} (amide II), 1220 cm^{-1} (amide III); imide group: 1780cm^{-1} (ν C=O), 1720 cm^{-1}(amide I), 1510 cm^{-1} (amide II), 1360 cm^{-1} (ν C-N), 720cm^{-1} (ν CO).

3.2 Morphology of the PUI membranes

The morphology of PUI was investigated by XRD as shown in Figure 4. The spectra of PUI membranes displayed a broad band around $2\theta=20°$ indicating the amorphous structures. It was known that pure PEA had a high crystallization. But the crystallization of PEA was absent when it became the soft segment of PUI due to the disrupted chain packing. An amorphous structure was advantageous to the permeability of small molecules.

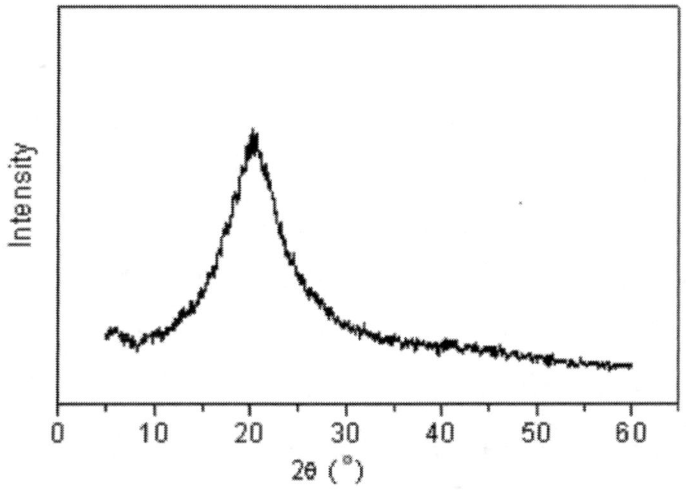

Figure 3 XRD spectra of PUI

3.3 Pervaperation performances

The pervaporation performance of PEA-based PUI membranes to benzene/cyclohexane mixture was shown as Table 1.

As expected, both results showed increased flux and decreased separation factor with feed temperature increasing. The increased flux was caused by the enhanced mobility of the polymer segments and the decrease selectivity was attributed to the increasing degree of swelling of the polyimide membranes with temperature.

Table 1 Effect of feed temperature on pervaporation performance of benzene/cyclohexane mixture

Temperature(°C)	J (kg·μm/m²·h)	Separation factor	Aromatic content in permeate(wt%)	$J_{benzene}$ (kg·μm/m²·h)	$J_{cyclohexane}$ (kg·μm/m²·h)
40	15.64	7.75	88.24	13.80	1.84
50	25.16	6.72	86.68	21.81	3.35
60	40.12	6.05	85.25	34.201	5.92
70	53.04	5.42	83.81	44.451	8.59
77	76.84	4.96	82.58	63.451	13.39

4.CONCLUSION

Polyurethane-imide (PUI) were successfully synthesized from the soft segment of poly(ethylene adipate)diol (PEA) and hard segments (MDI-PMDA) via a two-step method. the polymers had expected structures with urethane and imide groups,. Furthermore, the PUI membranes had a specific permeation flux of 15.64 $kg \cdot \mu m \cdot m^{-2} \cdot h^{-1}$ and separation factor of 7.75 in the benzene/cyclohexane mixture at 40°C. With the feed temperatures increasing, the flux of the PUI membranes was enhanced from 15.64$kg \cdot \mu m \cdot m^{-2} \cdot h^{-1}$ to 76.84 $kg \cdot \mu m \cdot m^{-2} \cdot h^{-1}$, but the separation factor was reduced from 7.75 to 4.96.

Acknowledgements

The authors greatly appreciate the financial supports of the Major State Basic Research Program of China (No.2003CB615701), National Natural Science Foundation of China (No. 20576059, No.20676067), SINOPEC Foundation (No. X505002) and CNPC Innovation Foundation (No.05051143).

REFERENCES

1. J.P.G. Villaluenga and A Tabe-Mohammadi, *J. Membr. Sci.,* 169, 159 (2000)
2. B. Smitha, D. Suhanya, S. Sridhar, and M. Ramakrishna, *J. Membr. Sci.*, 241, 1(2004)
3. N. Wynn, *Chem. Eng. Prog.*, 97, 66 (2001)
4. J. D. Li, C.X, Chen, B.B. Han, Y. .Peng, J. Zou and W.J. Jiang, *J. Membr. Sci.*, 203, 127(2002)
5. S.V. Satyanarayana and P.K. Bhattacharya, *J. Membr. Sci.*, 238, 103 (2004)
6. A. Jonquieres, R. Clement, P. Lochon, J. Neel, M. Dresch, and B.Chretien, *J. Membr. Sci.*, 206, 87(2002)
7. B.A. Koenitzer, US Patent 4929358 (1990)

Preparation of Poly (Phthalazinone Ether Ketone) Asymmetric Ultrafiltration Membrane via Phase Inversion: Composition of Casting Solutions

Weina Sun Cuixian Chen Jiding Li

Abstract

The effects of composition of poly (phthalazinone ether ketone)/NMP/PEG600 casting solutions on membrane structure and performance were studied. According to the method that the mass concentration of one component was constant, and the ratio of the other two components changed, the viscosity, the gelation rate, the membrane structure and performance of three groups of casting solutions were discussed. The experiments showd that PPEK concentration was the main factor of viscosity, while PEG600 concentration was the secondary factor; the viscosity and hydrophilicity of casting solutions had strong effects on gelation rate, which increased with the decrease of viscosity and the increase of hydrophilicity; and membrane structure and performance could be well explained by the viscosity and gelation rate of casting solutions.
Keywords: Poly (phthalazinone ether ketone); asymmetric ultrafiltration membrane; viscosity; gelation kinetics; structure and performance

Introduction

Poly (phthalazinone ether ketone) (PPEK), synthesized by Jian et al[1][2], was a novel polymer power with rigid heterocyclic configuration. It has high transition temperature (Tg>263℃) and high molecular weight. Therefore, ultrafiltration membrane was prepared using this material [3]~[6]。
Asymmetric PPEK ultrafiltration membrane was prepared by wet phase inversion; the effects of the composition of casting solutions on the apparent viscosity, the gelation rate, the membrane morphology and performance were studied in this paper.

Experimental

Materials

Poly (phthalazinone ether ketone) (PPEK) powder (see Figure 1) was purchased from Dalian Polymer new Material Co. Ltd., Liaoning Province, P.R China. N-methyl pyrrolidone (NMP) and polyethylene glycol 600 (PEG600), were purchased from Beijing Yili Fine Chemicals Co. Ltd., Beijing, P.R China.

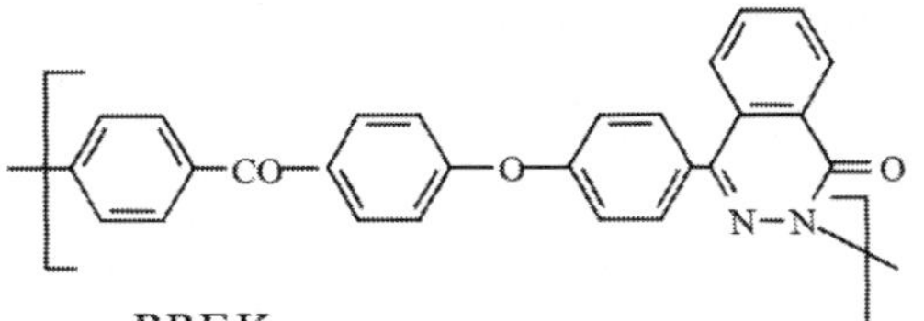

Figure 1 Structural of poly（phthalazinone ether ketone）

Composition of PPEK casting solutions

Three groups of experiments were employed, as illustrated in OA, OB, OC direction in Figure 2. The compositions of three casting solutions were listed in Table 1.

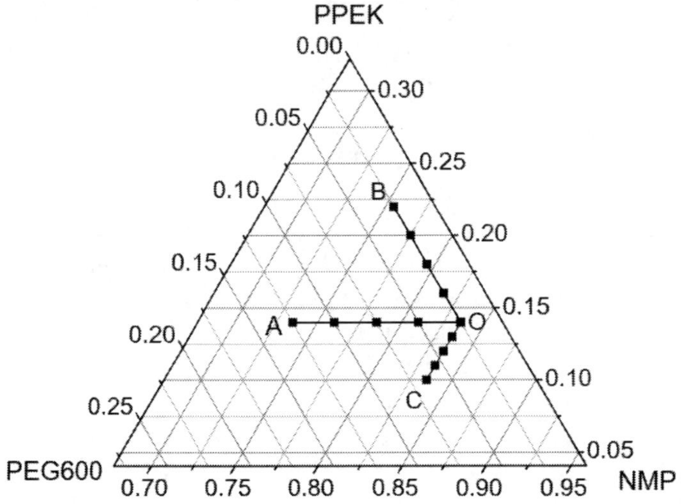

Figure 2 Phase diagram of PPEK/NMP/PEG600 casting solutions, O is the initial composition, A, B and C is the composition of three groups of casting solutions

Table 1 the composition of three groups of casting solutions

Casting solutions	P (wt%)	NS (wt%)	S (wt%)
O	14	2.5	83.5
A1	14	5	81
A2	14	7.5	78.5
A3	14	10	76
A4	14	12.5	73.5
B1	16	2.5	81.5
B2	18	2.5	79.5
B3	20	2.5	77.5
B4	22	2.5	75.5
C1	13	3.5	83.5
C2	12	4.5	83.5
C3	11	5.5	83.5
C4	10	6.5	83.5

Determination of apparent viscosity

The apparent viscosity of casting solutions was determined by Brookfield (DV-II+CP) viscometer.

Membrane preparation

PPEK and PEG600 were dissolved in NMP and stirred at least 40 hours, bubbles were removed in vacuum drying chamber, the dope solutions were cast onto non-woven fabrics by means of membrane-preparation machine, and membrane was formed by wet phase inversion.

Gelation kinetics experimental

The gelation process could be visually investigated in online optical microscope–CCD camera experimental system (OM–CC system) including an Olympus IX71 optical microscope, a CCD camera, a personal computer and a specially designed microscope slide (see Figure 2)[7]. From the recording gelation thickness of dope solution (X), we could distinguish the gelation rates of different dope solutions.

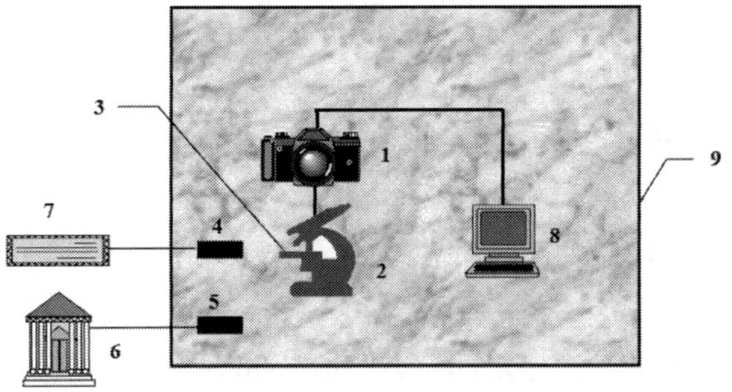

Figure 3 Experiment set up for the observation of the gelation process:
1,CCD camera; 2, optical microscope; 3, special microscope accessory; 4,temperature sensor; 5, humidity sensor; 6, dehumidifier; 7, air-condition; 8,personal computer; 9, clean room.

Characterization of membrane morphology and performance

The surface and cross-section images of membranes were observed by Scan Electron Microscope (SEM, Flotu JSM-7401F). For cross-section analysis the membrane samples are freeze fractured in liquid nitrogen. All samples are sputter-coated with gold before observed. Porosity of membrane surface was calculated by digital image processing program[8].

Membrane performance was characterized by pure water flux (J) and bovine serum albumin rejection (R) (the molecular weight of bovine serum albumin is 67,000 Da) by means of equipments made by Tsinghua University. J and R were calculated by the formula below:

$$J = \frac{V}{A \cdot t}, \quad R = \frac{C_f - C_p}{C_f}$$

Results and discussion

Effects of the composition of casting solutions on the apparent viscosity

The effects of PEG600/NMP on the viscosity of A, B and C casting solutions were illustrated in Figure 5. It can be seen that the viscosity of A and B increased with the increase of PEG600/NMP, but that of C decreased with the increase of PEG600/NMP. As a result, the concentration of PPEK was the main factor of the viscosity of casting solutions.

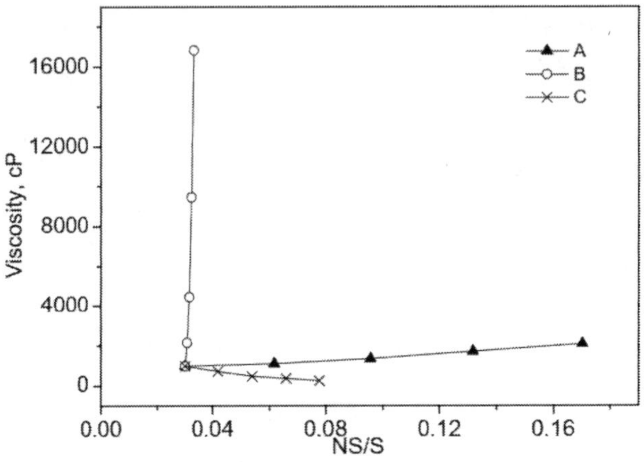

Figure 5 The viscosity of casting solutions A, B and C against NS/S ratio.

Effects of the composition of casting solutions on gelation rate

Gelation kinetic curves of A, B and C casting solutions were shown in Figure 6 (A), (B) and (C). In Figure 6 (A), the gelation rate increased gradually with increased PEG600 concentration, which was probably because that the interaction between casting solutions and gelation bath increased with increased PEG600, and the decrease of the dissolving capacity of NMP also resulted in the faster gelation rate.

In Figure 6 (B), the increase of PPEK concentration made the viscosity of casting solutions increase; therefore, the resistant of exchange rate between NMP and gelation bath inclined and the gelation rate inclined too.

In Figure 6 (C), the viscosity of casting solutions decreased because of decreased PPEK concentration, and the hydrophilicity of casting solutions increased because of increased PEG600 concentration; therefore, the gelation rate increased accordingly.

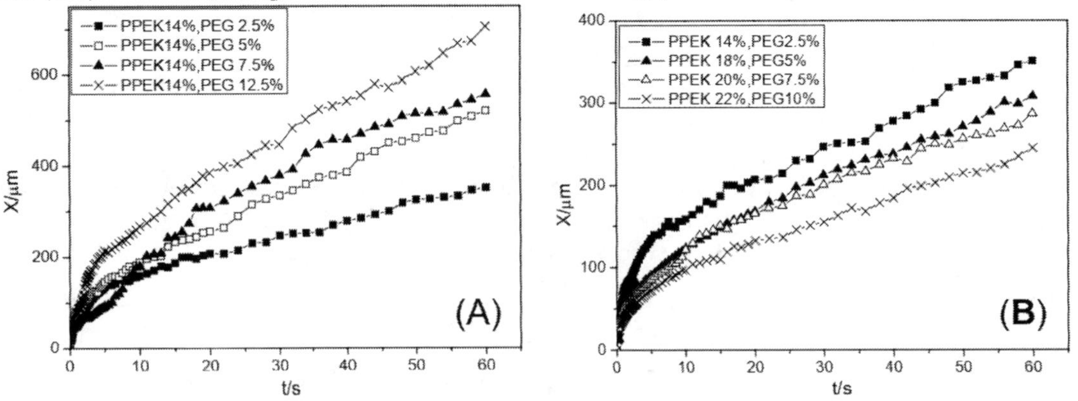

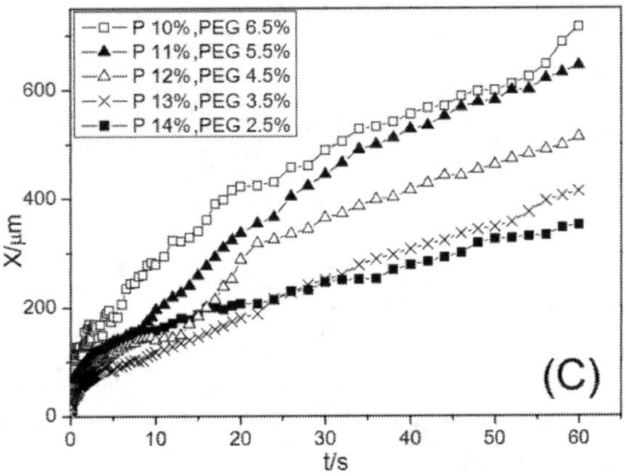

Figure 6 Gelation kinetic curves of three groups of casting solutions, (A), (B) and (C) represent A, B and C casting solutions respectively.

Effects of the composition of casting solutions on membrane morphology and performance

FESEM images of membrane surface and cross-section of A, B and C membranes were illustrated in Figure 7 (O), (A), (B) and (C). The porosity of membrane surface was listed in Table 2. Pure water flux (J) and bovine serum albumin rejection (R) were shown in Figure 8.

The pores of membrane surface presented as O distributed uniformly, and the membrane cross-section was finger pores. For group A, with the increase of PEG600 concentration, the membrane cross-section turned into sponge structure, the porosity of membrane surface increased accordingly; J decreased first and increased later, R increased first and decreased later.

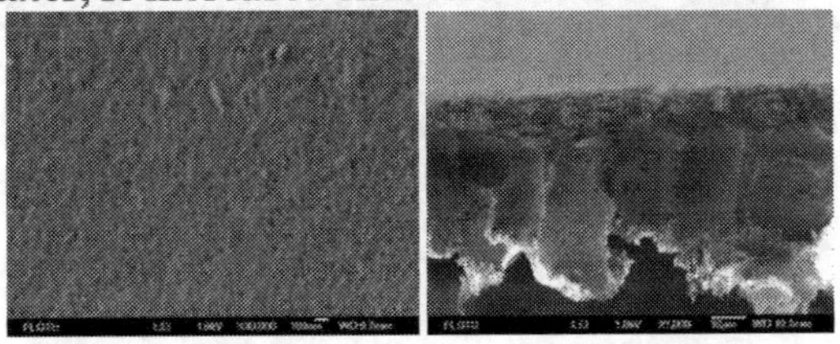

(O) initial point O composition: P is 14%, PEG6002.5%, NMP83.5%.

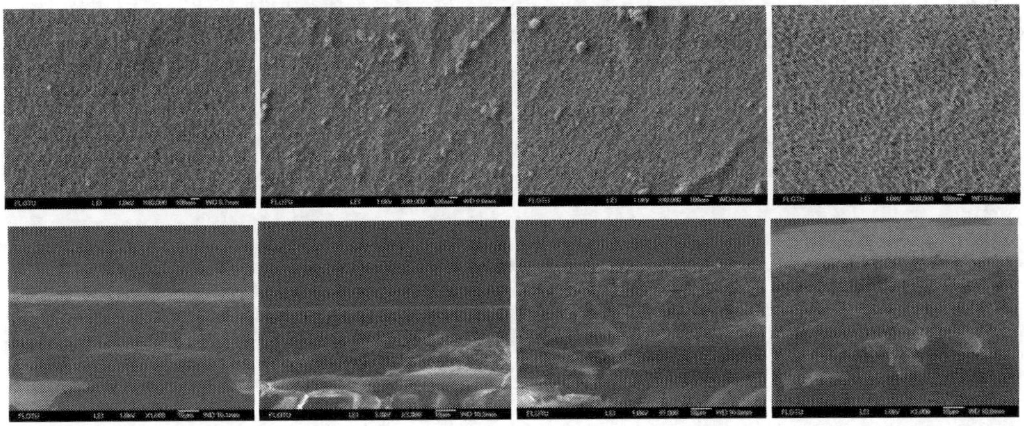

(A) group A composition: P is 14%, PEG600 is 5%, 7.5%, 10% and 12.5% respectively.

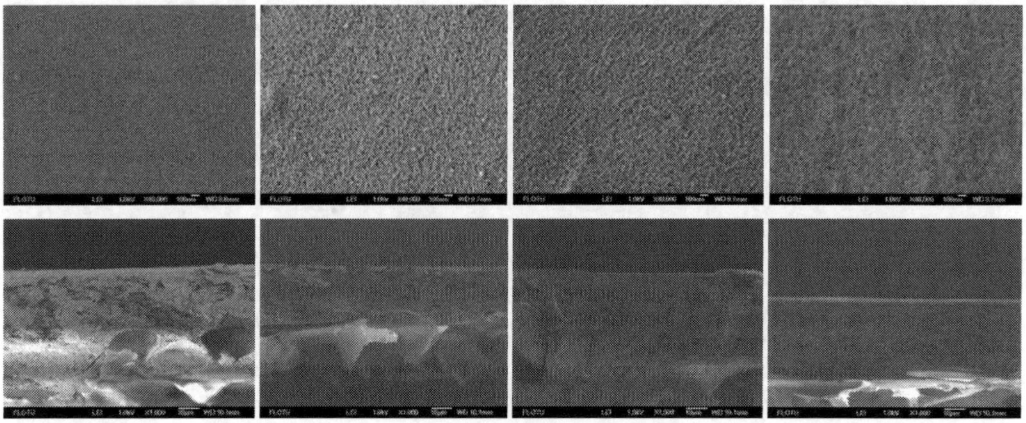

(B) group B composition: PEG600 is 2.5%, P is 16%, 18%, 20% and 22% respectively.

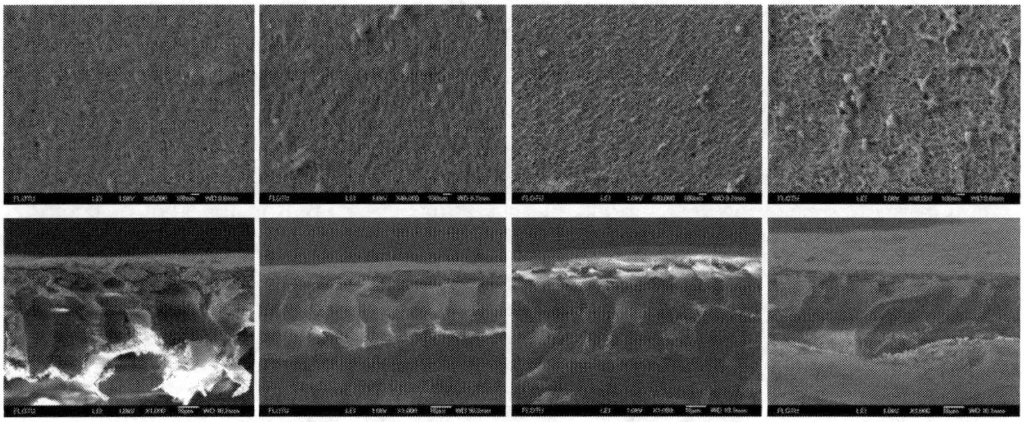

(C) group C composition: NMP is 83.5%, P is 13%, 12%, 11% and 10% respectively.

Figure 7 FESEM images of surface and cross-section of three groups of membranes.

For group B, the membrane cross-section turned into sponge structure with the increase of PPEK concentration; the increased PPEK concentration caused the membrane surface to become dense and made the gelation rate become slow, which resulted in the decrease of *J* and the increase of *R*.

For group C, because of the decrease of PPEK concentration and the increase of PEG600 concentration, the porosity of membrane surface increased gradually, the membrane cross-section remained the finger pores, but the sublayer became thin. The concentration of PPEK was too slow; therefore the viscosity of casting solutions also became too slow, which made the intensity of membrane cross-section weak, as shown in Figure 7 (C).

Table 2 Porosity of membrane surface of three groups of membranes

membranes	Porosity of membrane surface/%
Initial composition O	20.94
A1	23.68
A2	24.11
A3	24.70
A4	30.40

B1	20.53
B2	19.78
B3	18.70
B4	16.35
C1	19.46
C2	20.19
C3	21.73
C4	24.03

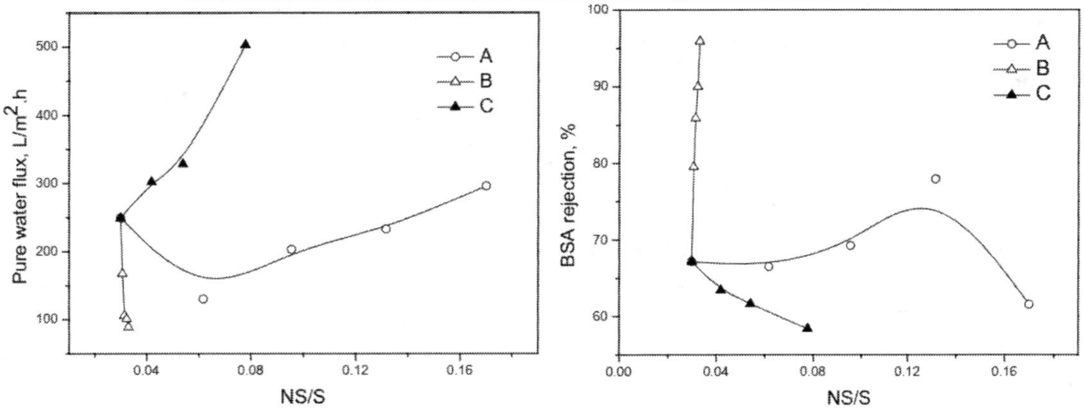

Figure 8 Pure water flux (a) and BSA rejection (b) of three groups of membranes.

Conclusions

(1) When PPEK concentration was constant and the ratio of PEG600/NMP increased, the viscosity, the gelation rate and the porosity of membrane surface increased, the morphology of membrane cross-section turned into sponge structure, *J* inclined first and increased later, *R* increased first and decreased later. When PEG600 concentration was constant and the ratio of PEG600/NMP increased, the viscosity increased, the gelation rate and the porosity of membrane surface decreased, the membrane cross-section turned into sponge structure, *J* inclined and *R* increased. When NMP concentration was constant and the ratio of PEG600/NMP increased, the viscosity decreased, the gelation rate and the porosity of membrane surface increased, the membrane cross-section turned into sponge structure, *J* increased and *R* inclined.

(2) The viscosity of casting solutions was mainly affected by polymer concentration; the gelation rate was affected by both the viscosity and the hydrophilicity of casting solutions; the faster gelation rate resulted in the larger *J*, but *R* was not only influenced by the porosity of membrane surface, but also the connectivity of pores.

Acknowledgment

The authors greatly appreciate the financial supports of National Natural Science Foundation of China (No. 20576059, No. 20676067).

References

[1] Liu YJ, Jian XG, Liu SJ, Synthesis and properties of novel poly (phthalazinone ether ketone ketone), *J Appl Polym Sci*, 2001, 82: 823
[2] Zhang SH, Jian XG, Wang GQ, *et al*. Synthesis of poly (phthalazinone ether ketone) containing sodium sulfonate groups via direct polymerization, *Chinese Chem Lett*, 2002, 13: 926
[3] Xia JQ, Zhang J, Liao GX, *et al*. Copolymerization and blending of poly (phthalazinone ether ketone)s to improve their melt processability, *J Appl Polym Sci*, 2007, 103: 2575
[4] Chen YL, Meng YZ, Li XH, *et al*. Poly (phthalazinone ether ketone) ionomers synthesized via N-C coupling reaction for fuel cell applications, *Macromolecules*, 2005, 38: 10007
[5] Sun YM, Wu TC, Lee HC, *et al*. Sulfonated poly (phthalazinone ether ketone) for proton exchange membranes in direct methanol fuel cells, *J Membr Sci*, 2005, 265: 108
[6] Lu ZQ, Liu YY, Qin JG, *et al*. Design and synthesis of a new nonlinear optical poly (phthalazinone ether ketone), *React Funct Polym*, 2004, 61: 379
[7] Qin PY, Chen CX, Han BB; *et al*. Preparation of poly (phthalazinone ether sulfone ketone) asymmetric ultrafiltration membrane - II. The gelation process, *J Membr Sci*, 2006, 268, 181
[8] Sun WN, Chen T, Chen CX, *et al*. A study on membrane morphology by digital image processing, *J Membr Sci*, 2007, in press

330e Water Uptake, Transport And Structural Characterization In Crosslinked N-Vinyl-2-Pyrrolidone/ N,N'-Methylene Bisacrylamide Membranes

Yuan-Hsuan Wu and Benny Freeman. Chemical Engineering, University of Texas at Austin, 10100 Burnet Rd. Building 133, Austin, TX 78758

The lack of sufficient amounts of pure water is one of the world's most serious concerns. Consequently, water reuse and management is becoming more and more important. Produced water is the single largest waste stream in oil and gas production, often containing salts, heavy metals, emulsified oil and other organics. If the organic content and salinity of produced water could be reduced to acceptable limits, produced water could represent a potential new water source with a wide variety of uses. Membranes may be an effective tool for treating water produced during oil and gas production. However, membrane fouling is a serious problem and limits the efficiency of water purification in this application.

The initial objective of this research was to examine potential fouling-reducing coating materials and determined fundamental structure/property relationships governing water transport, rejection of model foulants such as emulsified oil droplets or protein aggregates, and fouling characteristics. A systematic series of hydrogel-based polymers was prepared, and their permeation and fouling properties were studied and correlated with their structure. In a future study, composite membranes prepared via coating of commercial UF membranes will be performed.

In this presentation, the effects of UV intensity, water content in the prepolymer solution, and crosslinker content on film morphology, water transport and rejection properties are presented. Crosslinked free-standing films were prepared by UV-induced free radical polymerization at different UV intensities, using N,N'-methylene bisacrylamide (MBAA) as the crosslinker and N-vinyl-2-pyrrolidone (NVP) as the monomer. The water content in the prepolymer solution was varied from 30 wt % to 70 wt%, and the MBAA content of total solids was varied from 15 wt% to 30 wt%. The crosslinked NVP films show polymerization induced phase separation (PIPS). These PIPS films contain pores that scatter visible light and render them opaque. The films exhibited a porous morphology in CryoSEM studies. Liquid water uptake increased from 100 to 350 g water/100 g dry polymer as water content in the prepolymer solution increased. Dead end filtration of deionized water, oil/water solutions (1500 ppm), and bovine serum albumin (BSA) solutions (1g/l), was performed to explore the fundamental transport and fouling properties of these materials. Different internal structures of crosslinked NVP films at the same composition were made by polymerization at different UV intensities and characterized by molecular weight cutoff (MWCO) measurements. Polymerization under high intensity UV light (~3000 μW/cm2) leads to high water permeability, up to ~3000 L μm/(hr m2 bar) in film containing 70 wt% water and 30 wt% solids (NVP/MBAA: 85/15) and high MWCO, ~10,000 kDa. At the same composition with polymerization at low intensity (~1100 μW/cm2), the water permeability was ~1300 L μm/(hr m2 bar) and the MWCO was also lower, ~650 kDa. The total permeate flux of pure water was almost equal to that of the BSA containing solution and oil/water emulsion containing solution, indicating that crosslinked NVP films resist fouling by oil droplets and BSA under the conditions studied. The oil rejection was ~97% in all samples, while the protein rejection depended on the prepolymer water content, crosslinker concentration, and UV intensity.

330f Mixed Matrix Membranes For Gas Separations

Pipat Singha-In[1], Thirasak Rirksomboon[1], and Santi Kulprathipanja[2]. (1) The Petroleum and Petrochemical College, Chulalongkorn University, Chula 12, Phyathai Road, 10330, Bangkok, Thailand, (2) UOP, A Honeywell Company, 25 East Algonquin Road, Des Plaines, IL 60017-5016

One of the most attractive technologies for gas separations is a membrane separation owing to its compact unit and low energy consumption compared to the current energy-intensive cryogenic distillation, absorption, or adsorption. However, membranes have also a drawback such as plasticization phenomenon. The plasticization in glassy polymer membranes can occur by hydrocarbons or CO_2 acting as a plasticizer. Plasticization phenomenon in glassy polymer membranes results in increasing gas permeability which in turn results in reducing membrane selectivity. This is due to the segmental chain mobility enhancement and increase in free volume of the polymer matrices. Mixed matrix membranes (MMMs) have been developed and evaluated for the effects of plasticization. In this study, MMMs were fabricated by incorporating zeolites such as NaX, NaY, and silicalite into Matrimid® polymers. The olefin/paraffin separation performance and plasticizing characteristics of the membranes were carried out using pure gas permeability measurements. Preliminary results are encouraging. Detailed information on membrane preparation, gas permeation and selectivity will be presented and discussed.

330g Isotropic Microporous Mixed Matrix (Zeotips) Membranes

Caleb V. Funk, Bonnie L. Beavers, Courtney A. Taylor, and Douglas R. Lloyd. Chemical Engineering, The University of Texas at Austin, 1 University Station, Mailcode C0400, Austin, TX 78712

This presentation will discuss unique, isotropic, microporous mixed matrix membranes for liquid separation and gas separations. These membranes consist of a flexible microporous polymer matrix with nanoporous zeolite particles connecting the microporous voids of the matrix. These unique zeolite-containing membranes are formed via the thermally induced phase separation process; thus, they are referred to as ZeoTIPS membranes. In these membranes, the microporous cellular voids of the polymer matrix allow for rapid transport through the membrane while the nanoporous zeolite particles that connect the microporous cells provide the membrane with high selectivity capabilities. There are a variety of complicating factors occurring in the formation of ZeoTIPS membranes involving the choice of polymer-diluent-zeolite system and the processing conditions used. In addition to explaining the rationale behind the development of isotropic microporous mixed matrix membranes, this presentation will discuss in some detail the challenges associated with selecting a polymer and diluent for use in a ZeoTIPS membranes. Finally, the challenges related to zeolite surface chemistry in a TIPS system, including surface modification of the particles, will be presented.

Removing of dimethyl carbonate from dimethyl carbonate / methanol azeotrope by pervaporation

Luying Wang, Jiding Li, Yangzheng Lin, Cuixian Chen
Department of Chemical Engineering, Tsinghua University, Beijing, PR China

Introduction

In recent years, dimethyl carbonate (DMC) as a new kind of green chemical material is more and more important in chemical industry for its low toxicity, its quick biodegradation, and its versatility [1-3]. DMC can replace some virulent carcinogens to be used as a methylating, carbonylating, and methoxylating agent [4, 5], and can replace DMC methyl t-butyl ether (MTBE) and t-amyl methyl ether (TAME) to act as an oxygenation agent because of its high oxygen content (53%) [6, 7]. It is produced mainly by three methods, including (a) phosgenation of methanol (MeOH), (b) oxidative carbonylation of MeOH, and (c) transesterification of ethylene carbonate with MeOH [8-11]. However, DMC is always obtained as a mixture with MeOH, and this DMC/MeOH mixture is an azeotrope (30 wt.% DMC, 70 wt.% MeOH) under normal pressure. So it is difficulty to separation DMC/MeOH mixture by distillation under normal pressure. Some traditional methods used for separating DMC/MeOH azeotrope has been reported, such as low temperature crystallization [12, 13], high-pressure distillation [14, 15], azeotropic distillation [16, 17], extractive distillation [18, 19], and adsorption [20, 21].

Recently, membrane pervaporation is becoming one of the most promising candidates for its low-cost, energy saving, high efficient, and easy operation, especially for separation of azeotropic or close-boiling mixtures. Pervaporation is a membrane separation process where a liquid mixture is in contact with the upstream side of the dense membranes, and the product permeating through the membrane is removed as a vapor at the downstream side. Its applications can be found in: dehydration of water/organic mixtures [22], removal of organic components from water [23], and separation of organic/organic mixtures [24]. In our earlier work [25, 26], DMC/MeOH mixtures can be separated by pervaporation with PAA/PVA blend membranes and cross-linked PVA membranes, which membranes are both MeOH selective.

Polydimethylsiloxane (PDMS), as a membrane material for the preferential permeation of organic compounds, has been widely researched and developed [27]. The pervaporation performance of PDMS membrane can be enhanced by several approaches, for example, filling [28], grafting [29], crosslinking [30], and surface modification [31]. Compared other approaches mentioned above, filling hydrophobic or organophilic filler into the PDMS membrane is a simple and effective way for improving the separation properities. In this study, PDMS membranes and zeolite filled PDMS membranes were prepared in expectation of removing DMC from DMC/MeOH azeotrope by pervaporation. The effects of cross-linking agent content in PDMS membranes and the zeolite content in zeolite filled PDMS membrane on pervaporation performances were studied. Moreover, the effects of operation temperature on pervaporation performances of the homogeneous PDMS membrane were also examined.

Experimental

Materials

107 silicone rubber as PDMS prepolymer and HZSM-5 zeolite were supplied respectively from Beijing Huaer Co. Ltd. and Beijing Bowen Kefeng Co. Ethyl orthosilicate (EOS), dibutyltin dilaurate, and n-heptane were all purchased from Beijing Bei Hua Fine Chemicals Co. All chemicals were used as received.

Membrane preparation

107 silicone rubber and n-heptane as solvent were mixed by stirring for about 2h at ambient temperature. To prepare filled PDMS membranes, some amount of HZSM-5 zeolite was added to the rubber solution by stirring at ambient temperature until obtaining a homogeneous dispersion. Then EOS as cross-linking agent and dibutyltin dilaurate as catalyst were add to the homogeneous solution by stirring. After removing bubbles, the polymer solution was cast to the desired thickness on a glass plate. The resulting homogeneous membranes were allowed to dry in the air at room temperature at least 12hrs, and then were heated in a vacuum oven at about 60°C for 12h for evaporating n-heptane completely and cross-linking sufficiently. The membranes prepared in this study have a thickness of $120 \pm 5\,\mu m$.

Pervaporation experiments

The pervaporation apparatus used in this study was showed in Fig.1. The membrane was installed in a membrane cell, and the effective membrane area in this cell was 21.70cm^2. The feed was continuously circulated from a feed tank to the upstream side of the membrane at a desired temperature by a pump. The pressure at downstream side was maintained at about 200Pa respectively. Upon reaching steady-state conditions which was obtained after about 3hrs, permeates were collected in liquid nitrogen traps with certain intervals, and weighted to calculate the permeate flux. The composition of collected permeate was determined by GC (SHIMADZU,GC-14C) to calculate the separation factor.

The separation factor (α) was defined as

$$\alpha = \frac{Y_a/Y_b}{X_a/X_b} \tag{1}$$

where X and Y represent the weight fraction in the feed and permeate, respectively; indices a and b refer to the more permeable component and the less permeable component, respectively.

The permeate flux (J) was calculated from

$$J = \frac{w}{At} \tag{2}$$

where w is the weight of permeate, g; A is the effective area of membrane, m^2; and t is the time, h.

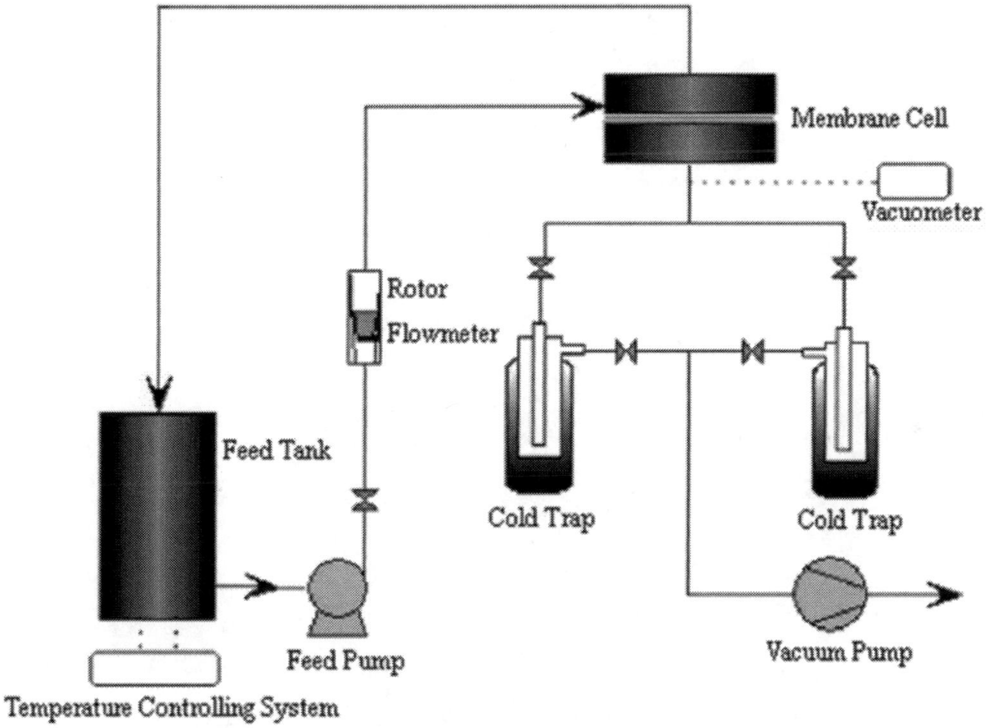

Fig.1. Schematic diagram of pervaporation apparatus

Results and Discussions

The effect of EOS content on pervaporation performances

The pervaporation performances of PDMS membranes with different EOS content (presented by EOS/PDMS, weight ratio) is shown in Fig.2, where the separation factor increases first and then decreases but the permeate flux decreases with increasing the EOS/PDMS from 0.1 to 0.5. As the EOS/rubber increases, the degree of cross-linking enhances, so the free volume in the PDMS membranes decreases leading to the permeate flux also decreases. With increasing the EOS/PDMS from 0.1 to 0.3, the selectivity of the PDMS membrane for DMC increasing, but when the EOS/PDMS is higher than 0.3, DMC is difficult to permeate the PDMS membrane because of the low free volume.

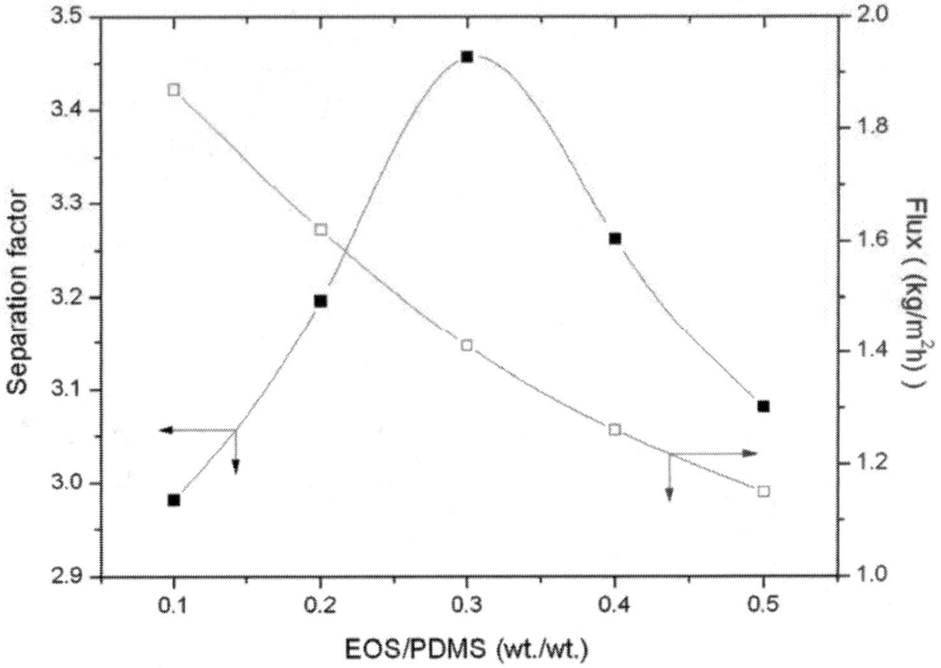

Fig.2. The effect of EOS content on pervaporation performances
(Operation temperature: 40°C)

The effect of HZSM-5 zeolite content on pervaporation performances

Fig.3 shows the effect of HZSM-5 zeolite content in PDMS membrane (EOS/PDMS: 0.3) on pervaoration performances. With increasing the zeolite content in the PDMS membrane to 40 wt.%, the separation factor increases but the permeate flux decreases. It was proved that the HZSM-5 zeolite could absorb DMC preferentially [32]. Therefore, the higher the zeolite content in the PDMS membrane is, the higher the separation factor is. On the other hand, because the PDMS content decreases with increasing the zeolite content, so the swelling of the membrane decreases leading to the permeate flux deceases.

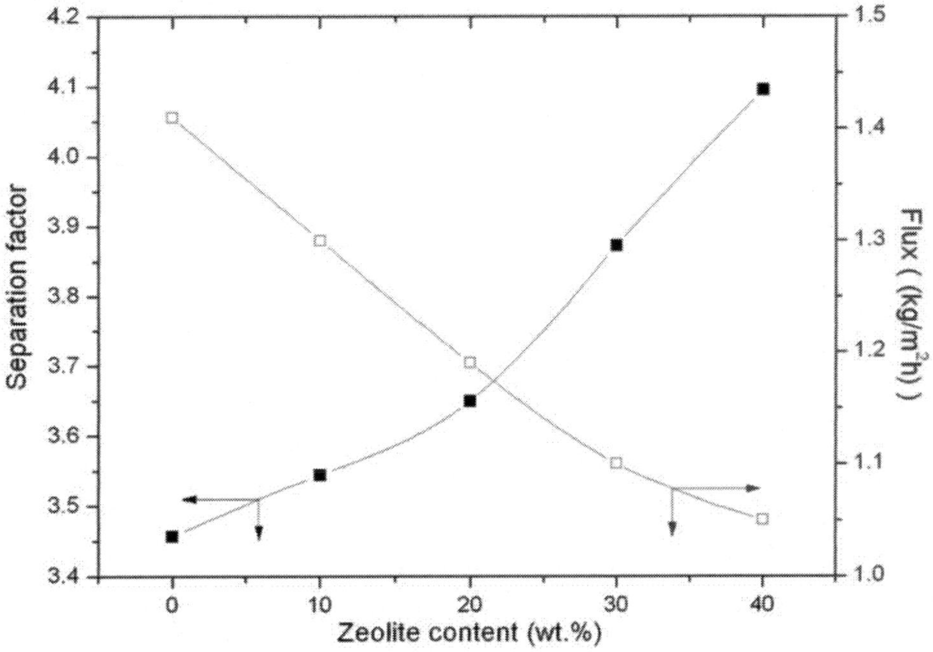

Fig.3. The effect of HZSM-% zeolite content on pervaporation performances
(Operation temperature: 40°C)

The effect of operation temperature on pervaporation performances

The effect of operation temperature on pervaporation performances of the PDMS membrane (EOS/PDMS: 0.3) and the HZSM-5 filled PDMS membrane (EOS/PDMS: 0.3, zeolite content: 40 wt.%) is shown in Fig.4. It can be seen the separation factors of these membranes both decrease but the permeate fluxes of these membranes both increases with increasing the operation factor from 40°C to 60°C. When the operation temperature increases, the free volume becomes larger resulting in the separation factor decreases but the permeate flux increases.

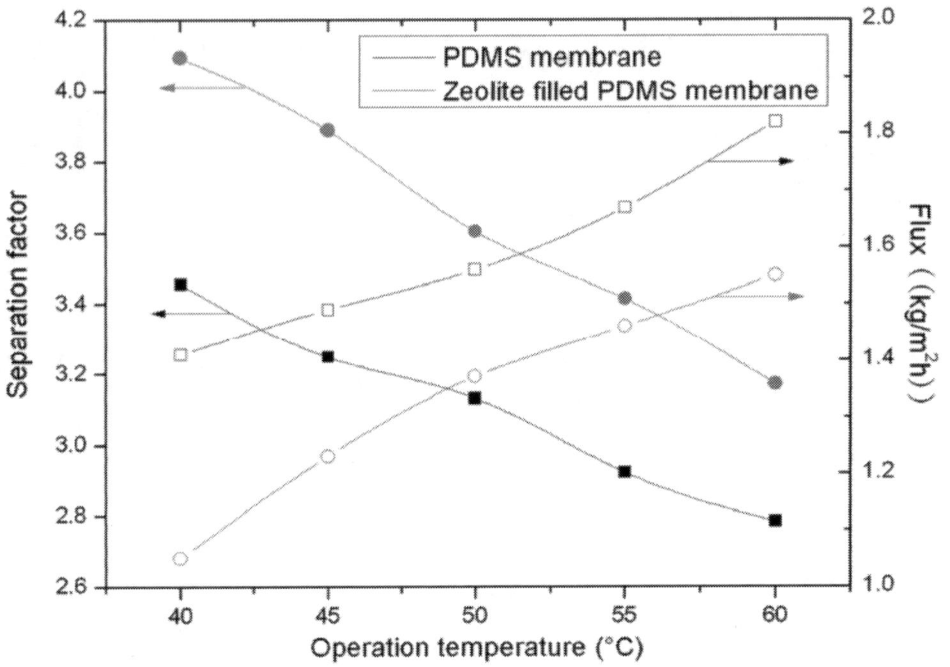

Fig.4. The effect of operation temperature on pervaporation performance

Conlusion

The pervaporation performances of the PDMS membrane and the HZSM-5 zeolite filled PDMS membrane for removing DMC from DMC/MeOH azeotrope were investigated. For the PDMS membrane, the separation factor increased first and then decreased but the permeate flux decreased with increasing the EOS content in the PDMS membrane. For the HZSM-5 zeolite filled PDMS membrane, the separation factor increased but the permeate flux decreased with increasing the zeolite content in the HZSM-5 zeolite filled PDMS membrane. As the operation temperature from 40°C to 60°C, the separation factors of these membranes both decreased, and the permeate fluxes of these membranes both increased. The membrane, in which the EOS/PDMS is 0.3 and the HZSM-5 zeolite content is 40 wt.%, has a highest separation factor of 4.1 and a permeate flux of 1.05 kg/(m^2h).

Acknowledgements

The authors greatly appreciate the financial supports of the Major State Basic Research Program of China (no. 2003CB615701), National Natural Science Foundation of China (nos. 20576059, 20676067), SINOPEC Foundation (no. X505002), and CNPC Innovation Foundation (no. 05051143).

References

1. P. Tundo, "New developments in dimethyl carbonate chemistry", *Pure Appl. Chem*, 73 (2001) 1117-1124.
2. Y. Ono, T. Baba, "Selective reactions over solid base catalysts", *Catal. Today*, 38 (1997) 321-337.
3. H. Hattori, "Heterogeneous basic catalysis", *Chem. Rev.*, 95 (1995) 537-558.

4. D. Delledonne, F. Rivetti, U. Romano, "Developments in the production and application of dimethyl carbonate", *Appl. Catal. A: Gen.*, 221 (2001) 241-251.
5. A.-A.G. Shaikh, S. Sivaram, "Organic Carbonates", *Chem. Rev.*, 96 (1996) 951-976.
6. M.A. Pacheco, C.L. Marshall, "Review of dimethyl carbonate (DMC) manufacture and its characteristics as a fuel additive", *Energy Fuels*, 11 (1997) 2–29.
7. K. Pan, C.S. Cheung, M.A. Liu and S.C. Lee, "Performance of DMC blended diesel in reducing pollutant emissions", *Paper presented at the conference: Better Air Quality in Asian and Pacific Rim Cities (BAQ 2002)*, Hong Kong SAR, 2002, Dec16–18.
8. N.S. lsaecs, B. O'Sullivan, C. Verhaelen, "High Pressure Routes to Dimethyl Carbonate from Supercritical Carbon Dioxide", *Tetrahedron*, 55 (1999) 11949-11956.
9. H.-J. Buysch, H. Krimm, S. Bohm, "Process for the preparation of dimethyl carbonate", *US Patent 4335051*, 1982.
10. Y. Ono. "Catalysis in the production and reactions of dimethyl carbonate, an environmentally benign building block", *Appl. Catal. A: Gen.*, 155 (1997) 133-166.
11. S. Uchiumi, K. Ataka, T. Matsuzaki, "Review: Oxidative reactions by a palladium–alkyl nitrite system", *J. Organometallic Chem.*, 576 (1999) 279-289.
12. J.A. Gilpin, A.H. Emmons, "Synthesis of dimethyl carbonate", *US Patent 3803201*, 1974.
13. G. A. Passoni, "Purification of dimethyl carbonate", *DE Patent 2450856*, 1973.
14. U. Romano, "Recovery of dimethyl carbonate from its azeotropic mixture with methanol", *DE Patent 2607003*, 1976.
15. J. Rechner, A. Klausener, H.-J. Buysch and P. Wagner, "Process for the preparation of dialkyl carbonates", *US Patent 5274163*, 1993.
16. B.H. Josef, K. Heindrich, R. Hans, G. Offen, "Process for separating off dimethyl carbonate", *DE Patent 2737265*, 1979.
17. N. Keigo, "Process for purifying dimethyl carbonate", *JP Patent 270249*, 1991.
18. H. Wechsler, L.P.B. Heindrich, R.L. Herbst, "Separation of alcohols from esters", *US Patent 2872478*, 1959.
19. H. Walter, "Purification of dimethyl carbonate", *DE Patent 2706684*, 1977.
20. M.S. Spencer, "Carbonate production", *US Patent 4582645*, 1984.
21. I. Janisch, H. Landscheidt, W. Struver, A. Klausener, "Process for separating off methanol from a mixture of dimethyl carbonate and methanol", *US Patent 5455368*, 1994.
22. J.M. Haack, W. Lenk, D. Lehmann, K. Lunkwitz, "Pervaporation separation of water/alcohol mixtures using composite membranes based on polyelectrolyte multiplayer assemblies", J. *Membr. Sci.*, 184 (2001) 233–243.
23. R.W. Baker, J.G. Wijmans, "Process for recovering organic components from liquid streams", *US Patent 5169533*, 1992.
24. B. Smitha, D. Suhanya, S. Sridhar, "M. Ramakrishna, Separation of organic–organic mixtures by pervaporation—a review", *J. Membr. Sci.*, 241 (2004) 1–21.
25. Luying Wang, Jiding Li, Yangzheng Lin, Cuixian Chen, "Separation of dimethyl carbonate / methanol mixtures by pervaporation with poly(acrylic acid)/ poly(vinyl alcohol) blend membranes", *J. Membr. Sci.*, In Press.
26. Luying Wang, Jiding Li, Yangzheng Lin, Cuixian Chen, "Cross-linked poly(vinyl alcohol) membranes for separation of dimethyl carbonate / methanol mixtures by pervaporation", Under Review.
27. J.E. Mark, "Some interesting things about polysiloxanes", *Acc. Chem. Res.*, 37 (2004) 946–953.

28. E. Favre, Q. T. Nguyen, S. Bruneau, "Extraction of 1-butanol from aqueous solutions by pervaporation", *J. Chem. Tech. Biotech.*, 65 (1996) 221-228.
29. S. Mishima, T. Nakagawa, "Plasma-grafting of fluoroalkyl methacrylate onto PDMS membranes and their VOC separation properties for pervaporation", *J. Appl. Polym. Sci.*, 73 (1999) 1835-1844.
30. T. Uragami, T. Ohshima, T. Miyata, "Removal of Benzene from an Aqueous Solution of Dilute Benzene by Various Cross-Linked Poly(dimethylsiloxane) Membranes during Pervaporation", *Macromolecules*, 36 (2003) 9430-9436.
31. T. Miyata, H. Yamada, T. Uragami, "Surface Modification of Microphase-Separated Membranes by Fluorine-Containing Polymer Additive and Removal of Dilute Benzene in Water through These Membranes", *Macromolecules*, 34 (2001) 8026-8033.
32. M. S. Spencer, "Carbonate production", *US patent*, 4582645.

Poly(dimethyl siloxane)/Poly(vinylidene fluoride) composite membrane for separation of *n*-butane/nitrogen mixture

Sheng Lu [1],Jianhua Cao, Jiding Li [1,] ,Yangzheng Lin[1], Cuixian Chen [1]*

[1]Membrane Technology & Engineering Research Center, Department of Chemical Engineering, Tsinghua University, Beijing 100084, P. R. China

1. Introduction

Membrane-based gas separation techniques have developed intensively during recent decades. This field is believed to be one of the main areas of growth for membrane technology [1].Membranes made by coating rubbery polymers on polymeric substrates have been used for separating volatile organic compounds(VOC's) from permanent gases in many commercial applications. These processes have recovered high-value hydrocarbons resulting in substantial savings of energy and raw materials. Membrane Technology Research (MTR), USA, GKSS, Germany and Dalian Institute of Chemical Physics (DICP), P.R. China, have successfully implemented many applications in a wide range of industries. These applications for recovering hydrocarbons and recycling permanent gas in the petroleum and polymer industries have mostly operated at sub-ambient temperature by solubility-selective polymeric membranes [2–7].Currently, PDMS is the most commonly used rubbery membrane material due to its preferential selectivity for hydrocarbons. Its glass transition temperature is the lowest amongst polymers (-129□), indicating a very flexible polymer backbone with long-range segmental motion, which is sufficiently active, even at very low temperatures [8,9]. The permeation properties of thick PDMS films have been evaluated in laboratory for the characteristics of various pure gases,

* Corresponding author, Tel.: +86-10-62782432; Fax: +86-10-62770304; E-mail: lijiding@mail.tsinghua.edu.cn

hydrocarbons and some hydrocarbons/permanent gas mixtures over a temperature range of 20–95℃ [10–15]. There are very few research articles dealing with the permeation of pure gases and especially gaseous mixtures through ultra thin asymmetric membranes for industrial separation purposes [16]. Recently Jiang and Kumar [17] reported that propylene and propane significantly plasticized the thin PDMS coating on polysulfone membrane at ambient temperature. Pfromm et al. [18] compared asymmetric polysulfone, polycarbonate and poly(ester carbonate)/PDMS membranes with their isotropic films, showing different plasticization and conditioning behavior for CO_2 permeation due to the fact that the skin layer morphology may contain a different distribution of free volume in asymmetric membranes. Wessling et al.[19] found that the plasticization behavior of polyimide/PDMS composite membranes is also thickness-dependent. Pinnau and He [20] recently reported the permeation properties for a multi component mixture of hydrogen, methane, ethane, propane and *n*-butane in a 150 μm thick, homogenous PDMS film for the temperaturerange of -20 to 35 □. The swelling, induced mainly by propane and *n*-butane sorption, significantly influenced the permeabilities and selectivities of each component in the mixture. An increase in penetrant diffusivity was believed to occur from increased polymer local segmental motion caused by the presence of condensable penetrant molecules in the polymer matrix at a constant temperature. As penetrant pressure and, therefore, the penetrant concentration in the polymer increases, the degree of swelling in a polymer matrix increases for those strongly sorbing penetrant, leading to significant increases in selectivities particularly with the decrease of temperature. It is apparent that there are advantages of combining the condensable pretreatment unit running at sub-ambient temperatures with membrane modules for the removal of extra hydrocarbons from feed gas. This approach could take the benefits of increased permeance and selectivity of hydrocarbons to reduce the required membrane area for the separation process. Moreover, residual gas could also be maintained at the pressure for recycling to the system. However, there is no published literature on the permeances and selectivities of C_2 and C_3 hydrocarbons and nitrogen mixturesusing the thin silicone coated polysulfone composite membranes at sub-ambient temperatures for designing such a process. For commercial applicability, the membrane must exhibit both high permeability to the desired gas (or the component of a gas mixture and high level of selectivity (or separation factor). The first demand requires the polymer films to be sufficiently thin, the second requires specific structure and composition of an active layer to provide desired selective properties. The present work reports permeation performances of *n*-butane and nitrogen through composite PDMS–PVDF membrane at a temperature range of 20–60 □.

2. Experimental

2.1 Materials

PDMS (107, viscosity 10000 Pa s), tetraethylorthosilicate(TEOS) , vinyltrimethoxysilicane(VTMS), triethoxyphenylsilane(TEPS), dibutyltin dilaurate (DBTL)were purchased from Beijing Chemical company. Polyvinylidene fluoride (PVDF, Solef®1010, average Mw211 000) was supplied from Solvay. Dimethyl acetamide (DMAc) and n-hexane were of analytic grade and obtained from the Beijing chemical reagent company. Deionized water was used as the nonsolvent for the PVDF. Nitrogen and propane (purity, 99.99%) were come from Beijing Huayuan Gas Chemical industry Co., Ltd. All chemicals were used as received.

2.2 Preparation of PVDF microporous membrane

The PVDF porous membranes were synthesized by means of the phase inversion process. Specific amount of PVDF was dissolved in DMAc by mechanical rotating the viscous solution at 60 ℃. After 24h, the solution was filtrated and degassed under vaccum condition. The transparent and viscous solution was obtained. The solution was cast on a polyester non-woven fiber using a casting knife set a thickness of 180μm. After 15s, the polyester non-woven fiber was immersed in deionized water (21□). Within a few minutes, solidification was observed and the membrane was removed from the coagulation bath after a certain time. The membrane was soaked in the deionized water every 2h interval. The membrane was dried in oven at 50□. The white membrane was obtained and the thickness was about 60 μm.

2.3 Preparation of PVDF/PDMS composite membrane

PDMS, cross-linking agent and catalyst dibutyltin dilaurate(DBTL) were dissolved in n-heptane by magnetic stirrer at ambient temperature. After degassed under vacuum, the solution was cast onto the PVDF membrane. The membrane was first vulcanized under room temperature to evaporate the solvent, and then introduced into a vacuum oven to complete crosslinking. The thickness of the top skin layer could be determined by means of SEM photographs. All experiments in this study were performed with the same membrane.

2.4 Characterization of composite membrane

2.4.1 SEM

Scanning electron microscopy was used to study the cross sectional morphology of the composite PDMS membrane and to measure the skin layer thicknesses. The composite PDMS membranes were fractured in liquid nitrogen. The fractured section was coated with a conductive layer of sputtered gold. The cross-section of the composite membrane was investigated using a Hitachi S-800 SEM.

2.4.2 FTIR

The chemical structure of PDMS pre-polymer and PDMS was studied by FTIR (Fourier transform infrared pectra) spectrometer (Nicolet, IR560). FTIR spectra were recorded within the range of 4000–400cm^{-1}.

2.5 Gas permeation test

The experimental setup is schematically shown in Figure 1. The feed gas at a given composition was obtained by two mass flow controllers and the gas composition was determined using a gas chromatograph (GC-14C, Shimadzu) equipped with a TCD. The effective membrane area for permeation was 22.55 cm2. The membrane was cut into circular coupons and fitted in the permeation cell. The permeation cell consisted of two detachable stainless steel parts. A sintered porous stainless steel plate was imbedded in one part of the cell to support the membrane. Double O-rings were used to provide a seal between the membrane and the cell. The residue flow was controlled such that the stage cut was always below 0.05. The pressure of the feed side was controlled by means of a needle valve, while the pressure on the permeate side was kept atmospheric. The permeate flow rate was measured by a bubble flow meter and the permeate composition was analyzed using the gas chromatograph. The permeability coefficient of a gas component i across the membrane, J_i, was obtained using the following equation:

$$J_i = \frac{Qy_i}{A\Delta P_i} 1 \tag{1}$$

where J_i is the permeability coefficient in Barrers (1 Barrer = 10−10 cm3 (STP) cm cm−2 s−1 cmHg−1), Q the overall gas permeation rate, A the effective area of the membrane (cm2), and ΔP_i is the partial pressure difference of component i across the membrane. $\Delta P_i = P_h x_i - P_l y_i$, where P_h and P_l are the gas pressures in the feed and permeate, respectively, and x_i is the mole fraction of component i on the feed side.

The perselectivity is given by the permeability ratio of the two components

$$\beta = \frac{J_A}{J_B} \tag{2}$$

Where subscript A represents the faster permeating component. The separation factor of the gas mixtures was calculated from

$$\alpha = \frac{y_A / y_B}{x_A / x_B} \tag{3}$$

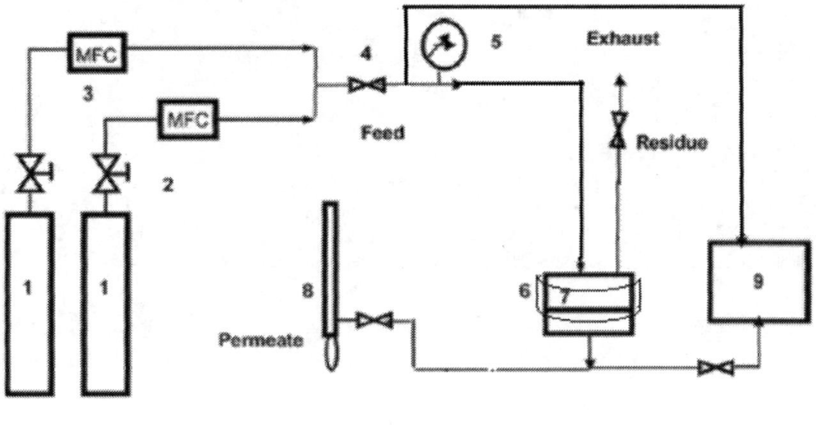

Figure 1 Schematic diagram of experimental setup for vapor permeation: (1) gas cylinder, (2) pressure regulator,(3) mass flow controller, (4) vavle, (5) pressure gauge, (6) heating tape (7) membrane cell, (8) bubble flow meter, (9) gas chromatograph.

3. Results and discussion

3.1 Structure of organic vapor separation membrane

3.1.1 Morphology

Figure 2 was the surface and cross-section SEM images of PDMS/PVDF composite membrane with TEPS as cross-linking agent. It is shown that the surface of composite membrane is very smooth and dense even the 40000 magnification. The composite membrane consists of dense PDMS thin film layer and microporous PVDF support layer. The thickness of the dense layer is about 24μm.

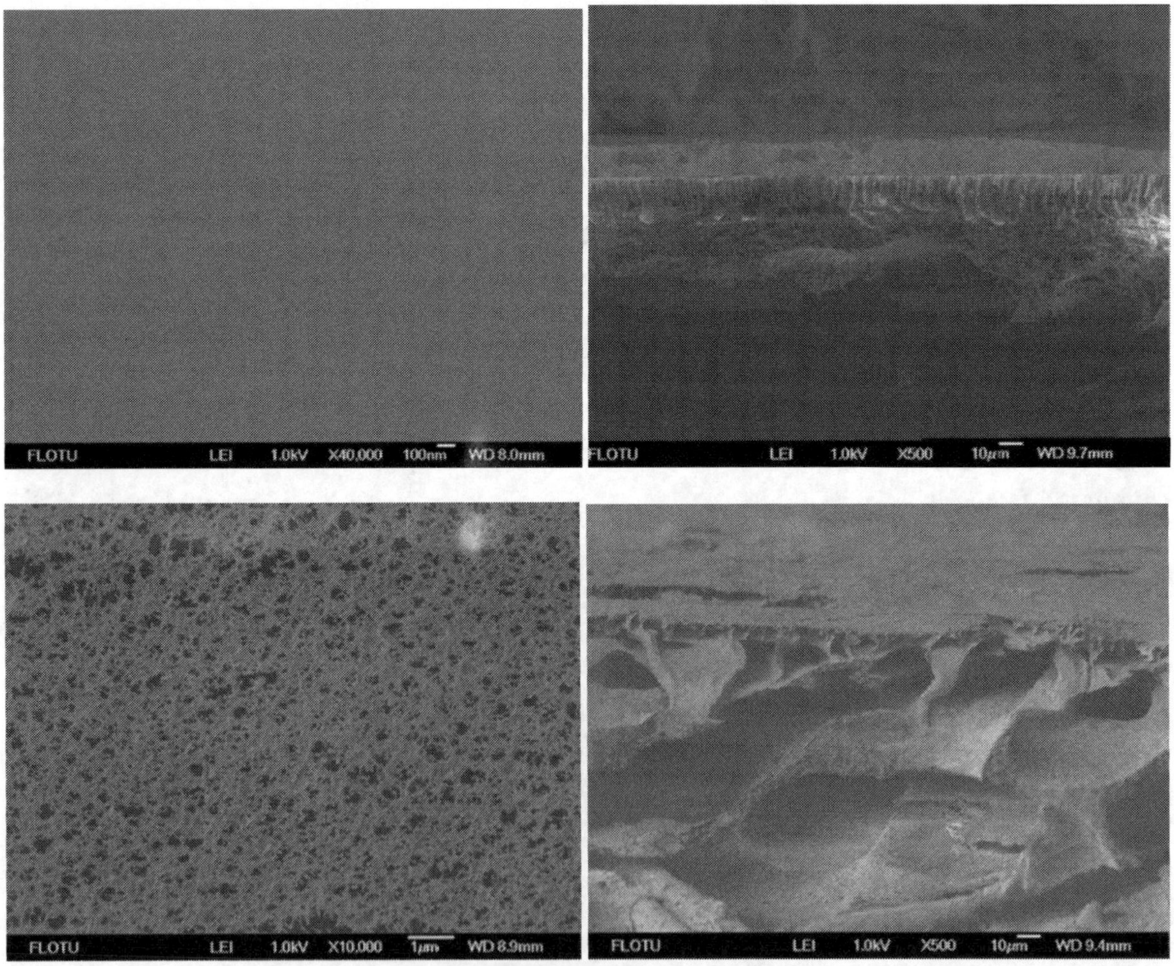

Figure 2 (Up)The surface (left) and cross-section (right) SEM images of PDMS/PVDF composite membrane with TEPS as cross-linking agent.(Down) The surface (left) and cross-section (right) SEM images of PVDF microporous membrane

3.1.2 FTIR analysis

As shown in figure 3, the characteristic peaks of Si-OH vibration are between 3000—2890cm^{-1}and 1150—1000cm^{-1}. It can be seen in figure 3 that the peaks in these areas weakened apparently. It proved that crosslinking occurred.

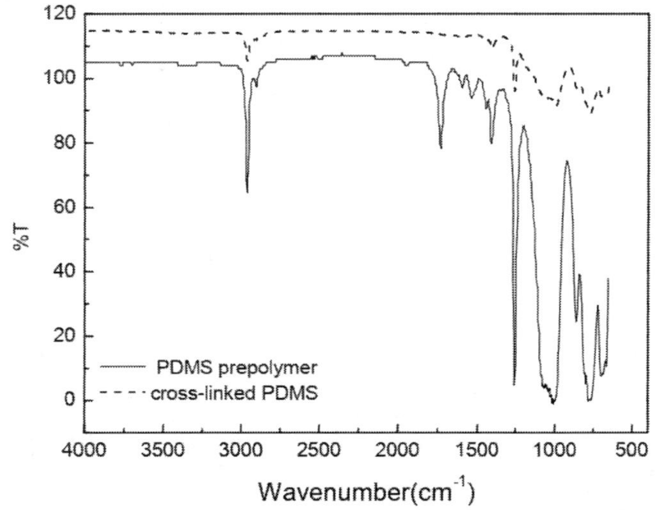

Figure 3 FTIR spectrum of PDMS pre-polymer and PDMS

3.2 The gas permeation

In this work, The PDMS/PVDF composite membranes were fabricated by casting a thin layer of PDMS on microporous PVDF substrates. The PDMS was cross-linked with three different solidified agents. Figure 3 shows the permeability coefficient of *n*-butane and N_2 and the permeability ratio as a function of the pressure difference. The permeability coefficient of n-butane and permeability ratio of n-butane and N_2 increased with the pressure difference increasing, however, the permeability coefficient of N_2 was independent of pressure variation for the three composite membranes. The permeability coefficient and the permeability ratio is a<b<c in the studied range. When the TEPS was used as cross-linking agent, the permeability coefficient of *n*-butane is 15500 Barrer and the permeability ratio was 81 at 0.4MPa pressure difference and 30□。

Figure 3 Permeability performance of pure propane(A) and nitrogen (B) and the permeability ratio of C3H6/N2 (C) through different cross-linked PDMS/PVDF composite membrane. (a, TEOS; b, VTMS; c, TEPS). Operating temperature:30℃

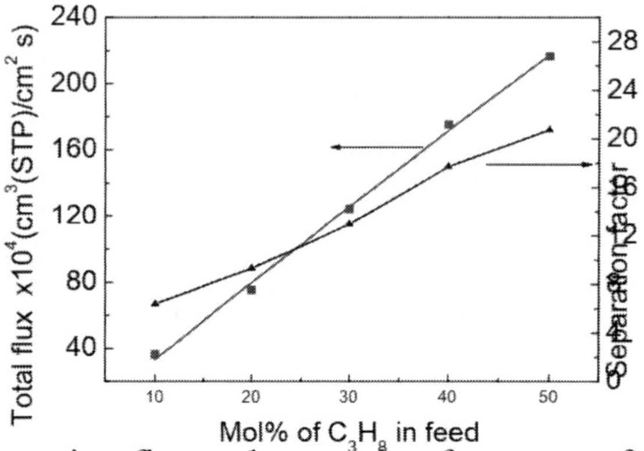

Figure 4 Total permeation flux and separation factor as a function of Mol% of C3H8 in the feed. Operating temperature 30℃, feed pressure 0.5MPa.

4. Conclusions

The TEPS is the preferential to prepare PDMS film in three investigated cross-linking agents. The permeability coefficient of *n*-C4H10 is 15500 Barrer and the permeability ratio was 81 at 0.1MPa pressure difference and 30℃. The Total permeation flux is 125cm3(STP)/cm2 •s and the separation factor is 13 when the mole percent of *n*-C4H10 in the gas mixture is 30 at the 0.1MPa feed pressure and 30□.

Acknowledgment

Financial support for this work was provided by China Postdoctoral Science Foundation (20060400436), the Major State Basic Research Program of China (No. 2003CB615701), National Natural Science Foundation of China (No. 20576059, No.20676067), SINOPEC Foundation (No. X505002) and CNPC Innovation Foundation (No. 05051143).

References

[1] D.R. Paul, Y.P. Yampol'skii, Polymeric Gas Separation Membranes, CRC Press, Boca Raton, FL, (1994).
[2] S.I. Semenova, Polymer membranes for hydrocarbon separation and removal, J. Membr. Sci. 231 (2004) 189.
[3] A. Jonquieres, R. Clement, P. Lochon, J. Neel, M. Dresch, B. Chretien, Industrial state-the-art of pervaporation and vapour permeation in the western countries, J. Membr. Sci. 206 (2002) 87.
[4] H. Strathmann, Membrane separation processes: current relevance and future opportunities, AIChE J. 47 (5) (2001) 1077.

[5] R.W. Backer, Future directions of membrane gas separation technology, Ind. Eng. Chem. Res. 41 (2002) 1393.
[6] T.K. Poddar, K.K. Sirkar, A hybrid of vapor permeation and membrane based absorption-stripping for VOC removal and recovery from gaseous emissions, J. Membr. Sci. 132 (1997) 229.
[7] R.W. Baker, J.G. Wijmans, J.H. Kaschemekat, The design of membrane vapor-gas separation system, J. Membr. Sci. 151 (1998) 55.
[8]R.W. Baker, N. Yoshioka, J.M. Mohr, A.J. Khan, Separation of organic vapors from air, J. Membr. Sci. 31 (1987) 259.
[9]B. Freeman, I. Pinnau, Separation of gases using solubility-selective polymers,TAIP 5 (1997) 167.
[10]T.C. Merkel, V.I. Bondar, K. Nagai, B.D. Freeman, I. Pinnau, Gas sorption, diffusion, and permeation in poly(dimethylsiloxane), J. Polym. Sci. Polym.Phys. 38 (2000) 415.
[11]M. Leemann, G. Eigenberger, H. Strathmann, Vapor permeation for the recovery of organic solvents from waste air streams: separation capacitiesand process optimization, J. Membr. Sci. 113 (1996) 313.
[12]C.K.Yeom, S.H. Lee, H.Y. Song, J.M. Lee,Vapor permeations of a series of VOCs/N2 mixtures through PDMS membrane, J. Membr. Sci. 198 (2002)129.
[13]S. Majumdar, D. Bhaumik, K.K. Sirkar, Performance of commercial size plasma polymerized PDMS-coated hollow fiber modules in removingVOCs from N2/air, J. Membr. Sci. 214 (2003) 323.
[14]A. Singh, B.D. Freeman, I. Pinnau, Pure and mixed gas acetone/nitrogen permeation properties of polydimethylsiloxane [PDMS], J. Polym. Sci. Polym. Phys. 36 (1998) 289.
[15]C.K. Yeom, S.H. Lee, J.M. Lee, Study of transport of pure and mixed CO2/N2 gases through polymeric membranes, J. Appl. Polym. Sci. 78(2000) 179.
[16]G.C. Kapantaidakis, G.H. Koops, M. Wessling, CO2 plasticization of polyethersulfone/polyimide gas-separation membranes, AIChE J. 49(2003) 1702.
[17]X. Jiang, A. Kumar, Performance of silicone-coated polymeric membrane in separation of hydrocarbons and nitrogen mixtures, J. Membr. Sci. 254(2005) 179.
[18]P.H. Pfromm, I. Pinnau, W.J. Koros, Gas transport through integral asymmetric membranes: a comparison to isotropic film transport properties,J. Appl. Polym. Sci. 48 (1993) 2161.
[19]M. Wessling, M. Lidon Lopez, H. Strathmann, Accelerated plasticizationof thin-film composite membranes used in gas separation, J. Membr. Sci.24 (2001) 223.
[20]I. Pinnau, Z.J. He, Pure- and mixed-gas permeation properties of polydimethylsiloxanefor hydrocarbon/methane and hydrocarbon/hydrogenseparation, J. Membr. Sci. 244 (2004) 227.

On the Use of Recycle in Reverse Osmosis Method for Desalination of Sea Water

Dr. Kal Renganathan Sharma PE
Adjunct Professor, Department of Chemical Engineering
Prairie View A & M University, PO Box 519; MS 2505
A. E. Greaux Building, Room 206C
Prairie View, TX 77446-0519
Tel: (936) 261 9413
Fax: (936) 261 9419
Email: jyoti_kalpika@yahoo.com

(Prepared for Presentation at the 99th AIChE Annual Meeting, Salt Lake City, UT, Nov' 4th – Nov' 9th 2007)

Abstract

The osmotic pressure is the primary contributor of the high pressure needed to desalinate sea water at a 3.6 % by weight concentration during the reverse osmosis operations using semi-permeable membranes with a good balance of salt rejection rates and water flux rates. Examples of such membranes are aromatic polyamide and thin film composite membranes. Compared with solutions of starch, sugar for the salt solutions the solute molecular size is small. On account of this, the osmotic pressure calculated by equating the fugacities of the two phases assumed to be in equilibrium on either side of the membrane increases with decrease in solute size exponentially. i.e.,

$$Pi = (Pw - Ps) = -RT/V_w \ln(X_w)$$

Thus for NaCl the osmotic pressure at the high pressure pump needed was found to be 16.0 atm compared with starch or sugar where it is a lot lower. The use of recycle was evaluated as shown in Figure 5.0. For example, 63 lakh litres of water at a salt concentration of 100ppm was generated from a feed with a salt concentration of 3.6 wt % NaCl. The size of the pump needed was 16 atm. Inorder to reduce the load of the high pressure pump recycling the treated water with the feed water was considered. The actual feed into the RO module will be lower in solute concentration, which will decrease the osmotic pressure need to be overcome (Figure 1.0). For a recycle ratio of 3.3 the pressure at the pump was found to be 3.7 atm. The residue to product ratio is 0.5. 82% reduction in pump size needed can be achieved in this example.

Introduction

Liquid solutions can be separated into solute and solvent using the methods of *Reverse osmosis*, ultrafiltration, molecular sieves and electrodialysis. The difference between the methods of Reverse osmosis and ultrafiltration differ in the solute size rejected. The conventional filtration reject particles of the size of 10 microns and above, the *ultrafiltration* rejects solute sizes of 10 nm to 10 microns and the reverse osmosis rejects solute sizes in the ionic range of 1 pico meter to 0.1 nm. The driving force for RO, ultrafiltration and molecular sieve operations come from a pressure difference. No filter cake is allowed to form. The driving force in *electrodialysis* is an induced electric field and polarization of the ions to the anode and cathode. The compartments are made from alternating anode and cathode and the ions get segregated from the diluate water in alternating compartments. It is used to desalinate brackish water and is found to be the most economical method at low salt concentration in the feed at little over 1 wt%. Any material that can exclude molecular species by size is referred to as *molecular sieve*. Inorganic materials that possess uniform pores with diameters in either less than 2 nm, micro- range or 2-20 nm, meso range.

In *reverse osmosis* the solvent from the solution is pumped across a semi-permeable membrane opposing the osmotic pressure difference with the solute largely rejected by the membrane. (Figure 1.0). For example in sea-water desalination by reverse osmosis, the osmotic pressure will cause a flow from the region of low solute concentration to a region of higher solute concentration. Pressure is supplied in order to overcome this, so that the solvent from the sea water flows across a semi-permeable membrane that permits only the solvent and not the solute. Although the membrane rejects the solute, some will diffuse across the pores.

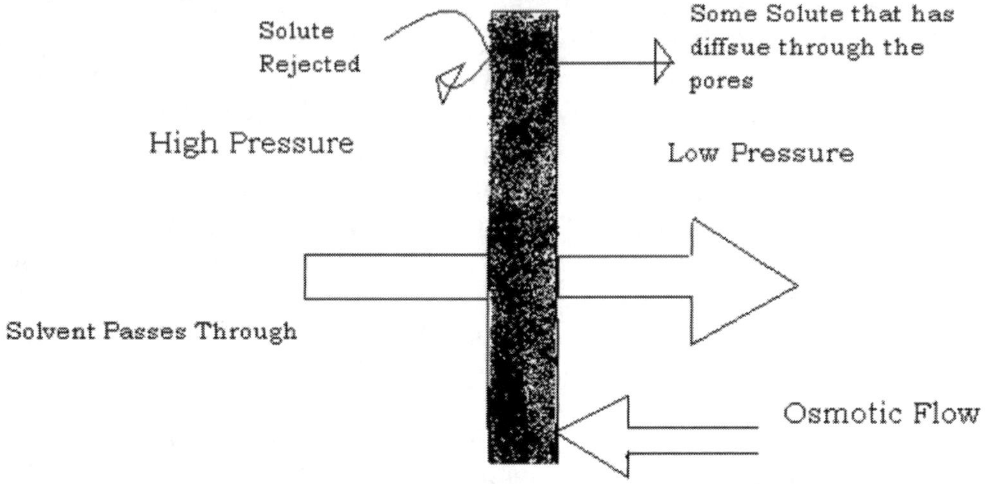

Figure 1.0 Transport Processes during Reverse Osmosis

The drinking water needs of major cities around the world can be met from the reverse osmosis method for desalination as the drought conditions in major cities have become a important concern. In the middle-east gulf region where the energy is cheap and river water scarce, sea water reverse osmosis, SWRO plants are increasingly relied upon to obtain potable drinking water (Figure 1.0) Reverse osmosis is also used in the food industry to concentrate fruit juices. Reverse osmosis is dominated by desalination that is by removing salt from water. It is used in the treatment of seawater, brackish water and reclamation of municipal waste water. It is used in boiler feed water pretreatment in order to avoid scale formation later on. The market for ultra pure water is growing and is used in injectable pharmaceuticals and for semi-conductors, water for domestic use, sweetener concentration, and fruit juice concentration and fermentation product recovery.

In the 1950s it was first shown that cellulose acetate RO membranes were capable of separating salt from water. But the water fluxes obtained were small to be feasible in practice. Since then the RO membrane technology has improved a lot. Recently du Pont patented a polyamide membrane made from interfacial polymerization and has a high solvent flux and turbulence is used to reduce the concentration polarization layer thickness formed during the operation of the RO during desalination. Advances in thin-film composite membranes and polymer materials have widened the applications of RO from desalination to treatment of hazardous wastes, material recovery in electroplating industries, production of ultra pure water, water softening and food processing, dairy and semi-conductor industry.

The advantages of these systems over traditional separation processes such as distillation, extraction, ion exchange and adsorption are because RO is a pressure driven process no energy intensive phase changes, or use of expensive solvents or adsorbents are needed. RO processes have simple design and easy to operate compared with other traditional methods.

Membrane properties play a pivotal role in the performance of RO technology. These properties depend on the chemical structure of the membrane. An ideal RO membrane is low in cost, resistant to chemical and microbial attack, posses high mechanical and structural stability over long period of operation, and wide range of temperature, and has the desired separation characteristics for the given system.

RO membranes are classified into asymmetric membranes (containing one polymer) and thin film composite membranes. *Asymmetric RO membranes* have a thin permselective skin layer of about 100 nm thickness supported on a more porous sublayer of the same polymer. The dense skin layer determines the fluxes and selectivities of these membranes. The porous layer serves as a mechanical support for the skin layer. Asymmetric membranes are formed by phase inversion polymer precipitation process. In this process, a polymer solution is precipitated into a polymer rich solid phase that forms the membrane and a polymer poor liquid phase that forms the membranes or void spaces. *Composite RO and ultrafiltration membranes* are thin films consisting of a thin polymer barrier layer formed on one or more porous support layers which is a different polymer compared with the barrier layer. The surface layer determines the flux and separation characteristics of the membrane. The porous backing is intended largely for the support for the barrier layer. The barrier layer is extremely thin thus allowing high water fluxes. The most important thin-film composite membranes are made by interfacial polymerization a process in which a highly porous membrane such as polysulfone is coated with an aqueous solution of a polymer or monomer and then reacted with a cross-linking agent in a water-immiscible solvent.

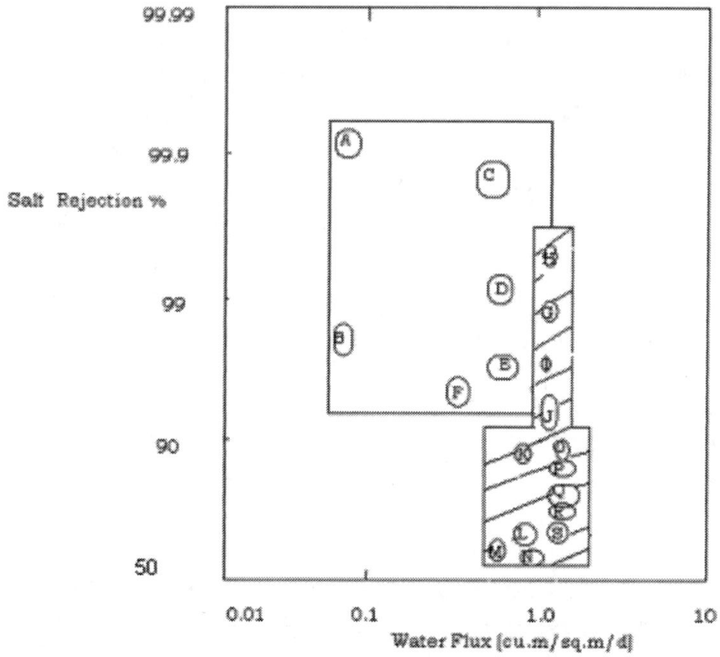

Figure 2.0 Water Flux and Salt Rejection for Various SWRO Semi-Permeable Membranes (A-F – Sea Water Membranes -operate at 5.5 Mpa and 25 ^{0}C. H-J Brackish water membranes – operate at 1500 mg/L NaCl feed, 1.5 Mpa and 25 ^{0}C. K-S – nanofiltration membranes – operate at 500 mg/L NaCl feed, 0.74 Mpa and 25 ^{0}C.

A- cellulose acetate/triacetate
B- linear aromatic polyamide
C- cross-linked polyether
D- cross-linked fully aromatic polyamide
I- Du Pont A-15
J,K, L-Nitto-Denko, NTR-739HF NTR-729HF, NTR-7250
O- Toray, UTC-40HF
E – other thin film composites
F -asymmetric membranes
G – FilmTec, BW 30
H – Toray, SU –700
Q- Toray, UTC 60
R –Toray, UTC 20HF
S – Filmtec, NF50
P- Filmtec, NF70

The most important RO membrane materials are, linear and cross-linked aromatic polyamide, aryl-alkyl polyetherurea and cellulose acetate/triacetate. Cellulose acetate materials hydrolyze and exhibits compaction upon application of higher pressure thus reducing the water flux and have a narrow temperature and pH application range. Polyamide and polyurea composite membranes have higher water fluxes and solute rejections (Figure 2.0) and operate at higher temperature and pH range and better resistance to biological attack, and mechanical compaction. These membrane are less resistant to chlorine attack.

The membrane module needs to offer mechanical support to the RO membrane upto 8 Mpa, offer low pressure drop, and be resistant to fouling and concentration polarization. The commercially available modules are plate and frame, tubular, spiral wound and hollow fiber elements. For large scale desalination plants, spiral wound modules are used. The RO process (Figure 4.0) is simple in design. It consists of feed water source, fed pretreatment, high pressure pump, RO membrane modules, and in some cases post-treatment steps.

In order to write the flux equations for reverse osmosis in a precise manner, three phenomena need to be accounted for (Figure 6.1). These are;

1) Osmotic Pressure
2) Permeability for Solvent Across Membrane and Starling's Law
3) Diffusion of Solute across the Membrane

In order to better quantitate the above concepts the systems in the human anatomy, the solute transport, plasma filtration and the relevant transport processes across the cell membrane are discussed below. The phenomena of osmotic pressure need to be understood. In SWRO, sea water reverse osmosis processes the osmotic pressure drop needs to be quantitated with the correct sign and added with the hydraulic pressure drop for reliable design of these systems. The concept of osmotic pressure is illustrated in Figure 6.3. When a balloon made out of a semi-permeable membrane filled with salt solution is immersed in a bath of pure water, the water will travel from the jar into the balloon and the size of the balloon will increase until equilibrium is reached in terms of the chemical potential on both sides of the membrane. The semi-permeable membrane can be chosen in a manner that it permits only water and not the solute in large measure. The flow of water is an example of the concept of osmotic pressure. *Osmosis* is the flow of solvent from a region of low solute concentration to a region of high solute concentration. The pressure difference that causes this flow is called as osmotic pressure. This pressure is caused by the presence of solutes. Hence it is called *colloid osmotic pressure*. For the human plasma in the blood the colloid oncotic pressure is about 28 mm Hg. The colloid osmotic pressure is small compared with the osmotic pressure developed when a human cell is placed in pure water. The total osmotic pressure of the intracellular fluid would be 5450 mm Hg at 37 ^{0} C.

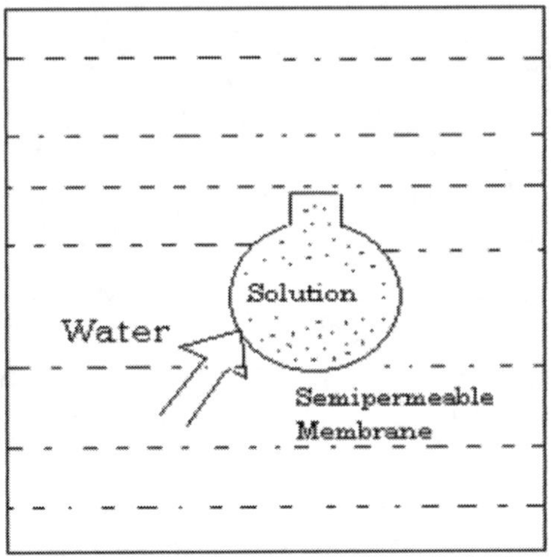

Figure 3.0 Concept of Osmotic Pressure

This is caused by the presence of solutes such as K^+ ions, phosphocreatine, Mg^{++}, Cl^-, HCO_3^-, HPO_4^{2-} ions, carnosine, amino acids, creatine, lactate, Na^+ ions, urea, ATP, hexose monophosphate, and others. A number of solute molecules contribute to the osmotic pressure. Some of them are dissociating. For example, NaCl dissociates to the Na^+ and Cl^- ions. Each ion particle exerts its own osmotic pressure. The charge of the ion has no bearing on the osmotic pressure. Substances such as glucose do not dissociate and the osmotic pressure exerted is based upon its concentration. The term *osmole* is introduced to account for the effect of a dissociating solute. One osmole is therefore defined as one mole of a nondissociating substance. One mole of dissociating NaCl is equivalent to two osmoles. *Osmolarity* defines the number of osmoles per liter of solution. For physiological solutions the units used for convenience, is mOsM. If a cell is placed within a solution that has a lower concentration of solutes or osmolarity, then the cell is in a *hypotonic solution* and establishment of osmotic equilibrium requires the osmotic flow of water into the cell. The influx of water into the cell results in swelling of the cell and a subsequent decrease in its osmolarity. On the other hand if the cell is placed in a solution with a higher concentration of solutes or osmolarity, i.e., *hypertonic solution*, then osmotic equilibrium requires osmotic flow of water out of the cell. An *isotonic solution*, is a fluid that has the same osmolarity of the cell. When cells are placed in an isotonic solution there is neither swelling not shrinkage of the cell. Examples of isotonic solutions are 0.9 % by weight NaCl in water solution and 5 % by weight of glucose solutions with respect to a human cell.

Vant' Hoff's law to determine the osmotic pressure in terms of the concentration of the solution can be derived from the concept that fugacity of different phases needs to be equal at equilibrium. Thus at equilibrium, in Figure

3.0 the temperature and fugacity of the water and that of the solution must be equal. Fugacity is a measure of the chemical potential of the system. As discussed above the difference in the chemical potential between the solution and solvent causes the osmotic flow from a region of low solute concentration to a region of high solute concentration.

$$f_w(T, P_w) = f_s(T, P_s) \quad (1.0)$$

The fugacity of solution can be written in terms of the pure component fugacity using the Poynting factor. The mole fraction of water and the activity coefficient of water are also needed. The Poynting factor corrects for the effect of pressure on the pure component fugacity where V_w is the molar volume.

$$f_w = \gamma_w x_w f_w \exp(-V_w(P_w - P_s)/RT \quad (2.0)$$

The osmotic pressure, is given by $(P_s - P_w) = \pi$ and can be solved for from Eq. [6.2] as;

$$\pi = (P_w - P_s) = -RT/V_w \ln(\gamma_w x_w) \quad (3.0)$$

For an ideal solution, the activity coefficient may be taken as 1. For dilute systems using Taylor series expansion the logarithmic functionality can be approximated as;

$$\ln(x_w) = \ln(1 - x_s) \approx -(x_s) \quad (4.0)$$

Eq. (3.0) becomes on substituting for Eq. (4.0) in Eq. (3.0);

$$\pi = RTx_s/V_w = RTC_s \quad (5.0)$$

Eq. (5.0)is called the van't Hoff's law to determine the osmotic pressure. Eq.(3.0] may be used where activity coefficient information is available. If the solution contains N ideal solutes, then the osmotic pressure can be obtained as a sum of the contributions from each solute;

$$\pi = RTx_s/V_w = RT\sum_{1}^{N} C_{si} \quad (6.0)$$

The number of molecules and not the absolute weight of the solute determine the osmotic pressure of the solution. The combined effect of osmotic pressure and hydrostatic pressure can be seen in ***Starling's Law*** which gives the relation between the flow of fluid across the capillary wall or a porous membrane and the pressure difference across the capillary. The volumetric fluid transfer rate, J, across the capillary membrane is given as;

$$J/L_pS = \Delta P_h - \Delta\pi = \overline{\Delta P_h} \qquad (7.0)$$

Where the effective pressure, $\overline{\Delta P_h}$ is the resultant ant of the hydrodynamic pressure, drop, ΔP_h, and osmotic pressure difference, $\Delta\pi$. L_p is the hydraulic conductance and S is the effective cross-sectional area through which the fluid flows. The hydraulic conductance is often determined by experiment. The hydraulic conductance varies from 1 E-9 m^2s/kg in capillaries in the kidney's glomeruli to 1 E-14 m^2s/Kg for endothelial cells found in the capillaries of the rabbit brain. One use of Eq. (7.0) is to better understand the flow of plasma across the capillary wall in human anatomy. This can be used in the SWRO systems used to desalinate sea water to drinking potable water. The hydraulic conductance can also be derived from the properties of the system. Thus should the membrane be viewed as a series of parallel cylindrical pores,

$$L_p = (r^2A_p/8\mu t_m S_c) \qquad (8.0)$$

Where A_p is the cross-sectional area of the pore of radius r in the capillary wall, S_c is the circumferential area, t_m is the wall thickness. The ratio A_p/S_c is the porosity of the capillary wall, ε.

Often times the solvent moving across the membrane will carry with it the solute molecules. Some molecules will be filtered on account of their large size. Even when the membrane is semi-permeable some solute diffusion will take place. The solute separation on account of the size can be accounted for by the introduction of the sieving coefficient, S_a. The sieving coefficient is defined as the ratio of the solute concentration in the filtrate, C_{sp} to the solute concentration of the feed solution, C_{sf}. Theoretical expressions based on the motion of a spherical solute moving through a cylindrical pore have been developed based on the motion of a spherical solute moving through a cylindrical pore have been developed to estimate the value of the sieving coefficient (Deen, 1987).

$$S_a = C_p/C_f = (1-\lambda)^2(2-(1-\lambda)^2)(1-2/3\lambda^2-0.163\lambda^3) \qquad (9.0)$$

Where λ is the ratio of the solute radius, a, to the capillary pore radius, r. *Concentration polarization* refers to the formation of a coat of retained solutes on the feed side of the membrane.

The solute diffusion across the membrane can be treated with the Fick's laws of diffusion at steady state and the generalized Fick's laws of diffusion in for transient applications. Different diffusivities, such as pore diffusion, diffusion in polymeric systems, convective effect can be added together. The size of the solute can be estimated from the Stokes Einstein relation. The diffusivity of the solute in the liquid needs to be known. If the diffusivity is not known, the solute size can first be estimated from the following equation that is written assuming that the solute of molecular weight, M_w is a sphere with a density of approximately 1 gm/cm^3 and is the same as that of the solute in the solid phase.

$$a = (3M_w/4\pi\rho A_N)^{1/3} \qquad (10)$$

Steric exclusion and hindered diffusion can also be accounted for in the expressions for diffusion of the solute depending on the problem at hand. Steric exclusion refers to the problem that only the volume in the pores other than that of the solute is available for diffusion. The hydrodynamic drag experienced by the solute is referred to as the hindered diffusion. The Renkin equation (1954) gives the ratio of the pore diffusivity to that of the bulk diffusivity. The solute flux can be written as;

$$J_s'' = DH/l\,(C_{sf} - C_{sp}) \qquad (11)$$

$$J_s = P_m S(C_{sf} - C_{sp}) \qquad (12)$$

Where P_m is the permeability of the solute. The combined effect of hydraulic pressure and osmotic pressure and the solute flux can be written from the application of irreversible thermodynamics (Kedem and Katchalsky, 1958). The cross coefficients that are from secondary effects are equal according to the Onsager relationships. The relative flow between the solvent and solute is capable of providing a separation of the solute and solvent. The sieving mechanism is called *ultrafiltration* and when the solute permeability is low it is referred to as *reverse osmosis*.

$$J = SL_p(\Delta P - \sigma RT\Delta C) \qquad (13)$$

$$J_s = SL_p(-\sigma(\Delta P + L_s/L_pRT\Delta C) \qquad (14)$$

The parameter $\sigma = -L_{sp}/L_p$ is called the ***Staverman reflection coefficient***. If the membrane is permeable to solvent and not to the solute, $\sigma = 1$. When $\sigma = 0$, the membrane is equally permeable to both solvent and solute. Eq. (13) gives the flux of solvent across the semi-permeable membrane and Eq.(14) gives the flux of solute.

Worked Example 1.0 SWRO Plant

Prepare a preliminary estimate of a SWRO plant where the inlet NaCl composition is 3.6% by weight and the outlet composition is 100 ppm. A typical RO process schematic is given in Figure 6.4. What ought to be the size of the pump. The RO membrane used is a linear aromatic polyamide. The expected water supply for the township is 63 lakh liters per day. Spiral wound modules are used with interfacial area of 5 m^2/gm How much membrane is needed ? The operating temperature is 25 ^{0} C.

Volumetric flow of water = 63 E 5 x 1 E-3/24/3600 = 0.07292 cu.m/s

Water flux for linear aromatic polyamide

(from Figure 6.2, Salt Rejection vs. Water Flux Diagram) = 0.08 cu.m/m^2/d

= 9.2593 E-7 cu.m/m^2/s

Amount of membrane needed = 0.07292/9.2593E-7/5000 = 1575 gm

= 605.77 cm^3

Radius of the cylindrical pores in the membrane module ,

$r = 2V_m/a$ = 605.8 cm^3

= 24.2 μm.

The effective length of the membrane module, L = 605.77/Pi/r^2 = 328.415 km.

The hydraulic conductance $SL_p = (r^2A_p/8\mu t_m)$ = 8.89 E-6 m^4s/kg

Hydraulic Pressure Drop at each stage =

$J/L_pS + \Delta\pi = \Delta P_{h1}$

0.104 + 2477.6* (615.4 – 1.71) = 15.0 atm

Pressure at the high pressure pump = 16.0 atm.

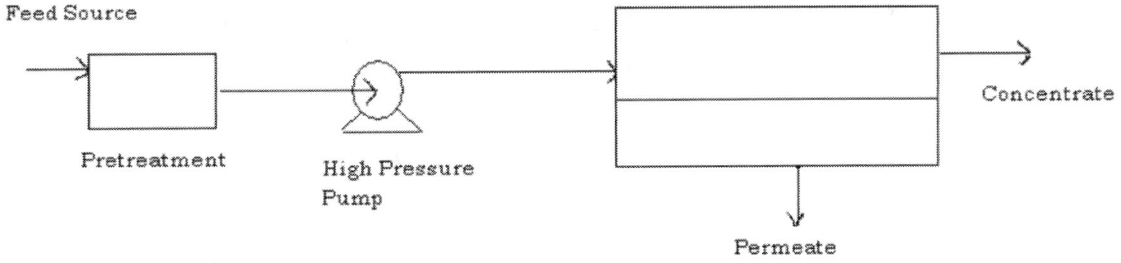

Figure 4 0 Schematic of a SWRO Semi-Permeable Membrane Process

Worked Example 2.0 Effect of Solute Concentration

van't Hoff's law as given by Eq. [5] was derived assuming that the solute concentration was small. For not so dilute systems what would be expression for osmotic pressure. What effect does it have in the solution in the worked example 6.1. b) Use the $\ln(x_w)$ in the van't Hoff's law instead of the Taylor series approximation. What effect does it have ?

In the Taylor series expansion of $\ln(1 - x_s)$ another two terms are taken;

$$\ln(x_w) = \ln(1 - x_s) \approx -(x_s) \quad + x_s^2/2 \; - x_s^3/6 \tag{15}$$

Eq. [6.3] becomes on substituting for Eq. (6.15] in Eq. [6.3);

$$\pi = RTx_s(1 - x_s/2)/V_w = RTC_s(1 - C_sV_w/2 + (C_sV_w)^3/6 \qquad (16)]$$

In worked example 1.0,

$$\pi = 2477.6^* \; 615.4\ (1 - (0.0111)^\wedge 2/2 - (.0111)^3/6)$$

$$= 2477.6^* \; 615.4\ (1 \; - 2.27\ E\text{-}7) = 15.0\ \text{atm.}$$

For the solute concentration of 3.6 wt % NaCl, Eq. [5] is a reasonable approximation of the logarithmic quantity.

b) The osmotic pressure difference when the filtrate concentration of solute is not negligible is as follows.

$$f_{sf} = f_{sp} \tag{17}$$

The fugacity of the solution can be written in terms of the pure component fugacity of water using the Poynting correction factor. For ideal solutions the activity coefficient may be taken as 1.

$$X_{wf}f_w\exp(-V_w(P_{sf} - P_w)/RT) = x_{wp}f_w\exp(-V_w(P_{sp} - P_w)/RT) \quad (18)$$

$$\text{or } -RT/V_w \ln(x_{wf}/x_{wp}) = P_{sp} - P_{sf} = \pi \quad (19)$$

In the worked example 1.0 if the Taylor series was not used to approximate for the logarithmic function of the mole fraction of the solute, Eq. [6.19] will be applicable.

Concentration of solute on the product side = 100 ppm = 3.156 E-5

$\pi = 2477.6/V_w \ln((1 - x_{sf})/(1 - x_{sp})) = 15.48$ atm

The error involved in using the Taylor series approximation is 3.2 % in the osmotic pressure estimation.

Worked Example 3.0 Multistage Sea Water Reverse Osmosis Operation

For the problem discussed in worked example 1.0, for some practical reason say due to mechanical compaction of the membrane under high pressure water from the pump, the same objectives can be met by a multistage operation. At each stage 10% of the solute is rejected. How many stages are needed ? What are the intermediate compositions. What is the net pressure drop required ?

Eq. [19] is used to obtain the pressure drop required to overcome the osmotic pressure at each stage. A MS excel spreadsheet on a Compaq X 86 desktop personal computer. At each stage 10% of the solute is rejected. At stage 9 the 19.99% of the solute is rejected.

	Xwf xwp Pi	atm
	1 0.98860.989776163630.981.614833	
	20.9897760.990912 157887.61.558153	
	30.9909120.992048 157706.71.556367	
	40.9920480.993184157526.211.554586	
	50.993184 0.99432157346.141.552809	
	6 0.994320.995456157166.471.551036	
	70.9954560.996592156987.221.549267	
	80.9965920.997728156808.371.547502	
0.000369	90.9977280.999886297444.642.935406	
369.24 ppm		15.41996

It can be seen that the outlet from the 9th stage is 369.2 ppm. The osmotic pressure difference that is needed to be overcome is given in column Pi in N/m^2 and as atmospheric pressure in the column atm. The pressure drop required at each stage is 1.6 atm except at the 9th stage at 2.9 atm. The total pressure drop is 0.4% lower than from a direct 1 stage operation. 9 pumps of a factor of 10 lower than that required by a single stage operation can be used in this design.

Worked Example 4.0 Use of Reflection Coefficient

A brackish water body is desalinated using the reverse osmosis method. The concentration of the NaCl feed is 15 gm/liter. A Filmtec NF40HF semi-permeable membrane is used. How much water flux can be handled when a spiral module with interfacial area of 8 m^2/gm is used and an exit water at a concentration of 100 ppm is produced. The operating temperature is 37 ^{0}C. How much membrane is needed. What is the pressure needed at the high pressure pump ?

From Figure 2.0, for Filmtech NF_40HF resin, the water flux can be seen to be 1.0 m^3/m^2.day and the salt rejection % is 51%.

The sieving coefficient, and the Staverman reflection coefficient is then 0.51.

The solute radius of NaCl, a = $(3M_w/4\pi\rho A_N)^{1/3}$ =

3*58.5/4/3.14)/2/6.023E23)^0.3333 = 22.7 nm

By using Eq. [6.9], which can be used as a relation between, σ and a/r, the pore radius r can be found by trial and error to be 50 nm.

The volumetric flow of water in cu.m/s, Water flux for Filmtech NF40HF resin in (1 cu.m/m^2/d), amount of membrane needed in gms and cm^3, radius of the cylindrical pores in the membrane module in nm and ($r = 2V_m/a$), and thee effective length of the membrane module, L for a 1 mm thick membrane are all related. These quantities were derived for the worked example 6.1 in MS excel spreadsheet. In this problem the volumetric flow rate was changed until the radius of the pore of the membrane was equal to 50 nm by trial and error. Here are the set of values for which this is the case;

Volumetric flux = 42 lakh/lit/day

= 1.15741 cu.m/s

Amount of membrane needed = 525 gm, 202 cc at 8 m^2/gm

$J/L_pS = 0.9$

The water flux is , $J = SL_p(\Delta P - \sigma RT\Delta C)$

and,

$\Delta P = 0.9 + 0.51*2577.34*(256 - 1.7)$

$= 0.9 + 6.48 = 7.38$ atm

The pressure at the pump = 8.4 atm.

2.0 Optimal Reverse Osmosis Design

Many mechanistic and mathematical models have been proposed to describe reverse osmosis membranes. Reliable models reduce the number of experimental data needed. The mathematical models that describe the RO process can be classified into three categories; models from irreversible thermodynamics (Kadem and Katchalsky); b) solute diffusion model and c) pore models.

Nanofiltration membranes can be explained using charged RO membrane theories. The principle of Donnan exclusion and extended Nernst – Planck equation include the electrostatic effects. Most models focus on the thin asymmetric layer that determines the water fluxes and selectivities. The homogeneous models and porous membrane models differ in the assumption of the porosity of the membrane. One assumes a pore size and certain geometry of pores, and the other assumes that there are no pores.

The use of stages can be made of once the equilibrium characteristics of the liquid-liquid systems on the feed and filtrate side is better quantitated. Some relations have been developed for the estimation of sieve coefficient. Some of the stages at lower level of solute concentration can be substituted with other methods for a minimum total cost. For instance RO and electrodialysis can be used in sequence in desalination of sea water to make potable drinking water at a minimum total cost.

The better understanding of the feed characteristics of membrane polymers and organics solute interactions, fundamental transport mechanisms, hazards, flux drop phenomena, concentration polarization layer formation needs to be better understood for an optimal design of RO based separations. Some experimental studies have shown that substituted phenolic compounds can cause reduction in solvent flux.

Liquid chromatography can be used to measure the membrane solute equilibrium characteristics. Equilibrium distribution coefficients, Gibbs free energy and surface energy for a number of different polymers can be measured. Quantum chemistry calculations, force field computer molecular simulations, and IR spectroscopy have been used to calculate the interactions. Small hydrocarbons occupy the void spaces in the membranes and reduce the water flux.

The ability of RO membranes to remove both inorganic and organic compounds, have made these attractive to treat drinking water supplies. RO processes can remove the hardness in water, various inorganics such as bromate, color, many kinds of bacteria and viruses, and organic contaminants such as agricultural chemicals and trihalomethane precursors. Natural organic matter, NOM, and disinfection byproducts, DBP can be removed using RO technology. The application of RO technology to treat municipal wastes has been successful. Dissolved solids that cannot be separated by other processes can be removed. Color, nitrate and organic levels can be reduced using RO membranes. The water factory 21 site in orange county, California, USA, is a large scale municipal wastewater treatment site. The membrane throughput is 190 lakh liters/day. The process reduces TDS and organics to levels that allowed the effluent to be injected into ground water aquifers used for water supplies. TDS solids of various slats rejection of greater than 99% and TOC rejections as high as 90% were found to be feasible. Losses in water flux over a period of time have been offset by periodic cleaning.

From the mid 90s the primary use of RO technology has been in desalination of seawater and brackish water. The industry is driven by the demand for potable water in some regions of the world. Desalination involves reduction of the TDS concentration to less than 200 ppm. RO technology was first applied to brackish water which has a 1 wt % salt concentration compared with 3.6 wt % in sea water. This difference in concentration as shown in the worked examples translates to large differences in pressure needed at the pump. RO membranes that can withstand 103 atm pressure were developed. The current desalination plants in operation around the world treat both brackish water and sea water.

The brackish water plant in Yuma, Arizona, has a capacity of 27.5 crore liters/day with a TDS intake of 0.3 wt %. Spiral-wound cellulose acetate membranes produce permeate water having 200 ppm of solids. Seawater desalination requires a much higher pressure to achieve a given separation than brackish water. The SWRO desalination plant in Jeddah, Saudi Arabia has a capacity of 56.8 crore liters/day. The Bhaba Atomic Research Center, BARC, at Trombay, Mumbai, India has set up a large desalination plant at the atomic power plant at Kalpakkam about 50 km near Chennai. This plant has two sections. One section produces 18 lakh liters of potable water a day from sea water using the reverse osmosis method and the other section, 45 lakh liters a day using the thermal method. Another desalination plant with two units at 50,000 liters per hour was inaugurated at Koodankukulum in Tirunelveli, Tamil Nadu using reverse osmosis technology. Some desalination plants combine distillation with RO to produce both power and water. MSF, multi-stage flash processes are used to produce both power and distilled water.

Costs for RO technology include both capital and recurring costs. Equipment costs in terms of capacity corresponding to 1000 liters/day range from Rs. 18,000 for brackish water plants and Rs. 40,500 per liter. Operating costs include energy, labor, chemicals and membrane replacement are 0.70 paise per liter for brackish water plants and 2.250 paise per liter for SWRO plants. Costs may vary depending on the pre-treatment and post treatment needed. RO membrane modules are available from Du Pont, Dow/FilmTech, UOP, Millipore, Nitto-Denko and Toray.

3.0 Use of Recycle to Reduce the Size of the High Pressure Pump

In worked example 1.0, 63 lakh liters of water at a salt concentration of 100ppm was generated from a feed with a salt concentration of 3.6 wt % NaCl. The size of the pump needed was 16 atm. In order to reduce the load of the high pressure pump consider recycling the treated water with the feed water. The actual feed into the RO module will be lower in solute concentration, which will decrease the osmotic pressure need to be overcome (Figure 6.5). a) What is the pressure at the pump for a recycle ratio of 3.3. The residue to product ratio is 0.5. What is the % reduction in pump size needed.

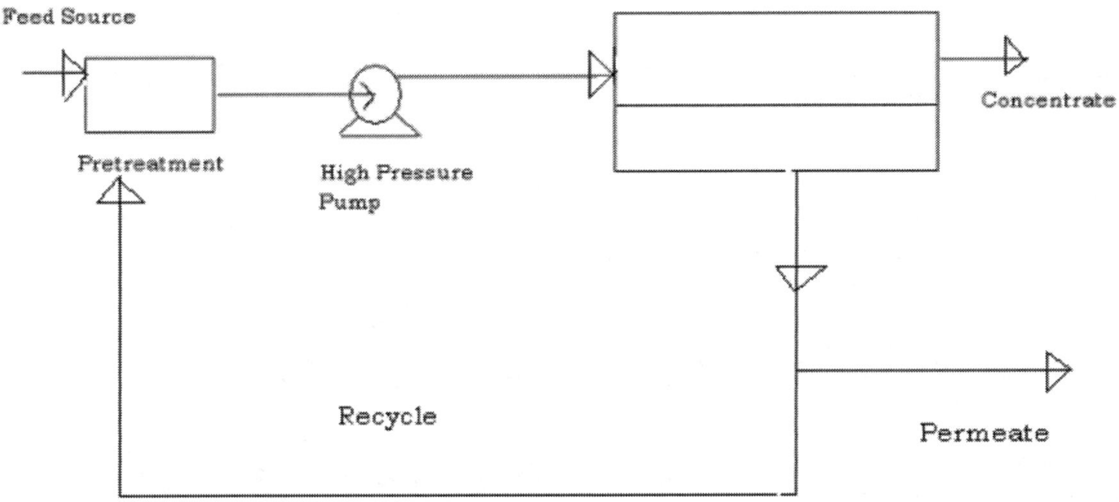

Figure 5.0 Use of Recycle to Reduce the Load on the High Pressure Pump

The product flow rate of water, P = 63 lakh liters/day

Feed flow rate F = P + W

Where W is the concentrate flow rate, R' is the rate of return to the feed tank

Let R be the recycle ratio of treated water, $R = R^{'}/P$ (20)

Then,
The concentration at the throat of the membrane module

$= (x_f F + R' x_{sp})/(F + R') = (0.0114*94.5 + 3.11E\text{-}5*207.9)/(207.9 + 94.5)$

x_f = 0.0114

x_{sp} = 3.11 E-5

R = 0.8

x_f' = 0.0036

The new osmotic pressure needed to be overcome, Eq. [6.19]

$$= -RT/V_w \ln((1-x_f')/(1-x_{sp})) = 2.67 \text{ atm} \quad (21)$$

Pressure needed at the pump = 3.7 atm.

This is a 82 % reduction in the pump size needed. However at this recycle ratio some fresh water is called for. Large recycle storage tanks and metering devices are needed. But the pressure at the pump can be lowered in this fashion.

References

Applegate, L., 1984, Chem. Eng., 91, 64.

Colton, C. K., 1091, AIChE J, 17, 772-780.

Daniekl, F. K., "*Dialysis and Electrodialysis*", in Encyclopedia of Chemical Technology, **5**, 1-20, USA.

Deen, W. M, 1987, *Hindered Transport of Large Molecules in Liquid-Filled Pores,* AIChE J, **33**, 1409-1425.

Fournier, R. L., 1999, *Basic Transport Phenomena in Biomedical Engineering,* Taylor & Francis, Philadelphia, PA, USA.

Ho, W. S., and Sirkar, K. K., 1992, *Membrane Handbook,* Van Nostrand-Reinhold, New York.

Katchalsky, A., and Curan, P. F., 1967, *Non-Equilibrium Thermodynamics in Biophysics,* Harvard university Press, Cambridge, MA, USA.

Kedem, O., and Katchalsky, A., 1958, *Thermodynamic Analysis of the Permeability of Biological Membranes to non-Electrolytes,* Biochmica et Biophysica Acta, **27**, 229.

Lonsdale, H., 1982, J. Membrane Sci., **10**, 81.

Noble, R. D., and Stern, S. A., 1993, *Handbook of Membrane Separations*, Marcel Dekker, New York.

Renkin, E. M., 1954, *Filtration, Diffusion and Molecular Sieving through Porous Cellulose Membranes,* J. Gen. Physiol., **38**, 225-243.

Sharma, K. R., 2005, *Damped Wave Transport and Relaxation,* Elsevier, Amsterdam.

Sharma, K. R., 2002, *Concentration Polarization Layer in Reverse Osmosis,* 94th AIChE Annual Meeting, Indianapolis, IN, USA, November.

Sharma, K. R., 2002, *Reverse Osmosis Unit for Water Removal for Blood Storage in the Army,* 54th Southeast Regional Meeting of the ACS, SERMACS, Charleston, SC, USA, November.

Sharma, K. R., 2002, *Ultrafiltration Device for Concentration of Transfusion Blood during Combat,* 54th Southeast Regional Meeting of the ACS, SERMACS, Charleston, SC, USA, November.

Sharma, K. R., 2003, *Water Processing using Polyamide Membrane by Reverse Osmosis and with Composite Resin in Ion Exchange Technology*, Chemistry Preprint Archive, Volume 2003, Issue 12, December, Pages 98-103.

Sharma, K. R., 2001, *Process Design for a Polyamide Membrane used in Water Purification,* 17th National Convention of Chemical Engineers, Kochi, India, October.

Sharma, K. R., 2003, *Molecular Sieve Design for Reuse of Army Vehicle Washeries under Mild Conditions,* AIChE Spring National Meeting, New Orleans, LA, USA, March.

Sharma, K. R., 2003, *Removal of Arsenic from Drinking Water by Molecular Sieve Adsorption,* 226th ACS National Meeting, New York, NY, USA, September.

Stein, W. D., 1984, *The Movement of Molecules across Cell Membranes,* Marcel Dekker, New York.

330k New Sulfonated Copolymer Membranes For High Temperature Proton-Exchange Membrane Fuel Cells

He Bai, Department of Chemical and Biomolecular Engineering, The Ohio State University, 125A Koffolt Labs, 140 West 19th Avenue, Columbus, OH 43210-1180 and W.S. Winston Ho, Department of Chemical and Biomolecular Engineering, The Ohio State University, 125A Koffolt Labs, 140 West 19th Avenue, Columbus, OH 43210-1180.

In recent years, great progress has been made on the development of proton-exchange membrane fuel cells (PEMFCs) for both mobile and stationary applications, particularly for fuel cell vehicles. Dupont's Nafion® or other perfluorinated sulfonic acid membranes are currently popular to use with low temperature PEMFCs due to their high proton conductivity, and desirable mechanical strength and chemical stability. However, some disadvantages, such as high cost and insufficient membrane performance at high temperatures and low relative humidities (RHs), have seriously limited the application of these membranes. High temperature operations can increase the anode's tolerable level of CO in the fuel and accelerate the reaction rates. Low humidity operations can facilitate the water management of the fuel cell system. Therefore, it is desirable for a PEMFC to operate at high temperatures (above 100oC) and low humidities (blow 50% RH). As a result, the development of competitive and less expensive PEMs that have good performance at high temperatures and low humidities is crucial for fuel cell applications.

Recently, we have synthesized sulfonated polyimide (SPI) copolymer membranes containing hydrophilic poly(ethylene oxide) (PEO) soft segments for water retention at high temperatures. The membrane showed similar fuel cell performance as Nafion® 112 at 70oC and 80% RH. However, the SPI membrane showed much better fuel cell performance than Nafion® 112 at 120oC and 50% RH, which was due to the improved water retention by PEO soft segments. We have also synthesized new sulfonated polybenzimidazole (SPBI)-based membranes. The membrane has exhibited a very high conductivity (> 0.1 S/cm) at high temperatures (> 120oC) and low humidities (even the anhydrous condition), which could essentially meet or exceed the DOE (Department of Energy) targets for PEM materials. Additionally, the newly synthesized SPBI-based membrane has possessed excellent thermal, oxidative, chemical, and hydrolytic stability even at high temperatures. Thus, it has the great potential for the high temperature and low humidity PEMFC application. All of these new membranes should be much more cost-effective since the starting materials are more than two orders of magnitude less expensive than those for Nafion® membranes.

330l Hydrogen Sulfide And Carbon Dioxide-Selective Facilitated Transport Membranes For Fuel-Cell Fuel Processing

He Bai, Jin Huang, and W.S. Winston Ho. Department of Chemical and Biomolecular Engineering, The Ohio State University, 125A Koffolt Labs, 140 West 19th Avenue, Columbus, OH 43210-1180

Hydrogen sulfide is a common contaminant in the hydrogen derived form fossil fuels, and it poisons the fuel cell anode catalyst. Since this poisoning mechanism is irreversible, even trace concentrations of H2S (> 10 ppb) in the hydrogen can significantly degrade fuel cell performance. So it must be almost completely removed prior to feeding hydrogen to fuel cells. In this study, hydrogen sulfide-selective facilitated transport membranes containing amino groups were investigated to remove hydrogen sulfide in hydrogen purification for proton-exchange membrane (PEM) fuel cells. Facilitated transport membranes can achieve high selectivity for the gas of interest due to the reversible reaction of the target gas with the active carriers that are incorporated in the membranes. Compared to the conventional amine scrubbing, hydrogen sulfide and carbon dioxide removal by membrane separation has many advantages, such as low energy consumption, high weight and space efficiency, and operational simplicity as well as no regeneration step required and no moving parts involved in the pressure driven separation.

We have synthesized new membranes for the removal of hydrogen sulfide and carbon dioxide from hydrogen-containing synthesis gas by incorporating amino groups into polymer networks. The membranes are selective to hydrogen sulfide and carbon dioxide preferentially versus hydrogen since the acid gases permeate through the amine-containing membranes via the facilitated transport mechanism due to their reversible reactions with the amine whereas hydrogen has no such reaction. The membranes synthesized have shown high hydrogen sulfide and carbon dioxide permeabilities and selectivities vs. hydrogen. This type of membranes has the potential for hydrogen purification for environmentally friendly fuel cells.

Hydrogen sulfide has much higher reaction rate with the amine than carbon dioxide as the former reacts with the amine via proton transfer and the latter reacts with the amine via carbamate formation primarily. Thus, hydrogen sulfide can permeate through the membrane much faster than carbon dioxide. In the absorption of hydrogen sulfide and carbon dioxide into aqueous amine solutions, the hydrogen sulfide/carbon dioxide selectivity is about 3. Our initial membrane data have also shown hydrogen sulfide with about 3 times permeability of carbon dioxide. The modeling based on hydrogen sulfide permeability data has shown that less than 10 ppb hydrogen sulfide is achievable from typical synthesis gas with small membrane area requirement. In addition, our initial experiments with a limited membrane area have shown a nearly complete removal of hydrogen sulfide from 50 ppm in the synthesis gas feed to about 10 ppb in the hydrogen product, which is good for fuel cell applications.

Preparation of multi-layer PDMS/PVDF composite membranes for ethanol recovery from water by pervaporation

*Xia Zhan, Jiding Li *, Cuixian Chen , Junqi Huang*

Membrane Technology & Engineering Research Center, Department of Chemical Engineering, Tsinghua University, Beijing 100084, People's Republic of China

Introduction

The production of renewable biofuels has been received great attention due to the energy crisis and environment crisis nowadays. And ethanol biofuel is one of the most important biofuels widely used in the world. Ethanol recovery from biomass fermentation processes is inevitably due to that microorganisms used in the fermentation process usually experienced strong ethanol inhibition. Pervaporation, advantageous for the separation of azeotropic mixtures such as ethanol/water mixtures with low energy consumption, has become an efficient, cost-effective manner.

The properties of the membrane material and feed species are the main factors dictating the separation performance achieved in pervaporation process. Many membrane materials have been developed for ethanol recovery from water by pervaporation, such as silicone-containing polymers, fluorine-containing polymers, and other hydrophobic polymers. Poly(dimethylsiloxane) (PDMS) is a typical ethanol perm-selective material due to its hydrophobic properties and large free volume. However, pure PDMS has low selectivity for ethanol recovery from water, with separation factor ranged from 4.4 to 10.8[1]. Much effort has been expended searching for attractive materials with excellent separation performance for the recovery of ethanol from water, and a great amount of investigation was carried out based on PDMS due to its low-cost and easiness to handle. And Leland M Vane [2] etc. proposed that PDMS will continue to be the main organic membrane material for the ethanol concentration in the near term.

In the present paper, pervaporation of ethanol/water mixtures was also conducted with pure PDMS. To surpass the limit of pure PDMS' selectivity for ethanol recovery, a multi-layer PDMS/PVDF composite membrane with an alternative PDMS/PVDF/non-wovon-fiber/PVDF/PDMS configuration was developed by curing PDMS cast on porous PVDF substrate. The multi-layer composite membrane showed much better separation performance of ethanol/water solutions compared with common PDMS/PVDF composite membrane

Experimental

Materials

α,ω-polydimethylsiloxanediol(PDMS) prepolymer was received from Beijing Chemical Reagents Corporation, with kinetic viscosity of 50,000mPa·s. N-heptane and di-n-butyltin dilaurate(DBTOL) were obtained from Beijing Jingyi Chemical Reagents Corporation, and was used as received. Phenyltrimethoxylsiane (PTMOS) was supplied by Debang New Chemical Materials Corporation of Yingcheng. Polyvinylidene fluoride (PVDF,1015) was used for the preparation of porous membrane support. Triethyl phosphate (TEP, reagent grade, Beijing Chemical Corporation) was used as the solvent for PVDF membrane formation. All reagents were used as received unless otherwise mentioned.

Preparation of bi-layer PVDF membrane

The bi-layer PVDF membrane was prepared by dissolving PVDF in TEP solvent to form a 15wt% solution. The PVDF/TEP solution was cast on one side of the non-woven fiber and immersed into water to induce polymer precipitation. The residual solvents was exchanged with alcohol for 5min and dried at room temperature. The other side of the non-woven fiber was also coated with PVDF following the procedure the same as mentioned above. The thickness of both PVDF layers was controlled in the range of 45μm~50μm.

Preparation of PDMS/PVDF composite membrane

Multi-layer PDMS/PVDF composite membrane

A prescribed amount of PDMS and cross-linking agent PTMOS were dissolved in n-heptane with vigorous stirring and then DBTOL catalyst was added into the solution. The resulting homogeneous PDMS solution was coated onto the both sides of the porous PVDF substrate in two steps and then the composite membrane was dried at room temperature for 48h. The thickness of both PDMS layers was controlled to 15±1μm.

Single layer PDMS/PVDF composite membrane

The TEP solution containing 15wt% PVDF was cast onto only one side of the non-woven fiber and other post treatment was all the same as mentioned above. The casting solution of PDMS was just coated onto the PVDF surface, and the PDMS layer was also controlled with the thickness of 15±1μm.

Pervaporation separation process

The pervaporation experiments were conducted by using an apparatus developed by our laboratory as shown in Fig.1. The experiments were carried out with low ethanol feed concentration ranging from 4.5wt% to 30wt%. And the operation temperature varied from 40℃ to 80℃ with receiving-side pressure of ca 200Pa.. The permeation vapor was condensed and collected using a liquid nitrogen trap. The composition of permeate collected in the trap was analyzed by gas chromatography.

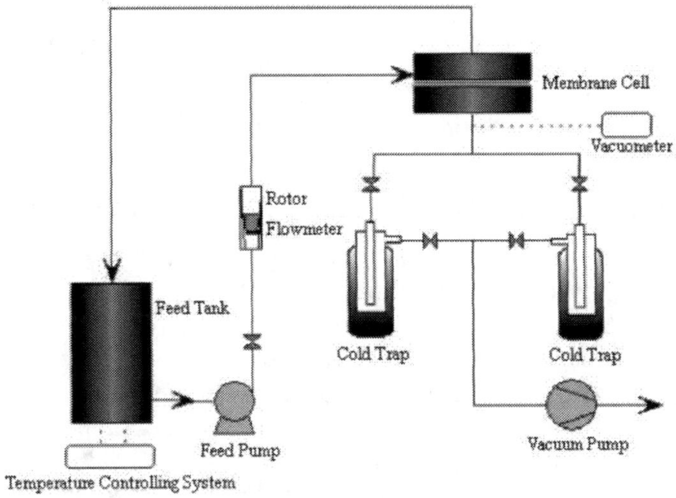

Fig.1 Schematic diagram of the Pervaporation apparatus

Results and discussions

Effect of ethanol feed concentration on pervaporation performance

The effect of ethanol feed concentration on the total flux and selectivity for separation of ethanol/water mixtures were represented in Figure 2. respectively. As the feed concentration increased from 4.5wt% to 30wt%, the separation factor decreased from 14.9 to 3.2, and the total flux increased from 208.6g/m2h to 407.2g/m2h with the operation temperature of 60℃.The effect of ethanol feed concentration on pervaporation performance of multi-layer PDMS/PVDF composite membrane was similar to the single layer PDMS/PVDF composite membrane reported by other researchers [3]. And the explanation of this phenomenon was explained almost the same as below. As the feed concentration increased from 4.5wt% to 30wt%, the swelling ratio of PDMS became much larger which would make the chain mobility more quickly. And hence the free volume in PDMS became much more, and the diffusivity increased. However, the increase in the diffusivity of water caused by an increase of free volume is much larger than that in the diffusivity of ethanol as the molecular size of water was smaller than that of ethanol. Therefore, the separation factor decreased as the feed concentration increased.

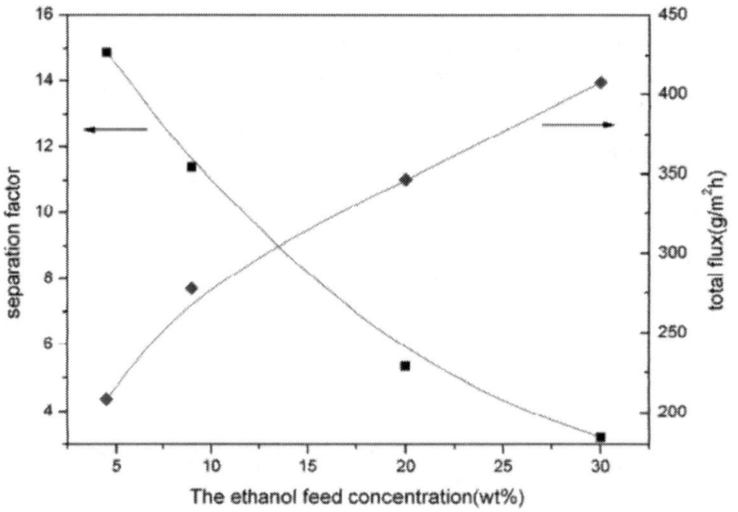

Fig 2 The effect of ethanol feed concentration on the pervaporation performance of multi-layer PDMS/PVDF composite membrane.

Effect of operation temperature on pervaporation performance

The effect of operation temperature on the pervaporation performance of multi-layer PDMS/PVDF composite membrane was shown in Figs. 4-5. The separation of ethanol aqueous solutions with four kinds of concentration was carried out through varying the operation temperature. With the ethanol feed concentration of 4.5wt%, 9wt% and 20wt%, a novel phenomenon appeared that the separation factor showed a maximum value at 60℃ with the total flux increasing all along, as the operation temperature increased from 40℃ to 80℃. Our results are quite evidently different from those reported in the literature. Lei Li. et al.[4] and Molina et al.[5] found that the separation factor of PDMS for ethanol/water mixtures increased with increasing temperature, while Vankelecom et al.[6] proposed that the separation factor of ethanol aqueous solution tended to decrease with increasing temperature using PDMS membranes. Many results and explanations were advanced, and most of them are based on solution-diffusion model. The significant changes of solubility and diffusion of penetrating components through varying temperature could be the main reason for the novel phenomenon. The novel phenomenon discovered in our experiment was also thought to be related with changes of permeation characteristics caused by varying temperature, and especially associated with the much different permeation process of the multi-layer structure of the composite PDMS membrane[7].As shown in figure 3-4, the influence of temperature on the separation factor became more significantly as the ethanol feed concentration decreased from 30wt% to 4.5wt%, while the total flux followed the reversed order. The reason for this phenomenon was also considered to be related with the much different permeation process of the multi-layer structure of the composite PDMS membrane. Further study is being carried on.

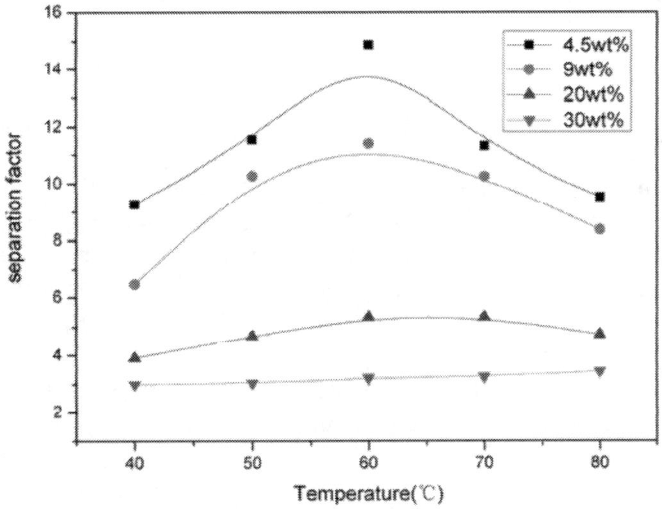

Fig 3 The effect of operation temperature on the separation factor of multi-layer PDMS/PVDF composite membrane.

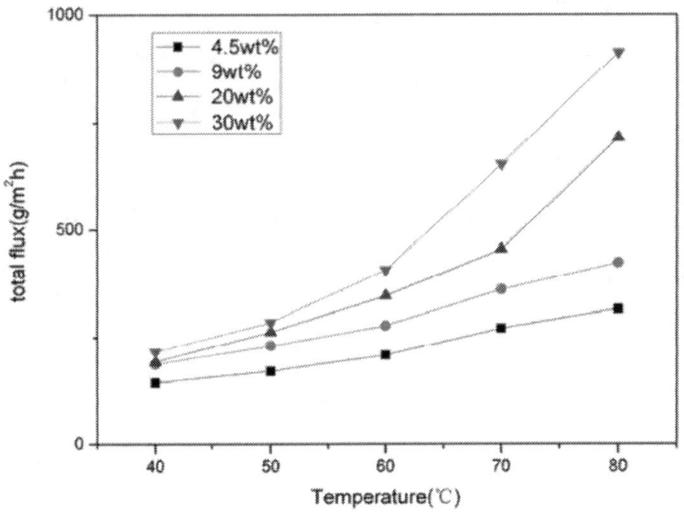

Fig 4 The effect of operation temperature on the total flux of multi-layer PDMS/PVDF composite membrane.

Comparison of pervaporation performance with single-layer PDMS/PVDF composite membrane

For comparison purpose, the pervaporation performance of single-layer PDMS/PVDF composite membrane developed with the similar method and same materials was also investigated. The thickness of the PDMS layer is 15μm determined by micro screw gauge. The comparison was conducted with the same operation conditions such as 9wt% ethanol feed concentration. The comparison of pervaporation performance was represented in Figs. 5-6. The separation factor of the multi-layer composite membrane was much higher than that of single layer PDMS/PVDF membrane in almost whole range of operation temperature except at 40℃. The reason of this phenomenon was not clear, and further investigation was needed. And the total flux followed the reversed order. It was clear that the total flux would be depressed as the mass transport resistance of the multi-layer membrane increased. Although the total flux was depressed a lot by the increase of mass transport resistance, we speculated that with higher selectivity, PDMS membrane with multi-layer configuration may also be expected to be commercially applied as ethanol perm-selective membranes by decreasing the thickness of both the PDMS layer. And experiment for improving the total flux by decreasing the thickness of the PDMS layer is being carried on.

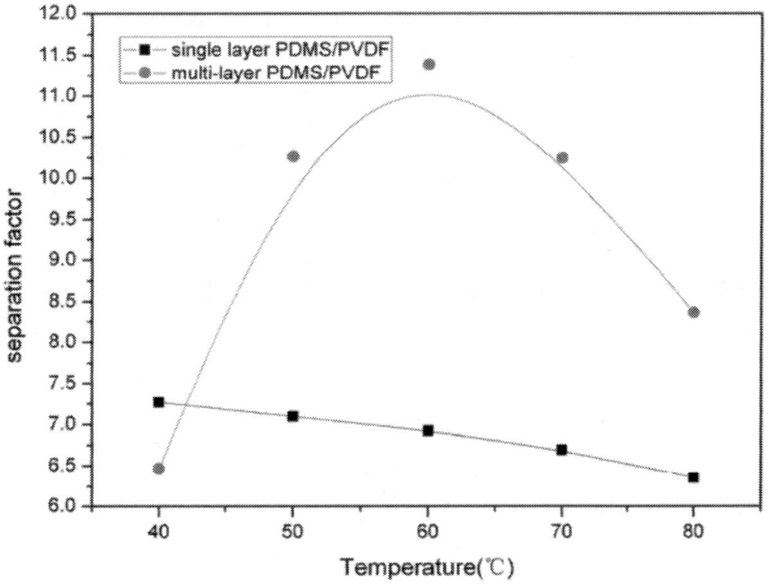

Fig 5 The effect of temperature on separation factor of the PDMS/PVDF composite membrane

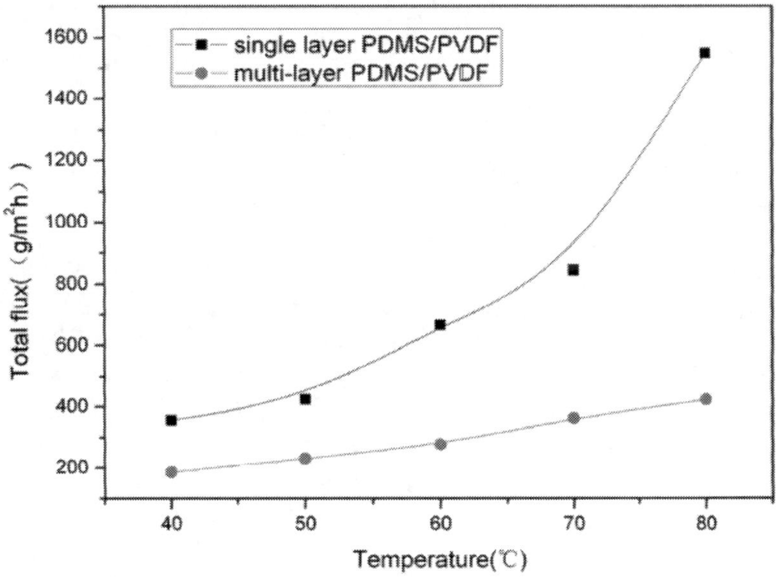

Fig 6 The effect of temperature on total flux of the PDMS/PVDF composite membrane

Conclusion

The multi-layer PDMS/PVDF composite membrane developed in our study surpassed the utmost of the selectivity of pure PDMS. A novel phenomenon appeared that the separation factor had a maximum value at 60℃ as the operation temperature increased, and the total flux increased all along, as the operation temperature increased from 40℃ to 80℃. The separation factor of the multi-layer composite membrane reached as high as 14.9 with the total flux of 208.6g/m^2h with the feed concentration of 4.5wt% at 60℃. Although the total flux of the multi-layer composite membrane was depressed due to the increase of mass transport resistance, the membrane surpassed the utmost of pure PDMS' selectivity remarkably for ethanol recovery. And the PDMS membrane with multi-layer configuration may open a door for its application in ethanol recovery from water in the future.

Acknowledgements

The authors greatly appreciate the financial supports of the Major State Basic Research Program of China (No. 2003CB615701), National Natural Science Foundation of China (No. 20576059, No.20676067), SINOPEC Foundation (No. X505002) and CNPC Innovation Foundation (No. 05051143).

References

1. Toshiki Aoki, Macromolecular design of permselective membranes. Prog. Polym. Sci. 24 (1999) 951-993.
2. Leland M Vane, A review of pervaporation for product recovery from biomass fermentation processes. Journal of Chemical Technology and Biotechnology, (2005) 80:603-629.
3. C.S.Slater, P.J.Hickey, and F.P.Juricic, Pervaporation of aqueous ethanol mixtures through poly(dimethyl siloxane) membrane. Separation Science and Technology, (1990)25(9&10), 1063-1077.
4. Lei Li, Zeyi Xiao, Shujuan Tan, Liang Pu, Zhibing Zhang, Composite PDMS membrane with high flux for the separation of organics from water by pervaporation. Journal of Membrane Science 243 (2004) 177-187.
5. J.M. Molina, G. Vatai, E.B. Molnar, Comparison of pervaporation of different alcohols from water on CMG-OM-010 and SULZER membranes, Desalination 149 (2002) 89-94.
6. I.F.J. Vankelecom, D. Depre, S.D. Beukelaer, J.B. Uytterhoeven, Influence of zeolites in PDMS membranes: pervaporation of water/alcohol mixtures, J. Phys. Chem. 99 (1995) 13193-13197.
7. Cheng-Liang Chang, Ming-Shih Chang, Preparation of multi-layer silicone/PVDF composite membranes for pervaporation of ethanol aqueous solutions. Journal of Membrane Science 238 (2004) 117-122.

330n Anomalous Diffusion of Water Vapor in Dense Polyimide Membranes Prepared from 3,3'4,4'-Benzophenonetetracarboxylic Dianhydride and Phenylene Diamine

Yexin Xu[1], Cuixian Chen[2], and Jiding Li[2]. (1) Chemical Engineering, Tsinghua University, Room 464, Beijing, 100084, China, (2) Department of chemical engineering, Tsinghua University, Room 464, Beijing, 100084, China

The dynamic sorption and diffusion of small solvent molecules in dense polymeric membrane are the prerequisite for membrane separation processes such as pervaporation, vapor permeation and gas separation. In this study, the dense polyimide membrane was prepared form 3,3'4,4'-benzophenonetetracarboxylic dianhydride (BTDA) and phenylenediamine (PDA) monomers. The dynamic sorption and diffusion behaviors at 298 K and 308 K of water vapor in BTDA + PDA polyimide membrane were investigated by experiments. According to the results, the diffusion behaviors did not obey the Fickian diffusion law and belonged to anomalous diffusion. The anomalous diffusion factors, calculated from the fractal diffusion theory, of water vapor in BTDA + PDA polyimide membrane are 2.38 at 298 K, and 2.26 at 308 K. It implies that the diffusion of water vapor in BTDA + PDA is more slowly than the Fickian diffusion.

330o Synthesis And Characterization Of Polycaprolactone Nanoporous Membranes For Implantable Drug Delivery Devices

Chi Yen, Hongyan He, L. James Lee, and W.S. Winston Ho. Chemical and Biomolecular Engineering, The Ohio State University, 140 WEST 19TH AVE, Columbus, OH 43210

In recent years, implantable drug delivery devices have been promising therapeutic systems for the treatment of chronic diseases. Chronic patients suffer from health problems as a result of permanent illness and/or organ failure. Therefore, medical methodology which can be effective for a long time has attracted a lot of attention. Implantable drug delivery devices are able to function in the human body for a long period. However, an improper amount of drug released from the device would cause uncomfortable side effect. To optimize the pharmacotherapy, drug release has to be well controlled in accordance with the remedy during a treatment course. An ideal drug delivery device is capable of releasing the desirable amount of drug to the targeted site for a required period of time. In order to achieve this goal, biodegradable polycaprolactone nanoporous membranes may be expected to control the drug release rate via tuning the porous structure.

In this study, biodegradable and biocompatible polycaprolactone (PCL) nanoporous membranes have been successfully prepared via the combination of thermally and nonsolvent induced phase separation techniques. The polymer solution with a specific composition was cast on a suitable plate. Then, the cast film of the solution was immersed into a lower temperature coagulation bath, i.e., both thermally and nonsolvent induced phase separations took place simultaneously. In the membrane preparation, the thermally induced phase separation could enhance porosity or avoid the formation of a dense top-layer. On the other hand, the nonsolvent induced phase separation could play a role on the formation and the dimension/size of nanoscale pores. Our results showed that the pore size reduced and the porosity increased as the coagulation bath temperature decreased. Also, the release rate of the model drug compound, lysozyme, into an aqueous solution was higher as a lower coagulation bath temperature was used. As a result, the drug release rate can be well controlled via adjusting the parameters for the membrane preparation, including coagulation bath temperature and casting solution composition. In addition, the nanoporous membrane provides a barrier for preventing the passage of cytotoxic cells, macrophages, antibodies and other immune substances.

330p A Finite Element Model Of A Gas Permeation Unit

William M. Clark and Jeremy Lebowitz. Chemical Engineering Department, Worcester Polytechnic Institute, 100 Institute Road, Worcester, MA 01609

A finite element simulation of a gas permeation unit has been developed using COMSOL Multiphysics™ software. The simulation provides the solution to momentum and mass balance differential equations with appropriate boundary conditions and allows visualization of the velocity, pressure, and concentration profiles within the unit. Input to the model includes only physical properties from the literature, permeate and retentate operating pressures, and the inlet flow rate. Model predictions for permeate and retentate flow rates and compositions for both co- and counter- current flow configurations compare favorably with experimental results for a pilot scale Monsanto Prism™ membrane unit for air separation with hollow fiber polysulphone membranes. The solution-diffusion mechanism of separation is clearly illustrated using solubility and diffusivity data from the literature as model parameters. Important trends in the experimental results including (1) the permeate flow rate increasing with increasing pressure, but nearly independent of inlet flow rate, and (2) the permeate concentration increasing with increasing pressure and increasing inlet flow rate are correctly predicted by the model which is shown to have advantages over the simple algebraic well-mixed model and other gas permeation models in the literature.

Modeling of Polymer Electrolyte Membrane and Proton Transport via Atomic Scale Molecular Dynamics Simulation

Pil Seung Chung, Parag Jain, Haigang Chen, Lorenz T. Biegler, and Myung S. Jhon
Carnegie Mellon University, Department of Chemical Engineering, Pittsburgh, PA

Fuel cells (FCs) have gained considerable attention recently as an alternative energy source for power generation, with applications in power plants, automotives, and portable electronics, possessing the attractive features of higher efficiency (more than twice the internal combustion engines), near zero greenhouse emissions, and simple as well as modular design. Amongst the various FCs, polymer electrolyte membranes (PEM) FCs are especially promising due to their higher efficiencies, and low temperature operation leading to shorter start-up times for varied applications. However, there are several limitations in the PEMFC technology which need to be overcome before its commercial entry, and some of the critical issues are related to the system component PEM.

A PEM is a semipermeable ionomer, designed to conduct protons, while being impermeable to electrons, and reactants such as oxygen and hydrogen. There are stringent requirements for the 'ideal' PEM: (i) to deliver high proton conductivity at high temperatures, (ii) to be impermeable to methanol and other reactants to prevent the undesirable cross-over phenomena, (iii) to be mechanically and chemically robust, and (iv) to be within the desirable cost. Although, the state-of-the-art PEM materials Gore®, and Nafion® posses some of these features, but their conductivity is a strong function of the hydration level which is difficult to control during a typical FC operation. Moreover, these PEMs are highly sensitive to high operating temperatures, and have a cost which is about an order higher than targeted [1]. Hence alternative PEMs are needed to be designed having high conductivity, and satisfying the constraints of durability and cost. This entails a detailed molecular level understanding of various interactions and transport processes within the PEM.

In this paper, we examined the proton transport phenomena in PEM using a molecular dynamics (MD) simulation to obtain a relationship between the PEM structure, its water uptake, and the macroscopic properties such as the conductivity and water diffusion coefficient. The goal was to build a tool which may be exploited later, to suggest molecular structures of PEM possessing the above mentioned desirable features.

Based on our previous study [2, 3], Nafion® molecules were modeled via a coarse-grained, bead-spring model, which neglects the detailed atomistic structural information while keeping the essence of the molecular structure. Lennard Jones and finitely extensible nonlinear elastic potential energy models were used to describe interaction among all the beads. In addition, to examine the water uptake in the PEM, long-range Coulombic interactions between water molecules and the beads were explicitly incorporated in our model. With the total potential energy constructed, MD simulations were performed using the Langevin equation.

In our preliminary study, we examined the water density distribution as well as proton diffusion in PEM. Specifically, we calculated the PEM density for different amounts of water uptake. According to our simulation, the density decreased with an increase in water uptake, which agrees with experimental data of Gierke *et al.* [4]. This means that swelling of the electrolyte can be qualitatively described by our MD simulation. To further examine this phenomenon in detail, we calculated the radius of gyration of Nafion® molecules. We found that the radius of gyration increases when the number of water molecules in the system increases indicating the swelling of Nafion® molecule, *i.e.*, a reduction in the density.

The radial distribution functions of proton-water, proton-polar group (sulfur), and proton-nonpolar group were also calculated. We found sharp peaks in proton-water and proton-polar group, which implies that almost all the counterions and the water molecules exist in a hydrophilic clustered region. However, we did not observe any clear peak in the proton-nonpolar group case. We made visualization for the water density distribution to illustrate this point (Fig. 1).

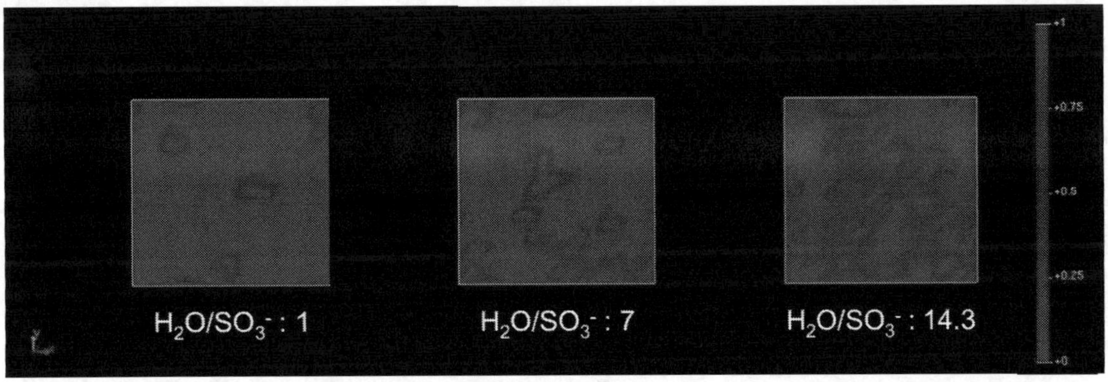

Fig. 1. Visualization of water distribution in a cross section of the system for three different water contents

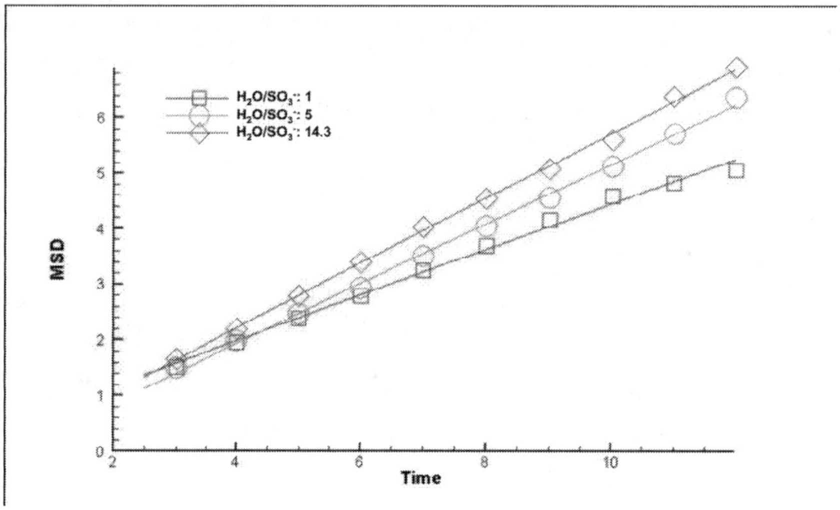

Fig. 2. Proton MSD dependence on water uptake for three different water contents

We also examined transport phenomena of protons and water molecules under various water uptake conditions. Specifically, we calculated the mean square displacement (MSD) of protons to examine self-diffusion phenomenon (Fig. 2). We found that the MSD increases (diffusion increases) as the water uptake increases.

Reference:

1. P Costamagna, S Srinivasan, *J. Power Sources*, **102**, 242 (2001).
2. Q. Guo, H. G. Chen, B. C. Smith, and M. S. Jhon, *J. Appl. Phys.*, **97**, 10P301 (2005).
3. Q. Guo, P. S. Chung, H. G. Chen, and M. S. Jhon, *J. Appl. Phys.*, **99**, 08N105 (2006).
4. T.D. Gierke, G.E. Munn, and F.C. Wilson, *J. Polym. Sci, Part A-2*, **19**, 1687 (1981).

330r Complex Fluids For Membrane-Based Separations

Stephen M. Martin, Chemical Engineering, Virginia Tech, 134 Randolph Hall, Blacksburg, VA 24061 and Sangil Han, Chemical Engineering, Massachusetts Institute of Technology, 77 Massachusetts Ave, 66-321, Cambridge, MA 02139.

Self-assembled, mesoscopically ordered materials possess a number of intriguing properties that suggest potential applications in the separation of chemical or biological species, including the separation of enantiomers. These materials include both lyotropic and thermotropic liquid crystals. There exists a large library of materials known to form thermotropic liquid crystalline (LC) phases. Unfortunately, the published reports for the majority of these materials concentrate exclusively on their suitability for use in optoelectronic devices and do not expand on their transport properties.

We have begun to explore the suitability of several thermotropic liquid crystalline materials for the separation of chemical components. Our initial studies have concentrated on the development of techniques to accurately measure the diffusion rates and solubilities of solutes in LC materials in order to examine the dependence of molecule size, structure and functionality on solute permeability.

330s Dehydration Separation of Tetrahydrofuran-Water Mixtures by Pervaporation with Commercially Available Membranes

Timothy Moroz, **Colleen McGinness**, David Schiavi, Mariano J. Savelski, and C. Stewart Slater. Department of Chemical Engineering, Rowan University, 201 Mullica Hill Road, Glassboro, NJ 08028-1701

Pervaporation technology can be successfully employed to dehydrate tetrahydrofuran-water mixtures found in the pharmaceutical and specialty chemical industry. The conventional azeotropic distillation of this separation requires a high amount of chemical entrainer to perform the separation and generates significant waste. Pervaporation does not require additional chemicals to push past the azeotrope making it a competitive unit operation.

In this study, a design of a hybrid pervaporation-distillation system has been compared to a conventional THF dehydration. Vendor specifications were been utilized to model the hybrid system and several membrane areas were studied. Dehydration studies were performed on a lab scale to study the effect of feed concentration, feed temperature and permeate-side pressure on membrane performance. A life cycle assessment of the process was performed to compare the environmental impacts of both processes.

This study shows that the addition of a pervaporation unit to a conventional system is a competitive separation. Process improvements that yield cost reduction, energy savings, and solvent reductions are all potential benefits of using pervaporation over existing practices. The use of pervaporation is well suited for separations in the pharmaceutical and specialty chemical industry and leads to a green drying process.

330t Functionalized Microfiltration Membranes For Bio-Separations

Saurav Datta[1], A. Nath[2], and D. Bhattacharyya[1]. (1) University of Kentucky, 177 Anderson Hall, Lexington, Kentucky 40506, (2) Johns Hopkins University, Department of Neurology and Neuroscience, Baltimore, MD 21287

Functionalized microfiltration (MF) membranes are becoming increasingly popular in biotechnology and pharmaceutical industries. Affinity based separation using functionalized MF membranes are emerging as a potential alternative to conventional bead-based column chromatography. Functionalization of MF membranes offer choices of attaching the active groups or moieties on different types of materials (hydrophilic, hydrophobic), environments (external surface, internal pore surface) and dimensions (pore size varying from 0.05 to 10 ƒÝm).This study deals with functionalization of microfiltration (MF) membranes within the pore structures and their applications in the field of bio-separations. Pore surface possesses the advantage of attaching more number of active groups or moieties on a single membrane as compared to the external surface of the membrane. Wide pore structure of MF membranes allows easy access of the active groups/moieties to the target molecules, under convective flow condition. Therefore, the mass transfer resistance, and hence the processing time per unit volume of feed is less.

This study describes different pathways to functionalize MF membranes with various active groups. Different functionalization techniques and experimental conditions are adopted for different situations. Once the pore surface of the membrane is activated, further attachments of groups, moieties or molecules are carried out depending on the area of application. For example, in this research work, avidin is attached to separate a biotin-tagged protein from mixture. Also, an enzyme is separated from mixture, and then enzyme-substrate reaction is studied within the membrane phase. Effect of spacer molecule on the efficiency of the process is studied. Characterizations of the functionalized membranes are also presented in this study. Characteristic fouling of the membrane matrix associated with bio-separations is also analyzed critically.

330u Transport Properties Of Supercritical Fluid Processed Proton Exchange Membranes

Juan C. Pulido, Chemical Engineering, University of Puerto Rico - Mayaguez, Mayagüez, PR 00681 and David Suleiman, Chemical Engr., University of Puerto Rico - Mayaguez, Mayagüez, PR 00681.

Text Not Available.

330v Transport Properties Of Supercritical Fluid Processed Proton Exchange Membranes: Methanol Permeability And Proton Conductivity

Juan C. Pulido, Alba Y. Marrero, and David Suleiman. Chemical Engineering, University of Puerto Rico - Mayaguez, Mayagüez, PR 00681

Proton exchange membranes (PEMs), commonly used in direct methanol fuel cells (DMFC), are typically limited by either high methanol permeability (also known as the cross-over limitation) or low proton conductivity. These membranes have phase segregated rubbery, glassy and ionic domains, which make processing challenging and limited by phase equilibria considerations. This study proposes to use supercritical fluids (SCFs) processing as an alternative, since the gas-like mass-transport properties of SCFs could allow for better penetration into the membranes and the use of co-solvents could also change the morphology of the PEMs, fine-tuning the physical and transport properties (e.g., fuel permeability and proton conductivity) needed for effective DMFC performance. This investigation focused on the effect of SCF processing on the physical and transport properties of tri-block copolymer ionomers commonly used as proton exchange membranes (PEMs). The PEMs selected included: Nafion®, sulfonated poly-(styrene-isobutylene-styrene) (SIBS) and sulfonated poly-(styrene-ethylene-ran-butylene-styrene) (SEBS). Processing was performed at 200 bars and 40 °C using SCF CO2 with and without different co-solvents. A variety of co-solvents were used including: acetic acid, acetone, acetonitrile, cyclohexanone, ethanol, isopropanol, methanol, methylene chloride, and tetrahydrofuran. They were selected based on their size, polarity, and chemical nature (e.g., protic or aprotic). Measurements of methanol permeabilities were made using an FT-IR technique, monitoring the methanol concentration with time through the membrane. Proton conductivity was measured using impedance spectroscopy. Measured permeabilities were of the order of 10-8 cm2/s, two orders of magnitude lower than unprocessed membranes for Nafion®. However, changes in proton conductivity were short of the required level (0.1 S/cm) and very sensitive to the processing orientation. These results coupled with thermogravimetric, swelling, and dynamic mechanical analyses have provided a better understanding of the structure property relation of these complex three-dimensional phase-segregated membranes.

330w Thermoporometric Investigation Of Solvent-Surface Interactions In Nanoporous Inorganic Membranes

Geoffrey D. Bothun, Chemical Engineering, University of Rhode Island, 205 Crawford Hall, Kingston, RI 02881

Thermoporometry provides a useful tool for characterizing nanoporous materials by relating the freeze/melt behavior of a confined probe liquid, typically water, to the pore size distribution using a relationship analogous to the Gibbs-Thompson equation. Near the pore wall, liquid-surface interactions arising from van der Waals, electrostatic, and structural forces, as well as adsorption, can yield an unfreezable liquid surface film that spans multiple molecular layers. In small nanopores (< 20 nm), the presence of this film reduces the measured pore size of the nanomaterial. In this work, a new technique is described where thermoporometry is used to detect liquid-surface interactions in nanoporous alumina, titania, and zirconia membranes using a host of solvent probe molecules including water, cyclohexane, and 1-octanol (i.e. a polar, an apolar cyclic, and an alkanol molecule, respectively). Particular emphasis is placed on characterizing these interactions and relating them to membrane solvent permeability to determine how surface forces influence fluid transport at the nanoscale.

330y Polymer and Inorganic Membrane Research Activities at the Ohio State University

W.S. Winston Ho, Department of Chemical and Biomolecular Engineering, The Ohio State University, 125A Koffolt Labs, 140 West 19th Avenue, Columbus, OH 43210-1180, Henk Verweij, The Ohio State University, 2041 College Road, Columbus, OH 43210-1178, and **He Bai**, Department of Chemical and Biomolecular Engineering, The Ohio State University, 125A Koffolt Labs, 140 West 19th Avenue, Columbus, OH 43210-1180.

Membrane applications include gas separation, water purification and fuel cell electrolytes. H2 purification may occur directly with inorganic micro-porous membranes, and indirectly with a CO2-selective polymer membrane. Water purification is achieved with a solution-diffusion mechanism in dense polymer membranes, or highly permeable mesoporous membranes that show space charge selectivity. Polymer electrolytes with –SO3H groups offer a cost-effective approach for proton-exchange membrane fuel cells. Zirconia, Zeolite L and Y membranes with optimum layer formation conditions offer promise for fuel cell and separation applications.

346b Adsorption And Phase Behavior Of Fluids In Novel Micro/mesoporous Silica Materials With Hierarchical Pore Structure

Matthias Thommes, Quantachrome Instruments, 1900 Corporate Drive, Boynton Beach, FL 33426, Ö. Sel, MPI of Colloids and Interfaces, Potsdam, Germany, Research Campus Golm, Potsdam, Germany, Peter I. Ravikovitch, Present Address: ExxonMobil, Clinton, NJ 08542, Bernd Smarsly, Max Planck Institute of Colloids and INterfaces, Am Muehlenberg 1, Golm, D-14476 Golm, Germany, and Alexander V. Neimark, Department of Chemical and Biochemical Engineering, Rutgers University, 98 Brett Road, Piscataway, NJ 08854-8058.

In contrast to the recent progress achieved in the understanding of the adsorption mechanism of fluids in highly ordered mesoporous materials with simple pore geometries (e.g. M41S materials), there are still many open questions concerning the phase and sorption behavior of fluids in more complex pore systems. In order to address this problem, we have performed systematic adsorption studies of nitrogen (77 K) and argon (at 87 K and 77 K) on a series of novel micro/mesoporous silica materials with hierarchical pore structures (KLE and KLE/IL silicas) [1,2]. These materials consist of spherical pores (mean pore diameters ranging from below 10 nm up to 35 nm) connected through narrow micro- and mesopores. The hierarchical pore structure of these samples has been also confirmed by small-angle neutron scattering (SANS) in combination with in-situ nitrogen adsorption at 77 K. The nitrogen and argon adsorption isotherms obtained on these samples reveal wide hysteresis loops of types H1/H2. An analysis of these data confirms that pore condensation occurs delayed due to the existence of metastable adsorption films and hindered nucleation of liquid bridges, while evaporation of the condensed pore fluid from the spherical cavities occurs by cavitation if the width of the connecting pores is smaller than a critical value („T 5 nm) [2]. Our data reveal that the size of the cavities has an appreciable effect on the cavitation pressure if the cavity size is smaller than ca. 14 nm, but not for larger cavity diameters. It appears further that the size of the connecting pores does not effect the cavitation pressure. Summarizing, our results lead to a better understanding of the adsorption and phase behavior of fluids in micro/mesoporous materials, which is not only important for an accurate textural characterization of adsorbents consisting of complex micro/mesoporous pore networks but also for optimizing their use in practical applications (e.g., catalysis, separation).

References [1] O. Sel, D, Kuang, M. Thommes, B. Smarsly, Langmuir 22, (2006) 2311 [2] M. Thommes, B. Smarsly, M. Groenewolt, P.I, Ravikovitch, A.V. Neimark, Langmuir 22, (2006) 756 [3] O. Sel, A.Brandt, D. Wallacher, M. Thommes, and B. Smarsly, Langmuir 23, (2007), 4724

346c Characterization Of The Structural And Vibrational Properties Of Phyllosilicate Clay Minerals

James P. Larentzos, Jeffery A. Greathouse, and Randall T. Cygan. Geochemistry Department, Sandia National Laboratories, P.O. Box 5800, MS 0754, Albuquerque, NM 87185

Phyllosilicates, including the clay minerals talc, pyrophyllite and montmorillonite, are associated with numerous environmental, agricultural and industrial processes. Due to the lack of well-crystallized materials, experimental determinations of the crystal structure and mineral behavior for most clay minerals are difficult to obtain. Molecular simulations, however, provide an opportunity to evaluate the detailed structure and dynamics for idealized models of these complex clay phases, usually as a supplement to the limited x-ray diffraction and vibrational spectroscopy data.

In this work, the structural and vibrational properties of charged and uncharged phyllosilicates are investigated with ab initio and classical molecular dynamics simulations and compared with IR and Raman spectroscopic experiments. The quantum mechanical simulations are based on density functional theory (DFT), which is sufficiently accurate to predict the structural and vibrational properties of clay minerals. However, to understand the nature of adsorption, diffusion, and other complex processes in clay minerals, simulation of these systems in a full quantum mechanical manner becomes infeasible due to the computational burden. The classical molecular dynamics (MD) simulation using an accurate force field (CLAYFF) is a viable alternative, where it has been shown to faithfully reproduce the crystal structures with relatively simple analytical functions that include primarily non-bonded interactions.

Critical to the understanding of adsorption properties in charged and uncharged clays is the disposition of the hydroxyl group in the octahedral sheets. Signature spectroscopic peaks that are characteristic of the hydroxyl group vibrations and bends are easily identified experimentally. With the assistance of molecular simulation, the relationship between hydroxyl group vibrational modes and the molecular-scale structure is explored. In particular, the orientation of the hydroxyl group is closely monitored to understand the effects of the dioctahedral vacancy and counterbalancing cations on the clay behavior. Overall, the structures obtained from ab initio and classical molecular dynamics simulations provide a detailed understanding of the important interactions that influence the adsorption properties of clay minerals.

High Pressure Flow Reactor Test and *In-Situ* FTIR Characterization of Ammonia Absorption/Desorption Performance Over Zeolites

Hanjing Tian[1], *Robert W. Stevens*[1], *Frank Thomas*[1] *and Ranjani V. Siriwardane*[2]

1. Parsons Infrastructure & Technology Group, Inc, Morgantown, WV 26507-0880
2. U.S. Department of Energy, National Energy Technology Laboratory, Morgantown, WV 26507-0880

This technical effort was performed in support of the U.S. Department of Energy /National Energy Technology Laboratory's on-going research in multi-contaminants removal from coal synthesis gas under the RDS contract DE-AC26-04NT4181

Introduction

The integrated gasification combined cycle (IGCC) process produces electricity from coal with high efficiency by converting the fuel to coal gas. Coal gas can be utilized for various applications, such as a gas turbine or fuel cells. These applications require contaminant-free gas. Hence, processes must be incorporated to remove impurities that are formed during gasification, such as sulfur, chorine, ammonia, and heavy metal constituents indigenous to the coal. One highly undesirable impurity is ammonia, a reactive, corrosive gas. Removal of ammonia species from hot coal gas will eliminate subsequent production of NOx during combustion. The ammonia concentration is about 50 to1000 ppm in typical coal-based synthesis gas, and the DOE removal target is set as less than 10 ppm.

Many efforts have been conducted on processes to remove ammonia. The initial attempt was catalytic decomposition ammonia to nitrogen and hydrogen. However, the nickel and molybdenum-based catalyst utilized during these studies deactivated rapidly when significant levels of H_2S were present in the fuel gas. Recently, researchers have conducted an ammonia removal by sorption process. SRI international has investigated the sorption performance of zeolite samples for ammonia removal. Various H-based zeolites were tested under 300 psi at 200 °C; the results indicated that Y-type zeolite would be a promising sorbent for ammonia adsorption at elevated temperature. The best performance (0.0045g NH_3/g sorbent) was obtained by CBV712 with a 300 °C activation. In addition, the changes in the silica-to-alumina ratio and the BET surface area of the Y-type zeolite did not have a significant effect on the ammonia adsorption capacity. However, these studies were conducted in the absence of steam, which is present in coal synthesis gas. Hence, it is necessary to understand the effect of steam on the ammonia sorption process.

In current study, we have investigated ammonia sorption over several zeolites at high temperature and pressure. Four reaction-regeneration cycles were conducted in a bench scale reactor to test the reactivity and stability of zeolites on ammonia sorption reaction in the presence of steam. High pressure FTIR was also applied to study the reaction and regeneration mechanism of ammonia over zeolites.

Experimental setup and conditions

Sample preparation:

Six zeolite samples (5A, 4A, H-ZSM5, 13X, Z-10-10 and Z-10-8) were obtained from Fisher and Zeochem. For comparison, pure alumina support (Sigma-Aldrich, 150 mesh and 155 m^2/g) was also tested.

Bench-scale screening test:

The sorption/regeneration tests of ammonia removal over zeolite samples were conducted on NETL high-pressure reactor (HPR) facility. 0.5 mg of zeolite sample were loaded in the reactor and treated at 350°C via N_2/ He mixture (50 sccm each) for 1.5 hour. Each sample was tested for 4 reaction /regeneration cycles in both the absence and presence of steam (0.05 ml/min). The reaction tests were conducted at 315°C with 1% NH_3 /N_2 with a total flow rate of 150 sccm until breakthrough saturation, which is defined as a 100 ppm total NH_3 concentration. Following breakthrough, the inlet gas was replaced by He (100 sccm each) and depressurized to 1 atm. Regeneration was then initiated by heating the samples to 350 °C in He for 1.5 hous. Subsequent to regeneration, the reactor unit was cooled back to 315 °C and for next sorption cycle.

Empty reactor bed tests were conducted, and the time corresponding to dead space for the reactor was measured as 4.5 minutes. For the reaction in the absence of steam, data collection was conducted by an online MS at the outlet of reactor. For the reaction with steam, the data collection was conducted by an online multi-gases analyzer with 10 times dilution by N_2. Adsorption capacities were calculated from the measured breakthrough data.

FTIR experiment:

Powered zeolite samples were used to maximize spectral quality via diffuse reflectance infrared Fourier transform spectroscopy (DRIFTS) analysis. For each test, approximately 20 mg of zeolite sample was charged to a DRIFTS reactor cell (Thermo-Nicolet model 0031-902). The DRIFTS cell is a cold-walled reactor that features temperature control and ZnSe windows for IR transmission within the 650-4000 cm^{-1} range. Infrared spectra were collected with a thermo-nicolet Nexus 670 FTIR instrument, which housed the DRIFTS reactor cell. Spectra were collected at the resolution of 4 cm^{-1} with 50 co-added scans.

2000 ppm NH_3 in N_2 gas was introduced to the cell at a total flow rate of 40 sccm. Prior to each experiment, the zeolite samples were pre-treated *in-situ* by heating to 350 °C and 1 atm in flowing N_2 for 1 hour. The pre-treated sample was then cooled to adsorption temperature (315 °C), and the cell was pressurized to 200 psi. The pre-treated zeolite sample at the desired adsorption temperature served as the IR background for each experiment. Adsorption of ammonia onto the zeolite samples was conducted using 2000 ppm NH_3/N_2. IR spectra were collected during selected intervals to observe changes in the sample due to the reaction. The adsorption phase of the test was conducted for about 30 minutes. After the adsorption reaction, the system was returned to the desorption condition (1 atm and 350 °C) and FTIR analysis was conducted during desorption.

Results and discussion

1. Optimization of experimental condition:

In order to select the best experimental condition for sorbent screening, zeolite 5A was tested at different conditions. In Table 1, the sorption capacity of zeolite 5A, tested at different reaction temperatures and pre-treatment temperatures, is listed. The results indicated that the sorbent performance (breakthrough time and sorption capacity) decreased significantly as the temperature increased. For a given reaction temperature, higher pre-treatment temperatures also had a negative influence on the sorbent performance.

Table 1. sorption performance of zeolite 5A under various reaction condition

Pretreatment Temperature (°C)	Reaction Temperature (°C)	Ammonia flow rate (sccm)	Breakthrough time (mins)	sorption capacity (mol NH_3/Kg absorbent)
350	150	150	58	7.03
350	250	150	53	6.82
350	315	150	48	5.88
500	315	150	36	4.75

2. Sorption performance in the absence of steam

The sorption performances of various zeolites are summarized in Table 2. Among the six zeolite samples tested, 5A, 4A and Z-10-10 indicated excellent sorption capacity in the first cycle. Zeolite 5A exhibited excellent sorption performance initially, but the sorption capacity decreased in cycles 3 and 4, though it still possessed better performance than other samples. Zeolite 4A also had good initial performance but rapidly lost the adsorption ability. Although the initial reactivity of zeolite Z-10-10 is not outstanding, its activity increased during the multi-cycle test and the performance did not change after multi-cycle tests. The sorption mechanism of zeolite Z-10-10 needs to be further investigated. In summary, zeolite 5A is identified as a good candidate of sorbent for ammonia removal.

Table 2. Ammonia sorption capacities (NH_3 moles/Kg sorbent) zeolites

Absorbent Samples	Cycle 1	Cycle 2	Cycle 3	Cycle 4
5A	5.88	6.00	4.75	4.16
4A	4.77	4.38	2.04	2.08
H-ZSM5	3.05	2.37	2.13	1.04
Z-10-10	3.35	4.34	3.35	3.45
13X	3.27	-	1.84	1.82
Z-10-8	2.29	1.08	3.72	3.37
Al_2O_3	3.44	3.26	3.33	3.36

3. Sorption performance in the presence of steam

A typical coal-based synthesis gas contains 6% water. Thus, it is very important to study the ammonia sorption on the presence of steam. During the flow reactor experiments, 6.2% steam was introduced.

The sorbent capacities of zeolite 5A and 4A during four-cycle reaction/regeneration tests in the presence and absence of steam are listed in Table 3. It was observed that the steam has a minor effect on the initial reactivity, but it significantly decreases the reaction performance in multi-cycle tests. After 4 cycle tests with steam, zeolite 5A lost 32.7% activity (compared to 29.3% in the reaction without steam), zeolite 4A lost 76.2 activity (compared to 56.4% in the reaction without steam). Therefore, it is very critical to develop water-resistant zeolite samples for ammonia removal from coal-based synthesis gas.

Table 3: ammonia sorption capacities with and without steam

Sorbent materials	Sorption capacity (without water)		Sorption capacity (with water)	
	1st cycle (mol/Kg)	4th cycle (mol/Kg)	1st cycle (mol/Kg)	4th cycle (mol/Kg)
5A	5.88	4.16	5.78	3.89
4 A zeolite	4.77	2.08	4.89	1.16

4. FTIR characterization:

NH_3-FTIR characterizations were conducted to investigate the sorption mechanism of ammonia over zeolite 5A under high pressure. The FTIR spectra of sorption reactions are shown in Figure 1a) and Figure 1b over different time intervals. A very small band at 3740 cm^{-1}, originating from terminal OH groups bound to silica framework atoms (called silanol group), was observed at all ammonia exposures as shown in Figure 1a. Three ammonia species contribute to ammonia sorption: the strong physisorbed ammonia species at 3485 cm^{-1}, chemisorbed ammonia on Bronsted acidic site at 1436cm^{-1}, and chemisorbed ammonia on Lewis acidic sites at 1276 cm^{-1}. During the ammonia sorption process with the zeolite 5A, the physisorbed species and chemisorbed ammonia associated with Bronsted sites were instantly formed as soon as the ammonia was introduced as shown in Figure 1b. With the increased reaction time, the band intensities of both physisorbed ammonia and chemisorbed ammonia species on Bronsted site increased. After introducing ammonia for 5 minutes, the amount of physisorbed ammonia and chemisorbed ammonia on Bronsted sites reached saturation and remained constant for further exposures. Moreover, the chemisorbed ammonia species associated with the Lewis site were first found after 3 minutes of exposure and reached maximum intensity after 8 minutes of exposure. During ammonia desorption, the FTIR intensities of these three species decrease immediately. After introduction of N_2 for 8 minutes, IR spectra showed trace amount of both physisorbed species and chemisorbed ammonia associated with Bronsted site, while the ammonia species with Lewis sites totally disappeared.

The FTIR study of ammonia sorption and desorption indicted that the acidic strength of Lewis sites are weaker than that of Bronsted sites over zeolite 5A since the ammonia is readily absorbed at Bronsted sites and desorbed at Lewis sites. However, the relative surface concentration of Bronsted to Lewis acid could not be accurately determined due to a broad IR band of NH_3 adsorbed on Lewis sites.

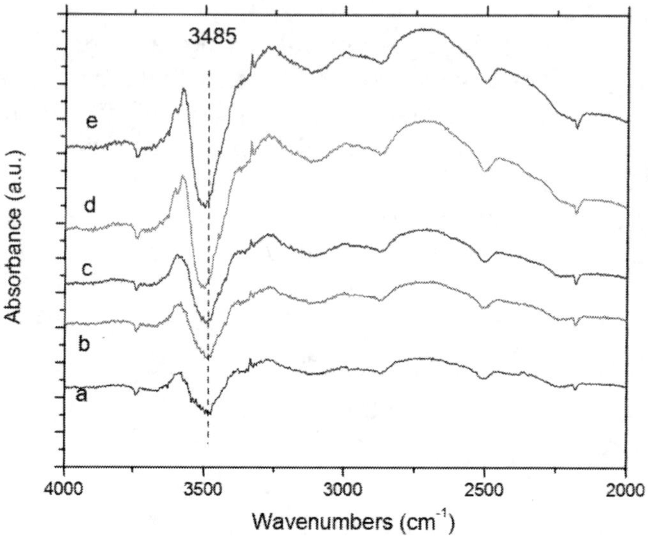

Figure 1a: IR spectra (region 4000-2000 cm^{-1}) of zeolite 5A at 315 °C and 200 psi after exposure to 2000 ppm for a) 0.5 mins., b) 2 mins., c) 3.5 mins., d) 5.5 mins., and e)8 mins.

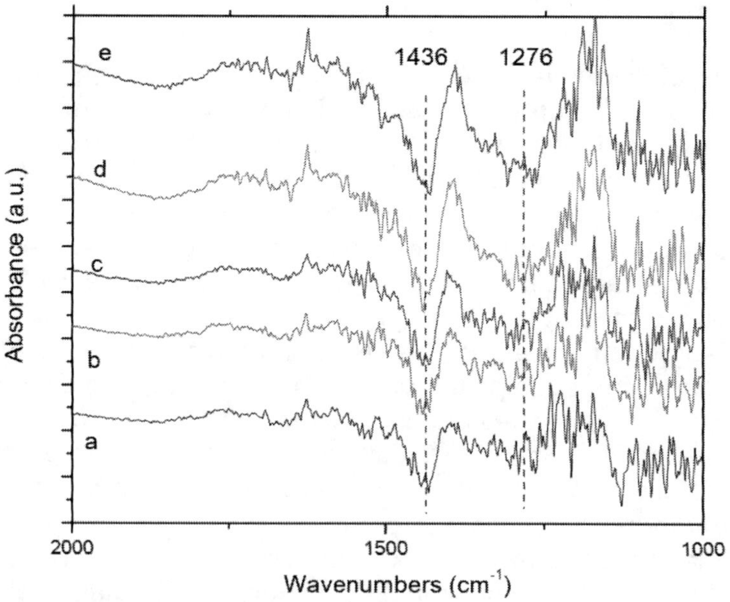

Figure 1b: IR spectra (region 2000-1000 cm^{-1}) of zeolite 5A at 315 °C and 200 psi after exposure to 2000 ppm for a) 0.5 mins., b) 2 mins., c) 3.5 mins., d) 5.5 mins., and e)8 mins.

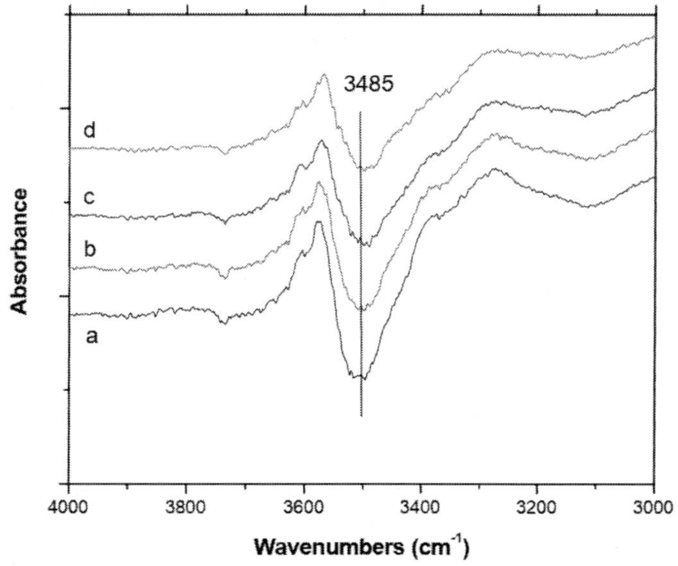

Figure 2a: IR spectra (region 4000-2000 cm^{-1}) of zeolite 5A at 350 °C and ATM after ammonia sorption and exposure to N_2 for a) 0.5 mins., b) 3 mins., c) 5 mins., and d) 10 mins.

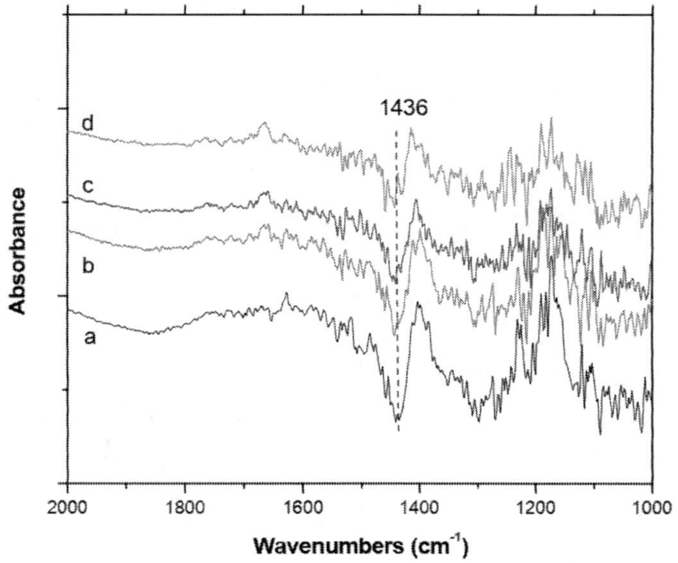

Figure 2b: IR spectra (region 2000-1000 cm^{-1}) of zeolite 5A at 350 °C and ATM after ammonia sorption and exposure to N_2 for a) 0.5 mins. b) 3 mins. c) 5 mins., and d)10 mins.

Conclusion

1. Six zeolites were successfully tested for ammonia sorption-regeneration both with and without steam, at high pressure and high temperature.
2. Zeolite 5A indicated good ammonia sorption performance in the first cycle, but deactivated significantly in the subsequent reactions cycles.
3. Addition of water to the gas steam could accelerate zeolite deactivation for ammonia sorption
4. FTIR spectra indicated the formation of three surface species for ammonia adsorption over zeolite 5A: physisorbed species, chemisorbed species on Bronsted sites, and chemisorbed species weakly interacting with Lewis sites.

346e Far Fourier-Transform Infrared Spectroscopy As A Characterization Tool For Zeolite-Like Porous Adsorbents

Tyler M. McGown[1], Xin Sun[2], **Jennifer L. Anthony**[2], and John R. Schlup[3]. (1) Burns & McDonnell, Kansas City, MO 64131, (2) Chemical Engineering, Kansas State University, 105 Durland Hall, Manhattan, KS 66506, (3) Department of Chemical Engineering, Kansas State University, 105 Durland Hall, Manhattan, KS 66506

Far-infrared (FIR) spectroscopy has been shown to be capable of identifying compositional and structural differences in zeolitic materials. The ability to observe differences between similar zeolites in the near- and mid-infrared regions can be limited; however, notable variations can become apparent in the far-infrared region. Although it is often an underused resource, this work illustrates the clear potential for using far-infrared spectroscopy to explore the inorganic and coordinated metal bonding sites in zeolite-like molecular sieves.

Diffuse reflectance Fourier transform infrared (DRIFT-IR) spectroscopy was employed to characterize a range of zeolites and zeolite-like molecular sieves in their powdered forms. Whiles spectra were collected from 8500 cm^{-1} to 150 cm^{-1}, the far-IR region (400 cm^{-1} - 10 cm^{-1}, but typically < 650 cm^{-1} in practice) was of particular interest as it can yield information on bonding of heavier elements, such as metals within the framework of the zeolite, tetrahedral atom-oxygen bending, and pore openings in zeolites. The advantage of these particular experiments is that diffuse reflection spectroscopy permits IR spectra of powders to be acquired with a minimum of sample preparation. Far-IR spectra result from vibrations involving elements heavier than those typically observed in mid-IR spectra or coordinated vibrations involving structures such as rings. Such species are expected with the zeolite-like materials being considered and in the products resulting from processes such as ion-exchange or calcination. Thus, these experiments are expected to address the questions often asked of these materials more directly than other spectroscopic techniques.

Results are presented for a series of zincosilicates with different framework structures (VPI-7, VPI-9, and VPI-10). Dramatic differences were observed across the far-IR region for each of the materials. Spectra are compared for as-made samples as well as calcined, ion-exchanged, and intentionally collapsed samples. Subtle differences in the far-IR region were identified for materials that had the same crystalline framework but different composition (zeolite beta (*BEA) and the zincosilicate CIT-6, treated to yield pure silicate, aluminosilicate, and zincosilicate forms).

346f Synthesis and Characterization of TiO2 Ion-Sieves for Lithium Adsorption

Qin-Hui Zhang, Chemical Engineering, East China University of Science and Technology, No. 130 Meilong Road, Xuhui District, Shanghai, 200237, China

TiO2 nanotubes (as in Fig. 1) were prepared by hydrothermal treatment of TiO2 nanoparticles in NaOH aqueous solution, and the TiO2 ion-sieves for lithium selective adsorption were successfully synthesized via a simple solid-state reaction and acid treatment process from the nanomaterials. The TiO2 nanomaterial have been characterized by XRD, TEM, HRTEM, SAED and BET methods, and the lithium ions adsorption property was tested via the adsorption isotherm and selectivity measurement. Results show that the ion-sieves prepared from different polymorphs have a remarkable selective adsorption capacity for lithium ions from brine or sea water (as in Fig. 2).

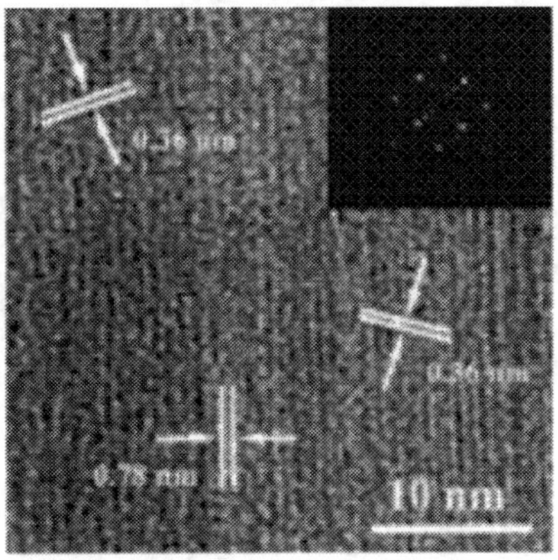

Fig. 1 HRTEM and SAED (inset) images of TiO_2 nanotube ion-sieve

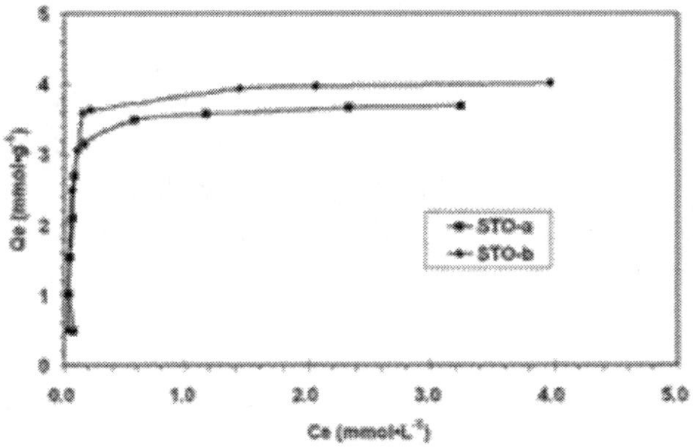

Fig. 2 Lithium ion adsorption isotherm of the TiO_2 nanomaterial

346g Adsorption Of Hydrogen And Argon On [Zn(Bdc)(Ted)0.5] ·2DMF·0.2H2O Metal-Organic Framework

Vaiva Krungleviciute[1], Kathleen Lask[1], Murat Bulut[1], Aldo D. Migone[1], Hyuna Lim[2], and Jing Li[2]. (1) Department of Physics, Southern Illinois University, Neckers 483A, Carbondale, IL 62901/4401, (2) Department of Chemistry and Chemical Biology, Rutgers University, 610 Taylor Road, Piscataway, NJ 08854

We performed volumetric adsorption isotherm measurements on a new[Zn(bdc)(ted)0.5]•2DMF•0.2H2O (bdc = benzenedicarboxylate, ted = triethylenediamine) metal-organic framework. Before starting our adsorption measurements, we outgassed this sample under vacuum at 373 K for ~15 hours. After the outgassing was completed the sample cell was transferred and reconnected to the adsorption setup without reexposing it to air. Hydrogen and argon adsorption isotherms were measured at 66 and 77 K. In addition to determining equilibrium adsorption isotherms for these adsorbates, we also measured the kinetics of their adsorption processes. The time needed to reach equilibrium after a dose of gas was admitted into the cell is around ~40,000 seconds for both gases in the low pressure, low coverage region.

350a Optimum Batch Distillation of Multicomponent Mixtures

Zbigniew T. Fidkowski, Air Products, 7201 Hamilton Boulevard, Allentown, PA 18195

Batch distillation is still widely used in numerous chemical plants that make multiple products at a relatively small scale. As in continuous distillation, various combinatorial sequences of batch separations can be derived. Each of these sequences can be optimized using commercial simulator and compared with the other sequences. The objective function in optimization of batch distillation is usually the cycle time that needs to be minimized. It corresponds to the minimum energy of separation in continuous distillation, because batch distillation columns operate usually at maximum boilup. Therefore, the shorter the cycle time, the lower the energy of separation. Optimization procedure has to take into account sizes of existing tanks and other available equipment, which may lead sometimes to a classical scheduling problem.

An example of selecting the optimum batch distillation sequence will be presented. The components in the feed may be lumped into five groups, ordered according to their volatilities: 1. Lights (L), 2. Product A, 3. Intermediates (I) , 4. Product B, 5. Heavies (H). Combining optimization with batch column model available in commercial simulator provides a useful tool for developing optimum batch distillation recipes.

350b A Rate-Based Process Modeling Study of Co2 Capture with Aqueous Mea Solution

Chau-Chyun Chen, Aspen Technology, Inc., 10 Canal Park, Cambridge, MA 02141 and Davy Zuo, Aspen Technology (Beijing), Ltd., 2966 Jin Ke Road, Pudong, Shanghai, China.

Recently available pilot plant data from U.Texas-Austin for carbon dioxide (CO2) capture with aqueous monoethanolamine (MEA) solution have been investigated and simulated with RateSep, a second generation rate-based absorption & stripping unit operation model in Aspen Plus. After a brief review on RateSep, thermodynamic and kinetic models, transport property models, and U.T.-Austin pilot plant data set, we present model predictions against the pilot plant data and we show the superior predictions of the rate-based model over the traditional equilibrium-stage models. We further examine the impacts on simulation results of various rate-based modeling constructs, i.e., film discretization, flow mixing, interfacial area correlations, etc. In addition, we report the sensitivities of the process performance with respect to liquid to gas ratio, CO2 concentration in gas stream, CO2 loading and MEA concentration in lean MEA solution, operating pressure, packing height and packing type. The RateSep model with proper parameters provides excellent predictive capability for the process and it should be very helpful for industrial process and plant design.

350c Rate-Based and Equilibrium Model Approach for Absorption of Aromatic Hydrocarbons in Multicomponent Mixtures: Simulation and Experiment

Diethmar Richter[1], Jens-Uwe Repke[2], Martin Kohrt[2], Holger Thielert[3], and Guenter Wozny[2]. (1) University of Technology Berlin, Straße des 17. Juni 135, Berlin, 10623, Germany, (2) Department of process dynamics and operation, Technical University of Berlin, Straße des 17. Juni 135, Berlin, 10623, Germany, (3) ThyssenKrupp Uhde / EnCoke GmbH, Friedrich-Uhde-Strasse 15, Dortmund, 44141, Germany

The determination of the absorption of aromatic hydrocarbons from coke-oven gases is strongly influenced by the complexity of the applied washing fluid. In the case of coal tar oil the number of components (400 up to 500) present in the mixture is inhibiting the application of conventional calculation techniques by binary interaction parameters. Investigations on an industrial plant (Schwelgern, Germany) have shown that the application of available gE models offered by process simulators are not able to calculate this process in an accurate manner and therefore great deviations occur between the measured process parameters from an industrial plant and the simulation results. For a better description of the regarded process comprehensive experiments were carried out in order to determine the phase equilibrium behaviour of the main aromatics, i.e. benzene, toluene and xylenes in coal tar oil whereby the washing oil mixture is treated as one pseudo component. Based on these studies an equilibrium as well as rate-based approach are developed and implemented in FORTRAN for simulating the regarded absorption process in the commercial simulation software ChemCAD. The calculation of diffusion coefficients are according to the methods of Fuller for the gas phase and Wilke-Chang for the liquid phase. Mass transfer coefficients are calculated by the methods of Billet and Schultes. The comparisons of the calculated profiles clarify that the assumption of temperature equilibrium between the present phases in the case of the equilibrium model is also leading to wide differences in the calculated concentration profiles.

On the basis of experiments carried out in a pilot plant the concentration as well as temperature profiles are evaluated for the recovery of benzene and toluene in coal tar oil. Therefore a pilot plant equipped with 2.6 m of structured packing Mellapak 350.Y with a diameter of 0.1 m was constructed at the laboratories of the University of Technology Berlin. In order to minimize the effects of maldistribution liquid collector and distributor systems were placed each 0.85 m along the column height. Gas samples are taken at four different points along the column and the temperature profile is measured at eight points between the gas and the liquid outlets. The operation conditions were varied for liquid loads between 6.5 and 11.0 m^3/m^2h as well as gas loads between F = 1.3 and 1.6 in order to obtain different rates of recovery. Gas concentrations at the inlet of the column were varied between 2.0 and 15 g/m^3 of benzene and 1.0 to 5 g/m^3 of toluene. The evaluation of the applied models is carried out by a great number of measurements. In the presentation both models and their applicability for the complex mixture will be discussed based on experimental investigations

As a result of these investigations the industrial process of the coke oven gas purification can be simulated in a more accurate manner and therefore turns out more efficiently in operation and process design.

Modelling, Simulation and Experimental Investigation of Reactive Hybrid Separation Processes

Carsten Buchaly, Peter Kreis(*), Andrzej Górak

University of Dortmund, Department of Biochemical and Chemical Engineering, Chair of Fluid Separation Processes, Emil-Figge-Strasse 70, 44227 Dortmund, Germany

Abstract

In this paper, the modelling, simulation and experimental investigation of a hybrid separation process, the combination of reactive distillation with membrane separation, is presented. The application is illustrated by the heterogeneously catalysed *n*-propyl propionate synthesis from 1-propanol and propionic acid. The membrane module is located in the distillate stream of the reactive distillation column in order to selectively remove the produced water without use of entrainers. For the stand-alone reactive separation process, the simulation results with a non-equilibrium model are in good agreement with experimental data obtained in a pilot scale column. Additionally, a comparison of the most common modelling depths, namely the non-equilibrium model with Maxwell-Stefan equations, the equilibrium model taking into account reaction kinetics and the equilibrium model assuming chemical equilibrium, is presented. Vapour permeation experiments using Sulzer Pervap™ 2201(D) have been performed in a pilot-scale membrane plant.

Keywords: modelling, hybrid processes, experimental investigation, reactive distillation, membrane separation, vapour permeation

1. Introduction

Reactive distillation is an important example of a reactive separation process. Especially for equilibrium reactions like esterifications, ester hydrolysis and etherifications, the combination of reaction and separation within one zone of a reactive distillation column is a well-known alternative to conventional processes with sequential reaction and separation steps [1, 2]. In several cases, non-ideal aqueous-organic mixtures are formed which tend to form azeotropes. They can be overcome using membrane separations like pervaporation and vapour permeation since they are very selective and not limited by vapour-liquid equilibrium [3]. Consequently, a hybrid process consisting of membrane-assisted reactive distillation contributes to sustainable process improvement due to arising synergy effects and allows for reduction of investment and operational costs.

In this work, the heterogeneously catalysed esterification of propionic acid (ProAc) with 1-propanol (POH) to *n*-propyl propionate (ProPro) and water (H2O) is investigated:

$$C_3H_8O + C_3H_6O_2 \Leftrightarrow C_6H_{12}O_2 + H_2O \qquad (1)$$

As catalyst, the strongly acidic ion exchange resin Amberlyst 46™ from Rohm & Haas is used. Amberlyst 46™ has acidic active sites (sulfonic-acid groups) only at the surface of the styrene-co-divinylbenzene matrix [4].

2. Process description

One possible process alternative for *n*-propyl propionate synthesis in one apparatus is the removal of the desired product (ProPro) at the bottom of the reactive distillation column while at the top, an almost azeotropic aqueous-organic mixture (POH/H2O) is obtained. A hydrophilic membrane unit is located in the distillate stream to remove water out of the process. The water depleted retentate is recycled back to the column. The coupling of the reactive distillation column with a membrane module results in a hybrid process (Figure 1).

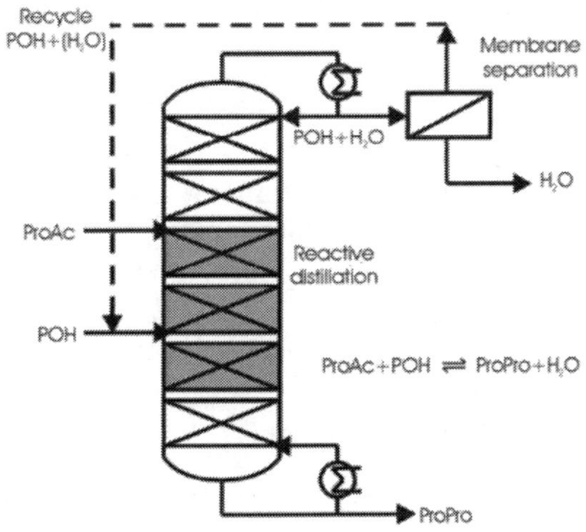

Figure 1: Reactive distillation column with a membrane separation located in the distillate stream.

3. Reactive distillation – modelling

Numerous papers about the modelling of reactive distillation processes can be found in the literature; comprehensive reviews about this topic are written by [5, 6, 7, 8]. They identify the key aspects for the proper description of reactive distillation processes being as follows:

- the description of thermodynamic and physical properties of the multicomponent system,
- the use of accurate reaction kinetics,
- the modelling of mass and heat transfer between liquid and vapour phase,
- the hydraulic characteristics of the column internals in terms of correlations for hold-up and pressure drop and
- the validation of the proposed process model with reliable experimental data.

3.1. Thermodynamic and physical properties

For the calculation of thermodynamic and physical properties, the software package Aspen Properties^{TM} (www.aspentech.com) is used. The UNIQUAC model for the calculation of the activity coefficients in the liquid phase is applied to take into account the liquid phase non-idealities. The Hayden-O'Connel equation of state is chosen to take into account the non-idealities in the vapour phase, e.g. the dimerisation of the carbonic acid.

3.2. Reaction kinetics

The use of accurate data for the reaction kinetics is crucial for a predictive column design. For the modelling of the heterogeneously catalysed synthesis of *n*-propyl propionate, a pseudo-homogeneous approach is applied:

$$r_i = m_{cat,dry} \cdot V_i \cdot c_{act} \cdot \left(k_1(T) \cdot a_{ProAc} \cdot a_{POH} - \frac{k_1(T)}{K_{eq}(T)} \cdot a_{ProPro} \cdot a_{H_2O} \right), \quad i = 1, \ldots, n_c. \qquad (2)$$

The temperature dependency on the activity-based equilibrium constant K_{eq} as well as the rate constant of the forward reaction k_1 are taken into account via the Arrhenius approach. Both parameters have been measured in the expected operating range of the reactive distillation column [9, 10].

$$k_1(T) = 7.381 \cdot 10^7 \cdot \exp\left\{\frac{-5.963 \cdot 10^4}{R_m \cdot T}\right\}; \quad K_{eq}(T) = 6.263 \cdot \exp\left\{\frac{4.519 \cdot 10^3}{R_m \cdot T}\right\} \qquad (3)$$

3.3. Reactive distillation model

In this work, a non-equilibrium stage model (NEQ model) is applied, which considers both the multicomponent mass transfer as well as heat transfer rates [11, 12, 13]. The influence of the hydrodynamic is taken into account by packing-specific correlations for mass-transfer coefficients in the vapour and liquid phase, interfacial area, liquid hold-up and pressure drop. The catalytic section of the reactive distillation column is equipped with the reactive structured packing Katapak-SP (type 11, Sulzer Chemtech); the structured packing Sulzer BX is used inside of the enrichment and stripping section. All model equations are implemented into the simulation environment Aspen Custom Modeler^{TM}. The peripherals, e.g. reboiler and condenser are assumed to be ideal.

4. Reactive distillation – experiments

For model validation purposes, a series of reactive distillation experiments have been carried out in a glass column with an inner diameter of 50 mm at atmospheric pressure. The set-up has a separation section below (0.5 m) and above (2.4 m) the reactive section (2.6 m). A total packing height of 5.5 m is realised. The mass of dry catalyst per meter of catalytic packing has been determined experimentally to $m_{cat,dry}$ = 0.205 $kg_{cat,dry}/m_{packing}$, whereas the activity of the catalyst has been measured to cact = 0.78 eq/$kg_{cat,dry}$ after the reactive distillation experiments have been successfully performed. A comparison between representative experimental results and model prediction is given in Figure 2, in which the liquid-phase molar fraction is shown along the packing height.

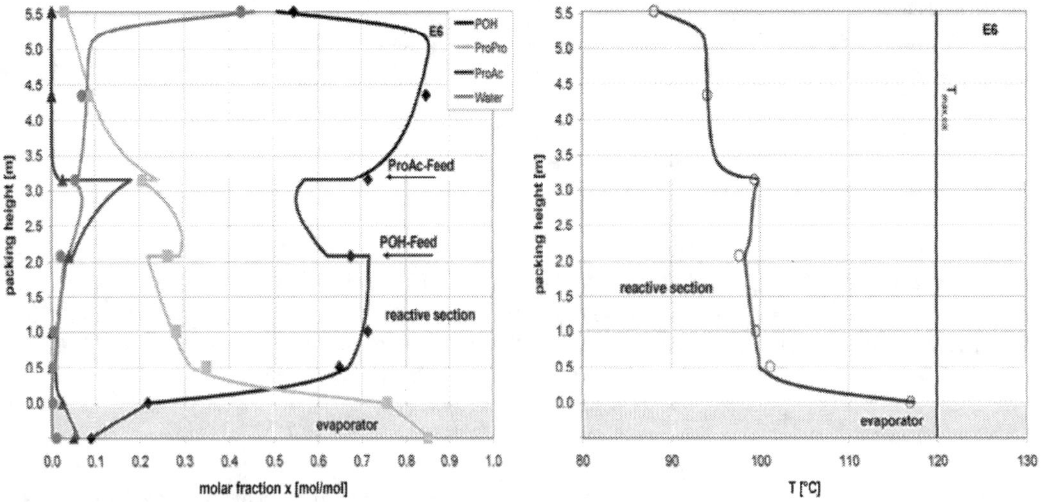

Figure 2: Liquid phase molar fraction (left) and temperature profile of the vapour phase (right) for experiment E6 (feed stream: 1.98 kg/h, RR: 2.49, D/Fmass: 0.424, molar feed ratio POH/ProAc: 2.068). The solid lines represent the simulation results, the symbols the experimental values [14].

5. Influence of modelling depth

The choice of an appropriate modelling depth, e.g. the use of the EQ or NEQ model for reactive distillation is not a trivial decision and has to be verified for each specific separation task. The most common modelling depths,

- the non-equilibrium model,
- the equilibrium model taking into account the reaction kinetics and
- the equilibrium model assuming chemical equilibrium,

have been applied in this paper. Since the desired product, *n*-propyl propionate, is removed via the bottom stream of the reactive distillation column, the impacts of the different modelling depths are investigated by a varying of the catalyst volume fraction, ψ_{cat}, and its influence on the acid conversion and therefore on the composition of the bottom stream (Figure 3). For the EQ model assuming chemical equilibrium the composition of the bottom stream is independent of ψ_{cat}, showing a product purity of $x_{ProPro,bottom}$ = 86.5%, while $x_{POH,bottom}$ = 13.3% and $x_{ProAc,bottom}$ = 0.2%.
The validated reference NEQ model shows a strong influence of the catalyst volume fraction on the composition of the bottom stream up to ψ_{cat} = 80%. With an increase in the amount of catalyst, which leads to a higher conversion of the propionic acid, an increase in the product purity can be achieved. Since an excess of alcohol is used, the 1-propanol composition is almost independent in the investigated range of catalyst volume fraction. The EQ model with reaction kinetics shows qualitatively similar results. Since the reaction is far from chemical equilibrium, the necessity of applying reaction kinetics for a proper description of the *n*-propyl propionate synthesis, which is a kinetically controlled reaction, is confirmed. To achieve the equilibrium conversion, an unrealistic catalyst volume fraction of ψ_{cat} = 160% is required for the EQ model with reaction kinetics.
It can be shown that especially for the actual catalyst amount of the pilot-scale column (DN 50, ψ_{cat} = 23% $\cong$ $m_{cat,dry}$ = 0.205 kg/m), in which the experimental validation has been performed, a lower product purity ($x_{ProPro,bottom}$ = 77%) is predicted by the EQ model in comparison to the NEQ model ($x_{ProPro,bottom}$ = 80%). In this range, the strongest influence of ψ_{cat} on conversion and thus the product purity is observed.

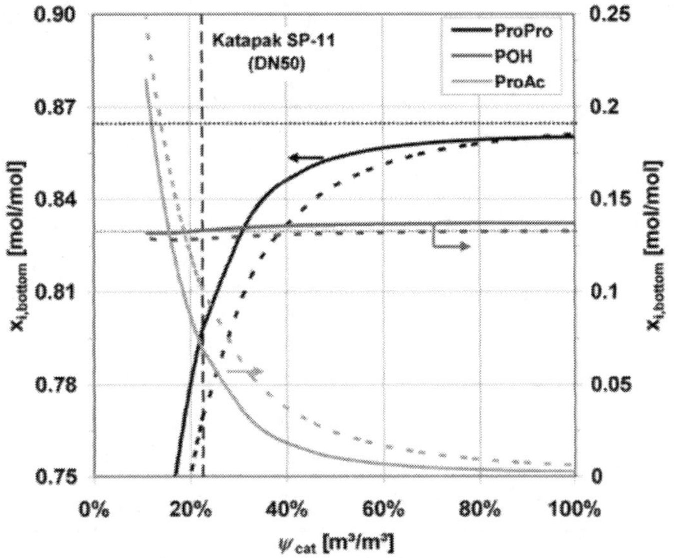

Figure 3: Influence of the modelling depth on the composition of the bottom stream of the reactive distillation column. Non-equilibrium model (solid lines), equilibrium model using reaction kinetics (dashed lines) and equilibrium model assuming chemical equilibrium (dotted lines) [10].

5.1. Vapour permeation

In vapour permeation, volatile components are separated by a non-porous membrane due to different sorption and diffusion properties. Consequently, the separation is very selective and not limited by the vapour-liquid equilibrium. The driving force is generated by lowering the partial pressure of the favourable permeating component on the permeate side by applying a vacuum. In vapour permeation, the feed is supplied as vapour, whereas in pervaporation, the feed stream enters the membrane module as liquid. For the coupling of a membrane separation unit with a reactive distillation column, the membrane is operated in vapour permeation mode to avoid polarisation effects in the boundary layer adjacent to the membrane surface at low feed streams.

5.2. Experimental setup

Vapour permeation experiments are carried out in a pilot-scale plant equipped with a plate and frame membrane module from Sulzer Chemtech and an effective membrane area of 0.5 m² to determine the separation characteristics of the used hydrophilic membrane. The plant is equipped with a process control system and can therefore be operated automatically to ensure steady-state conditions. In this work, the polymeric membrane Pervap™ 2201(D) from Sulzer Chemtech is applied. It consists of an active polyvinylalcohol (PVA) layer and a polyacrylnitrile (PAN) support layer to ensure its mechanical strength [15].

5.3. Experimental investigation

Experiments with a binary 1-propanol/water mixture are performed in order to investigate the separation behaviour of Sulzer Pervap™ 2201(D) with a membrane area of A_{memb} = 0.5 m² to deplete the obtained distillate stream of the column. The experimental data and operating conditions are summarised in Table 1.

Table 1: Experimental data for the operating conditions of the vapour permeation experiments [10]

$\dot{m}_{ret}$ [kg/h]	J_{total} [kg/m²h]	$w_{H20,feed}$ [kg/kg]	$w_{H20,ret}$ [kg/kg]	$w_{H20,perm}$ [kg/kg]	T_{feed} [°C]	□T_{sup} [°C]	p_{feed} [bar]	p_{perm} [mbar]
3.92	0.65	0.136	0.072	0.912	91.6	0.0	1.077	98
3.96	0.66	0.132	0.067	0.921	91.7	0.0	1.075	71
3.90	0.69	0.134	0.064	0.919	91.7	0.0	1.077	71
3.82	0.69	0.131	0.059	0.932	91.8	0.0	1.080	31
2.76	1.55	0.282	0.092	0.957	92.0	2.8	1.095	71
2.73	1.55	0.281	0.090	0.958	92.0	2.8	1.096	71
3.22	1.46	0.243	0.081	0.956	91.6	2.2	1.104	31
1.99	1.50	0.291	0.053	0.920	101.9	2.4	1.603	31
1.96	1.48	0.290	0.053	0.920	102.3	2.8	1.600	31
1.89	1.44	0.295	0.056	0.919	102.4	3.0	1.590	71
1.12	0.86	0.298	0.058	0.922	98.4	10.3	1.032	31
1.16	0.89	0.299	0.059	0.921	97.5	9.4	1.033	32
1.13	0.85	0.296	0.060	0.925	98.7	10.6	1.033	32

It can be seen that the separation characteristics show a strong dependency on the feed concentration.

5.4. Modelling of vapour permeation

For the modelling and simulation of the vapour permeation a detailed and flexible simulation model based on the "solution-diffusion theory" is applied. The model considers the influence of fluid dynamics, porous support layer and non-ideal effects like concentration and temperature polarisation on the process performance. For further details, refer to Kreis and Górak [16]. In this work, an empirical correlation based on two parameters for each component is applied as a transport equation to calculate the transmembrane flux as a function of the water concentration in the membrane feed stream:

$$J_i = Q_i^0 \cdot w_{i,Feed}^A \cdot \left(p_{i,feed} - p_{i,perm}\right)$$

The empirical model does not take into account the temperature dependency. All required parameters are determined by the minimum least square regression method and are listed in Table 2.

Table 2: Determined parameters for the empirical correlation used [10].

Component	Q_i^0	A
Water	897.8	0.74
1-propanol	1.22	-

6. Conclusion

The theoretical and experimental investigation shows that the heterogeneously catalysed synthesis of n-propyl propionate via a hybrid process consisting of reactive distillation and vapour permeation is feasible. For the simulation of the reactive distillation column, a NEQ model has been applied. Based on a series of reactive distillation experiments, a successful model validation has been performed. The most common modelling depths for the description of the reactive distillation process, namely the NEQ model, the EQ model taking into account reaction kinetics and the EQ model assuming chemical equilibrium, have been compared. One major result of this study is that the EQ model assuming chemical equilibrium is not sufficient for a proper description of the *n*-propyl propionate synthesis in a reactive distillation column at laboratory and pilot-scale. The NEQ model and the EQ model with reaction kinetics exhibit qualitatively similar results over a broad range of catalyst volume fractions. The simulation study shows that accurate reaction kinetic data is crucial for the investigation of the reactive distillation process in laboratory and pilot-scale columns. Industrial-scale columns with a high catalyst volume fraction show a lower sensitivity to the reaction kinetic data. Vapour permeation experiments with a binary 1-propanol/water mixture have been performed in a pilot-scale plant with a membrane area of 0.5 m² using the polymeric membrane Pervap™ 2201(D) from Sulzer Chemtech. The experiments proved the suitability of the membrane used and its stability at the operating conditions. In the measured concentration range, the membrane shows a high selectivity and high fluxes. The experimental results have been used to determine the required model parameters for the calculation of the overall transmembrane flux.

7. Acknowledgement

The authors wish to acknowledge the financial support provided by the European commission within the 6th Framework Programme, Project "INSERT-Integrating Separation and Reaction Technologies", Contract-No: NMP2-CT-2003-505862 and within the 5th Framework Programme, Marie Curie Training Site "Reactive Separations", Contract Nr. HPMT-CT-2001-00408. The support of "Fonds der Chemischen Industrie" with the required chemicals is highly appreciated.

References

[1] Hiwale, R.S., Mahajan, Y.S., Bhate, N.V., Mahajani, S.M., (2004) *Int. Journal of Chem. Reac. Eng.*, 2

[2] Kaibel, G., Miller, C., Holtmann, T., Schoenmakers, H., (2005) *Chem. Ing. Tech.*, 77, No.11, 1749-1758

[3] Rautenbach, R., *Membranverfahren- Grundlagen der Modul- und Anlagenauslegung*, Springer-Verlag, Berlin Heidelberg, 1st edn. (1997)

[4] E.G.Lundquist, Catalyzed esterification process, *U.S. Patent 5 426 199*, Rohm and Haas Company (1995)

[5] R. Taylor, R. Krishna, Modelling reactive distillation, *Chem. Eng. Sci.*, 55(2000) 5183-5229

[6] C. Noeres, E.Y. Kenig, A. Górak, Modelling of reactive separation processes: reactive absorption and reactive distillation, *Chem. Eng. Proc.*, 42(2003) 157-178

[7] E.Y. Kenig, A. Górak, in F.J. Keil (ed.), *Modelling of Process Intensification*, Wiley-VCH, Weinheim, 1st edn., 2007

[8] J. Richter, A. Górak, E.Y. Kenig, in H. Schmidt-Traub and A. Górak (eds.), *Reactive distillation*, Springer, Heidelberg, 2006

[9] C. Duarte, C. Buchaly, P. Kreis, J.M. Loureiro, Esterification of propionic acid with n-propanol catalytic and noncatalytic kinetic study, *Inz. Chem. Procesowa,* 27(2006) 273-286.

[10] C. Buchaly, P. Kreis, A. Górak, Hybrid separation processes-Combination of reactive distillation with membrane separation, *Chem. Eng. Proc.*, 46(2007) 790-799.

[11] M. Klöker, E.Y. Kenig, A. Hoffmann, P. Kreis, A. Górak, Rate-based modelling and simulation of reactive separations in gas/vapour-liquid systems, *Chem. Eng. Proc.*, 44(2005) 617-629

[12] A. Górak, A. Hoffmann, Catalytic Distillation in Structured Packings: Methyl Acetate Synthesis, *AIChE J.*, 47, No. 5(1999) 1067-1076

[13] R. Schneider, C. Noeres, L.U. Kreul, A. Górak, Dynamic Modelling and Simulation of Reactive Batch Distillation, *Comp. Chem. Eng.*, 23(1999) 423-426

[14] C. Buchaly, P. Kreis, A. Górak, in E. Sorensen (ed.), *Distillation & Absorption 2006*, IChemE Symposium Series No. 152, ISBN 100852955057, 2006, 373-383

[15] D. van Baelen, A. Reyniers, B. van der Bruggen, C. Vandecasteele, J. Degreve, Pervaporation of Binary and Ternary Mixtures of Water with Methanol and/or Ethanol, Sep.Sci Tec., 39(2004), 563-580

[16] P. Kreis, A. Górak, Process analysis of hybrid separation processes-combination of distillation and Pervaporation, *Chem. Eng. Res. Des.*, 84(2006) 595-600

350e Benchmarking Packing Performance Correlations With A Nonequilibrium Model

Harry Kooijman and Ross Taylor. Chemical Engineering, Clarkson University, Potsdam, NY 13699

Recently published correlations for predicting capacity, pressure drop and mass transfer were benchmarked on distillation test data of modern random and structured packings, using the nonequilibrium simulator ChemSep. The new models are compared to the established correlations and some recommendations are made for when to apply which correlation dependent on packing type. Packings include Sulzer Mellapak (Plus), Raschig Super-pak, Montz B(M), Flexipak (HC), Raschig Super-Ring, amongst others.

Web Page: www.chemsep.org/downloads/source.html

350f Modeling The Performance Of Conventional And High Capacity Structured Packings

Zarko Olujic, Separation Technology, Delft University of Technology, Leeghwaterstraat 44, Delft, 2628 CA, Netherlands

Recently, total reflux experiments have been carried out with a number of common and high performance corrugated sheet structured packings, using facilities available at SRP, University of Texas at Austin (0.17 to 4.14 bar pressure, cyclohexane/n-heptane, internal diameter 0.43 m, packed height up to 3.5 m), Delft University of Technology (atmospheric pressure, cyclohexane/n-heptane, internal diameter 0.45 m, packed height up to 2 m) and at Bayer Technology Services (0.1 bar and 1 bar pressure, chlorobenzene/ethylbenzene, internal diameter 0.6 m, packed height up to 4 m). As expected the pressure drop of the packing increases and capacity decreases with increasing operating pressure, irrespective of the test system. However, the mass transfer efficiency seems to be affected by the nature of the system. In case of the CH/nH system, the efficiency tends to increase with increasing operating pressure, while with CB/EB system there is no effect at all or there is an opposite effect. The latter suggests that packing size could also play a role, or, which is difficult to accept, presence of some peculiar (undetected) effect during that particular experiment. The observed test system dependence of mass transfer efficiency imposes certain difficulties with respect to applicability of a modeling approach striving to avoid use of any empirical, packing and/or system specific constants. The related problems will be discussed using a versatile predictive model developed in Delft to enable predicting the performance of structured packing based on variations in macro geometric characteristics of the corrugated sheets only, which proved to be capable of predicting geometry related effects on pressure drop side accordingly, but exhibits difficulties in following observed mass transfer trends.

353a Crystallization Research Past, Present And Future

Allan S. Myerson, Chemical Engineering, Illinois Institute of Technology, Chicago, IL 60616

Crystallization from solution is an important separation and purification process in the chemical, pharmaceutical and food industries. The past 40 years has seen great advances in our understanding of the fundamentals of crytallization. This talk with discuss major developments in crystallization research and with emphasis on the contributions of Don Kirwan and Ron Rousseau.

353b Thermodynamic Models For The Solubility And Purity Of Amino Acid Crystals In Multicomponent Systems

Angel Olivera-Toro, Ronald W. Rousseau, and **Amyn S. Teja**. School of Chemical & Biomolecular Engineering, Georgia Institute of Technology, Atlanta, GA 30332-0100

We present and evaluate several models for the correlation and prediction of crystal solubility and purity when solid solution formation is the dominant mechanism that leads to impure crystals. The models include our previously published relationship based on measurements of the purity of amino acid crystals obtained from aqueous solutions of two amino acids, and a new thermodynamic model based on dilute solution theory. An interesting result of our analysis is that the crystal phase exhibits significant nonidealities even when the dominant mechanism that leads to impure crystals is lattice substitution by isomorphic or near-isomorphic compounds. Potential causes for these nonidealities are explored.

353c Strategic Opportunities For Hyperthermophilic Microorganisms And Enzymes In Bioenergy Conversion Processes

Robert M. Kelly, Chemical and Biomolecular Engineering, North Carolina State University, 911 Partners Way, EB-1, Box 7905, Raleigh, NC 27695-7905

The conversion of biomass into ethanol and H_2 by biological routes has been touted as an answer to current problems with energy availability and global warming. In particular, α- and β-glucans represent promising energy feedstocks if bioprocesses based on efficient biocatalysts and microorganisms can be developed to achieve high energy yields and high rates. Biomass conversion at elevated temperatures can take advantage of favorable thermodynamics and accelerated processing rates. As such, hyperthermophilic microorganisms, those that grow optimally at 80°C and higher, represent a strategic opportunity for bioenergy production that merits close examination.

Discussed here will be two model hyperthermophilic microorganisms, *Pyrococcus furiosus* and *Thermotoga maritima*, which are being investigated as sources of thermostable and thermoactive enzymes for polysaccharide hydrolysis into fermentable sugars as well as for their capacity to produce biohydrogen from carbohydrate-based feedstocks. A functional genomics approach to biocatalyst discovery is being used to identify ORFs encoding new glucan-hydrolyzing enzymes and to provide clues as to the physiological role of these glycoside hydrolases. Biohydrogen production by these hyperthermophiles was strongly affected by the carbohydrate used as primary carbon/energy source. For *P. furiosus*, sugar linkage (cellobiose vs. maltose) had a profound effect on bioenergetics, hydrogen generation rates, as well as the transcriptome. For *T. maritima*, co-fermentation of glucose and xylose was studied and compared to fermentation patterns for the individual sugars. Significant interaction between hexose and pentose processing pathways was found to impact hydrogen production and this observation could be interpreted in terms of transcriptional response analysis. The results presented here point to the potential importance of hyperthermophilic microorganisms for the production of biofuels and the utility of functional genomics approaches for examining metabolic pathways for biohydrogenesis and for the discovery of biocatalysts.

Nucleation Kinetics Determination in High-Throughput Microfluidic Systems

L.M. Goh, K.J. Chen, G.H. He, V. Bhamidi, P.J.A. Kenis, C.F. Zukoski, and R.D. Braatz*

Department of Chemical and Biomolecular Engineering, University of Illinois at Urbana-Champaign, 600 South Matthews Avenue, Urbana, Illinois 61801
* Corresponding author: Phone: 217-333-5073, Fax: 217-333-5052, E-mail: braatz@uiuc.edu.

I. Introduction

Numerous studies have been carried out to obtain a better fundamental understanding of nucleation mechanisms (Blow et al, 1994; tenWolde & Frenkel, 1997; Larson et al, 1998; Baird, 1999; Aizenberg et al, 1999; Cacciuto et al, 2004). More recently, high-throughput microfluidic systems have been used to crystallize a variety of organic compounds including amino acids, proteins, and active pharmaceutical ingredients (APIs) (Sanjoh & Tsukihara, 1999; Zheng et al, 2003; Talreja et al, 2005; Squires et al, 2005; Hansen et al, 2006). The goals of such studies include the identification of conditions to produce protein crystals for X-ray and neutron crystallography, the high-throughput search for polymorphs for pharmaceutical compounds, and the exploration of the behavior of nucleation and crystallization kinetics at high supersaturation (Vekilov, 2004; Helliwell, 2005; Talreja et al, 2005). Potential applications include structure-function analysis, pharmaceuticals design, bioseparation, controlled drug delivery, treatment of protein condensation diseases, and study of human degenerative conditions (Bucciantini et al, 2002; Vekilov, 2004; Pan et al, 2005). Estimates of the nucleation kinetics can be used to revise crystallization conditions to produce higher quality protein crystals, for the design of kinetic processes such as separations, and for investigations into the fundamentals of nucleation. Microfluidic crystallization platforms enable the measurement of a large number of nucleation induction times for various supersaturation conditions with a small amount of starting material, and many researchers have looked into the estimation of nucleation kinetics from such data.

Although the estimation of crystallization kinetics for macro-scale crystallizers is well established (e.g., Rawlings et al, 1993; Fujiwara et al, 2005), nucleation kinetics estimation in microfluidic systems is still an active research area. Several methods have been used to estimate nucleation kinetics from microfluidic systems. Nucleation in nanoliter-volume drops ("nanodrops (includes microdrops)") is a stochastic phenomenon, which has motivated researchers to collect large experimental data sets for analysis (Galkin & Vekilov, 1999; Izmagilov et al, 1999; Knezic et al, 2004). Deterministic models such as population balance equations and method of moments as used by numerous researchers (Larson et al, 1998; Talreja et al, 2007) to estimate nucleation kinetics do not capture this randomness. Due to the small volume in a nanodrop, only a few crystals can form. Consequently, fluctuations in the measured nucleation induction times can be quite large and researchers have computed nucleation rates by averaging many induction times measured under identical conditions (e.g., Galkin & Vekilov, 1999). Galkin and Vekilov (1999) observed that the Poisson distribution provided a good fit to distributions in the number of lysozyme crystals nucleated at constant supersaturation in droplets using a temperature jump technique. Penkova et al (2004) built on the temperature jump technique by applying a supersaturation gradient, which the authors argued consumes less material. Izmagilov et al (1999) assumed that induction times could be fit by a gamma distribution. Knezic et al (2004) employed electrodynamic levitation and a two-step model which analyzed cluster size fluctuations and cluster rearrangements to study nucleation kinetics at high supersaturation. Various researchers have estimated nucleation kinetics via light scattering (Rosenberger et al, 1993; Kulkarni & Zukoski, 2002; Knezic et al, 2004) from measured mean induction times at different supersaturation levels, with supersaturation kept constant for each nucleation cycle.

The above references do not fully characterize the distribution of crystal nuclei in nanodrops in which the supersaturation varies with time, which occurs in some microfluidic platforms. For example, for high-throughput screening of crystallization conditions an evaporation-based microwell crystallization platform has the advantage that it can be designed to always force a phase change in every experiment (Talreja et al, 2005; He et al, 2006). An alternative evaporation-based platform utilizes nanotubes to form nanodrops of varying concentrations (Zheng et al, 2003). In either case, the supersaturation in each nanodrop monotonically increases in a controlled way as the evaporation proceeds.

Section II describes the analytical solution of the Master equation for crystal nucleation when the supersaturation is time-varying. The stochastic model describes the probability that the number of crystals in the system has the value *n* at any time *t*, for all values of *n*. This analytical solution enables the determination of nucleation kinetics within a nanodrop when multiple nuclei are detected at any time for any time-evolution of the supersaturation. Expressions are derived for the mean induction time and the most likely induction time, which are shown to be equal, and the variance in the measured induction times. Section III provides an example application of the analytical expressions to estimate the parameters in the classical nucleation expression from induction times measured in an evaporation-based crystallization platform (He et al, 2006).

II. Analytical Solution to the Master Equation

Assume the time for a nucleus to grow large enough to be observable is negligible and that the solution is spatially uniform (see Goh and Braatz (2007) for a detailed analysis that shows this latter assumption to be valid). The time-dependent function $P_n(t)$ is defined as the probability that the number of crystals, $N(t)$, at time t is equal to n where $n = 0, 1, 2, \ldots$ and κ is the transition probability of nucleation in units of number of crystals per unit time. That is, there is a chance of κdt that a critical nucleus will form during an infinitesimal time-interval of length dt. The Master equation that describes the time-evolution of the probability of the system to occupy each of the states is:

$$\frac{dP_0(t)}{dt} = -\kappa(t)P_0(t) \text{ for } n=0,\ P_0(0)=1, \text{ and}$$

$$\frac{dP_n(t)}{dt} = \kappa(t)(P_{n-1}(t) - P_n(t)) \text{ for } n>0,\ P_n(0)=0 \text{ for } n>0. \quad (1)$$

To solve this infinite set of differential equations, define the probability-generating function[1]

$$\varphi(z,t) = \sum_{n=0}^{\infty} P_n z^n = P_0 + P_1 z + P_2 z^2 + \ldots + P_n z^n. \quad (2)$$

[1] See Kendel (1949) for other examples of the application of probability-generating functions.

Multiplying the Master equation (1) for each state n by z^n and summing for all n results in a single differential equation

$$\frac{\partial \varphi(z,t)}{\partial t} = \kappa(t)\left(z-1\right)\varphi(z,t), \;\; \varphi(z,0)=1. \tag{3}$$

Redefining the time scale by $\tau = \int_0^t \kappa(s)ds$ allows (3) to be written as

$$\frac{\partial \varphi(z,\tau)}{\partial \tau} = \left(z-1\right)\varphi(z,\tau), \;\; \varphi(z,0)=1. \tag{4}$$

whose analytical solution is

$$\varphi(z,\tau) = e^{(z-1)\tau} = e^{(z-1)\int_0^t \kappa(s)ds} = e^{z\int_0^t \kappa(s)ds}\, e^{-\int_0^t \kappa(s)ds}. \tag{5}$$

Expanding $e^{z\int_0^t \kappa(s)ds}$ in a Taylor series and matching coefficients gives the probabilities:

$$P_n(t) = \frac{1}{n!}\left[\int_0^t \kappa(s)ds\right]^n e^{-\int_0^t \kappa(s)ds}, \;\; n=0,1,2,... \tag{6}$$

Figure 1ab shows the time-evolution of the probabilities and number of nuclei n for the case where κ is a constant:

$$P_n(t) = \frac{1}{n!}(\kappa t)^n e^{-\kappa t}, \;\; n=0,1,2,... \tag{7}$$

which is the expected Poisson distribution. The shapes of the curves in Figure 1b are identical to those observed experimentally for nucleation in nanodrops under constant supersaturation (e.g., see Fig. 2 of Galkin & Vekilov, 2000). The nucleation kinetics can be estimated by fitting (7) to the entire distribution of induction times for nanodrops with time-varying supersaturation.

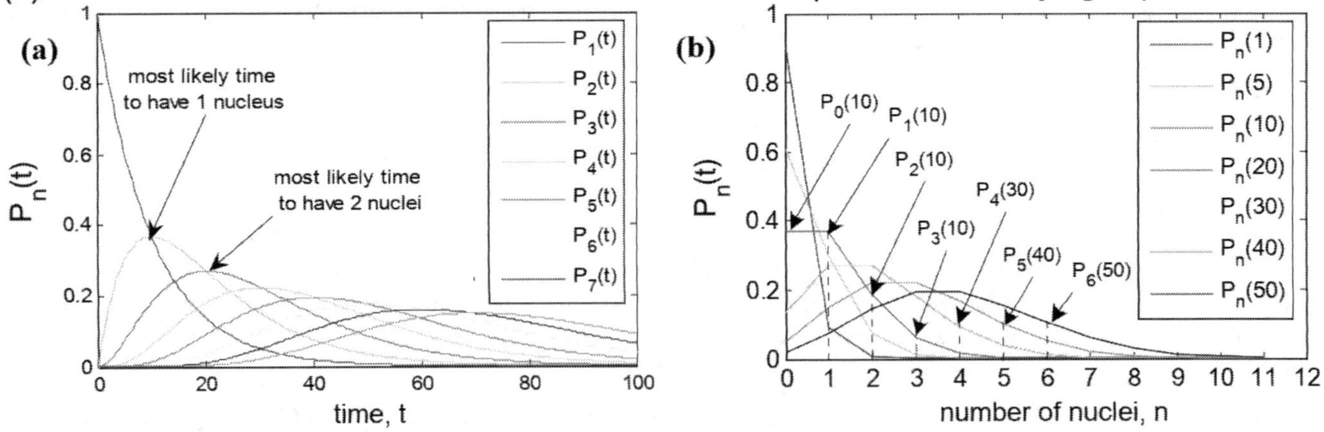

Figure 1. (a) Time-evolution of probabilities $P_n(t)$ for n = 0, 1, 2, …, 7 for κ = 0.1; (b) The corresponding $P_n(t)$ vs. n for different times.

Consider a large number of experiments carried out in a high-throughput microfluidic device. The mean number of nucleated crystals at time t (averaged over a sufficiently large number of experiments) is

$$\bar{N}(t) = \sum_{n=0}^{\infty} nP_n(t) = \sum_{n=0}^{\infty} \frac{n}{n!}\left[\int_0^t \kappa(s)ds\right]^n e^{-\int_0^t \kappa(s)ds} = \int_0^t \kappa(s)ds. \tag{8}$$

Hence the time when the mean number of crystals is equal to 1 is

$$\int_0^{t_{mean}} \kappa(s)ds = 1. \tag{9}$$

The *most likely* time of having exactly 1 crystal in the system is given by

$$\frac{dP_1(t)}{dt} = \kappa(t)e^{-\int_0^t \kappa(s)ds}\left[1 - \int_0^t \kappa(s)ds\right] = 0 \;\Rightarrow\; \int_0^{t_{ml}} \kappa(s)ds = 1. \tag{10}$$

Hence, the most likely time for there to be exactly one crystal in the system t_{ml} is equal to the time t_{mean} for the mean number of crystals to be equal to 1. Either (9) or (10) can be used to define the mean induction time:

$$\int_0^{t_{ind}} \kappa(t)dt = 1. \tag{11}$$

At this time, the probability distributions are given by (6)

$$P_n(t_{ind}) = \frac{1}{n!e},\; n = 0,1,2,... \tag{12}$$

and the variance in the number of crystals *n* is

$$\text{Var}[n(t_{ind})] = \sum_{n=0}^{\infty}(n-\bar{N})^2 P_n(t_{ind}) = \frac{1}{e}\sum_{n=0}^{\infty}\frac{(n-1)^2}{n!} = 1 \tag{13}$$

indicating that the number of crystals measured in each drop at the mean induction time varies widely. That is, the large value for the variance indicates a high probability in a particular experiment that there is either no nucleus or multiple nuclei at the mean induction time. This is also seen by the low probability for having exactly 1 crystal in the system at the mean induction time, $P_n(t_{ind}) = 1/e$. These conclusions hold irrespective of the time-evolution of the transition probability of nucleation, κ, and hence irrespective of the time-evolution of the supersaturation. The next section uses the above results to determine the distribution in induction times measured in nanodrops that undergo identical time-evolution of the supersaturation, for a specific application.

III. A Specific Application

This section illustrates the application of the above analysis to experimental data for the nucleation of paracetamol crystals in aqueous nanodrops in an evaporation-based microfluidic platform.[2] Define *S*(*t*) to be the relative supersaturation ($S(t) = (c(t) - c_{sat})/c_{sat}$ where *c*(*t*) is the concentration at time *t* and c_{sat} is the solubility), and define *J*(*S*) as the number of crystals formed per time per unit volume. The classical nucleation expression is:

$$J(S) = A(S+1)\exp(-B/(\ln(S+1))^2) \tag{14}$$

which corresponds to a transition probability of $\kappa(t) = J(S(t))V(t)$ where *V*(*t*) is the volume of the drop determined from the evaporation rate and a material balance on the solvent and the supersaturation is determined from a material balance on the solute. The optimal estimates of the nucleation parameters were determined by least-squares minimization (Beck & Arnold, 1977):

$$\min_{A,B} \frac{1}{19}\sum_{i=1}^{20}\left(t^i_{ind,model} - t^i_{ind,measured}\right)^2 \tag{15}$$

for 20 experiments with varying evaporation rates and initial solution concentrations (see Table 1), where the superscript refers to the experiment # and the model induction time was

[2] For details on the microfluidic platform and experimental setup, see (Talreja et al, 2005).

computed from (11). Figure 2a compares the experimentally measured induction times with the classical nucleation model for the 20 experiments.

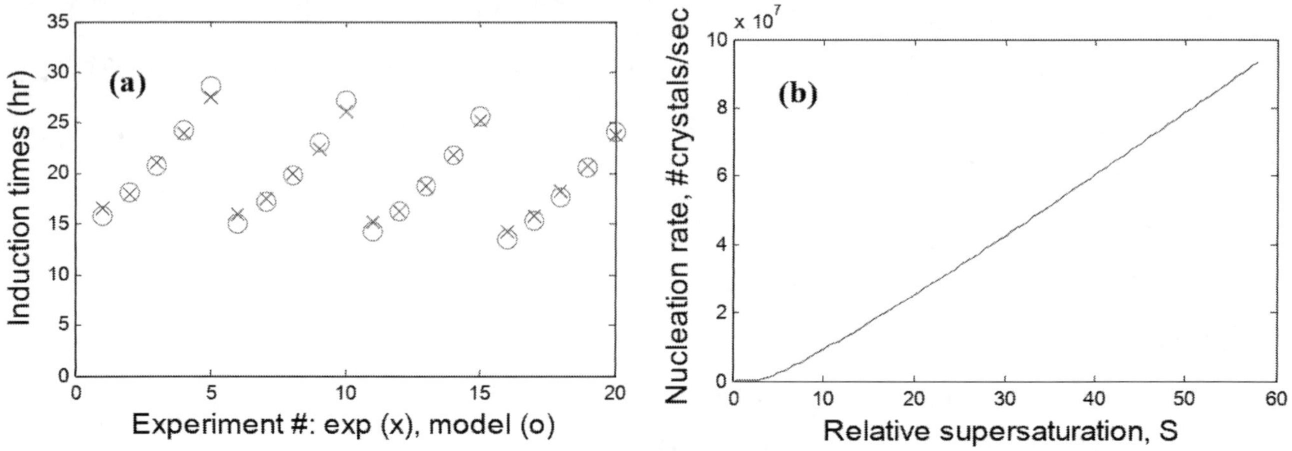

Figure 2. (a) Comparison of the induction times from experiments with the classical nucleation model with best-fit parameters $A = 2.2{\times}10^6$ and $B = 5.8$. (b) the estimated nucleation rate expression.

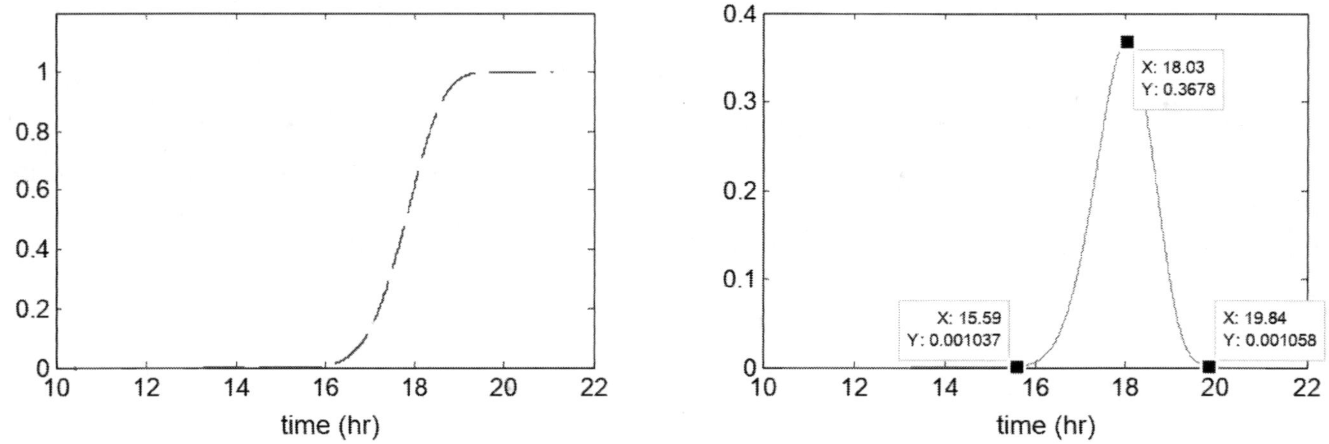

Figure 3. (a) Cumulative distribution for solution concentration for the conditions of Experiment 2. (b) Probability for having exactly *one* crystal in a drop.

The value for (15) was 0.36 hr^2, corresponding to a standard error in the induction time of 0.60 hr, which is small relative to the magnitude of the induction times. The fitted nucleation expression can be inserted into (6) to determine the variation in the measured induction times for each set of experimental conditions. The probability for having at least one crystal in the system is

$$1-P_0(t)=1-e^{-\int_0^t \kappa(s)ds} \qquad (16)$$

which is shown in Figure 3a for Experiment 2 in Table 1. The probability of forming at least one crystal increases from ~0 at 16 hr to ~1 at 19 hr, which is the range in induction times that would be measured for a large number of microfluidic nucleation experiments. The probability of having exactly 1 crystal in a drop is

$$P_1(t)=e^{-\int_0^t \kappa(s)ds}\int_0^t \kappa(s)ds \qquad (17)$$

Table 1. The measured induction times (last column) for paracetamol in water for different evaporation rates and initial solute concentrations, t_{lower} is the time at which $P_1(t)$ first becomes 0.001, t_{upper} is the time at which $P_1(t)$ drops to 0.001, and t_{ml} is the time that the probability of forming one crystal is highest (see Figure 3b for illustration).

No.	Evaporation rate (mg/hr)	c_0 (g/kg)	t_{lower} (hr)	t_{upper} (hr)	t_{ml} (hr)	$t_{ind,measured}$ (hr)
1	34.7	6.0	13.6	17.4	15.8	16.5
2	30.2	6.0	15.6	19.8	18.0	18.0
3	26.1	6.0	18.0	22.8	20.7	21.0
4	22.2	6.0	21.1	26.6	24.3	24.0
5	18.7	6.0	25.0	31.3	28.6	27.5
6	34.7	8.0	12.4	17.0	15.0	16.0
7	30.2	8.0	14.2	19.3	17.2	17.5
8	26.1	8.0	16.4	22.1	19.7	20.0
9	22.2	8.0	19.3	25.8	23.1	22.5
10	18.7	8.0	22.8	30.3	27.2	26.2
11	34.7	10.0	11.3	16.5	14.3	15.3
12	30.2	10.0	12.9	18.7	16.3	16.3
13	26.1	10.0	14.9	21.4	18.7	18.8
14	22.2	10.0	17.4	24.9	21.8	21.8
15	18.7	10.0	20.6	29.2	25.7	25.3
16	34.7	12.0	10.1	15.9	13.5	14.3
17	30.2	12.0	11.5	18.0	15.3	15.8
18	26.1	12.0	13.3	20.6	17.6	18.3
19	22.2	12.0	15.5	23.9	20.5	20.8
20	18.7	12.0	18.3	28.0	24.1	23.8

Figure 3b shows that the probability of having exact one crystal in a nanodrop has a similar spread, from t_{lower} = 15.6 to t_{upper} = 19.8 hr, with the highest probability of forming one crystal occurring at t_{ml} = 18.0 hr which is the same as the experimental induction time of 18 hr (Table 1). The model predictions agree with the measured induction times, within the calculated stochasticity of the experiments (compare $t_{ind,measured}$ with t_{lower} and t_{upper} in Table 1). The time-evolution of the probabilities for having n crystals in the nanodrop is shown in Figure 4. Although the probability of having one crystal in the nanodrop is the highest at t = 18.0 hr, the total probability of having from two to four crystals is equally high at that time.

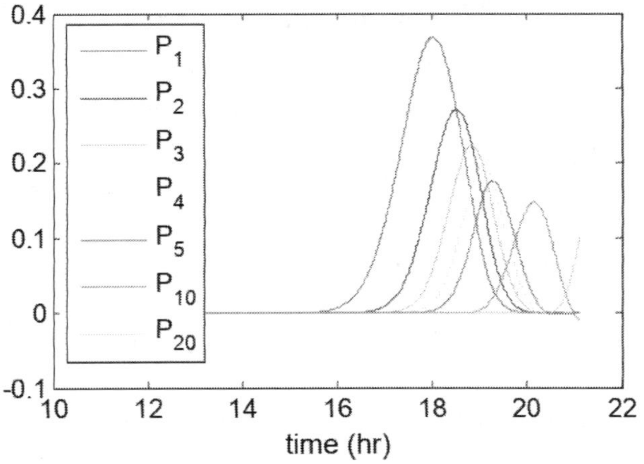

Figure 4. Probabilities for having n crystals in the microdrop at each time (hr).

The analysis for this specific application assumes that the nucleation occurs in the solution nanodrop and not on the liquid-gas or solid-liquid interface. The same analytical solution to the Master equation in Section II applies for such heterogeneous nucleation events as well, by substituting the expression for heterogeneous nucleation in for the nucleation transition probability κ. For example, for nucleation on the solid-liquid interface, the expression would be $\kappa(t) = J_{het}(S(t))A(t)$ where $A(t)$ is the contact area between the nanodrop and the solid surface.

References

- Aizenberg, J., Black, A.J., Whitesides, G.M. (1999) Control of crystal nucleation by patterned self-assembled monolayers, *Nature,* **398**, 495.
- Baird, J.K. (1999) Theory of protein crystal nucleation and growth controlled by solvent evaporation. *J. Crystal Growth,* **204**, 553.
- Beck, J.V., Arnold, K.J. (1977) Parameter Estimation in Engineering and Science, John Wiley, New York.
- Blow, D.M., Chayen, N.E., Lloyd, L.F., Saridakis, E. (1994) Control of nucleation of protein crystals. *Protein Sci.,* **3**, 1638.
- Bucciantini, M., Giannoni, E., Chiti, F., Baroni, F., Formigli, L., Zurdos, J., Taddei, N., Ramponi, G., Dobson, C.M., Stefani, M. (2002) Inherent toxicity of aggregates implies a common mechanism for protein misfolding diseases. *Nature,* **416**, 507.
- Cacciuto, A., Auer, S., Frenkel, D. (2004) Onset of heterogeneous crystal nucleation in colloidal suspensions, *Nature,* **428**, 404.
- Fujiwara, M., Nagy, Z.K., Chew, J.W., Braatz, R.D. (2005) First-principles and direct design approaches for the control of pharmaceutical crystallization. *J. of Process Control,* **15**, 493.
- Galkin, O., Vekilov, P.G. (1999) Direct determination of the nucleation rates of protein crystals. *J. Phys. Chem. B*, **103**, 10965.
- Goh, L.M., Braatz, R.D. (2007) Analysis of Concentration Field within an Evaporating Hanging Droplet. *AIChE Annual Meeting*, Paper 467e.
- He, G., Bhamidi, V., Tan, R.B.H., Kenis, P.J.A., Zukoski, C.F. (2006) Determination of critical supersaturation from microdroplet evaporation experiments. *Crystal Growth & Design*, **6**, 1175.
- Hansen, C.L., Classen, S., Berger, J., Quake, S.R. (2006) A microfluidic device for kinetic optimization of protein crystallization and in situ structure determination. *JACS*, **128**, 3142.
- Izmailov, A.F., Myerson., A.S., Arnold, S. (1999) A statistical understanding of nucleation. *J. Crystal Growth*, **196**, 234.
- Kendal, D.G. (1949) Stochastic processes and population growth – Symposium on stochastic processes. *J. Roy. Stat. Soc. Ser. B*, **11**, 230.
- Knezic, D., Zaccaro, Julien., Myerson, A.S. (2004) Nucleation induction time in levitated droplets. *J. Phys. Chem. B,* **108**, 10672.
- Kulkarni, A.M., Zukoski, C.F. (2002) Nanoparticle crystal nucleation: Influence of solution conditions. *Langmuir*, **18**, 3090.
- Larson, M.A., Saikumar, M.V., Glatz, C.E. (1998) Lysozyme crystal growth and nucleation kinetics. *J. Crystal Growth*, **187**, 277.
- Nyvlt, J., (1985) The kinetics of industrial crystallization. *Elsevier Science Pub. Co., Inc.*

- Pan, W., Kolomeisky, A.B., Vekilov, P.G. (2005) Nucleation of ordered solid phases of proteins via a disordered high-density stat: Phenomenological approach. *J. Chem. Phys.,* **122**, 174905.
- Penkova, A., Dimitrov, I., Nanev, C. (2004) Nucleation of insulin crystals in a wide continuous supersaturation gradient. *Annals New York Academy of Sciences,* **1027**, 56.
- Rawlings, J.B., Miller, S.M., Witkowski, W.R. (1993) Model identification and control of solution crystallization processes: A review. *Ind. Eng. Chem. Res.,* **32**, 1275.
- Rosenberger, F., Howard, S.B., Sowers, J.W., Nyce, T.A., (1993) Temperature dependence of protein solubility: determination and application to crystallization in X-ray capillaries. *J. Crystal Growth*, **129**, 1.
- Sanjoh, A., Tsukihara, T. (1999) Spatiotemporal protein crystal growth studies using microfluidic silicon devices. *J. Crystal Growth*, **196**, 691.
- Squires, T.M., Quake, S.R. (2005) Microfluidics: Fluid physics at the nanoliter scale. *Rev. Mod. Phys.,* **77**, 977.
- Talreja, S., Kenis, P.J.A., Zukoski, C.F. (2007). A kinetic model to simulate protein crystal growth in an evaporation-based crystallization platform. *Langmuir*, **23**, 4516.
- Talreja, S., Kim, D.Y., Mirarefi, A.Y., Zukoski, C.F., Kenis, P.J.A. (2005) Determination of critical supersaturation from microdroplet evaporation experiments. *J. Appl. Cryst.*, **38**, 988.
- tenWolde, P.R., Frenkel, D. (1997) Enhancement of protein crystal nucleation by critical density fluctuations. *Science,* **277**, 1975.
- Vekilov, P.G. (2004) Dense liquid precursor for the nucleation of ordered solid phases from solution. *Crystal Growth & Design*, **4**, 671.
- Zheng, B., Roach, L.S., Ismagilov, R.F. (2003) Screening of protein crystallization conditions on a microfluidic chip using nanoliter-size droplets. *JACS*, **125**, 11170.

Novel Design Strategies for Solution Crystallization Processes

Herman J.M. Kramer, Richard Lakerveld, Pieter J. Jansens, Johan Grievink,
Delft University of Technology, Delft, The Netherlands

Introduction

Crystallisation is one of the oldest and economically most important separation technologies in chemical industry. Many interacting physical phenomena occur in industrial crystallizers. This complicates model-based optimization studies and reduces flexible operation and design. The selection of crystallisation equipment is traditionally done in an early phase of the design from a limited number of state-of-art industrial crystallizers followed by optimization of that particular type of equipment. This reduces the design space and creative input of a designer. Furthermore, optimization of each individual physical phenomena is not possible because in present industrial crystallizers these phenomena are strongly interrelated.

Task based design approach

To improve on these drawbacks a more synthesis-focused design approach is proposed called task-based design. In the task-based design approach an attempt is made to conceptually construct the crystallization process from fundamental building blocks called physical tasks. Tasks can be connected in a network to accomplish the complete transition of a given feed into a desired product. The aim is to generate alternative ways of structuring the tasks, leading to a task superstructure that contains all possible and relevant tasks and their interconnectivity. Optimization of the task superstructure can be realized based on product quality, throughput or economic evaluation of the process and can be dependent on design constraints. The task based design approach is more flexible than traditional design approaches and allows for the optimization of each individual crystallisation task. In this way a much larger solution space is created which is needed to arrive at process intensification.

Experimental optimization of a single task

One of the key challenges of the task based design research is to control a certain crystallization task independently from the other tasks, which makes optimization of that particular task possible. In that perspective, a novel experimental crystallisation configuration that mainly facilities the task crystal growth is studied. The setup consists of a bubble column in which supersaturation is created by simultaneous cooling and evaporation of the solvent by sparging air. The crystals are kept in suspension by the upward velocity of the bubbles, eliminating the need for a stirrer or a circulation pump. In this way attrition caused by crystal-impeller collisions is absent. Moreover, the generation of supersaturation is nicely distributed along the complete length of the column. Experiments on lab-scale show the feasibility of the concept. The experimental results from the seeded experiments in the bubble column are summarized in table 1.

Table 1. Experimental results seeded experiments in bubble column

Exp. ref.	C_s [W_s / W_{th}]	L_s [µm]	L_p experimental [µm]	L_p ideal growth [µm]	X90 / X10 [µm / µm]
1311	0.067	223	601	562	2.5
1312	0.055	223	580	596	2.5

Figure 1 shows the measured final median size of two seeding experiments (for an ammonium sulphate water system) in a so-called seed chart introduced by Jagadesh et al. [1]. It can be seen that the final median size is on the ideal growth line which approximates a situation in which the total number of crystals do not change. It can therefore be concluded that nucleation is indeed minimal in the bubble column. As a reference, figure 1, also contains dynamic experimental results from Hojjati et al.,[2], indicated by the black line, who did seeding experiments on a lab scale for cooling crystallization of ammonium sulphate for seeds with two different median sizes in a stirred crystallizer. Furthermore, figure 1 contains experimental results, indicated by the dotted lines, from seeding experiments in a 75-liter draft tube crystallizer operated in evaporative fed-batch mode. The seeds that were used for those experiments were identical to the seeds that were used for the experiments with the bubble column. It can be concluded that optimization of the single task crystal growth in principle can be succesfull and allows for more flexible operation as the final product size can be manipulated directly by adjusting the seed mass.

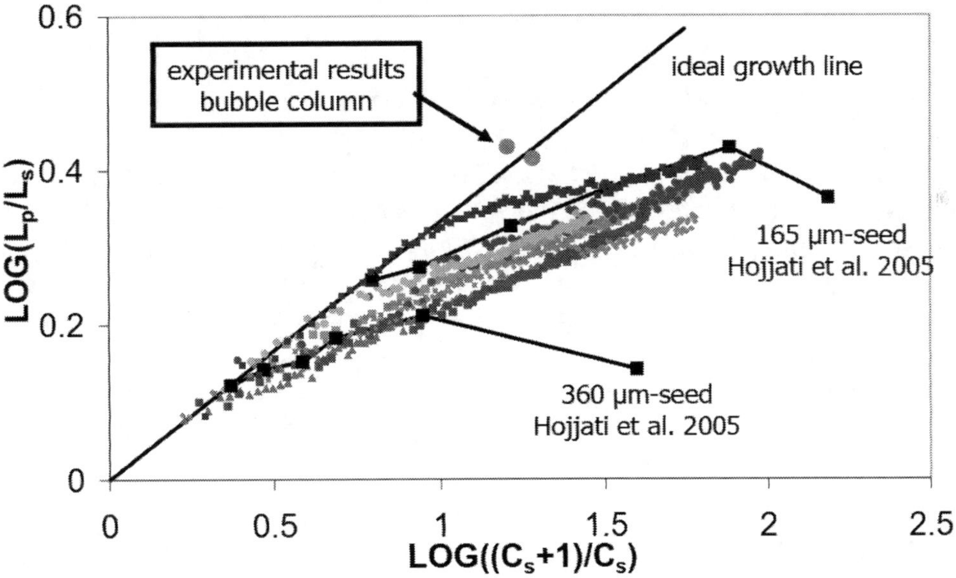

Figure 1: Seed chart showing results from ground seeds in 75-l DT crystallizer (dotted lines, initial seed mass between 0.4 and 1.4 kg, Ls ≈ 230 µm), results from literature [1] and results from seeded experiments in the bubble column (circles indicated by arrows, see Table 2 for operating conditions)

Experiments in which the bubble column, done in a continuous flow mode show that a device that isolates the task crystal growth can also be used to increase the median size of a slurry while maintaining the total number of particles. The aim is to connect the bubble column to devices that contain other isolated tasks. In that perspective, the use of membranes is interesting as they can be used to optimize the generation of supersaturation independently from the other tasks. Finally, the use of ultrasound is investigated for production of nuclei as has been shown by Virone et al [3]. In the end, the combined experimental setup is the translation of a task superstructure to real life equipment.

Conclusion

There is a need for a more synthesis-focused design approach for industrial crystallizers. A new design approach is proposed that aims to conceptually build-up the crystallization process from fundamental building blocks called physical tasks. In the framework of such a design approach experimental results from a bubble column crystallizer are discussed. The experimental setup isolates the physical task of crystal growth. The results show that the system is dominated by growth and nucleation is minimized, which allows for more flexible operation by for example choosing proper seeding conditions in case of batch-wise operation.

Nomenclature

Lp	[μm]	Product volume-based mean size
Ls	[μm]	Seed volume-based mean size
X10	[μm]	10-% quantile of the CSD
X90	[μm]	90-% quantile of the CSD
Cs	[kg kg-1]	Seed load defined as Ws / Wth
Ws	[kg]	Seed mass
Wth	[kg]	Theoretical crystal yield

References

1. Jagadesh, D., Kubota, N., Yokota, M., Doki, N. and Sato, A. "Seeding effect on Batch Crystallization of Potassium Sulfate under Natural Cooling Mode and a Simple Design Method of Crystallizer". *Journal of Chemical Engineering of Japan* **32**(4) (1999) 514-520.

2 Hojjati, H. and Rohani, S. "Cooling and seeding effect on supersaturation and final crystal size distribution (CSD) of ammonium sulphate in a batch crystallizer". *Chemical Engineering and Processing* **44** (2005) 949-957.

3 Virone, C., Kramer, H.J.M., Van Rosmalen, G.M., Stoop, A. H. and Bakker, T. W. "Primary nucleation induced by ultrasonic cavitation". *Journal of Crystal Growth* **294**(1) (2006) 9-15.

353f Developing The Final And Finishing Steps For R-411 Active Pharmaceutical Ingredient

Ram Narayan[1], Jay Lawrimore[1], Brandon Shealy[1], Rob Huber[1], Tom Cleary[1], and Michael Blocker[2]. (1) Chemical Development, Roche Carolina Inc, 6173 E. Old Marion Highway, Florence, SC 29501, (2) Pilot Plant, Roche Carolina, 6173 E. Old Marion Highway, Florence, SC 29501

This paper deals with an integrated approach between chemical reaction engineering, crystallization and solid-liquid separation to identify optimized conditions to improve the filterability of the API in the final and finishing steps. The simple chemical transformation from the free base to the hydrochloride salt had challenging issues related to the filterability of the salt. A Design of Experiments approach was used to study the problem. The specific cake resistance was measured in the laboratory under a variety of process conditions. The results were used to identify optimized conditions for successful scale-up in the plant.

361a Chromatography Membrane Reactor System for the Continuous Synthesis of Galactosyl-Oligosaccharides

Peter Czermak, Larisa Engel, and Mehrdad Ebrahimi. Department of Biotechnology, University of Applied Sciences Giessen Friedberg, Wiesenstrasse 14, Giessen, 35390, Germany

Beta-Galactosidase (EC 3.2.1.23), commonly known as lactase, catalyses not only the hydrolysis of lactose to the monosaccharides glucose and galactose but also the transgalactosylation reaction to produce galactosyl-oligosaccharides (GOS). GOS are non-digestible oligosaccharides which are recognized as prebiotics. Prebiotics have been found to reach the human colon without being hydrolyzed or absorbed in the upper part of the gastrointestinal tract. GOS selectively stimulate the growth of bifidobacteria in the lower part of the human intestine. Increase in the growth of bifidobacteria is usually accompanied by suppression of potentially harmful bacteria such as Clostridia and Bacteroides species in the intestine. GOS consist of a number of oligosaccharides with varying ß-glycosidic linkages depending on the enzyme source. There have been several investigations on the synthesis of GOS by ß-galactosidases from various sources. Some techniques have been developed for immobilization of ß-galactosidase including non-covalent adsorption, covalent binding, entrapment and encapsulation. Synthetic microporous membrane adsorbers as chromatographic media are an attractive alternative to traditionally used packed bed chromatography, which has several limitations. The adsorption of enzymes on ionic exchange resins is still the most popular, simplest and oldest technique for reversible immobilization of enzymes. The chromatography membranes are an ion exchange support containing functional quaternary amines or sulphopropyl groups e.g. supported cross-linked polyethersulfone (PES) or regenerated cellulose. The pores in the membranes are large enough to allow the biomolecules access to all the binding sites by direct fluid convection. This results in a very high capacity for large biomolecules. Pressure filtration forces the liquid through the micropores of the membrane, bringing target substances into direct contact with the binding sites. This direct convective transport to the binding sites minimizes the diffusion limitation of mass transfer without sacrificing capacity. The main applications of membrane adsorbers are separation and analysis of proteins. In our study, the membrane adsorbers have been investigated as an alternative support for immobilization of an enzyme to produce GOS. Some of the advantages of immobilizing enzymes via physical adsorption on an activated support are: the activated support is chemically inert, good stability during storage, controlled residence time on the enzyme, simple recovery of the enzyme after usage, the possibility of a continuous process in enzyme reactors and the elimination of carryover of the enzyme to the final product. This present study focuses on developing a method for immobilization of ß-galactosidase in a chromatography membrane, investigating the properties of the immobilized enzyme and the possibility of using the immobilized enzyme system for the synthesis of GOS from lactose. The ß-galactosidase from Kluyveromyces lactis was chosen as a model enzyme for this study. Different strongly basic anion exchange membranes were used here to achieve optimum conditions for the enzyme.

361b Pure Ester Produced Via A Water Removing Membrane Reactor

Juan C. Cruz[1], **Mary E. Rezac**[1], Stuart Nemser[2], and Sudipto Majumdar[2]. (1) Chemical Engineering, Kansas State University, Manhattan, KS 66506, (2) Research, Compact Membrane Systems, Inc, 325 Water Street, Wilmington, DE 19804

For thermodynamically-limited reactions, selective removal of a product can create an environment in which complete conversion of the substrates are possible. In this presentation, we'll discuss the esterification of an alcohol with an acid to produce an ester plus water. To promote the forward reaction, the water level in the reaction mixture should be minimized. The current study evaluates the use of membrane technology for water concentration in the reaction mixture. The process operates by contacting the reaction mixture with a synthetic membrane. On the opposite side of the membrane, a vacuum is applied to minimize the activity of water and provide a driving force for the mass transport. This presentation will evaluate the impact of applying this technology to the enzyme-catalyzed esterification of geraniol.

The synthesis of geranyl acetate by esterification of geraniol with acetic acid catalyzed by an immobilized lipase from Candida antarctica is the model reaction investigated. Geranyl acetate is an important flavor and fragrance compound. All reactions were completed in n-hexane solvent. Highly permeable, chemically and thermally resistant membranes manufactured by Compact Membrane Systems, Inc. were used to selectively remove water from the reaction mixture.

Conversion beyond the conventional equilibrium limitation was studied through the selective removal of water from the reaction mixture. Significant enhancement in reaction rate was observed. The impact of processing conditions and membrane properties on the reaction product spectrum will be discussed in this presentation.

Lipase Catalyzed Hydrolysis of Olive Oil in the Biphasic Enzyme Membrane Reactor: Modeling of Mass Transfer Effect

G. Pugazhenthi[1*] and Anil Kumar[2]
[*1] *Department of Chemical Engineering, Indian Institute of Technology Guwahati, Guwahati-781039, India. Email: pugal@iitg.ernet.in*
[2] *Department of Chemical Engineering, Indian Institute of Technology Kanpur, Kanpur – 208016, India. Email: anilk@iitk.ac.in*

Abstract

This article presents a model to explain the mass transfer phenomenon of the ions (OH^-, Na^+ and $RCOO^-$) formed during the hydrolysis reaction on the aqueous side of the reaction sites in the biphasic enzyme membrane reactor (EMR) using lipase immobilized on modified poly (methyl methacrylate-ethylene glycol dimethacrylate) (PMMA-EGDM) clay composite membrane. Our model divides the aqueous side of the system into two regions, an unstirred film (aqueous side) adjacent to the pore end (region I) and charged capillaries of the membrane (region II). The process involves transport of Na^+, OH^- and $RCOO^-$ ions in region I whereas the film is considered as an electro-neutral region. Space charge model (consisting of Nernst-Planck equation for ion transport and nonlinear Poisson-Boltzmann (PB) equation for charge distribution) is used to describe transport of ions inside charged capillaries (pores) in region II. The solution of the nonlinear PB equation is computationally intensive and to overcome this difficulty, a series solution of the PB equation for the charge distribution inside the capillaries has been developed. The suggested solution has been used to determine the rate of formation of oleic acid using the information on pore size (determined from molecular weight cut-off experiment) and diffusivity data from the literature. The average pore wall charge using the model has been determined to be 0.735 mV.

Keywords: PMMA-EGDM; Composite membrane; Hydrolysis; Olive oil

1. Introduction

The fats and oils are converted into its acids using an ultrafiltration membrane having immobilized enzyme in enzyme membrane reactors (EMR). The reactor has the advantages of integrating catalytic conversion and product separation into a single operation. In recent years, there have been a series of publications dealing with the modeling and simulation of catalytic membrane reactor to understand their behavior [1]. Membrane bioreactors have been similarly modeled using approaches of the conventional catalytic reactor. Completely effective models to describe the membrane bioreactor performance are yet to be published and the development of such models still remains as an important technical challenge [1, 2]. Many researchers have studied the kinetics of chemical steps of the enzyme hydrolysis of triglycerides [3, 4], the transport of reactants in the membrane, in the oil and aqueous phase compartments in the membrane reactors [5, 6]. Malcata and coworkers [7] have proposed the kinetic for lipase-catalyzed butter oil reaction using a hollow fiber membrane reactor. Paiva et al. [8] have reviewed the kinetics and mechanisms of reactions catalyzed by immobilized lipases. Ceynowa et al. [9, 10] have modeled the hydrolysis of olive oil to study the reactor performance by means of thermodynamic network representation of both the chemical reaction and the transport of the reactants. The parameters of the model were estimated based on experimental data and/or literature notations. Further, they have presented the simulation of the biphasic EMR performance at different limitations, such as kinetics of the chemical processes, diffusional permeabilities of the fatty acids and glycerides, rates of inhibition of the immobilized enzyme and concentration of the enzyme in the reaction zone. Their studies indicated that improved productivity in the membrane reactor could only be obtained for the case in which the reactor is kinetically limited. In our earlier paper [11], we have reported the immobilization of lipase on to the surface of chemically modified poly (methyl methacrylate-ethylene glycol dimethacrylate) (PMMA-EGDM) clay composite membrane. We have used these to study the effect of operating conditions (pH of aqueous phase, solvent used for olive oil, substrate (olive oil) concentration, aqueous phase circulation rate and temperature) on reaction rate for the hydrolysis of olive oil in biphasic EMR.

In this work, we propose a model to describe the mass transfer phenomenon of the ions (OH^-, Na^+ and $RCOO^-$) formed during the hydrolysis reaction on the aqueous side of the reaction sites (enzyme bonded at pore entrance). For this purpose, the aqueous side of the system is divided into two regions (Figure 1), the region I consisting of an electro-neutral unstirred film adjacent to the pore end and the region II comprising charged capillaries. The flux and the concentrations of Na^+ and $RCOO^-$ ions in region I have been determined from experimental results. In region II, the space charge model is used to describe the transport of ions inside the capillaries and to determine the average wall charge of the membrane used.

2. Mass Transfer Model

Biphasic EMR experimental setup and procedure has been described in our earlier work [11]. However, some information should be given here to get better understanding of the reactor performance. The pore size of the membrane used in this work (~ 9.5 nm) is very small, it is expected that the enzyme do not enter into the pores. Hydrolysis of olive oil occurs at the enzyme located on the oil side of the membrane (at pore mouth) and it generates fatty acid. The fatty acid diffuses through the capillaries of the hydrophilic membrane into the aqueous phase maintained at constant basic pH. For simplification of analysis, the fatty acid formed is assumed to be oleic acid alone. The extraction is facilitated by the reaction of acid molecule with the OH^- ions (given in equation 1 below) in the oil side of the membrane (where the enzyme is immobilized), which is assumed to be instantaneous.

$$RCOOH + OH^- \rightarrow RCOO^- + H_2O \qquad (1)$$

The schematic diagram depicting the mass transfer phenomenon is shown in figure 1. Here, AA′ represents the reaction front and the oleic acid molecule formed at enzyme sites (at pore entrance) diffuse towards the reaction front. The OH^- ions in the aqueous phase at the right hand side of AA′ migrate towards the reaction front where they are consumed and $RCOO^-$ ions are generated. The acid ions diffuse towards the bulk of aqueous phase through the pores of membrane due to concentration gradient. The NaOH is continuously added into the aqueous phase side to maintain a constant pH which leads to a buildup of Na^+ ions in the bulk. The region to the right of reaction front (AA′) contains Na^+ (hereafter referred to as ion 1), OH^- (hereafter referred to as ion 2) and $RCOO^-$ (hereafter referred to as ion 3) ions. Since the diffusivity of OH^- ions (5.29×10^{-9} m^2/s) is much larger (about 1000 times) than that of the RCOOH (1.2×10^{-12} m^2/s, evaluated using modified Wilke - Chang equation [12] and the reaction front is assumed to lie near the pore entrance for the short reaction time considered in this work. The mass transport phenomenon inside the membrane is assumed to be similar to that outside and so the entire system can be divided into two regions as shown in figure 1. The first region (region I) consists of an unstirred film from pore end to the bulk of aqueous phase. The second region is comprised of the capillaries of the membrane as shown and since the membrane is charged, the ion transport is described using space charge model with no fluid flow.

Figure 1 (b) shows the schematic diagram of region I and it consists of an unstirred film of thickness δ. The film thickness is related to mass transfer coefficient as

$$k = \frac{D}{\delta} \tag{2}$$

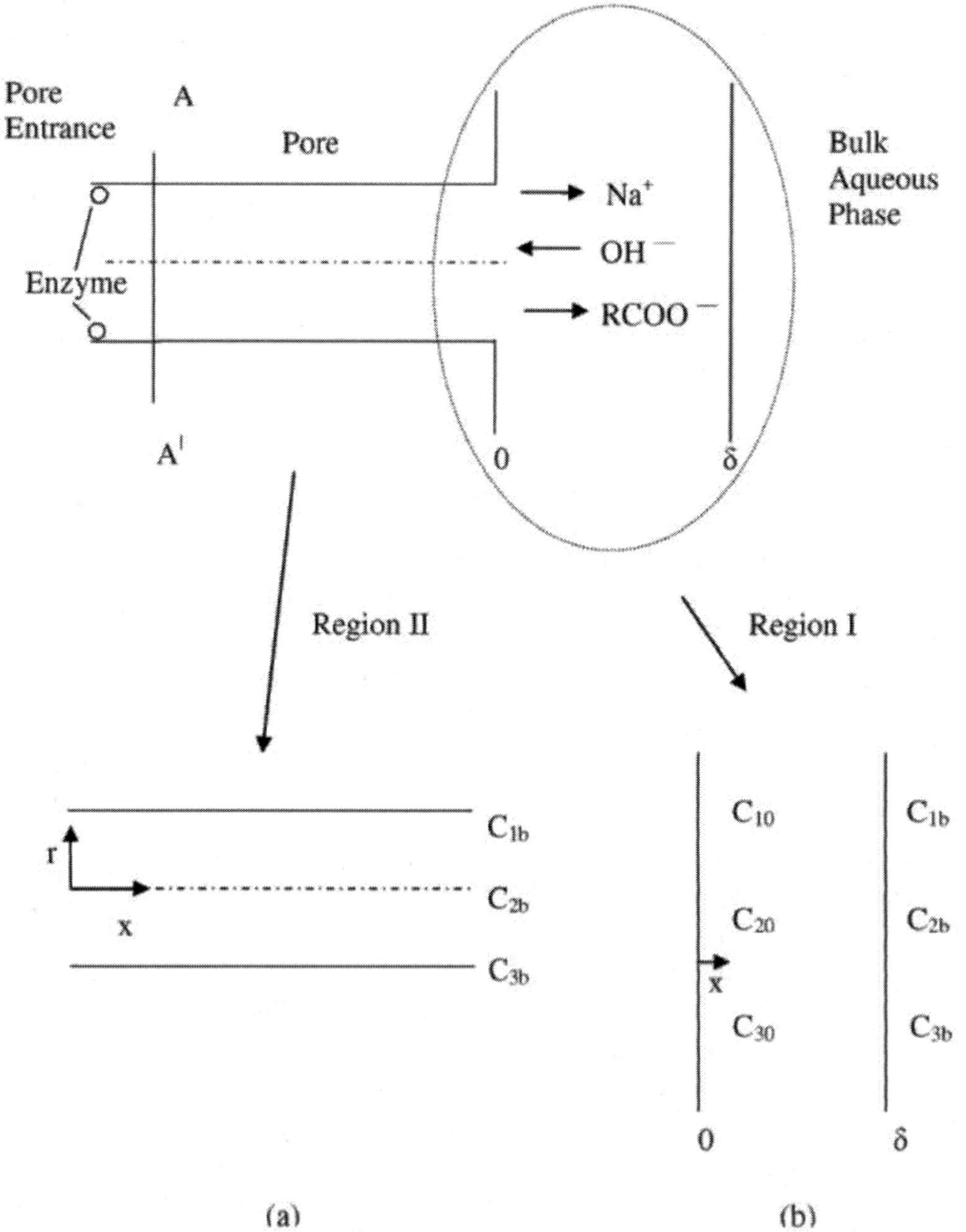

Figure 1. Schematic diagram showing the mass transfer phenomenon inside the membrane and on the aqueous side of the membrane.

and the mass transfer coefficient can be evaluated using the following relations given in literature [13]

$$Sh = \frac{kd}{D} = 0.023\,\mathrm{Re}^{7/8}\,Sc^{1/4} \qquad (3)$$

$$where \quad \mathrm{Re} = \frac{\rho u d}{\mu} \quad and \quad Sc = \frac{\mu}{\rho D}$$

where d is the diameter of the cylindrical compartment, ρ is fluid density, μ is the fluid viscosity and u is the velocity of water in the aqueous compartment. The thickness is evaluated using the diffusivity data of OH^- (5.29×10^{-9} m^2/s) ions.

The flux of ions is given as

$$J_i = -D_i\frac{dC_i}{dx} - \frac{D_i z_i F C_i}{\Re T}\frac{d\Phi}{dx} \qquad (4)$$

where C_i is concentration of ion i, D_i is the diffusivity, z_i is the ion valency, Φ is electrical potential, ℜ is universal gas constant and T is the absolute temperature. The electrical potential is related to charge density by the Poisson equation as

$$\frac{d^2\Phi}{dx^2} = -\frac{F\sum_{i=1}^{3} z_i C_i}{\varepsilon} \qquad (5)$$

The region I is assume to be electro-neutral i.e.,

$$\sum_{i=1}^{3} z_i C_i = 0 \qquad (6)$$

and substituting equation (6) in equation (5), we obtain

$$\frac{d\Phi}{dx} = \text{constant} = \mathrm{E}\ (\text{say}) \qquad (7)$$

where E is the electric field. It has to be self-generated in the absence of any externally applied field and it implies existence of unbalanced charge. The flux equation of the ions (assumed electro-neutral) may be written as

$$\left.\begin{aligned} J_1 &= -D_1\frac{dC_1}{dx} - \frac{D_1 z_1 F C_1}{\Re T}E && \text{(a)}\\ J_2 &= -D_2\frac{dC_2}{dx} - \frac{D_2 z_2 F C_2}{\Re T}E && \text{(b)}\\ J_3 &= -D_3\frac{dC_3}{dx} - \frac{D_3 z_3 F C_3}{\Re T}E && \text{(c)} \end{aligned}\right\} (8)$$

In absence of reaction and assuming constant diffusion coefficient, the continuity equation of the ions may be written as

$$\frac{\partial C_i}{\partial t} = -D_i\frac{\partial^2 C_i}{\partial x^2} - \frac{D_i F z_i E}{\Re T}\frac{\partial C_i}{\partial x} \qquad (9)$$

where t is time.

It is found experimentally that for the short reaction time (60-120 minutes) considered in this work [11], a constant rate of addition of NaOH is required to maintain the pH of aqueous compartment at a constant value. As the flux and bulk concentration of ion 2 is constant, it may be assumed that the flux and concentration profile of ion 2 in region I do not change with time. Thus, for ion 2 equation (9) may be solved to give its concentration profile as

$$C_2 = p_{21}\exp(-\alpha_2 x) + p_{22} \tag{10}$$

where p_{21}, p_{22} are constants and $\alpha_2 = \dfrac{z_2 FE}{\Re T}$.

The flux of ion 2 may be calculated from the rate of addition of NaOH (β) as

$$J_2 = \frac{\beta V_{aq}}{A_c} \tag{11}$$

where V_{aq} is the volume of the aqueous phase and A_c, the cross sectional area of the aqueous phase compartment. The total current flow is zero, which gives the following relation,

$$\sum_{i=1}^{3} z_i J_i = 0 \tag{12}$$

Taking partial derivative of equation 6 with respect to time and using $\dfrac{\partial C_2}{\partial t} = 0$, we obtain

$$\frac{\partial C_1}{\partial t} = \frac{\partial C_3}{\partial t} \tag{13}$$

Using equation (6), (9) and (13), we obtain

$$\left.\begin{aligned}
(D_1 - D_3)\frac{\partial^2 C_1}{\partial x^2} + \frac{FE(D_1 z_1 - D_3 z_3)}{\Re T}\frac{\partial C_1}{\partial x} &= -D_1\frac{\partial^2 C_2}{\partial x^2} - \frac{FED_1 z_1}{\Re T}\frac{\partial C_2}{\partial x} && \text{(a)}\\
(D_1 - D_3)\frac{\partial^2 C_3}{\partial x^2} + \frac{FE(D_1 z_1 - D_3 z_3)}{\Re T}\frac{\partial C_3}{\partial x} &= -D_3\frac{\partial^2 C_2}{\partial x^2} - \frac{FED_3 z_3}{\Re T}\frac{\partial C_2}{\partial x} && \text{(b)}
\end{aligned}\right\} \tag{14}$$

The above equation may be solved to give the concentration profile of ion 1 and 3 as

$$\left.\begin{aligned}
C_1 &= p_{11}\exp(-\alpha_1 x) + p_{12} + A_1\exp(-\alpha_2 x) && \text{(a)}\\
C_3 &= p_{31}\exp(-\alpha_3 x) + p_{32} + A_3\exp(-\alpha_2 x) && \text{(b)}
\end{aligned}\right\} \tag{15}$$

where p_{ij} are constants and

$$A_1 = \frac{p_{21}(D_3 z_3 FE/\Re T - D_3\alpha_2)}{\alpha_2(D_1 - D_3) - FE/\Re T(D_1 z_1 - D_3 z_3)}$$

$$A_3 = \frac{p_{21}(D_1 z_1 FE/\Re T - D_1\alpha_2)}{\alpha_2(D_1 - D_3) - FE/\Re T(D_1 z_1 - D_3 z_3)}$$

The governing equations have been developed for the region I and they are solved computationally. The transport of ions inside the capillaries of the membrane is considered to be at steady state and is described by space charge model. The fluid velocity inside the capillary is zero because it is found experimentally that there is no flow of aqueous phase into the oil compartment or of oil into the aqueous compartment. The transport of ions is described by the Nernst-Planck equation as (cylindrical coordinates). The non-linear Poisson-Boltzmann (PB) equation for charge distribution is solved analytically in the form of infinite power series which is given in our earlier publications [14-17].

3. Result and Discussion

In biphasic EMR, the reaction starts at the immobilized enzyme and the fatty acid is formed immediately at the olive oil interface. As a result of this, the pH of the aqueous phase changes continuously and to measure the rate of the fatty acid formation, we have kept its pH constant by adding the sodium hydroxide solution continuously to it. The flux of OH^- (ion 2) is calculated from the rate of addition of sodium hydroxide solution. So the fatty acid flux is dependent both on the diffusion of fatty acid and OH^- ions. The experimental data show that the rate of addition of sodium hydroxide solution is constant which implies that the flux of ion 2 is constant and is found to be 33.34×10^{-9} kmol/m^{2}s, 27.83×10^{-9} kmol/m^{2}s and 9.152×10^{-9} kmol/m^{2}s for the solvent used for olive oil are heptane, pentane and hexane respectively (the pH of aqueous phase solution as 9.5; olive oil concentration as 50 vol. %; organic and aqueous phase circulation rate as 400 ml/min and temperature as 30 °C).

Figure 2 shows the growth of the film thickness with time for different solvents used for olive oil. From this figure, it can be seen clearly that the film thickness grows slowly and was found to be constant after 120 minutes for all the cases. Initially when the reaction starts the film thickness is zero as there is no concentration gradient. As the carboxylic acid ions starts to diffuse, the region where the concentration gradient exist starts to increase. After sufficient time the region will grow to the extent where mixing is possible and film will stop to grow. The film thickness can also be determined by well known mass transfer correlation given in the literature [13] and is found to be considerably larger (~ 2 μm). Figure 3 shows the variation of ion 3 ($RCOO^-$) concentration at $x = 0$ with time for different solvent used for olive oil. It may be seen that the concentration increases with time which matches well with the experimental results reported in our previous work [11] that shows the rate formation of fatty acid increases with time. For heptane solvent (base case), the ion 3 concentration is higher compared to that of the other solvents (pentane and hexane) due to the higher rate of production of fatty acid with time. It is already reported in the literature [18] that the enzyme activity depends dramatically on the organic solvents used. The solubility of olive oil in heptane may be more compare to other solvents used in this study. This may be the reason for the higher rate of formation of fatty acid. Figure 4 shows the concentration profile of the three ions at three different times for the base case. It may be seen that the ion concentration varies in such a way that the sum of ion 2 and 3 concentration is equal to concentration of ion 1. Initially, the ion 1 and 2 concentrations are almost the same, as the time increases the difference between these two increases due to an increase in the concentration of ion 3.

The flux of these ions is extremely small compared to their concentration (C_1, C_2 and C_3) in the film and so the pseudo steady state assumption at the boundary of regions II and I can be made. To evaluate the wall potential of the membrane and mass transfer rate in region II, the pore radius of the membrane is taken as 9.5 nm, which is reported in our earlier publication [15] using the molecular weight cut-off experiment. The pore length is taken as equal to the thickness of the polymer film (40 μm) and the porosity is found to be 0.20 [15]. The solution scheme [14] requires the ion concentrations and their gradient only at the pore end (i.e., at the boundary of regions I and II). The ions concentrations are already determined from the solution of the region I. We assumed the pseudo steady state at the boundary of regions I and II which implies the flux of ions on both side are the same. The wall potential, determined this way, is found to be 0.735 mV for the various sets of experimental data and this low value is due to the extremely low concentration of the ions in the aqueous phase.

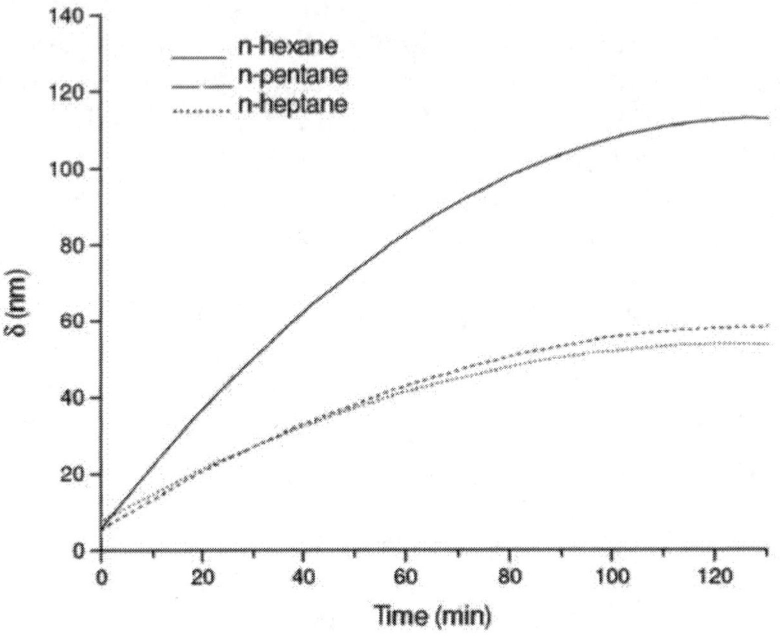

Figure 2. Variation of film thickness with time.

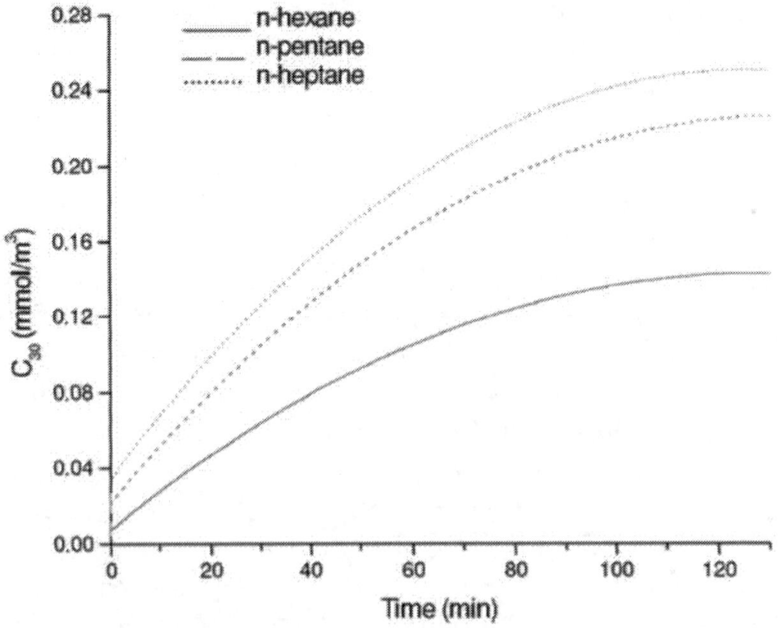

Figure 3. Variation of ion 3 ($RCOO^-$) concentration at x = 0 with time

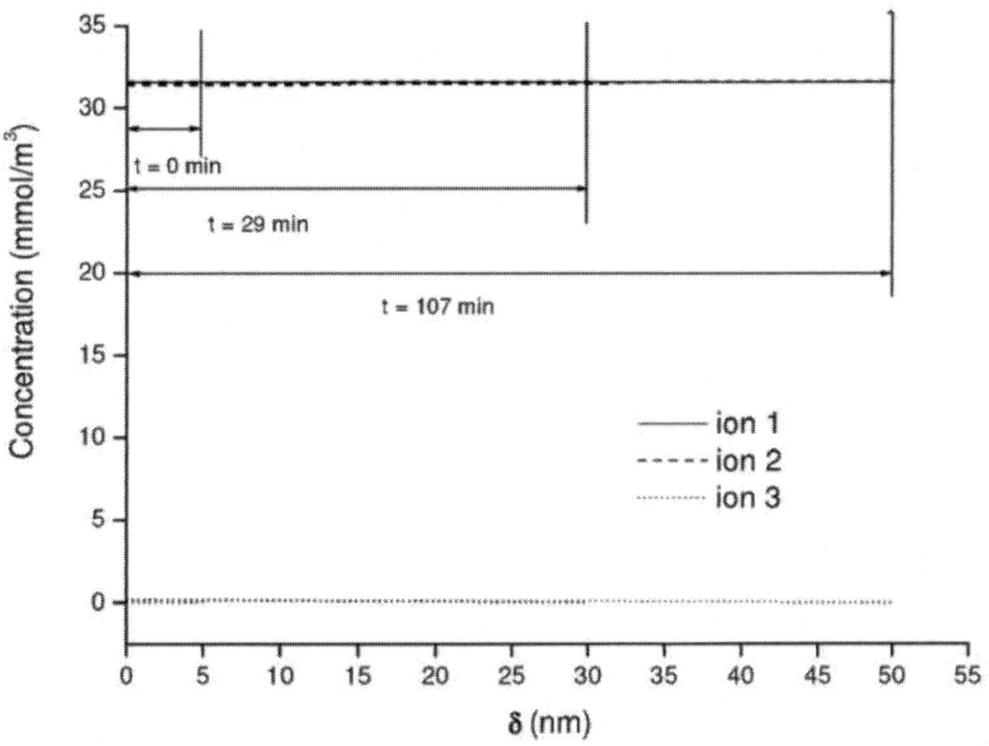

Figure 4. Concentration profile of the ions in region I at three different times t = 0 min, t = 29 min and t = 107 min. The profiles are determined using base case experimental data and for all times C_1-C_2-C_3 = 0

4. Conclusion

A model is presented to describe the mass transfer phenomenon of the ions on the aqueous side of the reaction sites (lipase located at the pore entrance). For this, the model divides the aqueous side of system into two regions, an unstirred film adjacent to the pore end (region I) and the charge capillaries of the membrane (region II). The transport of ions has been described using the film theory in the first region and space charge model in the second region. Due to low flow rate of ions, justifies the use of pseudo-steady state assumption and it enables the use of solution scheme for the space charge model which requires the ion concentrations and their gradient only at the pore end. The result shows that the film thickness increases with duration of reaction. The average wall potential has been determined from the space charge model and is found to be 0.735mV.

References

1. J.G.S. Marcano, T.T. Tsotsis, Catalytic Membranes and Membrane Reactors. Weinheim: Wiley-VCH Verlag GmbH, 2002.
2. G. Salzman, R. Tadmor, S. Guzy, S. Sideman, N. Lotan, Hollow fiber enzymic reactors for a two substrate process: analytical modeling and numerical simulations, Chem. Eng. Proc., 38 (1999) 289.

3. F.X. Malcata, H.R. Reyes, H.S. Garcia, C.G. Hill, C.H. Amundson, Kinetics and mechanisms of reactions catalyzed by immobilized lipases, Enzyme Microb. Technol., 14 (1992) 426-446.
4. F.X. Malcata, C.G. Hill, C.H. Amundson, Hydrolysis of butteroil by immobilized lipase using a hollow-fiber reactor: Part III. Multiresponse kinetic studies, Biotecnol. Bioeng., 39 (1992) 1002-1012.
5. P. Gemeiner, V. Stefuca, V. Bales, Biochemical Engineering of biocatalysis immobilized on cellulosic materials, Enzyme Microb. Technol., 15 (1983) 551-566.
6. S.W. Tsai, H.J. Wei, C.L. Chiang, Action of lipolytical enzymes in biphasic organic-aqueous systems: Dynamics of the irreversible Michaelis-Menten reaction, Biotechnol. Bioeng., 41 (1993) 603-611.
7. F.X. Malcata, C.G. Hill, C.H. Amundson, Use of a lipase immobilized membrane in a membrane reactor to hydrolyze the glycerides of butteroil, Biotecnol. Bioeng., 38 (1991) 853-868.
8. A.L. Paiva, V.M. Balcao, F.X. Malcata, Kinetics and mechanisms of reactions catalyzed by immobilized lipases: Review. Enzym.Micro.Tech., 27 (2000) 187-204.
9. J. Ceynowa, P. Adamczak, Staniszewski, Kinetics of olive oil hydrolysis in an enzyme membrane reactor bond graph network model of reactor performance (part I), Acta Biotechnologica., 17 (1997) 161–176.
10. J. Ceynowa, P. Adamczak, Analysis of the bond graph network model of the membrane reactor for olive oil hydrolysis, Sep. Purific. Tech., 22-23 (2001) 443-449.
11. G. Pugazhenthi and Anil Kumar, Enzyme membrane reactor for hydrolysis of olive oil using lipase immobilized on modified PMMA composite membrane. J. Membr. Sci., 228 (2004) 187-197.
12. R.C. Reid, J.M. Prausnitz, B.E. Poling, The properties of gases and liquids. New York: McGraw-Hill (1987).
13. M.C. Porter, Membrane filtration. In: Schweitzer PA, editor. Handbook of separation techniques for chemical engineers. New York: McGraw-Hil (1988) pp 2-41.
14. A. Shukla and A. Kumar, Analysis of separation of trivalent cations through analcime zeolite composite membranes using space charge model, J. Colloid Interf. Sci., 274 (2004) 204-215.
15. G. Pugazhenthi and Anil Kumar, Experimental and modeling of separation of Cr (VI) from its aqueous solution using PMMA-EGDM ultrafiltration clay composite anion exchange membrane, AIChE J, 51 (2005) 2001-2010.
16. G. Pugazhenthi, S. Sachan, Nanda Kishore and Anil Kumar, Separation of chromium (VI) using modified ultrafiltration charged carbon membrane and its mathematical modeling, J. Membr. Sci., 254 (2005) 229-239.
17. S. Sachan, S. Sachdeva, G. Pugazhenthi and Anil Kumar, Hydrolysis of olive oil using lipase bonded to modified carbon membrane, AIChE J, 52 (4) (2006) 1611-1620.
18. G. Carrea and S. Riva, Properties and synthetic applications of enzymes in organic solvents, Angew.Chem.Int.Ed., 39 (2000) 2227-2254.

361d Simulation of a Reactive Separation Application for CO2 Removal

Hyun Tae Hwang[1], Aadesh Harale[1], Paul K. T. Liu[2], Muhammad Sahimi[1], and Theodore Tsotsis[1]. (1) Mork Family Department of Chemical Engineering and Materials Science, University of Southern California, 925 Bloom Walk, HED 216, Los Angeles, CA 90089, (2) Media and Process Technology Inc, 1155 William Pitt Way, Pittsburgh, PA 15238

For long-duration human presence on Mars or the Moon one requires highly reliable and efficient systems which provide basic life support provisions, such as food, water and air for the crews. A key function of the space life-support systems is effective removal of metabolic CO2 from the environment in the living quarters, and in the cabin during space flights. Without continuous removal, the amount of CO2 in a closed environment would increase to unacceptable levels.

Several methods of CO2 removal have been studied for use in the space habitats and suits. Adsorption of CO2 by LiOH, for example, is the current standard approach. For long-term flights the method presents challenges, however, since LiOH is not easily regenerable. Other similar systems based on CO2 chemisorption by silver oxide and solid phase amines also face the same challenge. Molecular sieves, such as zeolites, can trap CO2 in their pores by physical adsorption instead, and are more amenable to regeneration, e.g., via pressure swing adsorption (PSA), at the cost, nevertheless, of a greater process complexity.

A promising approach is the use of the methanation (Sabatier) reaction, in which CO2 is catalytically activated in the presence of hydrogen, and is hydrogenated to methane with simultaneous production of water. The simplest life support systems are "open-loop" and provide all the required resources, from water and oxygen to food, from either storage or re-supply. For such systems the amount of provisions needed increases proportionally as the mission duration and crew size increase. More complex systems are "closed-loop", at least to a certain extent, and though they also require an initial supply of resources, they subsequently process the waste products generated, such as CO2, urine, and wastewater, to recover additional useful resources.

In this study, we investigate the application of a reactive separation technology, in which the catalytic and separation steps are coupled in situ through the use of CO2 permselective membranes. We also present an overview of our current experimental and modeling efforts in this area aiming to establish the feasibility of the proposed reactive separation application for "closed-loop" life support systems.

Key words: Membrane Reactor, Carbon dioxide, Methanation, Sabatier reaction, Modeling, Air revitalization system, Closed-loop system

361e Computer Aided Design, Analysis And Experimental Investigation Of Membrane Assisted Batch Reaction

Piotr Tomasz Mitkowski[1], Carsten Buchaly[2], Peter Kreis[2], Gunnar Jonsson[1], Rafiqul Gani[1], and Andrzej Górak[2]. (1) Department of Chemical Engineering, Denmark Technical University, Building. 227, Søltoft Plads, Lyngby, 2800, Denmark, (2) Dortmund University, Chair of Fluid Separation Processes, 44221 Dortmund, Germany

Equilibrium or kinetically controlled reactions are common in chemical and biochemical manufacturing. This type of reaction is usually characterized by low process yield or low selectivity towards the desired product, when parallel reactions occur. On-site removal of product(s) enhances the yield, suppresses undesired side reaction(s) and therefore leads to reduced processing times of batch processes. The integration of separation and reaction can save investing and operating resources leading to more sustainable process, however specific "knock-off" criteria with respect to the reaction phase, catalyst, residence time and operating temperature and pressure must be satisfied.

Combination of at least two unit operations, basing on different physical phenomena, is called a hybrid process since they jointly contribute to fulfil the process step. In general, two types of hybrid processes are distinguished: reaction-separation for example the combination of batch reaction and membrane-based separation, and separation-separation like coupling distillation with pervaporation. The design of hybrid processes need to take into account interdependency of combined processes and it is possible to identify reliable and feasible design alternatives through a model-based computer-aided technique, saving thereby valuable experimental resources.

In this work the propyl-propionate synthesis from propionic acid and n-propanol is investigated. The design algorithm consists of four mains steps. At step 1 of the methodology, concentration profiles published in [1] are used to determine a temperature dependent kinetic constants, this includes thermodynamic analysis and identifying the operating window of the reaction. The esterification reaction has a conversion of around 64% (equimolar). In step 2 the objective is defined in order to increase the process productivity and to simplify the complex downstream processing due to thermodynamic non-idealities of the reaction mixture. Pervaporation is selected as the membrane-based separation technique (step 3) and a hydrophilic membrane is applied to remove selectively the by-product water. In step 4 various process scenarios (batch reaction, membrane assisted batch reaction, etc.) are analysed with the ICAS-MoT modelling environment. The separation characteristics of the applied polymeric membrane (Sulzer Pervap 2201) for binary and quaternary mixtures were obtained from a previous work at a multipurpose PV/VP lab-scale plant at University of Dortmund [2]. The transmembrane flux is calculated with a semi-empirical approach. The operating window in terms of reactant ratio, temperature and ratio of mass of catalyst and mass of reaction medium is investigated experimentally. The experiments show good agreement with the simulations.

The results of this work show, that a reactor coupled with a pervaporation unit is a promising process configuration with respect to product yield and batch time by overcoming limitations of kinetically controlled reaction. By the application of the membrane reactor concept it is possible to increase the conversion up to 95% at a decreased operating time of 4 h. Moreover, this high conversion benefits in less complex downstream operation for the final purification of the ester.

Reference:

[1] Duarte, C., Buchaly, C., Kreis, P., Loureiro, J.M., Inz.Chem.Proc. (2006)

[2] Lauterbach, S., Kreis, P., Euromembrane 2006 Taormina, Italy (2006)

361f A Study On Steam Reforming Of Methanol In A Pd-Based Membrane Reactor Fabricated By Surfactant Induced Electroless Plating

Mohammad A. Islam, Mohammad H. Akanda, and Shamsuddin Ilias. Mechanical and Chemical Engineering Department, North Carolina A&T State University, Greensboro, NC 27411

Steam reforming of methanol was investigated in Pd-based stainless steel microporous membrane reactor. The Pd-composite membrane was fabricated on microporous stainless steel tubular support by novel electroless deposition technique in which surfactants were used to enhance the Pd deposition and its surface morphology. Membrane surface morphology was characterized by SEM (Scanning Electron Microscopy), EDS (Energy Dispersive Spectroscopy), XRD (X-Ray Diffraction) and AFM (Atomic Force Microscopy). Single and mixed gases were used to measure the H2-perm-selectivity. The membrane-reactor was configured in shell-tube structure. The reaction was carried out on the shell side and the permeate hydrogen was taken out in the tube side. The reaction was carried out over commercially available Cu/ZnO/Al2O3 catalyst. A two-dimensional, pseudo-homogeneous membrane-reactor model was developed to investigate the steam-methanol reforming reactions in a Pd-based membrane reactor. Radial diffusion was taken into consideration to account for the concentration gradient in the radial direction due to hydrogen permeation through the membrane wall. With appropriate reaction rate expressions, a set of partial differential equations was derived using the continuity equation for the reaction system. The equations were solved by finite difference method. The model was also capable of handling steam-methanol reforming reactions under non-membrane condition and equilibrium reaction conversions. Experimental results shows membrane reactors with relatively higher selectivity values, have considerably higher conversion than that of non membrane counter part. Experimental methanol conversions were found to be in good agreement with the model predictions. Some of our experimental and modeling results will be presented in this paper.

362a Network Structure Modification of Crosslinked Polyethylene Oxide for Reverse-Selective Carbon Dioxide Removal from Mixtures with Hydrogen

Victor A. Kusuma[1], Scott Matteucci[1], Benny D. Freeman[1], Douglass S. Kalika[2], Michael K. Danquah[2], and Matthew A. Borns[2]. (1) Chemical Engineering, The University of Texas at Austin, 10100 Burnet Rd., Bldg 133, Center for Energy and Environmental Resources, Austin, TX 78758, (2) Chemical and Materials Engineering, University of Kentucky, 177 F. Paul Anderson Tower, Lexington, KY 40506-0046

The removal of carbon dioxide, which is a major impurity in hydrogen streams, could benefit from application of reverse-selective membrane separation. While size-sieving polymers tend to favor the transport of H_2 over larger molecules such as CO_2, reverse-selective membranes preferentially remove CO_2 from mixtures with H_2. Since H_2 is often required to be at high pressure for further use, this approach minimizes or avoids the need for the expensive H_2 recompression associated with conventional membrane separation [1]. Reverse-selectivity can be achieved by increasing the membrane affinity for CO_2; for instance, introducing polar groups into the membrane will increase the solubility of CO_2, which is quadrupolar [2].

One of the more effective polar groups for increasing CO_2 affinity is ethylene oxide. However, high molecular weight poly(ethylene oxide) (PEO) has low permeability due to crystallization [3]. We have demonstrated that UV cross-linking of low molecular weight PEO diacrylates can produce a completely amorphous membrane with much higher permeability, close to the theoretical upper bound of Robeson's trade-off curve [4]. Not only does this membrane demonstrate reverse-selectivity for CO_2/H_2 separation, its diffusivity (and thus permeability) can be improved further by copolymerizing PEO diacrylates with certain mono-functionalized PEO [1]. This improvement has been attributed to an increase in fractional free volume of the polymer, as measured by various methods. The increase of fractional free volume leads to improved CO_2/H_2 selectivity due to a decrease in diffusivity selectivity.

In this study, we further examine diffusivity enhancement by strategic copolymerization of selected PEO diacrylates and monoacrylates. We are especially interested in assessing the influence of the monoacrylate end group on the free volume and transport properties of the resulting membranes. In addition to the measurement of transport properties via pure gas permeability and solubility experiments, the thermomechanical properties of the membranes have been characterized using dynamic mechanical analysis. The results underscore the strong correlation between network structure and gas transport in these crosslinked PEO copolymers.

[1] H. Lin, E. Van Wagner, B. D. Freeman, L. G. Toy, R. P. Gupta, Plasticization-Enhanced H_2 Purification Using Polymeric Membranes, Science 2006, 311, 639-642.

[2] H. Lin, B. D. Freeman, Materials Selection Guidelines for Membranes that Remove CO_2 from Gas Mixtures, Journal of Molecular Structure 2005, 739, 57-74.

[3] H. Lin, B. D. Freeman, Gas Solubility, Diffusivity and Permeability in Poly(ethylene oxide), Journal of Membrane Science 2004, 239, 105-117.

[4] H. Lin, T. Kai, B. D. Freeman, S. Kalakkunnath, D. S. Kalika, The Effect of Crosslinking on Gas Permeability in Crosslinked Poly(ethylene glycol diacrylate), Macromolecules 2005, 38, 8381-8393.

362b Enhancing High Temperature Hydrogen Separation On Mfi Zeolite Membranes By On-Stream Modification Of Zeolitic Pores

Xuehong Gu, Zhong Tang, and **Junhang Dong**. Department of Chemical & Materials Engineering, University of Cincinnati, 2600 Clifton Ave., Cincinnati, OH 45221

High temperature hydrogen (H2) separation membranes may play an important role in future cost-effective production of H2 from catalytic reforming of coal, natural gas, and biomass for use in the emerging fuel cell-based automobiles and clean power systems. The microporous inorganic membranes, particularly molecular sieve zeolite membranes, are potentially useful for high temperature H2 separation from catalytic reforming streams. Highly siliceous MFI zeolite (i.e. silicalite and ZSM-5) membranes with zeolitic pore size of ~0.56nm offer high permeance of H2 and extraordinary hydrothermal and chemical stabilities. However, the H2 selectivity over small gases like CO2 was found to be low at high temperatures because transport of molecules with sizes significantly smaller than the pore diameter may be primarily governed by Knudsen diffusion. Furthermore, nanometer scale intercrystalline pores and grain boundaries exist in the polycrystalline membranes which can severely reduce the membrane selectivity especially when the separation relies on molecular sieving or size and shape selectivity. To enhance the selectivity for H2 separation from small molecule gases, the MFI membranes must be modified to minimize the nonselective intercrystalline pores as well as reduce the intracrystalline zeolitic pore size.

In this study, alpha-alumina-supported MFI zeolite membranes were modified by catalytic cracking of methyldiethoxysilane (MDES) molecules inside the zeolitic channels on-stream during separation of H2/CO2 gas mixture at 450oC. The MDES vapor was carried by the H2/CO2 feed stream and the effect of modification was monitored continuously through analysis of the permeate stream. A dramatic increase in H2 selectivity over CO2 was achieved with a moderate loss in H2 permeance on the modified membrane. The modified membrane exhibited good stability in separation of H2/CO2 gas mixture containing water vapor at 450oC and atmospheric pressure.

First-Principles Approaches to Screening Multicomponent Metal Alloys for Hydrogen Purification Membranes

Lymarie Semidey-Flecha, **Chen Ling** and **David S. Sholl**, Carnegie Mellon University, Department of Chemical Engineering, Carnegie Mellon University, Pittsburgh, PA 15213

Dense metal membranes are a well established technology for purification of hydrogen from mixed gas streams. To improve upon this technology, there is a great need to identify new materials compositions that can enhance hydrogen permeation rates while simultaneously providing a membrane that is robust in the presence of common chemical contaminants. We will describe how first-principles calculations can be used to make quantitative predictions regarding the permeance of ternary Pd-based alloys in the absence of experimental data for these materials. Our calculations focus on the permeation of hydrogen from pure gas streams, since high permeation rates in this situation are a prerequisite for considering the complicated problem of how membranes perform in contaminated environments. The aim of our calculations is to identify additive metals that can be introduced into binary Pd-based alloys that are known to have useful properties in order to increase hydrogen permeation rates relative to the initial binary alloy.

362d Evaluation And Characterization Of Pd-Based Membrane Fabricated By Surfactant Induced Electroless Plating

Mohammad A. Islam and Shamsuddin Ilias. Mechanical and Chemical Engineering Department, North Carolina A&T State University, Greensboro, NC 27411

Deposition of Pd/Pd-alloy thin film on microporous ceramic and stainless steel substrates by electroless plating for fabricating H2-selective membrane for high temperature gas separations and membrane reactors have been studied extensively over the past two decades and continue to be studied in a laboratory scale. Initially, electroless plating route for thin film deposition promised significant advantages over other methods (e.g. Chemical Vapor Deposition, Physical Vapor Deposition, Radio Frequency Cathodic Sputtering, etc.), but there are technological issues that still need to be resolved. In the conventional electroless plating, oxidation-reduction reactions between Pd-complex and hydrazine (reducing agent) result in evolution of ammonia and nitrogen gas bubbles. When adhered to the substrate surface and in the pores, these gas bubbles hinder uniform Pd-film deposition. This is a major weakness of the conventional electroless plating bath recipe. Stirring or rotation of substrate during plating may lessen the problem but can not provide satisfactory resolution of the Pd-grains and subsequent agglomeration. To address the problem associated with the growing gas bubbles within the plating substrate vicinity, we explored the role of surface active agents with favorable structures in the electroless plating process. The surface morphology and compositional microstructures were examined using SEM (Scanning Electron Microscopy) coupled with in-situ EDS (Energy Dispersive Spectroscopy). Palladium deposition rate was measured by the weight gain as a function of time. The surfactants studied included Triton X-100 (non-ionic surfactant), dodecyltrimethyl-ammonium chloride (cationic surfactant) and sodium dodecyl-benzenesulfonate (anionic surfactant). The surfactant molecules play an important role in the plating process by removing tiny gas bubbles through adsorption at the gas-liquid interface. The water-soluble surfactants which have almost neutral HLB (hydrophilic-lipophilic balance) value are believed to have stronger capability of removing gas bubbles from the substrate surface. It was observed that addition of non-ionic surfactant increased the palladium deposition rate by about 20% with improved surface morphology and microstructures. These membranes were tested for H2-perm-selectivity using pure and mixed gases at elevated temperature. Some of these results will be presented in this paper.

Long-Term H_2 Characterization of Composite Pd and Pd/Ag Membranes Synthesized from Agitated Electroless Pd and Ag Plating Baths

M. Engin Ayturk and Yi Hua Ma

Center for Inorganic Membrane Studies (CIMS)
Department of Chemical Engineering
Worcester Polytechnic Institute
100 Institute Road, Worcester, MA 01609 USA

Introduction

Hydrogen separation technologies are of great importance due to the essential nature of H_2 as an alternate clean energy carrier and the increasing demand for hydrogen in various chemical and petrochemical industries. Composite Pd and Pd/alloy membranes prepared by the electroless plating on porous sintered metal supports have potential applications for high purity H_2 production in steam reforming reactors, as well as in H_2 separation and purification[1]. Membrane synthesis via electroless plating is a widely used technique to form uniform deposition on surfaces with complex geometries and large surface areas with the added advantages such as the hardness of the deposited film, good adhesion to the substrate, and simple equipment and ease of scale-up[2]. Although there are only few studies in the literature[3] that discuss the effects of plating parameters on the deposition rates from hydrazine-based plating baths, the effect of external mass transfer on the electroless plating rates of Pd has not been quantified yet.

The objective of this study was twofold: To investigate the effects of bath agitation to elucidate the effect of the external mass transfer on the electroless plating rates and the deposition morphology of Pd and Ag metals and to study the long-term H_2 permeation characteristics of composite Pd and Pd/Ag membranes synthesized from agitated electroless Pd and Ag plating baths.

Experimental

The porous sintered metal supports used in this study were consisted of 0.5 µm-media grade, ½" OD and 1" long porous stainless steel (PSS) cylindrical cups and 0.1 µm-media grade, ½" OD and 2" long porous Inconel tubes, which were purchased from Mott Metallurgical Corp. The PSS and Inconel supports were oxidized at 600°C and 800°C, respectively, for 12 hours in order to form an intermetallic diffusion barrier[4]. Following the oxidation and surface activation[1, 2, 4], the Inconel supports were modified with the $Al(OH)_3$ grading procedure, similar to that of Tong *et al.* [5, 6] and modified by Ayturk[7]. The PSS supports were used to study the effect of bath agitation on the electroless Pd and Ag plating rates by using atomic absorption (AA) flame analysis[8] for the measurement of the Pd and Ag ion concentrations in the plating baths, while the Inconel supports were used for the synthesis of a 20 µm thick Pd/Ag (10 wt% Ag) and a 5 µm thick pure-Pd membrane, which were designated as membranes 022 and 025, respectively. The thicknesses and the Ag content of the membranes were determined gravimetrically. The H_2 permeation tests for the composite Pd/Inconel and Pd/Ag/Inconel membranes were conducted using the experimental set-up and procedure described elsewhere[2, 9].

Results and Discussion

In order to assess the role of the external mass transfer on the electroless Pd and Ag plating rates, the effect of the bath agitation rate was studied over a range of 0-600 rpm and at a constant bath temperature of 60°C. The initial metal ion and the hydrazine concentrations were 8.16 mM and 5.41 mM for Pd and 3.05 mM and 3.61 mM for Ag, respectively. As shown in Figure 1, the bath agitation at a rate of 400 rpm resulted in a significant increase in the amount of the Pd and Ag deposits, suggesting that the electroless plating of both Pd and Ag was strongly affected by the external mass transfer. While the conversion of the reaction (% conversion) after ~40 minutes of electroless Pd plating was 54% in the absence of bath agitation (rpm=0), the agitation rate of 400 rpm resulted in a ~94% conversion under the same conditions (Figure 1a). In the case of Ag, a 100% conversion was achieved only after ~40 minutes of Ag plating at an agitation rate of 400 rpm, which was significantly higher than the 36% conversion achieved after 30 minutes of Ag plating in the absence of bath agitation (Figure 1b). SEM analysis of the surface microstructure indicated that the rate of agitation also had a substantial effect on both the Pd and Ag deposition morphologies as shown by the micrographs in Figure 2.

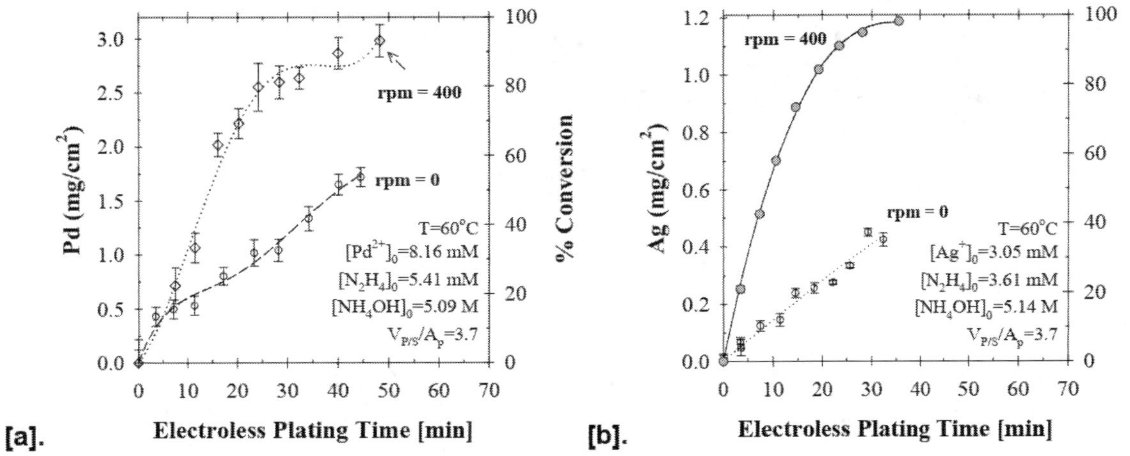

Figure 1. The effects of plating bath agitation on [a]. Pd and [b]. Ag plating rates

In comparison to the regular surface morphology of the tightly knitted Pd clusters formed in the absence of the bath agitation (rpm=0)[2, 9], the Pd morphology formed by the bath agitation at a rate of 400 rpm in Figure 2 was more uniform and smoother in appearance with relatively larger clusters, which consisted of nanosize Pd grains. The effect of bath agitation on the Ag deposition morphology was more pronounced at a rate of 400 rpm. In comparison to the regular Ag deposition morphology (rpm=0) consisting of randomly distributed clusters with different shapes and sizes, the transformation at an agitation rate of 400 rpm was quite significant resulting in a uniform, Pd-like surface topology in the form of large Ag clusters with an average size of ~2 μm in diameter (Figure 2).

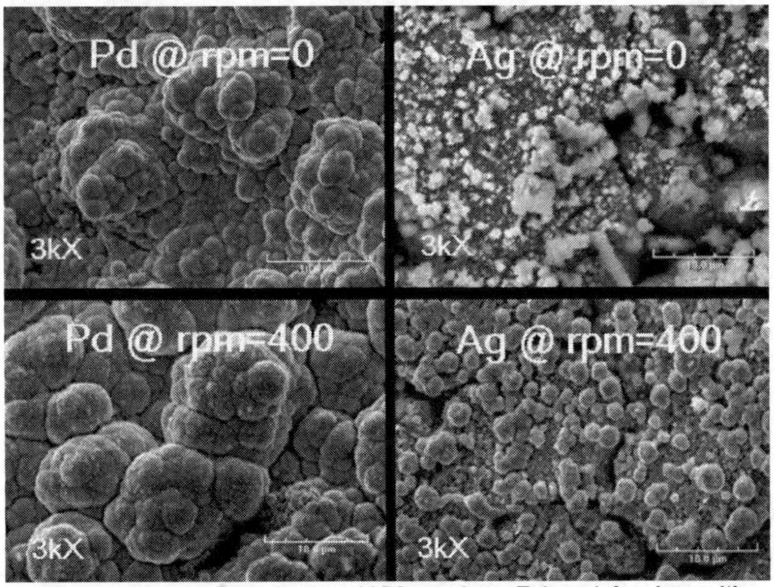

Figure 2. The effect of the bath agitation (400 rpm) on Pd and Ag deposition morphology

Since the rate of the agitation had no further effect on the electroless plating rates of both Pd and Ag at or above 400 rpm, indicating that the agitation rate of 400 rpm was high enough to minimize the external mass transfer limitations, the synthesis of membranes 022 and 025 were conducted at a constant agitation rate of 400 rpm and at the same temperature, initial metal ion and hydrazine concentrations reported for Figures 1a and 1b. Figure 3 shows the surface appearance of the Pd and Ag layers deposited on porous Inconel supports. In good agreement with the discussion in Figure 2, the agitation rate of 400 rpm resulted in the formation of extremely uniform Pd and Ag layers. The H_2 testing history for membranes 022 and 025 is shown in Figures 4a and 4b, respectively. The H_2 testing history for membranes 022 and 025 is shown in Figures 4a and 4b, respectively. For the 20 μm thick Pd/Ag (10 wt% Ag) membrane, the H_2 permeance and the H_2/He selectivity of membrane 022 at 300°C was 15 m^3/m^2-h-$atm^{0.5}$ and ~200, respectively.

Figure 3. Surface appearance of [a]. Pd and [b]. Ag membrane layers deposited on Inconel supports at an agitation rate of 400 rpm

The low value for the initial H_2/He selectivity was due to the partial peeling off of the $Al(OH)_3$ grading layer and a possible formation of a gradient in the membrane layer. However, it was interesting to note that the H_2/He selectivity of membrane 022 remained constant or even enhanced with the increasing temperature. The H_2 permeance and the H_2/He selectivity at 450°C were as high as 28 m^3/m^2-h-$atm^{0.5}$ and ~200, respectively, after a total testing period of 500 hours (Figure 4a). The slow leak growth for the 20 µm thick Pd/Ag (10 wt% Ag) membrane suggested that the uniform deposition morphology from an agitated plating bath, shown in Figures 2 and 3, might have an important effect on the coherent sintering of the as-synthesized Pd and Ag clusters. As can be depicted in Figure 4b, the H_2 permeance for the 5 µm thick pure-Pd membrane 025 at 300 and 400°C were 45 and as high as 63 m^3/m^2-h-$atm^{0.5}$. Furthermore, the H_2/He selectivity for the membrane 025 was over 310 after a total testing period of 690 hours in the temperature range of 300-400°C.

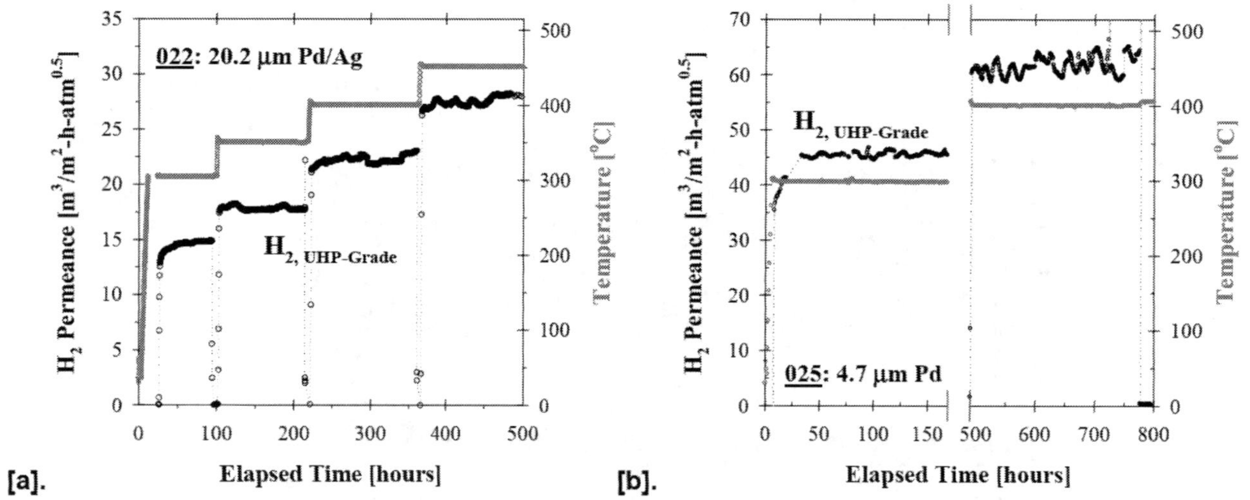

Figure 4. H_2 characterization of membranes 022 and 025

Conclusions

It was found that the electroless plating of both Pd and Ag were strongly affected by the external mass transfer in the absence of bath agitation. The external mass transfer limitations for both Pd and Ag deposition have been minimized at or above an agitation rate of 400 rpm, resulting in a maximum conversion of the plating reaction at 60°C and dramatically shortened plating times with an added advantage of uniform deposition morphology as evidenced by the SEI micrographs. The measurement of the metal ion concentrations in the plating baths via AA suggested that, for a commercial plating plant, successful utilization of the plating solutions through understanding of the electroless plating kinetics might have a substantial impact on reducing both labor and manufacturing costs and furthermore, eliminating the necessity for a metal recovery or waste treatment unit.

Furthermore, the bath agitation was used for the synthesis of composite Pd and Pd/Ag membranes on porous Inconel supports. The H_2 permeance for the 20 µm thick Pd/Ag (10 wt% Ag) and the 5 µm thick pure-Pd membranes (022 and 025, respectively) at 400°C was 22 and as high as 63 m^3/m^2-h-$atm^{0.5}$, respectively. In addition, after a total testing period of 500 and 690 hours in the temperature range of 300-450°C, the H_2/He selectivity for the membranes 022 and 025 were ~200 and over 310, respectively. The slow leak growth for both membranes was attributed to the uniform deposition morphology resulting from the agitated plating baths.

Acknowledgements

The authors gratefully acknowledge the financial support provided by Shell International Exploration and Production Inc., and Shell Hydrogen.

References

1. Mardilovich, P. P.; She, Y.; Ma, Y. H.; Rei, M.-H., Defect-Free Palladium Membranes on Porous Stainless-Steel Support. *AIChE Journal* **1998,** 44, (2), 310.
2. Ma, Y. H.; Akis, B. C.; Ayturk, M. E.; Guazzone, F.; Engwall, E. E.; Mardilovich, I. P., Characterization of Intermetallic Diffusion Barrier and Alloy Formation for Pd/Cu and Pd/Ag Porous Stainless Steel Composite Membranes. *Ind. Eng. Chem. Res.* **2004,** 43, (12), 2936-2945.
3. Yeung, K. L.; Christiansen, S. C.; Varma, A., Palladium composite membranes by electroless plating technique. Relationships between plating kinetics, film microstructure and membrane performance. *J. Membrane Sci.* **1999,** 159, 107-122.
4. Ma, Y. H.; Mardilovich, P. P.; She, Y. Hydrogen gas-extraction module and method of fabrication. US Patent 6,152,987, November 28, 2000.
5. Tong, J.; Matsumura, Y., Thin Pd membrane prepared on macroporous stainless steel tube filter by an in-situ multi-dimensional plating mechanism. *Chem. Commun.* **2004**, 2460-2461.
6. Tong, J.; Suda, H.; Haraya, K.; Matsumura, Y., A novel method for the preparation of thin dense Pd membrane on macroporous stainless steel tube filter. *Journal of Membrane Science* **2005,** 260, (1-2), 10-18.
7. Ayturk, M. E. Synthesis, Annealing Strategies and *in-situ* Characterization of Thermally Stable Composite Thin Pd/Ag Alloy Membranes for H_2 Separation. Worcester Polytechnic Institute, Worcester, 2007.
8. Beaty, R. B.; Kerber, J. D., *Concepts, Instrumentation and Techniques in Atomic Absorption Spectroscopy.* 2nd Edition ed.; Norwalk, CT, U.S.A., 1993.
9. Ayturk, M. E.; Mardilovich, I. P.; Engwall, E. E.; Ma, Y. H., Synthesis of composite Pd-porous stainless steel (PSS) membranes with a Pd/Ag intermetallic diffusion barrier. *Journal of Membrane Science* **2006,** 285, (1-2), 385.

Surface and fluid phase transport effects on hydrogen permeability through palladium-based membranes

Abdulrahman Y. Al Raisi and Tracy Q. Gardner, Colorado School of Mines, Golden, CO, USA

Introduction

Palladium and palladium alloy membranes are promising materials for hydrogen purification or removal processes. Hydrogen permeation through these materials has typically been modeled using the solution-diffusion approach in the past. When diffusion of hydrogen atoms through the bulk metal is the rate limiting step, hydrogen permeation flux at steady state can be described by Sieverts' Law. Sieverts' Law is widely applied in analyzing hydrogen permeation through Pd and Pd alloy membranes, even in some cases where the assumptions made in deriving Sieverts' Law are not valid. Recent advances in membrane synthesis, including thin metal films grown on porous supports, as well as considering hydrogen separation from mixtures where it is in low concentration, lead to situations in which Sieverts' Law is not valid, and the transport must be modeled in more detail. When Sieverts' Law is applied for hydrogen permeation through metal membranes, the exponent on the hydrogen partial pressure is often allowed to vary as if it were an empirical fit parameter. The closer this value is to 0.5, the better the membrane is said to behave as Sieverts' Law predicts. In cases of hydrogen purification or separation from other gases, concentration polarization can occur in the feed, and hydrogen transport through this fluid film can be rate limiting. We have modeled transport of hydrogen in metal membranes considering hydrogen transport through the fluid phase to assess concentration polarization effects on hydrogen permeation through metal membranes. Hydrogen permeation at 473-573 K through 25 μm Pd foils was measured from pure H_2 and H_2/He mixture feeds, with and without helium sweep gas. Transport through the fluid phase was modeled assuming constant total pressure on the feed side. Pure hydrogen permeation results followed Sieverts' Law, whereas mixture feed experiments did not. Hydrogen flux from feed mixtures was proportional to the hydrogen partial pressure difference rather than to the difference in the square roots of the hydrogen partial pressures as Sieverts' Law predicts.

Theory

Transport of hydrogen through metal membranes is often modeled using the solution-diffusion approach: the flux of hydrogen is assumed to be the product of its solubility in the metal, a measure of the amount of hydrogen absorbed in the membrane under equilibrium conditions, and its diffusivity, a measure of how fast hydrogen atoms transport through the membrane. The hydrogen flux through palladium films (foils and supported membranes) follows Sieverts' Law only if the hydrogen pressure exponent, n, is equal 0.5, which is usually valid for membranes that are at least 10 μm thick[1-3].

Pure Hydrogen Permeation

Hydrogen permeation through palladium membranes is a complex, multi-step process, as shown in Figure 1. Typically, diffusion of atomic hydrogen through bulk metal is the slowest step in the process and is considered to be rate limiting. Assuming Langmuir adsorption of only

H_2 on the surface, constant diffusivity in the metal, and taking the sorbed phase atomic hydrogen gradient as the driving force, the steady state hydrogen flux, J_{H2}, is:

$$J_{H2} = \frac{D}{2L}\left(\left(\frac{\sqrt{\frac{K_{H2,f}}{RT}P_{H2,f}}}{1+\sqrt{\frac{K_{H2,f}}{RT}P_{H2,f}}}\right)-\left(\frac{\sqrt{\frac{K_{H2,p}}{RT}P_{H2,p}}}{1+\sqrt{\frac{K_{H2,p}}{RT}P_{H2,p}}}\right)\right) \quad (1)$$

where D is the diffusion coefficient (m^2/s), L is the membrane thickness (m), $K_{H2,f}$ and $K_{H2,p}$ are the adsorption equilibrium constants on feed and permeate side (mol/m^3), and $P_{H2,f}$ and $P_{H2,p}$ are the feed and permeate side partial pressures of H_2 (Pa), respectively. In order to obtain the above equation the following assumptions were made:

- Diffusion of H through Pd metal is rate limiting
- Surface coverage of H at both the feed and permeate sides of the membrane is in equilibrium with the respective fluid phases and follows Langmuir adsorption behavior
- No hydrogen partial pressure gradient on the feed side
- Constant diffusion coefficient in Pd
- Linear H in metal phase concentration profile
- No species other than H on surface - either adsorbed molecules or contaminants

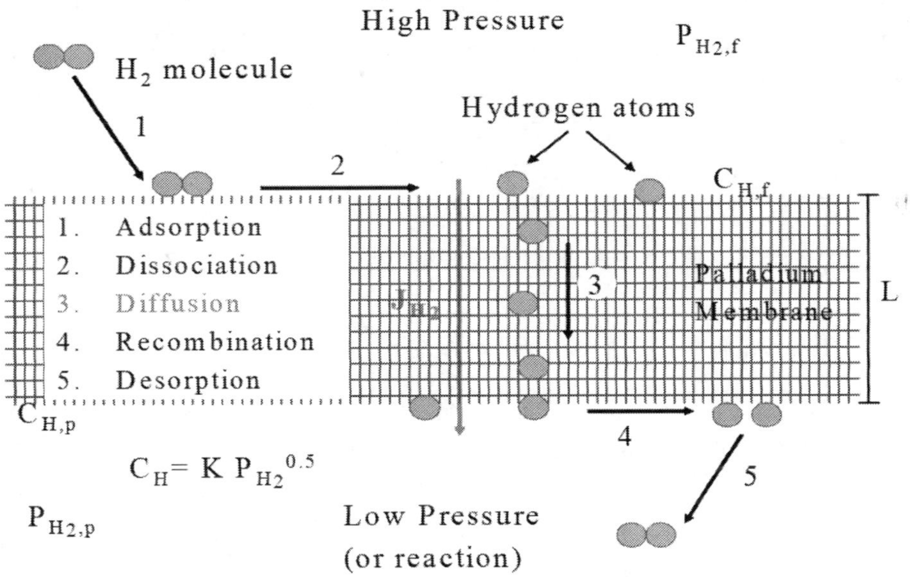

Figure 1 Schematic of hydrogen transport mechanism through Pd-based membranes

If adsorption on the surface is weak or the hydrogen partial pressure is low enough, then

$$\sqrt{\frac{K_{H2}}{RT}P_{H2}} \ll 1$$

and if the adsorption equilibrium constant is the same on both sides, then the expression can be simplified to Sieverts' Law:

$$J_{H2} = \frac{Q_{H2}}{L}\left(P_{H2,f}^{0.5} - P_{H2,p}^{0.5}\right) \quad (2)$$

where Q_{H2} is the permeability of H_2 through membrane ($mol/m\ s\ Pa^{0.5}$) [2-6].

Mixture Feed Permeation

When a mixture is fed to the membrane, for example hydrogen in helium, fast hydrogen depletion through the membrane can lead to an accumulation of the non-permeating specie near the membrane surface, and therefore a boundary layer is formed (concentration polarization). In such a case, hydrogen must first diffuse through helium before reaching the membrane surface. The steady state flux of molecular hydrogen through the boundary layer can be described as:

$$J_{H2} = \frac{-D_{H2,He} P_f}{RT\sigma}\left(y_{H2,s} - y_{H2,f}\right) \tag{3}$$

where $D_{H2,He}$ is diffusivity of H_2 in He, σ is boundary layer thickness, P_f is the feed pressure, and $y_{H2,s}$, $y_{H2,f}$ are H_2 mole fractions at the surface and in bulk feed, respectively (see Fig. 2).

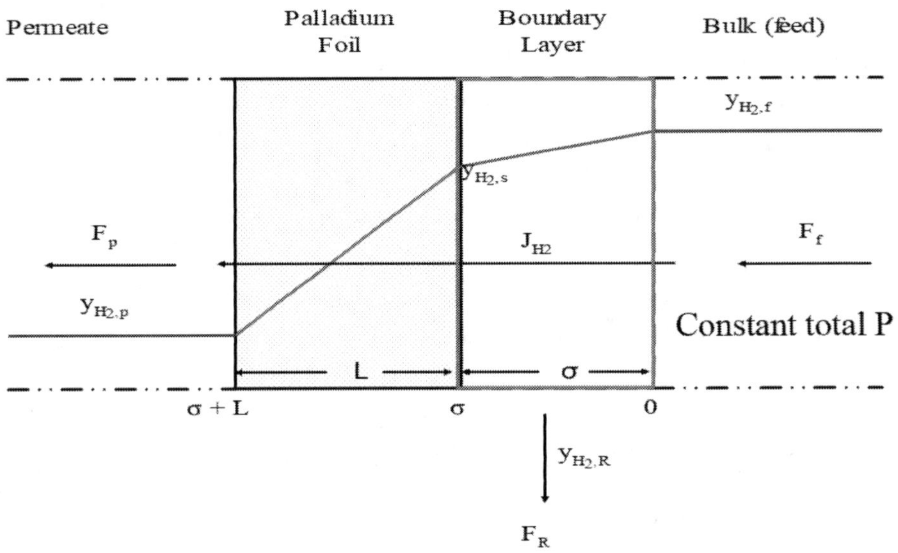

Figure 2 Schematic diagram showing concentration polarization in mixed feed

Then the steady state flux of hydrogen in palladium can be expressed in terms of the hydrogen partial pressure at the surface instead of that in the bulk feed, as was assumed for Sieverts' Law:

$$J_{H2} = \frac{Q_{H2}}{L}\left(P_f^{0.5} y_{H2,s}^{0.5} - P_p^{0.5} y_{H2,p}^{0.5}\right) \tag{4}$$

where $y_{H2,p}$ is the H_2 mole fraction on the permeate side. In Figure 2: F_f, F_p, and F_R are total molar flow rates at the feed, permeate, and retentate, respectively.

Experimental

Hydrogen permeation was measured through 25 μm thick, 40 mm diameter Pd foils. The foils were washed in acetone and thermally treated ex-situ in air for 30 minutes prior to hydrogen permeation measurements. The membranes were sealed with Kalrez o-rings in a 47-mm diameter Millipore flange fitted with two inlet and two outlet lines to allow for flow past the membrane surface on the feed and permeate sides (see Fig. 3). Measurements were made with ultra high purity hydrogen feed and with mixtures of hydrogen with ultra high purity helium, and with or without a helium sweep gas on the permeate side. Pure hydrogen and mixture feed permeation fluxes were measured at 473-573 K in a GC-8A oven and ΔP_{H2} up to 20 psi, with and without helium sweep. The permeate side total pressure was always atmospheric in Golden, CO (~84 kPa). Feed flow rates of hydrogen and helium were set with 5850E Brooks mass flow controllers, total volumetric flow rates were measured with a bubble flow meter, and for mixtures the concentration of hydrogen was measured using a Pfeiffer Vacuum QMS 200 mass spectrometer. Total pressures were measured with HEISE (CMM-121977) pressure gauges, and the temperature near the membrane surface was measured with Omega type J thermocouples.

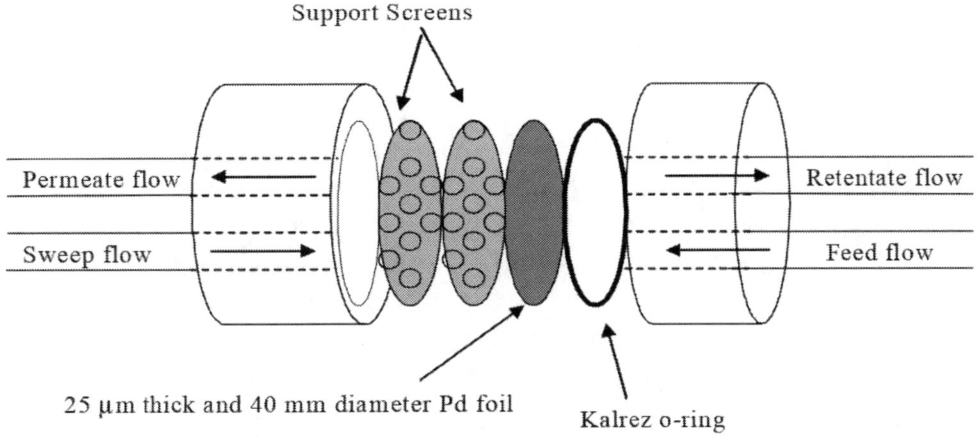

Figure 3 Membrane module

Results and Discussion

Pure Hydrogen Permeation

The hydrogen permeation fluxes through the 40 mm Pd foil (area = 12.6 cm^2) as a function of the pressure are presented in Figure 4. These results fit the pressure dependence given by Sievert's Law at all temperatures investigated. Some of the measurements were reproduced to be sure that the measured fluxes were at steady state. Fluxes were similar to those reported in the literature [7].

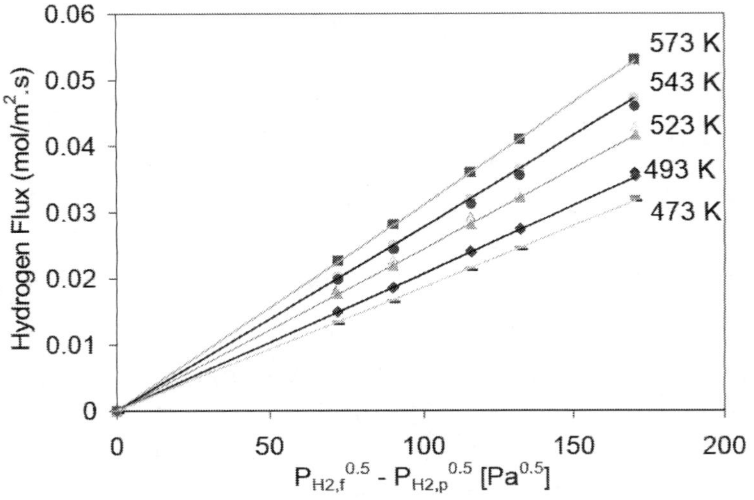

Figure 4 Pure hydrogen permeation fluxes through 25 μm thick 40 mm diameter Pd foil

From Arrhenius plots of the permeability, the activation energy, E_p, and pre-exponential factors, Q_o, for the hydrogen transport through the 40 mm diameter Pd foil were determined (see Table 1). Since pure hydrogen permeation fluxes followed Sievert's law, the assumptions made to derive Sieverts' Law appear to be valid for pure hydrogen permeation through these membranes.

Table 1 Hydrogen permeability parameters for a 25 μm thick and 40 mm diameter Pd foil

T (K)	Q_o (mol/m.s.$Pa^{0.5}$)	Q_o (literature) [7]	E_p (kJ/mol)	E_p (literature) [7]
473-573	8.61E-08	1.47E-08 – 3.82E-08	11.50	11.92 - 20.50

Mixture Feed Permeation

Normalized hydrogen permeation fluxes, from a 95%H_2-5%He (by volume) feed mixture at 523 K and hydrogen partial pressure drops of 3, 9, and 17 psi are shown in Figure 5. The fluxes are normalized by the hydrogen flux measured from a pure feed at the same hydrogen partial pressure drops. Mixture feed fluxes were lower than pure hydrogen fluxes, and approached the pure hydrogen flux as the total feed flow rate was increased. With the same helium mole fraction in the feed, lower fluxes were obtained at higher hydrogen partial pressure drops.

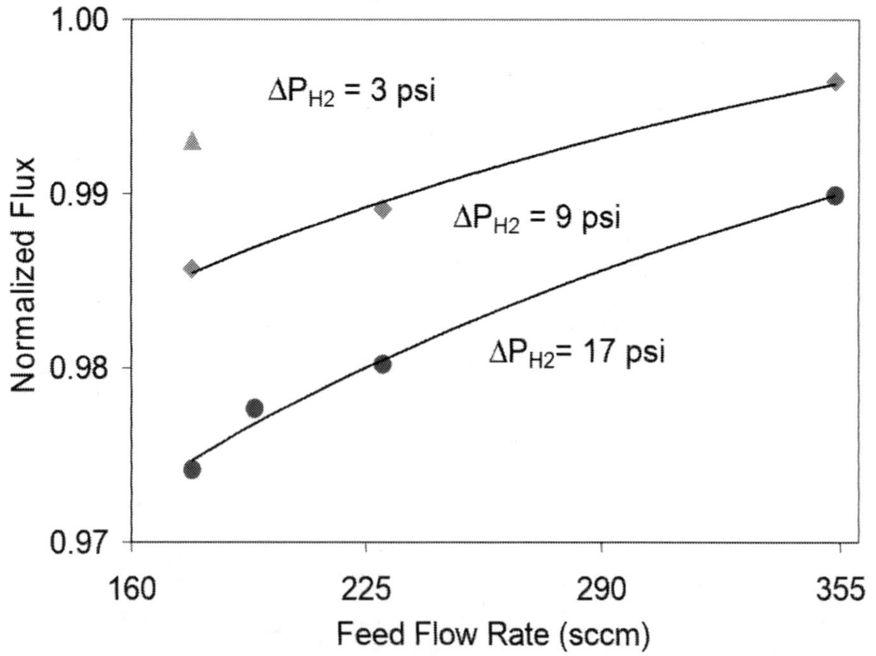

Figure 5 Hydrogen permeation fluxes through Pd foil for a 5%He feed stream at 523 K

Concentration polarization was more severe, as expected, when a 70%H_2-30%He (by volume) feed mixture was fed to the membrane at 523 K and a hydrogen partial pressure drop of 9 psi (see Fig. 6). Again as feed flow rate increases, concentration polarization decreases.

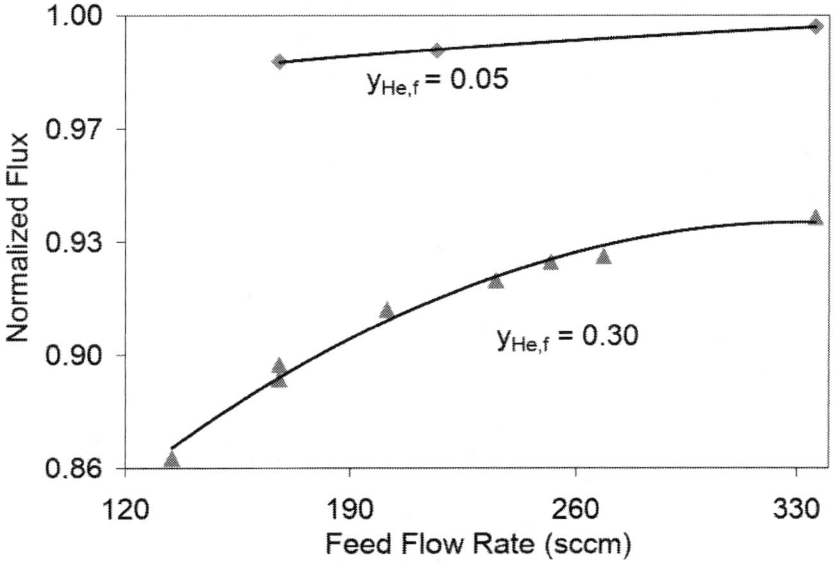

Figure 6 Hydrogen permeation fluxes through Pd foil for a 30%He feed stream at 523 K and ΔP_{H2} of 9 psi

Hydrogen permeation from mixture feeds was also measured at 523 K with ambient pressure on both the feed and permeate sides. Helium was used as a diluent in the feed and as a sweep on the permeate side. Helium sweep was maintained at 181 sccm, and the helium mole fraction was varied from 0.08 to 0.83 in the feed mixtures. The experiments were repeated three times at the same conditions, as shown in Figure 7. The fluxes with the mixture feeds did not vary with ($P_{H2,f}^{0.5} - P_{H2,p}^{0.5}$) as predicted by Sievert's Law. To get a linear plot of hydrogen flux vs. ($P_{H2,f}^{n} - P_{H2,p}^{n}$) for these data, n must be 1.

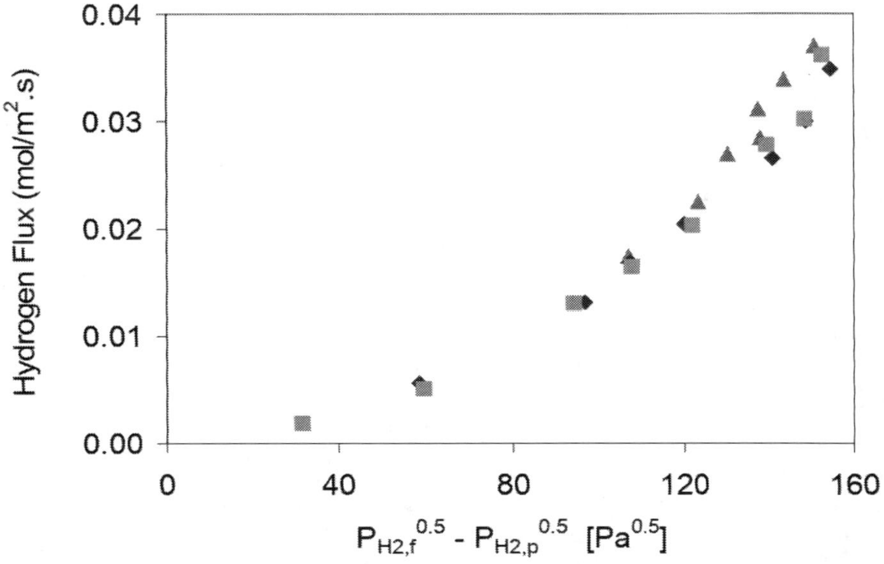

Figure 7 Mixed feeds permeation at 523 K and ambient feed and permeate pressures

Selective and fast hydrogen depletion through Pd leads to accumulation of helium near the Pd foil, which causes a concentration gradient across a boundary layer in the feed (concentration polarization), as represented in Figure 8. Due to this concentration polarization, the partial pressure of hydrogen in the bulk feed is not equal to the partial of pressure of hydrogen at the surface. Hydrogen diffusion through the membrane is, in this case, limited by the hydrogen diffusion through the boundary layer, leading to the n = 1 dependence on flux with hydrogen partial pressure difference in the gas phase.

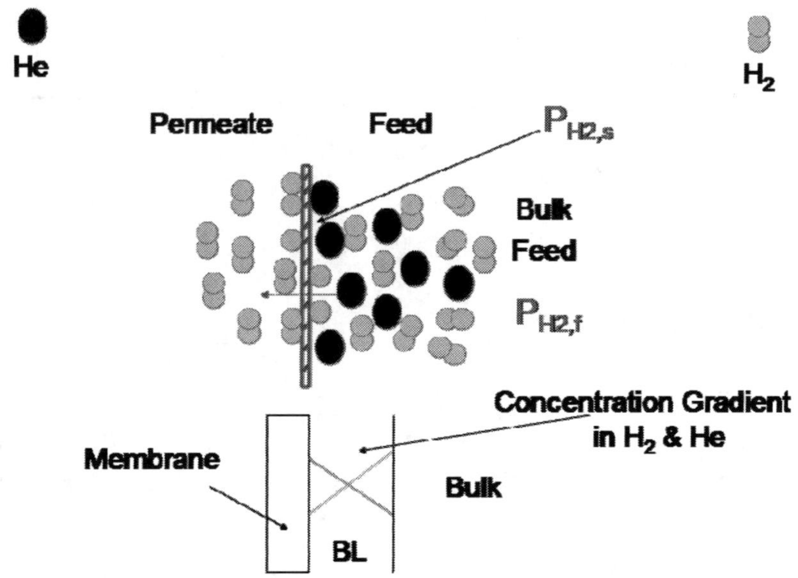

Figure 8 Hydrogen transport through Pd in existence of concentration polarization

The effective boundary layer thicknesses required to produce the measured fluxes were determined using equations (3) and (4), and are plotted in Figure 9 as a function of helium composition in the feed.

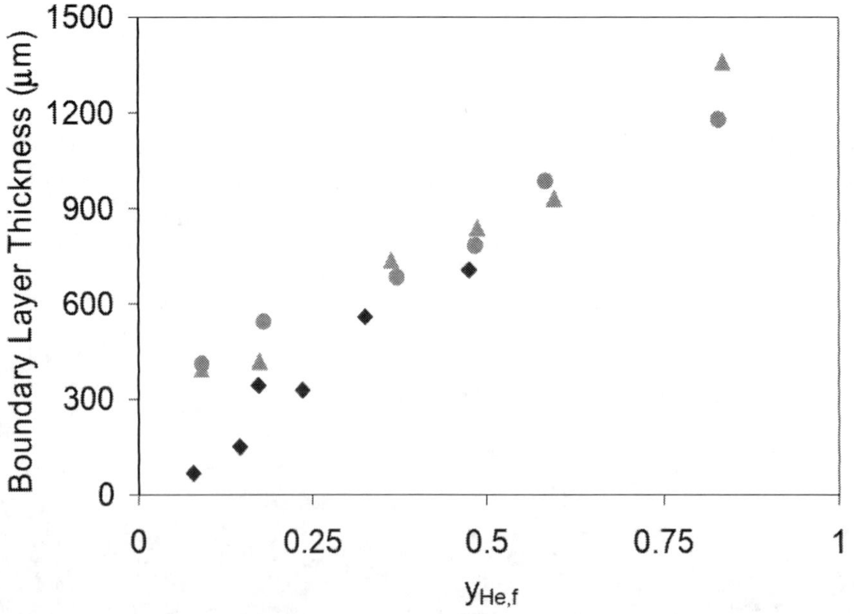

Figure 9 Boundary layer formation for hydrogen/helium mixture feeds

As the helium concentration in the feed increases, the estimated boundary layer thickness (which was 3 to 55 times the foil thickness) increases, as expected, because of the higher degree of concentration polarization with more helium. This effect can be minimized by increasing the feed flow rate.

Conclusions

Hydrogen permeation measurements from mixture feeds with hydrogen mole fractions as high as 0.95 showed significant deviation from Sieverts' Law. Fast and selective hydrogen permeation through Pd membranes can cause concentration polarization in mixture feeds, and diffusion of hydrogen through the consequent boundary layer can be rate limiting. It is important to account for concentration polarization in analyzing hydrogen transport from mixture feeds through metal membranes. Concentration polarization occurred even with only 5 mol% He in the feed and at feed flow rates as high as 353.5 sccm for a 12.6 cm^2 surface area disk membrane. Direct use of Sieverts' Law to these data led to a value of $n = 1$, treating n as an empirical fit parameter. A more detailed model of hydrogen permeation through metal membrane in which it is not necessary to treat n as an empirical fit parameter has been developed

Acknowledgements

A.Y. Al Raisi would like to thank Professor Douglas Way at Colorado School of Mines, for his advices and support. The authors also gratefully acknowledge the financial support provided by United Arab Emirates University (UAEU) and the Petroleum Institute (PI) in Abu Dhabi. Special thanks to the Center for Space Resources (CSR) at Colorado School of Mines for use of lab space.

References

1. Musket, R.G. (1976) "Effects of contamination on the interactions of hydrogen gas with palladium: A review", *J. Less-Common Met.*, 45, p. 173
2. Paglieri, S. N. and Way, J. D. (2002) "Innovations in Palladium Membrane Research", *Separation and Purification Methods*, 31(1), pp. 1-169
3. Mulder, M. (1991) "Basic Principles of Membrane Technology", *Kluwer Academic Publishers*
4. Baker, R.W. (2000) "Membrane Technology and Applications", *McGraw-Hill*
5. Shu, J.; Grandjean, B.P.A.; Van Neste, A.; Kaliaguine, S. (1991) "Catalytic palladium-based membrane reactors: a review", *The Canadian Journal of Chemical Engineering*, 69, pp. 1036-1060
6. Ward, T. L. and Dao, T. (1999) "Model of hydrogen permeation behavior in palladium membranes", *Journal of Membrane Science*, 153, pp. 211-231
7. Bryan D. Morreale, Michael V. Ciocco, Robert M. Enick, Badi I. Morsi, Bret H. Howard, Anthony V. Cugini and Kurt S. Rothenberger (2003) "The permeability of hydrogen in bulk palladium at elevated temperatures and pressures", *Journal of Membrane Science*, 212, pp. 87-97

362g Nanopore Membrane Based Reactors for Production of High Purity Hydrogen from Ammonia & Methanol

Sameer H. Israni, Chemical Engineering, University of Houston, S289, Engg Bldg-1, 4800 Calhoun, Houston, TX 77204 and Michael P. Harold, Department of Chemical Engineering, University of Houston, 4800 Calhoun Road, Houston, TX 77204.

Pd membrane based reactors have the potential to generate high purity H_2 in a single unit for stationary and mobile applications spanning power stations, soldier-power, and vehicles. The major challenges facing the deployment of these reactors include improvement of permeation fluxes and reduction of cost of the composite membranes. Previous studies have shown that these objectives can be achieved by depositing a thin Pd layer (< 5 microns) on a porous support, but in most cases this is accompanied by a drop in permselectivity due to defect formation. In our work we have synthesized a novel type of Pd membrane called the "Pd Nanopore" membrane which has yielded very encouraging results in terms of H_2 flux, permselectivity & long term stability of ultrathin membranes as compared to the conventional "top layer" membranes. We are currently applying the nanopore membrane in both high temperature separator and packed-bed membrane reactor (PBMR) configurations to generate high purity H_2. Catalytic ammonia decomposition on Ni/Al_2O_3 catalyst and methanol reforming on $Cu/ZnO/Al_2O_3$ catalyst are tests we are investigating which span a range of hydrogen volume productivities. Experimental & modeling results in the PBMR configuration for both systems will be discussed in detail in terms of reactor productivity, hydrogen utilization, and fuel conversion.

391a Microfluidic Platforms for Membrane Protein Crystallization

S.L. Perry, Griffin W. Roberts, Sameer Talreja, J.D. Tice, C.F. Zukoski, and P.J.A. Kenis. Department of Chemical & Biomolecular Engineering, University of Illinois at Urbana-Champaign, 800 S. Mathews, Urbana, IL 61801

The crystallization of membrane-type proteins remains a field of study that is more art than science. Membrane proteins are critical components of many fundamental biological processes such as ion regulation and transport through the cell membrane and catalyzing reactions such as charge separation and conversion of energy. As such they are often the targets of drug treatments and their malfunction has been linked to numerous diseases. Our understanding of such fundamental membrane-associated biological processes at a molecular level is limited due to the lack of knowledge of high-resolution structure of the membrane proteins as obtained from the analysis of crystalline materials by X-ray diffraction, electron microscopy, or solid state NMR. Numerous factors such as protein amphiphilicity, quantity, and non-rigid conformations increase the difficulty of determining suitable crystallization conditions. Two methods exist for the crystallization of membrane proteins. The in-surfo method utilizes detergents to solubilize membrane proteins, thereby creating a solution that can be screened against various precipitants to find crystal-forming conditions. In prior work we have shown that kinetic control over crystal growth can be achieved by variation of the evaporation rate. However, the in-surfo method suffers from limitations such as the choice of detergent for solubilization, denaturation due to the loss of the lateral membrane pressure, and applicability to various proteins. The in-meso method for protein crystallization allows for the crystallization of membrane proteins without their removal from a membrane-like environment. This alleviates concerns for both the solubilization of amphiphilic molecules and denaturation due to the loss of the lateral membrane pressure. However in-meso crystallization requires the homogeneous mixing of two materials with viscosities differing ~30x. For both of these methods, the determination of crystallization conditions is often done using high throughput sparse matrix screening methods. While traditional well plate technology is well established, such systems typically use large and expensive fluid handling equipment, and may be on a scale that is inaccessible given the quantities of protein available. Here we report on the design, fabrication, and testing of a microfluidic device designed for the on-chip formation and screening of in-meso crystallization conditions for membrane-type proteins at volumes of >20 nL per sample. This platform allows for the preparation of multiple mesophase samples at once, while also facilitating high throughput screening of multiple precipitant solutions on a single chip. In addition to screening for appropriate crystallization conditions, independent control over precipitant injection has the potential to allow for kinetic control of crystal growth done in-meso through the use of additives to control the rate of protein diffusion by affecting the curvature of the protein-containing mesophases. We also demonstrate kinetic control for in-surfo crystallization as achieved by the controlled evaporation of solvent within a microfluidic device. This device enables the rapid identification and optimization of crystallization conditions for membrane proteins and other molecules by precise control of the evaporation rate. In this manner we can independently control the rate of supersaturation and ensure a phase transition for every trial. The advantage of controlling the rate of supersaturation independently manifests in the improvement of crystal quality and an increase in crystal size in the case of soluble as well as membrane proteins. In addition, the microfluidic platform requires only ~5nL of sample per trial. These technology will help facilitate the crystallization of membrane proteins by providing multiple methods for crystal growth and easing the sample volumes necessary for screening. The membrane protein bacteriorhodopsin (bR) is our model system.

391c The Impact Of Mixing Times Scales And Process Parameters On The Optimisation And Scale Up Of Anti-Solvent Addition Crystallizations

Mark Barrett, Dr. Eoin Casey, and Dr. Brian Glennon. School of Chemical and BioProcess Engineering, University College Dublin, Engineering Building, Belfield Campus, Dublin, 4, Ireland

Previous work within this research group has highlighted the impact of addition location, addition rate, agitation and scale on the Meta Stable Zone (MSZ). The current research aims to assess the impact of macro, meso and micro mixing times scales on the induction time for the anti-solvent addition crystallization of benzoic acid form an ethanol-water solution. In addition to this the effect of nozzle addition and power law feed profiles on overcoming regions of poor addition (i.e. regions or scales of long mixing time scales) and on optimizing crystallization characteristics such as Metastable Zone Width (MSZW) and chord length distributions (CLDs) have been assessed. Anti-solvent addition crystallizations are mixing sensitive as they rely on the incorporation and blending of two liquids to produce a homogenous system. On scale up the incorporation and efficient blending of the anti-solvent to the bulk solution becomes more difficult, hence they are problematic to scale up. Vessel homogeneity and mixing times scales (macro, meso and micro) are all negatively impacted on scale up, thus leading to fluctuations in product yield / purity and in MSZW and chord length distributions (CLDs) produced. This paper develops a semi-quantitative method for the optimization and scale up of the anti-solvent addition crystallization of benzoic acid from 500 ml to 2L and finally to a 70 L crystallizer. The use of computational fluid dynamics (CFD) for understanding the mechanisms of importance, supersaturation nucleation and growth, is outlined in detail. The combination of this modeling tool with other in-situ tools such as Focused Beam Reflectance Measurement (FBRM), particle dimension characterization, Fourier Transform Infrared spectroscopy (FTIR), for on-line supersaturation and Particle Vision Measurement (PVM), for on-line particle imagery, allow for a more robust and strategic strategy for scale up.

The importance of macro, micro and meso-mixing time scales for successful scale up of anti-solvent crystallizations has been identified. Examining the effect of these times scales on premature nucleation events at the 500ml scale, due to excess supersaturation at the feed point, has been studied. The effect of feed point location and feeding mechanism (i.e. feed tube diameter) on macro and meso-mixing time scales have all been assessed for their impact on inconsistent crystallization characteristics, which include fluctuations in the FBRM statistics and premature nucleation events. Overall the dominant time scale for allowing premature and inconsistent chord length distributions (CLDs) was the meso mixing times scale. The effect of the process parameters, outlined above, on this time scale have been examined, thus allowing for problematic regions of addition (i.e. regions of long meso-mixing time scale) to be noted prior to scale up. This increase in process understanding through CFD minimized the experiments required to efficiently scale from a 500ml crystallizer to a 2L and finally to a pilot scale crystallizer of capacity 70L.

391d Kinetics And Size Distributions Resulting From Fractional Crystallization Of Multiple Solutes From Complex Solutions

Laurent Nassif, George Dumont, and Ronald W. Rousseau. School of Chemical & Biomolecular Engineering, Georgia Institute of Technology, Atlanta, GA 30332-0100

Crystallization has been used in a wide range of industrial applications; it is now being considered for pretreatment of nuclear waste. The Hanford Site tank farms have 53 million gallons of mixed liquid waste (waste with both hazardous and radioactive components) contained in 149 single-shell tanks (SST) and 28 double-shell tanks (DST). It is anticipated that tank waste will be partitioned to separate the high-level liquid waste (HLLW) constituents from the large volumes of low-activity liquid waste (LALW) in the tanks. A Pretreatment System for the LALW is required to ensure that these problematic waste components are diverted preferentially to the Bulk Vitrification facility, while radionuclides (primarily ^{137}Cs and ^{99}Tc) are diverted preferentially to the WTP.

Semi-batch fractional evaporative crystallization involves the removal of multiple solutes from a feed solution by the progressive achievement of supersaturation (through evaporation) and concomitant nucleation and growth of each species. The feed solution to the evaporator is expected to contain several sodium salts, among them sodium nitrate, sodium carbonate, sodium sulfate, sodium oxalate, sodium fluoride and sodium chloride. It is anticipated that evaporation from the feed solution produces burkeite (Na_2CO_3•$2Na_2SO_4$), sodium carbonate monohydrate (Na2CO3•H2O) and sodium nitrate ($NaNO_3$), as well as other sodium salts, which do not interfere with the crystallization kinetics of the three main species.

The present work explores the application of evaporative fractional crystallization on a multi-species solution leading to the nucleation of several sodium salts, each with distinct nucleation points, crystallization kinetics, and resultant crystal morphologies. Moreover, interactions among species are likely to influence the kinetics and crystalline habit of each.

Experiments were performed to investigate the influence of several operating variables on the characteristics of the final crystal product. Sampling was performed on a regular basis, and analysis of PLM images of the samples allowed determination of the nucleation point of the different species and comparing them to thermodynamic simulations. Results showed that the interactions between species modified the nucleation points of each of the main species (sodium nitrate, sodium carbonate and burkeite).

In addition, growth and nucleation rates were estimated for the overall crystal population along with those of specific crystalline species. An especially noteworthy observation was that the sodium nitrate-burkeite interactions had an effect on the crystal growth and potentially on the morphology of sodium nitrate crystals.

Finally, the evolution of the crystal population through the experiment was determined and presented as a series of histograms. Useful information was obtained concerning the correlation between population and mass distributions for multiple-species crystallization and the decomposition of overall distributions into single-species distributions. Results obtained on this very complex system represent the first step leading toward the modeling of complex multiple species crystallization with species interactions.

391e Rethinking Crystallization Processes

Daniel A. Green, DuPont Engineering Research & Technology, DuPont Experimental Station, Building E304-A222, Wilmington, DE 19880-0304 and Changzai Chi, DuPont Engineering Research, DuPont Engineering Research & Technology, DuPont Experimental Station, Building E304-A222, Wilmington, DE 19880-0304.

Most industrial crystallization process technology is well established, with many years experience. Consequently, there have been relatively few major improvements recently. Most industrial crystallization is done by creating supersaturation in stirred tanks with simultaneous crystal nucleation and growth. Since both nucleation and growth are strong functions of supersaturation, they can not in general be controlled individually in traditional crystallizers. In addition, new process technology is needed that can commercialize recent advances in the application of molecular and supramolecular control of crystallization, such as template directed crystal nucleation. We will review alternate crystallization process strategies that enable greater control of crystal properties. In particular, we will present research leading to a system that separates crystal nucleation and growth, and that can create templated nuclei.

391f Effect Of Crystal Aging And Agglomeration On Particle Breakage In Saturated Solutions

Devkant S. Gandhi, Dave C. Swalm School of Chemical Engineering, Mississippi State University, Box 9595, Mississippi State, MS 39762 and Priscilla Hill, Chemical Engineering, Mississippi State University, Box 9595, Mississippi State, MS 39762.

Particle breakage in stirred vessels such as crystallizers occurs due to collision of crystals with each other and with the mechanical parts of the crystallizer such as the stirrer, baffles, vessel wall etc. Researchers investigating the particle breakage phenomenon in crystallizers usually use a non-solvent in order to isolate breakage from other phenomena such as crystal aging and agglomeration. One type of crystal aging is Ostwald ripening, where large particles grow larger at the expense of smaller particles dissolving in a saturated solution. Ostwald ripening poses an obvious concern to the experimentalist researching particle breakage. If a significant quantity of small particles is formed, it could dissolve and significantly change the shape and size of the larger particles. Another type of aging is due to small temperature oscillations. It has been shown in the past that some crystals (for example, terephthalic acid, barium-lead sulfate) undergo shape changes in their saturated solutions when they are subjected to small temperature fluctuations. This type of aging poses a bigger concern because even in a temperature-controlled crystallizer there is a chance that temperature maybe slightly higher or lower than the set point at various locations in the vessel. For example, the temperature could be slightly higher at the tip of the impeller compared to the wall of the vessel. This could cause a suspended crystal in the saturated solution to experience random periods of high and low temperatures due to stirring and thereby accelerate the aging process and consequently change the shape and size of the crystals. While using non-solvents prevents other phenomena from occurring simultaneously, it doesn't always adequately represent real crystallization conditions. This is because of the difference in properties of saturated solutions and non-solvents. Only a few systems have been studied in the past using saturated solutions. Consequently, very little experimental data is available on crystal breakage behavior in saturated solutions. Although, it is common knowledge that phenomena like agglomeration, Ostwald ripening and aging occur in saturated solutions, the severity of such phenomena and their impact on breakage behavior is not known. The limited number of systems tested cannot be considered as representative of all crystallization processes. This work studies different systems in saturated solutions and quantitatively investigates the effects of Ostwald ripening, aging and agglomeration on crystal size and shape. Since it is impossible to follow one single particle in a stirred vessel, the particle is studied using a microscope hot-cold stage to control the temperature, and a microscope to determine changes in particle size and shape. Computer software is employed to measure various dimensional characteristics of particles to quantitatively analyze changes in shape and size. To investigate the effect of temperature fluctuations, the temperature of the hot-cold stage is varied randomly between 1 to 3 °C. Different sizes of crystals in saturated solutions and non-solvents are studied to determine if the original crystal size has a significant effect. Testing for agglomeration and Ostwald ripening is performed by adding small crystals to a saturated solution with a large crystal, and observing the large crystal. Images at regular time intervals are taken and analyzed. A variety of systems with different solubilities are tested to determine the effect solubility has on the results. Results of these experiments and comparisons between systems are presented.

391g Computationally Efficient Hybrid Method for the Inversion of Chord Length Distributions to Particle Size Distributions

Pavan Kumar Akkisetty[1], Michael Lasinski[2], Nandkishor Nere[2], Venkat Venkatasubramanian[2], Gintaras. V. Reklaitis[2], Doraiswami Ramkrishna[2], Willis V. Bell III[3], and Gary Blau[2]. (1) Chemical Engg, Purdue University, 480 Stadium Mall Drive, West Lafayette, IN 47907-2100, (2) Chemical Engineering, Purdue University, 480 Stadium Mall Drive, West Lafayette, IN 47907-2100, (3) Eli Lilly & Company, Tippecanoe Laboratories, 1650 Lilly Road, Lafayette, IN 47909

The product particle size distribution (PSD) and the associated properties are typically the important control objectives for crystallization operation. For the purpose of monitoring the PSD a focused beam reflectance measurement (FBRM) device is used to provide on-line and in-situ information of crystal size and particle concentration by means of a chord length distribution (CLD). Though used in general, CLD data are not the direct measure of the PSD

Present work focuses on the development of a computationally efficient tool to convert a CLD to a PSD. Conversion of CLDs to PSDs is usually done in a two step process: 1) Generation of a matrix that converts the PSD of a population of particles with given shape into the corresponding CLD using a 3 dimensional geometric model. 2) Computation of the solution of the resulting linear matrix equation for the PSD [1]. Since the inverse problem is usually mathematically ill-posed, the method of constrained least squares minimization (CLSM) used for the inversion requires regularization. The selection of the optimum regularization parameter is computationally expensive due to the requirement of huge number of repetitive minimizations to generate L-curve that forms a basis for the selection of an optimal regularization parameter.

In the present work we propose the use of a direct analytic model of the L-curve for Tikhonov regularization parameter selection proposed by Mc Carthy [2]. Implementation of a CLSM along with the direct analytic model reduces the computational time significantly

The tool developed is expected to speed up the process of the inversion of the CLD to PSDs which in turn are used in the extraction of the crystallization kinetics using the inverse problem approach of Mahoney et al (2002). A hybrid Population balance [3] - data driven modeling approach is underdevelopment for modeling the crystallization process.

References

1. Jorg Worlitschek (2003), "Monitoring, Modeling and Optimization of batch cooling crystallization"(PhD. Thesis), Zurich, Swiss Federal Institute of Technology.

2. P J Mc Carthy, "Direct analytic model of the L-curve for Tikhonov regularization parameter

selection," Inverse Problems, 19 (2003) 643-663.

3. Alan W. Mahoney, Francis J. Doyle III, Doraiswami Ramkrishna, Inverse problems in population balances: Growth and nucleation from dynamic data, AIChE J., 48(5), 981-990 (2002)

392a Effective Gas/liquid Contact Area

J. Christopher Lewis, A. Frank Seibert, and James R. Fair. Separations Research Program, University of Texas at Austin, 10100 Burnet Road, Bldg 133/CEER, Austin, TX 78758

Accurate measurement of the interfacial area of random and structured packing will allow better models to be created for use in designing distillation and absorber columns. Researchers at the University of Texas at Austin developed a procedure that allows the effective interfacial area of random and structured packing to be calculated directly from experimental results. These data are compared to existing interfacial area models and will be used to model the effective area in a forthcoming SRP model.

An Improved Model to Predict Liquid-Side Mass Transfer Coefficient in Stripping Applications

Jacinto Lopez-Toledo[1], Mahesh Sawant[2], Flor Castillo[1], Patrick Au-Yeung[1], and John Pendergast[1]. (1) The Dow Chemical Company, 2301 N Brazosport Blvd, B 1603, Freeport, TX 77541, (2) Dow Chemical International Pvt. Ltd, 6th Floor, Tower 'C',, Panchshil Tech Park, Pune, India

Abstract

Several correlations to predict mass transfer coefficients in separation towers exist in the literature (i.e., Onda et al [1], Billet and Schultes [2]). However, when these correlations are applied to existing columns using the rate-based methodology, they do not accurately predict the performance across the range of normally encountered operations. Thus, a more accurate correlation is needed for the gas side, the area for mass transfer, and for the liquid side, discussed in this paper.

A modified correlation for the liquid-side mass transfer correlation was developed using more than 400 unpublished stripping data points (screened from an original data set of more than 800 points), taking into consideration the influence of packing depth and gas velocity. The exponents of the parameters in the constituting correlation change with the operating regime of the packed tower (indicated by the *L/G* ratio). This correlation was tested with previously unpublished data to check the accuracy of the improved model using the system IPA/Water with 38 mm metal Pall rings in a 31.1 cm inside diameter column.

Background

The widespread use of packed towers with random and structured packing for distillation, stripping, and absorption operations across many industries calls for reliable and accurate mathematical models for predicting their hydrodynamic and mass transfer behavior. Decades of research and publications in both the academic and industrial community have been dedicated to this very topic of mass transfer in packed columns with only limited success. Dow itself has an extensive experimental database of random packing performance in stripping operations, and total reflux distillation operations.

The objective of this work was to take advantage of the packing performance database and build an improved mass transfer correlations that could be applied in RateSep™, the newer Aspen Plus® rate based modeling tool.

Experimental Data.

The experimental data which is evaluated to develop a modified liquid side mass transfer correlation consists of a total of 862 unpublished data points.

The evaluation of K_La is sensitive to experimental errors or limitations in two regions as indicated in Figure 1 [3]. For a particular tower height, cross-sectional area and mass transfer coefficient, the curved line indicates the calculated removal efficiency. The trends of this curve at low stripping factors (S) and very high removal efficiencies are important here. A small error in magnitude of the calculated stripping factor at low values ($S < 2$) can produce a large error in the calculated overall mass transfer coefficient. The magnitude of the calculated stripping factor is also susceptible to small errors related with the measurement of the air and water flowrate or the value of Henry's constant. Moreover the term $S/(S-1)$ in the expression to calculate K_La itself creates significant error when $S < 2$.

The second region of sensitivity to experimental variation [3] occurs when the effluent concentration is low and the fractional efficiency is very high. Even small analytical errors in determining the effluent concentration when it is small can lead to significant errors in the ratio of inlet to outlet VOC concentration and thus can lead to large errors in the calculated K_La.

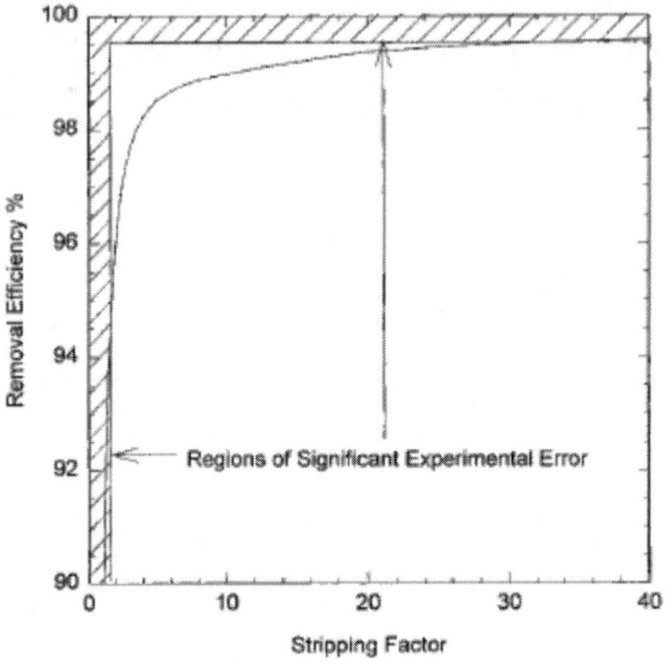

Figure 1 . Conditions where the calculated overall mass transfer coefficient is sensitive to experimental error. [3]

Due to the sensitivity of the calculated mass transfer coefficient in these two regions, a selection criterion was applied to the original set of experimental data. Thus, only those datasets with stripping factor greater than 2 and effluent concentration greater than 1 ppm were selected for further analysis, to eliminate any inherent error in the calculation of the overall mass transfer coefficient. Thus the experimental global data (862 data sets) set was whittled down to 553.

According to the penetration theory, the liquid-side mass transfer, mainly depending on liquid rate and contact time, is conventionally assumed to be independent of the gas rate, however early work by Sherwood and Holloway [4] have shown that, in the loading region, the gas rate increase can enhance the liquid-phase mass transfer. Kutzer and Mersmann [5] have also made similar observations where they confirm that the increase of gas flowrate reduces the overall height of a transfer unit (*HTU*) for the air stripping of perchloroethene, trichloroethene, 1,2-dichloroethane and 1,1-2 trichloroethane.

The influence of stripping factor on the overall *HTU* is shown in Figure 2. The respective trends represent the influence of the gas flow rate on the overall mass transfer efficiency wherein the individual points represent the experimental *HTU*, in meters, obtained for a particular packing. The gas flow rate was varied keeping all the other experimental parameters constant. As can be seen from the trends the increase in the gas flow reduces the *HTU*, or in other words increases the overall mass transfer coefficient.

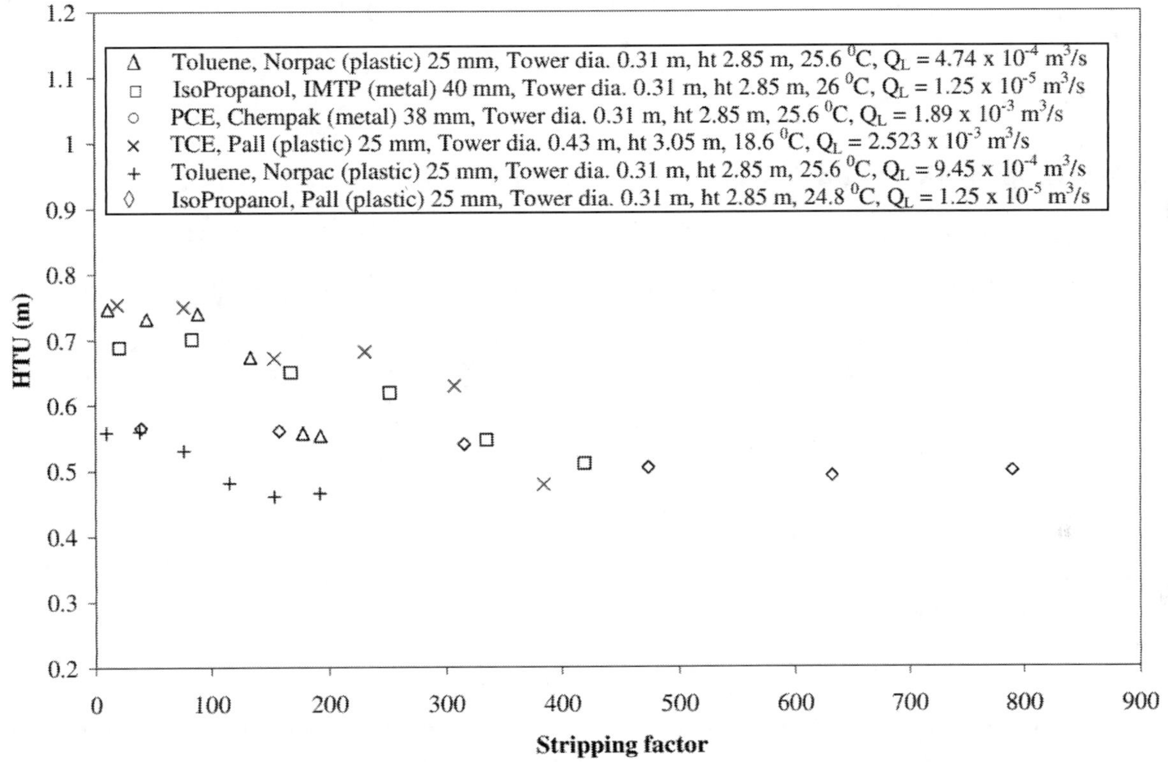

Thus, the *L/G* ratio which determines the operating regime of a packed column needs to be accounted for when developing a modified liquid side mass transfer coefficient.

Figure 2. Effect of gas flowrate on *HTU* (experimental data)

Channeling, end effects and liquid mal-distribution are some of the factors that are known to affect the transfer process, but they are either ill-considered or completely ignored by traditional mass transfer correlations. The mass transfer calculations in packed columns assume a constant mass transfer coefficient throughout the column height. This assumption however has been proven to be invalid at least at the entrance of a column up to a certain depth. Perry and Nutter [6] have showed that a certain height of the packed bed is required to develop a natural frequency of distribution beyond which the mass transfer coefficient is supposed to become constant. Djebbar and Narbaitz [7] have also demonstrated in their studies that the experimental k_L varied with depth.

The present work attempts to present an improved liquid side mass transfer correlation by including two dimensionless quantities:

1. the ratio of liquid to gas velocity (u_L/u_G) which would take into consideration the effect of the gas velocity on k_L
2. the ratio of the height of the packed bed to the packing diameter (L/d_p) to consider the effect of packed depth.

Correlation for the liquid side mass transfer coefficient:

The overall mass transfer coefficient K_La which is obtained from the experimental data needs to be separated into the individual phase mass transfer contributions before developing any phase-specific correlation. The following assumptions are made for further analysis.

1. The wetted surface area of the packing available for mass transfer was assumed to be equal to the total surface area of the packing.
2. The Onda gas phase mass transfer correlation was used to represent the gas phase resistance.

Further, the individual contribution of gas phase resistance to the overall mass transfer resistance was evaluated, and only those experimental data with the gas phase resistance contributing 10% or less of the overall resistance were included in the analysis. This ensured that the liquid side mass transfer resistance would dominate the overall mass transfer mechanism, which reduced the experimental data set to be analyzed from 553 to 417.

On the basis of two-film theory, the experimental liquid side mass transfer coefficient (k_L) is evaluated from the measured overall mass transfer coefficient (K_L) by the following equation:

$$\frac{1}{K_L a} = \frac{1}{k_L a} + \frac{1}{Hk_G a} \qquad (1)$$

where H is the Henrys' constant for the VOC in water.

The objective function for minimizing the error between the predicted mass transfer coefficient and the experimental value is given by the Average Absolute Relative Error (*AARE*) given below

$$\mathrm{AARE} = \frac{1}{n}\sum_{i=1}^{n}\frac{\left|(k_L)_{\exp} - (k_L)_{pred}\right|}{(k_L)_{\exp}} \qquad (2)$$

The expression for k_L includes the liquid Reynolds number based on the total packing surface area (as provided by the vendor), the Schmidt number, the ratio of liquid to gas velocity and the ratio of the length of the packed bed to the packing diameter. The exponent of the Schmidt number is forced at a value of -0.5 which strictly adheres to the fundamentals of the mass transfer theory. Bravo et. al. [8] have shown that the parameters used to represent k_L change from one operating region to another, which means that different liquid and gas flowrates, as operating regions, can lead to different exponent values in the same experimental conditions. Thus the experimental dataset is further classified according to the *L/G* ratio which is a prime indicator of the operating regime of a packed tower. The classification was made on the basis of *L/G* ratio less than or equal to 1 (37 points) and *L/G* ratio greater than 1 (380 points). Thus, these two different datasets yield different set of exponents due to the inherent difference in the operating conditions.

For *L/G* ratio greater than one, the AARE obtained after regressing the parameters was 16.60%., with over 85% of the regressed dataset lying within ±25%, as shown in Figure 3 .

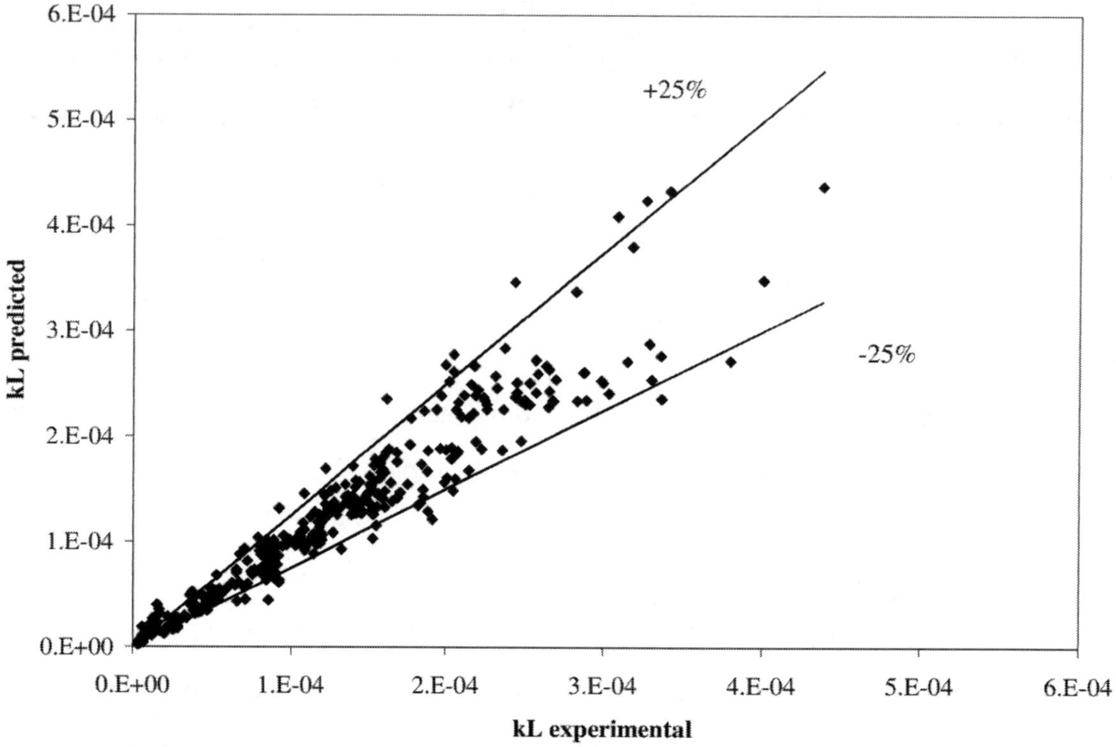

Figure 3. Parity plot for liquid side mass transfer coefficient k_L (m/s) when $L/G > 1$.

Whereas for $L/G < 1$ the AARE obtained after regressing the parameters was 18.4% with over 84% of the data points lying within ±20%, depicted in Figure 4. The correlations obtained for the two operating regions are given below.

L/G (mass/mass) < 1

$$k_L = 0.0944\left(\frac{\mu_L g}{\rho_L}\right)^{1/3} \mathrm{Re}_L^{0.953}\, Sc_L^{-0.5}\left(\frac{L}{d_p}\right)^{-1.0}\left(\frac{u_L}{u_G}\right)^{-0.063} \qquad (3)$$

L/G (mass/mass) > 1

$$k_L = 0.0027\left(\frac{\mu_L g}{\rho_L}\right)^{1/3} \mathrm{Re}_L^{0.88}\, Sc_L^{-0.5}\left(\frac{L}{d_p}\right)^{-0.014}\left(\frac{u_L}{u_G}\right)^{-0.05} \qquad (4)$$

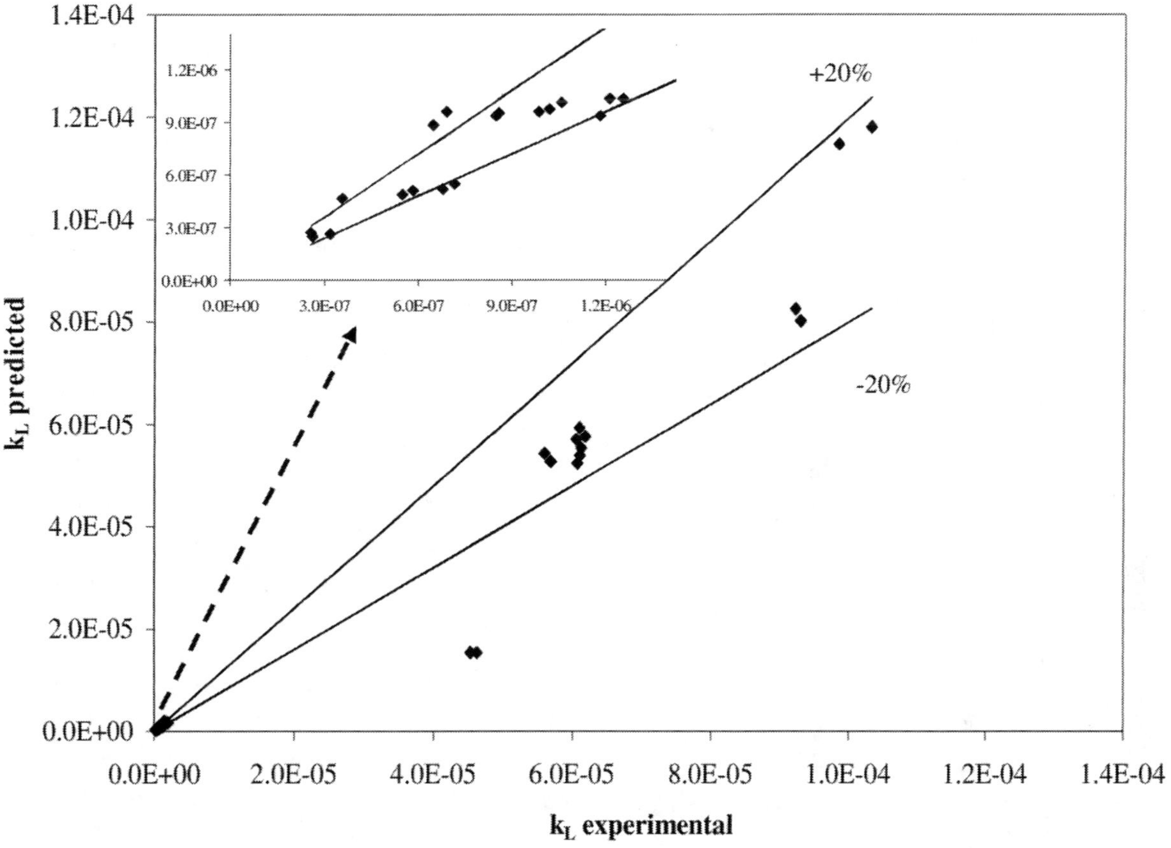

Figure 4. Parity plot for liquid side mass transfer coefficient k_L (m/s) when $L/G < 1$.

Test of New Correlations

The developed correlations were further tested against the experimental data-sets which were not included in developing the correlation ($S<2$) and shown in Table 1. The experimental runs were simulated in RateSep^{TM} (the correlation was implemented in a Fortran subroutine), using the improved correlations and compared with the predictions from Billet-Schultes, Onda, and Bravo-Fair correlation, which are available in RateSep^{TM}. The comparison is shown in Figure 5. It should be mentioned that the calculations were performed using Equation 1 as follows:

1. Gas side contribution as well as wetted area was still calculated using the original Onda correlation.
2. Liquid side contribution was calculated with new correlations and contact area was assumed to be total packing area.

3. Overall mass transfer coefficient was calculated as the sum of both resistances calculated in 1 and 2.

Table 1. Results of striping isopropanol (IPA) from water using air in a 0.31 m internal diameter column with 2.54 m of 38 mm metal Pall rings.

Liquid flow rate, kg/hr	IPA in, ppm	IPA out, ppm	Feed Temperature, °C	Air flow rate, kg/hr
11.34	7726	94	27.5	113.88
11.34	7634	755	27.6	40.67
16.10	4896	504	25.2	54.96
16.10	4883	579	25.5	59.36
16.10	4768	522	24.0	54.96
16.10	4851	25	25.3	407.82
22.44	7663	74	25.8	274.81
22.44	7661	306	26.5	108.17
29.02	4881	247	25.6	178.52
29.02	4850	314	25.8	209.52
29.02	4872	355	24.9	214.13
40.58	4869	315	24.9	426.95
40.58	4779	948	25.5	207.98
40.58	4867	1162	25.4	214.35
40.58	4792	244	25.4	415.96
44.66	7687	106	25.2	570.07
46.48	7463	1768	26.3	127.07
46.48	7610	464	26.4	295.70
90.23	7635	309	29.1	586.78
90.23	7589	2131	26.7	283.61
103.38	4850	258	25.5	821.36
103.38	4872	1285	25.5	426.95

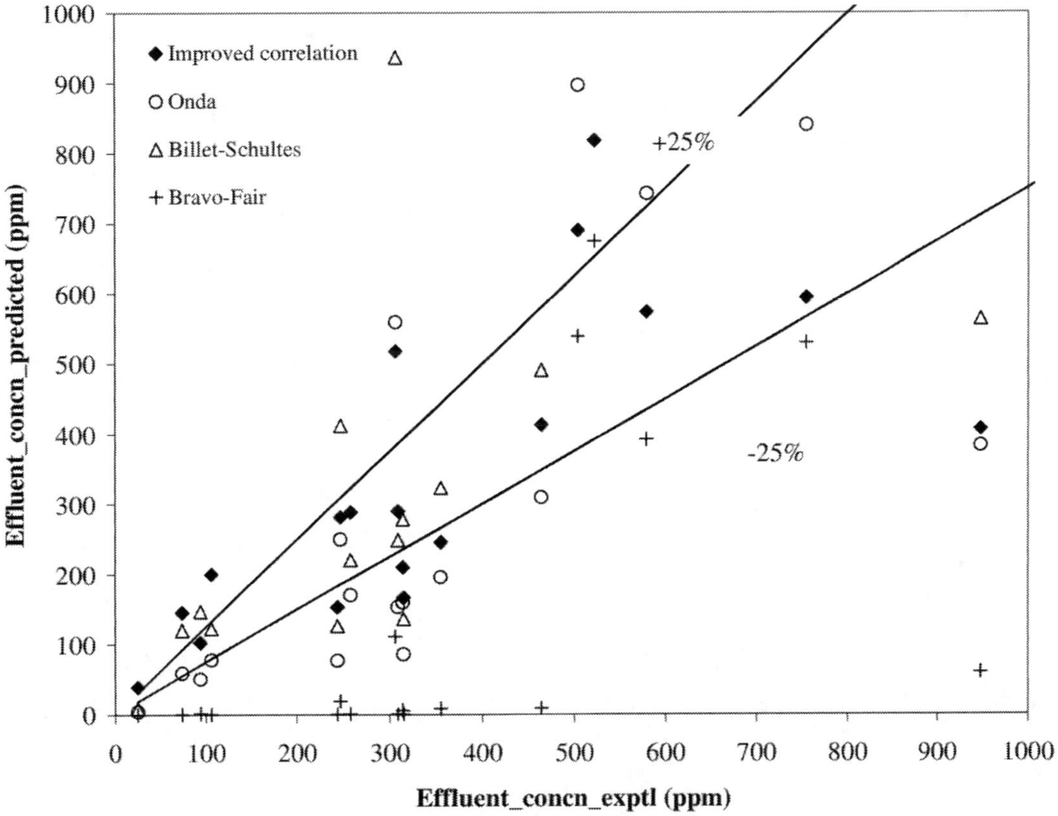

Figure 5. Comparison of improved correlation with Onda, Billet-Schultes and Bravo-Fair.

The modified correlation is slightly better than the original Onda. As mentioned previously, the original Onda correlation is still used for the gas resistance (for IPA/Water, there is still significant gas resistance, $S<2$). The new correlation ia also better than Billet and Schultes. In order to improve the accuracy, better models for interfacial area and for the gas side resistance are needed.

Conclusion

A modified correlation for the liquid-side mass transfer correlation was developed taking into consideration the influence of packing depth and gas velocity. The exponents of the parameters in the constituting correlation change with the operating regime of the packed tower (indicated by the L/G ratio). The developed correlation gives slightly better predictions than the Onda, Billet-Schultes and Bravo-Fair correlations.

Future Work

The inherent assumptions made while developing the correlation need to be corrected to represent both the liquid and gas phase mass transfer coefficients accurately. The Separations Research Program (SRP) at the University of Texas at Austin is developing a new expression for the effective interfacial area of structured as well as random packings. The liquid and gas (future work) phase mass transfer correlations will be re-evaluated with this new information which may yield slightly different exponents.

References

[1] Onda, K., Takeuchi, H., Okumuto,Y. J., Mass Transfer Coefficients between Gas and Liquid Phases in Packed Columns, J. Chem. Eng. Jpn. **1968**, 1 (1), 56.

[2] Billet, R., Schultes, M., Fluid Dynamics and Mass Transfer in the Total Capacity Range of Packed Columns up to the Flood Point, Chemical Engineering Technology, 18, **1995,** 371-379

[3] Dvorak, B., Lawler, D.; Fair, J., Handler N. Evaluation of the Onda Correlations for mass transfer with large random packings. *Enviro. Sci. Tech.*. **1996,** *30*, 945

[4] Sherwood, T. K., Holloway, F. A. L., Performance of Packed Towers Liquid Film Data for Several Packings. *Trans. Am. Inst. Chem. Eng*. **1940**, *36*, 39.

[5] Kutzer, S., Wintrich, H., Mersmann, A., Air Stripping – A Method for treatment of Wastewater, *Chem. Eng. Tech.* **1995**,*18*, 149.

[6] Perry, D., Nutter, D. E., Hale, A., Liquid Distribution for Optimum Packing Performance. *Chem. Eng. Prog.* **1990,** *86*, 30.

[7] Djebbar, Y., Narbaitz, M. R., Mass Transfer Correlations for Air Stripping Towers. *Environ. Prog*. **1995**, *14*, 137.

[8] Bravo, J. L., Patwardhan A. A., Edgar T. E., Influence of Effective Interfacial Areas in the Operation and Control of Packed Distillation Columns. *Ind. Eng. Chem. Res.*,**1982,** *31*, 604.

Abbreviations

HTU Height of a transfer unit, m

S Stripping factor, $S = H\frac{Q_G}{Q_L}$

K_La Overal liquid phase volumetric mass transfer coefficient (s^{-1})

k_L Liquid phase mass transfer coefficient (m/s)

L/G Liquid to Gas ratio

VOC Volatile Organic Compound

392c Uniqueness And Stability Of The Steady State In Flash Reactors

Yuan Xu, Department of Chemical Engineering, Carnegie Mellon University, Pittsburgh, PA 15213 and B. Erik Ydstie, Chemical Engineering, Carnegie Mellon University, 5000 Forbes Ave, Pittsburgh, PA 15213.

The uniqueness and stability of the steady state is of fundamental importance to the understanding of the operation of flash reactors. It has attracted the wide interest in the separation field. Doherty and Perkins (1982) showed that the differential equations governing the dynamic behavior of the distillation of a non-ideal multi-component mixture on a single theoretical CMO plate (flash) posses a globally asymptotically stable steady state. Lucia (1986) proved the uniqueness of the steady state in the isothermal isobaric flash reactors with the aid of Gibbs-Duhem relation and Caruchy interlace theorem. This analysis framework was extended to prove the uniqueness of the steady state for the binary homogeneous distillation (Sridhar and Lucia 1989) and the multi-component distillation (Sridhar and Lucia 1990).

In this work, we build a new uniqueness and stability proof on the geometric insight. The computationally challenging problem of finding the global minimum of the system Gibbs energy is equivalent to the geometrically intuitive problem of constructing the convex hull for all the Gibbs energy functions of candidate phases. Analytically, a convex hull is defined with Caratheodary theorem (Rockafellar 1970). The infimum operator in the theorem can be dropped under mild assumptions. The two-phase region on the convex hull is proved to a ruled surface. Its projection onto the component space is also endowed with the special structure, which allows the use of an extended level rule for the phase distribution relation. The uniqueness problem is transformed into the determination of intersection points of two trajectories. The stability proof is constructed with help of Lyapunov theorem. A ternary T-P flash is used as an example to demonstrate the idea of geometric analysis employed in this work. Future work includes the extension of the analysis in this work to reactive flash reactors and multi-staged systems.

Reference: Doherty, M. F., and J. D. Perkins (1982), On the dynamics of distillation processes – IV, Chem. Eng. Sci., 37, 381-392.

Lucia A. (1986), Uniqueness of solution to single-stage isobaric flash processes involving homogeneous mixtures, AIChE Journal, 32, 1761- 1770.

Rockafellar T. (1970), Convex analysis, Princeton University Press, Princeton, NJ.

Sridhar, L. and A. Lucia (1989), Analysis and algorithms for multi-stage separation processes. Ind. Eng. Chem. Res., 28, 793-803.

Sridhar, L., and A. Lucia (1990). Analysis of multi-component, multistage separation process: fixed temperature and fixed pressure profiles, Ind. Eng. Chem. Res., 29, 1668-1675.

Asymptotic Analysis of Effects of Bulk Liquid Reaction Disequilibria on Absorption in Packed Columns

Jerry H. Meldon and Ricardo Fernandes, Tufts University, Department of Chemical and Biological Engineering, Medford, MA 02155

INTRODUCTION

Analyses of absorption with reversible chemical reaction in packed columns generally assume that reaction equilibrium prevails in well-mixed bulk liquid. In most cases of practical interest this assumption is justified by high reaction rates and/or liquid holdup.

However, this justification collapses when reaction is sufficiently slow; and while slowness of reaction is generally considered undesirable, in certain important applications involving the simultaneous absorption of two or more gases, *selectivity* is based on the relative rates at which the absorbed gases react.

For example, the high H_2S/CO_2 selectivity of alkaline solutions – which is desirable – is attributable to the effective instantaneity of the reaction of dissolved H_2S and sluggishness of reaction of dissolved CO_2. In such cases, rational design requires estimation of small but potentially significant effects on absorption rates of slow reactions in bulk liquid.

As a first step towards rigorous analysis of acid gas and similar separations, we examine the effects of finite bulk liquid reaction rates on *single*-gas absorption. Since reaction kinetics are frequently nonlinear, we focus on absorption with the following reversible liquid-phase reaction:

$$A + B \rightleftharpoons C$$

where "A" is physically dissolved gas, "B" and "C" are nonvolatile solutes, and the reaction kinetics are assumed to be:

$$r = k_1[A][B] - k_2[C] \tag{1}$$

Our goals are:

- to identify, in terms of dimensionless parameters, operating conditions under which the assumption of bulk liquid reaction equilibrium breaks down, and

• to develop a methodlogy for estimating absorption rates under those conditions

MODEL FORMULATION

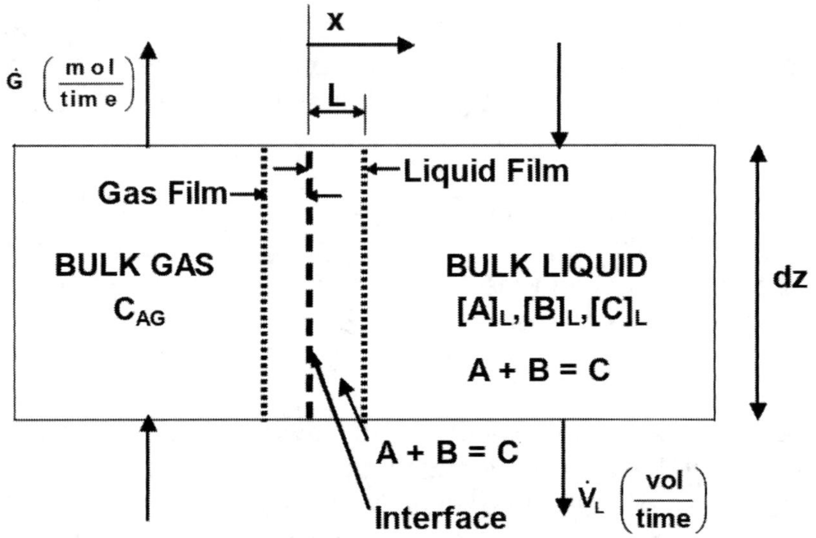

Figure 1: Film model of mass transfer in packed column with well-stirred bulk fluids; *z* is distance from *top* of column, *L* is liquid film thickness.

Referring to Fig. 1, a <u>steady-state</u> species *A* balance for bulk liquid within differential packed column volume $dV_{col} = \pi D_{col}^2 4/dz$ may be expressed as follows:

$$-D_A\left(\frac{d[A]}{dx}\right)_L\left(\frac{\pi D_{col}^2 a}{4}\right)-\left(1-\frac{aL}{h}\right)\left(\dot{V}^L\frac{d[A]_L}{dz}+\frac{\pi D_{col}^2 r_L}{4}\right)=0 \qquad (2)$$

where D_A is the diffusion coefficient of dissolved *A*; *a* is the interfacial area per unit of packed column volume; *L* is the liquid film thickness; subscript *L* denotes x = L, as well as bulk liquid; *h* is liquid holdup (liquid volume/column vol.ume); $\dot{V}_L$ is the volumetric flowrate of liquid (assumed constant); $\dot{G}$ is the molar flowrate of non-absorbing gas; and *r* is as given by Eq. 1.

The three terms in Eq. 2 acount for diffusion from the liquid film, convection and reaction. Equations analogous to Eq. 2 may be derived for species *B* and *C*.

Obtaining the first term in Eq. 2 requires a solution to the separate set of differential equations that express species balances *within* the liquid film, i.e.:

$$D_A \frac{d^2[A]}{dx^2} = D_B \frac{d^2[B]}{dx^2} = -D_C \frac{d^2[C]}{dx^2} = r \tag{3}$$

Eqs. 2 and 3 require simultaneous solution. Furthermore, the boundary conditions for Eqs. 3 include the gas-phase concentration of *A*, C_{AG}, and changes in C_{AG} are coupled to changes in the composition of bulk liquid according to:

$$\dot{V}_L\left(1-\frac{a\delta}{h}\right)\frac{d}{dz}\left([A]_L-[B]_L\right)=\dot{G}\frac{d}{dz}\left(\frac{C_{AG}}{C_G-C_{AG}}\right) \tag{4}$$

Accordingly, Eqs. 2-4 also require simultaneous solution. Moreover, the nonlinearity of the kinetics and of the term on the righthand side of Eq. 4 precludes an analytical solution. We linearized and solved Eqs. 2 and 3 using regular perturbation methods involving power series in terms of the "Hatta number," *Ha*, defined by:

$$Ha^2 \equiv \frac{k_1[B]_S L^2}{D_A} \tag{5}$$

where $[B]_S$ is the concentration of *B* in stripped liquid.

We are primarily interested in the behavior at low *Ha* values, when reaction is slow compared to diffusion.

RESULTS

Figure 2 depicts typical behavior at low Hatta numbers of the enhancement factor, *E*, defined as the ratio of the local absorption rates with and without reaction effects. "First" and "second" order results denote *E* values obtained by truncating the power series beyond the Ha^2 term and beyond the corsiderably more complex Ha^4 term, respectively.

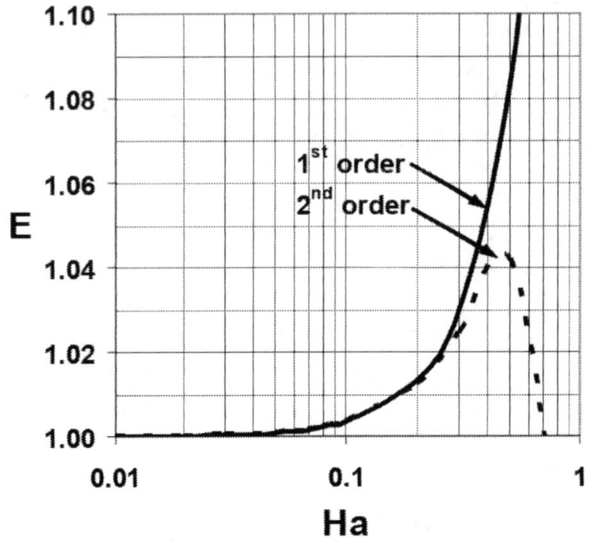

Figure 2: Typical enhancement factor *vs.* Hatta number curves based on series solutions truncated beyond the Ha^2 term ("first order') and the Ha^4 term ("second order").

As expected, reactive enhancement of absorption rates vanishes as *Ha* approaches zero. Enhancement only becomes significant when Ha > 0.1. Second order terms become significant when Ha > 0.2. It follows that the first order solution introduces negligible error, provided that Ha < 0.2.

Extensive additional data will be presented that explores the extent of bulk liquid reaction disequilibrium within packed columns at low Ha values, and the errors introduced by the assumption of bulk liquid equilibrium.

392e Side Reactor Configuration For The Production Of Ethyl Acetate

R. C. Tsai[1], **J. K. Cheng**[1], Hsiao-Ping Huang[1], Cheng-Ching Yu[1], Yi-Shan Shen[2], and Y. D. Chen[2]. (1) Chemical engineering, National Taiwan University, Taipei, 107-16, Taiwan, (2) Engineering Center, China petrochemical Development Corp., Ta-sheh, 815, Taiwan

Reactive distillation (RD) combines reaction and separation in a single unit to reduce energy consumption and capital investment. Despite of potential advantages, the reactive distillation may suffer from maintenance/design problems such as catalyst deactivation/replacement and hardware design (for catalyst packing), especially for heterogeneous catalyst such as ion exchange resin. In this work, an alternative design, side reactor configuration, is sought and the process of interest is the production of the ethyl acetate (EtAc) via esterification [1]. The reactive distillation study in [1] reveals that almost 90% conversion takes place in the column base of the RD and the rest of the 10% conversion occurs in the 10 reactive trays. This naturally leads to a coupled reactor/distillation configuration where all of the catalyst is packed in the bottoms base, denoted as Single Reactive Tray reactive distillation (SRT) hereafter. This mitigates the maintenance problem associated with conventional RD. However, simulation results show that, with the same amount catalyst loading (Wcat,RD), the SRT configuration cannot achieve the same performance as the RD (~93% conversion). In order to achieve the same conversion, the optimized design for SRT indicates that the catalyst loading should be 8.5 times of that for the reactive distillation while the heat input also increases by a factor of 24%. The cost is simply too much from maintenance perspective. Another alternative is then sought by adding external reactors to the Single Reactive Tray distillation column. This is termed as the Side Reactor Configuration (SRC) which has been proposed almost a decade ago [2-4]. Unlike conventional re active distillation design, the SRC design is less clear and design variables include: catalyst loading in the column base (Wcat, bot), % of equilibrium conversion for side reactor (X), sidestream withdrawn and return trays (Nss,r), side stream flow rate (Fss). A systematic design procedure is devised for the SRC design and the objective function to be minimized is the total annual cost (TAC). The results show that adding an external reactor decreases Wcat,bot substantially (from 8.5*Wcat,RD to 1.67*Wcat,RD) while the heat input is almost the same as that of the conventional reactive distillation and the TAC of the SRC only increases by a factor of 5% as compared to that of the RD. Considering the ease of catalyst replacement, the Side Reactor Configuration offers an attractive alternative to conventional reactive distillation.

References 1. Tang, Y. T.; Chen, Y. W.; Huang, H. P.; Yu, C. C.; Hung, S. B. ; Lee, M. J., "Design of reactive distillations for acetic acid esterification" AIChE Journal, 2005, 51, 6, 1683-1699. 2. Schoenmakers, H.; Buhler, W., "Distillation column with external reactors - an alternative to the reaction column" Chemie Ingenieur Technik, 1982, 54, 2, 163-163. 3. Jakobsson, K.; Pyhalahti, A.; Pakkanen, S.; Keskinen, K. ; Aittamaa, J., "Modelling of a side reactor configuration combining reaction and distillation" Chemical Engineering Science, 2002, 57, 9, 1521-1524. 4. Baur, R.; Krishna, R., "Distillation column with reactive pump arounds: An alternative to reactive distillation" Chemical Engineering and Processing, 2003, 43, 3, 435-445.

392f Modeling The Hydrodynamics Of Katapak-Sp Using The Particle Approach

Ricardo Macías-Salinas, ESIQIE-Chemical Engineering Department, Instituto Politécnico Nacional, UPALM, Av. IPN s/n, Col. Lindavista, México, D.F., Mexico

Various experimental efforts have been reported in the literature in an attempt to understand the hydrodynamic behavior of a relatively new family of catalytic structured packings: KATAPAK-SP. The only approach so far used to represent the experimental results obtained for KATAPAK-SP 11 & 12 relies in the channel model. The purpose of this work was to verify the suitability of a particle model originally developed for random and structured packings (Stichlmair et al., Gas Sep. Purif., 3, 19, 1989) by extending its applicability to catalytic structured packings, particularly for the KATAPAK-SP family. In doing that, a more suitable liquid holdup correlation for the open channels was proposed. The question on how liquid splits into the open channels and catalyst bags still remains partially unanswered. In an attempt to solve this, a new calculation procedure has been devised here to reasonably determine the liquid share between the open channels and the catalyst bags. Model predictions in terms of pressure drop and liquid holdup are presented and discussed in this work for the case of KATAPAK-SP 11 & 12.

403a Characterization Of Boron-Doped Carbon Synthesized Under Atmospheric Conditions And Its Influence On Pt Nanoparticles

Chethan Acharya and Heath Turner. Chemical and Biological Engineering, University of Alabama, Box 870203, Tuscaloosa, AL 35487-0203

The activity of nanometer-sized catalyst particles may be prolonged by stabilizing the catalysts on various supports. The stability may possibly be increased by strengthening the interaction between the nanoparticles and the support, thus decreasing the mobility and the agglomeration of the catalyst particles. Our density functional theory calculations of Pt and PtRu clusters on carbon supports have shown that the adsorption energies of these metal clusters increased substantially in the presence of substitutional boron defects in carbon lattice. To corroborate the simulation results with experiments, boron doped amorphous carbon powders were produced at 1000C and under 1 atmosphere of pressure using a simple tube furnace apparatus. Doping of the powders was monitored using XRD and Raman spectroscopy, and doping was confirmed by XPS. The XPS data showed a peak at 188 eV which coincides with the B1s energy when bound to carbon. XRD data showed a decreasing d002-spacing between graphitic planes, indicating increased graphitization. The full width at half maximum of this peak also decreased, owing to a more ordered powder. Raman microscopy was performed on the samples and showed a decrease in the G-band, which is due to the substitution of boron into the graphitic lattice. The intensity ratio of the G- to D-Bands increased, indicating larger grain sizes in the powders. Platinum nanoparticles were deposited on pure and boron-doped carbon and the stability of the nanoparticles were tested using XRD and HRTEM. The activity of the catalysts for reactions such as CO oxidation was also analyzed.

IS ^{129}Xe NMR A USEFUL TECHNIQUE FOR PROBING THE PORE STRUCTURE AND SURFACE PROPERTIES OF CARBONACEOUS SOLIDS

Konstantin V. Romanenko , Olga Lapina ,
Boreskov Institute of Catalysis, Academy of Sciences, Novosibirsk, Russia
Jean-Baptiste d'Espinose de Lacaillerie, Jacques P. Fraissard
Laboratoire de Physique Quantique, ESPCI, Paris, France

Abstract

Generally the interpretation of ^{129}Xe NMR data obtained for porous carbons is complicated. There are several reasons: structural disorder, distribution of the crystallite size or presence of amorphous domains, heterogeneity of surface properties, strong paramagnetic sites, etc. Thus it would appear that structural disorder, bad reproducibility of properties and various impurities make ^{129}Xe NMR ineffective when applied to porous carbon materials.

We have re-examined this problem for a series of microporous carbons with similar surface chemical properties. There is no relationship between the value, δ_S, of the chemical shift extrapolated to zero concentration and the pore diameter. However, the virial coefficient, $\delta_{Xe\text{-}Xe}$, arising from binary xenon collisions in the free volume, varies linearly with the micropore size.

Introduction

Besides the ordinary adsorption techniques, the use of ^{129}Xe NMR is generally accepted for the characterization of zeolites, mesoporous and amorphous silica. In contrast, ^{129}Xe NMR has been only rarely used to study porous carbon materials [1-14]. And a few of them aimed at finding some specific ^{129}Xe NMR data useful for the characterization of the carbon surfaces because the interpretation of ^{129}Xe NMR data obtained for porous carbon is complicated by several factors: structural disorder - distribution of the crystallite size or presence of amorphous domains; heterogeneity of surface properties - presence of various surface groups and surface structures (basal and edge); conductivity and strong paramagnetic sites. We have reinvestigated this problem for some relatively well defined microporous carbons, attempting to identify the most important parameters which perturb the determination of the porosity of carbons from the chemical shift of adsorbed xenon.

Experimental Section

Preparation of carbons

A series of activated carbons was synthesized by successive cyclic air oxidation/ pyrolysis treatments of the same precursor as described in reference [15]. The samples conserve the same basic carbonaceous structure and surface chemistry.

Characterization

The experimental conditions for the characterization of these solids are classical. The reader will find them in references [3].

Theory

The chemical shift of xenon adsorbed in a porous solid depends on its interaction with the pore surface and Xe-Xe interactions.

$$\delta = \delta_S + \delta_{Xe\text{-}Xe} \cdot \rho \qquad (1)$$

ρ *is* the density of adsorbed Xe and $\delta_{Xe\text{-}Xe}$ a coefficient characteristic of the interactions between two Xe atoms. δ_S is the shift extrapolated to zero concentration. It depends both on the interaction of the Xe atoms with the pore walls and on the pore dimensions [16-20]. However, in all studies, the results assume homogeneous surfaces consisting of uniformly distributed atoms of the same type, without the presence of strongly attracting isolated centres, which is not the case for carbon surfaces.

Until now little attention has been paid to the $\delta_{Xe\text{-}Xe} \cdot \rho$ term. This depends on the density of adsorbed xenon, ρ, but must also depend on the probability of Xe-Xe collisions, and therefore on the distribution of Xe atoms in the pores and their diffusion. It seems reasonable therefore to consider two extreme cases of Xe distribution, characterized by the volume xenon density in small micropores and the surface xenon density for large pores. We consider here only the first case.

Results and discussion

The characteristics of the samples have been published [3]. We recall only the useful data. The external surface area is considerably lower than the BET area. The total specific pore volume, W_t, and the specific micropore volume, W_0, are roughly similar for all cycles (Table 1).

Table 1. Textural properties of activated carbons obtained by air oxidation/pyrolysis cycles *D* - mean pore size, *Wt* - total specific pore volume, W_0 - specific micropore volume, S_{BET} - BET specific surface area, S_{EXT} - external specific surface area.

Cycle №	*D* (nm)	*Wt* ($cm^3 \cdot g^{-1}$)	W_0 ($cm^3 \cdot g^{-1}$)	S_{BET} (m^2g^{-1})	S_{EXT} (m^2g^{-1})
1	0.67	0.35	0.33	846	-
2	0.84	0.5	0.44	1157	8.00
3	0.97	0.48	0.42	1104	8.30
4	1.1	0.55	0.46	1236	16.65
5	1.32	0.6	0.48	-	-
6	1.45	0.68	0.5	1386	28.83
7	1.6	0.74	0.53	1464	36.05
8	1.75	0.82	0.56	1540	63.92
9	1.92	0.9	0.58	1522	71.11
10	2.26	0.89	0.47	1258	138.68

For all samples the chemical shift measured for these microporous carbons is a linear function of the xenon density ρ. As an example, Figure 1 represents five of these variations. The density ρ was calculated as the amount of adsorbed xenon divided by the total pore volume W_t of the sample considered. The chemical shift extrapolated to zero loading, δ_S, varies by 10 ppm over the pore size range 0.6-2.3 nm and without any monotonic behaviour (Figure 2); this despite the fact that the chemical nature of the pore surface can be considered identical for all samples. Thus, δ_S cannot be used as a pore size probe, except to give an idea of the order of magnitude.

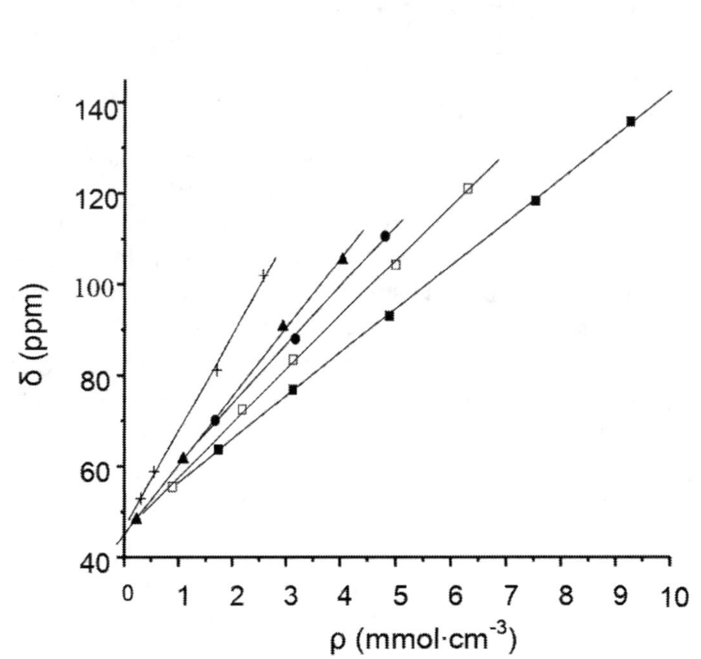

Fig. 1. [129]Xe NMR chemical shift vs. xenon density, ρ. (■) no.1 ; (□) no.3 ;(●) no.4 ;(▲) no.6 ;(+) no.10.

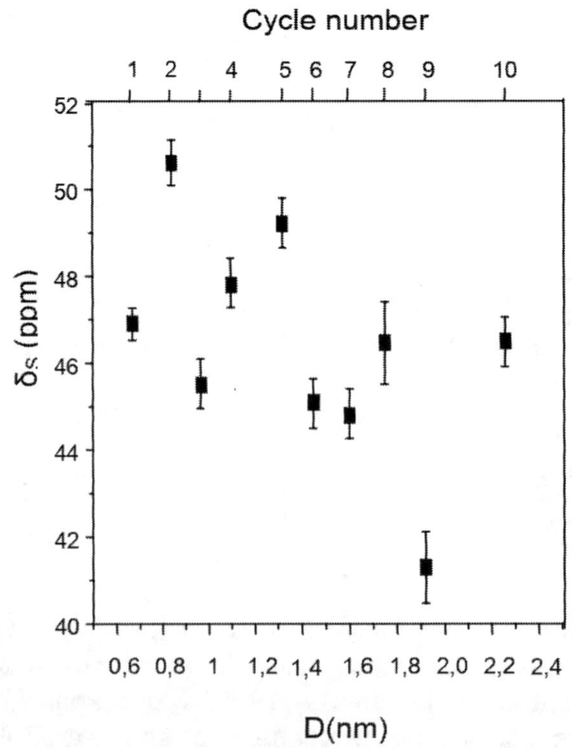

Fig. 2. [129]Xe NMR chemical shifts approximated to zero xenon density *vs* mean pore size. The treatment cycle numbers are given on the upper axis.

On the contrary, the slope of $\delta_{Xe\text{-}Xe}$ varies significantly and regularly with the pore size (Figure 3) according to the equation:

$$\delta_{Xe\text{-}Xe} = \eta + \kappa D \qquad (2)$$

$\delta_{Xe\text{-}Xe}$ is expressed in ppm.cm^3.mmol^{-1}; D is the mean pore size in nm, κ = 7±0.2 ppm.cm^3.mmol^{-1}.nm^{-1} and η = 5.1± 0.3 ppm.cm^3.mmol^{-1}.

It is, however, obvious that this equation, valid in the present case, cannot be generalized until other microporous carbons with the same surface homogeneity have been studied.

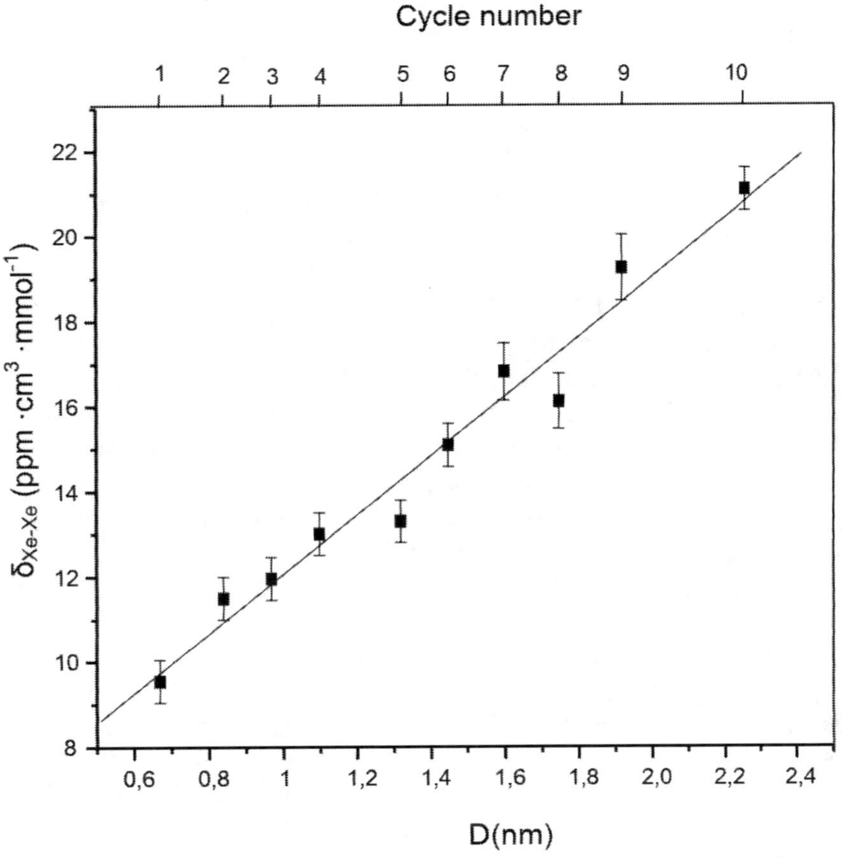

Fig. 3. Experimental $\delta_{Xe\text{-}Xe}$ values *vs.* mean pore size obtained for series of activated carbons (■).

Conclusion

^{129}Xe NMR is particularly well adapted to the characterization of zeolites and silica-based mesoporous solids. Its extension to the study of carbons is restricted for several reasons, such as structural disorder, heterogeneity of surface properties, paramagnetic impurities, etc. In particular, one does not find a simple relationship whereby the chemical shift approximated to zero density, δ_S, can be used for the characterization of pore diameters.

We have, however, shown that it is possible to adapt this technique to the study of carbon's microporosity by considering the virial coefficient $\delta_{Xe\text{-}Xe}$ arising from binary xenon collisions. There is a linear correlation between this term relative to the xenon collisions in the pore volume and the mean micropore size.

Acknowledgments

We thank Xavier Py (CNRS-PROMES, Technosud, Perpignan, France) for the microporous carbons.

References

[1] N. Bansal, H. C. Foley, D. S. Lafyatis, C. Dybowski, Catal. Today 14 (1992) 305.
[2] C. C. Tsiao, R. E. Botto, Energy & Fuels 5 (1991) 87.
[3] K. V. Romanenko, X. Py, J-B. d'Espinose de la Caillerie, O.B. Lapina, J. Fraissard, J. Phys. Chem. B 110 (2006) 3055.
[4] D. J. Suh, T. J. Park, S. K. Ihm, R. J. Ryoo, J. Phys. Chem. 95/9 (1991) 3767.
[5] P. C. Wernett, J. W. Larsen, O. Yamada, H. J. Yue, Energy and Fuels 4/4 (1990) 412.
[6] K. J. McGrath, Carbon 37 (1999) 1443.
[7] K. Saito, A. Kimura, H. Fujiwara, Magnetic Resonance Imaging 21 (2003) 401.
[8] P. A. Simonov, S.V. Filimonova, G. N. Kryukova, H. P. Boehm, E. M. Moroz, V. A. Likholobov, T. Kuretzky, Carbon 37 (1999) 591.
[9] K. V. Romanenko, J.-B. d'Espinose, J. Fraissard, T. V. Reshetenko, O. B. Lapina, Micropor. Mesopor. Mater. 81 (2005) 41.
[10] K. V. Romanenko, A. Fonseca, S. Dumonteil, J. B.Nagy, J.-B. d'Espinose, O. B. Lapina, J. Fraissard, Solid State NMR 28 (2005) 135.
[11] C. F. M. Clewett, T. Pietra, J. Phys. Chem. B 109 (2005) 17907.
[12] J. M. Kneller, R. J. Soto, S. E. Surber, J.-F. Colomer, A. Fonseca, J. B.Nagy, G. Van Tendeloo, T. Pietra, J. Am. Chem. Soc. 122 (2000) 10591.
[13] H. Ago, K. Tanaka, T. Yamabe, T. Miyoshi, K. Takegoshi, T. Terao, S. Yata, Y. Hato, S. Nagura, N. Ando, Carbon 35 (1997) 1781.
[14] D. Raftery, H. Long, T. Meersmann, P. J. Grandinetti, L. Reven, A. Pines, Phys. Rev. Lett. 66 (1991) 584.
[15] X. Py, A. Guillot, B. Gagnon, Carbon 41 (2003) 1533.
[16] J.L. Bonardet, A. Gédéon, M-A. Springuel-Huet and J. Fraissard, Molecular Sieves, Characterization II, 5 (2006) 155-248, Eds: H.G. Karge and J. Weitkamp, Springer, Berlin-Heidelberg.
[17] J.A. Ripmeester and C.I. Ratcliff, J. Phys. Chem. 94 (1990) 7652.
[18] J. Demarquay and J. Fraissard, Chem. Phys. Lett. 136 (1987) 314.
[19] M.-A. Springuel-Huet, J. Demarquay, T. Ito and J. Fraissard, Stud. Surf. Sci. Catal. 37 (1988) 183.
[20] T.T.P. Cheung, J. Phys.Chem. 99 (1995) 7089.

403g Kinetics Of H_2S, SO_2 And H_2O Desorption From Amorphous Carbon Films

J.B. Miller[1], E. Broitman[1], W. Michalak[1], A.J. Gellman[1], M.A. Alvin[2], and M. Mathur[2]. (1) Chemical Engineering, Carnegie Mellon University, Pittsburgh, PA 15213, (2) National Energy Technology Laboratory, US Department of Energy, Pittsburgh, PA 15236

Sulfur-compounds are ubiquitous in fossil fuels processing. In addition to their potential for adverse environmental impact, sulfur-compounds poison catalysts and foul process equipment. There is, therefore, significant interest in development of efficient catalyst-sorbents for their capture and conversion. Activated carbons are well known for their ability to adsorb sulfur-compounds; both morphology and surface chemistry contribute to their performance. However, primarily because control and characterization of activated carbon's surface properties are difficult, the interactions between sulfur-compounds and carbon surfaces are not understood at a fundamental level that enables rational design of new materials.

In this work we used ultra-high vacuum (UHV) techniques to carefully prepare and thoroughly characterize amorphous carbon (a-C) thin films as models of activated carbon sorbent-catalysts. Films with controlled and defined microstructure (sp2/sp3 bonding ratio), surface chemistry (dangling bonds, oxidation state), and morphology (roughness, porosity) were prepared by DC magnetron sputtering. Temperature programmed desorption (TPD) and x-ray photoelectron spectroscopy (XPS) were used to study H_2O, H_2S and SO_2 surface chemistry on the films. For comparison, we performed similar experiments on a highly oriented pyrolytic graphite (HOPG) surface.

Water desorbs from both the a-C and HOPG surfaces with zero-order kinetics and desorption energy, E_{des}, that increases with coverage, indicative of initial condensation via a cluster growth mechanism. On the a-C surface, SO_2 and H_2S both display TPD features consistent with first-order desorption from both mono- and multilayers. At sub-monolayer coverages, SO_2 ($E_{des} \sim 10$ kcal/mol) interacts more strongly with a-C than H2S ($E_{des} \sim 8$ kcal/mol). XPS analysis after TPD experiments on HOPG and a-C films provided no evidence of reaction between the carbon surfaces and the adsorbates. On the other hand, tightly bound residual sulfur was found on the surfaces of oxidized films, a-COx, after H_2S TPD, suggesting that chemical reaction between H_2S and the surface had occurred. This result illustrates the potential for chemical modification of the surface for altering adsorption/reaction pathways.

403d Characterization of Polysaccharide Based Chiral Sorbents Using Infrared Spectroscopy, X-Ray Diffraction, 13C Solid State NMR, and Density Functional Theory

Rahul B. Kasat, Nien-Hwa Linda Wang, and Elias I. Franses. School of Chemical Engineering, Purdue University, 480 Stadium Mall Drive, West Lafayette, IN 47907-2100

Polysaccharide based chiral sorbents are used widely in preparative scale chiral separations. The chiral recognition mechanisms of these sorbents are not yet completely elucidated. These polymers form helical rods with intra- and inter-rod hydrogen bonds (H-bonds). There are various nm-sized cavities formed between the polymer side-chains and rods. In the present work, we study the effects of the molecular structures of their backbone and side-chain on the molecular environments in the chiral cavities of three important sorbents, cellulose tris(3,5-dimethylphenylcarbamate) (CDMPC), amylose tris (3,5-dimethylphenylcarbamate) (ADMPC), and amylose tris((S)-á-methylbenzylcarbamate) (ASMBC), using attenuated total reflection infrared spectroscopy (ATR-IR), X-ray diffraction (XRD), 13C cross polarization/magic angle spinning (CP/MAS) and MAS solid state NMR, and DFT modeling [1]. ATR-IR is used to determine how the H-bonding states of the C=O and NH groups of the polymer depend on of the backbone and side-chain. The changes in the polymer crystallinity are characterized with XRD. The changes in the polymer helicity and molecular mobility for polymer-coated silica beads (commercially called Chiralcel OD, Chirapak AD, and Chiralpak AS) are probed with 13C CP/MAS and MAS solid-state NMR. The IR wavenumbers and the NMR chemical shifts for the polymer backbone monomers, dimers, and the side-chains are predicted using the DFT/B3LYP/6-311+g(d,p) level of theory. It is concluded that the molecular environments of the C=O, NH, and phenyl groups show significant differences in intramolecular and intermolecular interactions and in the nanostructures of the chiral cavities of these biopolymers. These results have implications for understanding how the molecular environments of chiral cavities of these polymers affect their molecular recognition mechanisms.

[1] Kasat, R. B.; Wang, N. H. L.; Franses, E. I. Biomacromolecules, 8, 2007, 1676-1685.

CHARACTERIZATION OF POWDERED ACTIVATED CARBON USING GASEOUS- AND AQUEOUS-PHASE INDICES

R. R. JAIN, D. K. LUDLOW and C. D. ADAMS*
Department of Chemical and Biological Engineering, University of Missouri-Rolla, Rolla, MO, 65409 USA

ABSTRACT - The adsorptive capacity of powdered activated carbon (PAC) is affected by the physicochemical characteristics of both the surface and sorbate as well as the pore-size distribution relative to sorbate size. The adsorptive capacities of PACs are commonly characterized using aqueous-phase indices such as iodine (I_2), methylene blue (MB), *p*-nitrophenol (PNP), molasses, and tannin. PACs may also be characterized using gas adsorption with nitrogen or carbon dioxide to develop detailed pore-size distribution information. The primary purpose of this research was to examine 12 different PACs to determine the relationship between common aqueous-phase indices and true pore-size distributions using gas-phase indices. The results showed that the internal surface areas of PACs tended to be ordered according to the source materials as: wood > bituminous coal > lignite coal. It was also found that the coal-based carbons had a majority of their internal pores in the microporous range (<2.0 nm), whereas wood-based carbons tended to have a much wider pore-size distribution, well into the mesoporous range. The molasses and tannin numbers correlated well with a total pore volume of greater than about 1.1 nm. The MB and I_2 numbers correlated well with the total surface area when micropores below 1.1 nm were also included. Surface functional groups were determined, using the Boehm titration methods, and were found to be less important in the sorptive capacity of PACs than the specific surface areas and pore-size distributions. It was proposed that the molasses-to-iodine ratio (MIR) and the inverse-tannin-to-iodine ratio (ITIR) may be useful as a means of estimating the pore-size distributions of PACs using only aqueous indices.

INTRODUCTION - Powdered activated carbon (PAC) is often used in water treatment plants to remove a wide variety of organic chemicals of varied molecular weights and sizes. These include synthetic organic chemicals (SOCs), taste and odor compounds such as methylisoborneol (MIB) and geosmin, and natural organic matter such as humic acid and fulvic acid (Lee et al., 1981). PACs are manufactured from materials, including bituminous coal, lignite coal, wood, coconut shells, peat, petroleum residue, bones, fruit seeds, and other sources (Gregg and Sing, 1982). The choice of source material, in large measure, dictates the producer's activation method (Naylor, 2001). Due to this, activated carbons may have different internal pore structures consisting of micropores (<2 nm), mesopores (2-50 nm), and macropores (>50 nm) (Webb and Orr, 1997). The progression of carbon reduction in nature under high temperature and pressure is (in order): wood, peat, lignite coal, bituminous coal, anthracite, graphite, and diamond. Thus, lignite coal contains greater volatile organics (and ash) than bituminous coal. Lignite coal may, therefore, be directly activated at a high temperature (i.e., establishment of extensive micro- and mesopores). Bituminous coal, on the other hand, is generally processed via a reagglomeration process involving crushing, mixing with a volatile (i.e., coat of tar pitch), being heat/pressure treated into brickets, crushing, and finally being activated at high temperatures. Wood-based carbons are usually treated in one of two ways: 1) steam activation, or 2) phosphoric acid treatment to dissolve cellulosic structure followed by high temperature activation (e.g., 500°C).

The chemical character of the carbon surface is affected by many parameters including source material, activation time, temperature, cooling rate, and environment, and others (Karanfil and Kilduff, 1999; Karanfil *et al.*, 1999; Greenbank, 2001). Pure carbon is hydrophobic and when the carbon precursor is activated using steam or carbon dioxide to increase the surface area and pore structure, the carbon surface is left with some residual oxide sites that give some hydrophilicity. The surface properties (e.g., degree of oxidation) can have a profound effect on both the adsorption of both target contaminants as well as the aqueous indices themselves (Switt, 2001; Boehm, 1994). Thus, jar testing still plays an important role in selecting and optimizing the use of PAC in water treatment applications.

PACs may be characterized experimentally using gas-phase and aqueous-phase methods. Gas-phase methods are particularly useful in characterizing the morphology of a PAC. They give information about pore-size distribution of the PAC as a function of surface area or pore volume because nitrogen and carbon dioxide molecules can move into very small pores, and are relatively inert with respect to interaction of the chemical functional groups on the surface of the carbon. Hence, characterization using gas-phase adsorption gives information about the surface area and morphology of the PAC.

The extent and nature of the oxide surface groups (acidic sites and basic sites) of the PACs can be characterized using neutralization with a variety of basic solutions and HCl. This information provides some insight into the chemical nature of the surface functional groups.

The more commonly used methods of PAC characterization are the standardized aqueous adsorption tests using *p*-nitrophenol (PNP), iodine (I_2), methylene blue (MB), molasses, and tannin as sorbate surrogates (Hassler, 1951; Galbraith *et al.*, 1958; Kasaoka *et al.*, 1989a). Aqueous-phase adsorption is complex because of the presence of surface functional groups and their interactions with different competing adsorbate molecules. Sontheimer et al. (1988) characterized eight AC in terms of several test methods (including surface area, total pore volume and some standard aqueous phase indices) and found that an overall ranking of the AC was not possible. Even though the relative capacity of a PAC for a particular adsorption system depends greatly on the total surface area and pore volume, it is also affected by accessibility to the surface area and by the chemical nature of the surface.

The purpose of this research was to examine how the common aqueous-phase indices used to characterize a PAC relate to the true pore-size distribution. A second purpose of this work was to examine the relative effect of the sorptive capacities of surface functional groups on a PAC versus their pore-size distribution.

In this study, we examined 12 PACs commonly used in the water treatment industry by using several gas-phase and common aqueous-phase techniques. In addition, we analyzed the morphology of the PAC pores (i.e., the surface area available and the pore size) and the PAC surface chemistry (basic and acidic oxide sites). These data were then compared statistically to assess the correlations between each parameter. An attempt was also made to establish the effect of the source of the material on the pore structure. There is a fundamental question, however, regarding what pore sizes are being characterized by the standardized aqueous index. Depending on the size of a target adsorbate molecule relative to the smallest pores, only a portion of a PAC's total surface area or pore volume may be available for adsorption (Romero *et al.* 2003). The correlation between a certain aqueous index and the pore structure was examined with reference to the adsorbate molecular size. Since these aqueous-phase characterization techniques, by themselves, have not been known to be suitable for all types of PACs, a new index is proposed that more effectively characterizes the pore-size distribution of a PAC. This information is useful for understanding the utility of commonly used indices. It may also provide the basis for improving the specification and selection approaches used by water utilities in choosing a PAC for a specific water treatment objective.

Gaseous Carbon Characterization Methods

Gas adsorption is a standard method used to characterize the internal surface area and structure of a porous material. Different adsorbate gases that have been used to characterize adsorbents include nitrogen, carbon dioxide, krypton, and argon (Gregg and Sing, 1982; Webb and Orr, 1997). Nitrogen is the most common adsorbate gas used to characterize surfaces because it is relatively inert and interacts with adsorbent surfaces primarily through a mechanism of physical adsorption at liquid nitrogen temperatures (77 K).

The *Brunauer, Emmett, and Teller (BET)* surface area method measures the area in the micro- and mesopores and is the most common method of quantifying the adsorptive capacity of porous material (Brunauer *et al.*, 1938; Ravikovitch *et al.*, 1998). Using nitrogen as the adsorbate, the BET method analyzes the moderate-pressure region of the nitrogen isotherm to predict surface area.

The *total pore volume* (TPV) method (Autosorb-1® Operating Manual) also measures micro- and mesopores. It is determined from the amount of vapor adsorbed at a relative pressure close to unity and assumes that the pores are filled with liquid adsorbate. The volume of nitrogen adsorbed is converted to the volume of liquid nitrogen contained in the pores.

The *Barrett, Joyner, and Halenda (BJH)* model measures mesopores (2 nm < pore diameter < 50 nm) and excludes micropores (Barrett *et al.*, 1951). The model assumes open-ended, cylindrical pores and uses the Kelvin equation to model the emptying of condensed adsorbate from the pores. Both adsorption and desorption isotherms can be used for this method. The BJH method is unsatisfactory over the microporous range, however, due to the failure of the Kelvin equation with smaller pores.

The *density functional theory (DFT)* method is used to develop pore-size distributions based on statistical thermodynamics and is one of the newer models used to analyze gas adsorption data (Ravikovitch *et al.*, 1998). To calculate the pore-size distribution, it is necessary to develop a model for pore filling which relates the pore width to the condensation pressure. Most theories related to isotherms tend to take a relatively macroscopic approach and may not be able to accurately predict the isotherms across their entire region. The DFT method takes a more microscopic approach and is based on statistical thermodynamics, taking the fluid-fluid and fluid-solid interactions into account as well as pore size, geometry, and temperature. When a gas is introduced into a pore space, the gas molecules experience an attractive force (van der Waal's forces) from the adsorbent's surface, which results in a higher average density of gas near the surface during equilibrium. Several researchers have used this model in their analyses of pore-size distributions of microporous materials (Lastoskie *et al.*, 1993; Krupa and Cannon, 1996; Ravikovitch *et al.* 1998).

The *Dubinin – Radushkevich (DR)* method is used to determine the surface area and pore volume in the microporous region using low-pressure nitrogen adsorption data (Dubinin –

Radushkevich, 1947). It is based on the Polanyi potential theory of adsorption which assumes that the fraction of the adsorption volume occupied by liquid adsorbate at various adsorption potentials can be expressed as a Gaussian function. The DR plot is linear and is used to compute the micropore surface area. The linear range of these plots is usually found at relative pressures of less than one percent of saturation pressure at 273K ($P/P_0<0.01$).

The *Horvath and Kawazoe (HK)* method is used to determine the micropore-size distribution from low-pressure nitrogen adsorption data (Horvath and Kawazoe, 1983). The HK model is appropriate for activated carbon since it is based on a slit-shaped pore geometry, which corresponds to the graphitic structure in activated carbon. The model expresses the adsorption potential function within slit-shaped micropores as a function of the effective pore width and assumes that the micropores are either completely full or completely empty, depending on the pressure of the adsorbate.

The CO_2 adsorption (CO_2) method measures adsorption at 273 K has been suggested by Marsh and Wynne-Jones (1964) and Garrido et al. (1987) as a good alternative when characterizing microporous carbon. At the low temperatures necessary for N_2 adsorption (77 K), diffusion of N_2 molecules may be kinetically limited in the narrow micropores, leading to excessive times required to reach equilibrium. To avoid this problem, CO_2 has been used as the adsorbate to characterize microporous carbon because: 1) its critical dimension is similar to nitrogen, and 2) the CO_2 adsorption temperature (273 K) would facilitate faster movement of the molecules down the narrow micropores than for N_2. The BET method can be used to determine the surface area using CO_2 isotherm data.

Grand canonical Monte Carlo (GC-MC) simulations use an analysis technique similar to the DFT approach and were used with CO_2 as the adsorbate. These and other adsorption models are incorporated in the standard software package (Autosorb-1 for Windows, Ver. 1.19) developed for the Autosorb-1C gas adsorption instrument (Quantachrome Corp., Boynton Beach, FL, USA).

Surface Functional Sites

When activated carbons are prepared from the precursor materials, various steps can be taken to “activate” the carbon. These process steps result in an increase in the surface area and the volume of the carbon, and also typically form various oxide sites that affect the sorptive capacity of the AC. The Boehm titration technique (Boehm, 1966; Bandosz *et al.*, 1993; Boehm, 1994, Bandosz *et al.*, 1996) is used to determine the acidic and basic nature of the carbon surface. There are three types of acidic sites on the carbon surface that can be measured (Boehm 1994). Some sites consist of a free carboxylic acid group, others consist of lactones and lactols (referred to as lactonic sites), while others sites have a phenolic nature. A given amount of washed and dried activated carbon is equilibrated with HCl, or NaOH, or Na_2CO_3, or $NaHCO_3$ for 24 hr that, in turn, neutralizes the various acidic or basic surface sites. The specific number of sites per mass of carbon (meq/g) is determined by titration of the supernatant.

Aqueous Indices

Another way of characterizing the pore structure of activated carbon is by the adsorption of molecules from an aqueous solution (Galbraith, *et al.*, 1958; Linares-Solano *et al.*, 1984; Kasaoka *et al*,. 1989a; Kasaoka *et al.*, 1989b; Krupa and Cannon, 1996; Chern and Huang, 1998; Pelekani and Snoeyink, 2000; Pelekani and Snoeyink, 2001). These methods of characterizing often use PNP, I_2, MB, molasses, and tannin. They are especially appropriate for charactering PACs because they specifically involve the aqueous-phase adsorption of sorbates from solution. Kasaoka *et al.*, 1989b, while examining adsorption on activated carbons, computed the area and volume occupied by each molecule to predict if the adsorption was multi-layer or monolayer. Their results indicated that dye and organic compounds adsorb in the activated carbon primarily as a monolayer and not by filling the pores. The specific methods are described below.

EXPERIMENTAL

Powdered Activated Carbons

The source materials of the PACs in this study were bituminous coal for five PACs, lignite coal for two PACs, and wood or wood-bark for four PACs (Table 1). Two different batches of Hydrodarco B were included in this research to study the variability of the activated carbon, as provided by the manufacturer.

Table 1. Powdered activated carbons (PACs) examined in study.

				Manufacturer Reported		
Name of PAC	**Symbol**	**Manufacturer**	**Carbon Source**	**I_2 Number (mg/g)**	**Molasses Number**	**Tannin Number (mg of PAC/L)**
WPL Pulverized	WPL	Calgon[1]	B. coal			
WPH Pulverized	WPH	Calgon[1]	B. coal			
PWA Pulverized	PWA	Calgon[1]	B. coal	900 (min)	210 (min)	
Colorsorb GP	COL	Calgon[1]	Wood	900 (min)	550	
Centaur Powder	CEN	Calgon[1]	B. coal	800 (min)		
PAC 20 B	20B	NORIT[2]	B. coal	900	200	360-400
HDW	HDW	NORIT[2]	Lignite coal	530		350
Hydrodarco B (Batch 1091 and 135-10C)	HD-A	NORIT[2]	Lignite coal	560	450-650	200
Watercarb PAC	WAT	Acticarb[3]	Wood Bark	600		
Nuchar SA-20	N-SA	Westvaco[4]	Hard wood	900 (min)		225
Nuchar SA	N-20	Westvaco[4]	Hard wood	900 (min)		225

[1] Calgon Carbon Corp., Pittsburgh, PA
[2] NORIT Americas Inc., Atlanta, GA
[3] Acticarb, Inc., Dunnellon, FL
[4] Westvaco Chemical Division, Westvaco Corp., Covington, VA

Determination of Standard Indices

For each gaseous or aqueous adsorbate studied (i.e., nitrogen, p-nitrophenol (PNP), methyl blue (MB), iodine (I_2), tannic acid and molasses)), molecular weight, dimensions, and average critical pore dimension are tabulated in Table 2. The dimensions were calculated using MacSpartan 3-D modeling software (Ver. 1.1, Wavefunction Inc., Irvine, CA, USA), except for nitrogen and PNP, which were from Linares-Solano *et al.* (1984). The MacSpartan software optimizes the initial molecular geometry submitted by the user and searches for a conformation that gives the lowest energy by using molecular mechanics principles to provide the estimated lowest-energy configuration.

Table 2. Molecular weights and sizes of the selected adsorbates.

Compounds		M.W.	Dimensions	Critical Dimension (nm)	Average Critical Pore Dimension (nm) ‡
Nitrogen		28	-	0.33†	-
p-Nitrophenol		139	0.76 x 0.5 x 0.19*	0.46†	0.8
Methylene Blue		320	1.6 x 0.84 x 0.47	0.84	1.52
Iodine		254	0.7 x 0.44 x 0.44	0.44	< 1.48
Tannic Acid		1701	2.68 x 2.48 x 1.32*	2.48	-
Molasses**		Mixture	-	-	-
Sucrose	35%	342	1.18 x 0.71 x 0.64*	0.71	-
Glucose	7 %	156	0.89 x 0.54 x 0.45*	0.54	-
Fructose	9 %	156	0.98 x 0.41 x 0.33*	0.41	-
Water	20 %	18	0.23 x 0.16 x 0.12*	0.16	-
Ash (K_2O, SiO_2, etc)	12 %	~64-94	Ex: SiO_2 0.43 x 0.26 x 0.26*	0.26	-

* Dimensions of the molecule and critical dimensions calculated using MacSpartan

** General composition of molasses, a mixture of several compounds.

† From (Linares-Solano et al, 1984)

‡ From (Kasaoka et al, 1989a)

The PNP capacity was determined for the PAC samples by using a procedure based on Hassler (1963), Linares-Solano *et al.* (1984), and Krupa and Cannon (1996). In this method, the PNP capacity is determined for a specific residual PNP concentration (100 mg/L) from isotherm data.

The iodine (I_2) number was determined for all PAC samples using the standard method (ASTM D4607-86, 1994) with five different masses of activated carbon, instead of the three required by the standard method. The I_2 number is determined as the amount of I_2 adsorbed per mass of PAC at a residual I_2 concentration of 0.02 N from a log-log plot ($0.95<R^2<0.99$).

The MB number is the adsorptive capacity corresponding to the residual concentration of 100 mg/L of MB (Galbraith *et al.*, 1958; Linares-Solano *et al.*, 1984; Krupa and Cannon, 1996). MB capacity was determined for all PAC samples by following two different methods. The first method involved developing an adsorption isotherm similar to the PNP method. The second method was developed by Hassler (1963) who proposed that a "single point" adsorption gives quantitative information for the MB adsorptive capacity of carbon as long as the amount of carbon added results in the removal of 60-99% of the initial dye. The method is reasonable since, in the adsorption isotherm, there is a relatively flat region where the amount adsorbed is not a strong function of the final equilibrium concentration.

The molasses number was determined for all PACs following the standard procedure (EPA 625/1-71-002a, 1973). This number is different from the other standard indices in that the decolorization capacity of a carbon is compared to that of a given standard carbon with a "known" molasses number.

The tannin number was determined for all PAC samples using the AWWA Standard for Powdered Activated Carbon (AWWA, 1996a). This number is different from the other tests mentioned above because of the fact that a higher tannin number value for a PAC signifies a lower tannin adsorptive capacity for the PAC. The tannin number denotes the concentration of PAC needed to achieve a certain amount of tannin removal. Thus, a lower concentration of carbon with a greater capacity is required to reach a desired final tannin concentration.

Gas Adsorption

In the nitrogen adsorption tests, the samples were analyzed with the Autosorb-1C gas sorption apparatus using a static volumetric technique for nitrogen adsorption. In this method, PAC samples are initially out-gassed under a high vacuum at 573 K for at least 6 hr to remove all moisture and volatile materials. The samples are then dosed with nitrogen at 77K to attain pressure equilibrium points during adsorption and desorption cycles to generate isotherms. These data are analyzed by a variety of micro- and meso-pore techniques built into the Autosorb software (Autosorb-1 for Windows, Ver. 1.19).

In the CO_2 adsorption tests, CO_2 adsorption was performed at 273 K using the same samples as were used for nitrogen adsorption with only the adsorption cycle data. These data were analyzed using the BET-CO_2 and GC-MC methods.

Determination of Surface Functional Groups

The Boehm titration method (Boehm, 1966; Boehm, 1994) was used to determine the number of surface oxygenated groups for each PAC. In this method, a several gram sample of PAC was mixed with 50 mL of distilled water in a mechanical shaker for 24 hr to be fully saturated. The supernatant was decanted off and the PAC again rinsed with distilled water. The PAC was then dried at 373 K. A 0.5 g sample of washed and dried PAC was placed in a vial with 50 mL of 0.05 N HCl solution, sealed, and shaken for 24 hr. The PAC/HCl slurry was then centrifuged to separate the PAC from the remaining HCl solution. A 5-mL aliquot of the HCl filtrate was titrated against 0.01N NaOH standard solution to determine the concentration of the HCl not neutralized by the basic sites on the carbon surface. From this, the number of basic sites (meq/g) was determined. Using a similar procedure, the acidic sites on the surface were determined by equilibrating the PAC with NaOH, Na_2CO_3, and $NaHCO_3$ for 24 hr and then titrating against 0.01 N HCl standard solution.

The basis of the titration method is that NaOH is strong enough to neutralize all acidic sites on the surface, Na_2CO_3 is strong enough to neutralize only the lactonic and free carboxylic sites, and $NaHCO_3$ is strong enough to neutralize only the free carboxylic sites. The difference between the number of sites determined from the NaOH and Na_2CO_3 neutralizations gives the number of phenolic sites, and the difference between the Na_2CO_3 and $NaHCO_3$ neutralizations gives the number of lactonic sites. The $NaHCO_3$ titration gives the number of free carboxylic sites.

RESULTS AND DISCUSSION

Pore-size Distribution and Surface Area.

For each PAC studied, three sets of gas adsorption data were generated. The first set, generated from atmospheric pressure nitrogen isotherms ($0.025<P/P_0<0.998$), was used to calculate BJH, BET-N_2, TPV, and DFT results (applicable toward mesopore, or combined micro- and mesopore analysis). The second set, generated from low pressure nitrogen isotherms ($2 \cdot 10^{-6}<P/P_0<0.12$), was used to calculate BET-N_2, DR-SA, DR-Vol, and HK Vol results (applicable toward micropore analysis only). The third set of data was generated from low pressure carbon dioxide isotherms ($0.001<P/P_0<0.030$) and was used to calculate GC-MC and BET microporous surface areas.

For each PAC studied, the source material, along with the results from the gas adsorption models, surface functional sites, and aqueous indices have been tabulated in Table 3. The WAT results were separated from the averages in Table 3 and in the following discussion because its source material was wood bark, rather than the hardwood itself.

Nitrogen Surface Area Analysis - The BET-N_2 results show that there are significant differences (α=0.05) between the BET-N_2 surface areas for wood-, bituminous-coal-, and lignite-coal-based PACs (excluding WAT) which were 1606(±86), 1070(±132), and 504(±114) m^2/g, respectively. Sorial *et al.* (1993) also noted that the lignite carbons had much lower BET surface areas than either bituminous carbon or wood-based carbons. Sorial *et al.* (1993) did not, however, observe significantly greater BET surface areas for wood-based versus bituminous-coal-based carbon.

The density functional theory (DFT) results (as used in this study) characterize the surface area for pores between 0.4 and 5.9 nm (thus including both micro- and meso-porous ranges). The results correlate relatively well with the BET-N_2 surface area results (r=0.939), and indicate that the average DFT surface areas were ordered: wood>bituminous carbon>lignite carbon (Table 3).

Table 3a. Summary of gas and standard indices for bituminous coal-based PACs. Specified ranges are 95%-confidence intervals.

PAC	20 B	WPL	WPH	CEN	PWA	Bitum. Coal Ave.
Gas Phase Adsorption						
Nitrogen Mesopore Analysis						
BET SA (N_2) (m^2/g)	1250 ± 0	1072 ± 120	1027 ± 90	1024 ± 115	976 ± 57	1070 ± 132
DFT SA(<59 Å) (m^2/g)	925	1050	756	1007	911	930 ± 140
BJH Vol. (<400 Å) (m^3/g)	0.45	0.28	0.3	0.24	0.24	0.30 ± 0.11
Total Pore Vol. (<400 Å) (m^3/g)	0.66	0.56	0.55	0.51	0.48	0.55 ± 0.08
Nitrogen Micropore Analysis						
DR SA (cm_2/g)	1394	1209	1080	1151	1032	1173 ± 175
DR Vol. (cm^3/g)	0.5	0.43	0.38	0.41	0.37	0.42 ± 0.06
HK Vol. (<15 Å) (cm^3/g)	0.48	0.43	0.39	0.4	0.37	0.41 ± 0.05
Carbon Dioxide Analysis						
GCMC SA (<12.5 Å) (m^2/g)	789	1005	747	815	1564	984 ± 421
BET SA (CO_2) (m^2/g)	877	953	769	911	1481	998 ± 346
Surface Chemistry						
Total Sites (meq/g)	0.49	0.86	0.54	0.87	0.73	0.70 ± 0.22
Basic sites (meq/g)	0.30	0.68	0.40	0.35	0.40	0.43 ± 0.18
Acidic (meq/g)	0.19	0.18	0.14	0.52	0.33	0.27 ± 0.19
Phenolic (meq/g)	0.18	0.04	0.04	0.22	0.16	0.13 ± 0.10
Lactonic (meq/g)	0.02	0.14	0.09	0.21	0.08	0.11 ± 0.09
Carboxylic (meq/g)	0.00	0.00	0.01	0.10	0.09	0.04 ± 0.06
Aqueous Indices						
PNP (mg/g)	202	168	182	138	170	172 ± 29
I_2 (mg/g)	1111 ± 129	802 ± 684	993 ± 84	597 ± 43	807 ± 885	862 ± 245
MB (mg/g PAC) Isotherm	369	249	320	220	278	287 ± 73
MB (mg/g PAC) Hassler	366 ± 73	235	324 ± 4	209	247	276 ± 82
Molasses	169 ± 11	170 ± 13	207 ± 108	139 ± 37	154 ± 9	168 ± 31
Tannin (mg PAC/L)	1173	1015	550	1782	1219	1148 ± 549

Table 3b. Summary of gas and standard indices for lignite coal-based PACs. Specified ranges are 95%-confidence intervals.

PAC	HD-B	HD-A	HDW	Lignite Coal Ave.
Gas Phase Adsorption				
Nitrogen Mesopore Analysis				
BET SA (N_2) (m^2/g)	546 ± 35	510 ± 33	455 ± 49	504 ± 114
DFT SA(<59 Å) (m^2/g)	515	481	453	483 ± 77
BJH Vol. (<400 Å) (m^3/g)	0.43	0.42	0.36	0.40 ± 0.09
Total Pore Vol. (<400 Å) (m^3/g)	0.48	0.48	0.41	0.46 ± 0.10
Nitrogen Micropore Analysis				
DR SA (cm_2/g)	594	562	524	560 ± 87
DR Vol. (cm^3/g)	0.21	0.2	0.19	0.20 ± 0.02
HK Vol. (<15 Å) (cm^3/g)	0.21	0.2	0.19	0.20 ± 0.02
Carbon Dioxide Analysis				
GCMC SA (<12.5 Å) (m^2/g)	485	467	503	485 ± 45
BET SA (CO_2)	405	380	418	401 ± 48
Surface Chemistry				
Total Sites (meq/g)	1.56	1.68	2.01	1.75 ± 0.59
Basic sites (meq/g)	1.35	1.44	1.79	1.53 ± 0.58
Acidic (meq/g)	0.21	0.24	0.22	0.22 ± 0.04
Phenolic (meq/g)	0.00	0.04	0.02	0.02 ± 0.05
Lactonic (meq/g)	0.12	0.17	0.16	0.15 ± 0.06
Carboxylic (meq/g)	0.09	0.04	0.05	0.06 ± 0.06
Aqueous Indices				
PNP (mg/g)	104	94	92	97 ± 16
I_2 (mg/g)	525 ± 128	415 ± 418	394 ± 286	445 ± 175
MB (mg/g PAC) Isotherm	171	162	148	160 ± 29
MB (mg/g PAC) Hassler	165	163	151	160 ± 19
Molasses	281 ± 1	223 ± 38	172 ± 1	225 ± 135
Tannin (mg PAC/L)	569	524	920	671 ± 539

Nitrogen Mesopore Analyses - The BJH pore volume characterizes primarily mesopores with little micropore contribution. The BJH results showed that at the 95-percent confidence level, there was no significant trend between the wood-, bituminous-coal-, and lignite-coal-based carbons (1.22(±0.68), 0.30(±0.11), and 0.40(±0.09) cm^2/g), respectively) (α=0.05) (Table 3). While the average BJH pore volume for the wood-based carbons was much greater than for the coal-based carbons, the variability in the data lead to a sufficiently large confidence interval to overlap with the other two carbons.

The total pore volume (TPV) results characterize both micro- and mesopores as less than 40 nm. These results indicate that the TPVs for wood (excluding WAT) were significantly greater than those for bituminous- or lignite-coal-based PACs (or 1.32(±0.39) versus 0.55(±0.08), and 0.46(±0.10) cm^2/g, respectively) (Table 3). While the average TPV for bituminous carbon was greater than that for lignite carbon (consistent with the BET-N_2 results), the difference was not significant (α=0.05). The BJH pore volume results correlated well with the total pore volume (TPV) (r=0.974; n=12) (Figure 1), although the TPVs were generally larger than the BJH values presumably due to the inclusion of a greater number of mesopores by the TPV method.

Table 3c. Summary of gas and standard indices for wood-based (and bark-based) PACs. Specified ranges are 95%-confidence intervals.

PAC	WAT (Bark-based)	N-20	N-SA	COL	Wood Ave (excluding WAT)
Gas Phase Adsorption					
Nitrogen Mesopore Analysis					
BET SA (N_2) (m^2/g)	604 ± 48	1633 ± 114	1567 ± 254	1619 ± 188	1606 ± 86
DFT SA(<59 Å) (m^2/g)	587	1132	1065	1151	1116 ± 112
BJH Vol. (<400 Å) (m^3/g)	0.09	1.53	1.1	1.02	1.22 ± 0.68
Total Pore Vol. (<400 Å) (m^3/g)	0.28	1.5	1.22	1.23	1.32 ± 0.39
Nitrogen Micropore Analysis					
DR SA (cm_2/g)	664	1634	1516	1705	1618 ± 237
DR Vol. (cm^3/g)	0.24	0.58	0.54	0.61	0.58 ± 0.09
HK Vol. (<15 Å) (cm^3/g)	0.23	0.57	0.54	0.61	0.57 ± 0.09
Carbon Dioxide Analysis					
GCMC SA (<12.5 Å) (m^2/g)	793	790	751	736	759 ± 69
BET SA (CO_2) (m^2/g)	623	821	875	874	857 ± 77
Surface Chemistry					
Total Sites (meq/g)	2.05	0.81	0.95	1.06	0.94 ± 0.31
Basic sites (meq/g)	1.10	0.31	0.39	0.00	0.23 ± 0.51
Acidic (meq/g)	0.95	0.50	0.56	1.06	0.71 ± 0.76
Phenolic (meq/g)	0.77	0.20	0.28	0.53	0.34 ± 0.43
Lactonic (meq/g)	0.18	0.30	0.17	0.03	0.17 ± 0.34
Carboxylic (meq/g)	0.00	0.00	0.11	0.50	0.20 ± 0.65
Aqueous Indices					
PNP (mg/g)	146	155	164	174	164 ± 24
I_2 (mg/g)	481 ± 279	926 ± 349	966 ± 427	977 ± 138	956 ± 67
MB (mg/g PAC) Isotherm	62	358	379	339	359 ± 50
MB (mg/g PAC) Hassler	61 ± 26	375 ± 121	395 ± 309	349 ± 185	373 ± 57
Molasses	113 ± 7	1047 ± 48	792 ± 1377	772 ± 91	870 ± 381
Tannin (mg PAC/L)	1804	275	296	232	268 ± 81

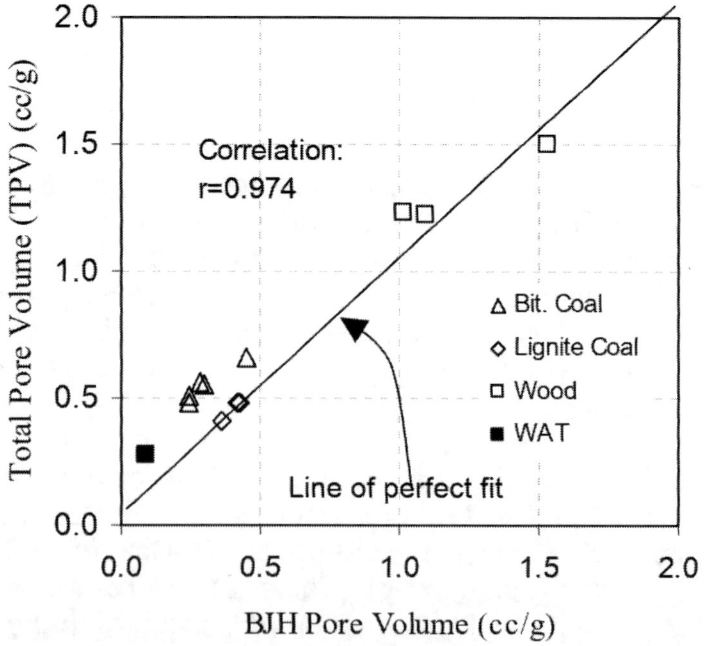

Figure 1. Comparison of BJH pore volume and total pore volume.

Nitrogen Micropore Analyses - The DR surface area, DR volume, and HK volumes are all representative of the microporous nature of activated carbons. Each of these models resulted in the same trend. Specifically, the surface area or volume trend was wood > bituminous coal > lignite coal (although the differences between the wood versus bituminous coal were not significant at $\alpha=0.05$). Comparison of the total pore volume (mesopores plus micropores) with the average of the DR and HK pore volumes showed that 43-44% of the pore volume was as micropores for the wood- and lignite-coal carbons. Based on DR and HK pore volumes for the bituminous-coal carbon, however, the micropores accounted for 75% of the combined mesopore and micropore pore volume.

Carbon Dioxide Micropore Analyses - The carbon dioxide adsorption (GC-MC) analysis was examined as an alternative means of characterizing the microporous nature of activated carbon. In general, the GC-MC results (<1.3 nm) contained greater variability than those obtained using nitrogen isotherms. However, as with the nitrogen results, the GC-MC surface areas of wood and bituminous coals were significantly larger ($\alpha=0.05$) than those of the lignite coal carbon.

Correlation Between the BET Nitrogen Surface Area and Other Gaseous Indices - Correlation coefficients (r) between the commonly used BET nitrogen surface area versus the other gaseous indices were determined using the data pairs for each PAC examined. The nitrogen BET SA correlated well with the SA determined by the DFT method (r=0.939). Correlations for the micropore analyses (DR Vol, HK Vol) and BET nitrogen SA were very strong (r >0.99) and only slightly weaker for mesopore analyses (BJH, TPV) ($0.767< r<0.888$) (Table 4). The BET-CO_2 SA and GCMC surface areas both exhibited much weaker correlations with the BET nitrogen SA results (r = 0.816 and 0.577, respectively). The reason for the poor correlation with the CO_2 data is not fully understood.

Table 4. Correlation coefficients (r) between BET(N_2) surface areas (m^2/g) and volumes or surface areas predicted by pore characterization models.

Analysis	Models	r
Mesopore	BJH Volume (<400 Å)	0.767
	Total Pore Volume (<400 Å)	0.888
	DFT SA (<59 Å)	0.939
Micropore	DR SA	0.992
	DR Volume	0.992
	HK Volume (<15 Å)	0.993
CO_2	BET SA (CO_2)	0.816
	GCMC SA (<12.5 Å)	0.577

Surface Functional Sites - The Boehm titrations were used to determine the specific number of acidic sites and basic sites (expressed as meq/g). The three wood-based carbons all had more acidic oxide surface groups than basic sites (Table 3). The lignite-based carbons had more basic sites, whereas the bituminous-based carbons generally had a similar number of acidic and basic sites. In terms of the total number of sites at the 95%-confidence level, there was no significant difference and the trend was lignite>wood>bituminous.

Aqueous Indices - Activated carbon capacities for small molecules are frequently characterized by use of the PNP, I_2 and MB methods, while capacities for large molecules are often characterized by the molasses and tannin methods (AWWA, 1996b). For the PNP sorption, the results show that bituminous carbon and wood-base carbon had significantly greater capacities than lignite carbons (Table 3) (α=0.05). For the I_2 number results, bituminous carbon and wood-based carbons also had significantly greater capacities than lignite coal. The WAT bark-based carbon had an I_2 number similar to that of the lignite coal carbon. The PNP and I_2 results agree with the gas adsorption results in that the bituminous and wood-based carbons tend to have much greater microporous structures than the lignite carbons.

MB capacities, determined using the isotherm methods versus the Hassler method, correlated very well (r=0.983) (Figure 2). Thus, these data provided justification for the use of the much quicker and easier Hassler method, versus the isotherm method. While the average MB values were ordered wood>bituminous coal>lignite coal, the differences were not significant (α=0.05).

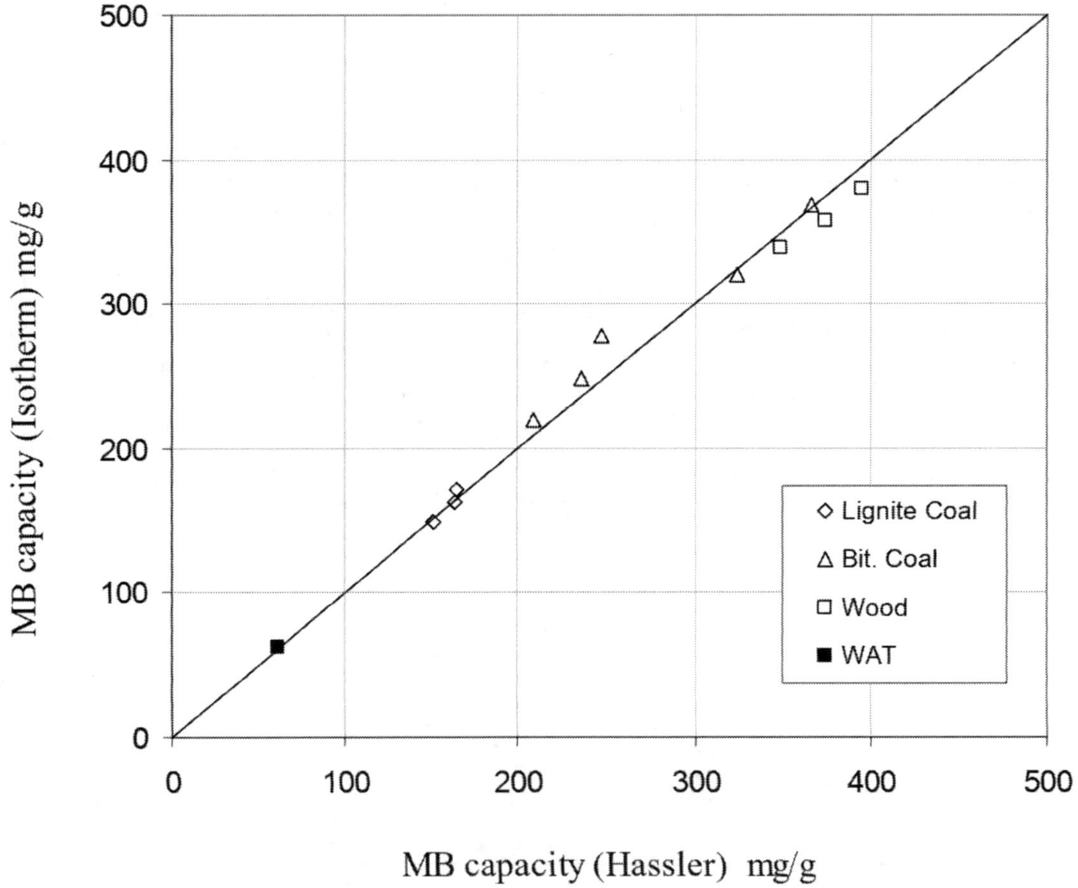

Figure 2. Comparison of the isotherm versus the Hassler methods for determining MB capacity.

Examination of the molasses and tannin number results shows that wood-based carbons (excluding bark) exhibit a greater capacity for these compounds than bituminous or lignite carbons (Table 3). Note that smaller tannin numbers indicate greater capacities. This result is consistent with the fact that the wood-based carbons tend to have greater mesoporous natures than coal-based carbons. Comparisons of the aqueous indices of the PACs in this study (Table 3) with the minimum values reported by producers (Table 1) were not always consistent.

Comparison of Gas-Phase and Aqueous-Phase Results - For the results from the 12 PACs studied, each gas-phase index was correlated with each aqueous-phase index and the correlation coefficients (*r*) were tabulated (Table 5). In the case of the tannin numbers, the correlations used the reciprocals of the tannin numbers.

In general, the molasses and tannin numbers correlated better than PNP, I_2, and MB with the gas-phase indices that characterize mesopores (i.e., BJH and TPV) (Table 6). Conversely, PNP, I_2, and MB correlated better than the molasses and tannin numbers did with the gas-phase indices that characterize micropores (i.e., DR-SA, DR-Vol., HK, BET-CO_2, and GC-MC) (Table 5).

Table 5. Correlation coefficients (*r*) between aqueous indices and volumes or surface areas predicted by pore characterization models.

Analysis	Models	PNP	I_2	MB	Molasses	Tannin
N_2 Mesopore	BET SA (N_2)	0.736	0.861	0.890	0.775	0.688
	DFT SA (<59 Å)	0.723	0.768	0.778	0.610	0.493
	BJH Vol. (<400 Å)	0.215	0.502	0.676	0.984	0.888
	TPV (<400 Å)	0.391	0.638	0.775	0.971	0.887
N_2 Micropore	DR SA	0.775	0.874	0.885	0.706	0.623
	DR Vol.	0.775	0.874	0.885	0.706	0.623
	HK Vol. (<15 Å)	0.774	0.874	0.885	0.709	0.632
CO_2	BET SA (CO_2)	0.851	0.775	0.689	0.312	0.192
	GCMC SA (<12.5Å)	0.769	0.568	0.388	0.096	0.067
Surface Chemistry	Total sites (meq/g)	0.802	0.864	0.853	0.219	0.160
	Basic sites (meq/g)	0.863	0.865	0.803	0.440	0.382
	Acidic sites (meq/g)	0.157	0.049	0.045	0.420	0.417

DFT Analysis - The DFT surface areas were computed for pores from 0.4 to 5.9 nm; that is, on the order of the size of smaller molecules (Table 2). The DFT surface area correlations with the molasses and tannin numbers are weaker than correlations with smaller molecule indices (PNP, I_2, and MB) even though the DFT characterizes the lower range of the mesopore sizes. The correlation coefficients with DFT are weaker for the larger molecules because the DFT does not account for pores greater than 5.9 nm. It should be noted that the DFT method can be expanded to much larger pore diameters than it was applied to in this study.

The pore-size distributions, based on surface areas, were determined for the 12 PACs using the DFT method (Figure 3). The cumulative surface area was calculated from these data and plotted in Figure 4. These data show an interesting anomaly with very few pores being indicated at both 1.5 and 2.2 nm (Figure 3). Similar anomalies have also been noted by other researchers studying six GACs (Krupa and Cannon, 1996). Because it is unlikely that few (or no) pores exist physically at these two distinct diameters, it has been suggested that the anomalies are a function of the DFT model itself. Krupa and Cannon (1996) suggested that the cause may be related to: 1) difficulties that the model has in transitioning from one to two molecular layers, 2) the model's presumption of a single adsorption energy, or 3) both.

These data showed that each bituminous carbon had very similar pore-size distributions, except below 1.5 nm, where CEN and WPL had significantly larger pore surface areas below 0.7 nm, than PWA, 20B or WPH (Figure 3). Each of the lignite and each of the wood-based carbons (with the exception of WAT) also had very similar pore-size distributions.

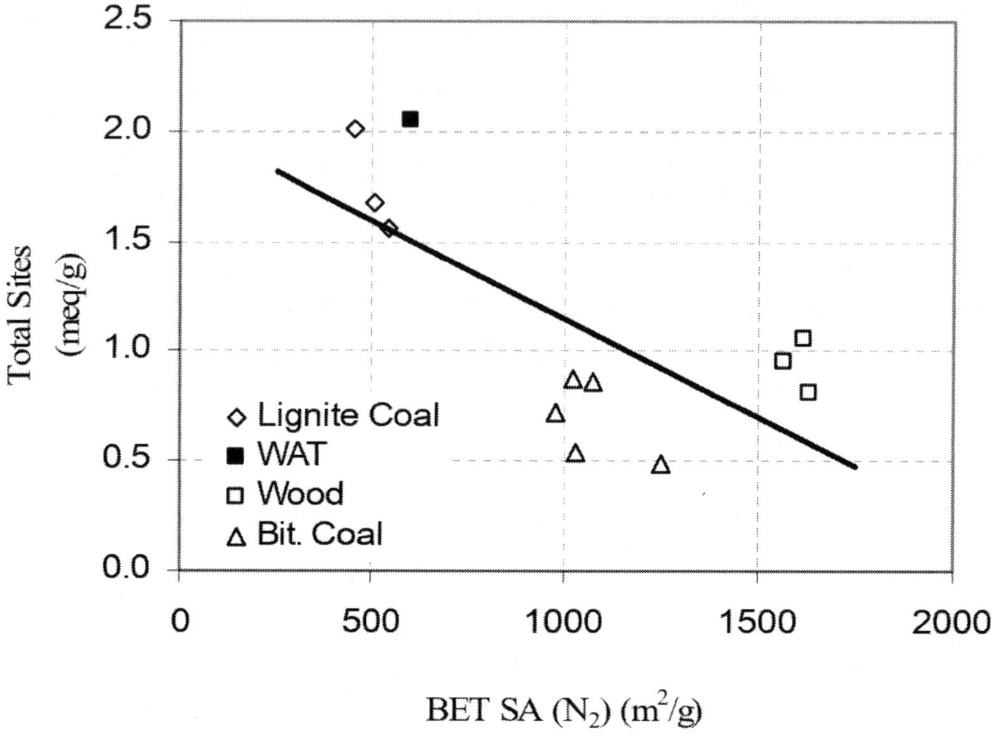

Figure 3. Total surface functional groups (sites) versus total surface area carbon source material for varied carbon types.

Based on DFT model data, bituminous and lignite carbons have a majority of their porous structures (>95%) in the microporous range (Figures 4 and 5). The wood-based carbons (excluding bark), however, have a much broader range of pore sizes, contributing to these cumulative pore surface areas (Figure 5). Overall, these data showed that the average cumulative pore areas ranged from 750-1050, 450-510, and 1050-1150 m^2/g for bituminous, lignite, and wood-based carbons (excluding bark), respectively. The cumulative pore area for WAT was 590 m^2/g. These results are consistent with work by Salame and Bandosz (2000) who stated that coal-based carbons (as well as petroleum pitch and coconut shell carbons) tend to be highly microporous in nature, and that wood-based carbons tend to be mesoporous in nature.

Comparison of Aqueous Indices and Pore-Size Distribution - The pore-size distributions, determined from the gas adsorption data, can be used to further elucidate the relationships between pore structures and standard aqueous indices. The pore-size distributions were used to calculate the total pore area for a given range of pore diameters for each carbon (i.e., between the given pore diameter and 5.9 nm), which was then correlated with the five aqueous indices. The correlation coefficients for this analysis were plotted versus the pore diameter for each of the 12 carbons (Figure 6). The data shows that the larger molecule indices (molasses and tannin) correlated better than the smaller molecules (PNP, I_2, and MB) everywhere except when the extremely small pore diameters were included (i.e., <0.9nm for I_2 and MB, and <0.6 nm for PNP) (Figure 6).

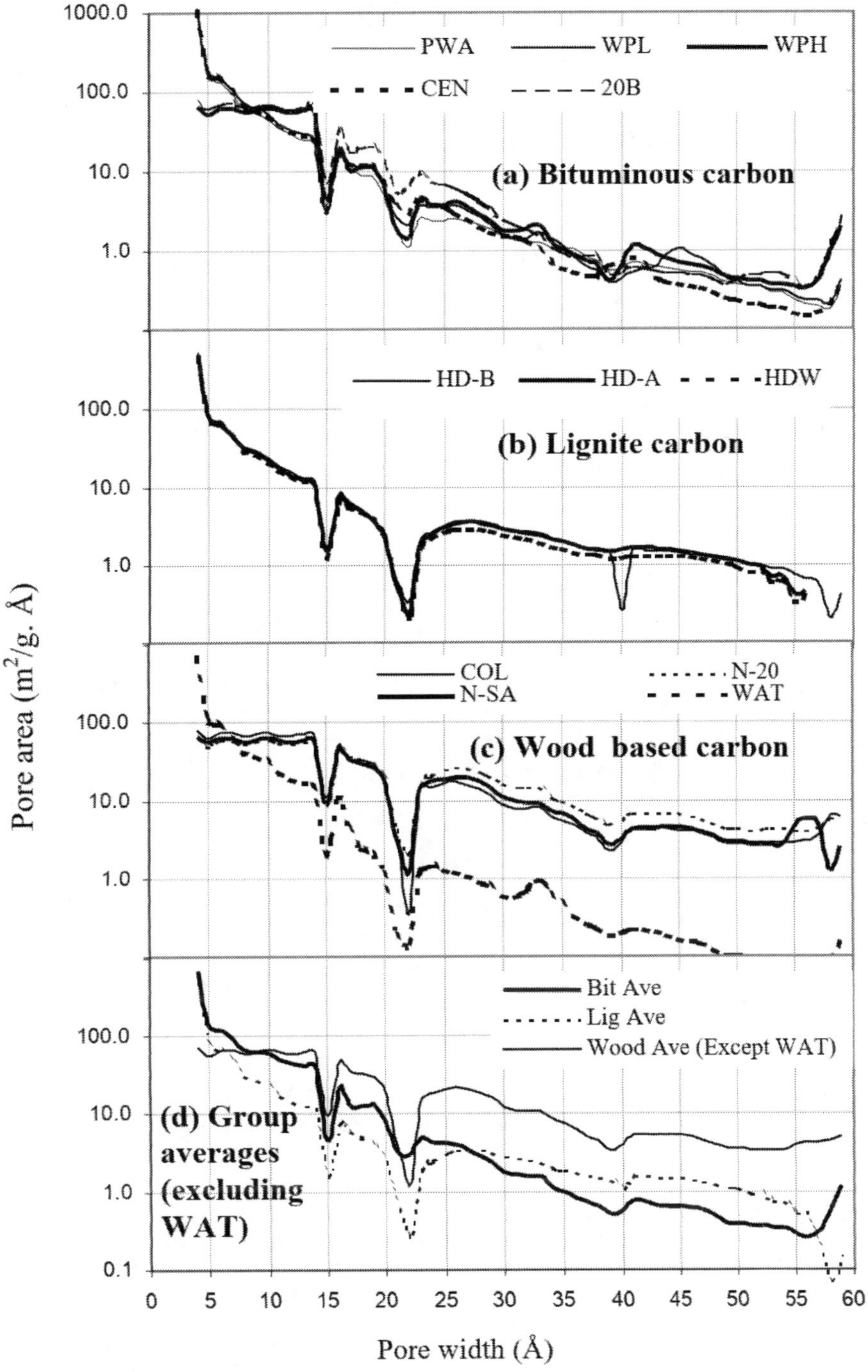

Figure 4. DFT pore-size distribution based on surface areas of 12 PAC groups by: (a) bituminous carbon, (b) lignite carbon, (c) wood based carbon, and (d) group average values.

The correlation coefficients for molasses and tannin decreased as smaller pore diameters were included in the cumulative surface area summation. The explanation for these observations is that the larger tannin molecules and the absorbing impurities in molasses are not readily diffused into these smaller pores.

Overall, the results of this study show that the commonly used I_2 and MB numbers are fairly good predictors of the volume of micropores plus mesopores within a PAC. Similarly, the molasses correlated well with the volume of mesopores within a PAC while the tannin number did not correlate as well.

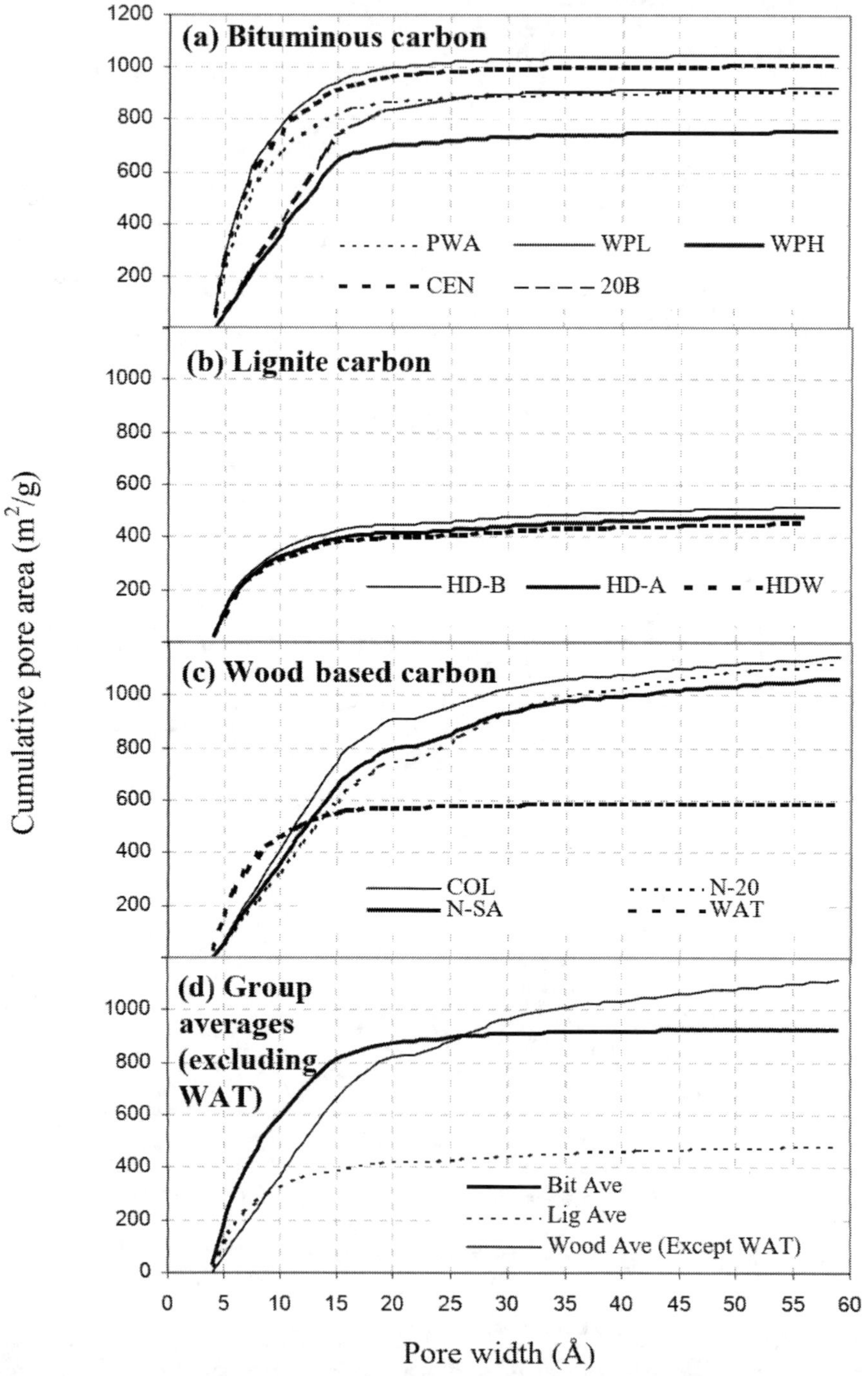

Figure 5. Cumulative DFT pore-size distribution based on surface areas of 12 PAC groups by: (a) bituminous carbon, (b) lignite carbon, (c) wood based carbon, and (d) group average values.

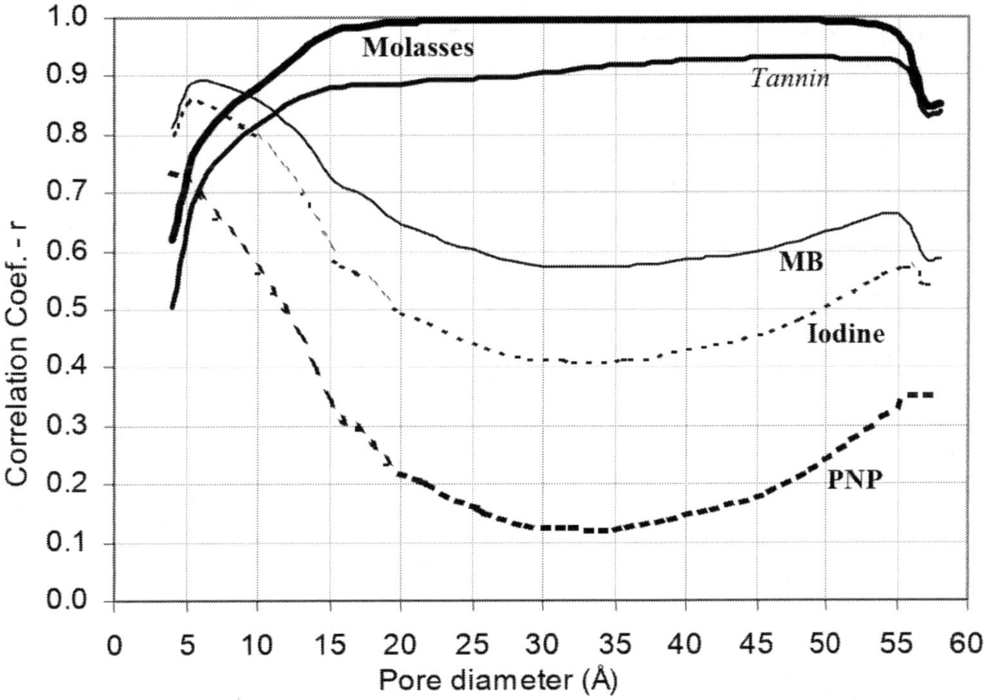

Figure 6. Correlation coefficients between the standard indices and the total pore surface areas, greater than the pore diameters plotted on the x- axis and <59 Å.

Proposed Aqueous Characterization Indices – We propose that ratios of commonly used indices can provide important information as to the micro- and meso-porous nature of activated carbons. First, it is proposed that the molasses-to-iodine ratio (MIR);

$$\text{MIR} = (\text{Molasses number})\cdot(\text{Iodine number})^{-1} \qquad (1)$$

is a useful parameter for characterizing the pore-size distribution of a PAC using aqueous indices. In Figure 7, MIR is plotted from each PAC versus the BET-N_2 surface area (top) and versus the iodine number (bottom). It may be seen that PACs from each carbon source (bituminous coal, lignite coal, and wood (excluding bark)) are tightly grouped with respect to their overall surface area and their MIR. A small MIR (e.g., <0.5) corresponds to a more microporous carbon, such as the bituminous carbons examined in this study (Figure 7). A larger MIR (e.g. >0.7) corresponds to wider pore-size distributions (including significant mesoporous natures, such as with the wood-based carbons in this study, but excluding bark) (Figure 7). The MIR may be a useful parameter for characterizing the pore-size distribution of PACs to assist in selection of the most appropriate PAC for a specific application.

Alternatively, it is proposed that the tannin (or specifically the inverse of the tannin number) and iodine numbers may be related as the inverse tannin-to-iodine ratio (ITIR):

$$\text{ITIR} = (\text{Tannin number})^{-1}\cdot(\text{Iodine number})^{-1} \qquad (2)$$

to provide information similar to that of the MIR (Figure 7). The tannin number is different from the other indices in that it is related to the amount of carbon needed to remove a given amount of tannin. Thus, a smaller tannin number indicates a greater surface area of a given size (and, hence, enhanced sorption). It is proposed that the ITIR can be used as an estimate of the pore-size distribution for an activated carbon. Similar to the proposed MIR, a lower ITIR indicates the more microporous nature of a carbon, whereas a greater ITIR would indicate a larger percentage of large pores. Because aqueous indices are commonly analyzed or specified for carbons, the use of these ratios may be appropriate for carbon specifications by utilities in selecting carbons with properties that are most suitable for specific objectives (e.g., color removal by sorption of large molecules, pesticide removal by carbons with large surface areas (in the microporous range), etc.).

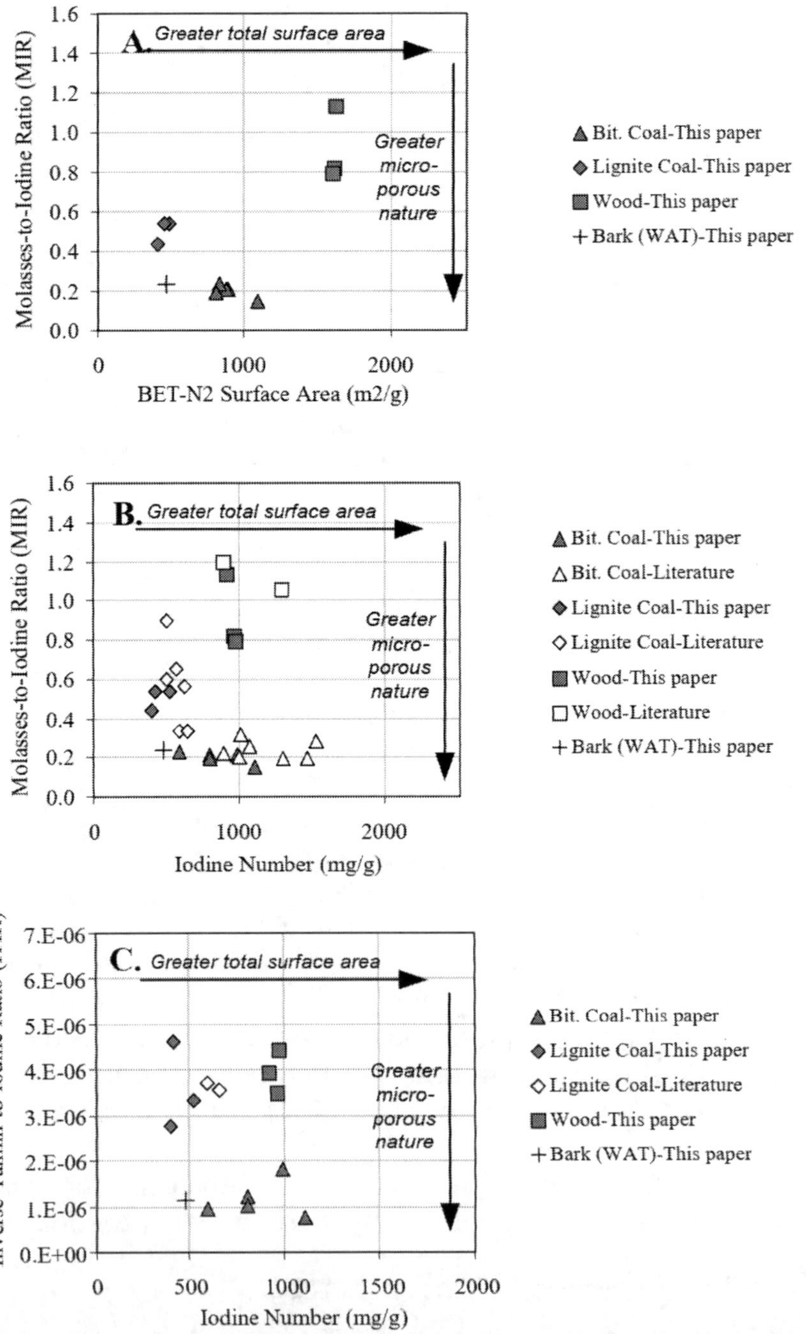

Figure 7. For 12 studied PACs and additional PACs from literature references, the Molasses-to-Iodine Ratios (MIR) are plotted versus the BET-N_2 surface areas (A (top)), and versus the Iodine numbers (B (middle)). In the bottom plot (C), the Inverse-Tannin-to-Iodine Ratios (ITIR) are plotted versus iodine numbers.

Additional common PACs (for which iodine, molasses, or tannin numbers were known) are also plotted in Figure 7, with designated source materials. These data show clear groupings and/or trends based on source materials. For example, bituminous coal has a consistent MIR of nominally 0.2 (indicating a microporous nature) and a range of iodine numbers (indicative of total surface area). Lignite carbons, however, have a fairly uniform iodine number (corresponding to uniform total surface areas) and a range of MIR (showing some variability in the degree of mesoporous character). The wood-based carbons tended to have a high iodine number (i.e., total surface areas) and a high MIR (a fairly mesoporous nature) (Figure 7). The results of this study suggest that the types of source material, and the associated activation methods, result in an activated carbon with a distinct total pore volume and pore-size-distribution.

Comparison of Surface Functional Groups and Aqueous-Phase Indices - The lignite-coal PACs had significantly (α=0.05) greater total exchange sites than either bituminous-coal or wood-based PACs (Table 3b). Similarly, the lignite-coal PACs also had significantly (α=0.05) greater basic exchange sites than either bituminous-coal or wood-based PACs (Table 3). These basic sites would include hydroxyl, carboxyl, and phenolic sites that increase surface polarity (and decrease sorptive capacity) when deprotonated at pH levels above their respective dissociation constants. There were no significant differences (α=0.05) between the numbers of acidic functional groups in carbons from different starting materials. These acidic functional groups might include amino-functionalities that may be protonated and cationic below their respective dissociation constants. While not statistically significant (α=0.05) due to variability between the wood-based PACs, the coal-based PACs had a lot fewer acidic functional groups than the wood-based carbons had.

The extent and types of surface oxide species are expected to have some effect on the sorptive capacity of the various aqueous adsorption probe molecules. The formation of the surface oxide groups, as part of the activation process, can have at least two effects on the sorptive nature of the surface. First, they change the polar nature (hydrophobicity) of the surface at those sites and, in some cases, can lead to an electrostatic attraction/repulsion between the surface site and the adsorbate. Iodine is a nonpolar molecule, so any addition of a polar surface oxide should have a negative effect on the amount adsorbed. The experimental results demonstrate a trend toward a strong inverse correlation between the number of surface oxide groups and the amount of iodine adsorbed (Table 6). The standard procedure for PNP adsorption is at a pH of 2.5, where the PNP probe will have a positive charge and, hence, it should be attracted to basic sites and repulsed from acidic sites. The standard procedure for MB adsorption is at a pH of 7.5, where MB will also have a positive charge. However, both PNP and MB demonstrated a strong inverse correlation with the total number of basic sites. A plot of the total surface sites, versus the surface area (Figure 3) can be used to explain this inverse relationship.

It is clear that the specific number of surface oxides depends on the precursor materials and that the lignite- and bark-based carbons have the smallest specific surface areas and the largest numbers of surface oxide sites. The surface areas of lignite-based carbons are approximately one third the size of the surface areas of wood-based carbons. This inverse correlation is due to the fact that the carbons with the most surface sites correspond to those with the smallest surface areas. Hence any enhancement in the sorptive capacity expected from the electrostatic attraction is overwhelmed by the decrease in surface area. When carbons from the same precursor are compared, some trends can be noted. For instance, for the bituminous-based carbons, CEN has the largest number of acidic sites and a corresponding smallest amount of PNP and MB adsorption (Table 3a). Tannin and molasses, which consist of larger adsorbate molecules with both polar and nonpolar portions, do not have a strong inverse correlation, with respect to the surface oxide sites. As a basis for our choice of adsorbate molecules, we observed that the amounts adsorbed depended more on the total surface area than on the surface chemistry.

CONCLUSIONS

To enhance the effectiveness of PAC application in water treatment plants, the intraporous nature of a PAC should be considered in the selection process. This study has shown that the I_2 and MB numbers are effective at characterizing total pore volume (and surface areas) of PACs, whereas the molasses and tannin numbers are effective in characterizing the mesopore volume (and surface areas) of PACs. The molasses-to-iodine ratio (MIR) and inverse tannin-to-iodine ratio (ITIR) are proposed as useful in characterizing the pore-size distribution of PACs by using commonly measured aqueous indices.

Acknowledgements – The authors would like to thank Mr. William Naylor (Norit Americas, Inc.), Dr. Michael Greenbank (Calgon Carbon Corp.), Ms. Kristine Switt (Acticarb), and Mr. Bob Beckler (Westvaco) for helpful discussions regarding the nature and processing of activated carbons, industrial applications of aqueous indices, and the results of this research. Acknowledgements are due to Kristopher Schumacher, Pechiwa Malope, Greg Scheidt, Goaletsa Seitiso, Jeanine Maxwell, and Sacha Patires, undergraduate researchers in the Department of Chemical Engineering, University of Missouri- Rolla, for help in performing the experiments.

REFERENCES

ASTM D4607-86 (1990) *Annual Book of ASTM Standards, vol.15.01*. American Society for Testing and Materials, Philadelphia, PA.

ASTM Standards (American Society for Testing and Materials) (1994) Standard test Method For Determination of Iodine Number of Activated Carbon. ASTM D 4607 - 94. Philadelphia, PA.

Autosorb-1® Operating Manual, II-7, Quantachrome Corp. Boynton Beach, FL, USA.

AWWA (1996a) AWWA Standard for Powdered Activated Carbon. ANSI/AWWA B600-96.

AWWA (1996b) Water Treatment Plant Design 3rd Ed., McGraw-Hill, New York.

AWWA, Water Quality and Treatment (1999) Fifth Edition. McGraw-Hill, Inc. New York.

Bandosz T. J., Jagiello J., Contescu C. and Schwarz J. A. (1993) Characterization of the surfaces of activated carbons in terms of their acidity constant distributions. *Carbon* **31(7)**, 1193-1202.

Bandosz T. J., Jagiello J. and Schwarz J. A. (1996) Effect of surface chemistry on sorption of water and methanol on activated carbons. *Langmuir* **12**, 6480-6486.

Barrett E. P., Joyner L. G. and Halenda P. P. (1951) The determination of pore volume and area distributions in porous substances. I. Computations from nitrogen isotherms. *Journal of American Chemical Society* **73(1),** 373-380.

Boehm, H. P. (1966) In *Advances in Catalysis,* Vol 16. Academic Press, New York.

Boehm, H. P. (1994) Some aspects of the surface chemistry of carbon blacks and other carbons. *Carbon* 32(5), 759-769.

Brunauer S., Emmett P. H. and Teller E. (1938) Adsorption of gases in multimolecular layers. *Journal of American Chemical Society* **60(2),** 309-319.

Chern J. and Huang S. (1998) Study of Nonlinear Wave Propagation Theory. 1. Dye Adsorption by Activated Carbon. *Ind. Eng. and Chem. Res.* **37(1)**, 253-257.

Dobbs R. A. and Cohen J. M. (1980) *Carbon adsorption isotherms for toxic organics.*

Dubinin M. M. and Radushkevich L. V. (1947) The equation of the characteristic curve of activated charcoal. *Proc Aced. Sci. USSR,* **55**, 327-329.

EPA 625/1-71-002a, “Process design manual for carbon adsorption” U.S. Environmental Protection Agency, October 1973.

Galbraith J.W, Giles C.H., Halliday A. G., Hassan A. S. A., McAllister D. C., Macaulay N., Macmillan, N.W. (1958) Adsorption at inorganic surfaces. III. The mechanism of adsorption of organic solutes, including dyes, by graphite. *Journal of App. Chem.*, **8**, 416-424.

Garrido J., Linares-Solano A., Martin-Martinez J.M., Molina-Sabio M., Rodriguez-Reinoso F. and Torregrosa R. (1987) Use of nitrogen vs. carbon dioxide in the characterization of activated carbons. *Langmuir* **3(1),** 76-81.

Greenbank, M. (2001) Calgon Carbon Corp., Personal communication.

Gregg S. J. and Sing K. S. W. (1982) *Adsorption, surface area and porosity* Second Edition, Academic Press Inc., New York.

Hassler J. W. (1963) *Activated carbon,* Chemical Publishing Co., Inc. New York, NY 324-326.

Hassler, J. W. (1951) *Active Carbon*. Chemical Publishing Co., Inc. Brooklyn, NY 325-354.

Horvath G. and Kawazoe K. (1983) Method for calculation of effective pore-size distribution in molecular sieve carbon. *Journal Chem. Eng. Japan* **16(6),** 470-475.

Karanfil, T., Kilduff, J. E. (1999) Role of Activate Carbon Surface Chemistry on the Adsorption of Organic Compounds. 1. Priority Pollutants, *Environ. Science and Technol.* **33**, 3217-3224.

Karanfil, T., Kitis, M., Kilduff, J. E. and Wigton, A. (1999) Role of Activate Carbon Surface Chemistry on the Adsorption of Organic Compounds. 1. Natural Organic Matter *Environ. Science and Technol.* **33**, 3225-3233.

Kasaoka S., Sakata Y., Tanaka E. and Naitoh R. (1989a) Design of molecular-sieve carbon. Studies on the adsorption of various dyes in the liquid phase. *Int. Chem. Eng.,* **29(4)**, 734-742.

Kasaoka S. Sakata Y., Tanaka E. and Naitoh R. (1989b) Preparation of activated fibrous carbon from phenolic fabric and its molecular-sieve properties. *Int. Chem. Eng.,* **29(1)**, 101-114.

Krupa N.E. and Cannon F. S. (1996) GAC: Pore structure versus dye adsorption. *Journ. American Water Works Assoc.*, **88(6)**, 94-108.

Lalezary S., Pirbazari M. and McGuire M. J. (1986) Evaluating activated carbons for removing low concentrations of taste-and odor-producing organics. *Journ. American Water Works Assoc.*, **78(11)**, 76-82.

Lastoskie C., Gubbins K. E. and Quirke N. (1993) Pore size heterogeneity and the carbon slit pore: A density functional theory model. *Langmuir* **9(10),** 2693-2702.

Lee M. C. , Snoeyink V. L. and Crittenden J. C. (1981) Activated carbon adsorption of humic substances. *Journ. American Water Works Assoc.,* **73** (8)**,** 440-446

Linares-Solano A., Rodriguez-Reinoso F., Molina-Sabio M., Lopez-Gonzalez J. D. (1984) Two-stage air-CO_2 activation in the preparation of activated carbons. II. Characterization by adsorption from solution. *Adsorption Sci. & Tech.,* **1**, 223-234.

Marsh H. and Wynne-Jones W. F. K. (1964) The surface properties of carbon. I. The effect of activated diffusion in the determination of surface area. *Carbon* **1,** 269-279.

Naylor, W. (2001), Norit America, Inc., Personal communication.

Pelekani C. and Snoeyink V. L. (2000) Competitive adsorption between atrazine and methylene blue on activated carbon: The importance of pore-size distribution. *Carbon* **38,** 1423-1436.

Pelekani C. and Snoeyink V. L. (2001) Kinetic and equilibrium study of competitive adsorption between atrazine and Congo red on activated carbon: The importance of pore size distribution. *Carbon* **39,** 25-37.

Perry J. H. (1950) *Chemical Engineers' Handbook*. Third Edition. McGraw-Hill Book Company, Inc. New York. 538-539.

Ravikovitch P. I., Haller G.L. and Neimark A. V. (1998) Density functional theory model for calculating pore size distributions: pore structure of nanoporous catalysts. *Adv. in Coll. and Interface Sci.***76-77**, 203-226.

Romero, L.C., Bonomo, A., Gonzo, E.E. (2003) Acid-activated Carbons from Peanut Shells: Synthesis, Characterization and Uptake of Organic Compounds from Aqueous Solutions. *Adsorption Science & Technology* **21(7)**, 617-626.

Salame I. I. and Bandosz T. J. (2000) Comparison of the surface features of two wood-based activated carbons. *Ind. Eng. Chem. Res.* **39**, 301-306.

Sorial G. A., Suidan M. T., Vidic R. D. and Brenner R. C. (1993) Effect of Gac characteristics on adsorption of organic pollutants. *Water Environment Research* **65(1)** 53-57.

Sontheimer, H., Crittenden, J., Summers, R. (1988) "Activated Carbon for Water Treatment," DVGW-Forschungsstelle, Engler-Bunte-Institut, Univ. Karlsruhe, Fed. Rep. of Germany.

Speth T. F. and Miltner R. J. (1990) Technical note. Adsorption capacity of GAC for synthetic organics. Journal American Water Works Association. **82(2)** 72-75.

Switt, K. (2001). Acticarb Corp., Personal Communication.

Webb P.A. and Orr C. (1997) Analytical methods in fine particle technology, Micrometrics Instrument Corporation, Norcross, GA.

Novel technique of estimating the fraction of open-ended carbon nanotubes

Sandeep Agnihotri,[1,] Yijing Zheng,[1] Paulo J. Mota,[2] Pyoungchung Kim[1]*

[1] Department of Civil and Environmental Engineering, University of Tennessee, Knoxville, TN 37996, [2] Requimte/CQFB, Departamento de Quimica, FCT/UNL, 2829-516, Caparica, Portugal. [*] Corresponding Author. Email: sagnihot@utk.edu Phone: (865) 974-7728

Abstract

We have developed an adsorption based technique to estimate the fraction of open-ended nanotubes in samples of single-walled carbon nanotubes (SWNTs). Our technique involves a practical approach for modeling heterogeneous bundles of SWNTs by sub-dividing the total adsorption on a bundle into two contributions: inside the nanotube in a bundle and on the external surface of the outermost nanotubes of the bundles. The technique integrates sample morphology, i.e., purity and sample characterization by Raman spectroscopy, into grand canonical Monte Carlo (GCMC) simulation of a small probe molecule, such as N_2 adsorption at 77 K, in nanotube bundles. We introduced a scaling parameter for endohedral adsorption which resulted in a near-perfect reproduction of experimental isotherms. We refer to this parameter as the fraction of open-ended nanotubes. We have tested several commercially available samples. We found that the fraction of open-ended nanotubes in samples ranged between 0 to 60% even for samples with high surface area of 100 m^2/g. Our methodology resulted in a perfect reproduction of experimental N_2 (77 K) adsorption isotherms for each of the tested samples. The details of our research are available in Agnihotri et al., *J Phys. Chem.* C (2007). The fraction of open-ended SWNTs cannot be determined by visual characterization of samples due to a large aspect ratio of the nanotubes and the spaghetti-like arrangement of the bundles. Therefore, gas adsorption could become an appropriate tool for quantifying this structural property of carbon nanotubes: much like estimating the surface area of a material using gas adsorption.

Introduction

The depiction of bundle heterogeneity is a known problem in developing a better understanding of adsorption in SWNTs. Heterogeneity refers to the non-uniform geometry of the cross-section of a nanotube bundle. It arises as a result of a range of nanotube sizes that are generated and high van der Waals interaction between individual nanotubes. Several attempts have been made to understand the effect of heterogeneity on adsorption in nanotubes. Striolo et al. used grand canonical Monte Carlo (GCMC) simulation in a bundle of three armchair nanotubes to study adsorption and transport of water. Zhao et al. explored exohedral and endohedral N_2 adsorption on side-walls of homogeneous SWNTs. Kuzmany et al. functionalized endohedral and exohedral sites of a nanotube sample. Of most significance is the work of Shi and Johnson, who simulated a heterogeneous bundle of SWNTs comprising several armchair tubes of varying chiralities. These authors showed that adsorption in the interstitial channels between adjacent nanotubes can occur for a wide diameter distribution. However, practical modeling of heterogeneous SWNT bundles with the same distribution and positional arrangement as that in an actual sample is not computationally tractable at present. Theoretical modeling of heterogeneity, therefore, could become sample specific; it may or may not be extended to the variety of samples that are now being synthesized worldwide.

Methodology

We have developed a practical approach for modeling the heterogeneity of SWNT bundles for adsorption applications. The method integrates sample morphology, especially tube-diameter distribution and sample purity, into GCMC simulation of a small probe molecule, such as nitrogen at 77 K, for comparison with commonly used standard experimental adsorption measurements. The methodology has been developed only recently,[22] and successfully tested for compatibility with other nanotube-sorbate combinations besides nitrogen.[23] Here, we are presenting a broader application of our method to several new commercially available samples, incorporating sample characterization by Raman scattering with multiple laser energies, and corroborate the need to take into account sample purity into GCMC modeling of experimental samples. Following is a brief description of the integrated experimental-simulation strategy for modeling of heterogeneous SWNT bundles.

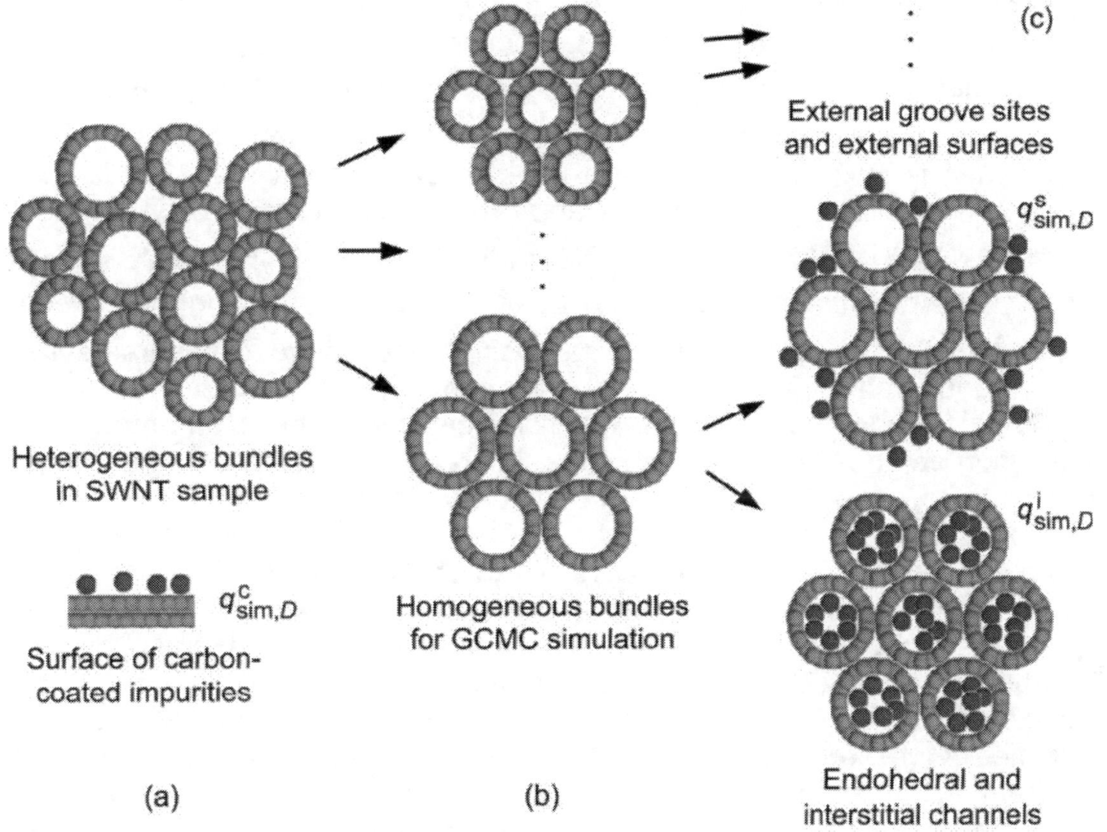

Figure 1. (a) Depiction of a typical heterogeneous SWNT bundle; the presence of impurities is approximated by an equivalent amount of planar nonporous carbon (q^{s}_{sim}). (b) Deconvolution of the heterogeneous bundle into several homogeneous bundles; the tube diameter, D, for each homogeneous bundle is determined from Raman scattering experiments. (c) GCMC simulation of adsorption on the external grooves and external surfaces (q^{s}_{sim}) and in the internal accessible volume (q^{i}_{sim}) of each homogeneous bundle. Total adsorption for the heterogeneous bundle is determined by weighted averaging of the individual contributions over the tube-diameter distribution obtained at each excitation energy λ.

GCMC calculations are carried out in three steps. *Firstly*, a heterogeneous bundle (Fig. 1a) is deconvoluted into several homogenous bundles (Fig. 1b). The tube-diameter distribution for a given SWNT sample can be obtained from the radial breathing mode (RBM) region of the spectrum at a given excitation energy λ, e.g. $\omega_{\mathrm{RBM}} = 12.5 + 223.5/D\,(\mathrm{nm})$ for λ = 785 nm, where *D* is the tube diameter and ω_{RBM} is the corresponding RBM frequency in the 100-300 cm^{-1} region. *Secondly*, each homogeneous bundle is further deconvoluted into two parts: the endohedral volume and the exohedral surface of the outermost SWNTs in the bundle (Fig. 1c). The former comprises adsorption inside the nanotubes and in the interstitial channels, whereas the latter represents adsorption in external groove sites and external surfaces, including the void space between the bundles. For each of these configurations, GCMC simulations of a small probe molecule are then carried out for comparison with experimental adsorption at same temperature and pressure conditions. For N_2 adsorption at 77 K, each nanotube can be approximated as a smooth structureless nanocylinder. The intertube distance for all simulation is kept fixed at 3.4 Å to mimic SWNTs adhering to each other via van der Waals forces forming bundles. Nitrogen is treated as a structureless spherical particle which interacts via dispersive forces only. The interaction between adsorbate molecules is modeled by a 12-6 Lennard-Jones (LJ) potential, $u_{ij} = 4\varepsilon_{ij}\left[(\sigma_{ij}/r)^{12} - (\sigma_{ij}/r)^{6}\right]$ (*r* is the intermolecular distance), as is the interaction between the carbon atoms of a nanotube and each adsorbate site. The well depths, ε_C/k_B and ε_{N2}/k_B (k_B is Boltzmann constant), used in calculations are 28.0 K and 100.4 K, respectively, and the collision diameters, σ_C and σ_{N2}, are 3.4 Å and 3.69 Å, respectively; the cross terms are obtained using standard Lorenz-Berthelot combination rules: $\varepsilon_{ij} = (\varepsilon_i \varepsilon_j)^{1/2}$ and $\sigma_{ij} = (\sigma_i + \sigma_j)/2$. The potential parameters for N_2 were obtained by fitting the simulated vapor-liquid coexistence curve to experimental data. Using these interaction parameters we have obtained very good agreement between the simulated N_2 adsorption isotherm and experimental data for nonporous carbon, which indicates that these parameters are a good first approximation to the adsorbate-nanotube interaction potential at the temperature of interest. The carbonaceous impurities are modeled as stacked graphene layers, with platelet spacing of 3.4 Å, using Steele's 10-4-3 potential. The GCMC simulations are carried out using established procedures;[25-29] the reader is referred to previous work[22-24,30] for further details. It should be pointed out that in our calculations, endohedral adsorption is expressed as amount adsorbed per unit weight (or volume) of bundle, whereas exohedral adsorption is calculated as amount adsorbed per unit external surface area of the bundle; if the latter value is multiplied by the specific external surface area of the bundle, S_p, the adsorption isotherm for the external surface becomes expressed per unit weight. Thus, the calculation of the total adsorptive contribution from the external surface of the bundle requires the estimation of S_p. In our previous work[22] a method for estimating S_p was developed. Briefly, plotting the experimental adsorption isotherm versus the simulated exohedral adsorption isotherm yields a curve, whose slope of the linear asymptote at high loading gives the value of S_p. *Thirdly*, once the endohedral and exohedral isotherms for each homogenous bundle have been calculated, they are weight-averaged according to the tube-diameter distribution obtained from the relative peak intensity at each RBM frequency. This provides theoretical isotherms, $q^{\mathrm{s}}_{\mathrm{sim}}$ and $q^{\mathrm{i}}_{\mathrm{sim}}$, incorporated with sample-specific heterogeneity:

$$\left\{q^{\mathrm{s}}_{\mathrm{sim}}, q^{\mathrm{i}}_{\mathrm{sim}}\right\} = \frac{1}{n_\lambda}\sum_{\lambda}\sum_{D} w_{\lambda,D}\left\{q^{\mathrm{s}}_{\mathrm{sim},D}, q^{\mathrm{i}}_{\mathrm{sim},D}\right\}, \qquad \sum_{D} w_{\lambda,D} = 1. \tag{1}$$

Here, $q^{s}_{sim,D}(P/P_o)$ and $q^{i}_{sim,D}(P/P_o)$ are the adsorption isotherms for the external surface (s) and internal volume (i) of an homogeneous bundle of tube diameter *D*, expressed as a function of relative pressure P/P_o; Σ_λ denotes summation over the n_λ excitation energies probed in the Raman scattering experiments; $w_{\lambda,D}$ is the tube-diameter distribution obtained at a particular excitation energy λ, which can be estimated from the relative ratio of peak heights at the corresponding ω_{RBM} in the Raman spectrum. The total adsorption isotherm for the sample, $q_{sim}(P/P_o)$, is then expressed as

$$q_{sim}(P/P_o) = \underset{\text{(endohedral)}}{(1-\eta)q^{i}_{sim}(P/P_o)} + \underset{\text{(exohedral)}}{S_p q^{s}_{sim}(P/P_o)} + \underset{\text{(impurities)}}{S_c \eta q^{c}_{sim}(P/P_o)} \quad (2)$$

where η is the weight-fraction of impurities in the sample, S_p is the external surface area of SWNTs in the sample (m^2/g); S_c is the surface area of impurities (m^2/g), and $q^{c}_{sim}(P/P_o)$ is the adsorption isotherm on nonporous carbon, which is assumed to be representative of the adsorptive contribution from carbon-coated impurities. In a related work,[24] it has been shown that most impurities present in SWNT samples can be modeled as the planar surface of a nonporous carbon, and S_c can be calculated by least-square fitting of Eq. 2 to the experimental adsorption isotherm. Notice that the first term on the r.h.s. of Eq. 2 is the total endohedral adsorption in the nanotube bundles; the other two terms account for adsorption outside of the internal accessible porous volume of the nanotubes.

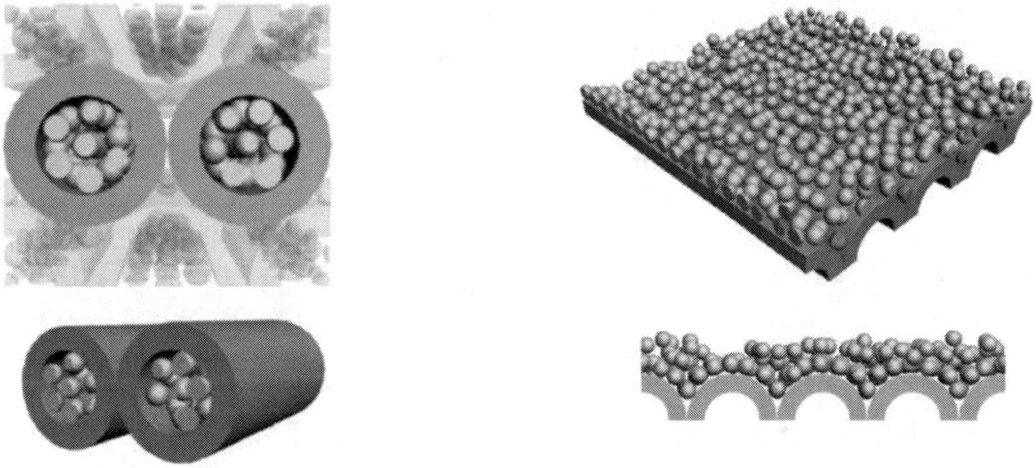

Figure 2. Snapshots of N_2 adsorbing on the inside (left) and outside (right) surface of homogenous arrays of 1.4 nm SWNTs at $P/P_o = 10^{-2}$ and T = 77 K.

Results and Discussion

The value of sample purity enters the GCMC calculations via the parameter η in eq 2. The experimental-simulation characterization method requires a reasonably good estimated of sample purity. Here, η was calculated from the purity value provided by the manufacturer. However, we specifically tested a > 95% pure sample by thermogravimetry. All samples were also analyzed for diameter distribution by Raman scattering at multiple excitation energies λ = 785 nm, 532 nm and 633 nm (Renishaw Raman Spectrometer, Configuration 1). These experiments were performed on solid samples. *D* was calculated by the relation, $\omega_{RBM} = 10 + 234.0 / D(nm)$ which accounts for the slight shift in RBM frequencies due to nanotubes present in bundles as opposed to dispersed individually. Figure below shows the results of sample characterization.

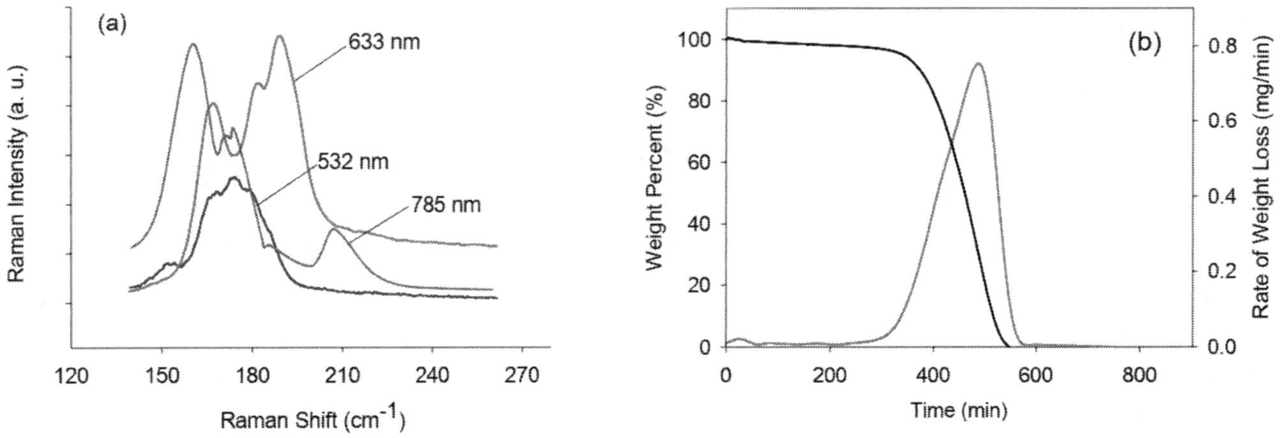

Figure 3. (a) RBM region of Raman spectra of the sample. (b) Sample purity analysis.

Total adsorption for a sample was estimated by Eq. 2. Figure 4 shows the total isotherm, $q_{\rm sim}(P/P_{\rm o})$, calculated for several samples. We are able to generate the characteristic type II adsorption isotherm for all samples. This isotherm shape is common for SWNT samples. It denotes the presence of mesoporosity (2 nm < pore size < 50 nm) in addition to microporosity (pore size < 2 nm). The typical diameter of an individual SWNT ranges between 0.7 nm to 2.0 nm. This is well within the microporous range; thus, the presence of any mesoporous characteristic would be counter-intuitive. The source for mesoporosity in SWNTs is the wide inter-bundle space between adjacent bundles; therefore, by calculating the external contribution, $q^{\rm s}_{\rm sim}(P/P_{\rm o})$, we are essentially modeling the adsorption in the void space between bundles. It is worth pointing out, however, that this mesoporous space has some microporosity attached to it, which is due to the external grooves. Once these sites get completely filled with adsorbate molecules, which happen at very low pressure, their influence is suppressed.

A striking feature is observed when the total adsorption isotherm determined by simulation, $q_{\rm sim}(P/P_{\rm o})$, is compared with the experimental one, $q_{\rm exp}(P/P_{\rm o})$. For all samples, the two isotherms are identical in shape, which corroborates our strategy of modeling heterogeneous bundles as smaller fractions of several homogeneous bundles. The simulated isotherms appear to be parallel to the corresponding experimental isotherms. However, the isotherms were found shifted vertically towards larger loading values, exhibiting more adsorption than experimentally measured at each $P/P_{\rm o}$. The explanation for this shift is as follows. Our calculations were carried out with the implicit assumption that all SWNTs are open or unblocked to allow unrestricted access for N_2 molecules. This assumption is not unrealistic given the fact that sonication and oxidation are routine practices in purification of as-produced nanotubes; these techniques are believed to cut the nanotubes to shorter lengths and oxidize their ends to allow easier access to their endohedral volume. Therefore, the simulated isotherm (left panel, Figure 4) and the corresponding surface area and pore volumes of any given sample represent its maximum achievable adsorption capacities and related physical properties.

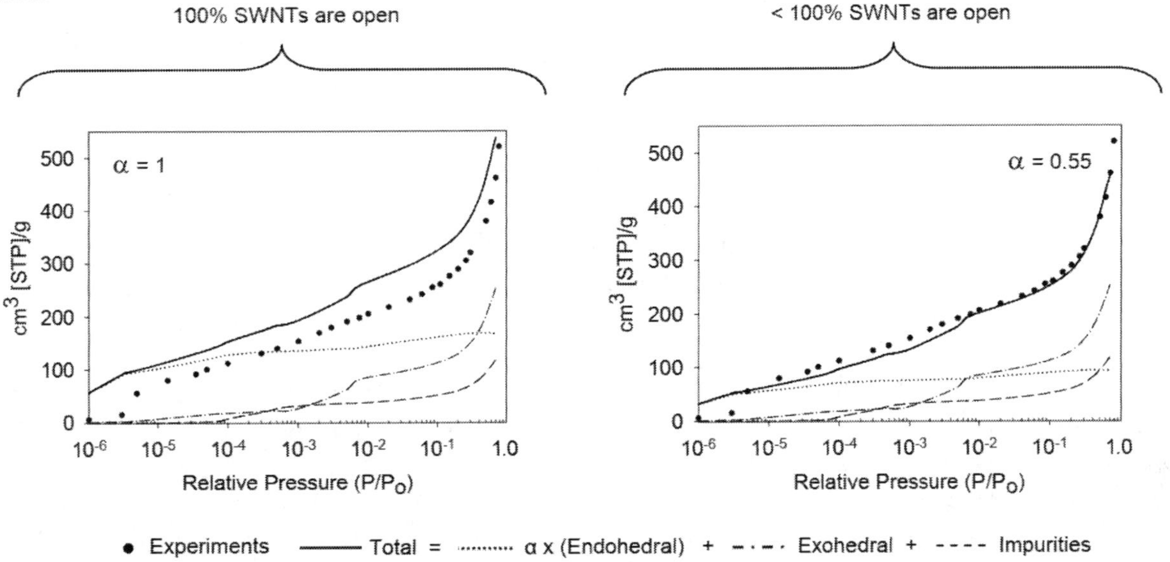

Figure 4. Experimental and simulated N_2 adsorption isotherms for sample a 70% pure SWNT sample as predicted by GCMC modeling of heterogeneous bundles. Left panel shows the isotherms for samples calculated with an implicit assumption that all nanotubes were open (α = 1). Right panel shows the same isotherms after endohedral adsorption was adjusted by a scaling parameter (α < 1.0) representing a fraction of open-ended nanotubes.

Volume Fraction of Open-Ended SWNTs

Assuming that not all SWNTs are open-ended or accessible for adsorption implies that the endohedral adsorption must be scaled down. Other adsorptive contributions remain unchanged, as adsorption on the external surface the bundles and on the impurities is unaffected by whether the nanotubes are open or closed. A scaling parameter, α, is introduced into our calculations, such that Eq. 2 is now modified as follows:

$$q_{\mathrm{sim}}(P/P_{\mathrm{o}}) = \underset{\text{(endohedral)}}{\alpha(1-\eta)q^{\mathrm{i}}_{\mathrm{sim}}(P/P_{\mathrm{o}})} + \underset{\text{(exohedral)}}{S_{\mathrm{p}}q^{\mathrm{s}}_{\mathrm{sim}}(P/P_{\mathrm{o}})} + \underset{\text{(impurities)}}{S_{\mathrm{c}}\eta q^{\mathrm{c}}_{\mathrm{sim}}(P/P_{\mathrm{o}})} \tag{3}$$

where the new parameter $0 \le \alpha \le 1$ is a structural property of the sample. It should be independent of *D* and P/P_{o} for most generality. It is particular only to a sample. This parameter can estimated by either trial-and-error to yield an isotherm, $q_{\mathrm{sim}}(P/P_{\mathrm{o}})$, that overlaps the experimental isotherm, or it can determined by least-square fitting of Eq. (3), or it can be taken as the ratio of the experimental micropore volume of the sample and the micropore volume calculated from simulations. The three methods result in comparable values. If the endohedral isotherm, $q^{\mathrm{i}}_{\mathrm{sim}}$, is scaled by $0 \le \alpha < 1$, it results in a near perfect fit between the total simulated isotherm and the experimentally measured one, for all samples (right panel, Figure

4). The surface area and pore volumes were recalculated, and were now comparable to the corresponding experimental values, which corroborates an accurate estimation of the *volume fraction of open-ended nanotubes* and the applicability of this concept to a variety of SWNT samples. Here, we define α as the *volume fraction of open-ended nanotubes*:

$$\alpha = \frac{q^{i}_{\exp}(P/P_o \rightarrow 1)}{(1-\eta)q^{i}_{\text{sim}}(P/P_o \rightarrow 1)} \tag{4}$$

Conclusions

We have developed an integrated experimental-simulation characterization technique for practical modeling of heterogeneous bundles of SWNTs for adsorption applications. Our approach resulted in a perfect reproduction of experimental N_2 adsorption isotherms at 77 K for several commercially available SWNTs. Typical experimental isotherms only provide total adsorption capacities without differentiating between adsorption on the various structural components of a sample. With our technique, we are able to deconvolute experimentally observed adsorption capacities into several components and accurately quantify adsorption inside the nanotubes, adsorption on the external surface of the bundles and adsorption on impurities. We are also able to estimate the fraction of open-ended nanotubes in several samples. It is a structural property which is specific to a sample. The fraction of open-ended SWNTs cannot be estimated by typical analytical techniques used in characterization of carbon nanotubes. Therefore, adsorption could become an appropriate method to quantify this structural property of nanotube samples, much like using gas adsorption to estimate the surface area of a material. This new information should convey that many advances are still needed in synthesis and processing of carbon nanotubes. It is also essential for a fair assessment of adsorption capabilities and limitations of carbon nanotubes for any practical use in adsorption and gas storage applications in competition with less expensive and more versatile activated carbons.

References

Striolo, A.; Chialvo, A. A.; Cummings, P. T.; Gubbins, K. E., *J. Chem. Phys.* 2006, 124.

Zhao, M. W.; Xia, Y. Y.; Ma, Y. C.; Ying, M. J.; Liu, X. D.; Mei, L. M., *Phys. Rev. B* 2002, 66.

Kuzmany, H.; Kukovecz, A.; et al., *Synthetic Metals* 2003, 135, 791-793.

Shi, W.; Johnson, J. K., *Phys. Rev. Lett.* 2003, 91.

Agnihotri, S.; Mota, J. P. B.; et al., *Langmuir* 2005, 21, 896-904.

Agnihotri, S.; Mota, J. P. B.; et al., *J. Phys. Chem. B* 2006, 110, 7640-7647.

Agnihotri, S.; et al., *Carbon* 2006, 44, 2376-2383.

Agnihotri, S.; et al., *J Phys Chem. C* 2007 (in press).

New method to determine micropore volume and surface area of single-walled carbon nanotubes

Yijing Zheng,[1] *Pyoungchung Kim,*[1] *Sandeep Agnihotri,*[1,*]

[1] *Department of Civil and Environmental Engineering, University of Tennessee, Knoxville, TN 37996*

Abstract

We have developed a microporosity and surface area estimation method that is specific to single-walled carbon nanotubes (SWNTs). Typically, micropore volume of adsorbents can be determined by the t-plot method, where the statistical thickness, *t*, of adsorbed layer is a function of the adsorbed amount, and is estimated by several equations such as de Boer equation, Halsey equation and Carbon Black equation. The t-plot method is one of the most common and versatile method. The estimated micropore volume is often specific to a model and to the segment of adsorption isotherm that is fitted to that model. SWNTs are relatively new adsorbents because of which applicability of these models to carbon nanotubes needs to be evaluated. We have developed a technique that compares experimental N_2 adsorption isotherm (77 K) of a sample with grand canonical Monte Carlo (GCMC) simulation isotherms of N_2 on the external surface of SWNT bundles ($10^{-6} < P/P_o < 0.99$). In this method, we plot the total experimental adsorption capacity vs. simulated external adsorption capacity from SWNT bundles. The plot is more or less a straight line for adsorption at $P/P_o \geq 10^{-3}$. The slope and intercept of the linear asymptote of the curve are interpreted as the external surface area of the bundles and the available micropore volume, respectively. We are also able to incorporate adsorption from impurities by modeling impurities as planar carbons and adding it to the total non-endohedral contributions. The main advantages of our method is that it is specific to SWNTs and, unlike other methods, it includes the entire experimental adsorption isotherm ($10^{-6} < P/P_o < 0.99$) in analysis, thus minimizing any experimental errors related to selecting an appropriate segment of an experimental isotherm to fit the data.

Introduction

The unique microporous structure of single-walled carbon nanotubes (SWNTs) is speculated to have remarkable adsorption properties. Theoretical studies have shown that adsorption surface area of arrays of SWNTs could be as high as 3,000 m^2/g. However, the reported BET surface area of nanotubes typically ranges between 150 m^2/g to 1587 m^2/g, which depends on the adsorbate gas used for analysis and the specific structure and composition of samples tested. Nevertheless, it is known that the adsorption surface area and porosity results of nanotubes are usually specific to a particular study and are generally not reproducible by other researchers. As a relatively new adsorbent, the pore volume and surface area of SWNTs needs to be evaluated. Several methods exist to determine the surface area and pore volume of adsorbents. V-T method is one of the most commonly used methods. It requires estimating the thickness of an adsorbate layer, such as N_2, on the adsorbent which is typically referred to as statistical thickness, *t*. The *t* values are calculated from relative pressure of the adsorbate using three different equations, such as the de Boer equation (eq.1), Carbon Black equation (eq. 2) and the Halsey Equation (eq. 3) presented below.

$$t(\overset{o}{A}) = \left[\frac{13.99}{log(P_o / P) + 0.034}\right]^{1/2} \tag{1}$$

$$t(\overset{o}{A}) = 0.88\,(P / P_o)^2 + 6.45\,(P / P_o) + 2.98 \tag{2}$$

$$t(\overset{o}{A}) = 3.54\left[\frac{5}{2.303\,log(P / P_o)}\right]^{1/3} \tag{3}$$

The t-plot equations are applicable to the adsorption data collected in the relative pressure range of $0.05 < P/P_o < 0.35$, where P is the actual vapor pressure and P_o is the saturation vapor pressure (= 760 torr for N_2 at 77 K). This pressure range is exactly the same as that used in estimation of the multi-point BET surface area. In the V-T plot, the adsorption capacity is plotted as the ordinate and t is plotted as the abscissa. Typically, the plot yields a straight line the intercept and slope of which are interpreted as the micropore volume and the external surface area, respectively.

$$V_{MP} = i \times 0.001547 (cm^3) \quad \text{and} \quad S_t\,(m^2/g) = 15.47\,x\,s$$

where, i is the intercept (cm^3[STP]/g) and *s* is the slope of the straight line fitted to the V-T plot. The V-T method was applied to two SWNT samples of > 95 wt% and 70 wt% purity. The samples are commercially available and procured from MER Corp and Carbon Solutions, respectively. We found that the micropore volume and external surface area of the samples were specific to the equation used in the estimation of *t*. Furthermore, the V-T plot did not always follow a straight line in the suggested pressure range of $0.05 < P/P_o < 0.35$. Representative values are provided in Table 1 and Figure 1. These results demonstrate the methods used for physical characterization of common adsorbents may not be applicable to carbon nanotubes.

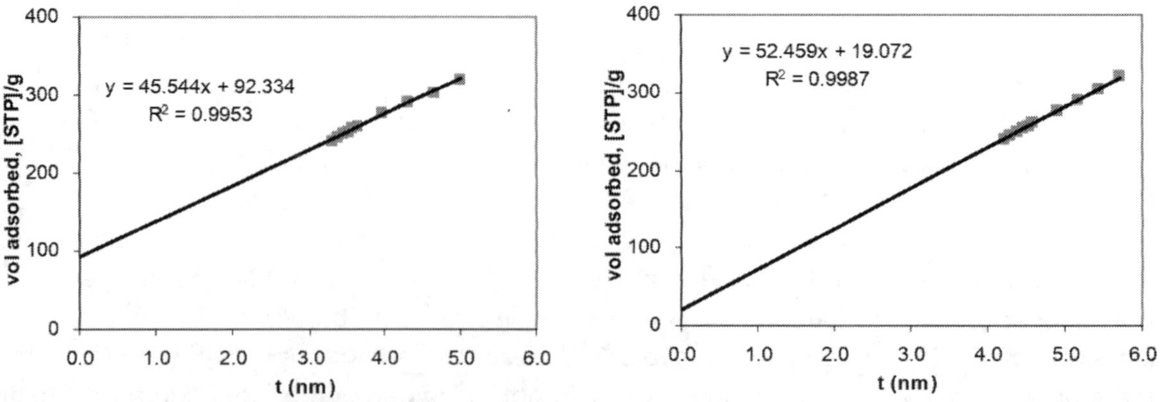

Figure 1. V-T plot for 70 wt% SWNT sample where t is estimated by Carbon Black equation (a) and Halsey equation (b).

Table 1. Micropore volume (cm^3/g) and external surface area (m^2/g) of SWNTs determined from three equations commonly used for calculating *t* in the V-T plot.

Sample	Micropore Volume (cm^3/g)			External Surface Area (m^2/g)		
	Eq. 1	Eq. 2	Eq. 3	Eq. 1	Eq. 2	Eq. 3
> 98% SWNT	0.155	0.151	0.101	304	314	359
70% SWNT	0.151	0.143	0.030	683	706	808

We have developed a new method of calculating the statistical thickness of N_2 at 77 K to estimate the micropore volume and external surface area of SWNT samples. The method was originally developed by Agnihotri et al., (2005) as part of practical modeling of the heterogeneity of SWNT bundles for adsorption applications. Here, we are presenting the application of this method to new samples, and incorporation of adsorption on impurities into the calculation of the statistical thickness.

Description of New Method

1. Calculation of Statistical Thickness by Molecular Simulation

Our method requires estimation of N_2 adsorption on the external surface of the SWNT bundles. This was performed by grand canonical Monte Carlo (GCMC) simulation (Figure 2).

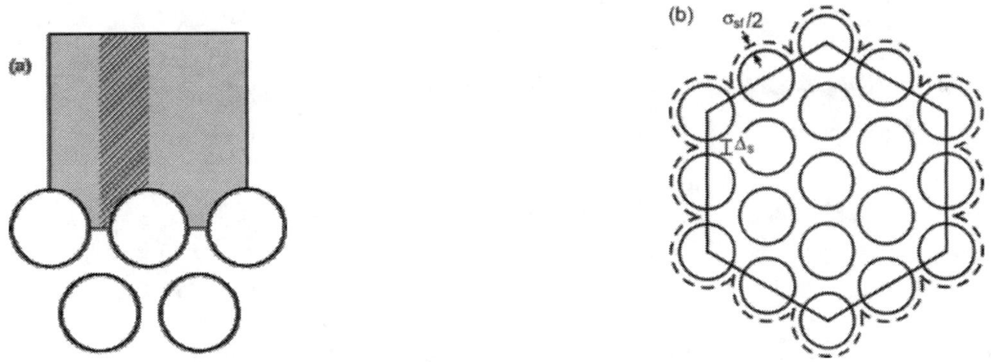

Figure 2. (a) Cross section of the unit simulation box for GCMC study of N_2 adsorption on the external groove sites and external surface of SWNT bundles at 77 K. (b) We define the external surface area of a bundle as the area of the plane boundaries of the prism with vertices on the center of the outermost shell of nanotubes (solid line). This is an unambiguous definition of external surface that is independent of sorbate size. It can be converted to that represented by the dashed line using the conversion formula $S_p' = S_p(\pi/2 - \theta)/\cos\theta$, where $\cos\theta = (D + \Delta_s)/(D + \sigma_{sf})$, *D* is the tube diameter, Δ_s = 3.4 Å is the intertube distance, and σ_{sf} is the collision diameter for sorbate-carbon interaction.

Each nanotube can be approximated as a smooth structureless nanocylinder at the temperature of interest. The intertube distance for all simulation is kept fixed at 3.4 Å to mimic SWNTs adhering to each other via van der Waals forces forming bundles. Each pseudoatom of the adsorbate molecule is treated as a structureless spherical particle which interacts via dispersive forces only. The interaction between adsorbate molecules is modeled by the 12-6 Lennard-Jones (LJ) potential. The cross terms are obtained using the standard Lorenz-Berthelot combination rules. For modeling of N_2 adsorption at 77 K, the well depths $\varepsilon_{C\text{-}C}/k_B$ and $\varepsilon_{N2\text{-}N2}/k_B$ used in calculations are 28.0 K and 100.4 K, respectively, and the collision diameters $\sigma_{C\text{-}C}$ and $\sigma_{N2\text{-}N2}$ are 3.4 Å and 3.69 Å, respectively. The GCMC simulation is carried out using established procedures. In normal GCMC, insertions are attempted uniformly throughout the volume of the simulation box. For nanotube bundles, however, it is known a *priori* that much of this volume is filled by carbon atoms. Furthermore, not all portions of the accessible space are equally favorable; there exist preferred regions in which sorbate molecules are localized. This information can be incorporated into a GCMC simulation if the insertions are not attempted randomly throughout the volume. This is achieved by using configurational-bias techniques. Due to the bias introduced into the MC insertion moves, the acceptance rules for insertions and deletions are altered to ensure that microscopic reversibility is satisfied and that the GC ensemble is still correctly sampled. The method has been described in detail by Agnihotri et al., (2005). Following is a brief description of their results.

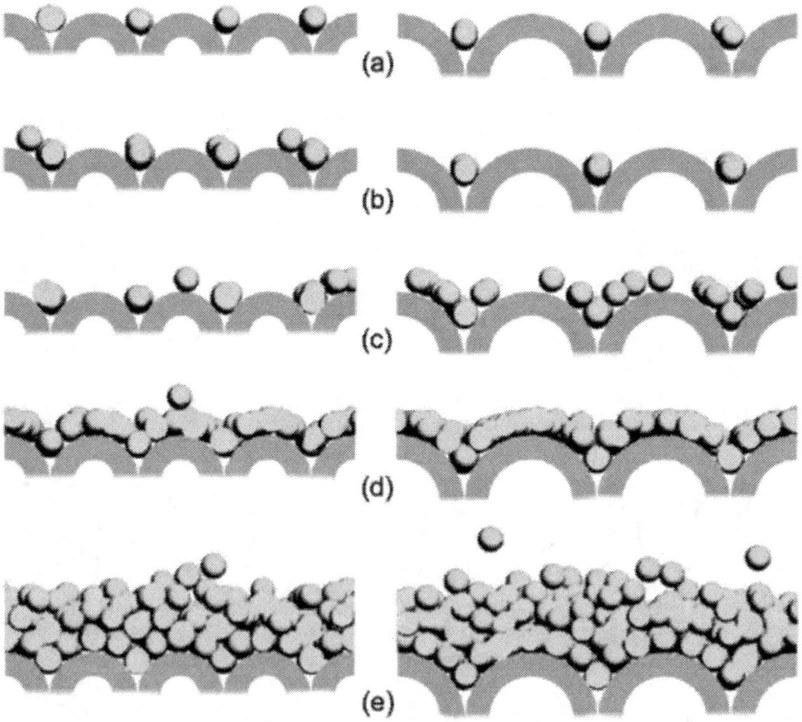

Figure 3. Snapshots of the simulation box for N_2 adsorption at 77 K on the external surface of homogenous SWNT bundles of tube diameter D = 9.0 Å (left) and D = 15.2 Å (right): (a) $P/P_0 = 10^{-6}$; (b) 10^{-4}; (c) 10^{-3}; (d) 10^{-2}; and (e) 0.9.

Simulations revealed that the peripheral grooves on nanotube bundles play an important role in adsorption on the surface of nanotubes (Figure 3), especially at low vapor pressures. Adsorption was observed at pressures as low as 10^{-6} p/p_o. At such pressures, some nitrogen molecules were noticed to adsorb on the external grooves sites. Adsorption continued on the grooves till they were saturated at about 10^{-4} p/p_o. Increasing the vapor concentration above 10^{-4} p/p_o resulted in adsorption on the curved surface of the nanotubes. At 10^{-3} p/p_o, partial coverage of the external surface was observed. Further increase in vapor concentration till 10^{-2} p/p_o resulted in complete monolayer formation following which adsorption proceeded rapidly with increasing vapor concentration. The mechanisms of nitrogen molecules adsorbing on the external surface of nanotubes are illustrated in Figure 3. It was also observed that the trends in external adsorption were independent of the nanotube size. Adsorption on an array of 15.2 Å diameter nanotubes proceeded in more or less the same fashion as that on 9.0 Å wide nanotube array, which showed that nanotube size has little or no effect on the total external adsorption capacity.

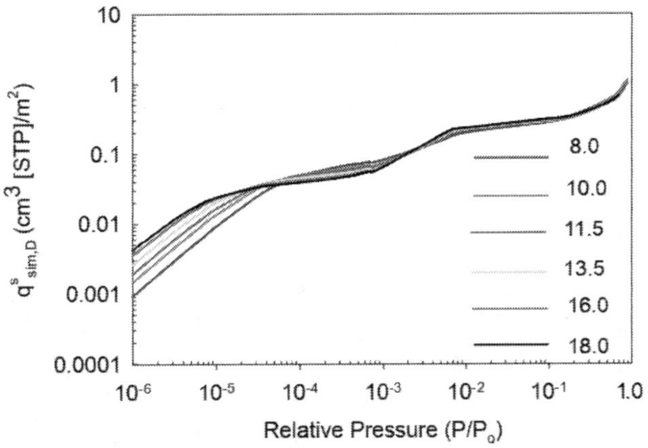

Figure 4. Exohedral N_2 adsorption isotherms at 77 K calculated for homogenous bundles of nanotubes sizes typically encountered in experimental samples.

The insensitivity of nanotube diameter towards the external adsorption capacity was evident from the isotherms (Figure 4). The isotherms followed similar trends for all nanotube sizes with minimal but perceptible adsorption till $\sim 10^{-3}$ p/p_o (< 10% of that at near saturation). The adsorption capacities of all arrays exhibited indistinguishable differences with respect to the nanotube diameter. Although, it appeared that at concentrations lower than 10^{-5} p/p_o, wider nanotubes had higher external adsorption capacities, the differences in absolute values of adsorption were too small to reach any such conclusion. Thus, it would be reasonable to say that adsorption on the external surface of SWNTs is insensitive to the diameter. The average of these values is presented hereafter as the statistical thickness, *t*.

2. Estimating Micropore Volume and External Surface Area

2a. Pure Sample

The statistical thickness calculated by simulations was then plotted against the adsorption capacity in a manner similar to the V-T plot. The external surface area of > 95 wt% SWNT sample was determined (Figure 5). The plot is interpreted to comprise of two regions: the first region shows a steep rise and the second region that nearly follows a straight line. The first region of the curve indicates that the experimental adsorption capacity surpassed the simulated values, which means that the internal adsorption in the sample was much higher than the adsorption on its external surface. The second region of the curve implies that the experimental adsorption capacity was linearly proportional to the simulated external adsorption capacity or, in other words, majority of the adsorption in this pressure range occurred on the external surface of the bundles. The intercept of the straight line through the linear part of the curve provides the micropore volume of the sample because the amount adsorbed at zero surface loading is adsorbed entirely inside the pores. Additionally, the slope of the second region of the curve, thus, represents the total external surface area of nanotubes in the sample. Furthermore, the relative pressure at the point of inflexion in the curve (i.e., meeting point of the two regions of the curve) is indicative of the maximum pressure above which the internal volume of nanotubes will be filled. We found that the external surface area of the sample was 160 m^2/g and the micropore volume was 0.16 cm^3/g.

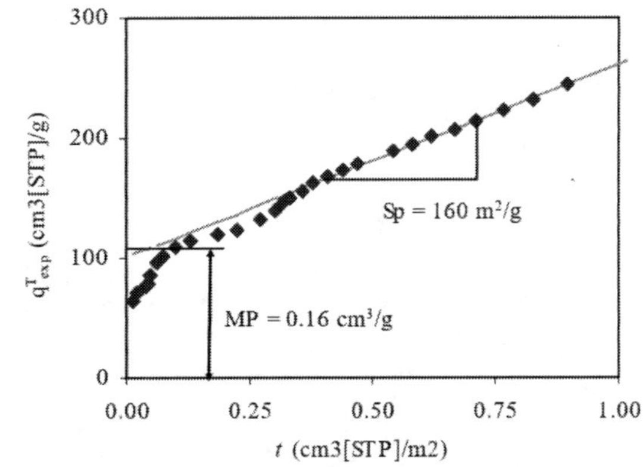

Figure 5. Total experimental adsorption capacity vs. surface adsorption capacity for 95 wt% pure SWNT sample. The intercept and slope of the straight line represents micropore volume and total external surface area of nanotubes of the sample.

2b. Impure Sample

The abovementioned approach was applied to a 70 wt% pure SWNT sample. The plot followed a straight line at $t > 0.3$ cm^3 (STP)/g. The intercept and slope of the line in this region could be taken as the micropore volume and the external surface area as presented in Figure 6a. However, when the fitted line was extrapolated to the entire data in the plot, it was observed that certain data points were above the line. This essentially means that the total adsorption on the sample was more than predicted by a simple directly proportional relationship with adsorption on the external surface of the bundles. We believe that this is because of the fact that this sample contains a substantial amount of impurities which are contributing to adsorption but are not accounted in the present analysis. Therefore, the statistical thickness (i.e., X axis) would need to be corrected for adsorption on impurities. Agnihotri et al., (2006) showed that for adsorption purposes impurities in SWNT samples can be modeled as planar carbon. They calculated the N_2 adsorption capacity of planar carbons using the GCMC technique described in the previous section. A representative figure of the N_2 isotherm and a snapshot is presented in Figure 7.

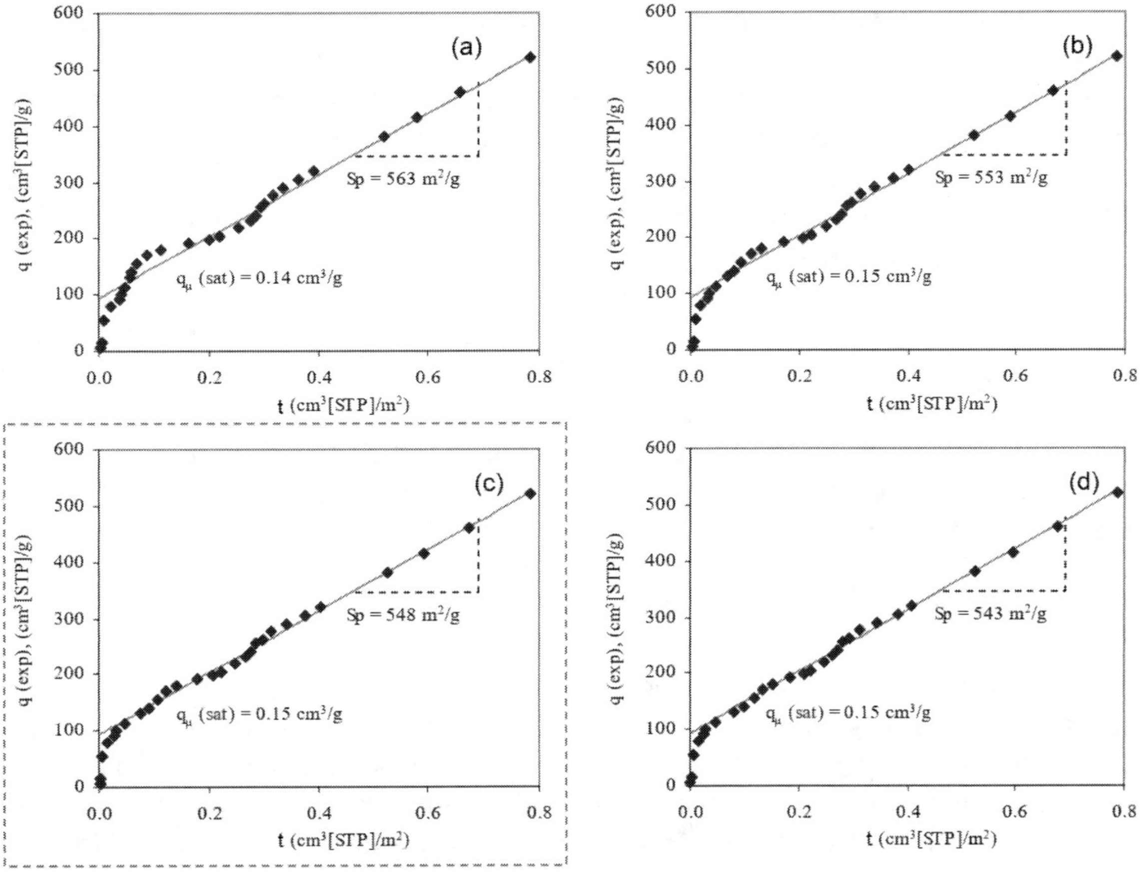

Figure 6. Total experimental adsorption capacity vs. simulated surface adsorption capacity for impure sample with f = (a) 1.0 (b) 0.8 (c) 0.7 and (d) 0.6. The intercept and slope of the straight line represents micropore volume and total external surface area of the sample.

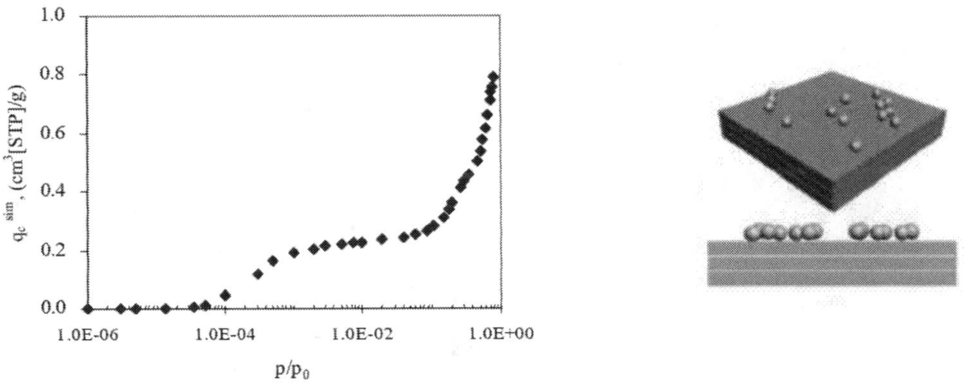

Figure 7. GCMC N_2 adsorption isotherm (left) at 77 K for a planar graphitic surface (right).

Therefore, the true statistical thickness can now be represented as a combination of adsorption on the external surface of the bundles and that on a planar carbon (Eq. 4).

$$t = f\,(q^{s}_{sim})_{SWNTs} \quad + \quad (1\text{-}f)\,(q^{c}_{sim})_{impurities} \tag{4}$$

Here, the parameter *f* is a floating parameter. It can be determined by trial-and-error. To find the correct value, we first started with an extreme case of *f* = 1 which resulted in an obvious overshoot of the curve same as that in Figure 6a. Then we selected a smaller value of *f* = 0.8 which seems to flatten the curve, thus reducing the difference between experimental values and fitted line representing the direct proportionality with the statistical thickness, *t* (Figure 6b). This trend seems to be complying with our hypothesis which forms the basis for Eq. 4. We then further reduced the value of *f* to 0.7 at which the fitted line superimposed the experimental data with no observable differences (Figure 6c). A further reduction in the value of *f* resulted in no appreciable changes in the data and the fitted line (Figure 6d). It was also noticed that the micropore volume and the surface area decreased slightly with decreasing values of *f*. We noticed that the *f* value at which the overshoot is nullified (Figure 6c) also corresponded to the purity of the sample when represented as a fraction (i.e., *f* = 0.7 for 70 wt% pure sample). Therefore, our method can be applied to a variety of samples with varying degrees of purity.

Summary and Conclusions

We are reporting a method for estimating the microporosity and external surface area of single-walled carbon nanotubes (SWNTs). The method is similar to the V-T plot approach; however, here the statistical thickness, t is estimated by GCMC simulations of N_2 at 77 K for an external corrugated surface of nanotubes. Furthermore, impurities could be modeled as planar carbon and their adsorptive contributions could be added into the statistical thickness. We plot the total experimental adsorption capacity vs. adjusted statistical thickness. The plot is more or less a straight line for adsorption at $P/P_o \geq 10^{-3}$. The slope and intercept of the linear asymptote of the curve are interpreted as the external surface area of the bundles and the available micropore volume, respectively.

424a Development of a Hybrid System to Improve the Anti-Fouling Performance of the Uf Membrane during the Specific Industrial Wastewater Treatment

Xi Wei, Hydrochem (S) Pte Ltd, a Hyflux Group Compan, 202 Kallang Bahru, Singapore 339339, Singapore, Singapore and **Tai-Shung Chung**, Chemical and biomolecular Engineering, National university of Singapore, 10 Kent Ridge Crescent, Singapore 119260, Singapore, Singapore.

Various foulants present in the industrial wastewater and affect the UF membrane filtration performance. To improve the UF membrane anti-fouling performance, it is important to get rid of the major foulants before the UF membrane filtration. In this study, an UF membrane which was the Hyflux newly developed UF Kristal 600 membrane was applied to three kinds of industry wastewater: the pulp and paper wastewater, the second effluent of industrial water and the mixture of sea water and biological wastewater. The Characteristics of these industrial wastewaters will be given during the presentation. The bacterial, organic matter and multivalent ions were the main foulants presented in these wastewaters, different in concentrations. As a result, this wastewater exhibited serious fouling tendency. The objectives of this study were 1) to filtrate the wastewater by means of the Hyflux newly developed UF Kristal 600 membranes and to specify the main foulants during the filtration, 2) to optimize the operating parameters to provide an efficient filtration process, and 3) to develop a hybrid system to prevent the high membrane fouling tendency and keep the UF membrane perform better anti-fouling performance. To achieve these targets, a pilot testing system was built on-site, and the membrane fouling studied was reported in this paper. It was found that bio-fouling and organic fouling were the main factors to affect the cake layer formation during the UF filtration process. To prevent this kind of fouling action occurred, the UF operating parameters were optimized and a hybrid system was set up. The foulants were removed step by step during the hybrid system and the whole system performed to produce quality water to the RO system at a high flow rate. It has been proved that a right filtration system could keep the UF membrane exhibit higher filtration performance constantly.

424b Hybrid Chitosan/poly(Ethylene Glycol) Hydrogels for Use as Highly Anti-Fouling Ultrafiltration Membrane Coatings for Synthetic Oily Water Treatment

Bryan D. McCloskey, Hao Ju, Ho-Bum Park, Victor A. Kusuma, and Benny D. Freeman. Chemical Engineering, The University of Texas at Austin, 10100 Burnet Rd. Bldg 133, Center for Energy and Environmental Resources, Austin, TX 78758

Membrane technology provides an attractive solution to wastewater treatment based on their relative versatility, small environmental and spatial footprint, and inherent economic advantages over other alternatives. However, the largest barrier to widespread use of water purification membranes is fouling, which is the deposition of matter in a membrane's pores (internal fouling) or on its surface (external fouling) that leads to a change in a membrane's flux characteristics, such as throughput and selectivity. Many different techniques can be used to combat membrane fouling, most of which focus on two general areas: introducing fluid instabilities in the feed stream, such as dean vortices, air sparging, and backflushing, and altering the surface properties of the membranes, either through surface grafting or coating. Although feed flow instabilities have been shown to increase flux in some MF and UF membrane applications, fouling is still a major concern, especially when internal fouling is the dominant fouling mechanism. Therefore, coating a thin, defect-free hydrophilic polymer on the surface of a porous membrane (which is usually made of a hydrophobic polymer) may provide the most effective way to eliminate internal membrane fouling while greatly reducing external fouling. A hydrophilic polymer coating, when made sufficiently thin, can also allow a high water throughput to maximize flux.

Here, we report the synthesis of a series of novel hydrophilic coating materials prepared by crosslinking chitosan with a bifunctional poly(ethylene glycol) macromer. The water permeability and chemical stability of these polymers have been found to be satisfactory for application as a potential membrane coating. Furthermore, the molecular weight cutoff (MWCO) using poly(ethylene glycol) feed solutions was found to be as low as 720 for these polymers, and generally varied inversely to the film's water permeability. The low MWCO indicates that high rejection can be expected for most organic contaminants, including emulsified oil droplets and most proteins. From a practical point of view, members of this polymer series were coated onto commercially-available polysulfone membranes to investigate anti-fouling characteristics using synthetic oily water. These composite membranes exhibited water flux values more than 3 times higher than uncoated membranes after one day of oily-water crossflow filtration, indicating that the hydrophilic polymer coating can significantly enhance anti-fouling properties. The organic rejection of the coated membranes was also higher than that of the unmodified polysulfone membranes.

424c Carrier Selection To Enhance Heavy Metal Ions Transmembrane Flux Through A Supported Liquid Membrane System

Junwen Lv, Qian Yang, Jianwen Jiang, and T.S. Chung. Chemical and Biomolecular Engineering, National University of Singapore, Kent Ridge, Singapore, 119260, Singapore

Extraction capabilities of three kinds of carriers, Aliquat 336, Kelex 100, LIX 54, for cadmium removal have been investigated in supported liquid membrane (SLM) systems. The maximum fluxes obtained using Aliquat 336, Kelex 100 and LIX 54 are1.12x10-9, 1.5x10-10 and 7.9x10-11 mol/(cm2•s), respectively. Theoretical calculation of these three carriers' extraction capabilities for cadmium in supported liquid membrane (SLM) systems using the quantum chemical computation method has also been carried out. The single point energy calculation results show that the energy changes in the complex formation process are in the order of Aliquat 336/Cd(II) > Kelex 100/Cd(II) > LIX 54/Cd(II), with energy changes of -1200.30, -329.19 and 4944.77 kcal/mol, respectively. This indicates the quantum chemical computation supports the experimental results well and can be proposed as an effective tool for carrier selection in the SLM system. FTIR results also agree with the computational results quite well. Investigation on the influence of stirring rate and strippant on the cadmium flux reveals that a stirring rate of 400 rpm and the use of 1 mM EDTA as the strippant constitute the optimal experimental conditions. It was also found that cadmium flux is a function of feed concentration at the low concentration stage and the cadmium flux is enhanced by appropriate addition of certain anion into the feed.

SURFACE MODIFICATION OF POLYETHYLENE HOLLOW FIBER MEMBRANE BY THE GRAFTING OF 2-HYDROXYETHYL METHACRYLATE

Jian Pan, Cuixian Chen, Yangzheng Lin, Jiding Li*
(Department of Chemical Engineering, Tsinghua University, Beijing, PR China)

Introduction

Polyethylenes (PE) are commercially available product, which occupy one-third plastic production amount. Polyethylene possesses excellent chemical stability and heat-insulation. Using melt-spinning stretching technique is an economic, convenient and cleaning route to produce mocroporous hollow fiber membrane. Polyethylene microfiltration membrane, when used in wastewater treatment, has better long-term stability than cellulose acetate and mixed cellulose membrane, even its low cost. However, because of its hydrophobic characteristics and low water absorbency, it is prevented from many purposes when used in cellulose and microfiltration membrane forms. So, when used in wastewater treatment, PE membranes must be processed by hydrophilic modification.

Numberous methods have been developed to modify polymer surfaces. Among them, corona treatment is widely used in industry. Corona discharge treatment results in an increase in surface energy by introducing polar groups on the surface, thus improving their adhesion and wetting properties. Oxidation plays a very important part in corona discharge treatment under different working gases to enhance a good adhesion and it is the main chemical mechanism for corona treatment. Corona treatment has a tendency to degrade over time. Aftertreatment, which may introduce polar groups to the macromolecular chains, is considered an effective way to reduce the aging effect. Among them, graft polymerization is mostly used in polymer surface modification.

In this work, Surface modification of PE hollow fiber microporous membrane was performed by the grafting of 2-hydroxyethyl methacrylate (HEMA) via corona treatment in order to improve the membrane hydrophilicity. Changes on the membrane surface were characterized by X-ray photoelectron spectroscopy (XPS), attenuated total reflection-Fourier transform infrared spectroscopy (ATR/FT-IR) and scanning electron microscopy (SEM). Moreover, the relationship of the treatment conditions and grafting degree of HEMA on the membrane surface was also studied.

Materials and Methods

Materials

PE hollow fiber microporous membrane was made by melt- spinning stretching technique. PE memberane was ultrasonic cleaned in absolute ethyl alcohol and dryed before treated with corona processing equipment (SDCD 16-2-10) for some time. Put the corona treated memberane into HEMA solution, then added a little $Fe(NH_4)_2(SO_4)_2$ solution, keep the grafting reaction in a nitrogen environment in some constant temperature. When the reaction finished, the membrane was cleaned with deionized water of 70°C, and then dried for measurements.

Corresponding author. Tel. +86 10 62782432; Fax: +86 10 62770304
E-mail: cxchen@mail.tsinghua.edu.cn

Measurements

Evidence for grafting was obtained from X-ray photoelectron spectroscopy (Perkin-Elmer Instruments, U.S.A.) and Fourier transform infrared (Nicolet Magna-IR spectroscopy 550, Japan). The morphology of membrane was observed by scanning electron microscopy (JEOL, JSM 7401). Differential scanning calorimeter (Seiko DSC 6200) was used to calculate the crystallinity of the membrane according to the following equation:

$$X_C = \frac{\Delta H}{\Delta H_C} \quad (1)$$

where the Xc is the calculated crystallinity of the membrane, ΔH is the measured melt-enthalpy, and the ΔHc is the melt-enthalpy of 100% crystalline PE. In this article, the ΔHc is considered 293.0 (J/g) according to reference. The heating scan up to 150°C was followed by cooling to 50°C at a 10°C/min scanning rate. Grafting degree was calculated according to the following equation:

$$\eta = \frac{m_g - m_0}{m_0} \times 100\% \quad (2)$$

where the η is the grafting degree, m_g is the weight of treated membrane, and the m_0 is the weight of original membrane.

Results and Discussion

Effect of Corona Treatment and Graft Polymerization

The ATR-FTIR spectra of PE membrane, showed in Fig.1, give clear dvidence of the reaction of HEMA and PE. The new peak at 1500~1730 cm^{-1} and 1180 cm^{-1} absent in PE is observed as a result of –C=O and –C-O, which were introduced to the surface of PE membrane by corona treatment and grafting polymerization. In addition, the peak intensity confirmed that the –C=O and –C-O were invited into the surface of membrane after corona treatment and enhanced after the following grafting polymerization.

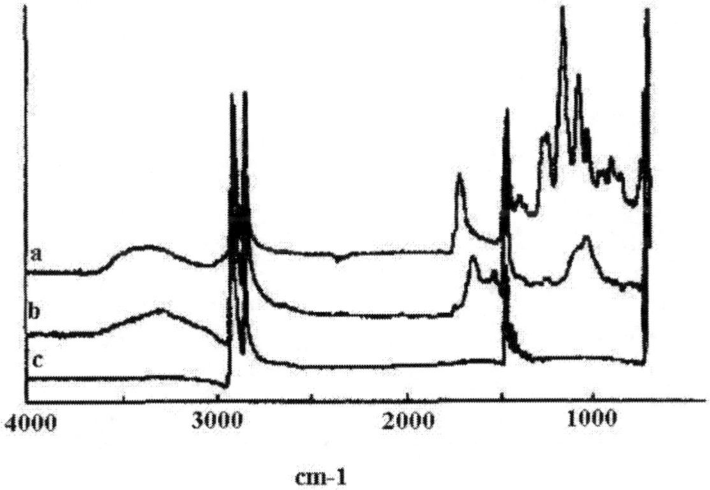

a. grafted b. corona treated c. original

Fig.1 ATR-FTIR spectra of PE membrane

Fig.2~Fig.4 show the XPS spectra of PE membrane before and after treatment. In the XPS spectra of the original membrane, only peak of C_{1s} (284.7eV) and weak O_{1s} (535eV) can be observed. After corona treatment, the peak of O_{1s} around 533.8eV turned out obviously showed in Fig.3, which indicate that the groups with oxygen have been invited to the surface of PE membrane. According to the XPS spectra of treated membrane (Fig.3, Fig.4), the main bonds are -C-C-, -C-O- and -C=O, this well matches the ATR-FTIR analyses. O_{1s}/ C_{1s} can characterize the oxygen content in the membrane surface. From Fig.3 and Fig.4 we can found that more -O-C=O turned out after graft polymerization.

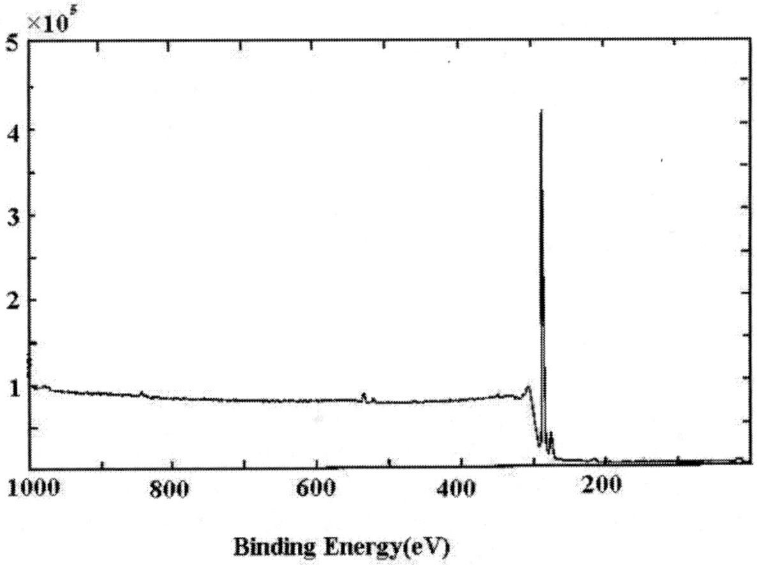

Fig.2 XPS spectra of the original membrane

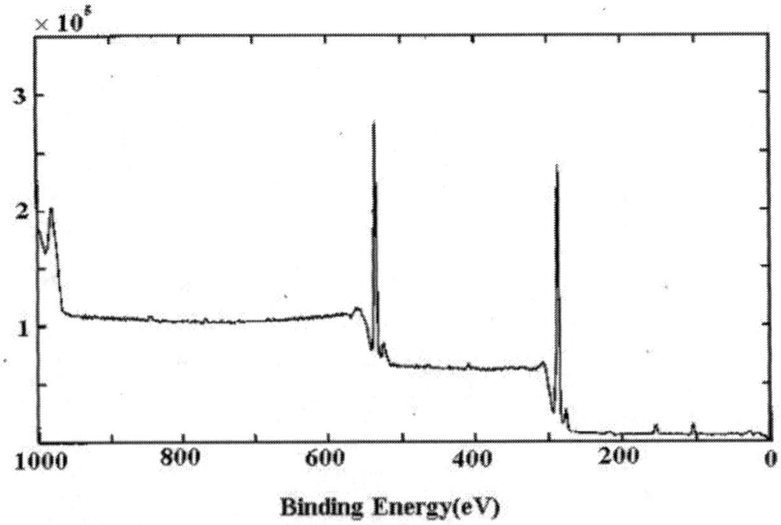

Fig.3 XPS spectra of the corona treated membrane

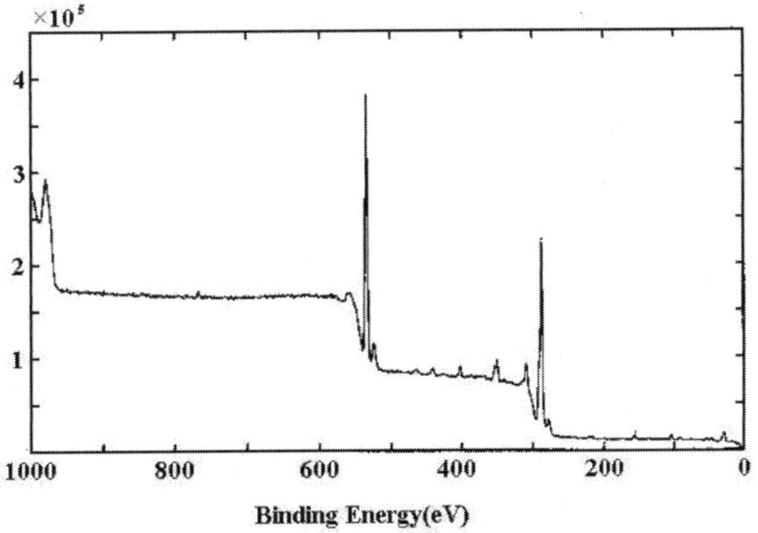

Fig.4 XPS spectra of the graft polymerization membrane

Fig.5 was SEM photographs of the PE membrane treated and not. In both of the photographs, regular lamellae structure can be observed. Voids were formed during initial stretching and these voids behaved like nucleation centers to form lager slit micropores by further stretch process. There are more crystal defects can be observed in the surface of the treated membrane, which may result in the decrease of the membrane porosity. Treated surface is more "smooth" than the original surface. But these will not influence the performance of the membrane much.

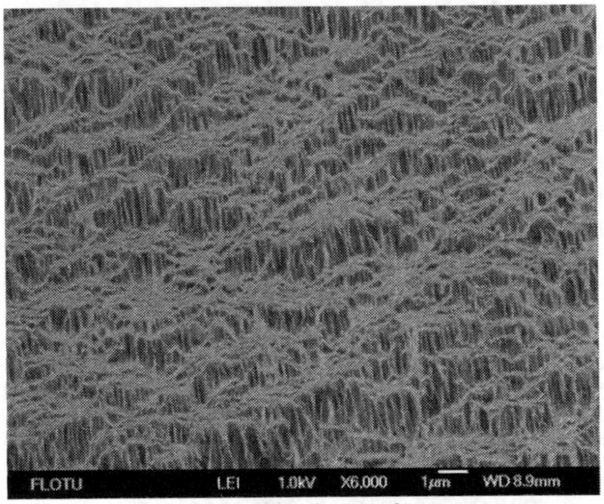

a. original memberane

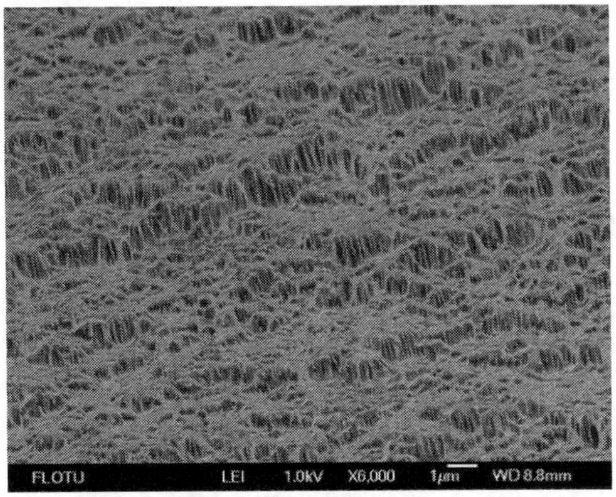

b. treated memberane

Fig.5 SEM photographs of PE membrane

The DSC curves of original, corona treated and grafted membrane are similar, as showed in Fig.6. This may be thought that the treatments did not change the main body of the PE membrane, the crystallinity was influenced little.

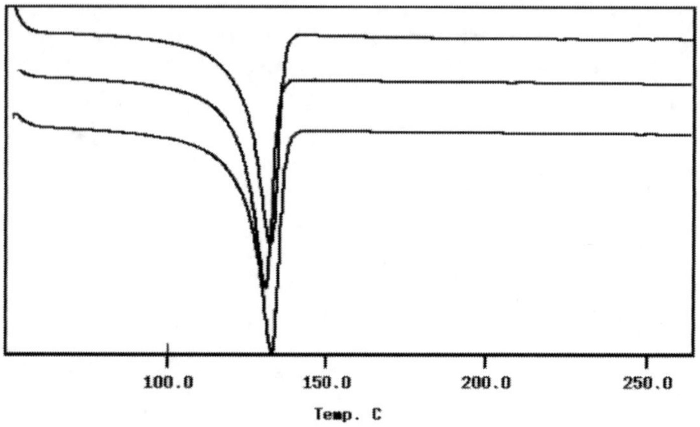

Fig.6 DSC curves of original and treated membrane

The Treatment Condition influence to the grafting degree

Many factors can change the grafting degree, such as the corona treatment time, the polymerization temperature, the grafting time, monomers' concentration, and so on. Fig.7, Fig. 8 and Fig.9 show the effects of the corona treatment time, the monomer concentration, and the grafting reaction time to the grafting degree.

Corona treated the PE membrane with different time, then grafting in the same monomer concentration for the same time. The grafting degree increased with the corona treatment time, as showed in Fig.7. The top grafting degree was reached when the corona treatment time is 80s, then the grafting degree begin to decrease with the time increases.

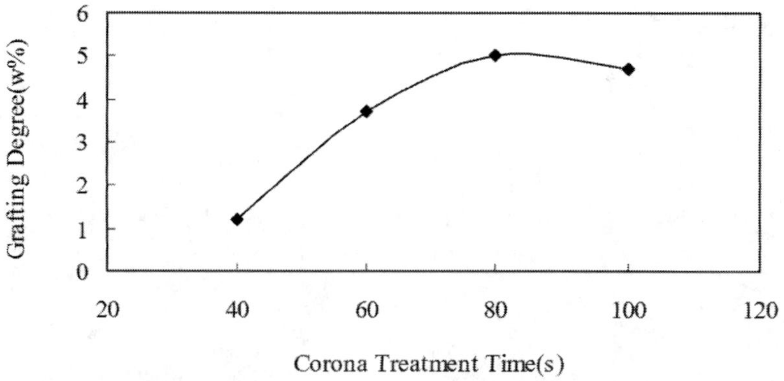

Fig.7 Effect of the corona treatment time

When monomer concentration is the only variable factor, the grafting degree changes as the Fig.8 show. The grafting degree increases with the monomer concentration when the monomer concentration changes between 10%~35%. Lower than 10%, the grafting effect can hardly measured; upper than 35%, the grafting degree changes little. So the monomer concentration should be kept in 10%~35% in this graft polymerization reaction.

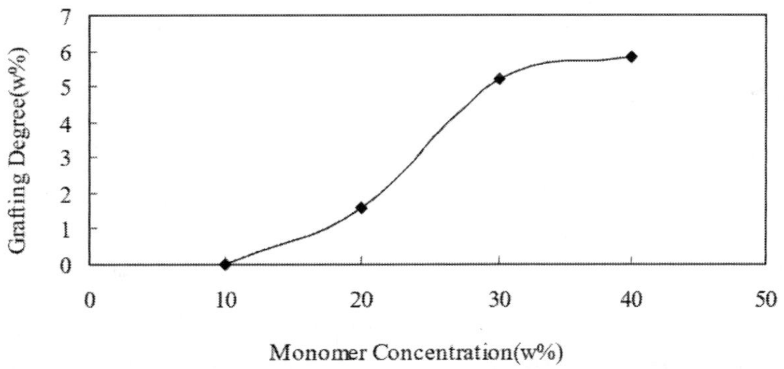

Fig.8 Effect of the monomer concentration

Keep the grafting reaction time as the only variable factor, the grafting degree changes showed in Fig.9. The grafting degree increases with the grafting reaction time within 3h, and not changes much after 3h. This indicates that the reaction trends to the balance after 3h. So the grafting reaction time should be kept within 1~2.5h.

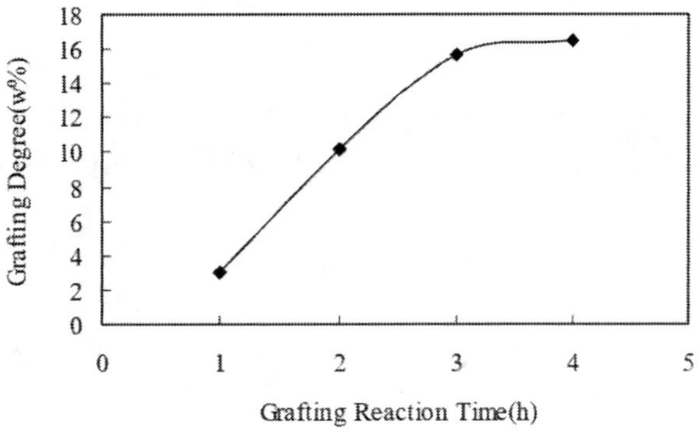

Fig.9 Effect of the grafting reaction time

Conclusions

Surface modification of polyethylene (PE) hollow fiber microporous membrane was successful performed by the grafting of 2-hydroxyethyl methacrylate (HEMA) via corona treatment. The results of ATR/FT-IR indicated the successful graft polymerization of HEMA onto the membrane surface. The grafting degree of HEMA could be modulated through the corona treatment time and monomer concentration. The modified PE hollow fiber membrane possessed better properties due to the improvement of the surface hydrophilicity by the grafting of HEMA.

Acknowledgement

The authors greatly appreciate the financial supports of the Major State Basic Research Program of China (No. 2003CB615701), National Natural Science Foundation of China (No. 20576059，

No.20676067), SINOPEC Foundation (No. X505002) and CNPC Innovation Foundation (No. 05051143).

References

1. Christine(Qin) Sun, Dong Zhang, Larry C Wadsworth(1999), "Corona Treatment of Polyolefin Films-A Review," in *Advances in Polymer Technology*, 18(2), pp. 171-180.

2. Lei J, Liao X(2001), "Surface graft copolymerization of acrylic acid onto LDPE film through corona discharge," *European Polymer Journal*, 37(4), pp. 771-779.

3. Wavhal D S, Fisher E R(2003), "Membrane surface modification by plasma-induced polymerization of acrylamide for improved surface properties and reduced protein fouling," *Langmuir*, pp.19-79.

4. Shi J L, Kang E T, Neoh K G, et al(1998), "Surface graft copolymerization of low density polyethylene films and its relevance to auto-adhesion," *Journal of Europe Polymer*, 34(10), pp.1429-1434.

5. Chen H, Belfort G(1999), "Surface modification of poly(ether sulfone) ultrafiltration membranes by low-temperature-plasma-induced graft polymerization," *Journal of Applied Polymer Science,* 72, pp.1699.

6. Olafesen K, Stori A, Tellefsen D A(1992), "Grafting of Acrylic Acid onto Corona-Treated Polyethylene Surfaces," *Journal of Applied Polymer Science,* 46, pp.1673

SYNTHESIS AND CHARACTERIZATION OF FUNCTIONALIZED COMPOSITE STYRENE-ACRYLONITRILE- DIVINYL BENZENE TERPOLYMER MEMBRANES FOR THE SEPARATION OF CHROMIC ACID

Anil Kumar* and Sonny Sachdeva
Department of Chemical Engineering
Indian Institute of Technology Kanpur
Kanpur-208016
India

anilk@iitk.ac.in

Introduction

Chromic acid is a highly toxic, carcinogenic oxidizing agent and is a major component of the effluents from leather tanning, chromium plating, galvanization, textile and dye industries [1]. In view of the environmental damage caused by it, chromium recovery and waste water recycling is one of the important problems of separation. Membrane separation processes have an edge over all the other separation methods (precipitation, exchange, evaporation, adsorption, and liquid-liquid extraction) as they do not require any addition of chemicals, require lesser energy, have a selective and nearly total recovery of chromates as well as produce purified water.

Membranes made from polystyrene and its copolymers are non cellulosic materials having high chemical stability and mechanical strength. Styrene can be copolymerized with many monomers like acrylonitrile and butadiene (ABS), acrylonitrile (SAN) and divinyl benzene (DVB) [2]. The SAN copolymers with 5-50% of acrylonitrile in styrene have greatly improved resistance to chemical attack by many reagents, better mechanical strength and excellent solvent resistance [2]. A number of membranes using SAN have been made for pervaporation [3, 4] due to their solvent stability but very little literature exists in the preparation of membranes from this polymer for nanofiltration separations [5] where these need to be used at high pressures for the separation of dyes.

The membrane reported in this work is a ceramic supported copolymer membrane of styrene and acrylonitrile which is tough and scratchproof. The dual initiator system (Benzoyl Peroxide and Azobisisobutyronitrile) along with the monomer (styrene and acrylonitrile) and accelerator dimethyl aniline (DMA) gives polymer syrup at room temperature having about 8% polymer in a clear solution. This result must be contrasted against a single initiator which starts polymerization at 50^0C and cannot be stopped and the polymer precipitates from the solution during polymerization. The polymer syrup prepared is spread uniformly on the clay support and heated further for crosslinking. The membranes so prepared are further modified by gas phase nitration followed by amination using hydrazine hydrate. After generating primary ammonium groups on the membrane surface, the aminated membrane is quaternized by reaction with dichloroethane and triethylamine. The membranes prepared in this way have high water flux at very low pressure drop without loss in rejection of chromic acid.

Experimental

Clay Support Preparation

Porous ceramic supports were prepared from a paste of a mixture of clays (kaolin, ball clay, pyrophallite, calcium carbonate, feldspar and quartz) in water and casted on a gypsum surface in the form of a circular disc followed by drying for 24h at room temperature, then 24h at $100^{0}C$ and finally for 24h at $250^{0}C$ to ensure the slow removal of water. The dried supports are then calcined at $900^{0}C$ for 8h and then finally polished using a silicon carbide abrasive paper (No. C-220) to get smooth and flat ceramic discs of diameter 64mm and 2-3mm thickness.

Polymer Preparation

A mixture styrene (6.073g, 0.0607gmole), divinyl benzene (3%) and acrylonitrile (10%, 0.607g, 3.067×10^{-3}gmole) using dual initiators consisting of Benzoyl Peroxide (0.082g, 3.388×10^{-4}), azobisisobutyronitrile (0.043g, 2.62×10^{-4}gmole) with an accelerator Dimethyl aniline (0.025g, 2.07×10^{-4}gmole) in presence of N_2 is copolymerized at $70^{0}C$ in a water bath for 1h to obtain a polymer syrup of about 8% conversion. A known amount (2g) of the polymer solution is spread over the wet support and is allowed to stand at controlled temperature and humidity ($33^{0}C$ and R.H 55%) for at least one hour. After the unreacted monomer and the solvent are evaporated, the composite membrane is heated at $70^{0}C$ for 3 hours in an oven to further crosslink the membrane polymer.

Chemical modification of composite SAN membrane

A mixture of NO and NO_2 (called NO_x) is generated in a 1.5 liter two neck flask by reacting sodium nitrate (10g, 0.14 gmole) with sulphuric acid (36N, sp.gr.1.18, 25 ml) in the presence of ferrous sulphate (5g, 0.072 gmole) [6, 7]. The composite membrane is placed inside the bottle and the NO_x (approximately 500ml) is introduced through the silicon septum of the reactor by a syringe. The reactor is kept into the oven at 80^{0} C and after the completion of the nitration reaction, the reactor is cooled to the room temperature.

The nitrated membrane is further refluxed with 50% hydrazine hydrate/water mixture for six hour in a water bath so as to reduce the NO_2 groups to amine (NH_2) groups as shown in the following reaction scheme and has also has been confirmed by FTIR. The aminated membrane is further modified by refluxing it with 2% v/v 1, 2 dichloroethane solution in ethanol for four hours. The modified membrane is quaternized by refluxing it with 5% solution of triethylamine (TEA) in ethanol for 4 h. This modified membrane is a strong anion exchange membrane having quaternary ammonium groups. The reaction scheme is presented as follows:

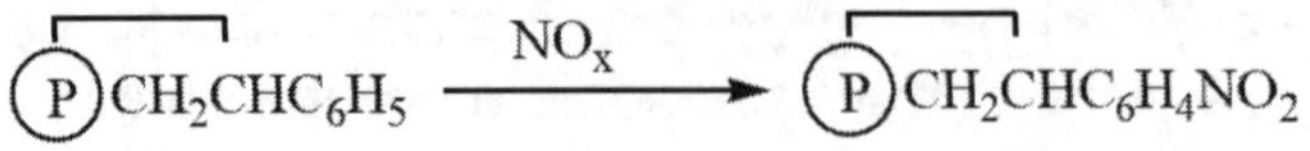

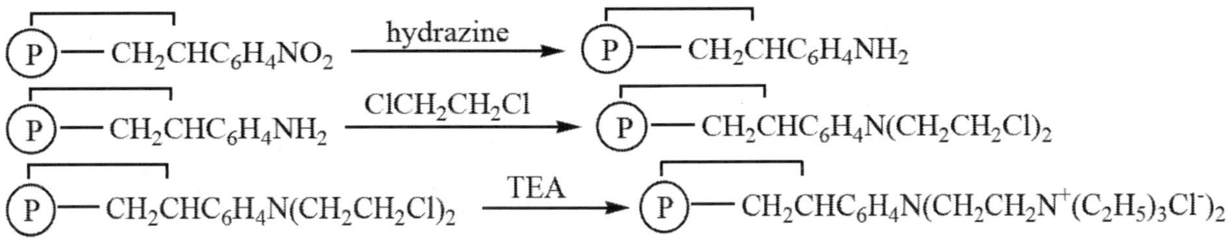

Characterization

The membranes have been fully characterized using Scanning Electron Microscopy (SEM), Fourier Transform Infrared Spectroscopy (FTIR), Atomic Force Microscopy (AFM), contact angle measurements, water content and molecular weight cut off experiments. The FTIR spectra confirmed the presence of presence of nitro and amine functional groups. The reduction in contact angle with each modification showed that the membrane became more hydrophilic. The pore diameter obtained using MWCO was found out to be 6.26, 8.32 and 11.2 nm for the unmodified, aminated and quaternized membranes, respectively. Figure 1 shows that polymer membrane is adhered to the support surface without penetrating the ceramic support.

Figure 1: Crossectional view of the SAN membrane. The polymer film is adhering to the top surface of the support without any interpenetration

Results and Discussion

Pure water permeation data has been obtained for the modified, nitrated, aminated and quaternized membranes. Figure 2 shows the pure water flux varies linearly with pressure drop for the unmodified and the modified membranes and obeys the Darcy's law ($J_v = L_p \Delta P$). From this figure, we determined the pure water permeability and found it to be 4.5365 X 10^{-8} (ms^{-1} KPa^{-1}) for the unmodified membrane, 1.06121 X 10^{-7} (ms^{-1} KPa^{-1}) for the nitrated membrane, 2.9978 X 10^{-7} (ms^{-1} kPa^{-1}) for the aminated membrane and 8.64124 X 10^{-7} (ms^{-1} KPa^{-1}) for the quaternized. We can infer from this figure that the water flux of nitrated and aminated membranes are much higher than the unmodified membrane due to the increased hydrophilicity and pore size of the modified membranes as has also been observed by contact angle measurements and morphological studies.

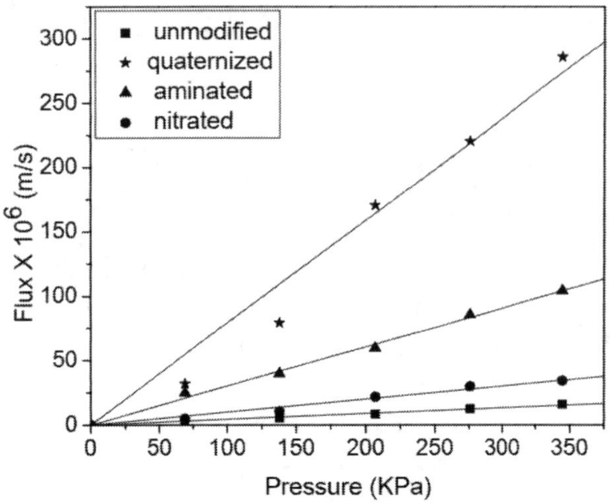

Figure 2: Water flux of SAN Membranes as a function of pressure

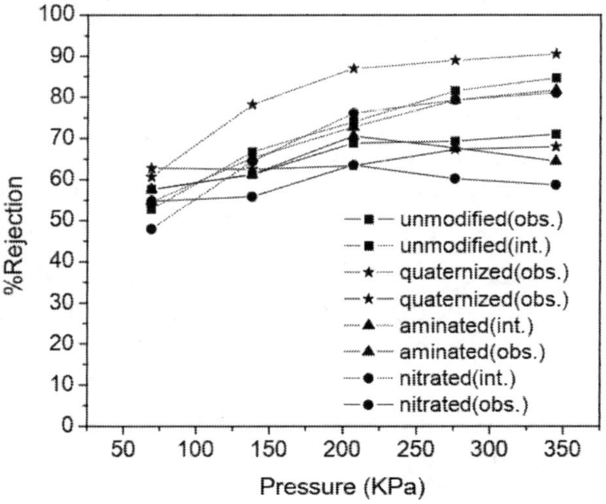

Figure3: Real and Observed rejection a of chromic acid as a function of pressure for concentration of 1000 ppm

Figure 3 shows that the chromium (VI) salt rejections as a function of pressure drop for the unmodified, nitrated, aminated and quaternized membranes. The observed rejection for the unmodified membrane increases with pressure drop but on modification the trend changes and the rejection at first increases and then falls with pressure. Therefore, observed rejection is no longer a proper parameter to express the membrane characteristics. The permeate flux was also found to be lower than the pure water flux which implies that the rejection results have to be corrected to include the effects of concentration polarization. Thus a procedure was developed as outlined in [8] to find the real rejection to characterize the membrane performance. It was found that the chromate concentration on the membrane surface was 2-3 times higher than the concentration in the bulk.

The intrinsic rejection always increased with pressure even though the trends obtained for observed rejection were much different. It was also found that the intrinsic rejection of the aminated membrane was lower than the unmodified and nitrated membranes as opposed to the observed rejection. This could be attributed to the much higher flux obtained for these membranes. It is also evident that the quaternized membrane had the maximum intrinsic rejection and maximum flux also. This can be explained on the basis of interaction of fixed membrane charge sites with ionic solutes. It can be seen that the intrinsic rejection increases with the increase in the applied pressure, which is typically found in the separation of electrolytes through charged membranes. For nitrated membrane the rejection is found to be less (determined in repeated experiments) compared to the unmodified membrane and this is likely to be due to an increase in the pore size.

Rejection data for different concentrations of chromic acid was obtained for the modified membranes. The experiments were conducted for the concentration of 1000ppm, 500 ppm and 100 ppm for the quaternized and the aminated membrane as shown in figure 4. It was found that the observed rejection increased with the decrease in concentration as has been reported in literature [9]. This is mainly because the effect of Donnnan exclusion reduces with the increase in concentration of the electrolyte. In the case higher electrolyte concentration, the membrane charge is shielded resulting in a lower effective charge of the membrane and hence lower retention.

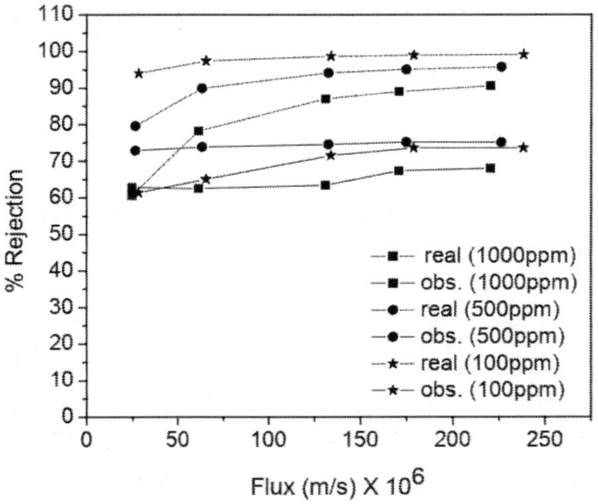

Figure 4: Observed rejection and real rejection of chromic acid for the quaternized SAN membranes as a function of concentration of chromic acid

Space Charge Model

The effective the wall potential of the membrane has been found out using space charge model [10] by the fitting of the experimental results including the concentration polarization effects. The space charge model uses Nernst-Planck equation to determine the flux of ions, Navier-Stokes equation to determine the fluid dynamics (bulk velocity/ volume flow) and Poisson-Boltzmann equation to relate the distribution of electrical potential to space charge density (concentration profile of solutes). The Poisson-Boltzman equation has been solved semianalytically using a series solution. The root mean square error between the predicted and the experimental value of the solute flux has been minimized to determine the values of wall potential which is found to be 110 mV for the quaternized membrane.

References

1. J. L. Kroschwitz, Kirk-Othmer Encyclopedia of Chemical Technology, Wiley, New York, 1991
2. C. A. Brighton, G. Pritchard, G. A. Skinner, Styrene Polymers: Technology and Environmental Aspects, Applied Scince Publishers Ltd., London
3. S. Mandal, V. G. Pangarkar, Pervaporative dehydration of 1-methoxy propanol with acrylonitrile based co-polymer membranes prepared through emulsion polymerization: a solubility parameter approach and study of structural impact, Journal of Membrane Science, 209 (2002) 53-66.
4. S. K. Ray, S. B. Sawant, J. B. Joshi, V. G. Pangarkar, Development of new synthetic membranes for the separation of benzene-cyclohexane meixtures by pervaporation: A solubility parameter approach, Industrial and Engineering Chemistry Research, 36 (1997), 5265-5276.
5. M. Joshi, A.K.Mukherjee, B. D. Thakur, Development of a new styrene copolymer membrane for recycling of polyester fibre dyeing effluent, Journal of Membrane Science, 189 (2001) 23-40
6. S. Roy Chowdhury, P. Kumar, P.K. Bhattacharya, A. Kumar, Separation characteristics of modified polysulfone ultrafiltration membrane using NO_x, Separation and Purification Technology, 24 (2001) 217-282
7. S. Sinha, A. Kumar, Preparation of high capacity chloromethylated strong base anion exchange resin using NO_x, Separation science and Technology, 37(2002) 895-919.
8. S. Ghose, C. Bhattacharjee, S. Datta, Simulation of unstirred batch ultrafiltration process based on reversible pore-plugging model, J. Membrane Science. 169 (2000) 29-38.
9. B. Chaufer, M. R.-Baudry, L. Guihard, G. Daufin, Retention of ions in nanofiltration at various ionic strength, Desalination, 104 (1996) 37-46
10. V. Sasidhar, E. Ruckenstein, Electrolyte osmosis through capillaries, Journal of Colloid and Interface Science, 82, 1981, 439-457

424f Sulfonated Block Copolymers For Water And Liquid Separation Applications

Scott R. Trenor and Dale L. Handlin Jr. Kraton Polymers, 3333 Hwy 6 S., Houston, TX 77082

Kraton Polymers is the inventor and world's leading producer of styrenic block copolymers which were first commercialized as part of Shell Elastomers in the 1960s. Kraton Polymers makes a wide range of specialty block copolymers in plants in the United States, Europe, South America, and Asia all with high precision polymerization control afforded by anionic polymerization. Kraton Polymers' use of anionic polymerization affords superior microstructure control and thus morphology management through precise control of block structures.

Numerous researchers have studied sulfonated styrenic block copolymers as semi-permeable membranes for various applications. Traditionally, the styrenic endblocks of triblock copolymers have been sulfonated, as was first practiced by Kraton Polymers as far back as the late 1960's. Under this process, once the polymer is hydrated, water plasticizes the endblocks which results in a significant degradation in the mechanical properties. Recently designed sulfonated block copolymer structures have been obtained that have continuous ionic domains and retain strength when fully hydrated. Utilizing 40+ years of experience in block copolymer structure and morphology, Kraton Polymers is now able to tailor the properties of the sulfonated polymer membranes via numerous methodologies including polymer structure, placement of sulfonic acid functionality, and membrane formation. The ability to tailor membrane properties provides the ability to manage the morphology of the membrane and decouple the degree of sulfonation and water swelling, as well as water swelling and transport properties. The initial characterization and the ability to manipulate the mechanical and transport properties of these polymers will be discussed.

430a Effect Of Bacteria And Virus Shape On Rejection By Microfiltration Membranes: Comparison Of Experiment With Hindered Transport Theory

Ruth E. Baltus[1], Wendong Xu[2], Appala Raju Badireddy[2], and Shankar Chellam[2]. (1) Chemical and Biomolecular Engineering, Clarkson University, Potsdam, NY 13699, (2) Department of Civil and Environmental Engineering, Department of Chemical Engineering, University of Houston, 4800 Calhoun Rd., Houston, 77204-4003

Microfiltration (MF) and ultrafiltration (UF) membranes are increasingly employed in the food processing, biotechnology, and pharmaceutical industries, for water and wastewater treatment, etc. Preventing microbial contamination of the filtered water (physical disinfection and sterile filtration) is integral to these applications. To date, several studies have empirically demonstrated the incomplete rejection of bacteria and viruses by microfilters. However, quantitative analyses of colloid passage across MF membranes have predominantly focused on spherical colloids, even though rod-shaped bacteria and non-spherical viruses are frequently encountered in membrane feed waters. Detailed information on removal of non-spherical colloids by membranes is not generally available. The objective of this research was to experimentally measure rejection of various organism with different aspect ratio (=length/diameter) from microfiltration membranes and to compare experimental observations to theoretical predictions. The rejection of two rod-shaped bacteria, *Brevundimonas* (formerly *Pseudomonas*) *diminuta* and *Serratia marcescens*, two bacterial viruses (PRD1 and T4), and spherical silica colloids from clean track-etched membranes was measured. The silica particles ranged in size from ~ 0.2 micron to 0.5 micron with a very narrow size distribution. The PRD1 bacteriophage was nearly spherical in shape, with diameter of 0.08 micron. The use of different growth periods provided us with two different *B. diminuta* samples, one with aspect ratio ~ 4; the other considerably longer with aspect ratio ~ 9. The *S. marcescens* bacteria were comparable in size to the *B. diminuta* , with an aspect ratio ~ 3.5. The T4 bacteriophage has six contractile tail fibers, a head and a tail. The aspect ratio of the head is ~ 2; considering the ratio of the head plus tail length to the head diameter gives an aspect ratio of ~ 2.7. Millipore polycarbonate membranes with pore sizes ranging from 0.1 micron to 5 micron were used for all rejection measurements. Pore sizes were determined using porometry as well as scanning electron microscope imaging. A stirred cell operating at high Peclet number conditions was used for rejection experiments, with samples collected before the membrane was fouled. Results obtained with spherical silica colloidal particles as well as with the PRD1 bacteriophage were in good agreement with hindered transport theory developed for rigid spherical particles1, confirming the validity of our pore size determinations and our experimental methods. Experimental results obtained with the non-spherical particles were quantitatively compared to predictions from hindered transport theory developed for rod shaped particles.2 This theory considers only configurational effects on particle transport, assuming that all possible particle orientations in the pore are equally probable. This theory predicts that the rejection coefficient for a rod shaped particle is greater than that for a spherical particle when particle size is described using a volume equivalent pore diameter, with rejection coefficient increasing as particle aspect ratio increases. In contrast to theoretical predictions, experimental results for the microorganisms with different sizes and aspect ratios were in general agreement with the prediction for a spherical particle when the volume equivalent sphere diameter is used to characterize particle size. When rod diameter is used to characterize particle size, rejection was found to be greater than the prediction for a sphere with the same diameter. Our results indicate that rod shaped particles are transported through the pore with a trajectory that is biased towards an 'end on' orientation, indicating that all particle orientations are not equally sampled. The unexpected conclusions are that the important particle size parameter appears to be the volume equivalent sphere diameter and that results show similar rejection behavior for particles with different aspect ratios. The 'take home' message from this work is that one can safely design a filtration system for microbial decontamination by using the rod diameter as the characteristic particle size parameter when selecting appropriate membrane pore size.

1 Bungay, P.M. and Brenner, H., "The Motion of a Closely Fitting Sphere in a Fluid-Filled Tube," *Int. J. Multiph. Flow* , 1, 25 (1973). 2 Anderson, J.L., "Configurational Effect on the Reflection Coefficient for Rigid Solutes in Capillary Pores," *J. Theor. Biol.* , 90, 405 (1981).

430b Temperature Effects On Sieving Characteristics Of Thin-Film Composite Nanofiltration Membranes: Pore Size Distributions And Transport Parameters

Shankar Chellam, Department of Civil and Environmental Engineering, Department of Chemical Engineering, University of Houston, 4800 Calhoun Rd., Houston, TX 77204-4003 and Ramesh Sharma, Trussell Tech. INnc., Pasadena, CA.

Even though polymeric nanofiltration (NF) and reverse osmosis (RO) membranes often operate on surface waters and surficial groundwaters whose temperature varies over time and with season, very little detailed mechanistic information on temperature effects on membrane selectivity is available to date. Hence, a study was undertaken to investigate the effects of operating temperature on the morphology and structure of two commercially available thin film composite NF membranes.

Crossflow filtration experiments were performed to measure transport of water and hydrophilic neutral organic solutes spanning a range of molecular sizes across two commercial thin-film composite nanofiltration membranes in the temperature range 5 – 41 °C. Non-viscous contributions to activation energies of pure water permeation across these polymeric membranes were calculated to be 3.9 and 6.4 kJ/mol. Analysis of solute rejection using a phenomenological model of membrane transport revealed that sizes of pores in the thin barrier layer of the thin-film composite membranes followed a lognormal distribution at any given temperature. Additionally, increasing temperature increased mean pore radii and the molecular weight cutoff suggesting changes in the structure and morphology of the polymer matrix comprising the membrane barrier layer.

Application of hydrodynamic models to experimental rejection of dilute solutions of hydrophilic neutral alcohols, sugars, and polyethylene glycols revealed changes in both the sieving coefficient and permeability of solutes below the membrane glass transition temperature. The vast majority of pores were smaller than 2 nm for both membranes (network pores) even though evidence for a small fraction of larger aggregate pores (~ 30 nm) was also obtained for one membrane. Increasing temperature appears to cause structural changes in network pores by increasing its pore size while simultaneously decreasing pore density. These increases in pore sizes partially explain reported reductions in contaminant (e.g. arsenic, salts, natural organic matter, hardness, etc.) removal by NF and RO membranes with increasing temperature.

Consistent with the free volume theory of activated gas transport, activation energies of neutral solute permeability in aqueous systems also increased with Stokes radius and molecular weight indicating their hindered diffusion in membrane pores. All activation energies for pore diffusion calculated in this study were greater than just the viscous contribution to bulk diffusion demonstrating hindered transport across the nanofiltration membranes. Finally, similar to gas transport across zeolites and rubbers, the activation energy and the Arrhenius pre-exponential factor for hindered diffusion coefficients increased with solute size and were highly correlated with each other.

430c Effects Of Supercritical CO2 Conditioning On Uncrosslinked Polyimide Membranes For Natural Gas Separations

Adam M. Kratochvil and William J Koros. Chemical & Biomolecular Engineering, Georgia Institute of Technology, 778 Atlantic Dr NW #450, Atlanta, GA 30332

The effects of supercritical CO2 (SCCO2) conditioning on a 6FDA-based polyimide membrane are investigated through gas permeability, sorption, and secondary characterization methods. At pressures above the supercritical point, CO2 permeability unexpectedly declines suggesting a structural reorganization of the polymer matrix facilitated by the presence of a highly soluble species. High pressure permeability isotherms for He and N2 do not exhibit a similar decline. Upon depressurization, a typical hysteresis is not observed; rather, CO2 permeability remains at reduced levels compared to the original pressurization. CO2/CH4 mixed gas testing of the sample prior to and after SCCO2 conditioning shows an enhancement in membrane separating performance after conditioning with a corresponding reduction in the membrane productivity. Repressurization of the same sample in CO2 to the supercritical region continues to exhibit reduced CO2 permeability confirming that the polymer chain packing has been altered to a new state. Wide angle x-ray diffraction and density measurements reveal a reduction in chain spacing and an increase in bulk density. Dual-mode modeling of the SCCO2 conditioned polymer suggests a large decrease in CH', or the Langmuir sorption capacity parameter. This reduction in CH' is consistent with the hypothesis that the reorganization observed in the supercritical region is a result of the removal of excess free volume through enhanced chain mobility coupled with high hydrostatic pressure.

430d Atomistic Packing Model For A Gas Seperation Membrane Polymer With Intrinsic Microporosity

Matthias Heuchel[1], Detlev Fritsch[1], Peter M. Budd[2], Peter M. McKeown[3], and Dieter Hofmann[1]. (1) Institute of Polymer Research, GKSS Research Center, Kanstr. 55, Teltow, D-14513, Germany, (2) Organic Materials Innovation Centre, School of Chemistry, The University of Manchester, Manchester, M13 9PL, United Kingdom, (3) School of Chemistry, Cardiff University, Cardiff, CF10 3AT, United Kingdom

Atomistic packing models have been built for PIM-1, a polymer belonging to a recently developed new type of "polymers of intrinsic microporosity" (PIMs). The molecular structure of PIM-1 contains sites of contortion (spiro-centres) within a rigid ladder polymer structure. Nitrogen sorption experiments have shown that the polymer, which traps in the glassy state a large amount of interconnected free volume, behaves in many respects like a microporous material. The models obtained by force field-based molecular mechanics and molecular dynamic methods allow a prediction of permeation properties of small gasses, the calculation of N_2 sorption isotherms at 77 K, and a determination of free volume distributions. The Gusev-Suter transition state theory (TST) was used to calculate gas solubility and diffusion coefficients for He, H_2, Ar, O_2, N_2, CH_4, and Xe. Good agreement with experimental data (factor 1-4) was found. The calculated nitrogen adsorption isotherm at 77 K shows properties typical for sorption on microporous materials. The obtained microporous volume is in good agreement with the respective experimental value. The size distribution function of free volume elements derived from the models agrees well in its range of pore width of 5 to 15 Å with results obtained from experimental sorption data and analyzed with the Horvath-Kawazoe method. Dihedral angle distributions in the backbone reveal a certain, unexpected flexibility of plane segments between the spiro-centres which act as connecting links.

The Modified Lévêque Limit for Spacer-Filled Channels

Abhishek Shrivastava, Satish Kumar, and E.L. Cussler
Chemical Engineering and Materials Science
University of Minnesota
Minneapolis, MN 55455

Introduction

The shortage of potable water will be a major problem of the next twenty years, probably compromising the lives of 40 percent of the world's population. This shortage will have a social impact on the same scale as that of increased energy prices. One major route to decreasing this water shortage will be membrane separations, especially because these processes remove impurities less than a micron in size. Ultrafiltration can remove viruses from groundwater; reverse osmosis can remove salt from sea water.

Both ultrafiltration and reverse osmosis are often compromised by concentration polarization. This polarization occurs when water passes quickly through the membrane and impurities accumulate on the feed (upstream) side of the membrane. In ultrafiltration, these impurities can foul the membrane; in reverse osmosis, the resulting concentrated solution at this surface compromises the water flux across the membrane. In either case, the membrane process has a lower productivity.

Because concentration polarization is a widely recognized problem, many have tried to reduce the effect with spacers or static mixers set close to the membrane's surface. The spacers try to disrupt the flow near the membrane, causing better mixing in the feed solution. Many shapes have been tried. Spacers shaped like ladders were a frequent early choice, later replaced by spacers shaped like diamonds. Designs for static mixers, normally aim at enhancing bulk mixing, should also work for this application. Asymmetric herringbone shapes have been successful for enhancing bulk mixing and so would be expected to reduce concentration polarization as well. In all these cases the results are less than would be expected from studies of turbulent heat transfer. The reason is that flow in membrane separations is almost always laminar, the consequence of the small channels chosen to get large membrane areas per process volume. These large areas are necessary to get large water fluxes.

In this paper, we report both electrochemical experiments and results of computational fluid mechanics to improve spacer design. We then use these experiments and computations to develop a predictive model for spacer design.

Results

We made four groups of experiments and simulations. First, we measured and calculated mass transfer coefficients in an empty rectangular channel, which lets us compare our results with a known analytical solution. Then we made experiments with three types of spacers: rectangular ridges placed perpendicular to the flow ("ladders"); similar asymmetric ridges placed at different angles to the flow ("herringbones"); and helically shaped spacers which tended to rotate the flow ("helices").

Comparisons with Analytical Results. Mass transfer in a thin channel has been very carefully studied, especially for short tubes. These studies are associated with Léveque, who made the earliest calculations for a cylindrical tube. The results for a rectangular slit whose width W is large relative to its height H are also well known:

$$\left(\frac{kH}{D}\right) = 1.47\left(\frac{H^2 v_o}{DL}\right)^{1/3} \quad (1)$$

where k is the average mass transfer coefficient at the channels walls, D is the diffusion coefficient in the fluids; v_o is the average fluid velocity; and L is the total channel length. The quantity in parentheses on the left-hand side of this equation is the Sherwood number; that on the right is the Graetz number. This result and its parallels for other cross sections have been carefully verified, both for heat transfer and for mass transfer. It is a secure standard for testing our experiments computations and our experimental results agree with this well.

With this basis in place, we now turn to results for different spacers (Figure 1)

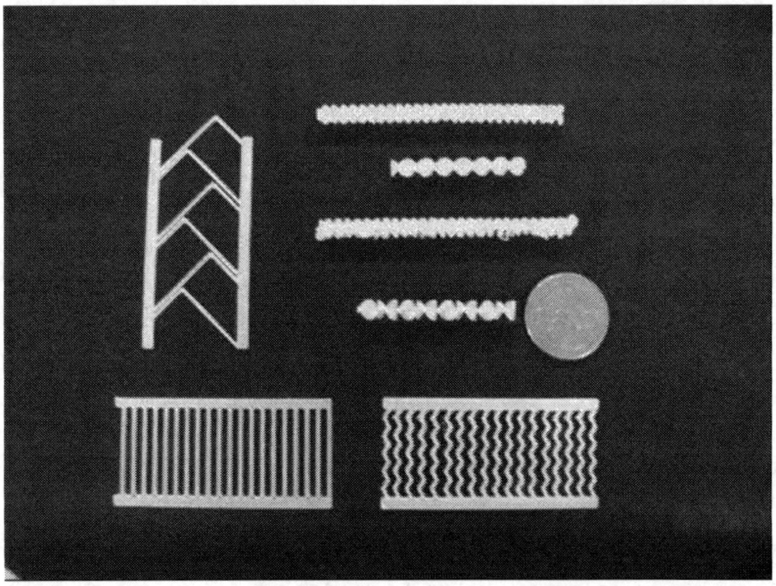

Figure 1. Static mixer designs

Ladders: We first consider spacers with rectangular barriers made by fused deposition modeling and glued to the bottom electrode. Electrochemical experiments and CFD simulations are conducted to determine the mass transfer coefficient due to these ladder type spacers on the top and bottom walls. Figure 2 represents the Sherwood number for different types of spacers including ladder type spacers.

Staggered Herringbone Spacers: One of the most interesting alternatives to the simple ladder spacers are those called "staggered herringbones" or "staggered chevrons." These spacers are known to significantly enhance mixing in small channels, and they may also improve interfacial heat transfer as well. Because they are asymmetric, they may generate a twisting, three-dimensional flow. The flow then depends not only on the height and spacing of the channel but also on the entire channel width perpendicular to the flow. This makes these flows different than those caused by ladder-type spacers. Electrochemical experiments were conducted at different flow rates with the two different staggered herringbone spacers shown in Figure 2.

Helical Spacers: Helical spacers take the idea of manipulating the flow in the third dimension a step further. Rather than allowing the fluid to meander across the width of the channel as asymmetric herringbones do, helical spacers force the fluid to twist along the width and the length of the channel. The result is a faster local velocity and a thinner mass transfer boundary layer. Helical spacers do not destroy the mass transfer boundary layer by obstructing the flow as ladders and herringbones rather helices sweep the fluid from one wall of the channel to the other. There is no physical obstruction to the flow, but there is a defined path that the fluid must follow.

Helical spacers like those in Figure 1 (top right), and constructed using fused deposition modeling, were installed in cylindrical channel 0.6 cm in diameter and 7 cm in length. The outer diameter of the spacers was the same as the inner diameter of the empty cylindrical channel. The diameter of the central rod in the channel determined the inner wall of the helical channel.
The data for the electro-chemical experiments conducted on different helical spacers are presented in Figure 2. The best helical spacer gives four times better mass transfer than an empty channel at the same flow rate. The data show that for the same flow rate, a smaller pitch and a smaller diameter give a higher limiting current. Again, the reason for these effects will become clearer from the analysis in the discussion section below.

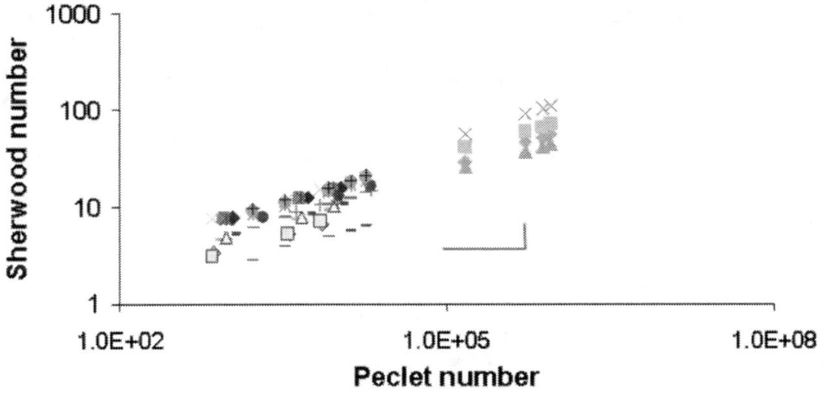

Figure 2: Simulations and Experiments for ladders, staggered herringbones and helices

Discussion

This work has two goals. First, we seek to show how specific spacer shapes can reduce concentration polarization and hence enhance ultrafiltration and reverse osmosis. Second, we hope to use results for specific spacers to develop a reliable guide for designing better spacers.

The results above show that we have succeeded in our first goal. We have made electrochemical measurements which give measurements of mass transfer accurate to within $\pm$ 5 percent. These mass transfer coefficients are an excellent measure of concentration polarization: large mass transfer coefficients mean small concentration polarization. We have shown that our measurements agree closely with well-established analytical solutions for mass transfer in a rectangular slit. We have also measured the enhanced mass transfer caused by spacers of different geometries placed within the slit. These enhancements are least for spacers shaped like ladders, greater for spacers shaped like asymmetric herringbones, and greatest for spacers shaped like helices. We have also calculated the mass transfer coefficients in flows around ladder-shaped spacers using two-dimensional computational fluid mechanics. The apparent agreement between these calculations and both analytical predictions and electrochemical measurements should encourage three-dimensional computations for the flows past herringbone and helical spacers.

However, the much more serious limitation of the research described so far is that we have not addressed our second goal: to provide an easy guide for improving spacer geometry. This guide should at least provide a rationale for better designs, and at most provide a means of evaluating them quantitatively. It should be simple to use giving results with a handheld calculator.

We have tried to develop such a guide by assuming that any channel is made of short segments, each of which can be described with the Léveque limit given in Eq. 1. By this assumption we imply that whenever the fluid encounters a new barrier, its concentration is instantly well mixed, and it behaves like a homogeneous solution entering a new tube. We make this assumption not because it seems an obvious approximation, but simply to start considering the effect of changes in cross-sectional area and resulting changes in velocity.

To test this assumption, we consider in detail the results for the top electrode of a channel containing a ladder-shaped spacer with n rungs glued to the bottom electrode. The top electrode now consists of $(2n + 1)$ channels connected end to end: one entrance channel; n channels where the bottom of the channel is a rung; $(n - 1)$ channels between rungs where the channel bottom is the bottom electrode, and one exit channel. In each channel, the mass transfer coefficient can be found by rearranging Eq. 1. For the entrance and exit regions and for the $(n - 1)$ regions between rungs, it is

$$k_i = 1.47\frac{D}{H}\left(\frac{HQ}{DL_i}\right)^{1/3} \qquad (2)$$

where H is the channel height, where Q is the volumetric flow per channel width, and L is the length of that particular channel. For the n regions above the rungs,

$$k_i = 1.47\frac{D}{H-h}\left(\frac{(H-h)Q}{Ds}\right)^{1/3} \qquad (3)$$

where $(H - h)$ is the new open channel height, and s is the distance between rungs. On this basis, the overall mass transfer coefficient is just the sum of these contributions

$$\begin{aligned} k_{avg} &= \frac{\int k_i d\ell}{\int d\ell} \\ &= \frac{1.47Q^{1/3}D^{1/3}}{L}\left[\frac{L_{ent}^{2/3}}{H^{2/3}} + \frac{nw^{2/3}}{\left(H-h\right)^{2/3}} + \frac{(n-1)s^{2/3}}{H^{2/3}} + \frac{L_{exit}^{2/3}}{H^{2/3}}\right] \end{aligned} \qquad (4)$$

where $L(= L_{ent} + nw + (n - 1)\,s + L_{exit})$ is the total length of the electrode.

The success of this idea is evaluated in Figure 3, which plots the predicted mass transfer coefficient *vs.* that found experimentally. In other words, Figure 3 uses Equation 4 to replot the ladder data in Figure 1 to give a single curve. No unknown adjustable parameters are used in this replotting. The agreement between the predictions and the experiments is within $\pm$ 4 percent.

We can test this assumption for other types of spacers simply by considering the changes in channel height and velocity. For a bottom electrode partly obscured by a ladder-shaped spacers we get the same mass transfer coefficient as that given in Eq. 5 and the same prediction as in Eq. 7 but without the term $\left[nw^{2/3}/(H - h)^{2/3}\right]$. The success of this prediction, accurate to $\pm$ 4 percent, is given in Figure 3 which replots the ladder data in Figure 2 for the bottom wall to give a single curve.

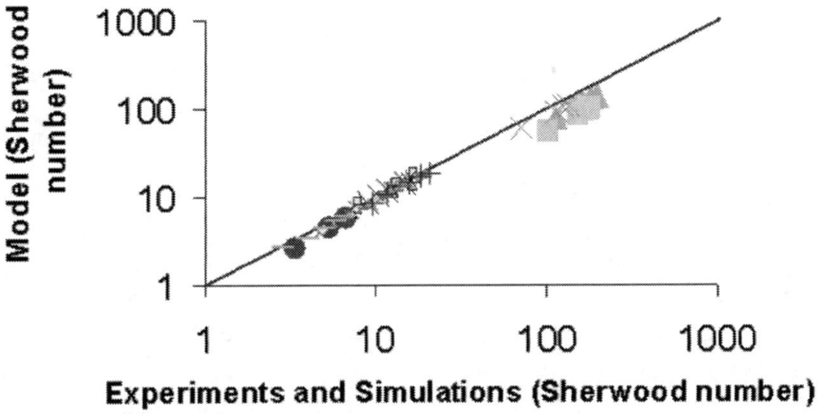

Figure 3: Predicted Sherwood number by the Modified Léveque limit model vs. the Sherwood number obtained by experiments and simulations. Experimental and Simulation data is same as in Figure 2.

The predictions of mass transfer with herringbone or helical spacers seem still more speculative because these spacers generate three-dimensional flows, and the Léveque equation which we are using describes two-dimensional flow in a slit. To make this jump, we must simply redefine the characteristic length and the inferred mass transfer coefficient. For the herringbones, the characteristic length is the repeat distance between chevrons L_{rot}, i.e., the distance which rotates the fluid completely. The height of the channel is that between the top electrode and the top of the chevrons $(H - h)$. Thus the mass transfer coefficient is

$$k = \frac{1.47\,Q^{1/3}D^{2/3}}{L_{rot}^{1/3}}\left(\frac{1}{H-h}\right)^{2/3} \tag{5}$$

Remember that Q is not the volumetric flow but that flow per channel width.

The results for helical spacers are analogous. We consider only the case where the helix turns many times down the length of the channel. Again the flow is rotated in a distance L_{rot}

$$L_{rot} = w \sec\theta \tag{6}$$

where w is the electrode width and θ is its pitch. If the pitch is zero, the fluid does not move forward but simply rotates in place; if the pitch is 90°, the fluid does not rotate but passes stright through the channel. (Is this correct?) The height of the channel is the difference between the channel radius $(d_{channel} / 2)$ and the radius of the case of the helix $(d_{case} / 2)$. Thus the mass transfer coefficient from rearranging Eq. 4 is

$$k = \frac{1.47\,Q^{1/3}D^{2/3}}{(W\sec\theta)^{1/3}}\left[\frac{2}{d_{channel} - d_{case}}\right]^{2/3} \tag{7}$$

Eqs. 5 and 7 are the basis for predicting behavior of herringbone and helical spacers, respectively.

These predictions are tested in Figure 3 and are within $\pm$ 6 percent of the experimental values shown uncorrelated in Figure 2. The differences between predicted and experimental values are larger for chevrons and helices than for ladder-shaped spacers, presumably a consequence of the three-dimensional, rotational flow caused by the non-ladder shapes.

The success of predictions based on these simple extensions of Leveque's result fulfill the second goal of this paper: we can now predict the effect of new spacer designs without major effort. We can simply use Eqs. 4, 5, 7 or their analogues and thus avoid large research efforts based either on electrochemical measurements or computational fluid mechanics.

The results raise two interesting questions for which we have no answers now. First, why do these extensions work so well? While we are not sure, we believe that the results imply that the most important aspect of mass transfer is a fresh surface. Thus our results can be considered a laminar parallel to arguments for the surface renewal theory for turbulent mass transfer. Three-dimensional flow doesn't matter anywhere nearly as much.

Second question is still more speculative: are there any spacer designs which might work better still? Again, while we don't know, we are intrigued by ideas of non-rigid, elastic spacers. Under some conditions, these spacers can move chaotically, randomly disrupting the flows near the wall. Such chaotic motion may reduce concentration polarization in that region. We look forward to learning about this possibility.

Experimental study and CFD modelling of an inorganic membrane module for H_2 purification

G. Montante, M. Giacinti Baschetti, D. Pizzi, J. Catalano, A. Paglianti, G.C. Sarti

Dipartimento di Ingegneria Chimica, Mineraria e delle Tecnologie Ambientali - Università di Bologna
Viale Risorgimento, 2 - 40136 Bologna - Italy

This work is aimed at investigating the capability of a fully predictive Computational Fluid Dynamics approach to reliably calculate the fluid dynamic and the separation performances of a lab-scale module based on inorganic membranes, for hydrogen purification.
The simulations are based on the numerical solution of the Navier-Stokes equations for a gas phase mixture on the three dimensional domain representing quite closely the module geometry. The tubular membrane is modelled as a selective layer, which allows the permeation of different components as a function of the transport mechanism and the driving force.
The computational strategy is strictly evaluated by comparing the results with experimental data obtained through permeation experiments. The simulation predictions show fairly good agreement with the experiments and allow to recognize the critical local fluid dynamic features of the separation module.

Introduction

Comprehensive knowledge of fluid dynamics and mass transfer mechanisms in industrial membrane separation processes is crucial for proper apparatus design and optimisation. To these purposes, Computational Fluid Dynamics (CFD) methods appear to be very promising, since their application could lead to detailed predictions of membrane separation operations in any geometrical configuration and scale of the modules. Nevertheless, to date development of proper modelling strategies and strict tests on their predictive capabilities are still required in order to reliable adopt CFD codes as a design tool in this field.
A membrane separation module can be modelled mainly considering two different and strictly linked phenomena: the fluid dynamic behaviour of the gas mixture in the whole module and the transport of each of the species in the mixture across the membrane.
One of the first application of CFD methods to membranes was devoted to the fluid dynamic analysis and geometrical optimisation of turbulence promoters (Cao et al., 2001). Afterwards, De Pinho et al. (2002) and Wiley and Fletcher (2003) performed initial attempts to couple simplified flow equations with mass transport mechanism models, focusing their attention on the concentration polarization phenomenon. Recently, a similar approach was adopted to desalination membranes by Ahmad et al. (2005). Liu et al. (2004, 2005) have modelled the mass transport across a pervaporation membrane as a pseudo-first-order chemical reaction, putting attention to the concentration polarization (Liu et al., 2004) and the baffle effects on the membrane separation efficiency (Liu et al., 2005).
For the case of gas mixture separation membranes, the available CFD simulations have been carried out so far in two-dimensional or simplified geometrical configurations by solving the momentum equations on the entire flow domain and modelling the mass transfer through species transport equations. Koukou at al (1999) found that the plug flow hypothesis does not allow to properly predict the experimental data. Takaba et al. (2005) have shown that the concentration polarization is more important the more membrane permeability and selectivity is increased. Moreover, as in the case of salt mixture, the

Reynolds number decrease, due to a reduction of the feed flow rate, worsens the mass fraction of the desired product in the permeate gas.
Overall, the simulations strategies developed up to now have shown that, although they still requires deep and broad validation tests, CFD methods are potentially very useful for overcoming the usually adopted simplified hypotheses on the apparatuses fluid dynamic behaviour and can provide detailed local information on the velocity and concentration distributions of the processed fluid mixtures, thus drawing attention on important aspects affecting the separation performances.
This work is aimed at providing a CFD simulation strategy for inorganic membrane apparatuses and its strict evaluation by comparison with experimental data.

Membrane module geometrical configuration

The experimental part of the investigation was carried out in a inorganic membrane module which consisted of a cylindrical shell of length equal to 185 mm and diameter equal to 54 mm, provided with one inlet and one outlet tube of 4 mm diameter each. The membrane was tubular in shape and its diameter was equal to 10.4 mm, which coincides with the fluid inlet diameter at the tube side. In particular, a ceramic membrane suitable for H_2 purification operations was considered. Permeation experiments were run under different operating conditions, thus providing the gas flow rates and compositions information required to strictly evaluate the purposely developed CFD model, which is described in the following.

Model equation and solution procedure

The mathematical modelling of the gas streams laminar flow in the separation module was started from the numerical solution of the steady-state continuity and momentum equation over the three-dimensional domain reproducing quite closely the experimental module geometrical details.
In particular, the following set of equations was considered:

$$\nabla \cdot (\rho \mathbf{U}) = 0 \tag{1}$$

$$\nabla \cdot (\rho \mathbf{U}\mathbf{U}) = -\nabla p + \nabla \cdot \bar{\bar{\tau}} + \rho \mathbf{g} \tag{2}$$

where ρ is the fluid density, **U** is the velocity vector, p is the static pressure, τ is the viscous stress tensor, **g** is the gravitational acceleration.
The mass transfer of gas species was modelled by a transport equation, based on the assumption that each species is distributed in the module by convection and diffusion. At the membrane boundary cells, source and sink terms are introduced, thus allowing to the species to disappear from one of the membrane sides and to come out on the other.
The following transport equation for the species i was considered:

$$\nabla \cdot (\rho \mathbf{U} \Phi_i) = \nabla (\rho D_{i,m} \nabla \Phi_i) + S_i \tag{3}$$

where Φ is the species i volumetric fraction, $D_{i,m}$ is its molecular diffusivity and S_i is the source term to account for the mass flow of the species across the membrane.
The source term equation has been based on the assumption that the driving force to mass transport in the inorganic membrane is the partial pressure difference and it is proportional to the species permeability. In particular, the following equation has been adopted:

$$S_{ji} = \frac{A_j}{V_j} Q_i \left(p_{shell} \phi_{i,j,shell} - p_{tube} \phi_{i,j,tube} \right) \tag{4}$$

where j is the cell index, A_j is its transfer area, V_j is its volume, Q_i is the permeability of the pure species i across the membrane, ϕ_i is the i molar fraction and p_{shell} and p_{tube} are the total pressure at the shell and tube side, respectively.

Extension of Eqn. (4) for including the contributions of Poiseuille flow and/or other mass transfer mechanisms through the membrane is possible, but they have been neglected as a first instance.

It is worth observing that the specific permeability is the only unpredictable variable of the present model. A single value for each species has been adopted in all the simulations, based on the permeability measurements of pure species. On the other end, a detailed description of the membrane structure and the consequent attempt to predict its permeability characteristics is not feasible, while investigating the full module fluid dynamic and transport characteristics.

One of the main advantages of the present modelling strategy is the possibility to adopt local, rather than, mean values of the driving forces, which feature should ensure a better accuracy and a deeper knowledge of the module behaviour.

The simulations were performed running the finite volume general purpose CFD code FLUENT 6.2. The computational grid, shown in Figure 1, consisted of a total of about 46000 tetrahedral cells over the full three dimensional domain. It reproduces quite closely the real module geometrical details, apart from the membrane that was assumed to be a zero-thickness wall for the above mentioned reasons. The grid was refined in the tubular membrane region (green surface in Figure 1) where the biggest variable gradients were expected.

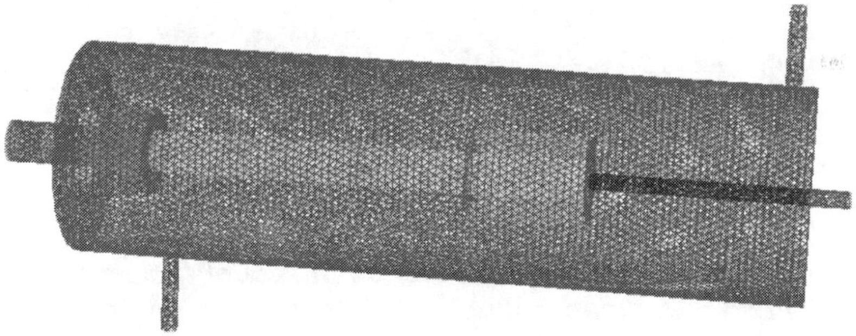

Figure 1. Meshed computational domain

A grid independency study was also performed by increasing of more than one order of magnitude its density, which confirmed that the selected grid was fine enough for obtaining accurate results.

The continuity and momentum equations as implemented in the code were numerically solved in a Cartesian coordinate system. To implement the mass flow terms (Eqn. 4), a user defined function provided by Fluent Inc. was written. The solution was obtained by the SIMPLEC algorithm to couple pressure and velocity and the second order upwind discretisation scheme was used for the convective terms. The solution convergence was carefully checked by monitoring the residuals of all variables as well as the species mass flow rate. Residuals were dropped to the order of 10^{-5} or less, which is at least two order of magnitude tighter than Fluent's default criteria. Typically, about 30,000 iterations were necessary in order to obtain the solution.

The solution of the model set of equations is a three-dimensional map of all of the gas species velocity and concentration in each of the computational cell. In the following, they will be compared with the available experimental data for a detailed and quantitative analysis and evaluation of the simulation strategy.

Results and discussion

In the experimental part of the work, the membrane performances were firstly investigated by feeding pure gas to the module shell side and measuring the gas flow rate at the shell and tube exit sections. In particular, the permeation of pure N_2 and H_2 was evaluated at a temperature T=500°C and fixed downstream total pressure (p_{tube}=1.05 bar) while varying the pressure from 2 to up to 5 bar in the shell side. These experiments allowed to obtain information about each gas permeability in the membrane and have been adopted as preliminary tests for the CFD computational strategy.
Before evaluating the mass transfer predictions, the hydrodynamics of the module, as obtained from the simulation, is analysed.
The importance to couple the fluid dynamic and the mass transfer modelling can be estimated from the velocity vector plots shown in Figure 2, where a particular shell portion is considered and the two different results obtained considering or neglecting the mass transfer across the membrane are shown . As can be observed, a different shape of the gas jet produced from the shell feed stream entrance is predicted by the two simulations, thus proving that the mass transfer produce visible effects on the present module hydrodynamics. Therefore, a simplified purely fluid dynamic modelling strategy is not viable for obtaining accurate predictions.

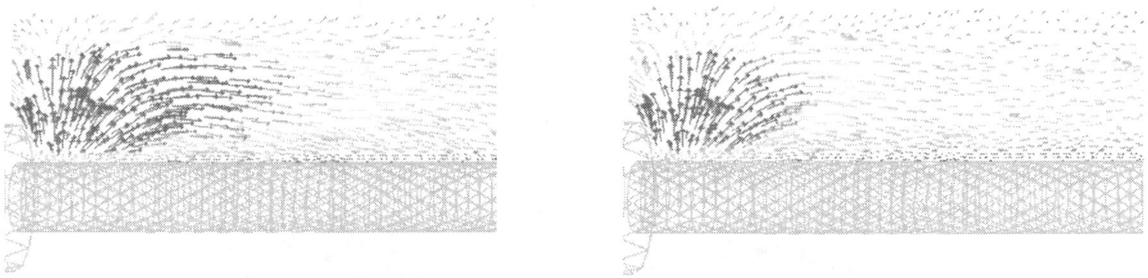

Figure 2 Velocity vector plots with species transport model included (left) and neglected (right). Shell side, portion of a plane orthogonal to that containing the shell inlet and outlet tubes.

The gas velocity map shown in Figure 3 reveals different velocity magnitude at the membrane shell side, thus suggesting that a constant driving force along the whole membrane active surface does not closely represent the real module behaviour. Indeed, the gas velocity considerably decreases along the membrane moving far from the shell feed tube region, which position and small section produces the strong jet shown in Figure 2.

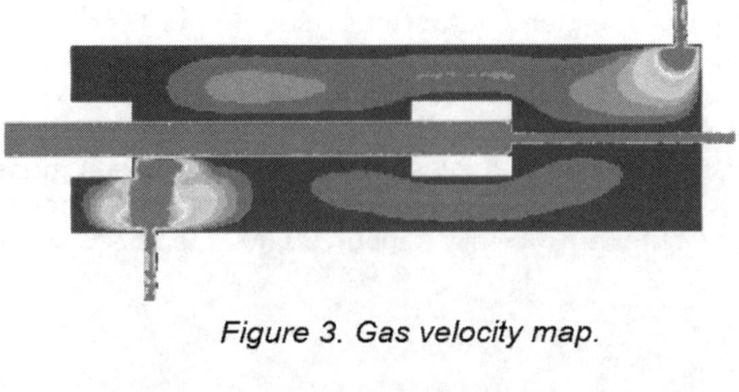

Figure 3. Gas velocity map.

For the particular geometrical configuration considered in this work, the flow field predictions have emphasized local effects of gas velocity, that could be responsible

of a non-homogeneous operation of the membrane surface.
A first evaluation of the computational approach accuracy can be performed by the comparison of the experimental and computed permeate flow rate shown in Figure 4 and 5, relevant to the pure H_2 and pure N_2 gas feed, respectively. As can be observed, for all the operating conditions a close agreement of the CFD results with the experimental values is obtained, thus confirming that in principle the computational strategy devised in this work can provide detailed and accurate information on the module behaviour, provided that information about the membrane permeability are available. In all the simulations, the H_2 permeability was assumed to be equal to 3.2 10^{-5} kg/s·m^2·bar, as obtained from the experiments assuming that the prevailing transport mechanism can be approximated by Eqn. (4), while the N_2 value, which was equal to 1.2 10^{-4} kg/s·m^2·bar was computed from the H_2 one by applying the Knudsen relationship.

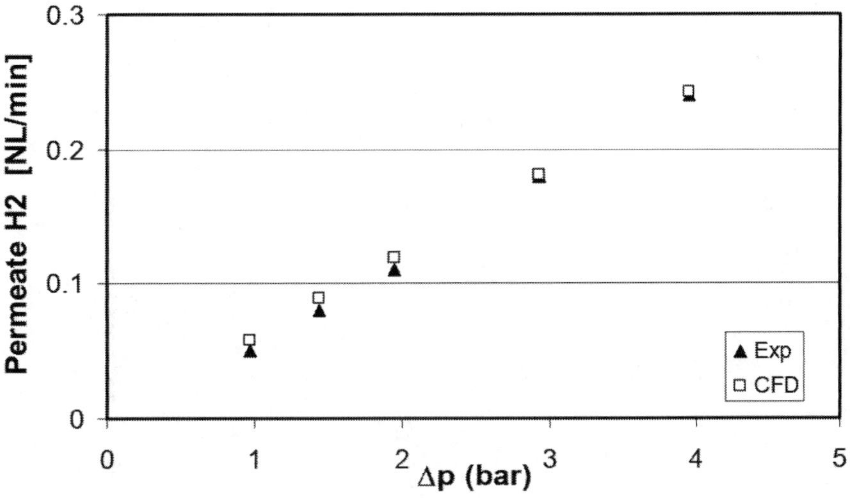

Figure 4. Comparison between experimental flow rate of H_2 permeate and CFD results at different operating conditions (gas feed on the shell side only).

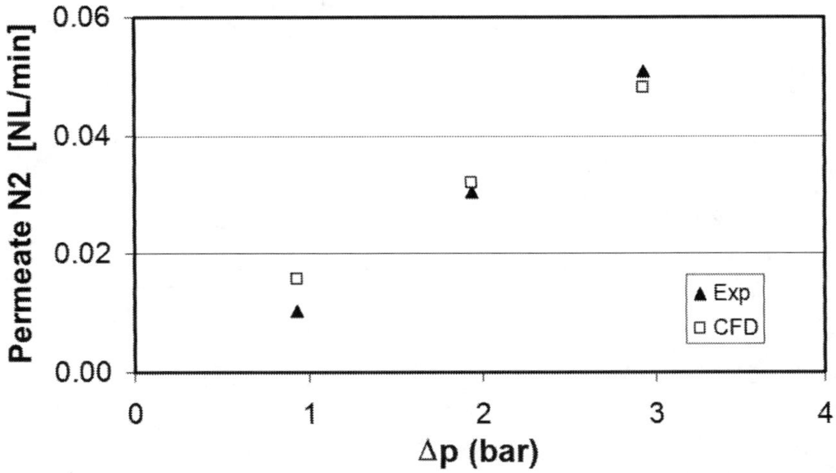

Figure 5. Comparison between experimental flow rate of N_2 permeate and CFD results at different operating conditions (gas feed on the shell side only).

Further preliminary experiments were subsequently performed by feeding pure H_2 at the shell side and pure N_2 from the tube inlet. In these cases, an equal total pressure was maintained at both module side and it was varied from 1.6 up to 5 bar in order to investigate the operating condition effect on the membrane behaviour.
The collected experimental data have provided a further stringent test for the simulations. In these cases, the membrane is permeated by two different gas in counter-current flow, following the same transport mechanism and different permeation properties, which are assumed to be independent from each other. Therefore, the same permeability values reported above have been adopted for all the cases.
The predicted H_2 and N_2 permeate flow rates (Figure 6 and 7 respectively) show a reasonable agreement with the relevant experimental data, but different accuracy levels are obtained at the different total pressure of the module.

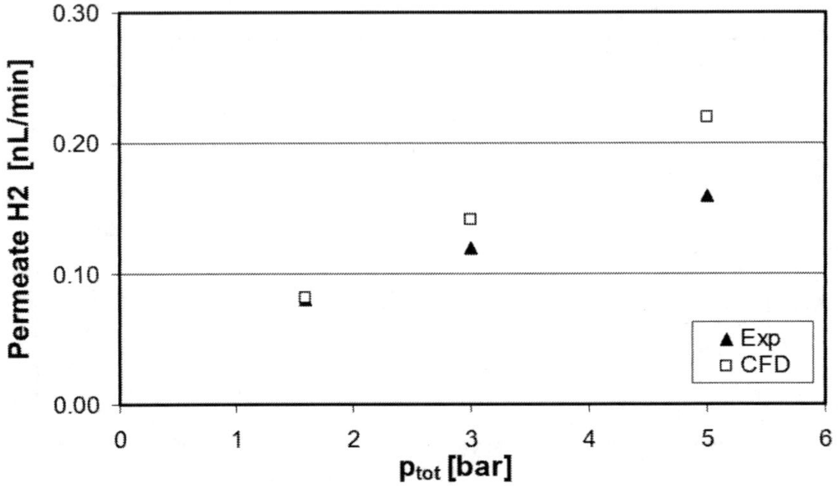

Figure 6. Comparison between experimental flow rate of H_2 permeate and CFD results at different operating conditions (gas feed on both shell and tube side).

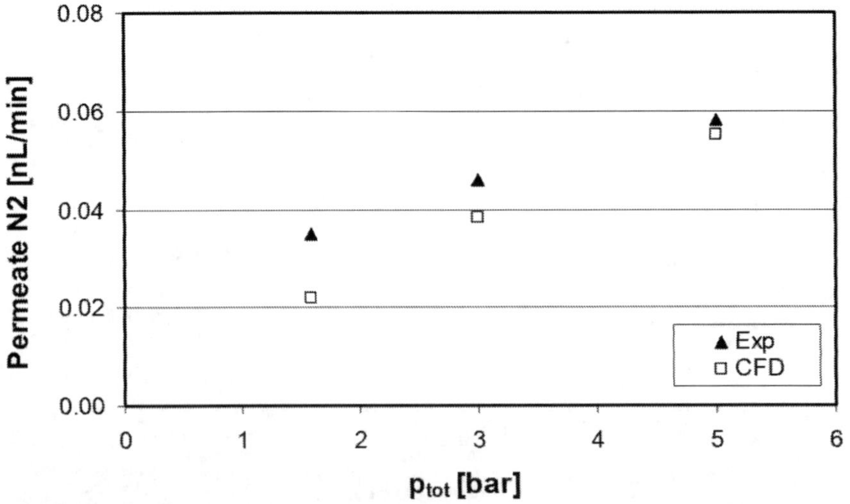

Figure 7. Comparison between experimental flow rate of N_2 permeate and CFD results at different operating conditions (gas feed on both shell and tube side).

In particular, the prediction accuracy decreases while increasing the total pressure for the H_2 permeate, while the opposite behaviour is obtained for N_2.
Probably, the simplified transport mechanism adopted in the present computational approach (Eqn. 4) can be further refined for better catching the pressure effect on the gas permeation and model developments are currently under study for gaining better insight on the membrane performances.
The present simulations also provided detailed information on the species distribution in the whole module volume, from which implications for mass transfer can be drawn putting attention to the H_2 and N_2 concentrations close to the membrane tubular surface. In particular, at the operating conditions considered in this work, radial and axial gradients for the species concentrations can be observed mainly on the shell side, as can be observed in Figure 8, where the species mass fractions close to the shell inlet tube are shown. Since the overall membrane efficiency can be affected by these features, preliminary knowledge could be very beneficial for efficient design and scale up of industrial separation apparatuses.

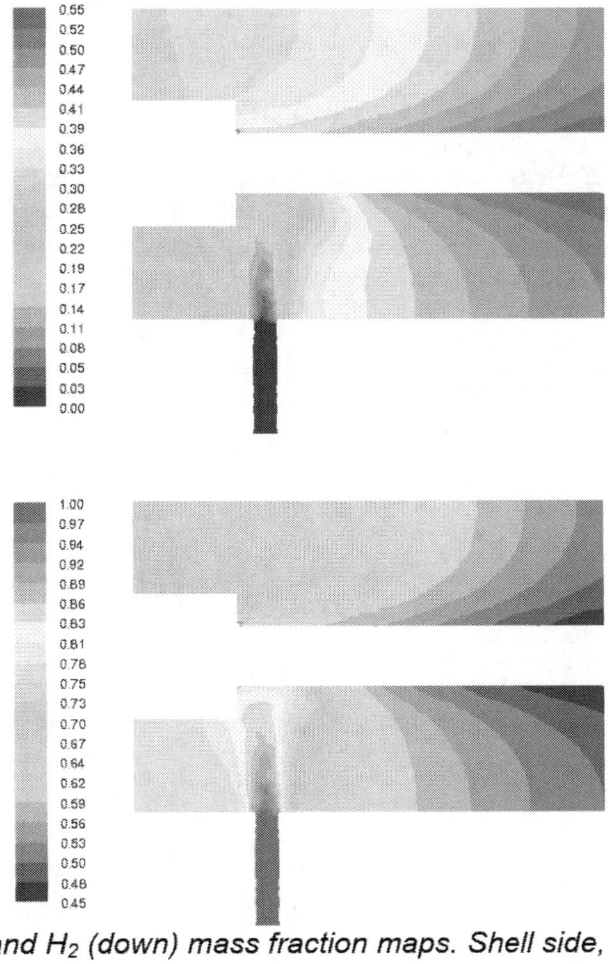

Figure 8. N_2 (up) and H_2 (down) mass fraction maps. Shell side, close to the feed inlet tube.

Overall, the results obtained so far suggest that the CFD model developed in the present work can provide an acceptable estimation of the membrane module behaviour. Further model developments and more extensive benchmark experiments are in progress for extending the model suitability to a wider range of applications.

Acknowledgements

This work was financially supported under "FISR DM 17/12/2002-anno 2001" project. The support of James Tibbatts, Ed Wynn (Fluent Europe) and André Bakker (Fluent Inc.) on the UDF implementation is gratefully acknowledged.

References

Ahmad A.L., Lau K.K., Abu Bakar M.Z., Abd. Shukor S.R., 2005. Integrated CFD simulation of concentration polarization in narrow membrane channel. Computers and Chemical Engineering, 29, 2087-2095.

Cao Z., Wiley D.E., Fane A.G., 2001. CFD simulations of net-type turbulence promoters in a narrow channel, J. of Membrane Sci., 185, 157-176.

De Pinho M. N., Semião V., Geraldes, V., 2002. Integrated modeling of transport processes in fluid/nanofiltration membrane systems. J. of Membrane Sci., 206, 189-200.

Koukou M.K., Papayannakos N., Markatos N.C., Bracht M., Van Veen H.M., Roskam A., 1999. Performance of ceramic membranes at elevated pressure and temperatures: effect of non-ideal flow conditions in a pilot scale membrane separator. J. of Membrane Sci., 155, 241-259.

Liu S.X., Peng M., Vane L.M., 2004. CFD modeling of pervaporative mass transfer in the boundary layer. Chem. Eng. Sci., 59, 5853-5857.

Liu S.X., Peng M., Vane L.M., 2005. CFD simulation of effect of baffle on mass transfer in a slit-type pervaporation module. J. of Membrane Sci., 265, 124-136.

Takaba H., Nakao S., 2005. Computational fluid dynamics study on concentration polarization in H2/CO separation membranes. J. of Membrane Sci., 249, 83-88.

Wiley D.E., Fletcher D.F., 2003. Techniques for computational fluid dynamics modelling of flow in membrane channels, J. of Membrane Sci., 211, 127-137.

430g A Novel Membrane Flux And Separation Factor Calculation Combined To Modified Unifac Model And Percolation-Fractal Theory

Xuechun Liu, Cuixian Chen, and Jiding Li. Chemical Engineering, Tsinghua University, Road Chengfu, District Haidian, Beijing, 100084, China

AICHE 2007 ABSTRACT

The study discusses the mass transfer models within membrane, which emphasizes particularly on the consideration to the transfer mechanism within membrane and membrane flux calculation. Solution-diffusion model was broadly applied in the separation mechanism of dense membrane, which includes solution (sorption) process, diffusion process and desorption process. It is generally suggested that the solution (sorption) process or diffusion process is the velocity-determined step. This solution process was more complex because of the plastication of components (membrane swelling) to polymer and the coupling effect between components.

Sorption models included UNIQUAC model, UNIFAC model, ASOG-FV model, Flory-Huggins theory, Entropic-FV model, Modified NRTL model and ENSIC model, etc. These models are related to statistic thermodynamics and physical chemistry. Among these models UNIFAC model is semi-empirical and needs less experimental data. Therefore UNIFAC model is often applied if experimental data is not enough.

Diffusion models included empirical models, dual sorption model and free volume theory. Because empirical models need too many parameters to apply in the complex (multi-component) system, the membrane flux can be calculated by the fitting and regression of mathematical methods. Dual sorption model might nevertheless be used as a foundation to understand the transfer process of high concentrations of liquids and vapors because it seems to be applicable to vapor permeation at low pressures. The essence of free volume model is that the small molecules transfer in the free volume to permeate from the membrane at last. The free volume can be described as temporary hollow spaces in the polymer structure as a result of density deviations of the polymer. The membrane flux can be obtained by reason of a set of complete calculation method in the free volume theory. However it is not excellent for these above-mentioned models to interpret the abnormal diffusion phenomenon within membrane.

The abnormal diffusion phenomenon is a kind of diffusion phenomena that does not conform to the Fick's law or are named as the non-Fick diffusion phenomena. In uniform Euclidean systems, the mean-square displacement of a random walker, , is proportional to the time t, , for any number of spatial dimensions d (Fick's Law). However, in disordered systems, this law is not valid in general. Rather, the diffusion law becomes anomalous , with , this indicates the slowing down of the diffusion. The slowing down of the diffusion is caused by the delay of the diffusing particles in the dangling ends, bottlenecks and backbends existing in the disordered structure. It is usually proposed that polymer membrane belongs to the disordered media system. Therefore the diffusion within membrane is a kind of abnormal diffusion. Very many researches deal with the abnormal diffusion phenomena, in which satisfactory interpretation can not be obtained.

Tiangang Shang and Cuixian Chen et al. put forward percolation-fractal model, in which anomalous diffusion phenomenon of small molecules transferring within membrane were systemically investigated. It is considered that the transfer process of small molecules is a random walk process in 'transfer path' within membrane. According to percolation theory, fractal geometry and random walk theory, the probability density expression of small molecules was deduced in the 'transfer path' of fractal region. It is not enough to consider the complex interaction among polymer molecules, among small molecules and between polymer molecules and small molecules in the percolation-fractal model.

According to the above-mentioned transfer mechanism within membrane, we suggested that it is necessary to sufficiently consider the various interactions among different components and between different components and membrane polymers. Firstly, sorption activities of small molecules can be calculated in feed and permeating side as the boundary values of the probability density expression in accordance with the modified UNIFAC model. The fractal dimension and the abnormal diffusion exponent can be determined by the renormalization and numerical calculation. The undetermined coefficients b and c in the probability density expression can be calculated adopting the boundary values from the modified UNIFAC model. Consequently the membrane flux and separation factor can be calculated adopting the probability density expression according to the integral transfer equation.

A novel membrane flux and separation factor calculation is proposed considering the interaction among polymer molecules, between polymer molecules and small permeating molecules, among small permeating molecules and probability density expression of small permeating molecules within membrane characteristic of stochastic fractal deducing from abnormal diffusion phenomenon within membrane. The continuous macroscopic membrane flux and separation factor can be calculated combined to the interaction between varied components in the system and the probability density that is microscopic discontinuous quantity.

A series of experiments are made in the ethanol, water and PVA system to obtain the experimental values of various feed ratio and substituting degree (sd). The theoretical calculations have been done adopting membrane flux and separation factor program in myself. Subsequently we compared the theoretical values with the experimental values and analysed to make an appropriate interpretation.

It is concluded from the comparing and analyzing of theoretical flux and experimental flux, calculation separation factor and experimental separation factor that the errors are usually below the permissible error 20%. This indicates that the membrane flux and separation factor calculation is appropriate to application.

We described the mass transfer behavior of small permeating molecules within PVA composite membrane and analysed the interaction among polymer molecules, between polymer molecules and small permeating molecules, among small permeating molecules. It is determined of the key factor to the mass transfer behavior of small permeating molecules. The substituting degree, feed and permeating concentration effect on membrane flux and separation factor were determined by the calculation method.

The authors made a thorough research on the probability density expression of small molecules in the mass transfer within membrane. The fractal dimension and abnormal diffusion exponent were determined by numerical calculation, series expansion and renormalization. The flux and separation factor were calculated adopting UNIFAC model and probability density expression of small molecules, which results in the application of this membrane flux and separation factor calculation.

In accordance with the thorough research on the transfer mechanism, the calculation was obtained to predict membrane flux and separation factor which are comparatively consistent with experimental. According to the comparing and analyzing of the above calculation and experimental, the various key factors to determine membrane flux and separation factor were obtained to be able to make a direction to membrane material choice, membrane preparation conditions and feed treatment conditions.

Energy Reduction in the Production of Ethanol by Membrane Dehydration

Pierre Côté, Christian Roy and Normand Bernier,
Vaperma Inc., St-Romuald, Quebec, Canada

Abstract

Distillation/dehydration represents the largest fraction of the energy used in the production of ethanol. The Siftek™ technology introduced in this paper carries the potential of reducing energy consumption of distillation/dehydration by up to 50% through the single pass removal of water from the water/ethanol stream at the beer column outlet, using a novel membrane process. For a modern corn-based ethanol plant, this is equivalent to 7.0 US¢/gal, or 1.8 US¢/L.

Siftek™ is a novel polymeric membrane that can operate continuously in the vapor phase and in presence of high water concentration. The hollow fiber membranes are made of a proprietary polyimide blend that offers high water/ethanol selectivity and water permeance. The water/ethanol vapor mixture is introduced at the beer column operating pressure on the bore side of the hollow fiber module. The water permeate stream is condensed and the steam latent heat is recovered. A vacuum provides the driving force between the bore side and the shell side of the membrane. The technology has been demonstrated in the field at a capacity of 1200 kg/day (2600 lbs/day) of vapor feed with an ethanol concentration varying between 75 and 90 wt% at one of the Greenfield Ethanol Inc plants in Tiverton, Ontario, since August 2006.

The presentation will include a detailed description of the technology, the results from the field demonstration unit, scenarios for retrofitting into existing or new green-field ethanol plants, and a summary of the energy savings and projected pay-back calculations.

Introduction

The use of ethanol as a fuel has been growing exponentially around the world, with global production now exceeding 13 billion gallons/year (50 million m^3/y). In 2006, the US production approached 5 billion gallons (19 million m^3) making it for the first time the largest producer in the world, ahead of Brazil. The US demand for ethanol in 2006 was stimulated by the mandatory replacement of methyl tertiary butyl ether (MTBE) as a gasoline additive. Going forward, ethanol is seen as a replacement for fossil fuels, decreasing the dependency on imports and reducing the emission of green house gases. While it is recognized that the amount of ethanol that can be produced from corn is limited (US estimates vary from 10-15 billion gallons/year), it is anticipated that the development of processes based on cellulosic materials will provide an almost infinite source of feedstock to meet the American Government objective of 20/10, i.e., replacing 20% of gasoline with bio-ethanol over the next 10 year.

After corn, steam is the most important cost in the production of ethanol. Based on an analysis of the dry milling process, Peschel et al (2006) estimated that steam represents approximately 15% of the total production cost (base case) estimated at US$1.61/gal (0.43 US$/L), before credits for any by-products.

The technology introduced in this paper carries the potential of reducing energy consumption for distillation/dehydration by up to 50% through the single pass removal of water from the water/ethanol stream at the beer column outlet, using a novel membrane process. For a large ethanol plant (42 million gal (160,000 ML) per year), this translates into a reduction of greenhouse gases (GHG) of 21,000 t/year, which is equivalent to over 2 million gallons of gasoline.

The data presented is for ethanol produced from corn, but the technology is also applicable, with similar benefits, to other feed stocks.

The use of membrane in the processing of ethanol is not new. Pervaporation (Huang, 1991) or vapor permeation (Ikeda and Watari, 2005) have been proposed as a replacement of the dehydration step (i.e., in replacement of the molecular sieves to break the azeotrope). These solutions carry limited potential for energy reduction as the bulk of the energy is spent in the distillation process itself.

The Siftek™ Membrane and System

Siftek™ is a hydrophilic polymer membrane that can be used to dry ethanol in the vapor phase, in a continuous process illustrated in Figure 1. The ethanol-water vapor mixture is fed to the bore side of a module containing thousands of fine polymeric hollow fiber membranes. As the mixture travels down the parallel channels, water vapor is drawn through the dense polymeric membrane (while ethanol is rejected) under a driving force established by a vacuum on the shell side of the module.

The membranes are used under rigorous conditions and must have exceptional thermal, mechanical and solvent resistance properties. The membrane is a proprietary formulation based on polyimide (US patent application 2006/0117955) which has been the object of 8 years of development at Laval University and at Vaperma (Huang, 2003a, 2003b). This membrane provides high flux and water/ethanol selectivity.

The hollow fiber membranes are assembled into an element which consist of several thousands hollow fibers potted at both ends into a tube sheet. An element is inserted into a stainless-steel pressure vessel to form a module. Modules are arranged in series-parallel in a system.

The key elements of a Siftek™ system are shown in Figure 2. The vapor feed to the membrane system should be available at a pressure of 6-7 psi gage (40-50 kPa) to account for flow control at the inlet, pressure losses through membrane modules and the fact that dry ethanol should be produced at a pressure slightly above atmospheric; for vacuum distillation columns, re-compression of the feed may be required. The pretreatment consists of unit operations to remove entrained particles and mist, and heat the ethanol-water vapor mixture to a temperature of about 9°F (5°C) above the dew point. Typically, the operating temperature is between 212-220°F (100-110°C).

A stage is defined by the method used to handle permeate as described below. The number of modules per stage normally decreases as water is withdrawn, to maintain a minimum vapor velocity in the hollow fibers.

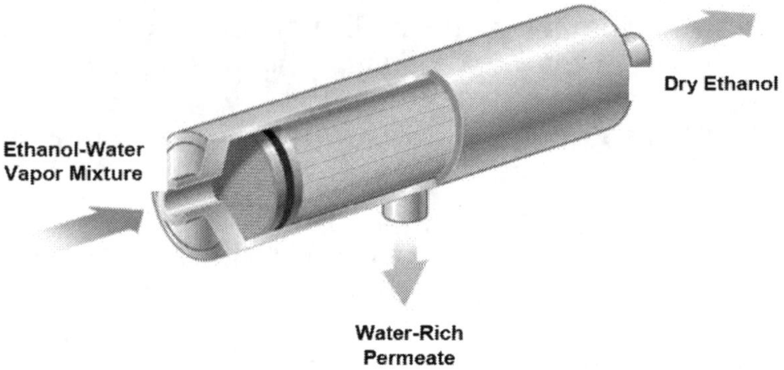

Figure 1 Conceptual operation of the Siftek™ membrane for the drying of ethanol

The basic options for permeate handling include mechanical vapor recompression (MVR) and condensation as illustrated for Stage I and Stage II in Figure 2. In both cases, the primary purpose is to establish a driving force for permeation by creating a vacuum on the permeate side of the membrane. A lower vacuum is required for Stage II as the feed water concentration near the product end is less than 1 wt%. Typical values are minus 10-12 psi gage (20-30 kPa abs) for Stage I and minus 13-14 psi gage (2-10 kPa abs) for Stage II. MVR allows to recompress the permeate vapor (mostly water) to the point where it can be reused as heating steam. In a condensation scheme, a vacuum pump is normally used to reach a lower vacuum and to evacuate non-condensable gases.

The degree of dryness of the ethanol product is controlled automatically by varying the power input to the compressors and vacuum pumps part of the system.

Ethanol plants are fully integrated from a mass and energy point of view. There are many options available to recycle the recovered vapor and liquid products to maximize the benefits of a membrane system; they are plant-specific and beyond the scope of this paper.

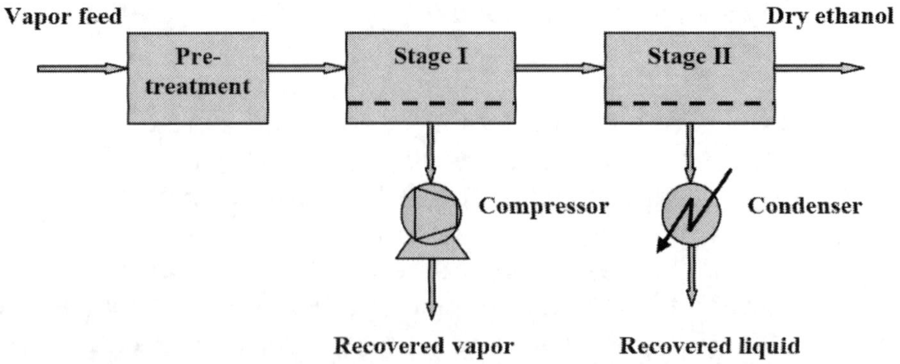

Figure 2 Components for the Siftek™ membrane system

Energy Consumption for Fuel-Ethanol Production

The process flow diagram for a most common dry-milling ethanol plant is presented in Figure 3 and consists of the following typical steps. The corn kernels are ground into flour; water and enzymes are added and the slurry is cooked to break down the starch and promote the formation of glucose. This mixture is transferred to fermenters where the yeast converts the sugars into ethanol. The beer, which contains about 10 wt% ethanol and a high concentration of suspended solids, is directed to distillation which is carried out in two steps. In the first step, called the beer column, the stillage (the solids and a large fraction of the water) is collected at the bottom. A water-ethanol mixture containing 40 to 70 wt% ethanol flows to a second column, called the rectification column, where the alcohol contents is increased to the maximum that can be reached by conventional distillation, 90-95 wt% (the azeotrope is 95.6 wt%). Final drying to fuel-grade ethanol is done with molecular sieves.

The stillage is centrifuged to produce distiller wet grain. The liquid fraction is concentrated by evaporation to produce condensed distiller solution which is mixed with the distiller wet grain to produce dry distiller grain with solubles (DDGS).

The total energy used in this process, including natural gas and electricity for non-denatured alcohol, varies between 38,900 and 50,000 BTU/gal (10.8-13.9 MJ/L), the lower figure representing the best available conventional technologies according to Kim and Dale (2005). The distillation and molecular sieve processes described above account for the largest fraction of energy used in a dry-milling plant, as shown in Table 1. Applying an average fraction of 50% from Table 1 to the total energy figures cited above, the energy needed for distillation/dehydration varies between 19,450 and 25,000 BTU/gal (5.4 – 6.9 MJ/L).

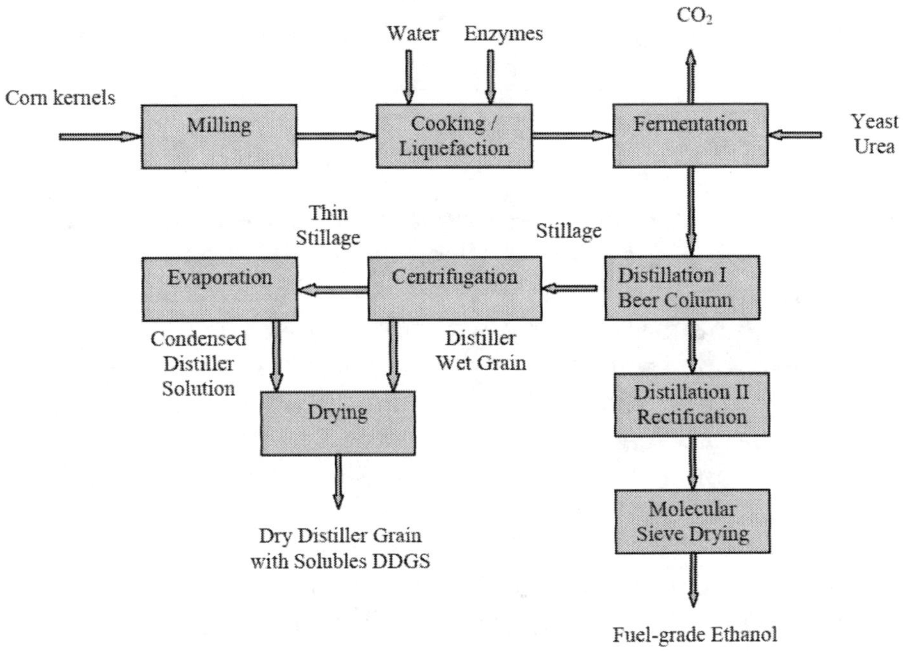

Figure 3 Conceptual process flow diagram for a corn-based, dry-milling ethanol plant

Based on these data, and recognizing that further improvements can be expected from conventional processes (Peschel et al, 2006), the values shown in Table 2 have been selected as reference for the estimation of the potential energy savings obtainable using the Siftek™ technology.

Table 1 Distribution of energy use in a dry milling ethanol plant

Process	Distribution (%)	
	McAloon et al (2004)	Kim and Dale (2005)
Milling	1.0	0.8
Cooking/liquefaction	19	29.6
Fermentation	1.0	3.5
Distillation/dehydration	45	56.5
DDGS recovery	34	9.6
Total	100	100

Table 2 Baseline energy for conventional ethanol distillation/dehydration (this work)

Technology	Total Energy for Distillation/dehydration	
	BTU/gal	MJ/L
Best available technology in a modern plant (New plant)	17,942	5.00
Older conventional technology in a existing plant (Retrofit plant)	25,632	7.15

Integration of Siftek™ into Ethanol Plants

Two case studies for New and Retrofit Plants were developed to illustrate the energy savings that can be derived from integration of the Siftek™ technology into ethanol plants. For both cases, the total capacity is 42 Mgal/y (160 ML/y). Full mass and energy balance were done using Hysis simulation software. Vaperma's proprietary model was used to design the membrane system. Scenarios were developed assuming the following ethanol concentrations: beer = 10 wt%; beer column outlet = 53 wt%; rectification column outlet = 90 wt%. Energy costs were estimated at 8.50$/GJ for natural gas, and .042 $/kWh for electricity (electricity costs considered were for running the compressors related to the Siftek™ system, in addition to other uses in the plants).

The results for the New Plants case studies are presented in Figure 4. When building a new plant, maximum benefits are obtained from integrating membranes at the core of the ethanol plant since the costs of conventional equipment can be avoided. However, for this case, it was assumed that membranes would compete with the efficient modern distillation process, i.e., 17,942 BTU/gal (5.0 MJ/L). Two application scenarios were considered, first using the membranes for drying only (New-Drying) and second using the membranes for dewatering (New-Dewatering).

In the New-Drying scenario, a single stage membrane system is used to replace the molecular sieves with a feed from the rectification column at 90 wt% ethanol. The permeate is condensed and returned to the beer column inlet. Results in Figure 4 show little energy benefit for this case. The bulk of the energy is consumed by distillation, not by the molecular sieves and the overall energy saving is only 3.8%.

In the New-Dewatering scenario, a two stage membrane system is used to replace both the rectification column and the molecular sieves with a feed from the beer column at 53 wt% ethanol. Permeate from the first stage is recompressed, condensed and the heat recovered is used to heat the beer column; the low ethanol liquid is recycled to the fermenters. Alternatively, this steam could be used to dry stillage with similar energy benefits. The permeate from the second stage is condensed and returned to the beer column inlet. Figure 4 shows that a significant reduction in steam energy is obtained at a relatively small electrical compression load. When converted into cost, savings of 43% are obtained, equivalent to 7.0 US¢/gal, or 1.8 US¢/L.

The results for the Retrofit Plants case studies are presented in Figure 5. In a retrofit situation the membrane equipment must be added to existing equipment with the potential benefit of de-bottlenecking the plant and increasing throughput. However, a higher baseline for older conventional distillation technology can be used, i.e., 25,632 BTU/gal (7.15 MJ/L). Two application scenarios were considered, first using the membranes for drying only, Retrofit-Drying, and second using the membranes for dewatering, Retrofit-Dewatering. In both cases, the Siftek™ systems were designed to handle 33% of the plant total flow rate.

		Conventional Technology	**Siftek™ Drying System**	**Siftek™ Dewatering System**
New Plants		Distillation I Beer column → Distillation II Rectification column → Molecular Sieves	Distillation I Beer column → Distillation II Rectification column → Siftek™ Membranes	Distillation I Beer column → Siftek™ Membranes I → Siftek™ Membranes II
Steam / Compression Energy	**BTU/gal**	**17,942 / 0**	**17,248 / 14.8**	**7,923 / 1652**
	kJ/L	**5,006 / 0**	**4,812 / 4.1**	**2,211 / 461**
Total Energy Cost	**US¢/gal**	**16.13**	**15.52**	**9.16**
	US¢/L	**4.26**	**4.10**	**2.42**
Savings			**3.8%**	**43%**

Figure 4 Integration of Siftek™ systems in New Plants

In the Retrofit-Drying scenario, a single stage membrane system is used on a 33% side-stream of the rectification column at 90 wt% ethanol. The permeate is condensed and returned to the beer column inlet. As was the case for the New Plants, results in Figure 5 show little energy benefit for this case.
In the Retrofit-Dewatering scenario, a two stage membrane system is used to treat a 33 % stream from the rectification column with an ethanol concentration of 60 wt% ethanol. The permeate streams are handled similarly to the New case. Figure 5 shows an overall energy savings of 28%, equivalent to 7.3 US¢/gal, or 1.9 US¢/L. It is worth noting that, in this scenario, the total energy consumption has been reduced to the level of the best available technology in a modern plant without membranes (Table 2).

The key reason a membrane can separate an ethanol-water mixture with little energy is that separation takes place based on the selectivity of the membrane, without phase change (as is the case for distillation). In a membrane system, the azeotrope has no influence over performance.

A fundamental difference between membranes and molecular sieves used for water removal is that the productivity of a membrane system <u>increases</u> with water concentration, while the productivity of molecular sieves <u>decreases</u> with water concentration. This is so because the flux through the membrane is proportional to the water concentration in the feed while the water holding capacity of a molecular sieve is finite.

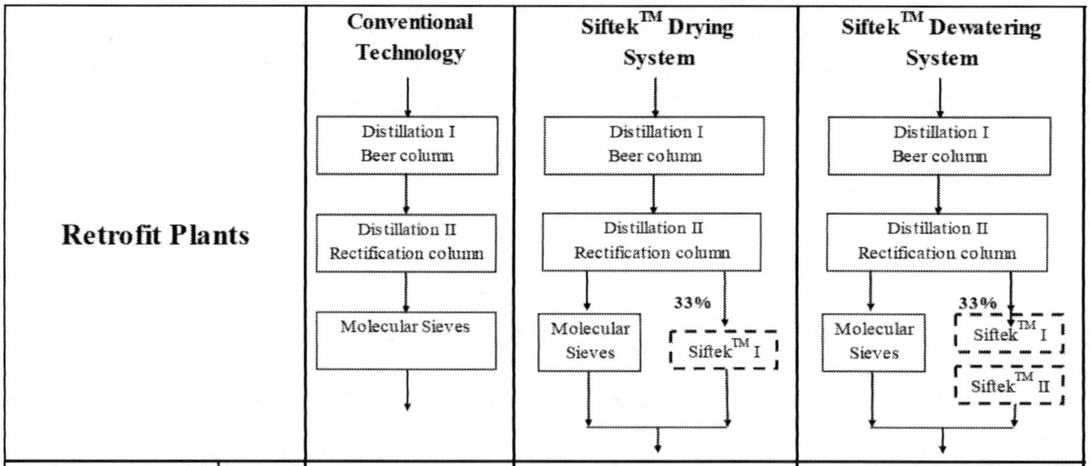

		Conventional Technology	Siftek™ Drying System	Siftek™ Dewatering System
Steam / Compression Energy	BTU/gal	25,632 / 0	17,246 / 5.0	17,486 / 412
	kJ/L	7,152 / 0	4,812 / 1.4	4,879 / 115
Total Energy Cost	US¢/gal	25.85	25.17	18.53
	US¢/L	6.82	6.64	4.89
Savings			2.6%	28%

Figure 5 Integration of Siftek™ systems in Retrofit Plants

The Siftek™ membrane process was designed to maximize energy savings by using membranes for both dewatering for partial replacement of distillation <u>and</u> drying, for replacement of molecular sieves.

The Tiverton Field Demonstration

The Siftek™ technology has been demonstrated in the field at the Greenfield Ethanol plant in Tiverton, Ontario, Canada. The vapor feed was a split stream from the rectification column with an ethanol concentration varying between 75 and 90 wt%. The unit, with a capacity of 1200 kg/day (2600 lbs/day) of vapor feed is fully automated, and operates at a total feed pressure of 120 to 140 kPa (3 to 6 psig) and temperature of 105 to 115 C° to produce fuel-grade ethanol (> 99 wt%).

Figure 6 illustrates the key elements of the field unit. The vaporous feed goes through a knock pot to remove any entrained particles or liquid droplets and flows to the Siftek™ membrane module. In a single-pass and continuous operation, the ethanol gets concentrated to the set-point value of 99.3 wt%. The vapor product is condensed, flows through a sampling reservoir and the liquid product is extracted by a pump.

To provide for a wide range of testing conditions, the equipment on the permeate side of the membrane in the field unit is more complex than in a typical plant. The permeate vapor can be cooled and recompressed before being condensed. A vacuum pump maintains the pressure required to achieve the target product quality and extract any non-condensable gases. The condensed permeate flows through a reservoir where the ethanol concentration is measured continuously.

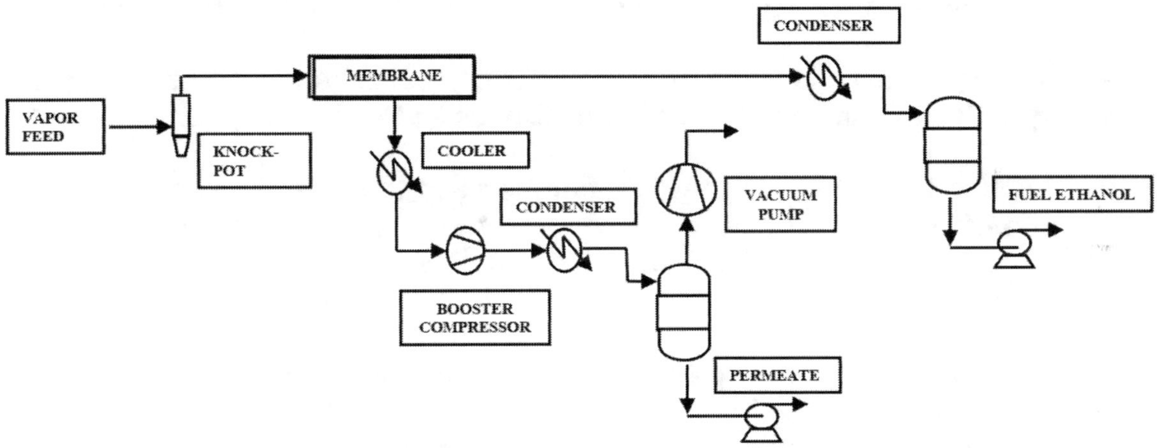

Figure 6 Process flow diagrams for the Siftek™ Tiverton field unit

The results of the field demonstration over of period of 140 days are shown in Figure 7. While there were a number of shut-down periods due to software issues and modifications to the equipment, the unit produced fuel-grade ethanol consistently.

Conclusions

Ethanol is booming, driven by policies to reduce dependency on foreign oil and the production of greenhouse gases. Ethanol producers cannot control the cost of raw materials, but they can reduce the processing cost with innovating technologies.

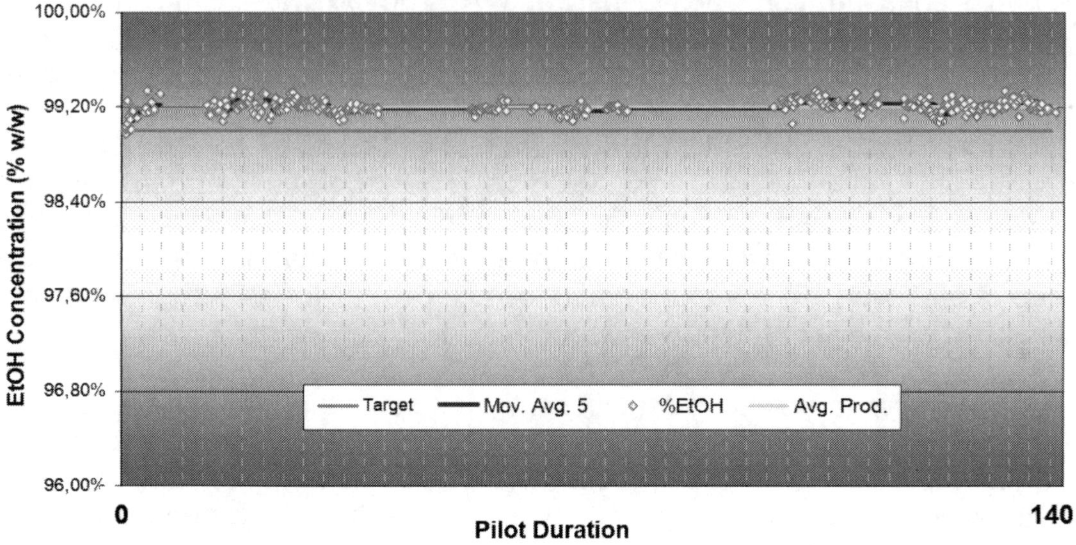

Figure 7 Results for the Siftek™ Tiverton field unit

After corn, the largest cost in the production of ethanol is energy and the largest amount of energy is consumed to purify the ethanol by distillation.

Siftek™ is a polymeric membrane that can be used to dry ethanol in the vapor phase. The membrane preferentially permeates water over ethanol in a continuous process. Energy reductions are obtained because this membrane is well suited to remove large quantities of water without phase change.

Siftek™ can be used for new plants or retrofits. In a typical New Plant, the technology can reduce distillation/dehydration energy by 43% as compared to the currently best available technology (equivalent to 7.0 US¢/gal, or 1.8 US¢/L). In a retrofit situation, Siftek™ can de-bottleneck the plant, increasing capacity while reducing energy.

For a large ethanol plant (42 million gal (160,000 ML) per year), the energy reduction obtained with the Siftek™ technology translates into a reduction of greenhouse gases (GHG) of 21,000 t/year, which is equivalent to over 2 million gallons of gasoline.

The pay-back for the Siftek™ technology is estimated to be less than 1 year for a new plant and between 1 and 2 years for a retrofit plant, based on the energy costs assumed in this paper.

The technology has been demonstrated in the field at the Greenfield Ethanol plant in Tiverton, Ontario, Canada.

References

J. Huang, R.J. Cranford, T. Matsuura, and C. Roy, "Water Vapor Permeation Properties of Aromatic Polyamides", Journal of Membrane Science, 215:129-140, 2003a.

J. Huang, R.J. Cranford, T. Matsuura, and C. Roy, "Water Vapor Sorption and Transport in Dense Polyimide Membranes", Journal of Applied Polymer Science, 87:2306-2317, 2003b.

R.Y.M. Huang, Editor, "Pervaporation Membrane Separation Processes", Elsevier Science Publishers B.V., Amsterdam, 1991.

S. Kim and B.E. Dale, "Environmental Aspects of Ethanol Derived from No-tilled Corn Grain: Non Renewable Energy Consumption and Greenhouse gas Emissions". Biomass and Bioenergy. 28:475-489, 2005.

S. Ikeda and M. Watari, "Ethanol Dehydration by Nanoporous Zeolite Membrane-based Vapor Permeation", Ethanol Producer Magazine, April 2005.

A. McAloon, W. Yee, and F. Taylor, "Production of Ethanol from Corn by the Dry Grind Process", USDA-ARS Eastern Regional Research Center, Wyndmoor, PA, 2004

A. Peschel, R. Karuppiah, M. Martin, I.E. Grossmann, L Zullo, and W. Martinson, "A Superstructure Optimization Approach for the Design of Corn-based Ethanol Plants", Paper 648e, AICHE Annual Meeting, San Francisco, November 12-17, 2006.

479b Fractionation of Pre-Treatment Hydrolysis Products from Lignocellulosic Biomass

Kendra Colyar[1], **John Pellegrino**[1], and Kiran L. Kadam[2]. (1) CEAE Department, University of Colorado, ECOT441, Boulder, CO 80309-0428, (2) PureVision Technology, 511 N. McKinley Ave., Fort Lupton, CO 80621

The commercial attractiveness of the enzymatic lignocellulosic based biorefinery for making transportation fuels relies on many factors, including the production of valuable byproducts; minimizing chemical consummables; and efficient water recovery and reuse. If the pre-treatment hydrolysis of woody or herbaceous feedstocks is done with acid and base in a novel sequential reactor design, one of the two final liquid effluents contains a variety of lignin species, alkalinity, and dissolved cellulose—similar to the black liquor produced in the pulp and paper industry. We will describe results from a two-step fractionation approach, pressure-driven ultrafiltration (UF) followed by electrodialysis (ED), with a focus on the ED step and the projected energy efficiency and material balance. The combined processes were run as batch separations. The overall process configuration indicates that >95% of the Klason lignin is recovered in the two streams (UF and ED retentate) and > 78% of the alkalinity can be recovered as NaOH for recycle back to the pre-treatment hydrolysis operation. The remaining $Na+$ is likely bound to the lignin recovered in the UF retentate. The presentation will focus on the influence of current density and ion exchange membranes' chemistry (all commercially available) on the overall mass fractionation balance and energy efficiency of the ED process.

Membrane steam reforming for H_2 production

*Erik Engwall, John Saukaitis[+], Mahendra Joshi, Andreas Matzakos, and Alan Del Paggio**

Shell International Exploration and Production/ [+]Shell Global Solutions
3333 Highway 6 South
Houston, TX 77082, USA

** CRI/Criterion Inc.*
PO Box 4489
Houston TX 77210-4489, USA

Composite palladium membrane technology has matured considerably in the past decade and is now poised for commercial applications in H_2 manufacture in the near future. The use of membrane reactors for H_2 manufacture is particularly attractive. Membrane reactors can combine steam reforming, water gas shift and hydrogen separation into a single unit operation. Pd-alloy membranes separate hydrogen from CO_2 at very high selectivity. The continuous withdrawal of hydrogen along the length of the reactor removes the thermodynamic limitations enabling conversions of 95 % or higher. The removal of equilibrium constraints on conversion enables the membrane reactor to operate at much lower temperatures (500°C) than conventional steam reforming reactors. In addition to pure hydrogen, the membrane reactor outputs a high pressure (30 Bara) retentate stream that is concentrated in CO_2, and is low in CO. This stream is advantaged for CO_2 sequestration as compared to low pressure CO_2 output from conventional SMR.

Shell has pursued a program in the development and commercialization of high temperature membranes, and membrane steam reforming (MSR) process technologies for H_2 production [1,2]. CRI/Criterion is in the process of commercializing Pd and Pd-alloy membranes on sintered porous metal supports. The scope of the CRI/Criterion effort includes the development of membranes for sale to third parties in addition to the support of the Shell MSR efforts. Figure 1 shows a 1"OD by 6"L Pd membrane from CRI/Criterion, which is of the type used in the MSR experiments presented below. The H_2 permeable section is the bright reflective region (center).

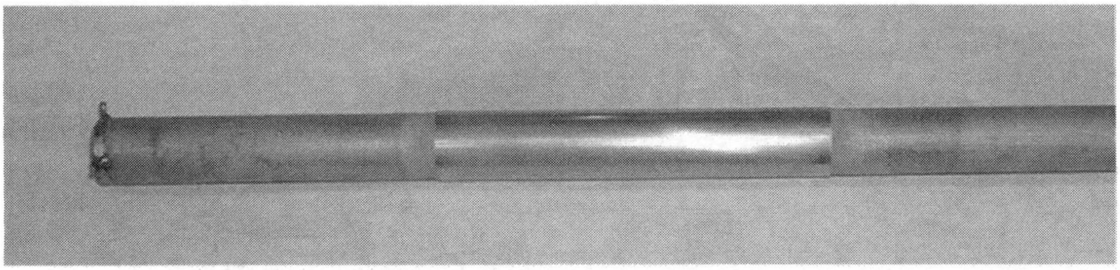

Figure 1: High temperature H_2 separation membrane manufactured by CRI-Criterion by deposition of Pd on a sintered porous metal support with permeable dimensions of 1"OD and 6"L.

Moreover, CRI has produced membranes of this type as large as 2"OD by 48"L, by welding two separate 24"L sections. These membranes can be produced with a hydrogen permeance in the range of 50-70 [$Nm^3/m^2.h.bar^{0.5}$]. Both the hydrogen permeance and the separation selectivity are stable at temperatures up to 500°C.

MSR experiments were conducted at 500°C with a reaction pressure of 30 Bara in an electrically heated shell and tube reactor. The annulus between the membrane and the reactor outer shell was packed with a commercial reforming catalyst. The hydrocarbon feed to the reactor was a blended gas that resembled natural gas (93 % CH_4, 4 % C_2H_6, 1% C_3H_8, 2% N_2) and feed steam to carbon ratio was 3.0. Figure 2 shows hydrogen output, purity and conversion data for an MSR run lasting **51 days**.

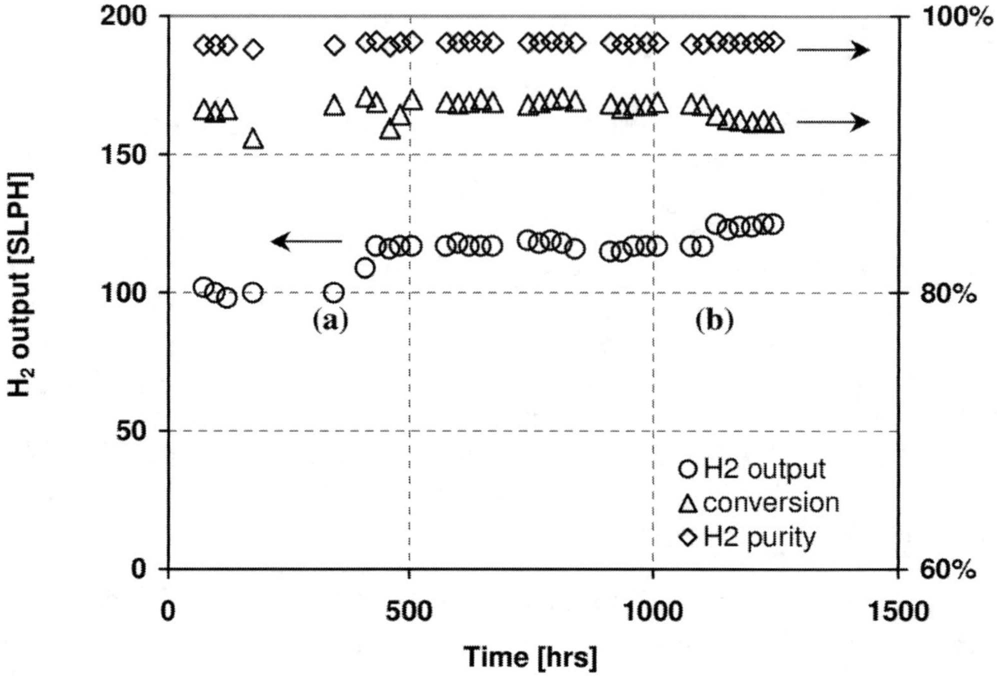

Figure 2. MSR with a CRI/Criterion composite Pd membrane like the one shown in above. The unit was operated at 500°C and a reaction pressure of 30 Bara.

The H_2 permeance was 56.1 [$Nm^3/m^2.h.bar^{0.5}$] at 500°C. The H_2 output of the unit was dependent on MSR feed flow rate. At the points labeled **a** and **b** the hydrocarbon feed was increased, resulting in increases in the H_2 flux and slight decreases in conversion. When operating conditions were held constant the H_2 flux and conversion were very stable. The highest output rate demonstrated for this trial was 125 SLPH and conversion varied between 94.0 and 92.4 %. The retentate stream composition from the reactor, measured at maximum H_2 output, was 91 mol% CO_2 and 0.7 mol% CO (dry basis). The membrane in this experiment yielded 98+ mol% hydrogen over the life of the test. The use of a high reaction pressure is advantageous because it increases the driving force for H_2 transport allowing less membrane area, and it provides advantages for CO_2 sequestration. Efforts are currently under way to improve the membranes to achieve higher permeate purities while retaining a high reaction pressure.

These results show that the CRI/Criterion membrane technology, is capable of maintaining stable H_2 flux and stable hydrogen product purity for extended periods at conditions of temperature and pressure which are both among the highest demonstrated for a **metal supported** Pd membrane. Further MSR evaluations are currently underway on a 2"OD by 24" long version of this membrane technology. Following the results shown in Figure 2, the membrane was provided as an evaluation sample to a third party with potential interest in purchasing membranes from CRI/Criterion.

---

[1] Matzakos, A.N.; Wellington, S.L.; Mikus, T.; Ward, J.M.. U.S. Patent 6,821,501 B2, Nov. 23, 2004.

[2] Andreas N. Matzakos and Scott L. Wellington. "Novel Membrane Steam Reforming (MSR) Reactor for Pure Hydrogen Production". 2004 Spring AIChE Meeting

479d The Separation Of Water From Steam/hydrogen Mixtures Using A Naa-Coated Alumina Membrane

Raymond Everson[1], Hein Neomagus[1], Henning Krieg[2], and Shawn Moodley[1]. (1) School of Chemical and Minerals Engineering, North-West University, Potchefstroom, Hoffmann Street, Potchefstroom. 2520, South Africa, (2) School of Chemistry, North-West University, Potchefstroom, Hoffmann Street, Potchefstroom. 2520, South Africa

The importance of coal-to-liquid and gas-to-liquid process technologies is being recognised throughout the world for the production of high value fuels and chemicals from coal and natural gas. At the heart of both these technologies are the Fischer-Tropsch reactors where purified synthesis gas, is converted over metal catalysts to yield a broad spectrum of hydrocarbons. In this reaction, excessive amounts of water are formed, which have a negative impact on the process, in terms of equilibrium conversion and kinetics of the reaction and the stability of the catalyst. It would therefore be beneficial to in-situ remove the water from the reaction mixtures, which could be facilitated by membranes. Since the harsh conditions of the Fischer-Tropsch reaction, ceramic membranes are preferred. It is the aim of this work to study the selective removal of water from steam/hydrogen mixtures. Hydrogen has been chosen, since it occurs as the largest fraction in the Fischer-Tropsch reaction product. For this purpose, a NaA composite zeolite tubular membrane was developed. The alumina support was prepared via centrifugal casting, and a NaA layer was deposited via a thermal deposition method. A shell and tube membrane module was used and both single component and binary mixtures were studied. The experimental results showed ideal selectivities less than unity, and decreasing rapidly from a temperature > 430 K. However, in binary mixtures, the selectivity of water was very large (> 1000) at low temperatures. As the temperature increases, a sharp decrease of the selectivity occurs at a temperature which was dependent on the partial pressure of the steam. This sudden decrease was explained by inter-crystalline diffusion, and the pore diameter of these inter-crystallines was estimated to be 2 nm.

Polybenzimidazole (PBI) Nanofiltration Hollow Fiber Membranes Applied in Forward Osmosis Process

Kai Yu Wang [a], Tai-Shung Chung [a, *], Jian-Jun Qin [b]

[a] Department of Chemical & Biomolecular Engineering
National University of Singapore
4 Engineering Drive 4
Singapore 117576

[b] Centre for Advanced Water Technology
Singapore Utilities International Pte Ltd
80/82 Toh Guan Road East
Singapore 608575

*Corresponding author
Email: chencts@nus.edu.sg
Tel: (65) 6516-6645
Fax: (65) 6779-1936

Keywords:
Forward osmosis, Polybenzimidazole, Nanofiltration, Membrane separation, Hollow fiber

1. Introduction

Recently, forward osmosis (FO) has draw much attention from researchers and its applications have been developed in various fields, such as wastewater treatment, pharmaceutical and juice concentration, desalinating seawater, even in the power generation and the potable water reuse in space. By employing the osmosis pressure as a driving force, forward osmosis can operate without high pressures and high temperatures which are necessary in the reverse osmosis and membrane distillation processes. As the barrier between draw solution and feed solution to transport water and reject solute, membrane plays an important role on the realization of forward osmosis. Presently, most available membranes used in the forward osmosis are dense semi-permeable membranes which are obtained from reverse osmosis industry except the commercialized FO membrane developed by Hydration Technologies Inc. In addition, almost all FO studies were based on the flat membrane configuration.

Because nanofiltration (NF) membranes have a much higher rejection to divalent ions but partly permit the monovalent ions get through, they can be employed to fractionate electrolyte mixtures. Nanofiltration membranes have been widely used for the separation of small neutral and charged molecules in aqueous solutions through their active nano-scale pores, with the estimated pore sizes of around 0.5 ~ 2 nm in diameter. As a pressure-driven process, nanofiltration has become an effective means for the removal of heavy metals ions and divalent anions from wastewater.

This paper explores if one can fabricate polybenzimidazole (PBI) NF hollow fiber membrane with desirable mean pore size and testing its performance in forward osmosis. PBI is a class of heterocyclic polymers that were commercially developed by the Celanese Corporation in 1983, with the chemical structure as shown bellow.

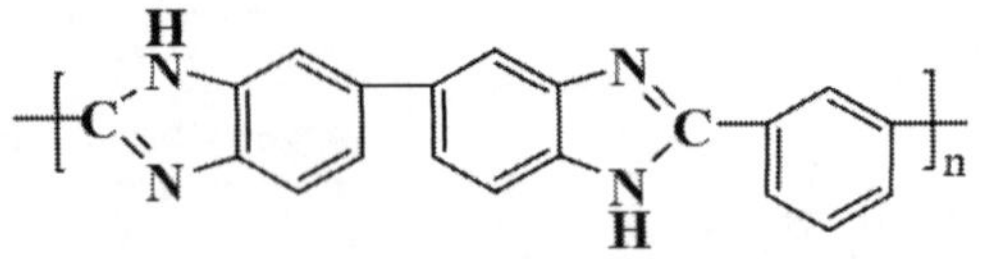

Moreover, PBI membranes have been fabricated to be employed in reverse osmosis and nanofiltration processes and as the ion exchange membrane in fuel cell for its excellent chemical and thermal stability. The heterocycle imidazole ring makes it possible for both inter and intramolecular hydrogen bonding to occur between PBI molecules. As a result, PBI may become self-charged in aqueous environments because the adjacent benzene ring delocalizes the proton of the imidazole group. The charged property makes PBI membrane applicable in water separation together with its high hydrophilicity for antifouling.

2. Fabrication of PBI nanofiltration hollow fiber membrane

The PBI hollow fibers were produced from the polymer dope of PBI/DMAc/LiCl (22.6/75.6/1.8 wt%). The selective layer was formed on the shell side of hollow fiber. After immersing in tap water for three days in order to remove the residual DMAc and LiCl, the fabricated hollow fiber membranes were soaked in a 50 wt% glycerol aqueous solution for 48 hrs and then air-dried at room temperature. SEM images illustrate the hollow fiber structure which consists of four types of morphologies, namely: asymmetric outer selective skin layer, finger-like macrovoids, sponge-like substructure, and porous inner skin layer. The outer and inner diameters of fiber are 273 and 137 microns, respectively. During PBI hollow fiber spinning, water is applied as the external coagulant to form the nano-porous outer selective layer. The formed sponge-like substructure near inner skin is due to the delayed phase separation induced by the solvent enriched bore fluid of DMAc/ethylene glycol mixture. This kind of microstructure makes PBI hollow fiber mechanically strong to sustain high trans-membrane pressure during nanofiltration except that the substructure may produce internal concentration polarization and seriously influence the water recovery during forward osmosis.

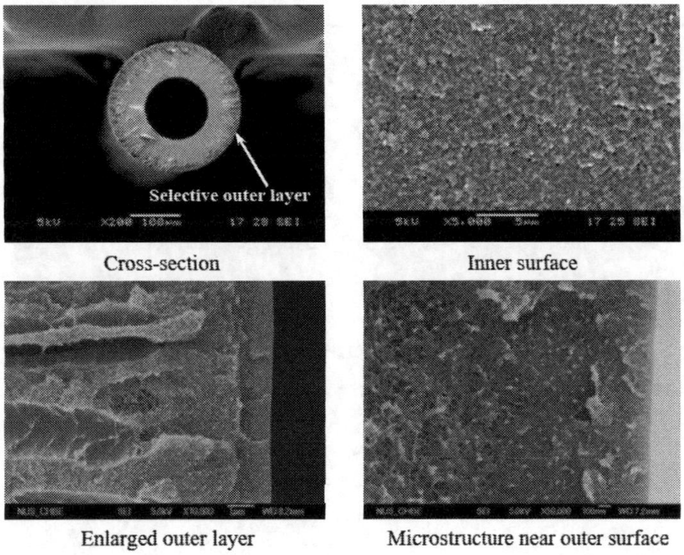

Cross-section — Inner surface

Enlarged outer layer — Microstructure near outer surface

3. Membrane characterization through nanofiltration

The pure water permeability is 0.5 liter m^{-2} bar^{-1} hr^{-1} measured at 10 bar. From the solute transport method, the mean effective pore radius of PBI NF hollow fiber membrane was 0.32 nm. The pore size distribution probability density curves show that the PBI NF membranes have a sharp pore size distribution which is essential for ion rejection. The newly developed PBI NF membrane has molecular-size pores throughout their outer selective layer that will determine the permeate flux and separation performance.

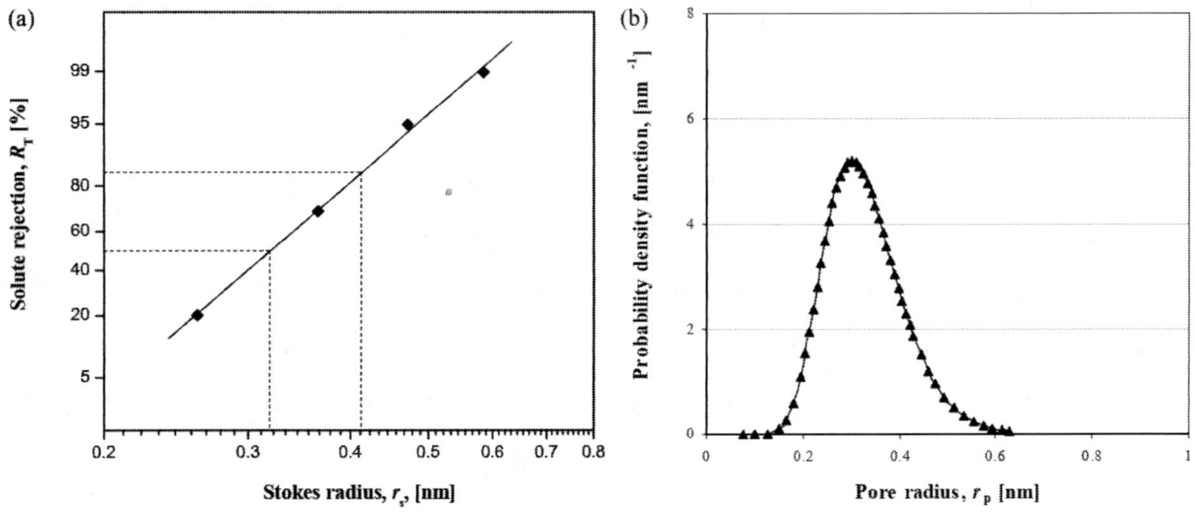

The salt rejection of the PBI hollow fiber as a function of trans-membrane pressure is shown below. This PBI NF membrane exhibits high rejection to divalent cations and anions whereas low rejection to monovalent ions in the diluted solution, which is attributed to the ion size exclusion and Donnan electrostatic effect. In addition, the solute rejection increases with increasing pressure. In nanofiltration, the salt rejection mechanisms include the steric (size exclusion) effect and electrostatic partitioning interaction (Donnan exclusion) between the membrane and external solution. Because divalent ions of Mg^{2+} and SO_4^{2-} have bigger ion size and higher change density than monovalent ions, the PBI membrane has higher rejections to divalent ions.

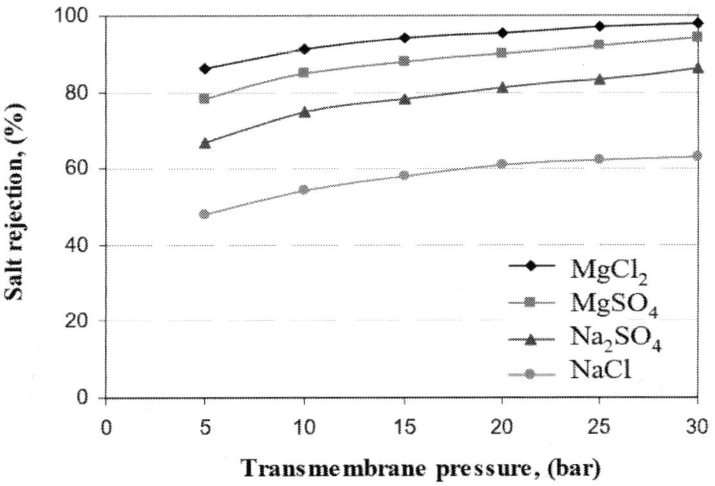

4,1. Forward osmosis based on the PBI NF membrane

Forward osmosis experiments were conducted on a lab-scale circulating filtration unit. Each membrane module had a filtration area of about 300 cm^2. The flowing channels of draw solution and DI water inside the membrane module can be interchanged to test the effect of membrane structure on the water permeation flux. The draw solution and DI water co-currently flowed through the module and were maintained at the same temperature by the water bath. The concentration of draw solution was always kept constant during operation (i.e., the draw solution never circulates to the original tank after flowing through the membrane), whereas DI water in the other side circulated. Two different membrane orientations were tested to investigate the effect of membrane structure and concentration polarization on water permeation flux; namely, the pressure retarded osmosis (PRO) mode for when the draw solution flows on (or against) the selective layer, and the forward osmosis (FO) mode for when the draw solution flows on the porous support layer. The PRO is equivalent to FO if the pressure in the draw solution side is negligible.

Water permeation flux was calculated from the weight changes of draw solution and DI water. Salt selectivity by the membrane was calculated from the concentration of salt in the DI water which permeating from the draw solution. The salt selectivity due to the salt transporting from the draw solution to DI water is defined as:

$$S_{\text{salt}} = (1 - \frac{n_{salt}}{t \times V_{DI} \times C_d}) \times 100\%$$

where n_{salt} is the amount of salt transporting from draw solution to DI water during a predetermined period, t, operating time, V_{DI}, the average volume of DI water at the begin and end of forward osmosis and C_d, the draw solution concentration.

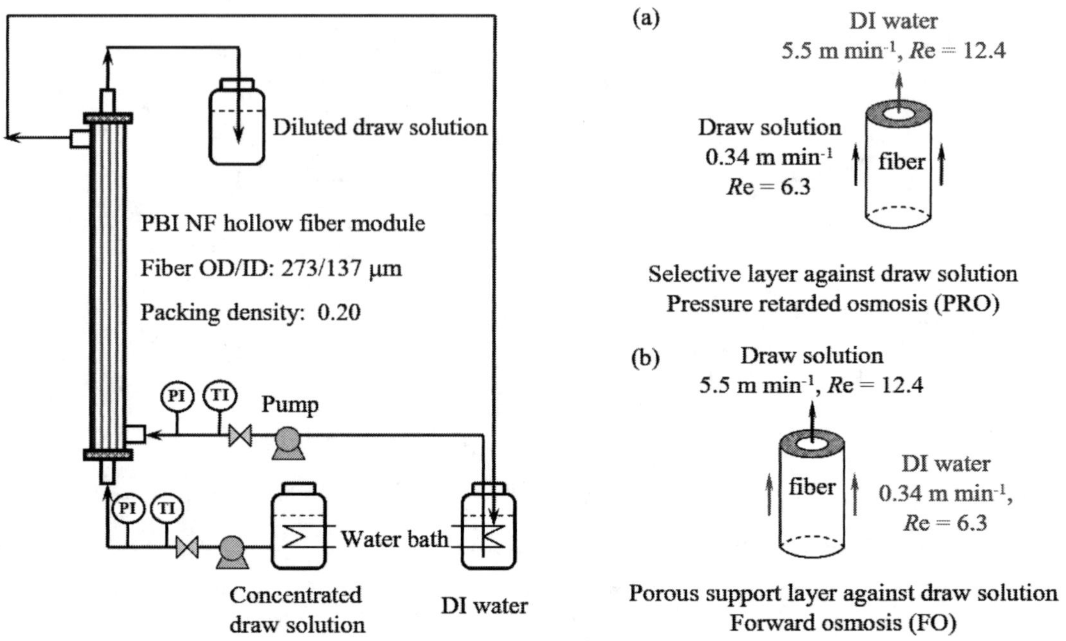

4.2. Effect of draw solution concentration on water permeation flux

The dependence of water permeation flux on draw solution concentration for two different flowing patterns inside the hollow fiber membrane is plotted below. It can be found that the water transport through the PBI membrane increases with an increase in draw solution concentration. The draw solution concentration dominates the membrane performance because the osmotic pressure is the driving force in the forward osmosis process. When draw solution flows against the outer selective layer of the hollow fiber (PRO mode), the water permeation is much higher than the other way that draw solution flows against the inner porous support layer of the hollow fiber (FO mode) although two streams are under laminar flow. This is attributed to the lack of internal concentration polarization because the feed solution of DI water flows against the inner porous layer. In addition, the high external concentration polarization may be reduced due to the cross-flow of draw solution along the hollow fiber outer selective layer. However, when draw solution flows against the porous support layer the dilutive internal concentration polarization due to the porous structure of asymmetric membrane seriously decreases the net driving force leading to reduced water permeation flux and there is no external concentration polarization because the dilute solution is the deionized water.

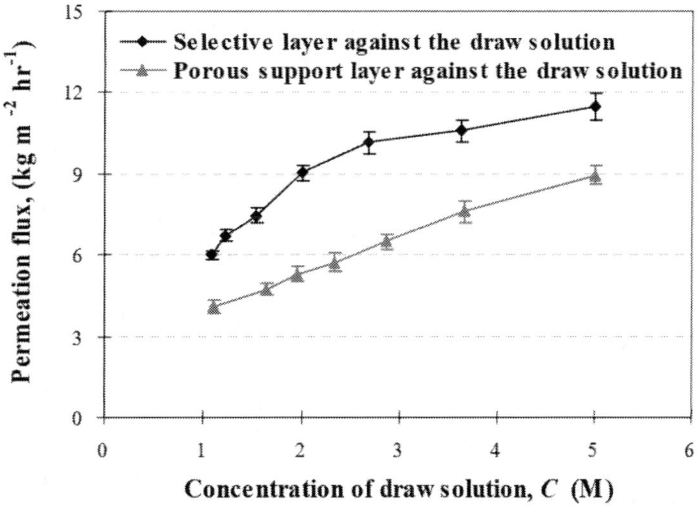

4.3. Effect of temperature on the water transport

Along with the increase in operation temperature, the water permeation flux gets a sharp increase. At the same time, the salt permeation through the membrane is effectively controlled at a quite lower level. High temperature can decrease the water viscosity and enhance the diffusion between the salt solution inside the porous membrane layer and the bulk salt solution in the lumen side, which partly lowers the effect of internal concentration polarization. Under the same temperature, the water permeation flux when draw solution flows against selective layer is much higher than that when draw solution flows against porous support layer. This is also attributed to the serious diluted internal concentration polarization in the porous substructure that is similar to the use of composite reverse osmosis membranes. The apparent activation energy of water permeation as a function of membrane orientations can be calculated based on the Arrhenius equation. E_J values of 18.74 and 17.12 kJ/mol are obtained for the draw solution flowing against the selective layer and porous support layer, respectively. The higher apparent activation energy value when the draw solution flowing against the porous support layer seems to demonstrate a higher internal concentration polarization in the porous support layer.

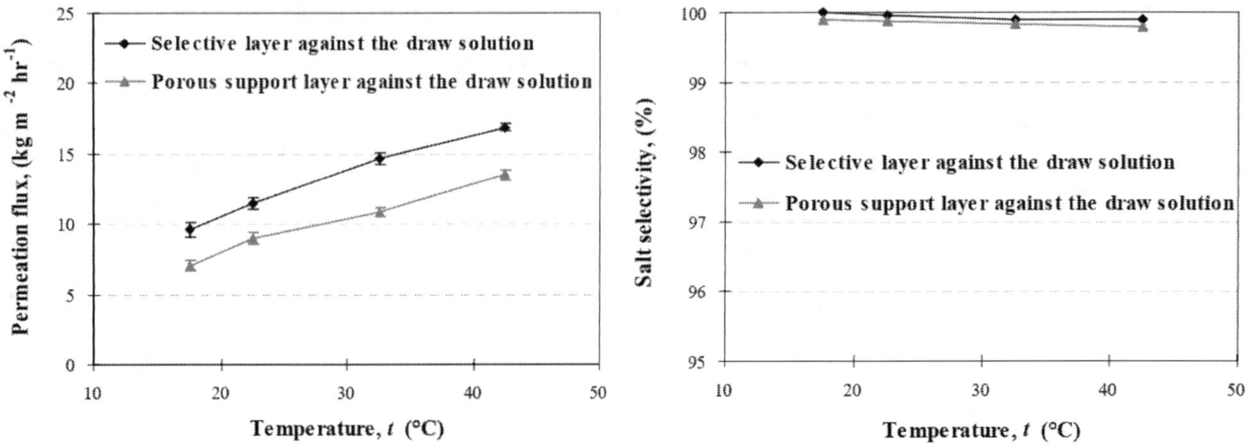

4.4. Effect of different salt solution on the water flux

The osmotic pressures are quite different for these four kinds of salt solutions although they have same molar concentrations because the osmotic pressure is a colligative property which is dependent on the number of ions in solution, so the $MgCl_2$ solution has the highest osmotic pressure, which produces the highest water permeation flux. Because the PBI NF membrane exhibits higher rejection to divalent anions and cations than monovalent ions due to the size exclusion under the condition of concentrated salt solutions, the salt selectivity of $MgSO_4$ by the membrane is highest.

Salt solution (2.0 M)	Osmotic pressure[(1)] (atm)	Permeation flux (kg m^{-2} hr^{-1})	Salt selectivity (%)
NaCl	100.5	3.84	97.31
$MgSO_4$	46.2	5.65	99.99
Na_2SO_4	-	7.74	99.98
$MgCl_2$	256.4	9.02	99.79

Draw solution against the selective layer, 0.34 m min^{-1}; DI water against the porous support layer, 5.5 m min^{-1}.

Moreover, the salt selectivity in the forward osmosis is much higher than the salt rejections in the nanofiltration process. This may be attributed to the fact that the mass transport during nanofiltration (hydraulic pressure driven process) and forward osmosis (osmotic pressure driven process) is different: in nanofiltration process, water transports from a high pressure side to a low pressure side, and salt goes through the membrane together with the water (convection); in the forward osmosis process, water transports from a low osmotic pressure side to a high osmotic pressure side but salt in the draw solution leaks into the low osmotic side (salt diffusion) which is in the opposite direction of water transporting. As a result, the salt rejection during nanofiltration is quite different from the salt leakage during forward osmosis.

In the case of $MgSO_4$ and NaCl, the order of their corresponding permeation fluxes is not consistent with the order of their osmotic pressures. This may be due the fact that the ion sizes of Na^+ and Cl^- are smaller than those of Mg^{2+} and SO_4^{2-}, respectively. When water transports from a low osmotic pressure side to a high osmotic pressure side during the forward osmosis, salt in the draw solution leaks into the low osmotic side (salt diffusion), which is in the opposite direction of water transporting. Under the FO system, the salt selectivity is lower when NaCl is used as the draw solution compared to that of $MgSO_4$. Similarly, under the NF system, the salt rejection is lower for NaCl compared to that of $MgSO_4$. These suggest that the PBI NF membrane has different reflect coefficients for NaCl and $MgSO_4$. As a consequent, the NaCl solution may likely have a less "effective osmotic pressure" than the $MgSO_4$ solution in the current PBI NF-based forward osmosis system.

5. Conclusions

This study has demonstrated that the asymmetric PBI nanofiltration hollow fiber membranes with a mean effective pore size around 0.32 nm in radius can be used effectively for forward osmosis. This newly developed PBI membrane exhibits desirable water permeation flux and high rejection to divalent ions. Therefore, as an environmentally beneficial and economically attractive means, the tailored PBI nanofiltration hollow fiber membrane is of great potential to be utilized in the forward osmosis process for water recover and concentrating various feed solutions. Experimental results verify that concentrated $MgCl_2$ solution can be considered as the draw solution for its high osmotic pressure. Future study will further optimize the PBI hollow fiber structure to decrease the internal concentration polarization and investigate their performance in the industrial applications.

480b Flux Enhancement in Membrane Distillation by Fabrication of Dual Layer Hydrophilic-Hydrophobic Hollow Fiber Membranes

Sina Bonyadi, Chemical & Biomolecular Engineering, National University of Singapore, 10 Kent Ridge Crescent, Singapore, Singapore and Tai Shung Chung, Department of Chemical and Biomolecular Engineering, National University of Singapore, 10 Kent Ridge Crescent,, Singapore, 119260, Singapore.

For the first time, co-extrusion was applied for the fabrication of dual layer hydrophilic-hydrophobic hollow fibers especially for the direct contact membrane distillation (DCMD) process. The effect of different nonsolvents on the morphology of the PVDF membranes was investigated and it was found that weak coagulants such as water/methanol (20/80 wt%) can induce a 3-dimensional porous structure on PVDF membranes with high surface and bulk porosities, big pore size, sharp pore size distribution, high surface contact angle and high permeability but rather weak mechanical properties. Hydrophobic and hydrophilic clay particles were incorporated into the outer and inner layer dope solutions, respectively, in order to enhance mechanical properties and modify the surface tension properties in the membrane inner and outer layers. Different membrane characterizations such as pore size distribution, gas permeation test, porosity and contact angle measurements were carried out as well. Ultimately, the fabricated hollow fibers were tested for the DCMD process and flux as high as 55 kg/m2hr at 90 °C was achieved in the test. This performance is much higher than most of the previous reports, indicating that the application of dual layer hydrophilic-hydrophobic hollow fibers may be a promising approach for MD.

480c Nasicon Membrane Electrolytic Technology for the Production of Sodium Hypochlorite

Justin Pendleton[1], Shekar Balagopal[2], Vinod Malhotra[3], and Alain Bergeron[3]. (1) NaSICON, Ceramatec Inc., 2425 S. 900 W., Salt Lake City, UT 84119, (2) Ceramatec, 2425 South 900 West, Salt Lake City, UT 84119, (3) Ampac, 10622 W 6400 N, Cedar City, UT 84721

A innovative and energy efficient, low temperature, cost-effective and environmental friendly electrochemical process using sodium ion conducting NaSICON ceramic membrane to continually produce sodium hypochlorite from sea water and low purity, un-softened water and sodium chloride is proposed. Ceramatec's proprietary membranes exhibit high selectivity towards sodium ions over hydronium and alkali ions to give the process an inherent advantage over technologies relying on organic membranes and un-divided cell methods to make sodium hypochlorite. Undivided and Organic membrane technologies have performance limitations from fouling and limited selectivity; however the NaSICON membranes operate at greater than 90 % of sodium transport efficiency, better ion transfer selectivity and not prone to fouling.

480d Sodium Removal From Contaminated Waste Streams By Nasicon Membrane Electrolysis

John J. Watkins, Justin Pendleton, and Shekar Balagopal. Ceramatec, 2425 South 900 West, Salt Lake City, UT 84119

Ceramatec has developed an electrolytic process for the separation of sodium from low-level waste and supplemental low activity waste (LAW) streams, which will allow for recycling of caustic and removal of sodium from waste prior to long term storage. The concept was proven in tests at the U.S. Savannah River Nuclear Site with NaSICON (Sodium Super Ionic CONductor) membrane containing laboratory-scale electrochemical cells using radioactive/contaminated solutions. The goal research was to complete demonstration to establish the ceramic NASICON membranes for the electrochemical separation process that will enable removal of sodium from low-level radioactively contaminated sodium containing waste at the Hanford site. We have successfully demonstrated the method to efficiently remove sodium from simulated waste streams and generate up to 50 weight % concentrated caustic with NASICON membrane-based laboratory-scale electrochemical cells. The electrolytic removal of sodium from low-level contaminated waste streams reduces the waste mass, caustic concentration and improves the process economics by recycling recovered caustic.

480e Theoretical Modeling of Phenolic Wastewater Treatment in a Hollow Fiber Biofilm Reactor

Haluk Beyenal, School of Chemical Engineering and Bioengineering, Washington State Univeristy, PO Box 642710, Pullman, WA 99164-2710

It has been demonstrated that biofilm reactors are effective in removal of phenol from wastewaters. As a new reactor type, hollow fiber biofilm reactors (HFBRs) have been applied to wastewater treatment and models predicting the activity of the biofilms in HBBRs would be beneficial in understanding the behavior of the reactor under varied conditions. In this study, we modeled phenol treatment in a hollow fiber biofilm reactor. We applied the maximum substrate consumption approach, which was previously developed by our research group, to predict biofilm density and to model phenol treatment in the hollow fiber biofilm reactor. The model assumes biofilms as a goal seeking system and hypothesis that biofilms arrange their structure to optimize phenol consumption rate within the active biofilm. The model solution predicted the biofilm density, active thickness, effectiveness factor, and phenol consumption rate using lumen phenol concentration as the only independent parameter. In addition, we evaluated the effect of radial flow velocity on the predicted parameters. The biofilm density exponentially decreased with increased lumen phenol concentration, while the active biofilm thickness exponentially increased. The maximum biofilm density was predicted as 110 g/L showing good correlation with literature data. The phenol consumption rate increased with the lumen concentration without being affected by phenol inhibition. However, the phenol consumption rate decreased sigmoidally with increased biofilm density. The results of this study showed that the performance of the proposed HFBR reactor depended on the lumen phenol concentration and radial flow velocity in the biofilm.

480f Membrane Bioreactor System For Purification Of Groundwaters Contaminated With Petroleum Hydrocarbons

Varadarajan Ravindran and Massoud Pirbazari. University of Southern California, Department of Civil and Environmental Engineering, Los Angeles, CA 90089-2531

The membrane bioreactor (MBR) process is a promising technology that provides a cost-effective alternative to conventional processes in water/wastewater treatment. The present research focuses on the application the MBR technology for the purification of groundwater contaminated with petroleum hydrocarbons exemplified by gasoline. These organic contaminants include several alkanes, iso-alkanes, and aromatics exemplified by benzene, toluene, ethylbenzene and xylenes, and oxygenated additives such as methyl-tert-butyl ether (MTBE). The MBR system employed a powder activated carbon (PAC) slurry with microorganisms acclimated to petroleum hydrocarbons. The process was operated as a closed system to prevent the volatilization of gasoline contaminants from the aqueous phase into the atmosphere.

A predictive model was employed for performance forecasting and design of the MBR process in the above application under a variety of operating and process conditions. The total organic carbon (TOC) was used as a lumped parameter for representing the total concentrations of petroleum hydrocarbons. The model involved the following essential features: (a) biological reaction in bulk liquid solution, (b) film transfer from bulk liquid phase to the biofilm-liquid interface, (c) diffusion with biological reaction inside biofilm, (d) adsorption equilibrium occurring at the biofilm-carbon interface, and (e) diffusion transport of contaminant within the activated carbon particles. The biological reaction in the bulk liquid phase as in the biofilm was represented by Monod kinetics, and the model assumed that biodegradation occurred in the bulk fluid phase as well as in the biofilm immobilized on the PAC particle. This modeling approach could be employed in upscaling the process from laboratory-scale to pilot-scale and eventually to full-scale using the techniques of dimensional analysis and similitude. The upscaling technique would involve dimensionless groups developed specifically for the adsorption and biodegradation sub-processes associated with the MBR process. The modeling and upscaling technique described above would serve as an efficient, time-saving, and cost-effective approach to design the MBR process.

The model parameters were determined from independent laboratory-scale experiments and correlation technique. The adsorption equilibrium and rate parameters for the model were determined by independent batch reactor studies. The biokinetic parameters for the degradation of gasoline components were determined from chemostat studies using continuous stirred tank reactors (CSTR). The laboratory-scale MBR experiments involved experiments in the presence and absence of activated carbon slurry for comparison purposes, and the organic contaminant removals as well as the permeate fluxes were measured in each case. These experiments established the effectiveness of the MBR technology in achieving nearly complete removals of organic contaminants under optimal conditions of PAC dosages and biomass concentrations. The results of these experiments also provided the means for model verification under different operating conditions. Model simulation studies were performed with respect to certain parameters including those representing liquid film mass transfer, biofilm transport, biological degradation kinetics, influent concentration, and reactor hydraulic retention time. These studies provided a qualitative evaluation of the parameters influencing the process dynamics under different conditions.

Key words: Membrane bioreactor, petroleum hydrocarbons, membrane fouling, modeling, biofilm transport, biodegradation kinetics, upscaling

MODELING OF A FLUIDISED BED BATCH CRYSTALLIZER

Akanksha[1], V.K.Srivastava[2] and Kuldeep[3]
[1,2,3] Department of Chemical Engineering, Indian Institute of Technology, Delhi
Hauz Khas, New Delhi-110016, INDIA

Introduction

Crystallization has been one of the important unit operations that combines both separation and purification step in a single operation. Many manufacturing industries rely on crystallization for the production of their materials. These industries include the pharmaceuticals, the agro-chemical, and general chemical manufacturers. Crystallization can be carried out from a vapour, from a melt or from a solution. Most of the industrial applications of the operation involve crystallization from solutions. The discussion henceforth deals with crystallization from solutions. The discussion henceforth deals with crystallization from solutions. Crystallization in a fluidized bed reactor (FBR) has been used in many water and wastewater treatment applications. The main objective of a crystallization operation is to produce crystals of uniform size with a narrow size distribution. This imparts aesthetic value for the product, which assumes significance when a particular product is marketed.

One such way of accomplishing the above mentioned objective has been by the use of Oslo crystallizers that operate on the fluidized-bed principle, in which a growing mass of crystals is suspended in an upward flow of the supersaturated solution through the crystallizer (Perry et al., 1984). As the supersaturated solution flows upward, the liquor contacting the bed relieves its supersaturation on the growing crystals and subsequently, the supersaturation decreases along the upward direction. As a result, the crystals segregate in the bed with the small ones at the top and the large ones at the bottom. Consequently, the behaviour of a fluidised bed of crystals exhibits a crystal size distribution (CSD), which is a prime objective in the design and operation of crystallizers. The important feature of the fluidised bed crystallizer is the absence of impellor, as opposed to the mixed-suspension, mixed-product removal crystallization, which effectively reduces nucleation thereby producing large and uniform size crystals.

Several attempts (Mullin and Nyvlt (1970), Al-Dibouni and Garside (1979), Tavare et al. (1990), Chianese et al. (1992) and Frances et al. (1994)) have been made to model fluidised-bed crystallizers. The models built were based on the assumption of size classification. proposed a model describing the crystallization behaviour in a continuous fluidized bed crystallizer. The other important simplifying assumption that is used in the modeling of fluidized bed crystallizers is the absence of nucleation.

All the above-mentioned studies have dealt in detail the modeling of fluidized bed crystallizers operated on a continuous basis, thus limiting the scope of application to systems in which seeds are continuously fed at a constant rate at one end and product crystals are withdrawn continuously at the other end. However, many industrial fluidized- crystallizers are operated in batch mode. Despite its importance, not much theoretical work has been done for this type of crystallizer. Shiau et al. in 1999 developed a model for batch mode fluidized bed crystallizer wherein seeds are initially charged in the bed and the final product crystals are removed entirely from the bottom of the crystallizer at the end of a run.

This paper is an attempt to model a fluidized-bed batch crystallizer. A rigorous model has been developed to describe the behaviour of crystal growth and the resulting CSD. It is assumed that liquid phase moves upward through the bed in plug flow and the solid phase in the fluidized bed is perfectly classified. The existing model is a modification of the Shiau et al. (1999) model, however, the basis has been changed to solute free solution flow rate that remains constant throughout the operation. This modification effectively reduces one equation from the original model, thereby reducing the computation time. The study presents significant information for crystallizer design and hence can be used for design purposes.

Model Development

The model is based on several simplifying assumptions as explained below:

1. The crystallizer is operated in isothermal mode.
2. Size independent crystal growth rate is exhibited by the system
   It is found that most of the systems exhibit size independent growth rate. Nevertheless, there are systems, which exhibit size dependent growth rate also. But in the present model, potash alum system has been chosen which is known to exhibit size independent crystal growth rate. The shape of the crystal is close to perfect octahedral (Mullin and Garside (1970)).
3. The supersaturated solution moves in plug flow pattern.
4. No nuclei formation takes place.
   Fluidized-bed crystallizes do not have impellers and therefore crystal-impeller contact is eliminated. This is one reason as to why there is no appreciable nucleation. Another notable reason is the near plug flow pattern of the supersaturated solution along the bed, thereby minimizing the crystal-crystal contact.
5. Crystal breakage is neglected.
   The absence of impeller is the reason as to why this assumption holds.
6. Uniform-size seeding is assumed
   The seeds are sieved in a narrow range before they are introduced in the crystallizer and hence they approximately follow Dirac delta function.

Performing an unsteady state mass balance on stage j, gives

$$\frac{d}{dt}\left(AH_j\varepsilon_j C_j\right)=Q_s\left(1+\frac{C_{j-1}}{\rho_P}\right)C_{j-1}-Q_s\left(1+\frac{C_j}{\rho_P}\right)C_j-\frac{1}{2}N_P\rho_P\beta L_j^2 G_j \qquad (1)$$

where j = 1,2,....,N

The modification done to the original model is the introduction of solute-free solution flow rate, Q_s that remains constant throughout the operation. In the original model, Shiau et al. (1999) considered the volumetric solution flow rate, Q. This gives rise to another equation describing the variation of volumetric flow rate, which happens to be time dependent, thereby increasing the computational time. Changing the basis to solute-free solution flow rate eliminates this equation consequently simplifying the problem. The above equation when simplified gives the following differential equation, which describes the rate of change of solution concentration in the fluidised-bed crystallizer.

$$\frac{dC_j}{dt} = \frac{1}{AH_j\varepsilon_j}\left[Q_s\left\{\left(1+\frac{C_{j-1}}{\rho_P}\right)C_{j-1} - \left(1+\frac{C_j}{\rho_P}\right)C_j\right\} - \frac{1}{2}N_P\rho_P\beta L_j^2\frac{dL_j}{dt} - AC_j\varepsilon_j\frac{dH_j}{dt} - AH_jC_j\frac{d\varepsilon_j}{dt}\right] \quad (2)$$

Growth rate, G_j, is given by a power-law expression as follows:

$$G_j = \frac{dL_j}{dt} = k_g\left(C_j - C^*\right)^n \quad (3)$$

In a fluidised-bed the superficial velocity, u_s, and the bed voidage, ε are related through the Richardson and Zaki (1954) equation given by:

$$u_s = u_i\varepsilon^z \quad (4)$$

where the value of u_i is nearly equal to the free-fall velocity of a single particle in an infinite fluid. Therefore the above equation can be written as:

$$u_s = u_{t,j}\varepsilon_j^{z_j} \quad (5)$$

The superficial velocity normally lies in between the terminal settling velocity and the minimum fluidization velocity. The exponent z varies with the particle Reynolds number and the ratio L_j/D and for the range of interest in crystallizer design is given by the following equations:

$$z = \left(4.4 + 18\frac{L_j}{D}\right)Re_{t,j}^{-0.1} \qquad 1<Re_{t,j}<200 \quad (6)$$

$$z = 4.4\,Re_{t,j}^{-0.1} \qquad 200<Re_{t,j}<500 \quad (7)$$

$$z = 2.4 \qquad Re_{t,j}>500 \quad (8)$$

$$Re_{t,j} = \frac{u_{t,j}\rho L_j}{\mu} \quad (9)$$

The value of terminal settling velocities at various Reynolds number ranges can be determined from the following equations (Levenspiel (1969)):

$$u_{t,j} = \frac{g(\rho_P - \rho)L_j^2}{18\mu} \qquad Re_{t,j}<0.4 \quad (10)$$

$$u_{t,j} = \left[\frac{4(\rho_P - \rho)^2 g^3}{225\rho\mu}\right]^{1/3} L_j \qquad 0.4<Re_{t,j}<500 \quad (11)$$

$$u_{t,j} = \left[\frac{3.1g(\rho_P - \rho)L_j}{\rho}\right]^{1/2} \qquad Re_{t,j}>500 \quad (12)$$

A solid balance on stage j yields the following equation for calculating the local bed height of stage j on equating the volume of the solid phase in stage j, $A(1-\varepsilon_j)H_j$, with the total volume of the crystals in stage j, $N_p\alpha L_j^3$:

$$H_j = \frac{N_P\alpha L_j^3}{A(1-\varepsilon_j)} \quad (13)$$

Differentiating the above equation with respect to t gives

$$\frac{dH_j}{dt} = \frac{1}{A(1-\varepsilon_j)}\left[AH_j\frac{d\varepsilon_j}{dt} + 3N_P\alpha L_j^2\frac{dL_j}{dt}\right] \qquad (14)$$

The model system consists of four ordinary differential equations (ODE's) with well-defined initial conditions (L_0, C_0, H_0/N, ε_0). ODE's that rule the evolution of solution concentration, crystal size, bed voidage and bed height are solved simultaneously by numerical methods as a function of operating times. The complexity of the model can be seen from the stiffness of differential equations. ODE15S, an inbuilt MATLAB routine based on Gear's algorithm is employed to solve the governing equations with a time step, $\Delta t = 0.002$ s.

Results and discussion

The performance of the model has been analysed for potassium alum system, for which the crystal shape and fluidization behaviour are all well known (Mullin and Garside (1970)) and exhibits size-independent crystal growth rate. The results obtained are exemplified in Fig. 1-8 together with corresponding published literature (Shiau et al. (1999)) results.

Fig. 1 shows the variation of solution concentration along the height of the crystallizer. It can be seen that the concentration drop is a monotonic decreasing function of H/H_t for different operating times. This is because, the supersaturation of the inlet solution decreases more rapidly as the crystals grow larger and the overall surface areas of the

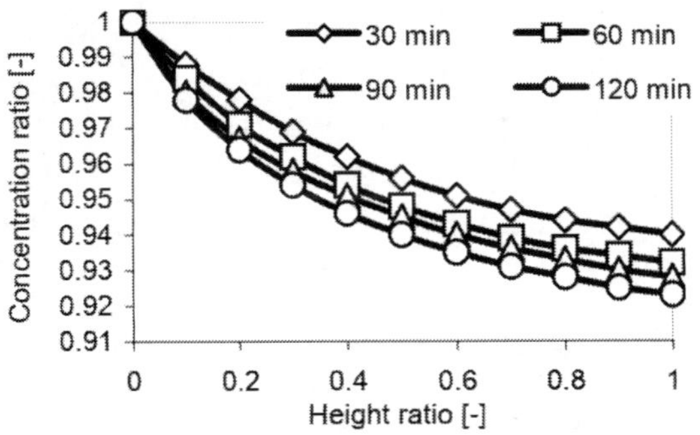

Fig. 1: Concentration change along the height of the crystallizer

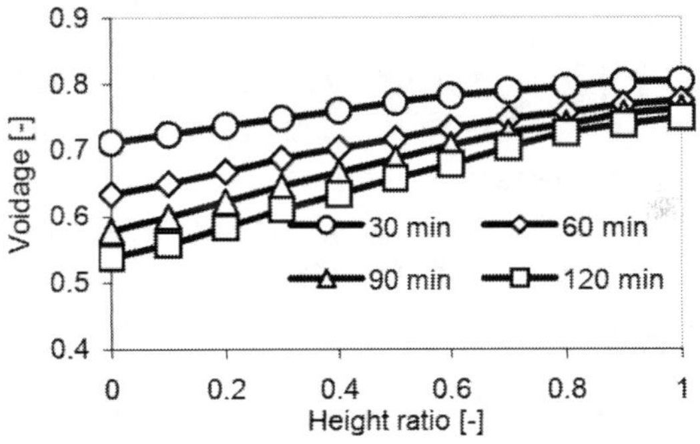

Fig. 2: Variation of bed voidage with bed height

crystals available for crystal growth become larger during the operation. It is found that the concentration decreases at a very slow rate and further it has also been observed that the change in concentration in a period of 2 hours at the top of the crystallizer is 1.8%. Fig. 2 depicts the variation of voidage in the bed at different operating times. Voidage was found to increase monotonically with the height ratio. The drop in ε is faster with operating times at the lower part (34%) than at the upper part (8%) of the bed. This is attributed to the presence of higher supersaturation at the bottom of the crystallizer.

Fig. 3 shows the variation in size ratio along the height of the crystallizer. The crystal size was found to decrease monotonically with the increase in height ratio.

Analysing the above results, it is observed that at the bottom of the crystallizer the size ratio increased from 1 to 3.751 after a 2 hour operation whereas the size ratio at the top of the crystallizer increased from 1 to 1.556 at the end of 2 hour operation. This reflects that a crystal size distribution is obtained at the end of the batch operation. Fig. 4 shows the cumulative size distribution of the crystals in the fluidised-bed crystallizer for four selected operating times; i.e. 30, 60, 90 and 120 min.

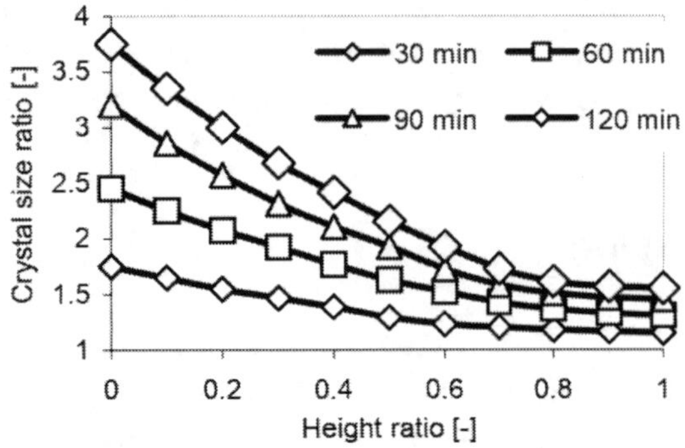

Fig. 3: Variation of crystal size ratio with bed height

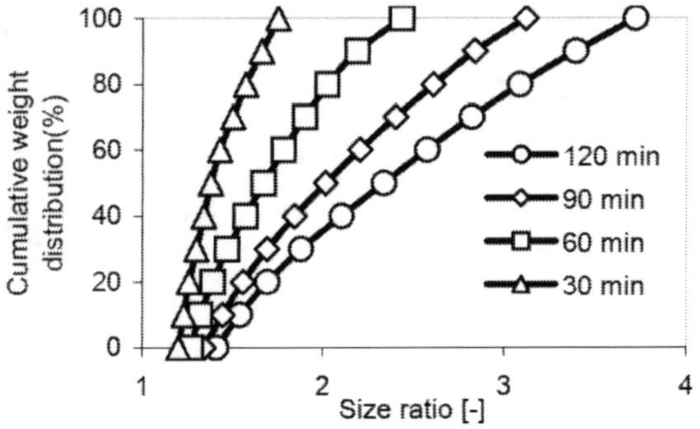

Fig. 4: Crystal size distribution within the bed

The weight fraction of crystal of size L_j in stage j can be expressed by

$$W_j = \frac{N_P \alpha L_j^3 \rho_P}{\sum_{j=1}^{N} N_P \alpha L_j^3 \rho_P} \quad (15)$$

A wider crystal size distribution was obtained with an increase in operating time.
The results obtained were also compared with the published data (Shiau et al. (1999)) and is represented in Figs. 5-8. The goodness of fit between predicted and published literature indicates the practical utility of the model described above in predicting actual behaviour of crystallization in a fluidised-bed crystallizer.

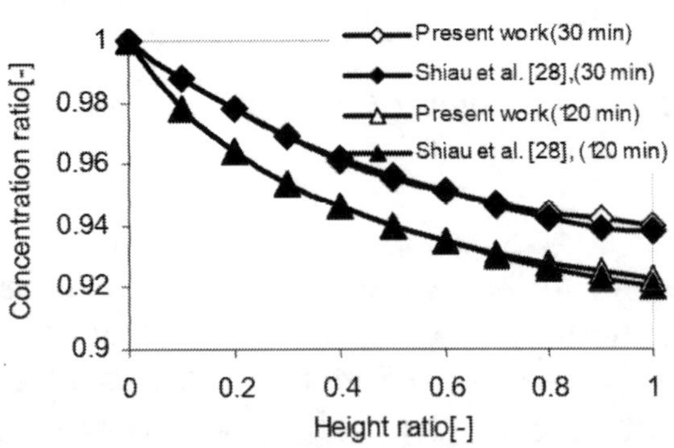

Fig. 5: Comparison of model results with published literature (variation of solution concentration with bed height)

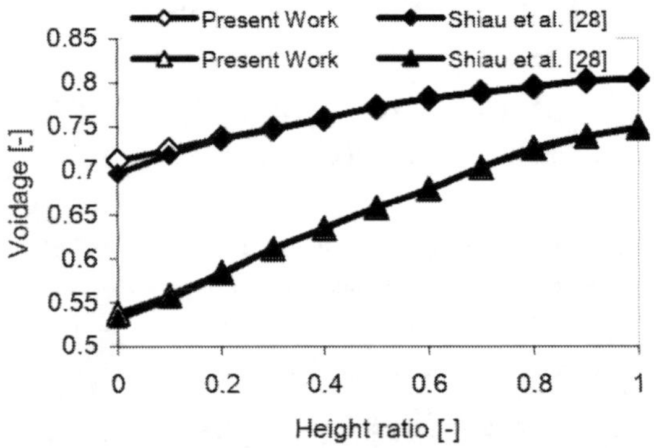

Fig. 6: Comparison of model results with published literature(variation of bed voidage with bed height)

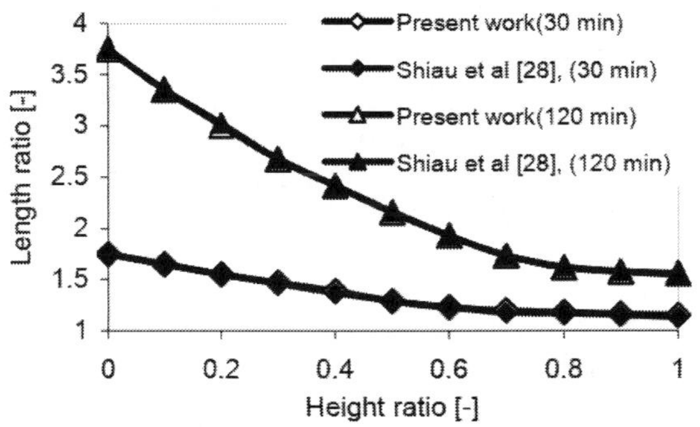

Fig. 7: Comparison of model results with published literature (variation of size with bed height)

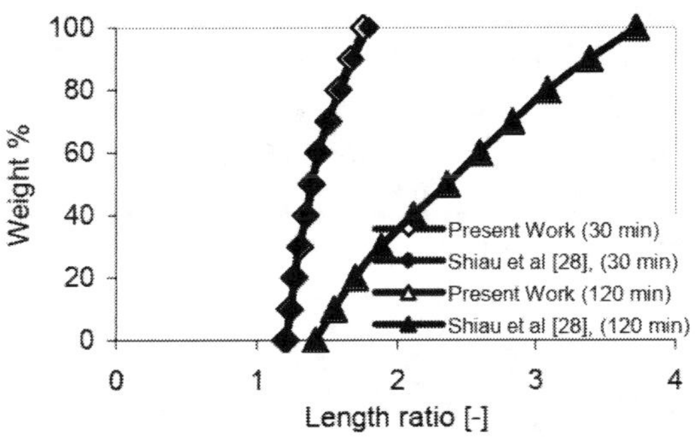

Fig. 8: Comparison of model results with published literature (cumulative crystal size distribution of crystals within the bed)

Conclusion

Potassium alum crystallization has been studied numerically in a fluidized-bed batch cystallizer for an isothermal case. An unsteady state mathematical model has been developed which is able to predict the variation of crystal size, the solution concentration, the bed voidage and the bed height within the cyrstallizer, the essential design parameters. The model is simple and time effective. Reasonably good agreement between the model predictions and published literature supports the general validity of the physical view underlying the model and thus can be extended to several relevant cases of practical significance. The study therefore provides a solution to the ultimate challenge of crystallization, which is control towards efficient separations and purifications.

Nomenclature

A	cross-section area of the bed, m^2
C_0	concentration of the influent solution, kg/m^3
C_j	concentration of the solution in jth stage, kg/m^3
C^*	saturated concentration of the solution, kg/m^3
D	diameter of the bed, m
g	gravity acceleration, m/s^2
G_j	growth rate of crystals in jth stage, m/s
H	vertical position from the bottom of the bed, m
H_0	total bed height at the start of a run, m
Hj	bed height of jth stage, m
k_g	constant in Eq. (3), $m^{3n+1}/s/kg^n$
L	crystal size, m
L_0	seed size, m
Lj	crystal size in jth stage, m
n	constant in Eq. (3), [-]
N	number of stages in the bed, [-]
N_P	number of crystals in each stage (= N_t/N), [-]

N_t total number of seeds, [-]
Q_j outlet flow rate of the solution from jth stage, m^3/s
Q_0 inlet flow rate of the solution, m^3/s
Q_s solute free solution flow rate ($Q_0 = Q_1 = \ldots = Q_j = \ldots = Q_N = Q_s$), m^3/s
$Re_{t,j}$ Reynolds number given by Eq. (9), [-]
t operating time, s
t_s duration of a batch operation, s
u_s superficial velocity, m/s
u_i velocity given by Eq. (4), m/s
$u_{t,j}$ terminal velocity of crystals in jth stage, m/s
W_0 total mass of seeds, kg
W_j weight fraction of crystal in jth stage, [-]
zj the expansion index given by Eq. (6-8), [-]

Greek letters

α volume shape factor of the crystal, [-]
β surface shape factor of the crystal, [-]
ε_0 initial bed voidage, [-]
ε_j bed voidage of jth stage, [-]
μ viscosity of the solution, kg/m/s
ρ density of the solution, kg/m^3
ρ_P density of the crystal, kg/m^3

References

1. Perry R.H., D.W. Green and J.O. Maloney (1984), Perry's Chemical Engineers' Handbook, sixth ed., McGraw-Hill, New York.
2. Mullin J.W. and J. Nyvlt (1970), "Design of classifying crystallizer", Trans. Instn Chem. Engrs., *48*, T7-T14.
3. Al Dibouni M.R. and J. Garside (1979), "Particle mixing and classification in liquid fluidized beds", Trans. Instn Chem. Engrs., *57*, 94-103.
4. Tavare N.S., M. Matsuoka and J. Garside (1990), "Modelling a continuous column crystallizer: dispersion and growth characteristics of a cooling section", J. Crystal Growth. *99,* 1151-1155.
5. Chianese A., C. Frances, F. Di Berardino and L. Bruno (1992), "On the behavior of a liquid fluidized bed of monosized sodium perborate crystals", Chem. Eng. J., *50*, 87-94.
6. Frances A., B. Biscans and C. Laguerie (1994), "Modelling of a continuous fluidized-bed crystallizer", Chem. Eng. Sci., *49*, 3269-3276.
7. Shiau L.D., S.H. Cheng and Y.C. Liu (1999), "Modelling of a fluidized bed crystallizer operated in a batch mode", Chem. Eng. Sci., *54*, 865-871.
8. Mullin J.W. and J. Garside (1970), "Velocity-voidage relationships in the design of suspended bed crystallizers", British Chem. Eng., *15*, 773-775.
9. Richardson J.F. and W.N. Zaki (1954), "Sedimentation and fluidization", Part I, Trans. Instn. Chem. Engrs., *32*, 35-52.
10. Kunii, O. Levenspiel (1969), Fluidization Engineering, Wiley, New York, pp.76-80.

512b Separation of M-Xylene from the Mixed Xylenes by Distillative Freezing

Lie-Ding Shiau and Chun-Ching Wen. Department of Chemical and Materials Engineering, Chang Gung University, 259 Wen-Hwa 1st Road, Taoyuan, 333, Taiwan

Xylenes have very broad applications in chemical industries. The mixed xylenes produced mainly consist of p-xylene (PX), m-xylene (MX), o-xylene (OX) and ethylbenzene (EB). Due to their similar molecular structures, these isomers have close boiling points. Therefore, it is very difficult to separate them from each other by distillation. The cost for separation and purification of these isomers is generally very high.

A new separation technique, called distillative freezing (DF), has been successfully applied to separate PX from a liquid mixture of 10% MX and 90% PX (Shiau et al., 2005). In principle, the DF process is operated at a triple point condition, in which the liquid mixture is simultaneously vaporized and crystallized due to the three-phase equilibrium. Thus, DF is a distillative crystallization technology, which combines distillation and crystallization to produce pure crystals. By lowering temperature and reducing pressure during the operation, the DF process results in the formation of pure crystals, and liquid phase and vapor phase of mixtures. The process is continued until the liquid phase is completely eliminated and only the pure crystals remain in the feed. Distillative freezing (DF) will be applied in this work to separate and produce MX crystals from the mixed xylene mixtures containing MX, PX, OX and EB. The experimental results indicate that, when the DF experiments are performed based on the simulated adiabatic DF operation, MX crystals can be produced and the purity of MX crystals can reach 99%~99.5% for the mixed xylene mixtures of MX, PX, OX and EB.

In essence, the DF operation is a combination of distillation and crystallization. There are several unique features of the DF operation: (1) The DF process is conducted under an adiabatic condition at a three-phase equilibrium, where the latent heat released in crystallization is removed by vaporization. Thus, it is an energy-conserving separation method compared to distillation. (2) The addition of solvent is not needed in DF compared to azeotropic or extractive distillation in the separation of the close-boiling mixture. Thus, it is a clean separation technique. (3) The DF process is continued until the liquid phase is completely eliminated and only the pure crystals remain in the feed. Compared to crystallization, subsequent filtration or centrifugation is not needed since no mother liquor is present with the pure crystals. (4) Crystal washing is not required since only the pure crystals remain in the feed and no impurities are adhered on the crystal surfaces at the end of the DF operation.

[1] Shiau, L. D.; Wen, C. C.; Lin. B. S.,Separation and purification of p-xylene from the mixture of m-xylene and p-xylene by distillative freezing" Ind. Eng. Chem. Res. 2005, 44, P2258-2265.

512c Optimization Of A Crystallization Process To Produce A High Purity Secondary Metabolite Intermediate

Rebecca A. Chmielowski, Jennifer M. Pollard, Mark Brower, and Kent Goklen. BioPurification Development, Merck & Company, Inc., 126 E. Lincoln Ave, P.O. Box 2000, Rahway, NJ 07065

The isolation of fermentation derived secondary metabolites is required for the synthesis of many anti-infective drugs. High purity (> 90 wt%) metabolites are required to support subsequent synthesis steps for conversion to the API. Purification of these compounds via crystallization is difficult due to the low purity of the initial product coupled with the production of various product analogs. A chromatography step is typically implemented to upgrade the product purity prior to performing crystallization. Since chromatography can negatively impact the process productivity and economics, the goal was to develop a crystallization utilizing a partially purified feed stream.

An investigation into the feasibility of crystallization from a crude, partially-purified product stream without using chromatography in the isolation process is presented. The motivation behind the development of a temperature cycling crystallization to produce high purity crystals (> 90 wt%) from crude extract will be described. Details of this crystallization process are presented, which include solubility curves, crystallization kinetics, and particle size distribution. The effect of system parameters, such as cooling rates, water content, and agitation, on yield and purity is also presented. The optimization of this preferential crystallization increased the product purity by 30-35 wt% with an increase in yield of 15-20% while also producing easily filterable crystals. This process was implemented on a 25 kg scale to produce material to support clinical development.

Web Page: None

512d Operation Alternative and Control of a Stirred Tank Continuous Isothermic Crystallizer

O. Velázquez-Camilo[1], E. Bolaños-Reynoso[2], J. J. Álvarez-Ramírez[1], and O. Xaca-Xaca[2]. (1) Departamento de Ingenieria Quimica, Universidad Autonoma Metropolitana, Av. San Rafael Atlixco N° 186, Col. Vicentina C.P. 09340, Iztapalapa., México D.F., Mexico, (2) Division de Estudios de Posgrado e Investigación, Instituto Tecnologico de Orizaba, Av. Ote 9 No. 852. Col. E. Zapata. Cp. 94320, Orizaba, Ver, Mexico

The crystallization is a unitary operation that consists on organic and inorganic chemical purifying by liquid - solid separation, where mass is transferred from a dissolved solute in a liquid phase to a solid phase (crystal pure); One of the key attributes of particulate systems is the co-presence of a continuous phase and a dispersed phase, which leads to the occurrence of physiochemical phenomena such as nucleation, growth and birth-death rates of the crystals. The driving force in order to occur the nucleation and the growth rates of the crystals is the supersaturation, therefore these phenomena has not any possibility to occur in a saturated or unsaturated solution.

In this work, an operation alternative was studied by simulation of a stirred tank continuous isothermic crystallizer through of its dynamic characterization. The process model is based on the moments technique, which has a less complex solution than to the population balance model (PBM); since the last one regularly presents complex distributed parameters models of solving mathematically. This technique has the advantage of being used by many investigators in crystallization in the last years. The methodological process consisted on the implementation of a SISO (single input-single output) feedback servocontrol algorithm and a PID (proportional-integral-derivative) controller. The tuning was carried out with the strategies of internal model control (IMC). The process variable to control was the average diameter in % volume D(4,3) = M4/M3, derivate of moments of the crystal size distribution (CSD) because is a useful variable in both commercial and industrial environment; the manipulate variable was the solute concentration in the feeder.

From obtained results, it is outstanding the stationary state analysis of the response from the process to different residence times (tao), where is observed that to tao < 0.6 hr., the crystallizer that generally has an oscillatory behavior (cycles limits); it has a trend toward a stable and controllable state. Simulation runs of the close-loop system clearly demonstrate that the implemented controller reached in smaller time (2 hrs) the set point value of (D(3,4) = 1,000 micron) than the open-loop process (approximately 10 hrs). With this control action, both the offset and the initial oscillations were eliminated; a good response was presented in the pursuit from the positive and negative step changes on set point. It is concluded that the system is functional; however it presents two inflection points, what makes it unstable in these zones.

512e A New Recrystallization Method for Nitroguanidine

Zhixiong Cha, Department of Civil and Environmental Engineering, University of Utah, 104 CMC, 122 South Central Campus Drive, Salt Lake City, UT 84112

The objective to crystallize nitroguanidine is to obtain high bulk-density nitroguanidine because the needle shape crystal of nitroguanidine has low bulk-density. Normally the recystallization process is realized by lowing the temperature with different rate in hot water or other solvent systems with the help of specific additives. Recently a new process, anti-solvent recrystallization, is used for recrystallization of nitroguanidine, HMX, and RDX. A good recrystallizing process can obtain specific shaped, much smaller and more uniform nitroguanidine crystals. The shape and size of nitroguanidine crystals are determined by recrystallizing additives and conditions. Normally spherical nitroguanidine crystal is use as explosive charge and short column crystal is used as propellant. Recently the best commercially available nitroguanidine crystals made in German can be smaller than 20 microns, but the distributions of nitroguanidine crystals are still not satisfied because the controlling parameters in a bulk recrystallizing vessel are always limited and indirect for size control, even for the anti-solvent recrystallization process.

Different from other processes, this technology has realized generation of high uniform and much smaller nitroguanidine crystals. The size of nitroguanidine crystals can be controlled between 1 micron to 50 micron and all crystals are identical under microscope. During crystallizing procedure, the shape and size of nitroguanidine crystals are adjusted by different additive concentrations and different temperature decreasing rate. After the nitroguanidine crystals are obtained, the additives are removed to purify the nitroguanidine crystals. Application of the new nitroguanidine crystals in automobile industry shows high reliability. Also applications of the new nitroguanidine crystals in propellant (M60 or missile) show high reliability and enhanced power. The cost of the new process is much lower and the size is more controllable than the anti-solvent recrystallization process.

512g Application Of Dynamic Regulation Profiles Of Vacuum Pressure In Batch Crystallization Of Cane Sugar

E. Bolaños-Reynoso[1], C. Santiago-de Jesus[1], O. Velázquez-Camilo[2], and P. A. Quintana-Hernández[3]. (1) Division de Estudios de Posgrado e Investigación, Instituto Tecnologico de Orizaba, Av. Ote 9 No. 852. Col. E. Zapata. Cp. 94320, Orizaba, Ver, Mexico, (2) Departamento de Ingenieria Quimica, Universidad Autonoma Metropolitana, Av. San Rafael Atlixco N° 186, Col. Vicentina C.P. 09340, Iztapalapa., México D.F., Mexico, (3) Departamento de Ingenieria Quimica, Instituto Tecnologico de Celaya, Av. Tecnológico y A.García Cubas S/N. C.P. 38010,, Celaya, Gto., Mexico

The crystallization as industrial process is important due to the great variety of materials that are commercialized in crystalline form. The product quality in a crystallization process is related to the crystal size distribution (CSD), the crystal morphology, the polymorphic effect and the purity degree. The main causes to control the product quality in batch crystallizers focus in the study of both phenomenological and process operation factors. The crystallization depends on thermodynamic and kinetic factors: nucleation rate, growth rate and production-reduction term (births and death rates of crystals), that are governed by variables like the local and average supersaturation levels, the mixed degree, impurities, additives, pH, dissolvent type, temperature and pressure of operation. It has been considered that both the crystal size distribution and the formed crystal mass (FCM) are important factors to the process and defines the final product characteristic. In this work, the dynamic regulation profiles of vacuum pressure implemented by means of a supervisory control system and data acquisition (SCADA) are introduced like novel operation technique in batch crystallization. This new operation technique represents an alternative for the manipulation of the average supersaturation, avoids agglomerate of formed crystalline product and eliminates the use of unsaturated liquor like means of dissolution in the batch process operation. This work has the objective of analyzing both the DTC and MCF behavior in the batch crystallization of cane sugar due to the application of dynamic regulation profiles of vacuum pressure.

Like a part of the methodological process, a strategy of dynamic regulation profiles of vacuum pressure was developed, which has a very important effect on the cooling by adiabatic evaporation into the batch process. The experimentation was carried out dissolving 9235.98 g of cane sugar and 2922.68 g of water into a stirred batch crystallizer, until reaching 75.96 °Brix, corresponding to the supersaturation temperature of 70 °C. It was included 6.33 g of crystal seed with average size of D(4,3) 195.30 µm (% volume) and standard deviation of S(4,3) 19.52 µm, during the first minute of process, approximately to 68 °C. A constant evaporation took place during the first 30 minutes of the batch, benefiting both the stabilization and the seeded crystal growth. Then, a dynamic regulation profile of vacuum pressure (called as maximum frequency, linear and cubic) was applied during 60 minutes, causing an abrupt cooling by adiabatic evaporation, which took to an increase of both crystal size and the formed mass. The final batch time was 90 minutes, with samplings every 15 minutes to gets experimental information of the process variables.

For the FCM analysis, the evaporated water amount was calculated and it depends on the dynamic regulation profile of vacuum pressure to follow during the crystallization, relating the amount of initial water of process and the obtained one by the temperature difference from 40 to 70 to °C. Later, the evaporated water fraction that retires every 15 minutes was determined. The sugar remnant amount is obtained by means of the relation of evaporated water and of the solution concentration based on the sampling times. By overall mass balance and based on the sugar remnant amount and the initial one, the FCM is obtained. This FCM is multiplied by a compensation of 2% (industrial parameter) corresponding to the sugar amount that is lost when unloading the crystallized solution, and at the centrifuging time (wastage). For the experimental concentration calculation, the °Brix obtained in each sampling are used, relating them to the sugar grams by each 100 milliliter of water. The supersaturation calculation (S) is obtained by means of the concentrations difference (theoretical and experimental).

An electronic microscopy technique was used to make the particles analysis, having like support the software of images acquisition system called IMAQ Vision Builder by National Instruments, with the purpose of capturing the crystal image in every sampling time. Later, CSD was determined (diameters average and the standard deviation) on the basis of the distribution moments: % number, % length, % surface and % volume, by means of the DTC Adq-Im software by Instituto Tecnológico de Orizaba, which receives like input data the crystals diameters obtained by IMAQ system in pixels.

The results corresponding to the "maximum frequency" dynamic regulation profile of vacuum pressure produce 1,128.70 g of evaporated water, a FCM of 4,910.38 g at the final batch time. With the °Brix measurement it was able to quantify and to give pursuit to the exhaustion of the sugar concentration in the batch available solution, from 3.1601 to 2.4364 g sugar/g water; this favored the formed crystal mass. At 15 minutes of process operation, it reached a supersaturation value of 0.038552 to finish with a decreasing behavior until 0.008635, as well as the relative supersaturation that it initiated in 0.012808 and it finished in 0.003557. This indicates that available supersaturation in the system has been exhausted in its majority. The error percentage between the experimental and theoretical FCM (overall mass balance) was of 0.922%, at the final batch time. For both "cubic" and "linear" dynamic regulation profiles of vacuum pressure with respect to the described profile as "maximum frequency", it determines that they produce minors amounts of evaporated water (717.65 g, and 842.41 g, respectively) and minor FCM at the final batch time (3,841.01 g and 4,155.68 g, respectively), because the adiabatic cooling in both cases are slower than in the "maximum frequency" dynamic regulation profile of vacuum pressure. The error percentage between the experimental and theoretical FCM for the "cubic" and "linear" dynamic regulation profiles of vacuum pressure was of 2.17% and 0.31%, respectively, at the final batch time.

The results of CSD analysis for the "maximum frequency" dynamic regulation profile of vacuum pressure produce a average crystal size of D(4,3) 732.72 μm (% volume) with standard deviation of S (4,3) 78.04 μm, being observed an almost linear growth of crystals during the process with variability in their size by hydrodynamic effects (agitation). The error between the base run and its repetition was of 6.07%. For the "cubic" and "linear" dynamic regulation profiles of vacuum pressure, it is obtained a D (4,3) of 567.79 and 585.18 μm with standard deviation of S(4,3) 64.07 and 69.44 μm, respectively. The error between the base runs and their repetitions were of 3.05% and 3.37%, respectively.

According to both CSD and FCM analysis, it is concluded that with the "maximum frequency" dynamic regulation profile of vacuum pressure, it is obtained the biggest crystal size distribution (D(4,3) of 732,72403 µm and S(4,3) of 78,042381 µm) and maximum amount of formed crystal mass (4.910,3820 g) with respect to the described profiles as "cubic " and "linear", contributing to the exhaustion of the processing molasses.

512h Evaporation-Based Microfluidic Platforms For Membrane Protein Crystallization

Sarah L. Perry, Sameer Talreja, C.F. Zukoski, and P.J.A. Kenis. Department of Chemical & Biomolecular Engineering, University of Illinois at Urbana-Champaign, 600 S. Mathews Ave., Urbana, IL 61801

Our research focuses on studying the physical chemistry governing various methods of membrane protein crystallization. Membrane proteins are critical components of many fundamental biological processes such as ion regulation and transport through the cell membrane and catalyzing reactions such as charge separation and conversion of energy. Membrane proteins also serve as a channel for signal transduction from one side of the cell membrane to the other and establish contact with the proteins of neighboring cells to form tissues such as muscles. Our understanding of such fundamental membrane-associated biological processes at a molecular level is limited due to the lack of knowledge of high-resolution structure of the membrane proteins which depend on obtaining X-ray quality crystals of proteins. While traditional well plate technology is well established, such systems typically use large and expensive fluid handling equipment, and may be on a scale that is inaccessible given the quantities of protein available. The goal of this work is to gain a deeper understanding of the phase behavior of the solutions/gels used for membrane protein crystallization. This study intends to address the challenge of membrane protein crystal growth by the development and validation of novel crystallization platforms. In this presentation we present a novel microfluidic platform that enables the rapid identification and optimization of crystallization conditions for membrane proteins and other molecules by precise control of the evaporation rate. In this manner we can independently control the rate of supersaturation and ensure a phase transition for every trial. The advantage of controlling the rate of supersaturation independently manifests in the improvement of crystal quality and an increase in crystal size in the case of soluble as well as membrane proteins. In addition, the microfluidic platform requires only ~5nL of sample per trial. This technology will help facilitate the crystallization of membrane proteins by easing the sample volume necessary for screening.

512i Pseudopolymorphic Forms Of Sodium Naproxen

Krystle J. Chavez and Ronald W. Rousseau. School of Chemical & Biomolecular Engineering, Georgia Institute of Technology, 311 Ferst Drive, Atlanta, GA 30332-0100

Solvates or hydrates (also known as pseudopolymorphs) crystallize in forms that differ from the unsolvated species, which should not be surprising because their stoichiometries are different. That is, they result from the incorporation of solvent bound in the lattice of the crystal in a specific stoichiometric ratio. Transformations among pseudopolymorphic forms often produce variations in physical properties such as density, solubility, dissolution rate, and bioavailability. These changes in physical form can interrupt downstream processing during manufacturing. Furthermore, the formation of pseudopolymorphs of pharmaceuticals can be of great importance because changes in the physical properties of the crystal can affect the safety and efficacy of a drug. Thus, predicting the occurrence of crystal pseudopolymorphs is of importance in developing robust, large-scale crystallization processes in the pharmaceutical industry.

Sodium naproxen is a non-steroidal, anti-inflammatory drug that has been shown to exhibit four pseudopolymorphic forms (anhydrous, monohydrated, dihydrated, and tetrahydrated). The purpose of the present study is to examine changes in the pseudopolymorphic states of sodium naproxen in mixed-solvent (e.g., methanol and water) systems with varying temperature. The work explores the transition from one pseudopolymorphic form to another using batch cooling crystallization, which can be seen clearly through a van't Hoff plot of solubility data. Specifically, the transition point between pseudopolymorphs occurs at decreasingly lower temperatures as the concentration of methanol is increased in the solvent. These data are being used to develop a capability that will allow prediction of the pseudopolymorphic state of sodium naproxen under a given set of conditions.

512j Investigating the Crystallization Mechanism of Amorphous Budesonide and Salbutamol Sulfate

Daniel J. Burnett and Frank Thielmann. Surface Measurement Systems, 2222 South 12th Street, Suite D, Allentown, PA 18103

PURPOSE: Many low molecular weight amorphous materials will relax to their thermodynamically stable, crystalline form. The timescale of this transition is dependent on both temperature and solvent concentration. Therefore, understanding the kinetics of this transition over a wide range of temperatures and solvent concentrations is vital for the processing and storage of materials containing amorphous phases. In this study the crystallization of both hydrophilic (salbutamol sulfate) and hydrophobic (budesonide) drugs are investigated via vapor sorption techniques.

METHODS: Salbutamol sulfate (SS) and budesonide (BD) were made (partially) amorphous via milling. Dynamic gravimetric Vapor Sorption (DVS) was used to monitor the crystallization behavior of the two drugs over a wide range of temperatures. This was observed as a sharp mass loss when exposed to high vapor concentrations. Water was used to crystallize SS, while ethanol was used to crystallize BD.

RESULTS AND DISCUSSION: When SS was exposed to water vapor concentrations above 42% relative humidity (RH) amorphous SS reverted to its crystalline form. This transition was monitored over a range of temperatures. Model-free mechanism analysis indicated SS crystallized in a one step, nth dimension Avrami-Erofeev process (n~0.5). Exposure to ethanol vapor above 55% P/Po at 25°C induced amorphous BD to crystallize. Ramping the ethanol vapor concentration from 0 to 95 % P/Po clearly showed vapor-induced glass transition and crystallization points.

CONCLUSIONS: The vapor-induced crystallization behavior of salbutamol sulfate (water vapor) and budesonide (ethanol vapor) was studied over a wide range of temperatures. Ethanol caused budesonide to crystallize above 55% P/Po at 25 °C. Mechanism modeling indicated a one-step Avrami-Erofeev process for water-induced salbutamol sulfate crystallization.

512k Effects Of Organic Compounds On The Fractional Evaporative Crystallization Of Hanford Waste Solution

George Dumont, Laurent Nassif, and Ronald W. Rousseau. School of Chemical & Biomolecular Engineering, Georgia Institute of Technology, Atlanta, GA 30332-0100

Fractional crystallization has been examined as a potential means of accelerating the remediation of mixed liquid waste at the Hanford Site. In the proposed process, a feed stream containing waste is partitioned by evaporating water and crystallizing a number of sodium salts. The salts are separated and ultimately treated in a bulk vitrification facility, while Cs and other radioactive species remain in solution and are sent to a waste treatment facility. The primary criterion by which performance was judged included the Cs content of the recovered crystals, the fraction of sodium recovered as crystals, and the fraction of feed sulfate that is diverted to the recovered crystals.

Previous work on fractional evaporative crystallization proved the feasibility of this method at the laboratory scale. Prior to pilot scale application, the robustness of the process was to be extended to include additional waste solutions that contained a number of organic compounds, including sodium glycolate ($NaC_2H_3O_3$), sodium acetate ($NaC_2H_3O_2$), tetrasodium EDTA dihydrate ($Na_4C_{10}H_{12}N_2O_8 \cdot 2H_2O$), and trisodium HEDTA dihydrate ($Na_3C_{10}H_{15}N_2O_7 \cdot 2H_2O$).

In the work to be discussed, the impact of organic compounds on the fractional evaporative crystallization process was explored. In particular, there were two major issues that were explored: (1) the effects of the organic compounds on the nucleation and growth kinetics of the inorganic sodium salts, and (2) whether or not the organic species remained in solution during the process. Either of these factors can alter the crystal size distribution and affect solid-liquid separation efficiency.

A methodology was developed in order to investigate potential interactions between organic species, along with the individual effects of each species on crystallization. Using this approach, it was determined that certain organics led to the crystallization of sodium sulfate instead of burkeite ($Na_2CO_3 \cdot 2Na_2SO_4$), which was the species crystallized in the absence of the organic compounds. In addition, organic species were responsible for process difficulties such as foaming during evaporation, film deposition on the crystallization equipment, and reduced solid-liquid separation efficiency.

541a High Zeolite Loading Mixed Matrix Membranes

Ryan T. Adams and William J. Koros. Chemical and Biomolecular Engineering, Georgia Institute of Technology, 311 Ferst Dr., Atlanta, GA 30332

Traditional gas separation membranes are polymer-based and are constrained by a fundamental "upper-bound" selectivity-permeability trade-off. Inorganic materials, such as zeolites, can exhibit desirable transport properties beyond the "upper-bound". Unfortunately, pure zeolite membranes are expensive, have poor mechanical properties, and cannot be made, defect-free, in useful amounts. Mixed matrix membranes (MMMs) combine the ease of processing polymers into large area, defect-free membranes with the superior transport properties of zeolites by blending the phases. Researchers have proven that MMMs can be successfully made with properties exceeding the "upper-bound" performance for loadings of up to 40 wt.% zeolite particles in a polymer matrix. Further increasing zeolite loadings will further increase the capabilities of MMMs as alternatives to polymer gas separation membranes.

Transport properties approaching those of zeolites are predicted as zeolite loadings in MMMs increase. However, observed transport properties fall short of these predictions. In fact, creating defect-free MMMs is challenging and becomes increasingly difficult as zeolite loading increases. Stress accumulation due to decreased polymer chain mobility, polymer-zeolite incompatibility, and the increased likelihood of forming defects with increasing filler content are the primary causes believed to be responsible for such defects.

Maximizing zeolite loading while maintaining polymeric processability to make defect-free MMMs with near-model predicted transport properties is the focus of this work. We pursue this goal using zeolite 4A and polyvinyl acetate (PVAc) as model materials and will then move to "upper-bound" polymers. Void-free, dense MMMs, with ≥ 50 vol.% zeolite 4A in PVAc, are made without exotic zeolite surface treatments or overly complex processing steps. Formation of void-free high loading MMMs is extremely sensitive to the total solids loading in the membrane casting solution. Strategies for forming void-free MMMs at this unprecedented zeolite loading are presented. Selected transport data as well as ways to avoid other undesired defects, like clogged zeolite pores, are also discussed.

Sorption and diffusion of organic vapors in mixed matrices based on Teflon AF2400 and fumed silica

M.C. Ferrari, L. Basile, M.G. De Angelis, T.C. Merkel°, G.C. Sarti
Dipartimento di Ingegneria Chimica, Mineraria e delle Tecnologie Ambientali
via Terracini 34, 40131 Bologna Italy
e-mail: grazia.deangelis@unibo.it
° MTR- Membrane Technology and Research, Menlo Park, CA (USA)

Abstract

The transport properties of two organic vapors in mixed matrix membranes based on amorphous Teflon® AF2400 and nanoscale fumed silica have been studied at 25°C. In particular, solubility and diffusivity of normal-butane and normal-pentane have been collected in pure AF2400, pure fumed silica and mixed matrices with weight fractions of filler equal to 0.25 and 0.40.
The experimental data show that there is no great effect of the filler content on the solubility, when the latter is evaluated over the total mass of the composite. On the other hand, by estimating the solubility in the polymeric fraction alone of the composite material, the solubility of the vapors increases monotonically with increasing filler content. The infinite dilution diffusivity evaluated from transient sorption data increases significantly with filler content, and the observed increase of diffusivity with penetrant concentration levels off with increasing filler content. The behavior of diffusivity is consistent with previous works and can be explained by the fact that the free volume increases after addition of filler. Such an increase in free volume is estimated by fitting the NELF model to the experimental solubility isotherms of composite matrices. The values of free volume estimated in this way increase with filler content, as expected, and correlate well also with the diffusivity data when a commonly used semiempirical law based on free volume theory is considered.

Introduction and background

Polymers are currently the dominant materials for gas separation membranes because their processability into high surface area hollow fibres enables economic production of modules for large scale separations. However, the segmental flexibility of polymers limits their separation ability compared to more selective, but unprocessable rigid molecular-sieving media. A step change over current technology may be achieved by dispersing selective inorganic materials in polymers that exhibit favourable selectivity. Composites of this kind are generally called mixed matrix membranes (MMM), with the highly selective rigid phase usually being represented by zeolites, carbon molecular sieves or silica particles dispersed in a continuous polymeric matrix. The polymers selected as matrices for MMM are generally chosen among glassy materials, whose gas separation performance is close to the trade-off curve in the selectivity/permeability plot for the mixture of interest; examples of such polymers include polyimides, poly(ether imide) (PEI), polyvinyl acetate (PVAc), polyethersulfone (PES), polysulfone (PSf), Amorphous Teflon (Teflon® AF), Poly(1-trimethylsilyl-1-propyne) (PTMSP) and poly(4-methyl-2-pentyne) (PMP).
The dispersed phase affects the surrounding polymer, leading to various and sometimes opposite effects: for instance, it may cause an undesirable void at the interface or introduce various degrees of rigidification in the polymer matrix.
Merkel et al. [1] have measured gas transport in the same material inspected here, namely amorphous Teflon® AF 2400 filled with nanoscale, nonporous, hydrophobic fumed silica (FS) particles, obtaining an increase in gas permeability. However, the permeability of large penetrants increases more than that of small molecules, thereby decreasing the size selectivity of this polymer, that, unlike PTMSP and PMP, is more permeable to small light gases than to larger condensable vapors. The same behavior is observed by adding FS to another stiff, high free volume glassy matrices such as PTMSP (PTMSP and AF2400

actually show different behavior – silica addition increases the relative permeability of large penetrants in AF2400 and small penetrants in PTMSP. We might say that in both size-selective AF2400 and vapor-selective PMP, silica addition increases the relative permeability of large penetrants. In contrast, in vapor-selective PTMSP, silica addition increases the relative permeability of small penetrants): [2] by loading silica particles the permeability of PTMSP to light gases increases more than the permeability of larger penetrants, thus leading to a reduction of vapor/permanent gas selectivity of PTMSP. Finally, an increase of both permeability and selectivity is obtained by adding fumed silica particles to PMP. [3]
The sorption of gases in PMP and PTMSP filled with FS particles was also considered in the above mentioned works, [1-3] and it was found that the vapor sorption isotherms reporting the amount of gas per unit composite mass remain almost unaffected by the addition of filler, that, based on a simple additive model, would be expected to depress the vapor uptake of the polymer. The same behavior is observed for the soprtion of n-butane in composites of AF2400 filled with fumed silica.
The diffusion coefficients, on the other hand, are significantly increased by the addition of filler, thus justifying the increase in permeability.
The gas and vapor sorption behavior of Teflon AF/FS, PMP/FS and PTMSP/FS MMMs could be explained considering three possible mechanisms: *i*) FS sorbs the same relative amount of gas/vapor per unit mass as the pure polymer does; *ii*) FS sorbs a smaller amount of gas/vapor with respect to the pure polymer, but modifies polymer chain packaging in a way that increases the polymer sorption capacity to compensate for the lower sorption of FS particles; *iii*) FS sorbs a higher amount of gas/vapor with respect to the pure polymer, but modifies polymer chain packaging in a way that decreases the polymer sorption capacity to compensate for the higher sorption of FS particles.
The first two mechanisms are believed to be the most reasonable because the silica particles are non porous, and therefore they can adsorb gas molecules only on their surface.
The study of the transport behavior of Teflon AF/FS, PMP/FS and PTMSP/FS MMMs has also been supported by a PALS characterization of the matrix free volume before and after the addition of fumed silica particles. As it is well known, in these high free volume polymers, two distinct distributions of free volume are present, represented by different characteristic dimensions. In Teflon AF 2400, PTMSP and PMP it was found that the addition of FS increases the size of large free volume elements, while it does not modify the size of the small free volume elements. The variation of the dimensions of larger free volume elements is rather small; however, it is believed to be the cause of the increased permeability and diffusivity of filled matrices. The increased free volume could also explain the behavior of the gas solubility, based on mechanism *ii*). The additional free volume that may be introduced when incorporating the filler must also be taken into account. The real dominating mechanism must be identified to guide the modelling activity.
With respect to modelling the transport properties in MMMs, only models for the prediction of permeability have been presented in the literature, to our knowledge; in particular, the Maxwell model, [4] originally developed to describe the permittivity of a dielectric medium, has been taken as a guideline by various authors due to the analogy between the constitutive equation governing electrical potential and the flux through membranes.[5] Different modifications of this model have been proposed to account for the various effects that may originate due to interactions between polymer and filler.[6-8] On the contrary, no specific models have been proposed for predicting either the solubility or the diffusivity of the composite material.

The NELF model extends the Lattice Fluid equation of state [9-11] valid for amorphous equilibrium phases, i.e. at temperatures above T_G for polymeric systems, to the non equilibrium state typical of glassy polymers. The model moves from the Lattice Fluid (LF) theory proposed by Sanchez and Lacombe, and uses the same characteristic parameters (P^*, T^*, ρ^*) to characterize the pure components behavior, and the same mixing rules to estimate the mixture properties.

For polymers, the pure components characteristic parameters are calculated by fitting the LF equation of state to pressure-volume-temperature (PVT) data above T_G, while for the penetrants either PVT data or vapor-liquid equilibrium data may be used.
In the LF model, the following relationship holds true for the number r_i^0 of lattice sites occupied by a molecule in its pure phase:[9]

$$r_i^0 = \frac{P_i^* M_i}{RT_i^* \rho_i^*} = \frac{M_i}{\rho_i^* v_i^*} \tag{1}$$

where M_i is the molecular weight of component i and v_i^* is the volume occupied by a lattice site of pure substance. The parameter r_i^0 is usually set to infinity for the polymer species. The mixing rules adopted to estimate the mixture characteristic parameters contain only one adjustable binary parameter, Ψ, appearing in the expression for P^*, the characteristic pressure for the mixture. Ψ affects the binary characteristic pressure P_{12}^*, representative of the characteristic energetic interactions per unit volume between gas and polymer molecules, labeled by subscripts 1 and 2, respectively:[11]

$$P_{12}^* = \Psi \sqrt{P_{11}^* \cdot P_{22}^*} \tag{2}$$

In the absence of specific binary data, a first-order approximation of the binary parameter is obtained by considering Ψ=1.
The extension to non-equilibrium states is given by the Non-Equilibrium Thermodynamics of Glassy Polymers (NET-GP) approach, in which the state of a glassy polymer-penetrant mixture can be described by the usual variables, e.g. temperature, composition and pressure, plus the polymer density ρ_2, which accounts for the departure from equilibrium frozen into the glass.[12-13] The model also assumes that ρ_2 is an internal state variable for the system, so that the specific thermodynamic relations for systems endowed with an internal state variable can be applied.[14]
The phase equilibrium between a pure external gas (g) and a solid glassy phase (s) requires:

$$\mu_1^{NE(s)}\left(T, p, \Omega_1^{PE}, \rho_1^{PE}\right) = \mu_1^{Eq(g)}(T, p) \tag{3}$$

where superscript PE labels the pseudo-equilibrium conditions asymptotically reached in the glassy phase, due to hindered long range mobility.
The evaluation of solubility thus requires the values of the independent characteristic parameters r_1^0, T_1^* and P_1^*, for the penetrants, and of the characteristic pressure, density and volume (P_2^*, ρ_2^* and v_2^*) for the polymer, which may be found in specific works and collections.[15-16]
Moreover, we need to know the value of the glassy polymer density ρ_2^0 at the experimental conditions of interest that may be different at the same temperature depending on the particular history experienced.
In general, at higher pressures for swelling penetrants, the value of ρ_2 varies appreciably during sorption and it can be obtained from data of polymer dilation during sorption: to this aim, a swelling coefficient k_{sw} is introduced to evaluate the polymer density at each pressure:

$$\rho_2(p) = \rho_2^0 (1 - k_{SW}\, p) \tag{4}$$

In Eq. (4) we have assumed, as many experimental works confirm, that the swelling isotherm curve representing volume variation *vs*. pressure has a linear behaviour.

In the absence of the specific dilation data for the system of interest, the parameter k_{sw} can be adjusted on one solubility datum at high pressure providing also a reliable estimate for the swelling behavior of the polymer matrix. The equilibrium condition becomes:

$$\mu_1^{NE(s)}\left(T,p,\Omega_1^{PE},\rho_2^0,k_{sw}\right)=\mu_1^{Eq(g)}(T,p) \tag{5}$$

For the following developments, it is useful to consider the fractional free volume of the polymer and its estimation. In the LF model, the fractional free volume of the polymer may be estimated as:

$$f=\frac{\rho_2^*-\rho_2^0}{\rho_2^*} \tag{6}$$

where ρ_2^*, the characteristic density of the polymer, is an estimate of the close packed density at 0 K; thus the fractional free volume calculated from eq. (6) measures the departure of the polymer volume from the hypothetical volume occupied by the crystal at absolute zero temperature. Use of eq. (6) is consistent with the theoretical framework considered, although it is different from what is often calculated based on Bondi's group contribution method leading to different values. We may as well evaluate the fractional free volume in the more common way:

$$FFV=\frac{V-1.3\cdot V_W}{V}=\frac{\rho_2^W-1.3\cdot\rho_2^0}{\rho_2^W} \tag{7}$$

where V_W is the van der Waals volume of the repeating unit of the polymer. This value may be calculated with group contribution methods and is available in the literature for the polymer under study ($\hat{V}_W$ for AF2400 is 0.301 L/kg).

The NELF model can be used to calculate sorption isotherms in organic/inorganic matrices, by considering separately the contribution of the two phases and applying mass additivity. To that aim one needs to consider that the presence of the inorganic rigid phase imposes constraints on the polymer matrix which affect *i*) its actual density and free volume, and *ii*) its swelling coefficient, which are relevant in the case of swelling penetrants. Changes in the polymer free volume, which usually are positive increments, are associated to phenomena occurring at the organic/inorganic phase boundary, and can be accounted for by comparing the actual average polymer density in the composite with the density of the pure polymer matrix alone.

In the cases of mixed matrices formed by high free volume glassy polymers and fumed silica, the actual solubility in the composite system is higher than what is expected on the basis of the pure additive model represented by the following equation:

$$C_M=\Phi_F\cdot C_F^0+\left(1-\Phi_F\right)\cdot C_P^0 \tag{8}$$

where Φ_F is the volume fraction of the filler, C_M the gas solubility in the mixed matrix, C_F^0 and C_P^0 the gas solubility in the pure filler and pure polymer, respectively.

We assumed that the non additive behavior experimentally observed can be explained by considering that filler addition modifies the properties of the polymeric phase alone, by generating additional free volume close to the interface between polymer and inorganic particles. The sorption capacity of the solid filler is considered unchanged. Clearly this assumption applies to the cases in which the adsorption onto the solid surfaces of the filler does not represent an important contribution. The gas solubility contribution of the inorganic phase of the mixed matrix is considered the same as that of the pure filler, while the contribution of the polymeric phase of the mixed matrix increases since it absorbs a higher amount of penetrant due to excess free volume induced by the filler. In the analysis, we consider that the sorptive capacity of the organic phase is increased from C_P^0 to C_P, so that the total amount of penetrant absorbed by the mixed matrix material can be expressed as:

$$C_M = \Phi_F \cdot C_F^0 + (1 - \Phi_F) \cdot C_P \tag{9}$$

In order to provide an estimate of the increase in free volume induced by the addition of filler and, possibly, the variations of the swellability of the matrix, we have used the NELF model.
On the basis of the experimental data for C_M and C_F^0, the contribution C_P is evaluated, isolating the solubility isotherm in the polymer phase alone of the MMM. By adjusting the NELF model isotherm to the C_P values, one calculates the actual values of polymer density and swelling coefficient of the polymer in its state present in the mixed matrix. Such new values can be used to calculate in a predictive way the solubility isotherms of other penetrants, different from the one used in the procedure above.

Estimation of diffusivity in mixed matrix membranes
In all the polymeric matrices here considered, penetrant diffusivity follows an exponential trend with concentration that may be expressed by the following relationship:

$$D_C = D_C(0) \cdot \exp(\gamma \cdot C) \tag{10}$$

where $D_C(0)$ is the infinite dilution diffusion coefficient in the composite material, and γ is a parameter characteristic of the gas-polymer couple, which depends on temperature. For the present analysis, we considered the infinite dilution diffusion coefficients $D_C(0)$ only that depends on the initial free volume of the material and should not be affected by the swelling induced by the penetrant.
The first assumption made is that the additional free volume induced by filler affects the properties of the polymeric phase only. Furthermore, we can reasonably assume that the following usual relationship, based on free volume theory, holds true between diffusivity at infinite dilution in the polymeric phase, $D_P(0)$, and its fractional free volume FFV:

$$\ln\left(D_p(0)\right) = A - \frac{B}{FFV} \tag{11}$$

where A and B are adjustable parameters depending on temperature and on the gas-polymer pair.
The gas diffusivity in mixed matrix materials is measured on the composite phase rather than on the polymeric phase only, thus the presence of the filler also has to be taken into account. In the case considered here, the dense filler is impermeable and its effect on gas transport is only associated to an increase in the mean path of a gas molecule diffusing through the material. $D_C(0)$ is therefore an apparent diffusivity, that may be simply related to the diffusivity in the polymeric phase, $D_P(0)$, by accounting for the tortuosity factor τ:

$$D_C(0) = \frac{D_P(0)}{\tau} \tag{12}$$

For the estimation of the tortuosity factor, we have assumed the validity of a Maxwell model in which the filler particles are considered to be spherical, a reasonable assumption in our case, thus one has:

$$\tau = 1 + \frac{1}{2}\Phi_F \tag{13}$$

By combining Eqs. 11 and 12 one obtains a relation correlating the apparent diffusivity data in the composite to the fractional free volume of the polymeric phase:

$$D_C(0) = \frac{1}{\tau} \cdot \exp\left(A - B \cdot {}^{1}\!/_{FFV}\right) \tag{14}$$

where the parameters A and B need to be adjusted on experimental data at various values or fractional free volume, or equivalently, at various amounts of filler.
For each mixed matrix, the FFV value was calculated from gas sorption data in the MMM, following the procedure illustrated above, then the value of apparent diffusivity in the composite material, $D_C(0)$, was estimated with Eqs. 12-14. The values of the parameters A and B have been adjusted to experimental diffusivity data in two matrices. The ratio between the diffusivity in the composite, $D_C(0)$, and the diffusivity in the pure polymer, $D_P^0(0)$, on the other hand, contains only one adjustable parameter, B:

$$\frac{D_C(0)}{D_P^0(0)} = \frac{1}{\tau} \exp\left[B\left(\frac{1}{FFV^P} - \frac{1}{FFV^C}\right)\right] \tag{15}$$

In order to test the validity of the approach illustrated so far, we have considered experimental apparent diffusivity values in the MMMs and have compared them with the values obtained by eq. (15) using experimental diffusivity data in Teflon AF2400, and the fractional free volume estimated from the NELF model, according to the procedure indicated above.

Experimental

Sorption isotherms were collected for two different organic vapors in mixed matrix membranes based on amorphous Teflon with a pressure decay apparatus.
Teflon® AF2400 is a glassy perfluorinated copolymer obtained by random copolymerization of 2,2-bis(trifluoromethyl)-4,5-difluoro-1,3-dioxole (BDD) and tetrafluoroethylene (TFE), at mole fractions of 87% and 13% respectively. Its glass transition temperature is 240°C and its density is equal to 1.74 g/cm^3; the polymer has an excellent chemical resistance and high gas permeability. The filler is non porous fumed silica TS-530 (Cabot Corporation) and has been chemically treated with hexamethyldisilazane to replace hydroxyl surface groups with hydrophobic trimethylsilyl surface groups. The particles have an average diameter of 12 nm and a density of 2.2 g/cm^3.
The films were obtained by casting a solution of 1% of Teflon in perfluoro-N-methylmorpholine on a glass plate. Fumed silica is added to the polymer solution in the case of mixed matrices and mixed at 18000 rpm before being cast. The final thickness of the films used is about 90 μm.
In particular, solubility and diffusivity of n-butane and n-pentane in pure Teflon and 2 mixed matrices with different filler content (25% wt and 40% wt content) were determined at 25°C, as well as adsorption on the pure filler.

Results and discussion

The sorption isotherms in pure Teflon AF2400, in two different mixed matrices and adsorption on the pure silica powder are reported in fig. 1a and 1b, plotted in terms of mass of penetrant per mass of total solid. In the case of n-butane (fig. 1a), the curves show the negative concavity that is characteristic of glassy polymers and the data are in very good agreement with literature values [1]. For n-pentane a broader activity range was inspected (fig. 1b), and at higher activities there is a change in the concavity that can be related to the plasticization effect of the matrix. In both cases, the penetrant mass uptake in the mixed matrix samples is similar to that in the pure polymer.

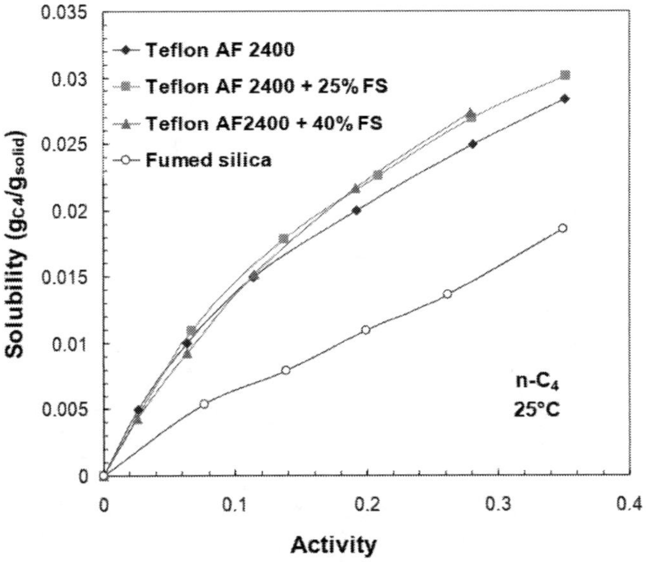

Figure 1a: Solubility of n-C_4 in pure AF2400, AF2400+25%FS andAF2400+40%FS, and adsorption onto pure fumed silica at 25°C, in grams of penetrant per gram of total solid

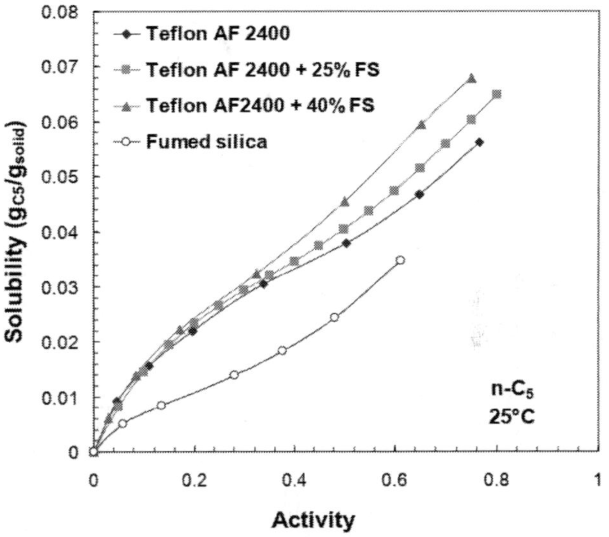

Figure 1b: Solubility of n-C_5 in pure AF2400, AF2400+25%F andAF2400+40%FS, and adsorption onto pure fumed silica ε 25°C, in grams of penetrant per gram of total solid

The solubility isotherm can also be reported in terms of the concentration of penetrant, C_p, absorbed in the polymeric phase alone, obtained from Eq. 9: the results are shown in figs. 2a (n-butane) and 2b (n-pentane). As expected, higher filler loading leads to higher mass uptake in the polymer, consistently with the fact that a higher free volume is present in the mixed matrix due to the filler addition.

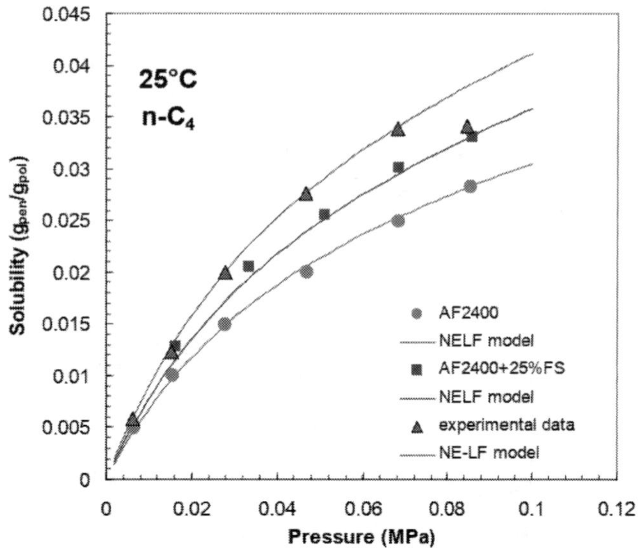

Figure 2a: Experimental solubility of n-C_4 in the polymeric phase of AF2400-based mixed matrices and comparison with NELF model.

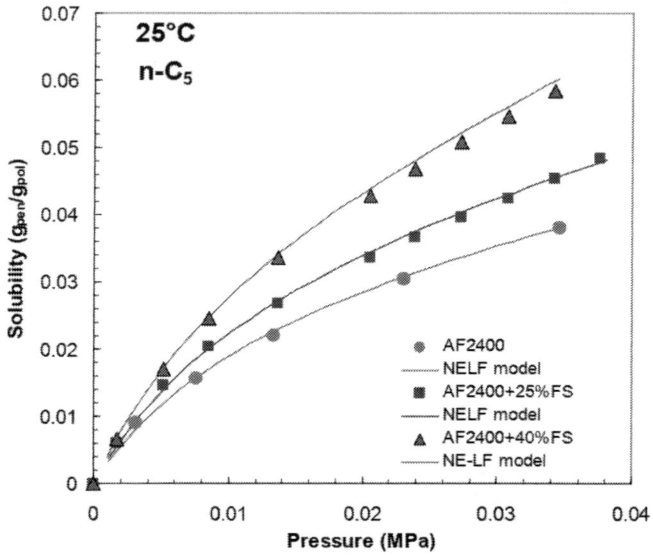

Figure 2b: Experimental solubility of n-C_5 in the polymeric phase of AF2400-based mixed matrices and comparison with NELF model.

The same figs. 2 also report the solubility curves calculated with the NELF model, by fitting the parameters to the experimental data. *N*-butane was used as reference penetrant. First, the *n*-C_4 solubility isotherm was used to obtain the binary parameter and the swelling coefficient from the NELF model, using the known value of the pure polymer density. For the mixed matrices, the polymer density and the swelling parameter were adjusted to the experimental solubility curve, using the value of the binary parameter calculated previously for the pure polymer. In the case of n-pentane only the first part of the isotherm was considered, before the concavity change due to the higher plasticization induced by *n*-C_5. The polymer density values and the binary parameters used are the same already determined for the case of *n*-C_4 sorption; only the swelling coefficients were modified and used as fitting parameters. In Table 1 we listed all the parameters used for the calculations.

Table 1: Model Parameters Ψ, *FFV* and k_{sw} for n-C_4 and n-C_5 in AF2400 Containing Various Amounts of FS

Penetrant	**Matrix**	**Ψ**	*Calculated* ρ_{pol} (g/cm^3)	*Calculated* *FFV*	*Calculated* k_{SW} (MPa^{-1})
n-C_4	AF 2400	0.14	1.740	0.320	0.13
n-C_4	AF 2400 +25% FS	0.14	1.714	0.330	0.20
n-C_4	AF 2400 +40% FS	0.14	1.680	0.344	0.21
n-C_5	AF 2400	0.14	1.740	0.320	0.85
n-C_5	AF 2400 +25% FS	0.14	1.714	0.330	1.20
n-C_5	AF 2400 +40% FS	0.14	1.680	0.344	2.00

The apparent diffusion coefficients were evaluated from the transient data of the sorption experiments and the results are plotted in figs. 3a and 3b versus the average concentration of penetrant within the film. The addition of filler greatly affects the diffusivity in the glassy matrices for both organic vapors. In particular, the diffusion coefficient increases with increasing filler content due to the higher free volume introduced by the fumed silica. In addition, the apparent diffusivity increases exponentially with penetrant concentration, with a weaker dependence the higher is the filler content.

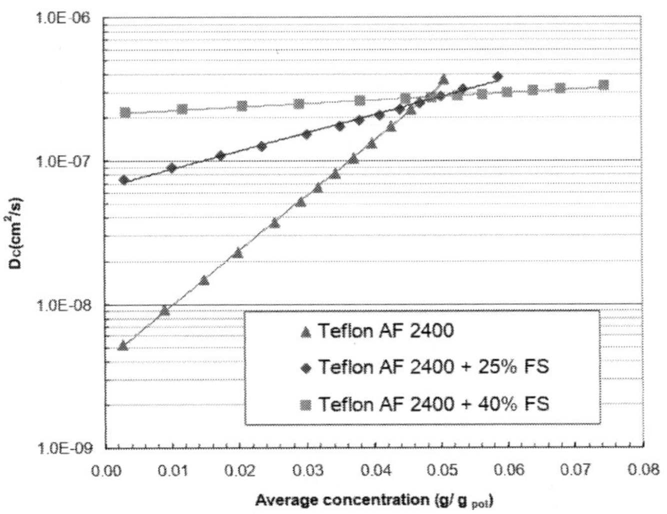

Figure 3a: Diffusivity of n-C_4 in AF2400-based mixed matrices at 25°C, versus average penetrant concentration in the step.

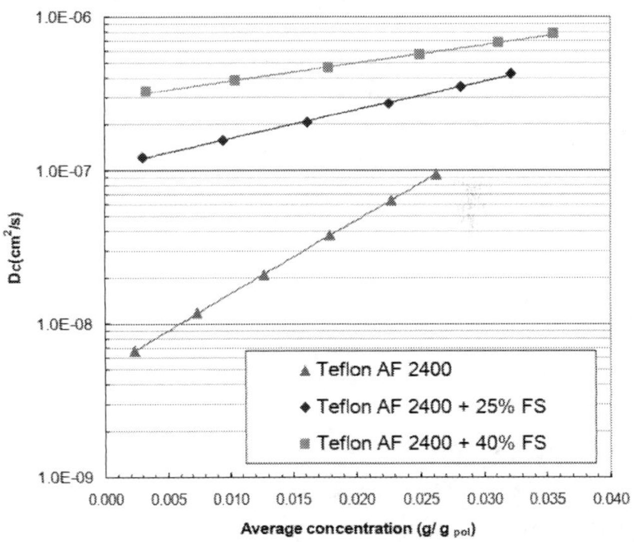

Figure 3b: Diffusivity of n-C_5 in AF2400-based mixed matrices at 25°C, versus average penetrant concentration in the step.

The experimental values of $D_C(0)$ in mixed matrices based on AF2400 reasonably follow eq. 14, and by reporting $ln(D_C(0)\cdot\tau)$ versus $1/FFV$ the data fall close to a straight line, as it can be seen in Figure 4a, with values of R^2 equal to 0.85 for butane and 0.88 for pentane. The experimental and calculated values of the diffusivity ratio (i.e., diffusivity in the MMM divided by diffusivity in the pure polymer) are reported in Figure 4b and in Table 2, from which it can be seen that the model allows prediction of the diffusivity increase due to filler addition; the agreement between experimental and simulated data is more satisfactory for the case of AF2400 loaded with 40% of FS but less satisfactory for the case of 25% FS.

Table 2 : Diffusivity values for n-C_4H_{10} and n-C_5H_{12} in Teflon AF2400-based mixed matrices at 25°C

	n-C_4H_{10} (B=18.2)		**n-C_5H_{12}** (B=18.0)	
	$D_C(0)/D_P(0)$ exp.	$D_C(0)/D_P(0)$ calc.	$D_C(0)/D_P(0)$ exp.	$D_C(0)/D_P(0)$ calc.
AF2400	1	1	1	1
AF2400/25%wt FS	20.4	4.3	16	4.2
AF2400/40%wt FS	56.6	46	52	44

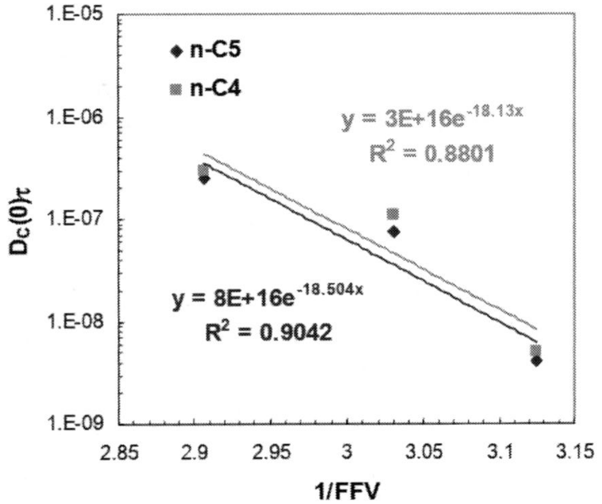

Figure 4a: Experimental diffusivity normalized for tortuosity versus 1/FFV

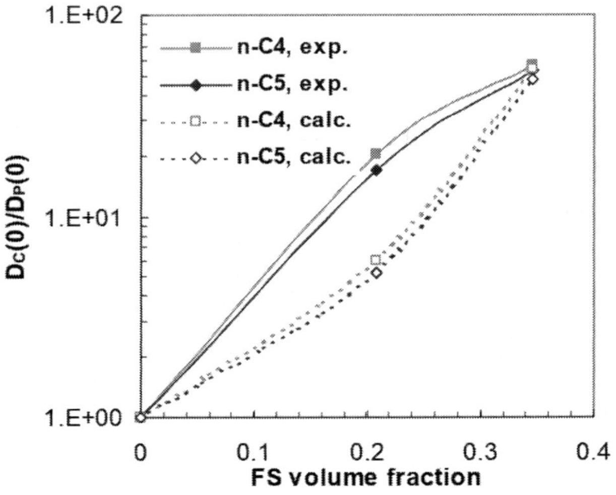

Figure 4b: Experimental and calculated values of $D_C(0)/D_P(0)$

Acknowledgements
The work was supported by the European Commission 6th Framework Program Project MultiMatDesign, Contract No. NMP3-CT-2005-013644

REFERENCES:

[1] Merkel, T. C.; He, Z.; Pinnau, I.; Freeman, B. D.; Meakin, P.; Hill, A. J. Sorption and Transport in Poly(2,2-bis(trifluoromethyl)-4,5-difluoro-1,3-dioxole-co-tetrafluoroethylene) Containing Nanoscale Fumed Silica" Macromolecules 2003; 36, 8406-8414.
[2] Merkel, T. C.; He, Z.; Pinnau, I.; Freeman, B. D.; Meakin, P.; Hill, A. J. Effect of Nanoparticles on Gas Sorption and Transport in Poly(1-trimethylsilyl-1-propyne) Macromolecules 2003; 36(18); 6844-6855.
[3] Merkel, T. C.; Freeman, B. D.; Spontak, R.J., He, Z.; Pinnau, I.; Meakin, P.; Hill, A. J. „Sorption, Transport, and Structural Evidence for Enhanced Free Volume in Poly(4-methyl-2-pentyne)/ Fumed Silica Nanocomposite Membranes" *Chem. Mater*. 2003, 15, 109-123.
[4] Maxwell, J.C. Treatise on electricity and Magnetism Vol.1 Clarendon Press, Oxford (1873).
[5] Moore, Mahajan, Vu, Koros, *AIChE J.* 50, 2004, 311-321.
[6] R.H.B. Bouma, A. Checchetti, G. Chidichimo and E. Orioli, Permeation through a heterogeneous membrane: the effect of the dispersed phase, *Journal of Membrane Science*, 128, 141-149, 1997.

[7] E.E. Gonzo, M.L. Parentis, J.C. Gottifredi Estimating models for predicting effective permeability of mixed matrix membranes, *Journal of Membrane Science*, 277, 2006, 46-54.
[8] Y.C. Chiew, E.D. Glandt, The effect of structure on the conductivity of a dispersion, *J. Colloid Interface Sci*. 94 (1983) 90–104.
[9] I.C. Sanchez, R. H. Lacombe, J. Phys. Chem., 1976, 80, 2352-2362
[10] R.H. Lacombe, I.C. Sanchez, J. Phys. Chem., 1976, 80, 2568-2580.
[11] I.C. Sanchez, R.H. Lacombe, Macromolecules, 1978, 11, 1145-1156.
[12] F. Doghieri, M. Quinzi, D.G. Rethwisch, G.C. Sarti, ACS Symposium Series 876 (2004) 74-90.
[13] M. Giacinti Baschetti, M. G. De Angelis, F. Doghieri, G. C. Sarti, in: M.A. Galan, E. Martin del Valle (Eds.), Chemical Engineering: Trends and Developments, Wiley, Chichester (UK), 2005, pp. 41-61.
[14] G. Astarita, Thermodynamics, Plenum Press, New York, USA, 1989.
[15] P. A. Rodgers, J. Appl. Polym. Sci. 48, 1993, 1061-1080.
[16] I.C. Sanchez and C. Panayiotou, in Models for Thermodynamic and Phase Equilibria Calculations, S. I. Sandler, Ed., Dekker, 1994.

541c Layered Silicate by Amh-3 Intercalation and Its Use in Nanocomposite Membranes for Gas Separation

Sunho Choi[1], Joaquin Coronas[2], Edgar Jordan[3], Weontae Oh[4], Sankar Nair[5], Frank Onorato[6], Daniel F. Shantz[3], and Michael Tsapatsis[1]. (1) Department of Chemical Engineering and Materials Science, University of Minnesota, 151 Amundson Hall, 421 Washington Avenue SE, Minneapolis, MN 55455-0132, (2) Chemical and Environmental Engineering Department, University of Zaragoza, Zaragoza, 50018, Spain, (3) Department of Chemical Engineering, Texas A&M University, College Station, TX 77843, (4) Nanotechnology Department, Dong-Eui University, Busan, 614-714, South Korea, (5) School of Chemical & Biomolecular Engineering, Georgia Institute of Technology, 311 Ferst Drive NW, Atlanta, GA 30332-0100, (6) Pall Corporation, 2200 Northern Blvd, East Hills, NY 11548

AMH-3 is a nanoporous layered material with unique three-dimensional porosity within the layer built of eight-membered rings. The swollen derivative of AMH-3 was prepared by a novel procedure that includes the sequential intercalation of organic surfactant following proton exchange. The increase of basal spacing, corresponding to the bilayer configuration of surfactant, allows the swollen AMH-3 to be incorporated and well-dispersed in polymer matrix. Nano-scale particles of exfoliated layers were used as a selective phase in nanocomposite membranes for gas separation. The hydrogen/carbon dioxide ideal selectivity of the nanocomposite membranes was improved by a factor of two compared to that of the pure polymer. Considering the morphology and the alignment of nanoparticles used in this study, as revealed by TEM and SAXS, the performance of these membranes has the potential to be improved further.

541d Formation Of Isotropic Microporous Mixed Matrix (Zeotips) Membranes

Caleb V. Funk, Bonnie L. Beavers, and Douglas R. Lloyd. Chemical Engineering, The University of Texas at Austin, 1 University Station, Mailcode C0400, Austin, TX 78712

This presentation will introduce a new concept in mixed matrix membranes for liquid separations and gas separations: nanoporous zeolite particles suspended in a flexible, isotropic, microporous polymer matrix. The cellular voids of the polymer matrix allow for rapid transport through the membrane while the nanoporous zeolite particles provide the membrane with high selectivity capabilities. These unique zeolite-containing membranes are formed via the thermally induced phase separation process; thus, they are referred to as ZeoTIPS membranes. This presentation will briefly explain the rationale behind the development of ZeoTIPS membranes and will then focus on the details of the membrane formation process. Specifically, the presentation will include a discussion of how zeolite particles influence the processes of droplet coarsening and pore formation in thermally induced phase separation by disrupting flow fields as well as changing local compositions and viscosities. The presentation will also discuss the issue of affinity between the zeolite particles and the polymer–diluent mixture as it relates to membrane morphology. Finally, the presentation will provide insight into the effects of the zeolite–polymer–diluent composition and the formation process conditions on the final membrane structure.

541e Hybrid Organic-Inorganic Opal Membranes For Chiral Separations

Ilya Zharov, Chemistry, University of Utah, 315 So. 1400 E., Salt Lake City, UT 84102

The surface of self-assembled silica colloidal crystals has been modified with chiral selector moieties, either as a monolayer or as a polymer brush. The chiral selectivity of these membranes to the transport of chiral amino acid derivatives has been studied using cyclic voltammetry. The factors affecting the chiral permselectivity will be discussed.

541f Organic-Mesoporous Inorganic Composite Membranes For Challenging Separations

Sukjoon Yoo[1], Seunguk Yeu[2], Achyuta Teella[1], Daniel F. Shantz[3], and David M. Ford[4]. (1) Chemical Engineering, University of Massachusetts, Amherst, 159 Goessmann Laboratory, North Pleasant Street, Amherst, MA 01003, (2) Department of Chemical Engineeing, Texas A&M University, TAMU 3122, College Station, TX 77843-3122, (3) Department of Chemical Engineering, Texas A&M University, TAMU 3122, College Station, TX 77843-3122, (4) Department of Chemical Engineering, University of Massachusetts, Amherst, 209 Goessmann Laboratory, 686 North Pleasant Street, Amherst, MA 01003

We synthesized organic-inorganic nanocomposite membranes by decorating the surfaces of mesoporous alumina or silica substrates with a selective organic material that is chemically anchored to the porous surfaces. The mesoporous substrates are either commercially available alumina membranes or custom-synthesized supported ordered mesoporous silica thin films. Linear alkyl molecules or hyperbranched triazine-based dendrimers with nanometer dimensions were covalently grown from the porous membrane surfaces as filling agents to give composite materials. The organic portion of the composite may be modified by varying the length and functionality of the terminating groups and by increasing the number of generations of the dendrimer grown on the surface. The uniformity of the mesoporous substrates and the various controls over chemistry enable the composite membranes to have a constant membrane performance with a high selectivity for target component. The separation focus of this research is the removal of organic molecules from light gas streams. The performance of our composite membranes compares very favorably to the best results reported in the literature for purely polymeric membranes.

593a Polystyrene Sacrificial Interlayers For The Preparation Of Silicon Carbide Membranes

Bahman Elyassi, Muhammad Sahimi, and **Theodore T. Tsotsis**. Mork Family Department of Chemical Engineering & Materials Science, University of Southern California, Los Angeles, CA 90089

Growing interest in the hydrogen economy is motivating research on inorganic, hydrogen-permselective membranes, which can be used in processes (related to H_2 production) that take place at high temperatures and pressures. A promising candidate material for a variety of inorganic membrane applications is silicon carbide due to its many unique properties, such as high thermal conductivity, thermal shock resistance, biocompatibility, resistance in acidic and alkali environments, chemical inertness, and high mechanical strength [1,2]. Previously, using a combination of slip-cating and dip-coating [1] we prepared SiC membranes typically showing an ideal H_2/CO_2 selectivity in the range of (42-96), and a H_2/CH_4 ideal selectivity in the range of (29-78). One of the challenges for applying membrane technology in reaction environment is to produce membranes with high steam stability. Steam stability experiments with our SiC membranes lasting 21 days, using an equimolar flowing mixture of He/H_2O at 200°C, indicated some initial decline in the permeance of He, after which the permeance became stable at these conditions. However, for a better perforamce higher permselectivity for hydrogen is needed. The next phase of our project was to increase the permeance of the membrane while keeping the separation factor at high levels. To do so, a novel method is presented for the preparation of microporous silicon carbide membranes. The strategy comprises periodic coating of sacrificial interlayers and SiC pre-ceramic layers. In order to keep in place the interlayers; a solvent has been selected for the pre-ceramic polymer which does not dissolve the sacrificial interlayer during the dip-coating process or afterwards. The new preparation method involves first dip-coating the slip-casted supports in polystyrene (PS) solution, and then drying at 100°C. The supports were subsequently dip-coated in allylhydridopolycarbosilane (AHPCS) solution, and then pyrolyzed at 750°C resulting in complete decomposition of the polystyrene, and the forming of a SiC membrane layer on the top. The procedure of coating, drying and pyrolyzing of PS and AHPCS layers was carried out for three additional times. Membranes prepared, so far, show single gas ideal separation factors of helium and hydrogen over argon in the range (176-420) and (100-200), respectively, with an increase of about two to three times in permeance compared to the our previously prepared membranes [1]. To get some insight of how these sacrificial interlayes promote membrane performance, SEM pictures were taken after the last layer coating of PS. Within our SEM resolution, no continuous distinguishable PS film was seen on the top. We speculate that the sacrificial interlayers fill the pores and prevent their blockage by the pre-ceramic polymer. Since the interlayer barrier can be decomposed in inert or oxidizing environments, the new method has promising potential for application to a variety of other inorganic membrane systems.

[1] Bahman Elyassi, Muhammad Sahimi, Theodore T. Tsotsis, Silicon carbide membranes for gas

separation applications, Journal of Membrane Science 288 (2007) 290–297.

[2] Richard J. Ciora, Babak Fayyaz, Paul K.T. Liu, Varaporn Suwanmethanond, Reyes Mallada, Muhammad Sahimi, Theodore T. Tsotsis, Preparation and reactive applications of nanoporous silicon carbide membranes, Chemical Engineering Science 59 (2004) 4957-4965.

593b Layer-By-Layer Deposition Of Permselective C-Oriented-Mcm-22/Silica Composite Films

Jungkyu Choi and Michael Tsapatsis. Chemical Engineering and Materials Science Dept., University of Minnesota, Department of Chemical Engineering and Materials Science, 421 Washington Avenue SE, Minneapolis, MN 55455-0132

Zeolite films can be synthesized by various methods, such as in situ, seeded (secondary) growth, vapor phase transport, etc. Most of these methods are based on the hydrothermal synthesis process which involves exposure of the bare or seeded substrates to the synthesis liquid or vapor at elevated temperature. Despite the success of these synthesis methods for many applications, the hydrothermal synthesis process is difficult to scale-up. Therefore, we have been investigating alternative synthesis methods that eliminate hydrothermal synthesis steps.

Here, we report an alternative film fabrication method, so called layer-by-layer deposition. The layer-by-layer deposition method combines deposition of plate-like MCM-22 zeolite crystals on substrates by sonication-assisted covalent attachment1, 2 with evaporation-induced-self-assembly of surfactant-templated silica.3 In this method, although MCM-22 zeolite particles are synthesized by hydrothermal procedure, no hydrothermal synthesis is involved during the film fabrication. The deposition cycles (particle deposition and silica coating) were repeated several times to make compact c-oriented-MCM-22/Silica films on porous á-alumina discs. The composite c-oriented-MCM-22/Silica films exhibited hydrogen ideal selectivities (e.g., H2/N2 ~ 7) above those expected by Knudsen diffusion indicating molecular sieving potential.

(1) Lee, J. S.; Ha, K.; Lee, Y.-J.; Yoon, K. B., Ultrasound-aided remarkably fast assembly of monolayers of zeolite crystals on glass with a very high degree of lateral close packing. Adv. Mater. (Weinheim, Ger.) 2005, 17, (7), 837-841. (2) Yoon, K. B., Organization of Zeolite Microcrystals for Production of Functional Materials. Acc. Chem. Res. 2007, 40, (1), 29-40. (3) Lu, Y.; Ganguli, R.; Drewien, C. A.; Anderson, M. T.; Brinker, C. J.; Gong, W.; Guo, Y.; Soyez, H.; Dunn, B.; Huang, M. H.; Zink, J. I., Continuous formation of supported cubic and hexagonal mesoporous films by sol-gel dip-coating. Nature (London) 1997, 389, (6649), 364-368.

593c Synthesis And Characterization Of Alumina-Ordered Mesoporous Silica Hybrid Membranes

Sukjoon Yoo, Chemical Engineering, University of Massachusetts, Amherst, 159 Goessmann Laboratory, North Pleasant Street, Amherst, MA 01003, David M. Ford, Department of Chemical Engineering, University of Massachusetts, Amherst, 209 Goessmann Laboratory, 686 North Pleasant Street, Amherst, MA 01003, Seunguk Yeu, Department of Chemical Engineeing, Texas A&M University, TAMU 3122, College Station, TX 77843-3122, and Daniel F. Shantz, Department of Chemical Engineering, Texas A&M University, TAMU 3122, College Station, TX 77843-3122.

Alumina-Ordered Mesoporous Silica (OMS) hybrid membranes were synthesized by using the evaporative induced self-assembly (EISA) process based on dip-coating of commercial macroporous alumina substrates with a Brij-56/TEOS/HCl/H2O solution. Two different commercial alumina membranes are used here as macroporous supports: Whatman Anopore and Pall Membralox. The former is scientifically interesting but not so practical for industrial use due to low mechanical strength, while the latter is a more practical route to a commercially useful product. Numerous analytical methods are used to probe both the hybrid material and the silica phase after dissolution of the alumina substrate. Most importantly, He/N2 permeation measurements show that the effective pore size of the membrane can be tuned from 20 to 5 nm based on the number of dip-coating cycles used. The observed He/N2 permselectivity is nearly identical to the theoretical Knudsen value, which is ascribed to the lack of pinhole defects in the materials. The silica obtained after dissolution of alumina substrate has been characterized using scanning and transmission electron microscopy. Those results indicate that the silica deposited in the membrane possesses ordered pores approximately 5 nm in size, consistent with the permeation studies. The current work presents an alternative approach to materials that possess many of the properties of mesoporous silica thin films (i.e. pores of controlled size and topology) without the difficulty of growing mesoporous silica thin films on porous supports.

593d Permeation In Mesoporous Silica-Anodized Alumina Composite Membranes

Aniket Sonalkar and Dr. Orhan Talu. Chemical and Biomedical Engineering, Cleveland State University, Cleveland, OH 44115

Ordered mesoporous silica (MPS) with 1-10 nm pores has been evaluated for numerous applications in adsorption. As synthesized MPS is in powder form not suitable for membrane applications. Anodized alumina (AA) with 50-200 nm pores has been evaluated for membrane applications such as ultrafilration. The pores of AA is quite large to separate in molecular scale. In this project, we are interested in growing MPS in AA pores for enhanced separation at molecular scale. To elucidate the transport mechanisms in suchj a composite membrane, several small gas molecules have been used. The permeation of Argon through these membranes before and after calcination was performed using a batch method at temperature range between 253 K – 353 K and at low pressures. The MPS-AA before calcination was almost impermeable to Argon because the mesopores are blocked by the surfactant molecules. Upon calcinations, the Argon flux is substantial and can be used to examine the diffusion/ permeation mechanism. The experimental flux through these membranes can be divided into different contributions and regimes such as surface diffusion, Knudsen diffusion, activated gaseous diffusion and viscous flow. A mathematical model is developed to interpret these contributions in terms of cross-sectional area of the pore channels or molecular mobility which can be suitably quantified by a physical property such as diffusivity. The temperature dependence of permeability is very useful to determine the mechanisms.

593f Synthesis And Properties Of High-Silica Sodalite And Deca-Dodecasil 3R Membranes Grown On α-Alumina Supports

Zhenkun Zheng, Chemical and Materials Engineering, University of Cincinnati, Cincinnati, OH 45221 and Vadim V. Guliants, Department of Chemical and Materials Engineering, University of Cincinnati, Cincinnati, OH 45221-0012.

Sodalite[1-3] and Deca-dodecasil 3R (DD3R) [4, 5] are promising membrane materials for hydrogen separation from larger gas molecules due to their 6-membered ring (0.28nm) and 8-membered ring pore (0.36×0.44 nm), respectively. The high-silica sodalite (SOD) membranes and deca-dodecasil 3R (DD3R) membranes were hydrothermally synthesized on porous α-alumina supports using tetramethylammonium hydroxide (TMAOH) and 1-adamantanamine (1-ADA) as the structure-directing agents (SDAs), respectively, by in-situ crystallization and seeded secondary growth methods. The XRD and SEM techniques were applied to investigate the phase composition and microstructure of the membranes and corresponding bulk phases. The H2, N2, CO2, and SF6 single gas permeances and their permselectivity of membranes were measured from 25oC to 600oC under different transmembrane pressures. The dependence of gas permeances on the temperature and the transmembrane pressure was investigated and the template removal behaviors of high-silica sodalite and DD3R were studied. For high-silica sodalite membranes, gas permeation studies showed the presence of small micropores after template removal. For DD3R membranes, the CO2/H2 permselectivity declined with increasing temperature and membranes became H2 selective at above 100oC because of the favorable adsorption for CO2 as compared to H2. H2/N2 permselectivity was about 9.0 at 25oC which is much higher than the corresponding Knudsen selectivity (~3.7) and CO2 has higher permeance than N2 at studied temperatures. It was shown that the permselectivities of the seeded membranes were greater than those of the unseeded membranes. It was revealed that the good quality DD3R membranes possessed molecular sieving function for gas separation. It was further demonstrated that repeated hydrothermal treatments improved the quality of these membranes.

Acknowledgements The financial support from Ohio Coal Development Office (OCDO) is gratefully acknowledged.

References 1. Motuzas, J., et al., Rapid synthesis of oriented silicalite-1 membranes by microwave-assisted hydrothermal treatment, Microporous and Mesoporous Materials, 2006. 92(1-3): p. 259-269. 2. Xu, X.C., et al., Synthesis and gas permeation properties of an NaA zeolite membrane, Chemical Communications, 2000(7): p. 603-604. 3. Zhenkun Zheng and Vadim V. Guliants, Synthesis and Properties of High-Silica Sodalite Membranes Grown on a-alumina Support, Thin Solid Films, submitted. 4. Tomita, T., K. Nakayama, and H. Sakai, Gas separation characteristics of DDR type zeolite membrane, Microporous and Mesoporous Materials, 2004. 68(1-3): p. 71-75. 5. Himeno, S., et al., Characterization and selectivity for methane and carbon dioxide adsorption on the all-silica DD3R zeolite, Microporous and Mesoporous Materials, 2007. 98(1-3): p. 62-69.

593g Screening Of Nanoporous Membranes For CO2 Separations Using Atomically-Detailed Simulations

Seda Keskin, Haibin Chen, Sang eun Jee, and David S. Sholl. Carnegie Mellon University, Department of Chemical Engineering, Carnegie Mellon University, Pittsburgh, PA 15213

Although large numbers of nanoporous materials exist that are potentially suitable for use as gas separation membranes, the experimental challenge of fabricating defect-free thin films makes screening of these materials a laborious process. Atomically-detailed simulations can offer a useful means to accelerate this screening process by making predictions of membrane properties that can be used to decide whether extensive efforts to fabricate membranes from new materials are justified. We will discuss the use of molecular simulations to simulate the transport of gas mixtures involving carbon dioxide through several types of ordered nanoporous materials, including small pore zeolites such as DDR, single-walled carbon nanotubes, and metal-organic framework materials. By using a multi-scale approach in our modeling, we are able to link atomically-detailed descriptions of each material with predictions of membrane performance under practical operating conditions.

626a New Insights Into The Ideal Cascade Theory For Membrane Separation Processes

Rugved Pathare and Rakesh Agrawal. School of Chemical Engineering and the Energy Center at Discovery Park, Purdue University, 480 Stadium Mall Drive, West Lafayette, IN 47907

Membrane separation processes have become increasingly popular in the last three decades. Most commercial membranes have finite permselectivity (ratio of the component permeabilities) values. The availability of modest permselectivity membranes has given impetus to the search for the most energy efficient design for a membrane process. Calculations show that considerable capital and energy savings can be made by cascading membrane stages together through recycle. To obtain specified high recoveries, the proper selection of cascade geometry and operating parameters, to minimize capital cost and energy consumption is an important problem. The analytical approaches to solve this problem were taken through the "Ideal cascade" theory [1-3] discussed by Cohen and others. An Ideal cascade insures that streams that are mixed are of the same compositions, thereby avoiding the energy penalty associated with mixing dissimilar streams.

The optimality condition for the 'Classical ideal cascade design', termed IC-1 here, was derived under a set of operating conditions in which the components' permselectivity was nearly unity. However, for contemporary applications we need to optimize configurations operating with higher permselectivities, under constraints dictating that only a few recycle compressors be used. For moderate membrane permselectivity, an additional degree of freedom, in the stage separation factor, is available, the proper choice of which can lead to up to 50% energy savings. With this insight, the present work demonstrates the mathematical relationships extending the scope of the theory to such applications. The present work identifies an optimal regime of operation for a membrane cascade, which can lead to energy savings of up to a factor of two, while still under the IC-1 design. Detailed investigations were also made into a second operating condition proposed by Herbst and McCandless [4]. This condition identifies another operating regime, still retaining no mixing condition and employing a classical scheme. It will be referred to as the IC-2 design in this work. Rigorous derivations are made, in the current work, for the stage separation factor limits within which the IC-2 design can be operated. The present work provides useful results into what parameter values are optimal with respect to both, energy considerations and number of stages, to perform a prespecified separation task, when still working in the classical scheme.

It is expected that the current work, will provide a base for the construction of a much needed systematic framework for selection of the energy optimal cascade configuration. Future work will focus on optimization of the cascade using alternative no mixing schemes and then extending the theory to handle multicomponent systems.

References

1. Cohen, K., The Theory of Isotope Separation as Applied to Large-Scale production of U235. 1st ed.

ed, ed. G.M. Murphy. 1951, New York: McGraw-Hill.

2. Benedict, M., T.H. Pigford, and H.W. Levi, Nuclear Chemical Engineering. 2nd ed. ed. 1981, New York: McGraw-Hill.

3. Pratt, H.R.C., Countercurrent Separation Processes. 1967, New York: Elsevier Publishing Co.

4. Herbst, R.S. and F.P. McCandless, No-Mix and Ideal Separation Cascades. Separation Science and Technology, 1994. 29(17): p. 2215 - 2226.

626b Energy Minimization Using Electrodeionization: Organic Acid Production and Separation

Jamie A. Hestekin, **Heath Himstedt**, Betty Phansiri, and Nicole Thomas. Ralph E. Martin Department of Chemical Engineering, University of Arkansas, 3202 Bell Engineering Center, Fayetteville, AR 72701

Bioproccessing offers advantages over traditional chemical processing including fewer by-products and milder operating conditions. One application of particular interest is biocatalyst-driven separation of organic acids for applications in industry and medicine. Currently these are quite expensive to produce due to the difficulty of separation from dilute bioprocessing solutions. Further, organic acids can have a negative effect on many fermentations (1) and the removal and\or control of the organic acids is key to efficiently designed processes. Our group has looked at a technology called wafer enhanced electrodeionization (WE-EDI), which uses ion exchange wafers and an electric field to selectively remove ions from uncharged species even at very low concentrations (2). WE-EDI is highly pursuable and desirable because of the large binding surface area, concurrent product removal and production, and high efficiency and ease of operations compared to loose resins. Further, as opposed to electrodialysis, WE-EDI can remove ions at extremely low concentrations, as low as 10 ppb. In this paper, we will talk about design aspects of a WE-EDI system, modeling of selective organic ion removal, and preliminary economics of such a process.

REFERENCES

1. J. A. Hestekin, S. Snyder, and B. Davison, “Direct Product Capture from Biotransformations”, ChemPlus Report, Chemical Vision 2020 (2002).

2. M. B. Arora and J. A. Hestekin, S. W. Snyder, E. J. St. Martin, M. I. Donnelly, C. Sanville-Millard, and Y. J. Lin, “The Separative Bioreactor: A Continuous Separation Process for Simultaneous Production and Direct Capture of Organic Acids”, Sep. Sci. Tech., in press (2007).

626c Gas Separation Membrane Opportunities to Capture Greenhouse Gases

Pushpinder S. Puri, Air Products and Chemicals, Inc., 7201 Hamilton Blvd., Allentown, PA 18195

Solar radiation reaching the earth heats the planet's surface (oceans, land masses and atmosphere), evaporates water, drives wind and weather, and powers plant photosynthesis. The earth's surface, in turn, radiates a portion of this energy at infrared (IR) wavelengths. Much of this energy escapes into space. However, a significant fraction of the energy carried by outgoing IR radiation is reflected back by to earth by particulates and aerosols in the atmosphere or is absorbed by atmospheric gases, such as carbon dioxide, water vapor, methane, ozone, nitrous oxide, chlorofluorocarbons (CFC's), etc., known as greenhouse gases. These gases (with the exception of man-made CFCs) originate from both natural and man-made processes. As concentration of greenhouse gases increases in the earth's atmosphere, retention of radiated energy also increases. This shifts the planet's overall energy balance – the balance between input from solar radiation absorbed and loss by radiation to space, contributing to global warming. This paper describes a variety of opportunities for membrane technologies to capture greenhouse gases. Increasing CO2 concentration in the atmosphere is responsible for approximately half of the human activity related contribution to global warming. Membranes' potential to contribute significantly to the control of the greenhouse effect, illustrated in this presentation for CO2 capture, is applicable to other greenhouse gases as well.

626d Design Optimization and Validation of A Solid State, Electrochemical Oxygen Compressor

Jie Bai, Merrill A. Wilson, Kent Hutchings, and Raymond Cutler. Ceramatec, Inc., 2425 South 900 West, Salt Lake City, UT 84119

This paper describes the design and validation of an ion transport membrane device which provides point of use ultra high purity oxygen at pressures as high as 2 MPa. This pressure is achieved via single stage electrochemical compression and is sufficient for distribution line and storage tank requirements. The benefits of this technology can greatly reduce energy required for multi-stage mechanical compression and the need and cost of cylinder systems.

Planar designs for SOFC's and oxygen separation devices typically involve fabrication of interconnects and electrolyte plates separately. After the components are manufactured they are typically joined together to form a stack. For pressurized oxygen production a mechanical endload must be applied to the stack to offset the internal oxygen pressure in order to prevent the development of leaks. Product gage pressures of 1 and 2 MPa could require 10 kN and 20 kN endloads respectively (slightly over 1 and 2 tons respectively). Alternatively, by co-sintering the electrochemical stack, these mechanical loads can be internally constrained eliminating the need for external mechanical loading systems.

This paper describes an innovative approach to develop a separation device capable of producing high pressure pure oxygen without the use of a counter balancing endload. In this approach all the functional layers are cosintered to form a mechanically robust device with excellent contact between the each layer enabling the production of pressurized oxygen without a mechanical endload. The oxygen pressure is limited by the design and strength of the bond between the porous supports and dense channeled layers.

626e Staged Membrane Systems for Hydrogen Recovery from H2/CO2 Mixtures

Santosh A. Sonalkar and **G. Glenn Lipscomb**. Chemical and Environmental Engineering, University of Toledo, 2801 West Bancroft Street, Mail Stop 305, Toledo, OH 43606

Steam reforming of various hydrocarbon feeds is the primary source of industrial hydrogen. The reforming process produces a stream containing primarily hydrogen and carbon dioxide which requires subsequent purification.

A variety of membrane separation processes are being considered for this separation. This work addresses the potential of polymeric membranes for hydrogen recovery. For most binary gas pairs, polymeric materials permeate one species preferentially over the over. However, for hydrogen-carbon dioxide materials are available that are selective for either hydrogen or carbon dioxide.

One may envision a number of different staging configurations based on these materials. For example, four different two-stage configurations exist in which each stage permeates either hydrogen or carbon dioxide. Moreover, one could use a single stage in which both types of membranes are intermingled – an internally staged configuration.

The performance of these configurations is evaluated theoretically for recovering 90% of the hydrogen from a gas stream containing 40%-60% hydrogen. The internally staged configuration produces the highest purity product. However, a two-stage configuration in which each stage preferentially permeates hydrogen performs nearly as well. Moreover, this two-stage design does not utilize inter-stage compression – the permeate pressure is reduced in two steps to its final, lowest value. The results are used to evaluate the economics of hydrogen recovery with polymeric membranes.

CO_2 selective membranes for carbon capture in water-gas-shift reactions

Introduction

During the last century, many new technologies have been developed and the industrialized world has grown fast. This has had and still has an enormous impact on the environment, especially through the greenhouse gas emissions associated with energy production in accordance with the Kyoto Protocol [1]. This global warming is mostly due to greenhouse effect through gases like CO_2. Apart from energy saving and durable sources, a way to reduce these emissions is to capture and store CO_2. Post-combustion capture technologies such as MEA scrubbing [2] and using polymeric membranes [2] already exist. Pre-combustion capture would have a lower capture penalty on the efficiency and takes place at higher temperatures where polymers do no longer function as a membrane. When starting out with a coal gasifier, a water gas shift membrane reactor can be used either with hydrogen selective membranes or CO_2 selective membranes. CO_2 selective membranes enable keeping hydrogen pressurized at a high temperature and is, therefore, an attractive alternative for noble metal alloy based hydrogen separating membranes [3,4,5,6]. Our research target within the Global Climate and Energy Project (GCEP [7]) is, therefore, to develop CO_2 separating membranes.

In the Sorption Enhanced Reforming Project (SERP) [8,9,10], as well as in other projects [11,12,13,14], hydrotalcite materials are investigated as CO_2 sorbents. These materials seem to be good candidates for adsorbing CO_2 although the desorption process is still rather slow. The presumption is that these materials might be applicable either as a dense or a porous membrane. Therefore, we decided to deeply study these materials and to look at all properties required for a viable membrane process.

Hydrotalcites are clay-like materials consisting of brucite-type layers separated by layers containing water and anions.

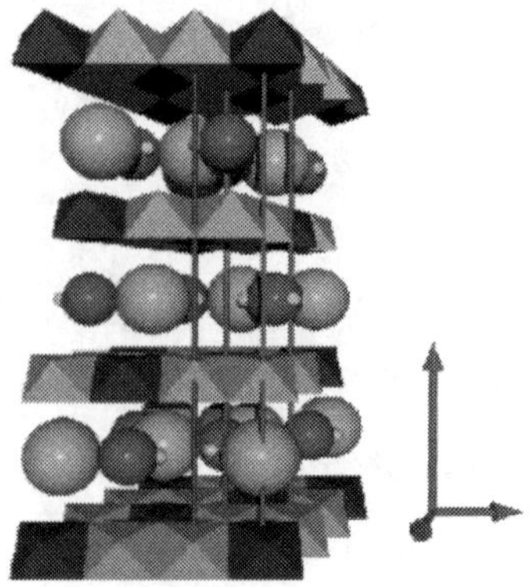

Fig.1: Hydrotalcite crystal

The hydrotalcite, with the general formula $\left(M_x^{2+}M_{1-x}^{3+}\right)(OH)_2(CO_3)_{\frac{1-x}{2}} \cdot yH_2O$ (with M^{2+} being Mg^{2+} and M^{3+} being Al^{3+} in this case), has a crystal structure in which, via its interlayer containing CO_3^{2-} anions and water, is presumed capable of having CO_2 molecules enter the structure on one side and get out of it on the other side; thus acting as a highly selective dense membrane. Porous membranes are presumed to transport CO_2 through the inter grain pores. Neither possibility has been proven yet.

It has been claimed in the literature that hydrotalcites could be synthesized with any Mg/Al ratio [15,16,17,18,19] and that it decomposes with increasing temperature, but so far no decomposition pathway has been described. We present our results here concerning hydrotalcite characterization and decomposition. We also propose a way to produce hydrotalcite membranes for pre-combustion CO_2 capture.

Materials and techniques

Commercial materials from Sasol and materials prepared in-house following the urea hydrothermal hydrolysis [20] with different Mg/Al ratios have been compared using X-Ray diffraction (XRD) and Scanning Electron Microscopy (SEM) with Emission Dispersive X-ray (EDX). Neutron Diffraction (ND) has been used to evaluate the exact Mg/Al ratio in hydrotalcite. The decomposition pathway has been established using ThermoGravimetric Analysis (TGA) attached to a Mass Spectrometer (MS), *in-situ* XRD and *in-situ* Diffuse Reflectance Infrared Fourier Transformed (DRIFT) spectroscopy.

Experimental

Characterization

A comparison between the commercial materials and the in-house produced ones has been performed with XRD. From this, it can be deduced that the commercial materials are poorly crystalline compared to the in-house produced ones. Furthermore, the closer the Mg/Al ratio is to 2, the more crystalline the materials are. Since the number of electrons of Mg^{2+} and Al^{3+} are equal, XRD can not be used to find the ratio between these ions, however, with ND one can. Three samples of different Mg/Al ratios have then been analyzed at the GEM beam line at ISIS.

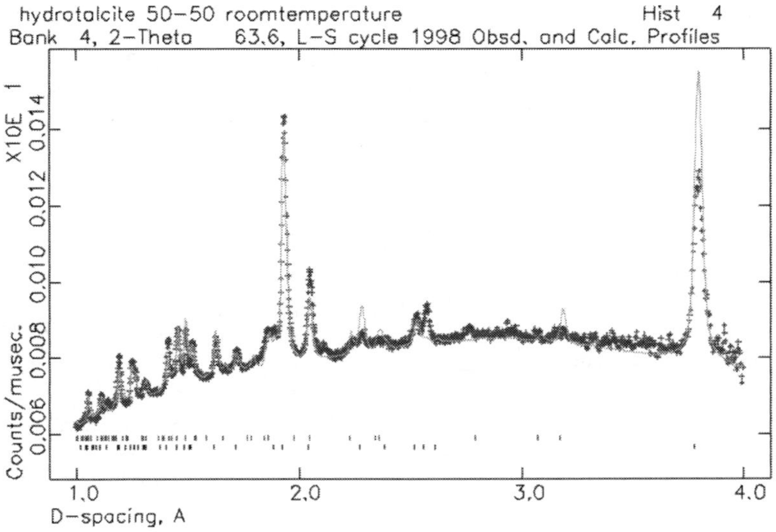

Fig.2: Mg50 (starting materials: Mg/Al=1) neutron patter

The pattern (fig.2) could be refined using the GSAS program suite, in which CO_3^{2-} and H_2O are treated as rigid bodies. The Mg/Al ratio could be determined to be about 1.8. Deviations from this value yielded either Al rich (Boehmite) or Mg rich (hydromagnesite) impurity phases.

Thermal stability

The water-gas-shift reaction takes place at about 400°C. TGA-MS has been performed on the commercial material MG61 (figure 3) in order to check the thermal stability.

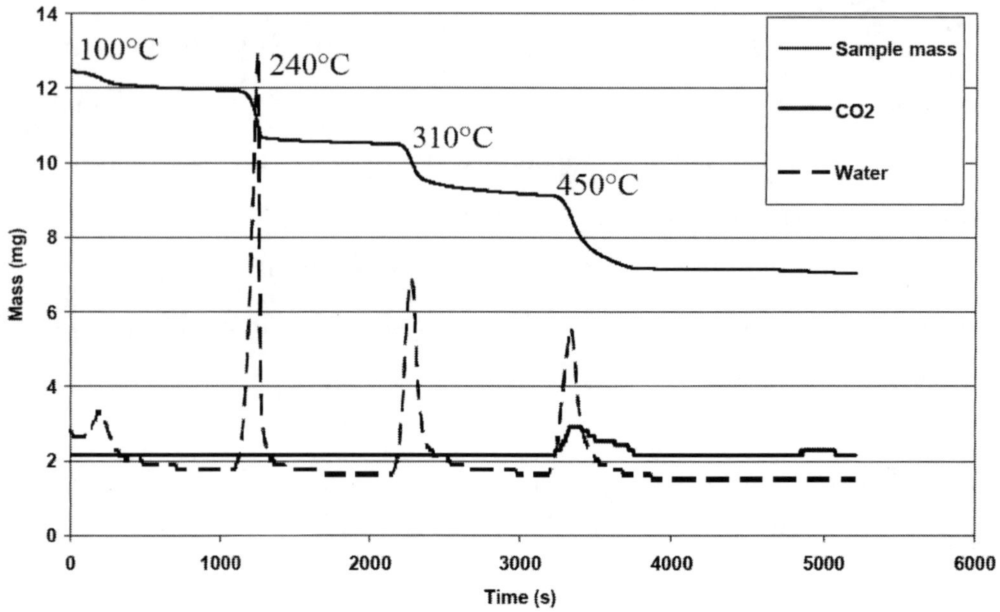

Figure 3: TGA-MS of MG61

In the figure 3, four decomposition steps can be seen. The first step at ≈100°C represents the loss of adsorbed water, the second at ≈240°C represents the interlayer water disappearing, the third at ≈310°C represents the hydroxides decomposing into oxides, and the fourth at ≈450°C represents the last hydroxides and carbonates decomposing into oxides. To characterize the structural evolution upon decomposition, *in-situ* XRD has been performed on the same material (fig.4).

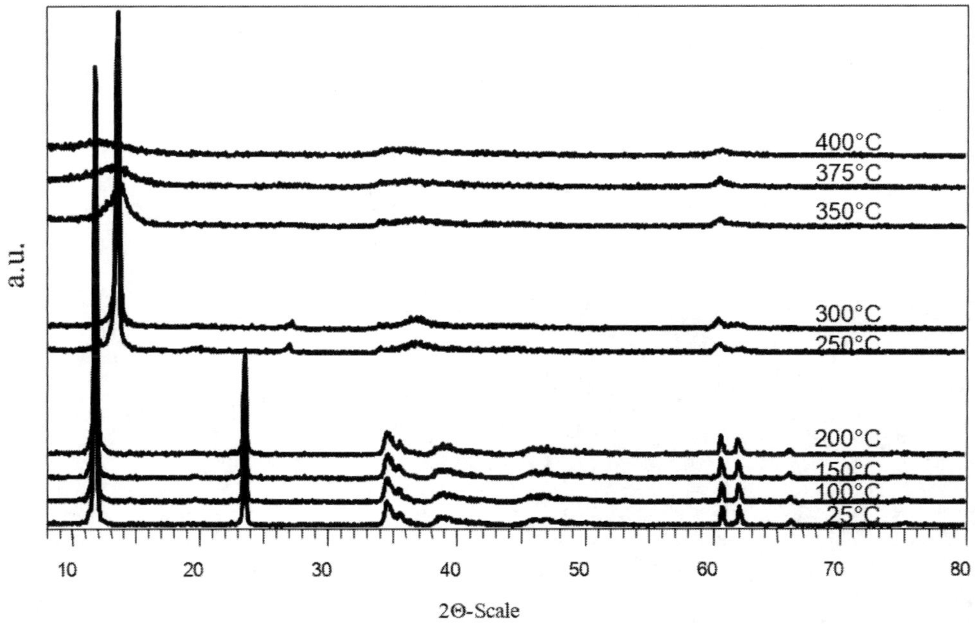

Fig.4: in-situ XRD pattern of MG61 as a function of temperature.

Figure 4 shows that the three first steps found with the TGA are confirmed by the XRD. Between 200°C and 250°C, there is a shift towards smaller d-spacing, indicative of the collapse of the interlayer at ≈240°C. Between 300°C and 350°C, the crystallinity of the material is strongly decreasing, while the hydroxides are decomposing into oxides.

More detailed information on the process can been obtained using DRIFT. This technique permits infrared spectroscopy under different gas atmospheres and temperature. Combining XRD, TGA-MS and DRIFT results yields the following decomposition pathway:

$$Mg_4Al_2(CO_3)(OH)_{12}.xH_2O.yH_2O_{ads} \xrightarrow{150^\circ C} Mg_4Al_2(CO_3)(OH)_{12}.xH_2O + yH_2O \quad (1)$$

$$Mg_4Al_2(CO_3)(OH)_{12}.xH_2O \xrightarrow{230^\circ C} Mg_4Al_2(CO_3)(OH)_{12} + xH_2O \quad (2)$$

$$Mg_4Al_2(CO_3)(OH)_{12} \xrightarrow{320^\circ C} MgCO_3.3Mg(OH)_2.Al_2O_3 + 3H_2O \quad (3)$$

$$MgCO_3.3Mg(OH)_2.Al_2O_3 \xrightarrow{440^\circ C} 3MgO.MgAl_2O_4 + CO_2 + 3H_2O \quad (4)$$

Membranes

The results of the previous paragraph clearly show that the dense membrane option is not feasible since the interlayer does not exist anymore at the operational temperature. A porous membrane is still an option. In this case, very small particles have to be produced that can be deposited on a support as thin layers. Using a conventional co-precipitation synthesis route it is possible to obtain nano-particles of 15nm diameter (fig.5).

Fig.5: hydrotalcite nano-particles

In order to have a homogeneous coating from a hydrotalcite sol, there is a need of having a very well dispersed sol particles. Sol-gel chemistry is very well known for providing homogeneous and stable sols [21]. In this study, the sol-gel route also provided very small particles but the particle size distribution was extremely broad, a.o. through aggregate formation, making any quantitative analysis of it impossible.

Conclusions and perspectives

Using XRD, TGA-MS, ND, DRIFT and SEM-EDX the decomposition pathway could be formulated. Furthermore the Mg/Al ratio in hydrotalcites could be determined.
It has been shown that hydrotalcite materials cannot be used as dense membranes. However, the option to make hydrotalcite based-porous membranes still seems viable.
It has been shown that small nano-particles needed to make such membranes can be produced. Still, there is a need of getting these particles better dispersed.

Acknowledgements

The authors thank the GCEP program (http://gcep.stanford.edu/) for financial support, NWO for access to the ISIS facilities and ISIS staff for measurements of the samples, and also Vera Smit-Groen, Andre van Zomeren and Sacha Valster for their help with XRD, TGA/MS and DRIFT experiments.

References

1. Convention-cadre des Nations Unies sur les changements climatiques (1992)

2. B. METZ, O. DAVIDSON, H. C. CONINCK DE, M. LOOS, and L. A. MEYER, IPCC special report on carbon dioxide capture and storage (2005)

3. Y. H. MA, in Proceedings of the 9th International Conference on Inorganic Membranes,Lillehammer, June 2006, p. 28.

4. Y. H. MA, I. P. MARDILOVICH, and E. E. ENGWALL, Annals of the New York Academy of Sciences 984 (2003) p. 346.

5. H. KLETTE, T. A. PETERS, A. MEJDELL, and R. BREDESEN, in Proceedings of the 8th International Conference on Greenhouse Gas Control Technologies, 2006

6. http://gcep.stanford.edu/

7. M. V. CIOCCO, B. D. MORREALE, O. IYOHA, R. M. ENICK, R. P. KILLMEYER, and B. H. HOWARD, in Proceedings of the International Technical Conference on Coal Utilization & Fuel Systems 30th, 2005, p. 303.

8. J. HUFTON, S. MAYORGA, T. GAFFNEY, S. NATARAJ, M. RAO, and S.SIRCAR, Sorption Enhanced Reaction Process (SERP) for the Production of Hydrogen (2005)

9. J. HUFTON, W. WALDRON, S. WEIGEL, M. RAO, S. NATARAJ, and S. SIRCAR, Sorption Enhanced Reaction Process (SERP) for the production of hydrogen PA 18195 (2000)

10. Y. DING and E. ALPAY, Chemical Engineering Science 55 (2000) p. 3929.

11. Y. DING and E. ALPAY, Trans IChemE Part B 79 (2001) p. 45.

12. Z. YONG, V. MATA, and A. E. RODRIGUES, Ind. Eng. Chem. Res. 40 (2001) p. 204.

13. Z. YONG and A. E. RODRIGUES, Energy Conversion and Management 43 (2002) p. 1865.

14. J.-I. YUNG and J.-N. KIM, Korean Journal of Chemical Engineering 23 (2005) p. 77.

15. M. TSUJI, G. MAO, T. YOSHIDA, and Y. TAMAURA, Material Research Society 8 (1993) p. 1137.

16. K. TAKEHIRA, T. KAWABATA, T. SHISHIDO, K. MURAKAMI, T. OHI, D. SHORO, M. HONDA, and K. TAKAKI, Journal of Catalysis 231 (2005) p. 92.

17. T. YAMAMOTO, T. KODAMA, N. HASEGAWA, M. TSUJI, and Y. TAMAURA, Energy Conversion and Management 36 (1995) p. 637.

18. R. MARCHIDAN, D. CRISAN, E. IVANA, M. GABROVSKA, R. EDREVA-KARDJIEVA, and A. ANDREEV, Bulgarian Chemical Communications 30 (1998) p. 217.

19. E. KANEZAKI, Mat. Res. Bull. 33 (1998) p. 773.

20. M. M. RAO, B. R. REDDY, M. JAYALAKSHMI, V. S. JAYA, and B. SRIDHAR, Materials Research Bulletin 40 (2005) p. 347.

21. C. J. BRINKER, G.W.SCHERER, Sol-Gel Science, The Physics and Chemistry of Sol-Gel Processing, Academic Press, Inc. (1990).

AUTHOR INDEX

AUTHOR INDEX

AUTHOR INDEX

AUTHOR INDEX

AUTHOR INDEX

AUTHOR INDEX

AUTHOR INDEX

AUTHOR INDEX

AUTHOR INDEX

AUTHOR INDEX

AUTHOR INDEX

AUTHOR INDEX

AUTHOR INDEX

American Institute of Chemical Engineers
3 Park Avenue
New York, New York 10016-5991

ISBN: 978-1-60560-000-0